$1s^2$
$2s^2$
$2p^6$

B^3

Main-Group Elements

						VIIIA
						2 He 4.00260

	IIIA	IVA	VA	VIA	VIIA	
	5 B 10.81	6 C 12.011	7 N 14.0067	8 O 15.9994	9 F 18.998403	10 Ne 20.179
	13 Al 26.98154	14 Si 28.0855	15 P 30.97376	16 S 32.06	17 Cl 35.453	18 Ar 39.948

IB	IIB					

28 Ni 58.69	29 Cu 63.546	30 Zn 65.38	31 Ga 69.72	32 Ge 72.59	33 As 74.9216	34 Se 78.96	35 Br 79.904	36 Kr 83.80
46 Pd 106.42	47 Ag 107.8682	48 Cd 112.41	49 In 114.82	50 Sn 118.69	51 Sb 121.75	52 Te 127.60	53 I 126.9045	54 Xe 131.29
78 Pt 195.08	79 Au 196.9665	80 Hg 200.59	81 Tl 204.383	82 Pb 207.2	83 Bi 208.9804	84 Po (209)	85 At (210)	86 Rn (222)

†Element synthesized, but no official name assigned

Inner-Transition Metals

63 Eu 151.96	64 Gd 157.25	65 Tb 158.9254	66 Dy 162.50	67 Ho 164.9304	68 Er 167.26	69 Tm 168.9342	70 Yb 173.04	71 Lu 174.967
95 Am (243)	96 Cm (247)	97 Bk (247)	98 Cf (251)	99 Es (252)	100 Fm (257)	101 Md (258)	102 No (259)	103 Lr (260)

General Chemistry

General Chemistry

Darrell D. Ebbing

Wayne State University

Consulting Editor:

Mark S. Wrighton

Massachusetts Institute of Technology

Houghton Mifflin Company   Boston

Dallas   Geneva, Illinois   Hopewell, New Jersey
Palo Alto

Fig. 1.2 Courtesy Mettler Instrument Corp., Hightstown, N.J. *Page 6* Cartoon by Sidney Harris. **Fig. 1.8** Photograph by James Scherer. **Fig. 1.9** U.S.D.A. Photograph by Forsythe. **Fig. 2.1** Courtesy F. P. Ottensmeyer, Ontario Cancer Institute, Toronto. From his "Scattered Electrons in Microscopy and Microanalysis," *Science,* 215:461, 29 January 1982. Copyright 1982 by the American Association for the Advancement of Science. **Fig. 2.3** Courtesy of Morton Salt, a division of Morton-Norwich. **Fig. 2.6** Photographs by James Scherer. **Fig. 3.1** Courtesy Finnigan Corporation. **Fig. 3.2** Photograph by James Scherer. **Fig. 5.2** Copyright Cavendish Laboratory, University of Cambridge. **Fig. 5.13** Wide World Photos. **Fig. 5.16** Courtesy Cambridge Instruments. **Fig. 6.6** Historical Pictures Service, Chicago. **Fig. 7.5** Adapted from Linus Pauling, *The Nature of the Chemical Bond.* Copyright 1939 and 1940, third edition © 1960 by Cornell University. Used by permission of Cornell University Press. **Fig. 9.4** Photographs by James Scherer. **Fig. 9.5** Fundamental Photographs, New York. **Fig. 10.1** Photographs by James Scherer. **Fig. 11.20** From Preston, *Proceedings of the Royal Society,* A, vol. 172, P1. 4, Fig. 5a. **Fig. 12.3** Courtesy James G. White, M.D., University of Minnesota. **Fig. 12.7** Fundamental Photographs, New York. **Fig. 13.3** Adapted from G. M. B. Dobson, *Exploring the Atmosphere.* Oxford University Press, 1968. **Fig. 13.9** Courtesy Contamination Control Incorporated. **Fig. 13.12** U.S.D.A. photo, courtesy American Chemical Society. **Fig. 14.3** United Press International Photo. **Fig. 15.11** Courtesy General Motors Corporation. **Fig. 16.4** Courtesy American Gas Association. **Fig. 17.1** Photograph by James Scherer. **Fig. 17.4** Courtesy Beckman Instruments, Inc. **Fig. 19.1** Photographs by James Scherer. **Fig. 19.2** Photographs by James Scherer. **Fig. 22.1** Courtesy Alan Davison, M.I.T. **Fig. 22.6** Courtesy of Bicron Corporation. **Fig. 22.12** Courtesy of Princeton University Plasma Physics Laboratory. **Fig. 23.5** Totem Poles in Giants Hall, Caverns of Luray, Virginia. **Fig. 24.1** Industrial diamond crystals manufactured by General Electric Company. **Fig. 24.2** Photograph by James Scherer. **Fig. 24.5** Wards Natural Science, Inc., Rochester, N.Y. **Fig. 24.13** Photograph by James Scherer. **Fig. 26.8** Atlantic Richfield Company. **Fig. 27.3** From *Biochemical Concepts* by Robert W. McGilvery, Copyright © 1975 by W. B. Saunders Company. Reprinted by permission of Holt, Rinehart and Winston, CBS College Publishing. **Fig. 27.4** From R. E. Dickerson in H. Neurath (ed.), *The Proteins,* Vol. 2, © 1964, Academic Press. **Fig. 27.7** Part a from E. Borek, *Trends in Biochemical Sciences* 2, 4 (1977). Part b from Nazar, Sitz, and Brusch (1975), *Journal of Biological Chemistry* 250:8591. Copyright The American Society of Biological Chemists, Inc. **Fig. 27.8** Photo by The Ealing Corporation. **Fig. 27.9** Photograph by Carolina Biological Supply Company. **Fig. 27.10** From *Experimental Biochemistry,* 2nd ed. by John M. Clark, Jr., and Robert L. Switzer. Copyright © 1977 by W. H. Freeman and Company. All rights reserved.

Color Section
Plate 1, *top,* Spectra-Physics; *bottom,* Bausch & Lomb. **Plate 2.** Wabash Instrument Corporation. **Plate 3,** *top,* James Hubbard, Commonwealth Photographers, Inc.; *bottom,* © Yoav Levy/Phototake. **Plate 4.** Photographs by James Scherer. **Plate 5.** Photographs by James Scherer. **Plate 6.** From *A Field Guide to Rocks and Minerals* by Frederick H. Pough. Copyright 1953, © 1955, 1960, 1976 by Frederick H. Pough. Reprinted by permission of Houghton Mifflin Company. **Plate 7.** Photographs by James Scherer. **Plate 8.** Csaba L. Martonyi, The University of Michigan.

Cover photograph by James Scherer.
Cover: Laser beams grazing the surface of laboratory glassware. The red beam (632.8 nm) is from a helium-neon laser; the blue-green beam (488.0) is from an argon ion laser.

Values for atomic weights listed in the periodic table on the inside front cover of this book are from the IUPAC report "Atomic Weights of the Elements 1981," *Pure and Applied Chemistry,* Vol. 55, No. 7 (July 1983), pp. 1107–1108.

Printed in the U.S.A.
Library of Congress Catalog Card Number: 81-83272
ISBN: 0-395-31489-5

DEFGHIJKLM-RM-898765

Foreword

Chemistry is the science that concerns the study of matter. Learning about chemistry is arguably necessary because the way we manipulate matter now has an important, practical impact in areas of medicine, energy, food production, and synthetic materials. Through our collective experience as instructors in the classroom and practicing chemists, Darrell Ebbing and I have gained valuable insight into aspects of chemistry that can both help and stimulate the learner. *General Chemistry* introduces today's students to the foundations and frontiers of chemical science, thus providing them with a path to a thorough understanding of the principles of chemistry. The book's presentation of these basic principles is enhanced by numerous examples, marginal notes, and problems that weave into the text the excitement and challenge of chemistry as it is practiced today. From research topics under study by the leading chemists in the world—such as catalysts for nitrogen fixation—to large-scale industrial processes involving the application of chemical principles—such as the electrochemical generation of chlorine—*General Chemistry* presents students with the chemistry worth knowing and encourages them to learn and to apply the basic concepts of the science.

As Consulting Editor for this project during the last four years, I have read and reread the successive drafts of the manuscript. My direct contributions to the development of the manuscript have been in the form of making suggestions for improvement, implementing many examples of modern chemical research, and adding new applications of basic principles. Further, I have gone over the numerous reviews of each chapter submitted by academic reviewers across the country and helped formulate strategies for bringing *General Chemistry* closer to the ideal text we all seek. Darrell Ebbing has created an excellent textbook. I and other members of the publishing team have worked to complement his achievement by heightening interest, providing consistency and balance throughout, and responding to the needs of today's students and instructors of chemistry. We have been uncompromising with respect to our goal of presenting the principles of chemistry so that they can be clearly understood and successfully applied to solving problems of practical and intellectual interest.

Chemistry is a contemporary science. This text incorporates recent examples of compounds, processes, and devices that illustrate basic principles of chemistry. From the *cisplatin* anticancer drug of the opening in Chapter 1 to the helium-neon and argon ion lasers used in our cover photograph or to the photoelectron spectroscopy described in Chapter 6, we have paid attention to identifying modern applications of basic principles. Balancing these modern applications on the timeline of chemistry is our coverage of key historical developments that reveal the basis for our current understanding of chemistry.

Chemistry is an experimental science. This text employs many examples of the use of qualitative observations and quantitative treatment of measurements. In Chapter 16 where equilibrium is considered, the pedagogical treatment includes the manipulation of real experimental data in coming to an appreciation of the meaning of equilibrium and quantitative relationships for chemical equilibrium systems. *General Chemistry* makes a point of developing principles from the experiments that generate them—both contemporary and historical firsts are used to describe the interplay between theory and experiment.

Chemistry is a fundamental part of the science curriculum. *General Chemistry* is written with the expectation that it will be *used* as a textbook. A major strength of this book is in the problem-solving approach. Worked-out example problems demonstrating specific operational skills are immediately followed by practice exercises and are further reinforced by problems at the end of the chapter. These exercises and problems contain many examples of descriptive chemistry and modern applications of chemistry. This text makes learning chemistry scientific—not "cookbook."

Chemistry is a timely science. In the 1980s the application of chemical principles has had technological consequences—for better and for worse. Part of the function of today's scientists is to establish scientifically and technologically sound solutions to major problems. The student of chemistry will be better prepared to discriminate among technological alternatives. To this end, the text relates technological issues of society that involve chemistry and emphasizes the relationships of chemistry with other sciences, with engineering, and with medicine.

Chemistry is a challenging science. *General Chemistry* provides a framework the student can use to learn chemistry and to prove it by solving problems. The spirit of the book should stimulate students to formulate original questions, answerable through research, where the answers may have intellectual and technological value. Not all of chemistry is known. In fact, the surprising thing is how fast the student using this book can generate fundamental, unanswered questions that stump the best workers in the field.

Mark S. Wrighton

Preface

As we proceed toward the twenty-first century, we see advancements in science and technology that outdistance the predictions of the most imaginative forecasters. Scientists delve into the molecular machinery of the biological cell and examine bits of material from the planets of the solar system. In these endeavors chemistry is fundamental. The challenge for the instructors of introductory chemistry is to capture the excitement of these discoveries while giving students a solid understanding of the basic principles and facts. The challenge for the students is to be receptive to a new way of thinking, which will allow them to be caught up in the excitement of discovery. In developing *General Chemistry* over the past several years, our goal has been to write a text that would make the fulfillment of these challenges to instructor and student seem as burdenless as possible. Keeping this goal constantly in mind, we set several demanding objectives: (1) to be as clear and lucid in our explanation of principles as possible by always relating abstract concepts to specific things in the real world; (2) to present topics in an order that is logical, yet flexible, and that avoids confusion or repetition; (3) to offer an abundance of meaningful instructional aids, particularly with respect to problem-solving.

Thoughtful Presentation of Principles

Considerable time has been spent in fine tuning all conceptual discussions and in providing explicit definitions of the many important terms that are a fundamental part of the chemist's lexicon. However, readability goes beyond pure use of language and clarity of expression, extending into the use of real-world information, concrete applications, and provocative ideas that infuse theory with genuine interest and relevance.

Standard Organization

The development of the sequence of chapters as a whole and the topics within individual chapters has been a lengthy process. The many reviewers of the six drafts of the manuscript were instrumental in developing the order of presentation that finally resulted. In spite of the considerable agreement among reviewers about the sequencing of topics, it is clear that most instructors do want the option of rearranging the order of chapters. Flexibility has been built into the presentation to accommodate rearrangement. Throughout the extensive reviewing and class-testing of the manuscript, alternative sequencing of chapters has been tested and proved to be pedagogically feasible and effective. For example, in an early draft of the manuscript, Chapter 4 on gases appeared immediately before the chapter on liquids and solids. Its present position, however, was shown in the class testing of the manuscript to provide a constructive follow-up to Chapter 3 on stoichiometry by illustrating the evolution of theory from experiment and to expand the scope of laboratory work.

Not surprisingly, the first three chapters of *General Chemistry* introduce basic concepts such as measurement, atomic theory, stoichiometry, and molar concentration. Chapter 4 on gases builds on this material, providing a model of the historical development of chemistry as well as demonstrating the interrelationship of experiment and theory. Chapters 5 through 8 develop the principles of atomic and molecular structure. Because the concepts are abstract, particular attention is given to relating the discussion to the real world. Thus, special care was given to making clear the experimental basis of atomic structure in Chapter 5. Then in Chapter 6, the relationship of electron configurations and the periodicity of elements is capped with a brief discussion of the characteristics of the main groups of elements. In developing specific skills, such as writing Lewis electron-dot formulas (Chapter 7), definite steps are outlined. These steps are carried through to include molecules that do not follow the octet rule. Chapter 8 uses the basic bonding concepts of the previous chapter as the framework for the discussion of the concept of molecular geometry. Chapter 9, reactions in aqueous solutions, draws on basic bonding concepts but is a welcome change of direction. This chapter deals with acid-base, precipitation, and oxidation-reduction reactions, the principal types of reactions in general chemistry.

With Chapter 10 on thermochemistry, we again begin a series of chapters on basic principles. Chapter 11 discusses the properties of liquids and solids, relating these to structure, and Chapter 12 discusses solutions. Chapters 13 and 14 have twin goals: to introduce the descriptive chemistry of oxygen, nitrogen, and hydrogen and to provide a review of basic principles by applying them to the chemistry of these elements.

Chapter 15, on rates of reactions, introduces a block of seven chapters (Chapters 15 through 21) dealing with dynamics, or rate, and equilibrium. Experience has shown that students often have difficulty assimilating the pertinent concepts and calculation skills. Again, we pay close attention here to relating the concepts to the real world. And, since the calculation skills are so central to the topic, we outline definite steps to be followed in the solving of problems throughout this block of chapters.

Chapter 22 covers nuclear chemistry, and Chapters 23 through 27 describe the chemistry of particular elements. Chapters 23 and 24 consider the chemistry of elements by their group in the periodic table. Chapter 25 describes the chemistry of selected transition elements and then discusses coordination compounds. Chapter 26 introduces organic chemistry, and Chapter 27 looks at compounds of biological interest.

Deliberate and Thorough Explanation of Problem-Solving

An area that instructors repeatedly cite as a number-one concern is that of problem-solving. The approach we have used in this text has met with consistent approval from reviewers of the manuscript. In every chapter each important problem-solving skill is introduced as an *Example*, in which the student is led through the reasoning involved in working out a particular type of problem. The skill featured in each solved Example has been intentionally selected to represent a specific category of problems encountered frequently by students in general chemistry. Each Example is accompanied by an *Exercise*, a similar problem that will immediately test the student's understanding of the skill presented in the Example. (Some Exercises are freestanding, unaccompanied by an Example because the calculations required are explained thoroughly within the narrative text and are not complex enough to justify a new formal Example.) At the end of each Exercise is a list of the end-of-chapter problems that will allow the student to develop mastery of the specific problem-solving skill. In addition, the review section at the back of each chapter includes *Operational Skills*, a list of all the problem-solving skills presented in the chapter that states what is needed and what is to be solved for in each skill and includes references to the Examples that illustrate the skills.

The *Review Questions* at the end of the chapter are designed to test the student's understanding of concepts and theory. The end-of-chapter problems that test the student's command of problem-solving are divided into two groups: *Problems* and *Additional Problems*. In the first group, the problems are arranged in matched pairs and are keyed to a particular skill or topic by headings. In the second group, Additional Problems, the problems are not keyed to skills or topics in the chapter in any way. This "mixed bag" of problems can be used by students as a self-test and sometimes provides a cumulative problem-solving review by picking up on necessary skills presented in chapters that would be covered earlier in the course. Answers to all odd-numbered problems are given in the text. And, importantly, there are three graded levels of difficulty for all the end-of-chapter problems: unstarred problems are for drill and practice of basic skills, single-starred problems require the use of more than one basic skill and are more challenging, and double-starred problems are the most difficult and thought-provoking problems involving a combination of skills.

Quantitatively, the problem-solving program in *General Chemistry* is extensive with 262 Examples, 368 within-chapter Exercises, 1116 Problems,

416 Additional Problems, and 508 Review Questions. With over 2400 problems, the instructor has great flexibility in assigning problems, and the student has tremendous opportunity to get thorough grounding in theory and calculation.

Student Orientation

My twenty years of experience teaching chemistry has made me aware of the need for the instructor to stimulate, support, and reinforce students in their learning of chemistry. The solid program of instructional aids in *General Chemistry* should be valuable in creating a comfortable context for the study of chemistry.

Within the body of the chapter, instructional aids are numerous. Each chapter opens with an outline of the contents to indicate to instructors and students the scope of material covered in the chapter. The actual narrative begins with the presentation of a "theme problem" that provides a concrete, tangible application of the principles covered in the chapter. The theme establishes a unifying thread that is returned to periodically throughout the chapter to clarify the abstract concepts with a practical, real-world example. Following the short statement of the theme is the *Chapter Overview*, which informs students in one or two paragraphs about the topical coverage in the chapter as a whole and the interrelationship of its parts. *Key terms*, set in boldface type, are always accurately and concisely defined when they are first introduced. *Key concepts and equations*, set in blue, direct the students' attention to the foundations of problem-solving. The twelve *Asides* interspersed throughout the text spotlight real-world applications, some higher-level material, and historical background, thus providing additional dimensions to chemistry for the student. Finally, within the chapter, important cross-references to pertinent discussions that appear in other parts of the book are handled in the marginal notes.

End-of-chapter instructional aids help the student review the material presented in the chapter. The list of *Important Terms* includes section numbers following each term to indicate where in the chapter it is defined and discussed. The *Summary of Facts and Concepts* summarizes the theory presented in the chapter and the *Operational Skills*, as we have already mentioned, stand as the summary of problem-solving skills. Likewise, whereas the *Review Questions* test the student's comprehension of the theory and concepts presented in the chapter, *Problems* and *Additional Problems* test the student's ability to solve problems and perform the necessary calculations.

Flexible Treatment of Descriptive Chemistry

Coverage of descriptive chemistry is systematically handled in three different ways to allow instructors flexibility in the degree of emphasis they place on this important aspect of chemistry. (1) Descriptive chemistry is incorporated within the narrative where appropriate and relevant. (2) Interesting descriptive chemistry is provided in the marginal notes to add relevance to what students sometimes find to be abstract conceptual discussions. (3) Descriptive chemistry is treated at length in eight separate chapters, which the instructor can use to the extent desired.

Complete Instructional Package

This textbook is complemented by a complete package of supplements that will meet the needs of the instructors and students in introductory chemistry.

Joan Senyk, James Braun, Larry Krannich: *Study Guide*

John Lorimer, Christopher Willis, Arnold Loebel: *Laboratory Experiments* (with accompanying Instructor's Manual)

Leonard Soltzberg: *Computer-Assisted Blackboard*

Solutions Manual

Instructor's Manual and Test Bank

Acknowledgments

The development of this book would have been impossible without the unflagging work of many people. Reviewers have helped immeasurably in shaping the final manuscript. I thank them for giving their time and their ideas to this project:

John J. Alexander, University of Cincinnati

J. Marshall Baker, Auburn University

Charles J. Barcelona, Community College of Allegheny County, Allegheny Campus

Ralph Barnhard, University of Oregon

Ernest Becker, University of Massachusetts, Boston

James R. Braun, Clayton Junior College

Clarence H. Breedlove, Montgomery Community College

John A. Chandler, University of Massachusetts, Amherst

Stanley Cherim, Delaware County Community College

Ronald J. Clark, Florida State University

Charles O. Cunningham, Long Beach City College

Geoffrey Davies, Northeastern University

Robert Desiderato, North Texas State University

Klaas Eriks, Boston University

Milton E. Fuller, California State University

Marian Hallada, University of Michigan, Ann Arbor

Henry Heikkinen, University of Maryland

F. Clyde Hentz, North Carolina State University

Robert F. Hollins, Chicago State University

Benjamin G. Hughes, Western Illinois University

Marc L. Kasner, Montclair State College

Robert C. Kerber, State University of New York at Stony Brook

Larry K. Krannich, University of Alabama in Birmingham

Norman Kulevsky, University of North Dakota

Arnold B. Loebel, Merritt College

Edward K. Mellon, Florida State University

Edward E. Mercer, University of South Carolina

Merle Pattengill, University of Kentucky

Theodore P. Perros, The George Washington University

Peter Politzer, University of New Orleans

Lucy T. Pryde, Southwestern College

James M. Purser, Millsaps College

William Reiff, Northeastern University

Patricia Samuel, Boston University

Jerry L. Sarquis, Miami University

Dan D. Scott, Middle Tennessee State University

Edward G. Senkbeil, Salisbury State College

Joan Senyk, University of Maryland

Bassam Z. Shakhashiri, University of Wisconsin, Madison

Donald Sink, Appalachian State University

Leonard Soltzberg, Simmons College

Kenneth Spitzer, Washington State University

Conrad Stanitski, Randolph-Macon College

Jack H. Stocker, University of New Orleans, Lakefront

Tamar Y. Susskind, Oakland Community College

Dwight C. Tardy, University of Iowa

Donald D. Titus, Temple University

Charles Wilkie, Marquette University

C. J. Willis, University of Western Ontario

George Woodbury, University of Montana

Special thanks are due Jim Braun, Milton Fuller, Henry Heikkinen, Marc Kasner, Arnold Loebel, Jimmy Purser, and Ken Spitzer who were instrumental in developing the features of the book and ensured consistency in its direction by giving detailed and thoughtful criticism of various drafts of the manuscript.

Mark Wrighton's role in determining the form and substance of this book was indeed extensive. As a result of Mark's participation as in-depth reviewer and consultant, many passages in the book will bear his stamp—even with my penchant for rewriting. My many thanks to Mark for his creativity, hard work, and support.

A number of people contributed in special ways. Joan Senyk class-tested a version of the principles chapters. Her incisive comments, as well as those of her students, significantly improved the presentation. In addition, she contributed to the writing of the glossary and provided answers to all of the Exercises. Ruth M. Doherty, with the able assistance of Henry Heikkinen of the University of Maryland, checked and worked the end-of-chapter Problems. Valerie Fassbender of Boston University double-checked the solutions and answers for all the Exercises and Problems. Al Boyd (University of Maryland) and Janice O. Tsokos Kuhn provided drafts of several descriptive chemistry chapters. Jed Harrison of M.I.T. contributed his time in setting up and supervising the cover photograph and the excellent demonstration photographs that appear throughout the book. My colleagues at Wayne State University were helpful in many ways, sometimes unknowingly, when their ideas became mine. Carl Johnson, Gene Reck, and George Schenk gave detailed reviews of chapters. Joe Oravek helped me with questions concerning lecture demonstrations. The manuscript was launched with the precision typing of Gloria Novak. She and her husband, Ed, also extended the hospitality of their mountain home in Asheville, North Carolina, to my wife and me at a time when the respite was much needed. I very much appreciate the contributions of all these people.

Throughout the lengthy process of writing, my wife, Jean, and my children, Julie, Linda, and Russell, have borne the brunt of my preoccupation and have still retained their good humor while encouraging me to finish. They all endured reading many drafts. Jean, an artist by profession, lent her skillful hand to rendering my word descriptions as drawings that eventually became figures in the book. Words are not enough to express my debt to them.

Darrell D. Ebbing

Contents

1. Chemistry and Measurement 1

Chapter Overview 2

Chemistry as a Quantitative Science 2
1.1 Development of Modern Chemistry 2
1.2 Experiment and Theory 5

Units of Measurement 7
1.3 Measurement and Significant Figures 7 Number of Significant Figures/ Significant Figures in Calculations/ Rounding
1.4 SI Units 11 Prefixes and Base Units/ Length, Mass, and Time/ Temperature
1.5 Working with Units: Dimensional Analysis 14
1.6 Derived Units 16 Volume/ Density/ Pressure/ Energy

A Checklist for Review 22
Important Terms/ Summary of Facts and Concepts/ Operational Skills

Review Questions 23

Problems 23

Additional Problems 26

2. Atomic Theory: Pure Substances and Mixtures 27

Chapter Overview 28

2.1 Atoms, Molecules, and Ions 28 Atoms/ Molecules and Molecular Substances/ Ions and Ionic Substances/ A Word on Naming Substances/ Chemical Reactions

2.2 Balancing Simple Chemical Equations 35
2.3 Classifications of Matter 37 Chemical Constitution —Element, Compound, or Mixture?/ Physical State—Solid, Liquid, or Gas?
2.4 Separation of Mixtures 41 Filtration/ Distillation/ Chromatography

A Checklist for Review 46
Important Terms/ Summary of Facts and Concepts/ Operational Skills

Review Questions 46

Problems 47

Additional Problems 49

3. Calculations with Chemical Formulas and Equations 50

Chapter Overview 51

Mass and Moles of Substance 51

3.1 Atomic Weights 54 *Aside: Atomic Weights and the Law of Combining Volumes*
3.2 Formula Weights 55
3.3 The Mole Concept 55 Definition of Mole/ Mole Calculations/ *Aside: Atomic Theory and Some Laws of Chemistry*

Determining Chemical Formulas 60

3.4 Mass Percentages from the Formula 60
3.5 Elemental Analysis: Percentages of Carbon, Hydrogen, and Oxygen 61

3.6 Determining Molecular Formulas 63   Empirical Formula from Elemental Composition/ Molecular Formula from Empirical Formula

Stoichiometry: Mole-Mass Relations in Chemical Reactions 66
3.7 Molar Interpretation of a Chemical Equation 66
3.8 Stoichiometry of a Chemical Reaction 67
3.9 Limiting Reactant; Theoretical and Percentage Yields 69

Calculations Involving Solutions 71
3.10 Molar Concentration 72
3.11 Diluting Solutions 74
3.12 Stoichiometry of Solution Reactions 76

A Checklist for Review 78   Important Terms/ Summary of Facts and Concepts/ Operational Skills

Review Questions 79

Problems 80

Additional Problems 84

4. The Gaseous State 88

Chapter Overview 89

Gas Laws 89
4.1 Measurement of Gas Pressure 89
4.2 Boyle's Law 91
4.3 Charles's Law 95   Variation of Volume with Absolute Temperature/ Charles's Law Calculations
4.4 Avogadro's Law 99
4.5 The Ideal Gas Law 101   Calculations Using the Ideal Gas Law/ Stoichiometry Problems with Gas Volumes/ Gas Density; Molecular-Weight Determination
4.6 Gas Mixtures; Law of Partial Pressures 107 Mole Fractions/ Collecting Gases over Water

Kinetic-Molecular Theory 111
4.7 Kinetic Theory of an Ideal Gas 111   Postulates of Kinetic Theory/ The Ideal Gas Law from Kinetic Theory
4.8 Molecular Speeds 115
4.9 Diffusion and Effusion 116
4.10 Real Gases 118

A Checklist for Review 121
Important Terms/ Summary of Facts and Concepts/ Operational Skills

Review Questions 122

Problems 122

Additional Problems 126

5. Atomic Structure 128

Chapter Overview 129

Basic Structure of Atoms 129
5.1 Discovery of the Electron 130   Cathode Rays/ Thomson's *m/e* Experiment/ Millikan's Oil Drop Experiment
5.2 The Nuclear Model of the Atom 133   Discovery of Radioactivity; Alpha Rays/ Alpha-Particle Scattering
5.3 Nuclear Structure 135
5.4 Mass Spectrometry and Atomic Weights 137

Electronic Structure of Atoms 139
5.5 The Wave Nature of Light 139
5.6 Quantum Effects and Photons 141   Planck's Quantization of Energy/ Photoelectric Effect/ *Aside: Photon Momentum*
5.7 The Bohr Theory of the Hydrogen Atom 144 Atomic Line Spectra/ Bohr's Postulates
5.8 Quantum Mechanics 149   de Broglie Waves/ Wave Functions
5.9 Quantum Numbers and Atomic Orbitals 153 Quantum Numbers/ Atomic Orbital Shapes

A Checklist for Review 158   Important Terms/ Summary of Facts and Concepts/ Operational Skills

Review Questions 159

Problems 159

Additional Problems 162

6. Electron Configurations and Periodicity 163

Chapter Overview 164

Electronic Structure of Atoms 164
6.1 Electron Spin and the Pauli Exclusion Principle 165   Electron Configurations and Orbital Diagrams/ Pauli Exclusion Principle

6.2 Building-Up Principle (Aufbau Principle) 167
Aside: X rays, Atomic Numbers, and Orbital Structure
6.3 Hund's Rule; Paramagnetism 173   Hund's Rule/
Paramagnetism

The Periodic Table 175
6.4 Periodic Classification of the Elements 176
Predictions from the Periodic Table/ Arrangement of the
Elements by Atomic Number/ Relationship to Electron
Configurations
6.5 Some Periodic Properties 180   Atomic Radius/
Ionization Energy/ Electron Affinity
**6.6 A Brief Description of the Main-Group
Elements** 187   Hydrogen: Configuration $1s^1$/ Group IA
Elements (Alkali Metals): Valence-Shell Configuration
ns^1/ Group IIA Elements (Alkaline Earth Metals):
Valence-Shell Configuration ns^2/ Group IIIA Elements:
Valence-Shell Configuration ns^2np^1/ Group IVA
Elements: Valence-Shell Configuration ns^2np^2/ Group
VA Elements: Valence-Shell Configuration ns^2np^3/ Group
VIA Elements: Valence-Shell Configuration ns^2np^4/
Group VIIA Elements (Halogens): Valence-Shell
Configuration ns^2np^5/ Group VIIIA Elements (Noble
Gases): Valence-Shell Configuration ns^2np^6

A Checklist for Review 191
Important Terms/ Summary of Facts and Concepts/
Operational Skills

Review Questions 192

Problems 193

Additional Problems 194

7. Ionic and Covalent Bonding 195

Chapter Overview 196

Ionic Bond 196
7.1 Describing Ionic Bonds 197   Lewis Electron-Dot
Symbols/ Energy Involved in Ionic Bonding
7.2 Some Common Ions 199   Monatomic Ions of the
Main-Group Elements/ Transition-Metal Ions/ Polyatomic
Ions/ Formulas of Ionic Compounds
7.3 Ionic Radii 204

Covalent Bond 206
7.4 Describing Covalent Bonds 207   Lewis Formulas/
Coordinate Covalent Bond/ Octet Rule/ Multiple Bonds
7.5 Polar Covalent Bond; Electronegativity 209
7.6 Writing Lewis Electron-Dot Formulas 212
Skeleton Structure of a Molecule/ Steps in Writing Lewis
Formulas

7.7 Exceptions to the Octet Rule 215
7.8 Delocalized Bonding; Resonance 218
7.9 Bond Length and Bond Order 220

Chemical Nomenclature 222
7.10 Oxidation Numbers 222
7.11 Naming Simple Compounds 224   Binary
Compounds/ Acids/ Ionic Substances

A Checklist for Review 228
Important Terms/ Summary of Facts and Concepts/
Operational Skills

Review Questions 229

Problems 230

Additional Problems 233

**8. Molecular Geometry and
Chemical Bonding Theory** 234

Chapter Overview 235

Molecular Geometry and Directional Bonding 236
**8.1 The Valence-Shell Electron-Pair Repulsion
(VSEPR) Model** 236   Two Electron Pairs (Linear
Arrangement)/ Three Electron Pairs (Trigonal Planar
Arrangement)/ Four Electron Pairs (Tetrahedral
Arrangement)/ Five Electron Pairs (Trigonal Bipyramidal
Arrangement)/ Six Electron Pairs (Octadedral
Arrangement)/ Summary of the VSEPR Model
8.2 Dipole Moment and Molecular Geometry 243
8.3 Valence Bond Theory 246   Basic Theory/ Hybrid
Orbitals
8.4 Description of Multiple Bonding 253

Molecular Orbital Theory 257
8.5 Principles of Molecular Orbital Theory 258
Bonding and Antibonding Orbitals/ Bond Order/ Factors
Determining Orbital Interaction
**8.6 Electron Configurations of Diatomic
Molecules** 261
8.7 Molecular Orbitals and Delocalized Bonding 264

A Checklist for Review 265
Important Terms/ Summary of Facts and Concepts/
Operational Skills

Review Questions 266

Problems 267

Additional Problems 268

9. Reactions in Aqueous Solutions 270

Chapter Overview 271

Ions in Solution; Ionic Equations 271

9.1 Electrolytes 271   A Note About the Hydrogen Ion/ Introduction to Chemical Equilibrium/ Strong and Weak Electrolytes
9.2 Ionic Equations 274
9.3 Types of Reactions 276

Metathesis Reactions 277

9.4 Solubility and Precipitation 277   Solubility Rules/ Precipitation Reactions
9.5 Reactions of Acids, Bases, and Salts 282
Neutralization/ Reactions of Salts; Formation of a Gas
9.6 Preparation of Acids, Bases, and Salts 289
Preparation of Acids/ Preparation of Bases/ Preparation of Salts

Oxidation-Reduction Reactions 292

9.7 Introduction to Oxidation-Reduction Reactions 292   Terminology/ Understanding Oxidation-Reduction Equations
9.8 Balancing Oxidation-Reduction Equations 297
Half-Reaction Method/ Oxidation-Number Method
9.9 Equivalents and Normality 303

A Checklist for Review 305
Important Terms/ Summary of Facts and Concepts/ Operational Skills

Review Questions 307

Problems 307

Additional Problems 310

10. Thermochemistry 311

Chapter Overview 312

Fundamentals of Thermochemistry 312

10.1 Heat of Reaction; Enthalpy 312   Heat of Reaction/ Enthalpy/ *Aside: Relating Enthalpy to Energy*
10.2 Calorimetry 319   Heat Capacity and Specific Heat/ Measurement of Reaction Heats
10.3 Stoichiometry of Reaction Heats 323
10.4 Hess's Law 325
10.5 Enthalpies of Formation 328

Applications of Thermochemistry 332
10.6 Bond Energy 332
10.7 Lattice Energies 336

A Checklist for Review 338
Important Terms/ Summary of Facts and Concepts/ Operational Skills

Review Questions 339

Problems 339

Additional Problems 343

11. Liquids and Solids; Changes of State 345

Chapter Overview 346

Liquids and Solids 346

11.1 Intermolecular Forces 347   London Forces/ Dipole–Dipole Forces/ Hydrogen Bonding
11.2 Properties of Liquids 351   Vapor Pressure/ Surface Tension/ Viscosity
11.3 Types of Solids 357
11.4 Crystalline and Amorphous Solids 358
Characteristics of Crystalline and Amorphous Solids/ The Crystalline Lattice/ Cubic Unit Cells/ Crystal Defects
11.5 The Structure of Some Crystalline Solids 363
Molecular Solids; Closest Packing/ Metallic Solids/ Ionic Solids/ Covalent Network Solids
11.6 X-ray Crystal Structure Determination 368

Changes of State 371

11.7 Phase Changes 371
11.8 Boiling Point and Melting Point 373   Boiling Point/ Melting Point/ Enthalpy Changes
11.9 Relating Physical Properties to Structure 376
Melting Point and Structure/ Boiling Point and Structure/ Hardness and Structure/ Electrical Conductivity and Structure
11.10 Phase Diagrams 380   Melting-Point Curve/ Vapor-Pressure Curve for the Liquid/ Vapor-Pressure Curve for the Solid/ Critical Temperature and Pressure

A Checklist for Review 383
Important Terms/ Summary of Facts and Concepts/ Operational Skills

Review Questions 384

Problems 384

Additional Problems 387

12. Solutions 389

Chapter Overview 390

Solubility; Colloid Formation 390

12.1 Types of Solutions 391   Gaseous Solutions/ Liquid Solutions/ Solid Solutions
12.2 The Solution Process 392   Factors Determining Solubility/ Molecular Solutions/ Ionic Solutions/ *Aside: Hemoglobin Solubility and Sickle Cell Anemia*
12.3 The Effects of Temperature and Pressure on Solubility 396   Temperature Change/ Pressure Change/ Henry's Law
12.4 Colloids 399   Tyndall Effect/ Types of Colloids/ Hydrophilic and Hydrophobic Colloids/ Coagulation/ Association Colloids

Colligative Properties 404

12.5 Ways of Expressing Concentration 404   Mass Percentage of Solute/ Molality/ Mole Fraction/ Conversion of Concentration Units
12.6 Vapor Pressure of a Solution 409
12.7 Boiling-Point Elevation and Freezing-Point Depression 411
12.8 Osmosis 415
12.9 Ionic Solutions 418

A Checklist for Review 419
Important Terms/ Summary of Facts and Concepts/ Operational Skills

Review Questions 420

Problems 421

Additional Problems 423

13. The Atmosphere: Oxygen, Nitrogen, and the Noble Gases 425

Chapter Overview 426

The Earth's Atmosphere 426

13.1 The Present Atmosphere 426   Regions of the Atmosphere/ Heating of the Thermosphere and the Stratosphere
13.2 Evolution of the Atmosphere 432   Lack of Noble Gases in the Atmosphere/ Formation of a Primitive Atmosphere/ Formation of Molecular Oxygen

Oxygen 436

13.3 Molecular Oxygen 437   Preparation of Oxygen in Small Quantities/ Production of Oxygen in Commercial Quantities/ Structure and Physical Properties of Oxygen/ Uses of Oxygen
13.4 Reactions of Oxygen 441   Reactions with Metals/ Reactions with Nonmetals/ Reactions with Compounds/ Basic and Acidic Oxides
13.5 Ozone 444   Properties of Ozone/ *Aside: Stratospheric Ozone Depletion*

Nitrogen 447

13.6 Molecular Nitrogen 447   Uses of Nitrogen/ Nitrogen Fixation
13.7 Compounds of Nitrogen 450   Ammonia and the Nitrides (Oxidation State −3)/ Hydrazine (Oxidation State −2)/ Hydroxylamine (Oxidation State −1)/ Nitrous Oxide (Oxidation State +1)/ Nitric Oxide (Oxidation State +2)/ Nitrogen Dioxide (Oxidation State +4)/ Dinitrogen Trioxide and Nitrous Acid (Oxidation State +3)/ Dinitrogen Pentoxide and Nitric Acid (Oxidation State +5)

Noble Gases and Other Constituents of Air 455

13.8 The Noble Gases 455   Discovery of the Noble Gases/ Compounds of the Noble Gases/ Preparation and Uses of the Noble Gases
13.9 Trace Constituents of Air 458   *Aside: The Earth's Thermal Moderators*

A Checklist for Review 461
Important Terms/ Summary of Facts and Concepts/ Operational Skills

Review Questions 462

Problems 463

Additional Problems 465

14. Hydrogen and Its Compounds with Oxygen: Water and Hydrogen Peroxide 466

Chapter Overview 467

Hydrogen, Water, and Hydrogen Peroxide 467

14.1 Discovery of Hydrogen; Composition of Water 467
14.2 Properties of Hydrogen 469   Physical Properties/ Preparation/ Hydrides/ Uses

14.3 Properties of Water 475   Hydrogen Bonding and the Physical Properties of Water/ Chemical Properties of Water
14.4 Hydrates and Hydrated Ions 477
14.5 Hydrogen Peroxide 478

Natural Waters 481
14.6 Physical Properties of Water and the Environment 481   Effect of Large Heat Capacity of Water/ Effect of Relative Densities of Ice and Water
14.7 Water Pollutants 482   Thermal Pollution/ Biological Oxygen Demand/ Inorganic Ions/ Other Pollutants
14.8 Water Treatment and Purification 484
14.9 Removing Ions from Water; Desalination 486

A Checklist for Review 488
Important Terms/ Summary of Facts and Concepts/ Operational Skills

Review Questions 489

Problems 489

Additional Problems 492

15. Rates of Reaction 493

Chapter Overview 494

Reaction Rates 494
15.1 Definition of Reaction Rate 495
15.2 The Experimental Determination of Rate 498
15.3 Dependence of Rate on Concentration 500
Reaction Order/ Determining the Rate Law
15.4 Change of Concentration with Time 504
Concentration–Time Equations/ Half-Life of a Reaction/ Graphical Plotting
15.5 Temperature and Rate; Collision and Transition-State Theories 511   Collision Theory/ Transition-State Theory/ Potential Energy Diagrams for Reactions
15.6 Arrhenius Equation 515

Reaction Mechanisms 517
15.7 Elementary Reactions 517   Molecularity/ Rate Equation for an Elementary Reaction
15.8 The Rate Law and the Mechanism 520   Rate-Determining Step/ Mechanisms with an Initial Fast Step
15.9 Catalysis 524

A Checklist for Review 528
Important Terms/ Summary of Facts and Concepts/ Operational Skills

Review Questions 529

Problems 530

Additional Problems 533

16. Chemical Equilibrium; Gaseous Reactions 535

Chapter Overview 536

Describing Chemical Equilibrium 537
16.1 Chemical Equilibrium—A Dynamic Equilibrium 537
16.2 The Equilibrium Constant 539   Definition of the Equilibrium Constant K_c/ Obtaining Equilibrium Constants for Reactions/ The Equilibrium Constant K_p
16.3 Heterogeneous Equilibria 544

Using an Equilibrium Constant 546
16.4 Qualitatively Interpreting an Equilibrium Constant 546
16.5 Predicting the Direction of Reaction 548
16.6 Calculating Equilibrium Concentrations 549

Changing the Reaction Conditions and the Application of LeChatelier's Principle 552
16.7 Adding a Catalyst 552
16.8 Removing or Adding Reactants or Products 555
16.9 Changing the Pressure and Temperature 558
Effect of Pressure Change/ Effect of Temperature Change/ Choosing the Optimum Conditions for Reaction

A Checklist for Review 562
Important Terms/ Summary of Facts and Concepts/ Operational Skills

Review Questions 563

Problems 564

Additional Problems 567

17. Acid–Base Concepts 569

Chapter Overview 570

17.1 Arrhenius Concept of Acids and Bases 570
17.2 Self-Ionization of Water 573
17.3 The pH of a Solution 575
17.4 Brønsted–Lowry Concept of Acids and Bases 578

Contents

17.5 Relative Strengths of Acids and Bases 581
17.6 Molecular Structure and Acid Strength 583
17.7 Acid–Base Properties of Salt Solutions 585
17.8 Lewis Concept of Acids and Bases 587

A Checklist for Review 589
Important Terms/ Summary of Facts and Concepts/ Operational Skills

Review Questions 590

Problems 590

Additional Problems 592

18. Acid–Base Equilibria 593

Chapter Overview 594

Solutions of a Weak Acid or Base or Salt 594
18.1 Acid Ionization Equilibria 594 Experimental Determination of K_a/ Calculations from K_a
18.2 Polyprotic Acids 601
18.3 Base Ionization Equilibria 603
18.4 Hydrolysis 605

Solutions of a Weak Acid or Base with Another Solute 608
18.5 Common-Ion Effect 609
18.6 Buffers 611
18.7 Acid–Base Titration Curves 615 Titration of a Strong Acid by a Strong Base/ Titration of a Weak Acid by a Strong Base/ Titration of a Weak Base by a Strong Acid

A Checklist for Review 619
Important Terms/ Summary of Facts and Concepts/ Operational Skills

Review Questions 620

Problems 621

Additional Problems 623

19. Solubility and Complex-Ion Equilibria 625

Chapter Overview 626

Solubility Equilibria 626
19.1 The Solubility Product Constant 626
19.2 Solubility and the Common-Ion Effect 629

19.3 Precipitation Calculations 633 Criterion for Precipitation/ Completeness of Precipitation/ Fractional Precipitation
19.4 Effect of pH on Solubility 638 Qualitative Effect of pH/ Separation of Metal Ions by Sulfide Precipitation

Complex-Ion Equilibria 642
19.5 Complex-Ion Formation 642 *Aside: Stepwise Formation Constants*
19.6 Complex Ions and Solubility 646

An Application of Solubility Equilibria 648
19.7 Qualitative Analysis of Metal Ions 648

A Checklist for Review 651
Important Terms/ Summary of Facts and Concepts/ Operational Skills

Review Questions 652

Problems 652

Additional Problems 654

20. Thermodynamics and Equilibrium 655

Chapter Overview 656

20.1 Enthalpy and Heat of Reaction 656

Spontaneous Processes and Entropy 657
20.2 Entropy and the Second Law 657 Entropy/ Second Law of Thermodynamics
20.3 Standard Entropies 660 Third Law of Thermodynamics/ Entropy of Reaction

Free Energy and Equilibrium 664
20.4 Free Energy and Spontaneity 664 Standard Free-Energy Change/ Standard Free Energies of Formation/ $\Delta G°$ as a Criterion for Spontaneity
20.5 Interpretation of Free Energy 668 Maximum Work/ Coupling of Reactions/ Free-Energy Change During Reaction
20.6 Calculation of Equilibrium Constants 671
20.7 Change of Free Energy with Temperature 673 Spontaneity and Temperature Change/ Calculation of $\Delta G°$ at Various Temperatures

A Checklist for Review 676
Important Terms/ Summary of Facts and Concepts/ Operational Skills

Review Questions 677

Problems 678

Additional Problems 680

21. *Electrochemistry* 682

Chapter Overview 683

21.1 Electrochemical Cells 683

Voltaic Cells 686

21.2 Some Commercial Voltaic Cells 687
21.3 Notation for a Voltaic Cell 691
21.4 Electromotive Force 693
21.5 Electrode Potentials 695   Cell emf/ Strength of an Oxidizing or Reducing Agent
21.6 Equilibrium Constants from emf's 698
21.7 Dependence of emf on Concentration 700
Nernst Equation/ Electrode Potentials for Nonstandard Conditions/ Determination of pH

Electrolysis 704

21.8 Aqueous Electrolysis 705
21.9 Stoichiometry of Electrolysis 708

A Checklist for Review 709
Important Terms/ Summary of Facts and Concepts/ Operational Skills

Review Questions 710

Problems 711

Additional Problems 714

22. *Nuclear Chemistry* 716

Chapter Overview 717

Radioactivity and Nuclear Bombardment Reactions 717

22.1 Radioactivity 718   Nuclear Equations/ Nuclear Stability/ Types of Radioactive Decay/ Radioactive Decay Series
22.2 Nuclear Bombardment Reactions 726
Transmutation/ Transuranium Elements
22.3 Radiations and Matter: Detection and Biological Effects 730   Radiation Counters/ Biological Effects and Radiation Dosage
22.4 Rate of Radioactive Decay   Rate of Radioactive Decay and Half-Life/ Radioactive Dating
22.5 Applications of Radioactive Isotopes 740
Chemical Analysis/ Medical Therapy and Diagnosis

Energy of Nuclear Reactions 744

22.6 Mass–Energy Calculations 744   Mass-Energy Equivalence/ Nuclear Binding Energy

22.7 Nuclear Fission and Nuclear Fusion 748
Nuclear Fission; Nuclear Reactors/ Breeder Reactor/ Nuclear Fusion

A Checklist for Review 753
Important Terms/ Summary of Facts and Concepts/ Operational Skills

Review Questions 755

Problems 756

Additional Problems 758

23. *The Main Group Elements: Groups IA to IIIA* 759

Chapter Overview 760

23.1 General Observations About the Main-Group Elements 760   Metallic–Nonmetallic Character/ Oxidation States/ Differences in Behavior of the Second-Row Elements
23.2 Group IA: The Alkali Metals 767   Properties of the Elements/ Preparation of the Elements/ Uses of the Elements and Their Compounds
23.3 Group IIA: The Alkaline Earth Metals 773
Properties of the Elements/ Preparation of the Elements/ Uses of the Elements and Their Compounds
23.4 Group IIIA: Boron and the Aluminum Family Metals 781   Properties of the Elements/ Preparation of the Elements/ Uses of the Elements and Their Compounds
23.5 Metallurgy 788   Preliminary Treatment/ Reduction/ Refining of a Metal

A Checklist for Review 791
Important Terms/ Summary of Facts and Concepts/ Operational Skills

Review Questions 792

Problems 794

Additional Problems 796

24. *The Main-Group Elements: Groups IVA to VIIA* 798

Chapter Overview 799

24.1 Group IVA: The Carbon Family 799   Properties of the Elements/ Preparation and Uses of the Elements/ *Aside: Zone Refining*/ Important Compounds

Contents

24.2 Group VA: The Phosphorus Family 812
Properties of the Elements/ Preparation and Uses of the Elements/ Important Compounds
24.3 Group VIA: The Sulfur Family 819   Properties of the Elements/ Preparation and Uses of the Elements/ Important Compounds
24.4 Group VIIA: The Halogens 826   Properties of the Elements/ Preparation and Uses of the Elements/ Important Compounds

A Checklist for Review 834
Important Terms/ Summary of Facts and Concepts/ Operational Skills

Review Questions 835

Problems 837

Additional Problems 838

25. The Transition Elements 841

Chapter Overview 842

Properties of the Transition Elements 842
25.1 Periodic Trends in the Transition Elements 843
Electron Configurations/ Melting Points, Boiling Points, and Hardness/ Atomic Radii/ Ionization Energies/ Oxidation States
25.2 The Chemistry of Selected Transition Metals 849   Chromium/ Iron/ Copper

Complex Ions and Coordination Compounds 856
25.3 Formation and Structure of Complexes 857
Basic Definitions/ Polydentate Ligands/ Discovery of Complexes; Formation of a Complex/ *Aside: On the Stability of Chelates*
25.4 Naming Coordination Compounds 862
25.5 Structure and Isomerism in Coordination Compounds 865   Structural Isomerism/ Stereoisomerism
25.6 Valence Bond Theory of Complexes 873
Octahedral Complexes/ Tetrahedral and Square Planar Complexes
25.7 Crystal Field Theory 878   Effect of an Octahedral Field on the *d* Orbitals/ High-Spin and Low-Spin Complexes/ Tetrahedral and Square Planar Complexes/ Visible Spectra of Transition-Metal Complexes/ *Aside: The Cooperative Release of Oxygen from Oxyhemoglobin*

A Checklist for Review 887
Important Terms/ Summary of Facts and Concepts/ Operational Skills

Review Questions 889

Problems 890

Additional Problems 892

26. Organic Chemistry 893

Chapter Overview 894

Hydrocarbons 894
26.1 Alkanes and Cycloalkanes 894   Methane, the Simplest Alkane/ The Alkane Series/ Nomenclature of Alkanes/ Cycloalkanes/ Sources of Alkanes and Cycloalkanes
26.2 Alkenes and Alkynes 901   Alkenes/ Alkynes
26.3 Aromatic Hydrocarbons 905   Derivatives of Benzene/ Sources and Uses of Aromatic Hydrocarbons
26.4 Reactions of Hydrocarbons 908   Oxidation/ Substitution Reactions of Alkanes/ Addition Reactions of Alkenes/ Substitution Reactions of Aromatic Hydrocarbons/ Petroleum Refining

Derivatives of Hydrocarbons 913
26.5 Organic Compounds Containing Oxygen 914
Alcohols and Ethers/ Aldehydes and Ketones/ Carboxylic Acids
26.6 Reactions of Oxygen-Containing Organic Compounds 920   Oxidation-Reduction Reactions/ Esterification and Saponification/ Polyesters
26.7 Organic Compounds Containing Nitrogen and Sulfur 925   Amines and Amides/ Thiols and Disulfides

A Checklist for Review 928
Important Terms/ Summary of Facts and Concepts/ Operational Skills

Review Questions 929

Problems 929

Additional Problems 934

27. Biochemistry 935

Chapter Overview 936

Introduction to Biological Systems 936
27.1 The Cell: Unit of Biological Structure 937
27.2 Energy and the Biological System 937

Biological Molecules 938
27.3 Biological Polymers 938

27.4 Proteins 940 Amino Acids/ Protein Primary
Structure/ Protein Conformations/ Enzymes
27.5 Carbohydrates 948 Monosaccharides/
Oligosaccharides and Polysaccharides
27.6 Nucleic Acids 953 Nucleotides/ Polynucleotides
and Their Conformations/ DNA and the Nature of the
Genetic Code/ RNA and the Transmission of the Genetic
Code/ Nucleotides and Metabolism
27.7 Lipids 961 Fats and Oils/ Biological Membranes

A Checklist for Review 964
Important Terms/ Summary of Facts and Concepts/
Operational Skills

Review Questions 966

Problems 967

Additional Problems 970

Appendix A Mathematical Skills A-1
**Appendix B Common Logarithms to Four
Places** A-10
**Appendix C Thermodynamic Quantities for
Substances at 25°C** A-12
**Answers to Exercises and Odd-Numbered
Problems** A-18
Glossary G-1
Index I-1

1. Chemistry and Measurement

Chemistry as a Quantitative Science

1.1 Development of Modern Chemistry
1.2 Experiment and Theory

Units of Measurement

1.3 Measurement and Significant Figures Number of Significant Figures/ Significant Figures in Calculations/ Rounding
1.4 SI Units Prefixes and Base Units/ Length, Mass, and Time/ Temperature
1.5 Working with Units: Dimensional Analysis
1.6 Derived Units Volume/ Density/ Pressure/ Energy

I n 1964 Barnett Rosenberg and his co-workers at Michigan State University were studying the effects of electricity on bacterial growth. They inserted platinum electrodes (electrical connections) into a live bacterial culture. Then they allowed an electrical current to pass through the culture. After 1 to 2 hours, they noted that cell division in the bacteria no longer occurred. The researchers were able to show that cell division was inhibited by a substance containing platinum, produced from the platinum electrodes by the electrical current. A substance such as this one, they thought, might be useful as an anticancer drug, because cancer involves runaway cell division. Later research confirmed their view, and in 1979 the Food and Drug Administration approved the marketing of *cisplatin*, a platinum-containing anticancer drug.

This story illustrates three significant reasons to study chemistry. First, chemistry has important practical applications. The development of life-saving drugs is one, and a complete list would touch upon most areas of modern technology.

Second, chemistry is an intellectual enterprise, a way of explaining our material world. When Rosenberg and his coworkers saw that cell division in their culture had ceased, they systematically looked for the chemical substance that caused it to cease. They sought a chemical explanation for the occurrence.

Finally, the concepts and thought processes you will encounter in the study of chemistry figure prominently in other fields. The above experiment began as a problem in biology; through the application of chemistry it led to an advance in medicine. Whatever your career plans, you will find that your knowledge of chemistry is a useful intellectual tool for making important decisions.

Chapter Overview

Chemistry, the science that studies materials, has a *quantitative aspect* that involves measurement and calculation. This quantitative aspect has been instrumental in the development of the science. Modern chemistry began with Lavoisier, who stressed the importance of measurements of mass (quantity of matter) in chemical research. Chemical research uses the scientific method, which depends on the interrelationship of *experiment* and *theory*. Because measurement and calculation are vital to chemistry, this chapter will concentrate on *units of measurement* and the calculation technique known as *dimensional analysis*.

Chemistry as a Quantitative Science

Chemistry is the science of the materials around us such as air, water, rocks, and plant and animal substances. Much of chemistry involves describing these materials and the changes they undergo. However, chemistry also has a quantitative side, which is concerned with measuring and calculating the characteristics of materials. This quantitative aspect has played, and continues to play, an important role in modern chemistry.

1.1 Development of Modern Chemistry

The origins of chemistry are ancient, and probably began with the use of natural materials for practical purposes. Modern chemistry emerged in the eighteenth century when the balance began to be used systematically as a

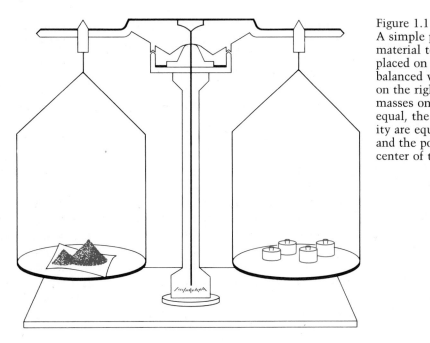

Figure 1.1
A simple pan balance. The material to be weighed is placed on the left pan and is balanced with weights placed on the right pan. When the masses on the two pans are equal, the forces due to gravity are equal and in balance, and the pointer rests at the center of the scale.

tool in research. Balances measure the **mass** or quantity of matter in a material (Figures 1.1 and 1.2). **Matter** is the general term for the material things around us and is often defined as whatever occupies space and can be perceived by our senses.

Antoine Lavoisier (1743–1794), a French chemist, insisted on the use of the balance in chemical research. His experiments demonstrated the **law of conservation of mass,** which states that mass remains constant during a chemical change (chemical reaction). ■ A flash bulb gives a convenient illustration of this law. The flash from such a bulb accompanies a chemical reaction triggered by an electrical current. But a flash bulb that weighs 11.2 grams before it is flashed still weighs 11.2 grams afterward; the mass (11.2 grams) remains constant.

In a series of experiments, Lavoisier showed that when a metal or any other substance burns, something in the air chemically combines with it. He called this component of air *oxygen*. For example, Lavoisier found that the liquid metal mercury was transformed in air to a red-colored substance. The substance had greater mass than the original mercury. This was due, he said, to the chemical combination of mercury with oxygen. Furthermore, the new substance (a mercury oxide) could be heated to recover the original mass of mercury (see Figure 1.3). Lavoisier's explanation of burning, or combustion, can be written

$$\text{mercury} + \text{oxygen} \longrightarrow \text{a mercury oxide}$$

The following exercise illustrates how the law of conservation of mass can be used to investigate chemical changes such as combustion.*

■ Heat or other forms of energy may be lost (or gained) during a chemical reaction. As Einstein's theory of relativity shows, mass and energy are equivalent: any substance having mass m has an energy mc^2, where c is the speed of light. Thus, when energy is lost as heat, mass is also lost. Changes of mass in chemical reactions (billionths of a gram) are too small to detect. Nuclear reactions, which are discussed in Chapter 22, however, involve enormous changes of energy and therefore detectable mass changes. In both kinds of reactions, any loss of mass that occurs in the reaction mixture is gained by the surrounding environment (from the energy released) so that the *total* mass always remains constant.

*An exercise unaccompanied by a worked-out example can be solved by using the ideas just discussed in the text. The problems mentioned after the exercise appear at the end of the chapter and reinforce the skills associated with the exercise.

Figure 1.2
Basic features of a modern single-pan balance. (a) The balance arm has a counterweight at one end, which just balances the weights and the pan on the other end. When an object is placed on the pan, weights equal to the mass of the object are removed by means of a mechanism (not shown) that is connected to dials at the front of the balance. (b) Front view of a single-pan balance. (c) Modern automatic electronic balance.

Exercise 1.1

When 2.53 grams of metallic mercury are heated in air, they are converted to 2.73 grams of red-colored residue. Assuming that the chemical change is due to the reaction of the metal with oxygen from the air and using the law of conservation of mass, determine the mass of the oxygen that has reacted. When the residue is strongly heated, it decomposes back to mercury, a silvery liquid. What is the mass of the oxygen that is lost when the residue is heated?

(See Problems 1.7 and 1.8.)

Measurements of mass before and after the combustion of various substances were necessary to convince chemists that Lavoisier's views were

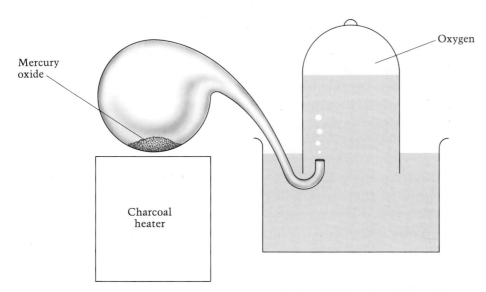

Figure 1.3
Apparatus used by Lavoisier in his experiments on the combustion of mercury.

correct. Thereafter, measurements of mass became indispensable to chemical research.

1.2 Experiment and Theory

Chemical research has two aspects, experiment and theory, which are closely interrelated. This interrelationship can be seen in the research described in the chapter opening. The scientists began with a tentative idea—that bacterial growth might be affected by an electrical current. During their experiment, they observed that the bacteria ceased dividing. They developed an explanation (a theory) for what had happened, and this prompted further experiments to test the explanation. In this section, we define the terms *experiment* and *theory* precisely, then show how they are related in scientific research.

An **experiment** is a well-defined, controlled procedure carried out so that others may duplicate the results. A scientist must record the exact situation under which the experiment was performed and identify any condition that might affect the results. He or she must report the results of the experiment as faithfully as possible, taking care that any interpretation of these results is clearly labeled. Others must be able to repeat the experiment from the published account in order to verify the results.■

After a series of experiments, one might be able to make a *generalization* about the results. A **law** (or **principle**) is a generalization about the behavior of nature that is sometimes limited in scope and that can be stated briefly, perhaps as a simple mathematical equation. An example is the law of conservation of mass. A **theory** is also a generalization about nature, usually of a wider scope. A law or theory often begins as a **hypothesis,** a new and untested explanation. The value to experimental science of any law or theory is that it helps us to organize our knowledge and to predict future events, which, in turn, allows us to plan new experiments.

■ Publishing the results of one's experiments is an important part of any scientific research. The fact that the published work will be studied carefully and criticized may serve to stimulate others to investigate new aspects of the research.

"I think you should be more explicit here in step two."

It is important to realize that a theory or law is always tentative. When the results of an experiment do not agree with it, the theory must give way. Sometimes a theory can be slightly modified to explain the results of a new experiment. In this way, a theory achieves wider application. However, if simple modification is not sufficient, the theory may have to be discarded; the history of science is littered with discarded theories. Scientists may replace the disproved theory with a new hypothesis, which is then tested by more experiments.

The creative process we have just described is the **scientific method.** Starting from a particular theoretical background, we set up experiments, recording the conditions and the results (see Figure 1.4). We then explain these results in terms of known theory, or we may have to devise a new explanation (a hypothesis). Experiments are set up to test this hypothesis. If it is confirmed, the hypothesis—now a new theory—joins the growing body of accepted understanding that makes up science. This theory suggests more experiments, and the cycle begins anew.

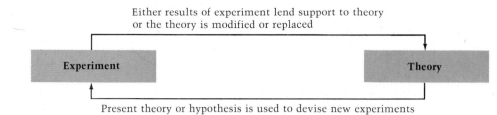

Units of Measurement

Cisplatin, the chemical substance featured in the chapter opening, was first prepared in 1845, over a hundred years before the discovery of its anti-cancer activity. It is a yellow, crystalline substance, known to chemists as *cis*-diamminedichloroplatinum(II).■ A like substance known as *trans*-diamminedichloroplatinum(II) was discovered the same year. Although these two substances are similar, we can easily distinguish between them by testing their solubilities in water, that is, by measuring the masses of the substances that will dissolve in a given quantity of water. The solubility of cisplatin is 0.252 grams in 100 grams of water at 25°C, whereas the solubility of the second substance is 0.037 grams in 100 grams of water.

■ The chemical name for cisplatin consists of parts written as one word: cis/di/ammine/di/chloro/platinum/(II). As we will discuss in Chapter 25, such names convey information about the structure of a substance.

This is only one illustration of the many uses of measurement in chemistry. In a modern chemical laboratory, complex measurements can be made with expensive instruments, yet many experiments begin with simple measurements of mass, volume, time, and so forth. In the next few sections, we will look at kinds and units of measurement.

1.3 Measurement and Significant Figures

Measurement is the comparison of a physical quantity to be measured with some fixed standard, called the **unit** of measurement. On a centimeter scale, the centimeter unit is the standard of comparison.■ A steel rod that measures 9.12 times the centimeter unit has a length of 9.12 centimeters. To record the measurement, we must be careful to give both the *measured number* (9.12) and the *unit* (centimeters).

■ 2.54 centimeters = 1 inch

In any series of measurements, the measured numbers are obtained to a certain degree of precision. **Precision** refers to the closeness of the set of values obtained from identical measurements of a quantity on the same instrument. **Accuracy** refers to the closeness of a single measurement to its true value.■ To illustrate the idea of precision, let us look at a simple measuring device, the centimeter ruler. In Figure 1.5, a steel rod has been placed next to a ruler subdivided into tenths of a centimeter. You can see that the rod measures just over 9.1 cm (cm = centimeter). With care, it is possible to estimate by eye to hundredths of a centimeter. Here we might give the measurement as 9.12 cm. Suppose we measure the length of this rod twice more. We find the values of these measurements to be 9.11 cm and 9.13 cm. Thus, we record the length of the rod as being somewhere between

■ Measurements that are of high precision are usually accurate. It is possible, however, to have a systematic error in a measurement. For example, suppose that in calibrating a ruler, the first centimeter is made too small by 0.1 centimeter. Then, although the measurements of length on this ruler are still precise to 0.01 centimeters, they are accurate to only 0.1 centimeter.

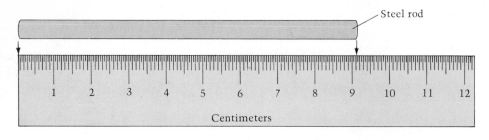

Figure 1.5
Precision of measurement with a centimeter ruler. The length of the rod is just over 9.1 centimeters. On successive measurements we estimate the length by eye at 9.12, 9.11, and 9.13. We record the length as between 9.11 cm and 9.13 cm.

9.11 cm and 9.13 cm. The spread of values indicates the precision with which a measurement can be made by this centimeter ruler.

To indicate the precision of a measured number, we write the number assuming that the last digit is uncertain and all other digits are not. In this case, it would be inappropriate to write 9.120 cm for the length of the rod, since this would say that the last digit (0) is uncertain but that the other digits (9.12) are not.

Number of Significant Figures

The phrase **number of significant figures** refers to the precision of a measured quantity, and equals the number of digits written, including the last one, even though its value is uncertain. Thus, there are three significant figures in 9.12 cm, whereas 9.123 cm has four. To count the number of significant figures in a given measured quantity, we observe the following rules.

1. All digits are significant except zeros at the beginning of the number and possibly terminal zeros (one or more zeros at the end of a number). Thus, 9.12 cm, 0.912 cm, and 0.00912 cm all contain three significant figures.
2. Terminal zeros ending at the right of the decimal point are significant. Each of the following has three significant figures: 9.00 cm, 9.10 cm, 90.0 cm.
3. Terminal zeros ending to the left of the decimal point may or may not be significant. When a measurement is given as 900 cm, we do not know whether one, two, or three significant figures were intended unless it is explicitly stated. Any uncertainty can be removed by expressing the measurement in *scientific notation*.

In **scientific notation,** a number is written in the form $A \times 10^n$, where A is a number with a single nonzero digit to the left of the decimal point, and n is an integer, or whole number. The measurement 900 cm precise to two significant figures can be written 9.0×10^2 cm. Scientific notation is also convenient for expressing very large or very small quantities. It is much easier (and it simplifies calculations) to write the speed of light as 3.00×10^8 (rather than as 300,000,000) meters per second.■

■ See Appendix A for a review of scientific notation.

Significant Figures in Calculations

Often, measurements are used in calculations. How do we report the significant figures in a calculation? Suppose we measure the volume of a gas and find it to be 77.1 liters at a pressure of 751.2 mmHg (mmHg = millimeters of mercury, a unit of pressure). In Chapter 4 we will see that we can obtain the volume of the gas at 760.0 mmHg (for a fixed temperature) from the following calculation.

$$\text{Volume} = 77.1 \text{ liters} \times \frac{751.2 \text{ mmHg}}{760.0 \text{ mmHg}}$$

Performing the arithmetic on a pocket calculator (Figure 1.6), we get 76.207263 for the numerical part of the answer (77.1 × 751.2 ÷ 760.0). But it would be incorrect to give this number as the final answer, since it is much more precise than the other numbers in the calculation. When multiplying or dividing measured quantities, there should be as many significant figures in the answer as there are in the measurement with the least number of significant figures. In the above calculation, 77.1 liters has the least number of significant figures (three). Hence, the answer should be reported to three significant figures, that is, 76.2 liters.

A different rule applies in the case of addition or subtraction of measured quantities. When adding or subtracting measured quantities, there should be the same number of decimal places in the answer as there are in the measurement with the least number of decimal places. Suppose we wish to add 184.2 grams and 2.324 grams. On a calculator, we find that 184.2 + 2.324 = 186.524. But since the quantity 184.2 grams has the least number of decimal places—one, whereas 2.324 grams has three—the answer is 186.5 grams. Note that any number whose value is known exactly does not affect the number of significant figures in a result. For example, there are 12 inches in 1 foot. The number 12 is known exactly, so 12 inches/foot × 3.16 feet = 37.9 inches. The number of significant figures is determined by 3.16, not 12.

Figure 1.6
Not all of the figures that appear on a calculator are significant. In performing the calculation 77.1 × 751.2 ÷ 760.0, the calculator shows 76.207263. We would report the answer as 76.2 because the factor 77.1 has the least number of significant figures (three).

Rounding

In reporting the sum of 184.2 grams and 2.324 grams as 186.5 grams, we *rounded* the number read off the calculator (186.524) by simply dropping the last two digits. The general procedure for **rounding** is as follows. Look at the leftmost digit to be dropped.

1. If this digit is greater than 5, or is 5 followed by nonzeros, add 1 to the last digit to be retained and drop all digits further to the right. Thus, rounding 1.2151 to three significant figures gives 1.22.

2. If this digit is less than 5, simply drop it and all digits further to the right. Rounding 1.2143 to three significant figures gives 1.21.

3. If this digit is simply 5 or 5 followed by zeros, and if the last digit to be retained is even, just drop the 5 and any zeros after it. If the last digit to be retained is odd, add 1 to it and drop the 5 and any zeros after it. For example, rounding 1.225, 1.22500, and 1.21500 to three significant figures gives 1.22 in each case.

In doing a calculation of two or more steps, it is desirable to retain additional digits for intermediate answers. This ensures that small errors from rounding do not appear in the final result. If you use a calculator, you can simply enter numbers one after the other, performing each arithmetic operation and rounding just the final answer. To keep track of the correct number of significant figures, you will need to record intermediate answers with a line under the last significant figures, as shown in the solution to part (c) of the following example.

Example 1.1

Perform the following calculations and round the answers to the correct number of significant figures (units of measurement have been omitted for clarity):

(a) $\dfrac{2.568 \times 5.8}{4.186}$ (b) $5.41 - 0.398$

(c) $4.18 - 58.16 \times (3.38 - 3.01)$

Solution

(a) The factor 5.8 has the fewest significant figures; therefore, the answer should be reported to two significant figures. Round the answer to 3.6. (b) The number with the least number of decimal places is 5.41. Therefore, the answer is rounded to two decimal places, to 5.01. (c) In any complicated arithmetic setup

such as this, proceed step by step. Carry out operations within parentheses as a separate calculation. By convention, multiplications and divisions are performed before additions and subtractions. It is best not to round intermediate answers, but you must keep track of the rightmost digit that would be retained after rounding (shown by an underline). The steps are:

$$4.1\underline{8} - 58.1\underline{6} \times (3.3\underline{8} - 3.0\underline{1}) = 4.1\underline{8} - 58.1\underline{6} \times 0.3\underline{7}$$
$$= 4.1\underline{8} - 2\underline{1}.5$$
$$= -1\underline{7}.32$$

The final answer is -17.

Exercise 1.2

Give answers to the following arithmetic setups. Round to the correct number of significant figures.

(a) $\dfrac{5.61 \times 7.891}{9.1}$ (b) $8.91 - 6.435$

(c) $6.81 - 6.730$ (d) $38.91 \times (6.81 - 6.730)$ (See Problems 1.15 and 1.16.)

1.4 SI Units

The first measurements were probably based on the human body (the length of the foot, for example). In time, fixed standards developed, but these varied from place to place. Each country or government (and often every trade) adopted its own units. As science became more quantitative in the seventeenth and eighteenth centuries, scientists found that the lack of standard units was a problem.■ They began to seek a simple, international system of measurement. In 1791, a study committee of the French Academy of Sciences devised such a system. Called the *metric system*, it became the official system of measurement for France and was soon used by scientists throughout the world. Most nations have since adopted the metric system or, at least, have set a schedule for changing to it.

■ In the system of units that Lavoisier used in the eighteenth century, there were 9216 grains to the pound (the *livre*). English chemists of the same period used a system in which there were 7000 grains in a pound— unless they were trained as apothecaries, in which case there were 5700 grains to the pound.

Prefixes and Base Units

One of the advantages of the metric system is that it is a decimal system. Larger and smaller units for a physical quantity are related by powers of 10 and are indicated by appropriate metric **prefixes.** For example, the unit of length in the metric system is the meter (somewhat longer than the yard), and a *centi*meter is 10^{-2} meters. Thus 2.54 centimeters are equal to 0.0254 meters. We convert centimeters to meters merely by moving a decimal point.

As quantitative science developed, a number of different metric units for the same quantity came into use. In an effort to simplify matters and to fix the standards, the General Conference of Weights and Measures adopted a particular choice of metric units in October 1960. The system of units selected as the standard is the **International System** of units (or **SI,** after the French *le Système International d'Unités*), which has seven base quantities and **base units** from which all other quantities and units are derived. Table 1.1 lists these base units and the symbols used to represent them. In this chapter we will discuss four base quantities: length, mass, time, and temperature.■ The prefixes used in the SI are listed in Table 1.2. Of these, only those shown in color will be used in this book: *mega-* (10^6), *kilo-* (10^3), *deci-* (10^{-1}), *centi-* (10^{-2}), *milli-* (10^{-3}), *micro-* (10^{-6}), *nano-* (10^{-9}), and *pico-* (10^{-12}).

■ The amount of substance is discussed in Chapter 3, and the electric current (ampere) is introduced in Chapter 21. Luminous intensity will not be used.

Length, Mass, and Time

The **meter (m)** is the SI base unit of length.■ By combining it with one of the SI prefixes, we can obtain a unit of appropriate size for any length measurement. For the very small lengths used in chemistry, the nanometer (1 nanometer = 10^{-9} meters) or the picometer (1 picometer = 10^{-12} meters) is an acceptable SI unit. A non-SI unit traditionally used by chemists is the **angstrom** (Å), which equals 10^{-10} meters.

For mass, the SI base unit is the **kilogram (kg).**■ (The kilogram is about 2.2 pounds.) This is an unusual base unit in that it contains a prefix. In forming other SI mass units, prefixes are added to the word *gram* to give units such as the *milli*gram (mg; 1 milligram = 10^{-3} grams).

■ The meter was originally defined in terms of a standard platinum-iridium bar kept at Sèvres, France. In the fall of 1983, the meter was defined as the length equal to the distance traveled by light in a vacuum in 1/299,792,458 seconds.

■ The present standard of mass is the platinum-iridium kilogram mass kept at the International Bureau of Weights and Measures in Sèvres, France.

Table 1.1
SI Base Units

Quantity	Unit	Symbol
Length	meter	m
Mass	kilogram	kg
Time	second	s
Temperature	kelvin	K
Amount of substance	mole	mol
Electric current	ampere	A
Luminous intensity	candela	cd

The base unit for time is the **second** (s). Combining this with prefixes such as *milli-*, *micro-*, *nano-*, and *pico-*, we create units appropriate for measuring very rapid events. For example, the time it takes to add two 10-digit numbers on a modern high-speed computer is roughly a nanosecond. The time required for the fastest chemical processes is about a picosecond. When we measure times much longer than a few hundred seconds, we revert to *minutes* and *hours,* an obvious exception to the prefix-base format of the International System.

Exercise 1.3

Express the following quantities using an SI prefix and a base unit. For instance, 1.6×10^{-6} m $= 1.6 \ \mu$m.

(a) 1.84×10^{-9} m (b) 5.67×10^{-12} s (c) 7.85×10^{-3} g
(d) 9.7×10^{3} m   (e) 2.34×10^{-1} m

(See Problems 1.17 and 1.18.)

Table 1.2
SI Prefixes

Multiple	Prefix	Symbol
10^{18}	exa	E
10^{15}	peta	P
10^{12}	tera	T
10^{9}	giga	G
10^{6}	mega	M
10^{3}	kilo	k
10^{2}	hecto	h
10	deka	da
10^{-1}	deci	d
10^{-2}	centi	c
10^{-3}	milli	m
10^{-6}	micro	μ
10^{-9}	nano	n
10^{-12}	pico	p
10^{-15}	femto	f
10^{-18}	atto	a

Temperature

Temperature is difficult to define precisely, but we all have an intuitive idea of what we mean by it. It is a measure of "hotness." A hot object placed next to a cold one becomes cooler, while the cold object becomes hotter. Later in the chapter, we will note that heat energy passes from a hot object to a cold one, and that the quantity of heat passed between the objects depends on the difference in temperature between the two. Therefore, temperature and heat are different, but related, concepts.

A thermometer is a device for measuring temperature. The common type consists of a glass capillary containing a column of liquid, whose length varies with temperature. A scale alongside the capillary gives a measure of the temperature. The **Celsius scale** (formerly the *centigrade* scale) is the working temperature scale for scientific use. On this scale, the freezing point of water is 0°C and the boiling point of water at normal barometric pressure is 100°C. However, the SI base unit of temperature is the **kelvin (K)**, a unit on an *absolute temperature* scale.■ On any absolute scale, the lowest point that can be attained theoretically is zero. Both the Celsius and Kelvin scales have equal size units, but 0°C is equivalent to 273.15 K. Thus it is easy to convert from one scale to the other, using the formula

$$K = °C + 273.15$$

A temperature of 20°C (about room temperature) equals 293 K.

The Fahrenheit scale is at present the common temperature scale in the United States. Figure 1.7 compares Kelvin, Celsius, and Fahrenheit scales. As the figure shows, 0°C is the same as 32°F (both exact) and 100°C corresponds to 212°F (both exact). Therefore, there are $212 - 32 = 180$ Fahrenheit degrees in the range of 100 Celsius degrees. That is, there are 1.8 Fahrenheit degrees for every Celsius degree. Knowing this, and that 0°C equals 32°F, we can derive a formula to convert degrees Celsius to degrees Fahrenheit. It is

$$°F = (1.8 × °C) + 32$$

This formula can be rearranged to give a formula for converting degrees Fahrenheit to degrees Celsius:

$$°C = \frac{°F - 32}{1.8}$$

■ Note that the degree sign (°) is not used with the Kelvin scale, and the unit of temperature is given in kelvins (not capitalized).

Example 1.2

The hottest place on record in North America has been Death Valley in California. A temperature of 134°F was reached there in 1913. What is this temperature reading in degrees Celsius? In kelvins?

Solution

Substituting, we find that

$$°C = \frac{°F - 32}{1.8} = \frac{134 - 32}{1.8} = 56.7$$

In kelvins,

$$K = °C + 273.15 = 56.7 + 273.15 = 329.8$$

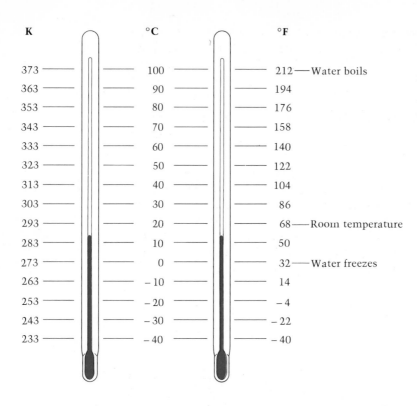

Figure 1.7
Comparison of thermometer scales. Room temperature is about 20°C, 293 K, and 68°F. Water freezes at 0°C, 273 K, and 32°F. Water boils at 100°C, 373 K, and 212°F.

Exercise 1.4

(a) A cooling mixture of Dry Ice and isopropyl alcohol has a temperature of −78°C. What is this in kelvins?   (b) A person has a fever of 102.5°F. What is this temperature in degrees Celsius?

(See Problems 1.19, 1.20, 1.21, and 1.22.)

1.5 Working with Units: Dimensional Analysis

In performing numerical calculations with physical quantities, it is a good habit to enter each quantity as a number with its associated unit. Both the numbers and the units are then carried through the indicated algebraic operations. The advantages of this are twofold:

1. The units for the answer will come out of the calculations automatically.
2. If you make an error in arranging factors in the calculation (for example, if you use the wrong formula), this will become apparent because the final units will be nonsense.

Dimensional analysis is the method of calculation in which one carries along the units for quantities. As an illustration, suppose we want to find the volume, V, of a cube, given s, the length of a side of the cube. Since $V = s^3$, if $s = 5.00$ cm, we find that $V = (5.00 \text{ cm})^3 = 5.00^3 \text{ cm}^3$. There is no guesswork about the unit of volume here; it is cubic centimeters (cm^3).

Suppose, however, that we wish to express the volume in liters (L), a metric unit that equals 10^3 cubic centimeters (approximately one quart). We can write this equality as

$$1L = 10^3 \text{ cm}^3$$

If we divide both sides of the equality by the right-hand quantity, we get

$$\frac{1L}{10^3 \text{ cm}^3} = \frac{\cancel{10^3 \text{ cm}^3}}{\cancel{10^3 \text{ cm}^3}} = 1$$

Observe that units are treated in the same way as algebraic quantities. Note too that the right-hand side now equals 1 and that there are no units associated with it. Since it is always possible to multiply any quantity by 1 without changing that quantity, we can multiply our previous expression for the volume by the factor $1L/10^3$ cm^3 without changing the actual volume. We are only changing the way in which we express this volume:

$$V = 5.00^3 \cancel{\text{cm}^3} \times \underbrace{\frac{1L}{10^3 \cancel{\text{cm}^3}}}_{\substack{\text{converts} \\ \text{cm}^3 \text{ to L}}} = 125 \times 10^{-3} \text{ L} = 0.125 \text{ L}$$

The factor $1L/10^3$ cm^3 is called a **conversion factor** because it converts an expression in one unit (cubic centimeters) to an expression in another unit (liters).■ Note that the numbers in this conversion factor are *exact*, because one liter equals exactly one thousand cubic centimeters. We do not use such exact conversion factors to determine the number of significant figures in an arithmetic result. In the previous calculation, the quantity 5.00 cm (the measured length of the side) determines or limits the number of significant figures.

The relationships between certain U.S. units of length, mass, and volume and corresponding metric units are given in Table 1.3. We can use this table to obtain conversion factors from one system to the other. For example, we see that 1 pound = 453.6 grams. Thus, the conversion factor from pounds to grams is 453.6 g/1 lb. This conversion factor is not exact; it has been written to only four significant figures (enough for most purposes). The exact definition of the U.S. pound is 1 pound = 0.45359237 kilograms.

■ It takes more room to explain conversion factors than it does to use them. With practice, you will be able to write the final conversion step without the intermediate algebraic manipulations outlined here. The use of conversion factors is helpful in solving many problems in chemistry. We will employ it throughout this book where appropriate.

Length	Mass	Volume
1 in = 2.54 cm (exact)*	1 lb = 0.4536 kg	1 qt = 0.9464 L
1 yd = 0.9144 m (exact)*	1 lb = 16 oz (exact)	4 qt = 1 gallon (exact)
1 mile = 1.609 km	1 oz = 28.35 g	
1 mile = 5280 ft (exact)		

Table 1.3
Relationships of Some U.S. and Metric Units

*These conversion equations are exact definitions of the inch and the yard.

Example 1.3

How many centimeters are there in 6.51 miles? Use the exact definitions 1 mi = 5280 ft, 1 ft = 12 in, and 1 in = 2.54 cm.

Solution

This problem involves several conversions, which can be done all at once. From the above definitions, we get conversion factors of

$$1 = \frac{5280 \text{ ft}}{1 \text{ mi}} \qquad 1 = \frac{12 \text{ in}}{1 \text{ ft}} \qquad 1 = \frac{2.54 \text{ cm}}{1 \text{ in}}$$

Then,

$$6.51 \text{ mi} = 6.51 \text{ mi} \times \underbrace{\frac{5280 \text{ ft}}{1 \text{ mi}}}_{\substack{\text{converts} \\ \text{mi to ft}}} \times \underbrace{\frac{12 \text{ in}}{1 \text{ ft}}}_{\substack{\text{converts} \\ \text{ft to in}}} \times \underbrace{\frac{2.54 \text{ cm}}{1 \text{ in}}}_{\substack{\text{converts} \\ \text{in to cm}}}$$

$$6.51 \text{ mi} = 1.05 \times 10^6 \text{ cm}$$

Since all of the conversion factors are exact, the number of significant figures in the result is determined by the number of significant figures in 6.51 miles.

Exercise 1.5

Using the definitions 1 in = 2.54 cm and 1 yd = 36 in (both exact), obtain the conversion factor for yards to meters. How many meters are there in 3.54 yards?

(See Problems 1.27, 1.28, 1.29, and 1.30.)

Exercise 1.6

A particular water droplet in a cloud has a diameter of 1.6×10^{-5} m. What is its diameter in millimeters? Convert this answer from millimeters to micrometers.

(See Problems 1.31 and 1.32.)

1.6 Derived Units

Once base units have been defined for a system of measurement, we can derive other units from them. We do this by using base units in an equation that defines a physical quantity. For example, *area*, which can be defined as length times length, has the SI unit of m × m, or m^2. *Speed*, defined as the rate of change of distance with time, has the SI unit of meters per second (that is, meters divided by seconds). This unit is symbolized as m/s or $m \cdot s^{-1}$. Thus SI units of area and speed are **derived units.** (Table 1.4 defines these and some other derived units.)

Table 1.4
Derived Units

Quantity	Definition of Quantity	SI Unit
Area	Length squared	m^2
Volume	Length cubed	m^3
Density	Mass per unit volume	kg/m^3
Speed	Distance traveled per unit time	m/s
Acceleration	Speed changed per unit time	m/s^2
Force	Mass times acceleration of object	$kg \cdot m/s^2$ (= newton, N)
Pressure	Force per unit area	$kg/(m \cdot s^2)$ (= pascal, Pa)
Energy	Force times distance traveled	$kg \cdot m^2/s^2$ (= joule, J)

Figure 1.8
Some laboratory glassware.
(a) A 600-mL beaker. (b) A
100-mL graduated cylinder.
(c) A 200-mL volumetric
flask. (d) A 250-mL Erlenmeyer flask. (e) A 5-mL pipet.
A graduated cylinder is used
to measure volumes of liquids. A pipet is calibrated to
deliver a specified volume of
liquid. It is filled to an
etched line with a suction
bulb and then allowed to
drain.

Volume

Volume is defined as length cubed and has the SI unit of cubic meter (m^3).
This is too large a unit for normal laboratory work, so we use either cubic
decimeters (dm^3) or cubic centimeters (cm^3, also written cc). Traditionally,
chemists have used the **liter** (**L**), which is equal to a cubic decimeter, as the
unit of volume. In fact, most laboratory glassware (Figure 1.8) is calibrated
in liters or milliliters (1000 mL = 1L). Because 1 cubic decimeter =
1000 cubic centimeters, a milliliter is equivalent to a cubic centimeter. ■

$$1L = 1\ dm^3 \qquad 1\ mL = 1\ cm^3$$

■ To show that 1 dm^3 =
1000 cm^3, convert 1 dm^3 to
cubic meters, then to cubic
centimeters:

$$1\ dm^3 \times \left(\frac{10^{-1}\ m}{1\ dm}\right)^3 \times \left(\frac{1\ cm}{10^{-2}\ m}\right)^3$$
$$= 10^3\ cm^3$$

Example 1.4

The world's oceans contain approximately 1.35×10^9
km^3 of water. What is this volume in liters?

Solution

Convert this value to cubic decimeters.

$$1.35 \times 10^9\ km^3 = 1.35 \times 10^9\ \cancel{km^3}$$
$$\times \underbrace{\left(\frac{10^3\ \cancel{m}}{1\ \cancel{km}}\right)^3}_{\substack{\text{converts} \\ \text{km}^3 \text{ to m}^3}} \times \underbrace{\left(\frac{1\ dm}{10^{-1}\ \cancel{m}}\right)^3}_{\substack{\text{converts} \\ \text{m}^3 \text{ to dm}^3}} = 1.35 \times 10^{21}\ dm^3$$

Since a cubic decimeter is equal to a liter, the volume
of the oceans is 1.35×10^{21} L.

Exercise 1.7

A crystal is constructed by the stacking of small, identical pieces of crystal, much as we construct a brick wall by stacking bricks. A unit cell is the smallest such piece from which a crystal can be made. The unit cell of a crystal of gold metal has a volume of 67.6 Å^3. What is this volume in cubic meters?

(See Problems 1.33 and 1.34.)

Density

The **density** of an object is its mass per unit volume. We can express this as

$$d = \frac{m}{V}$$

where d is the density, m is the mass, and V is the volume. Suppose an object has a mass of 15.0 g and a volume of 10.0 cm^3. Substituting, we find that

$$d = \frac{15.0 \text{ g}}{10.0 \text{ cm}^3} = 1.50 \text{ g/cm}^3$$

The density of the object is 1.50 g/cm^3 (or 1.50 g·cm^{-3}).

   Density is an important characteristic property of a material. Water, for example, has a density of 1.000 g/cm^3 at 4°C, and a density of 0.998 g/cm^3 at 20°C. Lead has a density of 11.3 g/cm^3 at 20°C. ■ Oxygen gas has a density of 1.33 × 10^{-3} g/cm^3 at normal pressure and 20°C. (Like other gases under normal conditions, oxygen has a density that is about 1000 times smaller than liquids and solids.) Because the density is characteristic of a substance, it can be helpful in identifying it. Exercise 1.8, which follows, illustrates this point. Density can also be useful in determining whether a substance is pure. Consider a gold bar whose purity is questioned. Any of the metals likely to be mixed with gold, such as silver or copper, have lower densities than gold. Therefore, an adulterated (impure) gold bar can be expected to be far less dense than pure gold.

■ The density of solid materials on earth ranges from about 1 g/cm^3 to 22.5 g/cm^3 (osmium metal). In the interior of certain stars, the density of matter is truly staggering. Black neutron stars, stars composed of neutrons, or atomic cores compressed by gravity to a super dense state, have densities of about 10^{15} g/cm^3.

Exercise 1.8

Calculate the density of a metal object with a volume of 20.2 cm^3 and a mass of 159 g. The object is made of manganese, iron, or nickel, which have densities of 7.21 g/cm^3, 7.87 g/cm^3, and 8.90 g/cm^3, respectively. From which metal is the object made?

(See Problems 1.35, 1.36, 1.37, and 1.38.)

Exercise 1.9

Ethanol (grain alcohol) has a density of 0.789 g/cm^3. What is the volume occupied by 30.3 g?

(See Problems 1.39, 1.40, 1.41, and 1.42.)

Pressure

Pressure is defined as force exerted per unit area. To obtain the SI unit of pressure, we must examine the SI unit of force. According to Newton,

$$\text{force} = \text{mass} \times \text{acceleration}$$

Acceleration is the rate of change of speed. The SI unit of speed is the meter per second (m/s), so the SI unit of acceleration is the meter per second per second (m/s^2). Thus,

SI unit of force = SI unit of mass × SI unit of acceleration

= kg × m/s^2

The SI unit of force, kg · m/s^2, is given the name *newton* (N).

Pressure, as we said, equals force per unit area. Therefore,

SI unit of pressure = SI unit of force/SI unit of area

= (kg · m/s^2)/m^2 = kg/(m · s^2)

The SI unit of pressure, kg/(m · s^2), is given the name **pascal (Pa)**. The pascal is an extremely small unit. The pressure exerted on a table by a flat coin the size of a penny is more than 100 pascals, and the pressure exerted by the atmosphere is approximately 100,000 pascals.■

Traditionally, chemists have used two other units of pressure, the **atmosphere (atm)** and **millimeters of mercury (mmHg)**, also called **torr**:

1 atm = 101.325 kPa   (exact definition)

1 atm = 760 mmHg   (exact definition)

We will have more to say about pressure in Chapter 4, on gases.

■ The pressure exerted by a flat coin is the force downward, or weight, divided by its cross-sectional area. The mass of a coin the size of a penny (not really flat) is about 3.0 g (3.0 × 10^{-3} kg). Its weight is the mass times acceleration due to gravity (9.8 m/s^2): 3.0 × 10^{-3} kg × 9.8 m/s^2 = 2.9 × 10^{-2} kg · m/s^2. The cross-sectional area is πr^2, where r is the radius of the coin (9.3 mm for a penny). This gives an area of 2.7 × 10^{-4} m^2. Hence, the pressure is (2.9 × 10^{-2} kg · m/s^2)/(2.7 × 10^{-4} m^2) = 1.1 × 10^2 kg/(m · s^2).

Example 1.5

A pressure of 794.0 mmHg equals how many kilopascals?

Solution

Because 760 mmHg = 101.3 kPa, we have

$$794.0 \text{ mmHg} = 794.0 \text{ mmHg} \times \frac{101.3 \text{ kPa}}{760 \text{ mmHg}}$$

$$= 105.8 \text{ kPa}$$

Exercise 1.10

The barometric pressure just before a storm is found to be 733 mmHg. What is this barometric pressure in atmospheres? What is it in pascals?

(See Problems 1.45 and 1.46.)

Energy

We can define **energy** briefly as the capacity to move matter.■ The energy associated with a moving object is called **kinetic energy.** An object of mass m and speed or velocity v has a kinetic energy, E_k, equal to

$$E_k = \tfrac{1}{2}mv^2$$

■ This definition includes heat. The atoms or molecules of a material move faster when heat is added.

Potential energy is the energy an object has by virtue of its position. For example, water at the top of a dam or chute has potential energy (in addition to whatever kinetic energy it may possess because the water is moving). This potential energy results from the relatively high position of the water in the

Figure 1.9
Energy as the capacity to
move matter. Water at the
top of the chute has potential
energy, which can be used to
move a water wheel as the
water flows over it.

gravitational field of the earth. It is converted to kinetic energy when the
water falls to a lower level. As the water falls, it moves faster. The potential
energy decreases, and the kinetic energy increases. In Figure 1.9, the potential energy of water is converted to useful energy as the water falls over a
chute and moves a water wheel. Recognizing that the potential energy of
water at the top of the chute does have the capacity to move matter (in this
case, the water wheel), we are brought back to our original definition of
energy.

The SI unit of energy may be obtained from the equation for the energy
needed to move an object by applying a force:

$$\text{Energy} = \text{force} \times \text{distance moved}$$

Thus the SI unit of energy equals the SI unit of force times the SI unit of
distance, or $(kg \cdot m/s^2) \times m$. This unit, symbolized $kg \cdot m^2/s^2$, is given the
name **joule (J)**.

To appreciate the smallness of the joule, note that the *watt* is a measure
of the quantity of energy used per unit time and equals 1 joule per second.
A 100-watt bulb, for example, uses 100 joules of energy every second that it
is lit.■ A kilowatt-hour, which is the unit by which electricity is sold,

■ An even smaller metric unit of
energy is the *erg*. This unit is
obtained when the base units
are the centimeter, the gram,
and the second (called the *cgs
system*). The erg is not an SI
unit, but it is related to the joule
by the formula

$$1 \text{ erg} = 10^{-7} \text{ J}$$

We will not use the erg in this
book.

equals 3600 kilowatt-seconds (because there are 3600 seconds in 1 hour), or 3.6 million joules. A household might use something like 1000 kilowatt-hours (3.6 billion joules) of electricity in a month.

A unit of energy commonly used by chemists is the **calorie** (**cal**). Originally the calorie was defined as the amount of energy required to raise the temperature of one gram of water by one degree Celsius, and this is still good to remember as a rough definition. ■ Since then, however, it has been discovered that the energy needed to heat water depends slightly on the temperature of the water. In 1925 the calorie was defined in terms of the joule:

$$1 \text{ cal} = 4.184 \text{ J} \quad \text{(exact definition)}$$

■ The "calorie" used in nutrition is actually a kilocalorie. A reasonable daily intake of energy for an average person is 2000 kcal to 2500 kcal.

Example 1.6

Recall that the units for a derived quantity can be obtained from the defining equation. Use the definition of kinetic energy ($E_k = \frac{1}{2}mv^2$) to derive the SI unit of energy. Compare your answer with the result given in the text.

Solution

The factor $\frac{1}{2}$ has no units. Therefore, the unit of energy equals

$$\text{unit of } m \times (\text{unit of } v)^2 = \text{kg} \times (\text{m/s})^2 = \text{kg} \cdot \text{m}^2/\text{s}^2$$

This is identical to the unit derived in the text.

Exercise 1.11

Any fluid resists the movement of objects through it. This resistance to motion depends on a property of the fluid called *viscosity*. Syrup, for example, has greater viscosity than water. The force exerted upward on a spherical object of radius r that is falling through a fluid of viscosity η with a speed v is $6\pi\eta rv$. Using the known SI units of force, and those for r and for v, obtain the SI unit of viscosity.

(See Problems 1.53 and 1.54.)

Heat is energy that has moved or has been transferred from one object to another because of a difference in temperature of the objects. When a hot object is placed next to a cold one, energy moves from the hot object to the cold one. We say that heat has been transferred from one object (the hot one) to another (the cold one).

When heat is transferred to a substance, its temperature is raised. **Specific heat** is the quantity of heat required to raise the temperature of one gram of a substance by one degree Celsius (or by one kelvin). To find the total heat that is transferred when the temperature of some mass of substance changes by several degrees Celsius, we use the following equation:

Heat transferred = specific heat × mass × temperature change

The temperature change is obtained by taking the final temperature and subtracting the initial temperature. When this temperature change is positive, the heat transferred is a positive quantity, which means that energy is *added* to the substance. When the temperature change is negative, the heat transferred is a negative quantity, which means that energy is *subtracted* from the substance. As an example of the use of this equation, let us calculate the quantity of heat that must be transferred to 15.0 g of water to raise its temperature from 20.0°C to 50.0°C. Water has a specific heat of 4.18 J/(g · °C). Therefore,

$$\text{heat transferred} = \text{specific heat} \times \text{mass} \times \text{temperature change}$$
$$= 4.18 \text{ J/(g} \cdot \text{°C)} \times 15.0 \text{ g} \times (50.0\text{°C} - 20.0\text{°C})$$
$$= 4.18 \text{ J/(} \cancel{\text{g}} \cdot \cancel{\text{°C}}) \times 15.0 \cancel{\text{g}} \times (+30.0\cancel{\text{°C}})$$
$$= 1.88 \times 10^3 \text{ J}$$

Thus, 1.88×10^3 J of energy (as heat) must be added to the substance.

As the illustration of the dam shows, the various forms of energy may be interconverted. During the conversion of energy, the total quantity of energy remains constant. This is a result of the **law of conservation of energy:** energy may be converted from one form to another, but the total quantity of energy remains constant.

A Checklist for Review

Important Terms   (The number in parentheses denotes the section in which the term was defined.)

mass (1.1)	**prefixes** (1.4)	**pressure** (1.6)
matter (1.1)	**International System (SI)** (1.4)	**pascal (Pa)** (1.6)
law of conservation of mass (1.1)	**base units** (1.4)	**atmosphere (atm)** (1.6)
experiment (1.2)	**meter (m)** (1.4)	**millimeters of mercury (mmHg)** (1.6)
law (principle) (1.2)	**angstrom (Å)** (1.4)	**torr** (1.6)
theory (1.2)	**kilogram (kg)** (1.4)	**energy** (1.6)
hypothesis (1.2)	**second (s)** (1.4)	**kinetic energy** (1.6)
scientific method (1.2)	**Celsius scale** (1.4)	**potential energy** (1.6)
unit (1.3)	**kelvin (K)** (1.4)	**joule (J)** (1.6)
precision (1.3)	**dimensional analysis** (1.5)	**calorie (cal)** (1.6)
accuracy (1.3)	**conversion factor** (1.5)	**heat** (1.6)
number of significant figures (1.3)	**derived units** (1.6)	**specific heat** (1.6)
scientific notation (1.3)	**liter (L)** (1.6)	**law of conservation of energy** (1.6)
rounding (1.3)	**density** (1.6)	

Summary of Facts and Concepts

Chemistry emerged as a *quantitative science* with the work of the eighteenth-century French chemist Antoine Lavoisier. He made use of the idea that the mass remains constant during a chemical reaction (*law of conservation of mass*).

Chemistry is an experimental science in that the facts of chemistry are obtained by *experiment*. These facts are systematized and explained by *theory*, and theory suggests more experiments. The *scientific method* involves this interplay, in which the body of accepted knowledge grows as it is tested by experiment.

A quantitative science requires one to make measurements. Any measurement has limited *precision*, which we convey by writing the measured number to a certain number of *significant figures*. There are many different systems of measurement, but scientists generally use the metric system. The International System (SI) uses a particular choice of metric units. It employs seven *base units* combined with *prefixes* to obtain units of various size. Units for other quantities are derived from these.

Dimensional analysis is a technique of calculating

with physical quantities in which units are included and treated in the same way as numbers. Using dimensional analysis, one can obtain the *conversion factor* needed to express a quantity in new units.

To obtain a *derived unit* in SI for a quantity such as the volume or density, we merely substitute base units into a defining equation for the quantity. For *pressure*, we get the derived unit $kg/(m \cdot s^2)$ (pascal). The derived unit of *energy* is $kg \cdot m/s^2$ (joule). Energy occurs in various forms (*potential, kinetic, heat*) that can be interconverted, though the total energy remains constant (*law of conservation of energy*).

Operational Skills

1. Given an arithmetic setup, report the answer to the correct number of significant figures and round it properly (Example 1.1).

2. Given a temperature reading on one scale, convert it to another scale, Celsius, Kelvin, or Fahrenheit (Example 1.2).

3. Given an equation relating one unit to another (or a series of such equations), convert a measurement expressed in one unit to a new unit (Examples 1.3, 1.4, and 1.5).

4. Given an equation involving physical quantities, and given the units of all quantities except one, find the units of that quantity (Example 1.6).

Review Questions

1.1 State the law of conservation of mass. Describe how you might demonstrate this law.

1.2 Define the terms *experiment* and *theory*. How are theory and experiment related? What is a hypothesis?

1.3 What is meant by the precision of a measurement? How is it indicated?

1.4 How does the International System (SI) obtain units of different size from a given unit? How does the International System obtain units for all possible physical quantities from only seven base units?

1.5 Define density, pressure, and energy. Cite examples to illustrate the size of the pascal and the joule.

1.6 Describe the interconversions of potential and kinetic energy in a moving pendulum. A moving pendulum eventually comes to rest. Has the energy been lost? If not, what has happened to it?

Problems

Key: The problems are arranged in matched pairs. The first problem of each pair (whose number is in color) appears in the left column, and its answer is given in the back of the book. Problems without stars require only basic problem-solving skills, which have usually been illustrated by Examples. Problems with one star are of moderate difficulty, and problems with two require more thought.

Conservation of Mass

1.7 An 8.4-gram sample of sodium hydrogen carbonate is added to a solution of acetic acid weighing 20.0 grams. The two substances react, releasing carbon dioxide gas to the atmosphere. After reaction, the contents of the reaction vessel weigh 24.0 grams. What is the mass of carbon dioxide given off during the reaction?

1.8 Some magnesium wire weighing 2.4 grams is placed in a beaker and covered with 15.0 grams of dilute hydrochloric acid. The acid reacts with the metal and gives off hydrogen gas, which escapes into the surrounding air. After reaction, the contents of the beaker weigh 17.2 grams. What is the mass of hydrogen gas produced by the reaction?

Significant Figures (Units of measurement have been omitted for clarity.)

1.9 How many significant figures are there in each of the following numbers?
    (a) 0.0058363   (b) 0.0600   (c) 701.00
    (d) 4.810×10^3 (e) 7001   (f) 0.508

1.10 How many significant figures are there in each of the following numbers?
    (a) 5280.0   (b) 0.097   (c) 0.0100
    (d) 9.736×10^5 (e) 1.000×10^4 (f) 0.15

1.11 If the number 1000 is accurate to as many figures as are given, how would you write it in scientific notation?

1.12 If the number 1800 is accurate to as many figures as are given, how would you write it in scientific notation?

1.13 Round the following numbers to three significant figures.
    (a) 8.73458   (b) 57.6501   (c) 3.56500
    (d) 8.75500   (e) 5.04500×10^4 (f) 8.53500×10^2

1.14 Round the following numbers to four significant figures.
    (a) 79.44321   (b) 8.964689
    (c) 0.007643500   (d) 0.05432500
    (e) 1.004500×10^3 (f) 2.863500×10^2

1.15 Do the indicated arithmetic and give the answer to the correct number of significant figures.
    (a) $\dfrac{8.71 \times 0.0301}{0.056}$
    (b) $0.71 + 81.8$
    (c) $934 \times 0.00435 + 107$
    (d) $(847.89 - 847.73) \times 14673$

1.16 Do the indicated arithmetic and give the answer to the correct number of significant figures.
    (a) $\dfrac{0.871 \times 0.23}{5.871}$
    (b) $8.937 - 8.930$
    (c) $8.937 + 8.930$
    (d) $0.00015 \times 54.6 + 1.002$

Measurement and Unit Conversions

1.17 Give the following measurements using the most appropriate SI prefix.
    (a) 2.31×10^{-12} m   (b) 5.43×10^{-9} s
    (c) 8.7×10^{-6} g   (d) 9.3×10^{-3} m

1.18 Give the following measurements using the most appropriate SI prefix.
    (a) 1.52×10^3 g   (b) 9.5×10^{-12} s
    (c) 2.961×10^{-1} m   (d) 7.9×10^{-3} s

1.19 Convert:
    (a) 32°F to degrees Celsius
    (b) −58°F to degrees Celsius
    (c) 68°F to degrees Celsius
    (d) −11°F to degrees Celsius
    (e) 37°C to degrees Fahrenheit
    (f) −77°C to degrees Fahrenheit

1.20 Convert:
    (a) 85°F to degrees Celsius
    (b) 121°F to degrees Celsius
    (c) 21°F to degrees Celsius
    (d) −15°F to degrees Celsius
    (e) −45°C to degree Fahrenheit
    (f) −96°C to degrees Fahrenheit

1.21 (a) What is 0 K in degrees Celsius? (b) What is theoretically the lowest attainable temperature in degrees Fahrenheit? (Give answers to the nearest degree.)

1.22 (a) A particular colony of bacteria thrives at 98°F. What is this temperature in degrees Celsius? (b) What is this temperature in kelvins?

1.23 One acre equals exactly 4840 yd^2. How many square yards are there in 6.40×10^2 acres?

1.24 One Btu (British thermal unit) equals 252 cal. How many calories are there in 2.500×10^3 Btu?

1.25 One U.S. gallon (liquid) equals exactly 231 in^3. How many gallons are there in 1.500×10^3 in^3?

1.26 One Imperial gallon equals 1.201 U.S. gal (liquid). An automobile gasoline tank that holds 16.0 U.S. gal holds how many Imperial gallons?

1.27 How many grams are there in 3.58 short tons? Note that 1 g = 0.03527 oz (ounces avoirdupois), 1 lb (pound) = 16 oz, and 1 short ton = 2000 lb. (These relations are exact.)

1.28 How many liters are there in 8.46 U.S. gal? Note that 1 qt (U.S. liquid) = 946.4 cm^3 and 4 qt = 1 gal (exact).

1.29 The first measurement of sea depth was made in 1840 in the central South Atlantic, where a plummet was lowered 2425 fathoms. What is this depth in meters? Note that 1 fathom = 6 ft, 1 ft = 12 in, and 1 in = 2.54×10^{-2} m. (These relations are exact.)

1.30 The estimated amount of recoverable oil from the field at Prudhoe Bay in Alaska is 9.6×10^9 barrels. What is this amount of oil in cubic meters? One barrel = 42 gal (exact), 1 gal = 4 qt (exact), and 1 qt = 9.46×10^{-4} m^3.

1.31 Convert:
  (a) 7.36 kg to milligrams
  (b) 655 μs to milliseconds
  (c) 57 km to nanometers
  (d) 25.4 mm to centimeters

1.32 Convert:
  (a) 518 Å to micrometers
  (b) 34.5 kg to milligrams
  (c) 16.5 cm to millimeters
  (d) 7.8 ns to microseconds

Derived Units

1.33 The total amount of fresh water on Earth is estimated to be 3.73×10^8 km^3. What is this in cubic meters? in liters?

1.34 A submicroscopic particle suspended in a solution has a volume of 1.4 μm^3. What is this volume in liters?

1.35 A certain sample of the mineral galena (lead sulfide) weighs 12.4 g and has a volume of 1.64 cm^3. What is the density of galena?

1.36 A flask contains 25.0 mL of diethyl ether weighing 17.84 g. What is the density of the ether?

1.37 A liquid with a volume of 10.7 mL has a mass of 9.42 g. The liquid is known to be either octane, ethanol, or benzene, the densities of which are 0.702 g/cm^3, 0.789 g/cm^3, and 0.879 g/cm^3, respectively. What is the identity of the liquid?

1.38 A mineral sample has a mass of 16.3 g and a volume of 2.3 cm^3. The mineral is either sphalerite (density = 4.0 g/cm^3), cassiterite (density = 6.99 g/cm^3), or cinnabar (density = 8.10 g/cm^3). Which is it?

1.39 Platinum has a density of 22.5 g/cm^3. What is the mass of 5.9 cm^3 of this metal?

1.40 What is the mass of a 51.6-mL sample of gasoline, which has a density of 0.70 g/cm^3?

1.41 Ethyl alcohol has a density of 0.789 g/cm^3. What volume of ethyl alcohol must be poured into a graduated cylinder to give 19.8 g of alcohol?

1.42 Toluene has a density of 0.867 g/cm^3. A sample of toluene weighing 12.6 g occupies what volume?

****1.43** The following procedure may be used to determine the density of a crystalline material such as table sugar (sucrose). A sample of sucrose weighing 9.35 g is placed in a 15.00-mL flask, and then the flask is completely filled with benzene (a liquid), the density of which is 0.879 g/cm^3. The sucrose and benzene together weigh 17.33 g. What is the density of sucrose? (Sucrose does not dissolve in benzene.)

****1.44** The density of a sample of olive oil was determined in the following way. A flask was first filled with ethanol, the mass of which was 8.02 g. Then the flask was emptied and filled with olive oil. The mass of this olive oil was 9.32 g. The density of ethanol is 0.789 g/cm^3. Calculate the density of the olive oil.

1.45 A barometer showed a pressure reading of 775 mmHg. What is this in atmospheres? in kilopascals?

1.46 What is a pressure of 125.5 kPa in mmHg? in atmospheres?

1.47 A piece of wood was burned and gave off 135.6 kcal of heat. What is this energy in kilojoules?

1.48 One gram of sulfur was burned in oxygen and released 9.28 kJ of heat. How many calories of heat were given off by this combustion?

****1.49** How many calories of heat are needed to raise the temperature of 4.5 g of water by 2.1°C?

****1.50** A vessel containing 7.9 cm^3 of water was heated from 70.1°C to 75.6°C. How much heat energy was required to heat the water? Give the answer in joules.

****1.51** Calculate the amount of kinetic energy in a 5.0-kg mass moving at 105 km/hr.

****1.52** How much heat will be released in braking to a stop an automobile weighing 2.2×10^3 lb that is moving at 55 mi/hr?

1.53 The momentum of a moving object is defined as the product of the mass of the object and its speed. What is the SI unit of momentum?

1.54 A steel spring that is extended a distance x beyond its normal length exerts a restoring force F according to the formula $F = -kx$. What is the SI unit of k?

Additional Problems

Key: Each odd-numbered additional problem and its following even-numbered problem are similar. Answers to odd-numbered additional problems (whose numbers appear in color) are given in the back of the book.

1.55 A person has 6.0 qt of blood containing 95 mg of glucose per 100.0 mL. How much glucose is there in this person's total blood supply?

1.56 A common disease of tropical aquarium fish, called "ick," is characterized by small white spots. One treatment consists of adding 0.125 grain of potassium permanganate to each gallon of water. How many grams of potassium permanganate should be added to a 30.0-gal tank? (One gram equals exactly 15 grains.)

1.57 Some bottles of colorless liquids were being labeled when the technicians accidentally mixed them up and lost track of their contents. A 15.0-mL sample withdrawn from one bottle weighed 22.3 g. The technicians knew that the liquid was either acetone, benzene, chloroform, or carbon tetrachloride (which have densities of 0.792 g/cm^3, 0.899 g/cm^3, 1.489 g/cm^3, and 1.595 g/cm^3, respectively). What was the identity of the liquid?

1.58 A solid will float on any liquid that is more dense than it is. The volume of a piece of calcite weighing 35.6 g is 12.9 cm^3. On which of the following liquids will the calcite float: carbon tetrachloride (density = 1.60 g/cm^3), methylene bromide (density = 2.50 g/cm^3), tetrabromoethane (density = 2.96 g/cm^3), methylene iodide (density = 3.33 g/cm^3)?

1.59 Liquid hydrogen peroxide has been used as a propellant for rockets. Hydrogen peroxide decomposes into oxygen and water, giving off heat energy equal to 686 Btu per pound of propellant. What is this energy in joules per gram of hydrogen peroxide? (One Btu equals 252 cal. Other necessary conversions can be found in the text.)

1.60 Industrial-quality diamonds have been made by melting graphite at 2000°C with metal catalysts (substances that increase the rate of conversion) under great pressure, for example, at 1.0×10^6 lb/in^2. What is this pressure in pascals? (One pound per square inch equals

51.7 mmHg. Other conversion factors are given in the text.)

**1.61 A sample of an ethanol–water solution has a volume of 54.2 cm^3 and a mass of 49.6 g. What is the percentage of ethanol (by mass) in the solution? (Assume that there is no change in volume when the pure compounds are mixed.) The density of ethanol is 0.789 g/cm^3 and that of water is 0.998 g/cm^3. Alcoholic beverages are rated in *proof*, which is a measure of the relative amount of ethanol in the beverage. Pure ethanol is exactly 200 proof; a solution that is 50% ethanol by volume is exactly 100 proof. What is the proof of the given ethanol–water solution?

**1.62 You have a piece of gold jewelry weighing 9.35 g. Its volume is 0.654 cm^3. Assume that the metal is an alloy (mixture) of gold and silver, which have densities of 19.3 g/cm^3 and 10.5 g/cm^3, respectively. Also assume that there is no change in volume when the pure metals are mixed. Calculate the percentage of gold (by mass) in the alloy. The relative amount of gold in an alloy is measured in *karats*. Pure gold is 24 karats; an alloy of 50% gold is 12 karats. State the proportion of gold in the jewelry in karats.

**1.63 Assume that the density of sea water is 1.025 g/cm^3. Calculate the depth at which an object would be subjected to a pressure of 2.00 atm (that is, to a pressure of 1.00 atm from air and to an additional 1.00 atm from the sea water). The constant acceleration of gravity is 9.81 m/s^2.

**1.64 The amount of water flowing over Niagara Falls that can be taken for hydroelectric power is regulated. The amount of water from Niagara Falls that is used for electric power averages 1.30×10^5 ft^3/s. If we assume a fall of 326 ft, what is the power generated in watts? The constant acceleration of gravity is 9.81 m/s^2; the density of water is 0.998 g/cm^3.

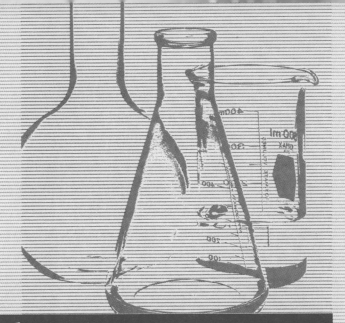

2. Atomic Theory: Pure Substances and Mixtures

2.1 Atoms, Molecules, and Ions Atoms/ Molecules and Molecular Substances/ Ions and Ionic Substances/ A Word on Naming Substances/ Chemical Reactions

2.2 Balancing Simple Chemical Equations

2.3 Classifications of Matter Chemical Constitution —Element, Compound, or Mixture?/ Physical State—Gas, Liquid, or Solid?

2.4 Separation of Mixtures Filtration/ Distillation/ Chromatography

S odium is a soft, bright silvery metal. Because it readily reacts with water, it is stored under kerosene. Chlorine is a toxic, pale green gas with a choking odor. A dramatic chemical reaction occurs when a pea-sized piece of sodium is placed in a spoon, heated, then lowered into a flask containing chlorine. The metal bursts into brilliant yellow flame and burns to a white, crystalline powder. Watching this violent reaction is fascinating enough, but examining the end product is more so. We find that this white, crystalline solid is sodium chloride—common table salt, an edible substance!

Sodium metal and chlorine gas have undergone a chemical reaction, a change in the character of the substances, giving us table salt. A striking change in the character of substances is one feature of a chemical reaction. How can we explain chemical reactions such as the combustion of sodium in chlorine? And how do we explain the differences between the substances we start with and the end products? These are basic questions in chemistry. In this chapter, atomic theory will provide some answers.

Chapter Overview

Atomic theory postulates the existence of atoms, or exceedingly small particles, from which the materials around us are composed. In this chapter, we see how we can *explain the great variety of substances* that exist in terms of a few different kinds of atoms. We also see how we can *explain chemical reactions* with this atomic concept. Moreover, we see how the atomic concept gives us *a way of classifying matter,* for example, into pure substances and mixtures. In the final section, we describe some *methods for separating mixtures* into pure substances.

2.1 Atoms, Molecules, and Ions

As we have seen, Lavoisier laid the experimental foundation of modern chemistry. But it was the British physicist and chemist John Dalton (1766–1844) who provided the basic theory: all matter—whether solid, liquid, or gas—is composed of small particles called atoms.

Atoms

Dalton had a lifelong interest in meteorology, the science of weather. This interest led him to speculate on the composition of air, and from this to put forward a hypothesis about the structure of matter. The main features of this hypothesis, now called the **atomic theory,** were developed by 1803.

The postulates of this theory, which are basically unchanged today, are

1. All matter is composed of small particles called **atoms** (Figure 2.1).
2. Elements are kinds of matter composed of only one chemically distinct type of atom. There are only a small number of different elements (108 are known at present).

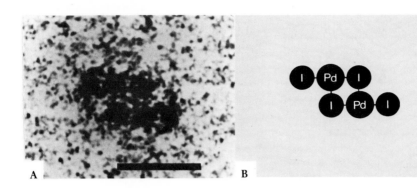

Figure 2.1
Image of iodine and palladium atoms. The image (a) is of a group of atoms (identified in b) within a large molecule (other atoms are not visible). It was obtained by a high-resolution electron microscope. The scale bar, added to the lower part of (a), measures 10Å.

3. Atoms of different elements have different properties. For example, the atoms of an element have definite mass.■

4. Compounds are kinds of matter composed of atoms of two or more elements. The relative number of atoms of each element in every sample of a compound is constant.

5. A chemical reaction, a change in chemical identities of matter, consists of rearranging the combinations of atoms present in the original kinds of matter to give new combinations present in the final kinds of matter. No atoms are created or destroyed in a chemical reaction.

■ Atomic masses are discussed in Chapter 3.

Some common elements and their **atomic symbols** (the letters used to designate their atoms) are listed in Table 2.1. These symbols were introduced by Jöns Jakob Berzelius (1779–1848), a Swedish contemporary of Dalton. Typically, the atomic symbol of an element is derived from the first one or two letters of its English name, though in some cases it is based on a foreign (usually Latin) name.

Molecules and Molecular Substances

Molecules are tightly connected (bonded) groups of two or more atoms of the same or different elements.■ In many cases, a piece of matter consists of a large number of molecules. (Later, we will describe another possibility, in which matter is composed of atoms having electrical charges.) The great variety of such chemical substances results from the many possible ways in which atoms can be bonded, or connected, to form molecules. Water is an example of a kind of matter composed of molecules. The water molecule consists of two atoms of hydrogen bonded to one atom of oxygen. Using atomic symbols, H for hydrogen and O for oxygen, we write H_2O to represent this molecule. The subscript 2 on the H indicates that the molecule contains two H atoms. A subscript 1 on any atomic symbol is not written, but is understood to be there.

■ The nature of the force of attraction between atoms that results in chemical bonding is discussed in Chapters 7 and 8.

The expression H_2O is called the formula of water. A **formula** gives the ratio of different kinds of atoms in a sample of substance. For example, a sample of water that has 90 H_2O molecules in it contains 2×90 H atoms and 1×90 O atoms. Ninety molecules is an exceptionally small sample.

Table 2.1
Some Common Elements

Name of Element	Atomic Symbol	Physical Appearance of Element*	Formula for Element[†]
Aluminum	Al	Silvery white metal	Al
Barium	Ba	Silvery white metal	Ba
Bromine	Br	Reddish brown liquid	Br_2
Calcium	Ca	Silvery white metal	Ca
Carbon	C		
(graphite)		Soft, black solid	C
(diamond)		Hard, colorless crystal	C
Chlorine	Cl	Greenish yellow gas	Cl_2
Chromium	Cr	Silvery white metal	Cr
Cobalt	Co	Silvery white metal	Co
Copper	Cu (from *cuprum*)	Reddish metal	Cu
Fluorine	F	Pale yellow gas	F_2
Hydrogen	H	Colorless gas	H_2
Iodine	I	Bluish black solid	I_2
Iron	Fe (from *ferrum*)	Silvery white metal	Fe
Lead	Pb (from *plumbum*)	Bluish white metal	Pb
Magnesium	Mg	Silvery white metal	Mg
Manganese	Mn	Gray white metal	Mn
Mercury	Hg (from *hydrargyrum*)	Silvery white liquid metal	Hg
Nickel	Ni	Silvery white metal	Ni
Nitrogen	N	Colorless gas	N_2
Oxygen	O	Colorless gas	O_2
Phosphorus (white)	P	Yellowish white waxy solid	P_4
Potassium	K (from *kalium*)	Soft, silvery white metal	K
Silicon	Si	Gray lustrous solid	Si
Silver	Ag (from *argentum*)	Silvery white metal	Ag
Sodium	Na (from *natrium*)	Soft, silvery white metal	Na
Sulfur	S	Yellow solid	S_8
Tin	Sn (from *stannum*)	Silvery white metal	Sn
Zinc	Zn	Bluish white metal	Zn

*Common form of the element under normal conditions.
[†]Formulas are discussed later in this section. Elements that exist as molecules are given molecular formulas. Otherwise, the atomic symbol is used to represent the element.

One gram of water (about one-fifth teaspoon) contains 3.3×10^{22} H_2O molecules. If you had a penny for every H_2O molecule in one gram of water, the stack of pennies would be over 300 million times the distance from the earth to the sun.

Example 2.1

A particular glass of wine contains 23.0 g of ethanol, C_2H_6O (ethyl alcohol). The number of C_2H_6O molecules in 23.0 g is 3.01×10^{23}. How many carbon atoms come from ethanol in this glass of wine?

Solution

Each C_2H_6O molecule contains two carbon atoms, so to form the conversion factor we can write

$$1 \ C_2H_6O \text{ molecule} \simeq 2 \text{ carbon atoms}$$

The symbol $\simeq$ means "is chemically equivalent to." It can be treated like an equals sign to form the conversion factor. Hence,

$$3.01 \times 10^{23} \ \overline{C_2H_6O \text{ molecules}} \times \frac{2 \text{ C atoms}}{1 \ \overline{C_2H_6O \text{ molecule}}}$$

$$\simeq 6.02 \times 10^{23} \text{ C atoms}$$

converts number of C_2H_6O molecules to number of C atoms

Exercise 2.1

A sample of methane, a major constituent of natural gas, contains 1.8×10^{24} CH_4 molecules. How many hydrogen atoms are there in the sample?

(See Problems 2.13 and 2.14.)

A formula, as we said, gives the relative number of different kinds of atoms in a substance. If this substance is composed of molecules, we usually write the **molecular formula,** which gives the exact number of different atoms in the molecule, as we did in writing H_2O for water. However, the atoms in a molecule do not combine randomly. Rather they are connected, or chemically bonded, in precise ways. A **structural formula** shows how the atoms in a molecule are bonded. The structural formula of water is H—O—H, the line joining atomic symbols representing a *chemical bond* between atoms. *Molecular models* are used to visualize the shapes and relative sizes of molecules. Figure 2.2 shows molecular and structural formulas and molecular models for several compounds.

Exercise 2.2

Propane is a gas that is easily liquefied. It is sold in metal canisters as a fuel for torches and stoves. The propane molecule has the structural formula

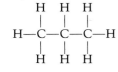

What is its molecular formula?

(See Problems 2.15 and 2.16.)

Ions and Ionic Substances

Although many substances are molecular, that is, composed of molecules, others are composed of ions. **Ions** are electrically charged atoms or groups of atoms. Table salt, for example, the substance described in the chapter opening as sodium chloride, is made up of ions and is an ionic substance.

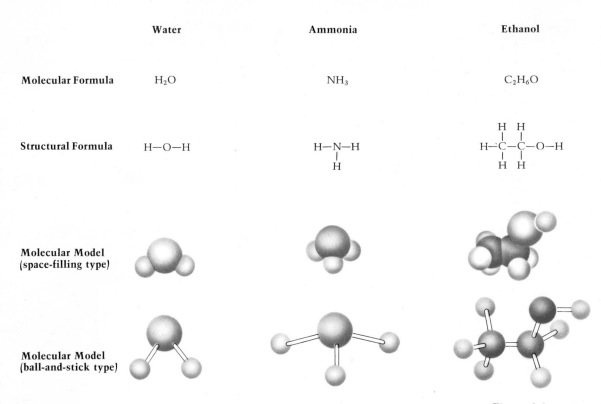

	Water	Ammonia	Ethanol
Molecular Formula	H_2O	NH_3	C_2H_6O

Structural Formula	H—O—H	H—N—H
		     |
		     H
			$H-\overset{\displaystyle H}{\underset{\displaystyle H}{C}}-\overset{\displaystyle H}{\underset{\displaystyle H}{C}}-O-H$

Molecular Model (space-filling type)			
Molecular Model (ball-and-stick type)			

Figure 2.2
Examples of molecular and structural formulas and molecular models. Three common molecules—water, ammonia, and ethanol—are shown.

To understand the relationship of ions to atoms, we must look briefly at the structure of an atom. An atom is composed of subatomic particles, some of which are electrically charged. *Electrons* are negatively charged subatomic particles. Although atoms are normally electrically neutral because their positive and negative charges are in balance, they become ions when electrons are transferred from one atom to another. An atom that picks up an extra electron becomes a negatively charged ion and is called an **anion.** An atom that loses an electron is left as a positively charged ion, called a **cation** (pronounced "cat-ion"). Sodium atoms, for example, tend to lose electrons to form sodium cations (denoted Na^+). Chlorine atoms tend to gain electrons to form chloride anions (denoted Cl^-).■

In sodium chloride, each Na^+ ion is surrounded by six Cl^- ions, and each Cl^- ion is in turn surrounded by six Na^+ ions. There are equal numbers of positive and negative charges from ions so that the whole arrangement is electrically neutral. Figure 2.3 shows this arrangement of ions, which is an example of a crystal. A **crystal** is a regular, repeating arrangement of ions, atoms, or molecules. The ions in sodium chloride have a cubic arrangement that shows up in the external shape of the crystal, which can be seen if you look closely at table salt.

The strong electrical attraction between positive and negative charges holds the ions together as a crystal. The size of a crystal varies, depending upon the total number of ions in it. Note that the sodium chloride crystal is not an array of NaCl molecules, but an array of equal numbers of Na^+ and

■ When we study chemical bonding in Chapter 7, we will see why some atoms tend to lose electrons and others tend to gain them. We will also learn to predict how many electrons are lost or gained by each atom in forming stable substances. We will see that sodium atoms tend to give Na^+ ions, whereas calcium atoms tend to give Ca^{2+} ions.

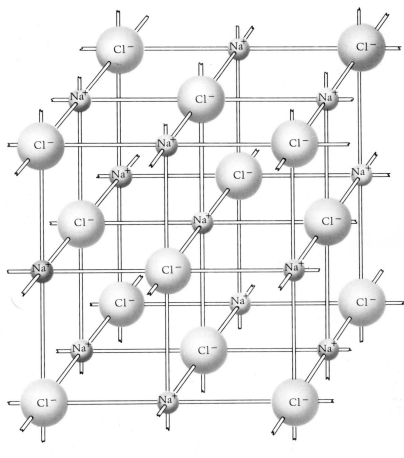

Figure 2.3
The sodium chloride crystal. (*Top*) A photograph showing the cubic shape of sodium chloride crystals. (*Right*) A model of a portion of the crystal detailing the regular arrangement of sodium ions and chloride ions. Each sodium ion is surrounded by six chloride ions, and each chloride ion by six sodium ions.

Cl^- ions. The formula of sodium chloride might be written Na^+Cl^- to indicate the *formula unit* or simplest ratio of the different ions in the crystal. However, a formula normally does not indicate whether a substance is molecular or ionic. The formula of sodium chloride is written without the ionic charges as NaCl. After we have discussed chemical bonding later in the text, you will be able to recognize when a substance is expected to be ionic and when it is molecular.

Some ions contain more than one atom, such as the sulfate ion, SO_4^{2-}. The superscript $2-$ (two minus) indicates an excess of two electrons on the ion. The formula unit of iron(III) sulfate contains two Fe^{3+} ions and three SO_4^{2-} ions. This formula is written $Fe_2(SO_4)_3$. The three SO_4^{2-} ions are shown by the subscript 3 after the parentheses enclosing the ion formula, which is written without charges. Note that each formula unit of $Fe_2(SO_4)_3$ contains 2 Fe atoms, 3 S atoms, and 3×4 O atoms.

Exercise 2.3

A crystal of sodium sulfate is composed of twice as many Na^+ ions as SO_4^{2-} ions. What is the formula of sodium sulfate?

(See Problems 2.17 and 2.18.)

Exercise 2.4

How many atoms of carbon are there for each atom of lead in the lead acetate crystal? The formula of lead acetate is $Pb(C_2H_3O_2)_2$.

(See Problems 2.19 and 2.20.)

A Word on Naming Substances

Chemical substances are given names that convey information about their formulas or molecular structures. The naming of chemical substances is called *chemical nomenclature.*

In naming ionic substances, the positive ion is given first. The same order is followed for the formula. For example, the substance containing Na^+ and SO_4^{2-} ions is named sodium sulfate, and the formula is Na_2SO_4. When more than one positive charge is possible for a metallic element, the charge is indicated in Roman numerals. For example, iron has ions Fe^{2+} and Fe^{3+}. The compound $Fe_2(SO_4)_3$ is composed of Fe^{3+} and SO_4^{2-} ions, and thus is named iron(III) sulfate. Similarly, $FeSO_4$ is composed of Fe^{2+} and SO_4^{2-} ions and is named iron(II) sulfate.■ Negative monatomic ions, or ions consisting of a single atom, such as Cl^- and O^{2-}, are named after the element, with the ending changed to *-ide.* Thus, Cl^- is chlor*ide* and O^{2-} is ox*ide.*

■ Chemical nomenclature, including this use of Roman numerals, is discussed more fully in Chapter 7.

Molecular substances composed of two elements can be named by using prefixes to indicate the number of atoms in the formula, such as *mono-* (one), *di-* (two), *tri-* (three), *tetra-* (four), *penta-* (five). For example, N_2O_4 is named *di*nitrogen *tetr*oxide. The suffix for the last-named element is *-ide.* (The element listed first in the formula and in the name is the one that is "most positive." Chapter 7 will show how this is decided.)

Chemical Reactions

Change is a familiar part of life. Plants grow, ice melts, a shiny new car eventually rusts. There are two kinds of change, physical change and chemical change. A **physical change** is a change in the form of a material but not in its composition or chemical identity. The melting of ice involves no change in the composition or chemical identity of water, so it is a physical change.

A **chemical change** or **chemical reaction** is the change of a substance into a new one that has a different chemical identity. It is usually accompanied by easily observed physical effects, such as the emission of light and heat, or color change.■ The burning of sodium in chlorine gas to produce salt is an example of a chemical change. So, too, is the rusting of iron, during which iron combines with oxygen in the air to form a new substance, rust. Plant growth involves many complex chemical changes, some of which are understood, many of which are not.

■ You are already familiar with many chemical reactions and their varied physical effects. A well-known antacid fizzes when the tablet is added to water because carbon dioxide gas is released. Fire is a dramatic chemical reaction that gives off heat energy and light. An explosion is even more dramatic; it is a fast chemical reaction accompanied by a sharp report.

An important feature of the atomic theory is that it accounts simply for chemical changes as rearrangement of the atoms of molecules. Consider the burning of natural gas, which is composed mostly of methane, CH_4. Using atomic theory, we describe this burning as the chemical reaction of one

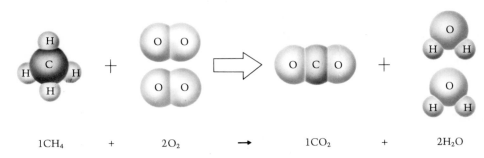

Figure 2.4
Molecular models representing the reaction of methane, CH_4, with oxygen, O_2, to give carbon dioxide, CO_2, and water, H_2O.

$$1CH_4 \qquad + \qquad 2O_2 \qquad \longrightarrow \qquad 1CO_2 \qquad + \qquad 2H_2O$$

molecule of methane, CH_4, with two molecules of oxygen, O_2, to form one molecule of carbon dioxide, CO_2, and two molecules of water, H_2O:

$$CH_4 \quad + \quad 2O_2 \quad \longrightarrow \quad CO_2 \quad + \quad 2H_2O$$

| one molecule of methane | + | two molecules of oxygen | react to form | one molecule of carbon dioxide | + | two molecules of water |

The reaction starts with one carbon atom, four hydrogen atoms, and four oxygen atoms arranged as one molecule of methane and two molecules of oxygen. During the reaction these atoms are rearranged to give one molecule of carbon dioxide and two molecules of water. *None of the atoms are lost, and no new atoms are created.* Figure 2.4 shows the reaction with molecular models.

2.2 Balancing Simple Chemical Equations

A **chemical equation** is the symbolic representation of a chemical reaction. For example, the burning of methane discussed in the previous section is represented

$$CH_4 + 2O_2 \longrightarrow CO_2 + 2H_2O$$

The formulas on the left side of the equation before the arrow represent the starting substances, or **reactants.** The arrow means "react to form" or "yield". The formulas on the right represent the substances that result from the reaction, or the **products.** The numbers in front of the formulas in the chemical equation give the relative number of molecules or formula units involved and are called **coefficients** (1's are usually not written). When the coefficients are correctly given, the numbers of each kind of atom are equal on both sides of the equation, which is then said to be *balanced.*

Before we can write the chemical equation for a reaction, we must determine *by experiment* those substances that are reactants and those that are products. We must also determine the formulas of each of these substances. Once we know these things, we can write the chemical equation. For example, by experiment we determine that propane reacts with oxygen to give carbon dioxide and water. We also determine from experiment that the formulas of propane, oxygen, carbon dioxide, and water are C_3H_8, O_2, CO_2, and H_2O, respectively. Now we can write

$$C_3H_8 + O_2 \longrightarrow CO_2 + H_2O$$

This equation is not balanced because the coefficients that give the relative number of molecules involved have not yet been determined. To balance the equation, select coefficients that will make the numbers of each kind of atom equal on both sides of the equation.

Since there are three carbon atoms on the left side of the above equation (C_3H_8), we must have three carbon atoms on the right. This is achieved by writing 3 for the coefficient of CO_2.

$$1C_3H_8 + O_2 \longrightarrow 3CO_2 + H_2O$$

(We have written the coefficient for C_3H_8 to show that it is now determined; normally when the coefficient is 1, it is omitted.) Similarly, there are eight hydrogen atoms on the left (in C_3H_8), so we write 4 for the coefficient of H_2O.

$$1C_3H_8 + O_2 \longrightarrow 3CO_2 + 4H_2O$$

The coefficients on the right are now determined, and there are ten oxygen atoms on this side of the equation: 6 from the three CO_2 molecules and 4 from the four H_2O molecules. We now write 5 for the coefficient of O_2. We drop the coefficient 1 from C_3H_8 and have the balanced equation.

$$C_3H_8 + 5O_2 \longrightarrow 3CO_2 + 4H_2O$$

It is a good idea to check your work by counting the number of each kind of atom on the left side and then on the right side of the equation.

The combustion of propane can be equally well represented by

$$2C_3H_8 + 10O_2 \longrightarrow 6CO_2 + 8H_2O$$

in which the coefficients of the previous equation have been doubled. Usually, however, it is preferable to *write the coefficients so that they are the smallest whole numbers possible.* When we refer to a balanced equation, we will assume that this is the case unless otherwise specified.

The method we have outlined—called *balancing by inspection*—is essentially a trial-and-error method.■ It can be made easier, however, by observing the following rule. Balance first those kinds of atoms that occur in only one substance on each side of the equation. The usefulness of this rule is illustrated in the next example. Remember that in any formula, such as $Fe_2(SO_4)_3$, a subscript to the right of the parentheses multiplies each subscript within the parentheses. Thus, $Fe_2(SO_4)_3$ represents 4×3 oxygen atoms. Remember, too, that you cannot change a subscript in any formula; only the coefficients can be altered to balance an equation.

■ Balancing equations this way is a relatively fast method for all but the most complicated equations. Special methods for these equations are described in Chapter 9.

Example 2.2

Balance the following equations:
(a) $H_3PO_3 \longrightarrow H_3PO_4 + PH_3$
(b) $Ca + H_2O \longrightarrow Ca(OH)_2 + H_2$
(c) $Fe_2(SO_4)_3 + NH_3 + H_2O \longrightarrow Fe(OH)_3 + (NH_4)_2SO_4$

Solution

(a) Only oxygen occurs in just one of the products (H_3PO_4). It is therefore easiest to balance O atoms first.

$$4H_3PO_3 \longrightarrow 3H_3PO_4 + PH_3$$

This equation is now also balanced in P and H atoms.

(Continued)

Thus the balanced equation is

$$4H_3PO_3 \longrightarrow 3H_3PO_4 + PH_3$$

(b) The equation is balanced in Ca atoms as its stands.

$$1Ca + H_2O \longrightarrow 1Ca(OH)_2 + H_2$$

O atoms occur in only one reactant and in only one product, so they are balanced next.

$$1Ca + 2H_2O \longrightarrow 1Ca(OH)_2 + H_2$$

The equation is now also balanced in H atoms. Thus, the answer is

$$Ca + 2H_2O \longrightarrow Ca(OH)_2 + H_2$$

(c) To balance the equation in Fe atoms, we write

$$1Fe_2(SO_4)_3 + NH_3 + H_2O \longrightarrow 2Fe(OH)_3 + (NH_4)_2SO_4$$

We balance the S atoms by placing the coefficient 3 for $(NH_4)_2SO_4$.

$$1Fe_2(SO_4)_3 + NH_3 + H_2O \longrightarrow 2Fe(OH)_3 + 3(NH_4)_2SO_4$$

Now we balance N atoms:

$$1Fe_2(SO_4)_3 + 6NH_3 + H_2O \rightarrow 2Fe(OH)_3 + 3(NH_4)_2SO_4$$

Finally, the O atoms are balanced. First count the number of O atoms on the right (18). Then count the number of O atoms in substances on the left with known coefficients. There are 12 O's in $Fe_2(SO_4)_3$; hence, the number of remaining O's (in H_2O) must be $18 - 12 = 6$.

$$1Fe_2(SO_4)_3 + 6NH_3 + 6H_2O \longrightarrow 2Fe(OH)_3 + 3(NH_4)_2SO_4$$

The answer is

$$Fe_2(SO_4)_3 + 6NH_3 + 6H_2O \longrightarrow 2Fe(OH)_3 + 3(NH_4)_2SO_4$$

Check each of the final equations by counting the number of each kind of atom on both sides of the equations.

Exercise 2.5

Find the coefficients that balance each of the following equations:
(a) $O_2 + PCl_3 \longrightarrow POCl_3$
(b) $P_4 + N_2O \longrightarrow P_4O_6 + N_2$
(c) $As_2S_3 + O_2 \longrightarrow As_2O_3 + SO_2$
(d) $Ca_3(PO_4)_2 + H_3PO_4 \longrightarrow Ca(H_2PO_4)_2$

(See Problems 2.25, 2.26, 2.27, 2.28, 2.29, 2.30, 2.31, and 2.32.)

2.3 Classifications of Matter

In chemistry, there are two fundamental ways of classifying materials. We can classify a material by its chemical constitution as an element, a compound, or a mixture. And, we can classify it by its physical state as solid, liquid, or gas.

Chemical Constitution—Element, Compound, or Mixture?

An element, as we mentioned earlier, is a kind of matter, or material, composed of only one chemically distinct kind of atom. The element oxygen is a gas composed of O_2 molecules. The physical appearance and formula of some common elements are given in Table 2.1. Compounds, you will recall, are materials composed of atoms of two or more elements. The composition of a compound is constant, that is, the relative numbers of atoms of each element in every sample are constant. Water (H_2O), methane (CH_4), and

Figure 2.5
Classification of matter by
its chemical constitution. A
material may be classified as
an element, a compound, a
homogeneous mixture, or a
heterogeneous mixture.

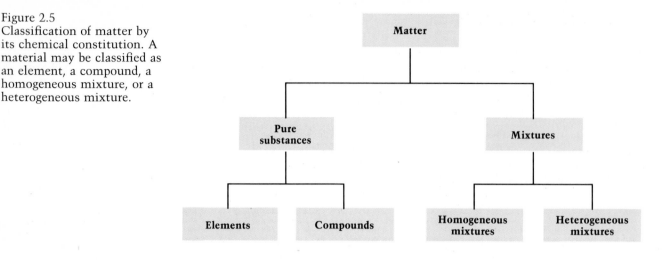

ethanol (C_2H_6O) are compounds. Ethanol, for example, is always composed of two carbon atoms and six hydrogen atoms for each oxygen atom.

Elements and compounds are pure substances. A **pure substance** is a material that has constant composition and definite *properties,* or distinctive traits. **Chemical properties** of a substance involve chemical changes. A chemical property of iron is that it rusts. **Physical properties** are properties that do not involve chemical changes. Examples are density and melting point (temperature at which melting occurs). A material that is a **mixture** of pure substances has a composition and properties that vary with the relative amounts of pure substances in the mixture. The properties of a mixture depend on its composition.

Mixtures are classified into two types. *Heterogeneous mixtures* consist of distinct parts. A mixture of sugar and salt is a heterogeneous mixture. Looking closely, we can see separate crystals of sugar and salt. *Homogeneous mixtures* are uniform throughout (every bit of the mixture appears the same). A homogeneous mixture is usually called a **solution.** Air is a gaseous solution, principally of N_2 and O_2. The mixture is uniform, so we cannot see separate parts by any physical means. The uniformity exists down to the molecular level; that is, the N_2 and O_2 molecules are completely mixed. Likewise, as sugar dissolves in water to give a solution, sugar molecules are dispersed thoroughly among the water molecules.

Figure 2.5 shows the relationship of elements, compounds, and mixtures.

Exercise 2.6

Classify each of the following as an element, compound, or mixture: (a) silver alloy (containing varying amounts of silver and copper)   (b) iodine (a black, lustrous solid with the formula I_2)   (c) baking powder (typically contains sodium hydrogen carbonate, $NaHCO_3$, and potassium hydrogen tartrate, $KHC_4H_4O_6$)   (d) mercury(II) oxide (a red, crystalline substance with the formula HgO)   (e) magnesium (a metal with the formula Mg).

(See Problems 2.33 and 2.34.)

Physical State—Solid, Liquid, or Gas?

Matter can exist in different forms—solid, liquid, or gas—under different conditions. Water, for example, exists as ice (solid water), as liquid water, and as steam (gaseous water). These different forms are referred to as the **states of matter.** They may be characterized by the physical properties of compressibility, fluidity, and density.

If a form of matter can be easily made to fit into a smaller volume by applying a slight pressure, it is *compressible.* Air is compressible. You can put more and more of it into a tire, which will increase only slightly in volume. A piece of steel is relatively *incompressible*.

If matter takes the shape of its container and can be poured, it is *fluid.* (Looking at matter like sand, we apply the test to the individual grains. They are not fluid.) Liquid water is fluid. Dry Ice is a solid form of carbon dioxide and is therefore not fluid, but *rigid.* But if we place this solid in a beaker, it will transform into the gaseous state and fill the beaker. The gas can then be poured over a candle flame, which is extinguished by the lack of oxygen (Figure 2.6). Thus we see that a gas is fluid.

Density, as we discussed earlier, is the quantity of matter (mass) per unit volume. Recall also that the solid and liquid states of a substance have relatively high densities, but that the gaseous state has much lower density. The densities of ice and liquid water are 0.92 g/cm^3 and 1.00 g/cm^3 at 0°C, respectively. The density of steam at 100°C, 1 atm is 5.9×10^{-4} g/cm^3.

Figure 2.6
The solid and gaseous states of carbon dioxide. (*Left*) Dry Ice is a solid form of carbon dioxide. It vaporizes to form a gas that is more dense than air and will remain for some time in a beaker. (*Right*) Carbon dioxide gas flows from the beaker and surrounds the candle flame, which is extinguished by the gas.

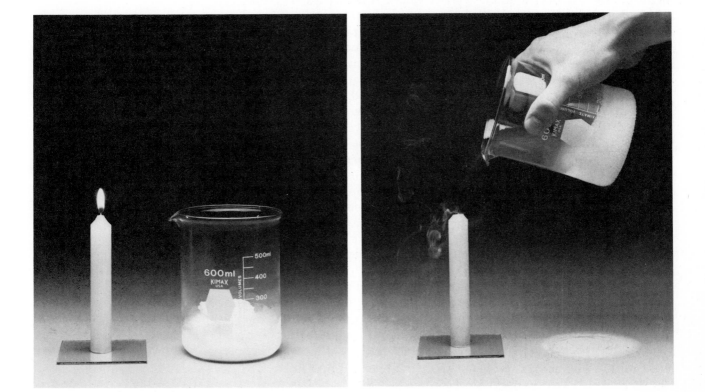

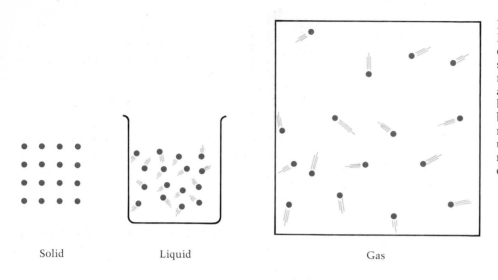

Figure 2.7
Representations of the states of matter. In a solid, the basic units (atoms, ions, or molecules) are closely spaced at fixed sites. In a liquid, the basic units are closely spaced but are in constant, random motion. In a gas, the basic units are in constant, random motion through largely empty space.

Solid   Liquid   Gas

These properties—compressibility or incompressibility, fluidity or rigidity, and low or high density—result from the structure of the material. A **solid** is characterized by rigidity; it does not flow, nor is it significantly compressible. Solids have definite volume and shape, and they are relatively dense. The structural units of a solid are atoms, ions, or molecules. These basic units are closely spaced, giving a relatively high density, and at approximately fixed sites, giving rigidity (see Figure 2.7, *left*).

Liquids have definite volume but, unlike solids, do not have definite shape. When poured from one vessel to another, a liquid flows to assume the contours of its container. Like a solid, however, a liquid is essentially incompressible, and its density is comparable to that of the corresponding solid. We explain these facts by saying that, like a solid, a liquid consists of atoms, ions, or molecules that are closely spaced. In a liquid, however, these units are not at fixed sites but are in constant motion, giving changing, random arrangements (Figure 2.7, *center*).

A **gas** is similar to a liquid in being fluid, but it differs in being compressible and having relatively low density.■ We explain these properties by saying that a gas, unlike a liquid or a solid, consists of molecules, atoms, or ions that are in constant, random motion through mostly empty space (Figure 2.7, *right*). These characteristics of the states of matter are summarized in Table 2.2.

The term **phase** is used to indicate the different homogeneous portions of a material, separated from other parts by observable boundaries. A phase may be either a pure substance or a solution in any one of the three states of matter. Thus, a glass of water with ice in it contains two phases, a liquid water phase and a solid ice phase (or three, if we include the gaseous phase over the water). A mixture of salt and sand contains two solid phases, the salt phase and the sand phase. A mixture of oil floating on water has two liquid phases, an oil phase and a water phase. Figure 2.8 illustrates these examples of phases.

■ Butane, C_4H_{10}, is an easily liquefied gaseous fuel derived from petroleum. It is the colorless liquid in some disposable cigarette lighters, where it is contained under a slight pressure. When you open the valve on a lighter, the liquid vaporizes, and the gas issues from the jet. The resulting gas occupies a volume 240 times that of the liquid at 0°C, 1 atm pressure.

Table 2.2
Characteristics of the
States of Matter

State of Matter	Compressibility	Fluidity or Rigidity	Relative Density	Structure
Gas	High	Fluid	Low	Atoms, ions, or molecules in constant, random motion through mostly empty space.
Liquid	Very low	Fluid	High	Atoms, ions, or molecules closely spaced and in constant motion.
Solid	Very low	Rigid	High	Atoms, ions, or molecules at closely spaced, approximately fixed sites.

2.4 Separation of Mixtures

The separation of mixtures into their components is of great commercial importance, as well as a necessary step in the preparation of a pure compound in the laboratory. A raw ore is often treated to obtain the pure metal compound before subsequent steps are undertaken to isolate the desired metal. The juice of sugar beets or cane is concentrated, crystallized, and centrifuged to give white table sugar.■ Steel objects are picked out by magnetic separators from other refuse at recycling plants. In this section, we will discuss three common methods for separating mixtures: filtration, distillation, and chromatography.

■ A centrifuge is a device used to separate large molecules or small crystals from a solution by spinning. The spin cycle of a clothes washer separates water from the clothes by centrifugation.

Filtration

Filtration is the process of separating solid particles suspended in a liquid solution by pouring the mixture through a filter. For example, photographic developing solution contains silver compounds. If sodium chloride is added

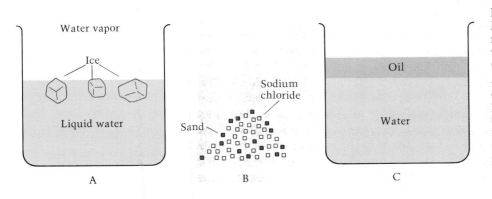

A   B   C

Figure 2.8
Materials that consist of more than one phase. (a) Ice water has three phases: liquid water, ice and vapor. (b) A mixture of sodium chloride and sand has two solid phases: a sand phase and a sodium chloride phase. (c) A mixture of oil and water has two liquid phases: an oil phase and a water phase.

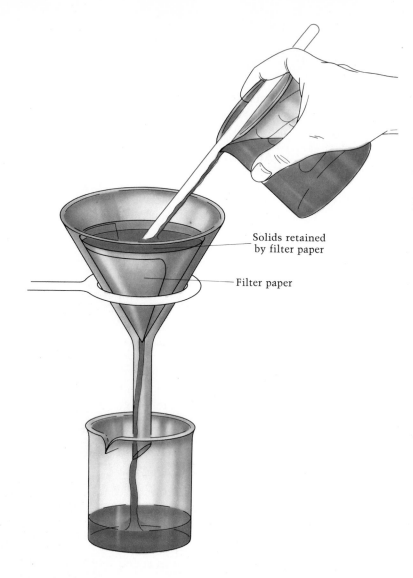

Figure 2.9
Separation of suspended solids by filtration. A filter-paper disk is folded in quarters, opened into a cone, and placed in the funnel. The solution containing the suspended solids is poured carefully from the beaker down a glass rod (to prevent splashing) and into the filter cone. The suspended solids are retained on the filter paper and may be washed of impurities by pouring water over them.

Solids retained
by filter paper

Filter paper

to the developing solution, fine crystals of silver chloride form. When this mixture is poured into a cone made of filter paper (Figure 2.9), the solution passes through, leaving the silver chloride behind. In this way, the silver can be recovered from the developing solution (as silver chloride).

Distillation

A liquid or solid that changes readily to the gaseous state is said to be *volatile.* When the temperature of a volatile liquid is increased sufficiently, the liquid boils, changing rapidly to the gaseous or vapor state. **Distillation** uses the differences in volatility of substances to separate a solution into its components. For example, if a solution of sodium chloride in water is heated, the water (which is fairly volatile) boils off, leaving behind sodium chloride (which is not volatile). Figure 2.10 shows a typical laboratory distillation

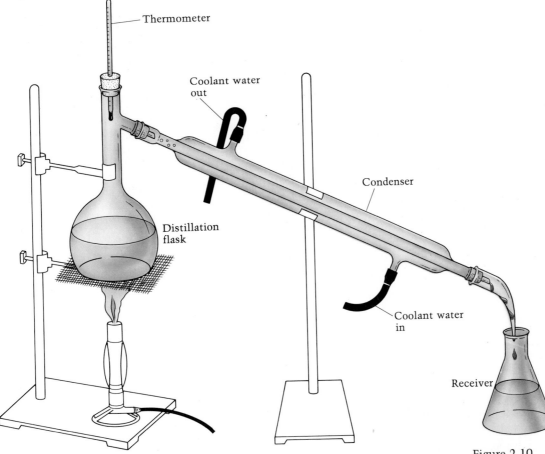

Thermometer

Coolant water
out

Condenser

Distillation
flask

Coolant water
in

Receiver

Figure 2.10
A simple distillation apparatus. Volatile components pass from the flask into the condenser. There the vapor changes back to liquid, which is collected in the receiver. Less volatile or nonvolatile components remain in the distillation flask.

apparatus. The sodium chloride solution is placed in the *distillation flask* and heated. The water vaporizes and passes into the *condenser*, leaving sodium chloride crystals in the flask. In the condenser, the water vapor is cooled and changes back to a liquid. The pure (distilled) water is collected in the *receiver*.

Fractional distillation is useful when the solution consists of two or more volatile components. The temperature at which a liquid boils is known as its *boiling point*. When a solution containing liquid substances of different boiling points is distilled, the substance with the lowest boiling point normally distills over first.■ An apparatus for fractional distillation is shown in Figure 2.11. A *fractionating column*, containing glass beads, is placed at the top of the distillation flask. The glass beads provide a surface on which the less volatile part of the vapor condenses. The temperature varies along the length of the column, being hotter at the bottom and cooler at the top. The substance with the lowest boiling point passes over into the condenser from the top of the column, whereas substances with higher boiling points condense on the beads in the column or remain in the distillation flask. As the substance of lowest boiling point is removed from the flask, the temperature of the boiling liquid increases and the next substance, now the one with the lowest boiling point, begins to distill.

■ Although solutions of volatile liquids usually increase in boiling point as the more volatile components distill off, some solutions exhibit a constant boiling point. For example, a solution of 95.6% ethanol (grain alcohol) and 4.4% water by mass boils at the constant temperature of 78.15°C; pure ethanol boils at 78.5°C. Such a constant-boiling mixture is called an *azeotrope*. It is impossible to separate ethanol and water completely by fractional distillation of simple ethanol–water mixtures.

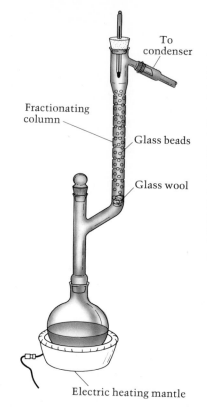

Figure 2.11
An apparatus for fractional distillation. The temperature varies up the length of the fractionating column, being hotter at the lower end and cooler at the top. The most volatile component passes over into the condenser, whereas less volatile components condense on the beads in the column or remain in the distillation flask. As the most volatile component distills over, the temperature of the boiling liquid increases until the next most volatile component distills over.

Fractional distillation is employed commercially to separate crude oil or petroleum into useful products. For this, the distillation is operated continuously, various fractions being taken off at different heights up the fractionating column. Thus, the lowest boiling fraction (20°–60°C), called petroleum ether, is obtained near the top of the column. Light naphtha or ligroin is a fraction with a slightly higher boiling point (60°–100°C), and it is taken off the column just below the petroleum ether. Gasoline boils between 50°C and 200°C, and kerosene between 175°C and 275°C; these come off even lower on the column. Furnace and diesel fuels boil at still higher temperatures. ■

Chromatography

Chromatography is the name given to a group of similar, very useful separation techniques that involve a stationary phase and a moving, or mobile, phase. The separation of dye pigments by chromatography can be seen if a drop of ink is placed on a sheet of paper. The spot grows outward by capillary action or wetting, each dye moving with the solution at a different, but characteristic, rate. The result is a dark disk surrounded by colored rings. Wetted paper fibers form the stationary phase, and the ink solution is the mobile phase. The rate at which each dye moves depends upon how strongly the dye is attracted to the wet fibers in the paper.

The Russian botanist Mikhail Tswett was the first to separate chemical substances by chromatography. He described the use of *column chro-*

■ Petroleum is a mixture of many compounds. Most of them are *hydrocarbons,* or compounds containing only hydrogen and carbon. Petroleum ether consists mostly of hydrocarbons with five or six carbon atoms. Pentane, C_5H_{12}, for example, has the structural formula

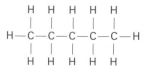

Ligroin contains hydrocarbon molecules of 6 to 7 carbon atoms, gasoline has molecules with 6 to 12 carbon atoms, and kerosene has molecules with about 12 to 18 carbon atoms.

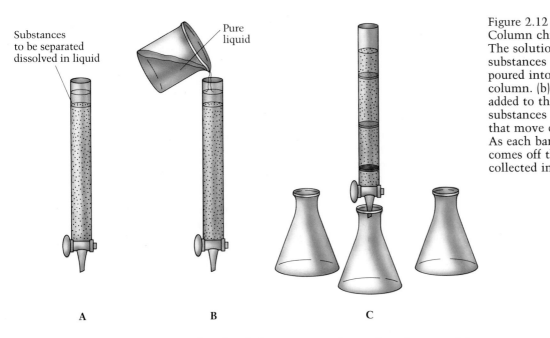

Figure 2.12
Column chromatography. (a) The solution containing the substances to be separated is poured into the top of the column. (b) Pure liquid is added to the column. (c) The substances separate as bands that move down the column. As each band of substance comes off the column, it is collected in a flask.

matography in 1906. Tswett dissolved the pigments from plant leaves with the liquid petroleum ether. After packing a glass tube or column with powdered chalk (calcium carbonate, $CaCO_3$), he poured the solution of plant pigments into the top of the column. When he washed, or *eluted*, the column by pouring in more pure petroleum ether after the solution, it began to show distinct yellow and green bands. These bands, each containing a pure pigment, became well separated as they moved down the column, so that the pure pigments could be obtained (see Figure 2.12).

Vapor phase chromatography (VPC), also called *gas chromatography* (GC), is a more recent separation method. Here the mobile phase consists of a mixture of gaseous or vaporized substances plus a gas such as helium, called the *carrier*, that acts as an eluting agent. The stationary phase consists of a solid or a liquid adhering to a solid, packed in a column. As the gas passes through the column, substances in the mixture are attracted differently to the stationary phase and thus are separated. ■

Vapor phase chromatography is a rapid method of separating mixtures and has become important in *chemical synthesis*, the preparation of complicated substances from simpler ones. A synthesis may require many chemical reactions, each one followed by the separation of products. With vapor phase chromatography, a synthesis that formerly took months might instead be completed in days. Vapor phase chromatography can also be used to detect minute quantities of substances. For example, as little as a picogram of amphetamine can be detected in urine using vapor phase chromatography with a special detector. ■

■ Automated VPC or GC units have been used in interplanetary space probes (such as *Viking* and *Pioneer Venus*) to analyze Martian soil and planetary atmospheres.

■ The detector used for such small quantities is the mass spectrometer, an instrument that measures the masses of atoms, molecules, and molecular fragments. It is described in Chapter 5.

Exercise 2.7

Briefly describe how you might separate each of the following mixtures into pure components: (a) sand and salt (b) sand, salt, and iron filings (c) benzene and toluene (Benzene, C_6H_6, is a liquid that boils at 80°C; toluene, C_7H_8, is a liquid that boils at 111°C.) (d) a gaseous mixture of ethane and ethylene.

(See Problems 2.37 and 2.38.)

A Checklist for Review

Important Terms

atomic theory (2.1)
atoms (2.1)
elements (2.1)
compounds (2.1)
chemical reaction (2.1)
atomic symbols (2.1)
molecules (2.1)
formula (2.1)
molecular formula (2.1)
structural formula (2.1)
ions (2.1)

anion (2.1)
cation (2.1)
crystal (2.1)
physical change (2.1)
chemical change (2.1)
chemical equation (2.2)
reactants (2.2)
products (2.2)
coefficients (2.2)
pure substances (2.3)
chemical properties (2.3)

physical properties (2.3)
mixture (2.3)
solution (2.3)
states of matter (2.3)
solid (2.3)
liquid (2.3)
gas (2.3)
phase (2.3)
filtration (2.4)
distillation (2.4)
chromatography (2.4)

Summary of Facts and Concepts

Atomic theory is central to chemistry. According to this theory, all matter is composed of small particles, or *atoms*. There are 108 chemically distinct atoms. Atoms are combined into tightly bonded groups (*molecules* or *ions*), which are responsible for the characteristics of a chemical substance. *Chemical reaction* occurs when the atoms in substances rearrange and combine into new substances. Chemists symbolize a reaction by a *chemical equation*, writing a chemical formula for each reactant and product. The coefficients in the equation indicate the relative numbers of reactant and product molecules or ions. Once the reactants and products, together with their formulas, have been determined by ex-periment, the coefficients are determined by *balancing* the numbers of each kind of atom on both sides of the equation.

Matter may be classified by its chemical constitution into *elements* (substances made up of only one chemically distinct kind of atom), *compounds* (substances made up of more than one element), or *mixtures* of substances. Matter may also be classified according to its physical state as a *gas, liquid,* or *solid.* Since any chemical reaction involves mixtures of substances, techniques of separating them—such as *filtration, distillation,* and *chromatography*—are very important.

Operational Skills

1. Given the number of molecules in a sample of compound of specified formula, calculate the number of atoms of a particular kind; or from the number of atoms, determine the number of molecules (Example 2.1).

2. Given the formulas of the reactants and products in a chemical reaction, obtain the coefficients of the balanced equation (Example 2.2).

Review Questions

2.1 Describe atomic theory and discuss how it explains the great variety of substances. How does it explain chemical reactions?

2.2 Ethane consists of molecules with two atoms of carbon and six atoms of hydrogen. Write the molecular formula for ethane.

2.3 What is the difference between a molecular formula and a structural formula?

2.4 What must be known from experiment and what can be done using pencil and paper to write a balanced chemical equation?

2.5 Consider the following chemical reaction, written symbolically:

$$N_2 + 3H_2 \longrightarrow 2NH_3$$

What does this mean in words? How many nitrogen atoms are there on the left side of the arrow? How many on the right? Are there the same number on the right as on the left? Answer the same questions for hydrogen.

2.6 Consider a mixture of 10 million O_2 molecules and 20 million H_2 molecules. In what way is this similar to 20 million water molecules? In what way is it dissimilar?

2.7 Give examples of an element, a compound, a heterogeneous mixture, and a homogeneous mixture.

2.8 Characterize gases, liquids, and solids in terms of compressibility, fluidity, and density. Describe these states of matter in molecular terms.

2.9 Give an example of a pure substance consisting of two phases. Give an example of a heterogeneous mixture consisting of two phases. What are the different phases in each example?

2.10 For each of the following methods of separation, give an example of a mixture that can be separated by the method and briefly describe how this separation is done: (a) filtration (b) distillation (c) chromatography.

Problems

Symbols and Formulas

2.11 Name each of the elements in the following formulas:
  (a) KSCN   (b) Na_3PO_4
  (c) $CuSO_4$ (d) CCl_4

2.13 One gram of toluene contains 6.5×10^{21} C_7H_8 molecules. How many carbon atoms are there in this mass of toluene? How many hydrogen atoms?

2.15 Give the molecular formulas for the following structural formulas:

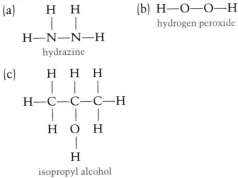

(a) hydrazine

(b) H—O—O—H
hydrogen peroxide

(c) isopropyl alcohol

2.12 For each atomic symbol, give the name of the element.
  (a) Ca   (b) Pb   (c) Hg   (d) Sn   (e) Mg

2.14 A sample of ammonia, NH_3, contains 3.3×10^{21} hydrogen atoms. How many NH_3 molecules are there in the sample?

2.16 What molecular formulas correspond to the following structural formulas?

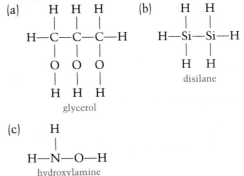

(a) glycerol

(b) disilane

(c) hydroxylamine

2.17 Aluminum sulfate contains Al^{3+} and SO_4^{2-} ions in the ratio 2:3. What is the formula of aluminum sulfate?

2.19 Iron(III) sulfate has the formula $Fe_2(SO_4)_3$. What is the ratio of iron atoms to oxygen atoms in this compound?

2.21 Given that an ionic compound is electrically neutral, what is the formula of barium phosphate, a compound of the ions Ba^{2+} and PO_4^{3-}?

2.18 Potassium chromium sulfate has the formula $KCr(SO_4)_2$. What is the ratio of chromium ions, Cr^{3+}, to sulfate ions, SO_4^{2-}, in the compound?

2.20 Ammonium phosphate, $(NH_4)_3PO_4$, has how many hydrogen atoms for each oxygen atom?

2.22 What is the formula of magnesium citrate, an ionic compound of magnesium ion, Mg^{2+}, and citrate ion, $C_6H_5O_7^{3-}$? Note that the compound must be electrically neutral.

Equations

2.23 How many total oxygen atoms are there in $As_4O_6 + 6H_2O$?

2.24 In the equation, $2PbS + O_2 \longrightarrow 2PbO + 2SO_2$, how many oxygen atoms are there on the right side? Is the equation balanced as written?

2.25 Balance the following equations:
(a) $Sn + NaOH \longrightarrow Na_2SnO_2 + H_2$
(b) $Al + Fe_3O_4 \longrightarrow Al_2O_3 + Fe$
(c) $CH_3OH + O_2 \longrightarrow CO_2 + H_2O$
(d) $P_4O_{10} + H_2O \longrightarrow H_3PO_4$
(e) $PCl_5 + H_2O \longrightarrow H_3PO_4 + HCl$

2.26 Balance the following equations:
(a) $Cl_2O_7 + H_2O \longrightarrow HClO_4$
(b) $MnO_2 + HCl \longrightarrow MnCl_2 + Cl_2 + H_2O$
(c) $Na_2S_2O_3 + I_2 \longrightarrow NaI + Na_2S_4O_6$
(d) $Al_4C_3 + H_2O \longrightarrow Al(OH)_3 + CH_4$
(e) $NO_2 + H_2O \longrightarrow HNO_3 + NO$

2.27 Find the correct coefficients for the following equations:
(a) $SbCl_5 + H_2O \longrightarrow SbOCl_3 + HCl$
(b) $MgO + Si \longrightarrow Mg + SiO_2$
(c) $CaCl_2 + Na_2CO_3 \longrightarrow CaCO_3 + NaCl$
(d) $C_6H_6 + O_2 \longrightarrow CO_2 + H_2O$
(e) $Al_2S_3 + H_2O \longrightarrow Al(OH)_3 + H_2S$

2.28 Find the correct coefficients for the following equations:
(a) $TiCl_4 + H_2O \longrightarrow TiO_2 + HCl$
(b) $Fe_3O_4 + H_2 \longrightarrow Fe + H_2O$
(c) $V_2O_5 + H_2 \longrightarrow V_2O_3 + H_2O$
(d) $(NH_4)_2Cr_2O_7 \longrightarrow Cr_2O_3 + H_2O + N_2$
(e) $C_4H_{10} + O_2 \longrightarrow CO_2 + H_2O$

2.29 Calcium phosphate, $Ca_3(PO_4)_2$, reacts with sulfuric acid, H_2SO_4, to give phosphoric acid, H_3PO_4, and calcium sulfate, $CaSO_4$, which comes out of the solution as a solid. Write a balanced equation for the reaction.

2.30 Sodium metal reacts with water, giving sodium hydroxide, $NaOH$, and releasing hydrogen, H_2. Write a balanced equation for the reaction.

2.31 Ammonium chloride, NH_4Cl, reacts with a solution of barium hydroxide, $Ba(OH)_2$, to give off ammonia gas, NH_3. Barium chloride, $BaCl_2$, and water are also products. Write a balanced equation for the reaction.

2.32 Lead is produced by heating lead sulfide, PbS, with lead sulfate, $PbSO_4$. Sulfur dioxide gas, SO_2, is the other product. Write a balanced equation for the reaction.

Classifications of Matter

2.33 State whether each of the following is an element, a compound, or a mixture:
(a) glucose (blood sugar; the formula is $C_6H_{12}O_6$)
(b) neon (a gas with the formula Ne)
(c) granite (a rock containing chiefly the substances quartz, orthoclase, and mica)
(d) bronze (a material made of varying amounts of copper and tin)
(e) isopropyl alcohol (a substance with the formula C_3H_8O)

2.34 Of the following, which are elements, which are compounds, and which are mixtures?
(a) aspirin (a substance with the formula $C_9H_8O_4$)
(b) selenium (a substance with the formula Se)
(c) milk of magnesia (a suspension of solid particles of magnesium hydroxide, $Mg(OH)_2$, in water)
(d) gasoline (a fuel containing various substances called hydrocarbons, such as C_6H_{14}, C_7H_{16}, and C_8H_{18})
(e) caustic soda (a substance with the formula $NaOH$)

2.35 Give the state of matter (gas, liquid, or solid) under normal conditions for each of the following:
(a) oxygen, O_2
(b) silver chloride, AgCl
(c) carbon dioxide, CO_2
(d) bromine, Br_2

2.36 Classify the following according to the state of matter under normal conditions:
(a) isopropyl alcohol, C_3H_8O (rubbing alcohol)
(b) sodium, Na
(c) methane, CH_4
(d) chlorine, Cl_2

Separation of Mixtures

2.37 Name or briefly describe a method you could use to separate each of the following:

(a) a mixture of the liquids heptane, C_7H_{16}, and octane, C_8H_{18}

(b) a mixture of sodium chloride, NaCl, and silver chloride, AgCl

(c) a powdered mixture of sulfur and iron

(d) the pigments from an extract of leaves

2.38 Name or briefly describe a method you could use to separate each of the following:

(a) the pigment in beet juice

(b) a mixture of the liquids benzene, C_6H_6, and toluene, C_7H_8

(c) the components of gunpowder—a mixture of charcoal, potassium nitrate (soluble in water), and sulfur (soluble in carbon disulfide)

(d) magnesium hydroxide, $Mg(OH)_2$, from milk of magnesia (solid magnesium hydroxide particles in water)

Additional Problems

2.39 A sample of iron(III) nitrate, $Fe(NO_3)_3$, contains 7.8×10^{23} Fe atoms. How many O atoms are there in the sample?

2.40 A sample of acetic acid, $HC_2H_3O_2$, contains 4.9×10^{22} acetic acid molecules. How many H atoms are there in the sample?

2.41 When ammonia, NH_3, burns in oxygen in the presence of a catalyst (a material that speeds up the reaction), nitric oxide, NO, and water are produced. Write a balanced equation for this reaction.

2.42 When lead sulfide, PbS, is heated in the presence of oxygen, it yields lead(II) oxide, PbO, and sulfur dioxide, SO_2. Write a balanced equation for this reaction.

2.43 Powdered iron and sulfur are mixed and then heated to produce iron(II) sulfide, according to the equation

$$8Fe + S_8 \longrightarrow 8FeS$$

The resulting mixture may contain unreacted iron and sulfur in addition to the product. Devise a scheme for separating this mixture into pure substances. Hint: sulfur dissolves in carbon disulfide (a liquid), but iron and iron sulfide do not. Also, iron is magnetic, and iron sulfide is not.

2.44 When barium carbonate, $BaCO_3$, is heated, it decomposes into barium oxide, BaO, and carbon dioxide:

$$BaCO_3 \longrightarrow BaO + CO_2$$

Assume that, after heating, the vessel contains a mixture of barium carbonate and barium oxide. Devise a method of separating the mixture into pure compounds. Hint: barium carbonate is insoluble in water, whereas the oxide reacts with water to give the soluble hydroxide:

$$BaO + H_2O \longrightarrow Ba(OH)_2$$

Strongly heating the solid barium hydroxide drives off water, leaving the oxide

$$Ba(OH)_2 \longrightarrow BaO + H_2O$$

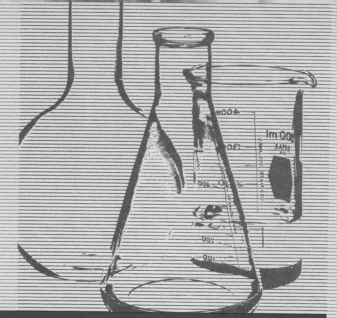

3. Calculations with Chemical Formulas and Equations

Mass and Moles of Substance

3.1 Atomic Weights Aside: Atomic Weights and the Law of Combining Volumes

3.2 Formula Weights

3.3 The Mole Concept Definition of Mole/ Mole Calculations/ Aside: Atomic Theory and Some Laws of Chemistry

Determining Chemical Formulas

3.4 Mass Percentages from the Formula

3.5 Elemental Analysis: Percentages of Carbon, Hydrogen, and Oxygen

3.6 Determining Molecular Formulas
Empirical Formula from Elemental Composition/ Molecular Formula from Empirical Formula

Stoichiometry: Mole–Mass Relations in Chemical Reactions

3.7 Molar Interpretation of a Chemical Equation

3.8 Stoichiometry of a Chemical Reaction

3.9 Limiting Reactant; Theoretical and Percentage Yields

Calculations Involving Solutions

3.10 Molar Concentration

3.11 Diluting Solutions

3.12 Stoichiometry of Solution Reactions

Butyric acid is a colorless liquid with the odor of rancid butter or sour milk. It occurs in small amounts on the skins of mammals and in the perspiration of humans. It is also one of several related compounds that give the characteristic odor to strong cheeses such as Limburger.

Interestingly, butyric acid is the commercial source of some compounds used to make artificial fruit flavors. Ethyl butyrate is one such compound. It has an odor like pineapples. Ethyl butyrate is made by boiling butyric acid with ethanol (grain alcohol, C_2H_6O).

The difference between the sour-milk odor of the reactant butyric acid and the fruity odor of the product ethyl butyrate is striking. Yet, as we saw in Chapter 2, striking differences are a characteristic feature of chemical change.

A chemist who has discovered a chemical reaction that produces a new compound is immediately faced with two important questions. What is the formula of this compound? What is the equation for the chemical reaction? As an illustration of how a chemist answers these questions, we will look at the reaction of butyric acid and ethanol. We will first obtain the formula of butyric acid. With similar data, you will be able to obtain the formula of ethyl butyrate. Then you can write the chemical equation for the reaction of butyric acid and ethanol. Once you have this equation, you will be able to answer the question, How many grams of ethanol are needed to react with 10.0 g of butyric acid? Questions like this arise constantly in chemical research and industry.

Chapter Overview

In Chapter 2 we saw that the central theory of chemistry is that all matter is composed of atoms (atomic theory). We mentioned the concept that each kind of atom has a particular mass. In this chapter we explore this concept of *atomic mass*. With it we can determine the mass of a substance that contains a certain number of atoms, molecules, or ions. Or we can calculate the number of atoms contained in a given mass of an element. This leads us to a way of *determining the formula* of a compound.

Once we have found the formulas of the reactants and products in a reaction, we can write the chemical equation. Then, from *calculations with the chemical equations*, we can determine the amounts of substances involved in the reaction. In order to look at reactions in solution, we define the *molar concentration*. This allows us to measure amounts of reactants by volume of solution.

Mass and Moles of Substance

You buy a quantity of groceries in several ways. If you buy eggs, you buy them by the dozen, that is, by number. Eggs are easy to count out. So are oranges and lemons. Other items, though countable, are more conveniently sold by mass. A dozen peanuts is too small a number to buy, and several hundred are too difficult to count. We buy peanuts by the pound or by the kilogram, that is, by mass. Chemists also are interested in quantity—the quantity of an element or a compound. As with grocery items, a chemist might measure a quantity by number or by mass. But though one can easily weigh a sample of a substance, the number of atoms or molecules in it is much too large to count. You may recall from Section 2.1 that a gram of water contains 3.3×10^{22} H_2O molecules. Nevertheless, chemists are interested in knowing numbers. How many atoms of carbon are there in one molecule of butyric acid? How many molecules of butyric acid react with one molecule of ethanol to give ethyl butyrate? Let us see how chemists solve this problem of measuring numbers of atoms, molecules, and ions.

3.1 Atomic Weights

According to Dalton, each chemically distinct kind of atom has a characteristic mass. Because he could not weigh individual atoms, Dalton measured the relative masses of the elements required to form a compound in order to deduce *relative atomic masses.* For example, one can determine by experiment that 1.0000 gram of hydrogen gas reacts with 7.9367 grams of oxygen gas to give water. If we know the formula of water, we can easily determine the mass of an oxygen atom relative to that of a hydrogen atom. Dalton did not have a way of determining this formula without knowing relative atomic masses, however. He merely assumed the simplest possibility, that the formula was HO. ■ From this assumption, it follows that any sample of water contains equal numbers of hydrogen atoms and oxygen atoms. Therefore, the oxygen atom would have a mass that was 7.9367 times that of a hydrogen atom. The formula for water was later correctly given as H_2O. Since a sample of water actually contains twice as many hydrogen atoms as oxygen atoms, the relative mass of an oxygen atom must be $2 \times 7.9367 = 15.873$ times that of a hydrogen atom.

Dalton's relative atomic mass scale, based on hydrogen, was eventually replaced by a relative scale based on oxygen, and then in 1961 by the present *carbon-12 atomic mass scale.* This new scale, which differs numerically only slightly from the oxygen scale, depends on measurements of atomic mass by an instrument called a *mass spectrometer.* Mass spectrometers, invented early in this century, determine atomic masses from the deflection of atomic ions in electric and magnetic fields. ■ (See Figure 3.1.)

Using the mass spectrometer, it was found that most elements consist of atoms of more than one characteristic mass, that is, they consist of isotopes. **Isotopes** are atoms of the same element having different masses. Naturally occurring carbon, for example, consists mainly of two isotopes. They have masses of 1.99268×10^{-23} g (called carbon-12) and 2.15929×10^{-23} g (carbon-13). ■

On the carbon-12 atomic mass scale, the carbon-12 isotope is arbitrarily assigned a mass of exactly 12 atomic mass units. One **atomic mass unit (amu)**, therefore, equals exactly one-twelfth the mass of a carbon-12 atom.

$$1 \text{ amu} = \frac{1}{12} \times 1.99268 \times 10^{-23} \text{ g} = 1.66057 \times 10^{-24} \text{ g}$$

Naturally occurring samples of an element usually have the same

■ Dalton's method requires the correct formula; but, as we will see later in the chapter, the determination of the formula of a compound requires knowledge of atomic masses. This was Dalton's dilemma. It was solved later by use of the law of combining volumes described in the Aside at the end of this section.

■ The mass spectrometer is discussed in Chapter 5.

■ ·A third isotope of carbon, carbon−14, is radioactive. It is in a natural sample of carbon as an extremely small fraction that varies with the time since it was part of living matter. This fact is used to date samples of charcoal, cloth, and so forth. This topic is discussed further in Chapter 22.

Figure 3.1
A mass spectrometer, which measures the masses of atoms, molecules, and molecular fragments. How a mass spectrometer works is discussed in Chapter 5.

composition wherever they are found on earth. A sample of naturally occurring carbon, for example, usually consists of 98.889% carbon-12 and 1.111% carbon-13. The average mass of a carbon atom in such a sample is 12.011 amu.■

The average atomic mass for a naturally occurring element, expressed in atomic mass units, is called its **atomic weight.** Atomic weights of some common elements are given in Table 3.1. A complete table is given on the inside back cover of this book.■

■ The method of calculating average masses will be explained in Chapter 5.

■ Atomic weights are sometimes defined without explicit mention of units. In that case, they are treated as *relative atomic masses* or ratios, where the unit of mass cancels. The standard of comparison (carbon-12) is implicitly understood.

Aside: Atomic Weights and the Law of Combining Volumes

Dalton's method of determining relative atomic weights requires one to know the formulas of simple compounds. How to find these without knowing atomic weights was eventually solved by using the *law of combining volumes,* discovered in 1808 by Joseph Louis Gay-Lussac (1778–1850). According to this law, gases at the same temperature and pressure react with one another in volume ratios of small whole numbers. For example, Gay-Lussac found that two volumes of hydrogen combine with one volume of oxygen to give water.

(Continued)

The simplest explanation of this law is that the volumes of the reacting elements represent the ratios of those atoms in the resulting compound. Using the then-accepted formulas for hydrogen and oxygen, one would write

$$2H \;+\; O \;\longrightarrow\; H_2O$$

     2 volumes   1 volume   1 volume ?

and the formula of water would be H_2O, instead of HO as Dalton had supposed. Gay-Lussac's law seemed to imply that equal volumes of gases (at the same temperature and pressure) contain equal numbers of molecules. Many scientists, including Dalton, resisted this conclusion. For one thing, one can show that *two* volumes of water vapor result from combining these volumes of hydrogen and oxygen, and not *one* volume of water as this equation implies.

Amedeo Avogadro (1776–1856), an Italian chemist, resolved this difficulty by assuming not only that equal volumes of gases contain the same number of molecules but also that the molecules of the gaseous elements, such as hydrogen and oxygen, each contain two atoms. He wrote

$$2H_2 \;+\; O_2 \;\longrightarrow\; 2H_2O$$

     2 volumes   1 volume   2 volumes

Avogadro's reasoning can be extended to obtain the formulas of other compounds. For example, one volume of nitrogen reacts with three volumes of hydrogen to give two volumes of ammonia. These results are explained by writing NH_3 for ammonia and N_2 for nitrogen:

$$N_2 \;+\; 3H_2 \;\longrightarrow\; 2NH_3$$

     1 volume   3 volumes   2 volumes

Unfortunately, Avogadro's work was ignored for nearly 50 years until it was resurrected by another Italian chemist, Stanislao Cannizzaro (1826–1910). In an 1858 paper, he painstakingly showed how Avogadro's interpretation of Gay-Lussac's law would lead to accurate atomic weights. Within a few years, the majority of chemists accepted Cannizzaro's views.

Table 3.1
Atomic Weights of
Some Common Elements
(Carbon-12 Scale)

Element	Symbol	Atomic Weight (amu)	Element	Symbol	Atomic Weight (amu)
Aluminum	Al	26.98154	Magnesium	Mg	24.305
Barium	Ba	137.33	Manganese	Mn	54.9380
Bromine	Br	79.904	Mercury	Hg	200.59
Calcium	Ca	40.08	Nickel	Ni	58.69
Carbon	C	12.011	Nitrogen	N	14.0067
Chlorine	Cl	35.453	Oxygen	O	15.9994
Chromium	Cr	51.996	Phosphorus	P	30.97376
Cobalt	Co	58.9332	Potassium	K	39.0983
Copper	Cu	63.546	Silicon	Si	28.0855
Fluorine	F	18.998403	Silver	Ag	107.8682
Hydrogen	H	1.00794	Sodium	Na	22.98977
Iodine	I	126.9045	Sulfur	S	32.06
Iron	Fe	55.847	Tin	Sn	118.69
Lead	Pb	207.2	Zinc	Zn	65.38

3.2 Formula Weights

The **molecular weight** (MW) of a substance is the sum of the atomic weights of all of the atoms in the molecule. It is, therefore, the average mass of a molecule of that substance, expressed in atomic mass units. If the molecular formula for the substance is not known, the molecular weight can be determined experimentally by means of a mass spectrometer. In later chapters, we will discuss simple, inexpensive methods of molecular weight determination.

The term **formula weight** (FW) refers to the average mass (in amu) computed from a given formula, whether the substance is molecular or not. Sodium chloride, NaCl, has a formula weight of 58.44 amu (22.99 amu from Na plus 35.45 amu from Cl). Since NaCl is ionic, the expression "molecular weight of NaCl" has no meaning. On the other hand, the molecular weight and the formula weight calculated from the molecular formula of a substance are identical.

Example 3.1

Calculate the formula weight of each of the following to three significant figures. Use a table of atomic weights (AW). (a) chloroform, $CHCl_3$ (b) iron(III) sulfate, $Fe_2(SO_4)_3$

Solution

(a) The calculation is

$$
\begin{array}{lll}
1 \times \text{ AW of C} = & & 12.01 \text{ amu} \\
1 \times \text{ AW of H} = & & 1.01 \text{ amu} \\
3 \times \text{ AW of Cl} = 3 \times 35.45 \text{ amu} = & \underline{106.35 \text{ amu}} \\
\text{FW of } CHCl_3 = & & 119.37 \text{ amu}
\end{array}
$$

The final answer rounded to three significant figures is 119. (b) The point to remember here is that every atomic symbol within the parentheses is multiplied by the subscript 3. The calculation is:

$$
\begin{array}{lll}
2 \times \text{ AW of Fe} = & 2 \times 55.8 \text{ amu} = & 111.6 \text{ amu} \\
3 \times \text{ AW of S} = & 3 \times 32.1 \text{ amu} = & 96.3 \text{ amu} \\
3 \times 4 \times \text{ AW of O} = & 12 \times 16.0 \text{ amu} = & \underline{192.0 \text{ amu}} \\
\text{FW of } Fe_2(SO_4)_3 = & & 399.9 \text{ amu}
\end{array}
$$

The answer rounded to three significant figures is 4.00×10^2 amu.

Exercise 3.1

Calculate the formula weights of the following compounds, using a table of atomic weights. Give the answers to three significant figures. (a) nitrogen dioxide, NO_2 (b) glucose, $C_6H_{12}O_6$ (c) sodium hydroxide, NaOH (d) magnesium hydroxide, $Mg(OH)_2$
(See Problems 3.15 and 3.16.)

3.3 The Mole Concept

When we prepare a compound industrially or even study a reaction in the laboratory, we deal with tremendous numbers of molecules or ions. Suppose we are preparing the flavoring agent ethyl butyrate, as described in the chapter opening. In this preparation, we use 10.0 g of butyric acid, which contains 6.8×10^{22} molecules, a truly staggering number. Imagine a device that counts molecules at the rate of one million per second. It would take over two billion years—nearly one-half the age of the earth—for this device to count out that many molecules! Chemists have adopted the *mole concept* as a convenient way to deal with the enormous numbers of molecules or ions in the samples they work with.

Definition of Mole

A **mole** (symbol **mol**) is defined as the quantity of a given substance that contains as many molecules or formula units as the number of atoms in exactly 12 g of carbon-12. One mole of ethanol, for example, contains the same number of ethanol molecules as there are carbon atoms in 12 g of carbon-12.

The number of atoms in a 12-g sample of carbon-12 is called **Avogadro's number** (to which we give the symbol N_A). We can calculate this number from the mass of an individual carbon-12 atom, which weighs 1.99268×10^{-23} g. The mass of one carbon-12 atom multiplied by Avogadro's number equals 12 g of carbon-12:

$$12 \text{ g} = 1.99268 \times 10^{-23} \text{ g} \times N_A$$

or

$$N_A = \frac{12 \text{ g}}{1.99268 \times 10^{-23} \text{ g}} = 6.02204 \times 10^{23}$$

Avogadro's number, to three significant figures, is

$$N_A = 6.02 \times 10^{23}$$

A mole of a substance contains Avogadro's number (6.02×10^{23}) of molecules (or formula units). The term *mole*, like a dozen or a gross, thus refers to a particular number of things. A dozen eggs equals 12 eggs, a gross of pencils equals 144 pencils, and a mole of ethanol equals 6.02×10^{23} ethanol molecules.■

In using the term *mole* for ionic substances, we mean the number of formula units of the substance. For example, a mole of sodium carbonate, Na_2CO_3, is a quantity containing 6.02×10^{23} Na_2CO_3 units.■ But, as we will see when we discuss ionic compounds in Chapter 7, each formula unit of Na_2CO_3 contains two Na^+ ions and one CO_3^{2-} ion. Therefore, a mole of Na_2CO_3 also contains $2 \times 6.02 \times 10^{23}$ Na^+ ions and $1 \times 6.02 \times 10^{23}$ CO_3^{2-} ions.

It is important when using the term *mole* to specify the formula of the unit to avoid any misunderstanding. For example, a mole of oxygen atoms (with the formula O) contains 6.02×10^{23} O atoms. A mole of oxygen molecules (formula O_2) contains 6.02×10^{23} O_2 molecules, that is, $2 \times 6.02 \times 10^{23}$ O atoms.

The **molar mass** of a substance is the mass of one mole of the substance. Carbon-12 has a molar mass of exactly 12 g, by definition. For all substances, the molar mass in grams is numerically equal to the formula weight in atomic mass units. (For this reason, the molar mass in grams is also called the *gram formula weight* or the *gram molecular weight*.) Thus, ethanol, C_2H_6O, has a molecular weight of 46.1 amu and a molar mass of 46.1 g/mol. Figure 3.2 shows molar amounts of different substances.

■ Avogadro's number of eggs would have a mass approximately 200 times that of the earth.

■ Sodium carbonate, Na_2CO_3, is a white, crystalline solid known commercially as soda ash. In the hydrated form (with water molecules in the crystal lattice), it is known as washing soda. Large amounts of soda ash are used in the manufacture of glass.

molar mass = formula wt

Figure 3.2
One mole each of various substances. From left to right, these substances are 1-octanol ($C_8H_{17}OH$), methanol (CH_3OH), sodium iodide (NaI), and mercury(II) iodide (HgI_2).

Example 3.2

Hydrogen chloride, HCl, is a colorless gas with a sharp, penetrating odor. It dissolves readily in water, producing a solution called hydrochloric acid. What is the mass in grams of an HCl molecule?

Solution

The molecular weight of HCl is 36.5 amu. Therefore, the molar mass is 36.5 g/mol. Since 1 mol HCl (36.5 g) contains Avogadro's number of HCl molecules, dividing 36.5 g by N_A gives

$$\text{Mass of an HCl molecule} = \frac{36.5 \text{ g}}{6.02 \times 10^{23}}$$

$$= 6.06 \times 10^{-23} \text{ g}$$

Exercise 3.2

Ethylene glycol has the structural formula

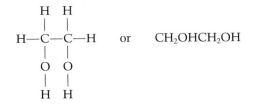

or CH_2OHCH_2OH

It is used as an automobile radiator antifreeze. What is the mass in grams of an ethylene glycol molecule?

(See Problems 3.19 and 3.20.)

Mole Calculations

Now that we know how to find the mass of one mole of substance, there are two important questions to ask. First, how much does a given number of moles of a substance weigh? Second, how many moles of a given formula unit does a given mass of substance contain? Both of these questions are easily answered using dimensional analysis, or the conversion-factor method.■

To illustrate, consider the conversion of grams of ethanol, C_2H_6O, to moles of ethanol. Since the molar mass of ethanol is 46.1 g/mol, we write

$$1 \text{ mol } C_2H_6O = 46.1 \text{ g } C_2H_6O$$

Thus, the factor converting grams of ethanol to moles of ethanol is 1 mol C_2H_6O/46.1 g C_2H_6O. To convert moles of ethanol to grams of ethanol, we would simply invert the conversion factor (46.1 g C_2H_6O/1 mol C_2H_6O).

Again, suppose we are preparing the flavoring agent ethyl butyrate in the laboratory. We start with a beaker containing 28.3 grams of ethanol. How many moles of C_2H_6O is this? We have ■

$$28.3 \text{ g } \cancel{C_2H_6O} \times \frac{1 \text{ mol } C_2H_6O}{46.1 \text{ g } \cancel{C_2H_6O}} = 0.614 \text{ mol } C_2H_6O$$

■ Alternatively, since the molar mass is the mass per mole, we can relate mass and moles by means of the formula: molar mass = mass/moles.

■ If you prefer to use the formula for molar mass, the calculation is

molar mass = mass/moles   or
        moles = mass/molar mass

Hence,

moles = 28.3 g/46.1 g/mol
            = 0.614 mol

Example 3.3

Nitric acid, HNO_3, is a colorless, highly corrosive liquid. It is used in the manufacture of fertilizers and explosives. Nitric acid turns skin a yellow color immediately on contact, and longer exposure gives a painful burn. An experiment calls for 0.345 mol HNO_3. What is the mass of this amount of nitric acid?

Solution

From a table of atomic weights, we find the molar mass of HNO_3 to be 63.0 g/mol. That is,

$$1 \text{ mol } HNO_3 = 63.0 \text{ g } HNO_3$$

Therefore,

$$0.345 \text{ } \cancel{\text{mol } HNO_3} \times \frac{63.0 \text{ g } HNO_3}{1 \text{ } \cancel{\text{mol } HNO_3}} = 21.7 \text{ g } HNO_3$$

Exercise 3.3

Hydrogen peroxide, H_2O_2, is a colorless liquid. A concentrated solution of it is used as a source of oxygen for rocket propellant fuels. Dilute aqueous (water) solutions are used as a bleach. Analysis of a solution shows that it contains 0.909 mol H_2O_2 in 1.00 L of solution. What is the mass of hydrogen peroxide in this volume of solution?

(See Problems 3.21 and 3.22.)

Example 3.4

The sales of sulfuric acid, H_2SO_4, are often used as a market indicator for the chemical industry. It is the chemical produced in greatest volume. The pure acid is a colorless, oily liquid. In an experiment, a chemist places 35.7 g of pure sulfuric acid in a beaker. How many moles of H_2SO_4 is this?

Solution

From atomic weights, we find that the molar mass of H_2SO_4 is 98.1 g/mol. That is,

$$1 \text{ mol } H_2SO_4 = 98.1 \text{ g } H_2SO_4$$

Therefore,

$$35.7 \text{ g } H_2SO_4 \times \frac{1 \text{ mol } H_2SO_4}{98.1 \text{ g } H_2SO_4} = 0.364 \text{ mol } H_2SO_4$$

Exercise 3.4

Ammonium nitrate, NH_4NO_3, is a white, crystalline solid used as a fertilizer. How many moles of NH_4NO_3 are there in 34.8 g of ammonium nitrate?

(See Problems 3.23 and 3.24.)

Example 3.5

How many HCl molecules are there in 3.46 kg of hydrogen chloride?

Solution

We first convert the mass of hydrogen chloride from kilograms to grams and then to moles of HCl. Finally, we convert moles of HCl to the number of HCl molecules. The conversions can be done one after the other.

$$3.46 \text{ kg } HCl \times \frac{1 \times 10^3 \text{ g } HCl}{1 \text{ kg } HCl} \times \frac{1 \text{ mol } HCl}{36.5 \text{ g } HCl}$$

$$\times \frac{6.02 \times 10^{23} \text{ HCl molecules}}{1 \text{ mol } HCl}$$

$$= 5.71 \times 10^{25} \text{ HCl molecules}$$

Note that we multiplied what is given (3.46 kg HCl) by a succession of conversion factors. The denominator of each conversion factor cancels the units of the previous term, finally leaving the desired units.

Exercise 3.5

Hydrogen cyanide, HCN, is a volatile, colorless liquid with the odor of certain fruit pits, such as bitter almonds, peaches, and cherries. The compound is highly poisonous. How many molecules are there in 56 mg HCN (the average toxic dose)?

(See Problems 3.25 and 3.26)

Aside: Atomic Theory and Some Laws of Chemistry

Once we accept the idea that matter is composed of atoms and that atoms have a definite mass, we can explain some laws of chemistry. Recall the law of conservation of mass, discussed in Section 1.1. Since a chemical reaction, ac-

(*Continued*)

cording to atomic theory, involves only a rearrangement of atoms with none created or destroyed, the total mass remains constant.

The *law of definite proportions* (also called the law of constant composition) states that a compound always contains a definite or constant proportion of elements by mass. This law was known to chemists before Dalton began work on his atomic theory. Consider carbon dioxide as an example. This compound always contains 2.6641 grams of oxygen for each gram of carbon, no matter how it is made or where it is found. Atomic theory can explain this if we say that each molecule of carbon dioxide contains two oxygen atoms (with mass 2×15.9994 amu $= 31.9988$ amu) and one carbon atom (with mass 12.0111 amu). Since these masses are constant, we expect any sample of carbon dioxide to contain oxygen and carbon in the ratio 31.9988 : 12.0111 or 2.6641 : 1.

The *law of multiple proportions,* first expressed by Dalton, states that different compounds of the same two elements have mass ratios of the elements that are simple multiples of one another. Carbon monoxide contains 1.3321 grams of oxygen for each gram of carbon. The ratio of oxygen to carbon (1.3321 : 1) is one-half that for carbon dioxide (2.6641 : 1). Using atomic theory, we explain this by saying that carbon monoxide contains only half as many oxygen atoms for each carbon atom as does carbon dioxide. The deduction of the law of multiple proportions from atomic theory was important in convincing chemists of the validity of the theory.

Determining Chemical Formulas

As mentioned in the chapter opening, when a chemist has discovered a new compound, the first question to answer is, What is the formula? To answer this, we begin by determining the *mass percentages,* or fractions by mass, of each element in the compound. The **elemental composition** of a compound is a list that gives the mass percentages of all the elements. We then determine the formula from this elemental composition. As we will see, we must also know the molecular weight of the compound in order to determine the molecular formula.

In the next section, we describe the calculation of mass percentages. Then in two following sections, we describe how to determine a formula. As an illustration, we use butyric acid, described in the chapter opening.

3.4 Mass Percentages from the Formula

Suppose that A is a part of something, that is, part of a whole. It could be an element in a compound, or one substance in a mixture. We define the **mass percentage** of A as follows:

$$\text{Mass \% } A = \frac{\text{mass of } A \text{ in the whole}}{\text{mass of the whole}} \times 100\%$$

We can look at the mass percentage of A as the number of grams of A in 100 grams of the whole.

To get practice with the concept of mass percentage, we will start with a compound (formaldehyde, CH_2O) whose formula we are given and obtain the elemental composition, or mass percentages of the element.

Example 3.6

Formaldehyde, CH_2O, is a toxic gas with a pungent odor. Large quantities are consumed in the manufacture of plastics, and a water solution of the compound has been used to preserve biological specimens. Calculate the mass percentages of each element in formaldehyde (give answers to three significant figures).

Solution

To calculate mass percentage, we need the mass of an element in a given mass of compound. We can get this information by interpreting the formula in molar terms, then converting moles to masses, using a table of atomic weights. Thus, 1 mol CH_2O has a mass of 30.0 g and contains 1 mol C (12.0 g), 2 mol H (2 × 1.01 g), and 1 mol O (16.0 g). Hence,

$$\% \ C = \frac{12.0 \ g}{30.0 \ g} \times 100\% \quad = 40.0\%$$

$$\% \ H = \frac{2 \times 1.01 \ g}{30.0 \ g} \times 100\% = 6.73\%$$

We can calculate the percentage of O in the same way, but it can also be found by subtracting the percentages of C and H from 100%:

$$\% \ O = 100\% - (40.0\% + 6.73\%) = 53.3\%$$

Exercise 3.6

Acetic acid, $HC_2H_3O_2$, is a colorless liquid with a strong, penetrating odor. Vinegar contains 4–5% acetic acid. Calculate the mass percentages of the elements in acetic acid to three significant figures.

(See Problems 3.33 and 3.34.)

If we determine the mass percentages of elements in glucose (blood sugar), $C_6H_{12}O_6$, in the same way, the values are found to be identical to those for formaldehyde, CH_2O. This is because the molecular formula of glucose is a multiple of that for formaldehyde; that is, $C_6H_{12}O_6 = (CH_2O)_6$. *Compounds whose molecular formulas are multiples of one another have the same mass percentages of elements.*

3.5 Elemental Analysis: Percentages of Carbon, Hydrogen, and Oxygen

Suppose we have a newly discovered compound, whose formula we wish to determine. The first step is to obtain an elemental composition (the mass percentages of all elements) by chemical analysis. As an example, consider the determination of the percentages of carbon, hydrogen, and oxygen in compounds containing only these three elements. The basic idea is this. We burn a sample of the compound of known mass and get CO_2 and H_2O. Then, we relate the masses of CO_2 and H_2O to masses of carbon and hydrogen. Once we know the masses of carbon and hydrogen in the sample, we calculate the mass percentages of C and H. We then find the mass percentage of O by subtracting the mass percentages of C and H from 100.

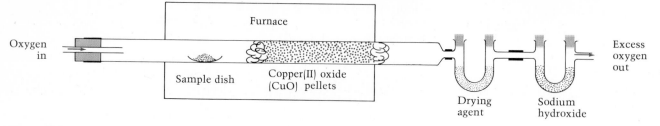

Figure 3.3
Combustion method for determining the percentages of carbon and hydrogen in a compound. The compound is placed in the sample dish and is heated by the furnace. Vapor of the compound burns in O_2 in the presence of CuO pellets, giving CO_2 and H_2O. The water vapor is collected by a drying agent, and CO_2 combines with the sodium hydroxide. Amounts of CO_2 and H_2O are obtained by weighing the U-tubes before and after combustion.

Figure 3.3 shows an apparatus used to find the amount of carbon and hydrogen in a compound. The compound is burned in a stream of oxygen gas. The vapor of the compound and its combustion products pass over copper pellets coated with copper(II) oxide, CuO, which supplies additional oxygen and ensures that the compound is completely burned. As a result of the combustion, every mole of carbon (C) in the compound ends up as a mole of carbon dioxide (CO_2), and every mole of hydrogen (H) ends up as one-half mole of water (H_2O). The water is collected by a drying agent, a substance that has a strong affinity for water. The carbon dioxide is collected by chemical reaction with sodium hydroxide, NaOH.■ By weighing the U tubes containing the drying agent and the sodium hydroxide before and after the combustion, it is possible to determine the masses of water and carbon dioxide produced. From these data, we can calculate the moles, then the masses, and finally the mass percentages of carbon and hydrogen. Subtracting these amounts from 100% gives the percentage of oxygen.

The chapter opened with a discussion of butyric acid. The next example shows how to determine the elemental composition of this substance from combustion data. We will use this elemental composition in Example 3.9 to determine the formula of butyric acid.

■ Sodium hydroxide is a white, crystalline solid, known commercially as caustic soda (very corrosive to skin). It is sold as a drain cleaner because it chemically dissolves fats and hair. The substance reacts with carbon dioxide according to the equations

$$NaOH + CO_2 \longrightarrow NaHCO_3$$
$$2NaOH + CO_2 \longrightarrow$$
$$Na_2CO_3 + H_2O$$

Example 3.7

Butyric acid contains only C, H, and O. A 4.24-mg sample of butyric acid is completely burned. It gives 8.45 mg of carbon dioxide and 3.46 mg of water. What is the mass percentage of each element in butyric acid?

Solution

We first convert the mass of CO_2 to moles of CO_2. Then we convert this to moles of C, noting that 1 mol $CO_2 \simeq 1$ mol C. (Recall that the symbol $\simeq$ means "is chemically equivalent to.") Finally, we convert to mass

of C. We write these conversions as follows:

$$8.45 \times 10^{-3} \text{ g } CO_2 \times \frac{1 \text{ mol } CO_2}{44.0 \text{ g } CO_2} \times \frac{1 \text{ mol } C}{1 \text{ mol } CO_2}$$

$$\times \frac{12.0 \text{ g C}}{1 \text{ mol C}} = 2.30 \times 10^{-3} \text{ g C} \quad \text{(or 2.30 mg C)}$$

For hydrogen, noting that 1 mol $H_2O \simeq 2$ mol H, we have the conversions

(Continued)

$$3.46 \times 10^{-3} \, \text{g} \, H_2O \times \frac{1 \, \text{mol} \, H_2O}{18.0 \, \text{g} \, H_2O} \times \frac{2 \, \text{mol} \, H}{1 \, \text{mol} \, H_2O}$$

$$\times \frac{1.01 \, \text{g} \, H}{1 \, \text{mol} \, H} = 3.88 \times 10^{-4} \, \text{g} \, H \quad \text{(or 0.388 mg H)}$$

The mass percentages of C and H in butyric acid can now be calculated.

$$\text{Mass \% C} = \frac{2.30 \, \text{mg}}{4.24 \, \text{mg}} \times 100\% = 54.2\%$$

$$\text{Mass \% H} = \frac{0.388 \, \text{mg}}{4.24 \, \text{mg}} \times 100\% = 9.15\%$$

The mass percentage of oxygen is found by subtracting the sum of these percentages from 100%:

$$\text{Mass \% O} = 100\% - (54.2\% + 9.15\%) = 36.6\%$$

Thus, the elemental composition of butyric acid is 54.2% C, 9.15% H, and 36.6% O.

Exercise 3.7

A 3.87-mg sample of ascorbic acid (vitamin C) gives 5.80 mg CO_2 and 1.58 mg H_2O on combustion. What is the elemental composition of this compound (the mass percentage of each element)? Ascorbic acid contains only C, H, and O.

(See Problems 3.37 and 3.38.)

3.6 Determining Molecular Formulas

Glyceraldehyde is a white, crystalline compound with the elemental composition 40.0% C, 6.7% H, and 53.3% O. Its molecular weight is 90.1 amu. What is its molecular formula? If you look back at the result in Example 3.6, you will note that the elemental composition of glyceraldehyde is identical to that of formaldehyde, CH_2O. Recall also that compounds whose molecular formulas are multiples of one another have the same elemental composition. We can conclude that the molecular formula of glyceraldehyde is either CH_2O or some multiple of it, such as $C_2H_4O_2$, $C_3H_6O_3$, $C_4H_8O_4$, and so forth. Of these formulas, however, only $C_3H_6O_3$ has a molecular weight of 90.1 amu. Therefore, the molecular formula of glyceraldehyde must be $C_3H_6O_3$.

We see that two pieces of information are needed to determine the molecular formula of a substance: (1) the elemental composition and (2) the molecular weight. If the molecular weight is not available, it is usual to give the **empirical formula** (or **simplest formula**) for a compound. The empirical formula is the one with the smallest integer subscripts. For glyceraldehyde (or glucose or formaldehyde), the empirical formula is CH_2O.

Empirical Formula from Elemental Composition

The empirical formula of a compound shows the ratios of numbers of atoms in the compound. We can find this formula from the elemental composition by converting from mass percentages of the elements to moles of the elements, which are proportional to the numbers of atoms. The next example shows these calculations in detail.

Example 3.8

Chromium forms compounds of various colors (the word *chromium* comes from the Greek *khroma*, color). Sodium dichromate is the most important commercial chromium compound, from which many other chromium compounds are manufactured. It is a bright orange, crystalline substance. An analysis of sodium dichromate gives the following mass percentages: 17.5% Na, 39.7% Cr, and 42.8% O. What is the empirical formula of this compound? (Sodium dichromate is ionic, so has no molecular formula.)

Solution

Assume for the purposes of this calculation that we have 100.0 g of sodium dichromate. Of this quantity, 17.5 g are Na (since the substance is 17.5% Na). Similarly, this sample contains 39.7 g Cr and 42.8 g O. We convert these amounts to moles.

$$17.5 \; \text{g Na} \times \frac{1 \; \text{mol Na}}{23.0 \; \text{g Na}} = 0.761 \; \text{mol Na}$$

$$39.7 \; \text{g Cr} \times \frac{1 \; \text{mol Cr}}{52.0 \; \text{g Cr}} = 0.763 \; \text{mol Cr}$$

$$42.8 \; \text{g O} \times \frac{1 \; \text{mol O}}{16.0 \; \text{g O}} = 2.68 \; \text{mol O}$$

These numbers are in the same ratio as the subscripts in the empirical formula. However, we would not write $Na_{0.761}Cr_{0.763}O_{2.68}$ because it is not possible to have fractional atoms. Therefore, we need to change these subscripts to integers.

To obtain the set of smallest integers from these mole numbers, we first divide each one by the smallest mole number.

$$\text{For Na:} \quad \frac{0.761 \; \text{mol}}{0.761 \; \text{mol}} = 1.00$$

$$\text{For Cr:} \quad \frac{0.763 \; \text{mol}}{0.761 \; \text{mol}} = 1.00$$

$$\text{For O:} \quad \frac{2.68 \; \text{mol}}{0.761 \; \text{mol}} = 3.52$$

We must decide whether these numbers are integers, within experimental error. If we round off the rightmost digit, which is subject to experimental error, we get $Na_{1.0}Cr_{1.0}O_{3.5}$. At this stage in the calculation, there are two possibilities. (1) The subscripts are all integers, in which case the empirical formula has been found. This is not the case here. (2) The subscripts are not all integers, but can be made into integers by multiplying each one by some whole number. If we multiply the subscripts that we have calculated by 2, we get $Na_{2.0}Cr_{2.0}O_{7.0}$. Thus, the empirical formula is $Na_2Cr_2O_7$. The sodium dichromate crystal consists of sodium ions, Na^+, and dichromate ions, $Cr_2O_7^{2-}$. Sodium ions are colorless, but dichromate ions give an orange color to the compound.

Exercise 3.8

Benzoic acid is a white, crystalline powder used as a food preservative. The compound contains 68.8% C, 5.0% H, and 26.2% O by mass. What is the empirical formula?

(See Problems 3.39, 3.40, 3.41 and 3.42.)

Molecular Formula from Empirical Formula

The molecular formula of a compound is a multiple of its empirical formula. Thus, the molecular formula of glyceraldehyde, $C_3H_6O_3$, is equivalent to $(CH_2O)_3$, or three times the empirical formula, CH_2O. It follows that the molecular weight is three times the empirical formula weight, which is obtained by summing the atomic weights from the empirical formula. For any molecular compound, we can write,

$$\text{molecular weight} = n \times \text{empirical formula weight}$$

where n is the number of empirical formula units in the molecule. We get the molecular formula by multiplying the subscripts of the empirical formula by n, which we calculate from the equation

$$n = \frac{\text{molecular weight}}{\text{empirical formula weight}}$$

Once we determine the empirical formula for a compound, we can calculate its empirical formula weight. If we have an experimental determination of its molecular weight, we can calculate n and then the molecular formula. The next example illustrates how we use the elemental composition and molecular weight to determine the molecular formula of butyric acid.

Example 3.9

In Example 3.7, we found the elemental composition of butyric acid to be 54.2% C, 9.2% H, and 36.6% O. Determine the empirical formula. The molecular weight of butyric acid was determined by experiment to be 88 amu. What is the molecular formula?

Solution

A sample of 100.0 g of butyric acid contains 54.2 g C, 9.2 g H, and 36.6 g O. Converting these masses to moles gives 4.51 mol C, 9.1 mol H, and 2.29 mol O. Dividing these mole numbers by the smallest one gives

1.97 for C, 4.0 for H, and 1.00 for O. The empirical formula of butyric acid is C_2H_4O. This has a formula weight of 44 amu. Dividing the formula weight into the molecular weight gives the number by which the subscripts in C_2H_4O must be multiplied.

$$n = \frac{MW}{FW} = \frac{88 \text{ amu}}{44 \text{ amu}} = 2.0$$

The molecular formula of butyric acid is $(C_2H_4O)_2$, or $C_4H_8O_2$.

Exercise 3.9

The elemental composition of ethyl butyrate is 62.0% C, 10.4% H, 27.6% O, and the molecular weight is 116 amu. Obtain the molecular formula.

(See Problems 3.45 and 3.46.)

Recall from the chapter opener that when butyric acid is boiled with ethanol, ethyl butyrate and water are obtained. We can now write the chemical equation for this reaction. From Example 3.9, we found that the formula for butyric acid is $C_4H_8O_2$. In Exercise 3.9, you obtained the formula of ethyl butyrate, $C_6H_{12}O_2$. The formula of ethanol, C_2H_6O, has already been given in the text. We first write the formulas for the reactants and products, then balance the equation.

$$C_4H_8O_2 \ + \ C_2H_6O \ \longrightarrow \ C_6H_{12}O_2 \ + \ H_2O$$

butyric acid   ethanol   ethyl butyrate   water

As it happens, the equation is already balanced with coefficients of 1. In the next sections, we are going to see how such a balanced equation can be used to calculate the amounts of reactants and products.

Stoichiometry: Mole–Mass Relations in Chemical Reactions

In Chapter 2, we described a chemical equation as a representation of what occurs when molecules (or ions) react. We will now study chemical reactions more closely to answer questions about the stoichiometry of reactions. **Stoichiometry** (pronounced stoi–key–om´–e–tree) is the calculation of the quantities of reactants and products involved in a chemical reaction. It is based on the chemical equation and on the relationship between mass and moles. Such calculations are fundamental to most quantitative work in chemistry. In the next sections, we will use the industrial Haber process to illustrate stoichiometric calculations.

3.7 Molar Interpretation of a Chemical Equation

In 1909, Fritz Haber, a German chemist, developed an industrial process for converting atmospheric nitrogen to ammonia. The Haber process has become the key to modern farming. Repeated harvesting of high-yield crops quickly depletes the soil of nitrogen compounds. Farmers maintain the fertility of the soil by adding nitrogen fertilizers, either as ammonia or as compounds derived from ammonia. In the Haber process, nitrogen reacts with hydrogen at high temperature and pressure.

$$N_2 + 3H_2 \longrightarrow 2NH_3$$

Hydrogen is usually obtained from natural gas or petroleum, and so is relatively expensive. For this reason, the price of hydrogen partly determines the price of ammonia. Thus, an important question to answer is, How much hydrogen is required to give a particular quantity of ammonia? How much hydrogen, for example, would be needed to produce one ton (907 kg) of ammonia? Similar kinds of questions arise throughout chemistry.

To answer this kind of question, we must first look at the chemical equation. The equation for the preparation of ammonia by the Haber process tells us that one N_2 molecule and three H_2 molecules react to produce two NH_3 molecules. A similar statement involving multiples of these numbers of molecules is also correct (Figure 3.4). For example, 6.02×10^{23} N_2 molecules react with $3 \times 6.02 \times 10^{23}$ H_2 molecules, giving $2 \times 6.02 \times 10^{23}$ NH_3 molecules. This last statement can be put in molar terminology: one mole of N_2 reacts with three moles of H_2 to give two moles of NH_3. We conclude that we may interpret a chemical equation either in terms of numbers of molecules (or ions or formula units) or in terms of numbers of moles, depending on our needs.

Since moles can be converted to mass, we can also give a mass interpretation of a chemical equation. The molar masses of N_2, H_2, and NH_3 are 28.0, 2.02, and 17.0 g/mol, respectively. Therefore, 28.0 g of N_2 react with 3×2.02 g of H_2 to yield 2×17.0 g of NH_3.

We summarize these three interpretations as follows:

N_2	+	$3H_2$	$\longrightarrow$	$2NH_3$	
1 molecule N_2	+	3 molecules H_2	$\longrightarrow$	2 molecules NH_3	(molecular interpretation)
1 mol N_2	+	3 mol H_2	$\longrightarrow$	2 mol NH_3	(molar interpretation)
28.0 g N_2	+	3×2.02 g H_2	$\longrightarrow$	2×17.0 g NH_3	(mass interpretation)

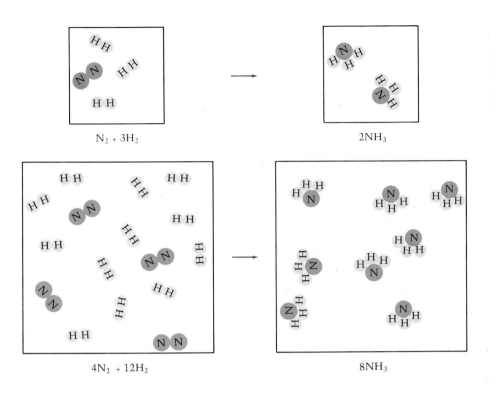

Figure 3.4
The synthesis of ammonia:
$N_2 + 3H_2 \longrightarrow 2NH_3$. Any multiple of the equation is also a correct representation of the reaction, for example, $4N_2 + 12H_2 \longrightarrow 8NH_3$.

Suppose we ask how many grams of atmospheric nitrogen will react with 6.0 g of hydrogen. We see from the last equation that the answer is 28.0 g N_2. We formulated this question for one mole of atmospheric nitrogen. Recalling the question posed earlier, suppose we asked, How much hydrogen is needed to yield 907 kg of ammonia in the Haber process? The solution to this problem depends on the fact that *the number of moles involved in a reaction is proportional to the coefficients in the balanced chemical equation*. In the next section, we will describe a procedure for solving similar problems.

Exercise 3.10

In an industrial process, hydrogen chloride, HCl, is prepared by burning hydrogen gas, H_2, in an atmosphere of chlorine, Cl_2. Write the chemical equation for the reaction. Below the equation give the molecular, molar, and mass interpretations.
(See Problems 3.47 and 3.48.)

3.8 Stoichiometry of a Chemical Reaction

The chemical equation for the Haber process
$$N_2 + 3H_2 \longrightarrow 2NH_3$$
tells us that three moles of H_2 produce two moles of NH_3. If we write this as a mathematical equation,
$$3 \text{ mol } H_2 \simeq 2 \text{ mol } NH_3$$
we can obtain a factor that converts from moles of hydrogen to moles of ammonia (2 mol NH_3/3 mol H_2) or from moles of ammonia to moles of hydrogen (3 mol H_2/2 mol NH_3).∎

■ These conversion factors simply express the fact that the mole ratio of NH_3 to H_2 in the reaction is 2 to 3.

Now we can see how to calculate the mass of hydrogen required to produce 907 kg ammonia (9.07×10^5 g NH_3). Because the chemical equation directly relates moles NH_3 to moles H_2, we first convert grams NH_3 to moles NH_3, then to moles H_2. Finally, we convert moles H_2 to grams H_2, which is what we are seeking. The factors for converting mass to moles or moles to mass come from molar masses, as described earlier.

$$9.07 \times 10^5 \text{ g NH}_3 \times \frac{1 \text{ mol NH}_3}{17.0 \text{ g NH}_3} \times \frac{3 \text{ mol H}_2}{2 \text{ mol NH}_3} \times \frac{2.02 \text{ g H}_2}{1 \text{ mol H}_2}$$

$$= 1.62 \times 10^5 \text{ g H}_2 \quad \text{(or 162 kg H}_2\text{)}$$

The following examples illustrate additional variations of this type of calculation.

Example 3.10

Hematite, Fe_2O_3, is an important ore of iron. (An ore is a natural substance from which the metal can be profitably obtained.) The free metal is obtained by reacting hematite with carbon monoxide, CO, in a blast furnace. Carbon monoxide is formed in the furnace by partial combustion of carbon. The reaction in which the metal is produced is

$$\text{Fe}_2\text{O}_3 + 3\text{CO} \longrightarrow 2\text{Fe} + 3\text{CO}_2$$

How many grams of iron can be produced from 1.00 kg of Fe_2O_3?

Solution

The calculation involves the conversion of a quantity of Fe_2O_3 to a quantity of Fe. An essential feature of this type of calculation is the conversion of moles of a given substance to moles of another substance. Therefore, we first convert the mass of Fe_2O_3 (1.00 kg Fe_2O_3 = 1.00 $\times$

10^3 g Fe_2O_3) to moles Fe_2O_3. Then, we convert this to moles Fe and to grams Fe. We use the following information to obtain conversion factors:

1 mol Fe_2O_3 = 160 g Fe_2O_3
(from the molar mass of Fe_2O_3)

1 mol Fe_2O_3 ≃ 2 mol Fe
(from the balanced chemical equation)

1 mol Fe = 55.8 g Fe
(from the molar mass of Fe)

The calculation is as follows:

$$1.00 \times 10^3 \text{ g Fe}_2\text{O}_3 \times \frac{1 \text{ mol Fe}_2\text{O}_3}{160 \text{ g Fe}_2\text{O}_3} \times \frac{2 \text{ mol Fe}}{1 \text{ mol Fe}_2\text{O}_3}$$

$$\times \frac{55.8 \text{ g Fe}}{1 \text{ mol Fe}} = 698 \text{ g Fe}$$

Exercise 3.11

Sodium is a soft, reactive metal. It is normally stored under a liquid like kerosene to keep it from reacting with moisture in the air. Sodium instantly reacts with water to give hydrogen gas and a solution of sodium hydroxide, NaOH. How many grams of sodium metal are needed to give 7.81 g of hydrogen by this reaction? (Remember to write the balanced equation first.)

(See Problems 3.49, 3.50, 3.51, and 3.52.)

Example 3.11

Today, chlorine is prepared from sodium chloride by electrochemical decomposition, discussed in Chapter 24. Formerly, chlorine was produced by heating hydrochloric acid with pyrolusite (manganese dioxide, MnO_2), a common manganese ore. Small amounts of

chlorine may be prepared in the laboratory by the same reaction:

$$4\text{HCl} + \text{MnO}_2 \longrightarrow 2\text{H}_2\text{O} + \text{MnCl}_2 + \text{Cl}_2$$

5.00 α

(*Continued*)

How many grams of HCl react with 5.00 g of manganese dioxide, according to this equation?

Solution

We write down what is given (5.00 g MnO_2) and convert this to moles, then to moles of what is desired (mol HCl). Finally, we convert this to mass (g HCl). We use the following information to obtain conversion factors:

$$1 \text{ mol } MnO_2 = 86.9 \text{ g } MnO_2$$
(from the molar mass of MnO_2)

$$1 \text{ mol } MnO_2 \simeq 4 \text{ mol HCl}$$
(from the balanced chemical equation)

$$1 \text{ mol HCl} = 36.5 \text{ g HCl}$$
(from the molar mass of HCl)

The calculations are

$$5.00 \text{ g } MnO_2 \times \frac{1 \text{ mol } MnO_2}{86.9 \text{ g } MnO_2} \times \frac{4 \text{ mol HCl}}{1 \text{ mol } MnO_2}$$

$$\times \frac{36.5 \text{ g HCl}}{1 \text{ mol HCl}} = 8.40 \text{ g HCl}$$

Exercise 3.12

Sphalerite is a zinc sulfide (ZnS) mineral and an important commercial source of zinc metal. The first step in the processing of the ore consists of heating the sulfide with oxygen to give zinc oxide, ZnO, and sulfur dioxide, SO_2. How many kilograms of oxygen gas combine with 5.00×10^3 g of zinc sulfide in this reaction? (You must first write out the balanced chemical equation.)

(See Problems 3.53 and 3.54.)

Exercise 3.13

British chemist Joseph Priestley prepared oxygen in 1774 by heating mercury(II) oxide, HgO. Mercury metal is the other product. If 6.47 g of oxygen are collected, how many grams of mercury metal are also produced?

(See Problems 3.55 and 3.56.)

3.9 Limiting Reactant; Theoretical and Percentage Yields

Often reactants are added to a reaction vessel in amounts different from the molar proportions given by the chemical equation. In that case, only one of the reactants is completely consumed at the end of the reaction. The **limiting reactant** (or **limiting reagent**) is the reactant that is entirely consumed when a reaction goes to completion. Other reactants are in excess and remain partially unreacted. Because any reactant that is added in excess to a reaction mixture does not completely react, its starting amount cannot be used to calculate the moles of product. The moles of product are always determined by the moles of limiting reactant. Consider the burning of hydrogen in oxygen.

$$2H_2 + O_2 \longrightarrow 2H_2O$$

If we add the reactants H_2 and O_2 to a vessel in the molar proportions given by the chemical equation, both will be used up when the reaction is complete. Thus, 2 mol H_2 and 1 mol O_2 react completely to give the product, H_2O. However, suppose we place only 1 mol H_2 in the reaction vessel with the same amount of O_2 (1 mol O_2). After the reaction is complete, all of the H_2 is consumed, but $\frac{1}{2}$ mol O_2 remains unreacted.

We cannot obtain the amount of H_2O produced in the reaction by converting from the moles O_2 added to the reaction vessel to moles H_2O because only part of the O_2 reacts. However, since the H_2 is entirely consumed, we can use the starting amount of this reactant (1 mol H_2) to obtain the amount of H_2O produced. From the chemical equation, we note that

$$2 \text{ mol } H_2 \simeq 2 \text{ mol } H_2O$$

This gives us the conversion factor from moles H_2 to moles H_2O. Therefore, the amount of H_2O produced is

$$1 \text{ mol } H_2 \times \frac{2 \text{ mol } H_2O}{2 \text{ mol } H_2} = 1 \text{ mol } H_2O$$

Let us summarize the situation. Suppose we are given the amounts of reactants added to a vessel, and we wish to calculate the amount of product obtained when the reaction is complete. Unless we know that the reactants have been added in the molar proportions given by the chemical equation, the problem is twofold. (1) We must first identify the limiting reactant. (2) We calculate the amount of product from the amount of limiting reactant. The next example illustrates both steps.

Example 3.12

Carbonate-containing compounds react with hydrochloric acid, HCl, to give carbon dioxide gas, CO_2. This is used as a test for limestone, which is mainly calcium carbonate, $CaCO_3$. Hydrochloric acid reacts with calcium carbonate, giving bubbles of carbon dioxide:

$$2HCl + CaCO_3 \longrightarrow CaCl_2 + H_2O + CO_2$$

If 4.50 g HCl are placed in a reaction vessel with 15.00 g $CaCO_3$, how many grams of CO_2 will be produced? Calculate the amount of excess reactant left over.

Solution

Step 1 We must first determine which of the two reactants, HCl or $CaCO_3$, is the limiting reactant. To do this, we take each reactant in turn and ask, How much product (CO_2) would be obtained if each were totally consumed? The reactant that gives the smaller amount of product is the limiting reactant. To do the calculations, we find the moles of each reactant and convert this to moles of CO_2.

Converting from 4.50 g HCl to moles CO_2:

$$4.50 \text{ g HCl} \times \frac{1 \text{ mol HCl}}{36.5 \text{ g HCl}} \times \frac{1 \text{ mol } CO_2}{2 \text{ mol HCl}}$$
$$= 0.0616 \text{ mol } CO_2$$

Converting from 15.00 g $CaCO_3$ to moles CO_2:

$$15.00 \text{ g } CaCO_3 \times \frac{1 \text{ mol } CaCO_3}{100.1 \text{ g } CaCO_3} \times \frac{1 \text{ mol } CO_2}{1 \text{ mol } CaCO_3}$$
$$= 0.1499 \text{ mol } CO_2$$

The reactant that gives the smaller amount of product (HCl) is the limiting reactant. The other reactant ($CaCO_3$) is in excess and is not all converted to CO_2 (therefore 0.1499 mol CO_2 is not actually produced).

Step 2 Once we know the limiting reactant (HCl), we convert the quantity of this reactant to grams of the product CO_2. We have already found that 4.50 g HCl gives 0.0616 mol CO_2. All that remains is to convert from this amount of CO_2 to grams CO_2:

$$0.0616 \text{ mol } CO_2 \times \frac{44.0 \text{ g } CO_2}{1 \text{ mol } CO_2} = 2.71 \text{ g } CO_2$$

To calculate the amount of the excess reactant ($CaCO_3$) left over, we must first calculate the amount of excess reactant consumed in the reaction. The mass of $CaCO_3$ consumed is obtained by converting moles CO_2 produced to moles $CaCO_3$ and then to grams $CaCO_3$:

$$0.0616 \text{ mol } CO_2 \times \frac{1 \text{ mol } CaCO_3}{1 \text{ mol } CO_2} \times \frac{100.1 \text{ g } CaCO_3}{1 \text{ mol } CaCO_3}$$
$$= 6.17 \text{ g } CaCO_3 \quad \text{(mass of } CaCO_3 \text{ consumed)}$$

To find the grams $CaCO_3$ that remain after the reaction, we subtract the mass consumed from the total available (15.00 g).

$$(15.00 - 6.17) \text{ g } CaCO_3 = 8.83 \text{ g } CaCO_3$$
$$\text{(mass of } CaCO_3 \text{ remaining)}$$

Exercise 3.14

In an experiment, 7.36 g of zinc were heated with 6.45 g of sulfur. Assume that these substances react according to the equation

$$8Zn + S_8 \longrightarrow 8ZnS$$

What amount of zinc sulfide will be produced by the reaction?

(See Problems 3.57 and 3.58.)

The **theoretical yield** of a product is the maximum amount of product that can be obtained in a reaction. It is the amount that we calculate from the stoichiometry based on the limiting reactant. In Example 3.12, the theoretical yield of CO_2 is 2.71 g. In practice, the **actual yield** of a product may be much less than this, for several possible reasons. First, some product may be lost during the process of separating it from the final reaction mixture. Then there may be other, competing reactions that occur simultaneously with the reaction on which the theoretical yield is based. Finally, many reactions appear to stop before they reach completion; they give mixtures of reactants and products.■

It is important to know the actual yield from a reaction in order to make economic decisions about a method of preparation. The reactants for a given method may not be too costly per kilogram, but if the actual yield is very low, the final cost can be very high. The **percentage yield** of product is the actual yield (experimentally determined) expressed as a percentage of the theoretical yield (calculated):

■ Such reactions reach chemical equilibrium, a topic introduced in Chapter 9.

$$\text{percentage yield} = \frac{\text{actual yield}}{\text{theoretical yield}} \times 100\%$$

To illustrate the calculation of percentage yield, recall that the theoretical yield of CO_2 calculated in Example 3.12 was 2.71 g. If the actual yield of CO_2 obtained in an experiment, using the amounts of reactants given in Example 3.12, is 2.01 g, then

$$\text{percentage yield of } CO_2 = \frac{2.01\,g}{2.71\,g} \times 100\% = 74.2\%$$

Exercise 3.15

In an experiment, 10.0 g of butyric acid, $C_4H_8O_2$, are heated with 3.14 g of ethanol, C_2H_6O, to produce ethyl butyrate, $C_6H_{12}O_2$. (Water is also a product.) What is the theoretical yield of ethyl butyrate? If the actual yield of ethyl butyrate is 5.3 g, what is the percentage yield?

(See Problems 3.59 and 3.60.)

Calculations Involving Solutions

Reaction between two solid reactants often proceeds very slowly or not at all. This is because the reactant molecules are in fixed positions in their crystals and cannot come in contact to react. For this reason, most reactions are run in liquid solutions. In this way, reactant molecules are free to move throughout the liquid, and reaction is much faster. When running reactions in liquid solutions, it is convenient to dispense the amounts of reactants by measuring out the volumes of reactant solution. In the next sections, we will discuss calculations involving volumes of solutions.

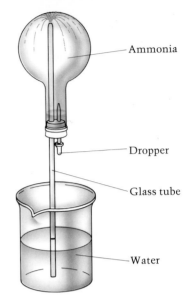

3.10 Molar Concentration

Ammonia gas dissolves readily in water (Figure 3.5), and aqueous ammonia solutions are often used in the laboratory. Solutions containing the maximum amount of ammonia are approximately 28% NH_3 by mass. If we want to measure out 28 g NH_3, we could weigh out 100 grams of this solution. But suppose we knew how many grams of NH_3 are in a liter of solution. Then we could simply measure out the volume containing 28 g NH_3. What we need to know is how volumes of solution and amounts of substances are related.

When we dissolve a substance in a liquid, we call the substance the *solute* and the liquid the *solvent*. In the case of aqueous ammonia, ammonia gas is the solute and water is the solvent.

Concentration is the general term used in referring to the quantity of solute in a standard quantity of solution. Qualitatively, we say that a solution is *dilute* when the solute concentration is low and that the solution is *concentrated* when the solute concentration is high. Usually these terms are used in a comparative sense and do not refer to a specific concentration. We say that one solution is more dilute or less concentrated than another. However, for commercially available solutions, the term *concentrated* refers to the maximum concentration available, or approximately this concentration. Thus, concentrated aqueous ammonia contains about 28% NH_3 by mass. Concentrated hydrochloric acid contains about 38% HCl by mass.

In these examples, we have expressed the concentration quantitatively by giving the mass percentage of solute, that is, the mass of solute in 100 g solution. However, our interest at the moment is in a unit of concentration that is convenient for doing reactions in solution.■ Molar concentration is such a unit. **Molar concentration** or **molarity (M)** is defined as the number of moles of solute dissolved in one liter (cubic decimeter) of solution:

$$\text{Molarity }(M) = \frac{\text{moles of solute}}{\text{liters of solution}}$$

An aqueous solution that is 0.15 M NH_3 (read this as "0.15 molar NH_3") contains 0.15 mol NH_3 per liter of solution.

If we want to prepare a solution that is, for example, 0.200 M NaOH, we place 0.200 mol NaOH in a 1.000 L volumetric flask (Figure 3.6), or a proportional amount in a flask of a different size. We add a small quantity of water to dissolve the NaOH. Then we fill the flask with additional water to the mark on the neck and mix the solution.

Figure 3.5
The "ammonia fountain." When we introduce a small amount of water from a dropper into the flask, the ammonia dissolves, reducing the pressure in the flask. Water spews from the glass tube, forced into the flask by atmospheric pressure.

■ Concentration can be expressed in various ways. It can be given in terms of mass or moles of solute in a standard mass of solvent or in a standard volume of solution. These different ways are described more completely in Chapter 12.

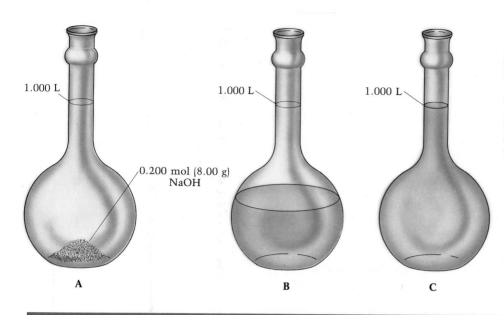

Figure 3.6
Preparing a 0.200 *M* NaOH solution. (a) 0.200 mol NaOH is added to a 1.000-L volumetric flask. (b) The NaOH is dissolved in a quantity of water. (c) The solution is filled to the mark on the neck of the flask, then mixed.

1.000 L

0.200 mol (8.00 g) NaOH

1.000 L

1.000 L

A

B

C

Example 3.13

A sample of NaOH weighing 0.38 g is placed in a 50.0-mL volumetric flask. The flask is then filled with water to the mark on the neck. What is the molarity of the resulting solution?

Solution

By definition, molarity is moles solute/liters solution. Therefore, we first convert grams of NaOH to moles. (We find that 0.38 g NaOH is 9.5×10^{-3} mol NaOH.) To obtain the molarity, we then divide the moles of solute by liters of solution (soln):

$$\text{Molarity} = \frac{9.5 \times 10^{-3} \text{ mol NaOH}}{50.0 \times 10^{-3} \text{ L soln}} = 0.19 \text{ } M \text{ NaOH}$$

Exercise 3.16

A sample of sodium chloride, NaCl, weighing 0.0678 g is placed in a 25.0-mL volumetric flask. Enough water is added to dissolve the NaCl, and then the flask is filled with water and carefully shaken to mix the contents. What is the molarity of the resulting solution?

(See Problems 3.61, 3.62, 3.63, and 3.64.)

Molarity may also be used as a factor for converting from liters of solution to moles of solute (or from moles of solute to liters of solution), as illustrated by the following example.

Example 3.14

How many moles of NaOH are contained in 31 mL of 0.15 *M* NaOH?

Solution

In a 0.15 *M* NaOH solution, we have

1 L soln ≏ 0.15 mol NaOH

(*Continued*)

> Therefore, the amount of NaOH in a 31 mL (or 31×10^{-3} ~~L soln~~ $\times \dfrac{0.15 \text{ mol NaOH}}{1 ~~\text{L soln}~~}$
>
> 31×10^{-3} L) solution is
>
> $$= 4.6 \times 10^{-3} \text{ mol NaOH}$$

Exercise 3.17

How many moles of sodium chloride should be put in a 50.0-mL volumetric flask to give a 0.15 M NaCl solution when the flask is filled with water? How many grams of NaCl is this?

(See Problems 3.65, 3.66, 3.67, and 3.68.)

Exercise 3.18

How many milliliters of 0.163 M NaCl are required to give 0.0958 g of sodium chloride?

(See Problems 3.69, 3.70, 3.71, and 3.72.)

3.11 Diluting Solutions

Commercially available aqueous ammonia (28.0% NH_3) is 14.8 M NH_3. Suppose, however, that we want a solution that is 1.00 M NH_3. We need to dilute the concentrated solution with a definite quantity of water. For this purpose, we must know the relationship between the molarity of the solution before dilution (the *initial molarity*) and that after dilution (the *final molarity*).

To obtain this relationship, first recall the equation defining molarity:

$$\text{Molarity} = \frac{\text{moles of solute}}{\text{liters of solution}}$$

This can be rearranged to give

$$\text{Moles of solute} = \text{molarity} \times \text{liters of solution}$$

The product of molarity and the volume (in liters) gives the number of moles of solute in the solution. Writing M_i for the initial molar concentration and V_i for the initial volume of solution we get

$$\text{Moles of solute} = M_i \times V_i$$

When the solution is diluted by adding more water, the concentration and volume change to M_f (the final molar concentration) and V_f (the final volume), and the moles of solute are

$$\text{Moles of solute} = M_f \times V_f$$

Since the number of moles of solute have not changed during the dilution,

$$M_i \times V_i = M_f \times V_f$$

(Note: *You can use any units for volume,* but both V_i and V_f must be in the same units.) Example 3.15 shows how to use this relationship in dilution problems.

Example 3.15

You are given 5.00 mL of 14.8 M NH_3. What will the final volume be after this solution is diluted with water to give 1.00 M NH_3?

Solution

We write the dilution formula just given.

$$M_i V_i = M_f V_f$$

Then, we rearrange it to give the final volume.

$$V_f = \frac{M_i V_i}{M_f}$$

After substituting the values given in the problem, we get

$$V_f = \frac{5.00 \text{ mL} \times 14.8\,\cancel{M}}{1.00\,\cancel{M}} = 74.0 \text{ mL}$$

Exercise 3.19

What is the final volume of a solution of sulfuric acid if you dilute 48 mL of 1.5 M H_2SO_4 to produce a solution of 0.18 M H_2SO_4?

(See Problems 3.73 and 3.74.)

Exercise 3.20

You have a solution that is 14.8 M NH_3. How many milliliters of this solution are required to give 50.0 mL of 1.00 M NH_3 when diluted?

(See Problems 3.75 and 3.76.)

The next example is somewhat more complicated than previous ones. We must put together several operational skills to find the answer.

Example 3.16

Commercially available concentrated hydrochloric acid is an aqueous (water) solution containing 38% HCl by mass. (a) What is the molarity of this solution? The density is 1.19 g/mL. (b) How many milliliters of concentrated HCl are required to make 1.00 L of 0.10 M HCl?

Solution

(a) To calculate the molarity, we select a convenient quantity (100.0 g) of the acid and calculate moles of HCl present and the volume of solution. Because the solution contains 38% HCl by mass, there are 38 g HCl in the quantity taken. We convert this mass to moles:

$$38 \text{ g HCl} \times \frac{1 \text{ mol HCl}}{36.5 \text{ g HCl}} = 1.04 \text{ mol HCl}$$

(We have retained an extra figure for the following calculations.) Now we calculate the volume of 100.0 g of the acid from its density, 1.19 g/mL. Recall that density = mass/volume. Therefore,

$$\text{volume} = \frac{\text{mass}}{\text{density}} = \frac{100.0 \text{ g}}{1.19 \text{ g/mL}}$$
$$= 84.0 \text{ mL (or 0.0840 L)}$$

Finally, we calculate the molarity by dividing the amount of HCl, 1.04 mol, by the volume just obtained.

$$\text{molarity} = \frac{\text{moles solute}}{\text{liters solution}} = \frac{1.04 \text{ mol HCl}}{0.0840 \text{ L soln}}$$
$$= 12.4 \text{ } M \text{ HCl}$$

Since the mass percentage of HCl was given to two significant figures, the answer should be reported as 12 M HCl. We will retain the extra figure for the next calculation. (b) We wish to dilute an initial volume of acid (not known) of initial molarity (12.4 M) to give a final volume (1.00 L = 1.00 × 10³ mL) and final molarity (0.10 M). We write the dilution formula

$$M_i V_i = M_f V_f$$

(Continued)

and solve for V_i

$$V_i = \frac{M_f V_f}{M_i}$$

Substituting, we find

$$V_i = \frac{0.10\,M \times 1.00 \times 10^3\,mL}{12.4\,M} = 8.1\,mL$$

Note that the solution of this problem requires an understanding of density, percentage composition, conversion of mass to moles, molarity, and dilution of solutions.

Exercise 3.21

Commercially available concentrated sulfuric acid is 95% H_2SO_4 by mass, and has a density of 1.84 g/mL. How many milliliters of this acid are needed to give 1.0 L of 0.15 M H_2SO_4?

(See Problems 3.77 and 3.78.)

3.12 Stoichiometry of Solution Reactions

Now that we see how the amount of solute can be measured out by volume of solution, let us look at the stoichiometry of solution reactions, that is, calculations of volumes involved in a reaction. The next example illustrates the method.

Example 3.17

Consider the reaction of sulfuric acid, H_2SO_4, with sodium hydroxide, NaOH.

$$H_2SO_4 + 2NaOH \longrightarrow 2H_2O + Na_2SO_4$$

Suppose a beaker contains 35.0 mL of 0.175 M H_2SO_4. How many milliliters of 0.250 M NaOH must be added just to react completely with the sulfuric acid?

Solution

We convert from 35.0 mL (or 35.0 × 10^{-3} L) H_2SO_4 solution to moles H_2SO_4 (using the molarity of H_2SO_4), then to moles NaOH (from the chemical equation).

Finally, we convert this to volume of NaOH solution (using the molarity of NaOH). The calculation is as follows:

$$35.0 \times 10^{-3}\,\text{L } H_2SO_4 \text{ soln} \times \frac{0.175\,\text{mol } H_2SO_4}{1\,\text{L } H_2SO_4 \text{ soln}}$$

$$\times \frac{2\,\text{mol NaOH}}{1\,\text{mol } H_2SO_4} \times \frac{1\,\text{L NaOH soln}}{0.250\,\text{mol NaOH}}$$

$$= 4.90 \times 10^{-2}\,\text{L NaOH soln} \quad \text{(or 49.0 mL NaOH soln)}$$

Thus, 35.0 mL of 0.175 M sulfuric acid solution react with 49.0 mL of .250 M sodium hydroxide solution.

Exercise 3.22

Nickel sulfate, $NiSO_4$, reacts with trisodium phosphate, Na_3PO_4, to give green crystals of nickel phosphate, $Ni_3(PO_4)_2$, and a solution of sodium sulfate, Na_2SO_4.

$$3NiSO_4 + 2Na_3PO_4 \rightarrow Ni_3(PO_4)_2 + 3Na_2SO_4$$

How many milliliters of 0.375 M $NiSO_4$ will react with 45.7 mL of 0.265 M Na_3PO_4?

(See Problems 3.79 and 3.80.)

Exercise 3.23

Sodium hydroxide, NaOH, may be prepared by reacting calcium hydroxide, $Ca(OH)_2$, with sodium carbonate, Na_2CO_3:

$$Ca(OH)_2 \ + \ Na_2CO_3 \rightarrow 2NaOH \ + \ CaCO_3$$

How many milliliters of 0.150 *M* $Ca(OH)_2$ will react with 2.55 g Na_2CO_3?

(See Problems 3.81 and 3.82.)

An important method for determining the amount of a particular substance is based on measuring volumes of reactant solutions. Suppose a substance *A* reacts in solution with substance *B*. If we know the volume and concentration of a solution of *B* that just reacts with substance *A* in a sample, we can determine the amount of *A*. **Titration** is a procedure for determining the amount of substance *A* by adding a carefully measured volume of a solution with known concentration of *B* until the reaction is just complete.

Figure 3.7 shows a flask containing a solution with an unknown amount of sodium hydroxide, NaOH, being titrated with hydrochloric acid, HCl, of known molarity. The reaction is

$$NaOH \ + \ HCl \longrightarrow NaCl \ + \ H_2O$$

To the NaOH solution are added a few drops of phenolphthalein indicator. An *indicator* is a substance that undergoes a color change as a reaction approaches completion.■ Phenolphthalein gives the NaOH solution a pink color. At the completion of the reaction with HCl, it becomes colorless. Hydrochloric acid with a concentration of 0.101 *M* is contained in a *buret*, a glass tube graduated to measure the volume of liquid delivered from the stopcock. The acid is added to the NaOH solution, rapidly at first, until the indicator color begins to lighten. Then the acid is added drop by drop until the solution is colorless. At this point, the reaction is complete and the

■ Indicators are discussed in Chapter 17.

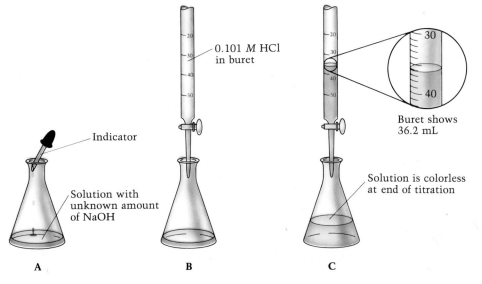

0.101 *M* HCl in buret

Buret shows 36.2 mL

Indicator

Solution with unknown amount of NaOH

Solution is colorless at end of titration

A   B   C

Figure 3.7
Titration of an unknown amount of NaOH. (a) A few drops of phenolphthalein indicator are added to the solution of NaOH. (b) The NaOH solution is then titrated with 0.101 *M* HCl. (c) At the end of the titration, the solution in the flask has turned from pink to colorless. The volume of HCl used (36.2 mL) is read on the buret. (Also see Color Plate 4, *top*.)

volume of HCl that reacts with the NaOH is read from the buret. This volume of hydrochloric acid is then used to obtain the mass of NaOH in the original solution.

Example 3.18

A flask contains an unknown amount of NaOH. Water is added to the flask to dissolve the NaOH. This solution is titrated with 0.101 M HCl. It takes 36.2 mL HCl to complete the reaction with NaOH. What is the mass of the NaOH?

Solution

We convert the volume of HCl (36.2 × 10⁻³ L HCl soln) to moles HCl (from the molarity of HCl), then to moles NaOH (from the chemical equation). Finally, we convert moles NaOH to grams NaOH. The calculation is as follows:

$$36.2 \times 10^{-3} \text{ L HCl soln} \times \frac{0.101 \text{ mol HCl}}{1 \text{ L HCl soln}}$$

$$\times \frac{1 \text{ mol NaOH}}{1 \text{ mol HCl}} \times \frac{40.0 \text{ g NaOH}}{1 \text{ mol NaOH}} = 0.146 \text{ g NaOH}$$

Exercise 3.24

A 5.00-g sample of vinegar is titrated with 0.108 M NaOH. If the vinegar requires 39.1 mL of the NaOH solution for complete reaction, what is the mass percentage of acetic acid, $HC_2H_3O_2$, in the vinegar? The reaction is

$$HC_2H_3O_2 + NaOH \longrightarrow NaC_2H_3O_2 + H_2O$$

(See Problems 3.83 and 3.84.)

A Checklist for Review

Important Terms

isotopes (3.1)
atomic mass unit (amu) (3.1)
atomic weight (3.1)
molecular weight (3.2)
formula weight (3.2)
mole (mol) (3.3)

Avogadro's number (3.3)
molar mass (3.3)
elemental composition (p. 60)
mass percentage (3.4)
empirical (simplest) formula (3.6)
stoichiometry (p. 66)

limiting reactant (limiting reagent) (3.9)
theoretical yield (3.9)
actual yield (3.9)
percentage yield (3.9)
molar concentration (molarity) (M) (3.10)
titration (3.12)

Summary of Facts and Concepts

The *atomic weight* of an element is the average mass of an atom of that element in *atomic mass units* (amu). An atomic mass unit equals one-twelfth the mass of an atom of carbon-12. A *formula weight* equals the sum of the atomic weights of the atoms in the formula of a compound. If the formula corresponds to that of a molecule, this sum of atomic weights equals the *molecular weight* of the compound. The mass of *Avogadro's number* (6.02 × 10²³) of formula units—that is, the mass of one *mole*

of substance—equals the mass in grams that corresponds to the numerical value of the formula weight in amu. This mass is called the *molar mass*.

The *empirical formula* (*simplest formula*) of a compound is obtained from the *elemental composition* of the substance, which is expressed as *mass percentages* of the elements. To calculate the empirical formula, mass percentages are converted to ratios of moles, which when expressed in smallest whole numbers give the subscripts in the formula. A molecular formula is a multiple of the empirical formula; this multiple is determined from the experimental value of the molecular weight.

A chemical equation may be interpreted in terms of moles of reactants and products, as well as in terms of molecules. Using this *molar interpretation*, we can convert from the mass of one substance in a chemical equation to the mass of another. The maximum amount of product from a reaction is determined by the *limiting reactant*, the reactant that is completely used up; the other reactants are in excess.

Molar concentration, or *molarity*, is the moles of solute in a liter of solution. Knowing the molarity allows us to calculate the amount of solute in any volume of solution. Because the moles of solute are constant during the *dilution of a solution*, we can determine to what volume to dilute a concentrated solution to give one of desired molarity. *Titration* is a method of chemical analysis by which we determine the volume of solution of known molarity that reacts with a compound of unknown amount.

Operational Skills

Note: A table of atomic weights is necessary for skills 1, 3–11, and 14.

1. Given the formula of a compound and a table of atomic weights, calculate the formula weight (Example 3.1).

2. Using the molar mass and Avogadro's number, calculate the mass of a molecule in grams (Example 3.2).

3. Given the moles of a compound with known formula, calculate the mass (Example 3.3). Or, given the mass of a compound of known formula, calculate the moles (Example 3.4).

4. Given the mass of a sample of a molecular substance and the formula, calculate the number of molecules in the sample (Example 3.5).

5. Given the formula of a compound, calculate the mass percentage of elements in it (Example 3.6).

6. Given the masses of CO_2 and H_2O obtained from the combustion of a known mass of a compound of C, H, and O, compute the mass percentages of each element (Example 3.7).

7. Given the mass percentages of all elements in a compound, obtain the empirical formula (Example 3.8).

8. Given the empirical formula and molecular weight of a substance, obtain the molecular formula (Example 3.9).

9. Given a chemical equation and the amount of one substance, calculate the amount of another substance involved in the reaction (Examples 3.10 and 3.11).

10. Given the amounts of reactants and the chemical equation, find the limiting reactant; then calculate the amount of a product (Example 3.12).

11. Given the mass of the solute and volume of solution, calculate the molarity (Example 3.13).

12. Given the volume and molarity of a solution, calculate the amount of solute. Or, given the amount of solute and molarity of a solution, calculate the volume (Example 3.14).

13. Calculate the volume of solution of known molarity required to make a specified volume of solution with different molarity (Example 3.15). Calculate this volume given a solution whose mass percentage of solute is known (Example 3.16).

14. Given the chemical equation, calculate the volume of solution of known molarity of one substance that just reacts with a given volume of solution of another substance (Example 3.17). Calculate the mass of one substance that reacts with a given volume of known molarity of solution of another substance (Example 3.18).

Review Questions

3.1 Describe how Dalton obtained relative atomic masses.

3.2 Define the term *atomic weight*. Why might values of atomic weights on Mars be different from those on Earth?

3.3 Define *atomic mass unit* (amu).

3.4 Describe in words how to obtain the formula weight of a compound from the formula.

3.5 One mole of N_2 contains how many N_2 molecules? How many N atoms are there in one mole of N_2? One mole of iron(III) sulfate, $Fe_2(SO_4)_3$, contains how many moles of SO_4^{2-} ions?

3.6 Explain what is involved in determining the composition of a compound of C, H, and O by combustion.

3.7 A substance has the molecular formula $C_6H_{12}O_2$. What is the empirical formula?

3.8 Explain in words what is involved in obtaining the empirical formula from the elemental composition.

3.9 Describe in words the meaning of the equation

$$CH_4 + 2O_2 \longrightarrow CO_2 + 2H_2O$$

using a molecular, a molar, and then a mass interpretation.

3.10 Explain in words how a chemical equation can be used to relate the amounts of different substances involved in a reaction.

3.11 What is a limiting reactant in a reaction mixture? Explain how it determines the amount of product.

3.12 From the definition of molar concentration, show that moles of solute equals the molar concentration times the volume of solution.

3.13 Why is it that the product of molar concentration and volume is constant for a dilution problem?

3.14 Describe in words how the amount of sodium hydroxide in a mixture can be determined by titration with hydrochloric acid of known molarity.

Problems

Formula Weights and Mole Calculations

3.15 Find the formula weights of the following substances to three significant figures:
   (a) methanol, CH_3OH
   (b) phosphorus trichloride, PCl_3
   (c) potassium carbonate, K_2CO_3
   (d) nickel phosphate, $Ni_3(PO_4)_2$

3.17 Sodium thiosulfate, $Na_2S_2O_3$, is used in photography to fix the negatives. What is the molar mass of $Na_2S_2O_3$?

3.19 Diethyl ether, commonly known as ether, is used as an anesthetic. It has the structural formula

$$\begin{array}{ccccc} H & H & & H & H \\ | & | & & | & | \\ H-C-C-O-C-C-H \\ | & | & & | & | \\ H & H & & H & H \end{array} \quad \text{or} \quad CH_3CH_2OCH_2CH_3$$

What is the mass in grams of a molecule of diethyl ether?

3.16 Find the formula weights of the following substances to three significant figures:
   (a) acetic acid, $HC_2H_3O_2$
   (b) phosphorus pentachloride, PCl_5
   (c) potassium sulfate, K_2SO_4
   (d) calcium hydroxide, $Ca(OH)_2$

3.18 Lead chromate, $PbCrO_4$, is a yellow pigment known as chrome yellow. Find the molar mass of $PbCrO_4$.

3.20 Glycerol, a sweet-tasting, syrupy liquid, has the structural formula

$$\begin{array}{ccc} H & H & H \\ | & | & | \\ H-C-C-C-H \\ | & | & | \\ O & O & O \\ | & | & | \\ H & H & H \end{array} \quad \text{or} \quad CH_2OHCHOHCH_2OH$$

It is used as a moistening agent (humectant) for candy and tobacco and is the starting material for nitroglycerine. Calculate the mass of a glycerol molecule in grams.

3.21 Boric acid, H_3BO_3, is a mild antiseptic and is often used as an eyewash. A sample contains 0.543 mol H_3BO_3. What is the mass of boric acid in the sample?

3.22 Carbon disulfide, CS_2, is a colorless, highly flammable liquid. The commercial product, which is used in the manufacture of rayon and cellophane, has a most disagreeable odor. A sample contains 0.0205 mol CS_2. Calculate the mass of carbon disulfide in the sample.

3.23 Calcium sulfate, $CaSO_4$, is a white, crystalline powder. Gypsum is a mineral, or natural substance, of calcium sulfate containing water molecules loosely bound in the crystal (such substances are known as hydrates). A 1.000-g sample of gypsum contains 0.791 g $CaSO_4$. How many moles of $CaSO_4$ are there in this sample? Assuming that the rest of the sample (0.209 g) is water, how many moles of H_2O are there in the sample?

3.25 Carbon tetrachloride, CCl_4, is a colorless, volatile, nonflammable liquid. Formerly, it was used as a dry-cleaning solvent, but it has been replaced by less toxic substances. Its principal use today is in the manufacture of fluorocarbons and as an industrial solvent. How many molecules are there in 8.53 mg of carbon tetrachloride?

3.24 Copper(II) sulfate, $CuSO_4$, is a white, crystalline compound that gives an aqueous solution from which bright blue copper(II) sulfate pentahydrate, $CuSO_4 \cdot 5H_2O$, crystallizes. (The dot in $CuSO_4 \cdot 5H_2O$ means that the H_2O is loosely bound in the crystal.) A 1.547-g sample of this substance is heated carefully to drive off the water. The white crystals of $CuSO_4$ that are left behind have a mass of 0.989 g. How many grams of water were in the original sample? How many moles of H_2O is this? From the mass of $CuSO_4$, calculate the moles of $CuSO_4$. Show that the relative molar amounts of $CuSO_4$ and H_2O agree with the formula of the hydrate.

3.26 Boron trifluoride, BF_3, is a colorless, nonflammable gas that fumes in moist air, producing dense white clouds. Calculate the number of BF_3 molecules in a 1.00-g sample.

Mass Percentage

3.27 A 1.836-g sample of coal contains 1.584 g C. Calculate the mass percentage of C in the coal.

3.29 A 43.4-mg sample of an alcohol contains 15.1 mg O. What is the mass percentage of O in the compound?

3.31 A fertilizer is advertised as containing 15.8% nitrogen (by mass). How much nitrogen is there in 4.15 kg of fertilizer?

3.28 A 5.69-g aqueous solution of isopropyl alcohol contains 4.87 g of isopropyl alcohol. What is the mass percentage of isopropyl alcohol in the solution?

3.30 Ethyl mercaptan is an odorous substance added to natural gas to make leaks easily detectable. A sample of ethyl mercaptan weighing 3.17 mg contains 1.64 mg of sulfur. What is the mass percentage of sulfur in the substance?

3.32 Sea water contains 0.0065% (by mass) of bromine. How many grams of bromine are there in 1.00 L of sea water? The density of sea water is 1.025 g/cm^3.

Determining Chemical Formulas

3.33 Cadaverine, one of the compounds responsible for the odor of decaying flesh, has the formula $NH_2CH_2CH_2CH_2CH_2CH_2NH_2$. What are the mass percentages of the elements in cadaverine? (Give answers to three significant figures.)

3.35 Acetylene is a gas with the formula C_2H_2 and benzene is a liquid with the formula C_6H_6. Calculate the mass percentages of C and H in each of these compounds to three significant figures. Compare and explain the results.

3.34 Ammonium nitrate, NH_4NO_3, and ammonium sulfate, $(NH_4)_2SO_4$, are both used as nitrogen fertilizers. Calculate the mass percentage of N in each of these compounds to three significant figures. Which compound contains more nitrogen on the basis of mass percentage?

3.36 Acetaldehyde, CH_3CHO, is formed in the human body during the metabolism of ethyl alcohol. Increased blood levels of acetaldehyde result in the condition known as acetaldehyde syndrome, or "hangover." Calculate the mass percentages of the elements in acetaldehyde. Also, obtain the mass percentages of the elements in ethyl acetate, $CH_3COOCH_2CH_3$. Give all percentages to three significant figures. Compare with those for acetaldehyde and explain.

3.37 Urea occurs in urine from the breakdown of proteins. It is manufactured commercially from ammonia and carbon dioxide for use as a nitrogen fertilizer and as a starting material for plastics. The compound contains C, H, N, and O. In a combustion analysis, 2.54 mg of urea gave 1.86 mg CO_2 and 1.53 mg H_2O. (Nitrogen in the compound has no effect on the results for C and H.) What are the mass percentages of C and H in the compound? In a separate analysis, urea was shown to contain 46.7% N. What is the percentage of O in the compound?

3.39 Potassium manganate is a dark green, crystalline substance whose composition is 39.7% K, 27.9% Mn, and 32.5% O, by mass. What is the empirical formula?

3.41 Acrylic acid, used in the manufacture of acrylic plastics, has the composition 50.0% C, 5.6% H, and 44.4% O. What is the empirical formula?

3.43 Putrescine, a substance produced by decaying animals, has the empirical formula C_2H_6N. Several determinations of molecular weight give values in the range of 87 to 90 amu. Find the molecular formula of putrescine.

3.45 Oxalic acid is a toxic substance used by laundries to remove rust stains. Its composition is 26.7% C, 2.2% H, and 71.1% O (by mass), and its molecular weight is 90 amu. What is the molecular formula?

3.38 Phenol, commonly known as carbolic acid, was used by Joseph Lister as an antiseptic for surgery in 1865. Its principal use today is in the manufacture of phenolic resins and plastics. Combustion of 5.23 mg of phenol yields 14.67 mg CO_2 and 3.01 mg H_2O. Phenol contains only C, H, and O. What is the percentage of each element in this substance?

3.40 Hydroquinone, used as a photographic developer, is 65.4% C, 5.5% H, and 29.1% O, by mass. What is the empirical formula of hydroquinone?

3.42 Malonic acid is used in the manufacture of barbiturates (sleeping pills). The composition of the acid is 34.6% C, 3.9% H, and 61.5% O. What is the empirical formula?

3.44 Compounds of boron with hydrogen are called boranes. One of these boranes has the empirical formula BH_3 and a molecular weight of 28 amu. What is the molecular formula?

3.46 Adipic acid is used in the manufacture of nylon. The composition of the acid is 49.3% C, 6.9% H, and 43.8% O (by mass), and the molecular weight is 146 amu. What is the molecular formula?

Stoichiometry: Mole–Mass Relations in Reactions

3.47 Ethylene, C_2H_4, burns in oxygen to give carbon dioxide, CO_2, and water. Write the equation for the reaction, giving molecular, molar, and mass interpretations below.

3.49 Tungsten metal, W, is used to make incandescent bulb filaments. The metal is produced from the yellow tungsten(VI) oxide, WO_3, by reaction with hydrogen:

$$WO_3 + 3H_2 \longrightarrow W + 3H_2O$$

How many grams of tungsten can be obtained from 4.81 kg of tungsten(VI) oxide?

3.51 Nitric acid, HNO_3, is manufactured by the Ostwald process, in which nitrogen dioxide, NO_2, reacts with water.

$$3NO_2 + H_2O \longrightarrow 2HNO_3 + NO$$

How many grams of nitrogen dioxide are required in the above reaction to produce 5.89×10^3 kg HNO_3?

3.48 Hydrogen sulfide gas, H_2S, burns in oxygen to give sulfur dioxide, SO_2, and water. Write the equation for the reaction, giving molecular, molar, and mass interpretations below.

3.50 White phosphorus, P_4, is prepared by fusing calcium phosphate, $Ca_3(PO_4)_2$, with carbon, C, and sand, SiO_2, in an electric furnace:

$$2Ca_3(PO_4)_2 + 6SiO_2 + 10C \longrightarrow P_4 + 6CaSiO_3 + 10CO$$

How many grams of calcium phosphate are required to give 1.00 kg of phosphorus?

3.52 Acrylonitrile, C_3H_3N, is the starting material for the production of a kind of synthetic fiber (acrylics). It can be made from propylene, C_3H_6, by reaction with nitric oxide, NO.

$$4C_3H_6 + 6NO \longrightarrow 4C_3H_3N + 6H_2O + N_2$$

(a) How many grams of acrylonitrile are obtained from 651 kg of propylene and excess NO?
(b) How many grams of nitrogen, N_2, are produced from 651 kg of propylene and excess NO?

3.53 The following reaction is used to make carbon tetrachloride, CCl_4, a solvent and starting material for the manufacture of fluorocarbon refrigerants and aerosol propellants:

$$CS_2 + 3Cl_2 \longrightarrow CCl_4 + S_2Cl_2$$

Calculate the number of grams of carbon disulfide, CS_2, needed for a laboratory-scale reaction with 43.2 g of chlorine, Cl_2.

3.55 A sample of calcium carbonate, $CaCO_3$, was decomposed by heating to give calcium oxide, CaO, and carbon dioxide, CO_2. If 6.79 g CaO was obtained, how many grams of CO_2 was produced?

3.54 Solutions of sodium hypochlorite, $NaClO$, are sold as a bleach (such as Clorox). They are prepared by the reaction of chlorine with sodium hydroxide:

$$2NaOH + Cl_2 \longrightarrow NaCl + NaClO + H_2O$$

If chlorine gas, Cl_2, is bubbled into a solution containing 60.0 g $NaOH$, how many grams of Cl_2 will eventually react?

3.56 Sodium nitrate, $NaNO_3$, was decomposed by heating to give sodium nitrite, $NaNO_2$, and oxygen, O_2. If 5.29 g of O_2 was collected from a sample of $NaNO_3$, how many grams of $NaNO_2$ was produced?

Limiting Reactant; Theoretical and Percentage Yields

3.57 Hydrogen chloride, HCl, can be prepared in the laboratory by heating sodium chloride, $NaCl$, with concentrated sulfuric acid, H_2SO_4:

$$NaCl + H_2SO_4 \longrightarrow NaHSO_4 + HCl$$

A reaction vessel contains 10.0 g of sodium chloride and 10.0 g of sulfuric acid. What is the limiting reactant? How many grams of hydrogen chloride gas are produced when the reaction is complete? How many grams of the excess reactant are left unconsumed?

3.59 Aspirin (acetylsalicylic acid) is prepared by heating salicylic acid, $C_7H_6O_3$, with acetic anhydride, $C_4H_6O_3$. The other product is acetic acid, $C_2H_4O_2$.

$$C_7H_6O_3 + C_4H_6O_3 \longrightarrow C_9H_8O_4 + C_2H_4O_2$$

What is the theoretical yield (in grams) of aspirin, $C_9H_8O_4$, when 2.00 g of salicylic acid are heated with 4.00 g of acetic anhydride? If the actual yield of aspirin is 2.21 g, what is the percentage yield?

3.58 Carbon disulfide, CS_2, burns in oxygen. Complete combustion gives the reaction

$$CS_2 + 3O_2 \longrightarrow CO_2 + 2SO_2$$

Calculate the grams of sulfur dioxide, SO_2, produced when a mixture of 15.0 g of carbon disulfide and 35.0 g of oxygen reacts. Which reactant remains unconsumed at the end of the combustion? How much of it remains?

3.60 Methyl salicylate (oil of wintergreen) is prepared by heating salicylic acid, $C_7H_6O_3$, with methanol, CH_3OH.

$$C_7H_6O_3 + CH_3OH \longrightarrow C_8H_8O_3 + H_2O$$

In an experiment, 1.50 g of salicylic acid is reacted with 11.20 g of methanol. The yield of methyl salicylate, $C_8H_8O_3$, is 1.24 g. What is the percentage yield?

Molarity

3.61 A sample of 0.0341 mol iron(III) chloride, $FeCl_3$, was dissolved in water to give 25.0 mL of solution. What is the molarity of the solution?

3.63 An aqueous solution is made from 0.834 g of potassium permanganate, $KMnO_4$. If the volume of solution is 50.0 mL, what is the molarity of $KMnO_4$ in the solution?

3.65 Heme, obtained from red blood cells, binds oxygen, O_2. How many moles of heme are there in 25 mL of 0.0015 M heme solution?

3.62 A 50.0-mL volume of $AgNO_3$ solution contains 0.0285 mol $AgNO_3$ (silver nitrate). What is the molarity of the solution?

3.64 A sample of oxalic acid, $H_2C_2O_4$, weighing 1.384 g is placed in a 100.0-mL volumetric flask, which is then filled to the mark on the neck with water. What is the molarity of $H_2C_2O_4$ in the solution?

3.66 Insulin is a hormone that controls the use of glucose in the body. How many moles of insulin are required to make up 28 mL of 0.0052 M insulin solution?

3.67 How many grams of potassium dichromate, $K_2Cr_2O_7$, should be added to a 50.0-mL volumetric flask to prepare 0.025 M $K_2Cr_2O_7$ when the flask is filled to the mark with water?

3.68 Describe how you would prepare 2.50×10^2 mL of 0.10 M NaOH (aqueous sodium hydroxide). What mass (in grams) of NaOH is needed?

3.69 What volume of 0.120 M NaOH is required to give 0.150 mol NaOH (sodium hydroxide)?

3.70 How many milliliters of 0.114 M $HClO_4$ (perchloric acid) are required to give 0.00752 mol $HClO_4$?

3.71 An experiment calls for 0.0353 g of potassium hydroxide, KOH. How many milliliters of 0.0176 M KOH are required?

3.72 What is the volume (in milliliters) of 0.215 M H_2SO_4 (sulfuric acid) containing 0.949 g H_2SO_4?

3.73 If you dilute 34 mL of 1.32 M $KMnO_4$ (potassium permanganate) to 0.28 M $KMnO_4$, what is the final volume of the solution?

3.74 To what final volume should 25 mL of 2.4 M $K_2Cr_2O_7$ (potassium dichromate) be diluted to give a solution that is 0.10 M $K_2Cr_2O_7$?

3.75 How many milliliters of 2.00 M $HC_2H_3O_2$ (acetic acid) are required to make 45.0 mL of a 0.18 M $HC_2H_3O_2$ solution?

3.76 Describe how you would prepare 1.00 L of 0.120 M NaOH from a 1.20 M NaOH solution.

3.77 How would you make up 255 mL of 0.150 M HNO_3 from nitric acid that is 68.0% HNO_3? The density of 68.0% HNO_3 is 1.41 g/mL.

3.78 A solution of aqueous ammonia contains 28.0% NH_3 by mass (density 0.898 g/mL). Describe the preparation of 425 mL of 0.320 M NH_3 from this solution.

3.79 What volume of 0.250 M HNO_3 (nitric acid) reacts with 53.1 mL of 0.150 M Na_2CO_3 (sodium carbonate) in the following reaction?

$$2HNO_3 + Na_2CO_3 \longrightarrow 2NaNO_3 + H_2O + CO_2$$

3.80 A flask contains 42.4 mL of 0.150 M $Ca(OH)_2$ (calcium hydroxide). How many milliliters of 0.350 M Na_2CO_3 (sodium carbonate) are required to just react with the calcium hydroxide by the following reaction?

$$Na_2CO_3 + Ca(OH)_2 \longrightarrow CaCO_3 + 2NaOH$$

3.81 How many milliliters of 0.150 M H_2SO_4 (sulfuric acid) are required to react with 1.87 g of sodium hydrogen carbonate, $NaHCO_3$, according to the following equation?

$$H_2SO_4 + 2NaHCO_3 \longrightarrow Na_2SO_4 + 2H_2O + 2CO_2$$

3.82 How many milliliters of 0.238 M $KMnO_4$ are needed to react with 2.16 g of iron(II) sulfate, $FeSO_4$? The reaction is

$$10FeSO_4 + 2KMnO_4 + 8H_2SO_4 \longrightarrow$$
$$5Fe_2(SO_4)_3 + 2MnSO_4 + K_2SO_4 + 8H_2O$$

3.83 A solution of hydrogen peroxide, H_2O_2, is titrated with a solution of potassium permanganate, $KMnO_4$. The reaction is

$$5H_2O_2 + 2KMnO_4 + 3H_2SO_4 \longrightarrow$$
$$5O_2 + 2MnSO_4 + K_2SO_4 + 8H_2O$$

It requires 46.9 mL of 0.145 M $KMnO_4$ to titrate 20.0 g of the solution of hydrogen peroxide. What is the mass percentage of H_2O_2 in the solution?

3.84 A 3.33-g sample of iron ore is transformed to a solution of iron(II) sulfate, $FeSO_4$, and this solution is titrated with 0.150 M $K_2Cr_2O_7$ (potassium dichromate). If it requires 41.4 mL of potassium dichromate solution to titrate the iron(II) sulfate solution, what is the percentage of iron in the ore? The reaction is

$$6FeSO_4 + K_2Cr_2O_7 + 7H_2SO_4 \longrightarrow$$
$$3Fe_2(SO_4)_3 + Cr_2(SO_4)_3 + 7H_2O + K_2SO_4$$

Additional Problems

3.85 Caffeine, the stimulant in coffee and tea, has the molecular formula $C_8H_{10}N_4O_2$. Calculate the mass percentage of each element in the substance. Give the answers to three significant figures.

3.86 Morphine, a narcotic substance obtained from opium, has the molecular formula $C_{17}H_{19}NO_3$. What is the mass percentage of each element in morphine (to three significant figures)?

3.87 A compound of nitrogen and oxygen contains 36.9% N (by mass). Determine the empirical formula of the compound.

3.88 Magnesium pyrophosphate contains 21.9% Mg, 27.8% P, and 50.3% O. What is the empirical formula of this substance?

3.89 A moth repellent, *para*-dichlorobenzene, has the composition 49.0% C, 2.7% H, and 48.2% Cl. The molecular weight of the compound is 147 amu. What is the molecular formula?

3.90 Sorbic acid is added to food as a mold inhibitor. Its elemental composition is 64.3% C, 7.2% H, and 28.5% O. The molecular weight of sorbic acid is 112 amu. What is the molecular formula?

*3.91 A water-soluble compound of gold and chlorine is treated with silver nitrate to convert the chlorine completely to silver chloride, AgCl. In an experiment, 328 mg of the compound gave 464 mg of silver chloride. Calculate the percentage of Cl in the compound. What is the empirical formula?

*3.92 A solution of scandium chloride was treated with silver nitrate. The chlorine in the scandium compound was converted to silver chloride, AgCl. A 58.9 mg-sample of scandium chloride gave 167.4 mg of silver chloride. What are the mass percentages of Sc and Cl in scandium chloride? What is the empirical formula?

**3.93 Dieldrin, like DDT, is a persistent insecticide (one that persists in the environment). Combustion of 29.72 mg of dieldrin gave 41.21 mg of carbon dioxide and 5.63 mg of water. In a separate analysis, all of the chlorine in 25.31 mg dieldrin was converted to 57.13 mg of silver chloride. The compound contains only C, H, Cl, and O. What is the empirical formula of dieldrin?

**3.94 Chloral hydrate is used as a hypnotic or sleep inducer (knockout drops) and as a starting material in the manufacture of the insecticide DDT. In a combustion analysis, 8.73 mg of chloral hydrate gave 4.65 mg CO_2 and 1.43 mg H_2O. In a separate experiment, 3.26 mg of chloral hydrate were treated to convert the chlorine in the substance to 8.47 mg of silver chloride, AgCl. The compound contains C, H, Cl, and O. What is the empirical formula?

**3.95 Thiophene is a liquid compound containing the elements carbon, hydrogen, and sulfur. A sample of thiophene weighing 7.96 mg was burned in oxygen, giving 16.65 mg CO_2. In a separate experiment, thiophene was subjected to a series of chemical reactions that transformed all of the sulfur in the compound to barium sulfate, $BaSO_4$. If 4.31 mg of thiophene gave 11.96 mg of barium sulfate, what is the empirical formula of thiophene? The molecular weight is 84 amu. What is the molecular formula?

**3.96 The combustion of a compound containing C, H, and N yields CO_2, H_2O, and N_2 as products. Aniline is a liquid compound consisting of these elements. If the combustion of 9.71 mg of aniline yields 6.63 mg H_2O and 1.46 mg N_2, what is the empirical formula? The molecular weight of aniline is 93 amu. What is the molecular formula?

**3.97 Hemoglobin is the oxygen-carrying molecule of red blood cells. It can be separated into a protein and a nonprotein substance. The nonprotein substance is called heme. A sample of heme weighing 35.2 mg contains 3.19 mg of iron. If a heme molecule contains one atom of iron, what is the molecular weight of heme?

**3.98 Penicillin V was treated chemically to convert sulfur to barium sulfate, $BaSO_4$. An 8.47-mg sample of penicillin V gave 5.46 mg $BaSO_4$. What is the percentage of sulfur in penicillin V? If there is one sulfur atom in the molecule, what is the molecular weight?

*3.99 A sample of limestone (containing calcium carbonate, $CaCO_3$) weighing 428 mg is treated with oxalic acid, $H_2C_2O_4$, to give calcium oxalate, CaC_2O_4:

$$CaCO_3 + H_2C_2O_4 \longrightarrow CaC_2O_4 + H_2O + CO_2$$

The mass of the calcium oxalate is 472 mg. What is the mass percentage of calcium carbonate in the limestone?

*3.100 Rutile is a natural source of titanium dioxide (TiO_2). The crude rock contains some iron oxide and silica. When it is heated with carbon in the presence of chlorine, titanium tetrachloride, $TiCl_4$, is formed:

$$TiO_2 + C + 2Cl_2 \longrightarrow TiCl_4 + CO_2$$

Titanium tetrachloride is a liquid and can be distilled from the mixture. If 35.4 g of titanium tetrachloride are recovered from 15.6 g of crude ore, what is the mass percentage of titanium dioxide in the ore (assuming all available TiO_2 reacts to form $TiCl_4$)?

*3.101 Sulfuric acid, H_2SO_4, is manufactured commercially by burning sulfur, S_8, to sulfur dioxide, SO_2, which is then reacted with more oxygen to produce sulfur trioxide, SO_3. Sulfur trioxide reacts with water to make sulfuric acid. How many grams of sulfur are needed to produce 5.00 kg of sulfuric acid?

*3.102 Carbon disulfide, CS_2, reacts with chlorine to give carbon tetrachloride, CCl_4:

$$CS_2 + 3Cl_2 \longrightarrow S_2Cl_2 + CCl_4$$

The disulfur dichloride, S_2Cl_2, reacts with carbon disulfide to produce additional carbon tetrachloride:

$$4CS_2 + 8S_2Cl_2 \longrightarrow 3S_8 + 4CCl_4$$

How many grams of chlorine, Cl_2, are needed to produce 5.00 kg of carbon tetrachloride?

3.103 Ethylene oxide, C_2H_4O, is made by the oxidation of ethylene, C_2H_4.

$$2C_2H_4 + O_2 \longrightarrow 2C_2H_4O$$

Ethylene oxide is used to make ethylene glycol for automobile antifreeze. In a pilot study, 10.6 g of ethylene gave 9.69 g of ethylene oxide. What is the percentage yield of ethylene oxide?

3.104 Nitrobenzene, $C_6H_5NO_2$, an important raw material for the dye industry, is prepared from benzene, C_6H_6, and nitric acid, HNO_3.

$$C_6H_6 + HNO_3 \longrightarrow C_6H_5NO_2 + H_2O$$

Using 20.3 g of benzene and an excess of HNO_3, what is the theoretical yield of nitrobenzene? If 28.7 g of nitrobenzene are recovered, what is the percentage yield?

*3.105 Zinc metal can be obtained from zinc oxide, ZnO, by reaction at high temperature with carbon monoxide, CO:

$$ZnO + CO \longrightarrow Zn + CO_2$$

The carbon monoxide is obtained from carbon:

$$2C + O_2 \longrightarrow 2CO$$

What is the maximum amount of zinc that can be obtained from 75.0 g of zinc oxide and 50.0 g of carbon?

*3.106 Hydrogen cyanide, HCN, can be made by a two-step process. First, ammonia is reacted with O_2 to give nitric oxide, NO:

$$4NH_3 + 5O_2 \longrightarrow 4NO + 6H_2O$$

Then, nitric oxide is reacted with methane, CH_4:

$$2NO + 2CH_4 \longrightarrow 2HCN + 2H_2O + H_2$$

Using 24.2 g of ammonia and 25.1 g of methane, how many grams of hydrogen cyanide can be produced?

*3.107 An aqueous solution contains 2.25 g of calcium chloride, $CaCl_2$, per liter. What is the molarity of $CaCl_2$? When calcium chloride dissolves in water, the calcium ions, Ca^{2+}, and chloride ions, Cl^-, in the crystal go into the solution. What is the molarity of each ion in the solution?

*3.108 An aqueous solution contains 3.45 g of iron(III) sulfate, $Fe_2(SO_4)_3$, per liter. What is the molarity of $Fe_2(SO_4)_3$? When the compound dissolves in water, the Fe^{3+} ions and SO_4^{2-} ions in the crystal go into solution. What is the molar concentration of each ion in the solution?

3.109 A stock solution of potassium dichromate, $K_2Cr_2O_7$, is made by dissolving 89.3 g of the compound in 1.00 L of solution. How many milliliters of this solution are required to prepare 1.00 L of 0.100 M $K_2Cr_2O_7$?

3.110 A 69.3-g sample of oxalic acid, $H_2C_2O_4$, was dissolved in 1.00 L of solution. How would you prepare 1.00 L of 0.150 M $H_2C_2O_4$ from this solution?

3.111 A solution contains 6.00% (by mass) NaBr (sodium bromide). The density of the solution is 1.046 g/cm^3. What is the molarity of NaBr?

3.112 An aqueous solution contains 4.50% NH_3 (ammonia) by mass. The density of the aqueous ammonia is 0.979 g/mL. What is the molar concentration of NH_3 in the solution?

**3.113 A 0.608-g sample of fertilizer contained nitrogen as ammonium sulfate, $(NH_4)_2SO_4$. It was analyzed for the percentage of nitrogen by heating with sodium hydroxide, NaOH, to drive off ammonia, NH_3:

$$(NH_4)_2SO_4 + 2NaOH \longrightarrow Na_2SO_4 + 2H_2O + 2NH_3$$

The ammonia was collected in 46.3 mL of 0.213 M HCl (hydrochloric acid), with which it reacted.

$$NH_3 + HCl \longrightarrow NH_4Cl + H_2O$$

This solution was titrated for excess hydrochloric acid with 44.3 mL of 0.128 M NaOH.

$$NaOH + HCl \longrightarrow NaCl + H_2O$$

What is the percentage of nitrogen in the fertilizer?

**3.114 An antacid tablet contains sodium hydrogen carbonate, $NaHCO_3$, and inert ingredients. A 0.500-g sample of powdered tablet was mixed with 50.0 mL of 0.190 M HCl (hydrochloric acid). The mixture was allowed to stand until it reacted.

$$NaHCO_3 + HCl \longrightarrow NaCl + H_2O + CO_2$$

The excess hydrochloric acid was titrated with 47.1 mL of 0.128 M NaOH (sodium hydroxide):

$$HCl + NaOH \longrightarrow NaCl + H_2O$$

What is the percentage of sodium hydrogen carbonate in the antacid?

3.115 Calcium carbide, CaC_2, used to produce acetylene, C_2H_2, is prepared by heating calcium oxide, CaO, and carbon, C, to high temperature:

$$CaO + 3C \longrightarrow CaC_2 + CO$$

If a reaction mixture contains 1.00 kg each of calcium oxide and carbon, how many grams of calcium carbide can be prepared?

3.116 A mixture consisting of 12.8 g of calcium fluoride, CaF_2, and 13.2 g of sulfuric acid, H_2SO_4, is heated to drive off hydrogen fluoride, HF:

$$CaF_2 + H_2SO_4 \longrightarrow 2HF + CaSO_4$$

What is the maximum number of grams of hydrogen fluoride that can be obtained?

*3.117 Alloys, or metallic mixtures, of mercury with another metal are called amalgams. Sodium in sodium amalgam reacts with water, giving off hydrogen:

$$2Na + H_2O \longrightarrow 2NaOH + H_2$$

If a 15.23-g sample of sodium amalgam evolves 0.108 g of hydrogen, what is the percentage of sodium in the amalgam? Mercury does not react.

*3.118 A type of sandstone consists of a mixture of silica, SiO_2 (silicon dioxide), and calcite, $CaCO_3$ (calcium carbonate). When the sandstone is heated, calcium carbonate decomposes into calcium oxide, CaO, and carbon dioxide:

$$CaCO_3 \longrightarrow CaO + CO_2$$

What is the percentage of silica in the sandstone if 21.8 mg of the rock yield 4.01 mg of carbon dioxide?

**3.119 Potassium permanganate, $KMnO_4$, is produced in a two-step process starting from manganese dioxide, MnO_2.

$$2MnO_2 + 4KOH + O_2 \longrightarrow 2K_2MnO_4 + 2H_2O$$

$$3K_2MnO_4 + 2CO_2 \longrightarrow 2KMnO_4 + 2K_2CO_3 + MnO_2$$

The manganese dioxide obtained as a product in the last step is recycled to the first step. When a chemical plant is in continuous operation, it can produce 76.5 kg $KMnO_4$ starting from 55.2 kg MnO_2. What is the theoretical yield of $KMnO_4$ based on this mass of MnO_2, assuming that the MnO_2 produced in the second step is continuously recycled to the first step? What is the percentage yield? (*Hint:* Since the MnO_2 is continuously recycled, all the manganese in it will eventually be incorporated into $KMnO_4$.)

**3.120 Nitric acid, HNO_3, is manufactured commercially from ammonia, NH_3, using the following series of steps (the Ostwald process):

$$4NH_3 + 5O_2 \longrightarrow 4NO + 6H_2O$$

$$2NO + O_2 \longrightarrow 2NO_2$$

$$3NO_2 + H_2O \longrightarrow 2HNO_3 + NO$$

The nitric oxide produced in the last step is recycled to the second step. In an experiment, 35.8 g of ammonia gave 125.3 g HNO_3. What is the theoretical yield based on this mass of ammonia, assuming that NO in the last step is continuously recycled to the previous step? What is the percentage yield? (*Hint:* Since the NO is continuously recycled, all the nitrogen in it will eventually be incorporated into HNO_3.)

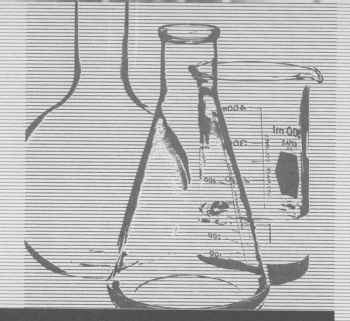

4. The Gaseous State

Gas Laws

4.1 Measurement of Gas Pressure
4.2 Boyle's Law
4.3 Charles's Law Variation of Volume with Absolute
Temperature/ Charles's Law Calculations
4.4 Avogadro's Law
4.5 The Ideal Gas Law Calculations Using the Ideal
Gas Law/ Stoichiometry Problems with Gas Volumes/ Gas
Density; Molecular-Weight Determination
4.6 Gas Mixtures; Law of Partial Pressures Mole
Fractions/ Collecting Gases over Water

Kinetic-Molecular Theory

4.7 Kinetic Theory of an Ideal Gas Postulates of
Kinetic Theory/ The Ideal Gas Law from Kinetic Theory
4.8 Molecular Speeds
4.9 Diffusion and Effusion
4.10 Real Gases

G ases are the simplest form of matter. A few simple equations relate pressure, volume, temperature, and molar amount for all gases. No such simple relationships exist for liquids and solids. Knowing these simple equations for gases is important for understanding and working with them. Say we want to determine the amount of compressed oxygen gas in a steel tank. We can conveniently measure the pressure of gas in the tank by attaching a pressure gauge. But to calculate the amount of oxygen, we also need to know the temperature and volume of gas (the volume of the tank). We would ask, "How many grams of oxygen are there in a 50.0-L tank at 21°C when the oxygen pressure is 15.7 atm?" Similarly, we can calculate the quantity of gas produced in a chemical reaction where the volume of gas is measured at a given pressure and temperature.

Kinetic-molecular theory describes a gas as composed of molecules in constant motion. This theory helps to explain how the physical properties of gases—pressure, volume, temperature, and amount—are related. It has also aided our understanding of the flow of fluids, the transmission of sound, and the conduction of heat. We can use this molecular concept to explain and predict the results of many different kinds of experiments. For example, suppose we want to find the average speed of an oxygen molecule in our 50.0-L tank at 21°C. The kinetic-molecular theory of gases, as we will see, predicts an average speed of 479 m/s (1071 mi/hr).

Chapter Overview

In this chapter, we will look at the relationships of pressure, volume, temperature, and amount of gas. We will begin by examining the experimental behavior of gases at low to moderate pressures and moderate temperatures. The experimental results are summarized by various *gas laws*. Next, we will look at the *kinetic-molecular theory* interpretation of the gas laws. According to this theory, a gas consists of molecules that are in constant random motion. The final section will discuss why gases deviate from the gas laws at high pressure and low temperature.

Gas Laws

All gases near room temperatures and at normal pressures show the same quantitative relationships among their physical properties of pressure, temperature, volume, and molar amount. The discovery of these quantitative relationships, called gas laws, occurred from the mid-seventeenth to the mid-nineteenth century. Before we discuss the gas laws, we need to understand how to measure the pressures of gases.

4.1 Measurement of Gas Pressure

A **barometer** is a device for measuring the pressure of the atmosphere. Recall from Section 1.6 that pressure is force per unit area. The mercury barometer, developed in 1643 by Evangelista Torricelli, is a simple but accurate instrument. It consists of a glass tube, about one meter long, that is filled with mercury and inverted in a dish of the same liquid metal (Figure 4.1). At sea level the mercury in the tube falls to a height of about 760 mm above the level in the dish.

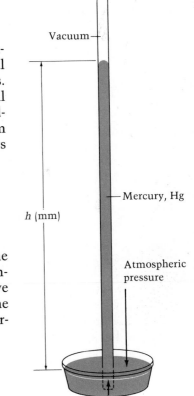

Figure 4.1
A mercury barometer. The height *h* is proportional to the barometric pressure. Thus, the pressure, *P*, can be given as the height of the mercury column, that is, mmHg.

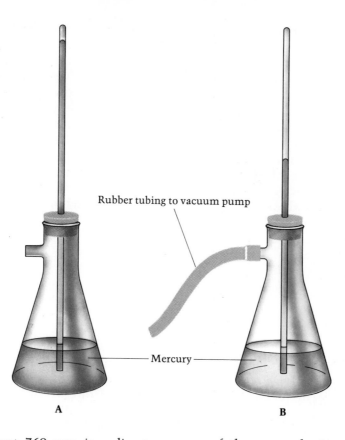

Rubber tubing to vacuum pump

Mercury

A **B**

Figure 4.2
Effect of reducing air pressure. When the side arm of the flask in (a) is connected to a vacuum pump as in (b), the mercury level in the tube falls.

This height of about 760 mm is a direct measure of the atmospheric pressure. Air pressure downward on the surface of the mercury in the dish is transmitted through the liquid, so that the same pressure is exerted upward at the base of the mercury column (see Figure 4.1). This upward pressure supports the mercury column.■

We can check this explanation of how a barometer works by reducing the air pressure over the mercury in the dish. In Figure 4.2a, the mercury tube is placed in a flask with a side arm. The mercury column is initially held up by the atmospheric pressure. However, when the side arm is connected to a vacuum pump that reduces the air pressure over the mercury in the flask, the level of the column in the glass tube begins to fall (Figure 4.2b).

A mercury column placed in a sealed flask, as in Figure 4.2b, no longer measures the pressure of the atmosphere. Instead, it measures the gas pressure in the flask. It is acting as a **manometer,** a device that measures the pressure of a gas or liquid in a sealed vessel.■ Figure 4.3 shows a flask equipped with a mercury manometer.

The exact relationship between pressure, P, and the height of a liquid column, h, in a manometer is

$$P = gdh$$

Here g is the constant acceleration of gravity (9.807 m/s^2) and d stands for the density of the liquid in the manometer. Since the density of mercury is 13.595 g/cm^3, or 13.595 × 10^3 kg/m^3, at 0°C, the pressure in SI units equivalent to a column of mercury 760.0 mm high is 1.013 × 10^5 Pa.■

■ Atmospheric pressure at a given location depends on the weather. A barometer is useful in weather forecasting because of the variation of atmospheric pressure with weather conditions. Atmospheric pressure also varies with height above the surface of the earth. At high altitudes, the pressure is lower, and in space the pressure is very low indeed—approximately 8.5 × 10^{-8} atm at 100 miles above the earth.

■ A sphygmomanometer is the familiar instrument used by doctors to measure blood pressure in the arteries.

■ Substituting the values of g, d, and h in $P = gdh$, we find that a column of mercury 760.0 mm high has a pressure of 9.807 m/s^2 × 13.595 × 10^3 kg/m^3 × 0.7600 m = 1.013 × 10^5 kg/m/s^2 (Pa).

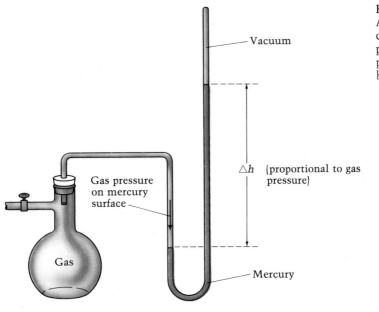

Figure 4.3
A flask equipped with a closed manometer. The gas pressure in the flask is proportional to the difference in heights of the liquid levels.

Vacuum

$\triangle h$ (proportional to gas pressure)

Gas pressure on mercury surface

Gas

Mercury

Because of the direct proportion between the height of a mercury column and pressure, the height of the column in millimeters is referred to as the pressure in units of mmHg (millimeters of mercury). This unit is also called the torr. A pressure of 760 mmHg equals one atmosphere (atm) and equals approximately 100 kPa. Table 4.1 summarizes the relationships of these various units of pressure.

4.2 Boyle's Law

One of the characteristic properties of a gas is its *compressibility*—its ability to be squeezed into a smaller volume by the application of pressure. By comparison, liquids and solids are relatively incompressible. Figure 4.4 shows a "Cartesian diver," a demonstration that depends on the difference in compressibility of air and water.

The compressibility of gases was first studied quantitatively by Robert Boyle in 1661.■ Figure 4.5 shows the type of apparatus he used to demonstrate the variation of volume with pressure of a gas at a given temperature. As mercury is poured into the open end of the J-shaped tube, the volume of

■ Robert Boyle (1627–1691) is noted for his contributions to the particle view of matter. In addition to his experiments on gases, he developed the concept of "primary" particles, a forerunner of our present concept of atoms and elements. He explained different kinds of matter in terms of the organization and motion of these primary particles.

Table 4.1
Important Units of Pressure

Unit	Relationship or definition
Pascal (Pa)	$kg/(m \cdot s^2)$ (SI unit)
Atmosphere (atm)	$1 \text{ atm} = 1.01325 \times 10^5 \text{ Pa} \simeq 100 \text{ kPa}$
mmHg or torr	$760 \text{ mmHg} = 1 \text{ atm}$

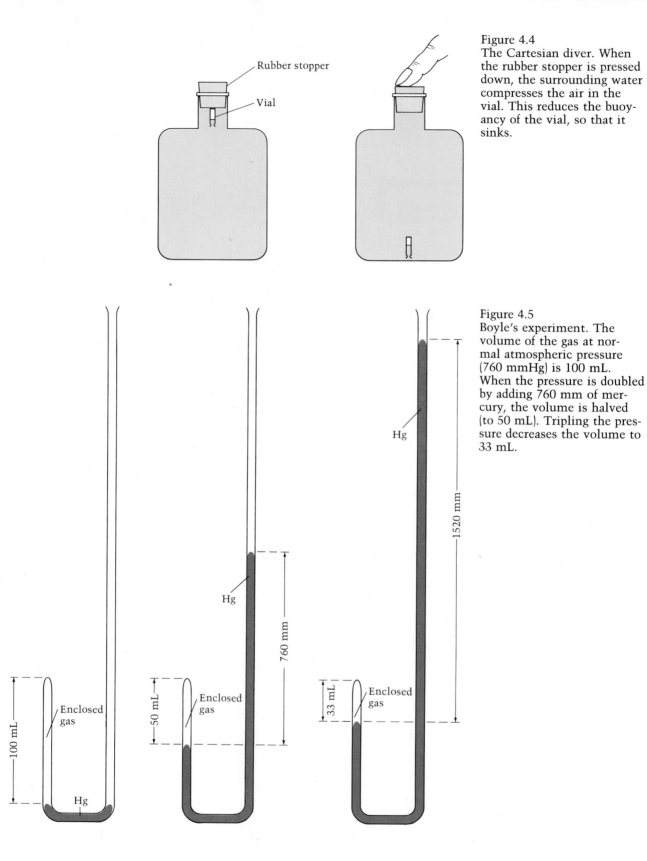

Rubber stopper

Vial

Figure 4.4
The Cartesian diver. When the rubber stopper is pressed down, the surrounding water compresses the air in the vial. This reduces the buoyancy of the vial, so that it sinks.

Figure 4.5
Boyle's experiment. The volume of the gas at normal atmospheric pressure (760 mmHg) is 100 mL. When the pressure is doubled by adding 760 mm of mercury, the volume is halved (to 50 mL). Tripling the pressure decreases the volume to 33 mL.

Hg

1520 mm

Hg

760 mm

Enclosed gas

50 mL

33 mL

Enclosed gas

Enclosed gas

100 mL

Hg

the enclosed gas decreases. *The volume of a sample of gas at a given temperature is found to vary inversely with the applied pressure.* (In Figure 4.5, the applied pressure, which is also the pressure of the gas in the tube, equals that of the atmosphere plus the pressure of the column of mercury that lies between the two liquid levels.) Thus, if the pressure is doubled, the volume is halved. Such an inverse relation can be written as a proportion, using ∝ as the symbol for "is proportional to":

$$V \propto \frac{1}{P} \quad \text{(at fixed temperature for a given amount of gas)}$$

where V is the volume occupied by the gas and P is the pressure. This relationship is known as **Boyle's law.**

Boyle's law can also be expressed in the form of an equation:

$$V = \frac{\text{constant}}{P} \quad \text{(at fixed temperature for a given amount of gas)}$$

Or, putting pressure and volume on the same side of the equation, we get

Boyle's law:
$$PV = \text{constant} \quad \text{(at fixed temperature for a given amount of gas)}$$

That is, for a given amount of gas at a fixed temperature, pressure times the volume equals a constant. Table 4.2 gives some pressure and volume data for 1.000 g O_2 at 0°C. Note that the product of the pressure and volume is nearly constant. By plotting the volume of the oxygen at different pressures (as shown in Figure 4.6), we can obtain a graph showing the inverse relationship of P and V.

Boyle's law is used to calculate the volume occupied by a gas when the pressure changes. Consider the 50.0-L tank of oxygen mentioned in the chapter opening. The pressure of gas in the tank is 15.7 atm at 21°C. What volume of oxygen can we get from the tank at 21°C if the atmospheric pressure is 1.00 atm? We can write P_i and V_i for the initial pressure (15.7 atm) and initial volume (50.0 L), and P_f and V_f for the final pressure (1.00 atm) and final volume (to be determined). Since the temperature does not change, the product of the pressure and volume remains constant. Thus, we can write

$$P_f V_f = P_i V_i$$

P (atm)	V (L)	PV
0.2500	2.801	0.7003
0.5000	1.400	0.7001
0.7500	0.9333	0.6999
1.000	0.6998	0.6998
2.000	0.3495	0.6991
3.000	0.2328	0.6984
4.000	0.1744	0.6977
5.000	0.1394	0.6969

Table 4.2
Pressure–Volume Data for 1.000 g O_2 at 0°C

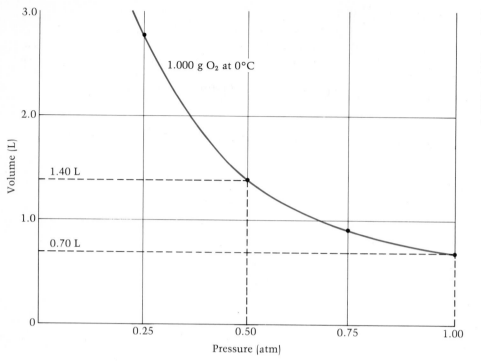

Figure 4.6
A plot of the volume of 1.000 g O_2 at 0°C for various pressures. Note how the volume decreases with increasing pressure. When the pressure is doubled (from 0.50 atm to 1.00 atm), the volume is halved.

Dividing both sides of the equation by P_f gives

$$V_f = V_i \times \frac{P_i}{P_f}$$

If we substitute into this equation, we get

$$V_f = 50.0 \text{ L} \times \frac{15.7 \text{ atm}}{1.00 \text{ atm}} = 785 \text{ L}$$

Note that the initial volume is multiplied by a ratio of pressures. Since we know that the oxygen gas is changing to a lower pressure and will therefore expand, this ratio will be greater than one. The final volume, 785 L, is that occupied by all of the gas at 1.00 atm pressure. However, since the tank is 50.0 L, this volume of gas remains in the tank. The volume that escapes is (785 − 50.0) L = 735 L.

Example 4.1

A volume of air occupying 12.0 dm³ at 98.9 kPa is compressed to a pressure of 119.0 kPa. The temperature remains constant. What is the new volume?

Solution

Putting the data for the problem in tabular form, we see what data we have and what we must find. This will suggest the method of solution.

$V_i = 12.0 \text{ dm}^3 \quad P_i = 98.9 \text{ kPa}$ ⎱ T and n (number
$V_f = \quad ? \quad P_f = 119.0 \text{ kPa}$ ⎰ of moles) remain constant.

Because P and V vary but T and n are constant, we use Boyle's law to obtain

(Continued)

$$V_f = V_i \times \frac{P_i}{P_f}$$

$$= 12.0 \text{ dm}^3 \times \frac{98.9 \text{ kPa}}{119.0 \text{ kPa}} = 9.97 \text{ dm}^3$$

Note that the pressure on the gas increases, so the gas is compressed (gives a smaller volume), and the ratio of pressures is less than 1. Thus, we could get the above result by simply multiplying the volume by the ratio of pressures, choosing the ratio so it is less than 1. (If we had expected the gas volume to increase, we would have chosen the ratio to be greater than 1.)

Exercise 4.1

A volume of carbon dioxide gas equal to 20.0 L was collected at 23°C and 1.00 atm pressure. What would be the volume of carbon dioxide if it was collected at 23°C and 0.830 atm?

(See Problems 4.25, 4.26, 4.27 and 4.28.)

Before leaving the subject of Boyle's law, we should note that the pressure–volume product for a gas is not precisely constant. We can see this from the PV data given in Table 4.2 for oxygen. In fact, all gases follow Boyle's law at low to moderate pressures, but deviate from this law at high pressures. The extent of deviation depends on the gas. We will return to this point at the end of the chapter.

4.3 Charles's Law

Temperature is another important variable affecting gas volume. Figure 4.7 shows that when a partially filled balloon is immersed in hot water, it becomes inflated. If the balloon is then placed in cold water, it quickly deflates. Thus, a gas expands when heated and contracts when cooled.

One of the first quantitative observations of gases at different temperatures was made by Jacques Alexandre Charles in 1787. Charles was a French physicist and a pioneer in hot-air and hydrogen-filled balloons. Later, John Dalton (in 1801) and Joseph Louis Gay-Lussac (in 1802) continued these kinds of experiments, which showed that a sample of gas at a fixed pressure increases in volume *linearly* with temperature, at a fixed pressure. By "linearly," we mean that if we plot the volume occupied by a given sample of gas at various temperatures, we get a straight line.

■ The first ascent of a hot-air balloon carrying people was made on November 21, 1783. A few days later, Jacques Alexandre Charles made an ascent in a hydrogen-filled balloon. On landing, the balloon was attacked and torn to shreds by terrified peasants armed with pitchforks.

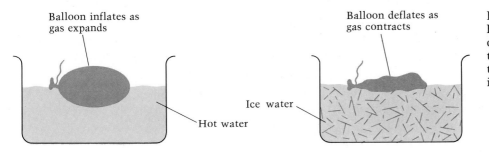

Balloon inflates as gas expands

Balloon deflates as gas contracts

Ice water

Hot water

Figure 4.7
Demonstration of the effect of temperature on a gas. Note that the balloon expands in the warm bath but contracts in the ice bath.

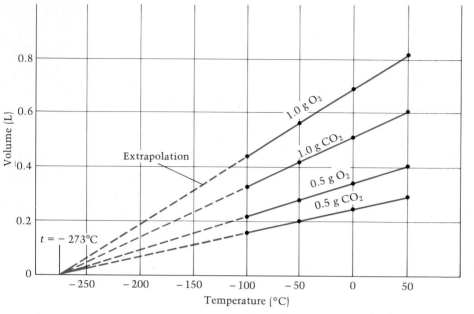

Figure 4.8
A graph showing that the volume of a gas at constant pressure varies linearly with temperature. Note that all lines extrapolate to $-273°C$ at zero volume.

Figure 4.8 plots the volumes at 1.00 atm pressure for samples of various gases at different temperatures. Note that each sample of gas shows a linear (straight line) variation of the volume with temperature. This linear variation is obtained independent of the amount or kind of gas.

Variation of Volume with Absolute Temperature

If the straight lines in Figure 4.8 are extended from the last experimental point toward lower temperatures—that is, if we *extrapolate* the straight lines backward—we find that they all intersect at a common point. This point occurs at a temperature of $-273.15°C$, where the graph indicates a volume of zero. This seems to say that if the substances remain gaseous, the volumes occupied will be zero at $-273.15°C$. This could not happen, however; all gases liquefy before they reach this temperature, and Charles's law would not apply. These extrapolations do show that we can express the volume variation of a gas with temperature more simply by choosing a different thermometer scale. Let us see how to do this.

The fact that the volume occupied by a gas varies linearly with degrees Celsius can be expressed mathematically by the following equation:

$$V = a + bt$$

where t is the temperature in degrees Celsius, and a and b are constants that determine the straight line.■

We can eliminate one of the constants (a) by observing that, on any one of the possible straight lines for gases, $V = 0$ at $t = -273.15$. Substituting

■ The mathematical equation of a straight line is discussed in Appendix A.

these values into the preceding equation, we get

$$0 = a + b(-273.15) \quad \text{or} \quad a = 273.15b$$

The equation for the volume can now be rewritten:

$$V = 273.15b + bt = b(t + 273.15)$$

Instead of following the variation of gas volume on the Celsius temperature scale, suppose we use a temperature scale equal to degrees Celsius plus 273.15. You may recognize that this gives the *Kelvin scale* (an absolute temperature scale), on which kelvins (K) are given by the formula ■

$$K = °C + 273.15$$

■ Absolute temperature and the Kelvin scale were discussed in Section 1.4.

Let us write T for the temperature on the Kelvin scale. Then we get

$$V = bT$$

This equation is a mathematical form of **Charles's law,** which we can state as follows: *The volume occupied by any sample of gas (at a constant pressure) is directly proportional to the absolute temperature.* Thus, doubling the absolute temperature of a gas doubles its volume.

Although most gases follow Charles's law fairly well, they deviate from it at high pressures and as the temperature is reduced—as we will see in the last section.

Charles's Law Calculations

The equation $V = bT$ for Charles's law can be rearranged into a form that is very useful for computation.

Charles's law:

$$\frac{V}{T} = \text{constant} \quad \text{(at a fixed pressure for a given amount of gas)}$$

It says that the volume occupied by a sample of gas at a fixed pressure, when divided by its absolute temperature, remains constant.

Consider a sample of gas at a fixed pressure, and suppose the temperature changes from its initial value T_i to a final value T_f. How does the volume change? Since the volume divided by absolute temperature is constant, we can write

$$\frac{V_f}{T_f} = \frac{V_i}{T_i}$$

Or, rearranging slightly,

$$V_f = V_i \times \frac{T_f}{T_i}$$

Note that the initial volume is multiplied by a ratio of absolute temperatures. The next example illustrates a calculation involving a change in temperature of a gas.

Example 4.2

In the previous section, we found that the total volume of oxygen that can be obtained from a particular tank at 1.00 atm and 21°C is 785 L. What would be the volume of oxygen if the temperature had instead been 28°C?

Solution

Before we can do the gas calculation, we must express the temperatures on the absolute temperature scale in kelvins:

$$T_i = (21 + 273) \text{ K} = 294 \text{ K}$$
$$T_f = (28 + 273) \text{ K} = 301 \text{ K}$$

Putting the data for the problem in tabular form:

$V_i = 785$ L $P_i = 1.00$ atm $T_i = 294$ K

$V_f = \quad ?$ $P_f = 1.00$ atm $T_f = 301$ K

Because V and T vary but P is constant, we use Charles's law.

$$V_f = V_i \times \frac{T_f}{T_i} = 785 \text{ L} \times \frac{301 \, K}{294 \, K} = 804 \text{ L}$$

Note that the temperature increases, so we expect the volume to increase. This means that the ratio of absolute temperatures is greater than 1. We could get the above result by simply multiplying the volume by the ratio of absolute temperatures, choosing the ratio so that it is greater than 1. (If we had expected the gas volume to decrease, we would have chosen the ratio to be less than 1.)

Exercise 4.2

If we expect a chemical reaction to produce 4.38 dm³ of oxygen (O_2) at 19°C and 101 kPa, what would be the volume at 25°C and 101 kPa?

(See Problems 4.31 and 4.32.)

Boyle's law ($V \propto 1/P$) and Charles's law ($V \propto T$) can be expressed in one statement: the volume occupied by a given amount of gas is proportional to the absolute temperature divided by the pressure.

$$V \propto \frac{T}{P} \quad \text{(for a given amount of gas)}$$

For calculation purposes we can write this as an equation.

$$V = \text{constant} \times \frac{T}{P} \quad \text{or} \quad \frac{PV}{T} = \text{constant} \quad \text{(for a given amount of gas)}$$

The last form of this equation shows that the pressure times volume divided by absolute temperature for a given amount of any gas remains constant.

Consider a problem in which we wish to calculate the final volume of a gas when the pressure and temperature are changed. Since PV/T is constant for a given amount of gas, we can write

$$\frac{P_f V_f}{T_f} = \frac{P_i V_i}{T_i}$$

which rearranges to

$$V_f = V_i \times \frac{P_i}{P_f} \times \frac{T_f}{T_i}$$

Thus, the final volume is obtained by multiplying the initial volume by ratios of pressures and absolute temperatures.

Example 4.3

In the Dumas method for determining the amount of nitrogen in a compound, the vapor of the substance is passed over hot copper(II) oxide. The compound is oxidized, and the nitrogen in the substance is converted to nitrogen gas (N_2). If 39.8 mg of caffeine give 22.9 cm³ of nitrogen gas at 23°C and 746 mmHg, what is the volume of nitrogen at 0°C and 760 mmHg?

Solution

We first express the temperatures in kelvins:

$$T_i = (23 + 273) \text{ K} = 296 \text{ K}$$
$$T_f = (0 + 273) \text{ K} = 273 \text{ K}$$

Now we put the data for the problem in tabular form:

$V_i = 22.9$ cm³	$P_i = 746$ mmHg	$T_i = 296$ K
$V_f = $?	$P_f = 760$ mmHg	$T_f = 273$ K

From the data, we see that we need to use Boyle's and Charles's laws combined to find how V varies as P and T change. Note that the increase in pressure decreases the volume, making the pressure ratio less than 1. The decrease in temperature also decreases the volume, and the ratio of absolute temperatures is less than 1.

$$V_f = V_i \times \frac{P_i}{P_f} \times \frac{T_f}{T_i}$$

$$= 22.9 \text{ cm}^3 \times \frac{746 \text{ mmHg}}{760 \text{ mmHg}} \times \frac{273 \text{ K}}{296 \text{ K}} = 20.7 \text{ cm}^3$$

Exercise 4.3

A balloon contains 5.41 dm³ of helium at 24°C and 101.5 kPa. Suppose the gas in the balloon is heated to 35°C. If the helium pressure is now 102.8 kPa, what is the volume?

(See Problems 4.35 and 4.36.)

4.4 Avogadro's Law

In 1808, Gay-Lussac published the results of some experiments on reactions of gases. His discovery that, when gases react, the volumes of reactant and product gases at a given temperature and pressure are in ratios of small whole numbers is known as Gay-Lussac's *law of combining volumes.* ■ For example, at some particular temperature and pressure, 3.00 L of carbon monoxide gas react with 1.50 L of oxygen gas to give 3.00 L of carbon dioxide gas. Thus, the ratio of volumes of carbon monoxide to oxygen is 3.00 : 1.50, which is equivalent to a ratio of 2 : 1. In other words, two volumes of carbon monoxide react with one volume of oxygen. By similar reasoning, we find that one volume of oxygen yields two volumes of carbon dioxide. Therefore, two volumes of carbon monoxide react with one volume of oxygen to give two volumes of carbon dioxide:

■ Gay-Lussac's law of combining volumes was discussed earlier in an Aside following Section 3.1.

$$2CO \quad + \quad O_2 \quad \longrightarrow \quad 2CO_2$$

3.00 L	1.50 L	3.00 L
(2 volumes)	(1 volume)	(2 volumes)

Three years later, Amedeo Avogadro interpreted Gay-Lussac's law of combining volumes in terms of what we now call **Avogadro's law:** *equal volumes of any two gases at the same temperature and pressure contain the same number of molecules.* ■ Using this principle, we can explain the law of combining volumes for the preceding reaction as follows. Suppose one volume of oxygen contains N O_2 molecules. Then two volumes of carbon monoxide contain $2N$ CO molecules, and two volumes of carbon dioxide

■ Although Avogadro's interpretation of the law of combining volumes was published in 1811, it did not receive general acceptance until about 1860. See the Aside following Section 3.1.

contain $2N$ CO_2 molecules. Thus, dividing by N, we get

$2N$ CO molecules + N O_2 molecules $\longrightarrow$ $2N$ CO_2 molecules

or   2 CO molecules + 1 O_2 molecule $\longrightarrow$ 2 CO_2 molecules

in agreement with the balanced chemical equation.

One mole of any gas contains the same number of molecules (Avogadro's number = 6.022×10^{23}) and by Avogadro's law must occupy the same volume at a given temperature and pressure. This volume is called the **molar gas volume.** The volumes of gases are often compared at standard conditions—that is, **standard temperature and pressure (STP).** These are chosen by convention to be 0°C and 1 atm pressure, respectively. At STP, the molar gas volume is found to be 22.41 L/mol. Thus, 22.41 L of oxygen at STP contain 6.022×10^{23} O_2 molecules. Similarly, 22.41 L of carbon monoxide at STP contain 6.022×10^{23} CO molecules (see Figure 4.9).

If we denote the molar gas volume at some temperature and pressure by V_m, the volume occupied by n moles of any gas at this temperature and pressure is

$$V = nV_m \quad \text{(any gas)}$$

This equation is equivalent to the earlier statement of Avogadro's law. To see this, consider samples of two different gases at the same temperature and pressure, with the same volume V. Each gas contains $n = V/V_m$ moles. Since the molar volume, V_m, is the same for every gas, the same volume, V, contains the same moles and therefore the same number of molecules.

Avogadro's law:
$V = nV_m$ where V_m has the same value for all gases
(at a fixed temperature and pressure)

Example 4.4

If 1.50 L of oxygen at some pressure and temperature contains 0.0588 mol O_2, how many moles of helium atoms are there in 2.25 L of helium at the same pressure and temperature? How many helium atoms are in this volume? (Helium is a gas consisting of He atoms.)

Solution

From oxygen, we can calculate the molar gas volume:

$$V_m = \frac{V}{n} = \frac{1.50 \text{ L}}{0.0588 \text{ mol}} = 25.5 \text{ L/mol}$$

This will also be the molar gas volume of helium at the same temperature and pressure, according to Avogadro's law. Since the total volume of helium

(2.25 L) will equal the moles, n, times the molar gas volume, 25.5 L/mol, we have

$$V = nV_m$$
$$2.25 \text{ L} = n \times 25.5 \text{ L/mol}$$

Hence, $n = \dfrac{2.25 \text{ L}}{25.5 \text{ L/mol}} = 0.0882 \text{ mol}$

Since 1 mol He contains 6.02×10^{23} He atoms, we obtain the number of atoms in 2.25 L He by converting the moles of helium to atoms:

$$0.0882 \text{ mol He} \times \frac{6.02 \times 10^{23} \text{ He atoms}}{1 \text{ mol He}}$$

$$= 5.31 \times 10^{22} \text{ He atoms}$$

Exercise 4.4

If the molar gas volume at 25°C and 1.000 atm pressure is 24.47 L, how many molecules of N_2 are there in 0.250 L of nitrogen gas?

(See Problems 4.39 and 4.40.)

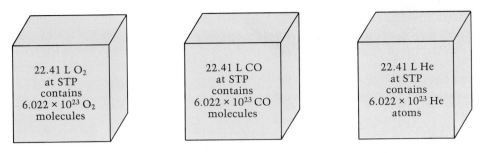

Figure 4.9
Avogadro's law. Equal volumes of any gas under the same conditions contain the same number of molecules.

4.5 The Ideal Gas Law

We have discussed three gas laws, each of which relates the volume occupied by a gas to another variable. These laws are

$$V \propto \frac{1}{P} \quad \text{at fixed } T, n \qquad \text{Boyle's law}$$

$$V \propto T \quad \text{at fixed } P, n \qquad \text{Charles's law}$$

$$V = nV_m \quad \text{at fixed } T, P \qquad \text{Avogadro's law}$$

These three laws can be conveniently summarized by one equation. How they lead to this equation is shown below.

Earlier we found that Boyle's and Charles's laws can be expressed as one equation:

$$V = \text{constant} \times \frac{T}{P} \quad \text{(for a given amount of gas)}$$

The "constant" in this equation has a definite value for a given amount of gas, but will have different values for other amounts. For one mole of gas, we can write

$$V_m = R \times \frac{T}{P}$$

where V_m is the volume occupied by one mole of gas (the molar gas volume) and R is the value of the constant in this equation for one mole of gas. Because the molar gas volume is the same for all gases, the value of R is the same for all gases. We call it the **molar gas constant (R).**

The preceding equation can be written for n moles of gas if we multiply both sides by n.

$$\underbrace{nV_m}_{V} = \frac{nRT}{P}$$

This equation, which combines all of the gas laws, is called the **ideal gas law.** If we write V for nV_m and rearrange the equation, we get

Ideal gas law:
$$PV = nRT$$

The limitations that apply to Boyle's, Charles's, and Avogadro's laws also apply to the ideal gas law. That is, the ideal gas law is most accurate for low to moderate pressures and for temperatures that are not too low.■

■ No gas is "ideal." But the ideal gas law has much practical utility even though it is only an approximation. The behavior of real gases is described at the end of this chapter.

To use this equation, we need to know the value of the molar gas constant, R. We can obtain it from the experimental value of the molar volume at STP (0°C, 1 atm). The data for the evaluation of R are

Variable	Value
P	1 atm
V	22.41 L
T	(0 + 273.2) K = 273.2 K
n	1 mol

Note that we have converted degrees Celsius (0°C) to kelvins (273.2 K).

Now we write the ideal gas law

$$PV = nRT$$

and solve for R:

$$R = \frac{PV}{nT}$$

Substituting from our data gives

$$R = \frac{1 \text{ atm} \times 22.41 \text{ L}}{1 \text{ mol} \times 273.2 \text{ K}}$$

$$= 0.08203 \text{ L} \cdot \text{atm}/(\text{K} \cdot \text{mol})$$

A more accurate value of R is 0.082057 L·atm/(K·mol). Table 4.3 lists values of the molar gas constant in various units. Thus, in SI units, $R = 8.31441$ J/(K·mol), which is equivalent to 8.31441 dm³·kPa/(K·mol).

The ideal gas law includes all the information contained in Boyle's, Charles's, and Avogadro's laws. It is therefore a more powerful law, since any gas calculation done with any one of the other laws can be done with the ideal gas law. In fact, starting with the ideal gas law, we can derive any of the other gas laws.

Table 4.3
Values of the Molar Gas Constant in Various Units

Value of R
0.082057 L · atm/(K · mol)
8.31441 J/(K · mol)
8.31441 kg · m²/(s² · K · mol)
8.31441 dm³ · kPa/(K · mol)
1.98719 cal/(K · mol)

Example 4.5

Prove the following statement: The pressure of a given amount of gas at a fixed volume is proportional to the absolute temperature. This is sometimes called *Amontons' law*. In 1702 Guillaume Amontons constructed a thermometer based on the measurement of the pressure of a fixed volume of air. The principle employed by Amontons is now used to construct gas thermometers, which are the accepted standards for calibrating other types of thermometers.

Solution

From the ideal gas law,

$$PV = nRT$$

Solving for P, we get

$$P = \left(\frac{nR}{V}\right)T$$

Note that everything in parentheses in this equation is constant. Therefore, we can write

$$P = \text{constant} \times T$$

Or, expressing this as a proportion, we get

$$P \propto T$$

Boyle's law and Charles's law follow from the ideal gas law by a similar derivation.

Exercise 4.5

Show that the moles of gas are proportional to the pressure for constant volume and temperature.

(See Problems 4.41 and 4.42.)

Calculations Using the Ideal Gas Law

The type of problem to which Boyle's and Charles's laws are applied involves a change in conditions (P, V, or T) of a gas. The ideal gas law allows us to solve another type of problem: given any three of the quantities P, V, n, and T, calculate the unknown quantity. Example 4.6 illustrates a typical problem.

Example 4.6

Answer the question asked in the chapter opening: How many grams of oxygen are there in a 50.0-L tank at 21°C when the oxygen pressure is 15.7 atm?

Solution

In asking for the mass of oxygen, we are in effect asking for moles of gas, n, since mass and moles are easily related. The data given in the problem are

Variable	Value
P	15.7 atm
V	50.0 L
T	(21 + 273) K = 294 K
n	?

Note that we must convert degrees Celsius (21°C) to kelvins (294 K).

From the data given, we see that we can use the ideal gas law to solve for n. The proper value to use for R depends on the units of V and P. Since these are in liters and atmospheres, respectively, we use (from Table 4.3)

$$R = 0.0821 \text{ L} \cdot \text{atm}/(\text{K} \cdot \text{mol})$$

We write the ideal gas law,

$$PV = nRT$$

and solve for n:

$$n = \frac{PV}{RT}$$

Substituting from the table of data gives

$$n = \frac{15.7 \text{ atm} \times 50.0 \text{ L}}{0.0821 \text{ L} \cdot \text{atm}/(\text{K} \cdot \text{mol}) \times 294 \text{ K}} = 32.5 \text{ mol}$$

and converting moles to mass of oxygen yields

$$32.5 \text{ mol } O_2 \times \frac{32.0 \text{ g } O_2}{1 \text{ mol } O_2} = 1.04 \times 10^3 \text{ g } O_2$$

Exercise 4.6

What is the pressure in a 50.0-L tank that contains 3.03 kg of oxygen, O_2, at 23°C?

(See Problems 4.45, 4.46, 4.47, and 4.48.)

Stoichiometry Problems with Gas Volumes

In Chapter 3, we learned how to find the mass of one substance in a chemical reaction from the mass of another substance in the reaction. Now we can extend our knowledge of stoichiometry to find the volumes of gases obtained from reactions. The ideal gas law allows us to calculate the moles of gas given the volume, or the volume given the moles of gas. Example 4.7 is a typical problem.

Example 4.7

How many liters of chlorine gas, Cl_2, can be obtained at 40°C and 787 mmHg from 9.41 g of hydrogen chloride, HCl, according to the following equation?

$$2KMnO_4 + 16HCl \longrightarrow$$
$$8H_2O + 2KCl + 2MnCl_2 + 5Cl_2$$

Solution

From the chemical equation, we can find how many moles of Cl_2 can be obtained from 9.41 g HCl using conversion factors. Then, from the moles of Cl_2, we can calculate the volume of Cl_2 under the specified conditions using the ideal gas law.

$$9.41 \text{ g HCl} \times \frac{1 \text{ mol HCl}}{36.5 \text{ g HCl}} \times \frac{5 \text{ mol } Cl_2}{16 \text{ mol HCl}}$$
$$= 0.0806 \text{ mol } Cl_2$$

Before doing the gas calculation, we list the available data:

Variable	Value
P	$787 \text{ mmHg} \times \dfrac{1 \text{ atm}}{760 \text{ mmHg}} = 1.036 \text{ atm}$
V	?
T	$(40 + 273) \text{ K} = 313 \text{ K}$
n	0.0806 mol

Note that we have converted the pressure in mmHg to atmospheres, and degrees Celsius to kelvins. Thus, we can use $R = 0.0821 \text{ L} \cdot \text{atm}/(K \cdot \text{mol})$, and the volume will come out in liters.

Now we write the ideal gas law,

$$PV = nRT$$

and solve for the volume:

$$V = \frac{nRT}{P}$$

Substituting from our data gives

$$V = \frac{0.0806 \text{ mol} \times 0.0821 \text{ L} \cdot \text{atm}/(K \cdot \text{mol}) \times 313 \text{ K}}{1.036 \text{ atm}}$$
$$= 2.00 \text{ L}$$

Exercise 4.7

Lithium hydroxide, LiOH, is used in spacecraft to absorb the carbon dioxide exhaled by astronauts. The reaction is

$$2LiOH + CO_2 \longrightarrow Li_2CO_3 + H_2O$$

Calculate the volume of carbon dioxide at 22°C and 748 mmHg absorbed by 1.00 g of lithium hydroxide.

(See Problems 4.49, 4.50, 4.51, and 4.52.)

Gas Density; Molecular-Weight Determination

Recall that density equals mass divided by volume. Because the volume of a gas depends on its temperature and pressure, the density of the gas must depend on these same variables. In the next example, we show how to find the density of a gaseous substance of known molecular weight at a given temperature and pressure.

Example 4.8

What is the density of oxygen, O_2, in grams per liter at 25°C and 0.850 atm?

Solution

Since density equals mass per unit volume, calculating the mass of 1 L of gas will give us the density of the gas.

(Continued)

$$n = \frac{m}{M_m} \quad \left(\frac{m}{V}\right) d = \frac{PM_m}{RT}$$

(The volume is treated as an exact quantity, so it is not used to obtain the number of significant figures in the calculation of mass.)

Variable	Value
P	0.850 atm
V	1 L
T	(25 + 273) K = 298 K
n	?

We start from the ideal gas law and solve for n:

$$PV = nRT \quad \text{or} \quad n = \frac{PV}{RT}$$

Substituting from the table,

$$n = \frac{0.850 \; \text{atm} \times 1 \; \text{L}}{0.0821 \; \text{L} \cdot \text{atm}/(\text{K} \cdot \text{mol}) \times 298 \; \text{K}} = 0.0347 \; \text{mol}$$

Now we convert moles oxygen to grams:

$$0.0347 \; \text{mol O}_2 \times \frac{32.0 \; \text{g O}_2}{1 \; \text{mol O}_2} = 1.11 \; \text{g O}_2$$

Therefore, the density of O_2 at 25°C and 0.850 atm is 1.11 g/L.

Exercise 4.8

Calculate the density of helium, He, in grams per liter at 21°C and 752 mmHg. The density of air under these conditions is 1.188 g/L. What is the difference in mass between 1 liter of air and 1 liter of helium? (This mass difference is equivalent to the buoyant or lifting force of helium per liter.)

(See Problems 4.55 and 4.56.)

■ Dumas obtained results for sulfur that were difficult to interpret. Sulfur, as we now know, exists as S_8 molecules in the solid and in the liquid. In the vapor, however, the molecules are not only S_8, but S_6 and smaller. Above 800°C, the vapor is mostly S_2, until about 2000°C, where sulfur exists mostly as a monatomic gas. These changes in molecular form can be correlated with the density measurements.

The density of a gas depends on its molecular weight. This suggests that we could use a measurement of gas density to determine the molecular weight. In fact, gas (vapor) density measurements provided one of the first methods of determining molecular weight. The method was worked out by the French chemist Jean-Baptiste André Dumas in 1826. It can be applied to any substance that can be vaporized without decomposing.■

Example 4.9

A quantity of carbon tetrachloride (a liquid), CCl_4, was put into a 200.0-mL flask. The open flask was then submerged in boiling water until all of the liquid evaporated and the vapor filled the flask (see Figure 4.10). (Any excess vapor escaped through the tube in the neck of the flask, so that the pressure of the vapor equaled the barometric pressure.) The flask was then cooled and weighed. Subtracting the mass of the empty flask gave the mass of CCl_4. The mass of the carbon tetrachloride vapor in the flask when the temperature of the boiling water was 99°C was found to be 0.970 g. The barometric pressure was 733 mmHg. What is the molecular weight of carbon tetrachloride?

Solution

We calculate the moles of vapor from the ideal gas law, then calculate the molar mass. Note that the temperature of the vapor is the same as the temperature of the boiling water, and the pressure of the vapor equals the barometric pressure. We have the following data for the vapor:

Variable	Value
P	$733 \; \text{mmHg} \times \dfrac{1 \; \text{atm}}{760 \; \text{mmHg}} = 0.964 \; \text{atm}$
V	200.0 mL = 0.2000 L
T	(99 + 273) K = 372 K
n	?

From the ideal gas law, $PV = nRT$, we have

$$n = \frac{PV}{RT}$$

Since we have the pressure in atmospheres and the volume in liters, we must use $R = 0.0821$ L · atm/(K · mol). Hence,

(Continued)

$$d = \frac{m}{V}$$

$$n = \frac{0.964 \text{ atm} \times 0.2000 \text{ L}}{0.0821 \text{ L} \cdot \text{atm}/(\text{K} \cdot \text{mol}) \times 372 \text{ K}}$$

$$= 0.00631 \text{ mol } CCl_4$$

Dividing the mass of the vapor by moles gives the mass per mole (the molar mass):

$$\text{Molar mass} = \frac{\text{grams vapor}}{\text{moles vapor}} = \frac{0.970 \text{ g}}{0.00631 \text{ mol}}$$

$$= 154 \text{ g/mol}$$

Thus the molecular weight is 154 amu.

Exercise 4.9

A sample of a gaseous substance at 25°C and 0.862 atm has a density of 2.26 g/L. What is the molecular weight of the substance? (*Hint:* Assume 1 L of gas with a mass of 2.26 g.)

(See Problems 4.59 and 4.60.)

An equation relating the density of a gas, d, and its molar mass, M_m, can be obtained from the ideal gas law. Recall that

$$M_m = \frac{\text{mass}}{n} \quad \text{or} \quad n = \frac{\text{mass}}{M_m}$$

We substitute mass/M_m for n in $PV = nRT$:

$$PV = \frac{\text{mass}}{M_m} RT$$

Rearranging, we get

$$PM_m = \frac{\text{mass}}{V} RT$$

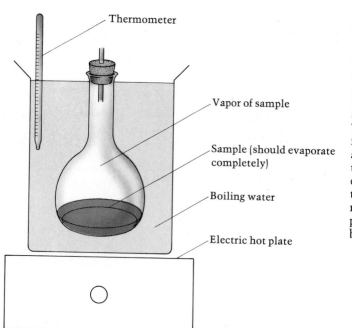

Thermometer

Vapor of sample

Sample (should evaporate completely)

Boiling water

Electric hot plate

Figure 4.10
Finding the vapor density of a substance by the Dumas method. After the flask is filled with vapor of the substance at the temperature of the boiling water, it is cooled so that the vapor condenses. The mass of the substance is found by weighing the flask and substance, then subtracting the mass of the empty flask. The pressure of the vapor equals the barometric pressure, and the temperature equals that of the boiling water.

or, since mass/V equals density,

$$PM_m = dRT$$

This equation gives us an alternative way to solve Examples 4.8 and 4.9. In Example 4.9, we first calculate the density of the carbon tetrachloride vapor in the flask.

$$d = \frac{0.970 \text{ g}}{0.2000 \text{ L}} = 4.85 \text{ g/L}$$

Rearranging the equation we just derived,

$$M_m = \frac{dRT}{P} = \frac{4.85 \text{ g/L} \times 0.0821 \text{ L} \cdot \text{atm/(K} \cdot \text{mol)} \times 372 \text{ K}}{0.964 \text{ atm}} = 154 \text{ g/mol}$$

4.6 Gas Mixtures; Law of Partial Pressures

While studying the composition of air, John Dalton concluded in 1801 that each gas in a mixture of unreactive gases acts, as far as its pressure is concerned, as though it were the only gas in the mixture. To illustrate, consider two 1-L flasks. One flask is filled with helium to a pressure of 152 mmHg at a given temperature. The other flask is filled with hydrogen to a pressure of 608 mmHg at the same temperature. Suppose all of the helium in the one flask is put in with the hydrogen in the other flask (see Figure 4.11).

Figure 4.11
A demonstration of Dalton's law of partial pressures. Flask A is filled with water so that the helium in it is pushed into Flask B, where it mixes with hydrogen. Each gas exerts the pressure it would exert if the other were not there.

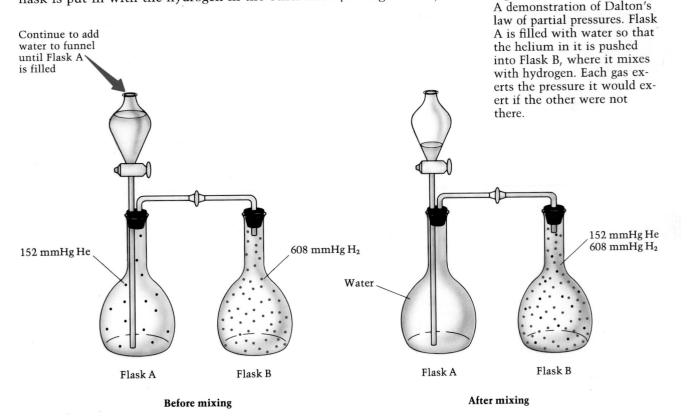

Continue to add water to funnel until Flask A is filled

152 mmHg He

608 mmHg H₂

Flask A

Flask B

Before mixing

Water

152 mmHg He
608 mmHg H₂

Flask A

Flask B

After mixing

After the gases are mixed in one flask, each gas occupies a volume of one liter, just as before, and has the same temperature. According to Dalton, each gas exerts the same pressure it would exert if it were the only gas in the flask. Thus, the pressure exerted by helium in the mixture is 152 mmHg. Similarly, the pressure exerted by hydrogen in the mixture is 608 mmHg. The total pressure exerted by the gases in the mixture is 152 mmHg + 608 mmHg = 760 mmHg.

The pressure exerted by a particular gas in a mixture is its **partial pressure.** The partial pressure of helium in the preceding mixture is 152 mmHg; the partial pressure of hydrogen in the mixture is 608 mmHg. According to Dalton, *the sum of the partial pressures of all the different gases in a mixture is equal to the total pressure of the mixture.* This is usually called **Dalton's law of partial pressures.**

If we let P be the total pressure and P_A, P_B, P_C, $\cdots$ be the partial pressures of the component gases in a mixture, the law of partial pressures can be written as

Dalton's law of partial pressures:
$$P = P_A + P_B + P_C + \cdots$$

The individual partial pressures, P_i, follow the ideal gas law

$$P_i V = n_i RT$$

where n_i is the number of moles of component i.

Example 4.10

A 1.000-L sample of dry air at 25.0°C (or 298.2 K) has the following composition: 0.8940 g N_2, 0.2740 g O_2, 0.0152 g Ar, and 0.0007 g CO_2. (a) What are the partial pressures of each component gas in the mixture? (b) What is the total pressure of the gas mixture?

Solution

(a) Each gas follows the ideal gas law. To calculate the partial pressure of N_2, we convert 0.8940 g N_2 to moles N_2.

$$0.8940 \text{ g } N_2 \times \frac{1 \text{ mol } N_2}{28.013 \text{ g } N_2} = 0.03191 \text{ mol } N_2$$

From the ideal gas law,

$$P_{N_2} = \frac{n_{N_2} RT}{V}$$

and substituting, using R to four significant figures,

$$P_{N_2} = \frac{0.03191 \text{ mol} \times 0.08206 \frac{L \cdot atm}{(K \cdot mol)} \times 298.2 \text{ K}}{1.000 \text{ L}}$$

$$P_{N_2} = 0.7808 \text{ atm}$$

The partial pressures of the other gases are calculated in a similar way. We obtain $P_{O_2} = 0.2095$ atm, $P_{Ar} = 0.0093$ atm, and $P_{CO_2} = 0.0004$ atm. (b) The total pressure of the mixture is

$$P = P_{N_2} + P_{O_2} + P_{Ar} + P_{CO_2}$$
$$= (0.7808 + 0.2095 + 0.0093 + 0.0004) \text{ atm}$$
$$= 1.0000 \text{ atm}$$

Exercise 4.10

A 10.0-L flask contains 1.031 g O_2 and 0.572 g CO_2 at 18°C. What are the partial pressures of oxygen and carbon dioxide? What is the total pressure?

(See Problems 4.61 and 4.62.)

Mole Fractions

The composition of a gas mixture is often described in terms of **mole fractions** of the component gases. The mole fraction of a component gas is simply the fraction of moles of that component in the total moles of gas mixture. If we let n_A be the moles of component A and n be the total moles of gas, then

$$\text{Mole fraction of } A = \frac{n_A}{n}$$

We can relate the mole fractions to the partial pressures. Let us obtain expressions for n_A and n. For the gas mixture, the ideal gas law is $PV = nRT$, so therefore $n = PV/RT$. A similar equation can be written for the moles of A, that is, $n_A = P_A V/RT$. If we substitute these expressions for n_A and n into the equation for the mole fraction of A, we get

$$\text{Mole fraction of } A = \frac{P_A V/RT}{PV/RT}$$

$$\text{Mole fraction of } A = \frac{n_A}{n} = \frac{P_A}{P}$$

Therefore, if we divide each partial pressure by the total pressure, we get the mole fractions.

Example 4.11

A sample of air has the following partial pressures of components: 593.4 mmHg N_2, 159.2 mmHg O_2, 7.1 mmHg Ar, and 0.3 mmHg CO_2. What is the composition of the air in mole fractions?

Solution

Since the total pressure is 760.0 mmHg,

$$\text{Mole fraction of } N_2 = \frac{P_{N_2}}{P} = \frac{593.4 \text{ mmHg}}{760.0 \text{ mmHg}} = 0.7808$$

$$\text{Mole fraction of } O_2 = \frac{P_{O_2}}{P} = \frac{159.2 \text{ mmHg}}{760.0 \text{ mmHg}} = 0.2095$$

$$\text{Mole fraction of Ar} = \frac{P_{Ar}}{P} = \frac{7.1 \text{ mmHg}}{760.0 \text{ mmHg}} = 0.0093$$

$$\text{Mole fraction of } CO_2 = \frac{P_{CO_2}}{P} = \frac{0.3 \text{ mmHg}}{760.0 \text{ mmHg}} = 0.0004$$

The sum of the mole fractions of all the components in a mixture should equal 1. Thus,

$$0.7808 + 0.2095 + 0.0093 + 0.0004 = 1.0000$$

Exercise 4.11

Natural gas from a gas field in Texas has the following composition (in mole fractions): methane (CH_4), 0.782; nitrogen (N_2), 0.131; ethane (C_2H_6), 0.041; propane (C_3H_8), 0.027; butane (C_4H_{10}), 0.013; and pentane (C_5H_{12}), 0.006. What would be the partial pressure of each gas when the total pressure is 782 mmHg?

(See Problems 4.63 and 4.64.)

Collecting Gases over Water

A useful application of the law of partial pressures arises when we collect gases over water. (To be collectable, the gas must not dissolve appreciably in

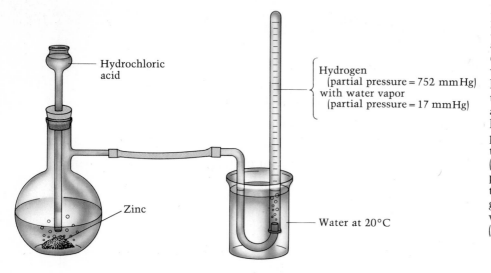

Hydrochloric acid

Zinc

Hydrogen (partial pressure = 752 mm Hg) with water vapor (partial pressure = 17 mmHg)

Water at 20°C

Figure 4.12
Hydrogen, prepared by the reaction of zinc with hydrochloric acid, is collected over water. When the gas collection tube is adjusted so that the level in the tube is at the same height as the level in the beaker, the gas pressure in the tube equals the barometric pressure (769 mmHg). The total gas pressure equals the sum of the partial pressure of hydrogen (752 mmHg) and the vapor pressure of water (17 mmHg).

water.) Figure 4.12 shows how a gas, produced by chemical reaction in the flask, is collected by leading it to an inverted tube, where it displaces water. As gas bubbles through the water, it picks up molecules of water vapor that mix with it. The partial pressure of water vapor in the gas mixture in the collection bottle depends only on the temperature. This partial pressure of water vapor is called the **vapor pressure** of water.■ Values of the vapor pressure of water at various temperatures are listed in Table 4.4. The following example shows how to find the partial pressure and then mass of the collected gas.

■ Vapor pressure is the maximum partial pressure of the vapor in the presence of the liquid. It is defined more precisely in Chapter 11.

Table 4.4
Vapor Pressure of Water at Various Temperatures

Temperature, °C	Pressure, mmHg	Temperature, °C	Pressure, mmHg
0	4.6	27	26.7
5	6.5	28	28.3
10	9.2	29	30.0
11	9.8	30	31.8
12	10.5	35	42.2
13	11.2	40	55.3
14	12.0	45	71.9
15	12.8	50	92.5
16	13.6	55	118.0
17	14.5	60	149.4
18	15.5	65	187.5
19	16.5	70	233.7
20	17.5	75	289.1
21	18.7	80	355.1
22	19.8	85	433.6
23	21.1	90	525.8
24	22.4	95	633.9
25	23.8	100	760.0
26	25.2	105	906.1

Example 4.12

Hydrogen gas is produced by the reaction of hydrochloric acid, HCl, on zinc metal:

$$2HCl + Zn \longrightarrow ZnCl_2 + H_2$$

The gas is collected over water. If 156 mL of gas are collected at 21°C and 769 mmHg total pressure, what is the mass of hydrogen collected?

Solution

The vapor pressure of water at 21°C is 18.7 mmHg. From Dalton's law of partial pressures, we know that the total gas pressure equals the partial pressure of hydrogen, P_{H_2}, plus the partial pressure of water, P_{H_2O}.

$$P = P_{H_2} + P_{H_2O}$$

Since we are interested in the partial pressure of hydrogen, we rearrange to obtain

$$P_{H_2} = P - P_{H_2O}$$

Substituting and solving for the partial pressure of hydrogen, we get

$$P_{H_2} = (769 - 18.7)\ \text{mmHg} = 750\ \text{mmHg}$$

Now we can use the ideal gas law to find the moles of hydrogen collected. The data are

Variable	Value
P	$750\ \text{mmHg} \times \dfrac{1\ \text{atm}}{760\ \text{mmHg}} = 0.987\ \text{atm}$
V	$156\ \text{mL} = 0.156\ \text{L}$
T	$(21 + 273)\ \text{K} = 294\ \text{K}$
n	?

From the ideal gas law, $PV = nRT$, we have

$$n = \frac{PV}{RT} = \frac{0.987\ \text{atm} \times 0.156\ \text{L}}{0.0821\ \text{L} \cdot \text{atm}/(\text{K} \cdot \text{mol}) \times 294\ \text{K}}$$

$$= 0.00638\ \text{mol}$$

We now convert moles of H_2 to grams of H_2:

$$0.00638\ \text{mol H}_2 \times \frac{2.02\ \text{g H}_2}{1\ \text{mol H}_2} = 0.0129\ \text{g H}_2$$

Exercise 4.12

Oxygen can be prepared by heating potassium chlorate, $KClO_3$, with manganese dioxide as a catalyst. (A catalyst merely speeds a reaction without being used up itself.) The reaction is

$$2KClO_3 \longrightarrow 2KCl + 3O_2$$

How many moles of O_2 would be obtained from 1.300 g of $KClO_3$? If this amount of O_2 was collected over water at 23°C and at a total pressure of 745 mmHg, what volume would it occupy?

(See Problems 4.65 and 4.66.)

Kinetic-Molecular Theory

In the following sections, we will see how the interpretation of a gas in terms of the *kinetic-molecular theory* (or simply kinetic theory) leads to the ideal gas law. According to this theory, a gas consists of molecules in constant random motion. (The word *kinetic* describes something in motion.)

4.7 Kinetic Theory of an Ideal Gas

Our present explanation of gas pressure is that it results from the continual bombardment of the container walls by constantly moving molecules. This kinetic interpretation of gas pressure was first put forth in 1676 by Robert

Hooke, who had earlier been an assistant of Boyle's. Hooke did not pursue this idea, however, so the generally accepted interpretation of gas pressure remained the one given by Isaac Newton, a contemporary of Hooke.

According to Newton, the pressure of a gas was due to the mutual repulsions of the gas particles (molecules). These repulsions pushed the molecules against the walls of the gas container, much as coiled springs packed tightly in a box would push against the walls of the box. This interpretation continued to be the dominant view of gas pressure until the middle of the nineteenth century.

Despite the dominance of Newton's view, however, some people followed the kinetic interpretation. In 1738, Daniel Bernoulli, a Swiss mathematician and physicist, gave a quantitative explanation of Boyle's law using the kinetic interpretation. He even suggested that molecules move faster at higher temperatures, in order to explain the experiments of Amontons on the temperature dependence of gas volume and pressure. However, Bernoulli's paper attracted little notice. A similar kinetic interpretation of gases was submitted for publication to the Royal Society of London in 1848 by John James Waterston. His paper was rejected as "nothing but nonsense."■

Soon afterward, the scientific climate for the kinetic view improved. The **kinetic theory of gases** was developed by a number of influential physicists, including James P. Joule (1848), Rudolf Clausius (1857), James Clerk Maxwell (1859), and Ludwig Boltzmann (in the 1870s). Throughout the last half of the nineteenth century, research continued on the kinetic theory, making it a cornerstone of our present view of molecular substances.

■ Waterston's paper, with the reviewers' comments, was discovered in the Royal Society's files by Lord Rayleigh in 1892. (Rayleigh, a physicist, codiscovered argon and received the 1904 Nobel Prize in physics).

Postulates of Kinetic Theory

Physical theories are often given in terms of *postulates*. These are the basic statements from which all conclusions or predictions of a theory are deduced. The postulates are accepted as long as the predictions from the theory agree with experiment. If a particular prediction did not agree with experiment, we would either limit the area to which the theory applies or modify the postulates. Or, we would start over with a new theory.

The kinetic theory of an ideal gas (a gas that follows the ideal gas law) is based on five major postulates.

Postulate 1 Gases are composed of molecules whose size is negligible compared to the distance between them. Most of the volume occupied by a gas is empty space. This means that we can ignore the volume occupied by the molecules. We can treat a gas to a very good approximation as if it consisted of point particles.

Evidence for Postulate 1 The fact that gases are highly compressible is in agreement with the idea that molecules take up very little of the gas volume. Note that a sample of liquid occupies much less volume than the corresponding vapor (gas), consistent with Postulate 1.

Postulate 2 Molecules move randomly in linear motion, that is, they are moving in straight lines in all directions and at various speeds. This means

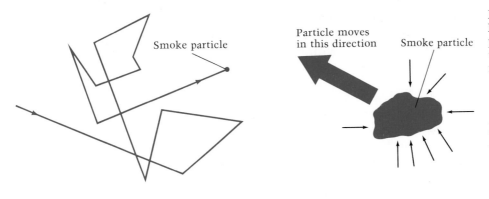

Figure 4.14 (*right*) Molecular collisions (small arrows) with a smoke particle. The particle moves in the direction of the large arrow due to unequal forces from molecular collisions at a given moment.

that properties of a gas, such as pressure, that depend on the motion of molecules will be the same in all directions.

Evidence for Postulate 2 In 1827, the botanist Robert Brown looked through a microscope at pollen grains in water. He saw the grains moving rapidly about in random fashion. We can see the same phenomenon by observing a particle of smoke in air (Figure 4.13). This so-called **Brownian motion** is explained by the random motion of air molecules buffeting the smoke particle (or water molecules hitting the pollen grains). The smoke particle is much larger than a single molecule, but it is small enough that the random collisions with molecules are detectable. If there are a few more collisions on one side than on the other, the particle will be jostled in the direction of the greater force (Figure 4.14).

Postulate 3 The forces of attraction or repulsion between two molecules (*intermolecular forces*) in a gas are very weak or negligible, except when they collide.■ This means that a molecule will continue moving in a straight line until it collides with another gas molecule or with the walls of the container.

Evidence for Postulate 3 The fact that gases fill any available volume, whereas liquids and solids have a relatively well-defined volume, can be taken as evidence for the very small intermolecular forces in the gaseous state. We explain the formation of a molecular liquid or solid in terms of the attractive forces between molecules. These forces hold the molecules together to produce the liquid or solid state. In a gas, where the molecules are far apart, these forces are weak, so gas molecules are free to move throughout the container.

The fact that the gases in a mixture behave independently of one another (Dalton's law of partial pressures) is also in agreement with Postulate 3. If there were significant molecular forces, one might expect the partial pressures to depend on what gases are in the mixture.

Postulate 4 When molecules collide with one another, the collisions are *elastic.* In an elastic collision, no kinetic energy is lost. To understand the difference between an elastic and an inelastic collision, compare the collision of two hard steel spheres with the collision of two masses of putty. The collision of steel spheres is nearly elastic (that is, the spheres bounce off

■ The statements that there are no intermolecular forces and that the volume of molecules is negligible are simplifications that lead to the ideal gas law. But intermolecular forces are needed to explain how we can get the liquid state from the gaseous state, since intermolecular forces hold molecules together in the liquid state.

each other and continue moving), but that of putty is not. Postulate 4 says that unless the kinetic energy of molecules is removed from a gas—for example, as heat—the molecules will forever move with the same average kinetic energy per molecule.

Evidence for Postulate 4 If kinetic energy were lost during a molecular collision, then the average speed of molecules would continuously decrease. A closed container insulated from heat exchanges would then show a steady drop of pressure (since pressure depends on the speed with which molecules hit the wall). We observe no such effect.

Postulate 5 The average kinetic energy of a molecule is proportional to the absolute temperature. This postulate establishes what we mean by temperature from a molecular point of view: the higher the temperature, the greater the molecular kinetic energy.

Evidence for Postulate 5 We can observe that Brownian motion increases with temperature. This indicates a relationship between molecular motion and temperature. And, as we will see, the ideal gas law is also in agreement with such a relationship.

The Ideal Gas Law from Kinetic Theory

One of the most important features of kinetic theory is its explanation of the ideal gas law. To show how we can get the ideal gas law from kinetic theory, we will first find an expression for the pressure of a gas.■

According to kinetic theory, the pressure of a gas is the result of molecules hitting a surface. Therefore, the pressure, P, will be proportional to the frequency of molecular collisions with the surface and to the average force exerted by a molecule in collision.

■ This description is essentially the one given by Daniel Bernoulli.

$$P \propto \text{frequency of collisions} \times \text{average force}$$

The average force exerted by a molecule depends on its mass, m, and its average speed, u, that is, on its average momentum, mu. In other words, the greater the mass of the molecule and the faster it is moving, the greater the force exerted during collision. The frequency of collisions also is proportional to the average speed, u, since the faster a molecule is moving, the more often it strikes the container walls. Frequency is inversely proportional to the gas volume, V, since the larger the volume, the less often a given molecule strikes the container walls. Finally, the frequency of collisions is proportional to the number, N, of molecules in the gas volume. Putting these factors together gives

$$P \propto \left(u \times \frac{1}{V} \times N \right) \times mu$$

If we bring the volume to the left side, we get

$$PV \propto Nmu^2$$

Since the average kinetic energy of a molecule of mass m and average speed u is $\frac{1}{2}mu^2$, we see that PV is proportional to the average kinetic energy of a molecule. ■ Moreover, the average kinetic energy is proportional to the absolute temperature, by Postulate 5. Noting that the number of molecules, N, is proportional to the moles of molecules, n, we have that

$$PV \propto nT$$

■ Kinetic energy is defined as $\frac{1}{2}m$ multiplied by (speed)^2. See Section 1.6.

We can write this as an equation by inserting a constant of proportionality, R, which we identify as the molar gas constant.

$$PV = nRT$$

The next two sections give additional deductions from kinetic theory, such as the average molecular speed.

4.8 Molecular Speeds

According to kinetic theory, in a given sample of gas, molecular speeds will vary over a range of values. Maxwell showed theoretically that they should be distributed as given in Figure 4.15. Note that molecules can have a wide range of speeds, though most are close to the average, with much smaller numbers at very high or low speeds. The distribution of speeds depends on the temperature, as Figure 4.15 shows. **Maxwell's distribution of molecular speeds** has since been demonstrated experimentally.

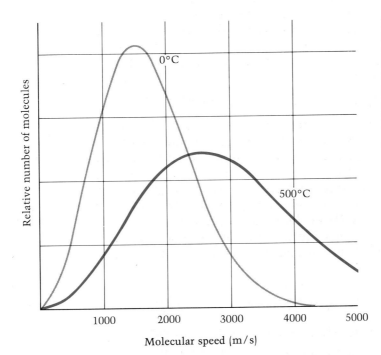

Figure 4.15
Maxwell's distribution of molecular speeds. The distributions of speeds of H_2 molecules are shown for 0°C and 500°C. Note that the speed corresponding to the maximum in the curve (this is the most probable speed) increases with temperature.

The **root-mean-square (rms) molecular speed,** u, a type of average molecular speed, is the speed of a molecule that has the average kinetic energy and can be shown to equal■

$$u = \sqrt{\frac{3RT}{M_m}}$$

where R is the molar gas constant, T is the absolute temperature, and M_m is the molar mass for the gas. Values of the average speed calculated from this formula indicate that molecular speeds are astonishingly high. For example, the average speed of H_2 molecules at 20°C are 1.90×10^3 m/s (over 4000 mi/hr).

■ According to detailed kinetic theory, the total kinetic energy of a mole of any gas equals $\frac{3}{2}RT$. We can get the equation for u from this. The average kinetic energy of a molecule, which by definition is $\frac{1}{2}mu^2$, would be obtained by dividing $\frac{3}{2}RT$ by Avogadro's number, N_A. Therefore, the kinetic energy is $\frac{1}{2}mu^2 = 3RT/(2N_A)$. Hence, $u^2 = 3RT/(N_A m)$. Or, noting that $N_A m$ equals the molar mass, M_m, we get $u^2 = 3RT/M_m$, from which we get the text equation.

Example 4.13

Calculate the rms molecular speed of O_2 molecules in a tank at 21°C and 15.7 atm. See Table 4.3 for the appropriate value of R.

Solution

The rms molecular speed is independent of pressure but does depend on the absolute temperature, which is (21 + 273) K = 294 K. To calculate u, it is best to use SI units throughout. In these units, the molar mass of O_2 is 32.0×10^{-3} kg/mol, and $R = 8.31$ kg·m²/(s²·K·mol). Hence

$$u = \left(\frac{3 \times 8.31 \text{ kg·m}^2/(\text{s}^2 \cdot \text{K} \cdot \text{mol}) \times 294 \text{ K}}{32.0 \times 10^{-3} \text{ kg/mol}}\right)^{1/2}$$

$$= 479 \text{ m/s}$$

Exercise 4.13

What is the rms speed (in m/s) of a carbon tetrachloride molecule at 22°C?

(See Problems 4.67 and 4.68.)

Exercise 4.14

At what temperature do hydrogen molecules, H_2, have the same rms speed as nitrogen molecules, N_2, at 455°C? At what temperature do hydrogen molecules have the same average kinetic energy?

(See Problems 4.69 and 4.70.)

4.9 Diffusion and Effusion

Gaseous **diffusion** is the process whereby a gas spreads out through another gas to occupy a space with uniform partial pressure. A gas or vapor having a relatively high partial pressure will spread out toward regions of lower partial pressure of that gas until the partial pressure becomes equal everywhere in the space. Thus, if we release an odorous substance at one corner of a room, minutes later it will be detected at the opposite corner. (It is well known, for example, how quickly the smell of baking apple pie draws people to the kitchen.)

We might consider the spread of an odor in minutes to be fast, yet when we think about our kinetic-theory calculations of molecular speed, we might ask why diffusion is not even much faster than it is. Why does it take minutes for the gas to diffuse throughout a room when the molecules are moving at perhaps a thousand miles per hour? This was in fact one of the first criticisms of kinetic theory. The answer is simply that a molecule never travels very far in one direction (at ordinary pressures) before it collides with another molecule and moves off in another direction. If we could trace out the path of an individual molecule, it would be a zigzagging trail reminiscent of Brownian motion (Figure 4.13). For a molecule to cross a room, it has to travel many times the straight-line distance.

Although the rate of diffusion certainly depends in part on the average molecular speed, the effect of molecular collisions makes the theoretical picture a bit complicated. *Effusion*, like diffusion, is a process involving the flow of a gas, but is theoretically much simpler.

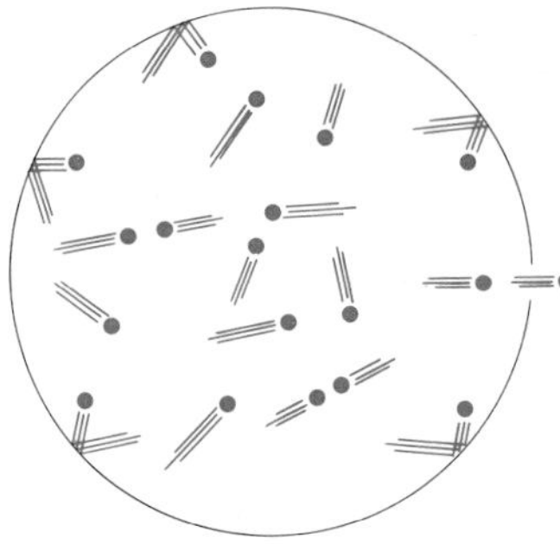

If we place a container of gas in a vacuum and then make a very small hole in the container, the gas escapes through the hole at the same speed it had in the container (Figure 4.16). The process is called **effusion.** It was first studied by Thomas Graham who, in 1846, discovered the law that bears his name. **Graham's law of effusion** states: *The rate of effusion of gas molecules from a particular hole is inversely proportional to the square root of the molecular weight of the gas at constant temperature and pressure.*

Let us consider a kinetic-theory analysis of an effusion experiment. Suppose the hole in the container is made small enough so that the gas molecules continue to move randomly (rather than moving together as they would in a wind). When a molecule happens to encounter the hole, it leaves the container. The collection of molecules leaving the container by chance encounters with the hole constitutes the effusing gas. All we have to consider for effusion is the rate at which molecules encounter the hole in the container.

The rate of effusion of molecules from a container depends on three factors: (1) the cross-sectional area of the hole (the larger this is, the more likely molecules are to escape); (2) the number of molecules per unit volume (the more crowded the molecules are, the more likely they are to encounter the hole); and (3) the average molecular speed (the faster the molecules are moving, the sooner they will escape).

If we compare the effusion of different gases from the same container, at the same temperature and pressure, factors (1) and (2) will be the same. The average molecular speeds will be different, however.

Since the average molecular speed essentially equals $\sqrt{3RT/M_m}$, where M_m is the molar mass, we see that the rate of effusion is proportional to $1/\sqrt{M_m}$. That is, the rate of effusion is inversely proportional to the square root of the molar mass (or molecular weight), as Graham's law states. The derivation of Graham's law from kinetic theory was considered a triumph of the theory, and greatly strengthened confidence in its validity.

Graham's law of effusion:
Rate of effusion of molecules $\propto 1/\sqrt{M_m}$
(for the same container at constant T and P)

Figure 4.16
Gas molecules effusing through a small hole in a container.

Example 4.14

Calculate the ratio of rates of effusion of molecules of carbon dioxide, CO_2, and sulfur dioxide, SO_2, from the same container and at the same temperature and pressure.

Solution

Since the two rates of effusion are inversely proportional to the square root of their molar masses, we can

(Continued)

write

$$\frac{\text{Rate of effusion of CO}_2}{\text{Rate of effusion of SO}_2} = \sqrt{\frac{M_m(\text{SO}_2)}{M_m(\text{CO}_2)}}$$

where $M_m(\text{SO}_2)$ is the molar mass of SO_2 (64.1 g/mol) and $M_m(\text{CO}_2)$ is the molar mass of CO_2 (44.0 g/mol). Substituting these molar masses into the formula gives

$$\frac{\text{Rate of effusion of CO}_2}{\text{Rate of effusion of SO}_2} = \sqrt{\frac{64.1 \text{ g/mol}}{44.0 \text{ g/mol}}} = 1.21$$

In other words, carbon dioxide effuses 1.21 times faster than sulfur dioxide (because CO_2 molecules move 1.21 times faster on average than SO_2 molecules).

Exercise 4.15

If it takes 4.67 times as long for a particular gas to effuse as it takes hydrogen, under the same conditions, what is the molecular weight of the gas? (Note that the rate of effusion is inversely proportional to the time it takes for a gas to effuse.)

(See Problems 4.71 and 4.72.)

Exercise 4.16

If it takes 3.52 s for 10.0 mL of helium to effuse through a hole in a container at a particular temperature and pressure, how long would it take for 10.0 mL of oxygen, O_2, to effuse from the same container at the same temperature and pressure? (Note that the rate of effusion can be given in terms of volume of gas effused per second.)

(See Problems 4.73 and 4.74.)

Graham's law has practical application in the preparation of fuel rods for nuclear fission reactors. Such reactors depend on the fact that the uranium-235 nucleus undergoes fission (splits) when bombarded with neutrons. When the nucleus splits, several neutrons are emitted and a large amount of energy is liberated. These neutrons bombard more uranium-235 nuclei, and the process continues with the evolution of more energy. However, natural uranium consists of 99.27% uranium-238 (which does not undergo fission) and only 0.72% uranium-235 (which does undergo fission). A uranium fuel rod must contain about 3% uranium-235 to sustain the nuclear reaction.

To increase the percentage of uranium-235 in a sample of uranium (a process called *enrichment*), one first prepares uranium hexafluoride, UF_6, a white, crystalline solid that is easily vaporized. Uranium hexafluoride vapor is allowed to pass through a series of porous membranes. Each membrane has many small holes through which the vapor can effuse. Since the UF_6 molecules with the lighter isotope of uranium travel about 0.4% faster than the UF_6 molecules with the heavier isotope, the gas that passes through first will be somewhat richer in uranium-235. When this vapor passes through another membrane, the uranium-235 vapor becomes further concentrated. It takes many effusion stages to reach the necessary enrichment; for complete separation of the isotopes, as required for bomb-grade uranium, about two million effusion stages are needed.

4.10 Real Gases

In Section 4.2, we found that, contrary to Boyle's law, the pressure–volume product for O_2 at 0°C was not quite constant, particularly at high pressures (see under PV in Table 4.2). Experiments have shown that the ideal gas law describes the behavior of a real gas quite well at moderate pressures and temperatures, but not so well at high pressures and low temperatures. We can see why this is so from the postulates of kinetic theory, from which the ideal gas law can be derived.

According to Postulate 1, the volume of space actually occupied by the molecules is small compared with the total volume that they occupy as a result of their motion through space. Further, according to Postulate 3, the molecules are sufficiently far apart on the average so the intermolecular forces can be disregarded. Both postulates are satisfied by a real gas when its density is low. The molecules are far enough apart to justify treating them as point particles with negligible intermolecular forces (such forces diminish as the molecules become further separated). But Postulates 1 and 3 apply less accurately as the density of a gas increases. As the molecules become more densely packed, the space occupied by molecules is no longer negligible compared to the total volume of gas. Moreover, the intermolecular forces become stronger as the molecules come closer together.

Figure 4.17 shows the behavior of the pressure–volume product at various pressures for several different gases. Note that the gases deviate from Boyle's law to different extents. This is because the two factors responsible for the deviation, the volume of the molecules and the magnitude of the intermolecular forces, vary with each gas. It turns out that the gases that deviate the most liquefy easiest. These gases have the strongest intermolecular forces.

When the ideal gas law is not sufficiently accurate for our purposes, we must use another of the many such equations available. A good alternative is **van der Waals equation,** which is particularly instructive and is reasonably accurate:

$$\left(P + \frac{n^2 a}{V^2}\right)(V - nb) = nRT$$

Here a and b are constants that must be determined for each kind of gas. Table 4.5 lists van der Waals constants for several gases.

We can obtain this equation from the ideal gas law, $PV = nRT$, by replacing P by $P + n^2 a/V^2$ and V by $V - nb$. To explain the change made to the volume term, we note that for the ideal gas law we assume that the molecules have negligible volume. Thus, any particular molecule is free to

Table 4.5
Van der Waals Constants[*]

Gas	a $L^2 \cdot atm/mol^2$	b L/mol
Helium, He	0.03412	0.02370
Hydrogen, H_2	0.2444	0.02661
Oxygen, O_2	1.360	0.03183
Carbon dioxide, CO_2	3.592	0.04267
Ethane, C_2H_6	5.489	0.06380

[*](Source: Reprinted with permission from Weast, R. C., Ed., in *CRC Handbook of Chemistry and Physics*, 59th ed., p. D-230. Copyright CRC Press Inc., Boca Raton, Florida.)

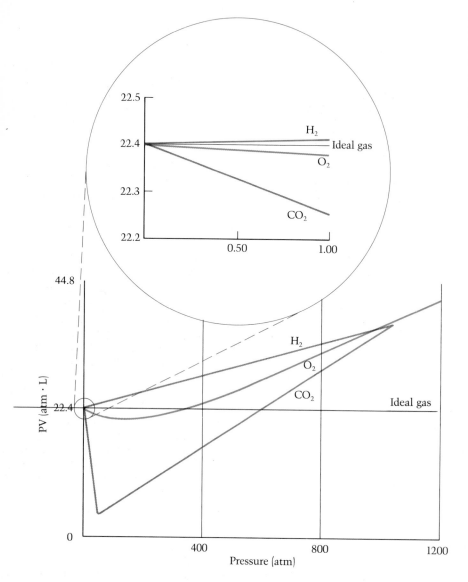

Figure 4.17
Pressure–volume product for
one mole of various gases at
0°C and at different pressures. Values at low pressure
are shown in the inset. For
an ideal gas, the pressure–
volume product should be
constant.

move throughout the entire volume V of the container. But if the molecules
do occupy a small but finite amount of space, the volume through which any
particular molecule can move is reduced. Thus, we subtract nb from V. The
constant b is a measure of molecular volume.

To explain the change made to the pressure term, we need to consider the
effect of intermolecular forces on the pressure the molecules exert. Such
forces will attract molecules to one another. A molecule about to collide
with the wall will be attracted by the other molecules, which will pull it
away from the wall. This attraction will reduce the force of the collision of
the molecule with the wall compared with that of the ideal gas. Therefore,
the actual pressure is less than that predicted by the ideal gas law, and we
must correct P to account for this effect of intermolecular forces. The correction is n^2a/V^2, where a is a measure of the intermolecular forces.

A Checklist for Review

Important Terms

barometer (4.1)
manometer (4.1)
Boyle's law (4.2)
Charles's law (4.3)
Avogadro's law (4.4)
molar gas volume (4.4)
standard temperature and
    pressure (STP) (4.4)
molar gas constant (R) (4.5)

ideal gas law (4.5)
partial pressure (4.6)
Dalton's law of partial
    pressures (4.6)
mole fractions (4.6)
vapor pressure (4.6)
kinetic theory of gases (4.7)
Brownian motion (4.7)

Maxwell's distribution of
    molecular speeds (4.8)
root-mean-square (rms) molecular
    speed (4.8)
diffusion (4.9)
effusion (4.9)
Graham's law of effusion (4.9)
van der Waals equation (4.10)

Summary of Facts and Concepts

Gases at low to moderate pressures and moderate temperatures follow the same simple relationships, or gas laws. Thus, for a given amount of gas at constant temperature, the volume varies inversely with pressure (*Boyle's law*). Also, for a given amount of gas at constant pressure, the volume is directly proportional to the absolute temperature (*Charles's law*). These two laws, together with *Avogadro's law* (equal volumes of any two gases at the same temperature and pressure contain the same number of molecules), can be formulated as one equation, $PV = nRT$ (*ideal gas law*). This equation gives the relationship between P, V, n, and T for a gas. It also relates these qualities for each component in a gas mixture. The total gas pressure, P, equals the sum of the partial pressures of each component (*law of partial pressures*).

The ideal gas law can be explained by *kinetic-molecular theory*. We define an ideal gas as consisting of molecules with negligible volume that are in constant random motion. In the ideal-gas model, there are no intermolecular forces between molecules, and the average kinetic energy of molecules is proportional to the absolute temperature. From kinetic theory, one can show that the *rms molecular speed* equals $\sqrt{3RT/M_m}$. Given this result, one can derive *Graham's law of effusion*: The rate of effusion under identical conditions is inversely proportional to the square root of the molecular weight.

Real gases deviate from the ideal gas law at high pressure and low temperature. From kinetic theory, we expect these deviations because real molecules do have volume, and intermolecular forces do exist. These two factors are partially accounted for in *van der Waals equation*.

Operational Skills

1. Given an initial volume occupied by a gas, calculate the final volume when the pressure changes at fixed temperature (Example 4.1); when the temperature changes at fixed pressure (Example 4.2); and when both pressure and temperature change (Example 4.3).

2. Given the molar volume of a gas (or equivalent information) at some temperature and pressure, calculate the number of moles in a given volume of another gas under the same conditions (Example 4.4).

3. Starting from the ideal gas law, derive the relationship between any two variables (Example 4.5).

4. Given any three of the variables, P, V, T, and n for a gas, calculate the fourth one from the ideal gas law (Example 4.6). Perform the same type of calculation combined with a stoichiometry problem (Example 4.7).

5. Given the molecular weight, calculate the density of a gas for a particular temperature and pressure (Example

4.8); or given the gas density, calculate the molecular weight (Example 4.9).

6. Given the masses of gases in a mixture, calculate the partial pressures and total pressure (Example 4.10).

7. Given the composition of a gas in partial pressures, obtain the composition in mole fractions, or vice versa (Example 4.11).

8. Given the volume, total pressure, and temperature of gas collected over water, calculate the mass of the dry gas (Example 4.12).

9. Given the molecular weight and temperature of a gas, calculate the rms molecular speed (Example 4.13).

10. Given the molecular weight of two gases, calculate the ratio of rates of effusion (Example 4.14). This type of problem can be rephrased to obtain the molecular weight of an unknown gas (as in Exercise 4.16).

Review Questions

4.1 What is the purpose of a manometer? How does it work?

4.2 What are the variables that determine the height of the liquid in a manometer?

4.3 Starting with Boyle's law (stated as an equation), obtain an equation for the final volume occupied by a gas from the initial volume when the pressure is changed at constant temperature.

4.4 The volume occupied by a gas depends linearly on degrees Celsius, at constant pressure, but is not directly proportional to degrees Celsius. However, it is directly proportional to kelvins. What is the difference between a linear relationship and a direct proportion?

4.5 Explain how you would set up an absolute temperature scale based on the Fahrenheit scale, knowing that absolute zero is −273.15°C.

4.6 Starting with Charles's law (stated as an equation), obtain an equation for the final volume of a gas from its initial volume when the temperature is changed at constant pressure.

4.7 State Avogadro's law in words. How does this law explain the law of combining volumes? Use the gas reaction $N_2 + 3H_2 \longrightarrow 2NH_3$ as an example in your explanation.

4.8 Starting from Boyle's, Charles's, and Avogadro's laws, obtain the ideal gas law, $PV = nRT$.

4.9 What variables are needed to describe a gas that obeys the ideal gas law? What are the SI units for each variable?

4.10 What is the value of R in units of $L \cdot mmHg/(K \cdot mol)$?

4.11 What are the standard conditions for comparing gas volumes?

4.12 What does the term *molar gas volume* mean? What is the molar gas volume (in liters) at STP for an ideal gas?

4.13 Give the postulates of kinetic theory, and state any evidence that supports them.

4.14 What is Brownian motion? How do we explain Brownian motion in terms of kinetic theory?

4.15 What is the origin of gas pressure, according to kinetic theory?

4.16 How does the rms molecular speed depend on absolute temperature? On molar volume?

4.17 Explain why a gas appears to diffuse more slowly than average molecular speeds might suggest.

4.18 What is effusion? Why does a gas whose molecules have smaller mass effuse faster than one whose molecules have large mass?

4.19 Under what conditions does a real gas begin to differ significantly from the ideal gas law?

4.20 What is the physical meaning of the a and b constants in the van der Waals equation?

Problems

Note: In these problems, the final zeros given in temperatures and pressures (for example, 20°C, 760 mmHg) are significant figures.

Pressure Measurements

*4.21 A closed manometer is filled with an oil with density of 0.776 g/mL. What is the pressure in pascals when the liquid column in the manometer stands at 85.7 mm?

*4.23 A lift pump operates by creating a vacuum above a column of liquid. Liquid is then pushed upward by the atmospheric pressure. The maximum height to which a liquid can be lifted equals the column height of liquid that can be supported by the barometric pressure. If the barometric pressure is 97.8 kPa, what is the maximum height to which water can be lifted? Assume a density of 1.000 g/cm³.

*4.22 (a) A manometer was checked against a known pressure. If the manometer column is 95.6 mm high, and the pressure is known to be 7.56 mmHg, what is the density of the manometric liquid? (b) What is the pressure when the manometer column reads 69.5 mm?

*4.24 What is the maximum height to which gasoline can be raised by a lift pump at 100.0 kPa (see Problem 4.23)? The density of the gasoline is 0.660 g/cm³.

Boyle's Law

4.25 A sample of helium at 0°C and 743 mmHg has a volume of 3.49 L. What would be the volume if the pressure were changed to 760 mmHg at the same temperature?

4.27 Nitrogen gas was contained in a tank at 21.6 atm pressure. If the gas were allowed to expand to a pressure equal to that of the normal atmosphere (1.00 atm) at 20°C, the total volume of gas would be 849 L. What would be the total volume of gas if it expanded from the tank to a pressure of 1.25 atm at the same temperature?

4.29 A McLeod gauge measures low gas pressures by compressing a known volume of the gas at constant temperature. If 345 cm³ of gas are compressed to a volume of 0.0457 cm³, under a pressure of 2.51 kPa, what was the original gas pressure?

4.26 A balloon contains hydrogen gas with a volume of 1.05 L at 20°C and 755 mmHg. What would be the volume of the hydrogen at a higher altitude, where the pressure of the gas in the balloon decreases to 625 mmHg? Assume that the temperature has not changed.

4.28 Water vapor above liquid water at 21°C has a pressure of 2.30×10^{-2} atm. If 1.00 L of this water vapor could be compressed to 1.00 atm at 21°C, what volume would you expect it to occupy?

4.30 If 456 dm³ of krypton at 101 kPa and 21°C are compressed into a 25.0-dm³ tank at the same temperature, what is the pressure of krypton in the tank?

Charles's Law

4.31 A sample of nitrogen gas collected at 18°C and 760 mmHg has a volume of 2.67 mL. What is the volume at standard temperature and pressure (0°C, 760 mmHg)?

4.33 A vessel containing 9.50 cm³ of helium at 25°C and 101 kPa was placed in dry ice. In order to maintain the same pressure of helium, ethyl alcohol was added to the vessel until the liquid occupied 3.28 cm³. What must be the temperature of dry ice?

4.35 A bacterial culture isolated from sewage produced 34.5 mL of methane, CH₄, at 30°C and 749 mmHg. What is the volume of methane at standard temperature and pressure (0°C, 760 mmHg)?

4.32 A mole of gas at 0°C and 760 mmHg occupies 22.41 L. What is the volume at 20°C and 760 mmHg?

4.34 A quantity of argon was contained in a J-shaped tube with mercury, as in Figure 4.5. The tube contained 46.0 cm³ of argon at 18°C and a total pressure of 150.0 kPa. A week later the volume of argon had changed. The mercury level was then adjusted in order to give the same pressure of argon. If the volume of argon was now 47.7 cm³, what was the temperature?

4.36 Pantothenic acid is a component of vitamin B complex. Using the Dumas method, we find that a sample weighing 81.9 mg gives 4.39 mL of nitrogen gas at 23°C and 785 mmHg. What is the volume of nitrogen at standard temperature and pressure (0°C, 760 mmHg)?

Avogadro's Law

4.37 In the presence of a platinum catalyst, ammonia, NH₃, burns in oxygen, O₂, to give nitric oxide, NO, and water vapor. How many volumes of nitric oxide are obtained from one volume of ammonia, at the same temperature and pressure?

4.39 At a particular temperature and pressure, 15.0 g of carbon dioxide, CO₂, occupy 7.16 L. What is the volume of 12.0 g of methane, CH₄, at the same temperature and pressure?

4.38 Methanol (wood alcohol, CH₃OH) is produced in some industrial plants by reacting carbon dioxide with hydrogen in the presence of a catalyst. Water is the other product. How many volumes of hydrogen are required for each volume of carbon dioxide, when each gas is at the same temperature and pressure?

4.40 At a particular temperature and pressure, 7.00 g of nitrogen, N₂, occupy 5.12 L. What would be the volume occupied by 1.00×10^{23} molecules of CO at the same temperature and pressure?

Ideal Gas Law

4.41 Starting from the ideal gas law, prove that the volume of a mole of gas is inversely proportional to the pressure at constant temperature (Boyle's law).

4.43 In a constant-volume thermometer, one measures the gas pressure and relates this to the temperature. If the gas pressure is 758 mmHg at 25°C, what is the temperature when the pressure is 784 mmHg?

4.45 Argon gas, Ar, is contained in a steel cylinder with a volume of 9.76 L. The temperature is 21°C, and the mass of the argon in the cylinder is 80.7 g. What is the pressure of the argon?

4.47 A carbonate mineral was treated with acid, which liberated 54.7 mL of carbon dioxide gas at 24°C and 758 mmHg. Calculate the moles of CO_2 obtained. How many grams of CO_2 is this?

4.49 Small amounts of hydrogen are conveniently prepared by reacting zinc with hydrochloric acid:

$$Zn + 2HCl \longrightarrow ZnCl_2 + H_2$$

How many grams of zinc are required to prepare 2.50 L H_2 gas at 765 mmHg and 22°C?

4.51 A person's daily caloric requirement can be obtained by ingesting about 726 g of glucose, $C_6H_{12}O_6$. Assume that the glucose reacts completely with oxygen to give carbon dioxide and water. Calculate the volume of oxygen at 20°C and 755 mmHg that must be breathed to react with this much glucose.

4.42 Starting from the ideal gas law, prove that the volume of a mole of gas is directly proportional to the absolute temperature at constant pressure (Charles's law).

4.44 If the pressure of nitrogen in a tank is 18.5 atm at 26°C, what is the pressure at 18°C?

4.46 If 0.843 g of oxygen, O_2, at 988 mmHg occupies a volume of 0.512 L, what is the temperature?

4.48 In an experiment, a student collected 195 mL of dry oxygen, O_2, at 753 mmHg and 23°C. The gas weighed 0.27 g. Obtain a value for the molar gas constant, R, from these data.

4.50 When carbon dioxide is bubbled into a solution of calcium hydroxide, fine crystals of calcium carbonate are formed:

$$Ca(OH)_2 + CO_2 \longrightarrow CaCO_3 + H_2O$$

During a particular experiment, 2.35 g of calcium carbonate were formed. How many liters of carbon dioxide at 15°C and 775 mmHg must have been bubbled into a solution of calcium hydroxide to give this quantity of calcium carbonate?

4.52 In the Solvay process, ammonia and carbon dioxide are passed under pressure into a solution of sodium chloride. Sodium hydrogen carbonate, $NaHCO_3$, crystallizes from solution.

$$NH_3 + CO_2 + H_2O + NaCl \longrightarrow NaHCO_3 + NH_4Cl$$

How many liters of NH_3 at 5°C and 2.00 atm are required to produce 1.00 kg $NaHCO_3$?

Gas Density

4.53 The density of dry air at 10°C and 760 mmHg is 1.247 g/L. What is the density of dry air at 25°C and 760 mmHg? (*Note:* The density of a gas is inversely proportional to the absolute temperature and directly proportional to the pressure.)

4.55 Butane, C_4H_{10}, is an easily liquefied gaseous fuel. Calculate the density of butane gas at 1.00 atm and 25°C. Give the answer in grams per liter.

4.54 The density of ozone gas, O_3, at 0°C and 760 mmHg is 2.144 g/L. What is the density of ozone at 20°C and 775 mmHg? (See note to Problem 4.53.)

4.56 Chloroform, $CHCl_3$, is a volatile (easily vaporized) liquid solvent. Calculate the density of chloroform vapor at 99°C and 745 mmHg. Give the answer in grams per liter.

$$PV = nRT$$

4.57 Ammonium chloride, NH_4Cl, is a white solid. When heated to 325°C, it gives a vapor that is a mixture of ammonia and hydrogen chloride:

$$NH_4Cl \longrightarrow NH_3 + HCl$$

Suppose someone contends that the vapor consists of NH_4Cl molecules, rather than a mixture of NH_3 and HCl. Could you decide between these alternative views on the basis of gas-density measurements? Explain.

4.59 The density of the vapor of a compound at 90°C and 720 mmHg is 1.434 g/L. Calculate the molecular weight of the compound.

4.58 Phosphorus pentachloride, PCl_5, is a white solid that sublimes (vaporizes without melting) at about 100°C. At higher temperatures, the PCl_5 vapor decomposes to give phosphorus trichloride, PCl_3, and chlorine, Cl_2:

$$PCl_5 \longrightarrow PCl_3 + Cl_2$$

How could gas-density measurements help to establish that PCl_5 vapor is decomposing?

4.60 A liquid compound was vaporized at 100°C and 755 mmHg to give 185 mL of vapor. The vapor weighed 0.427 g. What is the molecular weight of the compound?

Gas Mixtures

4.61 A 200.0-mL flask contains 1.03 mg O_2 and 0.41 mg He at 15°C. Calculate the partial pressures of oxygen and of helium in the flask. What is the total pressure?

4.63 The gas from a certain volcano had the following composition in mole percent (that is, mole fraction × 100): 65.0% CO_2, 25.0% H_2, 5.4% HCl, 2.8% HF, 1.7% SO_2, and 0.1% H_2S. What would be the partial pressure of these gases if the total pressure of volcanic gas were 760 mmHg?

4.65 Formic acid, $HCHO_2$, is a convenient source of small quantities of carbon monoxide. When warmed in the presence of sulfuric acid, formic acid decomposes to give carbon monoxide gas:

$$HCHO_2 \longrightarrow H_2O + CO$$

If 4.61 L of carbon monoxide were collected over water at 25°C and 689 mmHg, how many grams of formic acid were consumed? (Refer to Table 4.4.)

4.62 The atmosphere in a sealed diving bell contained oxygen and helium. If the gas mixture has a partial pressure of oxygen equal to 0.200 atm and a total pressure of 3.00 atm, calculate the mass of helium in 1.00 L of the gas mixture at 20°C.

4.64 In a series of experiments, the U.S. Navy developed an undersea habitat (submerged vessel) for the purpose of determining how well people can function at great depths. In one experiment, the composition in mole percent of the atmosphere in the undersea habitat was 79.0% He, 17.0% N_2, and 4.0% O_2. What would be the partial pressures of helium, nitrogen, and oxygen when the habitat is 58.8 m below sea level, where the pressure is 5.91 atm?

4.66 An aqueous solution of ammonium nitrite, NH_4NO_2, decomposes when heated to give off nitrogen, N_2:

$$NH_4NO_2 \longrightarrow 2H_2O + N_2$$

This reaction may be used to prepare pure nitrogen. How many grams of ammonium nitrite must have reacted if 3.40 dm^3 of nitrogen gas were collected over water at 19°C and 97.8 kPa? (Refer to Table 4.4.)

Molecular Speeds; Effusion

4.67 Uranium hexafluoride, UF_6, is a white solid that sublimes (vaporizes without melting) at 57°C under normal atmospheric pressure. What is the rms speed (in m/s) of a uranium hexafluoride molecule at 57°C?

4.69 If the rms speed of H_2 molecules at 20°C is 1.90 km/s, what is the rms speed of I_2 molecules at the same temperature? (Note that $u\sqrt{M_m} = \sqrt{3RT}$ = constant at a given temperature.)

4.68 For a spacecraft or a molecule to leave the moon, it must reach the escape velocity (speed) of the moon, which is 2.37 km/s. The average daytime temperature of the moon's surface is 365 K. What is the rms speed (in m/s) of a hydrogen molecule at this temperature? How does this compare with the escape velocity?

4.70 At what temperature would CO_2 molecules have an rms speed equal to that of H_2 molecules at 20°C? (See note in Problem 4.69.)

4.71 If 4.83 mL of an unknown gas effuse through a hole in a plate in the same time it takes 9.23 mL of argon, Ar, to effuse through the same hole under the same conditions, what is the molecular weight of the unknown gas?

4.73 If 0.10 mol I_2 vapor can effuse from an opening in a heated vessel in 52 s, how long will it take 0.10 mol H_2 to effuse under the same conditions?

4.72 A given volume of nitrogen, N_2, required 68.3 s to effuse from a hole in a chamber. Under the same conditions, another gas required 85.6 s for the same volume to effuse. What is the molecular weight of this gas?

4.74 Calculate the ratio of rates of effusion of $^{235}UF_6$ and ^{238}UF, where ^{235}U and ^{238}U are isotopes of uranium. The atomic masses are: ^{235}U, 235.04 amu; ^{238}U, 238.05 amu; ^{19}F (the only naturally occurring isotope), 18.998 amu. Carry five significant figures in the calculation.

Additional Problems

*4.75 A glass tumbler containing 243 cm³ of air at 1.00×10^2 kPa (the barometric pressure) and 20°C is turned upside down and immersed in a body of water to a depth of 20.5 m. The air in the glass is compressed by the weight of water above it. Calculate the volume of air in the glass, assuming the temperature and barometric pressure have not changed.

*4.76 The density of air at 20°C and 1.00 atm is 1.205 g/L. If this air were compressed at the same temperature to equal the pressure at 30.0 m below sea level, what would be its density? Assume the barometric pressure is constant at 1.00 atm. The density of sea water is 1.025 g/cm³.

4.77 A flask contains 183 mL of argon at 21°C and 738 mmHg. What is the volume of gas corrected to STP?

4.78 A steel bottle contains 12.0 L of a gas at 10.0 atm and 20°C. What is the volume of gas at STP?

4.79 A balloon containing 5.0 dm³ of gas at 14°C and 100.0 kPa rises to an altitude of 2000 m, where the temperature is 20°C. The pressure of gas in the balloon is now 79.0 kPa. What is the volume of gas in the balloon at this altitude?

4.80 A volume of air is taken from the earth's surface, at 15°C and 1.00 atm, to the stratosphere, where the temperature is -20°C and the pressure is 1.00×10^{-3} atm. By what factor is the volume increased?

4.81 A radioactive metal atom decays (goes to another kind of atom) by emitting an alpha particle (He^{2+} ion). The alpha particles are collected as helium gas. A sample of helium with a volume of 12.05 mL was obtained at 765 mmHg and 23°C. How many atoms decayed during the period of the experiment?

4.82 The combustion method used to analyze for carbon and hydrogen can be adapted to give percentage N by collecting the nitrogen from combustion of the compound as N_2. A sample of a compound weighing 8.75 mg gave 1.77 mL N_2 at 25°C and 749 mmHg. What is the percentage N in the compound?

4.83 A person exhales about 5.8×10^2 L of carbon dioxide per day (at STP). The carbon dioxide exhaled by an astronaut is absorbed from the air of a space capsule by reaction with lithium hydroxide, LiOH.

$$2LiOH + CO_2 \longrightarrow Li_2CO_3 + H_2O$$

How many grams of lithium hydroxide are required per astronaut per day?

4.84 Pyruvic acid, $HC_3H_3O_3$, is involved in cell metabolism. It can be assayed for (that is, the amount determined) by using a yeast enzyme. The enzyme makes the following reaction go to completion:

$$HC_3H_3O_3 \longrightarrow C_2H_4O + CO_2$$

If a sample containing pyruvic acid gives 20.3 mL of carbon dioxide gas, CO_2, at 343 mmHg and 30°C, how many grams of pyruvic acid are there in the sample?

4.85 Dry air at STP has a density of 1.2929 g/L. Calculate the average molecular weight of air from the density.

4.86 A hydrocarbon gas has a density of 1.22 g/L at 20°C and 1.00 atm. An analysis gives 80.0% C and 20.0% H. What is the molecular formula?

4.87 If the rms speed of NH_3 molecules is found to be 0.510 km/s, what is the temperature (in Celsius)?

4.88 If the rms speed of He atoms in the exosphere (highest region of the atmosphere) is 3.53×10^3 m/s, what is the temperature (in kelvins)?

*4.89 A 1.000-g sample of an unknown gas at 0°C gives the following data:

P(atm)	V(L)
0.2500	3.1908
0.5000	1.5928
0.7500	1.0601
1.0000	0.7930

Use these data to calculate the value of the molar mass at each of the given pressures from the ideal gas law (we will call this the "apparent molar mass" at this pressure). Plot the apparent molar masses against pressure and extrapolate to find the molar mass at zero pressure. Since the ideal gas law is most accurate at low pressures, this extrapolation will give an accurate value for the molar mass. What is the accurate molar mass?

**4.90 Plot the data given in Table 4.2 for oxygen at 0°C to obtain an accurate molar mass for O_2. (See Problem 4.89.)

4.91 Carbon monoxide, CO, and oxygen, O_2, react according to

$$2CO + O_2 \longrightarrow 2CO_2$$

Assuming that the reaction takes place and goes to completion, determine what substances remain and what their partial pressures are after the valve is opened in the apparatus represented below. Also assume that the temperature is fixed at 300 K.

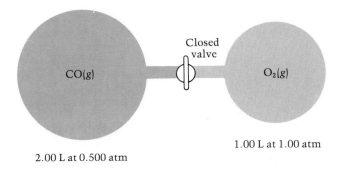

Closed valve

CO(g)

O_2(g)

1.00 L at 1.00 atm

2.00 L at 0.500 atm

**4.92 Consider the apparatus shown in Problem 4.91 that contains H_2 at 0.500 atm in the left vessel separated from O_2 at 1.00 atm in the other. If H_2 and O_2 react to give H_2O when the temperature is fixed at 533 K, what substances remain and what are their partial pressures after reaction?

**4.93 Calculate the molar volume of ethane at 1.00 atm and 0°C and at 10.0 atm and 0°C, using the van der Waals equation. The van der Waals constants are given in Table 4.5. To simplify, note that the term n^2a/V^2 is small compared with P. Hence, it may be approximated with negligible error by substituting $V = nRT/P$ for this term from the ideal gas law. Then the van der Waals equation can be solved for the volume. Compare the results with the ideal-gas-law values.

**4.94 Calculate the molar volume of oxygen at 1.00 atm and 0°C and at 10.0 atm and 0°C, using the van der Waals equation. The van der Waals constants are given in Table 4.5. (See the note on solving the equation given in Problem 4.93.) Compare the results with the ideal-gas-law values. Also compare with the values obtained from Table 4.2.

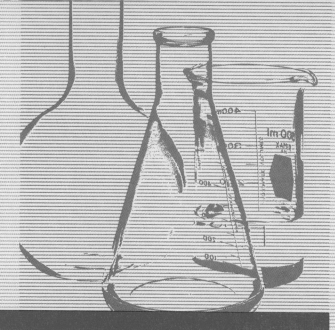

5. Atomic Structure

Basic Structure of Atoms

5.1 Discovery of the Electron Cathode Rays/ Thomson's *m*/*e* Experiment/ Millikan's Oil Drop Experiment

5.2 The Nuclear Model of the Atom Discovery of Radioactivity; Alpha Rays/ Alpha-Particle Scattering

5.3 Nuclear Structure

5.4 Mass Spectrometry and Atomic Weights

Electronic Structure of Atoms

5.5 The Wave Nature of Light

5.6 Quantum Effects and Photons Planck's Quantization of Energy/ Photoelectric Effect/ Aside: Photon Momentum

5.7 The Bohr Theory of the Hydrogen Atom Atomic Line Spectra/ Bohr's Postulates

5.8 Quantum Mechanics de Broglie Waves/ Wave Functions

5.9 Quantum Numbers and Atomic Orbitals Quantum Numbers/ Atomic Orbital Shapes

ireworks displays are fascinating to most people. The spectacular colors of fireworks are due to the presence of metal compounds in the burning material. Strontium nitrate, carbonate, or sulfate gives a deep red color; barium nitrate or chloride a green; and copper nitrate or sulfate a bluish green.

Chemists began studying colored flames in the eighteenth century and soon used "flame tests" to distinguish sodium from potassium compounds. A flame test is a simple way to identify an element. When sodium compounds are sprayed into a flame, the flame burns with a bright yellow color. Potassium gives a violet flame, and lithium and strontium give a red. (See Color Plate 3.)

Although the red flames of lithium and strontium appear similar, the light from each can be resolved (or separated) by means of a prism into distinctly different colors. This resolution easily distinguishes the two elements. A prism disperses the colors of white light just as small raindrops spread the colors of sunlight into a rainbow or spectrum. But the light from a flame, when passed through a prism, reveals something other than a rainbow. Instead of the continuous range of color from red to yellow to violet, the spectrum from a strontium flame, for example, shows a cluster of orange lines and a blue line against a black background. The spectrum of lithium is simpler, showing a red and an orange line against a black background.

Each element, in fact, has a characteristic line spectrum due to the emission of light from atoms in the hot gas. These spectra can be used to identify elements. Looking at such spectra, we can ask many questions about their meaning. How is it that each atom emits particular colors of light? What does a line spectrum tell us about the structure of an atom? If we know something about the structures of atoms, can we explain the formation of ions and molecules? We will answer these questions in this and the next few chapters.

Chapter Overview

In Chapter 2, we introduced atomic theory: matter consists of atoms. Now we want to explore the structure of these atoms. As far as the *basic structure* is concerned, we will see that an atom consists of a dense, positively charged core or nucleus, around which the negatively charged electrons move. In looking at the *electronic structure* of an atom, we will see that each electron can be found in a certain region of space about the nucleus and has a specific amount of energy. An electron in an atom can change energy, but only by definite amounts. It then "jumps," or undergoes a transition, to another region of space about the nucleus. As we will see, it is these transitions between definite energy states that produce the emission of particular colors of light.

Basic Structure of Atoms

An atom consists of a core or **nucleus,** around which move negatively charged particles called **electrons.** Each kind of atom has a nucleus (plural, *nuclei*) made up of a definite number of **protons** and **neutrons,** the building blocks of atomic nuclei. Protons and neutrons have approximately the same mass, but protons are positively charged and give a positive charge to the nucleus. Neutrons have no charge.

An electron differs from a proton or a neutron in being much lighter (about 1/2000 of the mass of a proton or neutron) and in having a negative charge.

The magnitude of the negative charge on an electron is the same as the magnitude of positive charge on the proton. Thus, an electrically neutral atom has the same number of electrons as protons.

In the next sections, we will look at some of the experiments that revealed this basic atomic structure, and we will discuss how the different atoms derive their characteristics from differences in this basic structure.

5.1 Discovery of the Electron

The neon sign with its reddish orange glow is a familiar part of modern life. It is a tube filled with neon gas at low pressure, through which a high-voltage electric current is discharged. A neon sign is an example of a *gas discharge tube.*■ Gas discharge tubes have played an important part in determining the structure of the atom. In particular, the study of such tubes led to the discovery of the electron.

Cathode Rays

The principle of the discharge tube—that gases at low pressure conduct electricity—was discovered in 1821.■ A typical discharge tube is shown in Figure 5.1. Electrical connections, or *electrodes,* are fused into the ends of a glass tube, and the tube is filled with a gas at low pressure. When several thousand volts of electricity are discharged through the tube, colored light characteristic of the gas filling the tube is emitted (neon emits an orange light). If the pressure is slowly reduced by pumping most of the gas out, the dark space that is initially near the *cathode* (negative electrode) gradually enlarges, and the colored glow vanishes.

It appears that some sort of rays leave the cathode and travel toward the *anode* (positive electrode). Because the rays are emitted by the cathode, they are called **cathode rays.** Where they come in contact with the glass, they give a greenish luminescence (glow).

■ The colored glow obtained from discharge tubes is caused by the characteristic emission of light from an atom, which we mentioned in the chapter opening. Other examples include fluorescent tubes and sodium vapor lamps (used for street lighting).

■ The English chemist Humphry Davy, who isolated the metallic elements sodium and potassium, discovered in 1821 that air becomes a better conductor of electricity at low pressure. Subsequently, gas discharge tubes were studied by many, particularly the English chemist William Crookes.

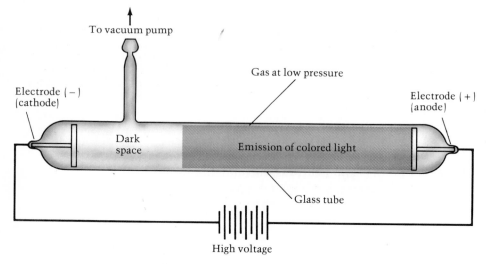

Figure 5.1
A typical electrical discharge tube containing a gas at low pressure. As an electric current passes through the gas, it emits a colored light. If the gas is pumped from the tube, the light diminishes.

To vacuum pump

Gas at low pressure

Electrode (−) (cathode)

Electrode (+) (anode)

Dark space

Emission of colored light

Glass tube

High voltage

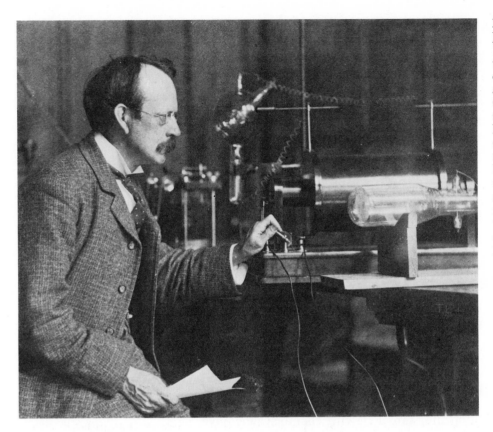

Figure 5.2
Joseph John Thomson (1856–1940). J. J. Thomson's scientific ability was recognized early with his appointment as Professor of Physics in the Cavendish Laboratory at Cambridge University when he was not quite twenty-eight years old. Soon after this appointment, Thomson began research on the discharge of electricity through gases. This work culminated in 1897 with the discovery of the electron. Thomson was awarded the Nobel Prize in physics in 1906.

Thomson's *m/e* Experiment

The identity of cathode rays remained controversial until experiments by J. J. Thomson (Figure 5.2) in 1897. Figure 5.3 shows the type of apparatus used by Thomson in one experiment. Here the anode, which had a hole near its center, was placed closer to the cathode. Cathode rays left the negative

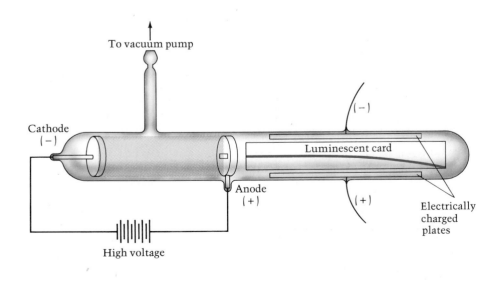

Figure 5.3
Bending of a cathode ray by electrically charged plates. The cathode ray is made visible by a luminescent card placed in the discharge tube.

electrode and passed through the hole in the anode, giving a beam (made visible where it fell on a luminescent card). This beam of cathode radiation then passed between two electrically charged plates, where it was bent, being repelled by the negative plate and attracted to the positive plate. Cathode rays were thus shown to be negatively charged.

Thomson's experiments also showed that cathode rays behave the same regardless of the type of gas filling the tube or the kind of material making up the electrodes. This suggested that cathode rays consist of particles contained in all kinds of matter. Thomson claimed that these negatively charged particles—now called *electrons*—are constituents of all atoms.

Cathode rays are also deflected by magnetic fields.■ By measuring the amount of deflection produced by electric and magnetic fields of known magnitude, Thomson was able to measure the ratio of the mass of an electron to its electrical charge. If the mass of an electron is denoted by m_e, and the size of its electrical charge by e, this mass-to-charge ratio is m_e/e and has the value 5.686×10^{-12} kg/C. (The SI unit of electrical charge is the *coulomb*, whose symbol is C.)

■ A television tube works through the deflection of a cathode ray beam by electromagnetic coils. The cathode ray beam is directed toward a coated screen, where it traces a luminescent image as the beam is bent by the varying magnetism of the deflection coils.

Millikan's Oil Drop Experiment

In 1909, the American physicist Robert Millikan performed a series of experiments in which, by observing the behavior of electrically charged oil drops, he determined the charge on an electron.■ The experimental apparatus is shown in Figure 5.4. A spray bottle produces a fine mist of oil droplets, some of which pass through an opening into a viewing chamber, where we can observe them with a microscope.

Often these droplets have an electrical charge, which is picked up from the friction of forming the mist (much as the friction of walking across a rug generates static electricity). A droplet may have one or more additional electrons on it, giving it a negative charge. If we observe the droplet while it is falling freely in air, we can determine its mass from its downward speed.■

As the droplet falls to the bottom of the chamber, it passes between two electrically charged plates. The droplet can be suspended between them if we adjust the voltage on the plates so that the electrical attraction upward just balances the force of gravity downward. We then use the voltage needed to establish this balance to calculate the mass-to-charge ratio for the droplet. Since we already know the mass of the droplet, we can find the charge on it.

A series of such determinations will reveal the smallest possible difference between the electrical charges on two oil drops. The difference will be the magnitude of the charge on a single electron.■ This value (e) is 1.602×10^{-19} C. It can be used with the value of m_e/e (5.686×10^{-12} kg/C) from Thomson's experiment to calculate the mass of the electron:

■ Water drops have been tried in this experiment, but evaporation causes the drops to continuously change mass.

■ A falling object at first accelerates by gravity, but soon reaches a constant speed as a result of air friction. This speed is related to the mass of the object.

■ We can also induce a given droplet to change charge by irradiating the surrounding air with x rays. The smallest possible change equals the charge of one electron.

$$m_e = (m_e/e) \times e$$
$$= 5.686 \times 10^{-12} \text{ kg/C} \times 1.602 \times 10^{-19} \text{ C} = 9.109 \times 10^{-31} \text{ kg}$$

Thus, the mass of an electron (to three significant figures) is 9.11×10^{-31} kg, nearly 2000 times smaller than the mass of the lightest atom (hydrogen).■

■ On the carbon-12 scale, the mass of the electron is 0.000549 amu.

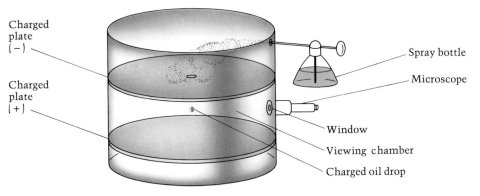

Charged plate (−)

Charged plate (+)

Spray bottle

Microscope

Window

Viewing chamber

Charged oil drop

Figure 5.4
Millikan's oil drop experiment. Oil is sprayed from a bottle to produce droplets. A droplet passes through the hole in the top plate and into the region between the charged plates, where it can be observed through a microscope. This droplet, when illuminated perpendicularly to the direction of view, appears in the microscope as a bright speck against a dark background.

5.2 The Nuclear Model of the Atom

The idea that the atom has a positively charged nucleus was developed to explain experiments involving the scattering of *alpha rays* from metal foils. Alpha rays, as we will now discuss, are given off by certain radioactive materials.

Discovery of Radioactivity; Alpha Rays

In 1896, the French physicist Antoine Henri Becquerel found that uranium and its compounds gave off a radiation that could blacken a photographic plate even when the plate was separated from the source of radiation by thick paper. This newly discovered phenomenon of spontaneous radiation from certain elements was called **radioactivity.**■

The radiation from uranium was shown to be separable by a magnet into two electrically charged beams (Figure 5.5), one called *alpha (α) rays* and the other called *beta (β) rays.* A third type of radiation found, called *gamma (γ) rays,* was unaffected by a magnet. Gamma rays are like visible light, but can easily pass through a layer of matter.

Ernest Rutherford, a British physicist, and others showed in 1903 that alpha rays were a stream of helium ions, He^{2+}. Beta rays, on the other hand,

■ Radioactivity is discussed in detail in Chapter 22.

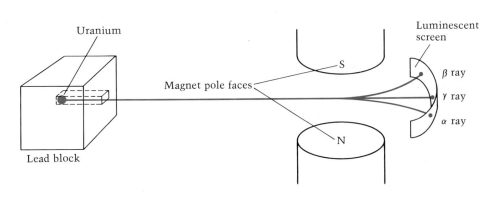

Uranium

Magnet pole faces

S

N

Lead block

Luminescent screen

β ray

γ ray

α ray

Figure 5.5
Separation of the radiation from a radioactive material (uranium). The radiation separates into alpha (α), beta (β), and gamma (γ) rays when it passes through a magnetic field.

Figure 5.6
Alpha-particle scattering from metal foils. Alpha radiation is produced by a radioactive source and formed into a beam by a lead plate with a hole in it. (Lead absorbs the radiation.) Scattered alpha particles are made visible by a zinc sulfide screen, which gives tiny flashes where particles strike it. A movable microscope is used for viewing the flashes.

proved to be a stream of high-speed electrons. Alpha radiation was very intriguing. It was clear evidence that uranium atoms had disintegrated to give off helium. No longer could atoms be considered unchangeable particles, as they had been since the time of Dalton.

Alpha-Particle Scattering

After the discovery of alpha particles, Hans Geiger and Ernest Marsden, working in Rutherford's laboratory, observed the effect of bombarding thin metal foils with these positively charged atomic ions (see Figure 5.6). First they tried gold foil, and later other metals. In each case, they found that most of the alpha particles passed on through the foil as if nothing were there. But just a few of the alpha particles (about 1 in 8000) scattered at large angles and

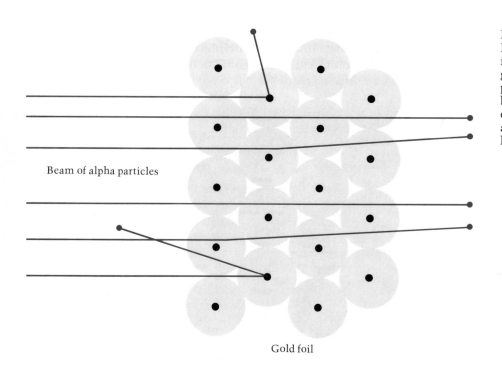

Figure 5.7
Representation of the scattering of alpha particles by a gold foil. Most of the alpha particles pass through the foil barely deflected. A few, however, collide with gold nuclei and are deflected through large angles.

sometimes almost backward. They behaved like bullets being shot at a wrought-iron fence. Usually the bullets would pass through undeflected, but occasionally one might hit an iron bar and ricochet backward.

In 1911, Rutherford successfully explained alpha-particle scattering from foils on the basis of the *nuclear model* of the atom. According to this model, most of the mass of an atom is concentrated in a positively charged center called the *nucleus,* around which negatively charged electrons move. Although most of the mass of an atom is in its nucleus (99.95% or more), the nucleus occupies only a very small portion of the atom. Nuclei have diameters of about 10^{-15} m (10^{-5} Å), whereas atomic diameters are about 10^{-10} m (1 Å)—a hundred thousand times larger. If we were to use a golf ball to represent the nucleus, the atom would be about three miles in diameter. Atoms are mostly empty space.

Thus, when the alpha particles (helium nuclei) are directed toward a metal foil, most pass through the empty space of the atoms more or less undeflected. But if an alpha particle happens to hit a nucleus, it is scattered backward by the repulsion of positive charges (see Figure 5.7).

5.3 Nuclear Structure

In 1919, Rutherford discovered that hydrogen nuclei appear to form when alpha particles collide with nitrogen atoms. Subsequently, it was shown that hydrogen nuclei form during the collision of alpha particles with other kinds of nuclei. These experiments clearly showed that atomic nuclei had structure and appeared to contain hydrogen nuclei, or *protons* as we now call them.

Then, in 1930, W. Bothe and H. Becker bombarded beryllium atoms with alpha particles and obtained a radiation that easily penetrated layers of matter. In 1932 James Chadwick showed that this radiation consisted of neutral particles, each with a mass approximately equal to that of a proton. Because they were neutral, these particles were called *neutrons.* Chadwick proposed that the neutron was a constituent of all nuclei. Later experiments showed that other nuclei gave off neutrons when bombarded with alpha particles.

It is now believed that atomic nuclei are composed of protons and neutrons. (The mass and charge of the proton and neutron are compared with those of the electron in Table 5.1.) Each different kind of nucleus is made up of different numbers of protons and neutrons. For example, the nucleus of a naturally occurring sodium atom contains 11 protons and 12 neutrons. The

Particle	Mass (kg)	Charge (C)	Mass (amu)	Charge (e)
Electron	9.10953×10^{-31}	-1.60219×10^{-19}	0.00055	-1
Proton	1.67265×10^{-27}	$+1.60219 \times 10^{-19}$	1.00728	$+1$
Neutron	1.67495×10^{-27}	0	1.00866	0

Table 5.1
Properties of the Electron, Proton, and Neutron

charge on a sodium nucleus is $+11e$, usually written as simply $+11$.■ Similarly, the nucleus of an aluminum atom contains 13 protons and 14 neutrons; the charge on the nucleus is $+13e$, or $+13$.

Often we characterize a nucleus by its atomic number (symbol Z) and mass number. The **atomic number (Z)** is the number of protons in the nucleus. The **mass number** is the total number of protons and neutrons in the nucleus.■

Mass number = number of protons + number of neutrons

Thus, a sodium nucleus has atomic number 11 and mass number 23 (11 + 12); an aluminum nucleus has atomic number 13 and mass number 27 (13 + 14). The complete notation or symbol for a nucleus consists of the atomic symbol for the element with the mass number written on the left as a superscript, and the atomic number as a subscript. For sodium, we have $^{23}_{11}Na$. The same information is conveyed by giving the name of the element and the mass number; $^{23}_{11}Na$ is also called sodium-23.

An atom is normally electrically neutral and has as many electrons moving about its nucleus as the nucleus has protons. A neutral sodium atom has a charge of $+11$ on its nucleus and has 11 electrons (with total charge -11) around this nucleus.

■ Recall that e is the magnitude of the charge on the electron. The actual charge is $-e$, since an electron has a negative charge. Therefore, the charge on a proton is $+e$.

■ The mass of an atom in atomic mass units is only approximately equal to the atom's mass number, partly because protons and neutrons do not have exactly unit mass (that is neither a proton nor a neutron mass equals exactly 1 amu). Even if their masses were exactly 1, however, nuclear masses would not be whole numbers because some mass is released as nuclear binding energy, according to Einstein's mass-energy equivalence. This is discussed in Chapter 22.

Example 5.1

How many electrons are there in the neutral helium atom? How many electrons are there in He^+ and in He^{2+} (the alpha particle)? The atomic number of He is 2.

Solution

Helium atoms have atomic number 2 and therefore a nuclear charge of $+2$. The neutral atom contains two electrons around the nucleus. The ion He^+ is a helium atom that has lost one electron, leaving a positive charge on the atom of $+1$. He^+ has one electron. Similarly, He^{2+} is a helium atom that has lost two electrons, leaving a positive charge of $+2$. He^{2+} has no electrons. Thus, the alpha particle is a bare helium nucleus.

Exercise 5.1

The atomic number of uranium is 92. How many electrons are there in the neutral atom? How many electrons are there in the U^{2+} ion?

(See Problems 5.23 and 5.24.)

Example 5.2

What is the complete symbol for a nucleus that contains 38 protons and 50 neutrons?

Solution

If you look at the table of atomic weights on the inside back cover of this book, you will note that the element with atomic number 38 (the number of protons in the nucleus) is strontium, symbol Sr. The mass number is $38 + 50 = 88$. The symbol is $^{88}_{38}Sr$.

Exercise 5.2

A nucleus consists of 17 protons and 18 neutrons. What is the complete symbol for this nucleus?

(See Problems 5.25 and 5.26.)

The atomic number of an atom characterizes an element, which always consists of atoms of the same atomic number. An element may, however, be a mixture of atoms of different mass number. **Isotopes** are atoms whose nuclei have the same atomic number but different mass numbers; that is, the nuclei have different numbers of neutrons.■ Sodium has only one naturally occurring isotope, as does aluminum, but most elements are mixtures of isotopes. Naturally occurring oxygen, for example, contains 99.759% $^{16}_{8}O$, 0.037% $^{17}_{8}O$, and 0.204% $^{18}_{8}O$. (These are percentages of the total number of atoms in any given sample of the element.)

■ Isotopes were first suspected in about 1912 when chemically identical elements with different atomic masses were found as radioactive decay products. The most convincing evidence for isotopes, however, came from the mass spectrometer (discussed in Section 5.4).

Exercise 5.3

Describe the structure of the carbon-14 atom (including the structure of the nucleus), according to the nuclear model.

(See Problems 5.27 and 5.28.)

5.4 Mass Spectrometry and Atomic Weights

It is possible to obtain the mass-to-charge ratio for positive ions in the same way that Thomson studied cathode rays. Since the charge on an ion is equal to the magnitude of the electron charge (or some multiple of it), these experiments can, in effect, measure the actual masses of atoms (and molecules). The instrument that is based on these principles and that separates ions by mass-to-charge ratio is called a **mass spectrometer.**

A simplified diagram of a mass spectrometer is shown in Figure 5.8, showing the mass analysis of neon. Neon (which consists of neon atoms, Ne) is

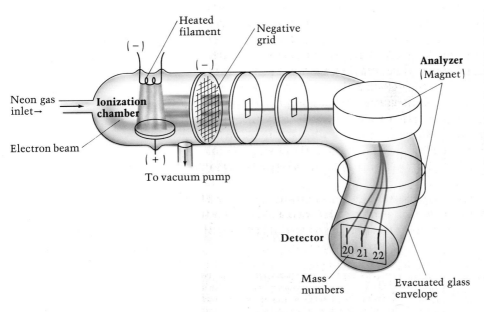

Figure 5.8
Diagram of a simple mass spectrometer, showing the separation of neon isotopes. Neon gas enters the ionization chamber, where neon atoms are ionized by a beam of electrons. Positive neon ions, Ne^+, are accelerated from the ionization chamber by the negative grid and pass to the analyzer, here shown as magnetic pole faces. The ion beam is split in three by the analyzer according to the mass-to-charge ratios. The three beams then travel to a detector. In most modern instruments, the detector registers an electrical signal for different positions, and these are recorded on a chart.

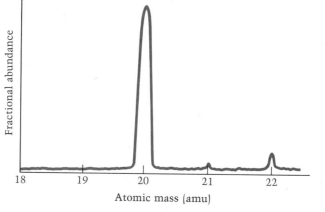

Figure 5.9
The mass spectrum of neon. Neon is separated into its isotopes Ne-20, Ne-21, and Ne-22. The height at each mass peak is proportional to the fraction of that isotope in the element.

introduced into an *ionization chamber*, where the gas stream is crossed by an electron beam. Electrons collide with neon atoms, producing neon ions, Ne^+. ■ Once these positive ions are formed, they are accelerated toward a negative grid and pass through slits to form a positive beam. This beam then travels through a magnetic field (the *analyzer*), which deflects the ions according to their mass-to-charge ratio. Just as a heavy car is buffeted less than a light car by a strong wind, the most massive ions are deflected the least by the magnetic field. Thus each ion arrives in a definite place at the *detector* (charge-measuring device) according to its mass-to-charge ratio. Note that the neon ions are split into three beams by the magnetic field because neon consists of a mixture of three isotopes: $^{20}_{10}Ne$, $^{21}_{10}Ne$, and $^{22}_{10}Ne$. ■ The mass of each positive ion can be obtained from the amount of deflection. Moreover, the percentage of the total number of neon atoms that are neon-20 atoms, for example, can be obtained from the relative number of $^{20}_{10}Ne^+$ ions reaching the detector. The mass spectrum of neon, that is, the chart recording that shows the relative numbers of ions for various masses, is shown in Figure 5.9.

A mass analysis of a substance gives two kinds of information. It gives the masses of atoms or molecules, and it gives the relative number or relative *abundance* of each atomic or molecular ion reaching the detector. Masses are compared with those of the carbon-12 atom, which is assigned a mass of exactly 12 *atomic mass units* (amu). On this carbon-12 scale, neon isotopes have the following masses: neon-20, 19.992 amu; neon-21, 20.994 amu; and neon-22, 21.991 amu.

In a naturally occurring element, the fraction of the total number of atoms composed of a particular isotope is called the **fractional (isotopic) abundance.** The fractional abundances for the isotopes of neon are: neon-20, 0.9051; neon-21, 0.0027; and neon-22, 0.0922.

In Chapter 3, *atomic weight* was defined as the relative atomic mass of a naturally occurring element (expressed in atomic mass units). We now see that a naturally occurring element usually has two or more isotopes. So, the *atomic weight* of an element can be defined as the weighted average of the masses (in atomic mass units) of these isotopes. As the following example illustrates, the weighted average is found by multiplying each isotopic mass by its fractional abundance, then summing.

■ All atoms except hydrogen (which has only one electron) can also produce doubly charged ions; for example, neon gives Ne^{2+}. Large atoms may lose even more electrons.

■ Neon was the first element to be separated into isotopes. In 1913, J. J. Thomson detected a less-abundant mass at 22 amu, as well as the principal mass at 20 amu, in a sample of neon. At first, he thought the mass at 22 amu might be a new element contaminating the neon sample. Later, with improved apparatus, F. W. Aston showed that most elements, including neon, were mixtures of isotopes.

Example 5.3

Chromium, Cr, has the following isotopic masses and fractional abundances:

Mass Number	Mass (amu)	Fractional Abundance
50	49.9461	0.0435
52	51.9405	0.8379
53	52.9407	0.0950
54	53.9389	0.0236

What is the atomic weight of chromium?

Solution

Multiply each of the isotopic masses by the respective fractional abundance, then sum:

$$49.9461 \text{ amu} \times 0.0435 = 2.17 \text{ amu}$$
$$51.9405 \text{ amu} \times 0.8379 = 43.52 \text{ amu}$$
$$52.9407 \text{ amu} \times 0.0950 = 5.03 \text{ amu}$$
$$53.9389 \text{ amu} \times 0.0236 = \underline{1.27 \text{ amu}}$$
$$51.99 \text{ amu}$$

The atomic weight of chromium is 51.99 amu.

Exercise 5.4

Chlorine consists of the following isotopes:

Isotope	Mass (amu)	Fractional Abundance
Chlorine-35	34.96885	0.75771
Chlorine-37	36.96590	0.24229

What is the atomic weight of chlorine?

(See Problems 5.31, 5.32, 5.33, and 5.34.)

Electronic Structure of Atoms

So far we have described the basic structure of atoms. To understand the formation of chemical bonds, however, we need to look at the electronic structure of atoms. Our present theory of electronic structure started with explanations for the colored light produced by atoms in hot gases and flames. Before we can discuss this, we need to describe the nature of light.

5.5 The Wave Nature of Light

If we drop a stone into one end of a quiet pond, the impact of the stone with the water starts an up-and-down motion of the water surface. This up-and-down motion travels outward from where the stone hit; it is a familiar example of a wave. A *wave* is a continuously repeating change or oscillation of matter or a physical field in which the oscillation travels. Light is also a wave. It consists of oscillating electric and magnetic fields that can travel through space. Visible light, as well as x rays and radio waves, are forms of *electromagnetic radiation.*

A wave is characterized by its wavelength and frequency. The **wavelength,** denoted by the Greek letter λ (pronounced "lambda"), is the distance between any two adjacent identical points of a wave. Thus, the wavelength is the distance between two adjacent high points (maxima). Figure 5.10 shows a cross section of a water wave at a given moment, with the wavelength

Figure 5.10
A water wave, with the
wavelength (λ) indicated.

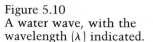

identified. Radio waves have a wavelength of approximately 100 m. Visible light has much shorter wavelengths, about 10^{-6} m. Wavelengths of visible light are often given in nanometers (1 nm = 10^{-9} m). Thus, light of wavelength 5.55×10^{-7} m, the greenish yellow light to which the human eye is most sensitive, equals 555 nm.

The **frequency** of a wave is the number of wavelengths of that wave that pass a fixed point in one unit of time (usually one second). For example, imagine you are anchored in a small boat on a pond when a stone is dropped into the water. Waves travel outward from this point and move past your boat. The number of wavelengths that pass you in one second is the frequency of that wave. Frequency is denoted by the Greek letter ν (pronounced "new"). The unit of frequency is /s, or s^{-1}, also called the *hertz* (Hz).

The wavelength and frequency of a wave can be related to each other. With a wave of frequency ν and wavelength λ, there will be ν wavelengths, each of length λ, that pass a fixed point every second. The product $\nu\lambda$ will be the total length of the wave that has passed the point in one second. This length of wave per second is the speed of the wave, c. Thus,

$$c = \nu\lambda$$

The speed of light waves depends on the medium through which they are passing. In a vacuum, the speed is 3.00×10^8 m/s, and the speed in air is only slightly less.■ For the calculations in this section, we will assume the light to be traveling in a vacuum.

■ To understand how fast the speed of light is, it might help to realize that it takes only 2.5 s for radar waves (which travel at the speed of light) to leave earth, bounce off the moon, and return—a total distance of 478,000 miles.

Example 5.4

What is the wavelength of the yellow sodium emission, which has a frequency of 5.09×10^{14}/s?

Solution

The frequency and wavelength are related by the formula $c = \nu\lambda$. This can be rearranged to give

$$\lambda = \frac{c}{\nu}$$

in which c is the speed of light (3.00×10^8 m/s). Substituting, we have

$$\lambda = \frac{3.00 \times 10^8 \text{ m/s}}{5.09 \times 10^{14}/\text{s}} = 5.89 \times 10^{-7} \text{ m, or } 589 \text{ nm}$$

Exercise 5.5

The frequency of the strong red line in the spectrum of potassium is 3.91×10^{14}/s. What is the wavelength of this light in nanometers?

(See Problems 5.39 and 5.40.)

Example 5.5

What is the frequency of violet light with a wavelength of 408 nm?

$$\nu = \frac{c}{\lambda}$$

Solution

The equation relating frequency and wavelength can also be rearranged to give

Substituting for λ (408 nm = 408×10^{-9} m) gives

$$\nu = \frac{3.00 \times 10^8 \text{ m/s}}{408 \times 10^{-9} \text{ m}} = 7.35 \times 10^{14}/\text{s}$$

Exercise 5.6

The element cesium was discovered in 1860 by Robert Bunsen and Gustav Kirchhoff, who found two bright blue lines in the spectrum of a substance isolated from a mineral water. One of these spectral lines of cesium has a wavelength of 456 nm. What is the frequency of this line?

(See Problems 5.41 and 5.42.)

The range of frequencies or wavelengths of electromagnetic radiation is called the **electromagnetic spectrum,** shown in Figure 5.11. Visible light extends from the violet end of the spectrum, which has a wavelength of about 400 nm, to the red end of the spectrum, with a wavelength of about 700 nm. Beyond these extremes, electromagnetic radiation is not visible to the human eye. Infrared radiation has wavelengths greater than 700 nm (greater than the wavelength of red light), and ultraviolet radiation has wavelengths less than 400 nm (less than the wavelength of violet light).

5.6 Quantum Effects and Photons

Isaac Newton, who studied the properties of light in the seventeenth century, believed that light consisted of a beam of particles. In 1801, however,

Figure 5.11
The electromagnetic spectrum.

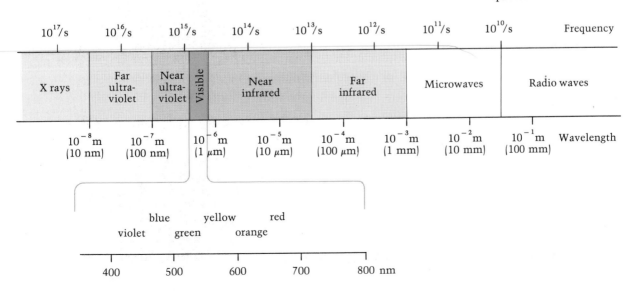

British physicist Thomas Young showed that light, like waves, could be diffracted. *Diffraction* is a property of waves in which the waves spread out when they encounter an obstruction or small hole about the size of the wavelength. We can observe diffraction by viewing a light source through a hole, for example, a street light through a mesh curtain. The image of the street light is blurred by diffraction. It was further shown in the early part of this century that light has particle properties, as well as wave properties. This wave–particle view of light was first postulated by Albert Einstein, starting from the work of the German physicist Max Planck.

Planck's Quantization of Energy

In 1900, Max Planck found a theoretical formula that exactly describes the intensity of light of various frequencies emitted by a hot solid at different temperatures.■ Earlier, others had shown experimentally that the light of maximum intensity from a hot solid varies in a definite way with temperature. A solid glows red at 750°C, then white as the temperature increases to 1200°C. At the lower temperature, chiefly red light is emitted. As the temperature increases, more yellow and blue light become mixed with the red, giving white light.■

According to Planck, the atoms of the solid oscillate, or vibrate, with a definite frequency, ν. But in order to reproduce the results of experiments on glowing solids, he found it necessary to accept a strange idea: that a vibrating atom could have only certain energies, E, those allowed by the formula

$$E = nh\nu, \; n = 1, 2, 3, \ldots$$

where h is a constant, now called **Planck's constant**, with the value 6.63×10^{-34} J · s. The value of n must be either 1 or 2 or some other whole number. Thus, the only energies a vibrating atom can have are $h\nu$, $2h\nu$, $3h\nu$, and so forth.

The numbers symbolized by n are called *quantum numbers*. The vibrational energies of the atoms are said to be *quantized;* that is, the possible energies are limited to certain values. Quantization of energy seems contradicted by everyday experience. Imagine the same concept being applied to the energy of a moving automobile. Quantization of a car's energy would mean that only certain speeds were possible. A car could travel at, say, 10, 20, or 30 miles per hour (mi/h), but not at 12 or 25 mi/h, or any other intermediate speeds. For car energies, this seems unreasonable. But for atoms, as we will see, quantization is the rule.

Photoelectric Effect

Planck himself was uneasy with the quantization assumption and tried (unsuccessfully) to eliminate it from his theory. Albert Einstein, on the other hand, boldly extended Planck's work to include the structure of light itself.■ Einstein reasoned that if a vibrating atom changed energy, say, from $3h\nu$ to $2h\nu$, it would decrease in energy by $h\nu$, and this energy would be emitted as

■ Max Planck was Professor of Physics at the University of Berlin when he did this research. He received the Nobel Prize in physics for it in 1918.

■ The radiation from the human body at normal temperatures is mostly infrared. Variation of the maximum intensity of infrared is used in medicine to measure temperatures of the body surface. It is possible to detect breast cancer (cancerous tissue is warmer) or potential strokes (narrowed arteries reduce the surface temperature of the head).

■ Albert Einstein obtained a Ph.D. from the University of Zurich in 1905. In the same year, he published five papers in *Annalen der Physik,* including one on the photoelectric effect (for which he received the Nobel Prize in physics in 1921) and two on special relativity.

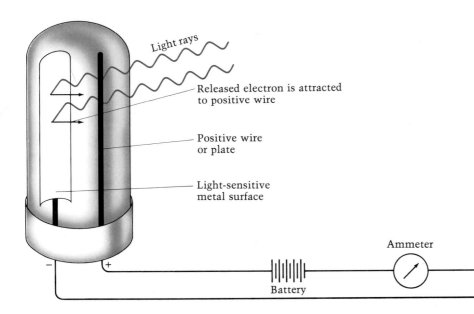

Light rays

Released electron is attracted to positive wire

Positive wire or plate

Light-sensitive metal surface

Ammeter

Battery

Figure 5.12
The photoelectric effect. Light shines on a metal surface, knocking out electrons. The metal surface is contained in an evacuated tube, so that the ejected electrons can be accelerated to a positively charged plate. As long as light of sufficient frequency shines on the metal, free electrons are produced, and a current flows through the tube. (The current is measured by an ammeter.) When the light is turned off, the current stops flowing.

a bit (or quantum) of light energy. He therefore postulated that light consists of particles (now called **photons**) with energy, E, proportional to the observed frequency of the light:

$$E = h\nu$$

In 1905, Einstein used this photon concept to explain the photoelectric effect.

The **photoelectric effect** is the ejection of electrons from the surface of a metal when light shines on it (see Figure 5.12). Electrons are ejected, however, only if the frequency of light is greater than a certain *threshold value* characteristic of the particular metal. For example, although violet light will cause potassium metal to eject electrons, no amount of red light (which has a lower frequency) has any effect.

To explain this dependence of the photoelectric effect on the frequency, Einstein assumed that an electron is ejected from a metal when it is struck by a single photon. Therefore this photon must have at least enough energy to remove the electron from the attractive forces of the metal. No matter how many photons strike the metal, if no single one has sufficient energy, an electron cannot be ejected.■ A photon of red light has insufficient energy to remove an electron from potassium. But a photon corresponding to the threshold frequency has just enough energy, and at higher frequencies it has more than enough energy. When the photon hits the metal, its energy, $h\nu$, is taken up by the electron. The photon ceases to exist as a particle; it is said to be *absorbed*.

The wave and particle pictures of light should be regarded as complementary views of the same physical entity. This is called the *wave–particle duality* of light. Neither wave nor particle view alone is a complete description of light.

■ As a rough analogy, consider five students, each with a nickel, who want to play a video game that takes quarters only. Collectively, there is enough money to play the game, but since no one student has a quarter, no one can play.

Example 5.6

The red spectral line of lithium occurs at 671 nm $(6.71 \times 10^{-7}\,\text{m})$. Calculate the energy of one photon of this light.

Solution

The frequency of this light is

$$\nu = \frac{c}{\lambda} = \frac{3.00 \times 10^8\,\text{m/s}}{6.71 \times 10^{-7}\,\text{m}} = 4.47 \times 10^{14}/\text{s}$$

Hence the energy of one photon is

$$E = h\nu = 6.63 \times 10^{-34}\,\text{J} \cdot \text{s} \times 4.47 \times 10^{14}/\text{s}$$
$$= 2.96 \times 10^{-19}\,\text{J}$$

Exercise 5.7

The following are representative wavelengths in the infrared, ultraviolet, and x-ray regions of the electromagnetic spectrum, respectively: $1.0 \times 10^{-6}\,\text{m}$, $1.0 \times 10^{-8}\,\text{m}$, and $1.0 \times 10^{-10}\,\text{m}$. What is the energy of a photon of each radiation? Which has the greatest amount of energy per photon? Which has the least?

(See Problems 5.43, 5.44, 5.45, and 5.46.)

Aside: Photon Momentum

Not only does the photon have energy, it also has momentum. (The *momentum* of a particle is its mass times its speed.) We can derive its value this way. According to Einstein's theory of relativity, energy is associated with mass. The amount of energy, E, equivalent to the mass, m, is

$$E = mc^2$$

where c is the speed of light. For a photon, whose energy equals $h\nu$, we have

$$mc^2 = h\nu$$

Dividing by c, we get

$$mc = \frac{h\nu}{c} = \frac{h}{\lambda}$$

In the last step, we used the relationship $\lambda = c/\nu$. This last equation relates the photon momentum, mc, to the wavelength of the associated light. Photon momentum is detected as radiation pressure. The orbit of an artificial satellite, for example, must be corrected while in motion for the pressure exerted by sunlight, which would otherwise deflect the satellite from its course.

Note that these two equations ($E = h\nu$ and $mc = h/\lambda$) relate particle properties (energy and momentum) to wave properties (frequency and wavelength).

5.7 The Bohr Theory of the Hydrogen Atom

According to Rutherford's nuclear model, the atom consists of a nucleus with most of the mass and a positive charge, around which move enough electrons to make the atom electrically neutral. But this model posed a

Figure 5.13
Niels Bohr (1885–1962). After Bohr developed his quantum theory of the hydrogen atom, he used his ideas to explain the periodic behavior of the elements. Later, when the new quantum mechanics was discovered by Schrödinger and Heisenberg, Bohr spent much of his time developing its philosophical basis. He received the Nobel Prize in physics in 1922.

dilemma. Using the then-current theory, one could show that an electrically charged particle (such as an electron) that revolves around a center would continuously lose energy as electromagnetic radiation. As an electron in an atom lost energy, it would spiral into the nucleus (in about 10^{-10} s, according to available theory). The stability of the atom could not be explained.

The solution to this theoretical dilemma was found in 1913 by Niels Bohr, a Danish physicist, who at the time was working with Rutherford (see Figure 5.13). Using the work of Planck and Einstein, Bohr applied a new theory to the simplest atom, hydrogen. Before we look at Bohr's theory, we need to consider the line spectra of atoms.

Atomic Line Spectra

As described in the previous section, a heated solid emits light. A heated tungsten filament in an ordinary light bulb is a typical example. We can spread out the light from a bulb with a prism to give a **continuous spectrum,** containing light of all wavelengths, like that of a rainbow (see Color Plate 1). The light emitted by a heated gas, however, yields different results. Rather than seeing a continuous spectrum, with all colors of the rainbow, we obtain a **line spectrum,** which shows only certain colors corresponding to specific wavelengths. When the light from a hydrogen gas discharge tube is analyzed into its components by a prism, it gives a spectrum of lines, each line corresponding to light of a given wavelength. The light produced in the discharge tube is emitted by hydrogen atoms. Color Plate 2 shows the line spectrum of the hydrogen atom, as well as line spectra of other atoms.■

The line spectrum of the hydrogen atom is especially simple. In the visible region, it consists of only four lines (a red, a blue green, a blue, and a violet), although others appear in the infrared and ultraviolet regions. In 1885, J. J. Balmer showed that the wavelengths, λ, in the visible spectrum of hydrogen could be reproduced by a simple formula.

$$\frac{1}{\lambda} = 1.097 \times 10^7/\text{m}\left(\frac{1}{2^2} - \frac{1}{n^2}\right)$$

Here n is some whole number (integer) greater than 2. By putting $n = 3$, for example, calculating $1/\lambda$ then λ, one finds $\lambda = 6.56 \times 10^{-7}$ m, or 656 nm, a wavelength corresponding to red light. The other lines in the H-atom spectrum are obtained by successively putting $n = 4$, $n = 5$, and $n = 6$.

■ The study of line spectra of gas phase atoms and ions is an important activity of chemists and physicists. *Lasers* are devices that emit light at a particular frequency, depending on the nature of the gas present in the discharge tube. The light beams shown on the cover of this book are from a helium-neon laser operated at 632.8 nm (red) and an argon ion laser operated at 488.0 nm (blue-green). Lasers now have practical use in surgical procedures and in scanning devices (for example, in some new cash registers).

Bohr's Postulates

Bohr set down the following two postulates to account for (1) the stability of the hydrogen atom (that the atom exists and its electron does not continuously radiate energy and spiral into the nucleus) and (2) the line spectrum of the atom.

1. Energy-level postulate An electron in an atom can have only specific energy values, which are called the **energy levels** of the electron in the atom. Thus, the atom itself can have only specific total energy values.

Bohr borrowed the idea of quantization of energy from Planck. Bohr, however, devised a rule for this quantization that could be applied to the motion of an electron in an atom. From this he derived the following formula for the energy levels of the electron in the hydrogen atom.

$$E = -\frac{R_H}{n^2}, \quad n = 1, 2, 3, \ldots \infty \quad \text{(for H atom)}$$

where R_H is a constant with the value 2.180×10^{-18} J.■ Different values of the possible energies of the electron are obtained by putting in different values of n, which can have only the integral values 1, 2, 3, and so forth (up to ∞). Here n is called the *principal quantum number*. The diagram in Figure 5.14 shows the energy levels of the electron in the H atom.

■ The energies have negative values because one takes the separated nucleus and electron to have zero energy. If the atom is to be stable, it must have less energy than this.

2. Transitions between energy levels An electron in an atom can change energy only by going from one energy level to another energy level. By so doing, the electron undergoes a *transition*. We explain the emission of light by atoms to give a line spectrum as follows. An electron in a higher energy level (initial energy level, E_i) undergoes a transition to a lower energy level (final energy level, E_f). (See Figure 5.14.) In this process, the electron loses energy, which is emitted as a photon. We obtain the energy of the photon by subtracting the lower energy of the electron, E_f, from the higher value, E_i.

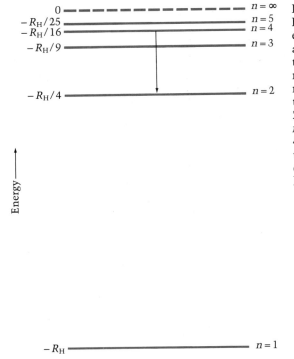

Figure 5.14
Energy-level diagram for the electron in the hydrogen atom. Energy is plotted on the vertical axis (in fractional multiples of R_H). The arrow represents an electron transition (discussed in Postulate 2) from level $n = 4$ to level $n = 2$. Light of wavelength 486 nm (blue green) is emitted. See Example 5.7 for the calculation of this wavelength.

This energy difference (the energy of the photon) is related to the frequency of the emitted light by Einstein's equation $E = h\nu$.

$$\text{Energy of emitted photon} = E_i - E_f = h\nu$$

In this postulate, Bohr used Einstein's photon concept to explain the line spectra of atoms. By substituting values of the energy levels of the electron in the H atom, which he had derived, into the above equation, he was able to exactly reproduce Balmer's formula. Moreover, he was able to predict all of the lines in the spectrum of the H atom, including those in the infrared and ultraviolet regions.

To show how Bohr obtained Balmer's formula, let us write n_i for the principal quantum number of the initial energy level, and n_f for the principal quantum number of the final energy level. Then, from Postulate 1,

$$E_i = -\frac{R_H}{n_i^2} \quad \text{and} \quad E_f = -\frac{R_H}{n_f^2}$$

Substituting these into the equation in Postulate 2,

$$E_i - E_f = \left(-\frac{R_H}{n_i^2}\right) - \left(-\frac{R_H}{n_f^2}\right) = h\nu$$

That is,

$$h\nu = R_H\left(\frac{1}{n_f^2} - \frac{1}{n_i^2}\right)$$

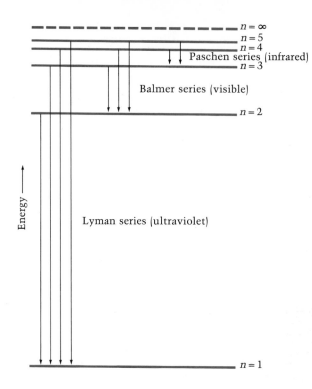

Figure 5.15
Energy-level diagram of the electron in the hydrogen atom, showing the Lyman, Balmer, and Paschen series of transitions that occur for n_f equal to 1, 2, and 3, respectively.

If we now recall that $\nu = c/\lambda$, we can rewrite this as

$$\frac{1}{\lambda} = \frac{R_H}{hc}\left(\frac{1}{n_f^2} - \frac{1}{n_i^2}\right)$$

By substituting $R_H = 2.180 \times 10^{-18}$ J, $h = 6.626 \times 10^{-34}$ J · s, and $c = 2.998 \times 10^8$ m, we find that $R_H/hc = 1.097 \times 10^7$/m, which is the constant given in the Balmer formula. The quantum number n_f, in Balmer's formula is 2. This means that Balmer's formula gives wavelengths that occur when electrons in H atoms undergo transitions from energy levels $n_i > 2$ to level $n_f = 2$. If we change n_f to other integers, we obtain a different series of lines (or wavelengths) for the spectrum of the H atom (see Figure 5.15).

Example 5.7

What is the wavelength of light emitted when the electron in a hydrogen atom undergoes a transition from energy level $n = 4$ to level $n = 2$?

Solution

We start with the formula for the energy levels of the H atom, $E = -R_H/n^2$, and obtain the energy change for the transition. This energy difference will equal the energy of the photon, from which we can calculate frequency, then wavelength, of the emitted light. (Although we could do this problem by "plugging into" the

equation for $1/\lambda$ previously derived, the method followed here requires us to remember only key formulas, shown in color.)

From the formula for the energy levels,

$$E_i = \frac{-R_H}{4^2} = \frac{-R_H}{16} \quad \text{and} \quad E_f = \frac{-R_H}{2^2} = \frac{-R_H}{4}$$

We subtract the lower value from the higher value (to get a positive result). Since this equals the energy of the

(Continued)

photon, we equate it to $h\nu$:

$$\left(\frac{-R_H}{16}\right) - \left(\frac{-R_H}{4}\right) = \frac{3\,R_H}{16} = h\nu$$

The frequency of the light emitted is

$$\nu = \frac{3\,R_H}{16\,h} = \frac{3}{16} \times \frac{2.180 \times 10^{-18}\,\cancel{J}}{6.63 \times 10^{-34}\,\cancel{J} \cdot s} = 6.17 \times 10^{14}/s$$

Since $\lambda = c/\nu$,

$$\lambda = \frac{3.00 \times 10^8\,m/\cancel{s}}{6.17 \times 10^{14}/\cancel{s}} = 4.86 \times 10^{-7}\,m \quad (486\ nm)$$

The color is blue green (see Figure 5.14).

Exercise 5.8

Calculate the wavelength of light emitted from the hydrogen atom when the electron undergoes a transition from level $n = 3$ to level $n = 1$.

(See Problems 5.47, 5.48, 5.49, and 5.50.)

Postulates 1 and 2 hold for atoms other than hydrogen, except that the energy levels cannot be obtained by a simple formula. However, if we know the wavelength of the emitted light, we can relate it to ν and then to the difference in energy levels of the atom. The energy levels of atoms have been experimentally determined in this way.

Exercise 5.9

What is the difference in energy levels of the sodium atom if emitted light has a wavelength of 589 nm?

(See Problems 5.53 and 5.54.)

5.8 Quantum Mechanics

Bohr's theory firmly established the concept of atomic energy levels. It was unsuccessful, however, in accounting for the details of atomic structure or in correctly predicting energy levels for atoms other than hydrogen. Further understanding of atomic structure required another theoretical development.

de Broglie Waves

According to Einstein, light has not only wave properties, which we characterize by frequency and wavelength, but also particle properties. Thus a particle of light, the photon, has a definite energy, $E = h\nu$. One can also show that the photon has momentum. (The momentum of a particle is the product of its mass and speed.) This momentum, mc, is related to the wavelength of the light, $mc = h/\lambda$ or $\lambda = h/mc$. ■

In 1923, the French physicist Louis de Broglie reasoned that if light (considered as a wave) exhibits particle aspects, then perhaps particles of matter show characteristics of waves under the proper circumstances.■ He therefore postulated that a particle of matter of mass m and speed v has an

■ The equation was derived in the Aside in Section 5.6.

■ De Broglie received the Nobel Prize in physics for this work in 1929.

associated wavelength, in analogy with light:

$$\lambda = \frac{h}{mv}$$

This is called the **de Broglie relation.**

If matter has wave properties, why are these not commonly observed? Using the de Broglie relation, calculation shows that a 1-kg mass moving at only 1 km/h has a wavelength of about 10^{-33} m, a value so incredibly small that such waves cannot be detected. On the other hand, electrons moving at experimentally accessible speeds have wavelengths the size of angstroms. Under the proper circumstances their wave character should be observable.

The wave property of electrons was first demonstrated in 1927 by C. Davison and L. H. Germer in the United States and by George Paget Thomson (son of J. J. Thomson) in England. They showed that a beam of electrons, just like x rays, can be diffracted by a crystal. This wave property is exploited in modern electron microscopes (Figure 5.16) to obtain high resolving power. The resolution (the ability to distinguish detail) of a microscope cannot be made smaller than the wavelength used. If we wish to resolve detail that is the size of angstroms (several hundred picometers), we need a wave whose wavelength is the size of angstroms. X rays have wavelengths in this range, but so far no practical means have been found for focusing them. Electrons, on the other hand, are readily focused with electric and magnetic fields. Figure 5.17 shows a photograph taken with an electron microscope.■

■ Electron waves are being investigated as a way to pack even more circuitry on miniature silicon chips for computers and other electronics. At present, ultraviolet light is used to focus an image on a chip coated with film. The ultraviolet light affects the coating so that it can be etched away, leaving the circuit pattern. Since an electron beam gives finer resolution, the size of circuitry might be reduced by a factor of one hundred.

Figure 5.16
Scanning electron microscope, with magnifying power up to 300,000 times. The sample is placed inside the chamber at the left and is viewed on the video screen in the center.

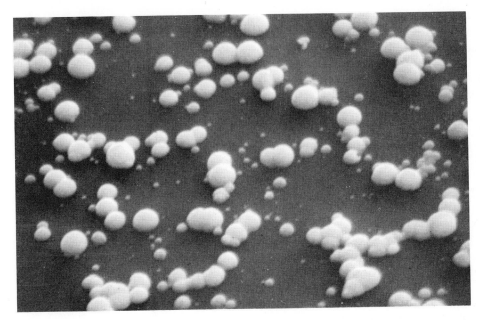

Figure 5.17
Scanning electron microscope image of elemental platinum, Pt (bright spots), deposited on a crystal of naturally occurring molybenite, MoS_2 (dark area). The image, magnified about 10,000 times, reveals that the surface is not uniformly coated. Instead, the platinum is found in islands.

Wave Functions

De Broglie's relation applies quantitatively only to moving particles not acted on by a force such as exists between an electron and a nucleus. It therefore cannot be directly applied to atoms and molecules. But in 1926, Erwin Schrödinger, guided by de Broglie's work, obtained a theory in which the quantum numbers for the hydrogen atom appear straightforwardly as part of the mathematical solution. Schrödinger developed the concept of matter waves into a complete theory that could be applied to other atoms and to molecules. The branch of physics that uses this theory to solve physical problems is called **quantum mechanics** or **wave mechanics.**■

    Without going into the mathematics of quantum mechanics here, we will discuss some of the most important conclusions of the theory. In particular, quantum mechanics alters the way we think about the motion of particles. Our usual concept of motion comes from what we see in the everyday world. We might, for instance, visually follow a ball that has been thrown. The path or orbit of the ball is given by its position at various times. We are therefore conditioned to think in terms of a continuous orbit for moving objects. In Bohr's theory, the electron was thought of as orbiting the nucleus in the way the earth orbits the sun. Quantum mechanics vastly changes this view of motion. We can no longer think of an electron as having an orbit in an atom. To describe such an orbit precisely, we would have to know the exact position of the electron at various times and exactly how long it would take it to travel to a nearby position in the orbit. That is, at any moment we would have to know not only the precise position, but also the precise speed, of the electron.

    In 1927, Werner Heisenberg showed from quantum mechanics that it was impossible to know simultaneously, with absolute precision, both the position and speed of a particle such as an electron. According to Heisenberg's

■ Schrödinger received the Nobel Prize in physics in 1933 for his wave formulation of quantum mechanics. Actually, Werner Heisenberg discovered quantum mechanics a few months before Schrödinger did. But Heisenberg's formulation employed matrix algebra, a mathematical discipline then seldom used by physicists. The results of Heisenberg's and Schrödinger's treatments are identical.

uncertainty principle, the product of the uncertainty in position and the uncertainty in momentum (mass times speed) of a particle can be no smaller than Planck's constant divided by 4π. Thus, letting Δx be the uncertainty in the x coordinate of the particle, and Δv_x the uncertainty in the speed in the x direction, we have

$$(\Delta x)(m\,\Delta v_x) \geq \frac{h}{4\pi} \qquad \text{Uncertainty principle}$$

where m is the mass of the particle. Note that, if m is large enough, the uncertainties Δx and Δv_x can be quite small and still satisfy the uncertainty principle. For objects large enough to be seen, the uncertainties are small, and we can easily describe their orbits. Thus, the orbit of a baseball or the orbit of the earth have meaning. But when the mass is that of an electron, the uncertainties become significant. (They become significant in just those cases where wave properties become appreciable.) The uncertainty in position of an electron in an atom is about the size of the atom. This means that it is impossible for us to describe or to know how the electron moves in an atom.

Although we cannot know how the electron moves in an atom, quantum mechanics does allow us to make *statistical* statements about the electron. For example, we can calculate the *average* speed of an electron in a hydrogen atom. (The average speed is 2.19×10^6 m/s.) We can also obtain the *probability* of finding an electron at a certain point in a hydrogen atom. Thus, although we cannot say that an electron will be at a particular position at a given time, we can say that the electron is likely (or not likely) to be found at this position.

Information about a particle associated with a given energy level (such as an electron in an atom) is contained in a mathematical expression or function. It is called a *wave function* and is often denoted by the Greek letter psi, ψ. The wave function is obtained by solving an equation of quantum mechanics (Schrödinger's equation). Its square, ψ^2, gives the probability of finding the particle within a region of space. More involved mathematical manipulations of ψ yield quantities such as the average speed of the particle.

Consider the wave function, ψ, for the lowest energy level of the hydrogen atom. This wave function has numerical values for each location about the nucleus. Suppose we ask: What is the probability of finding the electron in a hydrogen atom at a point P, if we were to look there for it? The answer is, the probability is ψ^2, where ψ is given its particular value at point P. We can ask this question about each of the points in space about the nucleus. Where we find the probability ψ^2 large, the electron is most likely to be found; where we find the probability small, the electron is less likely to be found. We will describe what these probability distributions for an electron in an atom look like in the next section. Generally, we can say that the wave functions for the hydrogen atom indicate that the electron in the lowest energy level is most likely to be found in a region near the nucleus. An electron in a higher energy level is most likely to be found in a region at a greater distance from the nucleus. Thus, although we cannot speak of electron orbits as we can of planetary orbits, we do know that the electrons are most likely to be in particular regions of space about the nucleus.

5.9 Quantum Numbers and Atomic Orbitals

According to quantum mechanics, each electron in an atom is described by four different quantum numbers. Three of the quantum numbers (n, l, and m_l) specify the wave function that gives the probability of finding the electron at various points in space.■ A wave function for an electron in an atom is called an **atomic orbital.** An *orbital* should not be confused with a Bohr *orbit*. An atomic orbital is pictured qualitatively by describing the region of space where there is high probability of finding the electrons. The atomic orbital so pictured has a definite shape. A fourth quantum number (m_s) refers to a magnetic property of electrons called *spin*. In this section we will first look at quantum numbers, then at atomic orbitals.

■ Three different quantum numbers are needed because there are three dimensions to space.

Quantum Numbers

The allowed values and general meaning of each of the four different quantum numbers of an electron in an atom are as follows:

1. Principal quantum number (n) This quantum number is the modern equivalent of the n quantum number in Bohr's theory. It can have any positive whole number value, 1, 2, 3, and so on.

The energy of an electron in an atom depends *principally* on n. The smaller n is, the lower the energy. In the case of the hydrogen atom or any other single-electron atom (such as Li^{2+} or He^+), it is the only quantum number determining the energy (which is given by Bohr's formula, discussed in Section 5.7). For other atoms, the energy also depends to a slight extent on the l quantum number. We will say more about orbital energies in Chapter 6.

The *size* of an orbital also depends on n. The larger the value of n is, the larger the orbital. Orbitals of the same quantum state n are said to belong to the same *shell*. Shells are sometimes designated by the following letters:

Letter	K	L	M	N...
n	1	2	3	4...

2. Angular momentum quantum number (l) (also called *azimuthal quantum number*) Within each shell of quantum number n, there are n different kinds of orbitals, each with a distinctive shape, denoted by the l quantum number. The allowed values of the l quantum number are the integers from 0 to $n - 1$. (Note that the value 0 is allowed for l.) For example, if an electron has a principal quantum number of 3, the possible values for l are 0, 1, and 2. Thus, within the M shell ($n = 3$), there are three kinds of orbitals, each of which has a different shape for the region where the electron is most likely found.

Orbitals of the same n, but different l, are said to belong to different *subshells* of a given shell. The different subshells are usually denoted by letters.

Letter	*s*	*p*	*d*	*f*	*g* ...
l	0	1	2	3	4 ...

To denote a subshell within a particular shell, we write the value of the n quantum number for the shell followed by the letter designation for the subshell. For example, $2p$ denotes a subshell with quantum numbers $n = 2$ and $l = 1$.■

3. Magnetic quantum number (m_l) Except for an s subshell, there is more than one orbital for each subshell. For a p subshell, for example, there are three different orbitals. There are always $2l + 1$ orbitals in each subshell of quantum number l. Thus, for $l = 2$ (the d subshell), there are $(2 \times 2) + 1 = 5$ orbitals. One way of designating these orbitals within a subshell is with the magnetic quantum number, m_l. The magnetic quantum number can have any whole number value from $-l$ to $+l$. For $l = 2$, the possible values of m_l are -2, -1, 0, 1, and 2.

Essentially, each of these orbitals of a subshell has the same shape, but a different *orientation*, or direction, in space. Each orbital of a particular subshell (no matter how it is oriented in space) has the same energy.

4. Spin quantum number (m_s) An electron acts as though it were spinning on its axis like the earth. Such an electron spin would give rise to a circulating electrical charge that would generate a magnetic field. Thus, an electron behaves like a small bar magnet, with a north and south pole. The spin quantum number, m_s, refers to the two possible orientations that are allowed in quantum mechanics for this magnet or spin. Values for the spin quantum number are $+\frac{1}{2}$ and $-\frac{1}{2}$.■

■ The rather odd choice of letter symbols for l quantum numbers survives from old spectroscopic terminology (describing the lines in a spectrum as *sharp, principal, diffuse, fundamental*).

■ Electron spin will be discussed further in Section 6.1.

Table 5.2 lists the permissible quantum numbers for all orbitals through the $n = 4$ shell. These values follow from the rules just given. Energies for these orbitals are shown in Figure 5.18 for the hydrogen atom. Note that all orbitals with the same principal quantum number, n, have the same energy. For atoms with more than one electron, however, only orbitals in the same subshell (denoted by a given n and l) have the same energy.

Table 5.2
Permissible Values of Quantum Numbers for Atomic Orbitals

n	l	m_l*	Subshell Notation	Number of Orbitals in the Subshell
1	0	0	$1s$	1
2	0	0	$2s$	1
2	1	$-1, 0, +1$	$2p$	3
3	0	0	$3s$	1
3	1	$-1, 0, +1$	$3p$	3
3	2	$-2, -1, 0, +1, +2$	$3d$	5
4	0	0	$4s$	1
4	1	$-1, 0, +1$	$4p$	3
4	2	$-2, -1, 0, +1, +2$	$4d$	5
4	3	$-3, -2, -1, 0, +1, +2, +3$	$4f$	7

*Any one of the m_l quantum numbers may be associated with the n and l quantum numbers on the same line.

Example 5.8

Using the rules for quantum numbers, state whether the following sets of quantum numbers are permissible for an electron in an atom. If not permitted, explain.
(a) $n = 1$, $l = 1$, $m_l = 0$, $m_s = +\frac{1}{2}$
(b) $n = 3$, $l = 1$, $m_l = -2$, $m_s = -\frac{1}{2}$
(c) $n = 2$, $l = 1$, $m_l = 0$, $m_s = +\frac{1}{2}$
(d) $n = 2$, $l = 0$, $m_l = 0$, $m_s = 1$

Solution

(a) Not permitted. The l quantum number is equal to n; it must be less than n. (b) Not permitted. The magnitude of the m_l quantum number (that is, the m_l value, ignoring its sign) must not be greater than l. (c) Permitted. (d) Not permitted. The m_s quantum number can only be $+\frac{1}{2}$ or $-\frac{1}{2}$.

Exercise 5.10

Explain why each of the following sets of quantum numbers is not permissible for an orbital.
(a) $n = 0$, $l = 1$, $m_l = 0$, $m_s = +\frac{1}{2}$
(b) $n = 2$, $l = 3$, $m_l = 0$, $m_s = -\frac{1}{2}$
(c) $n = 3$, $l = 2$, $m_l = 3$, $m_s = +\frac{1}{2}$
(d) $n = 3$, $l = 2$, $m_l = 2$, $m_s = 0$

(See Problems 5.61 and 5.62.)

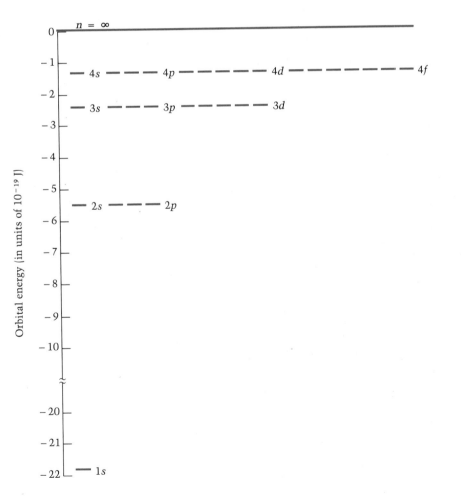

Figure 5.18
Orbital energies for the hydrogen atom. The lines for each subshell indicate the number of different orbitals. (Note break in energy scale.)

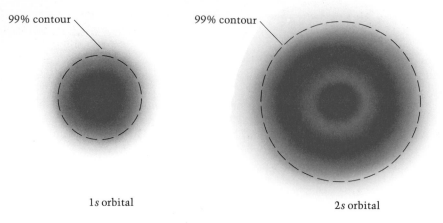

99% contour   99% contour

1s orbital   2s orbital

Figure 5.19
Cross-sectional representations of the probability distributions for a 1s orbital and a 2s orbital. In a 1s orbital, the probability distribution is largest near the nucleus. In a 2s orbital it is greatest in a spherical shell about the nucleus. Note the relative "size" of the orbitals, indicated by the 99% contour.

Atomic Orbital Shapes

An *s* orbital has a spherical shape, though specific details of the probability distribution depend on the value of *n*. Figure 5.19 shows cross-sectional representations of the probability distributions of a 1s and a 2s orbital. The color shading is darker where the electron is more likely to be found. In the case of a 1s orbital (Figure 5.19, left), the electron is most likely to be found near the nucleus. The shading becomes lighter as the distance from the nucleus increases, indicating that the electron is less likely to be found far from the nucleus.

The orbital does not abruptly end at some particular distance from the nucleus. An atom, therefore, has an indefinite extension or "size." We can gauge the "size" of the orbital by means of the *99% contour*. The electron has a 99% probability of being found within the space of the 99% contour (the dashed line in the diagram).

A 2s orbital differs in detail from a 1s orbital. As shown in Figure 5.19, right, the electron in a 2s orbital is likely to be found in two regions, one near the nucleus and the other in a spherical shell about the nucleus (the electron is most likely to be here). The 99% contour shows that the 2s orbital is larger than the 1s orbital.

A cross-sectional diagram cannot portray the three-dimensional aspect of the 1s and 2s atomic orbitals. Figure 5.20 shows cutaway diagrams, which better illustrate this three-dimensionality.

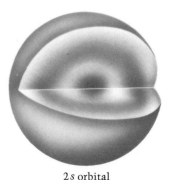

1s orbital   2s orbital

Figure 5.20
Cutaway diagrams showing the spherical shape of *s* orbitals. In both diagrams, the upper part of each orbital is cut away to reveal the electron distribution of the orbital.

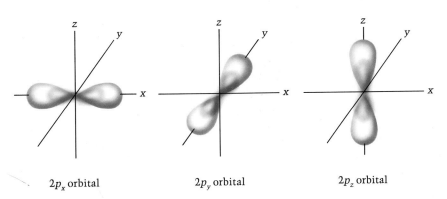

Figure 5.21
The three $2p$ orbitals. Each orbital consists of two lobes, with the lobes oriented along a given axis. The $2p_x$ orbital, for example, has its lobes along the x-axis.

$2p_x$ orbital $2p_y$ orbital $2p_z$ orbital

There are three p orbitals in each subshell, starting with the $2p$ subshell. All p orbitals have the same basic shape (two lobes arranged along a straight line with the nucleus between the lobes) but differ in their orientations in space. Since the three orbitals are set at right angles to each other, we can show each one as oriented along a different coordinate axis (Figure 5.21). We denote these orbitals as $2p_x$, $2p_y$, and $2p_z$. A $2p_x$ orbital has its greatest electron probability along the x-axis, a $2p_y$ orbital along the y-axis, and a $2p_z$ orbital along the z-axis. Other p orbitals, such as $3p$, have this same general shape, with differences in detail depending on n. We will discuss s and p orbital shapes again in Chapter 8 on chemical bonding.

There are five d orbitals, which have more complicated shapes than s and p orbitals. These are represented in Figure 5.22. We will need d orbitals in Chapter 25 in discussing the bonding in compounds of the transition elements.

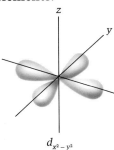

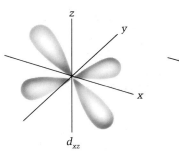

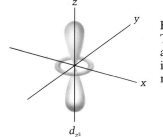

$d_{x^2-y^2}$ d_{xz} d_{z^2}

Figure 5.22
The five $3d$ orbitals. These are labeled by subscripts, as in d_{xy}, that describe their mathematical characteristics.

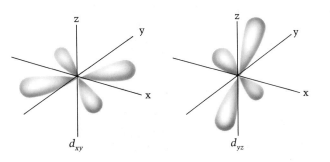

d_{xy} d_{yz}

A Checklist for Review

Important Terms

nucleus (p. 129)
electrons (p. 129)
protons (p. 129)
neutrons (p. 129)
cathode rays (5.1)
radioactivity (5.2)
atomic number (Z) (5.3)
mass number (5.3)
isotopes (5.3)
mass spectrometer (5.4)

fractional (isotopic) abundance (5.4)
wavelength (λ) (5.5)
frequency (ν) (5.5)
electromagnetic spectrum (5.5)
Planck's constant (5.6)
photons (5.6)
photoelectric effect (5.6)
continuous spectrum (5.7)
line spectrum (5.7)

energy levels (5.7)
de Broglie relation (5.8)
quantum (wave) mechanics (5.8)
uncertainty principle (5.8)
atomic orbital (5.9)
principal quantum number (n) (5.9)
angular momentum quantum number (l) (5.9)
magnetic quantum number (m_l) (5.9)
spin quantum number (m_s) (5.9)

Summary of Facts and Concepts

J. J. Thomson established that the cathode rays from an electrical discharge tube consist of negatively charged particles, called *electrons*, which are constituents of all atoms. Thomson measured the mass-to-charge ratio of the electron, and later Millikan measured its charge. From these measurements, the mass of the electron was found to be nearly 2000 times lighter than the lightest atom.

Rutherford proposed the nuclear model of the atom to account for the results of experiments in which alpha particles were scattered from metal foils. According to this model, the atom consists of a central core or *nucleus* around which the electrons move. The nucleus has most of the mass of the atom and consists of *protons* (with a positive charge) and *neutrons* (with no charge). Each chemically distinct atom has a nucleus with a specific number of protons, and around the nucleus in the neutral atom are an equal number of electrons. Atoms whose nuclei have the same number of protons but a different number of neutrons are called *isotopes*.

The atomic weight can be calculated from the atomic masses and *fractional abundances* of the isotopes in a naturally occurring element. These data can be determined by a *mass spectrometer*, which separates ions according to their mass-to-charge ratios.

One way to study the electronic structure of the atom is to analyze the electromagnetic radiation that is emitted from an atom. Electromagnetic radiation is characterized by its *wavelength* (λ), and frequency (ν), and these quantities are related to the speed of light, c ($c = \nu\lambda$).

Light was shown by Einstein to consist of particles (*photons*), each of energy $E = h\nu$, where h is Planck's constant. According to Bohr, electrons in an atom have *energy levels*, and when an electron in a higher energy level drops (or undergoes a transition) to a lower energy level, a photon is emitted. The energy of the photon equals the difference in energy between the two levels.

Electrons, and other particles of matter, have both particle and wave properties. For a particle of mass m and speed v, the wavelength is related to momentum, mv, by the *de Broglie relation*: $\lambda = h/mv$. The wave properties of a particle are described by a *wave function*, from which we can get the probability for finding the particle in different regions of space.

Each electron in an atom is characterized by four different quantum numbers. The distribution of an electron in space—its *atomic orbital*—is characterized by three of these quantum numbers: the principal quantum number, the angular momentum quantum number, and the magnetic quantum number. The fourth quantum number (spin quantum number) describes the magnetism of the electron.

Operational Skills

1. Given an atom or atomic ion and its nuclear charge, find the number of electrons (Example 5.1).

2. Given the number of protons and neutrons in a nucleus, write the complete nuclear symbol (Example 5.2).

3. Given the isotopic masses (in atomic mass units) and fractional isotopic abundances for a naturally occurring element, calculate its atomic weight (Example 5.3).

4. Given the wavelength of light, calculate the frequency, or vice versa (Examples 5.4 and 5.5).

5. Given the frequency or wavelength of light, calculate the energy associated with one photon (Example 5.6).

6. Given the initial and final principal quantum numbers for an electron transition in the hydrogen atom, calculate the frequency or wavelength of light emitted (Example 5.7). You need the value of R_H.

7. Given a set of quantum numbers n, l, m_l, m_s, state whether they are permissible for an electron (Example 5.8).

Review Questions

5.1 Explain the operation of a cathode ray tube. Describe the deflection of cathode rays by electrically charged plates placed within the cathode ray tube. What does this imply about cathode rays?

5.2 What is the evidence that cathode rays are a part of all matter?

5.3 Explain the oil drop experiment.

5.4 Describe the different radiations emitted by radioactive nuclei. How can they be distinguished?

5.5 Describe the nuclear model of the atom. How does this model explain the results of alpha-particle scattering from metal foils?

5.6 One theory of the atom current at the time Rutherford proposed his nuclear model was that the atom consisted of a positive distribution of electricity in which electrons were embedded. This is sometimes referred to as the "plum pudding" model, in which the positive electricity is the "pudding" and the electrons are the "plums." Shooting alpha particles at this "pudding" would be like shooting at a softwood board with lead bullets. The bullets would be slowed and perhaps stopped; some might pass on through. Contrast this with what is actually found.

5.7 What are the different kinds of particles in the nucleus? Compare their properties with each other and with those of an electron.

5.8 Briefly describe how a mass spectrometer works. What kinds of information does one obtain from the instrument?

5.9 Give a brief wave description of light. What are two characteristics of light waves?

5.10 Briefly describe the portions of the electromagnetic spectrum starting with shortest wavelengths and going to longer wavelengths.

5.11 In your own words, explain the photoelectric effect. How does the photon concept explain this effect?

5.12 Describe the wave–particle picture of light.

5.13 Physical theory at the time Rutherford described his nuclear model of the atom was not able to explain how this model could give a stable atom. Explain the nature of this difficulty.

5.14 Explain the main features of Bohr's theory. Do these features solve the difficulty alluded to in question 5.13?

5.15 Explain the process of emission of light by an atom.

5.16 What is the evidence for electron waves? Give a practical application.

5.17 What kind of information does a wave function give about an electron in an atom?

5.18 Give the possible values of (a) the principal quantum number, (b) the angular momentum quantum number, (c) the magnetic quantum number, and (d) the spin quantum number.

5.19 What is the notation for the subshell in which $n = 4$ and $l = 3$? How many orbitals are in this subshell?

5.20 What is the general shape of an s orbital? Of a p orbital?

Problems

Electrons, Protons, and Neutrons

5.21 A student has determined the mass-to-charge ratio for an electron to be 5.65×10^{-12} kg/C. In another experiment, using Millikan's oil drop apparatus, the charge on the electron was found to be 1.602×10^{-19} C. What would be the mass of the electron according to these data?

5.22 The mass-to-charge ratio for the sodium ion, Na^+, is 2.38×10^{-4} kg/C. Using the value of 1.602×10^{-19} C for the charge on the ion, calculate the mass of the sodium ion. (Since the electron mass is negligible by comparison with that of the ion, the ion mass is essentially the atomic mass.)

5.23 Compounds of europium, Eu, are used to coat the inside of color television tubes. If the europium nucleus has a charge of $+63$, how many electrons are there in the neutral atom? How many electrons are there in the Eu^{3+} ion?

5.25 What is the complete notation for the nucleus that contains 30 protons and 34 neutrons?

5.27 Naturally occurring chlorine is a mixture of the isotopes Cl-35 and Cl-37. How many neutrons are there in each isotope? (Chlorine's atomic number is 17.)

5.29 A nucleus of mass number 45 contains 24 neutrons. An atomic ion of this element has 18 electrons in it. Write the symbol for this atomic ion (give the symbol for the nucleus and give the ionic charge as a right superscript).

5.24 Cesium, Cs, is used in photoelectric cells ("electric eyes"). The cesium nucleus has a charge of $+55$. What is the number of electrons in the neutral atom? in the Cs^+ ion?

5.26 An atom contains seven protons and eight neutrons. What is the complete symbol for the nucleus?

5.28 Naturally occurring lithium is a mixture of ^6_3Li and ^7_3Li. Give the number of protons, neutrons, and electrons in the neutral atom of each isotope.

5.30 One isotope of a metallic element has mass number 208 and has 126 neutrons in the nucleus. An atomic ion has 80 electrons. Write the symbol for this ion (give the symbol for the nucleus and give the ionic charge as a right superscript).

Atomic Masses

5.31 Calculate the atomic weight of iridium, Ir, from the following data:

Isotope	Atomic Mass (amu)	Fractional Abundance
Ir-191	191.0	0.373
Ir-193	193.0	0.627

5.33 Magnesium has the following isotopes:

Isotope	Atomic Mass (amu)	Fractional Abundance
Mg-24	23.985	0.7870
Mg-25	24.986	0.1013
Mg-26	25.983	0.1117

What is the atomic weight of magnesium, calculated from these data?

**5.35 Boron has two naturally occurring isotopes, one of mass 10.0129 amu and the other of mass 11.00931 amu. Find the fractional abundances for these two isotopes.

5.32 An element has two naturally occurring isotopes, one of mass 68.93 amu and another of mass 70.92 amu. The fractional abundances of these isotopes are 0.604 and 0.396, respectively. Calculate the atomic weight of this element.

5.34 An element has isotopes with the following masses and abundances:

Atomic Mass (amu)	Fractional Abundance
27.977	0.9221
28.976	0.0470
29.974	0.0309

Calculate the atomic weight of this element. What is the identity of the element?

**5.36 Obtain the fractional abundances for the two naturally occurring isotopes of copper. The masses of the isotopes are: $^{63}_{29}Cu$, 62.9298 amu; $^{65}_{29}Cu$, 64.9278 amu.

Electromagnetic Waves

5.37 At its closest approach, Mars is 56 million km from Earth. How long would it take to send a radio message from a space probe of Mars to Earth, when the planets are at this closest distance?

5.39 Radio waves in the AM region have frequencies in the range 550 to 1400 kilocycles per second (550 to 1400 kHz). Calculate the wavelength corresponding to a radio wave of frequency 1.400×10^6/s (that is, 1400 kHz).

5.38 The space probe *Pioneer 11* was launched April 5, 1973, and reached Jupiter in December 1974, traveling a distance of 998 million km. How long did it take an electromagnetic signal to travel to Earth from *Pioneer 11* when it was near Jupiter?

5.40 Microwaves have frequencies in the range 10^9 to 10^{12}/s (cycles per second), equivalent to between 1 gigahertz and 1 terahertz. What is the wavelength of microwave radiation whose frequency is 1.000×10^{10}/s?

5.41 Light with a wavelength of 600 nm lies in the orange region of the visible spectrum. Calculate the frequency of this light (to three significant figures).

5.42 Calculate the frequency associated with light of wavelength 656 nm. (This corresponds to one of the wavelengths of light emitted by the hydrogen atom.)

Photons

5.43 What is the energy of a photon corresponding to radio waves of frequency $1.400 \times 10^6/s$?

5.44 What is the energy of a photon corresponding to microwave radiation of frequency $1.000 \times 10^{10}/s$?

5.45 The green line in the atomic spectrum of thallium has a wavelength of 535 nm. Calculate the energy of a photon of this light.

5.46 Indium compounds give a blue violet flame test. The atomic emission responsible for this blue violet color has a wavelength of 451 nm. Obtain the energy of a single photon of this wavelength.

Bohr Theory

5.47 Calculate the frequency of electromagnetic radiation emitted by the hydrogen atom in the electron transition $n = 4$ to $n = 3$.

5.48 An electron in a hydrogen atom in the level $n = 5$ undergoes a transition to level $n = 3$. What is the frequency of the emitted radiation?

5.49 The first line of the Lyman series of the hydrogen atom emission results from a transition from the $n = 2$ level to the $n = 1$ level. What is the wavelength of the emitted photon? Using Figure 5.11, describe the region of the electromagnetic spectrum in which this emission lies.

5.50 What is the wavelength of the electromagnetic radiation emitted from a hydrogen atom when the electron undergoes the transition $n = 5$ to $n = 4$? In what region of the spectrum does this line occur? See Figure 5.11.

*__5.51__ One of the lines in the Balmer series of the hydrogen atom emission spectrum is at 397 nm. It results from a transition from an upper energy level to $n = 2$. What is the principal quantum number of the upper level?

*__5.52__ A line of the Lyman series of the hydrogen atom spectrum has the wavelength 9.50×10^{-8} m. It results from a transition from an upper energy level to $n = 1$. What is the principal quantum number of the upper level?

5.53 What is the difference in energy between the levels that are responsible for the red emission line of the rubidium atom at 795 nm?

5.54 Calculate the difference in energy of the levels involved in the green calcium line at 554 nm.

Atomic Orbitals

5.55 If the n quantum number of an atomic orbital is 4, what are the possible values of l? If the l quantum number is 3, what are the possible values of m_l?

5.56 The n quantum number of an atomic orbital is 5. What are the possible values of l? What are the possible values of m_l if the l quantum number is 4?

5.57 How many subshells are there in the M shell? How many orbitals are there in the f subshell?

5.58 How many subshells are there in the N shell? How many orbitals are there in the g subshell?

5.59 Give the notation (using letter designations for l) for the subshells denoted by the following quantum numbers:
    (a) $n = 3, l = 1$
    (b) $n = 4, l = 2$
    (c) $n = 4, l = 0$
    (d) $n = 5, l = 3$

5.60 Give the notation for all of the subshells in the N shell.

5.61 From the following sets of quantum numbers, state which would be possible and which would be impossible for an electron in an atom:

(a) $n = 0$, $l = 0$, $m_l = 0$, $m_s = +\frac{1}{2}$
(b) $n = 1$, $l = 1$, $m_l = 0$, $m_s = +\frac{1}{2}$
(c) $n = 1$, $l = 0$, $m_l = 0$, $m_s = -\frac{1}{2}$
(d) $n = 2$, $l = 1$, $m_l = -2$, $m_s = +\frac{1}{2}$
(e) $n = 2$, $l = 1$, $m_l = -1$, $m_s = +\frac{1}{2}$

5.62 Explain why the following sets of quantum numbers would not be permissible for an electron, according to the rules for quantum numbers:

(a) $n = 1$, $l = 0$, $m_l = 0$, $m_s = +1$
(b) $n = 1$, $l = 3$, $m_l = 3$, $m_s = +\frac{1}{2}$
(c) $n = 3$, $l = 2$, $m_l = 3$, $m_s = -\frac{1}{2}$
(d) $n = 0$, $l = 1$, $m_l = 0$, $m_s = +\frac{1}{2}$
(e) $n = 2$, $l = 1$, $m_l = -1$, $m_s = +\frac{3}{2}$

Additional Problems

*5.63 In a series of oil drop experiments, the charges measured on the oil drops were -3.20×10^{-19} C, -6.40×10^{-19} C, -9.60×10^{-19} C, and -1.12×10^{-18} C. What is the smallest difference of charge between any two drops? If this is assumed to be the charge on the electron, how many excess electrons are there on each drop?

*5.64 In a hypothetical universe, an oil drop experiment gave the following measurements of charges on oil drops: -5.55×10^{-19} C, -9.25×10^{-19} C, -1.11×10^{-18} C, and -1.48×10^{-18} C. Assume that the smallest difference in charge equals the fundamental unit of negative charge in this universe. What is the value of this unit of charge? How many units of excess negative charge are there on each oil drop?

5.65 The blue line of the strontium atom emission has a wavelength of 461 nm. What is the frequency of this light? What is the energy of a photon of this light?

5.66 The barium atom has an emission with wavelength 554 nm (green). Calculate the frequency of this light and the energy of a photon of this light.

5.67 How many protons and neutrons are there in the nucleus of each of the following: ^9_4Be, $^{25}_{12}\text{Mg}$, $^{59}_{27}\text{Co}$?

5.68 Give the number of protons and neutrons in each of the following nuclei: $^{55}_{25}\text{Mn}$, $^{56}_{26}\text{Fe}$, $^{64}_{30}\text{Zn}$.

5.69 Write the complete symbol for the nucleus with atomic number 56 and mass number 138.

5.70 What is the complete symbol for the nucleus with atomic number 47 and mass number 107?

**5.71 There are two naturally occurring isotopes of silver, with isotope masses of 106.905 amu (^{107}Ag) and 108.905 amu (^{109}Ag). What is the fractional abundance of ^{109}Ag in silver?

**5.72 Europium is a silvery white metal. The element consists of two isotopes ^{151}Eu and ^{153}Eu, with masses of 150.92 amu and 152.92 amu, respectively. Calculate the fractional abundance of ^{151}Eu atoms in europium.

*5.73 Mass-to-charge ratios for some isotopic ions of sulfur are 3.3137×10^{-7} kg/C, 3.5205×10^{-7} kg/C, 1.6568×10^{-7} kg/C, and 1.7603×10^{-7} kg/C. What are the masses of each isotope in atomic mass units? What is the charge on each ion (in units of e)?

*5.74 The main isotopic ions of potassium give mass-to-charge ratios of 4.0383×10^{-7} kg/C, 4.1430×10^{-7} kg/C, 2.0192×10^{-7} kg/C, and 2.0715×10^{-7} kg/C. Obtain the isotope masses (in amu) and the charge on each ion (in units of e).

*5.75 The photoelectric work function of a metal is the minimum energy needed to eject an electron by irradiating the metal with light. For potassium, this work function equals 3.59×10^{-19} J. What is the minimum frequency of light for the photoelectric effect in potassium?

*5.76 The photoelectric work function for silver is 7.58×10^{-19} J. ("Work function" is defined in Problem 5.75.) Calculate the minimum frequency of light required to eject electrons from silver.

*5.77 A hydrogen-like ion has a nucleus of charge $+Ze$ and a single electron outside this nucleus. The energy levels of these ions are $-Z^2 R_H/n^2$ (where Z = atomic number). Calculate the wavelength of the transition from $n = 3$ to $n = 2$ for He$^+$, a hydrogen-like ion. In what region of the spectrum does this emission occur?

*5.78 What is the wavelength of the transition from $n = 4$ to $n = 3$ for Li^{2+}? (See Problem 5.77.)

*5.79 When an electron is accelerated by a voltage difference, the kinetic energy acquired by the electron equals the voltage times the charge on the electron. Thus, one volt imparts a kinetic energy of 1.602×10^{-19} volt-coulombs. (The volt-coulomb equals the joule; thus, this kinetic energy is 1.602×10^{-19} J.) What is the wavelength associated with electrons accelerated by 4.00×10^5 volts?

*5.80 What is the wavelength for electrons accelerated by 1.00×10^4 volts? (See Problem 5.79.)

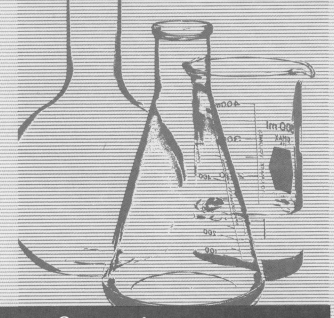

6. Electron Configurations and Periodicity

Electronic Structure of Atoms

6.1 Electron Spin and the Pauli Exclusion Principle Electron Configurations and Orbital Diagrams/ Pauli Exclusion Principle

6.2 Building-up Principle (Aufbau Principle) Aside: X Rays, Atomic Numbers, and Orbital Structure

6.3 Hund's Rule; Paramagnetism Hund's Rule/ Paramagnetism

The Periodic Table

6.4 Periodic Classification of the Elements Predictions from the Periodic Table/ Arrangement of the Elements by Atomic Number/ Relationship to Electron Configurations

6.5 Some Periodic Properties Atomic Radius/ Ionization Energy/ Electron Affinity

6.6 A Brief Description of the Main-Group Elements Hydrogen: Configuration $1s^1$/ Group IA Elements (Alkali Metals): Valence-Shell Configuration ns^1/ Group IIA Elements (Alkaline Earth Metals): Valence-Shell Configuration ns^2/ Group IIIA Elements: Valence-Shell Configuration ns^2np^1/ Group IVA Elements: Valence-Shell Configuration ns^2np^2/ Group VA Elements: Valence-Shell Configuration ns^2np^3/ Group VIA Elements: Valence-Shell Configuration ns^2np^4/ Group VIIA Elements (Halogens): Valence-Shell Configuration ns^2np^5/ Group VIIIA Elements (Noble Gases): Valence-Shell Configuration ns^2np^6

Marie Curie, a Polish-born French chemist, and her husband, Pierre, announced the discovery of radium in 1898. They had separated a very radioactive mixture from pitch-blende, an ore of uranium. This mixture was primarily a compound of barium. When the mixture was heated in a flame, however, it gave a new atomic line spectrum, in addition to the spectrum for barium. The Curies based their discovery of a new element on this finding of a new line spectrum. It took them four more years to obtain a pure compound of radium. Radium, like uranium, is a radioactive element. But in most of its chemical and physical properties, it is similar to the non-radioactive element barium. It was this similarity of radium to barium that made the final separation of the new element so difficult.

That groups of elements have similar properties was then well known to chemists. In 1869, Dmitri Mendeleev found that by arranging the elements in a particular way, they fell into columns, with elements in the same column displaying similar properties. Thus, Mendeleev placed beryllium, calcium, strontium, and barium in one column. Now, with the Curies' discovery, radium was added to this same column of elements.

Mendeleev's arrangement of the elements is called the *periodic table*. It was originally based on the observed chemical and physical properties of the elements and their compounds. We now explain this arrangement in terms of the electronic structure of atoms. In this chapter we will look at this electronic structure and its relationship to the periodic table of elements.

Chapter Overview

An electron configuration describes the arrangement of electrons in the subshells of an atom. In the first part of this chapter, the discussion centers on *electron configurations of atoms*. The chemical properties of elements are related to these configurations. If we arrange the elements by increasing atomic number (nuclear charge), we find that the configurations for the outer electrons of the atoms are repeated after a certain number of elements. This repetition of electron configurations is responsible for groups of elements having similar properties. We discuss this *periodicity of the elements* in the second part of the chapter.

Electronic Structure of Atoms

In Chapter 5, we found that an electron in an atom has four quantum numbers, n, l, m_l, and m_s, associated with it. The first three quantum numbers characterize the orbital (which mathematically describes the probability of finding the electron at places about the nucleus). We picture an orbital as the region of space where an electron is most likely found and say that the electron "occupies" this orbital. The spin quantum number, m_s, describes the spin orientation of an electron. We will look at electron spin in more detail in the next section. Then we will discuss how electrons are distributed among the possible orbitals.

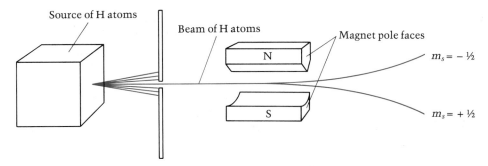

Figure 6.1
The Stern–Gerlach experiment. A beam of hydrogen atoms (in color) is split into two by a nonuniform magnetic field. One beam consists of atoms in which the electron has $m_s = +\frac{1}{2}$; the other consists of atoms in which the electron has $m_s = -\frac{1}{2}$.

6.1 Electron Spin and the Pauli Exclusion Principle

German physicists Otto Stern and Walther Gerlach first observed electron spin magnetism in 1921. They directed a beam of silver atoms into the field of a specially designed magnet. The same experiment can be done with hydrogen atoms. The beam of hydrogen atoms is split by the magnetic field into two, one half of the atoms being bent in one direction, and one half in the other (see Figure 6.1). The fact that the atoms are affected by the laboratory magnet shows that they themselves act as magnets.

The beam of hydrogen atoms is split into two because the electron in the atom behaves as a tiny magnet with only two possible orientations.■ We can picture the electron as a ball of spinning charge. Like an electric current in a coil of wire, the circulating charge creates a magnetic field. Electron spin, however, is subject to a quantum restriction on the possible directions of the spin axis. The resulting directions of spin magnetism, shown in Figure 6.2, correspond to spin quantum numbers $m_s = +\frac{1}{2}$ and $m_s = -\frac{1}{2}$.

■ Protons and many nuclei also have spin. Nuclear spin magnetism, however, is much smaller than that of electrons. The energy changes associated with changes of orientation of nuclear spins in a molecule depend on the electron environment of the nuclei. These energy changes are obtained as radio wave frequencies in nuclear magnetic resonance (NMR) spectroscopy. NMR spectroscopy is very useful in determining the structures of molecules.

Electron Configurations and Orbital Diagrams

The **electron configuration** of an atom is the particular distribution of electrons among available subshells. It is described by a notation that lists the subshell symbols, one after the other. Each symbol has a superscript on the right giving the number of electrons in that subshell. For example, a configuration of the lithium atom (atomic number 3) with two electrons in the $1s$ subshell and one electron in the $2s$ subshell is written $1s^2 2s^1$.

The notation for a configuration gives the number of electrons in each subshell but does not show how the orbitals of a subshell may be occupied by electrons. For this purpose, we use an **orbital diagram.** In such a diagram, an orbital is represented by parentheses (or a circle). Each group of orbitals in a subshell is labeled by its subshell notation. An electron in an orbital is shown by an arrow, with the arrow pointed upward when $m_s = +\frac{1}{2}$ and pointed downward when $m_s = -\frac{1}{2}$. The orbital diagram

$$(\uparrow \downarrow) \ (\uparrow \downarrow) \ (\uparrow \ )(\ \ )(\ \ )$$
$$1s \qquad 2s \qquad \qquad 2p$$

shows the electronic structure of an atom in which there are two electrons in the $1s$ subshell or orbital (one electron with $m_s = +\frac{1}{2}$, the other with

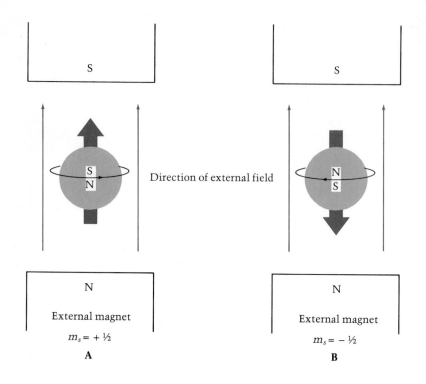

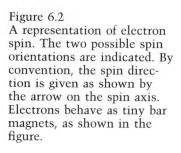

Figure 6.2
A representation of electron spin. The two possible spin orientations are indicated. By convention, the spin direction is given as shown by the arrow on the spin axis. Electrons behave as tiny bar magnets, as shown in the figure.

$m_s = -\frac{1}{2}$), two electrons in the $2s$ subshell $(m_s = +\frac{1}{2}, m_s = -\frac{1}{2})$, and one electron in the $2p$ subshell $(m_s = +\frac{1}{2})$. The electron configuration is $1s^2 2s^2 2p^1$.

Pauli Exclusion Principle

Not all of the conceivable arrangements of electrons among the orbitals of an atom are physically possible. The **Pauli exclusion principle**, which summarizes experimental observations, states that no two electrons in an atom can have the same four quantum numbers. If one electron in an atom has the quantum numbers $n = 1$, $l = 0$, $m_l = 0$, and $m_s = +\frac{1}{2}$, no other electron can have the same four quantum numbers. In other words, we cannot place two electrons with the same value of m_s in a $1s$ orbital. The orbital diagram

$$(\uparrow \ \uparrow)$$
$$1s$$

does not represent a possible arrangement of electrons.

Because there are only two possible values of m_s, an orbital can hold no more than two electrons, and then only if the two electrons have different spin quantum numbers. In an orbital diagram, an orbital with two electrons has to be written with arrows pointing in opposite directions. The two electrons are said to have opposite spins.

We can see that each subshell holds a maximum of twice as many electrons as the number of orbitals in the subshell. Thus, a $2p$ subshell, which has three orbitals (with $m_l = -1$, 0, and $+1$), can hold a maximum of six electrons. The maximum number of electrons in various subshells is given in the following table.

Subshell	Number of Orbitals	Maximum Number of Electrons
s $(l = 0)$	1	2
p $(l = 1)$	3	6
d $(l = 2)$	5	10
f $(l = 3)$	7	14

Example 6.1

Which of the following orbital diagrams or electron configurations are possible and which are impossible, according to the Pauli exclusion principle? Explain.

(a) $(\uparrow\downarrow)$ $(\uparrow\downarrow)$ $(\uparrow$ $)($ $)($ $)$
    1s   2s   2p

(b) $(\uparrow\downarrow)$ $(\uparrow\downarrow\uparrow)$ $($ $)($ $)($ $)$
    1s   2s   2p

(c) $(\uparrow\downarrow)$ $(\uparrow$ $)$ $(\uparrow\uparrow)($ $)($ $)$
    1s   2s   2p

(d) $1s^3 2s^1$

(e) $1s^2 2s^1 2p^7$

(f) $1s^2 2s^2 2p^6 3s^2 3p^6 3d^8 4s^2$

Solution

(a) Possible orbital diagram. (b) Impossible orbital diagram; there are three electrons in the $2s$ orbital. (c) Impossible orbital diagram; there are two electrons in a $2p$ orbital with the same spin. (d) Impossible electron configuration; there are three electrons in the $1s$ subshell (one orbital). (e) Impossible electron configuration; there are seven electrons in the $2p$ subshell (which can hold only six electrons). (f) Possible. Note that the $3d$ subshell can hold as many as ten electrons.

Exercise 6.1

Look at the following orbital diagrams and electron configurations. Which ones are possible and which ones are not, according to the Pauli exclusion principle? Explain.

(a) $(\uparrow$ $)$ $(\uparrow$ $)$ $($ $)($ $)($ $)$
    1s   2s   2p

(b) $(\uparrow$ $)$ $(\uparrow$ $)$ $(\uparrow\downarrow)(\uparrow\downarrow)(\uparrow\downarrow)$
    1s   2s   2p

(c) $(\uparrow\downarrow)$ $(\uparrow\downarrow)$ $(\uparrow\uparrow)(\uparrow\downarrow)(\uparrow\downarrow)$
    1s   2s   2p

(d) $1s^2 2s^2 2p^4$

(e) $1s^2 2s^4 2p^2$

(f) $1s^2 2s^2 2p^6 3s^2 3p^{10} 3d^{10}$

(See Problems 6.19, 6.20, 6.21, and 6.22.)

6.2 Building-up Principle (Aufbau Principle)

Every atom has an infinite number of possible electron configurations. The configuration associated with the lowest energy level of the atom corresponds to a quantum-mechanical state called the *ground state*. Other configurations correspond to *excited states*, associated with energy levels other than the lowest. For example, the ground state of the sodium atom is known from experiment to have the electron configuration $1s^2 2s^2 2p^6 3s^1$. Electron configuration $1s^2 2s^2 2p^6 3p^1$ represents an excited state of the sodium atom.■

■ The transition of the sodium atom from the excited state $1s^2 2s^2 2p^6 3p^1$ to the ground state $1s^2 2s^2 2p^6 3s^1$ is accompanied by the emission of yellow light at 589 nm. Thus, excited states are needed to describe the spectrum of an atom.

The chemical properties of an atom are related primarily to the electron configuration of its ground state. Table 6.1 lists the experimentally determined electron configurations of the ground states for neutral atoms.* In the abbreviated notation used in the table, [He], [Ne], and so on, stand for the configurations of the atom in brackets. Thus, "[Ne]" represents a "core" of electrons with the configuration $1s^2 2s^2 2p^6$. The configuration $1s^2 2s^2 2p^6 3s^1$ for the sodium atom is abbreviated [Ne]$3s^1$.

Most of the configurations in Table 6.1 can be explained in terms of the **building-up principle** (or **Aufbau principle**). Following this principle, we obtain the electron configuration of an atom by successively filling subshells in the following order (the *building-up order*): $1s, 2s, 2p, 3s, 3p, 4s, 3d, 4p,$ $5s, 4d, 5p, 6s, 4f, 5d, 6p, 7s, 5f$. This order was arrived at by noting that it reproduces the experimentally determined electron configurations (with some exceptions, which we will discuss later).

The building-up order corresponds for the most part to increasing energy of the subshells. We might expect this. By filling orbitals of lowest energy first, we usually get the lowest energy (ground state) of the atom. Recall that the energy of an orbital depends only on the quantum numbers n and l. ■ (The energy of the H atom, however, depends only on n.) Orbitals with the same n and l but different m_l, that is, different orbitals of the same subshell, have the same energy. The energy depends primarily on n, increasing with its value. Thus, a $3s$ orbital has greater energy than a $2s$ orbital because the value of n is greater. Except for the H atom, the energies of orbitals with the same n increase with the l quantum number. A $3p$ orbital has slightly greater energy than a $3s$ orbital because l is greater. The orbital of lowest energy is $1s$; the next higher ones are $2s$ and $2p$, then $3s$ and $3p$. The $3d$ subshell, however, has an energy just below but very close to that of the $4s$ orbital. Figure 6.3 shows the orbital energies calculated from theory for the scandium atom ($Z = 21$). Relative values of the orbital energies for other atoms are similar. Because the $3d$ energy is very close to the $4s$ energy, the total energy of atoms (which depends on both orbital energies and the energy of repulsion between electrons) turns out to be lower for those configurations obtained by filling the $4s$ orbital before the $3d$ subshell.

Let us see how we can reproduce the electron configurations of Table 6.1 using the building-up principle. Remember that the number of electrons in a neutral atom equals the atomic number Z (the nuclear charge is $+Z$). In the case of the simplest atom, hydrogen ($Z = 1$), we obtain the ground state by placing the single electron into the $1s$ orbital, giving the configuration $1s^1$. Now we go to helium ($Z = 2$). The first electron goes into the $1s$ orbital, as in hydrogen, followed by the second electron, since any orbital can hold two electrons. The configuration is $1s^2$. Filling the $n = 1$ shell gives a very stable configuration and, as we will discuss later, gives rise to a chemically unreactive atom.

We continue this way through the elements, each time increasing Z by one and adding another electron. The configuration of an atom is obtained from that of the preceding element by adding an electron into the next

■ The quantum numbers and characteristics of orbitals were discussed in Section 5.9.

*Configurations in Table 6.1, except those in parentheses, are taken from Charlotte E. Moore, *National Standard Reference Data Series*, National Bureau of Standards, 34 (September 1970). Those in parentheses were obtained on the basis of their assumed position in the periodic table. No official name has been adopted for elements beyond $Z = 103$.

Table 6.1 Electron Configurations of Atoms in the Ground State

Z	Element	Configuration	Z	Element	Configuration
1	H	$1s^1$	56	Ba	$[Xe]6s^2$
2	He	$1s^2$	57	La	$[Xe]5d^16s^2$
3	Li	$[He]2s^1$	58	Ce	$[Xe]4f^15d^16s^2$
4	Be	$[He]2s^2$	59	Pr	$[Xe]4f^36s^2$
5	B	$[He]2s^22p^1$	60	Nd	$[Xe]4f^46s^2$
6	C	$[He]2s^22p^2$	61	Pm	$[Xe]4f^56s^2$
7	N	$[He]2s^22p^3$	62	Sm	$[Xe]4f^66s^2$
8	O	$[He]2s^22p^4$	63	Eu	$[Xe]4f^76s^2$
9	F	$[He]2s^22p^5$	64	Gd	$[Xe]4f^75d^16s^2$
10	Ne	$[He]2s^22p^6$	65	Tb	$[Xe]4f^96s^2$
11	Na	$[Ne]3s^1$	66	Dy	$[Xe]4f^{10}6s^2$
12	Mg	$[Ne]3s^2$	67	Ho	$[Xe]4f^{11}6s^2$
13	Al	$[Ne]3s^23p^1$	68	Er	$[Xe]4f^{12}6s^2$
14	Si	$[Ne]3s^23p^2$	69	Tm	$[Xe]4f^{13}6s^2$
15	P	$[Ne]3s^23p^3$	70	Yb	$[Xe]4f^{14}6s^2$
16	S	$[Ne]3s^23p^4$	71	Lu	$[Xe]4f^{14}5d^16s^2$
17	Cl	$[Ne]3s^23p^5$	72	Hf	$[Xe]4f^{14}5d^26s^2$
18	Ar	$[Ne]3s^23p^6$	73	Ta	$[Xe]4f^{14}5d^36s^2$
19	K	$[Ar]4s^1$	74	W	$[Xe]4f^{14}5d^46s^2$
20	Ca	$[Ar]4s^2$	75	Re	$[Xe]4f^{14}5d^56s^2$
21	Sc	$[Ar]3d^14s^2$	76	Os	$[Xe]4f^{14}5d^66s^2$
22	Ti	$[Ar]3d^24s^2$	77	Ir	$[Xe]4f^{14}5d^76s^2$
23	V	$[Ar]3d^34s^2$	78	Pt	$[Xe]4f^{14}5d^96s^1$
24	Cr	$[Ar]3d^54s^1$	79	Au	$[Xe]4f^{14}5d^{10}6s^1$
25	Mn	$[Ar]3d^54s^2$	80	Hg	$[Xe]4f^{14}5d^{10}6s^2$
26	Fe	$[Ar]3d^64s^2$	81	Tl	$[Xe]4f^{14}5d^{10}6s^26p^1$
27	Co	$[Ar]3d^74s^2$	82	Pb	$[Xe]4f^{14}5d^{10}6s^26p^2$
28	Ni	$[Ar]3d^84s^2$	83	Bi	$[Xe]4f^{14}5d^{10}6s^26p^3$
29	Cu	$[Ar]3d^{10}4s^1$	84	Po	$[Xe]4f^{14}5d^{10}6s^26p^4$
30	Zn	$[Ar]3d^{10}4s^2$	85	At	$[Xe](4f^{14}5d^{10}6s^26p^5)$
31	Ga	$[Ar]3d^{10}4s^24p^1$	86	Rn	$[Xe]4f^{14}5d^{10}6s^26p^6$
32	Ge	$[Ar]3d^{10}4s^24p^2$	87	Fr	$[Rn](7s^1)$
33	As	$[Ar]3d^{10}4s^24p^3$	88	Ra	$[Rn]7s^2$
34	Se	$[Ar]3d^{10}4s^24p^4$	89	Ac	$[Rn]6d^17s^2$
35	Br	$[Ar]3d^{10}4s^24p^5$	90	Th	$[Rn]6d^27s^2$
36	Kr	$[Ar]3d^{10}4s^24p^6$	91	Pa	$[Rn](5f^26d^17s^2)$
37	Rb	$[Kr]5s^1$	92	U	$[Rn](5f^36d^17s^2)$
38	Sr	$[Kr]5s^2$	93	Np	$[Rn](5f^46d^17s^2)$
39	Y	$[Kr]4d^15s^2$	94	Pu	$[Rn]5f^67s^2$
40	Zr	$[Kr]4d^25s^2$	95	Am	$[Rn]5f^77s^2$
41	Nb	$[Kr]4d^45s^1$	96	Cm	$[Rn](5f^76d^17s^2)$
42	Mo	$[Kr]4d^55s^1$	97	Bk	$[Rn](5f^97s^2)$
43	Tc	$[Kr]4d^55s^2$	98	Cf	$[Rn](5f^{10}7s^2)$
44	Ru	$[Kr]4d^75s^1$	99	Es	$[Rn](5f^{11}7s^2)$
45	Rh	$[Kr]4d^85s^1$	100	Fm	$[Rn](5f^{12}7s^2)$
46	Pd	$[Kr]4d^{10}$	101	Md	$[Rn](5f^{13}7s^2)$
47	Ag	$[Kr]4d^{10}5s^1$	102	No	$[Rn](5f^{14}7s^2)$
48	Cd	$[Kr]4d^{10}5s^2$	103	Lr	$[Rn](5f^{14}6d^17s^2)$
49	In	$[Kr]4d^{10}5s^25p^1$	104	—	$[Rn](5f^{14}6d^27s^2)$
50	Sn	$[Kr]4d^{10}5s^25p^2$	105	—	$[Rn](5f^{14}6d^37s^2)$
51	Sb	$[Kr]4d^{10}5s^25p^3$	106	—	$[Rn](5f^{14}6d^47s^2)$
52	Te	$[Kr]4d^{10}5s^25p^4$	107	—	$[Rn](5f^{14}6d^57s^2)$
53	I	$[Kr]4d^{10}5s^25p^5$	108	—	$[Rn](5f^{14}6d^67s^2)$
54	Xe	$[Kr]4d^{10}5s^25p^6$	109	—	$[Rn](5f^{14}6d^77s^2)$
55	Cs	$[Xe]6s^1$			

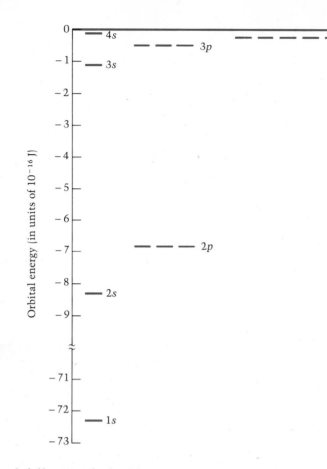

Figure 6.3
Orbital energies for the scandium atom ($Z = 21$). Note that unlike in the hydrogen atom, the subshells for each n are spread apart in energy. Thus the $2p$ energy is above the $2s$. Similarly, the $n = 3$ subshells are spread to give the order $3s < 3p < 3d$. The $3d$ subshell energy is now just below the $4s$. (Values for this figure were calculated from theory by Charlotte F. Fischer, Vanderbilt University.)

available orbital, following the building-up order. In lithium ($Z = 3$), the first two electrons give the configuration $1s^2$, like helium, but the third electron goes into the next higher orbital in the building-up order, since the $1s$ orbital is now filled. This gives the configuration $1s^2 2s^1$, or [He]$2s^1$. In beryllium ($Z = 4$), the fourth electron fills the $2s$ orbital, giving the configuration $1s^2 2s^2$, or [He]$2s^2$.

With boron ($Z = 5$), the electrons begin filling the $2p$ subshell. We get [He]$2s^2 2p^1$ for boron, [He]$2s^2 2p^2$ for carbon ($Z = 6$), and so forth, up to [He]$2s^2 2p^6$ for neon ($Z = 10$). Having filled the $2p$ subshell, we again find a stable configuration. Neon is chemically unreactive as a result. The configurations of the remaining elements are reproduced in similar fashion. Again, when a p subshell is filled, a stable configuration is obtained, giving a relatively unreactive element. These include, in addition to He and Ne, the elements Ar, Kr, Xe, and Rn. This group is called the *noble gases*.

A convenient memory aid will help to generate the building-up order. We write the symbols of the subshells for a given n on a separate line in order by l. Then we place the subshells with different n one under another in order by n (see Figure 6.4). Now we draw arrows diagonally, starting with the $1s$ orbital, then $2s$, and so forth, as Figure 6.4 shows. The order in which the symbols for the subshells are crossed by arrows, top to bottom, is the same as the building-up order. Example 6.2 illustrates the use of the building-up principle to obtain electron configurations of atoms.■

■ The building-up order can also be obtained from the following two rules. (1) Subshells fill in the order of $n + l$. For example, the $3d$ and $4s$ subshells have $n + l$ equal to 5 and 4, respectively. Therefore, $4s$ fills before $3d$. (2) If two subshells have the same $n + l$, the one of lower n fills first. For example, $2p$ and $3s$ both have $n + l = 3$, but $2p$ fills before $3s$.

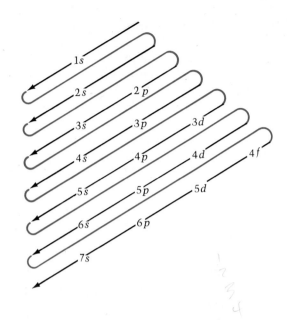

Figure 6.4
A memory aid for reproducing the building-up order.

Example 6.2

Use the building-up principle to obtain the configuration for the ground state of the gallium atom ($Z = 31$). Give the configuration in complete form (that is, do not abbreviate for the core).

Solution

Write the memory aid to obtain the building-up order of subshells.

This gives the order $1s$, $2s$, $2p$, $3s$, $3p$, $4s$, $3d$, $4p$. We fill these subshells with the 31 electrons present in the neutral gallium atom, starting with $1s$, then $2s$, and so forth. The configuration is $1s^2 2s^2 2p^6 3s^2 3p^6 3d^{10} 4s^2 4p^1$. We have written the orbitals in the configuration in the order of shells, with the subshells in order of l. This is a convention that we adopt in this book, although some prefer to arrange them in the building-up order.

Exercise 6.2

Use the building-up principle to obtain the electron configuration for the ground state of the manganese atom ($Z = 25$).

(See Problems 6.25, 6.26, 6.27 and 6.28.)

As mentioned, there are a number of exceptions to the building-up principle. Of these, three are common elements: chromium, copper, and silver. The configurations for these elements obtained from the building-up principle and the actual configurations observed experimentally are shown below.

Element	Building-up Configuration	Actual Configuration
Cr	$[Ar]3d^4 4s^2$	$[Ar]3d^5 4s^1$
Cu	$[Ar]3d^9 4s^2$	$[Ar]3d^{10} 4s^1$
Ag	$[Kr]4d^9 5s^2$	$[Kr]4d^{10} 5s^1$

We can remember the actual configurations by noting that *there is a tendency toward half-filled and filled* d *subshells*. A half-filled d subshell would be d^5, a filled subshell d^{10}. So when we obtain $[Ar]3d^44s^2$ for Cr using the building-up principle, we move one of the electrons from the $4s$ orbital to the $3d$ subshell to give $[Ar]3d^54s^1$ which has a half-filled d subshell.

Exercise 6.3

Use the building-up principle to obtain the electron configuration for the ground state of the molybdenum atom $(Z = 42)$. What might you expect for the actual configuration? Explain.   (See Problems 6.29 and 6.30.)

Aside: X Rays, Atomic Numbers, and Orbital Structure

In 1913, Henry G. J. Moseley, a student of Rutherford, used the technique of *x-ray spectroscopy* (just discovered by Max von Laue) to determine the atomic numbers of the elements. X rays are produced in a cathode ray tube when the electron beam (cathode ray) falls on a metal target. The explanation for the production of x rays is as follows. When an electron in the cathode ray hits a metal atom in the target, it can (if it has sufficient energy) knock an electron from an inner shell of the atom. A metal ion is produced with an electron missing from an inner orbital. Its electron configuration is unstable, and an electron from an orbital of higher energy drops into the half-filled orbital and a photon is emitted. The photon corresponds to electromagnetic radiation in the x-ray region.

The energies of the inner orbitals of an atom and the energy changes between them depend on the nuclear charge, $+Z$. Therefore, the photon energies, $h\nu$, and the frequencies, ν, of emitted x rays depend on the atomic number, Z, of the metal atom in the target. Figure 6.5 shows the x-ray spectra obtained by Moseley with various metal targets.

A related technique, *x-ray photoelectron spectroscopy*, experimentally confirms our theoretical view of the orbital structure of the atom. Instead of irradiating a sample with an electron beam and analyzing the frequencies of emitted x rays, we irradiate a sample with x rays and analyze the kinetic energies of ejected electrons. In other words, we observe the *photoelectric effect* on the sample (see Section 5.6).

As an example of photoelectron spectroscopy, consider a sample of neon gas (Ne atoms). Suppose the sample is irradiated with x rays of a specific frequency great enough to remove a $1s$ electron from the neon atom. Part of the energy of the x-ray photon, $h\nu$, is used to remove the electron from the atom (this is the *ionization energy, I.E.,* for that electron). The remaining energy appears as kinetic energy, E_K, of the ejected electron. From the law of conservation of energy, we can write

$$E_K = h\nu - I.E.$$

If we look at the electrons ejected from neon, they will have kinetic energies related to the ionization energies from all possible orbitals ($1s$, $2s$, and $2p$) in the atom. Thus, by scanning the various kinetic energies of ejected electrons, we will see a spectrum with peaks corresponding to the different occupied orbitals [see Figure 6.6(a)]. These ionization energies are approximately equal to the positive values of the orbital energies [Figure 6.6(b)], so this spectrum provides direct experimental verification of the discrete energy levels associated with the electrons of the atom.

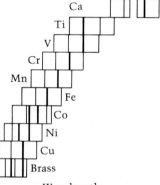

Wavelength→

Figure 6.5
X-ray spectra of the elements calcium to zinc, obtained by Moseley. Each line results from an emission of given wavelength. Because of the volatility of zinc, Moseley used brass (a copper–zinc alloy) to observe the spectrum of zinc. Note the copper lines in brass. (Figure taken from J. J. Lagowski, *The Structure of Atoms* (Boston: Houghton Mifflin Company, 1964), p. 80, Figure 26.)

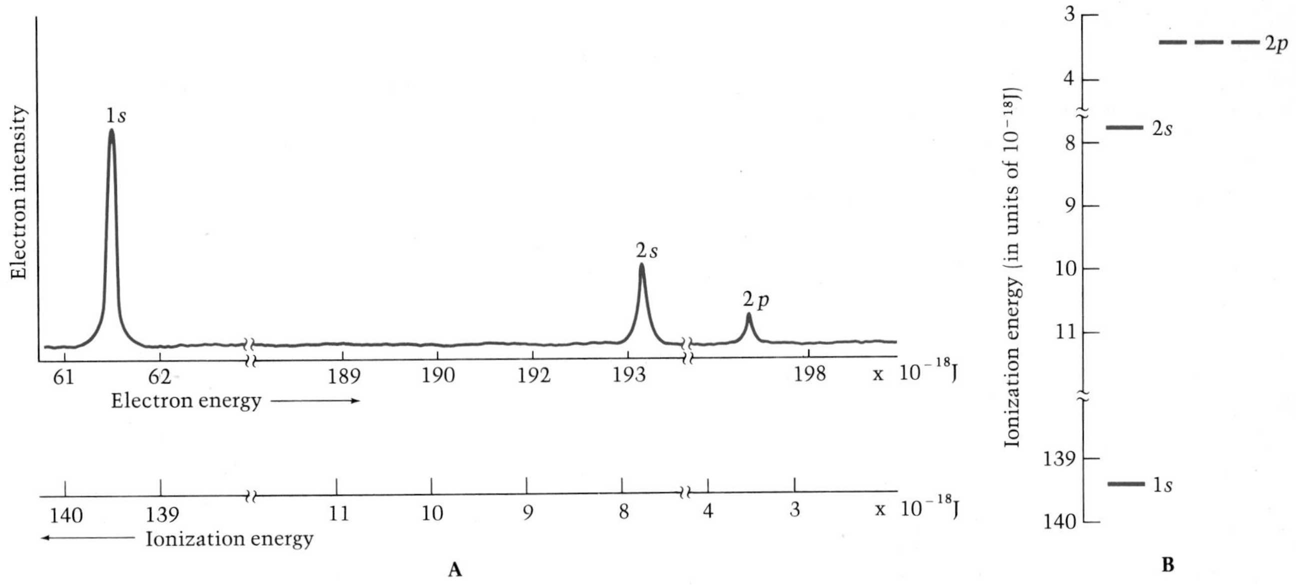

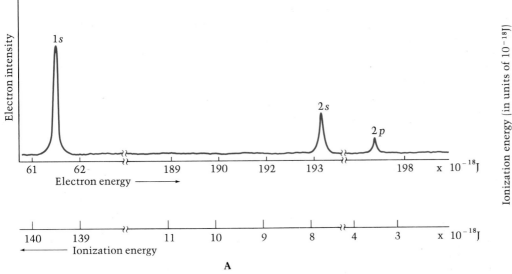

B

6.3 Hund's Rule; Paramagnetism

So far, in discussing the ground states of atoms, we have not described how the electrons are arranged in each subshell. There may be several different ways of arranging electrons in a particular configuration. Consider the carbon atom ($Z = 6$), with the ground-state configuration $1s^2 2s^2 2p^2$. Three possible arrangements are given in the following orbital diagrams.

Diagram 1: $(\uparrow\downarrow)$ $(\uparrow\downarrow)$ $(\uparrow\ )(\uparrow\ )(\ )$
$\quad\quad\quad\quad 1s\quad\quad 2s\quad\quad\quad 2p$

Diagram 2: $(\uparrow\downarrow)$ $(\uparrow\downarrow)$ $(\uparrow\ )(\ \downarrow)(\ )$
$\quad\quad\quad\quad 1s\quad\quad 2s\quad\quad\quad 2p$

Diagram 3: $(\uparrow\downarrow)$ $(\uparrow\downarrow)$ $(\uparrow\downarrow)(\ )(\ )$
$\quad\quad\quad\quad 1s\quad\quad 2s\quad\quad\quad 2p$

These orbital diagrams show different states of the carbon atom. Each state has a different energy and, as we will see, different magnetic characteristics.

Hund's Rule

In about 1927, German physicist Friedrich Hund experimentally discovered a rule determining the arrangement of electrons within a subshell. This rule describes the electron arrangement for the lowest energy level of an atom. According to **Hund's rule,** when electrons fill a subshell, every orbital in the subshell is occupied by a single electron before any orbital is doubly occupied, and all electrons in singly occupied orbitals have their spins in the same direction.

If we apply this rule to the carbon atom, the lowest energy is obtained when the two electrons of the $2p$ subshell go into separate orbitals with the same spin, as in Diagram 1 above. When the two electrons are in different

Figure 6.6
(a) X-ray photoelectron spectrum of neon. The energy of each x-ray photon is 200.9×10^{-18} J. (b) The energy-level diagram for neon shows the transitions that occur in the photo-electron spectrum.

Atom	Z	Configuration	Orbital Diagram		
			$1s$	$2s$	$2p$
Hydrogen	1	$1s^1$	(↑)	( )	( )( )( )
Helium	2	$1s^2$	(↑ ↓)	( )	( )( )( )
Lithium	3	$1s^2 2s^1$	(↑ ↓)	(↑)	( )( )( )
Beryllium	4	$1s^2 2s^2$	(↑ ↓)	(↑ ↓)	( )( )( )
Boron	5	$1s^2 2s^2 2p^1$	(↑ ↓)	(↑ ↓)	(↑)( )( )
Carbon	6	$1s^2 2s^2 2p^2$	(↑ ↓)	(↑ ↓)	(↑)(↑)( )
Nitrogen	7	$1s^2 2s^2 2p^3$	(↑ ↓)	(↑ ↓)	(↑)(↑)(↑)
Oxygen	8	$1s^2 2s^2 2p^4$	(↑ ↓)	(↑ ↓)	(↑ ↓)(↑)(↑)
Fluorine	9	$1s^2 2s^2 2p^5$	(↑ ↓)	(↑ ↓)	(↑ ↓)(↑ ↓)(↑)
Neon	10	$1s^2 2s^2 2p^6$	(↑ ↓)	(↑ ↓)	(↑ ↓)(↑ ↓)(↑ ↓)

Table 6.2
Orbital Diagrams for the Ground States of Atoms from $Z = 1$ to $Z = 10$

orbitals of the same energy, the mutual repulsion of the electrons is as low as possible. The arrangement of electrons shown in Diagram 3, with both electrons in the same $2p$ orbital, would give the atom a somewhat greater energy than that of Diagram 1.

The reason a lower energy is obtained when the electron spins are all in the same direction is more subtle. According to an extension of the Pauli exclusion principle, two electrons with the same spin avoid being in the same region of space. Thus, two electrons of the same spin avoid one another more effectively than do electrons of opposite spins. As a result, electron repulsion is lower when two electrons with the same spin occupy different orbitals than when electrons of opposite spin occupy different orbitals. Thus, the arrangement of electrons in Diagram 1 gives a lower energy than that of Diagram 2.

In the next example, Hund's rule is used to determine the orbital diagram for an atom. Orbital diagrams for the ground states of the first ten elements are shown in Table 6.2.

Example 6.3

Write an orbital diagram for the ground state of the iron atom.

Solution

From the building-up principle, we determine that the electron configuration of the iron atom is $1s^2 2s^2 2p^6 3s^2 3p^6 3d^6 4s^2$. All of the subshells but the $3d$ are filled. In placing the six electrons in the $3d$ subshell, we note that the first five go into separate $3d$ orbitals with their spin arrows in the same direction. The sixth electron must doubly occupy a $3d$ orbital.

The orbital diagram is

(↑ ↓)   (↑ ↓)   (↑ ↓)(↑ ↓)(↑ ↓)   (↑ ↓)
  $1s$ $2s$ $2p$ $3s$

(↑ ↓)(↑ ↓)(↑ ↓)   (↑ ↓)(↑)(↑)(↑)(↑)   (↑ ↓)
      $3p$ $3d$ $4s$

We can write this diagram in abbreviated form using [Ar] for the argon-like core of the iron atom:

[Ar]   (↑ ↓)(↑)(↑)(↑)(↑)   (↑ ↓)
           $3d$ $4s$

Exercise 6.4

Write an orbital diagram for the ground state of the phosphorus atom ($Z = 15$). Write out all orbitals.

(See Problems 6.31 and 6.32.)

Paramagnetism

When two electrons in an atom have opposite spins (they are said to be *paired*), their magnetisms cancel one another. Thus, a doubly occupied orbital does not show magnetism. However, an atom in which there is an excess of one kind of spin does exhibit magnetism. Such an atom (or molecule) is attracted by a magnetic field and is said to be **paramagnetic.**■ An atom (or molecule) with equal numbers of electrons of each kind of spin is not attracted by a magnetic field, but is weakly repelled by it. It is said to be **diamagnetic.**

The electron arrangements of the carbon atom shown in the diagrams given on page 173 exhibit different magnetic characteristics. Diagram 1 (the ground state) corresponds to a paramagnetic state. Diagrams 2 and 3 (excited states) correspond to diamagnetic states.

■ The strong, permanent magnetism seen in iron objects is called *ferromagnetism* and is due to the cooperative alignment of electron spins in many iron atoms. Paramagnetism is a much weaker effect. Nevertheless, paramagnetic substances can be attracted to a strong magnet. Liquid oxygen is composed of paramagnetic O_2 molecules. When poured over a magnet, the liquid is seen to cling to the poles. (See Figure 8.17.)

The Periodic Table

In 1869, the Russian chemist Dmitri Mendeleev (Figure 6.7) and the German chemist Lothar Meyer, working independently, had similar discoveries.

Figure 6.7
Dmitri Ivanovich Mendeleev (1834–1907). Mendeleev constructed a periodic table as part of his effort to systematize chemistry. He received many international honors for his work, but his reception in Czarist Russia was mixed. He had pushed for political reforms and made many enemies as a result.

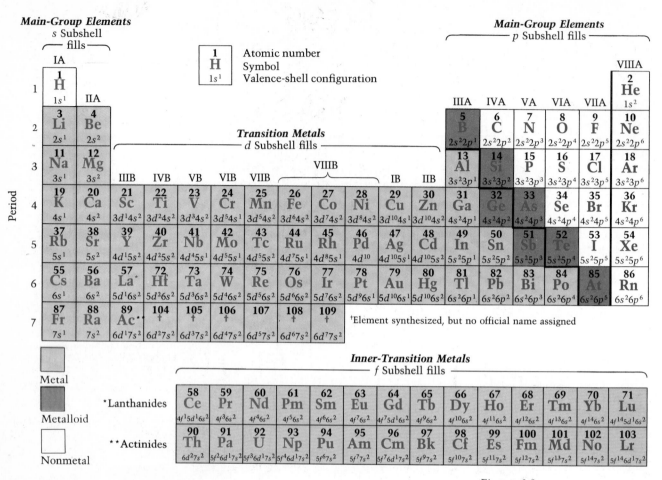

Figure 6.8
A modern form of periodic table. The position of the lanthanides and actinides are indicated in the main body of the table by an asterisk (*) and double asterisk (**), respectively. See also the inside front cover of the text.

They found that if they arranged the elements in order of their atomic weights, they could place them in horizontal rows, one row under the other, so that the elements in any one vertical column had similar properties. This arrangement of elements, highlighting the regular repetition of properties, is called a **periodic table.** The modern periodic table (Figure 6.8) is similar, differing primarily in its arrangement of elements by *atomic number*, rather than by atomic weight.

In the next sections, we will explain the structure of the periodic table and compare some of the properties of the elements in terms of this table.

6.4 Periodic Classification of the Elements

Mendeleev's periodic table was arranged in vertical columns, called *groups* (identified by Roman numerals), and horizontal rows (identified by Arabic numbers). See Figure 6.9. Elements in a given group have similar properties; for example, the elements of Group III (including boron, aluminum, and indium) have oxides with the general formula R_2O_3 (where R stands for a Group III element).

Reihen	Gruppe I. — R_2O	Gruppe II. — RO	Gruppe III. — R_2O_3	Gruppe IV. RH^4 RO_2	Gruppe V. RH^3 R_2O_5	Gruppe VI. RH^2 RO_3	Gruppe VII. RH R_2O_7	Gruppe VIII. — RO_4
1	H = 1							
2	Li = 7	Be = 9,4	B = 11	C = 12	N = 14	O = 16	F = 19	
3	Na = 23	Mg = 24	Al = 27,3	Si = 28	P = 31	S = 32	Cl = 35,5	
4	K = 39	Ca = 40	— = 44	Ti = 48	V = 51	Cr = 52	Mn = 55	Fe = 56, Co = 59, Ni = 59, Cu = 63.
5	(Cu = 63)	Zn = 65	— = 68	— = 72	As = 75	Se = 78	Br = 80	
6	Rb = 85	Sr = 87	?Yt = 88	Zr = 90	Nb = 94	Mo = 96	— = 100	Ru = 104, Rh = 104, Pd = 106, Ag = 108.
7	(Ag = 108)	Cd = 112	In = 113	Sn = 118	Sb = 122	Te = 125	J = 127	
8	Cs = 133	Ba = 137	?Di = 138	?Ce = 140	—	—	—	— — —
9	(—)	—	—		—	—	—	
10	—	—	?Er = 178	?La = 180	Ta = 182	W = 184	—	Os = 195, Ir = 197, Pt = 198, Au = 199.
11	(Au = 199)	Hg = 200	Tl = 204	Pb = 207	Bi = 208	—	—	— — —
12	—	—	—	Th = 231	—	U = 240	—	

Figure 6.9
Mendeleev's periodic table published in 1872.

Predictions from the Periodic Table

Mendeleev left spaces for what he felt were undiscovered elements. There are blank spaces in his row 5, for example, one directly under aluminum and another under silicon. By writing the known elements in this row with their atomic weights, he could determine approximate values (between the known ones) for the missing elements (values in parentheses):

Cu	Zn	—	—	As	Se	Br
63 amu	65 amu	(68 amu)	(72 amu)	75 amu	78 amu	80 amu

The element directly under aluminum, in Group III, Mendeleev called eka-aluminum, with the symbol Ea. (*Eka* is the Sanskrit word meaning "first"; thus, eka-aluminum is the first element under aluminum.) Since the known Group III elements have oxides of the form R_2O_3, Mendeleev predicted that eka-aluminum would have an oxide with the formula Ea_2O_3.

The physical properties of this undiscovered element could be predicted by comparing values for the neighboring known elements. For eka-aluminum Mendeleev predicted a density of 5.9 g/cm^3, a low *melting point* (the temperature at which a substance melts), and a high *boiling point* (the temperature at which a substance boils).

In 1874, the French chemist Lecoq de Boisbaudran found two previously unidentified lines in the atomic spectrum of a sample of sphalerite (a zinc sulfide, ZnS, mineral). Realizing he was on the verge of a discovery, Boisbaudran quickly prepared a large batch of the zinc mineral, from which he isolated a gram of a new element. He called this new element gallium. The properties of gallium were remarkably close to those predicted by Mendeleev for eka-aluminum.

Property	Predicted for Eka-aluminum	Found for Gallium
Atomic weight	68 amu	69.9 amu
Formula of oxide	Ea_2O_3	Ga_2O_3
Density of the element	5.9 g/cm^3	5.91 g/cm^3
Melting point of the element	Low	30.1°C
Boiling point of the element	High	1983°C

The predictive power of Mendeleev's periodic table was demonstrated again when scandium (eka-boron) was discovered in 1879 and when germanium (eka-silicon) was discovered in 1886. Both elements had properties remarkably like those predicted by Mendeleev.

Arrangement of the Elements by Atomic Number

In 1914, Henry G. J. Moseley (1887–1915) found that the elements could be ordered in the periodic table by their atomic numbers, which he determined from their x-ray spectra.■ In several cases, it had been found necessary to invert the order of two elements with respect to their atomic weights in order to place them correctly in the periodic table. Tellurium (atomic weight = 127.6 amu) and iodine (atomic weight = 126.9 amu) were prominent examples. Tellurium certainly belongs below selenium in Group VI, and iodine below bromine in Group VII, according to their chemical properties. Yet, arranging them by atomic weight would place iodine before tellurium. The atomic numbers, however, put tellurium ($Z = 52$) before iodine ($Z = 53$). *Once all elements were ordered by atomic number, such inversions disappeared.*

■ The relationship between x rays and atomic number is discussed in the Aside following Section 6.2.

Relationship to Electron Configurations

Elements are arranged in a modern periodic table (Figure 6.8) by increasing atomic number into horizontal rows called **periods** and vertical columns called **groups** (designated by Roman numerals and an *A* or *B*.) The first period consists of hydrogen ($1s^1$) and helium ($1s^2$), in which the *K*, or $n = 1$, shell is filled. In the second period, from lithium ([He]$2s^1$) to neon ([He]$2s^22p^6$), the *L*, or $n = 2$, shell is filled. In the third period, from sodium ([Ne]$3s^1$) to argon ([Ne]$3s^23p^6$), the 3s and 3p subshells of the *M*, or $n = 3$, shell are filled. Each period ends with the noble-gas element, in which the *p* subshell of the outer shell (the 1s shell in the case of He) has just filled.

The elements in these periods belong to vertical groups labeled *A*. All elements in the *A* groups are called **main-group,** or **representative, elements.** If we compare the configurations of the elements in any one group, we see similarities. For Groups IIA and IIIA, for example, we have

Group IIA		Group IIIA	
Beryllium, Be	[He]$2s^2$	Boron, B	[He]$2s^22p^1$
Magnesium, Mg	[Ne]$3s^2$	Aluminum, Al	[Ne]$3s^23p^1$
Calcium, Ca	[Ar]$4s^2$	Gallium, Ga	[Ar]$3d^{10}4s^24p^1$
Strontium, Sr	[Kr]$5s^2$	Indium, In	[Kr]$4d^{10}5s^25p^1$
Barium, Ba	[Xe]$6s^2$	Thallium, Tl	[Xe]$5d^{10}6s^26p^1$
Radium, Ra	[Rn]$7s^2$		

Each of the Group IIA elements consists of a noble-gas-like core and an ns^2 configuration of outer electrons. Group IIIA elements have an ns^2np^1 configuration of outer electrons, in addition to a noble-gas-like core. Starting with gallium, these elements also have a filled d subshell.

The noble-gas-like core and filled d subshell of Group IIIA elements appear to be relatively undisturbed by chemical reactions, whereas the outer ns and np electrons are directly involved.■ These outer electrons are referred to as **valence electrons,** meaning that they are important in describing the chemical bonding of the atom. It is the similarity of the configurations of valence electrons (the *valence-shell configurations*) that explains the like properties of the elements in a group. For this reason, we might expect barium and radium to behave similarly, which they do, as we saw in the chapter opening.

■ There is direct evidence that the core electrons are relatively undisturbed in most chemical reactions. The evidence comes from x-ray electron spectroscopy, which establishes that the energy required to remove a core electron (for example, a $1s$ electron from C) is almost independent of the compound. In contrast, the outer ns and np electrons are significantly affected depending on the compound.

Beginning with the fourth period, ten columns or vertical groups of elements are inserted between Groups IIA and IIIA (see Figure 6.8). These groups, labeled IIIB to VIIIB (with three columns) and then IB and IIB, are columns of **d-transition elements** or, simply, *transition elements.* In these elements, a d subshell is filling.

Then, in the sixth period after lanthanum $(Z = 57)$, the $4f$ subshell begins to fill, giving 14 **lanthanide elements.** Similarly, in the seventh period after actinium $(Z = 89)$, the $5f$ subshell begins to fill, giving 14 **actinide elements.** These are radioactive, and those beyond uranium $(Z = 92)$ are synthetic, that is, prepared in the laboratory from nuclear reactions.■ The lanthanides and actinides are collectively called **f-transition,** or **inner-transition, elements.** They are placed as separate rows at the bottom of the periodic table.

■ The nuclear reactions for the preparation of some of these synthetic elements are given in Chapter 22.

There are several known transition elements in the seventh period, including Element 109. These elements are all radioactive, having very short half-lives (the time in which one-half the nuclei disintegrate), and have been prepared by nuclear reactions. The period would presumably end with a noble-gas element $(Z = 118)$.

It is possible to determine the configuration of the outer electrons for a main-group (A-column) or transition (B-column) element knowing only its position in the periodic table, that is, its period and group numbers. Consider a main-group element. The period number gives the n quantum number of the outer shell, and the group number (the Roman numeral) gives the number of valence electrons. These electrons are put into the ns and np subshells, following the building-up principle described earlier. It gives a configuration of the form $ns^a np^b$ (b may be zero, and $a + b$ equals the group number).

For the transition elements, the determination of the configuration for the outer electrons from the position of the element in the periodic table is more complex. The period number gives the n quantum number of the highest-energy s orbital. The group number (Roman numeral) for Groups IIIB to VIIB gives the number of electrons in the ns orbital and in the $(n - 1)d$ subshell, which is filling in these elements. For Groups VIIIB, IB, and IIB elements, it is necessary to count, starting with 8 for the first column of VIIIB elements. Thus, the number of electrons in nickel is 10 (counting 8 for Fe, 9 for Co, and 10 for Ni). Following the building-up principle, the configuration for these electrons is $(n - 1)d^8 ns^2$. The next example illustrates the use of these ideas.

Example 6.4

What are the configurations for the outer electrons of the following atoms? (a) tellurium (Period 5, Group VIA) (b) cobalt (Period 4, Group VIIIB)

Solution

(a) Since tellurium is in Period 5, the subshells used by the outer electrons are $5s$ and $5p$. These subshells are filled by 6 electrons (because tellurium is in Group VIA). The valence-shell configuration is $5s^2 5p^4$.

(b) Since cobalt is in Period 4, the outer subshell is $4s$. Because cobalt is in a B column, it is a d-transition element. The d subshell is $(4-1)d$, or $3d$. The $3d$ and $4s$ subshells are filled with 9 electrons (counting 8 for Fe, then 9 for cobalt). Following the building-up principle, the configuration for the outer electrons is $3d^7 4s^2$.

Exercise 6.5

Arsenic is a Group VA element in the fourth period. From this information, write the valence-shell configuration of arsenic.

(See Problems 6.35, 6.36, 6.37, and 6.38.)

Exercise 6.6

The lead atom has the ground-state configuration $[Xe]4f^{14}5d^{10}6s^2 6p^2$. Knowing this, find the period and group for this element. From its position in the periodic table, would you classify lead as a main-group, d-transition, or inner-transition element?

(See Problems 6.39 and 6.40.)

6.5 Some Periodic Properties

The electron configurations of the atoms display a *periodic* variation with increasing atomic number (nuclear charge). That is, the pattern of configurations repeats after each period. As a result, the elements show periodic variations of physical and chemical behavior. We can state this **periodic law** thus: when the elements are arranged by atomic number, their physical and chemical properties vary periodically. In this section, we will look at three physical properties of an atom: atomic radius, ionization energy, and electron affinity. These three quantities, especially ionization energy and electron affinity, are important in discussions of chemical bonding, as we will see in Chapter 7.

Atomic Radius

An atom does not have a definite size, since the statistical distribution of electrons does not abruptly end but merely decreases to very small values as the distance from the nucleus increases. This can be seen in the plot of the electron distribution for the argon atom, shown in Figure 6.10. Consequently, atomic size must be defined in a somewhat arbitrary manner. The atomic radii plotted in Figure 6.11 are based on the distance from the nucleus to the position of the maximum in the electron distribution of the outermost orbital for each atom. (Other measures of atomic radii in chemically bonded situations will be discussed later.■)

■ One measure of atomic radius is obtained from metals. A *metal atom radius* is defined as one-half the distance between two nuclei in a crystal of the metal. Values of metal atom radii differ from the values of atomic radii described in the text, but both exhibit similar periodic trends.

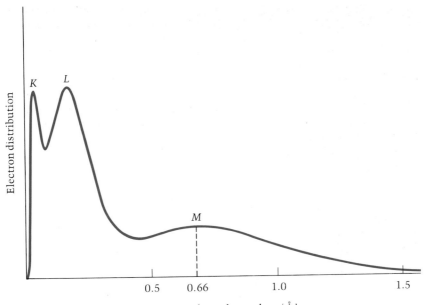

Figure 6.10
Electron distribution for the argon atom. The distribution shows three maxima, for the *K*, *L*, and *M* shells. The outermost maximum occurs at 0.66 Å (66 pm). After this maximum, the distribution falls steadily, becoming negligibly small after several angstroms.

Figure 6.11
A plot of atomic radii versus atomic number. Note that the curve is periodic (tends to repeat). Each period of elements begins with the Group IA atom (Li, Na, K, Rb, Cs, Fr), and the atomic radius tends to decrease until the Group VIIIA atom (He, Ne, Ar, Kr, Xe, Rn). (Data are from J. T. Waber and Don T. Cromer, "Orbital Radii of Atoms and Ions," *Journal of Chemical Physics* 42 (1965), 4116.)

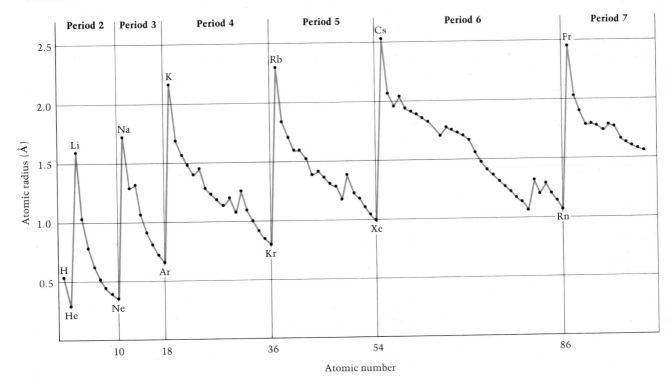

Looking at Figure 6.11 (also see Figure 6.12), we see the following general trends in atomic radius.

1. Within each period (horizontal row), the atomic radius tends to decrease with increasing atomic number (nuclear charge). Thus, the largest atom in a period is a Group IA element and the smallest atom is a noble-gas element.
2. Within each group (vertical column), the atomic radius tends to increase with the period number.

There is a large increase in atomic radius in going from any noble-gas element to the following Group IA element, giving the curve in Figure 6.11 a saw-tooth appearance.

These general trends in atomic radius can be explained if we look at the two factors that primarily determine the size of the outermost orbital. One of these is the principal quantum number, n, of the orbital; the larger n is, the larger the size of the orbital. The other is the **effective nuclear charge** acting on an electron in the orbital. Increasing the effective nuclear charge reduces the size of the orbital. The effective nuclear charge felt by an electron equals the nuclear charge less any shielding of this charge by intervening electrons. An electron in the core of an atom shields an outer electron from approximately $1e$ of nuclear charge (where $+e$ equals the charge on a single proton). For example, the effective nuclear charge felt by the $2s$ electron in the lithium atom (configuration $1s^2 2s^1$) is approximately $+3e - 2e = +1e$ (the nuclear charge $+3e$ minus e for the shielding from each of the two $1s$ electrons). The effective nuclear charge will be roughly equal to e times the number of valence electrons.

Consider a given period of elements. The principal quantum number of the outer orbitals remains constant. However, the effective nuclear charge increases, since the nuclear charge increases while the number of core electrons remains constant. Consequently, the size of the outermost orbital and thus the radius of the atom decreases with Z in any period.

Now consider a given column of elements. The effective nuclear charge remains nearly constant (approximately equal to e times the number of valence electrons), but n gets larger. Therefore, the atomic radius increases.∎

■ The sixth period has an additional 14 elements (the lanthanides) not present in preceding periods. There is an increase in effective nuclear charge and a corresponding decrease in atomic size throughout the lanthanides (called the *lanthanide contraction*). As a consequence, the elements just following the lanthanides have atomic sizes comparable to or less than those above them in the periodic table.

Example 6.5

Refer to a periodic table and use the trends noted for atomic radii to arrange the following in order of increasing atomic radius: Al, C, Si.

Solution

Note that C is above Si in Group IVA. Therefore, the radius of C is smaller than that of Si (the atomic radius increases in going down a group of elements). Note that Al and Si are in the same period. Therefore, the radius of Si is smaller than that of Al (radius decreases with Z in a period). Hence, the order of elements by increasing radius is C, Si, Al.

Exercise 6.7

Using a periodic table, arrange the following in order of increasing atomic radius: Na, Be, Mg.

(See Problems 6.41 and 6.42.)

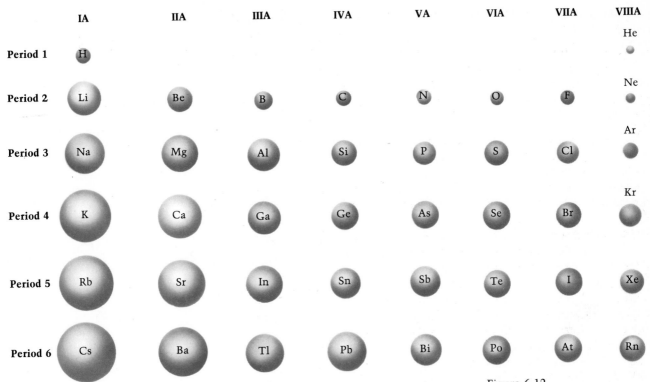

Figure 6.12
Representation of atomic radii of the main-group elements. Note the trends within each period and each group.

Ionization Energy

The **first ionization energy** or *ionization potential* of an atom is the minimum energy needed to remove the highest-energy (that is, the outermost) electron from the neutral atom in the gaseous state. (When the unqualified term *ionization energy* is used, it usually means first ionization energy.) For the lithium atom, the first ionization energy is the energy needed for the following process, called *ionization* (with electron configurations in parentheses):

$$\text{Li } (1s^2 2s^1) \longrightarrow \text{Li}^+ (1s^2) + e^-$$

Values of this energy are usually quoted for one mole of atoms (6.02×10^{23} atoms). The energy necessary to *ionize* (remove an electron from) the lithium atom is 521 kJ/mol. ■

Ionization energies display a periodic variation when plotted against atomic number, as Figure 6.13 shows. Within any period, values tend to increase with atomic number. Thus, the lowest values in a period are found for the Group IA elements (*alkali metals*). It is characteristic of reactive metals such as these to lose electrons easily. The largest ionization energies in any period occur for the noble-gas elements, which is a reflection of the stability of the noble-gas configuration. Because of this stability, these elements are rather unreactive.

This general trend—increasing ionization energy with atomic number in a given period—can be explained as follows. The energy needed to remove an electron from the outer shell is proportional to the effective nuclear

■ Ionization energies are often measured in electron volts (eV). This is the amount of energy imparted to an electron when it is accelerated through an electrical potential of one volt. One electron volt is approximately 100 kJ/mol (1 eV = 96.5 kJ/mol).

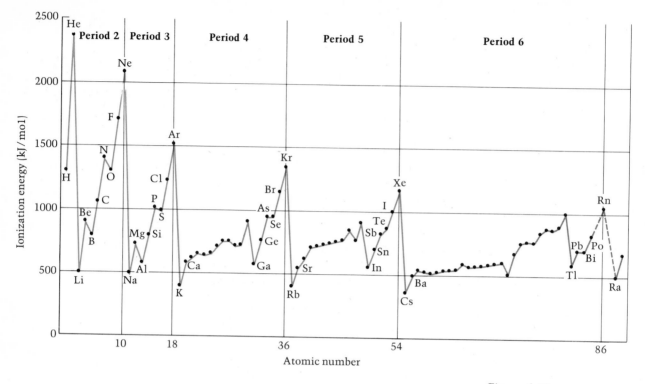

Figure 6.13
A plot of ionization energies versus atomic number. Note that the values tend to increase within each period.

charge divided by the average distance between electron and nucleus (which is inversely proportional to the effective nuclear charge). Hence, the ionization energy is proportional to the square of the effective nuclear charge and increases in going across a period.

Small discrepancies from this general trend can be seen. A IIIA element (ns^2np^1) has smaller ionization energy than the preceding IIA element (ns^2). Apparently the np electron of the IIIA element is more easily removed than one of the ns electrons of the preceding IIA element. Also note that a VIA element (ns^2np^4) has smaller ionization energy than the preceding VA element. As a result of electron repulsion, it is easier to remove an electron from the doubly occupied np orbital of the VIA element than an electron from a singly occupied orbital of the preceding VA element.

Ionization energies tend to decrease in going down any column of main-group elements. This results from the increasing atomic size in going down the column.

Example 6.6

Using a periodic table only, arrange the following elements in order of increasing ionization energy: Ar, Se, S.

Solution

Note that Se is below S in Group VIA. Therefore, the ionization energy of Se should be less than that of S.

Also, S and Ar are in the same period, with Z increasing from S to Ar. Therefore, the ionization energy of S should be less than that of Ar. Hence, the order is Se, S, Ar.

Exercise 6.8

The first ionization energy of the chlorine atom is 1251 kJ/mol. Without looking at Figure 6.13, state which of the following values would be the more likely ionization energy for the iodine atom. Explain.   (a) 1000 kJ/mol   (b) 1400 kJ/mol

(See Problems 6.43 and 6.44.)

The electrons of an atom can be removed successively. The energies required at each step are known as the *first* ionization energy, the *second* ionization energy, and so forth. Table 6.3 lists the successive ionization energies of the first ten elements. Note that the ionization energies for a given element increase as more electrons are removed. Thus, the first and second ionization energies of beryllium (electron configuration $1s^2\,2s^2$) are 899 kJ/mol and 1757 kJ/mol, respectively. The first ionization energy is the energy needed to remove a $2s$ electron from the Be atom. The second ionization energy is the energy needed to remove a $2s$ electron from the positive ion Be^+. Its value is greater than that of the first ionization energy because the electron is being removed from a positive ion, which strongly attracts the electron.

Note that there is a large jump in value from the second ionization energy (1757 kJ/mol) to the third ionization energy (14,848 kJ/mol). The second ionization energy corresponds to removing a valence electron (from the $2s$ orbital), which is relatively easy. The third ionization energy corresponds to removing an electron from the core of the atom (from the $1s$ orbital), that is, from a noble-gas configuration ($1s^2$). A line in Table 6.3 separates the energies needed to remove valence electrons from those needed to remove core electrons. For each element, a large increase in ionization energy occurs when crossing this line. The large increase results from the fact that once the valence electrons are removed, a stable noble-gas configuration is obtained. Further ionizations become much more difficult. We will see in Chapter 7 that metals often form compounds by losing valence electrons. However, core electrons are not involved in compound formation.

Table 6.3
Successive Ionization Energies of the First Ten Elements (kJ/mol)*

	First	Second	Third	Fourth	Fifth	Sixth	Seventh
H	1312						
He	2372	5250					
Li	520	7298	11,815				
Be	899	1757	14,848	21,006			
B	801	2427	3660	25,025	32,826		
C	1086	2353	4620	6222	37,829	47,276	
N	1402	2857	4578	7475	9445	53,265	64,358
O	1314	3388	5300	7469	10,989	13,326	71,333
F	1681	3374	6020	8407	11,022	15,164	17,867
Ne	2081	3952	6122	9370	12,177	15,238	19,998

*Ionization energies beyond the staircase line correspond to removal of electrons from the core of the atom.

Electron Affinity

When any neutral atom in the gaseous state picks up an electron to form a stable negative ion, energy is released. For example, a chlorine atom can pick up an electron to give a chloride ion, Cl^-, and 349 kJ/mol of energy are released. We write the process as follows (with electron configurations in parentheses):

$$Cl\ ([Ne]3s^23p^5)\ +\ e^-\ \longrightarrow\ Cl^-\ ([Ne]3s^23p^6)$$

In some atoms, the negative ion formed is unstable. For these atoms, energy is needed (rather than released) to add the electron to the atom.

The **electron affinity** is the energy involved when an electron is added to a neutral atom in the gaseous state to form a negative ion. When a stable negative ion is formed, the energy released is given as a negative number. When an unstable ion is formed, the energy needed is given as a positive number. Thus, the electron affinity of Cl is -349 kJ/mol; the electron affinity of He is $+21$ kJ/mol. Electron affinities of the main-group elements are given in Table 6.4.

If we plot electron affinities versus atomic number, we see a definite periodic variation (see Figure 6.14). Group VIIA elements (F, Cl, Br, I, and At), with valence-shell configurations ns^2np^5, have large negative electron affinities. By picking up an electron, they form very stable negative ions with noble-gas-like configurations. Group VIA elements (O, S, Se, Te, and Po) also have large negative electron affinities.

Figure 6.14
A plot of electron affinities versus atomic number. Negative values indicate the formation of a stable negative ion. Positive values indicate formation of an unstable negative ion. Elements indicated by open circles at zero indicate elements that form unstable negative ions. Their electron affinities are probably negative, but their exact values are uncertain.

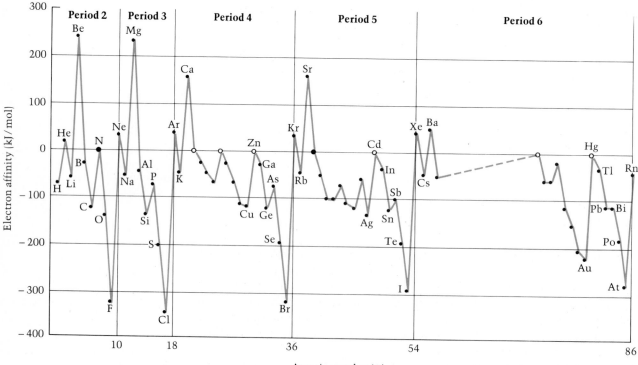

Table 6.4
Electron Affinities of the
Main-Group Elements
(kJ/mol)

	IA	IIA	IIIA	IVA	VA	VIA	VIIA	VIIIA
Period 1	H -73							He 21
Period 2	Li -60	Be 240	B -27	C -122	N 0	O -141	F -328	Ne 29
Period 3	Na -53	Mg 230	Al -44	Si -134	P -72	S -200	Cl -349	Ar 35
Period 4	K -48	Ca 156	Ga -30	Ge -120	As -77	Se -195	Br -325	Kr 39
Period 5	Rb -47	Sr 168	In -30	Sn -121	Sb -101	Te -190	I -295	Xe 41
Period 6	Cs -45	Ba 52	Tl -30	Pb -110	Bi -110	Po -180	At -270	Rn 41

Group IVA elements (C, Si, Ge, Sn, and Pb) have relatively large negative electron affinities as a result of the formation of a stable half-filled p subshell. For example, carbon ($[He]2s^2 2p^2$) gives the C^- ion ($[He]2s^2 2p^3$). On the other hand, Group VA elements, N, P, As, and Sb, with half-filled p subshells, have electron affinities that are higher than those of the preceding Group IVA elements. For example, nitrogen ($[He]2s^2 2p^3$) has a zero or perhaps small positive electron affinity, indicating an unstable negative ion.

Hydrogen ($1s^1$), Group IA elements (valence-shell configurations ns^1), and Group IIIA elements (valence-shell configurations $ns^2 np^1$) have small negative electron affinities. Elements with filled subshells (the noble gases, Group IIA and IIB elements) have positive electron affinities.

Exercise 6.9

Without looking at Figure 6.14, do you expect the electron affinity of barium to be positive or negative? Explain.

(See Problems 6.45 and 6.46.)

6.6 A Brief Description of the Main-Group Elements

The elements of the periodic table (Figure 6.8) are divided by a heavy "staircase" line into the metals on the left and the nonmetals on the right. **Metals** have a characteristic luster or shine and are generally good conductors of heat and electricity. Except for mercury, the metallic elements are solids at room temperature. They are more or less *malleable* (can be hammered into sheets) and *ductile* (can be drawn into wire). The ionization energies of the metallic elements are relatively low. ∎

Most of the **nonmetals** are gases (for example, nitrogen, N_2, and oxygen,

■ First ionization energies for the atoms of metallic elements vary from 380 kJ/mol for cesium to 1010 kJ/mol for mercury. Values for the nonmetals (excluding the noble gases) vary from 800 kJ/mol for boron to 1250 kJ/mol for fluorine.

O_2) or solids (for example, phosphorus, P_4, and sulfur, S_8). The solid non-metals are usually hard, brittle substances. Bromine, Br_2, is the only liquid nonmetal. Nonmetals have relatively high ionization energies and high electron affinities.

Elements bordering the staircase line in Figure 6.8 often have both metallic and nonmetallic properties. These elements, such as germanium (Ge), are called **metalloids** or **semimetals.**■

Going down a column of the periodic table, the elements normally show an increase in metallic characteristics. This is seen clearly in the Group IVA elements. The column starts with carbon, a nonmetal, and goes to germanium, a metalloid, and then to lead, a metal.

We will now look briefly at the properties of the main-group elements. As you read these descriptions, note the change in elements from nonmetallic (or semimetallic) to metallic characteristics in going down any column of Groups IIIA to VIIA. Also, note those groups of elements that are particularly reactive or unreactive. And note the formulas of oxides formed by groups of elements. Many of these oxides are common compounds. Some binary compounds of hydrogen and elements in Groups VIA and VIIA are also described.

■ The metalloids have properties that make them important in constructing solid-state electronic devices, such as transistors and solar cells.

Hydrogen: Configuration $1s^1$

Hydrogen is usually placed in a class by itself. Although its electron configuration is like that of the elements in Group IA, its chemical properties are quite different. The element hydrogen is a colorless, odorless gas composed of H_2 molecules. The gas burns rapidly in oxygen to produce water

$$2H_2 + O_2 \longrightarrow 2H_2O$$

Hydrogen also forms explosive mixtures with air.

Group IA Elements (Alkali Metals): Valence-Shell Configuration ns^1

Group IA elements (including sodium and potassium) are soft, reactive metals, usually referred to as **alkali metals.** They react violently with water, producing hydrogen gas. The reaction of sodium with water is

$$2Na + 2H_2O \longrightarrow 2NaOH + H_2$$

So much heat is released by this reaction that the hydrogen may be ignited, possibly causing an explosion. Because of their reactivity with air and moisture, these metals are stored under oil or kerosene. Alkali metals form oxides (compounds with oxygen) with the general formula R_2O.

Group IIA Elements (Alkaline Earth Metals): Valence-Shell Configuration ns^2

Group IIA elements—the **alkaline earth metals**—also are reactive metals. Like Group IA elements, calcium and the metals below it in Group IIA react with water, but much less vigorously:

$$Ca + 2H_2O \longrightarrow Ca(OH)_2 + H_2$$

Magnesium reacts slowly with hot water, and beryllium shows no reaction.

Calcium and the metals below it are too soft and reactive to use in structural applications (for building and fabricating). Magnesium is strong enough to use in the aircraft and missile industries, where its lightness is an advantage. It is also used to make ladders. Beryllium is an even harder metal. These elements form oxides with the general formula RO.

Group IIIA Elements: Valence-Shell Configuration ns^2np^1

The first element of Group IIIA, boron, B, is a metalloid with several allotropic forms. (**Allotropes** are forms of an element with different properties.) In one well-characterized form, boron is a black, crystalline substance with a metallic luster. Crystalline boron is very hard and brittle, and relatively unreactive.

Other elements in this group are metals. Aluminum, Al, is a common structural material. The next element in the group is gallium, Ga. Gallium melts at 30°C, so that it liquefies if placed in the hand (body temperature is 37°C); it boils, however, at 1983°C. Group IIIA elements form oxides of the general formula R_2O_3.

Group IVA Elements: Valence-Shell Configuration ns^2np^2

As mentioned earlier, Group IVA elements show a distinct trend from nonmetal (carbon, C) to metalloid (silicon, Si, and germanium, Ge) to metal (tin, Sn, and lead, Pb). Carbon exists in two well-known allotropic forms: graphite, a soft, black substance used in pencil leads, and diamond, a very hard, clear, crystalline substance. Both tin and lead are metals that were known to the ancients. Tin is mixed, or alloyed, with copper to make bronze.■

These elements form oxides of general formula RO_2. Examples are CO_2, a gas used to make Dry Ice (solid CO_2) and carbonated beverages; SiO_2, a solid that exists as quartz and white sand (particles of quartz); and SnO_2, the mineral cassiterite, principal ore of tin. Lead forms the dioxide, PbO_2, but the monoxide, PbO, is more stable. Carbon also forms a stable monoxide, CO.

■ Bronze, one of the first *alloys* (metallic mixtures) used in history, contains about 90% copper and 10% tin. It melts more easily than copper, but is much harder.

Group VA Elements: Valence-Shell Configuration ns^2np^3

Group VA elements also show the transition from nonmetal (nitrogen, N, and phosphorus, P) to metalloid (arsenic, As, and antimony, Sb) to metal (bismuth, Bi). Nitrogen occurs as a colorless, odorless gas, with N_2 molecules. It is a relatively unreactive element. White phosphorus is a white, waxy solid, with P_4 molecules. It is quite reactive and is normally stored under water. In air it bursts into flame, reacting with O_2 to give the oxides P_4O_{10} and P_4O_6. There are other allotropic forms, red and black phosphorus, that are much less reactive. Gray arsenic (the common form of the element) is a brittle solid with metallic luster. Yellow arsenic is an unstable, crystalline solid composed of As_4 molecules. Antimony is a brittle solid with a silvery, metallic luster. Bismuth is a hard, lustrous metal with pinkish tinge.

These elements form oxides with empirical formulas R_2O_5 and R_2O_3. In some cases, the molecular formulas are twice these formulas, that is, R_4O_{10} and R_4O_6. Nitrogen has the oxides N_2O_5 and N_2O_3, although it has other, better known oxides. Nitric oxide, NO, is a gas produced when nitrogen reacts with oxygen at high temperature (such as in a flame). This oxide combines readily with more oxygen to give nitrogen dioxide, NO_2, a brown gas.

$$N_2 + O_2 \longrightarrow 2NO$$
$$2NO + O_2 \longrightarrow 2NO_2$$

Phosphorus, as we noted earlier, forms the oxides P_4O_{10} and P_4O_6 when burned in air.

Group VIA Elements: Valence-Shell Configuration ns^2np^4

These elements show the transition from nonmetal (oxygen, O, sulfur, S, and selenium, Se) to metalloid (tellurium, Te) and metal (polonium, Po). Oxygen occurs as a colorless, odorless gas, with O_2 molecules. It also has an allotrope, ozone, of molecular formula O_3. Sulfur is a brittle, yellow solid, with molecular formula S_8. It burns in air to produce the gas sulfur dioxide, SO_2, and traces of sulfur trioxide, SO_3. Selenium exists in two allotropic forms, red and gray. Red selenium is a solid composed of Se_8 molecules. Gray selenium is a brittle solid that becomes electrically conductive when light shines on it. (It has been used to construct cells for measuring light intensity.) Selenium forms the oxide SeO_2 and SeO_3, though the dioxide is more stable. Selenium dioxide is a white solid that forms when selenium is burned in air. Tellurium is a white, metallic-looking substance; it is brittle and is a weak electrical conductor. Polonium is a radioactive metal.

Group VIA elements form compounds with hydrogen of the form H_2R, the best known of which is water, H_2O. Hydrogen sulfide, H_2S, is a poisonous gas with the odor of rotten eggs.

Group VIIA Elements (Halogens): Valence-Shell Configuration ns^2np^5

Group VIIA elements (usually called **halogens**) are reactive nonmetals with the general molecular formula X_2, where X symbolizes a halogen. Fluorine, F_2, is a pale yellow gas; chlorine, Cl_2, a pale green gas; bromine, Br_2, a reddish brown liquid; and iodine, I_2, a purplish black solid (it gives off a violet vapor). Little is known about astatine, At, because it is a synthetic element and is radioactive. It is more metallic than iodine and is perhaps a metalloid. Each of the halogens forms several compounds with oxygen, but these are rather unstable. The halogens form compounds with hydrogen of the general formula HX. Hydrogen fluoride is a liquid (b.p. 20°C); the other hydrogen halides are gases. All are corrosive and dissolve in water to give solutions known generally as *hydrohalic acids.* For example, hydrogen chloride (HCl) dissolves in water to give hydrochloric acid, denoted HCl(*aq*), where *aq* means aqueous (water solution).

Group VIIIA Elements (Noble Gases): Valence-Shell Configuration ns^2np^6

Group VIIIA elements exist as gases consisting of uncombined atoms (for example, helium, He). For a long time, these elements were thought to be chemically inert, since no compounds were known. In the early 1960s, several compounds of xenon were formed. Now compounds are also known for krypton and radon. These elements are known as **noble gases** because of their relative unreactivity. They are also called *rare gases* or *inert gases.* ■

■ Interestingly, some molecules involving noble gases exist only in excited electronic states but not in their ground states. When such molecules are composed of two different atoms, they are called exciplexes. Examples are ArF, KrF, XeF, and XeCl. They are used in newly developed lasers that emit ultraviolet light. These lasers are being used to separate isotopes and mixtures of compounds.

Exercise 6.10

(a) Write the formula for a compound of hydrogen and selenium.   (b) The formula of calcium sulfate is $CaSO_4$. Predict the formula of the similar compound of selenium, calcium selenate.

(See Problems 6.47 and 6.48.)

A Checklist for Review

Important Terms

electron configuration (6.1)
orbital diagram (6.1)
Pauli exclusion principle (6.1)
building-up (Aufbau) principle (6.2)
Hund's rule (6.3)
paramagnetic (6.3)
diamagnetic (6.3)
periodic table (p. 176)
periods (of periodic table) (6.4)
groups (of periodic table) (6.4)

main-group (representative) elements (6.4)
valence electrons (6.4)
d-transition elements (6.4)
lanthanide elements (6.4)
actinide elements (6.4)
f-transition (inner-transition) elements (6.4)
periodic law (6.5)
first ionization energy (6.5)
effective nuclear charge (6.5)

electron affinity (6.5)
metals (6.6)
nonmetals (6.6)
metalloids (semimetals) (6.6)
alkali metals (6.6)
alkaline earth metals (6.6)
allotropes (6.6)
halogens (6.6)
noble gases (6.6)

Summary of Facts and Concepts

To understand the similarities that exist among a group of elements, it is necessary to know the *electron configurations* for the ground states of atoms. Only those arrangements of electrons allowed by the *Pauli exclusion principle* are possible. The ground-state configuration of

an atom represents the electron arrangement that has the lowest total energy. This arrangement can be reproduced by the *building-up principle* (Aufbau principle), where electrons fill the subshells in a particular order (the building-up order) consistent with the Pauli exclusion principle. The arrangement of electrons in partially filled subshells is governed by *Hund's rule.*

Elements in the *periodic table* are arranged in rows (*periods*) and columns (*groups*) by atomic number. Those elements in the same group have similar *valence-shell configurations.* As a result, chemical and physical properties of the elements show periodic behavior. *Atomic radius,* for example, tends to decrease across any period (left to right) and increase down any group (top to bottom). *First ionization energies* tend to increase across a period and decrease down a group. *Electron affinities* are greatest for Group VIIA and VIA elements and are positive for those with filled subshells, that is, for Group IIA and IIB elements, and for the noble gases (the negative ions formed are unstable).

Operational Skills

1. Given an orbital diagram or electron configuration, decide whether it is possible or not, according to the Pauli exclusion principle (Example 6.1).

2. Given the atomic number of an atom, write the complete electron configuration for the ground state, according to the building-up principle for neutral atoms (Example 6.2).

3. Given the electron configuration for the ground state of an atom, write the orbital diagram (Example 6.3).

4. Given the period and group for an element, write the configuration of the outer electrons (Example 6.4.)

5. Using the known trends and by referring to a periodic table, arrange a series of elements in order by atomic radius (Example 6.5) or ionization energy (Example 6.6).

Review Questions

6.1 Describe the experiment of Stern and Gerlach. How are the results for the hydrogen atom explained?

6.2 Describe the model given in the text of electron spin. What are the restrictions on electron spin?

6.3 How does the Pauli exclusion principle limit the possible configurations of an atom?

6.4 What is the maximum number of electrons that can occupy a *g* subshell ($l = 4$)?

6.5 List the orbitals in order of increasing orbital energy up to and including the $3p$ orbital.

6.6 Give two different orbital diagrams for the $1s^2 2s^2 2p^4$ configuration of the oxygen atom, one of which should correspond to the ground state. Label the diagram for the ground state.

6.7 Define the terms *diamagnetic* and *paramagnetic.* Is the ground state of the oxygen atom diamagnetic or paramagnetic? Explain.

6.8 How was Mendeleev able to predict the properties of gallium before it was discovered?

6.9 What was Moseley's contribution to the modern form of periodic table?

6.10 What kind of subshell is being filled in Groups IA and IIA? In Groups IIIA to VIIIA? In the transition elements? In the lanthanides and actinides?

6.11 Describe the major trends seen when atomic radii are plotted versus atomic number. Describe the trends when first ionization energies are plotted versus atomic number.

6.12 What group in the periodic table has elements with the most negative electron affinities for each period? What electron configurations of neutral atoms give unstable negative ions?

6.13 Give some properties that are characteristic of a metal.

6.14 List the elements in Groups IIIA to VIA in the same order as in the periodic table. Label each element as a metal, metalloid, or nonmetal. Does each column of elements display the expected trend of increasing metallic characteristics?

6.15 Write an equation for the reaction of potassium metal with water.

6.16 From what is said in Section 6.6 about Group IIA elements, list some properties of barium.

6.17 Name and describe the two allotropes of carbon.

6.18 Match each description in the left column with the appropriate element in the right column.

(a) A waxy, white solid, normally stored under water   Sulfur

Sodium

(b) A yellow solid that burns in air   White phosphorus

(c) A reddish brown liquid   Bromine

(d) A soft, light metal that reacts vigorously with water

Problems

Pauli Exclusion Principle

6.19 Which of the following orbital diagrams are allowed and which are not allowed by the Pauli exclusion principle? Explain. For those that are allowed, write the electron configuration.

(a) (↑ ↓) (↑ ↓) (↑ ↑)(↑   )(↑   )
    1s   2s   2p

(b) (↑ ↓) (↑ ↓) (↑   )(↑   )(   )
    1s   2s   2p

(c) (↑ ↓) (↑   ) (↑ ↓ ↑)(   )(   )
    1s   2s   2p

(d) (↑ ↓) (↑ ↓) (↑ ↓)(↑   )(↑   )
    1s   2s   2p

6.20 Which of the following orbital diagrams are allowed by the Pauli exclusion principle? Explain how you arrived at this decision. Give the electron configuration for the allowed ones.

(a) (↑ ↓) (↑ ↑) (↓   )(↓   )(   )
    1s   2s   2p

(b) (↑ ↓) (↑ ↓ ↑) (↑ ↓ ↓)(   )(↓   )
    1s   2s   2p

(c) (↑ ↓) (↑ ↓) (↑   )(↑   )(↑   )
    1s   2s   2p

(d) (↑ ↓) (↑ ↓) (↑ ↓)(↑ ↑)(↓   )
    1s   2s   2p

6.21 Choose the electron configurations that are possible from the following. Explain why the others are impossible.

(a) $1s^1 2s^3 2p^5$ (b) $1s^2 2s^2 2p^5$
(c) $1s^2 2s^2 2p^6 3s^2 3p^7 3d^9$ (d) $1s^2 2s^2 2p^6 3s^2 3p^5 3d^{10}$

6.22 Which of the following electron configurations are possible ones? Explain why the others are not.
(a) $1s^2 2s^2 2p^7$ (b) $1s^2 2s^2 3d^{11}$
(c) $1s^1 2s^2 2p^5$ (d) $1s^2 2s^2 2p^6 3s^2 3p^6 3d^{12}$

6.23 Write out all of the possible orbital diagrams for the electron configuration $1s^2 2p^1$. (There are six different diagrams.)

6.24 Give the different orbital diagrams for the configuration $1s^1 2p^1$. (There are twelve different diagrams.)

Building-up Principle and Hund's Rule

6.25 Use the building-up principle to obtain the ground-state configuration for aluminum ($Z = 13$).

6.26 Give the electron configuration of the ground state of sulfur ($Z = 16$), using the building-up principle.

6.27 Use the building-up principle to obtain the electron configuration of the ground state of titanium ($Z = 22$).

6.28 Give the electron configuration for the ground state of germanium ($Z = 32$), using the building-up principle.

6.29 From the building-up principle, find the configuration of gold ($Z = 79$). Assuming that the tendency to half-filled and filled subshells operates here, what is the actual configuration for the ground state of the gold atom?

6.30 Use the building-up principle, together with the idea that there is a tendency toward half-filled and filled subshells, to write the electron configuration for the ground state of gadolinium ($Z = 64$).

6.31 Write the orbital diagram for the ground state of cobalt. The configuration is $[Ar]3d^7 4s^2$.

6.32 Write the orbital diagram for the ground state of cerium. The configuration is $[Xe]4f^1 5d^1 6s^2$.

6.33 Write an orbital diagram for the ground state of the magnesium atom ($Z = 12$). Is the atom diamagnetic or paramagnetic?

6.34 Write an orbital diagram for the ground state of the sulfur atom ($Z = 16$). Is the atom diamagnetic or paramagnetic?

Periodic Table

6.35 Indium is a Group IIIA element in Period 5. From this, deduce the valence-shell configuration of indium.

6.36 Antimony is a Group VA element in Period 5. With this information, write the valence-shell configuration of antimony.

6.37 Rhenium is a Group VIIB element in Period 6. What would you expect to be the configuration of outer electrons of rhenium, from its position in the periodic table?

6.39 Thallium has a ground-state configuration $[Xe]4f^{14}5d^{10}6s^26p^1$. Give the group and period for this element. Classify it as a main-group, transition, or inner-transition element.

6.38 Mercury is a Group IIB element in Period 6. Given its position in the periodic table, what would you give for the valence-shell configuration of mercury?

6.40 The configuration for the ground state of iridium is $[Xe]4f^{14}5d^76s^2$. What is the group and period for this element? Is it a main-group, d-transition, or f-transition element?

Periodic Trends

6.41 From your knowledge of periodic trends, arrange the following elements in order of increasing atomic radius: O, P, S.

6.43 Arrange the following elements in order of increasing ionization energy: Al, Ga, Si. Do not look at Figure 6.13.

6.45 From what you know in a general way about electron affinities, describe the following atoms as having a small positive or negative value or a large negative value of the electron affinity: Kr, Cd, S, F, Sr, P.

6.47 Write the simplest formulas expected for two oxides of arsenic.

6.42 Order the following elements by increasing atomic radius according to what you expect from periodic trends: Br, Cl, Se.

6.44 From your knowledge of periodic trends, arrange the following elements by increasing ionization energy: Ne, Li, F, B.

6.46 From your general knowledge, classify the following atoms as having a small positive or negative value or a large negative value of the electron affinity: As, Hg, Cl, O, Ar, Mg.

6.48 If potassium perchlorate has the formula $KClO_4$, what formula might be expected for sodium perbromate?

Additional Problems

6.49 For eka-lead, predict the electron configuration, whether it is a metal or nonmetal, and the formula of an oxide.

6.50 For eka-bismuth, predict the electron configuration, whether the element is a metal or nonmetal, and the formula of an oxide.

*__6.51__ From Figure 6.13, predict the first ionization energy of francium ($Z = 87$).

*__6.52__ From Figure 6.13, predict the first ionization energy of astatine ($Z = 85$).

6.53 Find the electron configuration of the element with $Z = 23$. From this, give its group and period in the periodic table. Classify the element as main-group, d-transition, or inner-transition.

6.54 Find the electron configuration for the element with $Z = 33$. From this, give its group and period in the periodic table. Is this a main-group, d-transition, or inner-transition element?

6.55 Write the orbital diagram corresponding to the ground state of Nb, whose configuration is $[Kr]4d^45s^1$.

6.56 Write the orbital diagram for the ground state of ruthenium. The configuration is $[Kr]4d^75s^1$.

6.57 Match the characteristics on the left with an element given in the column at the right.

   (a) A reactive nonmetal; the atom has a high electron affinity.   Sodium (Na)

   (b) A soft metal; the atom has low ionization energy.   Antimony (Sb)

   (c) A metalloid that forms an oxide of formula R_2O_3 Argon (Ar)

   (d) A chemically unreactive gas   Chlorine (Cl_2)

6.58 Match each of the elements on the right with a set of characteristics on the left.

   (a) A reactive, pale yellow gas; the atom has high electron affinity.   Oxygen (O_2)

   (b) A soft metal that reacts with water to produce hydrogen   Gallium (Ga)

   (c) A metal that forms an oxide of formula R_2O_3 Barium (Ba)

   (d) A colorless gas; the atom has moderately large negative electron affinity.   Fluorine (F_2)

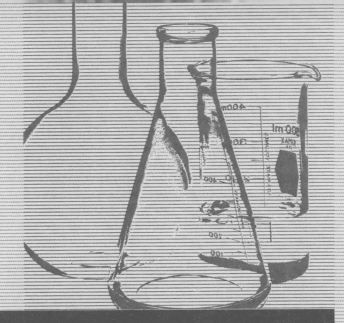

7. *Ionic and Covalent Bonding*

Ionic Bond

7.1 Describing Ionic Bonds Lewis Electron-Dot Symbols/ Energy Involved in Ionic Bonding
7.2 Some Common Ions Monatomic Ions of the Main-Group Elements/ Transition-Metal Ions/ Polyatomic Ions/ Formulas of Ionic Compounds
7.3 Ionic Radii

Covalent Bond

7.4 Describing Covalent Bonds Lewis Formulas/ Coordinate Covalent Bond/ Octet Rule/ Multiple Bonds
7.5 Polar Covalent Bond; Electronegativity
7.6 Writing Lewis Electron-Dot Formulas Skeleton Structure of a Molecule/ Steps in Writing Lewis Formulas
7.7 Exceptions to the Octet Rule
7.8 Delocalized Bonding; Resonance
7.9 Bond Length and Bond Order

Chemical Nomenclature

7.10 Oxidation Numbers
7.11 Naming Simple Compounds Binary Compounds/ Acids/ Ionic Substances

The properties of a substance are determined in part by the chemical bonds that hold the atoms together. A *chemical bond* is a strong attractive force that exists between certain atoms in a substance. In Chapter 2, we described how sodium (a silvery metal) reacts with chlorine (a pale green gas) to produce sodium chloride (table salt, a white solid). Each of the substances in this reaction is quite different, as are their chemical bonds. Sodium chloride, NaCl, consists of Na^+ and Cl^- ions held in a regular arrangement or crystal by *ionic bonds*. Ionic bonding results from the attractive force of oppositely charged ions.

A second kind of chemical bond is a *covalent bond*. In a covalent bond, two atoms share valence electrons (outer-shell electrons), which are attracted to the positively charged cores of both atoms, thus linking them. For example, chlorine gas consists of Cl_2 molecules. We would not expect the two Cl atoms in each Cl_2 molecule to acquire the opposite charges required for ionic bonding. Rather, a covalent bond holds the two atoms together. This is consistent with the equal sharing of electrons that we would expect between identical atoms. In most molecules, the atoms are linked by covalent bonds.

Metallic bonding, seen in sodium and other metals, represents another important type of bonding. A crystal of sodium metal consists of a regular arrangement of sodium atoms. The valence electrons of these atoms move throughout the crystal, attracted to the positive cores of all Na^+ ions. It is this attraction that holds the crystal together.

What determines the type of bonding in each substance? How do we describe the bonding in various substances? In this chapter, we will look at some simple, useful concepts of bonding that can help us answer these questions. We will be concerned with ionic and covalent bonds in particular.

Chapter Overview

The chapter begins with a description of the *ionic bonds* between atoms. We then look at the common ions encountered in compounds. The second part of the chapter describes how *covalent bonds* form between atoms. Following this, we discuss the bonding in molecules in terms of *Lewis electron-dot formulas*, which are simple representations of the valence-shell electrons. The last section of the chapter discusses *chemical nomenclature*, the systematic naming of compounds.

Ionic Bond

Chemical bonds hold different atoms together to give a great variety of known substances. What is the nature of the chemical bond? The first explanation of chemical bonding was suggested by the properties of *salts*, substances now known to be ionic. Salts are generally crystalline solids that melt at high temperatures. Sodium chloride, for example, melts at 801°C. A molten salt (the liquid after melting) conducts an electric current. And a salt dissolved in water gives a solution that also conducts an electric current. The electrical conductivity of the molten salt and the salt solution results from the motion of ions in the liquids.■

■ In 1812, Berzelius proposed that every atom has a particular positive or negative charge, and that compounds form due to the attraction of oppositely charged atoms. This idea became the basic principle of Berzelius's *dualistic theory*, the forerunner of our modern theory of ionic bonding.

7.1 Describing Ionic Bonds

An **ionic bond** corresponds to the electrostatic attraction between positive and negative ions. It forms between two atoms when one or more electrons are transferred from the valence shell of one atom to the valence shell of the other. The atom that loses electrons becomes a *cation* (positive ion), and the atom that gains electrons becomes an *anion* (negative ion). Any given ion tends to attract as many neighboring ions of opposite charge as possible. When large numbers of ions gather together, they form an ionic solid. The solid normally has a regular, crystalline structure that allows for the maximum attraction of ions.

To understand why ionic bonding occurs, consider the transfer of a valence electron from a sodium atom (electron configuration $[Ne]3s^1$) to the valence shell of a chlorine atom ($[Ne]3s^23p^5$). We can represent the electron transfer by the following equation:

$$Na\ ([Ne]3s^1) + Cl\ ([Ne]3s^23p^5) \longrightarrow Na^+\ ([Ne]) + Cl^-\ ([Ne]3s^23p^6)$$

As a result of the electron transfer, ions are formed, each of which has a noble-gas configuration. The sodium ion has lost its $3s$ electron and has taken on the neon configuration, $[Ne]$. The chloride ion has accepted the electron into its $3p$ subshell and has taken on the argon configuration, $[Ne]3s^23p^6$. Such noble-gas configurations and the corresponding ions are particularly stable. This stability of the ions accounts in part for the formation of the ionic solid NaCl. Once each cation or anion forms, it attracts ions of opposite charge. Within the sodium chloride crystal, NaCl, every Na^+ ion is surrounded by six Cl^- ions, and every Cl^- ion by six Na^+ ions. (Figure 2.3 on page 33 shows the arrangement of ions in the NaCl crystal.)

Lewis Electron-Dot Symbols

The preceding equation for the electron transfer between Na and Cl can be simplified by writing **Lewis electron-dot symbols** for the atoms and *monatomic ions* (ions formed from a single atom). In these symbols, the electrons in the valence shell of the atom or ion are represented by dots placed around the letter symbol of the element. Table 7.1 lists the Lewis symbol and corresponding valence-shell electron configurations for the atoms of the second and third periods. Note that the dots are placed one to each side of the letter symbol until all four sides are occupied. Then the dots

	IA ns^1	IIA ns^2	IIIA ns^2np^1	IVA ns^2np^2	VA ns^2np^3	VIA ns^2np^4	VIIA ns^2np^5	VIIIA ns^2np^6
Second Period	Li·	·Be·	·B·	·C·	:N·	:O·	:F·	:Ne:
Third Period	Na·	·Mg·	·Al·	·Si·	:P·	:S·	:Cl·	:Ar:

Table 7.1
Lewis Electron-Dot Symbols and Valence-Shell Electron Configurations for Atoms of the Second and Third Periods

are written two to a side until all valence electrons are accounted for. (The exact placement of the single dots is immaterial. For example, the single dot in the Lewis symbol for chlorine can be written on any one of the four sides.)

The equation representing the transfer of an electron from the sodium atom to the chlorine atom is written

$$Na\overset{\curvearrowright}{\cdot} + \cdot \ddot{Cl}: \longrightarrow Na^+ + [:\ddot{Cl}:]^-$$

The noble-gas configurations of the ions are apparent from the symbols. No dots are shown for the cation (all valence electrons have been removed, leaving the noble-gas core). There are eight dots shown in brackets for the anion (noble-gas configuration ns^2np^6).

Example 7.1

Use Lewis electron-dot symbols to represent the transfer of electrons from magnesium to fluorine atoms to form ions with noble-gas configurations.

Solution

The Lewis symbols for the atoms are $:\ddot{F}\cdot$ and $\cdot Mg\cdot$ (see Table 7.1). The magnesium atom loses two electrons to assume a noble-gas configuration. But since a fluorine atom can accept only one electron to fill its valence shell, two fluorine atoms must take part in the electron transfer. We can represent this electron transfer as follows:

$$:\ddot{F}\overset{\curvearrowleft}{\cdot} + \cdot Mg\overset{\curvearrowright}{\cdot} + \cdot \ddot{F}: \longrightarrow [:\ddot{F}:]^- + Mg^{2+} + [:\ddot{F}:]^-$$

Exercise 7.1

Represent the transfer of electrons from magnesium to oxygen atoms to assume noble-gas configurations. Use Lewis electron-dot symbols.

(See Problems 7.21 and 7.22.)

Energy Involved in Ionic Bonding

To gain some understanding of why certain atoms bond ionically and others do not, let us look at the energy involved in forming an ionic bond between two atoms. Consider again the formation of an ionic bond between a sodium atom and a chlorine atom. We will think of this as occurring in two steps. (1) An electron is transferred between the two separate atoms to give ions. (2) The ions then attract one another to form an ionic bond. In reality, the transfer of the electron and the formation of an ionic bond occur simultaneously as the atoms approach one another, rather than in discrete steps. But the *net* quantity of energy involved will be the same whether the steps occur one after the other or whether they occur at the same time.

The first step requires removal of the $3s$ electron from the sodium atom and the addition of this electron to the valence shell of the chlorine atom. Removing the electron from the sodium atom requires energy (the first ionization energy of the sodium atom, which equals 496 kJ/mol). Adding the electron to the chlorine atom releases energy (the electron affinity of the chlorine atom, which equals -349 kJ/mol).■ It requires more energy to remove an electron from the sodium atom than is gained when the electron is added to the chlorine atom. That is, the formation of ions from the atoms

■ Ionization energies and electron affinities of atoms were discussed in Section 6.5.

is not in itself energetically favorable. It requires additional energy equal to (496 − 349) kJ/mol, or 147 kJ/mol, to form ions.

However, the formation of an ionic bond (step 2) that results from the attraction of positive and negative ions also releases energy. When a sodium ion, Na^+, and a chloride ion, Cl^-, come together to form a bond, energy equal to 555 kJ/mol is released.■ This is more than enough to supply the additional energy required for the formation of ions in the first step (147 kJ/mol). Furthermore, when many ions come together to form the crystalline solid, even more energy is released.

We see that more energy is released in forming the NaCl crystal from ions than is required to form the ions from atoms. Overall, a net quantity of energy is released. This means that the ionic crystal is at lower energy than the separate atoms and is therefore more stable. Generally, we see that more energy will be released overall if the ionization energy of one atom is small and the electron affinity of the other is large and negative. Atoms with small ionization energy are classified metallic elements, and atoms with large negative electron affinity are classified nonmetallic elements. Therefore, we normally expect an ionic bond to form between a metal atom and a non-metal atom.

■ This energy analysis shows only why we expect sodium atoms and chlorine atoms to bond together. It does not describe the chemical reaction between the elements. Such a reaction involves the breaking of bonds in the elementary sub-stances to form new ones in the product, NaCl. We will look at the energy involved in chemical reactions in Chapter 10.

7.2 Some Common Ions

In the previous section, we described the formation of an ionic bond from atoms. Now we want to see what ions are found in common ionic sub-stances. These ions can be understood in terms of the electronic structure of the atoms.

Monatomic Ions of the Main-Group Elements

Table 7.2 shows common monatomic ions found in compounds of the main-group elements. For most of the metallic elements (those listed to the left of the "staircase" line), the ions have a positive charge corresponding to the loss of all valence electrons. In these instances, the cation charge is therefore equal to the group number. The exceptions will be discussed later.

Once the atoms of these metallic elements have lost their valence elec-trons, they have stable electron configurations. We can see the stability of

Table 7.2
Common Monatomic Ions of the Main-Group Elements*

	IA	IIA	IIIA	IVA	VA	VIA	VIIA
Period 1							H^-
Period 2	Li^+	Be^{2+}	B	C	N^{3-}	O^{2-}	F^-
Period 3	Na^+	Mg^{2+}	Al^{3+}	Si	P	S^{2-}	Cl^-
Period 4	K^+	Ca^{2+}	Ga^{3+}	Ge	As	Se^{2-}	Br^-
Period 5	Rb^+	Sr^{2+}	In^{3+}	Sn^{2+}	Sb	Te^{2-}	I^-
Period 6	Cs^+	Ba^{2+}	Tl^+, Tl^{3+}	Pb^{2+}	Bi^{3+}		

*Elements shown in color do not normally form compounds having monatomic ions.

Element	Successive Ionization Energies			
	First	Second	Third	Fourth
Na	496	4,562	6,912	9,543
Mg	738	1,451	7,733	10,540
Al	578	1,817	2,745	11,577

Table 7.3
Ionization Energies of Na, Mg, and Al (in kJ/mol)*

*Energies for the ionization of valence electrons lie to the left of the colored line.

these configurations by looking at the successive ionization energies of the atoms. Table 7.3 lists the first through fourth ionization energies of the atoms Na, Mg, and Al. The energy needed to remove the first electron from the Na atom is only 496 kJ/mol (first ionization energy). But the energy required to remove another electron (the second ionization energy) is nearly ten times greater (4562 kJ/mol). The electron in this case must be taken from Na^+, which has a neon configuration. Magnesium and aluminum atoms are similar. Their valence electrons are easily removed, but the energy needed to take an electron from either of the ions that result (Mg^{2+} and Al^{3+}) is extremely high. Thus, no compounds are found with ions having charges greater than the group number.

The ions of Group IA and IIA elements have noble-gas configurations. Similarly, Al^{3+} in Group IIIA has a noble-gas configuration. However, the other elements in Group IIIA have ions with filled d subshells (and a filled f subshell in the case of thallium) in addition to the noble-gas core. These ions with filled d and possibly f subshells, and a noble-gas core, are also quite stable. They are said to have **pseudonoble-gas configurations.**

The loss of successive electrons from an atom requires increasingly more energy. Consequently, Group IIIA elements show less tendency to form ionic compounds than do Group IA and IIA elements, which primarily form ionic compounds. Boron, in fact, forms no compounds with B^{3+} ions. The bonding is normally covalent, a topic discussed later in this chapter. However, the tendency to form ions becomes greater going down any column of the periodic table because of decreasing ionization energy. The remaining elements of Group IIIA do form compounds containing 3+ ions.

There is also a tendency for Groups IIIA to VA elements of higher periods, particularly Period 6, to form compounds with ions having a positive charge of two less than the group number. Thus, thallium in Group IIIA has compounds with 1+ ions and compounds with 3+ ions. Ions with charge equal to the group number minus two are obtained when the np electrons of an atom are lost but the ns^2 electrons are retained. For example,

$$Tl\ ([Xe]4f^{14}5d^{10}6s^26p^1) \longrightarrow Tl^+\ ([Xe]4f^{14}5d^{10}6s^2) + e^-$$

Few compounds of 4+ ions are known because a large quantity of energy is required to form the ions. The first three elements of Group IVA, C, Si, and Ge, are nonmetals (or metalloids) and usually form covalent rather than ionic bonds. Tin and lead, however, commonly form compounds with 2+ ions (ionic charge equal to the group number minus two). Bismuth, in Group VA, is a metallic element that forms compounds of Bi^{3+} (ionic charge equal to the group number minus two) where only the $6p$ electrons have been lost.

Group VIA and VIIA elements, whose atoms have the highest electron affinities, would be expected to form monatomic ions by gaining electrons to give noble-gas or pseudonoble-gas configurations. Group VIIA elements (valence-shell configuration ns^2np^5) pick up one electron to give $1-$ anions (ns^2np^6). Examples are F^- and Cl^-. (Hydrogen also forms compounds of the $1-$ ion, H^-. The hydride ion, H^-, has a $1s^2$ configuration like the noble-gas atom helium.) Group VIA elements (valence-shell configuration ns^2np^4) pick up two electrons to give $2-$ anions (ns^2np^6). Examples are O^{2-} and S^{2-}. Electron affinities of Group VA elements (valence-shell configuration ns^2np^3) are low, but the N^{3-} ion $(2s^22p^6)$ is stable in the presence of Li^+ and Group IIA ions such as Mg^{2+} and Ca^{2+}.

To summarize, the common monatomic ions found in the main-group elements fall into three categories.

1. Cations of Groups IA to IIIA having noble-gas or pseudonoble-gas configurations. The ion charges equal the group numbers.
2. Cations of Groups IIIA to VA having ns^2 electron configurations. The ion charges equal the group numbers minus two. Examples are Tl^+, Sn^{2+}, Pb^{2+}, and Bi^{3+}.
3. Anions of Groups VA to VIIA having noble-gas or pseudonoble-gas configurations. The ion charges equal the group numbers minus eight.

Example 7.2

Write the electron configuration and Lewis symbol for N^{3-}.

Solution

The electron configuration of the N atom is $[He]2s^22p^3$. By gaining three electrons, the atom assumes a $3-$ charge and the neon configuration $[He]2s^22p^6$. The Lewis symbol is

$$[:\ddot{N}:]^{3-}$$

Exercise 7.2

Write the electron configuration and Lewis symbol for Ca^{2+} and for S^{2-}.
                                                                  (See Problems 7.23 and 7.24.)

Exercise 7.3

Write the electron configurations of Pb and Pb^{2+}.
                                                                  (See Problems 7.25 and 7.26.)

Transition-Metal Ions

Most transition elements form cations of several different charges. For example, iron has the cations Fe^{2+} and Fe^{3+}. Neither of these ions has a noble-gas configuration; that would require the energetically impossible loss of eight electrons. The $2+$ ions are common for the transition elements and are obtained by the loss of the highest energy s electrons from the atom. Many transition elements also form $3+$ ions by losing one $(n-1)\, d$ electron in

Ion	Color in Water Solution	
Cr^{3+}	Violet	Table 7.4
Mn^{2+}	Pale pink	Common
Fe^{2+}	Pale green	Transition-Metal
Fe^{3+}	Colorless to pale yellow	Cations
Co^{2+}	Pink	
Ni^{2+}	Green	
Cu^{2+}	Blue	
Zn^{2+}	Colorless	
Ag^+	Colorless	
Cd^+	Colorless	
Hg^{2+}	Colorless	

addition to the two *ns* electrons. Ions with a partially filled *d* subshell exhibit various types and intensities of coloration in aqueous solutions. This is due to absorption of visible light involving *d* electrons.■ Table 7.4 lists common transition-metal ions.

■ Photons of this absorbed light provide the energy required to raise the *d* electrons from one energy level to a higher one. This is explained in detail in Chapter 25.

Example 7.3

Write the electron configurations of Fe^{2+} and Fe^{3+}.

Solution

The electron configuration of the Fe atom $(Z = 26)$ is $[Ar]3d^64s^2$. (This can be found using the building-up principle, as in Example 6.2, or from the position of Fe in the periodic table, as in Example 6.4.) To find the ion configurations, we first remove the *ns* electrons, and then the $(n - 1)$ *d* electrons. The number of units of positive charge on an ion indicates the number of electrons removed. Hence, the configuration of Fe^{2+} is $[Ar]3d^6$ and that of Fe^{3+} is $[Ar]3d^5$.

Exercise 7.4

Write the electron configuration of Mn^{2+}.

(See Problems 7.27 and 7.28.)

Polyatomic Ions

So far we have considered only monatomic ions, ions of only one atom. **Polyatomic ions** consist of several atoms linked together in a group by covalent bonds, with the group of atoms carrying a net electrical charge. There are few common polyatomic cations occurring in compounds. One of these is the ammonium ion, NH_4^+. It consists of an N atom surrounded by four H atoms, and has an overall positive charge. Another polyatomic cation is Hg_2^{2+}. Table 7.5 lists these and the most common polyatomic anions.

Formulas of Ionic Compounds

The formula of an ionic compound gives the simplest ratio of positive and negative ions present. Since a compound is always electrically neutral, this

Table 7.5
Some Common
Polyatomic Ions

Name	Formula	Name	Formula
Acetate	$C_2H_3O_2^-$	Hypochlorite	ClO^-
Ammonium	NH_4^+	Hydroxide	OH^-
Carbonate	CO_3^{2-}	Mercury(I)	Hg_2^{2+}
Chlorate	ClO_3^-	Monohydrogen phosphate	HPO_4^{2-}
Chlorite	ClO_2^-	Nitrate	NO_3^-
Chromate	CrO_4^{2-}	Nitrite	NO_2^-
Cyanide	CN^-	Oxalate	$C_2O_4^{2-}$
Dichromate	$Cr_2O_7^{2-}$	Perchlorate	ClO_4^-
Dihydrogen phosphate	$H_2PO_4^-$	Permanganate	MnO_4^-
Hydrogen carbonate (or bicarbonate)	HCO_3^-	Peroxide	O_2^{2-}
		Phosphate	PO_4^{3-}
Hydrogen sulfate (or bisulfate)	HSO_4^-	Sulfate	SO_4^{2-}
		Sulfite	SO_3^{2-}
Hydrogen sulfite (or bisulfite)	HSO_3^-	Thiosulfate	$S_2O_3^{2-}$

ratio of ions must be such that the electrical charges from the ions cancel. For example, magnesium nitrate is a compound of magnesium ions, Mg^{2+}, and nitrate ions, NO_3^-. There must be twice as many NO_3^- ions as Mg^{2+} ions to give a neutral compound. In a formula of an ionic compound, we write the cation before the anion and omit the charges. Subscripts are used to indicate the simplest ratio of ions present. Thus, the formula of magnesium nitrate is $Mg(NO_3)_2$, where parentheses enclose the formula of the polyatomic ion NO_3^-.

Example 7.4

(a) Aluminum oxide is a compound composed of Al^{3+} and O^{2-} ions. What is the formula of aluminum oxide?

(b) Strontium oxide is a compound composed of Sr^{2+} and O^{2-} ions. Write the formula of this compound.

Solution

(a) We can achieve electrical neutrality by taking as many cations as there are units of charge on the anion, and as many anions as there are units of charge on the cation. Thus, two Al^{3+} ions have a total charge of $6+$, and three O^{2-} ions have a total charge of $6-$, giving the combination a net charge of zero. The simplest ratio of Al^{3+} to O^{2-} is $2:3$, and the formula is Al_2O_3. Note that the charge (without its sign) on one ion becomes the subscript of the other ion.

$$Al^{③+} \quad O^{②-} \qquad \text{or} \qquad Al_2O_3$$

(b) We see that equal numbers of Sr^{2+} and O^{2-} ions will give a neutral compound. Thus, the formula is SrO. If we use the units of charge to find the subscripts, we get

$$Sr^{②+} \quad O^{②-} \qquad \text{or} \qquad Sr_2O_2$$

The final formula is SrO, since this gives the simplest ratio of ions.

Exercise 7.5

Write the formula of the compound composed of the ions K^+ and CrO_4^{2-} (potassium chromate).

(See Problems 7.29 and 7.30.)

7.3 Ionic Radii

A monatomic ion, like an atom, is a nucleus surrounded by a distribution of electrons. The **ionic radius** is a measure of the size of the spherical region around the nucleus within which the electrons of an ion are most likely to be found. Defining an ionic radius (just as with an atomic radius) is somewhat arbitrary because an electron distribution never abruptly ends. However, if we imagine ions to be spheres of definite size, we can obtain their radii from known distances between nuclei in crystals. (These distances can be determined accurately by observing how crystals diffract x rays.)■

To understand how to compute ionic radii, let us determine from experimental data the radius of an I^- ion in a lithium iodide (LiI) crystal. Figure 7.1 depicts a layer of ions in LiI. Note that the distance between adjacent iodine nuclei equals twice the I^- radius. From x-ray diffraction experiments, the iodine–iodine distance is found to be 4.26 Å. Therefore, the I^- radius in LiI is $1/2 \times 4.26$ Å = 2.13 Å. Other crystals give approximately the same radius for the I^- ion. Table 7.6, which lists average values of ionic radii for some main-group elements, gives 2.06 Å for the I^- radius.

That we can find values of ionic radii that agree with the known structures of many crystals is strong evidence for the existence of ions in the solid state. Moreover, these values of ionic radii compare with atomic radii in ways that we might expect. Thus, we expect a cation to be smaller and an anion to be larger than the corresponding atom (see Figure 7.2).

A cation formed when an atom loses all its valence electrons is smaller than the atom because it has one less shell of electrons. But even when only part of the valence electrons are lost from an atom the effective nuclear charge on the outer electrons is greater, because the nuclear charge is shielded by fewer electrons. The size of the outer orbitals, and thus the ionic

■ The study of crystal structure by x-ray diffraction is discussed in Chapter 11.

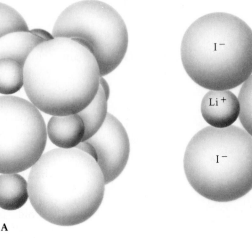

A   B

Figure 7.1
Determining the iodide ion radius in the lithium iodide (LiI) crystal. (a) A three-dimensional view of the crystal. (b) Cross section through a layer of ions. Iodide ions are assumed to be spheres in contact with one another. The distance between iodine nuclei (4.26 Å) is determined experimentally. One-half this distance (2.13 Å) equals the iodide ion radius.

Table 7.6
Ionic Radii (in Angstroms) of
Some Main-Group Elements*

	IA	IIA	IIIA	VIA	VIIA
Period 2	Li^+ 0.90	Be^{2+} 0.59		O^{2-} 1.26	F^- 1.19
Period 3	Na^+ 1.16	Mg^{2+} 0.86	Al^{3+} 0.68	S^{2-} 1.70	Cl^- 1.67
Period 4	K^+ 1.52	Ca^{2+} 1.14	Ga^{3+} 0.76	Se^{2-} 1.84	Br^- 1.82
Period 5	Rb^+ 1.66	Sr^{2+} 1.32	In^{3+} 0.94	Te^{2-} 2.07	I^- 2.06
Period 6	Cs^+ 1.81	Ba^{2+} 1.49	Tl^{3+} 1.02		

*Values are taken from R. D. Shannon, *Acta Crystallographica*, Vol. A32 (1976), p. 751.

radius, is smaller. Similarly, because an anion has more electrons than the atom, the nuclear shielding is greater. Since the effective nuclear charge on the outer electrons is less, the anion radius is larger.

Exercise 7.6

Which has the larger radius, S or S^{2-}? Explain.   (See Problems 7.31 and 7.32.)

The ionic radii of the main-group elements shown in Table 7.6 follow a regular pattern, just as atomic radii do. *Ionic radii increase down any column because of the addition of electron shells.*

Exercise 7.7

Without looking at Table 7.6, arrange the following ions in order of increasing ionic radius: Sr^{2+}, Mg^{2+}, Ca^{2+}. (You may use a periodic table.)
     (See Problems 7.33 and 7.34.)

The pattern across a period becomes clear if we look first at the cations, then at the anions. For example, in the third period we have

Cation	Na^+	Mg^{2+}	Al^{3+}
Radius (Å)	1.16	0.86	0.68

Na

$[He]2s^22p^63s^1$

Na^+

$[He]2s^22p^6$

Cl

$[Ne]3s^23p^5$

Cl^-

$[Ne]3s^23p^6$

Figure 7.2
Comparison of atomic and ionic radii. Note that the sodium atom loses its outer shell in forming the Na^+ ion. Thus, the cation is smaller than the atom. The Cl^- ion is larger than the Cl atom because the same nuclear charge holds a greater number of electrons less strongly.

All of these ions have the neon configuration $1s^2 2s^2 2p^6$, but have different nuclear charges. Ions such as these that have the same electron configuration are said to be **isoelectronic.** To understand the decrease in radius from Na^+ to Al^{3+}, imagine the nuclear charge (atomic number) of Na^+ to increase. With each increase of charge, the orbitals contract because of the greater attractive force of the nucleus. Thus, in any *isoelectronic sequence* of atomic ions, the ionic radius decreases with increasing atomic number (just as it does for the atoms).

If we look at the anions in the third-period elements, we notice that they are much larger than the cations in the same period. This abrupt increase in ionic radius is due to the fact that the anions S^{2-} and Cl^- have configurations with one more shell of electrons than the cations. And because these anions also constitute an isoelectronic sequence (with argon configuration), the ionic radius decreases with atomic number (as with the atoms):

Anion	S^{2-}	Cl^-
Radius (Å)	1.70	1.67

Thus, *in general, across a period, the cations decrease in radius. When we reach the anions, there is an abrupt increase in radius, and then the radii again decrease.*

Example 7.5

Without looking at Table 7.6, arrange the following ions in order of decreasing ionic radius: F^-, Mg^{2+}, O^{2-}. (You may use a periodic table.)

Solution

Note that F^-, Mg^{2+}, and O^{2-} are isoelectronic. If we arrange them by increasing nuclear charge, they will be in order of decreasing ionic radius. The order is O^{2-}, F^-, Mg^{2+}.

Exercise 7.8

Without looking at Table 7.6, arrange the following ions in order of increasing ionic radius: Cl^-, Ca^{2+}, P^{3-}. (You may use a periodic table.)

(See Problems 7.35 and 7.36.)

Covalent Bond

In the preceding sections, we looked at ionic substances, which are typically high-melting solids. Many substances, however, are molecular—gases, liquids, or low-melting solids consisting of molecules. A molecule is a group of atoms strongly linked by chemical bonds. Often the forces that hold the atoms together in a molecular substance cannot be understood solely on the basis of attraction of oppositely charged ions. An obvious example is the molecule H_2, where the two H atoms are held together tightly and no ions are present. In 1916, Gilbert Newton Lewis proposed that the strong attractive force between two atoms in a molecule resulted from a **covalent bond,** formed by the sharing of a pair of electrons between atoms. ■ In 1926

■ G. N. Lewis (1875–1946) was professor of chemistry at the University of California at Berkeley. Besides his work on chemical bonding, he was noted for his research in molecular spectroscopy and thermodynamics (the study of heat involved in chemical and physical processes).

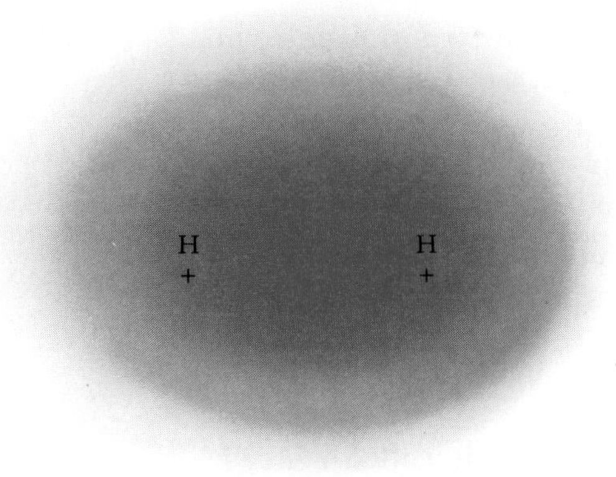

Figure 7.3
The electron probability distribution for the H_2 molecule. The electrons occupy the space around both atoms.

W. Heitler and F. London showed that the covalent bond in H_2 could be quantitatively explained by the newly discovered theory of quantum mechanics. We will discuss the descriptive aspects of covalent bonding in the following sections.

7.4 Describing Covalent Bonds

Consider the formation of a covalent bond between two H atoms to give the H_2 molecule. According to theory, as the atoms approach one another their $1s$ orbitals begin to overlap. Each electron can then occupy the space around both atoms (Figure 7.3). In other words, the two electrons can be shared by both atoms. The electrons are attracted simultaneously by the positive charges of the two hydrogen nuclei. This attraction that bonds the electrons to both nuclei is the force holding the atoms together. Thus, while ions do not exist in H_2, the force that holds the atoms together can still be regarded as arising from the attraction of oppositely charged particles, nuclei and electrons.

Lewis Formulas

We can represent the formation of the covalent bond in H_2 from atoms as follows:

$$H\cdot \; + \; \cdot H \longrightarrow H{:}H$$

Here we use the Lewis electron-dot symbol for the hydrogen atom (Table 7.1) and represent the covalent bond by a double dot. Recall that the two electrons from the covalent bond spend part of the time in the region of each atom. In this sense, each atom in H_2 has a helium configuration. We can draw a circle about each atom to emphasize this.

The formation of a bond between H and Cl to give an HCl molecule can be represented in a similar way.

$$\text{H}\cdot \; + \; \cdot\ddot{\text{C}}\ddot{\text{l}}: \; \longrightarrow \; \left(\text{H} :\ddot{\text{C}}\ddot{\text{l}}: \right)$$

As the two atoms approach, unpaired electrons on each atom pair up to form a covalent bond. The pair of electrons is shared by the two atoms. Each atom then acquires a noble-gas configuration of electrons, the H atom having two electrons about it (as in He), and the Cl atom having eight valence electrons about it (as in Ar).

A formula using dots to represent valence electrons is called a *Lewis electron-dot formula.* The electron pairs represented by double dots in such formulas are either **bonding pairs** (which are shared between two atoms) or **lone,** or **nonbonding, pairs** (electron pairs that remain on one atom and are not shared). For example,

lone pairs

bonding pair

Bonding pairs are often represented by dashes rather than by double dots.

Frequently, the number of covalent bonds formed by an atom equals the number of unpaired electrons shown in its Lewis symbol. Consider the formation of the molecules H_2O, NH_3, and CH_4 from atoms.

$$2\text{H}\cdot \; + \; \cdot\ddot{\text{O}}: \; \longrightarrow \; \text{H}:\ddot{\text{O}}: \\ \qquad\qquad\qquad\quad \ddot{\text{H}}$$

$$3\text{H}\cdot \; + \; \cdot\ddot{\text{N}}: \; \longrightarrow \; \begin{array}{c} \text{H} \\ \text{H}:\ddot{\text{N}}: \\ \ddot{\text{H}} \end{array}$$

$$4\text{H}\cdot \; + \; \cdot\dot{\underset{\cdot}{\text{C}}}\cdot \; \longrightarrow \; \begin{array}{c} \text{H} \\ \text{H}:\ddot{\text{C}}:\text{H} \\ \ddot{\text{H}} \end{array}$$

In each case, a bond is formed between an unpaired electron on one atom and an unpaired electron on another. In many instances, the number of bonds formed by an atom in Groups IVA to VIIA equals the number of unpaired electrons, which is eight minus the group number.■ For example, a nitrogen atom (Group VA) forms $8 - 5 = 3$ covalent bonds. We will discuss writing these "electron-dot" formulas of molecules in Section 7.6.

■ The numbers of unpaired electrons in the Lewis symbols for the atoms of Groups IA to IIIA elements equal the group numbers. But, except for the first elements of these groups, the atoms usually form ionic bonds.

Coordinate Covalent Bond

We can represent the formation of a bond between atoms that each donate an electron as follows:

$$\text{A}\cdot \; + \; \cdot\text{B} \longrightarrow \text{A}:\text{B}$$

A **coordinate covalent bond** is formed when both electrons of the bond are donated by one atom, as follows:

$$\text{A} \; + \; :\text{B} \longrightarrow \text{A}:\text{B}$$

A coordinate covalent bond is not essentially different from other covalent bonds, since it involves the sharing of a pair of electrons between two atoms. An example is the formation of the ammonium ion:

$$H^+ + :NH_3 \longrightarrow \left[\begin{array}{c} H \\ H:\overset{\cdot\cdot}{\underset{\cdot\cdot}{N}}:H \\ H \end{array} \right]^+$$

Octet Rule

In each of the molecules we have described so far, the atoms have obtained noble-gas configurations through the sharing of electrons. Atoms other than hydrogen have obtained eight electrons in their valence shells; hydrogen atoms have obtained two. This tendency of atoms in molecules to have eight electrons in their valence shells (two for hydrogen atoms) is known as the **octet rule.** Many of the molecules we will discuss follow the octet rule. However, as we will describe later, some do not.

Multiple Bonds

So far, the molecules we have described have had only **single bonds,** that is, covalent bonds in which a single pair of electrons is shared by two atoms. But it is possible for two atoms to share two electron pairs to give a **double bond** or three pairs to give a **triple bond.** As examples, consider ethylene, C_2H_4, and acetylene, C_2H_2. Their Lewis formulas are

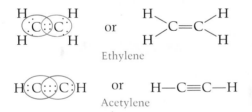

Ethylene

Acetylene

Note the octet of electrons on each C atom. Double bonds form primarily to C, N, O, and S atoms. Triple bonds form mostly to C and N atoms.

7.5 Polar Covalent Bond; Electronegativity

A covalent bond involves the sharing of a pair of electrons between two atoms. When the atoms are alike, as in the case of the H—H bond of H_2, the bonding electrons are equally shared. That is, the electrons spend equal time in the vicinity of each atom. But when the two atoms are of different elements, the bonding electrons need not be shared equally. A **polar covalent bond** is formed when the bonding electrons spend more time near one atom

Figure 7.4
The probability distribution
for the HCl bonding elec-
trons. Compare this with the
electron distribution for H_2
in Figure 7.3.

than the other. For example, in the case of the HCl molecule, we know that
the bonding electrons spend more time near the chlorine atom than the
hydrogen atom. Figure 7.4 shows the distribution of electrons in the H—Cl
bond. Thus the H—Cl bond is polar covalent.

We can look at the polar covalent bond as intermediate between a *nonpolar* covalent bond, as in H_2, and an ionic bond, as in NaCl. From this point
of view, an ionic bond is simply an extreme example of a polar covalent
bond. To illustrate, we can represent the bonding in H_2, HCl, and NaCl with
electron-dot formulas as follows:

$$H : H \qquad H : \overset{..}{\underset{..}{Cl}}: \qquad Na \quad :\overset{..}{\underset{..}{Cl}}:$$

Nonpolar covalent   Polar covalent   Ionic

As shown, the bonding pairs of electrons are equally shared in H_2, unequally
shared in HCl, and essentially not shared in NaCl. Thus we see that it is
possible to arrange different bonds to form a gradual transition from nonpolar covalent to ionic.

Electronegativity is a measure of the ability of an atom in a molecule to
draw bonding electrons to itself. Several electronegativity scales have been
proposed. In 1934, Robert S. Mulliken suggested on theoretical grounds that
the electronegativity (X) of an atom be given as one half its ionization energy
(I.E.) minus its electron affinity (E.A.).■

$$X = \frac{\text{I.E.} - \text{E.A.}}{2}$$

An atom such as fluorine that tends to pick up electrons easily (large negative E.A.) and to hold on to them strongly (large I.E.) will have a large
electronegativity. On the other hand, an atom such as lithium or cesium
that loses electrons readily (small I.E.) and has little tendency to gain electrons (small negative or positive E.A.) will have small electronegativity.
Until recently, only a few electron affinities had been measured. For this
reason, Mulliken's scale has had limited utility. A more widely used scale
was derived by Linus Pauling from values of bond energies, as described in
Chapter 10.■ Pauling's values are given in Figure 7.5. Because electronegativities depend somewhat on the bonds formed, these values are actually
average ones.

Fluorine, the most electronegative element, was assigned a value of 4.0 on
Pauling's scale. Lithium, at the left end of the same period, has a value of 1.0.
Cesium, in the same column but below lithium, has a value of 0.7. *In*

■ Robert S. Mulliken received
the Nobel Prize in chemistry in
1966 for his work on molecular
orbital theory (discussed in
Chapter 8).

■ Linus Pauling received the
Nobel Prize in chemistry in 1954
for his work on the nature of the
chemical bond. In 1962, he received the Nobel Prize for peace.

							H 2.1									

IA	IIA	IIIB	IVB	VB	VIB	VIIB	VIIIB	VIIIB	VIIIB	IB	IIB	IIIA	IVA	VA	VIA	VIIA
Li 1.0	Be 1.5											B 2.0	C 2.5	N 3.0	O 3.5	F 4.0
Na 0.9	Mg 1.2											Al 1.5	Si 1.8	P 2.1	S 2.5	Cl 3.0
K 0.8	Ca 1.0	Sc 1.3	Ti 1.5	V 1.6	Cr 1.6	Mn 1.5	Fe 1.8	Co 1.8	Ni 1.8	Cu 1.9	Zn 1.6	Ga 1.6	Ge 1.8	As 2.0	Se 2.4	Br 2.8
Rb 0.8	Sr 1.0	Y 1.2	Zr 1.4	Nb 1.6	Mo 1.8	Tc 1.9	Ru 2.2	Rh 2.2	Pd 2.2	Ag 1.9	Cd 1.7	In 1.7	Sn 1.8	Sb 1.9	Te 2.1	I 2.5
Cs 0.7	Ba 0.9	La–Lu 1.1–1.2	Hf 1.3	Ta 1.5	W 1.7	Re 1.9	Os 2.2	Ir 2.2	Pt 2.2	Au 2.4	Hg 1.9	Tl 1.8	Pb 1.8	Bi 1.9	Po 2.0	At 2.2
Fr 0.7	Ra 0.9	Ac–No 1.1–1.7														

Figure 7.5
Electronegativities of the elements.

general, *electronegativity increases from left to right and decreases from top to bottom in the periodic table.* Metals are the least electronegative elements (they are *electropositive*), and nonmetals are the most electronegative.

The absolute value of the difference in electronegativity of two bonded atoms gives us a rough measure of the polarity to be expected in a bond. When this difference is small, the bond is nonpolar. When it is large, the bond is polar, or if very large perhaps ionic. The electronegativity differences for the bonds H—H, H—Cl, and Na—Cl are 0, 0.9, and 2.1, respectively, following the expected order. Electronegativity differences explain why ionic bonds usually form between a metal atom and a nonmetal atom; the electronegativity difference would be largest between these elements. On the other hand, covalent bonds primarily form between two nonmetals because the electronegativity differences are small.■

■ The electronegativity difference in metal–metal bonding would also be small. This bonding frequently involves delocalized *metallic bonding*, briefly described in Section 7.8.

Example 7.6

Use electronegativity values (Figure 7.5) to arrange the following bonds in order by increasing polarity: P—H, H—O, C—Cl.

Solution
The absolute values of the electronegativity differences are P—H, 0.0; H—O, 1.4; C—Cl, 0.5. Hence, the order is P—H, C—Cl, H—O.

Exercise 7.9

Using electronegativities, decide which of the following bonds is most polar: C—O, C—S, H—Br.

(See Problems 7.43 and 7.44.)

We can use an electronegativity scale to predict the direction in which the electrons in a bond shift. Bonding electrons are pulled toward the more electronegative atom. For example, consider the H—Cl bond. Since the Cl atom $(X = 3.5)$ is more electronegative than the H atom $(X = 2.1)$, the bonding electrons in H—Cl are pulled toward Cl. Because the bonding electrons spend most of their time around the Cl atom, that end of the bond (and the molecule) acquires a partial negative charge (indicated $\delta -$). The H-atom end of the bond (and the molecule) has a partial positive charge $(\delta +)$. We can show this as follows:

$$\overset{\delta+ \quad \delta-}{\text{H—Cl}}$$

The HCl molecule is said to be a *polar molecule*. More will be said about polar molecules when we look at molecular structures in Chapter 8.

7.6 Writing Lewis Electron-Dot Formulas

The Lewis electron-dot formula of a molecule is similar to the structural formula in that it shows how atoms are bonded. Bonding electron pairs are indicated either by two dots or by a dash. In addition to the bonding electrons, however, an electron-dot formula shows the positions of lone pairs of electrons, whereas the structural formula does not. Thus, the electron-dot formula is a simple two-dimensional representation of the positions of electrons in a molecule. In the next chapter, we will see how to predict the three-dimensional shape of a molecule from the two-dimensional electron-dot formula.■ In this section, we will discuss the steps for writing the electron-dot formula for a molecule made from atoms of the main-group elements.

■ Lewis formulas do not directly convey information about molecular shape. For example, the Lewis formula of methane, CH_4, is written as the flat (two-dimensional) formula

$$\begin{array}{c} \text{H} \\ \ddot{\ } \\ \text{H:C:H} \\ \ddot{\text{H}} \end{array}$$

The actual methane molecule, however, is not flat; it has a three-dimensional structure, as explained in Chapter 8.

Skeleton Structure of a Molecule

Before we can write the Lewis formula of a molecule (or a polyatomic ion), we must know the *skeleton structure* of the molecule. The skeleton structure tells us which atoms are bonded to one another (without regard to whether the bonds are single or not). Normally, this information must be found by experiment. For simple molecules, we can often predict the skeleton structure from the following four rules. There are, of course, exceptions to these rules. *The only sure way to know the correct structure of a molecule is from experiment.*

Rule 1 Many small molecules or polyatomic ions consist of a central atom around which are bonded atoms of greater electronegativity, such as F, Cl, and O. Note the following structural formulas:

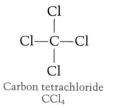

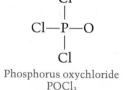

Carbon tetrachloride   Phosphorus oxychloride   Nitrosyl chloride
CCl_4 $POCl_3$ ClNO

Rule 2 Molecules or polyatomic ions with symmetrical formulas often have symmetrical structures. For example, disulfur dichloride, S_2Cl_2, is symmetrical, with the more electronegative Cl atoms around the S atoms: Cl—S—S—Cl.

Rule 3 *Oxyacids* are substances in which O atoms (and possibly other electronegative atoms) are bonded to a central atom, with one or more H atoms usually bonded to O atoms. Examples are

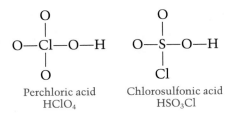

Perchloric acid
$HClO_4$

Chlorosulfonic acid
HSO_3Cl

Rule 4 Of several possible structural formulas, the one in which the atoms have their usual number of covalent bonds (8 − group number) is usually preferred. For example, hydrogen cyanide, HCN, has the Lewis formula H—C≡N:, not :C≡N—H. (Note that both of these formulas satisfy the octet rule, but in :C≡N—H the N atom has four covalent bonds rather than the normal three and the C atom has three rather than the normal four.)

Steps in Writing Lewis Formulas

Once we know the skeleton structure of a molecule, we can find the Lewis formula using the following four steps:

Step 1 Calculate the total number of valence electrons for the molecules by adding up the number of valence electrons (= group number) for each atom. If we are writing the Lewis formula of a polyatomic anion, we *add* the number of negative charges to this total. (Thus, for SO_4^{2-} we add 2 because the 2− charge indicates that there are 2 more electrons than are provided by the neutral atoms.) If we are writing the Lewis formula of a polyatomic cation, we *subtract* the number of positive charges from the total. (Thus, for NH_4^+ we subtract 1 because the 1+ charge indicates a loss of one electron from the group of neutral atoms.)

Step 2 Write the skeleton structure of the molecule, connecting every bonded pair of atoms by a double dot (or dash).

Step 3 Distribute electrons to the atoms surrounding the central atom (or atoms) to satisfy the octet rule for these surrounding atoms. (If the molecule contains only two atoms, skip this step.)

Step 4 Subtract the number of electrons so far distributed from the total found in Step 1. Distribute the remaining electrons as pairs to the central atom (or atoms). If there are less than eight on the central atom, this suggests that a multiple bond is present. (Two electrons less than an octet suggests a double bond; four less suggests a triple bond or two double bonds. As explained in the next section, atoms of Groups IA to IIIA elements should be left with less than eight valence electrons.) Atoms that often form multiple bonds are C, N, O, and S. Arrange the multiple bonds to satisfy the octet rule.

The next several examples illustrate how to write the Lewis electron-dot formulas for a small molecule, given the molecular formula.

Example 7.7

Thionyl chloride, $SOCl_2$, is a liquid with a corrosive vapor. Write the Lewis formula for the molecule.

Solution

We first calculate the total number of valence electrons. The number for each kind of atom (= group number) is S = 6, O = 6, and Cl = 7. Hence, the total is 6 + 6 + (2 × 7) = 26. Using Rule 1, we expect the skeleton structure to consist of an S atom surrounded by O and Cl atoms. After distributing electron pairs to these surrounding atoms to satisfy the octet rule, we

have

$$:\ddot{C}l: \\ :\ddot{C}l:\ddot{S}:\ddot{O}: \quad or \quad :\ddot{C}l—\ddot{S}—\ddot{O}:$$

This accounts for 12 pairs, or 24 electrons. Subtracting 24 from the total (26) gives 2 electrons, or 1 pair, which is placed on the S atom, giving an octet. The Lewis formula is

$$:\ddot{C}l: \\ :\ddot{C}l:\ddot{\underset{\cdot\cdot}{S}}:\ddot{O}: \quad or \quad :\ddot{C}l—\overset{\cdot\cdot}{S}—\ddot{O}:$$

Exercise 7.10

Dichlorodifluoromethane, CCl_2F_2, is a gas used as a refrigerant and aerosol propellant. (It is known commercially as one of the Freons.) Write the Lewis formula for CCl_2F_2.

(See Problems 7.47 and 7.48.)

Example 7.8

Carbonyl chloride, or phosgene, $COCl_2$, is a highly toxic gas used as a starting material for the preparation of polyurethane plastics. What is the electron-dot formula of $COCl_2$?

Solution

The total number of valence electrons is 4 + 6 + (2 × 7) = 24. From Rule 1, we expect the more electropositive atom, C, to be central, with the O and Cl atoms bonded to it. After distributing electron pairs to these surrounding atoms, we have

This accounts for all 24 valence electrons, leaving only 6 electrons on C. Since this is 2 less than an octet, a double bond is suggested. Because C and O atoms often form double bonds (and Cl atoms do not), we move a pair of electrons on the O atom to give a carbon–oxygen double bond. These electrons still count in the octet of the O atom, but are now shared with the C atom, giving it an octet. The electron-dot formula of $COCl_2$ is

$$:\ddot{C}l:C::\ddot{O} \quad or \quad :\ddot{C}l—C=\ddot{O} \\ \quad\ :\ddot{C}l: \qquad\qquad\quad |\quad\ \\ \qquad\qquad\qquad\qquad\quad :\ddot{C}l:$$

Check to see that each atom has an octet.

Exercise 7.11

Write the electron-dot formula of carbon dioxide, CO_2.

(See Problems 7.49 and 7.50.)

Example 7.9

Obtain the electron-dot formula of the sulfite ion, $SO_3{}^{2-}$.

Solution

Both S and O atoms have 6 valence electrons. Thus, the S atom and three O atoms provide 24 valence electrons. Because the anion has a charge of $2-$, it has 2 more electrons than are provided by the neutral atoms. Thus, the total number of available electrons is $24 + 2 = 26$. By Rule 1, we assign S as the central atom, placing the O atoms around it. After placing electron pairs on the O atoms to satisfy the octet rule for them, we have

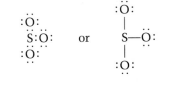

So far, we have used up 12 pairs, or 24 electrons. From the total number (26), this leaves 2 electrons, or 1 pair, which is placed on the S atom. Note that the octet rule is now satisfied for the S atom. The electron-dot formula for $SO_3{}^{2-}$ is

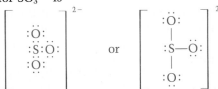

The charge on the entire ion is indicated by the superscript $2-$ to the square brackets enclosing the electron-dot formula.

Exercise 7.12

What is the electron-dot formula of (a) the hydronium ion, H_3O^+, (b) the chlorite ion, $ClO_2{}^-$?

(See Problems 7.51 and 7.52.)

7.7 Exceptions to the Octet Rule

Although most molecules composed of atoms of the main-group elements have electronic structures that satisfy the octet rule, a number of them do not. For example, the very common substance oxygen, O_2, does not obey this rule. A few molecules, such as NO, have an odd number of electrons, and so cannot satisfy the octet rule. The other exceptions to the octet rule fall into two groups. In one group are molecules with an atom having fewer than eight valence electrons around it. In the other group are molecules with an atom having more than eight valence electrons around it.

The first group consists mostly of molecules with boron atoms. For example, consider boron trifluoride, BF_3. The Lewis formula is believed to be

The boron atom in BF_3 has only six valence electrons around it. Many reactions of BF_3 result from this. Boron trifluoride reacts with molecules having a lone pair, such as with ammonia, NH_3, which gives the compound BF_3NH_3.

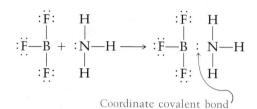

Coordinate covalent bond

In this reaction, a coordinate covalent bond forms between the boron and nitrogen atoms, and the boron atom achieves an octet.

Although the boron atom in BF_3 does not have an octet of valence electrons around it, it does have the normal number of covalent bonds. All the valence electrons of the boron atom have been paired with unpaired electrons from fluorine atoms. The formation of BF_3 from atoms can be written

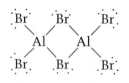

This suggests that other elements of Groups IA to IIIA might form molecules with an atom having less than eight electrons around it. Such molecules are uncommon, however, partly because these elements (particularly those in Groups IA and IIA) frequently form ionic bonds. Even when they form covalent bonds, the atoms tend to arrange themselves to give octets of electrons.

For example, consider aluminum bromide. Although the formula is often written $AlBr_3$, the solid substance consists of Al_2Br_6 molecules. The Lewis formula is

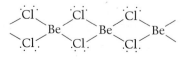

Each atom has an octet of electrons around it. Note that two of the Br atoms are in *bridge* positions, with each Br atom having two covalent bonds. Unlike the solid, aluminum bromide vapor does contain $AlBr_3$ molecules, with a Lewis formula similar to that of BF_3.

Exercise 7.13

Beryllium chloride, $BeCl_2$, is a solid substance consisting of long (essentially infinite) chains of atoms with Cl atoms in bridge positions.

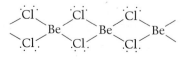

However, if the solid is heated, it forms a vapor of $BeCl_2$ molecules. Write the electron-dot formula of the $BeCl_2$ molecule.

(See Problems 7.53 and 7.54.)

The second group of exceptions to the octet rule consists of a variety of molecules having a central atom with more than eight valence electrons around it. Phosphorus pentachloride is an example. This is a pale yellow solid that readily forms a vapor of PCl_5 molecules. Each molecule consists of a phosphorus atom surrounded by five chlorine atoms. The phosphorus atom has ten valence electrons around it.■

■ Solid phosphorus pentachloride has the ionic structure $[PCl_4^+][PCl_6^-]$. The species within the brackets are covalently bonded. The phosphorus atom in PCl_6^- has 12 valence electrons around it.

The octet rule stems from the fact that the main-group elements in most cases employ only an ns and three np valence-shell orbitals in bonding, and these orbitals hold eight electrons. Elements of the second period are restricted to these orbitals, but from the third period on the elements also have unfilled nd orbitals, which may be used in bonding. For example, the valence-shell configuration of phosphorus is $3s^23p^3$. Using just these $3s$ and $3p$ orbitals, the phosphorus atom can accept only three additional electrons, forming three covalent bonds (as in PCl_3). However, more bonds can be formed if the empty $3d$ orbitals of the atom are used. If each of the five electrons of the phosphorus atom is paired with unpaired electrons of chlorine atoms, PCl_5 can be formed. Thus, phosphorus has both the trichloride (PCl_3) and the pentachloride (PCl_5). By contrast, nitrogen (which has no available d orbitals in its valence shell) has only the trichloride, NCl_3.

Example 7.10

Xenon, a noble gas, forms a number of compounds. One of these is xenon tetrafluoride, XeF_4, a white, crystalline solid first prepared in 1962. What is the electron-dot formula of the XeF_4 molecule?

Solution

Following the steps described in the previous section, we first calculate the total number of valence electrons for the molecule. There are 8 valence electrons from the Xe atom, and 7 from each F atom, giving a total of 36 valence electrons. For the skeleton structure, we draw the Xe atom surrounded by the electronegative F atoms. After placing electron pairs on the F atoms to satisfy the octet rule for them, we have

So far, this accounts for 16 pairs, or 32 electrons. Since a total of 36 electrons are available, we put an additional $36 - 32 = 4$ electrons (2 pairs) on the Xe atom. The Lewis formula is

Exercise 7.14

Sulfur tetrafluoride, SF_4, is a colorless gas. Write the electron-dot formula of the SF_4 molecule.

(See Problems 7.55 and 7.56.)

As we have seen, the *Lewis dot-octet rule approach* is powerful and generally applicable, save for exceptions such as those we have just considered. However, its failure to predict accurately certain aspects of the electron and molecular structure (magnetism and bond order) for something as common as molecular oxygen, O_2, is a serious problem. A more sophisticated theory will be discussed in Chapter 8.

7.8 Delocalized Bonding; Resonance

We have assumed up to now that the bonding electrons are localized to the region between two atoms. There are cases, however, where this assumption does not fit the experimental data. Suppose, for example, that we try to write an electron-dot formula for sulfur dioxide, SO_2. We find that we can write two formulas:

$$\ddot{S} \qquad \ddot{S}$$
$$:\!\ddot{O}\!:\!\quad:\!\ddot{O}\!: \qquad \text{and} \qquad :\!\ddot{O}\!:\!\quad:\!\ddot{O}\!:$$
$$\textbf{A} \qquad\qquad\qquad \textbf{B}$$

In formula A, the sulfur–oxygen bond on the left is a double bond and the sulfur–oxygen bond on the right is a single bond. In formula B, the situation is just the opposite. Experiment shows, however, that the two bonds are identical.■ Therefore, neither formula A nor formula B can be correct.

According to theory, one of the bonding pairs in sulfur dioxide is spread over the region of all three atoms, rather than localized to a particular sulfur–oxygen bond. This is called **delocalized bonding.** We might symbolically describe the delocalized bonding in sulfur dioxide as follows:

$$\underset{O}{\overset{S}{\diagup\!\!\!=\!\cdots\diagdown}} O$$

■ The lengths of the two sulfur–oxygen bonds (that is, the distances between the atomic nuclei) are each 1.43 Å.

(For clarity, only the bonding pairs are given.) The broken line indicates a bonding pair of electrons that spans three nuclei rather than only two. In effect, the sulfur–oxygen bond is neither a single bond nor a double bond, but an intermediate type.

A single electron-dot formula cannot properly describe delocalized bonding. Instead, a **resonance description** of delocalized bonding is often used. According to this, we describe the electron structure of a molecule by writing all possible electron-dot formulas. These are called the *resonance formulas* of the molecule. The actual electron distribution of the molecule is a composite of these resonance formulas.

Thus, the electron structure of sulfur dioxide can be described in terms of the two resonance formulas presented at the start of this section. By convention, one usually writes all of the resonance formulas connected by double-headed arrows. For sulfur dioxide, we would write

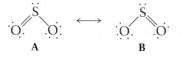
$$\textbf{A} \qquad\qquad\qquad \textbf{B}$$

Unfortunately, this notation can be misinterpreted. It does not mean that the sulfur dioxide molecule flips back and forth between two forms. There is only one sulfur dioxide molecule. The double-headed arrow notation means that you should form a mental picture of the molecule by fusing the various resonance formulas. Since the left sulfur–oxygen bond is double in formula A and the right one is double in formula B, you must picture an electron pair that actually encompasses both bonds.

In writing resonance formulas, it is important to realize that the nuclear positions must be the same in all electron-dot formulas. Remember that the resonance formulas describe one molecule, and every molecule has a definite nuclear arrangement. The following formula could not be a resonance formula of sulfur dioxide, since sulfur always has a central position in this molecule.

 (incorrect formula)

Attempting to write electron-dot formulas leads us to recognize that delocalized bonding exists in many molecules. Whenever we can write several plausible electron-dot formulas, which usually differ merely in their allocation of single and double bonds to the same kinds of atoms (as in sulfur dioxide), we can expect delocalized bonding.

Example 7.11

Describe the electron structure of the carbonate ion, CO_3^{2-}, in terms of electron-dot formulas.

Solution

One possible electron-dot formula for the carbonate ion is

Since we expect all carbon–oxygen bonds to be equivalent, we must describe the electron structure in resonance terms:

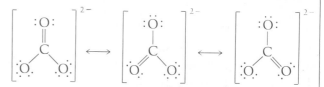

We expect that one electron pair will be delocalized over the region of all three carbon–oxygen bonds.

Exercise 7.15

Describe the bonding in NO_3^- using resonance formulas.

(See Problems 7.57, 7.58, 7.59, and 7.60.)

Metals are examples of extremely delocalized bonding. A sodium metal crystal, for example, can be regarded as an array of Na^+ ions surrounded by a "sea" of electrons (Figure 7.6). The valence or bonding electrons are delocalized over the entire metal crystal. The freedom of these electrons to move throughout the crystal is responsible for the electrical conductivity of a metal.

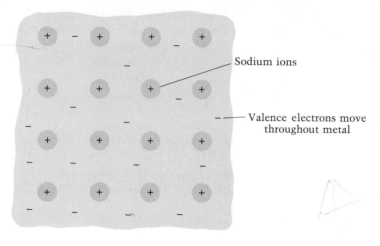

Figure 7.6
Delocalized bonding in sodium metal. The metal consists of positive sodium ions in a "sea" of valence electrons. Valence (bonding) electrons are free to move throughout the metal crystal (colored area).

7.9 Bond Length and Bond Order

The distance between the centers of the nuclei of the two atoms in a bond is called the **bond length** (or **bond distance**). Bond lengths are determined experimentally using x-ray diffraction or from analysis of molecular spectra. The value of a bond length in a particular molecule can sometimes give us a clue to the type of bonding present.

In many cases bond lengths for single covalent bonds in compounds can be predicted from covalent radii. **Covalent radii** are values assigned to atoms, found by taking one-half the single bond length between the nuclei of identical atoms. For example, the Cl—Cl bond length (in Cl_2) is 1.98 Å. Hence, the covalent radius of the Cl atom is $1/2 \times 1.98$ Å = 0.99 Å. Table 7.7 lists single-bond covalent radii for nonmetallic elements. The sum of the radii for two atoms in a bond reproduces known bond lengths in other compounds. For example, to predict the bond length of C—Cl, we add the covalent radii of the two atoms, C and Cl. We get (0.77 + 0.99) Å = 1.76 Å, which compares with the experimental value of 1.78 Å found in most compounds.

Exercise 7.16

Estimate the O—H bond length in H_2O, from the covalent radii listed in Table 7.7.
(See Problems 7.61, 7.62, 7.63, and 7.64.)

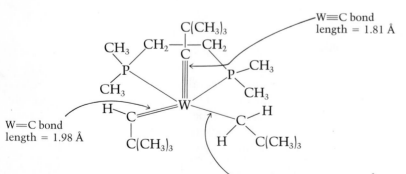

Figure 7.7
This unusual molecule contains tungsten-carbon single, double, and triple bonds and was synthesized by R. R. Schrock of M.I.T. The W—C bond distances were determined by x-ray crystallography by M. R. Churchill of State University of New York, Buffalo. The data show the tungsten-carbon distance decreases from single to double to triple bond (W—C > W=C > W≡C), just as carbon single, double, and triple bonds do in a series such as ethane, C_2H_6, ethylene, C_2H_4, and acetylene, C_2H_2.

				H
				0.30
B	C	N	O	F
0.88	0.77	0.70	0.66	0.64
	Si	P	S	Cl
	1.17	1.10	1.04	0.99
		As	Se	Br
		1.21	1.17	1.14
			Te	I
			1.37	1.33

Table 7.7
Single-Bond Covalent Radii
for Nonmetallic Elements
(in Å)

The **bond order,** defined in terms of the Lewis formula, is the number of pairs of electrons in the bond. For example, in C:C, the bond order is 1 (single bond); in C::C, the bond order is 2 (double bond). Bond length depends on bond order. As the order increases, the bond length decreases, and the strength of the bond increases. For example, look at carbon–carbon bonds. The average C—C bond length is 1.54 Å, whereas C=C is 1.34 Å, and C≡C is 1.20 A. Also see Figure 7.7.■

■ Although a double bond is stronger than a single bond, it is not necessarily less reactive than a single bond. Ethylene, CH_2=CH_2, for example, is more reactive than ethane, CH_3—CH_3, where carbon atoms are linked through a single bond. The reactivity of ethylene results from the simultaneous formation of a number of strong, single bonds.

Example 7.12

Consider the molecules N_2H_4, N_2, and N_2F_2. Which molecule has the shortest nitrogen–nitrogen bond? Which has the longest nitrogen–nitrogen bond?

Solution

First write the Lewis formulas

H—N̈—N̈—H   :N≡N: :F̈—N̈=N̈—F̈:
   | |
   H H
   N_2H_4 N_2 N_2F_2

The nitrogen–nitrogen bond should be the shortest in N_2, where it is a triple bond, and longest in N_2H_4, where it is a single bond. (Experimental values for the nitrogen–nitrogen bond lengths are 1.09 Å for N_2, 1.22 Å for N_2F_2, and 1.47 Å for N_2H_4.)

Exercise 7.17

Formic acid, isolated in 1670, is the irritant in ant bites. The structure of formic acid is

One of the carbon–oxygen bonds has a length of 1.36 Å, the other is 1.23 Å. What is the length of the C=O bond in formic acid?

(See Problems 7.65 and 7.66.)

The presence of electron delocalization in a molecule is often reflected in bond lengths. Bond lengths that are intermediate between those in single and double bonds or between those in double and triple bonds often indicate such electron delocalization. Consider the nitrous oxide molecule, N_2O. It consists of a central N atom, with another N atom on one side and an O atom on the other. The nitrogen–nitrogen bond length in N_2O is 1.13 Å. Compare this with the N≡N length in N_2 of 1.10 Å, the N=N length in N_2F_2 of 1.22 Å, and the N—N length predicted from Table 7.7 of 1.40 Å. We see that the nitrogen–nitrogen bond length in N_2O lies between the values for N≡N and N=N. This agrees with the resonance description of N_2O, which is

$$:N≡N—\ddot{O}: \longleftrightarrow \ddot{N}=N=\ddot{O}$$

According to this description, the length of the nitrogen–nitrogen bond in N_2O is between the length of a triple bond and that of a double bond.

Chemical Nomenclature

Chemical nomenclature is the systematic naming of chemical compounds. One system of naming compounds uses oxidation numbers, discussed in the next section.

7.10 Oxidation Numbers

We define the **oxidation number** (or **oxidation state**) of an atom in a substance as the charge it would have if all the bonding electrons belonged to the more electronegative atoms in the substance. In a compound such as sodium chloride that is composed of monatomic ions, all valence electrons belong to the more electronegative atom. Therefore, the ionic charge equals the oxidation number of the atom. Thus, in sodium chloride, the oxidation number of sodium is $+1$, and the oxidation number of chlorine is -1.

In a covalent compound or in an ion with covalent bonds, such as the polyatomic ion SO_4^{2-}, the electrons are shared between the atoms. Nevertheless, it is useful for bookkeeping to assume that the bonding electrons do belong to the more electronegative atoms. In the case of SO_4^{2-}, we assume that the bonding electrons belong to the oxygen atoms. The oxidation number is then the charge an atom would have after this electron assignment.■

If the two atoms of a bond are different, we assign both bonding electrons completely to the more electronegative atoms. In the hydrogen chloride molecule (HCl), for example, we assign both bonding electrons to chlorine, because it is more electronegative than hydrogen (3.0 for Cl, and 2.1 for H).

■ Besides their application in naming compounds, oxidation numbers are used to balance certain kinds of chemical equations, as shown in Chapter 9.

$$H : \ddot{C}l:$$

Since the H atom has no electrons assigned to it, the oxidation number is $+1$ (one electron less than that in the neutral atom). The chlorine atom has eight valence electrons, one more than that in the neutral atom. Therefore, its oxidation number is -1. If the atoms of a bond are of the same element, the two bonding electrons are assigned one to each atom. For example,

As a result of this electron assignment, the atoms in any elementary substance will have zero net charge, and the oxidation number will be 0.

The following rules can be applied to both ionic and covalent compounds to find the oxidation numbers of the atoms.

1. The oxidation number of an atom in an elementary substance is 0. Thus, the oxidation number of a chlorine atom in Cl_2 or of an O atom in O_2 is 0.
2. The oxidation number of a Group IA (alkali metal) atom in any compound is $+1$; the oxidation number of a Group IIA (alkaline earth) atom in any compound is $+2$.
3. The oxidation number of fluorine is -1 in all of its compounds.
4. The oxidation number of chlorine, bromine, and iodine is -1 in any compound containing only two elements, the halogen with a less electronegative element. ■
5. The usual oxidation number of oxygen in a compound is -2. (The major exceptions are peroxides and superoxides, discussed later in the text.)
6. The oxidation number of hydrogen in most of its compounds is $+1$. (The exceptions are *hydrides*, compounds in which hydrogen is bonded to metallic elements, where hydrogen has the oxidation number -1.)
7. The sum of the oxidation numbers of the atoms in a compound always equals zero. For a polyatomic ion, the oxidation numbers of the atoms add up to the charge on the ion.

■ For a compound such as BrCl, we obtain $+1$ for the bromine and -1 for the chlorine after following the rule of assigning all of the electrons of a covalent bond to the more electronegative atom.

Example 7.13

Use the preceding rules to obtain the oxidation number of chlorine in perchloric acid, $HClO_4$.

Solution

The oxidation numbers of H and O can be assigned immediately (rules 5 and 6). They are given above the atomic symbols.

$$\overset{+1}{H}\ \overset{?}{Cl}\ \overset{-2}{O_4}$$

Recall that the sum of the oxidation numbers for all the atoms in a compound equals zero (rule 7):

$$x_H + x_{Cl} + 4x_O = 0$$

(Here x_H is the oxidation number of the hydrogen atoms, and so forth.) Hence,

$$+1 + x_{Cl} + 4 \times (-2) = 0$$
$$x_{Cl} = +7$$

Exercise 7.18

Find the oxidation number of chromium in potassium dichromate, $K_2Cr_2O_7$.

(See Problems 7.67 and 7.68.)

Example 7.14

What is the oxidation number of sulfur in the sulfate ion, SO_4^{2-}?

Solution

The sum of the oxidation numbers equals the ion charge (rule 7). Hence,

$$x_S + 4x_O = -2$$
$$x_S + 4 \times (-2) = -2$$
$$x_S = +6$$

Exercise 7.19

What is the oxidation number of manganese in the permanganate ion, MnO_4^-?

(See Problems 7.69 and 7.70.)

7.11　Naming Simple Compounds

Before the structural basis of chemical substances became established, compounds were named after people, places, or particular characteristics. Examples are Glauber's salt (sodium sulfate, discovered by J. R. Glauber), sal ammoniac (ammonium chloride, named for the Egyptian deity Ammon from the temple near which the substance was made), and washing soda (sodium carbonate, used for softening of wash water). Today several million compounds are known and thousands of new ones are discovered every year. Without a system for naming compounds, coping with and understanding this multitude of substances would be a hopeless task. There are a number of systems for naming compounds, all based on the formulas or structures of substances. In this section, we will discuss the naming of **binary compounds** (compounds of only two elements), acids, and ionic compounds.

Binary Compounds

The formula of a binary compound made of a metal and a nonmetal lists the metal first. This agrees with the rule given in Section 7.2 for writing ionic compounds (the cation is written before the anion). The rule applies whether the bonding is ionic or covalent. If the binary compound is of two nonmetals (or metalloids) the nonmetal occurring first in the following sequence is written first in the formula:

Element	B	Si	C	Sb	As	P	N	H	Te	Se	S	I	Br	Cl	O	F
Group	IIIA	IVA		VA					VIA			VIIA				

Note that the nonmetals are arranged in order of the groups of the periodic table, listed from the bottom of the group upward. Oxygen, however, is placed just before fluorine, and hydrogen is placed between Groups VA and VIA.

IVA	VA	VIA	VIIA	Table 7.8 Names Used for Second Element in Binary Compounds
C carbide Si silicide	N nitride P phosphide	O oxide S sulfide Se selenide Te telluride	F fluoride Cl chloride Br bromide I iodide	

In writing binary compounds we name the elements in the same order as they appear in the formula. In NaCl we name sodium first. When two elements form only one compound, the compound is named by giving the name of the first element, followed by the name for the second element (normally a nonmetal) ending with the suffix *-ide* (see Table 7.8). Examples are:

NaCl	sodium chloride	HCl	hydrogen chloride
Al_2O_3	aluminum oxide	H_2S	hydrogen sulfide

When two elements can combine to form more than one compound, several different systems of naming are used. If the first-named element has two or more oxidation states, the *Stock system* may be used.■ According to this system, the oxidation state of the first-named element is given as a Roman numeral within parentheses, and this is appended to the element name. The Stock system is preferred for compounds in which the more electropositive element is a metal.

■ Among the main-group metallic elements with more than one oxidation state are Tl, Sn, Pb, and Bi, which have oxidation numbers equal to the group number and the group number minus two. Most transition elements have several oxidation states.

Cu_2O	copper(I) oxide	FeO	iron(II) oxide
CuO	copper(II) oxide	Fe_2O_3	iron(III) oxide

An older system uses suffixes with the stem name of the more electropositive element (which may be from the Latin) to differentiate oxidation states. The suffix *-ic* is used for the higher of two oxidation states, and *-ous* for the lower one.

Cu_2O	cuprous oxide	FeO	ferrous oxide
CuO	cupric oxide	Fe_2O_3	ferric oxide

The principal disadvantage of this system is its limitation to two oxidation states. Many transition elements have three or more common oxidation states. Another disadvantage is that the suffixes indicate only whether the oxidation state is the lower or higher of two values, but do not specify what these values are. (This system is not used in this text.)

The naming system usually preferred for binary compounds of two non-metallic elements uses Greek prefixes to indicate the number of atoms of each element in the formula. These prefixes are

one	*mono-*	five	*penta-*
two	*di-*	six	*hexa-*
three	*tri-*	seven	*hepta-*
four	*tetra-*	eight	*octa-*

The prefix *mono-* is omitted unless it is needed to distinguish between binary compounds of the same two elements. For example, CO and CO_2 are named carbon *mono*xide and carbon *di*oxide. (The prefix *mono-* before carbon is not used; it is understood that there is one atom of carbon in each formula.) Other examples of the use of Greek prefixes are

NO_2	nitrogen *di*oxide	PCl_5	phosphorus *penta*chloride
NCl_3	nitrogen *tri*chloride	SF_6	sulfur *hexa*fluoride
N_2O_4	*di*nitrogen *tetr*oxide	Cl_2O_7	*di*chlorine *hept*oxide

Acids

Acids are substances that give hydrogen ions, H^+, in water solution. An example is hydrogen chloride, HCl. It is a gaseous covalent compound that dissolves in water to give the ions H^+ and Cl^-. We distinguish the formula of the water solution from that of the pure compound by writing HCl(aq) for the solution (*aq* for *aqueous*, meaning water) and HCl(g) for the gaseous substance. The acidic solution obtained from a binary compound of hydrogen and another element (a nonmetal) is named by using the prefix *hydro-* and the suffix *-ic* with the stem name for the nonmetal, followed by the word *acid*. Examples are

HCl(aq)   *hydro*chloric acid $H_2S(aq)$ *hydro*sulfuric acid

Many acids are oxyacids. **Oxyacids** contain hydrogen and oxygen with another element (the characteristic or central element). The formulas are usually written with the H first, followed by the characteristic or central atom, then O. When there are two common oxyacids with the same central atom, they are labeled with *-ous* and *-ic* endings, indicating lower and higher oxidation states of this atom, respectively. Examples are

$\overset{+3}{H}NO_2$ nitr*ous* acid $\overset{+4}{H_2}SO_3$ sulfur*ous* acid
$\overset{+5}{H}NO_3$ nitr*ic* acid $\overset{+6}{H_2}SO_4$ sulfur*ic* acid

(Oxidation states of the central atom are indicated in color above the symbol of the element.) When there are three or four oxyacids of the same central element, the prefixes *hypo-* and *per-* are used with *-ous* and *-ic* to indicate a range of oxidation states of the central atom. The prefix *hypo-* with the suffix *-ous* corresponds to the lowest oxidation state, while *per-* with the suffix *-ic* signifies the maximum oxidation state. The latter is also the group number at the top of the periodic table.

$\overset{+7}{H}ClO_4$ *per*chlor*ic* acid
$\overset{+3}{H}ClO_2$ chlor*ous* acid
$\overset{+5}{H}ClO_3$ chlor*ic* acid
$\overset{+1}{H}ClO$ *hypo*chlorous acid ■

■ The formula for hypochlorous acid is often written as the structural formula HOCl, giving the atoms in the sequence in which they are bonded.

Ionic Substances

Names of ionic substances give the cation name followed by the anion name. Monatomic cations have the name of the element. When the cation can have several different charges, the charge is indicated by a Roman numeral (Stock system) or by the suffixes *-ous* and *-ic*, as described earlier for binary compounds. Monatomic anions are named using the stem name for the element and the suffix *-ide*, as in Table 7.8. (The common polyatomic anions OH^- and CN^- take the suffix *-ide*; OH^- is the hydroxide ion and CN^- is the cyanide ion.) A compound of a monatomic cation and a monatomic anion is a binary compound; it is named according to the rules for binary compounds.

Most polyatomic anions can be obtained from oxyacids by removing one or more hydrogen atoms. The name of the anion thus obtained is written by changing the *-ic* ending of the acid name to *-ate*, or the *-ous* ending to *-ite*. Examples are

Acid		**Anion**	
$HClO_4$	perchlor*ic* acid	ClO_4^-	perchlor*ate* ion
H_2SO_4	sulfur*ic* acid	SO_4^{2-}	sulf*ate* ion
H_2SO_3	sulfur*ous* acid	SO_3^{2-}	sulf*ite* ion
$HClO$	hypochlor*ous* acid	ClO^-	hypochlor*ite* ion

In an acid such as sulfuric acid, H_2SO_4, successive hydrogen ions can be removed to give the ions HSO_4^- and SO_4^{2-}. The *acid anion*, HSO_4^-, has an H atom that can be removed as H^+. It is distinguished from the normal anion by writing the word *hydrogen* before the anion name. (In an older system, not used in this text, one writes the prefix *bi-* in front of the anion name.)

HSO_4^- *hydrogen* sulfate (or *bi*sulfate) ion

When the acid has three or more hydrogen atoms that can be removed, a series of acid anions is possible. For example, phosphoric acid, H_3PO_4, gives the anions $H_2PO_4^-$, HPO_4^{2-}, and PO_4^{3-}. The number of hydrogen atoms in the acid anions are indicated by Greek prefixes in front of the word *hydrogen*. Thus,

$H_2PO_4^-$ *di*hydrogen phosphate ion
HPO_4^{2-} *mono*hydrogen phosphate ion

The following are names of ionic substances that have oxyacid anions.

$CuSO_4$ copper(II) sulfate
$KClO_4$ potassium perchlorate
$CaHPO_4$ calcium monohydrogen phosphate

Example 7.15

Give the names of the following: (a) Mg_3N_2, (b) $CrSO_4$, (c) N_2O_3. Use the Stock system or Greek prefixes, if necessary.

Solution

(a) Magnesium, a Group IIA element, is expected to form only a 2+ ion (the magnesium ion). Nitrogen

(Continued)

(Group VA) is expected to form an anion of charge equal to the group number minus eight (N^{3-}, the nitride ion). Named as an ionic compound, the substance is magnesium nitride. (b) Chromium is a transition element and like most such elements has more than one oxidation state. We can find the oxidation number for Cr if we know the formula of the anion. From Table 7.5, we see that the SO_4 in $CrSO_4$ refers to the anion SO_4^{2-} (the sulfate ion). Thus, the cation is Cr^{2+}, and the name of the compound is chromium(II) sulfate. (c) This is a compound between nonmetals, so we name it with Greek prefixes as dinitrogen trioxide.

Exercise 7.20

Write the names of the following compounds (use the Stock system or Greek prefixes where necessary): (a) CaO, (b) $PbCrO_4$, (c) SO_3.

(See Problems 7.71 and 7.72.)

Example 7.16

Write formulas for the following compounds: (a) ferrous phosphate, (b) titanium(IV) oxide.

Solution

(a) Ferrous ion is Fe^{2+}; phosphate ion is PO_4^{3-}. Hence, the formula is $Fe_3(PO_4)_2$. (b) The oxidation number of oxygen is -2, and that of titanium is $+4$. Since the sum of the oxidation numbers must be zero, the formula is TiO_2.

Exercise 7.21

A compound has the name thallium(III) nitrate. What is the formula?

(See Problems 7.73 and 7.74.)

A Checklist for Review

Important Terms

ionic bond (7.1)
Lewis electron-dot symbols (7.1)
pseudonoble-gas configurations (7.2)
polyatomic ions (7.2)
ionic radius (7.3)
isoelectronic (7.3)
covalent bond (p. 206)
bonding pairs (7.4)

lone (nonbonding) pairs (7.4)
coordinate covalent bond (7.4)
octet rule (7.4)
single bonds (7.4)
double bond (7.4)
triple bond (7.4)
polar covalent bond (7.5)
electronegativity (7.5)

delocalized bonding (7.8)
resonance description (7.8)
bond length (bond distance) (7.9)
covalent radii (7.9)
bond order (7.9)
oxidation number (oxidation state) (7.10)
binary compounds (7.11)
oxyacids (7.11)

Summary of Facts and Concepts

An *ionic bond* is a strong attractive force holding electrically charged atoms (ions) together. An ionic bond can form between two atoms by the transfer of electrons from the valence shell of one atom to the valence shell of the other. Many similar ions attract one another to form a crystalline solid, in which every positive ion is surrounded by negative ions and every negative ion is surrounded by positive ions (except at the outer surfaces of the crystal). Monatomic cations of the main-group elements have charges equal to the group number (or in some cases, the group number minus two). Monatomic anions of the main-group elements have charges equal to the group number minus eight.

A *covalent bond* is a strong attractive force holding two atoms together by the sharing of electrons. These bonding electrons are attracted simultaneously to both atomic nuclei and spend part of the time near one atom and part of the time near the other. If the electron pairs are not equally shared, the bond is *polar*. This polarity results from the difference in *electronegativities* of the atoms, that is, in their unequal abilities to draw bonding electrons to themselves.

Lewis electron-dot formulas are simple representations of the valence-shell electrons of atoms in molecules and ions and are useful for describing the covalent bonding in substances. In such formulas, the atoms satisfy the *octet rule*, that is, every atom has eight valence electrons about it (H has two). No single Lewis formula is sufficient to describe molecules with *delocalized bonding*. In these cases, a *resonance description* may be used. The electron distribution of the molecule is described as a composite of two or more Lewis formulas.

The last sections of the chapter discuss *chemical nomenclature*, the systematic naming of compounds based on their formulas or structures. One system uses *oxidation numbers*, the charges the atoms in a compound would have if the bonding electrons are assigned to the more electronegative atoms of each bond. Rules are given for naming binary compounds, acids, and ionic substances.

Operational Skills

Note: A periodic table is needed for skills 1, 2, 4, 6, and 7.

1. Given a metallic and a nonmetallic main-group element, use Lewis symbols to represent the transfer of electrons to form ions of noble-gas configurations (Example 7.1.).

2. Given an ion, write the electron configuration (Examples 7.2 and 7.3). For an ion of a main-group element, give the Lewis symbol.

3. Given the formulas of a cation and an anion, write the formula of the ionic compound (Example 7.4).

4. Given a series of ions, arrange them in order of increasing ionic radius (Exercise 7.7 and Example 7.5).

5. Given the electronegativities of the atoms, arrange a series of bonds in order by polarity (Example 7.6).

6. Given the molecular formula of a simple compound or ion, write the Lewis electron-dot formula (Examples 7.7, 7.8, 7.9, and 7.10).

7. Given a simple molecule with delocalized bonding, write the resonance description (Example 7.11).

8. Know the relationship between bond order and bond length (Example 7.12).

9. Given the formula of a simple compound or ion, write the oxidation numbers of the atoms, using the rules given in the text (Examples 7.13 and 7.14).

10. Given the name of a simple compound (binary or ionic), write the formula (Example 7.15), or vice versa (Example 7.16).

Review Questions

7.1 Describe the formation of a sodium chloride crystal from atoms.

7.2 Why does sodium chloride normally exist as a crystal rather than as a molecule composed of one cation and one anion?

7.3 Explain what energy terms are involved in the formation of an ionic solid from atoms. In what way should these terms change (become larger or smaller) to give the lowest energy possible for the solid?

7.4 Why do most monatomic cations of the main-group elements have a charge equal to the group number? Why do most monatomic anions of these elements have a charge (in units of e) equal to the group number minus eight?

7.5 Give examples of ions having pseudonoble-gas configurations. In what way do these differ from noble-gas configurations?

7.6 The 2+ ions of transition elements are common. Explain why this might be expected.

7.7 If the formulas of the cation and anion are known, the formula of the compound of these ions can be deduced. What is the basic idea involved in this?

7.8 Explain how ionic radii are obtained from known distances between nuclei in crystals.

7.9 Describe the trends shown by the radii of the monatomic ions for the main-group elements both across a period and down a column.

7.10 Describe the formation of a covalent bond in H_2 from atoms. What does it mean to say that the bonding electrons are shared by the two atoms?

7.11 Give an example of a molecule that has a coordinate covalent bond.

7.12 The octet rule correctly predicts the Lewis formula of many molecules involving main-group elements. Explain why this is so.

7.13 Describe the general trends in electronegativities of the elements in the periodic table both across a period and down a column.

7.14 What is the qualitative relationship between bond polarity and electronegativity difference?

7.15 Describe the kinds of exceptions to the octet rule that are met in compounds of the main-group elements. Give examples.

7.16 What is a resonance description of a molecule? Why is this concept required if we wish to retain Lewis formulas as a description of the electron structure of molecules?

7.17 What is the relationship between bond order and bond length? Use an example to illustrate it.

7.18 Give an example of a binary compound that is ionic. Give an example of a binary compound that is covalent.

Problems

Ionic Bonding

7.19 Write Lewis symbols for the following:
  (a) Ga (b) Ga^{3+}
  (c) Br (d) Br^-

7.20 Write Lewis symbols for the following:
  (a) Sr (b) Sr^{2+}
  (c) Se (d) Se^{2-}

7.21 Use Lewis symbols to represent the transfer of electrons from the following atoms to form ions with noble-gas configurations:
  (a) K and Br
  (b) K and O

7.22 Use Lewis symbols to represent the electron transfer from the following atoms to give ions with noble-gas configurations:
  (a) Ca and Br
  (b) Ca and S

7.23 For each of the following, write the electron configuration and Lewis symbol:
  (a) K^+ (b) Ar
  (c) S^{2-} (d) Ca^{2+}

7.24 For each of the following, write the electron configuration and Lewis symbol:
  (a) Sr^{2+} (b) Kr
  (c) Br^- (d) Se^{2-}

7.25 Write the electron configurations of Sn and Sn^{2+}.

7.26 Write the electron configurations of Bi and Bi^{3+}.

7.27 Give the electron configurations of Cr^{2+} and Cr^{3+}.

7.28 Give the electron configurations of Co^{2+} and Co^{3+}.

7.29 Write the formulas for the compounds of the following ions:
  (a) Fe^{3+} and Cl^- (b) K^+ and SO_4^{2-}
  (c) Li^+ and N^{3-} (d) Sr^{2+} and Cl^-

7.30 For each of the following pairs of ions, write the formula of the corresponding compound:
  (a) Co^{2+} and Br^- (b) NH_4^+ and SO_4^{2-}
  (c) Na^+ and PO_4^{3-} (d) Fe^{3+} and NO_3^-

Ionic Radii

7.31 Arrange the following pairs in the order of increasing radius and explain the order:
  (a) O^{2-}, O
  (b) Na^+, Na

7.32 Arrange the following pairs in the order of increasing radius and explain the order:
  (a) Br, Br^-
  (b) Sr, Sr^{2+}

7.33 Without looking at Table 7.6, arrange the following by increasing ionic radius: Se^{2-}, Te^{2-}, S^{2-}. Explain how you arrived at this order. (You may use a periodic table.)

7.34 Which has the larger radius, N^{3-} or P^{3-}? Explain. (You may use a periodic table.)

7.35 Arrange the following in order of increasing ionic radius: Sc^{3+}, S^{2-}, and K^+. Explain this order. (You may use a periodic table.)

7.36 Arrange the following in order of increasing ionic radius: Br^-, Se^{2-}, Sr^{2+}. Explain the order. (You may use a periodic table.)

Covalent Bonding

7.37 Use Lewis symbols to show the reaction of atoms to form phosphine, PH_3. Point out which electron pairs in the Lewis formula of PH_3 are bonding and which are nonbonding (lone pairs).

7.39 Assuming that the atoms form the normal number of covalent bonds, what is the molecular formula of the simplest compound of silicon and chlorine atoms?

7.38 Use Lewis symbols to show the reaction of atoms to form hydrogen selenide, H_2Se. Point out bonding pairs and lone pairs in the electron-dot formula of this compound.

7.40 Assuming that the atoms form the normal number of covalent bonds, what is the molecular formula of the simplest compound of arsenic and hydrogen atoms?

Polar Covalent Bonds; Electronegativity

7.41 With the aid of a periodic table (not Figure 7.5) arrange the following in order of increasing electronegativity:
   (a) Sr, Cs, Ba   (b) Ca, Ge, Ga
   (c) P, As, S

7.43 Decide which of the following bonds is least polar on the basis of electronegativities of atoms: H—Se, P—Cl, N—Cl.

7.45 Indicate the partial charges for the bonds given in Problem 7.43 (using the symbols δ^+ and δ^-).

7.42 Using a periodic table (but not Figure 7.5), arrange the following in order of increasing electronegativity:
   (a) P, O, N   (b) Na, Al, Mg
   (c) C, Al, Si

7.44 Arrange the following bonds in order by increasing polarity using electronegativities of atoms: Si—O, C—Br, As—Br.

7.46 Indicate the partial charges for the bonds given in Problem 7.44 (using the symbols δ^+ and δ^-).

Writing Lewis Formulas

7.47 Write Lewis formulas for the following molecules:
   (a) BrCl   (b) H_2S
   (c) $SeOF_2$ (d) H_3PO_4

7.49 Write Lewis formulas for the following molecules:
   (a) N_2 (b) COS
   (c) P_2H_4

7.51 Write Lewis formulas for the following ions:
   (a) ClO_4^- (b) PO_4^{3-}
   (c) O_2^{2-}

7.48 Write Lewis formulas for the following molecules:
   (a) Cl_2 (b) PH_3
   (c) SO_2Cl_2 (d) $HClO_2$

7.50 Write Lewis formulas for the following molecules:
   (a) CO   (b) CS_2
   (c) $COBr_2$

7.52 Write Lewis formulas for the following ions:
   (a) ClO_3^- (b) SeO_4^{2-}
   (c) CN^-

Exceptions to the Octet Rule

7.53 Write Lewis formulas for the following:
   (a) BCl_3
   (b) $TlCl_2^+$

7.55 Write Lewis formulas for the following:
   (a) XeF_2 (b) $TeCl_4$
   (c) SF_6 (d) ClF_5

7.54 Write Lewis formulas for the following:
   (a) BeF_2
   (b) BeF_3^-

7.56 Write Lewis formulas for the following:
   (a) AsF_5 (b) ClF_4^+
   (c) ICl_4^- (d) PCl_6^-

Resonance

7.57 Write a resonance description for the following:
   (a) FNO_2
   (b) SO_3

7.59 Give the resonance description of the formate ion. The skeleton structure is

7.58 Write a resonance description for the following:
   (a) NO_2^-
   (b) HNO_3

7.60 Use resonance to describe the electron structure of nitromethane, CH_3NO_2. The skeleton structure is

Bond Length and Bond Order

7.61 Use covalent radii (Table 7.7) to estimate the length of the H—S bond in hydrogen sulfide, H_2S.

7.63 Calculate the bond length for each of the following single bonds, using covalent radii (Table 7.7):
   (a) C—H   (b) S—Cl
   (c) Br—Cl   (d) Si—O

7.65 One of the following compounds has a carbon–nitrogen bond length of 1.16 Å, the other has a carbon–nitrogen bond length of 1.47 Å. Match a bond length with each compound.

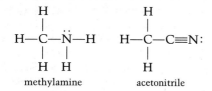

methylamine   acetonitrile

7.62 What do you expect for the B—F bond length in boron trifluoride, BF_3, on the basis of the covalent radii (Table 7.7)?

7.64 Calculate the C—H and C—Cl bond lengths in chloroform, $CHCl_3$, using values for the covalent radii from Table 7.7. How do these values compare with the experimental values: C—H, 1.07 Å; C—Cl, 1.77 Å?

7.66 Which of the following two compounds has the shorter carbon–oxygen bond?

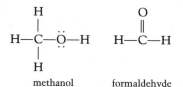

methanol   formaldehyde

Oxidation Numbers

7.67 Use the rules given in the text to assign the oxidation numbers of the underlined atoms:
   (a) $\underline{Mn}_2O_7$ (b) $\underline{Ga}_2O_3$
   (c) $\underline{Xe}F_2$ (d) $K\underline{Cl}O_3$

7.69 Write the oxidation numbers for the following:
   (a) B in BF_4^- (b) N in NO_2^-
   (c) Cr in CrO_4^{2-} (d) P in PO_4^{-3}

7.68 Write the oxidation numbers of the underlined atoms, using the rules described in the text:
   (a) $H_2\underline{S}$ (b) $\underline{B}_2O_3$
   (c) $K\underline{Cl}O_2$ (d) $\underline{Hg}_2Cl_2$

7.70 Write the oxidation numbers for the following:
   (a) Bi in BiO_3^- (b) As in AsO_3^{3-}
   (c) Mn in MnO_4^- (d) Al in $Al(OH)_4^-$

Naming Compounds

7.71 Name the following compounds:
   (a) Na_2SO_3 (b) SiO_2
   (c) $CuCl$ (d) Cr_2O_3

7.72 Name the following compounds:
   (a) CrO_3 (b) Mn_2O_7
   (c) NH_4HCO_3 (d) $Cu(NO_3)_2$

7.73 Write the formulas of:
  (a) lead(II) dichromate
  (b) barium bicarbonate
  (c) silicon tetrachloride
  (d) ferrous acetate

7.74 Write the formulas of:
  (a) sodium thiosulfate
  (b) cuprous oxide
  (c) calcium bicarbonate
  (d) tin(IV) fluoride

Additional Problems

7.75 For each of the following pairs of elements, state whether the binary compound formed is likely to be ionic or covalent. Give the formula and name of the compound.
  (a) Sr, O   (b) C, Br
  (c) Ga, F   (d) N, Br

7.76 For each of the following pairs of elements, state whether the binary compound formed is likely to be ionic or covalent. Give the formula and name of the compound.
  (a) Na, S   (b) Al, F
  (c) Ca, Cl   (d) Si, Br

7.77 Calcium phosphate is composed of Ca^{2+} and PO_4^{3-} ions. Write the formula of calcium phosphate.

7.78 Write the formula of aluminum sulfate. The compound is composed of Al^{3+} and SO_4^{2-} ions.

7.79 Knowing only that barium is a Group IIA element and that selenium is a Group VIA element, write the formula of barium selenide (an ionic compound of barium and selenium).

7.80 Write the formula of lithium nitride, an ionic compound of lithium (Group IA) and nitrogen (Group VA).

**7.81 Calculate a value for the electronegativity of chlorine using Mulliken's formula. Use values of the ionization energy and electron affinity in kilojoules per mole (see Figure 6.13 and Table 6.4). Divide this by 230 to get a value comparable with Pauling's scale.

**7.82 Calculate a value for the electronegativity of oxygen using Mulliken's formula. Use values of the ionization energy and electron affinity in kilojoules per mole (see Figure 6.13 and Table 6.4) and divide by 230 to get Pauling's scale.

7.83 Write formulas for compounds of iron(II) ion with each of the following anions:
  (a) acetate   (b) chloride
  (c) oxalate   (d) phosphate

7.84 Write formulas for compounds of iron(III) ion with each of the following anions:
  (a) acetate   (b) chloride
  (c) oxalate   (d) phosphate

7.85 What are the oxidation numbers of all the atoms in each of the following compounds?
  (a) $ZnSO_4$ (b) $CeOCl$
  (c) $CaWO_4$ (d) $NaAuCl_4$

7.86 What are the oxidation numbers of all the atoms in each of the following compounds?
  (a) $NaIO_4$ (b) $CaTeO_3$
  (c) Ag_2SO_4 (d) $NaOCl$

7.87 Iodic acid, HIO_3, is a colorless, crystalline compound. What is the electron-dot formula of iodic acid?

7.88 Selenic acid, H_2SeO_4, is a white, crystalline substance and a strong acid. Write the electron-dot formula of selenic acid.

7.89 Sodium amide, known commercially as sodamide, is used in preparing indigo, the dye used to color blue jeans. It is an ionic compound with the formula $NaNH_2$. What is the electron-dot formula of the amide anion, NH_2^-?

7.90 Lithium aluminum hydride, $LiAlH_4$, is an important reducing agent (an element or compound that generally has a strong tendency to give up electrons in its chemical reactions). Write the electron-dot formula of the AlH_4^- ion.

7.91 Nitronium perchlorate, NO_2ClO_4, is a reactive salt of the nitronium ion NO_2^+. Write the electron-dot formula of NO_2^+.

7.92 Solid phosphorus pentabromide, PBr_5, has been shown to have the ionic structure $[PBr_4^+][Br^-]$. Find the electron-dot formula of the PBr_4^+ cation.

7.93 Write electron-dot formulas for the following:
  (a) $SeOCl_2$ (b) CS_2
  (c) $GaCl_4^-$ (d) C_2^{2-}

7.94 Write electron-dot formulas for the following:
  (a) $POBr_3$ (b) Si_2H_6
  (c) $SeCN^-$ (d) NO^+

*7.95 Give a resonance description for the following:
  (a) N_2O_4
  (b) OCN^-

*7.96 Give a resonance description for the following:
  (a) $C_2O_4^{2-}$
  (b) N_3^-

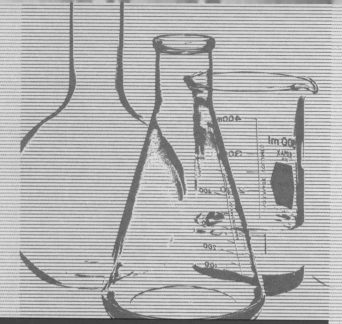

8. *Molecular Geometry and Chemical Bonding Theory*

Molecular Geometry and Directional Bonding

8.1 The Valence-Shell Electron-Pair Repulsion (VSEPR) Model
Two Electron Pairs (Linear Arrangement)/
Three Electron Pairs (Trigonal Planar Arrangement)/
Four Electron Pairs (Tetrahedral Arrangement)/
Five Electron Pairs (Trigonal Bipyramidal Arrangement)/
Six Electron Pairs (Octahedral Arrangement)/
Summary of the VSEPR Model
8.2 Dipole Moment and Molecular Geometry
8.3 Valence Bond Theory Basic Theory/
Hybrid Orbitals
8.4 Description of Multiple Bonding

Molecular Orbital Theory

8.5 Principles of Molecular Orbital Theory
Bonding and Antibonding Orbitals/ Bond Order/
Factors Determining Orbital Interaction
8.6 Electron Configurations of Diatomic Molecules
8.7 Molecular Orbitals and Delocalized Bonding

We know from various experiments that molecules have definite shapes or geometries. By definite shape, we mean that the atoms of a molecule occupy definite positions relative to one another in three-dimensional space. We also know that the shape of a molecule controls some of its chemical and physical properties. For example, the two structures shown below represent molecules having the same formula but different shapes:

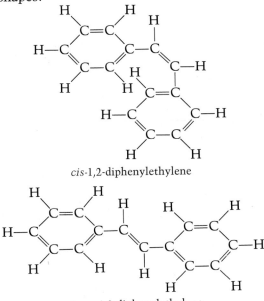

cis-1,2-diphenylethylene

trans-1,2-diphenylethylene

Trans-1,2-diphenylethylene is a solid at room temperature; *cis*-1,2-diphenylethylene is a liquid.

In this chapter we will be concerned with aspects of geometrical and electronic structure of molecules that can be understood by using simple, but powerful, models for the distribution of electrons in molecules. One of the most important issues to be dealt with concerns the shapes of small molecules. For example, when first encountering a molecule, how can one predict its shape? The issue can be illustrated by considering the molecules BF_3 and PF_3, which both contain three fluorine atoms attached to a central atom. The BF_3 molecule has a *trigonal planar* structure, where all four atoms are in a plane with the F atoms oriented in a regular array about the B atom. The PF_3 molecule has a *trigonal pyramidal* structure, with the P atom above a plane defined by the three F atoms.

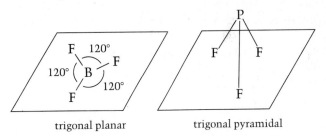

trigonal planar   trigonal pyramidal

We will see how to predict such structural differences.

Chapter Overview

In this chapter, we will extend our knowledge of bonding by considering spatial aspects. We will first look at *molecular geometry* and describe it in terms of the arrangement of electron pairs about an atom. Then we will touch on the relationship between *molecular geometry and dipole moment* (an important experimental measure of electrical polarity in molecules). In the final sections of the first part of this chapter, we will see how molecular geometry and bonding are related by *valence bond theory* to the directional character of orbitals.

In the second part of this chapter, we will look at *molecular orbital theory* as an alternative to valence bond theory. Molecular orbital theory uses the concept of electron configurations introduced in Chapter 6. This theory explains certain features of bonding that cannot be explained by valence bond theory.

Molecular Geometry and Directional Bonding

The spatial arrangement of atoms in a molecule is called the *molecular geometry* or *molecular shape*. Experimental determination of the precise positions of the nuclei in a molecule is a complicated problem. Yet, we can often predict the approximate geometry using a simple model. This model assumes that molecular shape is determined by the mutual repulsions of valence electrons about an atom in a molecule.

8.1 The Valence-Shell Electron-Pair Repulsion (VSEPR) Model

The **valence-shell electron-pair repulsion (VSEPR) model**■ is a model for predicting the shapes of molecules and ions. In this model, valence-shell electron pairs are arranged about an atom so that electron pairs are kept as far away from one another as possible, thus minimizing electron pair repulsions.■ For example, if there are only two electron pairs in the valence shell of an atom, these pairs tend to be at opposite sides of the nucleus, so that the repulsion is minimized. This gives a *linear* arrangement of electron pairs; that is, the electron pairs mainly occupy regions of space at an angle of 180° to one another (Figure 8.1).

If there are three electron pairs in the valence shell of an atom, they tend to be arranged in a plane directed toward the corners of a triangle of equal sides (equilateral triangle). This arrangement is called *trigonal planar*, in which the regions of space occupied by electron pairs are directed at 120° angles to one another (Figure 8.1).

Four electron pairs in the valence shell of an atom tend to have a *tetrahedral* arrangement. That is, if we imagine the atom at the center of a regular tetrahedron, each region of space in which an electron pair mainly lies will extend toward a corner or vertex (Figure 8.1). (A regular tetrahedron is a geometrical shape with four faces, each an equilateral triangle. Thus, it has the form of a triangular pyramid.) The regions of space mainly occupied by electron pairs are directed at approximately 109.5° angles to one another.

Five electron pairs tend to have a *trigonal bipyramidal* arrangement. The electron pairs tend to be directed to the corners of a trigonal bipyramid, a figure formed by placing the face of one tetrahedron onto the face of another tetrahedron (Figure 8.1).

Six electron pairs tend to have an *octahedral* arrangement. The electron pairs tend to be directed to the corners of a regular octahedron, a figure that has eight triangular faces and six vertexes or corners (Figure 8.1).

To predict the relative positions of other atoms around an atom using the VSEPR model, we first note the arrangement of valence-shell electron pairs. Some of these electron pairs will be bonding pairs and some will be lone pairs. *The direction in space of the bonding pairs gives us the molecular geometry.* For each arrangement of electron pairs, there are usually several possible molecular geometries (Figure 8.2). Let us look at these geometries.

■ The acronym VSEPR is pronounced "vesper."

■ The repulsion of electron pairs is partly electrostatic (due to electrical charges), but is mainly explained by the Pauli exclusion principle. Each electron pair has one electron with spin up and another with spin down. Since electrons of the same spin avoid being in the same region of space, each electron pair appears to strongly repel other pairs. Thus, each electron pair must occupy a different region about an atom.

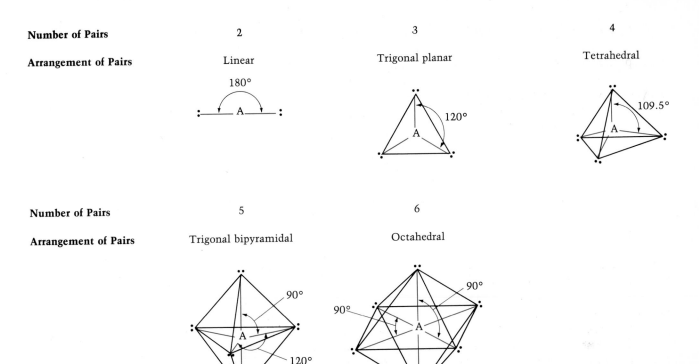

Figure 8.1
Arrangements of electron pairs about an atom, A.

Two Electron Pairs (Linear Arrangement)

Consider the geometry of a molecule with the general formula AX_n. Atom A is a central atom around which are arranged the X atoms (the X atoms need not be identical). To find the geometry, we first determine the number of valence-shell electron pairs around A. We can get this information from the electron-dot formula of AX_n. For example, following the rules given in Section 7.6 the BeF_2 molecule has the formula ■

■ Beryllium fluoride, BeF_2, is normally a solid. The BeF_2 molecule exists in the vapor phase at high temperature.

$$:\ddot{F}:Be:\ddot{F}:$$

There are two electron pairs in the valence shell for beryllium, and the VSEPR model predicts that they will have a linear arrangement (see Figure 8.1). Fluorine atoms are bonded in the same direction as the electron pairs. Hence the geometry of the BeF_2 molecule is linear, that is, arranged in a straight line (see Figure 8.2).

The VSEPR model can also be applied to molecules with multiple bonds. In this case, each multiple bond is treated as if it were a single electron pair (since all pairs of a multiple bond are required to be in approximately the same region). To predict the geometry of carbon dioxide, for example, we first write the electron-dot formula

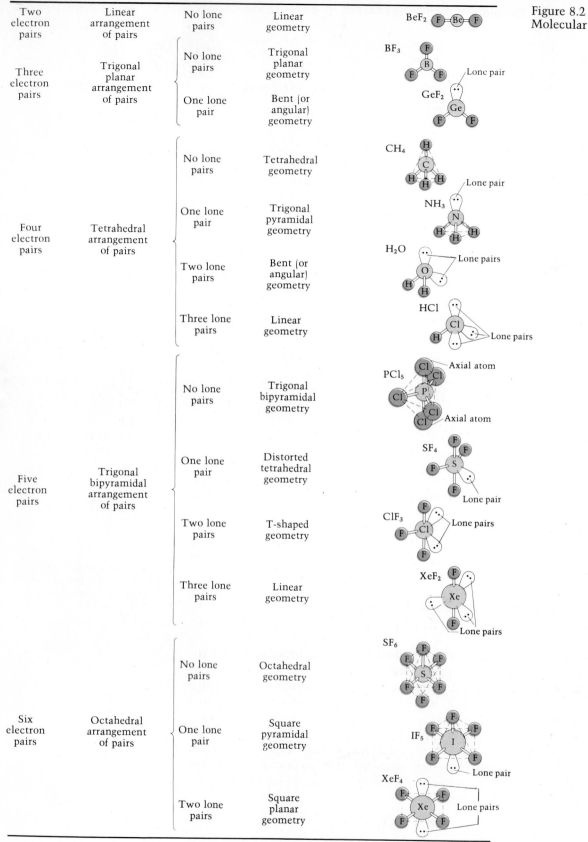

Two electron pairs	Linear arrangement of pairs	No lone pairs	Linear geometry	BeF_2
Three electron pairs	Trigonal planar arrangement of pairs	No lone pairs	Trigonal planar geometry	BF_3
		One lone pair	Bent (or angular) geometry	GeF_2
Four electron pairs	Tetrahedral arrangement of pairs	No lone pairs	Tetrahedral geometry	CH_4
		One lone pair	Trigonal pyramidal geometry	NH_3
		Two lone pairs	Bent (or angular) geometry	H_2O
		Three lone pairs	Linear geometry	HCl
Five electron pairs	Trigonal bipyramidal arrangement of pairs	No lone pairs	Trigonal bipyramidal geometry	PCl_5
		One lone pair	Distorted tetrahedral geometry	SF_4
		Two lone pairs	T-shaped geometry	ClF_3
		Three lone pairs	Linear geometry	XeF_2
Six electron pairs	Octahedral arrangement of pairs	No lone pairs	Octahedral geometry	SF_6
		One lone pair	Square pyramidal geometry	IF_5
		Two lone pairs	Square planar geometry	XeF_4

Figure 8.2
Molecular geometries

We have two electron groups about the carbon atom, and these are treated as if there were two pairs on carbon. Thus, according to the VSEPR model, the bonds will be arranged linearly, and the geometry of carbon dioxide is linear.

Three Electron Pairs (Trigonal Planar Arrangement)

Let us now predict the geometry of the boron trifluoride molecule, BF_3, introduced in the chapter opening. The electron-dot formula is

$$:\ddot{\underset{..}{F}}:$$
$$:\ddot{\underset{..}{F}}:B:\ddot{\underset{..}{F}}:$$

The three pairs on boron have a trigonal planar arrangement, and the molecular geometry, which is determined by the directions of the bonding pairs, is also trigonal planar (Figure 8.2).

Germanium difluoride, GeF_2, is an example of a molecule with three electron pairs, one of which is a lone pair.■ The electron-dot formula is

$$:\ddot{\underset{..}{F}}:Ge:$$
$$:\ddot{\underset{..}{F}}:$$

The three electron pairs have a trigonal planar arrangement, and the molecular geometry, which is determined by the direction of bonds to germanium, is *bent* or *angular* (Figure 8.2).

Note the difference between the *arrangement of electron pairs* and the *molecular geometry*, or the arrangement of nuclei. What we usually "see" by means of x-ray diffraction and similar methods are the nuclear positions, that is, the molecular geometry. Unseen, but nevertheless important, are the lone pairs, which occupy definite positions about an atom according to the VSEPR model.

To take a somewhat more complicated example, consider sulfur dioxide, SO_2. The electron-dot formula is

$$\ddot{\underset{..}{O}}::S:\qquad\text{(one of two possible resonance formulas)}$$
$$:\ddot{\underset{..}{O}}:$$

We need not be concerned that each bond is in fact an intermediate single–double bond. Rather, it is the three groups of electrons about sulfur (two bonds and one lone pair) that determine the geometry. These groups have a trigonal planar arrangement, and sulfur dioxide is a bent molecule.

Four Electron Pairs (Tetrahedral Arrangement)

The common and most important case of four electron pairs about the central atom (the octet rule) leads to four different molecular geometries, depending on the number of bonds formed. As examples, we have

H	H	H	
H:C:H	:N:H	:O:H	:Cl:H
H	H		
CH_4	NH_3	H_2O	HCl
Molecular geometry: tetrahedral	trigonal pyramidal	bent	linear

■ Germanium difluoride, GeF_2, is normally a white, crystalline solid with a polymeric structure similar to that of beryllium chloride (see Exercise 7.13). The GeF_2 molecule exists in the vapor phase.

In each case, the electron pairs are arranged tetrahedrally, and one or more atoms are bonded in these tetrahedral directions to give the different geometries (Figure 8.2).

When all electron pairs are bonding, as in methane, CH_4, the molecular geometry is tetrahedral. When three of the pairs are bonding and the other is nonbonding, as in ammonia, NH_3, the molecular geometry is *trigonal pyramidal*. The name comes from the shape formed by the connecting lines between the nuclei (see Figure 8.2). In this figure, the nitrogen in NH_3 is at the apex of the pyramid, with the three hydrogens extending downward to form the triangular base of the pyramid.

Now we can answer the question raised in the chapter opening: Why is the BF_3 molecule trigonal planar and the PF_3 molecule trigonal pyramidal? This difference is because in BF_3 there are three pairs of electrons around boron, and in PF_3 there are four pairs around phosphorus. As in NH_3, there are three bonding pairs and one lone pair in PF_3, which give rise to the trigonal pyramidal geometry.

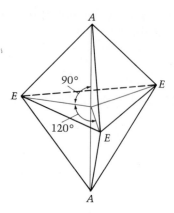

Figure 8.3
The trigonal bipyramidal arrangement. Electron pairs are directed along the colored lines to the vertexes of the trigonal bipyramid. Equatorial directions are labeled E, and axial directions are labeled A.

Five Electron Pairs (Trigonal Bipyramidal Arrangement)

Large atoms like phosphorus can accommodate more than eight valence electrons. The phosphorus atom in phosphorus pentachloride, PCl_5, has five electron pairs in its valence shell. With five electron pairs around phosphorus, all bonding, PCl_5 should have a trigonal bipyramidal molecular geometry (Figure 8.2). Note, however, that unlike the cases we have discussed so far the vertexes of the trigonal bipyramid are not all equivalent (that is, the angles between the electron pairs are not all the same). Thus, the directions to which the electron pairs point are not equivalent. Two of the directions, called *axial directions*, form an axis through the central atom (see Figure 8.3). They are 180° apart. The other three directions are called *equatorial directions*. These point toward the corners of the equilateral triangle that lies on a plane through the central atom, perpendicular (at 90°) to the axial directions. The equatorial directions are 120° from each other.

Other molecular geometries form when one or more of the five electron pairs are lone pairs. As an example, consider the sulfur tetrafluoride molecule, SF_4.■ The Lewis electron-dot formula is

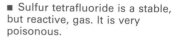

■ Sulfur tetrafluoride is a stable, but reactive, gas. It is very poisonous.

$$\begin{array}{c} \ddot{\text{F}}\text{:} \\ \ddot{\text{F}}\diagdown \text{ }| \\ \diagup\text{S:} \\ \ddot{\text{F}}\diagup\text{ }| \\ \text{:}\ddot{\text{F}}\text{:} \end{array}$$

The five electron pairs about sulfur should have a trigonal bipyramidal arrangement. Because the axial and equatorial positions of the electron pairs are not equivalent, we must decide in which of these positions the lone pair appears.

Since a lone pair is attracted to only one nucleus instead of two, the electrons are held less tightly and occupy more space. The lone pair is somewhat "larger" than a bonding pair. Therefore, the total repulsion between all pairs should be lower if the lone pair is in a position that puts it directly adjacent to the fewest number of other pairs.

An electron pair in an equatorial position is directly adjacent to only *two* other pairs—the two axial pairs at 90°. The other two equatorial pairs are further away at 120°. On the other hand, an electron pair in an axial position is directly adjacent to *three* other pairs—the three equatorial pairs at 90°. The other axial pair is much farther away at 180°.

Thus, we expect lone pairs to occupy equatorial positions in the trigonal bipyramidal arrangement. For sulfur tetrafluoride, this gives a *distorted tetrahedral molecular geometry* (Figure 8.2). Unlike a regular tetrahedron, a distorted tetrahedron has three different bond angles.

Chlorine trifluoride, ClF_3, has the electron-dot formula ■

■ Chlorine trifluoride is a colorless gas with a sweet, suffocating odor. It is very reactive and highly irritating to the eyes and skin.

The lone pairs on chlorine occupy two of the equatorial positions of the trigonal bipyramidal arrangement, giving a *T-shaped molecular geometry* (Figure 8.2). The four atoms all lie in one plane, with the chlorine nucleus at the intersection of the "T."

Xenon difluoride, XeF_2, has the electron-dot formula

The three lone pairs on xenon occupy the equatorial positions of the trigonal bipyramidal arrangement, giving a linear molecular geometry (Figure 8.2).

Six Electron Pairs (Octahedral Arrangement)

There are six electron pairs (all bonding pairs) about sulfur in sulfur hexafluoride, SF_6. Thus, it has an octahedral molecular geometry, with sulfur at the center of the octahedron and fluorine atoms at the vertexes (Figure 8.2).

Iodine pentafluoride, IF_5, has the electron-dot formula ■

■ The halogens form a number of molecules with one another. Except for those with only two atoms, these are mainly molecules with a chlorine, bromine, or iodine atom surrounded by an odd number of fluorine atoms.

The lone pair on iodine occupies one of the six equivalent positions in the octahedral arrangement, giving a *square pyramidal molecular geometry* (Figure 8.2). The name derives from the shape formed by drawing lines between atoms.

Xenon tetrafluoride, XeF_4, has the electron-dot formula ■

■ Xenon forms a series of fluorides, including XeF_2, XeF_4, and XeF_6. Xenon tetrafluoride, XeF_4, is a colorless, crystalline compound formed by the reaction of xenon, Xe, with fluorine, F_2.

The two lone pairs on xenon occupy opposing positions in the octahedral arrangement to minimize their repulsion. The result is a *square planar molecular geometry* (Figure 8.2).

Summary of the VSEPR Model

Let us summarize the steps to follow in order to predict the geometry of an AX_n molecule or ion by the VSEPR method (all X atoms of AX_n need not be identical).

1. Determine how many electron pairs are around the central atom, from the electron-dot formula. Count a multiple bond as one pair.
2. Arrange the electron pairs as shown in Figure 8.1.
3. Give the molecular geometry from the directions of bonding pairs, as shown in Figure 8.2.

The next examples illustrate the method.

Example 8.1

Predict the geometry of the following molecules, using the VSEPR method: (a) $BeCl_2$, (b) $SnCl_2$, (c) $SiCl_4$

Solution

(a) Following the steps outlined in Section 7.6 for writing Lewis formulas, we distribute the valence electrons to the skeleton structure of $BeCl_2$ as follows:

$$:\!\ddot{C}l\!:\!Be\!:\!\ddot{C}l\!:$$

This gives fewer than an octet of electrons about Be, but since Cl atoms do not normally form multiple bonds, we leave the formula as it is. (Moreover, with the dot symbol ·Be·, Be would be expected to form two single bonds.) The two pairs on Be have a linear arrangement, leading to a linear molecular geometry for $BeCl_2$. (See Figure 8.2 under 2 electron pairs, 0 lone pairs.) (b) Distributing the valence electrons to the skeleton struc-

ture of $SnCl_2$, we get

Again, the Cl atoms do not normally form multiple bonds, so we leave this electron-dot formula as it is. The three electron pairs around the Sn atom have a trigonal planar arrangement and thus the molecular geometry of $SnCl_2$ is bent. (See Figure 8.2 under 3 electron pairs, 1 lone pair.) (c) The electron-dot formula of $SiCl_4$ is

$$\begin{array}{c} :\!\ddot{C}l\!: \\ :\!\ddot{C}l\!:\!\underset{}{Si}\!:\!\ddot{C}l\!: \\ :\!\ddot{C}l\!: \end{array}$$

The molecular geometry is tetrahedral. (See Figure 8.2 under 4 electron pairs, 0 lone pairs.)

Exercise 8.1

Use the VSEPR method to predict the geometry of the following ion and molecules:
(a) ClO_3^- (b) OF_2 (c) SiF_4

(See Problems 8.17, 8.18, 8.19, and 8.20.)

Example 8.2

What do you expect for the geometry of tellurium tetrachloride, $TeCl_4$?

Solution

First, we distribute the valence electrons to the Cl atoms to satisfy the octet rule. Then we allocate the remaining ones to the central atom, Te (following the steps outlined in Section 7.6). The electron-dot formula is

Thus, there are five electron pairs in the valence shell of Te in $TeCl_4$. Of these, four are bonding pairs and one is a lone pair. The arrangement of electron pairs is trigonal bipyramidal. We expect the lone pair to occupy an equatorial position, so $TeCl_4$ has a distorted tetrahedral molecular geometry. (See Figure 8.2 under 5 electron pairs, 1 lone pair.)

Exercise 8.2

According to the VSEPR method, what molecular geometry would you predict for iodine trichloride, ICl_3?

(See Problems 8.21, 8.22, 8.23, and 8.24.)

8.2 Dipole Moment and Molecular Geometry

The VSEPR model provides a simple procedure for predicting the geometry of a molecule. However, predictions must be verified by experiment. Information about the geometry of a molecule can sometimes be obtained from an experimental quantity called the **dipole moment,** which is related to the polarity of the bonds in a molecule. It is a quantitative measure of the degree of charge separation in a molecule. The polarity of a bond, such as that in HCl, is characterized by a separation of electrical charge. We can represent this in HCl by indicating partial charges, δ^+ and δ^-, on the atoms.

$$\overset{\delta^+}{H}-\overset{\delta^-}{Cl}$$

Any molecule that has a net separation of charge, as in HCl, has a dipole moment. A molecule in which the distribution of electrical charge is equivalent to charges $+\delta$ and $-\delta$ separated by a distance d has a dipole moment equal to δd. Dipole moments are usually measured in units of *debyes* (D). In SI units, dipole moments are measured in coulomb-meters (C·m), and $1\ D = 3.34 \times 10^{-30}$ C·m.

Measurements of dipole moments are based on the fact that polar molecules (molecules having a dipole moment) can be oriented by an electric field. Figure 8.4 shows an electric field generated by charged plates. Note that the polar molecules tend to align themselves so that the negative ends of the molecules point toward the positive plate, and the positive ends point toward the negative plate. This orientation of the molecules will affect the *capacitance* of the charged plates (the capacity of the plates to hold a charge).

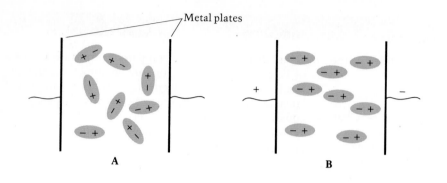

Figure 8.4
Alignment of polar molecules
by an electric field. (a) A polar substance is placed between metal plates. (b) When
these plates are connected to
an electrical voltage (whose
direction is shown by the +
and − signs), the polar molecules align so that their negative ends point toward the
positive plate.

Consequently, measurements of the capacitance of plates separated by different substances can be used to obtain the dipole moments of those substances.

We can sometimes relate the presence or absence of a dipole moment in a molecule to its molecular geometry. For example, consider the carbon dioxide molecule. Each carbon–oxygen bond has a polarity in which the more electronegative oxygen atom has a partial negative charge:

$$\overset{\delta^-}{O}=\overset{2\delta^+}{C}=\overset{\delta^-}{O}$$

We will denote the dipole-moment contribution from each bond (the bond dipole) by an arrow ($+\!\!\longrightarrow$) with a positive sign at one end. The dipole-moment arrow points from the positive partial charge toward the negative partial charge. Thus, we can rewrite the formula for carbon dioxide as

$$O\overset{\longleftarrow}{=}C\overset{\longrightarrow}{=}O$$

Each bond dipole, like a force, is a *vector* quantity; that is, it has both magnitude and direction. Like forces, two bond dipoles of equal magnitude but opposite direction cancel each other. (Think of two groups of people in a tug of war. As long as each group pulls on the rope with the same force but in the opposite direction, there is no movement—the net force is zero.) Since the two carbon–oxygen bonds in CO_2 are equal but point in opposite directions, they give a net dipole moment of zero for the molecule.

For comparison, let us consider the water molecule. The bond dipoles point from the hydrogen atoms toward the more electronegative oxygen:

$$\overset{\nearrow \; O \; \nwarrow}{\underset{H \qquad H}{\diagup \quad \diagdown}}$$

Here, however, the two bond dipoles do not point directly toward or away from each other. As a result, they add together to give a nonzero dipole moment for the molecule:

The dipole moment of H_2O has been observed to be 1.94 D.

Formula	Molecular Geometry	Dipole Moment	Table 8.1 Relationship between Molecular Geometry and Dipole Moment
AX	Linear	Usually nonzero	
AX_2	Linear	Zero	
	Bent	Usually nonzero	
AX_3	Trigonal planar	Zero	
	Trigonal pyramidal	Usually nonzero	
	T-shaped	Usually nonzero	
AX_4	Tetrahedral	Zero	
	Square planar	Zero	
	Distorted tetrahedral	Usually nonzero	
AX_5	Trigonal bipyramidal	Zero	
	Square pyramidal	Usually nonzero	
AX_6	Octahedral	Zero	

The fact that the water molecule has a dipole moment is excellent experimental evidence for a bent geometry. If the H_2O molecule were linear, the dipole moment would be zero.

The analysis we have just made for two different geometries of AX_2 molecules can be extended to other AX_n molecules (in which all X atoms are identical). Table 8.1 summarizes the relationship between geometry and dipole moment. Those geometries in which A—X bonds are directed symmetrically about the central atom (for example, linear, trigonal planar, and tetrahedral) will give molecules of zero dipole moment, that is, the molecules will be *nonpolar*. Those geometries in which the X atoms tend to be on one side of the molecule (for example, bent and trigonal pyramidal) will usually have nonzero dipole moments; that is, they will give *polar* molecules.

Example 8.3

Each of the following molecules has a nonzero dipole moment. Select the molecular geometry that is consistent with this information. Explain your reasoning.

(a) SO_2 linear, bent
(b) PH_3 trigonal planar, trigonal pyramidal

Solution

(a) In the linear geometry, the S—O bond contributions to the dipole moment would cancel, giving a zero dipole

moment. This would not happen in the bent geometry; hence, this must be the geometry for the SO_2 molecule. (b) In the trigonal planar geometry, the bond contributions to the dipole moment would cancel, giving a zero dipole moment. This would not occur in the trigonal pyramidal geometry; hence, this is a possible molecular geometry for PH_3.

Exercise 8.3

Bromine trifluoride, BrF_3, has a nonzero dipole moment. Which of the following geometries are consistent with this information? (a) trigonal planar (b) trigonal pyramidal (c) T-shaped

(See Problems 8.25 and 8.26.)

Exercise 8.4

Which of the following molecules would be expected to have a dipole moment of zero on the basis of symmetry? Explain.   (a) $SOCl_2$ (b) SiF_4 (c) OF_2

(See Problems 8.27 and 8.28.)

In Example 8.3, we used the fact that certain molecular geometries necessarily imply a zero dipole moment, so we could eliminate these geometries when considering a molecule with a nonzero dipole moment. The reverse argument, however, does not follow. For example, suppose a molecule of the type AX_3 is found to have no measurable dipole moment. It may be that the geometry is trigonal planar. But there are two other possibilities: the geometry is trigonal pyramidal (or T-shaped), but the bonds are nonpolar; or, the geometry is trigonal pyramidal (or T-shaped), but the lone pairs on the central atom offset the polarity of the bonds.■

We can see this effect of lone pairs on the dipole moment of nitrogen trifluoride, NF_3. Each N—F bond should be quite polar, judging by the electronegativity difference. Yet, the dipole moment of nitrogen trifluoride is only 0.2 D. (By contrast, ammonia has a dipole moment of 1.47 D.) The explanation for the small dipole moment in NF_3 is shown in Figure 8.5. The lone pair on nitrogen has a dipole-moment contribution that is directed outward from the nucleus, since the electrons are offset from the nuclear center. This dipole-moment contribution thus opposes the N—F bond moments.■

■ Why is the trigonal pyramidal geometry for AX_3 listed in Table 8.2 as "usually nonzero" and not simply as "nonzero"?

■ Can you explain why NH_3 has such a large dipole moment compared with NF_3?

8.3 Valence Bond Theory

The VSEPR model is usually a satisfactory method for predicting molecular geometries. To understand bonding and electronic structure, however, one must look to quantum mechanics. We will consider two theories stemming from quantum mechanics, valence bond theory and molecular orbital theory. Both theories use the methods of quantum mechanics but make different simplifying assumptions. In this section, we will look in a qualitative way at the basic ideas involved in valence bond theory.

Basic Theory

According to **valence bond theory,** a bond forms between two atoms when the following conditions are met:

1. An orbital on one atom comes to occupy a portion of the same region of space as an orbital on the other. The two orbitals are said to *overlap.*
2. The total number of electrons in both orbitals is no more than two.

As the orbital of one atom overlaps the orbital of another, the electrons in the orbitals begin to move about both atoms. Since they are attracted to both nuclei at once, they pull the atoms together. Strength of bonding depends on the amount of overlap; the greater the overlap, the greater the bond strength. The two orbitals cannot contain more than two electrons because a given

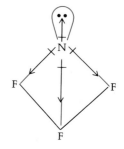

Figure 8.5
The lone-pair and bond contributions to the dipole moment of NF_3. Note that these tend to offset one another.

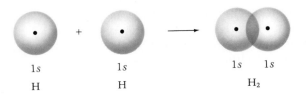

Figure 8.6
Formation of H_2 by the overlap of 1s orbitals on each
atom.

region of space can hold only two electrons (and then only if the spins of the electrons are opposite).

For example, consider the formation of the H_2 molecule from atoms. Each atom has the electron configuration $1s^1$. As the H atoms approach, their 1s orbitals begin to overlap and a covalent bond forms (Figure 8.6). At close approach, the strength of bonding from the overlap is countered by the strong repulsion of the nuclei for one another. Thus, the atoms eventually reach an optimum distance (the *bond distance* or *bond length*), where the total energy is a minimum (Figure 8.7).

Valence bond theory explains why two He atoms (each with electron configuration $1s^2$) do not bond. Suppose that two He atoms approach one another and that their 1s orbitals begin to overlap. Since each orbital is doubly occupied, the sharing of electrons between atoms would place the four valence electrons from the two atoms in the same region. This, of course, could not happen. As the orbitals begin to overlap, each electron pair strongly repels the other. The atoms come together, then fly apart.

Because the strength of bonding depends on orbital overlap, orbitals other than s orbitals bond only in given directions. Orbitals bond in the directions in which they protrude or point, to obtain maximum overlap. For example, consider the bonding between a hydrogen atom and a chlorine atom to give the HCl molecule. A chlorine atom has the electron configuration $[Ne]3s^23p^5$. Of the orbitals in the valence shell of the chlorine atom, three are doubly occupied by electrons and one (a 3p orbital) is singly occupied.

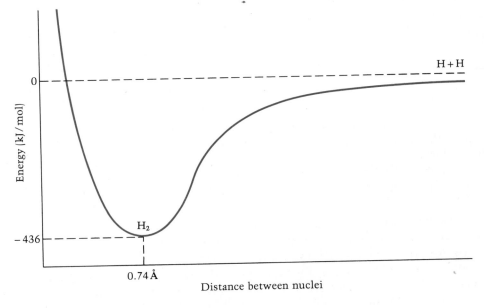

Distance between nuclei

Figure 8.7
Energy change during for-
mation of H_2 from atoms.
Separated atoms are shown at
the far right, corresponding
to a large distance between
nuclei. As the atoms come
together, the energy de-
creases. At closer approach,
repulsion from the nuclei
makes the energy increase.
The minimum energy occurs
at 0.74 Å, the bond length in
H_2.

The bonding of the hydrogen atom will have to occur with the singly oc-cupied $3p$ orbital of chlorine. For the strongest bonding to occur, the max-imum overlap of orbitals is required. The $1s$ orbital of hydrogen must overlap along the axis of the singly occupied $3p$ orbital of chlorine (Figure 8.8).

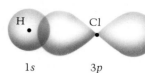

Figure 8.8
Bonding in HCl. The bond is formed by overlap of a hydro-gen $1s$ orbital along the axis of a chlorine $3p$ orbital (only the main lobes of the $3p$ or-bital are shown).

Hybrid Orbitals

From what has been said, one might expect the number of bonds formed by a given atom to equal the number of unpaired electrons in its valence shell. Chlorine, whose orbital diagram is

$$(\uparrow\downarrow)\ (\uparrow\downarrow)\ (\uparrow\downarrow)(\uparrow\downarrow)(\uparrow\downarrow)\ (\uparrow\downarrow)\ (\uparrow\downarrow)(\uparrow\downarrow)(\uparrow\ \ )$$
$$1s\quad\ \ 2s\qquad\qquad 2p\qquad\quad\ 3s\qquad\quad 3p$$

has one unpaired electron and forms one bond. Oxygen, whose orbital dia-gram is

$$(\uparrow\downarrow)\ (\uparrow\downarrow)\ (\uparrow\downarrow)(\uparrow\ \ )(\uparrow\ \ )$$
$$1s\quad\ \ 2s\qquad\qquad 2p$$

has two unpaired electrons and forms two bonds.

However, consider the carbon atom, whose orbital diagram is

$$(\uparrow\downarrow)\ (\uparrow\downarrow)\ (\uparrow\ \ )(\uparrow\ \ )(\ \ \ )$$
$$1s\quad\ \ 2s\qquad\qquad 2p$$

We might expect this atom to bond to two hydrogen atoms to form the CH_2 molecule. Although this molecule is known to be present momentarily during some reactions, it is very reactive and cannot be readily isolated. But methane, CH_4, in which the carbon atom bonds to four hydrogen atoms, is well known. In fact, a carbon atom usually forms four bonds.

We can explain this with the concept of electron *promotion* (electrons raised to a higher orbital level). Four unpaired electrons are formed when an electron from the $2s$ orbital of the carbon atom is promoted (excited) to the vacant $2p$ orbital:

C atom (ground state) $(\uparrow\downarrow)\ (\uparrow\downarrow)\ (\uparrow\ \ )(\uparrow\ \ )(\ \ \ )$
$$\qquad\qquad\qquad\qquad\quad 1s\qquad 2s\qquad\qquad 2p$$

$$\downarrow\ \text{promotion}$$

C atom (promoted) $(\uparrow\downarrow)\ (\uparrow\ \ )\ (\uparrow\ \ )(\uparrow\ \ )(\uparrow\ \ )$
$$\qquad\qquad\qquad\quad 1s\qquad 2s\qquad\qquad 2p$$

It requires some energy to promote the carbon atom this way, but more than enough energy is obtained from the formation of two additional covalent bonds (see Figure 8.9). Thus, we can expect energy to be released during the overall bonding process and the formation of four bonds to carbon to be energetically favorable.

According to this view, the bonding in methane would use one $2s$ and the three $2p$ orbitals of the valence shell of carbon. If C—H bonds were to form by direct overlap of each one of these orbitals with $1s$ orbitals from hydrogen atoms, two kinds of bonds would result. One bond would form from the overlap of the carbon $2s$ orbital with a hydrogen $1s$ orbital. Each of the other three bonds would form from a carbon $2p$ orbital and a hydrogen $1s$ orbital.

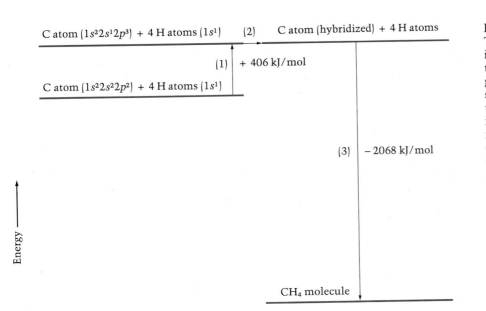

Figure 8.9
The steps in bond formation in CH_4. It requires 406 kJ/mol to promote carbon from the ground state to the $1s^2 2s^1 2p^3$ state (1). After the promotion, the orbitals are hybridized (2). Formation of bonds in CH_4 releases 2068 kJ/mol of energy (3). These three steps occur simultaneously as bonds form.

Experiment shows, however, that the four C—H bonds in methane are identical. This implies that the carbon orbitals involved in bonding are also equivalent. For this reason, valence bond theory assumes that the four valence orbitals of the carbon atom combine during the bonding process to form four new, but equivalent, orbitals. These equivalent orbitals obtained from atomic orbitals are called **hybrid orbitals.** Since, in this case, a set of these particular hybrid orbitals are constructed from one s orbital and three p orbitals, they are called sp^3 hybrid orbitals. Calculations from theory show that each sp^3 hybrid orbital has a large lobe pointing in one direction and a small lobe pointing in the opposite direction. The set of four sp^3 hybrid orbitals point in tetrahedral directions. Figure 8.10 shows the shape of a single sp^3 orbital and a set of four orbitals pointing in tetrahedral directions.

The C—H bonds in methane, CH_4, are described by valence bond theory as the overlapping of each sp^3 hybrid orbital of the carbon atom with $1s$ orbitals of hydrogen atoms (see Figure 8.10). Thus, the bonds are arranged tetrahedrally, which is the arrangement predicted by the VSEPR model. We can represent the hybridization of carbon and the bonding of hydrogen to the carbon atom in methane as follows:

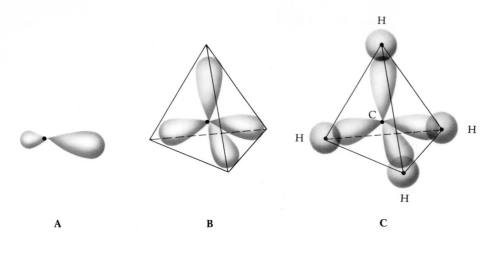

Figure 8.10
The spatial arrangement of sp^3 hybrid orbitals. (a) The shape of a single sp^3 hybrid orbital. (b) The set of four hybrid orbitals are arranged tetrahedrally in space (small lobes are omitted for clarity). (c) Bonding in CH_4. Each C—H bond is formed by the overlap of a $1s$ orbital from hydrogen with an sp^3 hybrid orbital of the carbon atom.

Here, the colored arrows represent electrons originally belonging to hydrogen atoms.

Hybrid orbitals can be formed from various numbers of atomic orbitals. The number of hybrid orbitals that are formed always equals the number of atomic orbitals used. For example, if we combine an s orbital and two p orbitals to get a set of equivalent orbitals, we get three hybrid orbitals (they are called sp^2 hybrid orbitals). A set of hybrid orbitals always has definite directional characteristics. Thus, all three sp^2 hybrid orbitals lie in a plane and are directed at 120° angles to one another. That is, they have a trigonal planar arrangement. Some possible hybrid orbitals and their geometrical arrangements are listed in Table 8.2 and shown in Figure 8.11. Note that the geometrical arrangements of hybrid orbitals are the same as those for electron pairs under the VSEPR model.

Only two of the three p orbitals are used to form sp^2 hybrid orbitals. The unhybridized p orbital is perpendicular to the plane of the sp^2 hybrid orbitals. Similarly, only one of the three p orbitals is used to form sp hybrid orbitals. The two unhybridized p orbitals are perpendicular to the axis of the sp hybrid orbitals and are perpendicular to each other. We will use these facts in discussing multiple bonding in the next section.

Usually hybrid orbitals are used in valence bond theory as a simple way to explain the bonding in a molecule with known geometry. (Computer cal-

Table 8.2
Kinds of Hybrid Orbitals

Hybrid Orbitals	Geometric Arrangement	Number of Orbitals	Example
sp	Linear	2	Be in BeF_2
sp^2	Trigonal planar	3	B in BF_3
sp^3	Tetrahedral	4	C in CH_4
sp^3d	Trigonal bipyramidal	5	P in PCl_5
sp^3d^2	Octahedral	6	S in SF_6

Linear arrangement

180°

sp hybrid orbitals

Trigonal planar arrangement

120°

sp² hybrid orbitals

Tetrahedral arrangement

109.5°

sp³ hybrid orbitals

Trigonal
bipyramidal
arrangement

sp³d hybrid orbitals

Octahedral
arrangement

sp³d² hybrid orbitals

Figure 8.11
Diagrams of hybrid orbitals
showing their spatial arrangement. Each lobe shown is
one hybrid orbital (small
lobes omitted for clarity).

culations can bypass the use of hybrid orbitals to predict the molecular geometry.) The next examples illustrate how we can use hybrid orbitals to describe bonding.

Example 8.4

Describe the bonding in H_2O according to valence bond theory. Assume that the molecular geometry is the same as given by the VSEPR model (in other words, assume that hybrid orbitals are used).

Solution

The Lewis formula for H_2O is

$$H : \overset{\displaystyle ..}{\underset{\displaystyle H}{O}} :$$

Note that there are four pairs of electrons about the oxygen atom. These, according to the VSEPR model, are directed tetrahedrally. We will assume that these four pairs of electrons about oxygen can be described in

terms of a set of four hybrid orbitals. From Table 8.2, we see that sp^3 hybrid orbitals occur as a set of four (pointing in tetrahedral directions). Therefore, we assume that sp^3 hybrids are used on the oxygen atom.

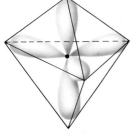

O atom (ground state)   [He] (↑↓) (↑↓)(↑)(↑)
                              2s   2p

                              ↓ hybridization

O atom (hybridized)   [He] (↑↓)(↑↓)(↑)(↑)
                                   sp^3

Each O—H bond is formed by the overlap of a $1s$ orbital of a hydrogen atom with one of the singly occupied sp^3
(Continued)

hybrid orbitals of the oxygen atom. The angle between the two bonds, according to this description, is 109.5° (the tetrahedral angle). Each lone pair of electrons occupies one of the remaining two sp^3 hybrid orbitals (see Figure 8.12). We can represent the bonding to the oxygen atom in H_2O as follows:

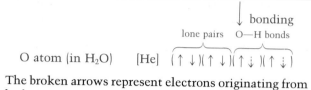

O atom (in H_2O)   [He]   (↑↓)(↑↓)(↑↓)(↑↓)

The broken arrows represent electrons originating from hydrogen atoms.

Exercise 8.5

Describe the bonding in NH_3 using hybrid orbitals

(See Problems 8.31 and 8.32.)

The actual angle between O—H bonds in H_2O has been experimentally determined to be 104.5°. Because this is quite close to the tetrahedral angle (109.5°), we believe the description of bonding given in Example 8.4 to be essentially correct. To explain why the actual bond angle is smaller than the tetrahedral value, we can say that the hybrid orbitals for bonds and lone pairs are not exactly equivalent. Qualitatively, we note that lone pairs are somewhat larger than bonding pairs. Because they take up more space, the lone pairs push the bonding pairs closer together than they are in the tetrahedral case.

As the next example illustrates, hybrid orbitals are useful for describing bonding when the central atom of a molecule is surrounded by more than eight valence electrons.

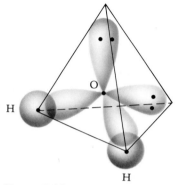

Figure 8.12
Bonding in H_2O (see Example 8.4).

Example 8.5

Describe the bonding in XeF_4 using hybrid orbitals.

Solution

The Lewis formula of XeF_4 is

$$:\ddot{F}: \quad :\ddot{F}:$$
$$:Xe:$$
$$:\ddot{F}: \quad :\ddot{F}:$$

The xenon atom has four single bonds and two lone pairs. It will require six orbitals to describe the bonding. This suggests that we use sp^3d^2 hybrid orbitals on xenon (according to Table 8.2 for a set of six hybrid orbitals.) The promotion and hybridization steps are shown below.

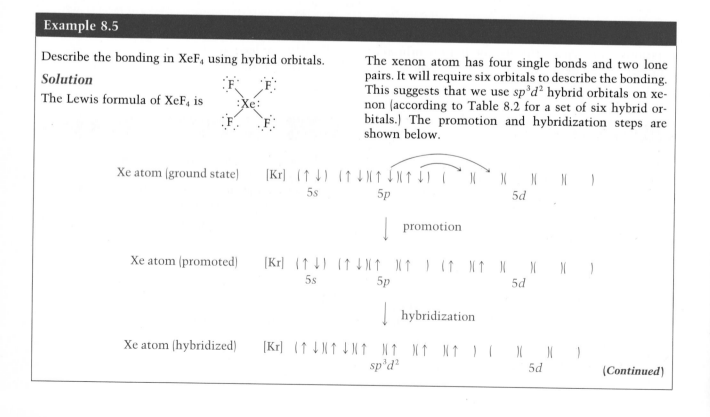

(Continued)

Since each fluorine atom (valence-shell configuration $2s^2 2p^5$) has one singly occupied $2p$ orbital, we will assume that this orbital is used in bonding. Each Xe—F bond is formed by the overlap of a xenon $sp^3 d^2$ hybrid orbital with a singly occupied fluorine $2p$ orbital. See the last line, in color, following.

$\downarrow$ bonding

lone pairs   Xe—F bonds

Xe atom (in XeF$_4$)   [Kr] (↑↓)(↑↓)(↑ ↓)(↑ ↓)(↑ ↓)(↑ ↓) ( )( )( )( )
                                        $sp^3 d^2$ $5d$

Exercise 8.6

Describe the bonding in PCl$_5$ using hybrid orbitals.

(See Problems 8.33 and 8.34.)

8.4 Description of Multiple Bonding

In the previous section, we described bonding as the overlap of *one* orbital from each of the bonding atoms. Now, however, we want to consider the possibility that *more than one* orbital from each of the bonding atoms might overlap, resulting in a multiple bond.

As an example, let us look at the ethylene molecule:

Hybrid orbitals are needed for each bond and for each lone pair. Because each carbon atom is bonded to three other atoms and there are no lone pairs, we will assume that the valence orbitals of each carbon atom are sp^2 hybrids (see Table 8.2). Thus, the $2s$ orbital and two of the $2p$ orbitals of each carbon atom form three hybrid orbitals having trigonal planar orientation. A third $2p$ orbital on each carbon atom remains unhybridized and is perpendicular to the plane of the three sp^2 hybrids.

C atom (ground state)   (↑↓) (↑↓) (↑)(↑)( )
                          $1s$ $2s$ $2p$

$\downarrow$ promotion

C atom (promoted)   (↑↓) (↑) (↑)(↑)(↑)
                      $1s$ $2s$ $2p$

$\downarrow$ hybridization

C atom (hybridized)   (↑↓) (↑)(↑)(↑) (↑)
                        $1s$ sp^2 $2p$

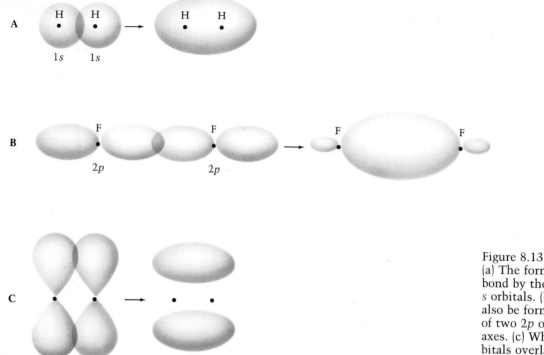

Figure 8.13
(a) The formation of a σ bond by the overlap of two s orbitals. (b) A σ bond can also be formed by the overlap of two $2p$ orbitals along their axes. (c) When two $2p$ orbitals overlap sidewise, a π bond is formed.

To describe the multiple bonding in ethylene, we will need to distinguish between two kinds of bonds. A **σ (sigma) bond** has a cylindrical shape about the bond axis. It is formed either when two s orbitals overlap, as in H_2 (Figure 8.13a), or when an orbital with directional character, such as a p orbital or a hybrid orbital, overlaps *along its axis*. Figure 8.13b illustrates the formation of a σ bond by the overlap of two $2p$ orbitals along their axes. The bonds we discussed in the previous section are σ bonds.

A **π (pi) bond** is formed by the *sidewise* overlap of two parallel p orbitals (see Figure 8.13c). A sidewise overlap will not give as strong a bond as an along-the-axis overlap of two p orbitals. A π bond occurs when two parallel orbitals are still available after strong σ bonds have formed. It has an electron distribution above and below the bond axis, as Figure 8.13c shows.

Now imagine that the separate atoms of ethylene move into their normal molecular positions. Each sp^2 hybrid carbon orbital overlaps a $1s$ orbital of a hydrogen atom or an sp^2 hybrid orbital of another carbon atom to form a σ bond (Figure 8.14a). Together, the σ bonds give the molecular framework of ethylene.

As we see from our orbital diagram for the hybridized C atom, a single $2p$ orbital still remains on each carbon atom. These orbitals are perpendicular to the plane of the hybrid orbitals, that is, perpendicular to the —CH_2 plane. Note that the two —CH_2 planes can rotate about the carbon–carbon axis without affecting the overlap of the hybrid orbitals. As these planes rotate, the $2p$ orbitals also rotate. When the —CH_2 planes rotate so that the $2p$ orbitals become parallel, the orbitals overlap to give a π bond (Figure 8.14b).

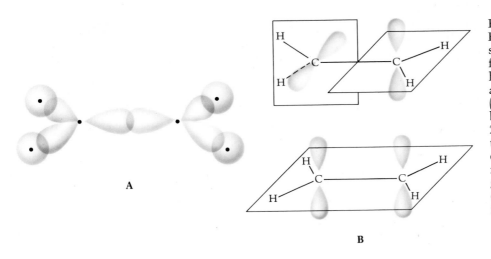

Figure 8.14
Bonding in ethylene. (a) The sigma bond framework formed by the overlap of sp^2 hybrid orbitals on C atoms and $1s$ orbitals on H atoms. (b) The formation of the pi bond in ethylene. When the $2p$ orbitals are perpendicular to one another, there is no overlap and no bond formation. When the two —CH_2 groups rotate so that the $2p$ orbitals are parallel, a π bond is formed.

Thus, we describe the carbon–carbon double bond as one σ bond and one π bond. Note that when the two $2p$ orbitals are parallel, the two —CH_2 ends of the molecule lie in the same plane. Thus, the formation of a π bond "locks" the two ends into a flat, rigid molecule.

We can describe the triple bonding in acetylene, H—C≡C—H, in similar fashion. Since each carbon atom is bonded to two other atoms and there are no lone pairs, two hybrid orbitals are needed. This suggests sp hybridization (see Table 8.2).

C atom (promoted) (↑↓) (↑) (↑)(↑)(↑)
                             1s   2s   2p

↓ hybridization

C atom (hybridized) (↑↓) (↑)(↑) (↑)(↑)
                              1s   sp   2p

These sp hybrid orbitals have a linear arrangement, so the H—C—C—H geometry is linear. Bonds formed by the overlap of these hybrid orbitals are σ bonds. The two $2p$ orbitals not used to construct hybrid orbitals are perpendicular to the bond axis and to each other. They are used to form two π bonds. Thus, the carbon–carbon triple bond consists of one σ bond and two π bonds (see Figure 8.15).

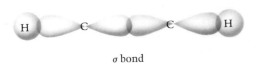

σ bond

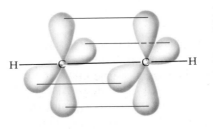

Two π bonds

Figure 8.15
Description of the bonding in acetylene. (a) The σ bond framework. (b) The two π bonds.

Example 8.6

Describe the bonding on a given N atom in dinitrogen difluoride, N_2F_2, using valence bond theory.

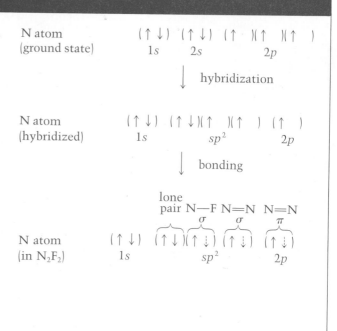

Solution

The electron-dot formula of N_2F_2 is

Note that a double bond is described as a π bond plus a σ bond. Hybrid orbitals are needed to describe each σ bond and each lone pair (a total of three hybrid orbitals for each N atom). This suggests sp^2 hybridization (see Table 8.2). According to this description, one of the sp^2 hybrid orbitals is used to form the N—F bond, another to form the σ bond of N=N, and the third to hold the lone pair on the N atom. The $2p$ orbitals on each N atom overlap to form the π bond of N=N. Hybridization and bonding of the N atoms are shown as follows:

Exercise 8.7

Discuss the bonding on the carbon atom in carbon dioxide, CO_2, using valence bond theory.

(See Problems 8.35 and 8.36.)

The π bond description of the double bond agrees well with experiment. The *geometrical*, or *cis–trans, isomers* of the compound 1,2-dichloroethylene can illustrate this. *Isomers* are compounds of the same molecular formula but with different arrangements of the atoms. (The numbers in the name 1,2-dichloroethylene refer to the positions of the chlorine atoms. Thus, one chlorine atom is attached to carbon atom 1 and the other to carbon atom 2.) The structures of these isomers of 1,2-dichloroethylene are

cis-1,2-dichloroethylene   *trans*-1,2-dichloroethylene

In order to transform one isomer into the other, one end of the molecule must be rotated as the other remains fixed. To do this, the π bond must be broken. Since breaking the π bond would require considerable energy, the *cis* and *trans* compounds are not easily interconverted. The *cis*- and *trans*-1,2-

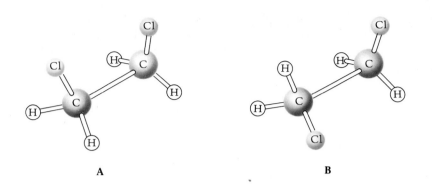

A   B

Figure 8.16
Because of rotation about the carbon–carbon bond in 1,2-dichloroethane, geometrical isomers are not possible. Note that the molecule pictured in (a) can be twisted easily to give (b).

diphenylethylene molecules shown in the chapter opening do not readily interconvert for the same reason.■ Contrast this with 1,2-dichloroethane (Figure 8.16), in which the two ends of the molecule can rotate without breaking any bonds. Here, isomers corresponding to different spatial orientations of the two chlorine atoms cannot be obtained.

Exercise 8.8

Dinitrogen difluoride (see Example 8.6) exists as *cis* and *trans* isomers. Write structural formulas for these isomers and explain (in terms of the valence bond theory of the double bond) why they exist.

(See Problems 8.37 and 8.38.)

■ The conversion of a *cis* isomer to a *trans* isomer is the central reaction in the visual process. A compound called 11-*cis*-retinal, which exists in the rod cells of the retina (in the eye), is converted to 11-*trans*-retinal when a photon of light is absorbed. This molecule triggers a series of reactions that generate a nerve impulse. The 11-*trans*-retinal is then reconverted to 11-*cis*-retinal to reset the visual process.

Molecular Orbital Theory

In the preceding sections, we looked at valence bond theory. Although this theory satisfactorily describes most of the molecules we encounter, it does not apply to all molecules. For example, under valence bond theory any molecule with an even number of electrons should be diamagnetic (not attracted to a magnet), since we assume the electrons to be paired and to have opposite spins. In fact, a few molecules with an even number of electrons are paramagnetic (attracted to a magnet), indicating that some of the electrons are not paired. The best-known example of such a paramagnetic molecule is O_2. Because of the paramagnetism of O_2, liquid oxygen sticks to a magnet when poured over it (Figure 8.17).■

Molecular orbital theory provides an alternative way to describe the electronic structure of molecules. This theory views the electronic structure of molecules to be much like the electronic structure of atoms. It assumes the existence of molecular orbitals that describe the region in which electrons in a molecule move. Each molecular orbital may be spread over several atoms (perhaps over the entire molecule) and has a definite energy. To obtain the ground state of a molecule, electrons are put into molecular orbitals of lowest energy, consistent with the Pauli exclusion principle, just as in atoms.

■ Valence bond theory can be extended to explain the electron structure of O_2.

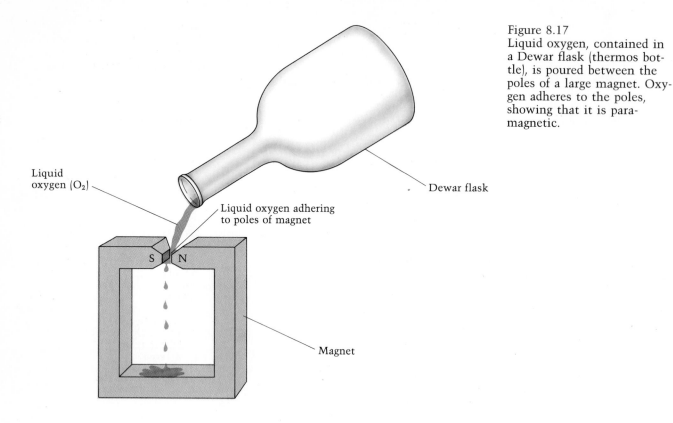

Liquid
oxygen (O_2)

Liquid oxygen adhering
to poles of magnet

Dewar flask

Magnet

S   N

Figure 8.17
Liquid oxygen, contained in
a Dewar flask (thermos bottle), is poured between the
poles of a large magnet. Oxygen adheres to the poles,
showing that it is para-
magnetic.

8.5 Principles of Molecular Orbital Theory

We can think of a molecular orbital as being formed from a combination of
atomic orbitals. As atoms approach each other and their atomic orbitals
overlap, molecular orbitals are formed.

Bonding and Antibonding Orbitals

Consider the H_2 molecule. As the atoms approach to form the molecule,
their $1s$ orbitals overlap. One molecular orbital is obtained by adding the two
$1s$ orbitals together (see Figure 8.18). Note that where the atomic orbitals
overlap, their values add together to give a larger result. This means that in
this molecular orbital the likelihood that an electron will be found in the
region between the two nuclei is high. Electrons in such a molecular orbital
tend to be in the region where they can hold the nuclei together. This
molecular orbital is called a **bonding orbital** and is denoted σ_{1s}. The σ (sigma)
means that the molecular orbital has a cylindrical shape about the bond axis.
The subscript $1s$ tells us that it was obtained from $1s$ atomic orbitals.

Another molecular orbital is obtained by subtracting the $1s$ orbital on one
atom from the $1s$ orbital on the other (Figure 8.18). When the orbitals are
subtracted, the resulting values in the region of the overlap are close to zero.
This means that in this molecular orbital the electrons spend little time

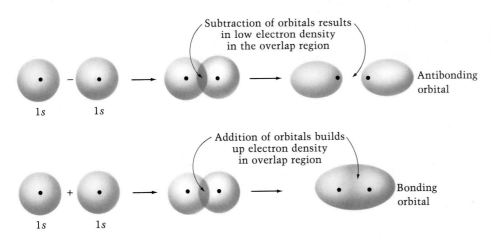

Figure 8.18
Formation of bonding and antibonding orbitals from $1s$ orbitals of hydrogen atoms.

between the nuclei. We call it an **antibonding orbital** and denote it $\sigma_{1s}^{\star}$. The asterisk tells us that the orbital is antibonding.

Figure 8.19 shows the energies of the molecular orbitals σ_{1s} and $\sigma_{1s}^{\star}$ relative to the atomic orbitals. Energies of the separate atomic orbitals are represented by heavy colored lines at the far left and right. The energies of the molecular orbitals are shown by heavy colored lines in the center. (These are connected by light gray lines to show which atomic orbitals were used to obtain the molecular orbitals.) Note that the energy of a bonding orbital is less than that of the separate atomic orbitals, whereas the energy of an antibonding orbital is higher.

We obtain the electron configuration for the ground state of H_2 by placing the two electrons (one from each atom) into the lower-energy orbital (see Figure 8.19). The orbital diagram is

$$(\uparrow \downarrow)(\quad)$$
$$\sigma_{1s} \quad \sigma_{1s}^{\star}$$

and the electron configuration is $(\sigma_{1s})^2$. Since the energy of the two electrons is lower than their energies in the isolated atoms, the H_2 molecule is stable.

Configurations involving the $\sigma_{1s}^{\star}$ orbital describe excited states of the molecule. As an example of an excited state, we have the orbital diagram

$$(\uparrow \quad)(\uparrow \quad)$$
$$\sigma_{1s} \quad \sigma_{1s}^{\star}$$

The corresponding electron configuration is $(\sigma_{1s})^1(\sigma_{1s}^{\star})^1$.

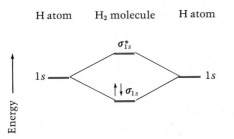

Figure 8.19
Relative energies of the $1s$ orbital of the H atom and the σ_{1s} and $\sigma_{1s}^{\star}$ molecular orbitals of H_2. Arrows denote occupation of the σ_{1s} orbital by electrons in the ground state of H_2.

We obtain a similar set of orbitals when we consider the approach of two helium atoms. To obtain the ground state of He_2, we note that there are four electrons, two from each atom, that would fill the molecular orbitals. Two of these electrons go into the σ_{1s} orbital and two go into the σ_{1s}^* orbital. The orbital diagram is

$$(\uparrow \downarrow)(\uparrow \downarrow)$$
$$\sigma_{1s} \quad \sigma_{1s}^*$$

and the configuration is $(\sigma_{1s})^2 (\sigma_{1s}^*)^2$. The energy lowering from the bonding electrons is offset by the energy increase from the antibonding electrons. Hence, He_2 is not a stable molecule. Molecular orbital theory explains, therefore, why the element helium exists as a monatomic gas, whereas hydrogen is diatomic.

Bond Order

The term *bond order* refers to the number of bonds that exist between two atoms. In molecular orbital theory, the bond order of a diatomic molecule is defined as one-half the difference between the number of electrons in bonding orbitals, n_b, and the number of electrons in antibonding orbitals, n_a:■

$$\text{Bond order} = \tfrac{1}{2}(n_b - n_a)$$

■ For a Lewis formula, the bond order equals the number of electron pairs shared between two atoms (see Section 7.9).

For H_2, which has two bonding electrons, we have

$$\text{Bond order} = \tfrac{1}{2}(2 - 0) = 1$$

That is, H_2 has a single bond. For He_2, which has two bonding and two antibonding electrons, we have

$$\text{Bond order} = \tfrac{1}{2}(2 - 2) = 0$$

Bond orders need not be whole numbers; bond orders of $\tfrac{1}{2}$, $\tfrac{3}{2}$, and so forth, are also possible. For example, the H_2^+ molecular ion, which is formed in mass spectrometers, has the configuration $(\sigma_{1s})^1$ and a bond order of $\tfrac{1}{2}(1 - 0) = \tfrac{1}{2}$.

Factors Determining Orbital Interaction

Suppose two lithium atoms approach to form the Li_2 molecule. In this case, we would consider the interaction of $1s$ and $2s$ orbitals on each atom. The strength of the interaction of two atomic orbitals to form molecular orbitals is determined by two factors: (1) the energy difference between the interacting orbitals and (2) the magnitude of their overlap. *In order for the interaction to be strong, the energies of the two orbitals must be similar, and the overlap large.*

We expect the $1s$ orbital of one lithium atom to interact with the $1s$ orbital of the other atom (since they have the same energy), and not with the $2s$ orbital (which has much greater energy). Similarly, the $2s$ orbital of one lithium atom interacts with the $2s$ orbital of the other atom. Thus, the $1s$ orbitals interact to give σ_{1s} and σ_{1s}^*, and the $2s$ orbitals interact to give σ_{2s} and

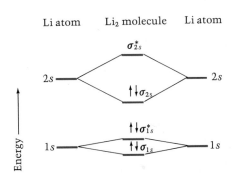

Figure 8.20
The energies of the molecular orbitals of Li$_2$. Arrows indicate the occupation of these orbitals by electrons in the ground state.

σ_{2s}^*. Further, as the atoms approach, the outer orbitals (the $2s$ orbitals) interact more strongly than the $1s$ orbitals, because they overlap more. Thus, the difference in energy between σ_{2s} and σ_{2s}^* is greater than that between σ_{1s} and σ_{1s}^*. These energy differences reflect the different strengths of interaction.∎ Figure 8.20 shows that the $1s$ orbitals and the $2s$ orbitals interact separately to give molecular orbitals.

We obtain the ground-state configuration of Li$_2$ by putting electrons into the orbitals whose energies are displayed in Figure 8.20. Each lithium atom contributes three electrons, giving a total of six. Putting two electrons into each of the lowest-energy orbitals gives the configuration

$$\text{Li}_2 \qquad (\sigma_{1s})^2(\sigma_{1s}^*)^2(\sigma_{2s})^2$$

Note that there are four bonding electrons and two antibonding electrons. Thus, the bond order is $\frac{1}{2}(4-2)=1$. We find that the molecule should exist and should have a single bond.

In Be$_2$, the energy diagram for molecular orbitals is similar to the one given in Figure 8.20 for Li$_2$ (involving $1s$ and $2s$ orbitals). Since each atom contributes four electrons, giving a total of eight, the ground-state configuration would be

$$\text{Be}_2 \qquad (\sigma_{1s})^2(\sigma_{1s}^*)^2(\sigma_{2s})^2(\sigma_{2s}^*)^2$$

However, we note that the configuration has four bonding and four antibonding electrons. Thus, the bond order is $\frac{1}{2}(4-4)=0$. No bond is formed; the Be$_2$ molecule, like He$_2$, is unstable.

- There is only a small energy difference between the σ_{1s} and σ_{1s}^* orbitals in Li$_2$ because core electrons are only slightly affected by bonding. Generally, only the valence orbitals are significantly affected when bonds are formed with other atoms.

8.6 Electron Configurations of Diatomic Molecules

In the previous section, we looked at the electron configurations of some simple molecules: H$_2$, He$_2$, Li$_2$, and Be$_2$. These are **homonuclear diatomic molecules,** that is, molecules with two like nuclei. (**Heteronuclear diatomic molecules** are composed of two different nuclei, for example, CO and NO.) To find the electron configurations of other homonuclear diatomic molecules, we need to know how additional molecular orbitals fill, just as we needed to know the building-up order to find the configurations of atoms.

We have already looked at the formation of molecular orbitals from s atomic orbitals. Now we need to consider the formation of molecular orbitals from p atomic orbitals. There are two different ways in which $2p$

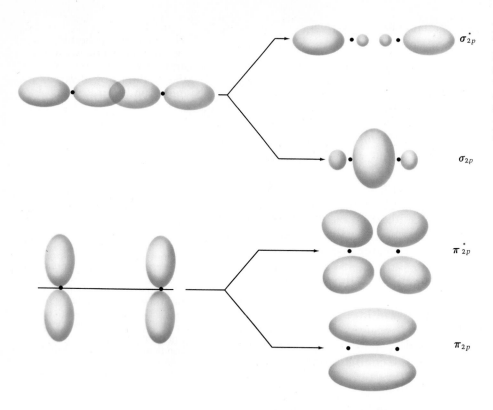

Figure 8.21
The different ways in which $2p$ atomic orbitals can interact. If they overlap along their axes, they form σ_{2p} and $\sigma_{2p}^{\star}$ molecular orbitals. If they overlap sidewise, they form π_{2p} and $\pi_{2p}^{\star}$ molecular orbitals.

atomic orbitals can interact. One set of $2p$ orbitals can overlap along their axes to give one bonding and one antibonding σ orbital (σ_{2p} and $\sigma_{2p}^{\star}$). The other two sets of $2p$ orbitals will then overlap sidewise to give two bonding and two antibonding π orbitals (π_{2p} and $\pi_{2p}^{\star}$). See Figure 8.21.

Figure 8.22 shows the relative energies of the molecular orbitals obtained from $1s$, $2s$, and $2p$ atomic orbitals. This order of molecular orbitals reproduces the known electron configurations of homonuclear diatomic mol-

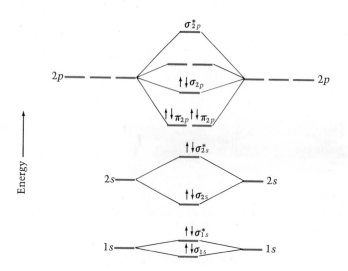

Figure 8.22
Relative energies of molecular orbitals of homonuclear diatomic molecules. The arrows show the occupation of molecular orbitals by electrons in N_2.

ecules composed of elements in the first and second rows of the periodic table.■ The order of filling is

$$\sigma_{1s} \quad \sigma_{1s}^{\star} \quad \sigma_{2s} \quad \sigma_{2s}^{\star} \quad \pi_{2p} \quad \sigma_{2p} \quad \pi_{2p}^{\star} \quad \sigma_{2p}^{\star}$$

Note that there are two orbitals in the π subshell and two orbitals in the $\pi^{\star}$ subshell. Since each orbital can hold two electrons, a π or $\pi^{\star}$ subshell can hold four electrons. The next example shows how the order of filling can be used to obtain the orbital diagram, magnetic character, electron configuration, and bond order of a homonuclear diatomic molecule.

■ The order (by energy) of the π_{2s} and σ_{2p} orbitals changes in the series of homonuclear diatomic molecules. In B_2, C_2, and N_2, π_{2p} is below σ_{2p}. In O_2 and F_2, σ_{2p} is below π_{2p}. However, the order in the text gives the correct configurations for all of the homonuclear diatomic molecules of the first- and second-period elements.

Example 8.7

Give the orbital diagram of the O_2 molecule. Is this molecule diamagnetic or paramagnetic? What is the electron configuration? What is the bond order of O_2?

Solution

There are 16 electrons in O_2 (8 from each atom), which occupy the molecular orbitals as shown in the following orbital diagram.

$$(\uparrow\downarrow)(\uparrow\downarrow)(\uparrow\downarrow)(\uparrow\downarrow)\underbrace{(\uparrow\downarrow)(\uparrow\downarrow)}(\uparrow\downarrow)\underbrace{(\uparrow)\ \ (\uparrow)}(\ \ )$$

$$\sigma_{1s} \quad \sigma_{1s}^{\star} \quad \sigma_{2s} \quad \sigma_{2s}^{\star} \quad\quad \pi_{2p} \quad\quad \sigma_{2p} \quad\quad \pi_{2p}^{\star} \quad\quad \sigma_{2p}^{\star}$$

Note that the two electrons in the $\pi_{2p}^{\star}$ subshell must go into different orbitals with their spins in the same direction (Hund's rule). Because there are two unpaired electrons, the molecule is paramagnetic. The electron configuration is

$$(\sigma_{1s})^2(\sigma_{1s}^{\star})^2(\sigma_{2s})^2(\sigma_{2s}^{\star})^2(\pi_{2p})^4(\sigma_{2p})^2(\pi_{2p}^{\star})^2$$

There are ten bonding electrons and six antibonding electrons. Therefore,

$$\text{Bond order} = \tfrac{1}{2}(10 - 6) = 2$$

Exercise 8.9

The C_2 molecule exists in the vapor phase over carbon at high temperature. Describe the molecular orbital structure of this molecule; that is, give the orbital diagram and electron configuration. Would you expect the molecule to be diamagnetic or paramagnetic? What is the bond order for C_2?

(See Problems 8.39 and 8.40.)

When the atoms in a heteronuclear diatomic molecule are close to one another in the periodic table, the molecular orbitals have the same relative order of energies as those for homonuclear diatomic molecules. In this case we can obtain the electron configurations in the same way, as the next example illustrates.

Example 8.8

Write the orbital diagram for nitric oxide, NO. What is the bond order of NO?

Solution

We assume that the order of filling of orbitals is the same as for homonuclear diatomic molecules. There are 15 electrons in NO. Thus, the orbital diagram is

$$(\uparrow\downarrow)(\uparrow\downarrow)(\uparrow\downarrow)(\uparrow\downarrow)\underbrace{(\uparrow\downarrow)(\uparrow\downarrow)}(\uparrow\downarrow)\underbrace{(\uparrow)\ \ (\ \ )}(\ \ )$$

$$\sigma_{1s} \quad \sigma_{1s}^{\star} \quad \sigma_{2s} \quad \sigma_{2s}^{\star} \quad\quad \pi_{2p} \quad\quad \sigma_{2p} \quad\quad \pi_{2p}^{\star} \quad\quad \sigma_{2p}^{\star}$$

Since there are ten bonding and five antibonding electrons, we have

$$\text{Bond order} = \tfrac{1}{2}(10 - 5) = \tfrac{5}{2}$$

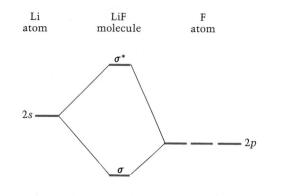

Figure 8.23
Molecular orbitals in LiF.
Bonding and antibonding orbitals are formed by the interaction of the lithium 2*s*
orbital and a fluorine 2*p* orbital (aligned along the bond
axis).

Exercise 8.10

Give the orbital diagram and electron configuration for the carbon monoxide
molecule, CO. What is the bond order of CO? Is the molecule diamagnetic or
paramagnetic?

(See Problems 8.41 and 8.42.)

When the two atoms in a heteronuclear diatomic molecule differ appreciably, we can no longer use the scheme appropriate for homonuclear diatomic molecules. The LiF molecule is an example. A sigma bonding orbital
is formed by combining the 2*s* orbital on the Li atom with a 2*p* orbital on the
F atom. (The 2*p* orbital must lie along the Li—F bond axis.) However, the
energy of the 2*p* fluorine orbital is much lower than that of the 2*s* lithium
orbital (see Figure 8.23), because the fluorine electrons are more tightly
held.■ As a result, the bonding orbital is made up primarily of the 2*p* fluorine
orbital, with very little of the 2*s* lithium orbital. This means that the
electrons occupying the bonding orbital lie mostly in the region of the
fluorine atom, corresponding to the ionic nature of the Li—F bond
(Li$^+$:F:$^-$). This result agrees with the greater electronegativity of the fluorine atom.

■ Recall that the ionization energy, or energy needed to remove an electron, increases in a
row of the periodic table from
left to right (Section 6.5). In the
second row, the ionization energy increases from lithium to
fluorine.

8.7 Molecular Orbitals and Delocalized Bonding

One of the advantages of molecular orbital theory is the simple way in which
it describes molecules with delocalized bonding. Whereas valence bond
theory has to use two or more resonance formulas, molecular orbital theory
describes the bonding in terms of a single electron configuration.

Consider the sulfur dioxide molecule as an example. Previously, we described this molecule in terms of resonance as

The part that is the same in both resonance formulas is

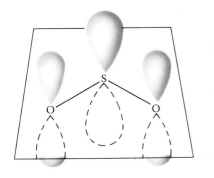

Antibonding π orbital

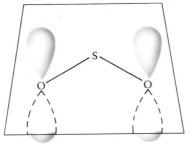

Nonbonding π orbital

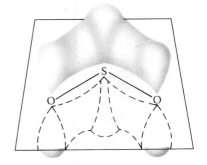

Bonding π orbital

Figure 8.24
The electron distributions of the π molecular orbitals of SO_2.

We can describe the molecular orbitals for this part of the electron structure as localized. The other two electron pairs have a delocalized molecular-orbital description.

Since each atom in the localized framework has three electron pairs about it, we will assume that each atom uses sp^2 hybrid orbitals (see Table 8.1). The overlap of a hybrid orbital from sulfur with one from oxygen gives an S—O bond. Another hybrid orbital from sulfur overlaps a hybrid orbital of the oxygen atom to form the other S—O bond. This leaves one hybrid orbital on the sulfur atom and two on each oxygen to describe the lone pairs on these atoms.

After hybrid orbitals are formed, one unhybridized p orbital still remains on each atom. These three p orbitals are perpendicular to the plane of the molecule and parallel to one another. Thus they can overlap sidewise to give three pi molecular orbitals (since three interacting atomic orbitals produce the same number of molecular orbitals). All three will span the entire molecule (see Figure 8.24). One of these orbitals turns out to be bonding, one is nonbonding (that is, it has an energy equal to that of the isolated atoms), and the third is antibonding.

The bonding and nonbonding π orbitals are both doubly occupied. This agrees with the resonance description, in which one delocalized pair is bonding and the other is a lone pair on oxygen.

A Checklist for Review

Important Terms

valence-shell electron-pair repulsion (VSEPR) model (8.1)
dipole moment (8.2)
valence bond theory (8.3)

hybrid orbitals (8.3)
σ (sigma) bond (8.4)
π (pi) bond (8.4)
molecular orbital theory (p. 257)

bonding orbital (8.5)
antibonding orbital (8.5)
homonuclear diatomic molecules (8.6)
heteronuclear diatomic molecules (8.6)

Summary of Facts and Concepts

Molecular geometry refers to the spatial arrangement of atoms in a molecule. The *valence-shell electron-pair repulsion (VSEPR) model* is a simple model for predicting molecular geometries. It is based on the idea that the valence-shell electron pairs are arranged symmetrically about an atom to minimize electron-pair repulsion. The geometry about an atom is then determined by the directions of the bonding pairs. Information about the geometry of a molecule can sometimes be obtained from the experimentally determined presence or absence of a *dipole moment.*

The bonding and geometry in a molecule can be described in terms of *valence bond theory.* In this theory, a bond is formed by the overlap of orbitals from two atoms. *Hybrid orbitals*, a set of equivalent orbitals formed by combining atomic orbitals, are often needed to describe this bond. Multiple bonds occur from the overlap of atomic orbitals to give σ *bonds* and π *bonds*. *Cis–trans* isomers result from the molecular rigidity imposed by the formation of a π bond.

Molecular orbital theory can also be used to explain bonding in molecules. According to this theory, electrons in a molecule occupy orbitals that may spread over the entire molecule. We can think of these molecular orbitals as constructed from atomic orbitals. Thus, when two atoms approach to form a diatomic molecule, the atomic orbitals interact to form *bonding* and *antibonding* molecular orbitals. The configuration of a diatomic molecule such as O_2 can be predicted from the order of filling of the molecular orbitals. From this configuration, we can predict the *bond order* and whether the molecule is diamagnetic or paramagnetic.

Operational Skills

1. Given the formula of a single molecule, predict the geometry, using the VSEPR model (Examples 8.1 and 8.2).

2. State what geometries of a molecule AX_n are consistent with the information that the molecule has a nonzero dipole moment (Example 8.3).

3. Given the formula of a simple molecule, describe the bonding using valence bond theory (Examples 8.4, 8.5, and 8.6).

4. Given the formula of a diatomic molecule obtained from first- or second-period elements, deduce the molecular-orbital configuration, bond order, and whether the molecule is diamagnetic or paramagnetic (Examples 8.7 and 8.8).

Review Questions

8.1 Describe the main features of the VSEPR model.

8.2 What are the arrangements of two, three, four, five, and six valence-shell electron pairs about an atom, according to the VSEPR model?

8.3 Why is a lone pair expected to occupy an equatorial position instead of an axial position in the trigonal bipyramidal arrangement?

8.4 Why is it possible for a molecule to have polar bonds yet have a dipole moment of zero?

8.5 Explain why nitrogen trifluoride has a small dipole moment even though it has polar bonds in a trigonal pyramidal arrangement.

8.6 Explain in terms of valence bond theory why the orbitals used on an atom give rise to a particular geometry about that atom.

8.7 What is the angle between two sp^3 hybrid orbitals?

8.8 What is the difference between a sigma and a pi bond?

8.9 Describe the bonding in ethylene, C_2H_4, in terms of valence bond theory.

8.10 How does the valence bond description of a carbon–carbon double bond account for *cis–trans* isomers?

8.11 What are the differences between a bonding and an antibonding molecular orbital of a diatomic molecule?

8.12 What are the factors that determine the strength of interaction of two atomic orbitals to form a molecular orbital?

8.13 Describe the formation of bonding and antibonding molecular orbitals resulting from the interaction of two $2s$ orbitals.

8.14 Describe the formation of molecular orbitals resulting from the interaction of two $2p$ orbitals.

8.15 How does molecular orbital theory describe the bonding in the LiF molecule?

8.16 Describe the bonding in SO_2 using molecular orbital theory. Compare with the resonance description.

Problems

The VSEPR Model

8.17 Predict the shape or geometry of each of the following molecules, using the VSEPR model:
  (a) PbF_2 (b) HCN (c) H_2S (d) $HgCl_2$
  (e) $POCl_3$ (f) $SnCl_4$ (g) NCl_3 (h) SOF_2

8.19 Predict the geometry of each of the following ions, using the electron-pair repulsion model:
  (a) $TlCl_2^+$ (b) AsF_2^+ (c) NO_3^-
  (d) PCl_4^+ (e) ClO_3^-

8.21 What geometry is expected for the following molecules, according to the VSEPR model?
  (a) TeF_4 (b) AsF_5 (c) BrF_5

8.23 Predict the geometries of the following ions, using the VSEPR model:
  (a) $SnCl_5^-$ (b) ICl_2^- (c) IF_4^-

8.18 Use the electron-pair repulsion model to predict the geometry of the following molecules:
  (a) $SOCl_2$ (b) $PbCl_2$ (c) C_2H_2 (d) $SbCl_3$
  (e) SCl_2 (f) SiH_4 (g) $NOCl$ (h) Hg_2Cl_2

8.20 Use the VSEPR model to predict the geometry of the following ions:
  (a) H_3O^+ (b) BeF_3^- (c) NO_2^+
  (d) CO_3^{2-} (e) ClO_4^-

8.22 From the electron-pair repulsion model, predict the geometry of the following molecules:
  (a) ClF_5 (b) $SeCl_4$ (c) PF_5

8.24 Name the geometries expected for the following ions, according to the electron-pair repulsion model:
  (a) ClF_2^- (b) ICl_4^- (c) ClF_4^+

Dipole Moment and Molecular Geometry

8.25 For each of the molecules, choose the geometry that is consistent with the information given about the dipole moment.

Molecule	Dipole Moment	Geometry
(a) AsF_3	2.59 D	Trigonal planar Trigonal pyramidal T-shaped
(b) H_2S	0.97 D	Linear Bent

8.27 Which of the following molecules would be expected to have zero dipole moments on the basis of their geometry?
  (a) $AsCl_3$ (b) $HgBr_2$ (c) SiF_2 (d) SF_6

8.26 For each of the molecules, choose the geometry that is consistent with the information given about the dipole moment.

Molecule	Dipole Moment	Geometry
(a) BrF_3	1.19 D	Trigonal planar Trigonal pyramidal T-shaped
(b) $TeCl_4$	2.54 D	Tetrahedral Distorted tetrahedral Square planar

8.28 Which of the following molecules would be expected to have a dipole moment of zero because of symmetry?
  (a) SO_3 (b) PH_3 (c) IF_5 (d) $SnBr_4$

Valence Bond Theory

8.29 What hybrid orbitals would be expected for the central atom in each of the following?
  (a) BeH_2 (b) SiF_6^{2-} (c) TeF_4 (d) CI_4

8.31 (a) Mercury(II) chloride dissolves in water to give poorly conducting solutions, indicating that the compound is largely nonionized in solution—it dissolves as $HgCl_2$ molecules. Describe the bonding of the $HgCl_2$ molecule using valence bond theory. (b) Phosphorus trichloride, PCl_3, is a colorless liquid with a highly irritating vapor. Describe the bonding in the PCl_3 molecule using valence bond theory. Use hybrid orbitals.

8.30 What hybrid orbitals would be expected for the central atom in each of the following?
  (a) SCl_2 (b) NCl_3 (c) $AlCl_6^{3-}$ (d) ICl_2^+

8.32 (a) Nitrogen trifluoride, NF_3, is a relatively unreactive, colorless gas. How would you describe the bonding in the NF_3 molecule in terms of valence bond theory? Use hybrid orbitals. (b) Silicon tetrafluoride, SiF_4, is a colorless gas formed when hydrofluoric acid attacks silica (SiO_2) or glass. Describe the bonding in the SiF_4 molecule using valence bond theory.

8.33 Phosphorus pentachloride is normally a white solid. It exists in this state as the ionic compound $[PCl_4^+][PCl_6^-]$. Describe the electron structure of the PCl_6^- ion in terms of valence bond theory.

8.35 (a) Formaldehyde, H_2CO, is a colorless, pungent gas used to make plastics. Give the valence bond description of the formaldehyde molecule. (Both hydrogen atoms are attached to the carbon atom.) (b) Nitrogen, N_2, makes up about 80% of the earth's atmosphere. Give the valence bond description of this molecule.

8.37 The hyponitrite ion, $^-O—N{=}N—O^-$, exists in solid compounds as the *trans* isomer. Using valence bond theory, explain why *cis*–*trans* isomers might be expected for this ion. Draw structural formulas of the two *cis*–*trans* isomers.

8.34 Iodine, I_2, dissolves in an aqueous solution of iodide ion, I^-, to give the triiodide ion, I_3^-. This ion consists of one iodine atom bonded to two others. Describe the bonding of I_3^- in terms of valence bond theory.

8.36 (a) The molecule $HN{=}NH$ exists as a transient species in certain reactions. Give the valence bond description of this species. (b) Hydrogen cyanide, HCN, is a very poisonous gas or liquid with the odor of bitter almonds. Give the valence bond description of HCN. (Carbon is the central atom.)

8.38 Fumaric acid, $C_4H_4O_4$, occurs in the metabolism of glucose in the cells of plants and animals. It is used commercially in beverages. The structural formula of fumaric acid is

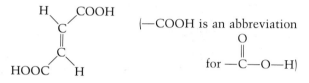

Maleic acid is the *cis* isomer of fumaric acid. Using valence bond theory, explain why these isomers are possible.

Molecular Orbital Theory

8.39 Describe the electronic structure of each of the following using molecular orbital theory. Calculate the bond order of each and decide whether or not it should be stable. For each, state whether it is diamagnetic or paramagnetic.

    (a) B_2 (b) B_2^+

8.41 Assume that the cyanide ion, CN^-, has molecular orbitals similar to those of a homonuclear diatomic molecule. Write the configuration and bond order of CN^-. Is the ion diamagnetic or paramagnetic?

8.40 Use molecular orbital theory to describe the bonding in the following. For each one, find the bond order and decide if it is stable. Is it diamagnetic or paramagnetic?

    (a) C_2^+ (b) Ne_2

8.42 Write the molecular-orbital configuration of the diatomic molecule BN. What is the bond order of BN? Is the molecule diamagnetic or paramagnetic? Use the order of energies that was given for homonuclear diatomic molecules.

Additional Problems

8.43 Which of the following molecules or ions are linear?

    (a) $BeCl_2$ (b) NH_2^- (c) CS_2 (d) ICl_2^+

8.44 Which of the following molecules or ions are trigonal planar?

    (a) $SnCl_3^-$ (b) BCl_3 (c) GaF_3 (d) PH_3

8.45 Describe the hybrid orbitals used by each carbon atom in the following molecules.

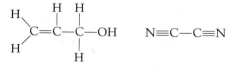

8.46 Describe the hybrid orbitals used by each nitrogen atom in the following molecules.

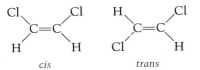

8.47 Explain how the dipole moment could be used to distinguish between the *cis* and *trans* isomers of 1,2-dichloroethylene:

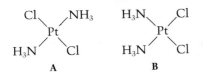

8.48 There are two compounds of the formula $Pt(NH_3)_2Cl_2$. (Compound B is *cisplatin*, mentioned in the opening to Chapter 1.) They have square planar structures:

One is expected to have a dipole moment, the other is not. Which one would have a dipole moment?

8.49 What is the molecular-orbital configuration of HeH^+? Do you expect the ion to be stable?

8.50 What is the molecular-orbital configuration of He_2^+? Do you expect the ion to be stable?

8.51 Calcium carbide, CaC_2, consists of Ca^{2+} and C_2^{2-} (acetylide) ions. Write the molecular-orbital configuration and bond order of the acetylide ion, C_2^{2-}.

8.52 Sodium peroxide, Na_2O_2, consists of Na^+ and O_2^{2-} (peroxide) ions. Write the molecular-orbital configuration and bond order of the peroxide ion, O_2^{2-}.

**8.53 The oxygen–oxygen bond in O_2^+ is 1.12 Å and in O_2 is 1.21 Å. Explain why the bond length in O_2^+ is shorter than in O_2. Would you expect the bond length in O_2^- to be longer or shorter than that of O_2? Why?

**8.54 The nitrogen–nitrogen bond distance in N_2 is 1.09 Å. On the basis of bond orders, would you expect the bond distance in N_2^+ to be less than or greater than 1.09 Å? Answer the similar question for N_2^-.

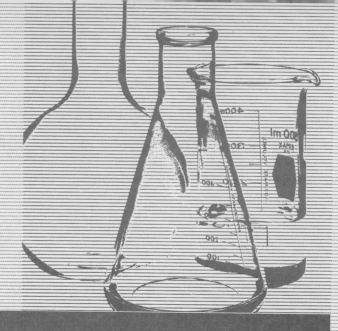

9. *Reactions in Aqueous Solutions*

Ions in Solution; Ionic Equations

9.1 Electrolytes A Note About the Hydrogen Ion/
Introduction to Chemical Equilibrium/ Strong and Weak
Electrolytes

9.2 Ionic Equations

9.3 Types of Reactions

Metathesis Reactions

9.4 Solubility and Precipitation Solubility Rules/
Precipitation Reactions

9.5 Reactions of Acids, Bases, and Salts
Neutralization/ Reactions of Salts; Formation of a Gas

9.6 Preparation of Acids, Bases, and Salts
Preparation of Acids/ Preparation of Bases/ Preparation
of Salts

Oxidation-Reduction Reactions

**9.7 Introduction to Oxidation-Reduction
Reactions** Terminology/ Understanding Oxidation-
Reduction Equations

9.8 Balancing Oxidation-Reduction Equations
Half-Reaction Method/ Oxidation-Number Method

9.9 Equivalents and Normality

Some of the world's most spectacular caves, such as Carlsbad Caverns in New Mexico, were formed in limestone by the chemical action of underground water. Water that has been exposed to air contains dissolved carbon dioxide. When this water flows through limestone deposits, which are primarily calcium carbonate, $CaCO_3$, the rock dissolves, leaving caverns. We can represent the reaction by the following equation:

$$CaCO_3 + H_2O + CO_2 \longrightarrow Ca^{2+} + 2HCO_3^-$$

As a result of the reaction, an aqueous (water) solution of ions Ca^{2+} and HCO_3^- is formed.

The dissolving of limestone by underground water is an example of a chemical reaction in aqueous solution, one that produces ions. Reactions in aqueous solution that involve ions as the products or reactants occur widely in geological processes. Such reactions also occur in biological cells. And, many industrial and laboratory processes use reactions that involve ions in aqueous solution.

Although a great many ionic reactions are possible in aqueous solution, most can be simply classified. Once we recognize the types of reactions possible, we can often predict other similar reactions. For example, we might look for reactions like the one responsible for the formation of limestone caves. Carbon dioxide dissolved in water is an acidic solution. Thus, we see that the dissolving of limestone by underground water is the reaction of a carbonate mineral with an acid. By extension, we would expect other carbonate minerals to be weathered by water containing carbon dioxide. As we will see, the property that makes carbonates react with acidic solutions is present in other compounds, such as sulfides. Thus, the weathering of a sulfide mineral such as sphalerite, ZnS, can be explained in a similar way.

Chapter Overview

In this chapter, we will discuss reactions in aqueous solution that involve ions as reactants or products. Many of the reactions we will encounter later in this book are of this kind, and such reactions occur in many everyday situations and in natural processes. These reactions are conveniently classified into two main types. One includes reactions in which there are no changes of oxidation states, or oxidation numbers, of any atoms. (Oxidation numbers were discussed in Section 7.10.) Within this main type, we will look at *precipitation reactions*, that is, reactions between ions that produce a solid product, and *reactions of acids and bases*. The other main type of reaction includes *oxidation-reduction* reactions, those in which some atoms change oxidation state. Our goal will be to recognize and understand these types of ionic reactions.

Ions in Solution; Ionic Equations

Before we can look in detail at reactions in aqueous solutions, we need to understand some basic concepts concerning ionic solutions. We also need to develop some skills in writing ionic equations.

9.1 Electrolytes

Pure water is a poor conductor of electricity. But if sodium chloride crystals, NaCl, are dissolved in it, the resulting aqueous solution becomes elec-

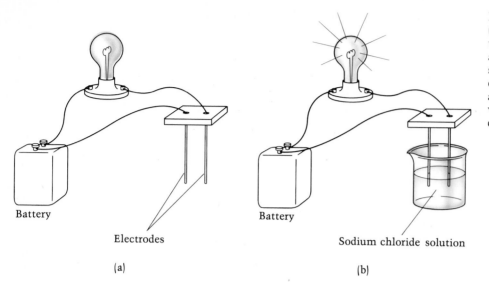

Battery

Electrodes

(a)

Battery

Sodium chloride solution

(b)

trically conducting. Figure 9.1 shows an apparatus used to demonstrate the conductivity of solutions. It consists of two electrodes, or electrical connections, one wired directly to a battery, the other connected to a bulb, which is in turn connected to the other side of the battery. As long as there is no electrical connection between the electrodes, the bulb cannot light; the electric circuit is not complete. The difference in electrical conductivities between pure water and a solution of NaCl can be seen if we first dip the electrodes into pure water. Nothing happens. But when the electrodes are dipped into a solution of NaCl, the bulb lights. Thus, the solution of NaCl is an electrical conductor and allows electricity to flow from the battery.

A substance, such as sodium chloride, that dissolves in water to give an electrically conducting solution is called an **electrolyte.** Other substances, such as sucrose, or table sugar $(C_{12}H_{22}O_{11})$, dissolve in water but give nonconducting or very poorly conducting solutions. These substances are called **nonelectrolytes.**

The conductivity of aqueous sodium chloride is explained by the ionic theory of solutions, proposed in 1884 by the Swedish chemist Svante Arrhenius (1859–1927). According to this theory, an electrolyte produces ions when it dissolves in water. Sodium chloride dissolves in water as Na^+ and Cl^- ions. Suppose we dip the electrodes of the apparatus described earlier (Figure 9.1) into this solution. One electrode is positively charged by the battery and attracts the negatively charged Cl^- ions. The other electrode is negatively charged and attracts Na^+ ions. Thus, the ions move in the solution. This movement of ions, or electrical charge, is responsible for the electric current that flows in the solution. A nonelectrolyte like ethanol dissolves in water as molecules. These are not electrically charged and are not attracted to the electrodes. Thus, no current flows.

Most soluble ionic substances dissolve in water as ions and are therefore electrolytes. Some molecular substances also dissolve to give ions. For example, hydrogen chloride gas, HCl, reacts with water to give the ions H_3O^+ and Cl^-. Since the aqueous solution of hydrogen chloride (called hydrochloric acid) contains ions, hydrogen chloride is an electrolyte. We represent the reaction by the equation

$$HCl(g) + H_2O(l) \longrightarrow H_3O^+(aq) + Cl^-(aq)$$

We use the labels *g*, *l*, and *aq* to denote gas, liquid, and aqueous solution, respectively (*s* denotes a solid). We will use these labels often to make clear the state or phase of a substance or ion. (The labels can be omitted if the phase is understood.)

A Note About the Hydrogen Ion

The free hydrogen ion, H^+, is a hydrogen nucleus (proton) and would strongly attract negative particles to itself. For this reason, a proton does not exist in aqueous solution as a bare particle. Rather, it is strongly associated with water molecules. There is evidence in certain crystals for the existence of a species H_3O^+ (the *hydronium ion*), in which a proton is bonded to one water molecule. But there is also evidence that in solution the proton is bonded to four water molecules, $(H_2O)_4H^+(aq)$; other water molecules are loosely associated with this species. The symbol $H^+(aq)$ for the hydrogen ion in aqueous solution leaves the details of this association of the proton with water unspecified, and we will usually use this symbol. In cases where we wish to show that the proton is transferred from one molecule to water, it is convenient to write the hydrogen ion as $H_3O^+(aq)$. Thus, both $H^+(aq)$ and $H_3O^+(aq)$ will be used to represent the hydrogen ion in aqueous solution.

Introduction to Chemical Equilibrium

To understand the electrical conductivities of certain substances, we need an understanding of *chemical equilibrium*. ■ As an example, consider the reaction of ammonia, NH_3, in aqueous solution. The gas dissolves in water to give a solution of NH_3 molecules. These molecules then react with water to produce NH_4^+ and OH^- ions:

■ Chemical equilibrium is discussed in more depth in Chapter 16 and in subsequent chapters.

$$NH_3(aq) + H_2O(l) \longrightarrow NH_4^+(aq) + OH^-(aq)$$

We call this the forward reaction. The ions produced by the forward reaction subsequently react to produce the original reactants, NH_3 and H_2O:

$$NH_4^+(aq) + OH^-(aq) \longrightarrow NH_3(aq) + H_2O(l)$$

We call this the reverse reaction.

Equilibrium occurs when the speed or rate of the reverse reaction ($NH_4^+ + OH^- \longrightarrow NH_3 + H_2O$) becomes equal to the rate of the forward reaction ($NH_3 + H_2O \longrightarrow NH_4^+ + OH^-$). When the rates are equal, the rate at which NH_3 and H_2O molecules are used up by the forward reaction equals the rate at which these molecules are replenished by the reverse reaction. Thus, the concentrations of all *species* (ions and molecules) in the mixture remain constant. Externally, nothing appears to be happening. But because the forward and reverse reactions are always occurring, though at the same rate, we say that chemical equilibrium is a *dynamic equilibrium*. We represent it by a double arrow:

$$NH_3(aq) + H_2O(l) \rightleftharpoons NH_4^+(aq) + OH^-(aq)$$

Strong and Weak Electrolytes

When electrolytes dissolve in water, they produce ions, but to varying extents. A **strong electrolyte** exists in solution almost entirely as ions. When HCl dissolves in water, it reacts almost completely to give the ions H_3O^+ and Cl^-. Thus, HCl is a strong electrolyte. The reverse reaction would be $H_3O^+(aq) + Cl^-(aq) \longrightarrow HCl(aq) + H_2O(l)$. At equilibrium, however, the concentration of HCl molecules in the solution is very small. For all practical purposes, the reaction of HCl goes completely to ions and we can ignore the reverse reaction. When the reverse reaction has no important effect, we write the equation with a single arrow instead of a double arrow:

$$HCl(aq) + H_2O(l) \longrightarrow H_3O^+(aq) + Cl^-(aq)$$

Most ionic substances, such as NaCl, dissolve in water as ions; that is, the ionic substance exists in aqueous solution almost completely as ions. Thus, such ionic substances are strong electrolytes.

A **weak electrolyte** dissolves in water to give an equilibrium between a molecular substance and a small concentration of ions. Ammonia, NH_3, is an example of a weak electrolyte. When NH_3 dissolves in water, it gives the equilibrium

$$NH_3(aq) + H_2O(l) \rightleftharpoons NH_4^+(aq) + OH^-(aq)$$

The reverse reaction, $NH_4^+ + OH^- \longrightarrow NH_3 + H_2O$, is important. At equilibrium, the concentrations of ions remain small because of this reverse reaction. For example, if we dissolve 0.100 mol NH_3 in water to give a liter of solution, it gives an equilibrium mixture containing 0.099 mol NH_3, 0.001 mol NH_4^+, and 0.001 mol OH^-. Only about 1% of the NH_3 molecules at any given moment have reacted to give ions. To emphasize the importance of the reverse reaction and the equilibrium, we write the equation with a double arrow.

9.2 Ionic Equations

If an aqueous solution of calcium chloride, $CaCl_2$, is added to an aqueous solution of sodium carbonate, Na_2CO_3, fine white crystals of calcium carbonate, $CaCO_3$, are formed. We can write the equation for the reaction as follows:

$$CaCl_2(aq) + Na_2CO_3(aq) \longrightarrow CaCO_3(s) + 2NaCl(aq)$$

We call this the **molecular equation** for the reaction, because the substances are written as if they were molecular substances, even though they actually exist in solution as ions. The molecular equation is useful because it is explicit about what solutions have been added and what products are obtained. Moreover, stoichiometric calculations are easily performed with the molecular equation.

The molecular equation for the reaction of $CaCl_2$ and Na_2CO_3 solutions, however, does not tell us the full story—that the reaction actually involves ions. When one mole of $CaCl_2(s)$ dissolves in water, it goes into the solution as one mole of Ca^{2+} ions and two moles of Cl^- ions. Thus, it would be more descriptive to write $[Ca^{2+}(aq) + 2Cl^-(aq)]$ in place of $CaCl_2(aq)$. An **ionic equation** for the reaction shows each of the aqueous solutions of ionic substances in terms of ions. Thus, the reaction of $CaCl_2(aq)$ with $Na_2CO_3(aq)$ would be

$$[Ca^{2+}(aq) + 2Cl^-(aq)] + [2Na^+(aq) + CO_3^{2-}(aq)] \longrightarrow$$
$$CaCO_3(s) + 2[Na^+(aq) + Cl^-(aq)]$$

Square brackets around the ions remind us of where these ions came from. Of course, once the ions are in solution, their origin is immaterial. The ions are mixed throughout the solution and roam freely about the liquid.

To transform a molecular equation into an ionic equation, we write an aqueous solution of a strong electrolyte as ions in aqueous solution. An aqueous solution of a weak electrolyte is left as the molecular formula; ionic solids are written as the formula for the solid, as we did for $CaCO_3$.

As an example, consider the reaction of the weak electrolyte hydrogen cyanide, HCN, with an aqueous solution of the ionic compound sodium hydroxide, NaOH, to produce the ionic substance sodium cyanide, NaCN, and water. The molecular equation is

$$HCN(aq) + NaOH(aq) \longrightarrow NaCN(aq) + H_2O(l)$$

For the corresponding ionic equation, we write the ionic substances NaOH and NaCN as ions, since the compounds are strong electrolytes. The formulas HCN and H_2O are left as they are, since these compounds are weak electrolytes. Thus,

$$HCN(aq) + [Na^+(aq) + OH^-(aq)] \longrightarrow [Na^+(aq) + CN^-(aq)] + H_2O(l)$$

Notice that in the two ionic equations we have written, some ions appear on both sides of the arrow. These ions, called **spectator ions,** do not take part in the reaction. Because spectator ions do not take part in the actual reaction, we can cancel them from both sides of the equation to leave the **net ionic equation.** This kind of equation tells us exactly which ions react. Again consider the previous equation.

$$\underbrace{Ca^{2+}(aq) + \cancel{2Cl^-(aq)}}_{\text{from } CaCl_2} + \underbrace{\cancel{2Na^+(aq)} + CO_3^{2-}(aq)}_{\text{from } Na_2CO_3} \longrightarrow$$
$$CaCO_3(s) + \cancel{2Na^+(aq)} + \cancel{2Cl^-(aq)}$$

The net ionic equation is

$$Ca^{2+}(aq) + CO_3^{2-}(aq) \longrightarrow CaCO_3(s)$$

In other words, the reaction consists of Ca^{2+} and CO_3^{2-} ions coming together to form solid calcium carbonate.

We can see from the net ionic equation that mixing any solution of calcium ion with any solution of carbonate ion will give the same reaction. For example, calcium nitrate, $Ca(NO_3)_2$, dissolves readily in water. Because it is an ionic compound, it goes into solution as ions (Ca^{2+} and NO_3^-). Similarly, potassium carbonate, K_2CO_3, dissolves in water to give K^+ and CO_3^{2-}. The molecular equation is

$$Ca(NO_3)_2(aq) + K_2CO_3(aq) \longrightarrow CaCO_3(s) + 2KNO_3(aq)$$

The ionic equation is

$$[Ca^{2+}(aq) + \cancel{2NO_3^-(aq)}] + [\cancel{2K^+(aq)} + CO_3^{2-}(aq)] \longrightarrow$$
$$CaCO_3(s) + \cancel{2[K^+(aq)} + \cancel{NO_3^-(aq)]}$$

and the net ionic equation is

$$Ca^{2+}(aq) + CO_3^{2-}(aq) \longrightarrow CaCO_3(s)$$

The net ionic equation is identical to the one obtained from the reaction of $CaCl_2$ and Na_2CO_3. The value of the net ionic equation is its generality. Thus, seawater contains Ca^{2+} and CO_3^{2-} ions from various sources. Regardless of what these sources are, we still know that sediments of calcium carbonate, and eventually limestone, could form from these ions.

Exercise 9.1

Write a net ionic equation for each of the following molecular equations. Molecular substances are labeled as strong or weak electrolytes.

(a) $\underset{\text{strong}}{HCl(aq)} + NaOH(aq) \longrightarrow NaCl(aq) + \underset{\text{weak}}{H_2O(l)}$

(b) $Pb(NO_3)_2(aq) + Na_2SO_4(aq) \longrightarrow PbSO_4(s) + 2NaNO_3(aq)$

(See Problems 9.19 and 9.20.)

9.3 Types of Reactions

Most of the ionic reactions we will discuss in this book are of two major types: metathesis (or double-replacement) and oxidation-reduction. In a **metathesis,** or **double-replacement, reaction,** the positive and negative parts of the two reactants are exchanged to give the products. If the substances are ionic, these positive and negative parts are the cations and anions. If the substances are molecular and form ions in aqueous solution, the positive and negative parts are identified with these ions.

For example, consider the metathesis reaction in aqueous solution of calcium chloride, $CaCl_2$, with sodium carbonate, Na_2CO_3:■

$$CaCl_2 + Na_2CO_3 \longrightarrow$$

We get the products by simply interchanging the anions (or by interchanging the cations). After making the interchange, we write the correct formulas for the new substances in a balanced equation:

$$CaCl_2 + Na_2CO_3 \longrightarrow CaCO_3 + 2NaCl$$

In a metathesis reaction, cations and anions are exchanged, but the oxidation states of the atoms are unchanged. In an **oxidation-reduction** (or

■ We may find it convenient to think of the reaction of $CaCl_2$ and Na_2CO_3 as a metathesis (double-replacement) reaction. But, in writing the net ionic equation, $Ca^{2+} + CO_3^{2-} \longrightarrow CaCO_3$, we see the reaction as simply the combination of ions. Thus, the classification of a reaction depends to some extent on how we look at it.

redox) reaction, however, atoms do change oxidation states. Consider the reaction of copper metal with nitric acid:

$$\overset{0}{3Cu}(s) + \overset{+5}{8HNO_3}(aq) \longrightarrow \overset{+2}{3Cu(NO_3)_2}(aq) + \overset{+2}{2NO}(g) + 4H_2O(l)$$

Note that the oxidation state of the Cu atom changes from 0 to +2 (given in color above the atoms). Nitrogen changes oxidation state from +5 to +2. Therefore, this is an oxidation-reduction reaction. Various kinds of metathesis and oxidation-reduction reactions will be discussed in later sections.

Exercise 9.2

For each of the following, decide whether the reaction is a metathesis or oxidation-reduction type. Explain.

(a) $H_2SO_4(aq) + Ca(OH)_2(aq) \longrightarrow CaSO_4(s) + 2H_2O(l)$
(b) $2H_2S(g) + SO_2(g) \longrightarrow 3S(s) + 2H_2O(l)$
(c) $2Na(s) + 2H_2O(l) \longrightarrow 2NaOH(aq) + H_2(g)$
(d) $AgNO_3(aq) + KCl(aq) \longrightarrow AgCl(s) + KNO_3(aq)$

(See Problems 9.21 and 9.22.)

Metathesis Reactions

The general equation for a metathesis reaction can be written

$$A\overgroup{B + C}D \longrightarrow AD + CB$$

A metathesis reaction consists of an exchange of parts of the two reactants. Why does the reaction go in a particular direction? We will find in all such reactions that one of the products is effectively removed from the reaction mixture. Consider a *precipitation reaction.* In these reactions, one of the products is a solid that comes out of the solution. Such reactions depend on the lack of *solubility* of one product; that is, one product does not dissolve readily in water. In the next section, we will look first at the concept of solubility and then at precipitation reactions.

9.4 Solubility and Precipitation

On a microscopic level, stirring a soluble ionic substance such as sodium chloride in water is a dynamic process. Suppose we stir 40.0 grams of sodium chloride crystals into 100 mL of water at 20°C. Sodium ions and chloride ions leave the surface of the crystals and go into the aqueous phase. The ions move about at random in the solution and may by chance collide with a crystal and stick, thus returning to the crystalline phase. As the sodium chloride continues to dissolve, more ions enter the solution, and the rate at which they return to the crystalline phase increases (the more ions in solution, the more likely ions are to collide with the crystals and stick). Eventually, a dynamic equilibrium is reached in which the rate of ions leaving the

Figure 9.2
Solubility equilibrium. The solid crystalline phase is in dynamic equilibrium with ions in a saturated solution. The rate of ions leaving the crystals equals the rate at which ions return to the crystals.

crystals equals the rate at which ions return to the crystals (see Figure 9.2). We write the dynamic equilibrium this way:

$$NaCl(s) \rightleftharpoons Na^+(aq) + Cl^-(aq)$$

At equilibrium, no more sodium chloride appears to dissolve. We find that 36.0 grams have gone into solution, leaving 4.0 grams of crystals at the bottom of the vessel. The solution is said to be **saturated**, that is, unable to dissolve more NaCl. The **solubility** of sodium chloride in water (the amount that dissolves in a given quantity of water at a given temperature to give a saturated solution) is given as 36.0 g/100 mL at 20°C. Note that if we had mixed 30.0 grams of sodium chloride with 100 mL of water, all of the crystals would have dissolved. Such a solution is called **unsaturated**, since it could dissolve more sodium chloride. (Saturated and unsaturated solutions are compared in Figure 9.3.)

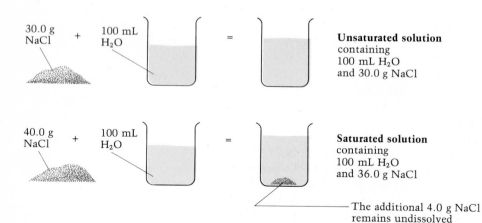

Figure 9.3
Comparison of unsaturated and saturated solutions. (*Top*) The 30.0-g pile of NaCl will dissolve completely in 100 mL of water at 20°C, giving an unsaturated solution. (*Bottom*) When 40.0 g of NaCl are stirred into 100 mL of water, only 36.0 g will dissolve at 20°C, leaving 4.0 g of the crystalline solid on the bottom of the beaker. This solution is saturated.

Figure 9.4
Crystallization from a super-saturated solution of sodium acetate. (a) Crystallization begins to occur where a small crystal of sodium acetate was added. (b) Within seconds, crystal growth spreads from the original crystal throughout the solution.

Sometimes it is possible to obtain a **supersaturated** solution, which contains more dissolved substance than does a saturated solution. For example, the solubility of sodium thiosulfate, $Na_2S_2O_3$, in water at 100°C is 231 g/100 mL. But at room temperature, the solubility is much less—about 50 g/100 mL. Suppose we prepare a solution saturated with sodium thiosulfate at 100°C. We might expect that as the water solution is cooled, sodium thiosulfate would crystallize out. In fact, if the solution is slowly cooled to room temperature, this does not occur. Instead, the result is a solution in which 231 grams of sodium thiosulfate are dissolved in 100 mL of cold water, as compared with the 50 grams that we would normally expect to find dissolved.

Supersaturated solutions are not in equilibrium with the solid phase. Thus, if a small crystal of sodium thiosulfate is added to a supersaturated solution, the excess immediately crystallizes out. Crystallization is usually quite fast and dramatic (see Figure 9.4). ■

■ Adding small crystals to a supersaturated solution to crystallize out the excess is called "seeding."

Solubility Rules

Table 9.1 summarizes the solubilities of common ionic substances. We have listed the solubility of a compound as *soluble, slightly soluble,* or *insoluble.* A "soluble" compound dissolves to the extent of 1 gram or more per 100 mL. The solubility of a "slightly soluble" compound is less than 1 gram, but more than 0.1 gram, per 100 mL. The solubility of an "insoluble" compound

Substance	Solubility	Some Important Exceptions
Alkali metal and ammonium (NH_4^+) compounds	Soluble	
Nitrates (NO_3^-), acetates ($C_2H_3O_2^-$), chlorates (ClO_3^-), and perchlorates (ClO_4^-)	Soluble	
Chlorides (Cl^-), bromides (Br^-), and iodides (I^-)	Soluble	Silver and mercury(I) halides,* and lead and mercury(II) iodides are insoluble; lead chloride, lead bromide, and mercury(II) bromide are slightly soluble.
Sulfates (SO_4^{2-})	Soluble	Strontium, barium, lead, and mercury(I) sulfates* are insoluble; calcium and silver sulfate are slightly soluble.
Hydroxides (OH^-)	Insoluble	Alkali metal and barium hydroxides are soluble; strontium and calcium hydroxides are slightly soluble.
Sulfides (S^{2-}), carbonates (CO_3^{2-}), and phosphates (PO_4^{3-})	Insoluble	Alkali metal and ammonium compounds are soluble.

*The mercury(I) ion is Hg_2^{2+}.

Table 9.1
Empirical Rules for the Solubility of Ionic Solids in Water

is less than 0.1 gram per 100 mL. We will find these categories useful for looking at reactions that yield insoluble products.

Precipitation Reactions

Among the simplest metathesis reactions are precipitation reactions, that is, reactions in which ions react to form a solid, or precipitate. A **precipitate** is a solid formed in solution by a reaction (Figure 9.5). When we mix together two solutions of soluble ionic compounds, we initially get a solution of four different ions, consisting of a cation and an anion from each compound. For example, mixing solutions of nickel(II) sulfate and sodium hydroxide gives a solution of Ni^{2+}, SO_4^{2-}, Na^+, and OH^-. We can see from Table 9.1, however, that nickel(II) hydroxide, $Ni(OH)_2$, is insoluble. This means that the ions Ni^{2+} and OH^- will react to form the solid nickel(II) hydroxide, which will come out of solution, or precipitate. We can filter off the nickel(II) hydroxide precipitate, leaving a *filtrate* (the solution that passes through the filter) of Na^+ and SO_4^{2-} ions. If the filtrate is evaporated, sodium sulfate crystals are left. ■

Let us write a molecular equation for this precipitation reaction using the formulas for the compounds we have mixed and the compounds we have recovered:

$$NiSO_4(aq) + 2NaOH(aq) \longrightarrow Ni(OH)_2(s) + Na_2SO_4(aq)$$

The reaction consists of two soluble compounds ($NiSO_4$ and $NaOH$) in an aqueous solution, giving a mixture of four ions (Ni^{2+}, SO_4^{2-}, Na^+, and OH^-) from which $Ni(OH)_2$ precipitates. Thus, Ni^{2+} and OH^- ions are removed,

■ The reactants $NiSO_4$ and NaOH must be added in amounts that exactly react. Otherwise, either $NiSO_4$ or NaOH will be in excess and will occur with the product Na_2SO_4 in the filtrate.

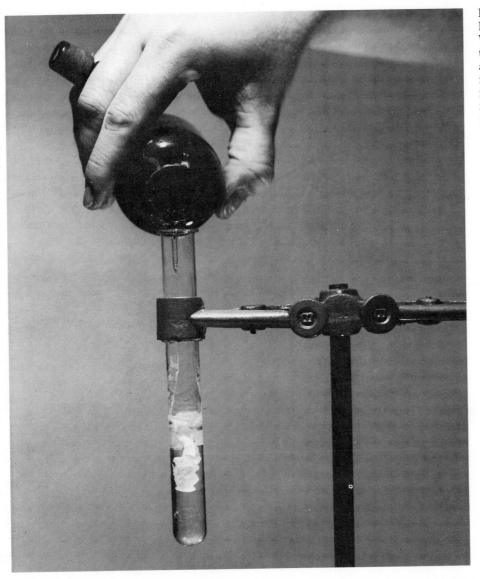

Figure 9.5
Formation of a precipitate.
When a solution of silver nitrate, AgNO$_3$, is poured into
a solution of sodium chloride, NaCl, a white precipitate of silver chloride, AgCl,
forms. Can you write the net
ionic equation?

leaving a solution of Na$^+$ and SO$_4^{2-}$ ions. *Note that it is the formation of a
precipitate that "drives" this metathesis reaction in a particular direction.*
   We obtain the ionic equation for this reaction by writing ions for all of the
soluble ionic compounds, leaving the insoluble nickel(II) hydroxide in terms
of the formula for the solid, Ni(OH)$_2$(s).

$$[Ni^{2+}(aq) + \cancel{SO_4^{2-}(aq)}] + 2[\cancel{Na^+(aq)} + OH^-(aq)] \longrightarrow$$
$$Ni(OH)_2(s) + [\cancel{2Na^+(aq)} + \cancel{SO_4^{2-}(aq)}]$$

If we cancel the spectator ions, we get the net ionic equation

$$Ni^{2+}(aq) + 2OH^-(aq) \longrightarrow Ni(OH)_2(s)$$

This equation represents the essential reaction that occurs: Ni^{2+} and OH$^-$
ions in aqueous solution react to form the solid compound Ni(OH)$_2$.

Example 9.1

For each of the following, decide if a precipitation reaction occurs. If it does, write the balanced molecular equation. Then write the net ionic equation. If no reaction occurs, write NR after the arrow.

(a) $NaCl + Fe(NO_3)_2 \longrightarrow$

(b) $Al_2(SO_4)_3 + NaOH \longrightarrow$

Solution

(a) To obtain the possible metathesis reaction, we write down the reactants and interchange anions to get the products. (Remember to write the correct formulas for the compounds after making this interchange. Example 7.4 in Section 7.2 discusses how to write formulas of ionic compounds.)

$$NaCl + Fe(NO_3)_2 \longrightarrow NaNO_3 + FeCl_2$$

(not balanced)

Referring to Table 9.1, we see that all of the compounds in this equation are soluble (for example, all chlorides except those of Ag^+, Hg_2^{2+}, and Pb^{2+} are soluble). Thus, no reaction occurs.

$$NaCl(aq) + Fe(NO_3)_2(aq) \longrightarrow NR$$

(b) To obtain the possible metathesis reaction, we write down the reactants and interchange anions to get the products:

$$Al_2(SO_4)_3 + NaOH \longrightarrow Al(OH)_3 + Na_2SO_4$$

(not balanced)

According to Table 9.1, sulfates are soluble, except for those with Sr^{2+}, Ba^{2+}, Pb^{2+}, and Hg_2^{2+}. Hydroxides are insoluble, except for those with alkali metal ions (Group IA ions), NH_4^+, and Ba^{2+}. Thus, $Al(OH)_3$ is insoluble, and we should get a precipitation reaction. We write the balanced equations with labels *aq* and *s* to make it clear which substances are soluble and which precipitate. The balanced molecular equation is

$$Al_2(SO_4)_3(aq) + 6NaOH(aq) \longrightarrow$$
$$2Al(OH)_3(s) + 3Na_2SO_4(aq)$$

Check to see that the formulas of ionic compounds are correctly written.

To get the net ionic equation, we write the soluble compounds as ions in aqueous solution and cancel spectator ions:

$$[2Al^{3+}(aq) + 3SO_4^{2-}(aq)] + 6[Na^+(aq) + OH^-(aq)] \longrightarrow$$
$$2Al(OH)_3(s) + 3[2Na^+(aq) + SO_4^{2-}(aq)]$$

The net ionic equation is

$$2Al^{3+}(aq) + 6OH^-(aq) \longrightarrow 2Al(OH)_3(s)$$

or, in simplest terms,

$$Al^{3+}(aq) + 3OH^-(aq) \longrightarrow Al(OH)_3(s)$$

Exercise 9.3

Aqueous solutions of sodium iodide and lead acetate are mixed. If a reaction occurs, write the molecular equation and the net ionic equation. If no reaction occurs, write NR after the arrow.

(See Problems 9.25 and 9.26.)

9.5 Reactions of Acids, Bases, and Salts

Many metathesis reactions involve substances called acids, bases, and salts. These are common substances, some of which are listed in Table 9.2. Before we can look at any reactions, we must first define what we mean by acids, bases, and salts. The definitions we will use for these substances were first developed by Arrhenius. They are based on the properties of water and aqueous solutions. Although pure water is a very poor conductor of electricity, it does produce ions to a very small extent (about $2 \times 10^{-7}\%$ of the molecules present react to give ions). ∎ The reaction is

$$H_2O(l) + H_2O(l) \rightleftharpoons H_3O^+(aq) + OH^-(aq)$$

∎ We can visualize the smallness of the concentration of ions in pure water by calculating the volume of water that holds one mole each of H^+ and OH^-. The concentration of H^+ (and of OH^-) is 1.0×10^{-7} M. Therefore, the volume of water with 1 mol H^+ (and 1 mol OH^-) is 1.0×10^7 L. This is about the volume carried by 300 gasoline tanker trucks.

Table 9.2
Common Acids, Bases, and Salts

Name	Formula	Remarks
Acids:		
Acetic acid	$HC_2H_3O_2$	Found in vinegar
Acetylsalicylic acid	$HC_9H_7O_4$	Aspirin
Ascorbic acid	$H_2C_6H_6O_6$	Vitamin C
Citric acid	$H_3C_6H_5O_7$	Found in lemon juice
Hydrochloric acid	HCl	Found in gastric juice (digestive fluid in stomach)
Sulfuric acid	H_2SO_4	Battery acid
Bases:		
Ammonia	NH_3	Aqueous solution used as a household cleaner
Calcium hydroxide	$Ca(OH)_2$	Slaked lime (used in mortar for building construction)
Magnesium hydroxide	$Mg(OH)_2$	Milk of magnesia (antacid and laxative)
Sodium hydroxide	NaOH	Drain cleaners, oven cleaners
Salts:		
Magnesium sulfate	$MgSO_4$	Epsom salts (laxative)
Potassium chloride	KCl	Salt substitute
Sodium carbonate	Na_2CO_3	Washing soda
Sodium chloride	NaCl	Table salt

Since the H_3O^+ ion is equivalently represented as $H^+(aq)$, it is common to write this reaction as

$$H_2O(l) \rightleftharpoons H^+(aq) + OH^-(aq)$$

This equilibrium gives special significance to the hydrogen ion, H^+, and hydroxide ion, OH^-, in aqueous solutions. Substances dissolved in water can alter the concentrations of these ions.

We will define a **salt** as an ionic substance containing any anion but OH^- or O^{2-}. An **acid** is any substance other than a salt that increases the concentration of hydrogen ion in aqueous solution.■ Hydrogen chloride, for example, is a gas that dissolves readily in water, where it reacts:

$$HCl(aq) + H_2O(l) \longrightarrow H_3O^+(aq) + Cl^-(aq)$$

This equation can also be written

$$HCl(aq) \longrightarrow H^+(aq) + Cl^-(aq)$$

Because HCl produces H^+ ion in water, it is an acid. The aqueous solution is called hydrochloric acid.

A **base** is any substance other than a salt that increases the concentration of hydroxide ion in aqueous solution. Sodium hydroxide, NaOH, is an ionic solid that dissolves in water to give hydroxide ions:

$$NaOH(s) \longrightarrow Na^+(aq) + OH^-(aq)$$

■ This definition of an acid could be relaxed to include salts. Even broader definitions are discussed in Chapter 17. Which definition is used depends on the use we wish to make of it.

Thus, sodium hydroxide is a base. Ammonia gas, NH_3, is also a base. It dissolves readily in water, reacting to the extent of about 1% to give OH^-:

$$NH_3(aq) + H_2O(l) \rightleftharpoons NH_4^+(aq) + OH^-(aq)$$

Because all acids increase the hydrogen ion concentration in water, they have certain common properties. Sour taste is one of these properties. The sour taste of vinegar, for example, is due to acetic acid, $HC_2H_3O_2$, which dissociates into ions to the extent of about 1% in water:∎

$$HC_2H_3O_2(aq) \rightleftharpoons H^+(aq) + C_2H_3O_2^-(aq)$$

∎ The formula for acetic acid is also written CH_3COOH, which is an abbreviation of its structural formula

Basic solutions, such as those of ammonia, have a soapy feel and a bitter taste. (But do not taste laboratory chemicals.)

Acids and bases also cause color changes in certain dyes. **Acid-base indicators** are dyes that are used to indicate the acidity or basicity of a solution. Such dyes are common in natural substances and mixtures. The amber color of tea, for example, is perceptibly lightened by the addition of lemon juice (citric acid). Various fruit and vegetable juices such as those of beets, red cabbage, and black cherries show color changes with acids. *Litmus* is a common laboratory acid–base indicator. It is a dye, extracted from certain species of lichens (a scaly or crustlike growth on rocks and trees), that turns red in acidic solution and blue in basic solution.

Acids and bases are classified as strong or weak. **Strong acids** are strong electrolytes. They are molecular substances that ionize completely in water (that is, they react completely to give ions). For example,

$$HCl(aq) \longrightarrow H^+(aq) + Cl^-(aq)$$
$$HNO_3(aq) \longrightarrow H^+(aq) + NO_3^-(aq)$$

Table 9.3 lists the common strong acids. Most of the other acids we will discuss are **weak acids,** that is, acids that are weak electrolytes. They only partially dissociate into ions in aqueous solution and are present in equilibrium with the acid molecule. One example is hydrocyanic acid, HCN.

$$HCN(aq) \rightleftharpoons H^+(aq) + CN^-(aq)$$

Strong bases are strong electrolytes. They are present in aqueous solution entirely as ions. The ionic compound sodium hydroxide, NaOH, is an example. It dissolves in water to give only ions. We represent the process of dissolving NaOH in water by the equation:

$$NaOH(s) \longrightarrow Na^+(aq) + OH^-(aq)$$

Strong Acids	Strong Bases
Perchloric acid, $HClO_4$	*Soluble*
Sulfuric acid, H_2SO_4	Group IA metal hydroxides ($NaOH$, KOH)
Hydriodic acid, HI	Barium hydroxide, $Ba(OH)_2$
Hydrobromic acid, HBr	*Slightly soluble*
Hydrochloric acid, HCl	Strontium hydroxide, $Sr(OH)_2$
Nitric acid, HNO_3	Calcium hydroxide, $Ca(OH)_2$
	Magnesium hydroxide, $Mg(OH)_2$

Table 9.3
Strong Acids and Bases

The common strong bases are hydroxides of Group IA and IIA elements, except for Be (see Table 9.3). **Weak bases,** like weak acids, are present in solution in equilibrium with a molecular species. Ammonia, NH_3, is an example:

$$NH_3(aq) + H_2O(l) \rightleftharpoons NH_4^+(aq) + OH^-(aq)$$

Neutralization

Neutralization is a reaction of acids and bases to produce salts. The reaction of a strong base with an acid yields a salt and water and can be written as a metathesis reaction.

Consider the neutralization of the strong acid $HCl(aq)$ with the strong base $NaOH(aq)$. The molecular equation is

$$HCl(aq) + NaOH(aq) \longrightarrow H_2O(l) + NaCl(aq)$$
$$\text{an acid} \qquad \text{a base} \qquad \text{water} \qquad \text{a salt}$$

Since the acid, base, and salt are strong electrolytes, the ionic reaction is

$$[H^+(aq) + \cancel{Cl^-(aq)}] + [\cancel{Na^+(aq)} + OH^-(aq)] \longrightarrow$$
$$H_2O(l) + [\cancel{Na^+(aq)} + \cancel{Cl^-(aq)}]$$

The net ionic equation is

$$H^+(aq) + OH^-(aq) \longrightarrow H_2O(l)$$

Thus, out of the mixture of ions produced by HCl and NaOH, the H^+ and OH^- ions are removed by reaction to produce the stable molecule H_2O (it ionizes only slightly). Compare this with a precipitation reaction. There we found that it was the formation of a precipitate that "drives" the reaction, or makes it go in a particular direction. *In a neutralization reaction, it is the formation of a weak electrolyte (H_2O) that "drives" the reaction.* The H_2O molecule is very stable, that is, it holds together and ionizes only slightly as a result.

The net ionic equation for the neutralization of a strong acid by a strong base is simply the reaction of H^+ and OH^- to give H_2O. But as shown in the next example, the net ionic equation for the neutralization of a weak acid by a strong base is the reaction of the acid with OH^-.

Example 9.2

Write the net ionic equation for the neutralization of acetic acid, $HC_2H_3O_2$, with sodium hydroxide, NaOH, in aqueous solution.

Solution

The molecular equation for the neutralization is

$$HC_2H_3O_2(aq) + NaOH(aq) \longrightarrow$$
$$H_2O(l) + NaC_2H_3O_2(aq)$$

Acetic acid is a weak acid (note that it is not one of the strong acids in Table 9.3). Both NaOH and $NaC_2H_3O_2$ are ionic compounds and strong electrolytes. Therefore, the ionic equation is

$$HC_2H_3O_2(aq) + [\cancel{Na^+(aq)} + OH^-(aq)] \longrightarrow$$
$$H_2O(l) + [\cancel{Na^+(aq)} + C_2H_3O_2^-(aq)]$$

The net ionic equation is

$$HC_2H_3O_2(aq) + OH^-(aq) \longrightarrow H_2O(l) + C_2H_3O_2^-(aq)$$

Exercise 9.4

Write the molecular equation and net ionic equation for the neutralization of hydrocyanic acid, HCN, by potassium hydroxide.

(See Problems 9.27 and 9.28.)

The neutralization of a base like NH_3 by a strong acid, and by a weak acid, are reactions where only a soluble salt is formed. They are shown in the following equations:

$$NH_3(aq) + H^+(aq) + Cl^-(aq) \longrightarrow NH_4^+(aq) + Cl^-(aq)$$
strong acid

$$NH_3(aq) + HC_2H_3O_2(aq) \longrightarrow NH_4^+(aq) + C_2H_3O_2^-(aq)$$
weak acid

The corresponding net ionic equations are

$$NH_3(aq) + H^+(aq) \longrightarrow NH_4^+(aq)$$

$$NH_3(aq) + HC_2H_3O_2(aq) \longrightarrow NH_4^+(aq) + C_2H_3O_2^-(aq)$$

Note that we can regard each of these as a reaction in which a proton is transferred to NH_3 to give NH_4^+. ■ The proton forms a covalent bond to the lone pair on the nitrogen atom of NH_3. It is the lone pair on nitrogen that makes NH_3 and similar compounds basic. Table 9.4 lists some weak bases having nitrogen atoms with lone pairs.

■ The concept of proton transfer in acid-base reactions is amplified in Chapter 17.

The acids HCl and $HC_2H_3O_2$ are *monoprotic acids*. These are acids that give up one mole of H^+ per mole of acid during neutralization. **Polyprotic acids** can give up two or more moles of H^+ per mole of acid. Sulfuric acid, H_2SO_4, and phosphoric acid, H_3PO_4, are examples of polyprotic acids. Specifically, H_2SO_4 is a *diprotic acid*, and H_3PO_4 is a *triprotic acid*. By reacting one mole of a polyprotic acid with one or two (or more) moles of base, we can obtain a series of salts. Thus,

$$H_3PO_4(aq) + NaOH(aq) \longrightarrow NaH_2PO_4(aq) + H_2O(l)$$
$$H_3PO_4(aq) + 2NaOH(aq) \longrightarrow Na_2HPO_4(aq) + 2H_2O(l)$$
$$H_3PO_4(aq) + 3NaOH(aq) \longrightarrow Na_3PO_4(aq) + 3H_2O(l)$$

Salts, such as NaH_2PO_4 and Na_2HPO_4, that have acidic hydrogen atoms and can undergo neutralization with bases are called **acid salts.**

Exercise 9.5

Write molecular equations for the successive neutralizations of each acidic hydrogen of H_2SO_4 with KOH.

(See Problems 9.29 and 9.30.)

Reactions of Salts; Formation of a Gas

A weak acid is a relatively stable molecule. It dissociates to ions, but only to a limited extent. For this reason, the anion corresponding to the weak acid tends to react with hydrogen ion and sources of hydrogen ion (acids) to form

Table 9.4
Weak Bases with
Nitrogen Atoms

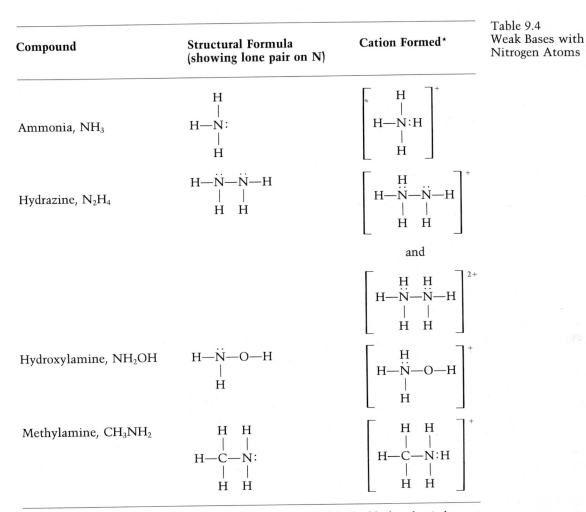

Compound	Structural Formula (showing lone pair on N)	Cation Formed*

*The bond formed with the proton to give the cation is denoted by double dots, but it does not differ from other N—H bonds.

the weak acid. This explains why limestone (calcium carbonate, $CaCO_3$) dissolves in water containing carbon dioxide. Such water is an acidic solution and contains carbonic acid, H_2CO_3.

$$CO_2(aq) + H_2O(l) \rightleftharpoons H_2CO_3(aq)$$

Carbonate ion, CO_3^{2-}, reacts with acids to form hydrogen carbonate ion, HCO_3^-, which is a weak acid. Thus, limestone dissolves because the CO_3^{2-} ion in $CaCO_3(s)$ reacts with the acidic water to give HCO_3^-, which goes into solution with Ca^{2+}.

For example, consider the reaction of calcium carbonate with a strong acid. Hydrogen ions react with $CaCO_3$ to form HCO_3^-:

$$CaCO_3(s) + H^+(aq) \longrightarrow Ca^{2+}(aq) + HCO_3^-(aq)$$

Since we have added a strong acid, there will be sufficient hydrogen ion to react with HCO_3^- to give carbonic acid. Carbonic acid decomposes easily

Salt	Gas	Example
Carbonate (CO_3^{2-})	CO_2	$Na_2CO_3 + 2HCl \longrightarrow 2NaCl + H_2O + CO_2$
Sulfide (S^{2-})	H_2S	$Na_2S + 2HCl \longrightarrow 2NaCl + H_2S$
Sulfite (SO_3^{2-})	SO_2	$Na_2SO_3 + 2HCl \longrightarrow 2NaCl + H_2O + SO_2$
Ammonium (NH_4^+)	NH_3	$NH_4Cl + NaOH \longrightarrow NaCl + H_2O + NH_3$

into CO_2 and H_2O:

$$HCO_3^-(aq) + H^+(aq) \longrightarrow H_2CO_3(aq) \longrightarrow H_2O(l) + CO_2(g)$$

Carbon dioxide bubbles from the reaction mixture. This reaction is the basis of a test for carbonate minerals. Any mineral that is a carbonate will fizz from bubbles of CO_2 if treated with hydrochloric acid.

The release, or *evolution*, of CO_2, with its subsequent loss from the reaction vessel, effectively drives the reaction of carbonate salts and strong acids in the direction in which gas is produced. Other salts, such as sulfides and sulfites, behave in a similar manner (see Table 9.5). Sulfides contain the S^{2-} ion, which in strong acid solution gives H_2S gas. Sulfites contain the SO_3^{2-} ion, which in strong acid gives sulfurous acid, H_2SO_3, that readily breaks down to give H_2O and SO_2 gas. *In the reaction of these salts with a strong acid, it is the formation and loss of a gas from the mixture that "drives" the reaction.*

Ammonium compounds react with strong bases to evolve ammonia gas, particularly if the vessel is warmed to reduce the solubility of the gas. In this case, ammonia is a weak base and is formed by the reaction of NH_4^+ with OH^-:

$$NH_4^+(aq) + OH^-(aq) \longrightarrow NH_3(g) + H_2O(l)$$

Once NH_3 is removed from the mixture, it is not available for the reverse reaction, $NH_3 + H_2O \longrightarrow NH_4^+ + OH^-$. As a result, the reaction is driven to products, NH_3 and H_2O.

Example 9.3

Write the balanced molecular equation for the reaction of zinc sulfide with hydrochloric acid. Also write the net ionic equation.

Solution

Consider the metathesis reaction:

$$ZnS + HCl \longrightarrow ZnCl_2 + H_2S \quad \text{(not balanced)}$$

Note that ZnS is insoluble and $ZnCl_2$ is soluble (Table 9.1). H_2S is a gas. The balanced molecular equation is

$$ZnS(s) + 2HCl(aq) \longrightarrow ZnCl_2(aq) + H_2S(g)$$

The ionic equation is

$$ZnS(s) + 2[H^+(aq) + Cl^-(aq)] \longrightarrow$$
$$[Zn^{2+}(aq) + 2Cl^-(aq)] + H_2S(g)$$

and the net ionic equation is

$$ZnS(s) + 2H^+(aq) \longrightarrow Zn^{2+}(aq) + H_2S(g)$$

Exercise 9.6

Write the molecular equation and net ionic equation for the reaction of sodium sulfite with hydrochloric acid.

(See Problems 9.31 and 9.32.)

Many acids are volatile compounds, that is, they easily vaporize when heated. If a salt of a volatile acid is heated with an acid of relatively high boiling point (concentrated sulfuric acid, which boils at 338°C, is commonly used), the volatile acid leaves the reaction mixture. For example, consider the reaction of calcium fluoride with concentrated sulfuric acid:

$$CaF_2(s) + H_2SO_4(l) \xrightarrow{\Delta} CaSO_4(s) + 2HF(g)$$

The symbol Δ over the arrow means that the reaction mixture is heated. Hydrogen fluoride, which boils at 20°C, leaves the reaction vessel as the vapor (gas). Although this is not an aqueous reaction, we can regard it as a metathesis reaction in which a gas is evolved.

9.6 Preparation of Acids, Bases, and Salts

The metathesis reactions we discussed in the preceding sections can be useful in preparation of acids, bases, and salts. The three kinds of reactions that we discussed were:

1. *Precipitation* Two soluble compounds react in aqueous solution to give a precipitate of an ionic compound plus a solution of another compound.
2. *Neutralization* An acid and base react to give a solution of a salt.
3. *Reaction of certain salts with a strong acid or base to form a gas or volatile compound* The product solution contains a salt; the volatile compound may be an acid or base.

Let us see how we can use these reactions in the preparation of compounds.

Preparation of Acids

As we noted in the previous section, many acids are volatile compounds. The only common acids that are relatively nonvolatile are sulfuric acid, H_2SO_4, and phosphoric acid, H_3PO_4. They are used in the commercial preparation of a number of acids. Hydrofluoric acid, HF, is made commercially from CaF_2, as described in the previous section. Until recently, hydrochloric acid was prepared commercially by heating sodium chloride with concentrated sulfuric acid:■

$$NaCl(s) + H_2SO_4(l) \xrightarrow{\Delta} NaHSO_4(s) + HCl(g)$$

■ Most HCl is now obtained as a by-product in the chlorination of certain organic (carbon-containing) compounds. For example,

$$CH_4 + Cl_2 \longrightarrow CH_3Cl + HCl$$

Sulfuric acid cannot be used to prepare HBr and HI from their salts because concentrated H_2SO_4 reacts with Br^- and I^- to give the elements, Br_2 and I_2. However, phosphoric acid may be used in its place. For example,

$$NaBr(s) + H_3PO_4(l) \xrightarrow{\Delta} NaH_2PO_4(s) + HBr(g)$$

Precipitation can be used to advantage in preparing an acid that might be decomposed by the heating required in the previous method. For example, chlorous acid, $HClO_2$, is an unstable acid. But an aqueous solution of the acid can be prepared from a solution of barium chlorite, $BaClO_2$, and aqueous sulfuric acid, $H_2SO_4(aq)$:

$$Ba(ClO_2)_2(aq) + H_2SO_4(aq) \longrightarrow BaSO_4(s) + 2HClO_2(aq)$$

This method depends on the fact that the products are a soluble acid and an insoluble salt. The insoluble salt can be filtered from the aqueous solution of the acid. If the aqueous chlorous acid must be relatively free of Ba^{2+} and SO_4^{2-} ions, it is necessary to add the stoichiometric amounts of reactants.

Example 9.4

(a) Nitric acid was at one time prepared commercially from the mineral Chile saltpeter, $NaNO_3$. The preparation depended on the volatility of the acid. Describe a possible reaction by writing the molecular equation. (b) How could you prepare an aqueous solution of nitrous acid, HNO_2, free of extraneous ions? Nitrous acid decomposes on heating to give NO, NO_2, and H_2O. Give the molecular equation for the preparation.

Solution

(a) Nitric acid can be distilled from a mixture of $NaNO_3$ and concentrated sulfuric acid:

$$NaNO_3(s) + H_2SO_4(l) \xrightarrow{\Delta} NaHSO_4(s) + HNO_3(g)$$

(b) Distillation cannot be used because HNO_2 decomposes. However, precipitation can be used. We look for a nitrite salt and an acid that will react to give an insoluble salt of the acid and an aqueous solution of HNO_2. Since barium sulfate is an insoluble salt, the following reaction is one possibility:

$$Ba(NO_2)_2(aq) + H_2SO_4(aq) \longrightarrow BaSO_4(s) + 2HNO_2(aq)$$

If equal molar amounts of $Ba(NO_2)_2$ and H_2SO_4 are added, essentially all Ba^{2+} and SO_4^{2-} ions precipitate as $BaSO_4$, which can be filtered from the solution of HNO_2.

Exercise 9.7

Describe preparations of the following: (a) perchloric acid, $HClO_4$, from $KClO_4$, based on the volatility of $HClO_4$; (b) an aqueous solution of perchloric acid from $AgClO_4$ by precipitation.

(See Problems 9.35, 9.36, 9.37, and 9.38.)

Preparation of Bases

Hydroxides of Group IA elements can be prepared by precipitation reactions. A commercial method of preparing lithium hydroxide uses the following reaction:

$$Li_2CO_3(aq) + Ca(OH)_2(aq) \longrightarrow CaCO_3(s) + 2LiOH(aq)$$

Filtering off the precipitate of calcium carbonate leaves a solution of lithium hydroxide.■ Note that the method uses a strong base and a salt to give an insoluble salt and the new base.

Ammonia and related bases, such as methylamine, CH_3NH_2, that are gaseous or volatile substances can be prepared from their salts by heating

■ At one time, NaOH was prepared in a similar way from Na_2CO_3. Now it is prepared by electrolysis (electrical decomposition) of NaCl(aq).

with a strong base. For example,

$$2NH_4Cl(aq) + Ca(OH)_2(aq) \xrightarrow{\Delta} 2NH_3(g) + H_2O(l) + CaCl_2(aq)$$

Example 9.5

Describe the following preparations: (a) KOH from $Ba(OH)_2$ by precipitation, (b) methylamine, CH_3NH_2, from the salt CH_3NH_3Cl (an ionic compound of $CH_3NH_3^+$ and Cl^-). Use molecular equations.

Solution

(a) Note that we start with a strong base, $Ba(OH)_2$, as one reactant. The preparation depends on the precipitation of a barium salt, such as $BaCO_3$ or $BaSO_4$. As the other reactant, we could use K_2SO_4. The reaction is

$$K_2SO_4(aq) + Ba(OH)_2(aq) \longrightarrow 2KOH(aq) + BaSO_4(s)$$

If equal molar amounts of the reactants are used, essentially all of the Ba^{2+} and SO_4^{2-} ions are precipitated as $BaSO_4$. After removing $BaSO_4$ by filtration, the filtrate can be evaporated to recover crystals of KOH. (b) If the salt CH_3NH_3Cl is heated with an aqueous solution of strong base, such as NaOH, methylamine gas will evolve:

$$CH_3NH_3Cl(aq) + NaOH(aq) \longrightarrow \\ CH_3NH_2(g) + H_2O(l) + NaCl(aq)$$

Exercise 9.8

Using molecular equations, describe the preparation of (a) NaOH from Na_2CO_3, (b) NH_3 from $(NH_4)_2SO_4$.

(See Problems 9.39, 9.40, 9.41, and 9.42.)

Preparation of Salts

Salts are often prepared by neutralization of the appropriate acid and base. Calcium acetate is obtained from calcium hydroxide and acetic acid:

$$Ca(OH)_2(aq) + 2HC_2H_3O_2(aq) \longrightarrow Ca(C_2H_3O_2)_2(aq) + 2H_2O(l)$$

Salts are also obtained from carbonates, sulfides, and sulfites by reaction with a strong acid to evolve a gas. For example, barium chloride is obtained commercially from barium sulfide:■

$$BaS(s) + 2HCl(aq) \longrightarrow BaCl_2(aq) + H_2S(g)$$

In some cases, precipitation is also useful. Silver iodide, which is used to make photographic film, is obtained from the soluble salts $AgNO_3$ and KI:

$$AgNO_3(s) + KI(aq) \longrightarrow AgI(s) + KNO_3(aq)$$

Silver iodide precipitates and can be filtered from the solution.

■ Barium sulfide is prepared from the mineral barite, $BaSO_4$:

$$BaSO_4(s) + 4C(s) \xrightarrow{\Delta} \\ BaS(s) + 4CO(g)$$

Barium sulfide is soluble in water and is used to prepare other barium salts, including purified $BaSO_4$.

Example 9.6

Give molecular equations for each of the following preparations: (a) $MgSO_4$ by neutralization, (b) NaF from Na_2CO_3, (c) $BaCO_3$ by precipitation.

Solution

(a) Looking at the formula of the salt, we see that the base must be $Mg(OH)_2$, and the acid H_2SO_4. From the

(*Continued*)

solubility rules, we see that $Mg(OH)_2$ is insoluble and $MgSO_4$ is soluble. Thus, the molecular equation is

$$Mg(OH)_2(s) + H_2SO_4(aq) \longrightarrow MgSO_4(aq) + 2H_2O(l)$$

(b) An acid reacts with Na_2CO_3 to give a salt, NaF, and $H_2O(l) + CO_2(g)$. The acid must be HF(aq). Hence,

$$Na_2CO_3(aq) + 2HF(aq) \longrightarrow$$
$$2NaF(aq) + H_2O(l) + CO_2(g)$$

(c) For reactants, we require two soluble salts, one supplying Ba^{2+}, the other CO_3^{2-}. One of the products is $BaCO_3$, which according to Table 9.1 is insoluble. The other product must be soluble so that the two products can be separated by filtration. Note that $BaCl_2$ and Na_2CO_3 are soluble (Table 9.1). Also, the product NaCl is soluble. Hence, one possible equation is

$$BaCl_2(aq) + Na_2CO_3(aq) \longrightarrow BaCO_3(s) + 2NaCl(aq)$$

Exercise 9.9

Give molecular equations to describe these preparations: (a) $Ca(ClO_4)_2$ by neutralization, (b) $ZnCl_2$ from ZnS, (c) $PbSO_4$ by precipitation.

(See Problems 9.43 and 9.44.)

Oxidation-Reduction Reactions

Oxidation-reduction reactions are those reactions in which some of the atoms change oxidation states. As we will see, some atoms are oxidized (that is, they increase in oxidation state), while other atoms are reduced (that is, they decrease in oxidation state). We will also see that oxidation and reduction occur together in such reactions, which is why they are called oxidation-reduction.

9.7 Introduction to Oxidation-Reduction Reactions

In this section, we will first discuss the terminology used to describe oxidation-reduction reactions. Then, we will look at some oxidation-reduction equations with a view toward gaining an understanding of them. At first glance, many such equations may appear complicated, but we will see how to find the important aspects of the equation.

Terminology

Oxidation-reduction reactions are common. Any combustion in oxygen, for example, is an oxidation-reduction reaction, since oxygen starts as the element (oxidation number 0) and ends up in a combined state (oxidation number -2). For example, if iron wire is heated and plunged into a vessel of oxygen gas, it burns, giving iron(III) oxide:

$$\overset{0}{4Fe} + \overset{0}{3O_2} \longrightarrow \overset{+3\ -2}{2Fe_2O_3}$$

(Oxidation numbers are shown above each atom.) A very similar reaction occurs when a heated iron wire is plunged into a vessel of chlorine gas. The iron burns, giving iron(III) chloride.

$$\overset{0}{Fe} + 3\overset{0}{Cl_2} \longrightarrow 2\overset{+3 \; -1}{FeCl_3}$$

Because Fe and Cl atoms change oxidation numbers, this is an oxidation-reduction equation.

Originally "oxidation" meant only reaction with oxygen. Now the term has a much more general meaning. In any oxidation-reduction reaction, the atom that increases in oxidation number (and thereby loses electrons) is said to undergo **oxidation,** or to be oxidized. Thus, iron in both of the preceding reactions goes from the element (oxidation number 0) to the compound (oxidation number +3). The oxidation number has increased. Also, in going from the element to the compound, iron atoms effectively lose electrons:

$$Fe \longrightarrow Fe^{3+} + 3e^-$$

Therefore, iron is oxidized. The atom that is reduced in oxidation number (and thereby gains electrons) is said to undergo **reduction,** or be reduced. Thus, in the first reaction, oxygen atoms are reduced from oxidation number 0 to oxidation number -2. In effect, oxygen atoms gain electrons:■

$$O + 2e^- \longrightarrow O^{2-}$$

In any oxidation-reduction reaction, oxidation and reduction occur together. If one atom loses electrons, another gains them. In other words, whenever one atom increases in oxidation number, another atom decreases. An **oxidizing agent** is the reactant that acts on another substance to oxidize one of its atoms (and has one of its own atoms reduced). Thus, since O_2 and Cl_2 oxidize iron metal, they function as oxidizing agents in these reactions. A **reducing agent** is the reactant that acts on another substance to reduce one of its atoms (and has one of its own atoms oxidized). In the reactions of iron with oxygen and with chlorine, iron reduces O_2 and Cl_2. Therefore, iron functions as a reducing agent in these reactions.

To summarize this terminology, consider another example. When a zinc metal strip is dipped into a blue solution of copper(II) sulfate, the zinc becomes coated with the reddish tinge of metallic copper (Figure 9.6). The reaction is

$$\overset{0}{Zn}(s) + \overset{+2}{Cu^{2+}}(aq) \longrightarrow \overset{+2}{Zn^{2+}}(aq) + \overset{0}{Cu}(s)$$

In this reaction, zinc increases in oxidation number (it is oxidized), and copper decreases in oxidation number (it is reduced). Or, in terms of electrons gained and lost, we have

$$Zn \longrightarrow Zn^{2+} + 2e^- \quad \text{(electrons lost, so Zn is oxidized)}$$
$$Cu^{2+} + 2e^- \longrightarrow Cu \quad \text{(electrons gained, so Cu is reduced)}$$

Thus, zinc metal is the reducing agent and Cu^{2+} is the oxidizing agent. This

■ Oxidation and reduction are approximate ways of describing changes in electron distribution in a reaction, just as oxidation numbers are approximate ways of assigning electrons to atoms. Only in certain cases can we say that a definite number of electrons have been totally lost or gained during oxidation or reduction.

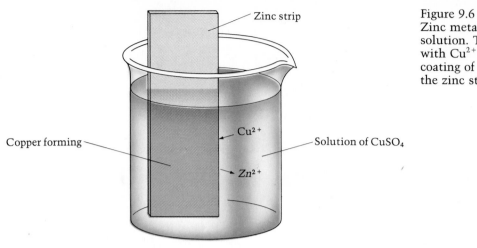

Figure 9.6
Zinc metal reacts in a $CuSO_4$ solution. The metal reacts with Cu^{2+} ion, giving a thin coating of copper metal over the zinc strip.

is summarized as follows:

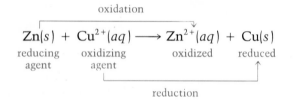

Understanding Oxidation-Reduction Equations

A standard test for bromide ion in aqueous solution is to oxidize the ion to bromine with potassium permanganate, $KMnO_4$, in the presence of acid. Bromine can be extracted from the aqueous solution with methylene chloride, CH_2Cl_2, a liquid to which bromine gives a yellow or orange color.■ The net ionic equation for the oxidation of Br^- by MnO_4^- in acidic solution is as follows:

■ Bromine in carbon tetrachloride, CCl_4, gives a more vivid color but CCl_4 is a toxic substance.

$$10Br^-(aq) + 2MnO_4^-(aq) + 16H^+(aq) \longrightarrow$$
$$5Br_2(aq) + 2Mn^{2+}(aq) + 8H_2O(l)$$

This is a typical oxidation-reduction equation. At first glance, such equations may look a bit intimidating. It is helpful to have a way of reducing such an equation to something easily understandable.

To do this, we first obtain the oxidation numbers of the atoms in the equation. Then we note the species containing atoms whose oxidation numbers change. If we write down just these species, ignoring coefficients, we get a *skeleton equation*. For the previous equation, we get

$$\overset{-1}{Br^-} + \overset{+7}{MnO_4^-} \longrightarrow \overset{0}{Br_2} + \overset{+2}{Mn^{2+}}$$

This shows us quickly what is being oxidized (Br^-) and what is the product

			Group				
	IA	IIA	IIIA	IVA	VA	VIA	VIIA
Period 2	**Li**	**Be**	**B**	**C**	**N**	**O**	**F**
	+1 Li^+	+2 Be^{2+}	+3 B_2O_3	+4 CO_2	+5 HNO_3	0 O_2	0 F_2
	0 Li	0 Be	0 B	+2 CO	+4 NO_2	−1 H_2O_2	−1 F^-
				0 C	+3 HNO_2	−2 H_2O	
				−4 CH_4	+2 NO		
					+1 N_2O		
					0 N_2		
					−1 NH_2OH		
					−2 N_2H_4		
					−3 NH_3		
Period 3	**Na**	**Mg**	**Al**	**Si**	**P**	**S**	**Cl**
	+1 Na^+	+2 Mg^{2+}	+3 Al^{3+}	+4 SiO_2	+5 H_3PO_4	+6 H_2SO_4	+7 $HClO_4$
	0 Na	0 Mg	0 Al	0 Si	+3 H_3PO_3	+4 SO_2	+5 $HClO_3$
				−4 SiH_4	0 P_4	0 S_8	+3 $HClO_2$
					−3 PH_3	−2 H_2S	+1 $HClO$
							0 Cl_2
							−1 Cl^-

*Relatively stable oxidation states are shown in color. Uncommon oxidation states have been omitted.

Table 9.6
Oxidation States and
Common Species of
Some Main-Group
Elements*

of this oxidation (Br_2). We also see what is the oxidizing agent (MnO_4^-) and to what it is reduced (Mn^{2+}).

Further understanding comes from knowing the various oxidation states (oxidation numbers) of the elements and common species with those oxidation states. For bromine, we have the following:

Bromine Species	Oxidation Number		
$HBrO_4$	+7	↑	
$HBrO_3$	+5		
$HBrO_2$	+3	oxidation	reduction
$HBrO$	+1		
Br_2	0		
Br^-	−1		↓

The most stable oxidation state is −1 (shown in color) and is represented by Br^-. Like most of the nonmetals, bromine displays a series of oxidation states (all odd), from a low value equal to the group number minus eight (−1) to a high value equal to the group number (+7). Oxidation of Br^- would give a species with higher oxidation number, which from the diagram of oxidation states would be Br_2 or perhaps an oxyacid. The actual product, we know, is Br_2.

Table 9.6 lists the oxidation states of the second and third rows of the main-group elements. Metals usually have oxidation states in compounds equal to the group number. Nonmetals, on the other hand, usually have a series of oxidation states extending from a low value equal to the group

Cr		Mn		Fe		Co		Ni	
+6	CrO_4^{2-}, $Cr_2O_7^{2-}$	+7	MnO_4^-	+3	Fe^{3+}	+3	Co^{3+}	+2	Ni^{2+}
+3	Cr^{3+}	+4	MnO_2	+2	Fe^{2+}	+2	Co^{2+}	0	Ni
+2	Cr^{2+}	+2	Mn^{2+}	0	Fe	0	Co		
0	Cr	0	Mn						

Cu		Ag		Zn		Cd		Hg	
+2	Cu^{2+}	+1	Ag^+	+2	Zn^{2+}	+2	Cd^{2+}	+2	Hg^{2+}
+1	Cu^+	0	Ag	0	Zn	0	Cd	+1	Hg_2^{2+}
0	Cu							0	Hg

Table 9.7
Oxidation States and
Common Species of
Some Transition Elements*

*Stable oxidation states are shown in color. Uncommon oxidation states have been omitted.

number minus eight to a high value equal to the group number. The exceptions are oxygen and fluorine. Usually, the series of oxidation states for compounds are all even or all odd (as for Br). Nitrogen is an exception, having compounds in all oxidation states from -3 to $+5$.

Table 9.7 lists oxidation states of some transition elements. From this, we see that the most stable oxidation state of manganese is $+2$. Manganese species in higher oxidation states would function as oxidizing agents, since there would be a tendency for them to go to the $+2$ state. We see this in the reaction of MnO_4^- with Br^-. Here MnO_4^- acts as the oxidizing agent and is converted in acid solution to the stable Mn^{2+} ion. In basic solution, MnO_4^- is reduced only to MnO_2. The MnO_2 precipitates in basic solution, which concludes the reaction.

Example 9.7

Copper metal, unlike most metals, does not react with H^+ ion from acids. (Zinc, for example, gives $Zn + 2H^+ \longrightarrow Zn^{2+} + H_2$.) Copper, however, does react with nitric acid and with hot, concentrated sulfuric acid. In hot, concentrated sulfuric acid, the reaction is

$$Cu(s) + 4H^+(aq) + SO_4^{2-}(aq) \xrightarrow{\Delta}$$
$$Cu^{2+}(aq) + SO_2(g) + 2H_2O(l)$$

Write the skeleton equation for this oxidation-reduction equation. Then draw a diagram indicating oxidation numbers of atoms that change their numbers, the reducing agent, the oxidizing agent, the oxidized species, and the reduced species. Also, with arrows, indicate the oxidation part of the reaction and the reduction part of the reaction. Finally, comment on the

reaction by referring to the oxidation states of the elements (Tables 9.6 and 9.7).

Solution

The oxidation numbers of all atoms are

$$\overset{0}{Cu} + \overset{+1}{4H^+} + \overset{+6}{S}\overset{-2}{O_4^{-2}} \longrightarrow \overset{2+}{Cu^{2+}} + \overset{+4}{S}\overset{-2}{O_2} + \overset{+1}{2H_2}\overset{-2}{O}$$

Thus, the atoms that change oxidation numbers are Cu and S. Hence, the skeleton equation is

$$\overset{0}{Cu} + \overset{+6}{SO_4^{2-}} \longrightarrow \overset{+2}{Cu^{2+}} + \overset{+4}{SO_2}$$

Since Cu increases in oxidation number, it is oxidized. Similarly, S decreases in oxidation number, so is re-

(Continued)

duced. We can now draw the diagram for the reaction:

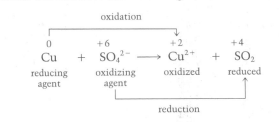

From Table 9.7, we see that the oxidation states of copper are 0, +1, and +2. The oxidation state 0 could be oxidized to either +1 (Cu^+) or +2 (Cu^{2+}). This reaction actually yields the +2 oxidation state, which is relatively stable. (Although Table 9.7 does not indicate it, the Cu^+ ion is normally unstable in aqueous solution and is formed only in the presence of certain substances that stabilize this oxidation state.) We would expect sulfur to have oxidation states from −2 (group number − 8) to +6 (group number). The ones commonly observed are −2, 0, +4, and +6 (see Table 9.6). Thus, if H_2SO_4 is reduced, we expect species in either +4, 0, or −2 oxidation states. At an oxidation state of +4, the compound H_2SO_3 might form; this, however, is unstable and would decompose to H_2O and $SO_2(g)$, which are the products actually observed.

Exercise 9.10

We can identify sulfite ion in aqueous solution by reacting it with aqueous bromine. The resulting sulfate ion gives a white precipitate of $BaSO_4$ with barium chloride solution. The reaction of sulfite ion with bromine is

$$SO_3{}^{2-}(aq) + Br_2(aq) + H_2O(l) \longrightarrow SO_4{}^{2-}(aq) + 2Br^-(aq) + 2H^+(aq)$$

Write the skeleton equation. Then draw a diagram indicating oxidation numbers of atoms that change their numbers, oxidizing and reducing agents, oxidized and reduced species, and the oxidation and reduction parts of the equation. Comment on the reaction by referring to the oxidation states commonly observed for the atoms.

(See Problems 9.49 and 9.50.)

9.8 Balancing Oxidation-Reduction Equations

Oxidation-reduction equations are often too difficult to balance by the inspection method we used for simpler equations.■ Two methods of balancing oxidation-reduction equations are described here. The first method (*half-reaction method*) is useful for balancing net ionic equations. The second (*oxidation-number method*) is useful for balancing molecular equations.

■ Balancing by inspection was discussed in Section 2.2.

Half-Reaction Method

The *half-reaction method* of balancing oxidation-reduction equations is based on the separation of the equation into two **half-reactions,** with each half-reaction representing either the oxidation or the reduction part of the reaction. These half-reactions are balanced, then recombined to get the balanced oxidation-reduction equation.

For example, consider the reaction of permanganate ion, $MnO_4{}^-$, with Fe^{2+} in acidic solution. The skeleton equation is

+7   +2   +2   +3
$$MnO_4{}^-(aq) + Fe^{2+}(aq) \longrightarrow Mn^{2+}(aq) + Fe^{3+}(aq) \quad \text{(acid) skeleton equation}$$

Note that this is a reaction in which MnO_4^- (purple color) acts as an oxidizing agent in acidic solution and is reduced to Mn^{2+} (pale pink to colorless). Iron(II) ion (pale green) is oxidized to Fe^{3+} (pale yellow to colorless). Note from Table 9.7 that Fe^{2+} and Fe^{3+} are common oxidation states of iron. This reaction may be used to obtain the amount of iron in a sample of iron ore (see Figure 9.7). The iron ore is first converted to a solution of Fe^{2+}. When this solution is titrated with $KMnO_4(aq)$, the purple color of MnO_4^- ion disappears because the ion is reduced to Mn^{2+}. When all of the Fe^{2+} ion has reacted, the purple color is retained. This marks the endpoint of titration.■

■ Titration was discussed in Section 3.12.

   To obtain a balanced equation for this reaction, we first split the skeleton equation into half-reactions. We write an incomplete half-reaction for manganese by copying species containing Mn from the skeleton equation:

$$MnO_4^- \longrightarrow Mn^{2+} \qquad \text{(incomplete half-reaction)}$$

Similarly, we copy the species containing Fe from the skeleton equation to get the other half-reaction:

$$Fe^{2+} \longrightarrow Fe^{3+} \qquad \text{(incomplete half-reaction)}$$

   Note that the manganese half-reaction is balanced in Mn atoms. If it had not been, we would balance it by adjusting coefficients of reactant and product species. We balance the oxygen in this half-reaction by adding $4H_2O$

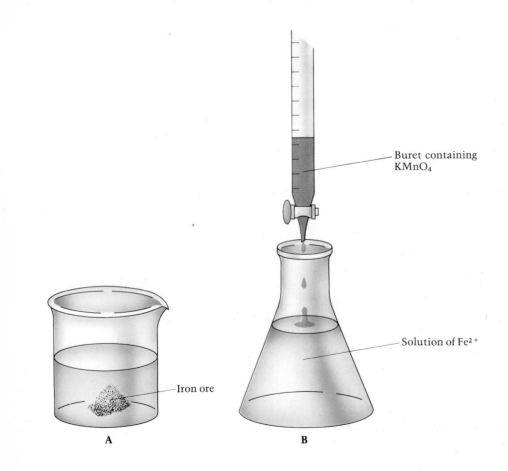

Figure 9.7
Analysis of iron ore. (a) Iron ore, containing Fe_2O_3, is first dissolved in an acid, forming Fe^{3+}. Iron in the $+3$ state is reduced to Fe^{2+} by reaction with zinc metal. (b) The solution of Fe^{2+} is analyzed (titrated) with potassium permanganate, $KMnO_4$, which has an intense purple color. Permanganate ion reacts with Fe^{2+}, giving a nearly colorless solution. When all of the Fe^{2+} is used up, another drop of potassium permanganate solution gives a magenta color to the contents of the flask. From the amount of potassium permanganate used, one can calculate the amount of iron in the sample.

Buret containing $KMnO_4$

Solution of Fe^{2+}

Iron ore

A

B

to the right side:

$$MnO_4^- \longrightarrow Mn^{2+} + 4H_2O$$

Now we add $8H^+$ to the left side to balance the hydrogen. (Since the reaction occurs in acidic solution, we can assume H^+ is available.)

$$MnO_4^- + 8H^+ \longrightarrow Mn^{2+} + 4H_2O$$

This half-reaction is not yet balanced in electrical charge (there is a $7+$ charge on the left and a $2+$ charge on the right). We add five electrons to the left side of the equation to achieve charge balance:

$$MnO_4^- + 8H^+ + 5e^- \longrightarrow Mn^{2+} + 4H_2O \quad \text{(reduction half-reaction)}$$

Check that this half-reaction is balanced in each kind of atom and that it is balanced in charge. Note that electrons are gained, as expected for reduction.

The half-reaction for iron is balanced in iron, but not in charge. Thus, we add one electron to the right side:

$$Fe^{2+} \longrightarrow Fe^{3+} + e^- \quad \text{(oxidation half-reaction)}$$

Check that this half-reaction is balanced. Note that electrons are lost, as expected for oxidation.

The last step is to multiply each half-reaction by a factor so that when the two half-reactions are added the electrons cancel. This is required, since free electrons cannot appear in the final equation.

$$\begin{aligned} 1 \times (MnO_4^- + 8H^+ + 5e^- &\longrightarrow Mn^{2+} + 4H_2O) \\ + 5 \times (Fe^{2+} &\longrightarrow Fe^{3+} + e^-) \\ \hline MnO_4^- + 8H^+ + \cancel{5e^-} + 5Fe^{2+} &\longrightarrow Mn^{2+} + 4H_2O + 5Fe^{3+} + \cancel{5e^-} \end{aligned}$$

The final balanced equation is

$$MnO_4^-(aq) + 5Fe^{2+}(aq) + 8H^+(aq) \longrightarrow Mn^{2+}(aq) + 5Fe^{3+}(aq) + 4H_2O(aq)$$

The half-reaction method for balancing a skeleton oxidation-reduction equation can be summarized in the following steps:

1. Split the skeleton equation into unbalanced half-reactions. Usually, it is obvious how this is to be done. If there is any difficulty, follow this procedure:

a. Obtain the oxidation numbers of the atoms in the skeleton equation in order to decide which atoms change oxidation numbers.

b. Write an incomplete half-reaction for reduction by copying just those species from the skeleton equation in which an atom decreases in oxidation number. Similarly, write an unbalanced half-reaction for oxidation by copying those species in which an atom increases in oxidation number. If a half-reaction involves O_2 on one side, the other side may require H_2O. Also, in some cases one species may occur in both half-reactions (one atom is both oxidized and reduced).

2. Now balance each half-reaction.

a. Balance each half-reaction in the atom being oxidized or reduced by adjusting coefficients of the reactant and product species.

b. Balance each half-reaction in O atoms by adding H_2O's to one side of the half-reaction.

c. Balance each half-reaction in H atoms by adding H^+ ions to one side of the half-reaction.

d. Balance each half-reaction in electrical charge by adding electrons (e^-) to one side of the half-reaction.

3. Combine the half-reactions to get the final balanced oxidation-reduction equation.

    a. Multiply each half-reaction by a factor so that when the half-reactions are added to get the balanced equation the electrons cancel.

    b. Simplify the balanced equation by canceling species that occur on both sides and reduce the coefficients to the smallest whole numbers.

To balance a half-reaction in basic solution, we can assume OH^- is available but H^+ is not. In some cases, a half-reaction in basic solution is easily balanced in O and H atoms by simply adding OH^-. For example, the incomplete half-reaction

$$Zn \longrightarrow Zn(OH)_4^{2-}$$

is balanced by adding OH^- and e^-:

$$Zn + 4OH^- \longrightarrow Zn(OH)_4^{2-} + 2e^-$$

In most cases, the balancing of O's and H's requires both OH^- and H_2O. One way to do this is to balance the half-reaction as if it were in acidic solution. Then, after Step 2d, insert the following step. Add enough OH^- to both sides of the half-reaction to combine with all H^+ ions according to the reaction $H^+ + OH^- \longrightarrow H_2O$. The effect of this can be stated as Step 2e:

2. e. If a reaction occurs in basic solution, then change all H^+ ions in a half-reaction to H_2O's and add the same number of OH^- ions to the other side of the equation. Cancel any H_2O's that occur on both sides of the half-reaction.

The half-reaction method is illustrated for reactions in basic solution in Example 9.8.

Example 9.8

Permanganate ion oxidizes sulfite ion according to the following skeleton equation:

$MnO_4^-(aq) + SO_3^{2-}(aq) \longrightarrow$
     $MnO_2(s) + SO_4^{2-}(aq)$ (basic) skeleton equation

Use the half-reaction method to write the balanced equation.

Solution

Before we begin balancing the equation, it is instructive to look at the oxidation states of sulfur and manganese (Tables 9.6 and 9.7). Sulfites are salts of sulfurous acid, H_2SO_3, which is obtained by dissolving SO_2 in water. If any of these species are oxidized, we

expect corresponding species in the $+6$ oxidation state of sulfur. The sulfites should be oxidized to sulfates (SO_4^{2-}). Permanganate ion, we know, is an oxidizing agent. In acidic solution, it goes all the way to the $+2$ oxidation state. But in basic solution, MnO_2 precipitates.

Following Step 1, we write the incomplete half-reactions for manganese and sulfur:

$$MnO_4^- \longrightarrow MnO_2$$
$$SO_3^{2-} \longrightarrow SO_4^{2-}$$

These are balanced in Mn and S (Step 2a).

(Continued)

We balance the oxygen in the manganese equation by adding H_2O's to the right (Step 2b):

$$MnO_4^- \longrightarrow MnO_2 + 2H_2O$$

Then we balance this in hydrogen by adding H^+ ions to the opposite side (Step 2c):

$$MnO_4^- + 4H^+ \longrightarrow MnO_2 + 2H_2O$$

Now we balance the charge by adding electrons (Step 2d):

$$MnO_4^- + 4H^+ + 3e^- \longrightarrow MnO_2 + 2H_2O$$

We can convert this half-reaction to one in basic solution by adding four OH^- ions to both sides of the equation (Step 2e):

$$MnO_4^- + 4H_2O + 3e^- \longrightarrow MnO_2 + 2H_2O + 4OH^-$$

Note that on the left we have added the four OH^- ions to four H^+ ions, giving four H_2O molecules. Canceling the extra water molecules gives

$$MnO_4^- + 2H_2O + 3e^- \longrightarrow MnO_2 + 4OH^-$$
$$\text{(reduction half-reaction)}$$

We can find the sulfur half-reaction in a similar manner:

$$SO_3^{2-} + 2OH^- \longrightarrow SO_4^{2-} + H_2O + 2e^-$$
$$\text{(oxidation half-reaction)}$$

Finally, the half-reactions are multiplied through by factors so that when they are added together the electrons cancel (Step 3a):

$$2 \times (MnO_4^- + 2H_2O + 3e^- \longrightarrow MnO_2 + 4OH^-)$$
$$+\ 3 \times (SO_3^{2-} + 2OH^- \longrightarrow SO_4^{2-} + H_2O + 2e^-)$$

$$\overset{1}{2MnO_4^- + \cancel{4}H_2O + \cancel{6e^-} + 3SO_3^{2-} + \cancel{6}\cancel{OH^-}} \longrightarrow$$

$$\overset{2}{2MnO_2 + \cancel{8}OH^- + 3SO_4^{2-} + \cancel{3}H_2O + \cancel{6e^-}}$$

After simplification (Step 3b), the balanced equation is

$$2MnO_4^- + 3SO_3^{2-} + H_2O \longrightarrow$$
$$2MnO_2 + 3SO_4^{2-} + 2OH^-$$

Exercise 9.11

Balance the following equation using the half-reaction method:

$$Zn + NO_3^- \longrightarrow Zn^{2+} + NH_4^+ \quad \text{(acidic)}$$

(See Problems 9.53 and 9.54.)

Oxidation-Number Method

One advantage of the half-reaction method is that we see how to balance half-reactions. This is useful for our later study of electrochemistry (Chapter 21). However, redox equations may also be balanced by the *oxidation-number method*. This method depends on the fact that the increase in oxidation number for the atoms that are oxidized must equal the absolute value of the decrease in oxidation number for the atoms that are reduced.

As an example of the oxidation-number method, consider the following molecular equation:

$$HNO_3(aq) + Cu_2O(s) \longrightarrow$$
$$Cu(NO_3)_2(aq) + NO(g) + H_2O(l) \quad \text{not balanced}$$

In order to discover what atoms change oxidation numbers, we find the oxidation numbers of all atoms. From this, we find that the N and Cu atoms change oxidation numbers:

$$\overset{+5}{HNO_3} + \overset{+1}{Cu_2O} \longrightarrow \overset{+2}{Cu(NO_3)_2} + \overset{+2}{NO} + H_2O$$

Notice that nitric acid, HNO_3, functions as an oxidizing agent. Since the nitrogen atom is in its highest oxidation state (see Table 9.6), there are a variety of species to which it might be reduced. In practice one finds that the reduction product depends on the concentration of HNO_3, as well as on the nature of the reducing agent. Here the product is NO. Copper(I) oxide is oxidized to Cu^{2+}, the usual ion in aqueous solution.

To balance the previous equation, we first balance each of the atoms that change oxidation number. Nitrogen atoms that change oxidation number are balanced. (N atoms that remain in oxidation state $+5$ are not, but we will balance them in the last step.) Copper atoms are not balanced, but can be by adjusting coefficients. We get

$$\overset{+5}{HNO_3} + \overset{+1}{Cu_2O} \longrightarrow 2\overset{+2}{Cu}(\overset{+2}{N}O_3)_2 + NO + H_2O$$

We now calculate the change in oxidation number for the copper atom. Since this is an increase from $+1$ to $+2$, it represents a loss of $2 - 1 = 1$ electron. However, there are two copper atoms, so a total of 2 electrons are lost. For the nitrogen atom, the oxidation number decreases from $+5$ to $+2$, which represents a gain of $5 - 2 = 3$ electrons. We can summarize this information as follows:

$$\overset{+5}{HNO_3} + \overset{+1}{Cu_2O} \longrightarrow 2\overset{+2}{Cu}(\overset{+2}{N}O_3)_2 + NO + H_2O$$

(3e⁻ gained; 2e⁻ lost)

Since the electrons lost by copper are gained by nitrogen, we multiply copper species by 3 and nitrogen species by 2, so that the number of electrons lost equals the number of electrons gained. We have

$$2\overset{+5}{HNO_3} + 3\overset{+1}{Cu_2O} \longrightarrow 6\overset{+2}{Cu}(\overset{+2}{N}O_3)_2 + 2NO + H_2O$$

(6e⁻ gained; 6e⁻ lost)

The remaining atoms are balanced by inspection. Although the nitrogen atoms that change oxidation number have been balanced, other nitrogen atoms are present as NO_3^- ions in $6Cu(NO_3)_2$. For this reason, we add $12HNO_3$ to the left side:

$$14HNO_3 + 3Cu_2O \longrightarrow 6Cu(NO_3)_2 + 2NO + H_2O$$

Now copper and nitrogen atoms have been balanced, and all coefficients except the one for H_2O have been fixed. We can obtain the coefficient of H_2O by balancing hydrogen atoms:

$$14HNO_3 + 3Cu_2O \longrightarrow 6Cu(NO_3)_2 + 2NO + 7H_2O$$

The oxidation-number method for balancing a molecular equation can be summarized in the following steps:

1. Obtain the oxidation numbers of all atoms. From this, determine which atoms change in oxidation number. Balance the equation in each of these atoms by adjusting the coefficients of species containing these atoms.
2. Determine the total number of electrons lost by oxidation and the total number of electrons gained by reduction. Make these numbers equal by multiplying the species involved in oxidation by one factor and those involved in reduction by another factor.
3. Balance other atoms in the equation by inspection.

The oxidation-number method can also be applied to balancing a skeleton ionic equation. In this case, we would follow Steps 1 and 2 for the molecular equation. Then we complete the equation by adding H_2O's to one side of the equation to balance oxygen, and adding H^+ ions to one side to balance hydrogen. If the reaction occurs in basic solution, we convert H^+ ions to H_2O by adding OH^- ions. We then add the same number of OH^- ions to the opposite side of the equation. The charge should balance and serves as a check.

Exercise 9.12

Iodic acid, HIO_3, can be prepared by reacting iodine, I_2, with concentrated nitric acid. The equation is

$$I_2 + HNO_3 \longrightarrow HIO_3 + NO_2 + H_2O \quad \text{(not balanced)}$$

Balance this equation by the oxidation-number method.

(See Problems 9.55 and 9.56.)

9.9 Equivalents and Normality

The concepts of *equivalents* and *normality* were devised to simplify calculations in chemical analyses. An **equivalent** (eq) of a reactant is the mass that exactly reacts with an equivalent of another reactant. In a neutralization, for example, an acid gives H^+ ion to a base. An equivalent of the acid is the mass that yields 1 mol H^+ in the neutralization. An equivalent of the base is the mass that accepts 1 mol H^+. For example, in the neutralization

$$H_2SO_4 + 2NaOH \longrightarrow Na_2SO_4 + 2H_2O$$

each mole of H_2SO_4 supplies 2 mol H^+. Since the molar mass is 98.1 g, an equivalent of H_2SO_4 in this reaction equals 98.1 g/2 = 49.0 g. Because each mole of NaOH accepts 1 mol H^+ (each OH^- ion reacts with an H^+ ion), an equivalent of NaOH equals the molar mass (40.0 g). One equivalent of H_2SO_4 reacts with one equivalent of NaOH.

However, in the reaction

$$H_2SO_4 + NaOH \longrightarrow NaHSO_4 + H_2O$$

each mole of H_2SO_4 supplies only 1 mol H^+. An equivalent of H_2SO_4 in this reaction equals the molar mass (98.1 g).

An equivalent of an oxidizing or reducing agent is the mass that effectively provides or uses one mole of electrons. Thus, one equivalent of oxidizing agent reacts with one equivalent of reducing agent. For example, the permanganate ion, MnO_4^- gains five electrons when it is reduced to Mn^{2+} in acidic solution:

$$MnO_4^- + 8H^+ + 5e^- \longrightarrow Mn^{2+} + 4H_2O$$

An equivalent of potassium permanganate, $KMnO_4$, is obtained by dividing its molar mass (158.0 g) by the number of electrons gained (5), that is, 158.0 g/5 = 31.61 g.

Example 9.9

A sample of a substance A is oxidized by $KMnO_4$ in acidic solution. If this sample reacts with 0.4863 g of $KMnO_4$, how many equivalents of the substance A are contained in the sample? If the sample is $FeCl_2$, which is oxidized to Fe^{3+} in acidic solution of $KMnO_4$, how many grams of $FeCl_2$ are in the sample?

Solution

We obtain the number of equivalents in 0.4863 g $KMnO_4$ by dividing by the mass of one equivalent, which we calculated above to be 31.61 g/eq (grams per equivalent). This will also give the number of equivalents of A.

$$\frac{0.4863 \text{ g}}{31.61 \text{ g/eq}} = 0.01538 \text{ eq}$$

Since Fe^{2+} in $FeCl_2$ is oxidized to Fe^{3+},

$$Fe^{2+} \longrightarrow Fe^{3+} + e^-$$

the mass of one equivalent of $FeCl_2$ is its molar mass, 126.8 g, divided by 1. The mass of $FeCl_2$ in the sample is the number of equivalents of $FeCl_2$ (0.01538 eq) times the mass per equivalent (126.8 g/eq):

$$0.01538 \text{ eq} \times \frac{126.8 \text{ g}}{\text{eq}} = 1.949 \text{ g}$$

Exercise 9.13

A sample of an unknown acid A is titrated by a volume of NaOH solution containing 1.032 g NaOH. If the mass of this sample of acid is 0.581 g, what is the mass of an equivalent of the acid? If this acid is diprotic and is completely neutralized in this titration, what is its molar mass?

(See Problems 9.57 and 9.58.)

The **normality** of a solution is the number of equivalents of a substance dissolved in a liter of solution. It equals the molarity of the solution times the number of equivalents per mole. For example, if the reaction is an oxidation by $KMnO_4$ in acidic solution, then the normality of 0.10 M $KMnO_4$ is

$$\text{normality} = \underbrace{\frac{\text{moles substance}}{\text{liters solution}}}_{M} \times \frac{\text{equivalents substance}}{\text{moles substance}}$$

$$= 0.10 \frac{\text{mol KMnO}_4}{\text{L soln}} \times \frac{5 \text{ eq KMnO}_4}{\text{mol KMnO}_4} = 0.50 \text{ N KMnO}_4$$

The solution is 0.50 N (read 0.50 normal) $KMnO_4$. The normality of a solution is useful for doing chemical analyses by titrations.

Example 9.10

A sample of 0.256 g of iron alloy (a mixture of iron with other metallic elements) was dissolved in hydrochloric acid to give a solution of Fe^{2+} ion. This solution was titrated to the endpoint with 35.6 mL of 0.100 N $KMnO_4$. What is the percentage of iron in the sample?

Solution

We first determine the amount of Fe^{2+} ion in the solution from the equivalents of $KMnO_4$ needed in the titration. This amount of Fe^{2+} equals the amount of Fe in the sample, from which we can calculate the percentage of iron in the alloy.

The equivalents of Fe^{2+} equals the equivalents of $KMnO_4$. This equals the volume of $KMnO_4$ in liters times the normality of $KMnO_4$.

$$0.356 \; \text{L KMnO}_4 \times 0.100 \; \frac{\text{eq KMnO}_4}{\text{L KMnO}_4} =$$
$$0.00356 \; \text{eq KMnO}_4 = 0.00356 \; \text{eq Fe}^{2+}$$

The mass of 1 eq Fe^{2+} equals the molar mass divided by the number of electrons lost in the oxidation to Fe^{3+} by $KMnO_4$. The oxidation half-reaction is $Fe^{2+} \longrightarrow Fe^{3+} + e^-$. Since the molar mass of Fe^{2+} is 55.8 g and since one electron is lost in the oxidation, the mass of an equivalent of Fe^{2+} is 55.8 g/1; that is, the mass per equivalent of Fe^{2+} is 55.8 g Fe^{2+}/eq Fe^{2+}. The mass of Fe^{2+} in the solid is

$$0.00356 \; \text{eq Fe}^{2+} \times \frac{55.8 \; \text{g Fe}^{2+}}{\text{eq Fe}^{2+}} = 0.199 \; \text{g Fe}^{2+}$$

This is also the mass of Fe in the sample of alloy. The percentage of iron in the alloy is obtained by dividing the mass of Fe by the mass of the sample:

$$\frac{0.199 \; \text{g}}{0.256 \; \text{g}} \times 100\% = 77.7\%$$

Exercise 9.14

In the laboratory, an impure sample of oxalic acid, $H_2C_2O_4$ (a diprotic acid), weighing 0.130 g was titrated with 0.100 N NaOH until it was completely neutralized. If the titration required 25.8 mL NaOH(aq), what is the percentage of oxalic acid in the sample?

(See Problems 9.59 and 9.60.)

A Checklist for Review

Important Terms

electrolyte (9.1)
nonelectrolytes (9.1)
strong electrolyte (9.1)
weak electrolyte (9.1)
molecular equation (9.2)
ionic equation (9.2)
spectator ions (9.2)
net ionic equation (9.2)
metathesis (double-replacement) reaction (9.3)
oxidation-reduction (redox) reaction (9.3)
saturated (9.4)

solubility (9.4)
unsaturated (9.4)
supersaturated (9.4)
precipitate (9.4)
salt (9.5)
acid (9.5)
base (9.5)
acid–base indicators (9.5)
strong acids (9.5)
weak acids (9.5)
strong bases (9.5)

weak bases (9.5)
neutralization (9.5)
polyprotic acids (9.5)
acid salts (9.5)
oxidation (9.7)
reduction (9.7)
oxidizing agent (9.7)
reducing agent (9.7)
half-reactions (9.8)
equivalent (9.9)
normality (9.9)

Summary of Facts and Concepts

Substances that dissolve in water to give ions are called *electrolytes*. Electrolytes that exist in solution completely as ions are called *strong electrolytes*. Electrolytes that exist in solution as an equilibrium between molecules and ions are called *weak electrolytes*.

A *molecular equation* for a reaction in aqueous solution is written as an *ionic equation* by writing the strong electrolytes as ions. By canceling the *spectator ions*, we get the *net ionic equation*. Ionic reactions are of two major types: *metathesis*, in which positive and negative parts of the reactants are interchanged; and *oxidation-reduction*, in which there are changes of oxidation numbers of some atoms. The metathesis reactions discussed in this chapter are *precipitation* (formation of a solid product from solution), *neutralization* (reaction of an *acid* and a *base*), and reactions of *salts* to produce a gas. These reactions are illustrated by preparations of acids, bases, and salts.

In an oxidation-reduction reaction, there is effectively a transfer of electrons from one atom to another. The atom that loses electrons and thereby increases in oxidation number is said to undergo *oxidation*. Similarly, the atom that gains electrons and thereby decreases in oxidation number is said to undergo *reduction*. Oxidation and reduction must occur together in a reaction. A balanced oxidation-reduction equation frequently looks complicated because of large coefficients and the presence of species such as H^+, OH^-, and H_2O. One can better understand these equations by looking at the "skeleton" equation for the reaction and the oxidation states of the elements. Oxidation-reduction equations can be balanced by either the *half-reaction method* or the *oxidation-number method*.

Equivalents and *normality* are defined in order to simplify calculations in chemical analyses. An equivalent of one reactant combines with an equivalent of another. A one normal solution contains one equivalent per liter of solution.

Operational Skills

1. Using solubility rules, decide whether two soluble ionic compounds will or will not react to form a precipitate; if they will, write the net ionic equation (Example 9.1).

2. Given an acid and base, write the molecular equation and then the net ionic equation for the neutralization (Example 9.2).

3. Given a reaction between a carbonate, sulfide, or sulfite and a strong acid, or the reaction between an ammonium compound and a strong base, write the molecular and net ionic equations (Example 9.3).

4. Given the salt of a volatile acid, write the molecular equation for the preparation of the acid based on its volatility (Example 9.4a). Using the solubility rules, give a molecular equation for the preparation of an aqueous solution of an acid based on precipitation (Example 9.4b).

5. Given a strong base, write the molecular equation for a possible preparation of a Group IA hydroxide using a precipitation reaction (Example 9.5a). Write a molecular equation for the preparation of a volatile base from one of its salts (Example 9.5b).

6. Given a salt, write the molecular equation for its preparation by neutralization (Example 9.6a). Given a carbonate, sulfide, or sulfite, write the molecular equation for the preparation of a particular salt from it (Example 9.6b). Write a molecular equation for the preparation of a given salt by a precipitation reaction (Example 9.6c).

7. Given a balanced oxidation-reduction equation, write the skeleton equation. Label the oxidizing and reducing agents, the oxidized and reduced species, and the oxidation and reduction parts of the equation. Comment on the reaction by referring to the commonly observed oxidation states (Example 9.7).

8. Given an oxidation-reduction equation (an unbalanced or a skeleton equation), complete and balance it. (Example 9.8 illustrates the half-reaction method; the oxidation-number method is described in the text.)

9. Given the mass of a reactant in a neutralization or oxidation-reduction reaction, calculate the number of equivalents. Also calculate the mass of the other reactant (Example 9.9).

10. Given the volume and normality of a solution used to titrate a substance, obtain the equivalents and mass of the substance (Example 9.10).

Review Questions

9.1 Explain the electrical conductivity of an electrolyte solution. Use an example to illustrate.

9.2 Explain why chemical equilibrium is called a dynamic equilibrium. Use an example to illustrate.

9.3 Define the terms *strong electrolyte* and *weak electrolyte*. Give an example of each.

9.4 What are the advantages and disadvantages of using a molecular equation to represent an ionic reaction?

9.5 What is a *spectator ion?* Illustrate with an ionic reaction.

9.6 What are the two major types of ionic reactions? Define and give an example of each type.

9.7 What is meant by the terms *unsaturated, saturated,* and *supersaturated?* What is meant by the *solubility* of a substance?

9.8 In words, describe how you would prepare pure crystalline AgCl and $NaNO_3$ from solid $AgNO_3$ and solid NaCl. The reaction is

$$AgNO_3(aq) + NaCl(aq) \longrightarrow AgCl(s) + NaNO_3(aq)$$

9.9 Define the following terms and give an example of each: salt, acid, base.

9.10 In what ways do we represent the hydrogen ion in solution? Describe a possible structure of this ion.

9.11 Give examples of the following metathesis reactions: a precipitation, a neutralization, a reaction that evolves a gas. Give molecular and net ionic equations.

9.12 Explain why NH_3 gives basic solutions. Use Lewis formulas to illustrate your answer.

9.13 Give an example of a polyprotic acid and write equations for the successive neutralizations of the acidic hydrogen atoms to produce a series of salts.

9.14 The mineral magnesite fizzes when treated with hydrochloric acid. The gas released is odorless. To the extent possible, explain what reaction is occurring.

9.15 Define *oxidation* and *reduction* in terms of electrons transferred and in terms of oxidation numbers.

9.16 Why must oxidation and reduction occur together in a reaction?

9.17 Define *equivalent* for a neutralization and for an oxidation-reduction reaction.

9.18 Barium hydroxide, $Ba(OH)_2$, is a strong base. In a neutralization, both OH^- ions from the base react. Describe how you would prepare a 0.10 N solution.

Problems

Ionic Reactions

9.19 Write net ionic equations for the following molecular equations. Molecular substances are labeled as strong or weak electrolytes.
(a) $HF(aq) + KOH(aq) \longrightarrow KF(aq) + H_2O(l)$
   weak   weak

(b) $AgNO_3(aq) + NaBr(aq) \longrightarrow AgBr(s) + NaNO_3(aq)$

9.21 Classify each of the following as a metathesis or oxidation-reduction reaction. Explain your answers.
(a) $2HBr(aq) + Cl_2(aq) \longrightarrow 2HCl(aq) + Br_2(aq)$
(b) $2HBr(aq) + Ca(OH)_2(aq) \longrightarrow CaBr_2(aq) + 2H_2O(l)$
(c) $Ba(ClO_2)_2(aq) + H_2SO_4(aq) \longrightarrow$
$$BaSO_4(s) + 2HClO_2(aq)$$
(d) $3NaClO(aq) \longrightarrow NaClO_3(aq) + 2NaCl(aq)$

9.20 Write net ionic equations for the following molecular equations. Molecular substances are labeled as strong or weak electrolytes.
(a) $HBr(aq) + NH_3(aq) \longrightarrow NH_4Br(aq)$
   strong   weak

(b) $Na_2CO_3(aq) + Ca(OH)_2(aq) \longrightarrow$
$$CaCO_3(s) + 2NaOH(aq)$$

9.22 Decide whether the reaction given is a metathesis or oxidation-reduction type. Explain your answers.
(a) $PbO_2(s) + 4HCl(aq) \longrightarrow$
$$PbCl_2(s) + Cl_2(g) + 2H_2O(l)$$
(b) $3HNO_2(aq) \longrightarrow HNO_3(aq) + 2NO(g) + H_2O(l)$
(c) $2NaNO_2(aq) + H_2SO_4(aq) \longrightarrow$
$$2HNO_2(aq) + Na_2SO_4(aq)$$
(d) $2HNO_3(aq) + Ba(OH)_2(aq) \longrightarrow$
$$Ba(NO_3)_2(aq) + 2H_2O(l)$$

Solubility and Precipitation

9.23 Using the solubility rules, decide whether the following ionic solids are soluble or insoluble in water:
(a) AgBr   (b) $Pb(NO_3)_2$ (c) $SrSO_4$ (d) Na_2CO_3

9.25 Write the molecular equation and net ionic equation for each of the following. If no reaction occurs, write NR after the arrow.
(a) $FeSO_4 + NaCl \longrightarrow$
(b) $Na_2CO_3 + CaCl_2 \longrightarrow$
(c) $MgSO_4 + NaOH \longrightarrow$
(d) $NiCl_2 + NaBr \longrightarrow$

9.24 Using the solubility rules, decide whether the following ionic solids are soluble or insoluble in water:
(a) $(NH_4)_2SO_4$ (b) $Ca(NO_3)_2$ (c) $BaCO_3$ (d) $PbSO_4$

9.26 Write the molecular equation and net ionic equation for each of the following. If no reactions occurs, write NR after the arrow.
(a) $AgClO_3 + NaI \longrightarrow$
(b) $Ba(NO_3)_2 + K_2SO_4 \longrightarrow$
(c) $Mg(NO_3)_2 + K_2SO_4 \longrightarrow$
(d) $CaCl_2 + Al(NO_3)_3 \longrightarrow$

Acids and Bases

9.27 Write the molecular equation and net ionic equation for the following neutralization reactions:
(a) nitric acid with sodium hydroxide
(b) acetic acid with potassium hydroxide

9.29 Oxalic acid, $H_2C_2O_4$, is a diprotic acid. Write molecular equations for the successive neutralizations of each acidic hydrogen of $H_2C_2O_4$ with NaOH.

9.31 Write the molecular equation and net ionic equation for the reaction of iron(II) sulfide with hydrochloric acid.

9.33 How could you prepare sulfur dioxide from calcium sulfite?

9.28 Write the molecular equation and net ionic equation for the following neutralization reactions:
(a) perchloric acid with sodium hydroxide
(b) hydrochloric acid with aqueous ammonia

9.30 Sulfurous acid, H_2SO_3, is a diprotic acid. Write a molecular equation for the neutralization of each acidic hydrogen of H_2SO_3 with KOH.

9.32 Write the molecular equation and net ionic equation for the reaction of ammonium sulfate with barium hydroxide.

9.34 How could you prepare carbon dioxide from magnesium carbonate?

Preparation of Acids, Bases, and Salts

9.35 Describe by a molecular equation the preparation of hydrogen iodide (hydriodic acid), HI, based on the volatility of HI. See note on the preparation of HI in the text (p. 289).

9.37 Chloric acid, $HClO_3$, is stable only in aqueous solutions. Give a molecular equation for the preparation of chloric acid from its barium salt using precipitation.

9.39 Write a molecular equation for the preparation of LiOH from Li_2SO_4.

9.41 Write a molecular equation for the preparation of hydrazine, N_2H_4, from its salt, $N_2H_6Cl_2$. (See Table 9.4 for the structure of the $N_2H_6^{2+}$ ion.)

9.43 Give molecular equations for the preparation of (a) aluminum sulfate by neutralization, (b) copper(II) sulfate from copper(II) sulfide, (c) calcium phosphate by precipitation.

*9.45 Use precipitation reactions to prepare
(a) $MgCl_2$ from $MgSO_4$ (b) $Sr(NO_3)_2$ from SrI_2

9.36 Hydrogen cyanide (hydrocyanic acid), HCN, boils at 26°C. It can be distilled from aqueous solution. Give a molecular equation for the preparation of HCN based on this volatility.

9.38 Give a molecular equation for the preparation of nitric acid from its silver salt using precipitation.

9.40 Write a molecular equation for the preparation of CsOH from Cs_2CO_3.

9.42 Aniline, $C_6H_5NH_2$, is a base similar to methylamine and ammonia. With a molecular equation, describe the preparation of aniline from its sulfate salt, $(C_6H_5NH_3)_2SO_4$.

9.44 Give molecular equations for the preparation of (a) calcium phosphate by neutralization, (b) magnesium chloride from magnesium carbonate, (c) lead bromide by precipitation.

*9.46 Use precipitation reactions to prepare
(a) $NaNO_3$ from NaCl   (b) NH_4Cl from $(NH_4)_2SO_4$

Oxidation-Reduction Equations

9.47 In the following, label the oxidizing agent and the reducing agent:
(a) $2Al + 3F_2 \longrightarrow 2AlF_3$
(b) $Hg^{2+} + NO_2^- + H_2O \longrightarrow Hg + 2H^+ + NO_3^-$

9.49 Cobalt(II) sulfide reacts with nitric acid:

$$3CoS(s) + 8H^+(aq) + 2NO_3^-(aq) \longrightarrow$$
$$3Co^{2+}(aq) + 2NO(g) + 3S(s) + 4H_2O(l)$$

(The molecular formula of sulfur is S_8, but sulfur is often written for simplicity as S.) Write the skeleton equation. Then draw a diagram indicating oxidation numbers of atoms that change their numbers, oxidizing and reducing agents, oxidized and reduced species, and the oxidation and reduction parts of the equation. Comment on the reaction by referring to the oxidation states commonly observed.

9.51 What are the half-reactions for the following reaction?

$$Ni + Cu^{2+} \longrightarrow Ni^{2+} + Cu$$

Which is the oxidation half-reaction, and which is the reduction half-reaction?

9.53 Balance the following oxidation-reduction equations using the half-reaction method (reactions occur in aqueous solution):
(a) $S^{2-} + NO_3^- \longrightarrow NO_2 + S_8$ (acidic)
(b) $NO_3^- + Cu \longrightarrow NO + Cu^{2+}$ (acidic)
(c) $MnO_4^- + SO_2 \longrightarrow SO_4^{2-} + Mn^{2+}$ (acidic)
(d) $Bi(OH)_3 + Sn(OH)_3^- \longrightarrow Sn(OH)_6^{2-} + Bi$ (basic)

9.55 Balance the following oxidation-reduction equations by the oxidation-number method. Identify the oxidizing agent.
(a) $AsH_3 + KClO_3 \longrightarrow H_3AsO_4 + KCl$
(b) $SnCl_2 + O_2 + HCl \longrightarrow H_2SnCl_6 + H_2O$
(c) $MnO_2 + HBr \longrightarrow Br_2 + MnBr_2 + H_2O$
(d) $P_4 + NaOH + H_2O \longrightarrow NaH_2PO_2 + PH_3$

9.57 An antacid tablet contains $Mg(OH)_2$, in addition to an inert binding ingredient. The mass of $Mg(OH)_2$ in a tablet was determined by titration with $HCl(aq)$.

$$Mg(OH)_2(s) + 2HCl(aq) \longrightarrow MgCl_2(aq) + H_2O(l)$$

What is the mass of one equivalent of $Mg(OH)_2$? If 0.375 g HCl are required to completely neutralize one tablet, how many grams of $Mg(OH)_2$ are there in this tablet?

9.59 A solution of phosphoric acid, H_3PO_4, was titrated with $NaOH(aq)$ to give Na_2HPO_4. What is the mass of one equivalent of H_3PO_4 in this reaction? If the titration required 42.1 mL of 0.0973 N NaOH to react with a volume of solution, how many equivalents of NaOH are used? How many grams of H_3PO_4 are in the solution?

9.48 In the following, label the oxidizing agent and the reducing agent:
(a) $Fe_2O_3 + 3CO \longrightarrow 2Fe + 3CO_2$
(b) $PbS + 4H_2O_2 \longrightarrow PbSO_4 + 4H_2O$

9.50 The first step in a common procedure for the detection of the nitrate ion consists in transforming it to NO:

$$3Fe^{2+}(aq) + NO_3^-(aq) + 4H^+(aq) \longrightarrow$$
$$3Fe^{3+}(aq) + NO(g) + 2H_2O(l)$$

Write the skeleton equation. Then draw a diagram indicating oxidation numbers of atoms that change their numbers, oxidizing and reducing agents, oxidized and reduced species, and the oxidation and reduction parts of the equation. Comment on the reaction by referring to the oxidation states commonly observed.

9.52 What are the half-reactions for the following reaction?

$$Zn + 2Ag^+ \longrightarrow Zn^{2+} + 2Ag$$

Which is the oxidation half-reaction, and which is the reduction half-reaction?

9.54 Balance the following oxidation-reduction equations using the half-reaction method (reactions occur in aqueous solution):
(a) $Hg_2^{2+} + H_2S \longrightarrow Hg + S_8$ (acidic)
(b) $S^{2-} + I_2 \longrightarrow SO_4^{2-} + I^-$ (basic)
(c) $Al + NO_3^- \longrightarrow Al(OH)_4^- + NH_3$ (basic)
(d) $MnO_4^- + C_2O_4^{2-} \longrightarrow MnO_2 + CO_2$ (basic)

9.56 Balance the following oxidation-reduction equations by the oxidation-number method. Identify the reducing agent.
(a) $P_4 + HNO_3 + H_2O \longrightarrow H_3PO_4 + NO$
(b) $K + KNO_3 \longrightarrow N_2 + K_2O$
(c) $MnSO_4 + PbO_2 + H_2SO_4 \longrightarrow$
$$HMnO_4 + PbSO_4 + H_2O$$
(d) $NO_2 + H_2O \longrightarrow HNO_3 + NO$

9.58 A solution of hydrogen peroxide, H_2O_2, was determined by titration with $KMnO_4$ in acidic solution in which H_2O_2 was oxidized to O_2. If a 34.5-g sample of hydrogen peroxide solution required 1.92 g $KMnO_4$ for complete reaction, how many equivalents of H_2O_2 are contained in the solution? What is the mass percentage of H_2O_2 in the solution?

9.60 The mass of Fe^{2+} in a solution was determined by oxidation to Fe^{3+} by $K_2Cr_2O_7$ in acidic solution. The $K_2Cr_2O_7$ was reduced to Cr^{3+}. What is the mass of Fe^{2+} in a solution that required 22.1 mL of 0.0500 N $K_2Cr_2O_7$?

Additional Problems

9.61 Decide if a reaction occurs for each of the following. If it does not, write NR after the arrow. If it does, write the balanced molecular equation; then write the net ionic equation.
(a) $LiOH + HCN \longrightarrow$ (b) $Li_2CO_3 + HNO_3 \longrightarrow$
(c) $LiCl + AgNO_3 \longrightarrow$ (d) $LiCl + MgSO_4 \longrightarrow$

9.62 Decide if a reaction occurs for each of the following. If it does not, write NR after the arrow. If it does, write the balanced molecular equation; then write the net ionic equation.
(a) $Al(OH)_3 + HNO_3 \longrightarrow$ (b) $FeS + HClO_4 \longrightarrow$
(c) $CaCl_2 + NaNO_3 \longrightarrow$ (d) $MgSO_4 + Ba(NO_3)_2 \longrightarrow$

9.63 Describe how you could do the following preparations. Use either precipitation, neutralization, or a reaction of a salt with acid to produce a gas as one product.
(a) $CuCl_2$ from $CuSO_4$ (b) $Ca(C_2H_3O_2)_2$ from $CaCO_3$
(c) $NaNO_3$ from Na_2SO_3 (d) $MgCl_2$ from $Mg(OH)_2$

9.64 Describe how you could do the following preparations. Use either precipitation, neutralization, or a reaction of a salt with acid to produce a gas as one product.
(a) $MgCl_2$ from $MgCO_3$ (b) $NaNO_3$ from $NaCl$
(c) $Al(OH)_3$ from $Al(NO_3)_3$ (d) H_2SO_4 from HCl

9.65 Use the half-reaction method to balance the following skeleton equations:
(a) $MnO_4^- + S^{2-} \longrightarrow MnO_2 + S$ (basic)
(b) $IO_3^- + HSO_3^- \longrightarrow I^- + SO_4^{2-}$ (acidic)
(c) $P_4 \longrightarrow PH_3 + H_2PO_2^-$ (basic)
(d) $Cl_2 \longrightarrow Cl^- + ClO^-$ (basic)

9.66 Use the half-reaction method to balance the following skeleton equations:
(a) $Fe^{2+} + Cr_2O_7^{2-} \longrightarrow Fe^{3+} + Cu^{3+}$ (acidic)
(b) $Zn + NO_3^- \longrightarrow Zn^{2+} + N_2O$ (acidic)
(c) $MnO_4^- + NO_2^- \longrightarrow MnO_2 + NO_3^-$ (basic)
(d) $Br_2 \longrightarrow Br^- + BrO_3^-$ (basic)

9.67 Iron(II) hydroxide is a greenish precipitate that is formed from iron(II) ion by the addition of a base. This precipitate gradually turns to the yellow brown iron(III) hydroxide by oxidation from O_2 in the air. Write a balanced equation for this oxidation.

9.68 A sensitive test for bismuth(III) ion consists of shaking a solution suspected of containing bismuth(III) ion with a basic solution of sodium stannite, Na_2SnO_2. If black precipitate of bismuth metal is formed, this is a positive test for bismuth(III). Stannite ion is oxidized to stannate ion, SnO_3^{2-}. Write a balanced equation for the reaction.

9.69 The normality of a solution of iodine, I_2, can be determined by titrating a solution containing a known mass of arsenious acid, H_3AsO_3, with the iodine solution. What is the normality of a solution of I_2 if 0.840 g H_3AsO_3 reacts with exactly 25.4 mL of $I_2(aq)$? In this reaction, H_3AsO_3 is oxidized to arsenic acid, H_3AsO_4, and I_2 is reduced to I^-.

9.70 The normality of a solution of potassium dichromate, $K_2Cr_2O_7$, can be determined by titration with potassium iodide solution. In acidic solution, $Cr_2O_7^{2-}$ is reduced to Cr^{3+} and I^- is oxidized to I_2. What is the normality of a solution of $K_2Cr_2O_7$ if 48.5 mL of the solution require 38.3 mL of 0.0500 N KI for complete reaction?

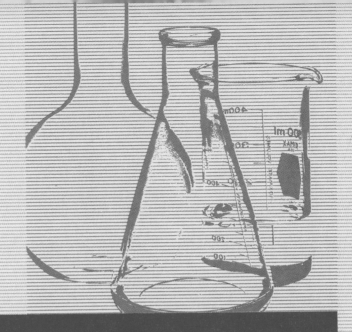

10. Thermochemistry

Fundamentals of Thermochemistry

10.1 Heat of Reaction; Enthalpy Heat of Reaction/ Enthalpy/ *Aside: Relating Enthalpy to Energy*

10.2 Calorimetry Heat Capacity and Specific Heat/ Measurement of Reaction Heats

10.3 Stoichiometry of Reaction Heats

10.4 Hess's Law

10.5 Enthalpies of Formation

Applications of Thermochemistry

10.6 Bond Energy

10.7 Lattice Energies

Nearly all chemical reactions involve either the release or absorption of heat. The burning of coal or gasoline is a dramatic example of chemical reactions in which a great deal of heat is released. Such reactions are important sources of warmth and power.

Chemical reactions that absorb heat are usually less dramatic, although the reaction of barium hydroxide octahydrate, $Ba(OH)_2 \cdot 8H_2O$, and ammonium thiocyanate, NH_4SCN, is an exception. If crystals of barium hydroxide octahydrate are mixed with crystals of ammonium thiocyanate in a flask, the solids first form a slush, then a liquid. Because the reaction mixture absorbs heat from the surroundings, the flask feels cool. It soon becomes so cold that if it is set in a puddle of water on a board, the water freezes; the board can then be inverted with the flask frozen to it (Figure 10.1).

In this chapter, we will be concerned with the quantity of heat released or absorbed in a chemical reaction. The questions we want to address are: How do we measure the quantity of heat released or absorbed by a chemical reaction? To what extent can we relate the quantity of heat involved in a given reaction to the quantities of heat in other reactions? Finally, how can we use this information?

Chapter Overview

You may wish to review the brief discussion of energy given in Section 1.6 before reading the present chapter. We begin here by defining some important terms needed for discussing the *fundamentals of thermochemistry* (study of heat of reaction). After describing the experimental determination of heats of reaction (calorimetry), we describe how the heat for one reaction can be related to the heats of other reactions. In the final sections, we look at some *applications of thermochemistry*.

Fundamentals of Thermochemistry

Thermochemistry is the study of the quantity of heat absorbed or evolved (given off) by chemical reactions. An example of a heat-evolving reaction, as mentioned, is the burning of fuel. There may be practical reasons why we want to know the quantity of heat evolved during the burning of a fuel: we could calculate the cost of the fuel per unit of heat energy produced; we could calculate the quantity of heat obtained per unit mass of rocket fuel; and so forth. But there are more basic, theoretical reasons for wanting to know the quantity of heat involved in a reaction. Knowing such values, we are able to calculate the amount of energy needed to break a particular kind of chemical bond and so learn something about the strength of that bond. And, from the quantity of energy released in forming an ionic solid, we can perhaps explain some of the properties of the crystal. We will see later that heat measurements also provide data needed to answer the central questions: does a particular chemical reaction occur? and if so, to what extent? ■

■ These questions concern chemical equilibrium and will be discussed in Chapter 20.

10.1 Heat of Reaction; Enthalpy

In the chapter opening, we used the term *heat*, relying on your everyday knowledge to give it meaning. Before we proceed, we will define this term and others more precisely.

Figure 10.1
An example of an endo-
thermic reaction. Two crystalline substances, barium
hydroxide octahydrate and
ammonium thiocyanate, are
mixed thoroughly in a flask.
Soon the solid mixture
liquefies and evolves ammonia:

$$Ba(OH)_2 \cdot 8H_2O(s)$$
$$+ 2NH_4SCN(s) \longrightarrow$$
$$Ba(SCN)_2(aq) + 2NH_3(g)$$
$$+ 10H_2O(l)$$

Then the flask, which feels
quite cold to the touch, is set
in a puddle of water on a
board. In a couple of
minutes, flask and board are
frozen solidly together.

Heat of Reaction

Suppose that we are interested in studying some physical or chemical
change. The substance or mixture of substances in which this change occurs
is called the **system.** The **surroundings** are everything in the vicinity of this
system. For example, if we are interested in the reaction of barium hydroxide
octahydrate with ammonium thiocyanate, which was described in the chapter opening, this reaction mixture is our system, and the reaction vessel and
everything outside it are the surroundings (see Figure 10.2). **Heat** is defined
as the energy that flows into or out of a system because of a difference in
temperature between the system and its surroundings. As long as a system
and its surroundings are in thermal contact (that is, not thermally insulated
from each other), energy—heat—flows between them to establish temperature equality or **thermal equilibrium.** Heat flows from a system at high
temperature to one at low temperature; but once thermal equilibrium is
established, heat flow stops.

   Suppose, for example, that a system and its surroundings are both initially
at 25°C. They are in thermal equilibrium and no heat flows. Imagine that we
initiate a chemical reaction in the system so that the temperature rises. Heat
begins to flow from the system to the surroundings. It continues to flow
until the reaction stops and the temperature of the system equals the temperature of the surroundings. If, on the other hand, the temperature of the
system falls because of the reaction, heat flows from the surroundings to

Figure 10.2
Illustration of a thermo-
dynamic system. The *system*
consists of the portion of the
universe that we choose to
study; in this case, it is a
solution of Ba(OH)$_2$ and
NH$_4$SCN. Everything else,
including the reaction vessel,
comprises the *surroundings*.

the system—the system absorbs heat. It will continue to absorb heat until
the reaction stops and the temperature of the system equals the temperature
of the surroundings (see Figure 10.3).

The total amount of heat that is evolved or absorbed by a system at a
particular temperature (25°C in the previous example) because of chemical
reaction is called the **heat of reaction.** A chemical reaction or physical
change is **exothermic** when heat is evolved, and **endothermic** when heat is
absorbed. Experimentally, we note that in the exothermic case the reaction
flask warms; in the endothermic case, the reaction flask cools.■

We will use the symbol q to denote heat. When heat is absorbed by a
system, energy is *added* to it; by convention, the number assigned to q is
positive. Thus, for an endothermic reaction, the heat, q, is a positive quantity. On the other hand, when heat is evolved during a reaction, energy is

■ A number of common physical processes are endothermic.
The cooling of a volatile liquid
by evaporation is one example.
Cooling from the evaporation of
moisture from the skin is an important mechanism for body-
temperature regulation.

Temperature of surroundings is 25°C
(arrows indicate heat flow)

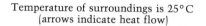

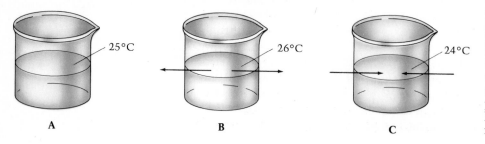

25°C

26°C

24°C

A   B   C

Figure 10.3
Heat flow. (a) The system
(solution in the beaker) is in
thermal equilibrium with the
surroundings; no heat flows.
(b) Reaction raises the temperature of the system, and
heat flows (as indicated by
the arrows) out of the system
to the surroundings. (c) Reaction decreases the tem-
perature of the system, and
heat flows from the sur-
roundings into the system.

subtracted from the system. The number assigned to q for an exothermic reaction is *negative*. The sign conventions for q are summarized below.

Type of Reaction	Experimental Effect Noted	Result on System	Sign of q
Endothermic	Reaction vessel cools (heat is absorbed)	Energy added	+
Exothermic	Reaction vessel warms (heat is evolved)	Energy subtracted	−

For example, suppose that in an experiment one mole of methane burns in oxygen,

$$CH_4(g) + 2O_2(g) \longrightarrow CO_2(g) + 2H_2O(l)$$

and evolves 890 kJ of heat. The reaction is exothermic: energy (as heat) is *subtracted* from the system. Therefore, the heat of reaction, q, is −890 kJ.

Or, consider the reaction described in the chapter opening, in which crystals of barium hydroxide octahydrate, $Ba(OH)_2 \cdot 8H_2O$, react with crystals of ammonium thiocyanate, NH_4SCN.■

$$Ba(OH)_2 \cdot 8H_2O(s) + 2NH_4SCN(s) \longrightarrow 2NH_3(g) + 10H_2O(l) + Ba(SCN)_2(aq)$$

When 1 mol $Ba(OH)_2 \cdot 8H_2O$ reacts with 2 mol NH_4SCN, the reaction mixture absorbs 167.2 kJ of heat. The reaction is endothermic: energy (as heat) is *added* to the system. Therefore, the heat of reaction, q, is +167.2 kJ.

■ The water molecules in $Ba(OH)_2 \cdot 8H_2O$ are contained in the crystal, loosely bound to the ions. Such crystalline compounds that contain loosely bound water are called *hydrates*. They are discussed in Chapter 14.

Exercise 10.1

Ammonia burns in the presence of a platinum catalyst to give nitric oxide, NO. (A catalyst only speeds up the reaction. Because it has no effect on the heat absorbed or released, it does not occur in the balanced equation.)

$$4NH_3(g) + 5O_2(g) \longrightarrow 4NO(g) + 6H_2O(l)$$

In an experiment, 4 mol NH_3 are burned and evolve 1170 kJ of heat. Is the reaction endothermic or exothermic? What is the value of q?

(See Problems 10.13 and 10.14.)

Enthalpy

When one mole of methane burns in oxygen to give carbon dioxide and liquid water, the amount of heat that is released depends on how the experiment is carried out. If the reaction occurs in the atmosphere, where the pressure is constant (equal to that of the atmosphere), the quantity of heat evolved at 1 atm and 25°C is 890 kJ ($q = -890$ kJ). On the other hand, if this same reaction is done in a closed container, the heat evolved at 25°C is 883 kJ ($q = -883$ kJ). In this case, although the volume of the system is constant (equal to the volume of the container), the pressure changes during the reaction.

Most reactions are carried out in beakers or flasks that are open to the atmosphere. In these cases, the volume changes but the pressure is constant, since it must equal that of the surrounding atmosphere. The heat of reaction at constant pressure, q_p, is related to a property of the reactants and products called **enthalpy** (or *heat content*). Enthalpy (denoted H) is a property of a substance, like mass and volume. Every substance has a definite quantity of enthalpy, just as it has definite mass and definite volume. The enthalpy of a substance depends on the amount of substance. Two moles of substance have twice as much enthalpy as one mole of substance. The quantity of enthalpy in a mole of substance depends only on those variables, such as temperature and pressure, that determine the state of the substance. Therefore, enthalpy is said to be a *state function*, or a state property.

Using the symbol Δ (meaning "change in"), we write the change in enthalpy as ΔH. The change in enthalpy for a reaction (called the **enthalpy of reaction**) is obtained by subtracting the enthalpy of the reactants from the enthalpy of the products:

$$\Delta H = H(\text{products}) - H(\text{reactants})$$

The heat of reaction at constant pressure equals the enthalpy of reaction: ■

$$q_p = \Delta H$$

To understand the relationship between the enthalpies of substances and the heat of reaction, consider the combustion (burning) of methane, CH_4:

$$CH_4(g) + 2O_2(g) \longrightarrow CO_2(g) + 2H_2O(l)$$

The enthalpies of $CH_4(g)$, $O_2(g)$, $CO_2(g)$, and $H_2O(l)$ are -75, 0, -393, and -286 kJ/mol, respectively. ■ Before reaction, the total enthalpy is the enthalpy of 1 mol CH_4 plus 2 mol O_2, which is -75 kJ $+ 2(0$ kJ$) = -75$ kJ. After reaction, the total enthalpy is that of 1 mol CO_2 plus 2 mol H_2O, or -393 kJ $+ 2(-286$ kJ$) = -965$ kJ. The change of enthalpy, ΔH, for the reaction is

$$\Delta H = H(\text{products}) - H(\text{reactants})$$
$$= -965 \text{ kJ} - (-75 \text{ kJ}) = -890 \text{ kJ}$$

This equals the heat of reaction at constant pressure. In other words, when 1 mol $CH_4(g)$ and 2 mol $O_2(g)$ react to produce 1 mol $CO_2(g)$ and 2 mol $H_2O(l)$, the enthalpy, or heat content, of the system decreases by 890 kJ, and 890 kJ of heat are released. The changes in enthalpy are shown in Figure 10.4.

Values of enthalpies of reaction are often given with the chemical equation. For example,

$$CH_4(g) + 2O_2(g) \longrightarrow CO_2(g) + 2H_2O(l); \Delta H = -890 \text{ kJ}$$

In writing $\Delta H = -890$ kJ, we mean that 890 kJ of heat are evolved when the reaction occurs using the molar amounts shown in the chemical equation (that is, when 1 mol CH_4 reacts with 2 mol O_2 to yield 1 mol CO_2 and 2 mol H_2O). Unless otherwise specified, values of ΔH given in this book are for 25°C and 1 atm pressure.

We were careful to give the state of each substance in the last equation. This is important because, in general, ΔH will be different if the state of any reactant or product changes. Consider the equation

$$CH_4(g) + 2O_2(g) \longrightarrow CO_2(g) + 2H_2O(g); \Delta H = -802 \text{ kJ}$$

Comparing the enthalpy of reaction with the one given before, we note that less heat is evolved for the second reaction (802 kJ versus 890 kJ). In the second reaction, water is obtained as the vapor, $H_2O(g)$, instead of as the

■ This relationship between enthalpy change and heat gave rise to the term *heat content*. Since the enthalpy of a substance or mixture can be changed in ways other than by absorbing or evolving heat, the term *heat content* is misleading and *enthalpy* is now preferred. However, the symbol H (from heat content) has been retained for enthalpy.

■ Absolute values of enthalpy cannot be measured. Only values relative to an arbitrary reference can be given. Since we are interested only in the differences (ΔH), the choice of this reference is immaterial. The conventional choice, as described in Section 10.5, leads to values known as enthalpies of formation. Enthalpy values given here are enthalpies of formation.

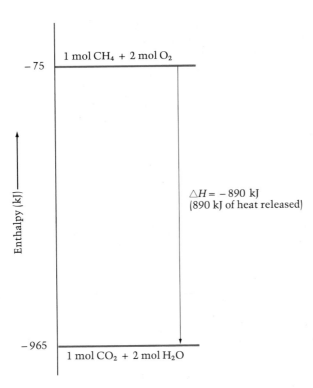

Figure 10.4
An enthalpy diagram. When 1 mol CH_4 and 2 mol O_2 react to give 1 mol CO_2 and 2 mol H_2O, the enthalpy decreases 890 kJ, and 890 kJ of heat are released.

liquid, $H_2O(l)$. The lower enthalpy of reaction is due to the fact that some heat is used to vaporize liquid water, so that less heat is released to the surroundings. ■

■ It takes 44.0 kJ of heat to vaporize one mole of liquid water.

Example 10.1

Chlorine gas, Cl_2, reacts with methane gas, CH_4, to give liquid carbon tetrachloride, CCl_4, and hydrogen chloride gas, HCl. For each mole of carbon tetrachloride produced, 433 kJ of heat are evolved at constant pressure. Write the complete equation for this reaction, including the heat of reaction and labels for the states of all reactants and products.

Solution
The balanced equation is

$$4Cl_2 + CH_4 \longrightarrow CCl_4 + 4HCl$$

Since the heat evolves at constant pressure and since one mole of CCl_4 is produced, the heat of reaction is $\Delta H = -433$ kJ. Including labels for the states and the value of ΔH, the equation is

$$4Cl_2(g) + CH_4(g) \longrightarrow CCl_4(l) + 4HCl(g);$$
$$\Delta H = -433 \text{ kJ}$$

Exercise 10.2

A propellant for rockets is obtained by mixing the liquids hydrazine, N_2H_4, and dinitrogen tetroxide, N_2O_4. These react to give gaseous nitrogen, N_2, and water vapor, evolving 1.15×10^3 kJ of heat at constant pressure when 1 mol N_2O_4 reacts. Write the balanced equation for this reaction, including the states of the compounds and the heat of reaction.

(See Problems 10.15 and 10.16.)

Aside: Relating Enthalpy to Energy

In Section 10.1, we discussed enthalpies of substances in terms of their relationship to heats of reactions. However, the enthalpy of a substance is actually defined in terms of other properties of the substance, namely energy, pressure, and volume. To see how the enthalpy is related to these properties, we must consider all of the energy changes taking place during a chemical reaction that occurs at constant pressure (say, in a vessel open to the atmosphere).

A chemical system changes energy if heat is absorbed or released. However, if the system changes in volume, this also involves energy. For instance, if the volume increases, the system expands and must push back the atmosphere. This requires energy. It can be shown that if the system increases in volume ΔV against a pressure P, the energy needed is $P\Delta V$. This energy is called *pressure-volume work*. ■ Since this energy comes from the system, the system changes in energy by the amount $-P\Delta V$. The negative sign means the energy of the system decreases when the volume increases (that is, when the system does work by pushing back the atmosphere).

The net change of energy of the system, ΔE, equals the energy added (or subtracted) because heat is absorbed (or released) at constant pressure, q_p, minus the energy used for pressure-volume work, $P\Delta V$.

$$\Delta E = q_p - P\Delta V$$

This equation is a mathematical statement of the *first law of thermodynamics,* which relates the net change of energy of a system to the sum of the various energy changes (see Figure 10.5).

By rearranging this equation, we can see that the heat of reaction equals the net change of energy of the system plus the pressure-volume work.

$$q_p = \Delta E + P\Delta V$$

Note that if the system does not change in volume ($\Delta V = 0$), the heat of reaction equals the difference in energy of the products and reactants.

For those reactions in which there is significant change of volume, the $P\Delta V$ term cannot be ignored; the heat of reaction does not equal ΔE. Let us write E_r for the energy of reactants and V_r for the volume of reactants. Similarly, E_p and V_p are the energy and volume of the products. Since $\Delta E = E_p - E_r$ and $\Delta V = V_p - V_r$, the preceding equation can be written

$$q_p = \Delta E + P\Delta V = (E_p - E_r) + P(V_p - V_r)$$

We can rearrange the terms on the right to give

$$q_p = (E_p + PV_p) - (E_r + PV_r)$$

Note that the heat of reaction, q_p, equals the difference in the quantity $E + PV$ for the products and reactants. That is, q_p equals the change in the property $E + PV$ when the reaction occurs:

$$q_p = \Delta(E + PV)$$

The property $E + PV$ is given the name enthalpy, H, so that this equation becomes the one we gave in the text:

$$q_p = \Delta H$$

■ The expression for pressure-volume work can be derived as follows. Assume that the system is contained in a cylinder of cross section A, and suppose the system expands and moves the atmosphere up by a distance d. The change in volume of the system, ΔV, is $d \times A$. From physics, we know that the energy used to move anything equals the force applied, F, times the distance the object is moved, d. That is, the energy used $= F \times d$. Note that $F \times d = (F/A) \times (d \times A)$. But F/A equals the pressure, P, and $d \times A = \Delta V$. Therefore, the energy used by the system in expanding is $P\Delta V$.

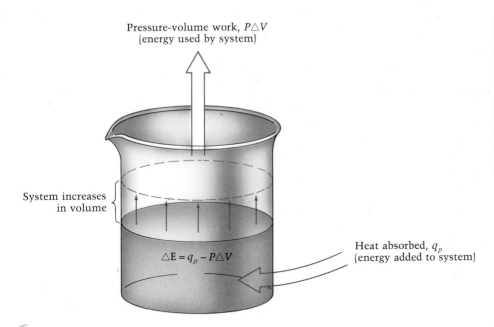

Pressure-volume work, $P\triangle V$
(energy used by system)

System increases
in volume

$\triangle E = q_p - P\triangle V$

Heat absorbed, q_p
(energy added to system)

Figure 10.5
Energy changes during a reaction. If the reaction is endothermic, energy is added as heat is absorbed. If the reaction mixture increases in volume, energy is taken from the system and used to push back the atmosphere. These energy changes are shown by the arrows.

10.2 Calorimetry

If a chemical reaction goes to completion without any other reactions occurring at the same time, we can directly determine the heat of reaction. A **calorimeter** measures the heat flow to or from a system. This device can be as simple as the apparatus sketched in Figure 10.6, which consists of an insulated container (such as a polystyrene coffee cup) with a thermometer in

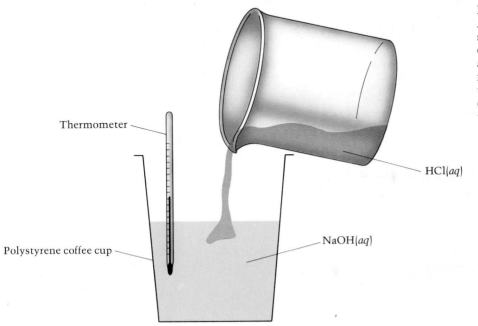

Thermometer

Polystyrene coffee cup

HCl(aq)

NaOH(aq)

Figure 10.6
A simple calorimeter, consisting of a polystyrene coffee cup. When reactants are added, the cup is covered to reduce heat loss by convection. The heat of reaction is determined by noting the temperature rise or fall.

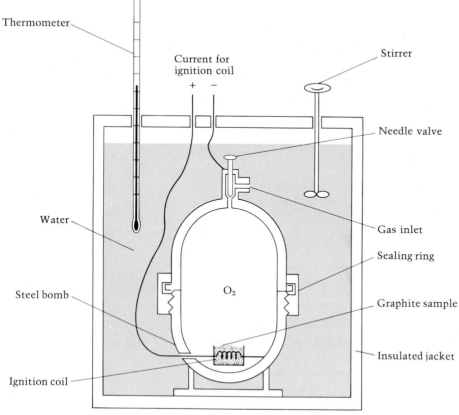

Thermometer

Current for
ignition coil

Stirrer

Needle valve

Water

Gas inlet

Sealing ring

Steel bomb

Graphite sample

Ignition coil

Insulated jacket

O_2

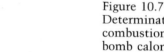

Figure 10.7
Determination of the heat of
combustion of graphite in a
bomb calorimeter. The reaction is started by an ignition
coil running through the
graphite sample.

it. More elaborate calorimeters are employed in precise research (Figure 10.7).

We will look at two types of heat measurement in this section: (1) the measurement of the specific heat of an individual substance and (2) the measurement of the heat released or absorbed from a reaction mixture.

Heat Capacity and Specific Heat

The **heat capacity**, C, of a sample of a substance is the quantity of heat needed to raise its temperature one degree Celsius (or one kelvin). Changing the temperature of the sample from an initial temperature, t_i, to a final temperature, t_f, requires heat equal to

$$q = C \Delta t$$

where Δt is the change of temperature and equals

$$\Delta t = t_f - t_i$$

For example, a piece of iron that takes 6.70 J of heat to raise its temperature by one degree Celsius has a heat capacity of 6.70 J/°C. The quantity of heat required to raise the temperature of the piece of iron from 25.0°C to 35.0°C is

$$q = C\Delta t = 6.70 \text{ J} \times (35.0°C - 25.0°C) = 67.0 \text{ J}$$

The molar heat capacity of a substance is its heat capacity per mole.

The *specific heat* of a substance is the quantity of heat required to raise the temperature of one gram of that substance by one degree Celsius (or one kelvin). Water has a specific heat of 4.184 J per gram per °C, that is, 4.184 J/(g·°C).■ Iron has a specific heat of 0.447 J/(g·°C); aluminum has a specific heat of 0.903 J/(g·°C).

■ The specific heat of water is 1.00 calorie per gram per degree.

Exercise 10.3

A 20.0-g sample of a metal has a heat capacity of 18.1 J/°C. What is the specific heat of the metal? Given that the metal is either aluminum or iron, which is it? (See the preceding text for data on aluminum and iron.)

(See Problems 10.17 and 10.18.)

To find total quantity of heat needed to change the temperature of a sample, we multiply the specific heat (sp. ht.) by the mass of the sample and the change of temperature, Δt:

$$q = \text{sp. ht.} \times \text{mass} \times \Delta t$$

Thus, the quantity of heat required to raise the temperature of 20.0 g of water from 25.0°C to 30.0°C is

$$q = \text{sp. ht.} \times \text{mass} \times \Delta t$$
$$= 4.184 \text{ J/(g·°C)} \times 20.0 \text{ g} \times (30.0°C - 25.0°C)$$
$$= 418 \text{ J}$$

Exercise 10.4

A 23.1-g sample of ethanol, C_2H_6O, was heated from 25.00°C to 27.34°C by 132 J of heat. What is the specific heat of ethanol? What is the molar heat capacity?

(See Problems 10.19, 10.20, 10.21, and 10.22.)

Example 10.2

A 10.40-g sample of silver was heated to 100.00°C. It was then added to 28.00 g of water in an insulated cup. The water temperature rose from 25.00°C to 26.48°C. What is the specific heat of silver?

Solution

When the heated silver is dropped into the cup of water, heat flows from the silver to the water. (Heat cannot flow to the surroundings because the calorimeter is insulated.) Therefore, the quantity of heat released by the silver, $q(Ag)$, equals the quantity of heat absorbed by the water, $q(H_2O)$; that is, $-q(Ag) = q(H_2O)$.

Let us calculate the quantity of heat absorbed by the water. The specific heat of water is 4.184 J/(g·°C). Since there are 28.0 g of water, and the temperature changed from 25.00°C to 26.48°C, the heat absorbed is

$$q(H_2O) = \text{sp. ht. } (H_2O) \times \text{mass } (H_2O) \times (t_f - t_i) \text{ for } H_2O$$
$$= 4.184 \text{ J/(g·°C)} \times 28.0 \text{ g} \times (26.48°C - 25.00°C)$$
$$= 173 \text{ J}$$

Note that $q(H_2O)$ is a positive quantity, indicating that heat is absorbed. And, because $q(H_2O) = -q(Ag)$,

(*Continued*)

$$q(Ag) = -173 \text{ J}$$

Now we can obtain the specific heat of silver from the equation

$$q(Ag) = \text{sp. ht. (Ag)} \times \text{mass (Ag)} \times (t_f - t_i) \text{ for Ag}$$

We have 10.40 g of Ag, which changes temperature from 100.00°C to 26.48°C (the final temperature of the water and silver). Substituting into the equation,

$$-173 \text{ J} = \text{sp. ht. (Ag)} \times 10.40 \text{ g} \times (26.48°C - 100.00°C)$$

Solving for the specific heat of silver, we get

$$\text{sp. ht. (Ag)} = \frac{-173 \text{ J}}{10.40 \text{ g} \times (26.48°C - 100.00°C)}$$

$$= 0.226 \text{ J/(g} \cdot °C)$$

Exercise 10.5

A 15.63-g sample of mercury is heated to 98.56°C. The metal is then poured into 20.18 g of water in an insulated cup. The water temperature rises from 25.00°C to 28.26°C. What is the specific heat of mercury?   (See Problems 10.23 and 10.24.)

Measurement of Reaction Heats

We measure the heat of reaction in a calorimeter, as we did for specific heat. We could use the coffee-cup calorimeter shown in Figure 10.6 or a more accurate calorimeter. The *bomb calorimeter* pictured in Figure 10.7 is often used to measure reactions that involve gases, such as a combustion. To measure the heat released when graphite burns in oxygen, for example, a sample of graphite is placed in a small cup surrounded by oxygen, which is sealed in a steel vessel. (An electrical circuit is activated to burn the graphite.) The heat of reaction is calculated from the change in water temperature caused by the reaction.

Note that when a reaction is carried out in a closed vessel such as in a bomb calorimeter, the pressure is not necessarily constant. Under these conditions, the heat of reaction does not equal ΔH, and a small correction is needed to get the enthalpy of reaction. However, the correction is negligible whenever the reaction involves no gases or if the number of moles of reactant gas equals the number of moles of product gas, as is the case in the next example.

Example 10.3

Suppose 0.562 g of graphite (a form of carbon) is placed in a calorimeter with an excess of oxygen at 25.00°C and 1 atm pressure. Excess O_2 ensures that all carbon burns to form CO_2. The graphite is then ignited, and it burns according to the equation:

$$C(graphite) + O_2(g) \longrightarrow CO_2(g)$$

On reaction, the water temperature rises from 25.00°C to 25.89°C. The heat capacity of the calorimeter and its contents was determined in a separate experiment to be 20.7 kJ/°C. What is the heat of reaction at 25.00°C and 1 atm pressure?

Solution

The heat released by the reaction, q_{rxn}, is the heat absorbed by the calorimeter and its contents, which is $C_{cal}\Delta t$ (C_{cal} is the heat capacity of the calorimeter and its contents.) Hence,

$$q_{rxn} = -C_{cal}\Delta t$$

Substituting, we get

$$q_{rxn} = -C_{cal}\Delta t = -20.7 \text{ kJ/°C} \times (25.89°C - 25.00°C)$$

$$= -18.4 \text{ kJ}$$

(Continued)

An extra figure has been retained in q_{rxn} for further computation. The negative sign indicates that the reaction is exothermic, as expected for a combustion.

Thus, ΔH for the combustion of 0.562 g of carbon equals -18.4 kJ. We can write

$$0.562 \text{ g C} \simeq -18.4 \text{ kJ}$$

To obtain the value for the combustion of one mole of carbon, we perform the following conversions to express one mole of carbon as kilojoules of heat: mol C $\longrightarrow$ g C $\longrightarrow$ kJ heat. Thus, rounding to two

significant figures,

$$1 \text{ mol C} \simeq 1 \text{ mol C} \times \frac{12.0 \text{ g C}}{1 \text{ mol C}} \times \frac{-18.4 \text{ kJ}}{0.562 \text{ g C}}$$

$$\simeq -3.9 \times 10^2 \text{ kJ}$$

When one mole of carbon burns, 3.9×10^2 kJ of heat are released. We can summarize the results by the equation

$$C(\text{graphite}) + O_2(g) \longrightarrow CO_2(g);$$
$$\Delta H = -3.9 \times 10^2 \text{ kJ/mol}$$

Exercise 10.6

Suppose 33 mL of 1.20 M HCl are added to 42 mL of a solution containing excess sodium hydroxide, NaOH, in a coffee-cup calorimeter. The solution temperature, originally 25.0°C, rises to 31.8°C. Give the enthalpy change, ΔH, for the reaction

$$HCl(aq) + NaOH(aq) \longrightarrow NaCl(aq) + H_2O(l)$$

For simplicity, assume that the heat capacity and the density of the final solution in the cup are those of water. (In more accurate work, these values must be determined.) Also assume that the total volume of the solution equals the sum of the volumes of HCl(aq) and NaOH(aq).

(See Problems 10.25 and 10.26.)

10.3 Stoichiometry of Reaction Heats

The method we described in Chapter 3 to calculate the masses of reactants and products from a chemical equation can be extended to problems involving the quantity of heat. For example, suppose 15.0 grams of O_2 react completely according to the following equation:

$$CH_4(g) + 2O_2(g) \longrightarrow CO_2(g) + 2H_2O(l); \ \Delta H = -890 \text{ kJ}$$

How much heat is evolved for the given amount of oxygen?

The calculation involves the following conversions:

$$\text{Grams of } O_2 \longrightarrow \text{moles of } O_2 \longrightarrow \text{kilojoules of heat}$$

The relevant information is

$$1 \text{ mol } O_2 \simeq 32.0 \text{ g } O_2 \quad \text{(from the molecular weight)}$$
$$2 \text{ mol } O_2 \simeq -890 \text{ kJ} \quad \text{(from the chemical equation)}$$

Hence,

$$15.0 \text{ g } O_2 \simeq 15.0 \text{ g } O_2 \times \frac{1 \text{ mol } O_2}{32.0 \text{ g } O_2} \times \frac{-890 \text{ kJ}}{2 \text{ mol } O_2} \simeq -209 \text{ kJ}$$

■ If we wanted the result in kilocalories, this answer can be converted as follows:

$$209 \text{ kJ} \times \frac{1 \text{ kcal}}{4.184 \text{ kJ}} = 50.0 \text{ kcal}$$

That is, 15.0 grams of O_2 react with sufficient methane to evolve 209 kJ of heat.■

Example 10.4

How much heat is evolved when 907 kg of ammonia are produced according to the following equation? (Assume that the reaction occurs at constant pressure.)

$$N_2(g) + 3H_2(g) \longrightarrow 2NH_3(g); \Delta H = -91.8 \text{ kJ}$$

Solution

The calculation involves converting grams of NH_3 to moles of NH_3 and then to kilojoules of heat. Note that

907 kg equals 9.07×10^5 g. Hence,

$$9.07 \times 10^5 \text{ g NH}_3 \times \frac{1 \text{ mol NH}_3}{17.0 \text{ g NH}_3} \times \frac{-91.8 \text{ kJ}}{2 \text{ mol NH}_3} \simeq$$
$$-2.45 \times 10^6 \text{ kJ}$$

Thus, 2.45×10^6 kJ of heat evolve.

Exercise 10.7

How much heat evolves when 10.0 g of hydrazine react according to the reaction described in Exercise 10.2 on page 317?

(See Problems 10.27, 10.28, 10.29, and 10.30.)

Suppose we want to write the equation given in Example 10.4 to show what happens when twice as many moles of nitrogen and hydrogen react to produce ammonia. Since double the amounts of substances are present, the heat evolved (the change in enthalpy) will be doubled. Similarly, if we wanted to write the equation to express the formation of only one mole of ammonia (instead of two), the quantities would be halved and the heat evolved would be halved. When an equation with a given ΔH is multiplied by any factor, the value of ΔH for the new equation is obtained by multiplying the value of ΔH in the original equation by that same factor. In the case of the equation for the synthesis of ammonia, we would obtain the following values.■

■ If we use the molar interpretation of a chemical equation, there is nothing unreasonable about using coefficients like $\frac{1}{2}$ and $\frac{3}{2}$.

$$N_2(g) + 3H_2(g) \longrightarrow 2NH_3(g); \Delta H = -91.8 \text{ kJ}$$
$$2N_2(g) + 6H_2(g) \longrightarrow 4NH_3(g); \Delta H = -183.6 \text{ kJ}$$
$$\tfrac{1}{2}N_2(g) + \tfrac{3}{2}H_2(g) \longrightarrow NH_3(g); \quad \Delta H = -45.9 \text{ kJ}$$

If we would like to know the enthalpy change when NH_3 dissociates into N_2 and H_2, we have only to reverse the equation for the synthesis of ammonia. When a chemical equation is reversed, the value of ΔH is reversed in sign. We get

$$NH_3(g) \longrightarrow \tfrac{1}{2}N_2(g) + \tfrac{3}{2}H_2(g); \Delta H = +45.9 \text{ kJ}$$

The dissociation of ammonia into its elements is an endothermic reaction, absorbing 45.9 kJ per mole of ammonia.

Example 10.5

When two moles of $H_2(g)$ and one mole of $O_2(g)$ react to give liquid water, 259 kJ of heat evolve:

$$2H_2(g) + O_2(g) \longrightarrow 2H_2O(l); \Delta H = -259 \text{ kJ}$$

Write this equation for one mole of liquid water. Give the reverse equation, in which one mole of liquid water dissociates into hydrogen and oxygen.

(Continued)

Solution

We multiply the coefficients and ΔH in the equation by one-half:

$$H_2(g) + \tfrac{1}{2}O_2(g) \longrightarrow H_2O(l); \ \Delta H = -130 \text{ kJ}$$

If we reverse the equation, we get

$$H_2O(l) \longrightarrow H_2(g) + \tfrac{1}{2}O_2(g); \ \Delta H = +130 \text{ kJ}$$

Exercise 10.8

(a) Write the equation described in Exercise 10.2 on page 317 for 1 mol N_2H_4. (b) Write the reverse of the equation described in Exercise 10.2.

(See Problems 10.31, 10.32, 10.33, and 10.34.)

10.4 Hess's Law

Although heats of reaction can in principle be determined by directly observing the heat evolved or absorbed, not all chemical reactions can be studied this way in practice. The reaction may go too slowly, or it may not go to completion (leaving a mixture of reactants and products), or it may not go "cleanly" (other products may be produced by competing reactions). This is what we would encounter if we were to try to obtain the enthalpy change for the reaction

$$2C(\text{graphite}) + O_2(g) \longrightarrow 2CO(g) \qquad \textbf{(1)}$$

by direct calorimetric measurement. Once carbon monoxide forms, it is impossible to keep it from reacting further to give carbon dioxide:

$$2CO(g) + O_2(g) \longrightarrow 2CO_2(g) \qquad \textbf{(2)}$$

At the same time, graphite may react in one step to give carbon dioxide:

$$C(\text{graphite}) + O_2(g) \longrightarrow CO_2(g) \qquad \textbf{(3)}$$

Since we cannot run Reaction 1 without Reactions 2 and 3, the heat we measure calorimetrically is some mixture of values for all three reactions.

Fortunately, there is a way out of this difficulty, through **Hess's law of heat summation.** This law was discovered empirically in 1840 by Germain Henri Hess, who was a professor of chemistry at the University of St. Petersburg, in Russia. According to Hess's law, we can show that it is possible to calculate ΔH for Reaction 1 from the values for Reactions 2 and 3. Enthalpies of reaction for Reactions 2 and 3 are readily determined by calorimetry. Carbon monoxide is burned completely in an excess of oxygen to give carbon dioxide:

$$2CO(g) + O_2(g) \longrightarrow 2CO_2(g); \ \Delta H = -566.0 \text{ kJ} \qquad \textbf{(2)}$$

Similarly, graphite burns in an excess of oxygen to give only carbon dioxide:

$$C(\text{graphite}) + O_2(g) \longrightarrow CO_2(g); \ \Delta H = -393.5 \text{ kJ} \qquad \textbf{(3)}$$

To obtain the enthalpy change for Reaction 1, let us *imagine* that this reaction proceeds in two steps. First, we burn *two* moles of graphite to carbon dioxide:

$$2C(graphite) + 2O_2(g) \longrightarrow 2CO_2(g); \Delta H = -787.0 \text{ kJ}$$

The value of ΔH is twice that given for Reaction 3, that is, 2×-393.5 kJ $= -787.0$ kJ. Now we *imagine* that all of the carbon dioxide that has formed can be taken to carbon monoxide and oxygen:

$$2CO_2(g) \longrightarrow 2CO(g) + O_2(g); \Delta H = 566.0 \text{ kJ}$$

This reaction normally goes in the opposite direction; we obtain it by reversing the direction of the equation for Reaction 2 and switching the sign of ΔH. The net change in these two steps is

$$2C(graphite) + 2O_2(g) \longrightarrow 2CO(g) + O_2(g)$$

which is identical to Reaction 1, after canceling one mole of O_2. The total heat for this two-step process equals the sum of the heats of both steps:

$$\Delta H = (-787.0 + 566.0) \text{ kJ}$$
$$= -221.0 \text{ kJ}$$

Thus, the ΔH for Reaction 1 is -221.0 kJ.

What we have done is this: we have found two reactions—Reactions 2 and 3—that when multiplied by the appropriate factors and added together give Reaction 1. Multiplying the enthalpies of reaction by the same factors and adding gives ΔH for Reaction 1.

It is worthwhile looking at Hess's law in a slightly different way. Suppose we represent the values of the enthalpies of substances on an enthalpy diagram (Figure 10.8). Since we are only interested in *changes* of enthalpies, we can arbitrarily put 2 mol C(graphite) plus 2 mol $O_2(g)$ at enthalpy zero. In the reaction in which 2 mol C(graphite) + 2 mol $O_2(g)$ go to 2 mol $CO_2(g)$, the change of enthalpy is -787.0 kJ. (This change is depicted by the arrow labeled $2\Delta H_3$.) So, 2 mol $CO_2(g)$ must be at enthalpy -787.0 kJ. We place

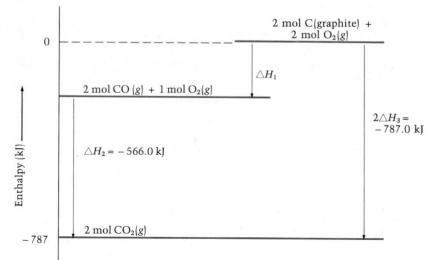

Figure 10.8
Enthalpy diagram for partial and complete combustion of graphite.

2 mol CO(g) + 1 mol O$_2$(g) at an enthalpy that is 566.0 kJ above that for 2 mol CO$_2$(g), because in the reaction in which 2 mol CO(g) + 1 mol O$_2$(g) go to 2 mol CO$_2$(g) (Reaction 2), the change of enthalpy is −566.0 kJ. (This change is depicted by the arrow labeled ΔH_2.) Now it is clear from the diagram that the length of the arrow labeled ΔH_1, which represents the enthalpy change for the reaction of 2 mol C(graphite) + 2 mol O$_2$(g) to give 2 mol CO(g) + 1 mol O$_2$(g), can be obtained by subtracting ΔH_2 from $2\Delta H_3$. This result is precisely what we got before, of course. But it shows graphically why we should expect that enthalpies of reaction will not all be independent of one another. Some values will be related to others, which is the essence of Hess's law.

Example 10.6

What is the enthalpy of reaction, ΔH, for the formation of tungsten carbide, WC, from the elements? (Tungsten carbide is very hard and is used to make cutting tools and rock drills.)

$$W(s) + C(\text{graphite}) \longrightarrow WC(s)$$

The enthalpy change for this reaction is difficult to measure directly, since the reaction occurs at 1400°C. However, the heats of combustion of the elements and of tungsten carbide can be measured:

$$2W(s) + 3O_2(g) \longrightarrow 2WO_3(s);$$
$$\Delta H = -1680.6 \text{ kJ} \qquad \textbf{(1)}$$

$$C(\text{graphite}) + O_2(g) \longrightarrow CO_2(g);$$
$$\Delta H = -393.5 \text{ kJ} \qquad \textbf{(2)}$$

$$2WC(s) + 5O_2(g) \longrightarrow 2WO_3(s) + 2CO_2(g);$$
$$\Delta H = -2391.6 \text{ kJ} \qquad \textbf{(3)}$$

Solution

The general procedure is to multiply each equation by a factor, perhaps reversing some, so that when they all are added together the desired equation results. The proper factor can usually be guessed by looking at the desired final equation. The desired reaction in this problem has W(s) on the left. Hence, multiply Reaction 1 by $\frac{1}{2}$. (Also multiply the ΔH for Reaction 1 by $\frac{1}{2}$.)

$$W(s) + \tfrac{3}{2}O_2(g) \longrightarrow WO_3(s);$$
$$\Delta H = \tfrac{1}{2} \times -1680.6 \text{ kJ} = -840.3 \text{ kJ}$$

Since the desired reaction has C(graphite) on the left, leave Reaction 2 as it is. The desired reaction has WC(s) on the right. Hence, reverse Reaction 3 and multiply it by $\frac{1}{2}$:

$$WO_3(s) + CO_2(g) \longrightarrow WC(s) + \tfrac{5}{2}O_2(g);$$
$$\Delta H = -\tfrac{1}{2} \times -2391.6 \text{ kJ} = 1195.8 \text{ kJ}$$

Note that the ΔH is obtained by multiplying the value for Reaction 3 by $-\frac{1}{2}$. Now these three reactions and the corresponding ΔH's are added together:

	ΔH, kJ
$W(s) + \tfrac{3}{2}O_2(g) \longrightarrow WO_3(s)$	−840.3
$C(\text{graphite}) + O_2(g) \longrightarrow CO_2(g)$	−393.5
$WO_3(s) + CO_2(g) \longrightarrow WC(s) + \tfrac{5}{2}O_2(g)$	1195.8

$$W(s) + \tfrac{3}{2}\cancel{O_2(g)} + C(\text{graphite}) + \cancel{O_2(g)} + \cancel{WO_3(s)}$$
$$+ \cancel{CO_2(g)} \longrightarrow \cancel{WO_3(s)} + \cancel{CO_2(g)} + WC(s) + \tfrac{5}{2}\cancel{O_2(g)};$$
$$\Delta H = -38.0 \text{ kJ}$$

The answer can be expressed in kcal, if desired:

$$\Delta H = -38.0 \text{ k\cancel{J}} \times \frac{1 \text{ kcal}}{4.184 \text{ k\cancel{J}}} = -9.08 \text{ kcal}$$

Exercise 10.9

Calculate the enthalpy change for breaking the four C—H bonds in methane, CH$_4$:

$$CH_4(g) \longrightarrow C(g) + 4H(g)$$

Use the following information:

$$C(\text{graphite}) \longrightarrow C(g); \Delta H = 715.0 \text{ kJ}$$
$$H_2(g) \longrightarrow 2H(g); \Delta H = 436.0 \text{ kJ}$$

$$C(graphite) + O_2(g) \longrightarrow CO_2(g); \Delta H = -393.5 \text{ kJ}$$

$$2H_2(g) + O_2(g) \longrightarrow 2H_2O(l); \Delta H = -571.7 \text{ kJ}$$

$$CH_4(g) + 2O_2(g) \longrightarrow CO_2(g) + 2H_2O(l); \Delta H = -890.3 \text{ kJ}$$

(See Problems 10.35, 10.36, 10.37, and 10.38.)

10.5 Enthalpies of Formation

Using calorimetric measurements and Hess's law, we can find the enthalpy change for almost any reaction. But how shall we store all of these data for future use? There is clearly no need to list all of these reactions, since Hess's law can be used to relate some ΔH values to others. A convenient solution to this problem is to list only the enthalpy changes for formation reactions, that is, reactions in which compounds are formed from the elements. Then, the enthalpy change of any other reaction would be obtained from the formation reactions using Hess's law. Table 10.1 lists **standard enthalpies of formation** (also called *standard heats of formation*). The standard enthalpy of formation, ΔH_f°, of a substance is the enthalpy change for the formation of one mole of the substance from its elements, at standard pressure (1 atm) and a specified temperature (25°C unless otherwise noted). These standard conditions for ΔH are indicated by a superscript degree sign, °, with a subscript f indicating a formation reaction. The standard enthalpy of formation of liquid water, for example, is the enthalpy change for the reaction at 1 atm pressure and 25°C:

$$H_2(g) + \tfrac{1}{2}O_2(g) \longrightarrow H_2O(l); \Delta H_f^\circ = -285.8 \text{ kJ}$$

Because there is often more than one form of an element, one allotrope is used as the reference to which the formation reactions apply.■ Usually, the stablest form of the element at the given conditions is chosen to be the reference. Graphite, for example, is the stablest form of carbon at 1 atm and 25°C and is generally taken as the reference for the element. Thus, the enthalpy of formation of methane, $CH_4(g)$, refers to the equation

■ Allotropes are forms of an element with different properties. For example, carbon exists as diamond and as graphite.

$$C(graphite) + 2H_2(g) \longrightarrow CH_4(g); \Delta H_f^\circ = -74.9 \text{ kJ}$$

and not to■

$$C(diamond) + 2H_2(g) \longrightarrow CH_4(g); \Delta H^\circ = -76.8 \text{ kJ}$$

The enthalpy of formation of diamond is

$$C(graphite) \longrightarrow C(diamond); \Delta H_f^\circ = 1.9 \text{ kJ}$$

■ Note that we have written ΔH_f° for the reaction involving graphite, but ΔH° for the one involving diamond.

but that for graphite is zero, since there is no change:

$$C(graphite) \longrightarrow C(graphite); \Delta H_f^\circ = 0$$

Before you use any table of enthalpies of formation, it is helpful to look over the reference states given for each listing. The reference state for an element will have $\Delta H_f^\circ = 0$.■

Now let us see how to use Table 10.1 to find the enthalpy change for a reaction. Consider the equation

■ Look at Table 10.1 and pick out the reference states for the elements.

$$C_2H_4(g) + H_2(g) \longrightarrow C_2H_6(g)$$
$$\text{ethylene} \qquad\qquad\qquad \text{ethane}$$

Formula	ΔH_f° (kJ/mol)	Formula	ΔH_f° (kJ/mol)
$e^-(g)$	0	*Nitrogen*	
		$N(g)$	473
Hydrogen		$N_2(g)$	0
$H^+(aq)$	0	$NH_3(g)$	−45.9
$H(g)$	218.0	$NH_4^+(aq)$	−132.8
$H_2(g)$	0		
		Oxygen	
Sodium		$O(g)$	249.2
$Na^+(g)$	609.8	$O_2(g)$	0
$Na^+(aq)$	−239.7	$O_3(g)$	143
$Na(g)$	107.8	$OH^-(aq)$	−229.9
$Na(s)$	0	$H_2O(g)$	−241.8
$NaCl(s)$	−411.1	$H_2O(l)$	−285.8
$NaHCO_3(s)$	−947.7		
$Na_2CO_3(s)$	−1130.8	*Sulfur*	
		$S(g)$	279
Calcium		$S_2(g)$	129
$Ca^{2+}(aq)$	−543.0	$S_8(\text{rhombic})$	0
$Ca(s)$	0	$S_8(\text{monoclinic})$	2
$CaCO_3(s)$	−1206.9	$SO_2(g)$	−296.8
		$H_2S(g)$	−20
Carbon			
$C(g)$	715.0	*Fluorine*	
$C(\text{graphite})$	0	$F^-(g)$	−255.6
$C(\text{diamond})$	1.9	$F^-(aq)$	−329.1
$CO(g)$	−110.5	$F_2(g)$	0
$CO_2(g)$	−393.5	$HF(g)$	−273
$HCO_3^-(aq)$	−691.1		
$CH_4(g)$	−74.9	*Chlorine*	
$C_2H_4(g)$	52.5	$Cl^-(aq)$	−167.5
$C_2H_6(g)$	−84.7	$Cl(g)$	121.0
$C_6H_6(l)$	49.0	$Cl_2(g)$	0
$HCHO(g)$	−116	$HCl(g)$	−92.3
$CH_3OH(l)$	−238.6		
$CS_2(g)$	117	*Bromine*	
$CS_2(l)$	87.9	$Br^-(g)$	−218.9
$HCN(g)$	135	$Br^-(aq)$	−120.9
$CCl_4(g)$	−96.0	$Br_2(l)$	0
$CCl_4(l)$	−139		
$CH_3CHO(g)$	−166	*Iodine*	
$C_2H_5OH(l)$	−277.6	$I^-(g)$	−194.7
		$I^-(aq)$	−55.9
Silicon		$I_2(s)$	0
$Si(s)$	0		
$SiO_2(s)$	−910.9	*Silver*	
$SiF_4(g)$	−1548	$Ag^+(g)$	1026.4
		$Ag^+(aq)$	105.9
Lead		$Ag(s)$	0
$Pb(s)$	0	$AgF(s)$	−203
$PbO(s)$	−219	$AgCl(s)$	−127.0
$PbS(s)$	−98.3	$AgBr(s)$	−99.5
		$AgI(s)$	−62.4

Table 10.1
Standard Enthalpies of
Formation (at 25°C)

From Table 10.1 we pick out the enthalpies of formation for $C_2H_4(g)$ and for $C_2H_6(g)$:

$$2C(\text{graphite}) + 2H_2(g) \longrightarrow C_2H_4(g); \Delta H_f^\circ = 52.5 \text{ kJ} \quad (1)$$
$$2C(\text{graphite}) + 3H_2(g) \longrightarrow C_2H_6(g); \Delta H_f^\circ = -84.7 \text{ kJ} \quad (2)$$

We now apply Hess's law to these equations. Since we want C_2H_4 to appear on the left, we reverse Reaction 1, then add Reaction 2:

	ΔH°, kJ
$C_2H_4(g) \longrightarrow 2C(\text{graphite}) + 2H_2(g)$	-52.5
$2C(\text{graphite}) + 3H_2(g) \longrightarrow C_2H_6(g)$	-84.7
$C_2H_4(g) + H_2(g) \longrightarrow C_2H_6(g)$	-137.2 kJ

Note that the reference elements, unless they are to appear in the final equation (as H_2 does here), drop out.

Once we realize that these elements do not appear in the results, we see that it is not necessary to write down the equations and manipulate them. Instead, we manipulate only the values of ΔH_f°. As an example, let us calculate the enthalpy of reaction for

$$CH_4(g) + 2Cl_2(g) \longrightarrow CH_2Cl_2(l) + 2HCl(g); \Delta H^\circ = ?$$

Enthalpies of formation (in kilojoules) of the substances in the equation are: CH_4, -75; Cl_2, 0; CH_2Cl_2, -117; HCl, -92. Let us write these beneath the equation, multiplying each by its stoichiometric coefficient (since ΔH_f° values are per mole of substance).

$$CH_4(g) + 2Cl_2(g) \longrightarrow CH_2Cl_2(l) + 2HCl(g)$$
$$-75 \quad 2 \times 0 \quad -117 \quad 2 \times (-92) \quad (\text{kJ})$$

Now we add the values for the products and subtract the values for the reactants.

$$\Delta H^\circ = [-117 - (2 \times 92) - (-75 + 0)] \text{ kJ} = -226 \text{ kJ}$$

To summarize: The value of ΔH° for a reaction is obtained by writing a balanced chemical equation, summing the values of ΔH_f° for the products (multiplying each by corresponding stoichiometric coefficients), and subtracting the values of ΔH_f° for the reactants (times stoichiometric coefficients).

Example 10.7

Use values of ΔH_f° to calculate the heat of vaporization, ΔH_{vap}°, of carbon disulfide at 25°C. The vaporization process is

$$CS_2(l) \longrightarrow CS_2(g)$$

Solution

The vaporization process can be treated just like a chemical reaction. Write the equation with values of ΔH_f° beneath reactant and product.

$$CS_2(l) \longrightarrow CS_2(g)$$
$$88 \quad 117 \quad (\text{kJ})$$

Then,

$$\Delta H_{vap}^\circ = (117 - 88) \text{ kJ} = 29 \text{ kJ}$$

Exercise 10.10

Calculate the heat of vaporization, ΔH°_{vap}, of water, using standard enthalpies of formation (Table 10.1).

(See Problems 10.39 and 10.40.)

Example 10.8

The commercial preparation of carbon disulfide is given by the equation

$$CH_4(g) + 2S_2(g) \longrightarrow CS_2(g) + 2H_2S(g)$$

(Solid sulfur consists of S_8 molecules, but the vapor at the temperature of the commercial preparation of CS_2 is mainly S_2.) Calculate the enthalpy of reaction from standard enthalpies of formation.

Solution

Write the equation with values of ΔH°_f beneath every reactant and product. Multiply each value by the corresponding stoichiometric coefficient:

$$CH_4(g) + 2S_2(g) \longrightarrow CS_2(g) + 2H_2S(g)$$
$$-75 \quad\ 2 \times 129 \qquad\quad 117 \quad 2 \times (-20) \quad (kJ)$$

Note that sulfur is given for the gas phase, so that ΔH°_f for it is not zero. The commercial reaction is actually run at 700°C, where sulfur is gaseous. Our calculated value of ΔH° is for 25°C; but it can be used as an approximate value for 700°C. The calculation for the enthalpy of reaction is

$$\Delta H^\circ = [117 + 2(-20) - (-75) - (2 \times 129)] \text{ kJ}$$
$$= -106 \text{ kJ}$$

Be very careful with arithmetical signs—they are the most likely source of mistakes. Also pay particular attention to the state of each substance. Here, for example, we use the value of ΔH°_f for $CS_2(g)$, not that for $CS_2(l)$.

Exercise 10.11

Calculate the enthalpy change for breaking the four bonds in methane, CH_4:

$$CH_4(g) \longrightarrow C(g) + 4H(g)$$

Use standard enthalpies of formation, ΔH°_f (Table 10.1). Compare with the solution of Exercise 10.9.

(See Problems 10.41, 10.42, 10.43, and 10.44.)

Enthalpies of formation can also be defined for ions. In this case, since it is not possible to take thermal measurements on individual ions, we must arbitrarily define the standard enthalpy of formation of one ion as zero. Then, values for all other ions can be deduced from calorimetric data. By convention, the standard enthalpy of formation of $H^+(aq)$ is taken as zero. Values of ΔH°_f for some ions are given in Table 10.1.

Example 10.9

When an aqueous solution of silver nitrate is added to an aqueous solution of sodium chloride, a precipitate of silver chloride forms (that is, solid silver chloride comes out of the solution). Both silver nitrate, $AgNO_3$, and sodium chloride, $NaCl$, dissolve in water as ions. We can write the equation in ionic form as follows:

$$[Ag^+(aq) + \cancel{NO_3^-(aq)}] + [\cancel{Na^+(aq)} + Cl^-(aq)] \longrightarrow$$
$$AgCl(s) + [\cancel{Na^+(aq)} + \cancel{NO_3^-(aq)}]$$

Ions that appear on both sides of the equation can be

(Continued)

omitted to give the equation for the precipitation:

$$Ag^+(aq) + Cl^-(aq) \longrightarrow AgCl(s)$$

What is $\Delta H°$ for this precipitation reaction? Use values of $\Delta H_f°$ from Table 10.1.

Solution

We write the equation for the formation of the precipitate from ions, putting values of $\Delta H_f°$ below each spe-

cies in the equation:

$$Ag^+(aq) + Cl^-(aq) \longrightarrow AgCl(s)$$
$$105.9 -167.5 -127.0 \quad (kJ)$$

The calculation of the enthalpy of reaction is

$$\Delta H° = [-127.0 - (105.9 - 167.5)] \text{ kJ}$$
$$= -65.4 \text{ kJ}$$

Exercise 10.12

Calculate the standard enthalpy change for the reaction of an aqueous solution of barium hydroxide, $Ba(OH)_2$, with an aqueous solution of ammonium thiocyanate, NH_4SCN. (Figure 10.1 illustrated this reaction using solids instead of solutions.)

$$[Ba^{2+}(aq) + 2OH^-(aq)] + 2[NH_4^+(aq) + SCN^-(aq)] \longrightarrow$$
$$2NH_3(g) + 2H_2O(l) + [Ba^{2+}(aq) + 2SCN^-(aq)]$$

(See Problems 10.45 and 10.46.)

Applications of Thermochemistry

The next two sections describe some applications of thermochemistry to problems in chemical bonding. We will look first at covalent *bond energies*, then at the energies involved in forming ionic solids (*lattice energies*).

10.6 Bond Energy

The heat of reaction is surely related in some way to the structure of the reactants and products. Thus, we should be able to use enthalpies of reaction to tell us something about structure. Conversely, knowing something about structure, we ought to be able to say something about enthalpies of reaction. The bond-energy concept is a simple idea devised as a link between our knowledge of heats of reaction and of molecular structure.

Consider the experimentally determined enthalpy changes for the breaking or dissociation of a C—H bond in methane, CH_4, and in ethane, C_2H_6, in the gaseous phase:

$$\text{H--C--H} \longrightarrow \text{H--C} + \text{H} \qquad \Delta H = 435 \text{ kJ}$$

$$\text{H--C--C--H} \longrightarrow \text{H--C--C} + \text{H} \qquad \Delta H = 410 \text{ kJ}$$

Note that the ΔH values are approximately the same in the two cases. This suggests that the enthalpy change for a particular kind of bond dissociation may be about the same in other molecules. Comparing different kinds of localized bonds in other molecules bears out this conclusion.

We define the A—B **bond energy** (denoted BE) as the *average* enthalpy change for the breaking of an A—B bond in a molecule in the gaseous phase. For example, to calculate a value for the C—H bond energy, or $BE(C—H)$, we might look at the experimentally determined enthalpy change for the breaking of all C—H bonds in methane:

$$CH_4(g) \longrightarrow C(g) + 4H(g); \Delta H = 1662 \text{ kJ}$$

Since four C—H bonds are broken, we obtain an average value for one C—H bond by dividing the enthalpy change for the reaction by four:

$$BE(C—H) = \tfrac{1}{4} \times 1662 \text{ kJ} = 416 \text{ kJ}$$

Similar calculations with other molecules, such as ethane, would yield approximately the same values for the C—H bond energy. Table 10.2 lists representative values of some bond energies. Because it takes energy to break a bond, bond energies are always positive numbers. When a bond is formed, energy is released in an amount equal to the negative of the bond energy.

Bond energy is a measure of the strength of a bond: the larger the bond energy, the stronger the chemical bond. Note from Table 10.2 that the bonds C—C, C=C, and C≡C have energies of 346, 602, and 835 kJ, respectively. These numbers indicate that the triple bond is stronger than the double bond, which in turn is stronger than the single bond.

Table 10.2
Bond Energies (in kJ/mol)*

| | **Single Bonds** | | | | | | | | |
	H	C	N	O	S	F	Cl	Br	I
H	432								
C	411	346							
N	386	305	167						
O	459	358	201	142					
S	363	272	—	—	226				
F	565	485	283	190	284	155			
Cl	428	327	313	218	255	249	240		
Br	362	285	—	201	217	249	216	190	
I	295	213	—	201	—	278	208	175	149

Multiple Bonds						
C=C	602	C=N	615	C=O	799
C≡C	835	C≡N	887	C≡O	1072
N=N	418	N=O	607	S=O (in SO_2)	532
N≡N	942	O_2	494	S=O (in SO_3)	469

*Data are taken from J. E. Huheey, *Inorganic Chemistry*, 2nd ed. (New York: Harper and Row, 1978), pp. 842–850.

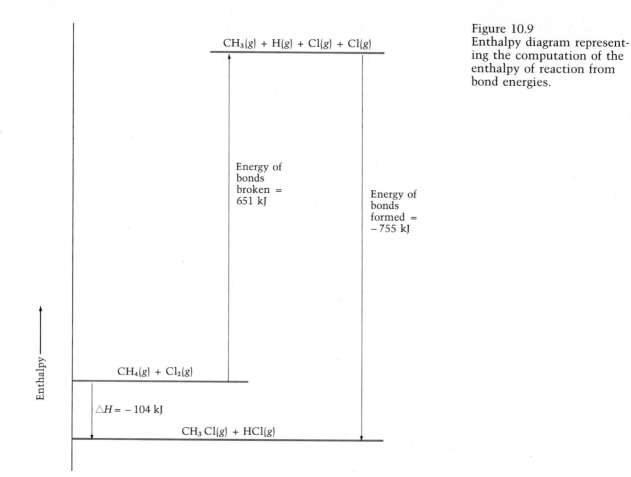

Figure 10.9
Enthalpy diagram representing the computation of the
enthalpy of reaction from
bond energies.

We can use this table of bond energies to estimate enthalpy changes for gaseous reactions. To illustrate this, let us find ΔH for the reaction

$$CH_4(g) \; + \; Cl_2(g) \longrightarrow CH_3Cl(g) \; + \; HCl(g)$$

We can *imagine* that the reaction takes place in steps involving the breaking and forming of bonds (see Figure 10.9). Starting with the reactants, we suppose that one C—H bond and the Cl—Cl bond break:

$$\begin{array}{c} \text{H} \\ | \\ \text{H}-\text{C}-\text{H} + \text{Cl}-\text{Cl} \longrightarrow \text{H}-\text{C} + \text{H} + \text{Cl} + \text{Cl} \\ | \qquad\qquad\qquad\qquad\quad | \\ \text{H} \qquad\qquad\qquad\qquad\quad \text{H} \end{array}$$

The enthalpy change for this is $BE(\text{C—H}) \; + \; BE(\text{Cl—Cl})$. Now the fragments are reassembled to give the products:

$$\begin{array}{c} \text{H} \qquad\qquad\qquad\qquad\qquad \text{H} \\ | \qquad\qquad\qquad\qquad\qquad | \\ \text{H}-\text{C} + \text{H} + \text{Cl} + \text{Cl} \longrightarrow \text{H}-\text{C}-\text{Cl} + \text{H}-\text{Cl} \\ | \qquad\qquad\qquad\qquad\qquad | \\ \text{H} \qquad\qquad\qquad\qquad\qquad \text{H} \end{array}$$

In this case, C—Cl and H—Cl bonds are formed, and the enthalpy change equals the negative of the bond energies $-BE(\text{C—Cl}) - BE(\text{H—Cl})$. Substituting bond-energy values from Table 10.2, we get the enthalpy of reaction

$$\Delta H \simeq BE(\text{C—H}) + BE(\text{Cl—Cl}) - BE(\text{C—Cl}) - BE(\text{H—Cl})$$
$$= (411 + 240 - 327 - 428) \text{ kJ}$$
$$= -104 \text{ kJ}$$

Since the bond-energy concept is only approximate, this value is only approximate. The experimental value is -101 kJ.

In general, the enthalpy of reaction is (approximately) equal to the sum of the bond energies for bonds broken minus the sum of the bond energies for bonds formed.

Example 10.10

Polyethylene is formed by linking many ethylene molecules into long chains. Estimate the enthalpy change per mole of ethylene for this reaction (shown below), using bond energies.

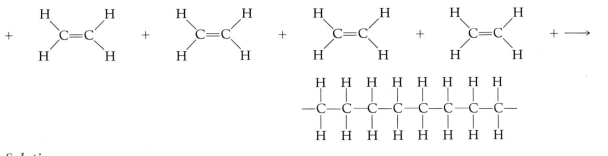

Solution

Imagine the reaction to involve the breaking of the carbon–carbon double bonds and the formation of carbon–carbon single bonds. For a very long chain, the net result is that for every C=C bond broken, two C—C bonds are formed:

$$\Delta H \simeq [602 - (2 \times 346)] \text{ kJ} = -90 \text{ kJ}$$

Exercise 10.13

Use bond energies to estimate the enthalpy change for the combustion of ethylene, C_2H_4, according to the equation

$$C_2H_4(g) + 3O_2(g) \longrightarrow 2CO_2(g) + 2H_2O(g)$$

(See Problems 10.47 and 10.48.)

As another example of the use of bond-energy values, let us look at Pauling's method for determining electronegativities. According to Pauling, the A—B bond energy is related to the average of the A—A and B—B bond energies, plus an energy contribution from the polar character of the bond that depends on the electronegativity difference:

$$BE(\text{A—B}) = \tfrac{1}{2}[BE(\text{A—A}) + BE(\text{B—B})] + k(X_A - X_B)^2$$

Here k is a constant that Pauling has determined; its value is 98.6 kJ. Also, X_A and X_B are the electronegativities of atoms A and B. Suppose we let atom A be oxygen, and atom B hydrogen. If we substitute values from Table 10.2 for the O—H, O—O, and H—H bond energies, we can calculate the O—H electronegativity difference. Thus,

$$459 \text{ kJ} = \tfrac{1}{2}[142 + 432] \text{ kJ} + 98.6 \text{ kJ} (X_O - X_H)^2$$

Note that $\tfrac{1}{2}[142 + 432]$ equals 287. Hence,

$$(X_O - X_H)^2 = \frac{459 - 287}{98.6} = 1.74$$

Taking the square root, we get $X_O - X_H$ equals either $+1.32$ or -1.32. But since oxygen is more electronegative than hydrogen, we choose the positive sign.

$$X_O - X_H = 1.32$$

Other differences can be calculated in a similar manner. Then, once the value for one element is assigned, those for the other elements can be obtained. This is essentially the method used to compute the values given in Figure 7.5 on page 211.

Exercise 10.14

Calculate the electronegativity difference for the H—S bond, using bond energies from Table 10.2. Compare this with the value computed from Figure 7.5.

(See Problems 10.49 and 10.50.)

10.7 Lattice Energies

The **lattice energy** of an ionic solid is the energy required to break the solid into isolated ions in the gas phase. It is a measure of the strength of bonding in an ionic solid. For sodium chloride, this is the enthalpy change of the reaction

$$NaCl(s) \longrightarrow Na^+(g) + Cl^-(g)$$

Direct experimental determination of the lattice energy is difficult. However, Hess's law enables us to relate the lattice energy to other quantities that have been measured experimentally. The method was devised by Max Born and Fritz Haber in 1919 and is called the *Born-Haber cycle.*

Following the method of the Born-Haber cycle, we think of solid sodium chloride being formed from the elements by two different routes. These are shown in Figure 10.10. In one route, NaCl(s) is formed directly from the elements, Na(s) and $\tfrac{1}{2}Cl_2(g)$. The enthalpy change for this is ΔH_f°, which is given in Table 10.1 as -411 kJ per mole of NaCl. The second route consists of five steps, marked 1 to 5 in Figure 10.10. These steps are

1. *Sublimation of sodium.* Metallic sodium is vaporized to a gas of sodium atoms. (*Sublimation* is the transformation of a solid to a gas.) The enthalpy change for this process has been measured experimentally and is equal to 108 kJ per mole of sodium.

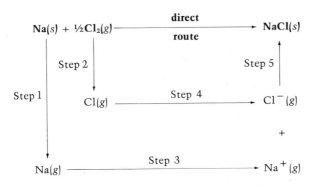

Figure 10.10
Born-Haber cycle for NaCl. The formation of NaCl(s) from the elements accomplished by two different routes. The direct route is the formation reaction (shown in color) and the enthalpy change is ΔH_f°. The indirect route occurs in five steps, numbers 1 to 5.

2. *Dissociation of chlorine.* Chlorine molecules are dissociated to atoms. The enthalpy change for this equals the Cl—Cl bond energy, which is 240 kJ per mole of bonds or 120 kJ per mole of Cl atoms.

3. *Ionization of sodium.* Sodium atoms are ionized to Na^+ ions. The enthalpy change is essentially the ionization energy of atomic sodium and equals 502 kJ per mole of Na.

4. *Formation of chloride ion.* The electrons from the ionization of sodium atoms are transferred to chlorine atoms. The enthalpy change for this is essentially the electron affinity of atomic chlorine and equals −349 kJ per mole of Cl atoms.

5. *Formation of NaCl(s) from ions.* The ions Na^+ and Cl^- formed in steps 3 and 4 combine to give solid sodium chloride. Since this process is just the reverse of the one corresponding to the lattice energy (breaking the solid into ions), the enthalpy change will be the negative of the lattice energy. If we let U be the lattice energy, the enthalpy change for step 5 is $-U$.

Let us write out these five steps and add them together. We also add the corresponding enthalpy changes, following Hess's law.

Step 1	$Na(s)$	$\longrightarrow Na(g)$	$\Delta H_1 =$	108 kJ
Step 2	$\frac{1}{2}Cl_2(g)$	$\longrightarrow Cl(g)$	$\Delta H_2 =$	120 kJ
Step 3	$Na(g)$	$\longrightarrow Na^+(g) + e^-(g)$	$\Delta H_3 =$	502 kJ
Step 4	$Cl(g) + e^-(g)$	$\longrightarrow Cl^-(g)$	$\Delta H_4 =$	−349 kJ
Step 5	$Na^+(g) + Cl^-(g)$	$\longrightarrow NaCl(s)$	$\Delta H_5 =$	$-U$
	$Na(s) + \frac{1}{2}Cl_2(g)$	$\longrightarrow NaCl(s)$	$\Delta H_f^\circ =$	381 kJ − U

In summing the equations, we have canceled terms that appear on both the left and right sides of the arrows. The final equation is simply the formation reaction for NaCl(s). Adding the enthalpy changes for steps 1 to 5, we find the enthalpy change for this formation reaction is 381 kJ − U. But the enthalpy of formation has been determined calorimetrically and equals −411 kJ. Equating these two values, we get

$$381 \text{ kJ} - U = -411 \text{ kJ}$$

Solving for U gives the lattice energy of NaCl:

$$U = (381 + 411) \text{ kJ} = 792 \text{ kJ}$$

It is instructive to compare the lattice energy of NaCl (792 kJ) with that of MgO (determined to be 3934 kJ). Clearly, MgO has much stronger bonding than NaCl. Using the ionic theory of bonding, one can calculate values for lattice energies that are close to those obtained by the Born-Haber cycle from experimental data. This gives strength to the ionic theory of bonding. Without doing a detailed calculation, we can see that the lattice energies of NaCl and MgO are approximately as we would expect from the theory. The energy released in bringing a cation and an anion together is proportional to the product of the ion charges divided by the distance between the centers of the ions. For NaCl, the product of ion charges is $(+1) \times (-1) = -1$, whereas for MgO it is $(+2) \times (-2) = -4$. If the distance between ions in these solids were about the same, we would expect their lattice energies to be proportional to the negative of these products of charges. That is, we would expect the lattice energy of MgO to be four times that of NaCl. The distance between ions is not the same, of course. The distance is actually smaller in MgO, so that we expect its lattice energy to be more than four times greater than in NaCl, as is the case.

A Checklist for Review

Important Terms

system (10.1)
surroundings (10.1)
heat (10.1)
thermal equilibrium (10.1)
heat of reaction (10.1)

exothermic (10.1)
endothermic (10.1)
enthalpy (10.1)
enthalpy of reaction (10.1)
calorimeter (10.2)

heat capacity (10.2)
Hess's law of heat summation (10.4)
standard enthalpies of formation (10.5)
bond energy (10.6)
lattice energy (10.7)

Summary of Facts and Concepts

Reactions absorb or evolve definite quantities of heat under given conditions. At constant pressure, this heat of reaction is the *enthalpy of reaction*, ΔH. One measures the heat of reaction in a *calorimeter*. Direct calorimetric determination of the heat of reaction requires a reaction that goes to completion without other reactions occurring at the same time. Otherwise, the heat or enthalpy of reaction is determined indirectly from other enthalpies of reaction by using *Hess's law of heat summation*. Thermochemical data are conveniently stored as *enthalpies*

of formation. By knowing values for each reactant and product in an equation, we can easily compute the enthalpy of reaction.

The A—B *bond energy* is the average enthalpy change when an A—B bond is broken. We can use values of the bond energy to estimate enthalpy changes for gaseous reactions. The Born-Haber cycle is an application of Hess's law to ionic solids. It allows us to obtain the *lattice energy* in terms of experimentally determined quantities.

Operational Skills

1. Given a chemical equation, states of substances, and the quantity of heat absorbed or evolved, write the complete equation, including labels for states and ΔH (Example 10.1).

2. A given amount of heated substance is added to a given amount of water. From the temperature changes of the substance and the water, calculate the specific heat of the substance (Example 10.2).

3. Given the amounts of reactants and the temperature change of a calorimeter of specified heat capacity, calculate the heat of reaction (Example 10.3).

4. Given the value of ΔH for a chemical equation, calculate the heat of reaction for a given mass of a reactant or product (Example 10.4). Atomic weights are needed.

5. Given a chemical equation including ΔH, write the equation for different multiples of the coefficients or write the reverse equation (Example 10.5).

6. Given a set of reactions with enthalpy changes, calculate ΔH for a specified reaction using Hess's law (Example 10.6).

7. Given a table of standard enthalpies of formation, calculate the enthalpy change for a specified reaction (Examples 10.7, 10.8, and 10.9).

8. Given a table of bond energies, estimate the enthalpy change for a specified reaction (Example 10.10).

Review Questions

10.1 Define an *exothermic* and an *endothermic* reaction. Give an example of each.

10.2 How does the enthalpy change for an endothermic reaction occurring at constant pressure?

10.3 Why is it important to give the states of the reactants and products when giving an equation for ΔH?

10.4 Define the heat capacity of a substance.

10.5 Describe a simple calorimeter. What are the measurements needed to determine the heat of reaction?

10.6 If an equation for a reaction is doubled and then reversed, how is the value of ΔH changed?

10.7 What is the essential idea behind Hess's law of heat summation?

10.8 What is the standard enthalpy of formation of a substance?

10.9 Write the chemical equation for the formation reaction of $H_2S(g)$.

10.10 Is the following reaction the appropriate one to use in determining the enthalpy of formation of methane, $CH_4(g)$? Why or why not?

$$C(g) + 4H(g) \longrightarrow CH_4(g)$$

10.11 Define bond energy. Explain in words how you can use bond energies to estimate the enthalpy change for a gaseous reaction.

10.12 Define lattice energy. Why do you expect the lattice energy to decrease in the following series of ionic compounds?

$$LiF, NaF, KF, RbF, CsF$$

Problems

Heat of Reaction

10.13 The gaseous reaction

$$CH_4 + N_2 \longrightarrow HCN + NH_3$$

absorbs 164 kJ of heat per mole of ammonia produced. Is the reaction endothermic or exothermic? What is the value of q?

10.15 When one mole of mercury(II) oxide crystals, HgO, decomposes into its elements at constant temperature and pressure, 90.8 kJ of heat are absorbed. Write an equation for this reaction, including the value of ΔH and labels for the states of reactants and products.

10.14 The gaseous reaction

$$2NO + O_2 \longrightarrow 2NO_2$$

evolves 57 kJ of heat per mole of nitric oxide, NO, consumed. Is this an endothermic or an exothermic reaction? Give the value of q.

10.16 When one mole of zinc metal reacts with hydrogen ion, H^+, in aqueous solution, it gives zinc ion, $Zn^{2+}(aq)$, and evolves hydrogen. At constant temperature and pressure, 152 kJ of heat are given off. Write a complete equation for the reaction, including labels for states and ΔH.

Calorimetry

10.17 A piece of copper with a mass of 21.2 g has a heat capacity of 8.16 J/°C. What is the specific heat of copper?

10.19 How much heat is required to raise the temperature of 15.0 g of iron from 10.2°C to 99.8°C? The specific heat of iron is 0.447 J/(g · °C).

10.21 What is the specific heat of carbon tetrachloride, CCl_4, if it requires 77.1 J of heat to raise the temperature of a 6.43-g sample from 21.1°C to 35°C? What is the molar heat capacity of carbon tetrachloride?

10.23 A 50.0-g sample of water at 100.00°C was placed in an insulated cup. Then 25.3 g of zinc metal at 25.00°C were added to the water. The temperature of the water dropped to 96.68°C. What is the specific heat of zinc?

10.25 When 15.3 g of sodium nitrate, $NaNO_3$, were dissolved in water in a calorimeter, the temperature fell from 25.00°C to 21.56°C. If the heat capacity of the solution and the calorimeter is 1071 J/°C, what is the enthalpy change when one mole of sodium nitrate dissolves in water? The reaction is

$$NaNO_3(s) \longrightarrow Na^+(aq) + NO_3^-(aq); \Delta H = ?$$

10.18 A metal was known to be either chromium, molybdenum, or tungsten. If a 5.42-g sample of metal had a heat capacity of 0.735 J/(g · °C), what is the metal? The specific heats of chromium, molybdenum, and tungsten are 0.449 J/(g · °C), 0.244 J/(g · °C), and 0.136 J/(g · °C), respectively.

10.20 An 80.0-g sample of nickel was heated from 25°C to 295°C. What is the heat that was required for this? The specific heat of nickel is 0.442 J/(g · °C).

10.22 A 4.35-g sample of a metal is heated from 25.0°C to 75.0°C. If this requires 53.7 J of heat, what is the specific heat of the metal? According to the law of Dulong and Petit, the molar heat capacity of a solid element is approximately 25 J/(mol · °C). Assuming the metal is a pure element, what is its approximate atomic weight?

10.24 A 19.6-g sample of a metal was heated to 61.67°C. When the metal was placed into 26.7 g of water in a calorimeter, the temperature of the water increased from 25.00°C to 30.00°C. The metal is known to be either titanium or aluminum. Which is it? The specific heat of titanium is 0.515 J/(g · °C), and that of aluminum is 0.903 J/(g · °C).

10.26 When 0.786 g of sulfur were burned in a calorimeter, the temperature rose from 25.014°C to 25.683°C. The reaction is

$$\tfrac{1}{8}S_8(s) + O_2(g) \longrightarrow SO_2(g)$$

What is the enthalpy change per mole of sulfur dioxide? The heat capacity of the calorimeter and contents is 10.87 kJ/°C.

Stoichiometry of Reaction Heats

10.27 Sulfur burns in oxygen to give sulfur dioxide:

$$S_8(s) + 8O_2(g) \longrightarrow 8SO_2(g); \Delta H = -2374 \text{ kJ}$$

Calculate the heat evolved per gram of sulfur burned.

10.29 A sample of calcium carbonate, $CaCO_3$, decomposes to give 12.8 g of calcium oxide, CaO. What is the enthalpy change to form this amount of CaO? The reaction is

$$CaCO_3(s) \longrightarrow CaO(s) + CO_2(g); \Delta H = 178 \text{ kJ}$$

10.28 Nitroglycerine, $C_3H_5(NO_3)_3$, is a common explosive, which decomposes according to the equation

$$4C_3H_5(NO_3)_3(l) \longrightarrow 12CO_2(g) + 10H_2O(g) + O_2(g)$$
$$+ 6N_2(g); \Delta H = -5.720 \times 10^3 \text{ kJ}$$

Calculate the heat for the decomposition of 10.6 g of nitroglycerine.

10.30 Iodine monochloride, ICl, can be prepared by direct interaction of the elements:

$$I_2(s) + Cl_2(g) \longrightarrow 2ICl(g); \Delta H = 36 \text{ kJ}$$

Calculate the amount of heat absorbed per gram of iodine monochloride formed at constant pressure.

10.31 Calcium carbide, CaC_2, reacts with water to give acetylene, C_2H_2, and calcium hydroxide, $Ca(OH)_2$:

$$CaC_2(s) + 2H_2O(l) \longrightarrow C_2H_2(g) + Ca(OH)_2(s);$$
$$\Delta H = -411 \text{ kJ}$$

What is ΔH for the following equation?

$$C_2H_2(g) + Ca(OH)_2(s) \longrightarrow CaC_2(s) + 2H_2O(l)$$

10.33 Phosphoric acid, H_3PO_4, can be prepared by the reaction of phosphorus(V) oxide, P_4O_{10}, with water:

$$\tfrac{1}{4}P_4O_{10}(s) + \tfrac{3}{2}H_2O(l) \longrightarrow H_3PO_4(aq); \ \Delta H = -107.6 \text{ kJ}$$

What is ΔH for the reaction involving one mole of P_4O_{10}?

$$P_4O_{10}(s) + 6H_2O(l) \longrightarrow 4H_3PO_4(aq)$$

10.32 A laboratory preparation of oxygen consists in decomposing potassium chlorate, $KClO_3$, using a catalyst of manganese dioxide, MnO_2:

$$2KClO_3(s) \longrightarrow 2KCl(s) + 3O_2(g); \ \Delta H = -44.7 \text{ kJ}$$

What is ΔH for the following reaction?

$$KCl(s) + \tfrac{3}{2}O_2(g) \longrightarrow KClO_3(s)$$

10.34 With a platinum catalyst, ammonia will burn in oxygen to give nitric oxide, NO:

$$4NH_3(g) + 5O_2(g) \longrightarrow 4NO(g) + 6H_2O(g);$$
$$\Delta H = -905 \text{ kJ}$$

What is the enthalpy change for the following reaction?

$$NO(g) + \tfrac{3}{2}H_2O(g) \longrightarrow NH_3(g) + \tfrac{5}{4}O_2(g)$$

Hess's Law

10.35 The reaction

$$H_2O(g) \longrightarrow 2H(g) + O(g)$$

corresponds to the breaking of two O—H bonds. Obtain the enthalpy change, ΔH, for this reaction from the following data:

$$H_2(g) \longrightarrow 2H(g); \ \Delta H = 218.0 \text{ kJ}$$
$$O_2(g) \longrightarrow 2O(g); \ \Delta H = 498.4 \text{ kJ}$$
$$H_2(g) + \tfrac{1}{2}O_2(g) \longrightarrow H_2O(g); \ \Delta H = -241.8 \text{ kJ}$$

10.37 Compounds with carbon–carbon double bonds, such as ethylene, C_2H_4, add hydrogen in a reaction called hydrogenation. Thus,

$$C_2H_4(g) + H_2(g) \longrightarrow C_2H_6(g)$$

Calculate the enthalpy change for this reaction, using the following combustion data:

$$C_2H_4(g) + 3O_2(g) \longrightarrow 2CO_2(g) + 2H_2O(l);$$
$$\Delta H = -1401 \text{ kJ}$$
$$C_2H_6(g) + \tfrac{7}{2}O_2(g) \longrightarrow 2CO_2(g) + 3H_2O(l);$$
$$\Delta H = -1550 \text{ kJ}$$
$$H_2(g) + \tfrac{1}{2}O_2(g) \longrightarrow H_2O(l); \ \Delta H = -286 \text{ kJ}$$

10.36 Find the enthalpy change for the reaction in which the carbon–oxygen bonds in carbon dioxide are broken:

$$CO_2(g) \longrightarrow C(g) + 2O(g)$$

Use the following data:

$$C(graphite) \longrightarrow C(g); \ \Delta H = 715.0 \text{ kJ}$$
$$O_2(g) \longrightarrow 2O(g); \ \Delta H = 249.2 \text{ kJ}$$
$$C(graphite) + O_2(g) \longrightarrow CO_2(g); \ \Delta H = -393.5 \text{ kJ}$$

10.38 Acetic acid, CH_3COOH, is contained in vinegar. Suppose acetic acid was formed from its elements,

$$2C(graphite) + 2H_2(g) + O_2(g) \longrightarrow CH_3COOH(l)$$

Find the enthalpy change, ΔH, for this reaction, using the following data:

$$CH_3COOH(l) + 2O_2(g) \longrightarrow 2CO_2(g) + 2H_2O(l);$$
$$\Delta H = -871 \text{ kJ}$$
$$C(graphite) + O_2(g) \longrightarrow CO_2(g); \ \Delta H = -394 \text{ kJ}$$
$$H_2(g) + \tfrac{1}{2}O_2(g) \longrightarrow H_2O(l); \ \Delta H = -286 \text{ kJ}$$

Enthalpies of Formation

10.39 Carbon tetrachloride, CCl_4, is a liquid used as an industrial solvent and in the preparation of fluorocarbons (Freons). What is the heat of vaporization of carbon tetrachloride?

$$CCl_4(l) \longrightarrow CCl_4(g); \ \Delta H = ?$$

Use enthalpies of formation (Table 10.1).

10.40 The cooling effect of alcohol on the skin is due to its evaporation. Calculate the heat of vaporization of ethanol (ethyl alcohol), C_2H_5OH:

$$C_2H_5OH(l) \longrightarrow C_2H_5OH(g); \ \Delta H = ?$$

The enthalpy of formation of $C_2H_5OH(l)$ is -277.6 kJ/mol and that of $C_2H_5OH(g)$ is -235.4 kJ/mol.

10.41 Use enthalpies of formation (Table 10.1) to calculate the enthalpy change, ΔH, involved in breaking the hydrogen–sulfur bonds in hydrogen sulfide, H_2S:

$$H_2S(g) \longrightarrow 2H(g) + S(g); \Delta H = ?$$

10.43 The first step in the preparation of lead from its ore (galena, PbS) consists of roasting of the ore:

$$2PbS(s) + 3O_2(g) \longrightarrow 2SO_2(g) + 2PbO(s)$$

Calculate the enthalpy change for this reaction using enthalpies of formation (see Appendix C).

10.45 Hydrogen chloride gas dissolves in water to form hydrochloric acid (an ionic solution):

$$HCl(g) \longrightarrow H^+(aq) + Cl^-(aq)$$

Find ΔH for this reaction. Data are given in Table 10.1.

10.42 Ozone, O_3, is produced in the upper atmosphere by the reaction of molecular oxygen with oxygen atoms:

$$O_2(g) + O(g) \longrightarrow O_3(g)$$

Oxygen atoms are formed when ultraviolet radiation interacts with molecular oxygen. Use enthalpies of formation to calculate the enthalpy change for the production of ozone by the above reaction (Table 10.1).

10.44 Iron is obtained from iron ore by reduction with carbon monoxide. The overall reaction can be described by the equation

$$Fe_2O_3(s) + 3CO(g) \longrightarrow 2Fe(s) + 3CO_2(g)$$

Calculate the enthalpy change for this equation using enthalpies of formation. The ΔH_f° for $Fe_2O_3(s)$ is -825.5 kJ/mol; other data are given in Table 10.1.

10.46 Carbon dioxide from the atmosphere "weathers," or dissolves, limestone ($CaCO_3$) by the reaction

$$CaCO_3(s) + CO_2(g) + H_2O(l) \longrightarrow$$
$$Ca^{2+}(aq) + 2HCO_3^-(aq)$$

Obtain ΔH for this reaction. Data are given in Table 10.1.

Bond Energy

10.47 Use bond energies (Table 10.2) to estimate ΔH for the following gas-phase reaction:

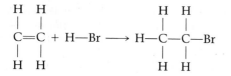

This is called an "addition" reaction, because a compound (HBr) is added across the double bond.

10.48 A commercial process for preparing ethanol (ethyl alcohol), C_2H_5OH, consists in passing ethylene gas, C_2H_4, and steam over an acid catalyst (to speed up the reaction). The gas-phase reaction is

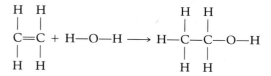

Estimate ΔH for this reaction, using bond energies (Table 10.2).

10.49 Using bond-energy values (Table 10.2), calculate the electronegativity difference between hydrogen and chlorine. Assume that the electronegativity of hydrogen is 2.1, and calculate the electronegativity of chlorine.

10.50 Because known compounds with N—I bonds tend to be unstable, there are no thermodynamic data available to calculate the N—I bond energy. However, we can estimate a value from Pauling's formula relating electronegativity to bond energies. Obtain electronegativities from Figure 7.5 and necessary bond energies from Table 10.2, then calculate the N—I bond energy.

Additional Problems

**10.51 A lead bullet is traveling at 241 m/s. Suppose the bullet is stopped by a wall, where the kinetic energy of the bullet is transformed into heat. If all of the heat goes to raise the temperature of the lead, which initially was at 25°C, what is the final temperature of the lead? The specific heat of lead is 0.1289 J/(g·°C). Kinetic energy is $\frac{1}{2}mv^2$, where m is the mass and v is the speed.

**10.52 Horseshoe Falls (the Canadian side of Niagara Falls) are 49 m high. Assume that all of the potential energy of the water is converted to heat, and that this heat raises the temperature of the water. What is the difference in temperature of the water between the top and bottom of the falls? The potential energy of a mass m, raised to a height h, is mgh, where g (acceleration of gravity) equals 9.81 m/s.

**10.53 For the reaction

$$N_2(g) + 3H_2(g) \longrightarrow 2NH_3(g); \Delta H = -45.9 \text{ kJ}$$

calculate the pressure-volume work (in joules) against a pressure of 1.00 atm when 1 mol N_2 reacts with 3 mol H_2 at 298 K. What is ΔE for the reaction?

**10.54 For the reaction

$$2H_2(g) + O_2(g) \longrightarrow 2H_2O(l); \Delta H = -571.7 \text{ kJ}$$

calculate the pressure-volume work (in joules) against a pressure of 1.00 atm when 2 mol H_2 react with 1 mol O_2 at 298 K. What is ΔE for the reaction?

10.55 In a calorimetric experiment, 6.48 g of lithium hydroxide, LiOH, were dissolved in water. The temperature of the calorimeter rose from 25.00°C to 36.66°C. What is ΔH for the solution process?

$$LiOH(s) \longrightarrow Li^+(aq) + OH^-(aq)$$

The heat capacity of the calorimeter and its contents is 547 J/°C.

10.56 When 21.45 g of potassium nitrate, KNO_3, were dissolved in water in a calorimeter, the temperature fell from 25.00°C to 14.14°C. What is the ΔH for the solution process?

$$KNO_3(s) \longrightarrow K^+(aq) + NO_3^-(aq)$$

The heat capacity of the calorimeter and its contents is 682 J/°C.

10.57 A 10.00-g sample of acetic acid, CH_3COOH, was burned in a bomb calorimeter in an excess of oxygen:

$$CH_3COOH(l) + 2O_2(g) \longrightarrow 2CO_2(g) + 2H_2O(l)$$

The temperature of the calorimeter rose from 25.00°C to 35.81°C. If the heat capacity of the calorimeter and its contents is 10.81 kJ/°C, what is the enthalpy change for the reaction?

10.58 The sugar arabinose, $C_5H_{10}O_5$, is burned completely in oxygen in a calorimeter:

$$C_5H_{10}O_5(s) + 5O_2(g) \longrightarrow 5CO_2(g) + 5H_2O(l)$$

Burning a 0.548-g sample caused the temperature to rise from 20.00°C to 20.54°C. The heat capacity of the calorimeter and its contents is 15.8 kJ/°C. Calculate ΔH for the combustion reaction per mole of arabinose.

10.59 Hydrogen sulfide, H_2S, is a very poisonous gas with the odor of rotten eggs. The enthalpy change for the reaction in which hydrogen sulfide is formed from the elements cannot be determined directly. However, the enthalpy changes for the combustion of H_2S, H_2, and S_8 can be easily determined:

$$H_2S(g) + \tfrac{3}{2}O_2(g) \longrightarrow H_2O(g) + SO_2(g); \Delta H = -519 \text{ kJ}$$
$$H_2(g) + \tfrac{1}{2}O_2(g) \longrightarrow H_2O(g); \Delta H = -242 \text{ kJ}$$
$$\tfrac{1}{8}S_8(s) + O_2(g) \longrightarrow SO_2(g); \Delta H = -297 \text{ kJ}$$

Obtain the enthalpy of formation of $H_2S(g)$ from these data.

10.60 Sucrose, $C_{12}H_{22}O_{11}$, is common table sugar. The enthalpy of combustion (that is, the ΔH for the complete burning in O_2) of sucrose is 5641 kJ per mole of sucrose. The products of combustion are $CO_2(g)$ and $H_2O(l)$. From this and from other data in Table 10.1, calculate the enthalpy of formation of sucrose.

10.61 Hydrogen, H_2, can be prepared by the *steam-reforming process*, in which hydrocarbons are reacted with steam in the presence of a catalyst. For CH_4

$$CH_4(g) + H_2O(g) \longrightarrow CO(g) + 3H_2(g)$$

Calculate the enthalpy change, ΔH, for this reaction, using enthalpies of formation.

10.62 Hydrogen is prepared from natural gas (mainly methane, CH_4) by partial oxidation:

$$2CH_4(g) + O_2(g) \longrightarrow 2CO(g) + 4H_2(g)$$

Calculate the enthalpy change, ΔH, for this reaction, using enthalpies of formation.

10.63 Calcium oxide, CaO, is prepared by heating calcium carbonate (from limestone and seashells):

$$CaCO_3(s) \longrightarrow CaO(s) + CO_2(g)$$

Calculate the enthalpy of reaction, ΔH, using enthalpies of formation. The enthalpy of formation of CaO(s) is −635 kJ/mol; other values are given in Table 10.1.

10.64 Sodium carbonate, Na_2CO_3, is used to manufacture glass. It is obtained from sodium hydrogen carbonate, $NaHCO_3$, by heating:

$$2NaHCO_3(s) \longrightarrow Na_2CO_3(s) + H_2O(g) + CO_2(g)$$

Calculate the enthalpy of reaction, using enthalpies of formation (Table 10.1).

*10.65 Obtain the average C—Cl bond energy from carbon tetrachloride, CCl_4, by the same method as used in the text to obtain the C—H bond energy from methane, CH_4. Use data from Table 10.1.

*10.66 Using data from Table 10.1, find a value for the average N—H bond energy, from ammonia, NH_3.

*10.67 Calculate the enthalpy of reaction for

$$HCN(g) \longrightarrow H(g) + C(g) + N(g)$$

from enthalpies of formation. Given that the C—H bond energy is 411 kJ, obtain a value for the C≡N bond energy. Compare with the value given in Table 10.2.

*10.68 Assume values for the C—H and C—C bond energies given in Table 10.2. Then, using data given in Table 10.1, calculate the C=O bond energy in acetaldehyde,

Compare with the value given in Table 10.2.

*10.69 Calculate the electronegativity difference for the N—H bond, using bond energies (Table 10.2).

*10.70 Calculate the electronegativity difference for the C—O bond, using bond energies (Table 10.2).

**10.71 Electron affinities were first obtained from the Born-Haber cycle. Calculate the electron affinity of the fluorine atom from the following data:

$$Na(s) + \tfrac{1}{2}F_2(g) \longrightarrow NaF(s); \ \Delta H_f = -575.4 \text{ kJ}$$
$$Na(s) \longrightarrow Na(g); \ \Delta H_{sub} = 107.8 \text{ kJ}$$
$$Na(g) \longrightarrow Na^+(g) + e^-(g); \ \Delta H_I = 502.1 \text{ kJ}$$
$$NaF(s) \longrightarrow Na^+(g) + F^-(g); \ \Delta H_U = 915.0 \text{ kJ}$$
$$F_2(g) \longrightarrow 2F(g); \ \Delta H_D = 157.8 \text{ kJ}$$

**10.72 The electron affinity of the oxygen atom to give O^{2-} cannot be determined directly because the isolated O^{2-} ion is unstable. The value can be obtained from the Born-Haber cycle, however. Calculate the electron affinity of the oxygen atom to give O^{2-} (that is, obtain ΔH for the reaction $O(g) + 2e^-(g) \longrightarrow O^{2-}(g)$). Use the following data:

$$Ca(s) + \tfrac{1}{2}O_2(g) \longrightarrow CaO(s); \ \Delta H_f = -635.1 \text{ kJ}$$
$$Ca(s) \longrightarrow Ca(g); \ \Delta H_{sub} = 192.6 \text{ kJ}$$
$$Ca(g) \longrightarrow Ca^{2+}(g) + 2e^-(g); \ \Delta H_I = 1741.5 \text{ kJ}$$
$$CaO(s) \longrightarrow Ca^{2+}(g) + O^{2-}(g); \ \Delta H_U = 3524 \text{ kJ}$$
$$O_2(g) \longrightarrow 2O(g); \ \Delta H_D = 498.4 \text{ kJ}$$

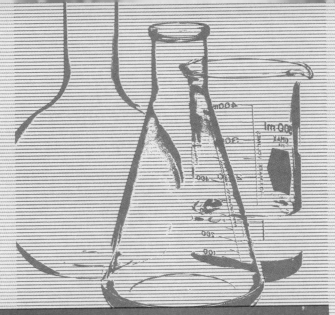

11. Liquids and Solids; Changes of State

Liquids and Solids

11.1 Intermolecular Forces London Forces/ Dipole–Dipole Forces/ Hydrogen Bonding

11.2 Properties of Liquids Vapor Pressure/ Surface Tension/ Viscosity

11.3 Types of Solids

11.4 Crystalline and Amorphous Solids Characteristics of Crystalline and Amorphous Solids/ The Crystalline Lattice/ Cubic Unit Cells/ Crystal Defects

11.5 The Structure of Some Crystalline Solids Molecular Solids; Closest Packing/ Metallic Solids/ Ionic Solids/ Covalent Network Solids

11.6 X-ray Crystal Structure Determination

Changes of State

11.7 Phase Changes

11.8 Boiling Point and Melting Point Boiling Point/ Melting Point/ Enthalpy Changes

11.9 Relating Physical Properties to Structure Melting Point and Structure/ Boiling Point and Structure/ Hardness and Structure/ Electrical Conductivity and Structure

11.10 Phase Diagrams Melting-Point Curve/ Vapor-Pressure Curve for the Liquid/ Vapor-Pressure Curve for the Solid/ Critical Temperature and Pressure

Dry Ice is solid carbon dioxide. It has an interesting property: at normal pressures it passes directly to the gaseous state without first melting to the liquid. This property, together with the fact that the vaporizing solid is at $-78°C$, makes solid carbon dioxide an excellent refrigerant. It cools other objects, but doesn't leave a liquid residue as ordinary ice does.

Liquid carbon dioxide exists, but only at pressures greater than 5 atm. In fact, carbon dioxide is usually transported in the liquid state either at room temperature in steel tanks at high pressure or at low temperature (about $-18°C$) in insulated vessels at moderate pressure. It too is an excellent refrigerant. When liquid carbon dioxide evaporates (changes to vapor), it absorbs large quantities of heat, cooling to as low as $-57°C$. A major use of liquid carbon dioxide is to freeze foods for grocery stores and fast-food restaurants.

Liquid carbon dioxide is obtained by compressing the gas to high pressure. If the compressed gas from the evaporating liquid is allowed to expand through a valve, carbon dioxide "snow" is obtained. (Most gases cool when they expand in this manner; for example, air cools when it escapes from the valve of an inflated tire.) The carbon dioxide snow is compacted into blocks and sold as Dry Ice.

Thus we see that carbon dioxide, like water and most other pure substances, exists in solid, liquid, and gaseous states and can undergo changes from one state to another. Substances change state under a variety of conditions of temperature and pressure. Can we obtain some understanding of these conditions for a change of state? Why, for example, is carbon dioxide normally a gas, whereas water is normally a liquid? What are the conditions under which carbon dioxide gas changes to a liquid or to a solid?

These are the kinds of questions that we will examine in this chapter. They concern certain physical properties that, as we will see, often can be related to the bonding and structure of the liquid and solid states.

Chapter Overview

We briefly described the three states of matter earlier in the book (Section 2.3). There we noted that gases are compressible, but solids and liquids are relatively incompressible. This relative incompressibility results from the fact that the structural units (ions, atoms, or molecules) are held closely together by attractive forces.

Now we will look in detail at *liquids and solids*. First, we will explore the relationship between intermolecular forces, the weak forces between molecules, and certain properties of liquids. Then we will examine the structure of solids. In the last part of the chapter, we will look at *changes of state* and discuss the conditions for such changes.

Liquids and Solids

We discussed the physical properties of gases in Chapter 4. There we found that the compressibility of a gas is accounted for by the large amount of empty space between molecules. Because gas molecules are relatively far apart, the forces between one molecule and another, called *intermolecular forces*, are quite weak.

In molecular liquids and solids, by contrast, molecules are close together, and intermolecular forces can be appreciable. Since these forces determine many of the properties of such liquids and solids, let us begin by discussing the nature of intermolecular forces.

11.1 Intermolecular Forces

Intermolecular forces, the attractive forces between molecules, are much weaker than the chemical bonding forces that exist between atoms in a molecule.■ We can classify intermolecular forces into two types: *van der Waals forces* and *hydrogen bonding*. Table 11.1 lists the types of intermolecular forces and compares their energies with those of chemical bonds.

■ Chemical bonding was discussed in Chapters 7 and 8.

London Forces

Van der Waals forces are one kind of attractive force between molecules. They arise from *London forces*, which exist between any two molecules, and from *dipole–dipole forces*, which are present only between polar molecules.■ For example, consider a simple substance like neon, which consists of molecules that are single atoms. Even though a chemical bond does not form between two neon atoms, a weak attractive force exists, and it becomes stronger as the two atoms are brought near each other. When two neon atoms are about 3 Å apart, this intermolecular force reaches a maximum. To separate the atoms requires an energy of about 0.3 kJ/mol. In order to appreciate how small this amount is, recall that the H—H bond energy is about 400 kJ/mol. That is, it requires about 400 kJ/mol to separate the hydrogen atoms in H_2. Thus, the attraction between neon atoms is about a thousand times smaller than the attraction between atoms in a chemical bond. But although it is small, it is responsible for the fact that at low temperature neon gas liquefies (changes to liquid). This attractive force holds the atoms together to form the liquid.

■ The Dutch physicist Johannes van der Waals (1837–1923) was the first to suggest the importance of intermolecular forces and used the concept to derive his equation for gases (see van der Waals equation, Section 4.10).

What is the nature of this attractive force that exists between two neon atoms, or between any two molecules? In 1930, Fritz London found that he could account for such a weak attraction in terms of small, *instantaneous* dipoles that occur as a result of the varying positions of the electrons during their motion about their nuclei. Because the theory was first developed by London, these attractive forces are called **London forces** (they are also known as *dispersion forces*).

To illustrate the origin of London forces, consider first a single neon atom. The electrons of this atom move about the nucleus so that over a very short period of time they have a spherical distribution. The electrons spend as

Table 11.1

Types of Intermolecular and Chemical Bonding Forces

Type of Force	Approximate Energy (kJ/mol)
Intermolecular	
van der Waals (London, dipole–dipole)	0.1 to 10
Hydrogen bonding	10 to 40
Chemical Bonding	
Ionic	100 to 1000
Covalent	100 to 1000

much time on one side of the nucleus as on the other. Experimentally, we find that the atom has no dipole moment (a measure of separation of charges).■ But at any instant, it can happen that more electrons are on one side of the nucleus. The atom becomes a small, instantaneous dipole, with one side having a partial negative charge, δ^-, and the other side having a partial positive charge, δ^+ (see Figure 11.1a).

■ Dipole moments were discussed in Section 8.2.

Now imagine another neon atom to be close to this first atom, say, next to its partial negative charge that has appeared at that instant. This partial negative charge repels the electrons in the second atom, so that at this instant it too becomes a small dipole (see Figure 11.1b). Note that these two dipoles are oriented with the partial negative charge of one atom next to the partial positive charge of the other atom. Therefore, an attractive force exists between the two atoms.

The electrons will be in constant motion, but the motion of electrons on one atom will affect the motion of electrons on the other atom. As a result, the *instantaneous dipoles* of the atoms will change together so as to give a net attractive force (compare b and c of Figure 11.1). Such instantaneous changes of electron distributions occur in all molecules. For this reason, London forces exist between any two molecules and are always attractive.

London forces tend to increase with molecular weight for the following reasons. First, molecules with larger molecular weight usually have more electrons, and London forces increase in strength with the number of electrons. Second, larger molecular weight often means larger atoms, which are more easily distorted to give instantaneous dipoles. The relationship of London forces to molecular weight is only approximate, however.

Figure 11.1
Origin of the London force. (a) At some instant, there are more electrons on one side of a neon atom. (b) If this atom is near another neon atom, the electrons on that atom are repelled. The result is two instantaneous dipoles that are oriented δ^- to δ^+, which gives an attractive force. (c) Later, the electrons on both atoms have moved, but the motions change together, so that there is always an attractive force between the atoms.

Dipole–Dipole Forces

Substances composed of *polar* molecules also have attractive dipole–dipole forces between molecules in addition to the London forces. The **dipole–dipole force** results from the tendency of polar molecules to align themselves so that the positive end of one molecule is near the negative end of another. Recall that a polar molecule has a dipole moment as a result of the electron structure of the molecule. Hydrogen chloride, HCl, is a polar molecule because of the difference in electronegativities of the H and Cl atoms:

$$\overset{\delta^+}{H}—\overset{\delta^-}{Cl}$$

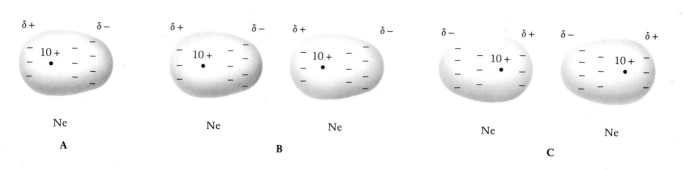

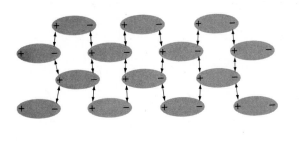

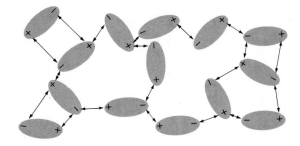

A

B

Figure 11.2
Alignment of polar molecules of HCl. (a) Molecules tend to line up in the solid so that positive ends point to negative ends. (b) The normal random motion of molecules in a liquid only partially disrupts this alignment of polar molecules.

A polar molecule has a *permanent* dipole (which we experimentally measure as its dipole moment) in addition to a small *instantaneous* dipole arising from the motion of the electrons.

Figure 11.2 shows the alignment of polar molecules in the case of HCl. Hydrogen chloride molecules are attracted to one another by both London forces and dipole–dipole forces. These forces account for the fact that this normally gaseous substance liquefies if cooled to $-85°C$. At this temperature, the molecules have slowed enough for the intermolecular forces to hold the molecules together as the liquid.

Hydrogen Bonding

Hydrogen bonding is a weak to moderate attractive force that exists between certain covalently bonded hydrogen atoms and the lone pairs of electrons of another atom. Hydrogen bonding occurs when the hydrogen atom is covalently bonded to a highly electronegative atom, particularly fluorine, oxygen, and nitrogen. Water provides a common example of this kind of intermolecular force (Figure 11.3a). The electrons of a hydrogen–oxygen bond are attracted to the electronegative atom (oxygen), leaving the positive charge of the proton (the hydrogen nucleus) partially exposed. As a result, this proton is strongly attracted to the lone pairs of other atoms. In the case of water, the proton of one water molecule is attracted to the lone pairs on the oxygen atom of another water molecule. A hydrogen bond is usually represented by a series of dots, for example, O···H (Figure 11.3b).

Although this type of interaction is called hydrogen bonding, the strength of the interaction is usually much less than that of normal chemical bonding. The energies of most hydrogen bonds lie in the range of 10 to 40 kJ/mol. Thus, the hydrogen-bonding interaction is stronger than van der Waals forces, but less than one-tenth that of covalent bonds. It is usual to think of the hydrogen-bonding interaction as a strong intermolecular force, rather than as another form of chemical bonding.■

Formic acid, HCOOH, provides another example of hydrogen bonding. Two molecules are strongly attracted to one another by two hydrogen bonds

■ Hydrogen bonding is crucial in determining the structure and function of biologically important molecules, such as DNA. We will discuss this in Chapter 27.

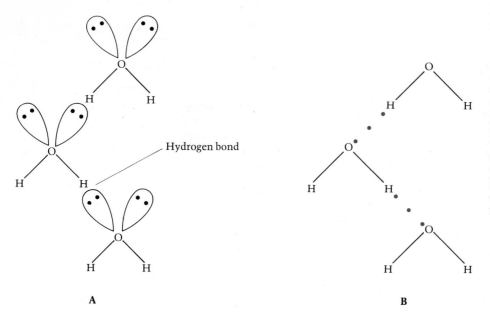

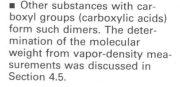

Figure 11.3
Hydrogen bonding in water.
(a) The electrons in the
O—H bond of H_2O mole-
cules are attracted to the oxygen atoms, leaving the positively charged protons
partially exposed. A proton
on one molecule will be attracted to a lone pair on an
oxygen atom in another water molecule. (b) Hydrogen
bonding between water molecules is represented by a series of dots.

A   **B**

of the type O—H$\cdots$O formed between the carboxyl groups (—COOH) of the
formic acid molecules:

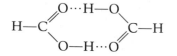

These molecule pairs are stable enough to be present in the vapor at the
boiling point of the liquid (101°C). This is shown by vapor-density measurements, which give a molecular weight between that of HCOOH and
$(HCOOH)_2$. Such stable molecule pairs as $(HCOOH)_2$ are called *dimers.* ■

Ammonia, NH_3, provides an example of an N—H$\cdots$N type of hydrogen
bond. This hydrogen bond is only about half as strong as the O—H$\cdots$O bond
in water, partly because of the smaller electronegativity of nitrogen compared with that of oxygen.

■ Other substances with carboxyl groups (carboxylic acids)
form such dimers. The determination of the molecular
weight from vapor-density measurements was discussed in
Section 4.5.

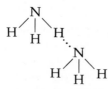

As you might expect from the large electronegativity of fluorine, there is
strong hydrogen bonding in hydrogen fluoride, HF. Hydrogen bonding in this
case appears to be directional (as in covalent bonding), giving the zigzag
arrangement:

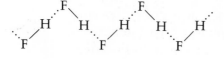

Example 11.1

What kinds of intermolecular forces (London, dipole–dipole, hydrogen bonding) are expected in the following substances? (a) methane, CH_4 (b) trichloromethane (chloroform), $CHCl_3$ (c) methanol (methyl alcohol), CH_3OH

Solution

(a) Methane is a nonpolar molecule. Hence, the only intermolecular attractions are London forces. (b) Tri-chloromethane is an unsymmetrical molecule with polar bonds. Thus, we expect dipole–dipole forces, in addition to London forces. (c) Methanol has a hydrogen atom attached to an oxygen atom. Therefore, we expect hydrogen bonding. Since the molecule is polar (from the O—H bond), we also expect dipole–dipole forces. London forces exist too, because such forces exist between all molecules.

Exercise 11.1

List the different intermolecular forces you would expect for each of the following compounds: (a) methylamine, CH_3NH_2 (b) carbon dioxide, CO_2 (c) sulfur dioxide, SO_2.

(See Problems 11.19 and 11.20.)

11.2 Properties of Liquids

Most common liquids are composed of molecules. The molecules are relatively close to one another and interact through intermolecular forces. In this section, we will look at three different properties of liquids that depend on intermolecular forces: vapor pressure, surface tension, and viscosity.

Vapor Pressure

The **vapor pressure** of a liquid is the partial pressure of the vapor over the liquid, measured at *equilibrium*.■ To understand what we mean by equilibrium, let us look at a simple method for measuring vapor pressure.

The method uses a mercury barometer. To measure the vapor pressure of water, we introduce a few drops of the liquid from a medicine dropper into the mercury column (Figure 11.4a). The water, being less dense than mercury, rises in the tube to the top, where it can evaporate or vaporize.

We can describe the process of *vaporization*, or change from liquid to gas, in this way. Water molecules are held in the liquid state by intermolecular forces. However, whenever a molecule at the liquid surface receives a sufficient jolt from its neighbors, it may leave the liquid and go into the space at the top of the barometer column (see Figure 11.4b). More and more water molecules begin to fill this space, and water vapor forms. As the number of molecules in the vapor state increases, more and more gaseous molecules collide with the liquid water surface, exerting pressure on it. So, as vaporization of water proceeds, the mercury column falls.

Some of the molecules in the vapor collide with the liquid surface and stick. This process, the change of gas to liquid, is the reverse of vaporization and is called *condensation*. The rate of condensation steadily increases as the number of molecules in the volume of vapor increases until the rate at

■ The development of a vapor pressure over a liquid is an example of a change of state (liquid to gas). Changes of state are discussed in detail later in the chapter.

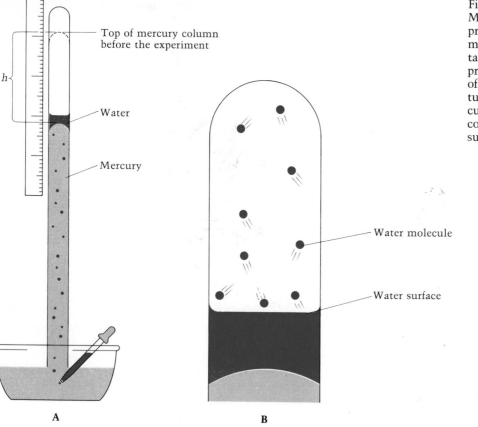

Figure 11.4
Measurement of the vapor pressure of water. (a) The mercury column drops a distance *h* due to the vapor pressure of water. (b) A detail of the upper portion of the tube showing water molecules evaporating from and condensing on the water surface.

which molecules are condensing on the liquid equals the rate at which molecules are vaporizing.

When the rates of vaporization and condensation have become equal, the number of molecules in the vapor at the top of the column stops increasing and remains unchanged. The partial pressure of the vapor (as measured by the fall in the mercury column) reaches a steady value, which is its vapor pressure.■ We say that the liquid and vapor are in *equilibrium*. Although the partial pressure of the vapor is unchanging, molecules are still leaving the liquid and coming back. For this reason, we speak of this as a *dynamic equilibrium*, that is, one in which the molecular processes (in this case, vaporization and condensation) are continuously occurring. We represent this dynamic equilibrium for the vaporization and condensation of water by the equation

$$\text{H}_2\text{O}(l) \rightleftharpoons \text{H}_2\text{O}(g)$$

The vapor pressure of a liquid will depend on intermolecular forces, since the ease or difficulty of a molecule leaving the liquid phase will depend on its strength of attraction to other molecules. When intermolecular forces are strong, we expect the vapor pressure to be low. Also, as Figure 11.5 shows, vapor pressure depends on temperature. As the temperature increases and the kinetic energy of molecular motion becomes greater, the vapor pressure increases.

■ Strictly speaking, the fall in the mercury column is a measure of the sum of the vapor pressures of water and mercury. Mercury, however, has a much lower vapor pressure (0.0012 mmHg at 20°C) than water (17.5 mmHg at 20°C), and we can ignore its effect on the vapor-pressure measurement of water.

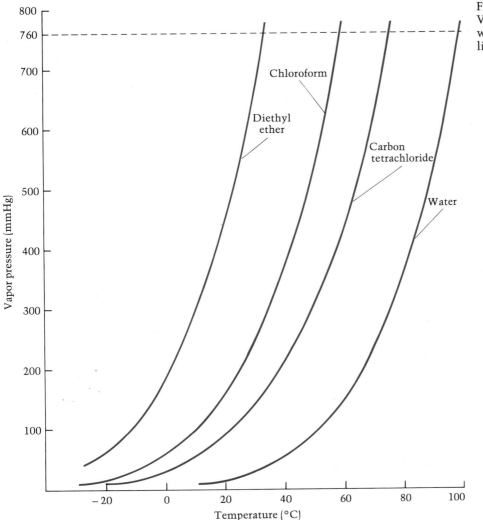

Figure 11.5
Variation of vapor pressure
with temperature for some
liquids.

Table 11.2 lists the vapor pressures of some liquids at 20°C. (Surface tension and viscosity, which we will discuss later in this section, are also listed.) The liquids are listed in order of decreasing vapor pressure, and are therefore in order of increasing intermolecular forces.

Recall that London forces increase with molecular weight. Therefore, if London forces are the principal intermolecular attractions in a liquid, we should expect the molecular weight to increase in the table from the top to the bottom. In fact, this is the case except for water and glycerol. In these substances, hydrogen bonding exists. Glycerol has the structure

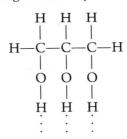

Substance	Molecular Weight (amu)	Vapor Pressure (mmHg)	Surface Tension (J/m²)	Viscosity (kg/m · s)
Carbon dioxide, CO_2	44	4.3×10^4	1.2×10^{-3}	7.1×10^{-5}
Isopentane, C_5H_{12}	72	5.8×10^2	1.4×10^{-2}	2.4×10^{-4}
Chloroform, $CHCl_3$	119	1.7×10^2	2.7×10^{-2}	5.8×10^{-4}
Carbon tetrachloride, CCl_4	154	8.7×10^1	2.7×10^{-2}	9.7×10^{-4}
Water, H_2O	18	1.8×10^1	7.3×10^{-2}	1.0×10^{-3}
Mercury, Hg	201	1.2×10^{-3}	4.4×10^{-1}	1.6×10^{-3}
Glycerol, $C_3H_8O_3$	92	1.6×10^{-4}	6.3×10^{-2}	1.5

Table 11.2
Properties of Some Liquids
at 20°C

where triple dots indicate hydrogen bonds to other glycerol molecules. There is, therefore, substantial hydrogen bonding in this substance. Mercury has a relatively low vapor pressure for a liquid due to strong metallic bonds.

Liquids with relatively high vapor pressure at normal temperatures are said to be *volatile*. Chloroform and carbon tetrachloride are both volatile liquids. Solids may also have a high vapor pressure. Naphthalene, $C_{10}H_8$, and *para*-dichlorobenzene, $C_6H_4Cl_2$, are solids with appreciable vapor pressures. Both are used to make moth balls.

Example 11.2

For each of the following pairs, choose the substance you expect to have the lower vapor pressure at a given temperature: (a) carbon dioxide (CO_2) or sulfur dioxide (SO_2); (b) dimethyl ether (CH_3OCH_3) or ethanol (CH_3CH_2OH).

Solution

(a) London forces increase roughly with molecular weight. The molecular weights for SO_2 and CO_2 are 64 amu and 44 amu, respectively. Therefore, the London forces between SO_2 molecules should be greater than the London forces between CO_2 molecules. Moreover, as SO_2 is polar but CO_2 is not, there are dipole–dipole forces between SO_2 molecules, but not between CO_2 molecules. We conclude that sulfur dioxide has the lower vapor pressure. Experimental values of the vapor pressure at 20°C are: CO_2, 56.3 atm; SO_2, 3.3 atm.

(b) Dimethyl ether and ethanol have the same molecular formulas but different structural formulas. The structural formulas are

$$
\begin{array}{ccccc}
 & H & & H & \\
 & | & & | & \\
H\!-\!\!&C&\!\!-\!O\!-\!&C&\!\!-\!H \\
 & | & & | & \\
 & H & & H & \\
\end{array}
\qquad
\begin{array}{ccccc}
 & H & H & & \\
 & | & | & & \\
H\!-\!\!&C&\!\!-\!\!&C&\!\!-\!O\!-\!H \\
 & | & | & & \\
 & H & H & & \\
\end{array}
$$

dimethyl ether   ethanol

The molecular weights are equal, and therefore the London forces are approximately the same. However, there will be strong hydrogen bonding in ethanol but not in dimethyl ether. We expect ethanol to have the lower vapor pressure. Experimental values of vapor pressure at 20°C are: CH_3OCH_3, 4.88 atm; CH_3CH_2OH, 0.056 atm.

Exercise 11.2

Arrange the following hydrocarbons in order of increasing vapor pressure: ethane, C_2H_6; propane, C_3H_8; and butane, C_4H_{10}. Explain your answer.

(See Problems 11.25 and 11.26.)

Exercise 11.3

Methyl chloride, CH_3Cl, has a vapor pressure of 1490 mmHg, and ethanol has a vapor pressure of 42 mmHg. Explain why you might expect methyl chloride to have a higher vapor pressure than ethanol even though methyl chloride has a somewhat larger molecular weight.

(See Problems 11.27 and 11.28.)

Surface Tension

As Figure 11.6 illustrates, a molecule within the body of a liquid tends to be attracted equally in all directions, so that it experiences no net force. On the other hand, a molecule at the surface of a liquid feels a net attraction by other molecules toward the interior of the liquid. As a result, there is a tendency for the surface area of a liquid to be reduced as much as possible. This explains why falling raindrops are nearly spherical. (The sphere has the smallest surface area for a given volume of any geometrical shape.)

Energy is needed to reverse the tendency toward reduction of surface area in liquids. The **surface tension** of a liquid is the energy required to increase the surface area by a unit amount. Table 11.2 lists values of the surface tension of some liquids.

Since surface tension depends on intermolecular forces, we might expect it to increase for a series of liquids as the vapor pressure decreases. This is roughly true, as Table 11.2 shows. The situation is complicated by the fact that unsymmetrical molecules can take an advantageous position at the surface so that the loss of energy for a surface molecule is not as great as it might otherwise be. Glycerol molecules, for example, can orient so that most of the hydrogen bonding is not broken for a surface molecule. Therefore, the surface tension of glycerol is not as high as we might expect from its low vapor pressure.

Capillary rise is a phenomenon that is related to surface tension. When a small-diameter glass tube, or capillary, is placed upright in water, a column of the liquid rises in the tube (Figure 11.7a). This capillary rise can be

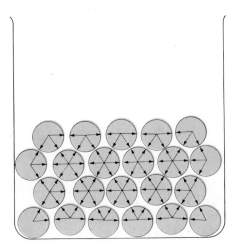

Figure 11.6
Diagrammatic representation of the forces between molecules in a liquid. Note that the molecules at the surface experience a net force toward the interior of the liquid.

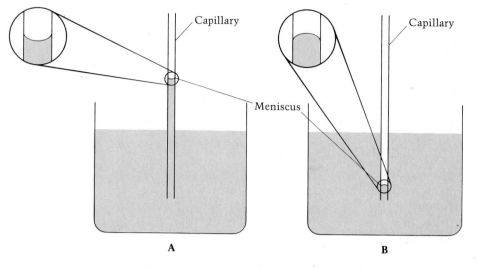

Figure 11.7
(a) Capillary rise, due to the attraction of water and glass. The final water level in the capillary is a balance between the force of gravity and the surface tension of water. (b) Depression, or lowering, of mercury level in a glass capillary. Unlike water, mercury is not attracted to glass.

explained in the following way. Water molecules happen to be attracted to glass (partly as a result of hydrogen bonding to oxygen atoms in the glass). Because of this attraction a thin film of water starts to move up the inside of the glass capillary. But in order to reduce the surface area of this film, the water level begins to rise also. The final water level is a balance between the surface tension and the potential energy required to lift the water against gravity. Note that the *meniscus*, or liquid surface within the tube, is curved downward, that is, it has a concave shape.

In the case of mercury, the liquid level in the capillary is lower than it is for the liquid outside (Figure 11.7b). Also, the meniscus is curved upward, that is, it has a convex shape. Compare Figure 11.7b with Figure 11.7a. The difference in behavior results from the fact that the attraction between two mercury atoms is greater than the attraction of mercury for glass, in contrast to the situation described for water and glass.

The surface tension of a liquid can be affected by dissolved substances. Soaps and detergents, in particular, drastically decrease the surface tension of water. An interesting, but simple, experiment shows the effect of soap on surface tension. Because of surface tension, a liquid behaves as if it had a skin. Water bugs seem to skitter across this skin as if they were ice-skating. You can actually float a needle on water, if you carefully lay it across the surface (Figure 11.8).■ If you now put a drop of soap solution onto the water, the soap spreads across the surface, and the needle sinks. Water bugs also sink in soapy water.

■ The easiest way to float a needle is by placing a square of tissue paper on the prongs of a fork, with the needle on top of the tissue. When the fork is lowered into the water, the tissue and needle float on the surface. However, the paper soon becomes water-logged and sinks, leaving the needle floating.

Viscosity

All liquids show some **viscosity,** or resistance to flow. Thus, syrup is a more viscous liquid than water; that is, it has a higher viscosity. Glass can be thought of as a liquid of very high viscosity. Although window glass flows very slowly, it has been found that glass panes from old homes are measurably thicker at the base, due to the downward flow caused by gravity. Table 11.2 lists the viscosities of some liquids.■

■ Lubricating oils are classified by viscosity, with each oil assigned an SAE (Society of Automotive Engineers) number. For example, SAE number 40 oil is more viscous than SAE number 10 oil.

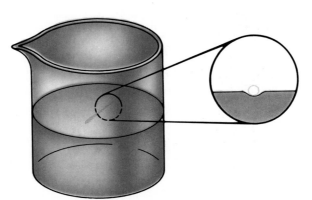

Figure 11.8
A needle will float on the surface of water because of surface tension. As the inset shows, the water surface is depressed and slightly stretched by the needle; but even more surface would have to be created for the needle to submerge—and this takes energy.

Since the resistance to flow is in part due to molecular attraction, viscosities should also depend on intermolecular forces. Glycerol has a very high viscosity, which is a result of hydrogen bonding. Sugars, such as glucose and sucrose, have many O—H groups in their molecules. Aqueous solutions of a sugar are called syrups and have high viscosities because of hydrogen bonding to these O—H groups. However, viscosity also depends on other factors, such as the possibility of molecules tangling together. Liquids with long molecules that tangle together are expected to have high viscosity.

11.3 Types of Solids

A solid consists of structural units, either atoms, molecules, or ions, that are attracted to one another strongly enough to give a rigid substance. One way to classify solids is by the type of force holding the structural units together. In some cases, these forces are intermolecular and hold molecules together. In other cases, these forces are chemical bonds, either metallic, ionic, or covalent, and hold atoms together. Viewed this way, we get four different types of solids: molecular, metallic, ionic, and covalent.

Molecular solids consist of atoms or molecules held together by intermolecular forces. Many solids are of this type. Examples: solid neon, solid water (ice), solid carbon dioxide (Dry Ice).

Metallic solids consist of atoms held together by metallic bonding.■ In this kind of bonding, positively charged atomic cores are surrounded by delocalized electrons. Examples: iron, copper, silver.

■ Metallic bonding was discussed briefly in Section 7.8.

Ionic solids consist of cations and anions held together by the electrical attraction of opposite charges (ionic bonds). Examples: cesium chloride, sodium chloride, zinc chloride.

Covalent network solids consist of atoms held together in large networks or chains by covalent bonds. Diamond is an example of a three-dimensional network solid. Every carbon atom in diamond is covalently bonded to four others, so that an entire crystal might be considered an immense molecule, or *macromolecule*. It is possible to have two-dimensional "sheet" and one-dimensional "chain" macromolecules, with atoms held together by covalent bonds. Examples: diamond (three-dimensional network), graphite (sheets), asbestos (chains).

Type of Solid	Structural Units	Attractive Forces Between Structural Units	Examples
Molecular	Atoms or molecules	Intermolecular forces	Ne, H_2O, CO_2
Metallic	Atoms (positive cores surrounded by electron "sea")	Metallic bonding (extreme delocalized bond)	Fe, Cu, Ag
Ionic	Ions	Ionic bonding	CsCl, NaCl, ZnS
Covalent network (macromolecular)	Atoms	Covalent bonding	Diamond, graphite, asbestos

Table 11.3
Types of Solids

Table 11.3 summarizes these four types of solids. Representative crystal structures are described in Section 11.5. Later in the chapter, after we have discussed changes of state, we will examine the relationship between the structure of a solid and some of its physical properties.

Example 11.3

Which of the four basic types of solids would you expect the following substances to be? (a) solid ammonia, NH_3 (b) cesium, Cs (c) cesium iodide, CsI (d) silicon, Si

Solution

(a) Ammonia has a molecular structure; therefore it freezes as a molecular solid. (b) Cesium is a metal; it is a metallic solid. (c) Cesium iodide is an ionic substance; it exists as an ionic solid. (d) Silicon atoms might be expected to form covalent bonds to other silicon atoms, as carbon does in diamond. A covalent network solid would result.

Exercise 11.4

Classify each of the following solids according to the forces of attraction that exist between the structural units: (a) zinc (b) sodium iodide, NaI (c) silicon carbide, SiC (d) methane, CH_4

(See Problems 11.29 and 11.30.)

11.4 Crystalline and Amorphous Solids

Solids may be either crystalline or amorphous. **Crystalline solids** have well-defined, regular shapes, but **amorphous solids** do not. Sodium chloride (table salt) and sucrose (table sugar) are examples of crystalline substances. If you look at table salt, you will see the cubic-shaped crystals. Metals are also crystalline, but since they are usually compact masses of crystals, this may not be obvious. Asphalt and paraffin wax are examples of amorphous solids.

Characteristics of Crystalline and Amorphous Solids

Looking at a crystalline solid, you can see its definite shape. If you break (cleave) it, you will find it gives flat, well-defined faces. When heated, the

Kind of Solid	Characteristics	Examples
Crystalline	Consist of individual crystals, each with well-defined shape. Cleave to give flat, well-defined faces. Melt at definite temperature.	Sodium chloride, sucrose, metals
Amorphous	Are noncrystalline, so have no well-defined shape. Break to give curved or irregular faces. Soften, then melt over a temperature range.	Asphalt, paraffin, window glass, obsidian, glassy form of glycerol

Table 11.4
Characteristics of Crystalline and Amorphous Solids

crystal melts at a definite temperature (unless it decomposes before melting). The change from solid to liquid occurs abruptly at the melting point.

When an amorphous solid is broken, it gives curved or irregular faces. Window glass, which is an amorphous solid, shatters into irregularly shaped pieces. An amorphous solid does not melt at a definite temperature. Rather, it softens over a broad temperature range, changing slowly from solid to liquid.

An amorphous solid is similar to a liquid in having a random arrangement of structural units. *Glasses*, in fact, are amorphous solids that are essentially highly viscous liquids, that is, liquids that flow very slowly. Ordinary window glass is an example.■ Obsidian is a natural glass formed after volcanic eruptions when molten rock solidifies too quickly for crystals to form. A normal liquid will freeze to form a glass if its molecules become "frozen" in random positions before they can assume an ordered crystalline arrangement. Glycerol, $CH_2OHCHOHCH_2OH$, is easily cooled to a glass, but is difficult to crystallize. Because of the high viscosity of the liquid, glycerol molecules move into an ordered arrangement with difficulty. Table 11.4 summarizes the characteristics of crystalline and amorphous solids.

■ When quartz crystals (SiO_2) are melted, then cooled rapidly, they form a glass called silica glass. Silica glass, although ideal for some purposes, is difficult to work with because of its high temperature of softening. This temperature can be lowered by adding various oxides, such as Na_2O and CaO. In practice, ordinary glass is formed by melting sand (quartz crystals) with sodium carbonate and calcium carbonate. The carbonates decompose to the oxides plus carbon dioxide.

The Crystalline Lattice

The regularity of a crystalline solid is due to the arrangement of structural units into an orderly array or *lattice*. We can imagine this crystalline lattice to be formed by stacking together identical small portions of the crystal, much as bricks are stacked to form a brick wall. A **unit cell** is the smallest repeating unit of a crystal. As such it is the smallest portion from which we can imagine constructing the crystal. A unit cell is composed of atoms, molecules, or ions arranged in a definite way. Therefore the unit cell has a definite shape, which accounts for the well-defined appearance of the crystal.

There are seven basic shapes possible for unit cells, which give rise to the seven **crystal systems** used to classify crystals. A crystal belonging to a given crystal system will have a unit cell with one of the seven shapes shown in Figure 11.9. Each unit-cell shape can be characterized by the angles between the edges and the relative lengths of the edges of the unit cell. Table 11.5 summarizes the relationships between the angles and edge lengths for the unit cell of each crystal system.

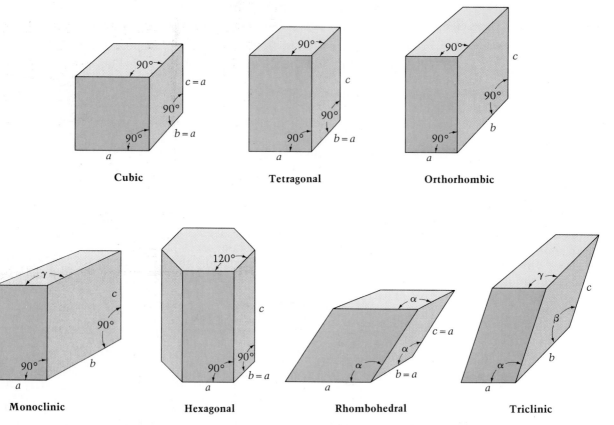

Cubic Unit Cells

Certain crystal systems have more than one type of unit cell, differing by the way in which structural units are arranged within the unit-cell shape. For example, the cubic crystal system has three possible cubic unit cells, called *simple cubic, body-centered cubic,* and *face-centered cubic.*

To see the different possible cubic unit cells, suppose that a cubic crystal is composed of atoms as the structural units (certain metals or solid noble-gas elements are examples). If the crystal has a **simple cubic** unit cell, the atoms are arranged only at the corners of the unit cell (Figure 11.10, *left*). A **body-centered cubic** unit cell has an atom at the center of the unit cell, as

Table 11.5
The Seven Crystal Systems

Crystal System	Edge Length	Angles	Examples
Cubic	$a = b = c$	$\alpha = \beta = \gamma = 90°$	NaCl, Cu
Tetragonal	$a = b \neq c$	$\alpha = \beta = \gamma = 90°$	TiO_2 (rutile), Sn (white tin)
Orthorhombic	$a \neq b \neq c$	$\alpha = \beta = \gamma = 90°$	$CaCO_3$ (aragonite), $BaSO_4$
Monoclinic	$a \neq b \neq c$	$\alpha = \beta = 90°, \gamma \neq 90°$	$Na_2B_4O_7 \cdot 10H_2O$ (borax), $PbCrO_4$
Hexagonal	$a = b \neq c$	$\alpha = \beta = 90°, \gamma = 120°$	C (graphite), ZnO
Rhombohedral	$a = b = c$	$\alpha = \beta = \gamma \neq 90°$	$CaCO_3$ (calcite), HgS (cinnabar)
Triclinic	$a \neq b \neq c$	$\alpha \neq \beta \neq \gamma \neq 90°$	$K_2Cr_2O_7$, $CuSO_4 \cdot 5H_2O$

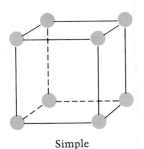
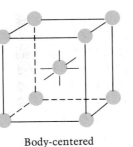
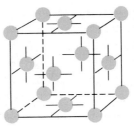

Simple   Body-centered   Face-centered

Figure 11.10
Cubic unit cells: simple,
body-centered, and face-
centered.

Figure 11.11
A face-centered cubic lattice.
The diagram shows how the
unit cells are stacked to give
a lattice. For clarity the
atomic positions are given
for one unit cell only.

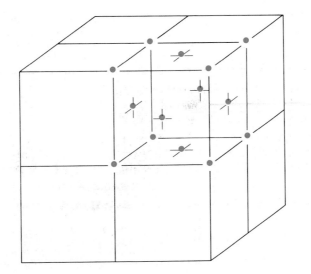

well as those at the corners (Figure 11.10, *center*). In a **face-centered cubic**
unit cell, there are atoms at the center of each face of the unit cell, in
addition to those at the corners (Figure 11.10, *right*).

A crystal lattice consists of many identical unit cells stacked together.
Thus, copper metal consists of face-centered cubic unit cells stacked together to give a crystal belonging to the cubic system. Figure 11.11 shows a
portion of a crystal lattice with a face-centered cubic unit cell.

Other crystal systems may also have more than one possible type of unit
cell. Altogether, fourteen different unit cells are possible.

Example 11.4

How many atoms are there in the face-centered cubic
unit cell of an atomic crystal? Remember that atoms at
the corners and faces of the unit cell are shared with
adjoining unit cells (see Figure 11.12).

Solution

Each atom at the center of a face is shared with one
other unit cell. Hence, only one-half of the atom be-
(Continued)

longs to a particular unit cell. But there are six faces on a cube. This gives three whole atoms from the faces. Each corner atom is shared among eight unit cells. Hence, only one-eighth of a corner atom belongs to a particular unit cell. But there are eight corners on a cube. Therefore, the corners contribute one whole atom. Thus, there are a total of four atoms in a face-centered cubic unit cell. To summarize, we have

$$6 \text{ faces} \times \frac{1/2 \text{ atom}}{\text{face}} = 3 \text{ atoms}$$

$$8 \text{ corners} \times \frac{1/8 \text{ atom}}{\text{corner}} = \underline{1 \text{ atom}}$$

$$\text{Total} \quad 4 \text{ atoms}$$

Exercise 11.5

Figure 11.13 shows solid dots ("atoms") forming a two-dimensional lattice. A unit cell is marked off by the dashed line. How many "atoms" are there per unit cell in such a two-dimensional lattice?

(See Problems 11.31 and 11.32.)

Figure 11.12
A representation of the face-centered cubic unit cell. Only that portion of each atom belonging to the unit cell is shown. Note that a face atom is shared between two unit cells, and a corner atom is shared with eight.

Crystal Defects

So far we have assumed that crystals have perfect order. In fact, real crystals have various defects or imperfections. These are principally of two kinds, chemical impurities and defects in the formation of the lattice.

Ruby is an example of a crystal with a chemical impurity. The crystal is mainly aluminum oxide, Al_2O_3, but occasional aluminum ions, Al^{3+}, are replaced by chromium(III) ions, Cr^{3+}.

Various lattice defects occur during crystallization. Crystal planes may be misaligned or sites in the crystal lattice may remain vacant. For example, there might be an equal number of sodium ion vacancies and chloride ion vacancies in a sodium chloride crystal. It is also possible to have an unequal number of cation and anion vacancies in an ionic crystal. For example, iron(II) oxide, FeO, usually crystallizes with some iron(II) sites left unoccupied. Enough of the remaining iron atoms are in the +3 oxidation state to give an electrically balanced crystal. As a result, there are more oxygen atoms than iron atoms in the crystal. Moreover, the exact composition of the crystal can vary. Therefore, the formula FeO is only approximate. Such a

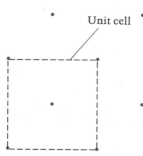

Unit cell

Figure 11.13
A two-dimensional lattice.

compound whose composition varies slightly from its idealized formula is said to be *nonstoichiometric*. Other examples of nonstoichiometric compounds are Cu_2O and Cu_2S. Each usually has less copper than expected from the formula.

11.5 The Structure of Some Crystalline Solids

We described the structure of crystals in a general way in Section 11.4. Now we want to look in detail at the structure of several crystalline solids representative of the different types: molecular, metallic, ionic, and covalent network.

Molecular Solids; Closest Packing

The simplest molecular solids are the frozen noble gases, for example, solid neon. In this case, the molecules are single atoms, and the intermolecular interactions are London forces. These forces are nondirectional (in contrast to covalent bonding, which is directional), and the maximum attraction will be obtained when each atom is surrounded by the largest possible number of other atoms. The problem, then, is simply to find how identical spheres can be packed as tightly as possible into a given space. Let us look at how to do that.

To form a layer of closest-packed spheres, place each row of spheres in the crevices of the adjoining rows. This is illustrated in Figure 11.14a. (You may find it helpful to stack identical coins in the manner described here for spheres.) Place the next layer of spheres in the hollows in the first layer. There are twice as many hollows, or sites to place spheres in, as spheres that can be placed in a layer. Once you have placed a sphere in a given hollow, it partially covers three neighboring hollows (Figure 11.14b) and completely determines the pattern of the layer (Figure 11.14c).

When we come to fill the third layer, we find that we have a choice of sites for spheres, labeled *x* and *y* in Figure 11.14c. Note that each of the *x* sites has a sphere in the first layer directly beneath it, but the *y* sites do not. When the spheres in the third layer are placed in the *x* sites, so that the third layer

Figure 11.14
Closest packing of spheres.
(a) Each layer is formed by placing spheres in the crevices of the adjoining row.
(b) When a sphere is placed in a hollow on top of a layer, three other hollows are partially covered. (c) There are two types of hollows on top of the second layer, labeled *x* and *y*. The *x* sites are directly above the spheres in the first layer; the *y* sites are not. Both are possible sites for the spheres of the third layer.

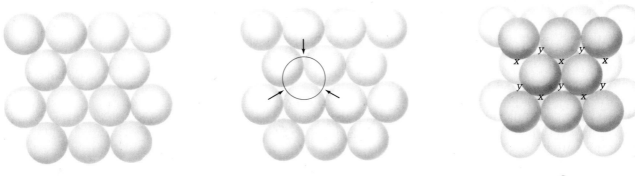

A   B   C

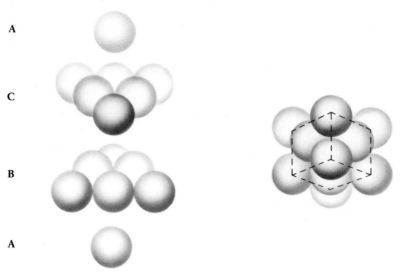

Figure 11.15
The cubic closest-packed lattice is identical to the face-centered cubic lattice. (*Left*) An "exploded" view of portions of four layers of the cubic closest-packed lattice. (*Right*) These layers form a face-centered cubic unit cell.

repeats the first layer, we label this stacking *ABA*. If successive layers are placed so that the spheres of each layer are directly over a layer that is one layer away, we get a stacking that we label *ABABABA* ⋯. (This notation indicates that the *A* layers are directly over other *A* layers, while the *B* layers are directly over other *B* layers.) This results in a **hexagonal closest-packed structure** (hcp). A crystal with this structure has a hexagonal unit cell.

When the spheres in the third layer are placed in *y* sites, so that the third layer is over neither the first nor the second layer, we get a stacking that we label *ABC*. The fourth layer must be over either the *A* or the *B* layer. If subsequent layers are stacked so that they are over the layer two layers below, we get the stacking *ABCABCABCA* ⋯, which results in a **cubic closest-packed structure** (ccp).∎

The cubic closest-packed lattice is identical to the lattice having a face-centered cubic unit cell. To see this, we take portions of four layers from the cubic closest-packed array (Figure 11.15, *left*). When these are placed together, they form a cube, as shown in Figure 11.15, *right*.

In any closest-packed arrangement, each interior atom is surrounded by twelve nearest-neighbor atoms. The number of nearest-neighbor atoms is called the **coordination number.** Thinking of atoms as hard spheres, one can calculate that the spheres occupy 74% of the space of the crystal. There is no way of packing identical spheres so that an atom has a coordination number greater than 12 or the spheres occupy more than 74% of the space of the crystal. All of the noble-gas solids have cubic closest-packed crystals except for helium, which has a hexagonal closest-packed crystal.

The crystal lattices of many polyatomic substances have closest-packed structures although precise descriptions of the crystals are complicated by the fact that the molecules may assume various orientations within the lattice. Solid carbon dioxide (Dry Ice), for example, forms cubic crystals with the carbon atoms of each carbon dioxide molecule at the sites of a face-centered cubic lattice. The molecules are oriented in the lattice to give the most compact structure (see Figure 11.16).

■ If we pack into a crate as many perfectly round oranges of the same size as possible, they will be closest-packed. However, they need not be either hexagonal or cubic closest-packed, since the *A*-, *B*-, and *C*-type layers can be placed at random (except that similar layers cannot be adjacent). A possible arrangement might be *ABCABABABC* ⋯. It is the forces between structural units in a closest-packed crystal that determine whether one of the ordered structures, either cubic or hexagonal, will occur.

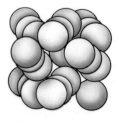

Figure 11.16
The crystal structure of solid carbon dioxide (Dry Ice).

Table 11.6
Crystal Structures of Metals*

Li	Be												
bcc	hcp												

Na	Mg										Al		
bcc	hcp										ccp		

K	Ca	Sc	Ti	V	Cr	Mn	Fe	Co	Ni	Cu	Zn	Ga	
bcc	ccp	hcp	hcp	bcc	bcc	bcc	bcc	hcp	ccp	ccp	hcp	o	

Rb	Sr	Y	Zr	Nb	Mo	Tc	Ru	Rh	Pd	Ag	Cd	In	Sn
bcc	ccp	hcp	hcp	bcc	bcc	hcp	hcp	ccp	ccp	ccp	hcp	o	o

Cs	Ba	La	Hf	Ta	W	Re	Os	Ir	Pt	Au	Hg	Tl	Pb	Bi
bcc	bcc	hcp	hcp	bcc	bcc	hcp	hcp	ccp	ccp	ccp	—	hcp	ccp	o

*hcp = hexagonal closest packing, ccp = cubic closest packing, bcc = body-centered cubic,
o = other.

Metallic Solids

If we were to assume that metallic bonding is completely nondirectional, we would expect metals to crystallize in one of the closest-packed structures, as the noble-gas elements do. Indeed, many of the metallic elements do have either cubic or hexagonal closest-packed crystals. Copper and silver, for example, are cubic closest-packed (that is, face-centered cubic). However, a body-centered cubic arrangement of spheres occupies almost as great a percent of space as the closest-packed ones. In a body-centered arrangement, 68% of the space is occupied by spheres, compared with the maximum value of 74% in the closest-packed arrangements. Each atom in a body-centered lattice has a coordination number of 8, rather than the 12 of a closest-packed lattice.

Table 11.6 lists the structures of the metallic elements. With few exceptions, they have either a closest-packed or a body-centered lattice. Iron has a body-centered cubic structure, as do the alkali metals.

Ionic Solids

The description of ionic crystals is complicated by the fact that we must give the positions in the crystal lattice of both the cations and the anions. We will look at the structures of three different cubic crystals with the general formula MX (where M is the metal and X is the nonmetal): cesium chloride (CsCl), sodium chloride (NaCl), and zinc blende (ZnS).

The cesium chloride lattice is shown in Figure 11.17. If you look just at the cesium ions, Cs^+, you will see that they are arranged in a simple cubic lattice. Similarly, the chloride ions, Cl^-, also have a simple cubic lattice. Thus, you can think of the cesium chloride structure as two interpenetrating simple cubic lattices. One choice for the unit cell is shown by the dotted lines. Chloride ions occur at the corners of the cubic unit cell, with a cesium ion at the center of the cube. Alternatively, one could put the cesium ions at the corners of the unit cell, with a chloride ion at the center. Note that

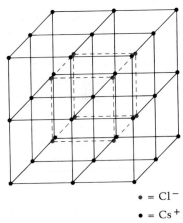

• = Cl^-

• = Cs^+

Figure 11.17
The cesium chloride crystal lattice. Each Cl^- ion is at the center of a cube of Cs^+ ions. One choice for the unit cell is shown by the dotted lines. Chloride ions occupy the corners of the unit cell, with a cesium ion at the center. Alternatively, the cesium ions can occupy the corners, with a chloride ion at the center.

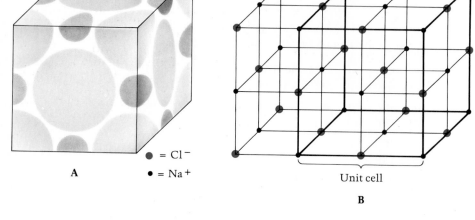

A

• = Cl⁻
• = Na⁺

Unit cell

B

Figure 11.18
The structure of sodium chloride. (a) The unit cell, showing chloride ions in a cubic closest-packed arrangement (face-centered cubic), with sodium ions in octahedral holes. (b) The crystal lattice, showing the face-centered cubic arrangement of sodium ions.

each ion is surrounded by eight ions of opposite charge. Ammonium chloride also has this type of structure.

In cesium chloride, the Cs^+ and Cl^- ions are of similar size. But when one kind of ion in a crystal is much larger than the other, that ion is often arranged in a closest-packed lattice. The smaller ions occupy cavities or holes in the closest-packed lattice of larger ions.

The sodium chloride structure can be described in this way. Figure 11.18a shows the sodium chloride unit cell, with the larger chloride ions, Cl^-, in a cubic closest-packed (face-centered cubic) arrangement. The sodium ions, Na^+, occupy what are called the *octahedral holes* in this closest-packed arrangement. They are called octahedral holes because if the centers of the six chloride ions surrounding a sodium ion are connected by lines, an octahedral figure is formed.

We can also describe the sodium chloride structure as interpenetrating face-centered lattices of sodium ions and chloride ions. The face-centered lattice of sodium ions can be seen by extending the unit cell, as shown in Figure 11.18b. Note that each ion is surrounded by six ions of the opposite charge. Some other compounds that crystallize in this structure are potassium chloride, KCl, calcium oxide, CaO, and silver chloride, AgCl.

Zinc blende (ZnS), also called sphalerite, consists of sulfide ions, S^{2-}, arranged in a cubic closest-packed (face-centered cubic) arrangement, with the zinc ions, Zn^{2+}, in alternate *tetrahedral holes* of the anion lattice.■ The tetrahedral holes can be located by dividing the unit cell into eight cubes (see Figure 11.19a). There is a hole or cavity formed by the four sulfide ions on the alternate corners of each of these smaller cubes. The cavity is called a tetrahedral hole because the figure formed by connecting the centers of these four sulfide ions is a tetrahedron.

The zinc blende structure can also be described as interpenetrating face-centered lattices of zinc ions and sulfide ions. Figure 11.19b shows the relative positions of these two lattices. Note that each ion is surrounded by four ions of the opposite charge. Zinc oxide, ZnO, and beryllium oxide, BeO, also have the zinc blende structure.

■ Zinc sulfide occurs naturally as zinc blende and as wurtzite. Wurtzite consists of hexagonal closest-packed sulfide ions with zinc ions in alternate tetrahedral holes of the anion lattice. Thus, zinc sulfide may have either of two different crystal structures, one cubic, the other hexagonal. It is said to be *polymorphic*.

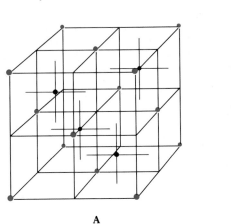
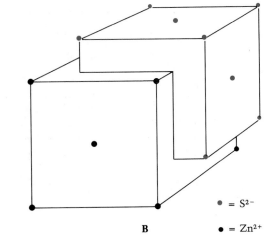

Figure 11.19
The structure of zinc blende (ZnS), or sphalerite. (a) The unit cell, showing sulfide ions in a closest-packed arrangement, with zinc ions in alternate tetrahedral holes. (b) The crystal lattice, showing the interpenetrating face-centered arrangement of zinc ions and sulfide ions.

A   B

$\bullet = S^{2-}$

$\bullet = Zn^{2+}$

Covalent Network Solids

In diamond, each carbon atom is covalently bonded to four other carbon atoms in tetrahedral directions to give a three-dimensional covalent network. The structural arrangement is similar to that of zinc blende, with each zinc ion and each sulfide ion replaced by a carbon atom (see Figure 11.19).

Graphite is another form of the element carbon.■ It consists of large sheets of the form

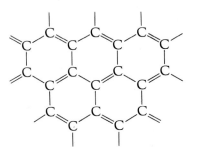

(one resonance formula)

Each carbon–carbon bond distance within a sheet is 1.42 Å (142 pm), which is almost midway between the length of a C—C bond (1.54 Å) and that for a C=C bond (1.34 Å). The intermediate bond length can be explained by resonance. Various resonance formulas can be drawn with the double and single bonds in different positions. The π electrons are therefore delocalized over the two-dimensional sheet of carbon atoms. These two-dimensional sheets are stacked one on top of the other, with bonding between sheets due to weak London forces. The between-layer distance is 3.35 Å (335 pm). The reason the distance between layers is greater than the bond distance is that London forces are weaker than covalent bonding forces.

Quartz (SiO_2) is a three-dimensional covalent network solid consisting of silicon atoms covalently bonded in tetrahedral directions to four oxygen atoms. Each oxygen atom is in turn covalently bonded to two silicon atoms.■

■ Diamond and graphite are allotropes of carbon (different forms of the same element) in the same state. They are also *polymorphs* (different crystal forms of the same substance) —see previous marginal note.

■ The structures of CO_2 and SiO_2 differ because carbon forms π bonds and silicon usually does not (probably because the silicon atom is too large, and π bonds to it are too weak). Thus, carbon can form molecular CO_2, with the structure O=C=O. Silicon, on the other hand, forms only single bonds, giving a network structure for SiO_2.

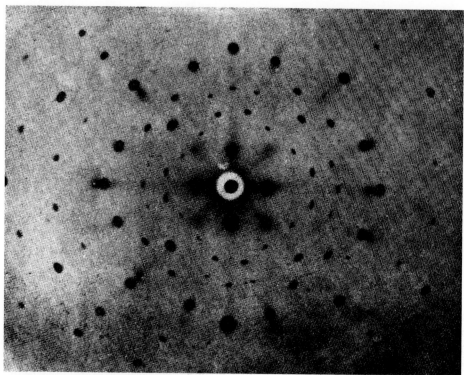

Figure 11.20
A diffraction pattern obtained by diffracting x rays from a sodium chloride crystal.

11.6 X-ray Crystal Structure Determination

X-ray diffraction is one of the most important ways of determining the structure of molecules.■ A crystal, because of its ordered structure, consists of repeating planes of the same kind of atom. These planes can act as reflecting surfaces for x rays. When x rays are reflected from these planes, they show a *diffraction pattern*, a series of spots on a photographic plate (see Figure 11.20). By analyzing this diffraction pattern, we can determine the positions of the atoms in the crystal. To understand how such a diffraction pattern occurs, let us first look at the diffraction of two waves.■

Suppose two waves of the same wavelength come together in phase. By *in phase* we mean that the waves come together with their maxima at the same points and their minima at the same points (see Figure 11.21, *top*). Two waves that come together in phase form a wave with higher maxima and lower minima. We say that the waves reinforce or undergo *constructive interference*. In the case of x rays, the intensity of the resultant ray is increased.

Now suppose two waves of the same wavelength come together out of phase. By *out of phase* we mean that where one wave has its maxima, the other has its minima (see Figure 11.21, *bottom*). When two waves come together out of phase, each maximum of one wave combines with a minimum of the other so that the wave is destroyed. We say that the waves undergo *destructive interference*.

Now consider the scattering, or reflection, of x rays from a crystal. Figure 11.22 shows two rays, one reflected from one plane and the other reflected

■ X-ray diffraction has been used to obtain the structure of proteins, which are large molecules essential to life. John Kendrew and Max Perutz received the Nobel Prize in chemistry in 1962 for their x-ray work on determining the structure of myoglobin (MW = 17,600 amu) and hemoglobin (MW = 66,000 amu). Hemoglobin is the oxygen-carrying protein of the red blood cells, and myoglobin is the oxygen-carrying protein in muscle tissue.

■ Diffraction of light was discussed briefly in Section 5.6.

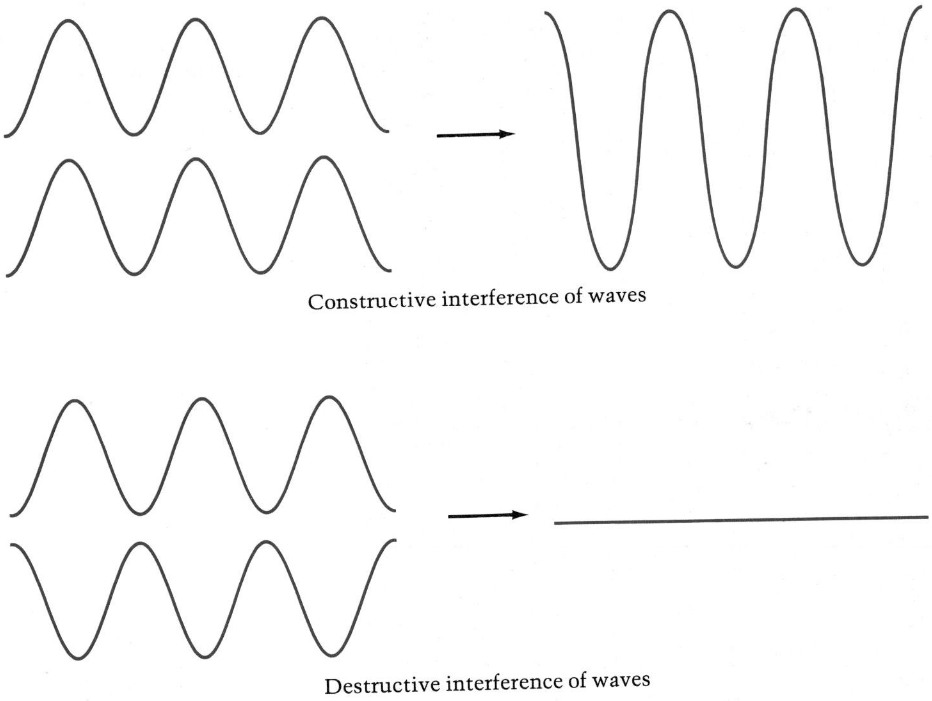

Constructive interference of waves

Figure 11.21
Wave interference. Two waves constructively interfere if they are in phase (that is, if their peaks match). Two waves destructively interfere if they are out of phase (that is, if the peaks of one wave match the valleys of the other).

Destructive interference of waves

Figure 11.22
Diffraction of x rays from crystal planes. At most angles, reflected waves will be out of phase and will destructively interfere, as in (a). However, at certain angles reflected waves will be in phase and will constructively interfere, as in (b).

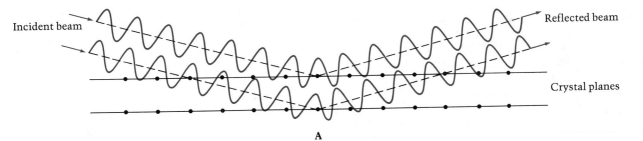

Incident beam

Reflected beam

Crystal planes

A

Incident beam

Reflected beam

Crystal planes

B

from another plane. The two rays start out in phase, but because one ray travels farther, they may end up out of phase after reflection. Only at certain angles of reflection do the two rays remain in phase. At some angles the rays will constructively interfere (giving dark areas on a negative photographic plate, as in Figure 11.20). At other angles the rays will destructively interfere (giving light areas on a photographic plate).■

We can relate the resulting pattern produced by the diffraction to the structure of the crystal. We can determine the type of unit cell and its size, and if the solid is composed of molecules, we can determine the position of each atom in the molecule.

> ■ X rays interact with the electrons in atoms, not the nuclei. Therefore, heavy atoms scatter best (that is, produce brighter spots) because of their large number of electrons.

Example 11.5

X-ray diffraction from crystals provides one of the most accurate ways of determining Avogadro's number. Silver crystallizes in a face-centered cubic lattice. The length of an edge of the unit cell can be determined by x-ray diffraction to be 4.086 Å (408.6 pm). The density of silver is 10.50 g/cm^3. Calculate the mass of a silver atom. Then, using the known value of the atomic weight, calculate Avogadro's number.

Solution

Knowing the edge length of a unit cell, we can calculate the unit-cell volume. Then, from the density we can find the mass of the unit cell and hence the mass of a silver atom. The volume, V, of the unit cell is obtained by cubing the length of an edge:

$$V = (4.086 \times 10^{-10} \text{ m})^3 = 6.822 \times 10^{-29} \text{ m}^3$$

The density, d, of silver in grams per cubic meter is

$$10.50 \frac{\text{g}}{\text{cm}^3} \times \left(\frac{10^2 \text{ cm}}{1 \text{ m}}\right)^3 = 1.050 \times 10^7 \text{ g/m}^3$$

Density is mass per volume; hence, the mass of a unit cell equals the density times the volume of the unit cell:

$$m = dV$$
$$= 1.050 \times 10^7 \text{ g/m}^3 \times 6.822 \times 10^{-29} \text{ m}^3$$
$$= 7.163 \times 10^{-22} \text{ g}$$

Since there are four atoms in a face-centered unit cell (see Example 11.4), the mass of a silver atom is

$$\text{Mass of 1 Ag atom} = \tfrac{1}{4} \times 7.163 \times 10^{-22} \text{ g}$$
$$= 1.791 \times 10^{-22} \text{ g}$$

The known atomic weight of silver is 107.87 amu. Thus, the molar mass (Avogadro's number of atoms) is 107.87 g/mol, and Avogadro's number, N_A, is

$$N_A = \frac{107.87 \text{ g/mol}}{1.791 \times 10^{-22} \text{ g}} = 6.023 \times 10^{23}/\text{mol}$$

Exercise 11.6

Lithium metal has a body-centered cubic structure with a unit-cell length of 3.509 Å. Calculate Avogadro's number. The density of lithium is 0.534 g/cm^3.

(See Problems 11.33 and 11.34.)

If we know or assume the structure of an atomic crystal, we can calculate the length of the unit-cell edge from the density of the substance. This is illustrated in the next example. Agreement of this value with that obtained from x-ray diffraction confirms that our view of the structure of crystals is correct.

Example 11.6

Platinum crystallizes in a face-centered cubic lattice. It has a density of 21.45 g/cm³ and an atomic weight of 195.08 amu. From these data, calculate the length of a unit-cell edge. Compare with the value of 3.924 Å obtained from x-ray diffraction.

Solution

The mass of an atom, hence the mass of a unit cell, can be calculated from the atomic weight. Knowing the density and the mass of the unit cell, we can calculate the volume and the edge length of a unit cell.

We can use Avogadro's number (6.022×10^{23}/mol) to convert the molar mass of platinum (195.08 g/mol) to the mass per atom:

$$\frac{195.08 \text{ g Pt}}{1 \text{ mol Pt}} \times \frac{1 \text{ mol Pt}}{6.022 \times 10^{23} \text{ Pt atoms}}$$

$$= \frac{3.239 \times 10^{-22} \text{ g Pt}}{1 \text{ Pt atom}}$$

Since there are four atoms per unit cell, the mass per unit cell can be calculated as follows:

$$\frac{3.239 \times 10^{-22} \text{ g Pt}}{1 \text{ Pt atom}} \times \frac{4 \text{ Pt atoms}}{1 \text{ unit cell}} = \frac{1.296 \times 10^{-21} \text{ g}}{1 \text{ unit cell}}$$

The volume of the unit cell is

$$V = \frac{m}{d} = \frac{1.296 \times 10^{-21} \text{ g}}{21.45 \text{ g/cm}^3} = 6.042 \times 10^{-23} \text{ cm}^3$$

If the edge length of the unit cell is denoted as l, the volume is

$$V = l^3$$

Hence, the edge length is

$$l = \sqrt[3]{V}$$
$$= \sqrt[3]{6.042 \times 10^{-23} \text{ cm}^3}$$
$$= 3.924 \times 10^{-8} \text{ cm } (3.924 \times 10^{-10} \text{ m})$$

Thus, the edge length is 3.924 Å or 392.4 pm, which is in excellent agreement with the x-ray diffraction value.

Exercise 11.7

Potassium metal has a body-centered cubic structure. The density of the metal is 0.856 g/cm³. Calculate the edge length of a unit cell.

(See Problems 11.35 and 11.36.)

Changes of State

In the chapter opener, we discussed the change of solid carbon dioxide (Dry Ice) directly to a gas. Such a change in a substance from one state to another is called a **change of state** or **phase change.** In the next sections of this chapter, we will look at the different kinds of phase change and the conditions under which they occur.

11.7 Phase Changes

In general, each of the three states of a substance can change into either one of the other states. Table 11.7 lists the six possible kinds of phase changes.

Melting is the change of a solid to the liquid state. For example,

$$\text{H}_2\text{O}(s) \longrightarrow \text{H}_2\text{O}(l) \qquad \text{(melting)}$$
$$\text{ice, snow} \qquad \text{liquid water}$$

Table 11.7
Kinds of Phase Change

Phase Change	Name	Examples
Solid ⟶ liquid	Melting	Melting of snow and ice
Solid ⟶ gas	Sublimation	Sublimation of Dry Ice, freeze-drying of coffee
Liquid ⟶ solid	Freezing	Freezing of water or a liquid metal
Liquid ⟶ gas	Vaporization	Evaporation of water or refrigerant
Gas ⟶ liquid	Condensation, liquefaction	Formation of dew, liquefaction of carbon dioxide
Gas ⟶ solid	Condensation	Formation of frost and snow

Freezing is the change of a liquid to the solid state. The freezing of liquid water to ice is a common example.

$$H_2O(l) \longrightarrow H_2O(s) \quad \text{(freezing)}$$
$$\text{liquid water} \qquad \text{ice}$$

The casting of a metal object involves the melting, then the freezing, of the metal in a mold.

Vaporization is the change of a solid or a liquid to the vapor. For example,

$$H_2O(l) \longrightarrow H_2O(g) \quad \text{(vaporization)}$$
$$\text{liquid water} \quad \text{water vapor}$$

The change of a solid to the vapor is specifically referred to as **sublimation.** Although the vapor pressure of many solids is quite low, some (usually molecular solids) have appreciable vapor pressure. Ice, for instance, has a vapor pressure of 4.7 mmHg at 0°C. For this reason, a pile of snow will slowly disappear in winter even though the temperature is too low for it to melt. The snow is being changed directly to water vapor:

$$H_2O(s) \longrightarrow H_2O(g) \quad \text{(sublimation)}$$
$$\text{ice, snow} \quad \text{water vapor}$$

Sublimation can be used to purify solids that readily vaporize. Figure 11.23 shows a simple apparatus for the purification of iodine by sublimation. Impure iodine is heated in a beaker so that it vaporizes, leaving nonvolatile impurities behind. The vapor crystallizes on the bottom surface of a dish containing ice. Freeze-drying of foods is a commercial application of sublimation. Brewed coffee, for example, is frozen and placed in a vacuum to remove water vapor. The ice continues to sublime until it is all gone, leaving freeze-dried coffee. Most freeze-dried foods are easily reconstituted by adding water.

Condensation is the change of a gas to either the liquid or solid state. Dew is liquid water formed by condensation of water vapor from the atmosphere. Frost is solid water formed by direct condensation of water vapor from the atmosphere without first forming liquid water. Snow is formed by a similar process in the upper atmosphere.

$$H_2O(g) \longrightarrow H_2O(l) \quad \text{(condensation)}$$
$$\text{water vapor} \qquad \text{dew}$$

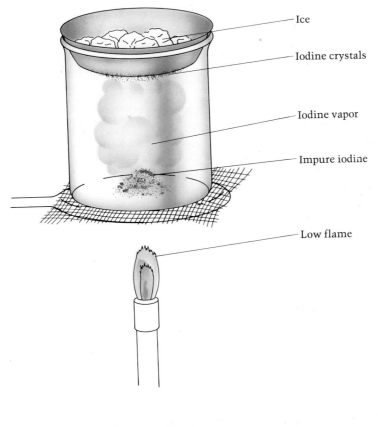

Figure 11.23
Purification by sublimation. Iodine is volatile, but the impurities are not. Iodine vapor crystallizes on the cool underside of the dish.

$$H_2O(g) \longrightarrow H_2O(s) \quad \text{(condensation)}$$

water vapor   frost, snow

When a substance that is normally a gas, such as carbon dioxide, changes to the liquid state, the phase change is often referred to as *liquefaction*.

11.8 Boiling Point and Melting Point

The boiling point and melting point of a substance are properties that, like vapor pressure, involve phase changes. In fact, the boiling point is closely related to the vapor pressure of the liquid.

Boiling Point

The temperature at which the vapor pressure of a liquid equals the atmospheric pressure is called the **boiling point** of the liquid. As the temperature of a liquid is raised, the vapor pressure increases. When the vapor pressure equals the atmospheric pressure, stable bubbles of vapor form within the body of the liquid. This process is called boiling. Once boiling starts, the temperature of the liquid remains at the boiling point (as long as sufficient heat is supplied) until no more liquid remains.

Since the atmospheric pressure can vary, the boiling point of a liquid can vary. For instance, at 1.00 atm (the average atmospheric pressure at sea level), the boiling point of water is 100°C. But at 0.83 atm (the average atmospheric pressure in Denver, Colorado, at 1609 m above sea level), the boiling point of water is 95°C. The *normal boiling point* of a liquid is the boiling point at 1 atm. ■

Melting Point

A pure liquid changes to a crystalline solid, or freezes, at a definite temperature called the **freezing point.** This crystalline solid changes to the liquid, or melts, at a definite temperature called the **melting point,** which is identical to the freezing point. The melting or freezing occurs at the temperature where the liquid and solid are in dynamic equilibrium: ■

$$\text{Solid} \rightleftharpoons \text{liquid}$$

Impure substances usually melt over a temperature range, instead of at a definite temperature.

Both the melting point and boiling point are characteristic properties of a substance and may be used to help in identifying it. Table 11.8 gives melting points and boiling points of several substances.

Enthalpy Changes

Any change of state involves the addition or removal of energy as heat from the substance. For example, heat is needed to melt a solid or to vaporize a liquid; these are endothermic processes.

The enthalpy change for the melting of a solid is called the **heat of fusion**

■ If you want to cook a stew in less time than is possible at atmospheric pressure, you can use a pressure cooker. This is especially valuable at high altitudes, where water boils at a lower temperature. The steam pressure inside the cooker is allowed to build up to a given value before being released by a valve. The stew then boils at a higher temperature and is cooked more quickly.

■ Unlike boiling points, melting points are affected only by large pressure changes. Small variations in the barometric pressure have negligible effect. The difference arises from the relative incompressibility of liquids and solids compared to gases.

Table 11.8
Melting Points and
Boiling Points (at 1 atm)

Name	Type of Substance	Melting Point, °C	Boiling Point, °C
Neon, Ne	Molecular	−249	−246
Hydrogen sulfide, H_2S	Molecular	−86	−61
Chloroform, $CHCl_3$	Molecular	−64	62
Water, H_2O	Molecular	0	100
Acetic acid, CH_3COOH	Molecular	17	118
Mercury, Hg	Metallic	−39	357
Sodium, Na	Metallic	98	883
Tungsten, W	Metallic	3410	5660
Cesium chloride, CsCl	Ionic	645	1290
Sodium chloride, NaCl	Ionic	801	1413
Magnesium oxide, MgO	Ionic	2800	3600
Quartz, SiO_2	Covalent network	1610	2230
Diamond, C	Covalent network	3550	4827

(or enthalpy of fusion) and is denoted ΔH_{fus}. For ice, the heat of fusion is 6.01 kJ per mole:

$$H_2O(s) \longrightarrow H_2O(l); \quad \Delta H_{fus} = 6.01 \text{ kJ/mol}$$

The enthalpy change for the vaporization of a liquid is called the **heat of vaporization** (or enthalpy of vaporization). At 100°C, the heat of vaporization of water is 40.7 kJ per mole: ■

$$H_2O(l) \longrightarrow H_2O(g); \quad \Delta H_{vap} = 40.7 \text{ kJ/mol}$$

Note that much more heat is required for vaporization than is required for melting. Melting needs only enough energy for the molecules to escape from their sites in the crystal lattice. For vaporization, enough energy must be supplied to break most of the intermolecular attractions.

A refrigerator relies on the cooling effect accompanying vaporization. Its mechanism contains an enclosed gas, such as ammonia or dichlorodifluoromethane (Freon), CCl_2F_2, that can be liquefied under pressure. As the liquid is allowed to evaporate, it absorbs heat and thus cools its surroundings (the interior space of the refrigerator). Gas from the evaporation is recycled to a compressor, where it is liquefied again.

■ That heat is required for vaporization can be demonstrated by evaporating a dish of water inside a vessel attached to a vacuum pump. With the pressure kept low, water evaporates quickly enough to freeze the water remaining in the dish.

Example 11.7

A particular refrigerator cools by evaporating liquefied dichlorodifluoromethane, CCl_2F_2. How many kilograms of this liquid must be evaporated to freeze a tray of water at 0°C to ice at 0°C? The mass of the water is 525 g, the heat of fusion of ice is 6.01 kJ/mol, and the heat of vaporization of dichlorodifluoromethane is 17.4 kJ/mol.

Solution

The heat that must be removed to freeze 525 g of water at 0°C is

$$525 \text{ g } H_2O \times \frac{1 \text{ mol } H_2O}{18.0 \text{ g } H_2O} \times \frac{-6.01 \text{ kJ}}{1 \text{ mol } H_2O} = -175 \text{ kJ}$$

The minus sign indicates that heat energy is taken away from the water. Consequently, the vaporization of dichlorodifluoromethane absorbs 175 kJ of heat.

$$175 \text{ kJ} \times \frac{1 \text{ mol } CCl_2F_2}{17.4 \text{ kJ}} \times \frac{121 \text{ g } CCl_2F_2}{1 \text{ mol } CCl_2F_2}$$
$$= 1.22 \times 10^3 \text{ g } CCl_2F_2$$

Thus, 1.22 kg of dichlorodifluoromethane must be evaporated.

Exercise 11.8

The heat of vaporization of ammonia is 4.80 kJ/mol. How much heat is required to vaporize 1.00 kg of ammonia? How many grams of water at 0°C could be frozen to ice at 0°C by the evaporation of this amount of ammonia?
(See Problems 11.43, 11.44, 11.45, and 11.46.)

The heat of vaporization of a liquid can be obtained from vapor-pressure data. The vapor pressure is needed for two different temperatures. Suppose the vapor pressure of a liquid is P_1 at absolute temperature T_1 and P_2 at absolute temperature T_2. According to the *Clausius-Clapeyron equation*:

$$\log \frac{P_2}{P_1} = \frac{\Delta H_{vap}}{2.303 \, R} \left(\frac{T_2 - T_1}{T_2 T_1} \right)$$

Here, R is the gas constant, 8.31 J/(K · mol). As an example of the use of this equation, consider the calculation of the heat of vaporization of water from the following data. The vapor pressure of water at 90°C (363 K) is 526 mmHg and at 100°C (373 K) is 760 mmHg. We identify one vapor pressure and its absolute temperature as P_1 and T_1. We will choose P_1 = 526 mmHg and T_1 = 363 K. Then, P_2 = 760 mmHg and T_2 = 373 K. Now we substitute into the Clausius-Clapeyron equation:

$$\log \frac{760 \text{ mmHg}}{526 \text{ mmHg}} = \frac{\Delta H_{vap}}{2.303 \times 8.31 \text{ J/(K·mol)}} \times \left(\frac{373 \text{ K} - 363 \text{ K}}{373 \text{ K} \times 363 \text{ K}}\right)$$

Solving for ΔH_{vap} gives

$$\Delta H_{vap} = 2.303 \times 8.31 \text{ J/mol} \times \left(\frac{373 \times 363}{373 - 363}\right) \times \log \frac{760}{526}$$

$$= 4.1 \times 10^4 \text{ J/mol, or 41 kJ/mol}$$

Within the precision of the calculation, this value is equal to the one given earlier for ΔH_{vap} for water.

11.9 Relating Physical Properties to Structure

Many physical properties of a substance can be directly related to the structure of the substance. Let us look at several of these properties.

Melting Point and Structure

For a solid to melt, the forces holding the structural units in the lattice sites must be broken, at least partially. In a molecular solid, these forces are weak intermolecular attractions. Thus, molecular solids tend to have low melting points (usually below 300°C). At room temperature, many molecular substances are either liquids (such as water and ethanol) or gases (such as carbon dioxide and ammonia). By contrast, for a covalent network solid to melt, covalent bonds must be broken. These substances have very high melting points. Quartz, for example, melts at 1610°C; diamond melts at above 3550°C.■

Ionic solids have high melting points for the same reason: chemical bonds must be broken. For example, sodium chloride melts at 801°C and magnesium oxide melts at 2800°C.

The difference between the melting points of sodium chloride and magnesium oxide can be explained in terms of the charges on the ions. We expect the melting point to rise with lattice energy.■ Lattice energies, however, depend on the product of the ionic charges. For sodium chloride (Na^+Cl^-) this product is $(+1) \times (-1) = -1$, whereas for magnesium oxide ($Mg^{2+}O^{2-}$) this product is $(+2) \times (-2) = -4$. Thus, the lattice energy is greater for magnesium oxide, and this is reflected in the much higher melting point of magnesium oxide.

Metals often have high melting points, but there is considerable variability. Mercury, which is a liquid at room temperature, melts at -39°C.

■ Both molecular and covalent network solids have covalent bonds, but none of the covalent bonds are broken during the melting of a molecular solid.

■ Lattice energy is the energy needed to separate a crystal into isolated ions in the gas phase. It represents the strength of attraction of ions in the solid.

Tungsten melts at 3410°C, the highest melting point of any metallic element. In general, melting points are low for the elements at the left side of the periodic table (Groups IA and IIA), but increase as we move right to the transition metals. Those elements in the center of the transition series (such as tungsten) have high melting points. Then, moving further to the right, the melting points decrease and are again low for Group IIB elements.

Boiling Point and Structure

The boiling points of liquids, like the melting points of solids, reflect the strength of the forces between the structural units. Molecular solids have low boiling points because the molecules are held in the liquid state by weak intermolecular forces. The boiling points of molecular substances usually increase with molecular weight because the strength of London forces usually increases with molecular weight. Figure 11.24 shows that the boiling points of the noble gases increase with atomic weight, as expected.

Figure 11.24 also shows the boiling points of the binary compounds of Group VIA elements with hydrogen. The boiling points of hydrogen sulfide (H_2S), hydrogen selenide (H_2Se), and hydrogen telluride (H_2Te) increase with molecular weight as expected.■ They are all gases at normal temperatures. Water, which has an even smaller molecular weight, however, is normally

■ Both London forces and dipole-dipole forces are present. London forces predominate, however, and increase with molecular weight.

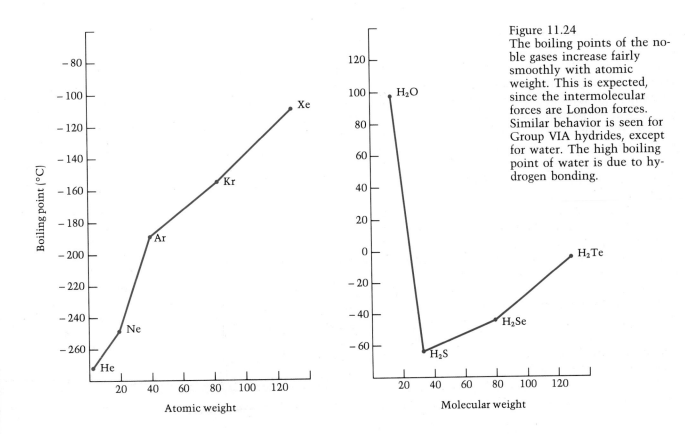

Figure 11.24
The boiling points of the noble gases increase fairly smoothly with atomic weight. This is expected, since the intermolecular forces are London forces. Similar behavior is seen for Group VIA hydrides, except for water. The high boiling point of water is due to hydrogen bonding.

a liquid. Its much higher boiling point is due to hydrogen bonding, which is negligible in H_2S, H_2Se, and H_2Te.

If you look again at Table 11.8, you will see the relationship described here between type of substance and melting and boiling points.

Example 11.8

Arrange the following elements in order of increasing melting point: silicon, hydrogen, lithium. Explain your reasoning.

Solution

Hydrogen is a molecular substance (H_2) with a very low molecular weight (2 amu). Thus, we expect it to have a very low melting point. Lithium is a Group IA ele-ment and is expected to be a relatively low-melting metal. Except for mercury, however, all metals are solids below 25°C. So the melting point of lithium is well above that of hydrogen. Silicon might be expected to have a covalent network structure like carbon and to have a high melting point. Therefore, the order of the elements in order of increasing melting point is hydrogen, lithium, and silicon.

Exercise 11.9

Decide the type of solid that is formed for each of the following substances: C_2H_5OH, CH_4, CH_3Cl, $MgSO_4$. On the basis of the type of solid and the expected magnitude of intermolecular forces (for molecular crystals), arrange these substances in order of increasing melting point. Explain your reasoning.

(See Problems 11.53 and 11.54.)

Hardness and Structure

Hardness depends on how easily the structural units can be moved. Thus, molecular crystals are rather soft in comparison with ionic and covalent network crystals. Three-dimensional covalent network solids are usually quite hard because of the rigidity given to the structure by strong covalent bonds throughout it. Diamond and silicon carbide (SiC), which are three-dimensional covalent networks, are among the hardest substances known.

If a covalent network solid consists of sheets, as in graphite, the sheets tend to separate easily to show this layer structure. The ease with which layers of graphite slide over one another accounts for some of the applications of this substance. The "lead" in a lead pencil is graphite. As the pencil is used for writing, layers of graphite are rubbed off. Powdered graphite is also used to lubricate locks and other closely fitted metal surfaces, since the graphite particles easily move over one another as the layers slide.

Molecular and ionic crystals are brittle. Any shifting of layers of molecules or ions, as happens when the crystal is struck, drastically weakens the forces of attraction. This is especially easy to see in the case of an ionic solid. When a layer shifts with respect to the next one, adjacent layers have identical charges in contact (see Figure 11.25, *top*), and the layers are no longer bonded. Hence, the crystal shatters.

Metallic crystals, by contrast, are *malleable*, that is, they can be shaped by hammering. When one layer shifts over another, the bonding does not change significantly (Figure 11.25, *bottom*).

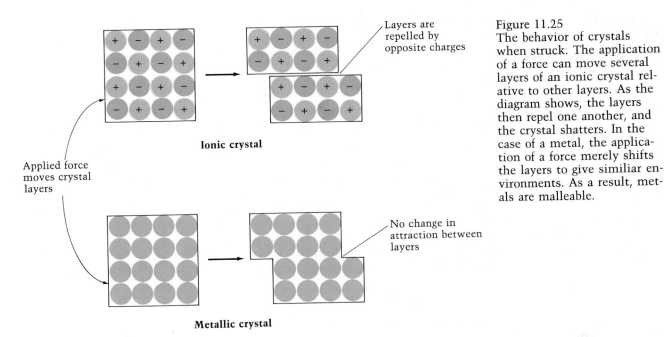

Ionic crystal

Metallic crystal

Applied force moves crystal layers

Layers are repelled by opposite charges

No change in attraction between layers

Figure 11.25
The behavior of crystals when struck. The application of a force can move several layers of an ionic crystal relative to other layers. As the diagram shows, the layers then repel one another, and the crystal shatters. In the case of a metal, the application of a force merely shifts the layers to give similiar environments. As a result, metals are malleable.

Electrical Conductivity and Structure

One of the characteristic properties of metals is their good electrical conductivity. The delocalized valence electrons are easily moved by an electrical field and are responsible for carrying the electric current. Graphite is a moderately good conductor because delocalization leads to mobile electrons within each two-dimensional sheet. In contrast, most covalent and ionic solids are nonconductors, since the electrons are localized to particular atoms or bonds. Ionic substances do become conducting in the liquid state, however, because the ions can move. In an ionic liquid, it is the ions that carry the electric current.

Table 11.9 summarizes the properties expected of different types of solids and liquids.

Table 11.9
Properties of the Different Types of Solids and Liquids

Types of Substance	Melting Point of Solid	Boiling Point of Liquid	Hardness and Brittleness	Electrical Conductivity
Molecular	Low	Low	Soft and brittle	Nonconducting
Metallic	Variable	Variable	Variable hardness; malleable	Conducting
Ionic	High to very high	High to very high	Hard and brittle	Nonconducting solid; conducting liquid
Covalent network	Very high	Very high	Very hard	Usually nonconducting

11.10 Phase Diagrams

As we mentioned in the chapter opening, the solid, liquid, and gaseous states of carbon dioxide exist under different temperature and pressure conditions. A **phase diagram** is a convenient way to summarize the conditions under which the different states of a substance are stable.

Melting-Point Curve

Figure 11.26 is a phase diagram for water. It consists of three curves that divide the diagram into regions labeled "solid," "liquid," and "gas." In each given region, the indicated state is stable. Every point on each of the curves indicates experimentally determined temperatures and pressures at which two states are in equilibrium. Thus, the curve AB, dividing the solid region from the liquid region, represents the conditions under which the solid and liquid are in equilibrium:

$$\text{Solid} \rightleftharpoons \text{liquid}$$

This curve gives the melting points of the solid at various pressures.

Usually, the melting point is only slightly affected by pressure. For this reason, the melting-point curve AB is nearly vertical. If a liquid is more dense than the solid, as is the case for water, the melting point decreases with pressure.■ The melting-point curve in such cases leans slightly to the

■ The following is a dramatic demonstration of the effect of pressure on the melting of ice. Suspend a block of ice between two chairs. Then loop a length of wire over the block and place weights on the ends of the wire. The pressure of the wire will melt the ice under it. Liquid water will flow over the top of the wire and freeze again, since it is no longer under the pressure of the wire. As a result, the wire will cut through the ice and fall to the floor, but the block will remain as one piece.

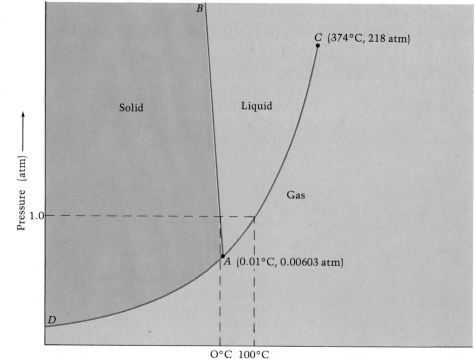

Figure 11.26
Phase diagram for water (not to scale). The curves AB, AC, and AD divide the diagram into regions that give the temperatures and pressures for which only one phase is stable. Along any curve, the two phases from the adjoining regions are in equilibrium.

left. In the case of ice, the decrease is indeed slight—only 1°C for a pressure increase of 133 atm. Usually, the liquid state is less dense than the solid. In that case, the melting-point curve leans slightly to the right.

Vapor-Pressure Curve for the Liquid

The curve AC that divides the liquid region from the gaseous region gives the vapor pressures of the liquid at various temperatures. It also gives the boiling points of the liquid for various pressures. The boiling point of water at 1 atm is shown on the phase diagram (Figure 11.26).

Vapor-Pressure Curve for the Solid

The curve AD that divides the solid region from the gaseous region gives the vapor pressures of the solid at various temperatures. This curve intersects the other curves at the point A, called the **triple point.** All three states of the substance are in equilibrium at the triple point, which for water occurs at 0.01°C, 0.00603 atm (4.58 mmHg).■

Suppose a solid is warmed at a pressure below the pressure at the triple point. In a phase diagram, this corresponds to moving along a horizontal line below the triple point. We can see from Figure 11.26 that such a line will intersect curve AD, which is the vapor-pressure curve for the solid. Thus, the solid will pass directly into the gas, that is, the solid will sublime. Freeze-drying of a food is accomplished by placing the frozen food in a vacuum (below 0.00603 atm) so that the ice in it sublimes.

The triple point of carbon dioxide is at -57°C and 5.1 atm (Figure 11.27). Therefore, the solid sublimes if warmed at any pressure below 5.1 atm. This is why solid carbon dioxide sublimes at normal atmospheric pressure (1 atm). Above 5.1 atm, however, the solid melts if warmed.

■ Because the triple point for water occurs at a definite temperature, it is used to define the Kelvin thermometer scale. The temperature of water at its triple point is defined to be 273.16 K or 0.01°C.

Critical Temperature and Pressure

Imagine an experiment in which liquid and gaseous carbon dioxide are sealed into a thick-walled glass vessel at 20°C. At this temperature, the liquid is in equilibrium with its vapor at a pressure of 57 atm. We see that the liquid and vapor are separated by a well-defined boundary or meniscus. Now suppose the temperature is raised. The vapor pressure will increase, and at 30°C it is 71 atm. Then, as the temperature approaches 31°C, a curious thing happens. The meniscus becomes fuzzy and less well defined. At 31°C, the meniscus disappears altogether. Above this temperature, there is only one fluid phase, carbon dioxide gas.

The temperature above which the liquid state of a substance no longer exists is called the **critical temperature.** For carbon dioxide, this is 31°C. The pressure at the critical temperature is called the **critical pressure** (73 atm for carbon dioxide). It is the minimum pressure that must be applied to a gas at the critical temperature to liquefy it.

On a phase diagram, the preceding experiment corresponds to following

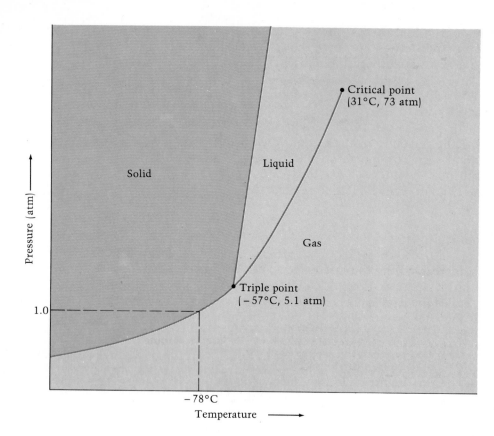

Figure 11.27
Phase diagram for carbon dioxide (not to scale).

the vapor-pressure curve, where the liquid and vapor are in equilibrium. Note that this curve in Figure 11.27 ends at a point at which the temperature and pressure have their critical values. This is the *critical point*. If you look at the phase diagram for water, you will see that the vapor-pressure curve for the liquid similarly ends, at point C, which is the critical point for water. In this case, the critical temperature is 374°C and the critical pressure is 218 atm.

Many important gases cannot be liquefied at room temperature. Nitrogen, for example, has a critical temperature of −147°C. This means the gas cannot be liquefied until the temperature is below −147°C.

Example 11.9

The critical temperature of ammonia and nitrogen are 132°C and −147°C, respectively. Explain why ammonia can be liquefied at room temperature by merely compressing the gas to a high enough pressure, whereas the compression of nitrogen requires a low temperature as well.

Solution

The critical temperature of ammonia is well above room temperature. If ammonia gas at room temperature is compressed, it will liquefy. Nitrogen, however, has a critical temperature well below room temperature. It cannot be liquefied by compression unless the temperature is below the critical temperature.

Exercise 11.10

Describe how you could liquefy the following gases: (a) methyl chloride, CH_3Cl (critical point, 144°C, 66 atm); (b) oxygen, O_2 (critical point, -119°C, 50 atm).

(See Problems 11.57 and 11.58.)

A Checklist for Review

Important Terms

intermolecular force (11.1)
van der Waals forces (11.1)
London force (11.1)
dipole–dipole force (11.1)
hydrogen bonding (11.1)
vapor pressure (11.2)
surface tension (11.2)
viscosity (11.2)
molecular solids (11.3)
metallic solids (11.3)
ionic solids (11.3)
covalent network solids (11.3)
crystalline solids (11.4)

amorphous solids (11.4)
unit cell (11.4)
crystal systems (11.4)
simple cubic (11.4)
body-centered cubic (11.4)
face-centered cubic (11.4)
hexagonal closest-packed structure (11.5)
cubic closest-packed structure (11.5)
coordination number (11.5)
change of state (phase change) (p. 371)
melting (11.7)
freezing (11.7)
vaporization (11.7)

sublimation (11.7)
condensation (11.7)
boiling point (11.8)
freezing point (11.8)
melting point (11.8)
heat of fusion (11.8)
heat of vaporization (11.8)
phase diagram (11.10)
triple point (11.10)
critical temperature (11.10)
critical pressure (11.10)

Summary of Facts and Concepts

Liquids and solids are relatively incompressible states of matter. In a molecular liquid or solid, the molecules are held closely together by *intermolecular forces.* Such properties of liquids as *vapor pressure, surface tension,* and *viscosity* depend on the strength of these intermolecular forces.

In solids, the forces of attraction are sufficient to give a rigid structure. A *crystalline solid* has an ordered arrangement or crystal lattice, which we may think of as constructed from *unit cells.* An *amorphous solid* is noncrystalline. Solids may be classified also according to the type of force between structural units. Thus, we have *molecular, metallic, ionic,* and *covalent network solids.*

Any given state of a substance may change to another state. Melting and freezing are examples of such *changes of state* or *phase changes.* The boiling point and melting point are characteristic properties of a substance that involve phase changes. These properties and others, such as hardness, can be related to the structure of the substance. The conditions under which a given state of a substance exists can be summarized by a *phase diagram.*

Operational Skills

1. Given the molecular structure, state the kinds of intermolecular forces expected for a substance (Example 11.1).

2. Given two liquids, decide on the basis of the intermolecular forces which has the higher vapor pressure at a given temperature (Example 11.2).

3. From what you know about the bonding in a solid, classify it as molecular, metallic, ionic, or covalent network (Example 11.3).

4. Given the description of a unit cell, find the number of atoms per cell (Example 11.4).

5. Given the edge length of the unit cell and the density of a metal, calculate the mass of a metal atom (Example 11.5).

6. Given the type of unit cell, the density, and the atomic weight for an element, calculate the edge length of the unit cell (Example 11.6).

7. Given the heat of fusion (or vaporization) of a substance, calculate the amount of heat required to melt (or vaporize) a given quantity of substance (Example 11.7).

8. Given a list of substances, arrange them in order of increasing melting point or boiling point from what you know of their structures (Example 11.8).

9. Given the critical temperature and pressure of a substance, describe the conditions necessary for liquefying the gaseous substance (Example 11.9).

Review Questions

11.1 Explain the origin of the London force that exists between two molecules.

11.2 Describe vapor pressure in molecular terms. What is meant by saying it involves a dynamic equilibrium?

11.3 Why does the vapor pressure of a liquid depend on the intermolecular forces?

11.4 Explain the surface tension of a liquid in molecular terms. How does the surface tension make a liquid act as if it had a "skin"?

11.5 Describe the distinguishing characteristics of a crystalline solid and an amorphous solid.

11.6 Describe the face-centered cubic unit cell.

11.7 Describe the structure of thallium(I) iodide, which has the same structure as cesium chloride.

11.8 What is the coordination number of Cs^+ in CsCl? of Na^+ in NaCl? of Zn^{2+} in ZnS?

11.9 Explain the production of an x-ray diffraction pattern by a crystal in terms of the interference of waves.

11.10 List the different phase changes that are possible and give examples of each.

11.11 Explain why 15 g of steam at 100°C will melt more ice than 15 g of liquid water at 100°C.

11.12 Why is the heat of fusion of a substance smaller than its heat of vaporization?

11.13 Explain why evaporation leads to cooling of the liquid.

11.14 Describe how you could purify iodine by sublimation.

11.15 Why do molecular substances have relatively low melting points?

11.16 Describe the behavior of a liquid and its vapor in a closed vessel as the temperature increases.

11.17 Gases that cannot be liquefied at room temperature merely by compression are called "permanent" gases. How could you liquefy such a gas?

11.18 The pressure in a cylinder of nitrogen continuously decreases as gas is released from it. On the other hand, a cylinder of propane maintains a constant pressure as propane is released. Explain this difference in behavior.

Problems

Intermolecular Forces

11.19 For each of the following substances, list the kinds of intermolecular forces expected:
   (a) boron trifluoride, BF_3
   (b) isopropyl alcohol, $CH_3CHOHCH_3$
   (c) hydrogen iodide, HI
   (d) krypton, Kr

11.21 Arrange the following substances in order of increasing magnitude of the London forces: $SiCl_4$, CCl_4, $GeCl_4$.

11.20 Of the following compounds, identify those you expect to exhibit only London forces:
   (a) carbon tetrachloride, CCl_4
   (b) methyl chloride, CH_3Cl
   (c) phosphorus trichloride, PCl_3
   (d) phosphorus pentachloride, PCl_5

11.22 Arrange the following substances in order of increasing magnitude of the London forces: Ar, He, Kr.

11.23 The molecular weight of acetic acid, CH_3COOH, calculated from its vapor density at 110°C is 91 amu. Explain why this result is much higher than that obtained from the formula CH_3COOH.

11.24 Benzoic acid was dissolved in benzene. For this solution at room temperature, a molecular-weight determination of benzoic acid gives a value corresponding closely to the formula $(C_6H_5COOH)_2$. Draw the structural formula for this species in solution.

Vapor Pressure

11.25 Methane, CH_4, reacts with chlorine, Cl_2, to produce a series of chlorinated hydrocarbons: methyl chloride (CH_3Cl), methylene chloride (CH_2Cl_2), chloroform $(CHCl_3)$, and carbon tetrachloride (CCl_4). Which of these compounds has the lowest vapor pressure at room temperature? Explain.

11.26 Which of the following hydrogen halides do you expect to have the highest vapor pressure at room temperature: HCl, HBr, HI? Explain.

11.27 Trimethylamine, $(CH_3)_3N$, and propylamine, $CH_3CH_2CH_2NH_2$, have odors that are ammonia-like and fishy. Explain why propylamine might be expected to have a much lower vapor pressure at a given temperature than trimethylamine.

11.28 Hydrogen iodide, HI, has a vapor pressure of 760.0 mmHg at −35.1°C. Hydrogen fluoride, however, has a vapor pressure of 69.6 mmHg at the same temperature. Explain why the vapor pressure of hydrogen fluoride is lower than that of hydrogen iodide when we might have expected the opposite result from their molecular weights.

Types of Solids

11.29 Classify each of the following solid elements as molecular, metallic, ionic, or covalent network:
- (a) tin, Sn
- (b) germanium, Ge
- (c) sulfur, S_8
- (d) iodine, I_2

11.30 Which of the following are expected to be molecular solids?
- (a) sodium hydroxide, NaOH
- (b) solid ethane, C_2H_6
- (c) nickel, Ni
- (d) solid silane, SiH_4

Crystal Structure

11.31 How many atoms are there in a simple cubic unit cell of an atomic crystal?

11.32 How many atoms are there in a body-centered cubic unit cell of an atomic crystal?

11.33 Metallic iron has a body-centered cubic lattice with a unit cell whose edge length is 2.866 Å. The density of iron is 7.87 g/cm^3. What is the mass of an iron atom? Compare this value with the value you obtain from the molar mass.

11.34 Nickel has a face-centered unit cell with an edge length of 3.524 Å. The density of metallic nickel is 8.91 g/cm^3. What is the mass of a nickel atom? From the atomic weight, calculate Avogadro's number.

11.35 Copper metal has a face-centered cubic structure and a density of 8.93 g/cm^3. The atomic weight is 63.5 amu. Calculate the edge length of the unit cell.

11.36 Barium metal has a body-centered cubic lattice; the density is 3.51 g/cm^3. From these data and the atomic weight, calculate the edge length of a unit cell.

11.37 Gold has cubic crystals whose unit cell has an edge length of 4.079 Å. The density of the metal is 19.3 g/cm^3. From these data and the atomic weight, calculate the number of gold atoms in a unit cell. What type of cubic lattice does gold have?

11.38 Chromium forms cubic crystals whose unit cell has an edge length of 2.885 Å. The density of the metal is 7.20 g/cm^3. Use these data and the atomic weight to calculate the number of atoms in a unit cell. What type of cubic lattice does chromium have?

11.39 Tungsten has a body-centered cubic lattice. The edge length of the unit cell is 3.165 Å. The atomic weight of tungsten is 183.8 amu. Calculate its density.

**11.41 Metallic magnesium has a hexagonal closest-packed structure and a density of 1.74 g/cm^3. Assume magnesium atoms to be spheres of radius r. Since magnesium has a closest-packed structure, 74.1% of the space is occupied by atoms. Calculate the volume of each atom; then find the atomic radius, r. Note that the volume of a sphere is $4\pi r^3/3$.

11.40 Lead has a face-centered cubic lattice with a unit-cell edge length of 4.950 Å. The atomic weight is 207.2 amu. What is the density of lead?

**11.42 Metallic barium has a body-centered cubic structure and a density of 3.51 g/cm^3. Assume barium atoms to be spheres. The spheres in a body-centered array occupy 68.0% of the total space. Find the atomic radius of barium. (See Problem 11.41.)

Heat of Phase Change

11.43 How much heat energy must be removed from benzene, C_6H_6, at its freezing point to freeze 15.0 g of benzene? The heat of fusion of benzene is 9.95 kJ/mol.

11.45 Water at 0°C was placed in a dish inside a vessel maintained at low pressure by a vacuum pump. After a quantity of water had evaporated, the remainder froze. If 7.83 g of ice at 0°C was obtained, how much liquid water must have evaporated? The heat of fusion of water is 6.01 kJ/mol and the heat of vaporization is 44.9 kJ/mol.

11.47 A quantity of ice at 0°C is added to 50.0 g of water in a glass at 55°C. After the ice melted, the temperature of the water in the glass was 15°C. How much ice was added?

11.49 White phosphorus, P_4, has a vapor pressure of 400.0 mmHg at 251.0°C and 760.0 mmHg at 280.0°C. What is the heat of vaporization of this substance?

*11.51 Chloroform, $CHCl_3$, boils at 61.7°C and has a heat of vaporization of 31.4 kJ/mol. What is the vapor pressure at 20.0°C?

11.44 Liquid butane, C_4H_{10}, is stored in cylinders to be used as a fuel. Suppose 25.5 g of butane gas are removed from a cylinder. How much heat energy must be provided to vaporize this much gas? The heat of vaporization of butane is 21.3 kJ/mol.

11.46 A quantity of ice was added to 50.0 g of water at 21.0°C to give water at 0.0°C. How much ice was added? The heat of fusion of water is 6.01 kJ/mol and the specific heat is 4.18 J/(g · °C).

11.48 Steam at 100°C was passed into a flask containing 175 g of water at 21°C, where the steam condensed. How many grams of steam must have condensed if the temperature of the water in the flask was raised to 85°C?

11.50 Carbon disulfide, CS_2, has a vapor pressure of 400.0 mmHg at 28.0°C and 760.0 mmHg at 46.5°C. What is the heat of vaporization of this substance?

*11.52 Methanol, CH_3OH, boils at 65.0°C and has a heat of vaporization of 37.4 kJ/mol. What is the vapor pressure at 20.0°C?

Structure and Physical Properties

11.53 List the following substances in order of increasing boiling points: methanol, CH_3OH; ethylene glycol, CH_2OHCH_2OH; ethane, C_2H_6; methane, CH_4.

11.55 Associate each type of solid in the left-hand column with two of the properties listed in the right-hand column. Each property may be used more than once.

(a) molecular solid	low-melting
(b) ionic solid	high-melting
(c) metallic solid	brittle
(d) covalent network	malleable
solid	hard
	electrically conducting

11.54 Explain the trend in the boiling points of the hydrogen halides HCl, HBr, and HI. Explain why HF does not follow this trend. The boiling points are: HF, 20°C; HCl, −85°C; HBr, −67°C; HI, −51°C.

11.56 On the basis of the description given, classify each of the following solids as molecular, metallic, ionic, or covalent network. Explain your answers.
   (a) A lustrous, yellow solid that conducts electricity
   (b) A hard, black solid melting at 2350°C to give a nonconducting liquid
   (c) A nonconducting, pink solid melting at 650°C to give an electrically conducting liquid
   (d) Red crystals having a characteristic odor and melting at 171°C

Phase Diagrams

11.57 Which of the following substances can be liquefied by applying pressure at 25°C? For those that cannot, describe the conditions under which they can be liquefied.

Substance	Critical Temperature	Critical Pressure
Sulfur dioxide, SO_2	158°C	78 atm
Acetylene, C_2H_2	36°C	62 atm
Methane, CH_4	−82°C	46 atm
Carbon monoxide, CO	−140°C	35 atm

11.58 A tank of gas at 21°C has a pressure of 1.0 atm. Using the data in the table, answer the following questions. Explain your answers.

(a) If the tank contains carbon tetrafluoride, CF_4, is the liquid state also present?

(b) If the tank contains butane, C_4H_{10}, is the liquid state also present?

Substance	Boiling Point at 1 atm	Critical Temperature	Critical Pressure
CF_4	−128°C	−46°C	41 atm
C_4H_{10}	−0.5°C	152°C	38 atm

*11.59 Bromine, Br_2, has a triple point at −7.3°C and 44 mmHg and a critical point at 315°C and 102 atm. The density of the solid is 3.4 g/cm³, and the density of the liquid is 3.1 g/cm³. Sketch a rough phase diagram of bromine, labeling all important features. Circle the correct word in each of the following sentences (and explain your answers):

(a) Bromine vapor at 40 mmHg will condense to the (liquid, solid) when cooled sufficiently.

(b) Bromine vapor at 400 mmHg will condense to the (liquid, solid) when cooled sufficiently.

*11.60 Krypton, Kr, has a triple point at −169°C and 133 mmHg and a critical point at −63°C and 54 atm. The density of the solid is 2.8 g/cm³, and the density of the liquid is 2.4 g/cm³. Sketch a rough phase diagram of krypton. Circle the correct word in each of the following sentences (and explain your answers):

(a) Solid krypton at 130 mmHg (melts, sublimes without melting) when the temperature is raised.

(b) Solid krypton at 760 mmHg (melts, sublimes without melting) when the temperature is raised.

Additional Problems

11.61 Iridium metal, Ir, crystallizes in a face-centered cubic structure. The edge length of the unit cell was determined by x-ray diffraction to be 3.839 Å. The density of iridium is 22.42 g/cm³. Calculate the mass of an iridium atom. Use Avogadro's number to calculate the atomic weight of iridium.

11.62 The edge length of the unit cell of tantalum metal, Ta, is 3.306 Å; the unit cell is body-centered cubic. Tantalum has a density of 16.69 g/cm³. What is the mass of a tantalum atom? Use Avogadro's number to calculate the atomic weight of tantalum.

**11.63 Use the answer from Problem 11.35 to calculate the radius of the copper atom. To do this, assume that copper atoms are spheres. Then note that the spheres on any face of a unit cell touch along the diagonal.

**11.64 Rubidium metal has a body-centered cubic structure. The density of the metal is 1.532 g/cm³. From this information and the atomic weight, calculate the edge length of the unit cell. Now, assume that rubidium atoms are spheres. Each corner sphere of the unit cell touches the body-centered sphere. Calculate the radius of a rubidium atom.

**11.65 Calculate the percent of volume that is actually occupied by spheres in a body-centered cubic lattice of identical spheres. You can do this by first relating the radius of a sphere, r, to the length of an edge of a unit cell, l. (Note that the spheres do not touch along an edge, but they do touch along a diagonal passing through the body-centered sphere.) Then calculate the volume of a unit cell in terms of r. The volume occupied by spheres equals the number of spheres per unit cell times the volume of a sphere ($4\pi r^3/3$).

**11.66 Calculate the percent of volume that is actually occupied by spheres in a face-centered cubic lattice of identical spheres. You can do this by first relating the radius of a sphere, r, to the length of an edge of a unit cell, l. (Note that the spheres do not touch along an edge, but they do touch along the diagonal of a face.) Then calculate the volume of a unit cell in terms of r. The volume occupied by spheres equals the number of spheres per unit cell times the volume of a sphere ($4\pi r^3/3$).

11.67 Decide which substance in each of the following pairs has the lower melting point. Explain how you made each choice.

(a) Potassium chloride, KCl, or calcium oxide, CaO
(b) Carbon tetrachloride, CCl₄, or hexachloro-ethane, C_2Cl_6
(c) Zinc, Zn, or chromium, Cr
(d) Acetic acid, CH_3COOH, or ethyl chloride, C_2H_5Cl

11.68 Decide which substance in each of the following pairs has the lower melting point. Explain how you made each choice.
(a) Magnesium oxide, MgO, or hexane, C_6H_{14}
(b) 1-Propanol, C_3H_7OH, or ethylene glycol, CH_2OHCH_2OH
(c) Silicon, Si, or sodium, Na
(d) Methane, CH_4, or silane, SiH_4

11.69 Ethylene glycol (CH_2OHCH_2OH) is a slightly viscous liquid that boils at 198°C. Pentane (C_5H_{12}), which has approximately the same molecular weight, is a mobile (nonviscous) liquid that boils at 36°C. Explain the differences in physical characteristics of these two compounds.

11.70 Benzoic acid, C_6H_5COOH, has a formula weight of 122 amu and a melting point of 122°C. Strontium oxide, SrO, has a formula weight of 104 amu and a melting point of 2430°C. Why does strontium oxide have a much higher melting point than benzoic acid?

*11.71 The percent relative humidity of a sample of air equals: (partial pressure of water vapor/vapor pressure of water) × 100. A sample of air at 23°C was cooled to 15°C, where moisture began to condense as dew. What was the relative humidity of the air at 23°C?

*11.72 A sample of air at 21°C has a relative humidity of 58%. At what temperature will water begin to condense as dew? (See Problem 11.71.)

**11.73 The vapor pressure of benzene is 100.0 mmHg at 26.1°C and 400.0 mmHg at 60.6°C. What is the boiling point of benzene at 760.0 mmHg?

**11.74 The vapor pressure of water is 17.5 mmHg at 20.0°C and 355.1 mmHg at 80.0°C. Calculate the boiling point of water at 760.0 mmHg.

11.75 Describe the behavior of carbon dioxide gas when compressed at the following temperatures:
(a) 20°C   (b) −70°C   (c) 40°C
The triple point of carbon dioxide is −57°C and 5.1 atm, and the critical point is 31°C and 73 atm.

11.76 Describe the behavior of iodine vapor when cooled at the pressures given:
(a) 120 atm   (b) 1 atm   (c) 50 mmHg
The triple point of iodine is 114°C and 90.1 mmHg, and the critical point is 512°C and 116 atm.

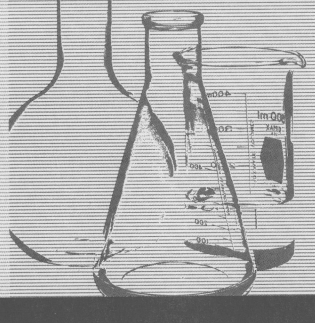

12. Solutions

Solubility; Colloid Formation

12.1 Types of Solutions Gaseous Solutions/ Liquid Solutions/ Solid Solutions

12.2 The Solution Process Factors Determining Solubility/ Molecular Solutions/ Ionic Solutions/ *Aside: Hemoglobin Solubility and Sickle Cell Anemia*

12.3 The Effects of Temperature and Pressure on Solubility Temperature Change/ Pressure Change/ Henry's Law

12.4 Colloids Tyndall Effect/ Types of Colloids/ Hydrophilic and Hydrophobic Colloids/ Coagulation/ Association Colloids

Colligative Properties

12.5 Ways of Expressing Concentration Mass Percentage of Solute/ Molality/ Mole Fraction/ Conversion of Concentration Units

12.6 Vapor Pressure of a Solution

12.7 Boiling-Point Elevation and Freezing-Point Depression

12.8 Osmosis

12.9 Ionic Solutions

There are various practical reasons for preparing solutions. For instance, most chemical reactions are run in solution. Also, solutions have particular properties that are useful. When gold is used for jewelry, it is mixed or alloyed with a small amount of silver. Gold–silver alloys are not only harder than pure gold, but they melt at lower temperatures and are therefore easier to cast.

Solubility varies with temperature and possibly pressure. (Solubility is the equilibrium amount of one substance that dissolves in another.) The variation of solubility with pressure can be a useful property. Acetylene gas (C_2H_2), for example, is used as a fuel in welding torches. It is transported under pressure in cylinders, in solution with acetone (CH_3COCH_3). Acetone is a liquid, and at 1 atm pressure one liter of the liquid dissolves 27 grams of acetylene. But at 12 atm, the pressure in a full cylinder, the same quantity of acetone dissolves 320 grams of acetylene, so that more can be transported. When the valve on a cylinder is opened and the pressure is reduced, acetylene gas comes out of solution.

Another useful property of solutions is their lower melting or freezing points compared with those of the major component. We have already mentioned the lowering of the melting point of gold when a small amount of silver is added. The use of ethylene glycol, CH_2OHCH_2OH, as an automobile antifreeze depends on the same property. Water containing ethylene glycol freezes at temperatures below the freezing point of pure water.

What is it that determines how much the freezing point of a solution is lowered? How does the solubility of a substance change when conditions such as pressure and temperature change? These are some of the questions that we will address in this chapter.

Chapter Overview

In the present chapter, we will examine some general characteristics of solutions. For example, we will look at the *factors determining solubility* and how solubility is affected by temperature and pressure changes. In the last part of the chapter, we will discuss *colligative properties* of solutions. These are properties, such as freezing-point depression, or lowering, that depend on the concentration of molecules or ions in solution, rather than on particular characteristics of the molecular or ionic substance.

Solubility; Colloid Formation

When sodium chloride dissolves in water, the resulting uniform dispersion of ions in water is called a *solution*. ■ In general, a solution is a homogeneous mixture of two or more substances, consisting of ions or molecules. A *colloid* is similar, in that it appears to be homogeneous like a solution. In fact, it consists of comparatively large particles of one substance dispersed throughout another substance or solution.

■ Ionic solutions were discussed in Chapter 9.

From the examples of solutions we mentioned in the chapter opening, we see that they may be quite varied in their characteristics. To begin our discussion, let us look at the different types of solutions that we might encounter.

Solution	State of Matter	Description
Air	Gas	Homogeneous mixture of gases
Soda water	Liquid	Gas (CO_2) dissolved in a liquid (H_2O)
Ethanol in water	Liquid	Liquid solution of two completely miscible liquids
Brine	Liquid	Solid (NaCl) dissolved in a liquid (H_2O)
Potassium–sodium alloy	Liquid	Solution of two solids (K + Na)
Hydrogen in platinum	Solid	Gas (H_2) dissolved in a solid (Pt)
Dental-filling alloy	Solid	Solution of a liquid (Hg) in a solid (Ag plus other metals)
Gold–silver alloy	Solid	Solution of two solids (Au + Ag)

Table 12.1
Examples of Solutions

12.1 Types of Solutions

Solutions may exist in any of the three states of matter, that is, they may be gases, liquids, or solids. Some examples are listed in Table 12.1.

Gaseous Solutions

In general, nonreactive gases or vapors can mix in all proportions to give a gaseous mixture. Two fluids that mix with or dissolve in each other are said to be **miscible.** Gases are thus completely miscible, that is, they mix in all proportions. Air, which is a mixture of nitrogen, oxygen, and smaller amounts of other gases, is an example of a gaseous solution.

Liquid Solutions

Most liquid solutions are obtained by dissolving a gas, liquid, or solid in some liquid. Soda water, for example, consists of a solution of carbon dioxide gas in water. Ethanol, C_2H_5OH, in water is an example of a liquid–liquid solution.■ Brine is water with sodium chloride (a solid) dissolved in it. Sea water contains both dissolved gases (from air) and solids (mostly sodium chloride).

It is also possible to make a liquid solution by mixing two solids together. Potassium–sodium alloy is an example of this. Both potassium and sodium are solid metals at room temperature, but a liquid solution results when the mixture contains 10 to 50% sodium.■

When a gas or solid dissolves in a liquid, the gas or solid is referred to as the **solute** and the liquid is called the **solvent.** For example, in an aqueous solution of sodium chloride, sodium chloride is the solute and water is the solvent. In the case of liquid–liquid solutions, the smaller amount of substance is called the solute, and the larger amount of substance is called the solvent.

■ Ethanol is completely miscible in water, so that solutions with varying amounts of ethanol and water can be prepared.

■ Sodium metal melts at 98°C and potassium metal melts at 63°C. When potassium is mixed with 20% sodium, for example, the melting point is lowered to −10°C. This is another example of the lowering of the freezing or melting point of solutions, which we will discuss later in the chapter. Sodium–potassium alloy is used as a heat-transfer medium in nuclear reactors.

Solid Solutions

Solid solutions are also possible. In the chapter opening, we mentioned gold–silver alloys. Dental-filling alloy is a solution of mercury (a liquid) in silver, with small amounts of other metals. Gases can also dissolve in solids. Hydrogen, H_2, for example, dissolves in platinum.

Exercise 12.1

Give an example of a solid solution prepared from a liquid and a solid.

(See Problems 12.15 and 12.16.)

12.2 The Solution Process

The *solubility* of a substance in a solvent is the maximum amount that can be dissolved at equilibrium at a given fixed temperature. A solution containing this maximum amount of substance is said to be *saturated*. For instance, the solubility of ammonium nitrate, NH_4NO_3, in water at 0°C is 118 grams per 100 mL of water. Any additional ammonium nitrate that is added to this saturated solution appears to remain undissolved at the bottom of the vessel.

In fact, the undissolved crystals are in *dynamic equilibrium* with the ions in solution:

$$NH_4NO_3(s) \rightleftharpoons NH_4^+(aq) + NO_3^-(aq)$$

As some ions leave the solid phase, other ions leave the solution and deposit on the crystals at the bottom of the vessel. The rates at which ions leave the crystals and deposit back onto the crystals are equal. Thus, the crystals appear to be unchanged, although they are in dynamic equilibrium with the ions in solution.■

■ The solubility equilibrium was discussed in Section 9.4. Equilibrium will be discussed further in Chapter 16.

Factors Determining Solubility

The solubility of one substance in another is determined by two factors. One of these is the natural inclination toward disorder, reflected in the tendency of substances to mix, as ink and water do when poured together.■ The other factor is the strength of the forces of attraction between species (molecules and ions). These forces, for example, may favor the unmixed solute and solvent, while the natural tendency to mix favors the solution. In such a case, it is the balance between these two factors that determines the solubility of the solute. Let us examine the solution process for several examples.

■ *Entropy* is a measure of disorder. Thus, the tendency toward disorder can be expressed as a tendency toward increasing entropy. The mixing of substances increases entropy. This increase is one of the primary factors leading to the spontaneous formation of a solution.

Molecular Solutions

If the process of dissolving one molecular substance in another were nothing more than the simple mixing of molecules, we would not expect a limit on solubility. As with the mixing of grains of salt and sand, the two substances would be completely miscible. It is the presence of intermolecular forces that gives rise to limited solubility. In the case of gases, where intermolecular forces are small, molecules mix freely. As a result, gases are completely miscible.

Substances may be completely miscible even when the intermolecular forces are not negligible. Consider the solution of the two similar liquid hydrocarbons heptane, C_7H_{16}, and octane, C_8H_{18}. The intermolecular attractions are due to London forces, and those between heptane and octane molecules are nearly equal to those between octane and octane and heptane and heptane molecules. Since the different intermolecular attractions are about the same strength, there are no favored attractions. Therefore, the tendency of molecules to mix results in complete miscibility of the substances.

As a counterexample, consider the mixing of octane with water. There are strong hydrogen-bonding forces between water molecules. In order for octane to mix with water, hydrogen bonds must be broken and replaced by much weaker London forces between water and octane. In this case, the maximum forces of attraction among molecules (and therefore the lower energy) are obtained if the octane and water remain unmixed. As a result, octane and water are nearly immiscible in one another. (Also see Figure 12.1.)

The statement that "like dissolves like" succinctly expresses these observations. That is, substances with similar intermolecular attractions are usually soluble in one another. Thus, the two similar hydrocarbons heptane and octane are completely miscible, whereas octane and water (with dissimilar intermolecular attractions) are immiscible.

For the series of alcohols (organic, or carbon-containing, compounds with an —OH group) listed in Table 12.2, the solubility in water decreases from completely miscible to slightly soluble. Water and alcohols are alike in having —OH groups through which strong hydrogen-bonding attractions arise.

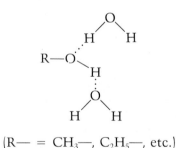

(R— = CH_3—, C_2H_5—, etc.)

Figure 12.1
The immiscibility of liquids. Suppose an A molecule moves from liquid A into liquid B. If the intermolecular attraction between two A molecules is much stronger than the intermolecular attraction between an A molecule and a B molecule, the net force of attraction will tend to pull the A molecule back into liquid A. Thus, liquid A will be immiscible with liquid B.

Table 12.2
Solubilities of Alcohols in Water

Name	Formula	Solubility in H_2O (g/100 g H_2O at 20°C)
Methanol	CH_3OH	Completely miscible
Ethanol	CH_3CH_2OH	Completely miscible
1-Propanol	$CH_3CH_2CH_2OH$	Completely miscible
1-Butanol	$CH_3CH_2CH_2CH_2OH$	7.9
1-Pentanol	$CH_3CH_2CH_2CH_2CH_2OH$	2.7
1-Hexanol	$CH_3CH_2CH_2CH_2CH_2CH_2OH$	0.6

However, as the hydrocarbon end, R—, of the alcohol becomes the more prominent portion of the molecule, the alcohol becomes less like water, and its solubility decreases.

Exercise 12.2

Which of the following compounds is likely to be more soluble in water: C_4H_9OH or C_4H_9SH? Explain.

(See Problems 12.17, 12.18, 12.19, and 12.20.)

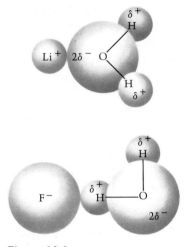

Figure 12.2
Attraction of water molecules to ions due to the ion–dipole force.

Ionic Solutions

Ionic substances differ markedly in their solubilities in water. For example, sodium chloride, NaCl, has a solubility of 36 grams per 100 mL of water at room temperature, whereas calcium phosphate, $Ca_3(PO_4)_2$, has a solubility of only 0.002 grams per 100 mL of water. In most cases, these differences in solubility can be explained in terms of the different energies of attraction between ions in the crystal and between ions and water.

The energy of attraction between an ion and a water molecule is due to an ion–dipole force. Water molecules are polar, and therefore they tend to orient with respect to nearby ions. In the case of a positive ion (Li^+, for example), water molecules orient with their oxygen atoms (the negative ends of the molecular dipoles) toward the ion. In the case of a negative ion (for instance, F^-), water molecules orient with their hydrogen atoms (the positive ends of the molecular dipoles) toward the ion (see Figure 12.2).

The attraction of ions for water molecules is called **hydration.**■ Hydration of ions favors the dissolving of an ionic solid in water. If the hydration of ions were the only factor in the solution process, we would expect all ionic solids to be soluble in water.

■ Hydration of ions also occurs in crystalline solids. For example, copper(II) sulfate pentahydrate, $CuSO_4 \cdot 5H_2O$, contains five water molecules for each $CuSO_4$ formula unit in the crystal. Substances like this, called hydrates, are discussed in Chapter 14.

The ions in a crystal, however, are very strongly attracted to one another. The solubility of an ionic solid, therefore, depends not only on the energy of hydration of ions, the energy associated with the attraction between ions and water molecules, but also on *lattice energy,* the energy holding ions together in the crystal lattice. Lattice energy works against the solution process, so that an ionic solid with relatively large lattice energy is usually insoluble.

Lattice energies depend on the charges on the ions (as well as on the distance between the centers of neighboring positive and negative ions). The greater the magnitude of ion charge, the greater the lattice energy.■ For this reason, we might expect substances with singly charged ions to be comparatively soluble, and those with multiply charged ions to be less soluble. This is borne out by the fact that compounds of the alkali metal ions (such as Na^+ and K^+) and ammonium ions (NH_4^+) are generally soluble, but those with phosphate ions (PO_4^{3-}), for example, are generally insoluble.

■ The energy of attraction of two ions, according to Coulomb's law, is proportional to the product of the ion charges and inversely proportional to the distance between the centers of the ions.

Lattice energy depends on the distance between neighboring ions, as we mentioned, and this distance in turn depends on the radii of the ions. Thus, the lattice energy of magnesium hydroxide, $Mg(OH)_2$, is inversely proportional to the sum of the radii of Mg^{2+} and OH^-. In the series of alkaline earth hydroxides—$Mg(OH)_2$, $Ca(OH)_2$, $Sr(OH)_2$, $Ba(OH)_2$—the lattice energy decreases as the radius of the alkaline earth ion increases (from Mg^{2+} to Ba^{2+}). If the lattice energy alone determines the trend in solubilities, we should expect the solubility to increase from magnesium hydroxide to barium hydroxide. In fact, this is what we find. Magnesium hydroxide is insoluble in water and barium hydroxide is soluble. But this is not the whole story. The energy of hydration also depends on ionic radius. A small ion has a concentrated electrical charge and a strong electric field that attracts water molecules. Therefore, the energy of hydration is greatest for a small ion like Mg^{2+} and least for a large ion like Ba^{2+}. If the energy of hydration of ions alone determined the trend in solubilities, we would have expected the solubilities to decrease from magnesium hydroxide to barium hydroxide, rather than increase. (We should also add that the energy of hydration increases with the charge on the ion. Thus, energy of hydration is greater for Mg^{2+} than for Na^+. Magnesium ion has a larger charge, as well as being a smaller ion than Na^+.)

The explanation for the observed solubility trend in the alkaline earth hydroxides is that the lattice energy decreases more in the series $Mg(OH)_2$, $Ca(OH)_2$, $Sr(OH)_2$, and $Ba(OH)_2$ than does the energy of hydration in the series of ions Mg^{2+}, Ca^{2+}, Sr^{2+}, and Ba^{2+}. For this reason, the lattice-energy factor dominates this solubility trend.

We see the opposite solubility trend when the energy of hydration decreases more, so that it dominates the trend. Consider the alkaline earth sulfates. Here the lattice energy depends on the sum of the cation radius and the sulfate ion radius. Because the sulfate ion, SO_4^{2-}, is much larger than the hydroxide ion, OH^-, the percent change in lattice energy in going through the series of sulfates from $MgSO_4$ to $BaSO_4$ is smaller than in the hydroxides. Thus, the lattice energy changes less, and the energy of hydration of the cation decreases by a greater amount. Now the energy of hydration dominates the solubility trend, and the solubility decreases from magnesium sulfate to barium sulfate. Magnesium sulfate is soluble in water and barium sulfate is insoluble.

Exercise 12.3

Which of the following ions has the larger hydration energy, Na^+ or K^+?
(See Problems 12.21 and 12.22.)

Aside: Hemoglobin Solubility and Sickle Cell Anemia

Sickle cell anemia was the first inherited disease shown to have a specific molecular basis. In people with the disease, the red blood cells tend to become elongated (sickle-shaped) when the concentration of oxygen (O_2) is low, as for example in the venous blood supply (see Figure 12.3). Once the red blood cells have sickled, they can no longer function in their normal capacity as oxygen carriers and they often break apart. Moreover, the sickled cells clog capillaries, interfering with the blood supply to vital organs.

In 1949, Linus Pauling showed that people with sickle cell anemia have abnormal hemoglobin. Hemoglobin is the substance in red blood cells responsible for carrying oxygen. It is normally present in solution within the red blood cells. But

(Continued)

in people with sickle cell anemia, the unoxygenated hemoglobin readily comes out of solution. It produces a fibrous precipitate that deforms the cell, giving it the characteristic sickle shape.

The normal and abnormal hemoglobins have been shown to differ very slightly. They are large molecules, with molecular weights of about 64,000 amu, and they are alike in structure except in one place. In this place, the normal hemoglobin has the group

$$\begin{array}{c} | \\ CH_2 \\ | \\ CH_2 \\ | \\ HO-C=O \end{array}$$

which helps confer water solubility on the molecule. The abnormal hemoglobin has the hydrocarbon group

$$\begin{array}{c} | \\ H-C-CH_3 \\ | \\ CH_3 \end{array}$$

instead. This small change makes the molecule less water soluble.

12.3 The Effects of Temperature and Pressure on Solubility

In general, the solubility of a substance depends on temperature. For example, the solubility of ammonium nitrate in 100 mL of water is 118 grams at 0°C and 811 grams at 100°C. Pressure may also have an effect on solubility, as we will see.

Figure 12.3
(a) Normal and (b) sickled red blood cells.

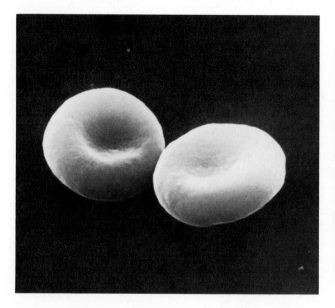

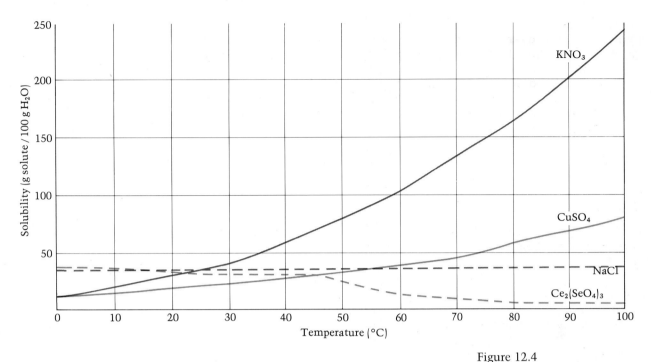

Figure 12.4
Variation of solubility with temperature. The solubility of the salts NaCl, KNO_3, and $CaSO_4$ rise with increasing temperature, as is the case with most ionic solids. The solubility of $Ce_2(SeO_4)_3$, however, falls with increasing temperature.

Temperature Change

Solubilities of substances usually vary with temperature. Most gases become less soluble in water at higher temperatures. The first bubbles that appear when tap water is heated are not bubbles of water vapor. They are bubbles of air released as the increasing temperature reduces the solubility of air in water. On the other hand, the usual behavior for ionic solids is to increase in solubility in water with rising temperature.

The variations of the solubilities of the salts KNO_3, $CuSO_4$, NaCl, and $Ce_2(SeO_4)_3$ are shown in Figure 12.4. Three of these salts show the usual behavior; their solubilities increase with rising temperature. Potassium nitrate, KNO_3, changes solubility dramatically from 14 g/100 g H_2O at 0°C to 245 g/100 g H_2O at 100°C. Copper sulfate, $CuSO_4$, shows a moderate increase in solubility over this temperature interval. Sodium chloride, NaCl, increases only slightly in solubility with temperature.

A number of ionic compounds decrease in solubility with increasing temperature. Calcium sulfate, $CaSO_4$, and calcium hydroxide, $Ca(OH)_2$, are common examples. They are slightly soluble compounds that become even less soluble at higher temperatures. Cerium selenate, $Ce_2(SeO_4)_3$, whose solubility variation with temperature is shown in Figure 12.4, is very soluble at 0°C, but much less soluble at 100°C.

Heat is usually released or absorbed when ionic substances are dissolved in water. In some cases, the effect of this heat release or absorption is quite noticeable. When sodium hydroxide is dissolved in water, the solution becomes hot (the solution process is exothermic). On the other hand, when ammonium nitrate is dissolved in water, the solution becomes very cold (the solution process is endothermic). This cooling effect from the dissolving of ammonium nitrate in water is the basis for instant cold packs used in hospitals and elsewhere. An instant cold pack consists of a bag of NH_4NO_3 crystals inside a bag of water. When the inner bag is broken, NH_4NO_3 dissolves in the water. Heat is absorbed, so the bag feels cold.

Pressure Change

In general, pressure change has little effect on the solubility of a liquid or solid in water. On the other hand, the solubility of a gas is very much affected by pressure. The qualitative effect of a change in pressure on the solubility of a gas can be predicted from **LeChatelier's principle.** According to LeChatelier's principle, if a system in physical or chemical equilibrium is altered by a change of some condition, a physical or chemical change will occur to shift the equilibrium composition in a way that attempts to reduce that change of condition. Let us see how LeChatelier's principle can predict the effect on the equilibrium composition when there is a change in pressure.

Imagine a cylindrical vessel, fitted with a movable piston, that contains carbon dioxide gas over its saturated water solution (Figure 12.5). Suppose the partial pressure of carbon dioxide gas is momentarily increased by pushing down the piston. This pressure increase represents a change of condition. According to LeChatelier's principle, more gas should now dissolve because this will decrease the pressure of carbon dioxide, and so reduce the effect of the pressure increase by the piston.

Thus, we predict that carbon dioxide is more soluble at higher pressures. Conversely, if the partial pressure of carbon dioxide gas is reduced, its solubility is decreased. A bottle of carbonated beverage fizzes when the cap is removed: as the partial pressure of carbon dioxide is reduced, gas comes out of solution.

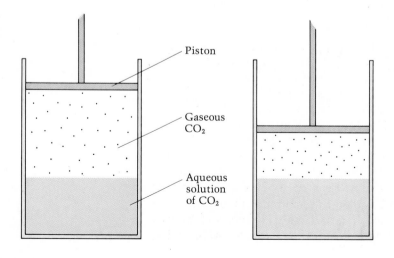

Piston

Gaseous CO_2

Aqueous solution of CO_2

Figure 12.5
If the piston is pushed down, increasing the pressure of carbon dioxide, more gas will dissolve.

The argument given for carbon dioxide holds for any gas. Therefore, all gases become more soluble in a liquid at a given temperature if the partial pressure of the gas over the solution is increased.

Henry's Law

The effect of pressure on the solubility of a gas can be predicted quantitatively. According to **Henry's law,** the solubility of a gas is directly proportional to the partial pressure of the gas above the solution. Expressed mathematically, the law is

$$S = k_H P$$

where S is the solubility of the gas (expressed as mass of solute per volume of solvent), k_H is Henry's law constant for the gas, and P is the partial pressure of the gas. The next example shows how this formula is used.

Example 12.1

In the chapter opening, we noted that 27 g of acetylene, C_2H_2, dissolve in 1 L of acetone at 1.0 atm pressure. If the partial pressure of acetylene is increased to 12 atm, what is its solubility in acetone?

Solution

Let S_1 be the solubility of the gas at partial pressure P_1 and S_2 be the solubility at partial pressure P_2. Then we can write Henry's law for both pressures:

$$S_1 = k_H P_1$$
$$S_2 = k_H P_2$$

If we divide the second equation by the first, we get

$$\frac{S_2}{S_1} = \frac{\cancel{k_H} P_2}{\cancel{k_H} P_1}$$

or

$$\frac{S_2}{S_1} = \frac{P_2}{P_1}$$

At 1.0 atm partial pressure of acetylene (P_1), the solubility S_1 is 27 g C_2H_2 per L of acetone. For a partial pressure of 12 atm (P_2), we get

$$\frac{S_2}{27 \text{ g } C_2H_2/\text{L acetone}} = \frac{12 \cancel{\text{ atm}}}{1.0 \cancel{\text{ atm}}}$$

Hence,

$$S_2 = \frac{27 \text{ g } C_2H_2}{\text{L acetone}} \times \frac{12}{1.0} = \frac{3.2 \times 10^2 \text{ g } C_2H_2}{\text{L acetone}}$$

That is, at 12 atm partial pressure of acetylene, 1 L of acetone will dissolve 3.2×10^2 g of acetylene.

Exercise 12.4

A liter of water at 25°C dissolves 0.0404 g O_2 when the partial pressure of the oxygen is 1.00 atm. What is the solubility of oxygen from air, in which the O_2 partial pressure is 159 mmHg?

(See Problems 12.25 and 12.26.)

12.4 Colloids

A **colloid** is a dispersion of particles of one substance (the dispersed phase) throughout another substance or solution (the continuous phase). Fog is an example of a colloid: it consists of very small water droplets (dispersed phase) in air (continuous phase). A colloid differs from a true solution in that

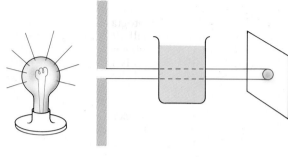

A　　　**B**

Figure 12.6
Demonstration of the Tyndall effect in a colloid.
(a) Light passes through a sodium chloride solution without scattering. (b) Light is scattered by the colloid.

the dispersed particles are larger than normal molecules, though too small to be seen in a microscope. The particles are from about 10 Å to about 2000 Å in size.

Tyndall Effect

Although a colloid appears to be homogeneous because the dispersed particles are quite small, it can be distinguished from a true solution by its ability to scatter light. The scattering of light by colloidal-sized particles is known as the **Tyndall effect.**■ For example, the atmosphere appears to be a clear gas, but a ray of sunshine against a dark background shows up many fine dust particles by light scattering. Similarly, if a beam of light is directed through clear gelatin (a colloid, not a true solution), the beam becomes visible by the scattering of light from colloidal gelatin particles. The beam appears as a ray passing through the solution (Figures 12.6 and 12.7). If the same experiment is performed with a true solution, for example, an aqueous solution of sodium chloride, the beam of light is not visible.

■ Although all gases and liquids scatter light, the scattering from a pure substance or true solution is quite small and usually not detectable. However, because of the considerable depth of the atmosphere, the scattering of light by air molecules can be seen. The blue color of the sky is due to the fact that blue light is scattered more easily than red light.

Types of Colloids

Colloids are characterized according to the state (solid, liquid, or gas) of the dispersed phase and the state of the continuous phase. Table 12.3 lists various types of colloids and some examples of each. Fog and smoke are examples of **aerosols** which are liquid droplets or solid particles dispersed throughout a gas. An **emulsion** consists of liquid droplets dispersed throughout another liquid (as particles of cream are dispersed through homogenized milk). A **sol** consists of solid particles dispersed in a liquid.

Hydrophilic and Hydrophobic Colloids

Colloids in which the continuous phase is water are also divided into two major classes: hydrophilic colloids and hydrophobic colloids. In a **hydrophilic colloid** there is a strong attraction between the dispersed phase and

Figure 12.7
A photograph showing the Tyndall effect. A light beam is visible perpendicular to its path only if light is scattered toward the viewer. The vessel on the left contains a colloid, which scatters light. The vessel on the right contains a true solution, which scatters negligible light.

the continuous phase (water). Many such colloids consist of macromolecules (very large molecules) dispersed in water. These are like normal solutions, except for the large size of the dispersed molecules. Protein solutions, such as gelatin in water, are hydrophilic colloids. Gelatin molecules are attracted to water molecules by London forces and hydrogen bonding.

In a **hydrophobic colloid** there is a lack of attraction between the dispersed phase and the continuous phase (water). Hydrophobic colloids are basically unstable. Given sufficient time, the dispersed phase comes out of solution by aggregating into larger particles. In this behavior, they are quite unlike true solutions and hydrophilic colloids. The time taken to separate may be extremely long, however. A colloid of gold particles in water prepared by Michael Faraday in 1857 is still preserved in the British Museum in Lon-

Table 12.3
Types of Colloids

Continuous Phase	Dispersed Phase	Name	Example
Gas	Liquid	Aerosol	Fog, mist
Gas	Solid	Aerosol	Smoke
Liquid	Gas	Foam	Whipped cream
Liquid	Liquid	Emulsion	Mayonnaise (oil dispersed in water)
Liquid	Solid	Sol	$AgCl(s)$ dispersed in H_2O
Solid	Gas	Foam	Pumice, plastic foams
Solid	Liquid	Gel	Jelly, opal (mineral with liquid inclusions)
Solid	Solid	Solid sol	Ruby glass (glass with dispersed metal)

don. This colloid is hydrophobic, as well as a sol (solid particles dispersed in water).

Hydrophobic sols are often formed when a solid crystallizes rapidly from a chemical reaction or a supersaturated solution. When crystallization occurs rapidly, many centers of crystallization (called *nuclei*) are formed at once. Ions are attracted to these nuclei and very small crystals are formed. These small crystals are prevented from settling out by the random thermal motion of the solvent molecules, which continue to buffet them.

You might expect these very small crystals to aggregate into larger crystals because the aggregation would bring ions of opposite charge into contact. However, sol formation appears to happen when for some reason each of the small crystals gets a preponderance of one kind of charge on its surface. For example, iron(III) hydroxide forms a colloid due to the presence of an excess of iron(III) ion (Fe^{3+}) on the surface, giving each crystal an excess of positive charge. These positively charged crystals repel one another, so that aggregation to larger particles is prevented.

Coagulation

An iron(III) hydroxide sol can be made to aggregate by the addition of an ionic solution, particularly if the solution contains anions with multiple charges (such as phosphate ions, PO_4^{3-}). **Coagulation** is the process by which a colloid is made to come out of solution by aggregation. We can picture what happens in the following way. A positively charged colloidal particle of iron(III) hydroxide gathers a layer of anions around it. The thickness of this layer is determined by the charge on the anions. The greater the magnitude of the negative charge, the more compact the layer of charge. Phosphate ions, for example, will gather more closely to the positively charged colloidal particles than would chloride ions (see Figure 12.8). If the ion layer is gathered close to the colloidal particle, the overall charge is effectively neutralized. In that case, two colloidal particles can approach close enough to aggregate.

The curdling of milk when it sours is another example of coagulation. Milk is a colloidal suspension in which the particles are prevented from aggregating because they are charged with the same sign. The ions re-

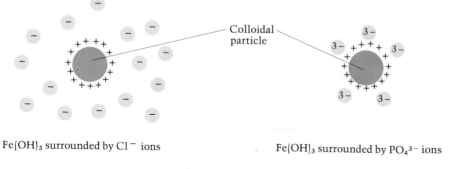

Fe(OH)₃ surrounded by Cl⁻ ions

A

Fe(OH)₃ surrounded by PO_4^{3-} ions

B

Figure 12.8
Layers of ions surrounding charged colloidal particles. (a) A positively charged colloidal particle (iron(III) hydroxide) surrounded by chloride ions; (b) the same colloidal particle surrounded by phosphate ions. Because these ions gather more closely to the colloidal particles, coagulation is more likely to occur.

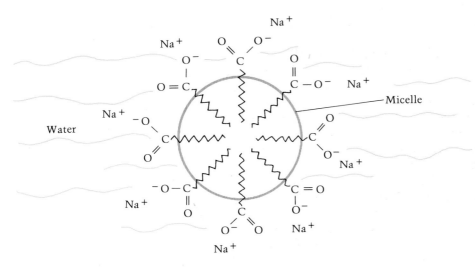

Figure 12.9
A stearate micelle formed in water solution.

sponsible for the coagulation (curdling) are formed when lactose (milk sugar) ferments to lactic acid. A third example is the coagulation of a colloidal suspension of soil in river water when the water meets the concentrated ionic solution of an ocean. The Mississippi delta was formed in this way.

Exercise 12.5

Colloidal sulfur particles are negatively charged with thiosulfate ions, $S_2O_3^{2-}$, and other ions on the surface of the sulfur. Which of the following would be most effective in coagulating colloidal sulfur: $NaCl$, $MgCl_2$, $AlCl_3$?

(See Problems 12.29 and 12.30.)

Association Colloids

When molecules that have both a hydrophobic and a hydrophilic end are dispersed in water, they associate or aggregate to form colloidal-sized particles, or **micelles.** The colloid that is formed is called an **association colloid.**

Ordinary soap in water provides an example of an association colloid. Soap consists of compounds like sodium stearate, $C_{17}H_{35}COONa$. The stearate ion has a long hydrocarbon end that is hydrophobic and a carboxyl group (COO^-) that is hydrophilic:

$$CH_3CH_2CH_2CH_2CH_2CH_2CH_2CH_2CH_2CH_2CH_2CH_2CH_2CH_2CH_2CH_2CH_2C{\overset{\displaystyle O}{\underset{\displaystyle O^-}{}}}$$

hydrophobic end   hydrophilic end

In water solution, the stearate ions associate into micelles in which the hydrocarbon ends point inward toward one another and away from the water, while carboxyl groups are on the outside of the micelle facing the water (Figure 12.9).

The cleaning action of soap occurs because oil and grease can be absorbed into the hydrophobic centers of the soap micelles and washed away (Figure 12.10). Synthetic detergents are also substances that form association colloids. Sodium lauryl sulfate is a synthetic detergent present in laundry soaps, toothpastes, and shampoos:

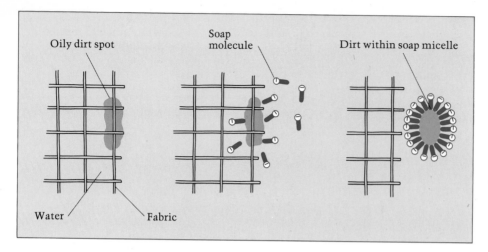

Figure 12.10
The cleaning action of soap.
The hydrocarbon ends of
soap molecules gather around
an oil spot, forming a micelle
that can be washed away in
the water.

■ Detergents may be anionic
(that is, the polar end may be
negatively charged, as in sodium lauryl sulfate), nonionic
(with hydrogen-bonding groups,
such as —OH), or cationic. Cationic detergents have germicidal
properties. They are used in certain mouthwashes.

$$CH_3CH_2CH_2CH_2CH_2CH_2CH_2CH_2CH_2CH_2CH_2CH_2OSO_3^- Na^+$$

sodium lauryl sulfate

It has a hydrophilic sulfate group ($—OSO_3^-$) and a hydrophobic dodecyl
group $C_{12}H_{25}—$ (the hydrocarbon end).■

Colligative Properties

In the chapter opening, we mentioned that the addition of ethylene glycol,
CH_2OHCH_2OH, to water lowers the freezing point of water below 0°C. For
example, if 0.010 mol of ethylene glycol is added to one kilogram of water,
the freezing point is lowered to −0.019°C. The magnitude of freezing-point
lowering is directly proportional to the number of ethylene glycol molecules
added to a quantity of water. Thus, if we add 0.020 (= 0.010 × 2) mol of
ethylene glycol to one kilogram of water, the freezing point is lowered to
−0.038°C (= −0.019°C × 2). Moreover, the same lowering is observed with
the addition of other molecular substances. For example, an aqueous solution of 0.0200 mol of urea, $(NH_2)_2CO$, in one kilogram of water also freezes
at −0.038°C.

Freezing-point lowering is a **colligative property.** Such properties of solutions depend on the concentration of solute molecules (that is, on the number of solute molecules or ions in a given quantity of solvent or solution), but
do not depend on the nature of the substance (whether it is ethylene glycol
or urea, for instance). In the remainder of the chapter, we will discuss several
colligative properties, which are expressed quantitatively in terms of various
concentration units. We will first look into the various ways of expressing
the concentration of a solution.

12.5 Ways of Expressing Concentration

The *concentration* of a solute is the amount of solute dissolved in a given
quantity of solvent or solution. The quantity of solvent or solution may be

expressed in terms of volume or in terms of mass or molar amount. Thus, there are several ways of expressing the concentration of a solution.

The *molarity* of a solution is the moles of solute in a liter of solution:■

■ Molarity was discussed in Section 3.10.

$$\text{Molarity} = \frac{\text{moles of solute}}{\text{liters of solution}}$$

For example, 0.20 mol ethylene glycol dissolved in water to give 2.0 L of solution has a molarity of

$$\frac{0.20 \text{ mol ethylene glycol}}{2.0 \text{ L solution}} = 0.10 \text{ } M \text{ ethylene glycol}$$

The solution is 0.10 molar (denoted M). This unit is especially useful when we wish to dispense a given amount of solute, since the amount of solute will be directly related to the volume of solution.

Other concentration units are defined in terms of the mass or molar amount of solvent or solution. The most important of these are mass (or weight) percentage of solute, molality, and mole fraction.

Mass Percentage of Solute

Solution concentration is sometimes expressed in terms of the **mass percentage of solute,** that is, the percent by mass of solute contained in the solution.

$$\text{Mass percentage of solute} = \frac{\text{mass of solute}}{\text{mass of solution}} \times 100\%$$

For example, an aqueous solution that is 3.5% sodium chloride by mass contains 3.5 grams of NaCl in 100.0 grams of solution. It could be prepared by dissolving 3.5 grams of NaCl in 96.5 grams of water (100.0 − 3.5 = 96.5).

Example 12.2

How would you prepare 425 g of an aqueous solution containing 2.40% by mass of sodium acetate, $NaC_2H_3O_2$?

Solution

The mass of sodium acetate in this quantity of solution is

Mass of $NaC_2H_3O_2$ = 425 g × 0.0240 = 10.2 g

The quantity of water in the solution is

Mass of H_2O = mass solution − mass $NaC_2H_3O_2$
= 425 g − 10.2 g = 415 g

Thus, you would prepare the solution by dissolving 10.2 g of sodium acetate in 415 g of water.

Exercise 12.6

An experiment calls for 35.0 g of 20.2% hydrochloric acid, HCl. How many grams of HCl is this? How many grams of water?

(See Problems 12.31 and 12.32.)

Molality

The **molality** of a solution is the moles of solute per kilogram of solvent:

$$\text{Molality} = \frac{\text{moles of solute}}{\text{kilograms of solvent}}$$

For example, 0.20 mol of ethylene glycol dissolved in 2.0×10^3 grams ($= 2.0$ kilograms) of water has a molality of

$$\frac{0.20 \text{ mol ethylene glycol}}{2.0 \text{ kg solvent}} = 0.10 \; m \text{ ethylene glycol}$$

That is, the solution is 0.10 molal (denoted m). The units of molality and molarity are sometimes confused. Note that molality is defined in terms of *mass of solvent* rather than *volume of solution*.

Example 12.3

Glucose, $C_6H_{12}O_6$, is a sugar that occurs in fruits. It is also known as "blood sugar" because it is found in blood and is the body's main source of energy. What is the molality of a solution containing 4.57 g of glucose dissolved in 25.2 g of water?

Solution

The moles of glucose (MW = 180.2 amu) in 4.57 g are found as follows:

$$4.57 \text{ g } C_6H_{12}O_6 \times \frac{1 \text{ mol } C_6H_{12}O_6}{180.2 \text{ g } C_6H_{12}O_6}$$

$$= 0.0254 \text{ mol } C_6H_{12}O_6$$

The molality is obtained by dividing the moles of solute (glucose) by the mass of solvent (water) in kilograms. The mass of water is 25.2 g or 25.2×10^{-3} kg.

$$\text{Molality} = \frac{0.0254 \text{ mol } C_6H_{12}O_6}{25.2 \times 10^{-3} \text{ kg solvent}} = 1.01 \; m \; C_6H_{12}O_6$$

Exercise 12.7

Toluene, $C_6H_5CH_3$, is a liquid compound similar to benzene, C_6H_6. It is the starting material for other substances, including trinitrotoluene (TNT). Find the molality of toluene in a solution that contains 35.6 g of toluene and 125 g of benzene.

(See Problems 12.33, 12.34, 12.35, and 12.36.)

Mole Fraction

The **mole fraction** of substance A (X_A) in a solution is defined as the number of moles of substance divided by the total moles of solution (that is, total moles of solute plus solvent):

$$X_A = \frac{\text{moles of substance A}}{\text{total moles of solution}}$$

For example, if a solution is made up of 1 mol of ethylene glycol and 9 mol of water, the total moles of solution are 1 mol + 9 mol = 10 mol. The mole fraction of ethylene glycol is 1/10 = 0.1, and the mole fraction of water is

9/10 = 0.9. Multiplying mole fractions by 100 gives *mole percent*. Hence, this solution is 10 mole percent glucose and 90 mole percent water. We can also say that 10% of the molecules in the solution are ethylene glycol and 90% are water. The sum of the mole fractions of all the components of a solution equals one.

Example 12.4

What are the mole fractions of glucose and water in a solution containing 4.57 g of glucose, $C_6H_{12}O_6$, dissolved in 25.2 g of water?

Solution

This is the glucose solution described in Example 12.3. There we found that 4.57 g of glucose equal 0.0254 mol of glucose. The moles of water in the solution are

$$25.2 \text{ g } H_2O \times \frac{1 \text{ mol } H_2O}{18.0 \text{ g } H_2O} = 1.40 \text{ mol } H_2O$$

Hence, the total moles of solution are

$$1.40 \text{ mol} + 0.0254 \text{ mol} = 1.425 \text{ mol}$$

(retaining an extra figure in the answer for further computation). Finally, we get

$$\text{Mole fraction glucose} = \frac{0.0254 \text{ mol}}{1.425 \text{ mol}} = 0.0178$$

$$\text{Mole fraction water} = \frac{1.40 \text{ mol}}{1.425 \text{ mol}} = 0.982$$

The sum of the mole fractions is 1.000.

Exercise 12.8

Calculate the mole fractions of toluene and benzene in the solution described in Exercise 12.7.

(See Problems 12.37 and 12.38.)

Conversion of Concentration Units

It is relatively easy to interconvert concentration units when they are expressed in terms of mass or moles of solute and solvent, as the following examples show.

Example 12.5

An aqueous solution is 0.120 *m* in glucose, $C_6H_{12}O_6$. What are the mole fractions of each component in the solution?

Solution

A 0.120 *m* glucose solution contains 0.120 mol of glucose in 1.00 kg of water. The number of moles in 1.00 kg of water is

$$1.00 \times 10^3 \text{ g } H_2O \times \frac{1 \text{ mol } H_2O}{18.0 \text{ g } H_2O} = 55.6 \text{ mol } H_2O$$

Hence,

$$\text{Mole fraction glucose} = \frac{0.120 \text{ mol}}{(0.120 + 55.6) \text{ mol}} = 0.00215$$

$$\text{Mole fraction water} = \frac{55.6 \text{ mol}}{(0.120 + 55.6) \text{ mol}} = 0.998$$

Exercise 12.9

A solution is 0.120 m methanol dissolved in ethanol. Calculate the mole fractions of methanol, CH_3OH, and ethanol, C_2H_5OH, in the solution.

(See Problems 12.39 and 12.40.)

Example 12.6

A solution is 0.150 mole fraction glucose, $C_6H_{12}O_6$, and 0.850 mole fraction water. What is the molality of glucose in the solution?

Solution

One mole of solution contains 0.150 mol of glucose and 0.850 mol of water. The mass of this amount of water is

$$0.850 \text{ mol H}_2\text{O} \times \frac{18.0 \text{ g H}_2\text{O}}{1 \text{ mol H}_2\text{O}} = 15.3 \text{ g H}_2\text{O}$$

$$(= 0.0153 \text{ kg H}_2\text{O})$$

Therefore, the molality of glucose, $C_6H_{12}O_6$, in the solution is

$$\text{Molality of C}_6\text{H}_{12}\text{O}_6 = \frac{0.150 \text{ mol C}_6\text{H}_{12}\text{O}_6}{0.0153 \text{ kg solvent}}$$

$$= 9.80 \ m \ \text{C}_6\text{H}_{12}\text{O}_6$$

Exercise 12.10

A solution is 0.250 mole fraction methanol, CH_3OH, and 0.750 mole fraction ethanol, C_2H_5OH. What is the molality of methanol in the solution?

(See Problems 12.41 and 12.42.)

$d = \frac{m}{v}$

$v = \frac{m}{d}$

To convert molality to molarity and vice versa requires that we know the density of the solution. The calculations are described in the next two examples.

Example 12.7

An aqueous solution is 0.273 m KCl. What is the molar concentration of potassium chloride, KCl? The density of the solution is 1.011×10^3 g/L.

Solution

There are 0.273 mol KCl per kilogram of water. To calculate the molarity, we must first find the volume of solution for a given mass of solvent. Let us consider an amount of solution containing a kilogram of water $(1.000 \times 10^3 \text{ g H}_2\text{O})$. The mass of potassium chloride in this quantity of solution is

$$0.273 \text{ mol KCl} \times \frac{74.6 \text{ g KCl}}{1 \text{ mol KCl}} = 20.4 \text{ g KCl}$$

The total mass of the solution equals the mass of water plus the mass of potassium chloride:

$$1.000 \times 10^3 \text{ g} + 20.4 \text{ g} = 1.020 \times 10^3 \text{ g}$$

The volume of solution equals the mass divided by the density of the solution:

$$\text{Volume of solution} = \frac{1.020 \times 10^3 \text{ g}}{1.011 \times 10^3 \text{ g/L}} = 1.009 \text{ L}$$

Hence, the molarity of the solution is

$$\frac{0.273 \text{ mol KCl}}{1.009 \text{ L solution}} = 0.271 \ M \ \text{KCl}$$

Note that the molarity and molality of this solution are approximately equal. This happens in cases where the solutions are dilute and the density is about 1 g/mL.

Exercise 12.11

Urea, $(NH_2)_2CO$, is used as a fertilizer. What is the molar concentration of an aqueous solution that is 3.42 m urea? The density of the solution is 1.045 g/mL.

(See Problems 12.43 and 12.44.)

Example 12.8

An aqueous solution is 0.907 M $Pb(NO_3)_2$. What is the molality of lead nitrate, $Pb(NO_3)_2$, in this solution? The density of the solution is 1.252 g/mL.

Solution

There are 0.907 mol $Pb(NO_3)_2$ per liter of solution. Let us consider 1 L ($= 1.000 \times 10^3$ mL) of solution and calculate its mass. We can then calculate the mass of lead nitrate and find the mass of water by difference.

$$\begin{aligned}
\text{Mass of solution} &= \text{density} \times \text{volume} \\
&= 1.252 \text{ g/mL} \times 1.000 \times 10^3 \text{ mL} \\
&= 1.252 \times 10^3 \text{ g}
\end{aligned}$$

The mass of lead nitrate is

$$0.907 \text{ mol } Pb(NO_3)_2 \times \frac{331.2 \text{ g } Pb(NO_3)_2}{1 \text{ mol } Pb(NO_3)_2}$$
$$= 3.00 \times 10^2 \text{ g } Pb(NO_3)_2$$

The mass of the water in this solution is

$$\begin{aligned}
\text{Mass of } H_2O &= \text{mass of solution} - \text{mass of } Pb(NO_3)_2 \\
&= 1.252 \times 10^3 \text{ g} - 3.00 \times 10^2 \text{ g} \\
&= 9.52 \times 10^2 \text{ g} \quad (= 0.952 \text{ kg})
\end{aligned}$$

Hence, the molality of lead nitrate in this solution is

$$\frac{0.907 \text{ mol } Pb(NO_3)_2}{0.952 \text{ kg solvent}} = 0.953 \text{ } m \text{ } Pb(NO_3)_2$$

Exercise 12.12

An aqueous solution is 2.00 M urea. The density of the solution is 1.029 g/mL. What is the molal concentration of urea in the solution?

(See Problems 12.45 and 12.46.)

12.6 Vapor Pressure of a Solution

In this and the next two sections, we will discuss the following colligative properties: vapor pressure lowering, boiling-point elevation, freezing-point depression, or lowering, and osmosis. The vapor pressure of a solvent is lowered by the addition of a nonvolatile solute (a substance with negligible vapor pressure). Consider the addition of ethylene glycol to water at 20°C. Ethylene glycol has a very low vapor pressure at 20°C, whereas pure water at this temperature has a vapor pressure of 17.54 mmHg. When ethylene glycol is added to water, the vapor pressure of the water is lowered. For instance, an aqueous solution containing 0.0100 mole fraction of ethylene glycol has a vapor pressure of 17.36 mmHg.

The **vapor-pressure lowering** of a solution is the vapor pressure of the pure solvent minus the vapor pressure of the solution. For the ethylene glycol solution just given,

Vapor-pressure lowering = 17.54 mmHg − 17.36 mmHg = 0.18 mmHg

Vapor-pressure lowering is a colligative property. It depends only on the concentration of solute molecules or ions in solution, not on their identity. If the mole fraction of ethylene glycol is doubled to 0.0200, the vapor-pressure lowering is doubled. Moreover, the same vapor-pressure lowering is obtained with other nonvolatile solutes. An aqueous solution that is 0.0100 mole fraction in urea, $(NH_2)_2CO$, has the same vapor pressure as one that is 0.0100 mole fraction in ethylene glycol.

To understand vapor-pressure lowering, think of a volatile solvent to which a small amount of nonvolatile solute has been added. The environment of each solvent molecule is nearly the same in such a dilute solution as it is in the pure solvent. Thus, the tendency of a solvent molecule to escape is nearly the same as in the pure solvent. However, the number of volatile molecules at the surface available for evaporation is reduced from that in the pure solvent and is proportional to the fraction of solvent molecules. Therefore, the rate of evaporation (and hence the vapor pressure) is directly proportional to the fraction of solvent molecules in solution (see Figure 12.11).

Raoult's law expresses this algebraically. The vapor pressure of the solution, P, is equal to the vapor pressure of the pure solvent, P^0, multiplied by the mole fraction of solvent in the solution, X:

$$P = P^0X$$

Since X is always less than 1 for a solution, the vapor pressure of a solution containing a nonvolatile solute, according to Raoult's law, is less than that for the pure solvent.

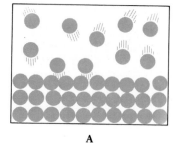

A

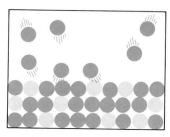

B

Figure 12.11
Vapor-pressure lowering.
(a) Vaporization of solvent molecules from pure solvent.
(b) The number of volatile molecules at the surface available for vaporization of solvent is reduced by adding a nonvolatile solute; hence, the vapor pressure decreases.

Example 12.9

Use Raoult's law to calculate the vapor pressure at 25°C of a solution containing 4.57 g of glucose, $C_6H_{12}O_6$, dissolved in 25.2 g of water. The vapor pressure of water at 25°C is 23.8 mmHg. Glucose has negligible vapor pressure. What is the vapor-pressure lowering?

Solution

This is the glucose solution described in Example 12.4. According to calculations there, the solution is 0.982

mole fraction of water. Therefore, the vapor pressure is

$$P = P^0X$$
$$P = 23.8 \text{ mmHg} \times 0.982 = 23.4 \text{ mmHg}$$

The vapor-pressure lowering is

$$P^0 - P = 23.8 \text{ mmHg} - 23.4 \text{ mmHg} = 0.4 \text{ mmHg}$$

Exercise 12.13

Naphthalene, $C_{10}H_8$, is used to make moth balls. Calculate the vapor pressure at 20°C of a solution containing 0.515 g of naphthalene in 60.8 g of chloroform, $CHCl_3$. The vapor pressure of chloroform at 20°C is 156 mmHg. Naphthalene has a much lower vapor pressure and may be assumed to be nonvolatile for this calculation.
(See Problems 12.47 and 12.48.)

The measurement of the vapor-pressure lowering of a solution can be used to obtain the mole fraction and hence the number of moles of a solute in solution. In effect, the amount of vapor-pressure lowering by a nonvolatile solute is an indirect measure of the number of solute molecules in a given amount of solvent. Freezing-point depression and boiling-point elevation, which are also colligative properties, are similarly a measure of the number of solute molecules and are usually easier to determine experimentally. We discuss these properties in the next section.

12.7 Boiling-Point Elevation and Freezing-Point Depression

The *normal* boiling point of a liquid is the temperature at which its vapor pressure equals 1 atm. Since the addition of a nonvolatile solute to a liquid will reduce its vapor pressure, the temperature must be increased to a value greater than the normal boiling point to achieve a vapor pressure of 1 atm. Figure 12.12 shows the vapor-pressure curve of a solution. The curve is below, but parallel to, the vapor-pressure curve of the pure liquid solvent. The **boiling-point elevation** is indicated in the diagram.

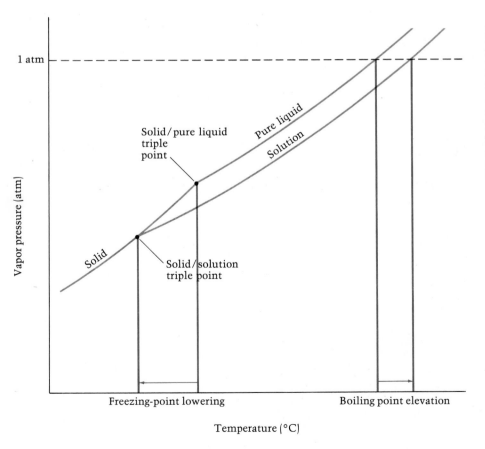

Figure 12.12
Vapor-pressure curves of pure liquid and solution. The boiling point of a solution with nonvolatile solute is elevated and the freezing point is lowered, compared with those for the pure liquid. (The normal freezing points differ slightly from the triple points; see text and also Figure 11.26 on page 380.)

Solvent	Formula	Melting Point (°C)	Boiling Point (°C)	K_f (°C/m)	K_b (°C/m)
Acetic acid	$HC_2H_3O_2$	16.60	118.5	3.59	3.08
Benzene	C_6H_6	5.455	80.2	5.065	2.61
Camphor	$C_{10}H_{16}O$	179.5	—	40	—
Carbon disulfide	CS_2	—	46.3	—	2.40
Ethanol	C_2H_5OH	—	78.3	—	1.07
Water	H_2O	0.000	100.000	1.858	0.521

*Data are taken from Landolt-Börnstein, 6th ed., *Zahlenwerte und Functionen aus Physik, Chemie, Astronomie, Geophysik, und Technik,* © Springer-Verlag, Heidelberg, Vol. II, part IIa (1960), pp. 844–849 and pp. 918–919.

Table 12.4
Boiling-Point-Elevation and
Freezing-Point-Depression
Constants*

The boiling-point elevation, ΔT_b, is found to be proportional to the molal concentration, c_m, of the solution (for dilute solutions):

$$\Delta T_b = K_b c_m$$

The constant of proportionality, K_b (called the *boiling-point-elevation constant*), depends only on the solvent. Table 12.4 lists values of K_b, as well as boiling points, for some solvents. Benzene, for example, has a boiling-point-elevation constant of 2.61°C/m. This means that a 1 m solution will boil at 2.61°C above the boiling point of pure benzene. Since pure benzene boils at 80.2°C, a 1 m solution boils at 80.2°C + 2.61°C = 82.8°C.

Figure 12.12 also shows the effect on the freezing point. Usually it is the pure solvent that freezes out of solution. Sea ice, for example, is almost pure water. For that reason, the vapor-pressure curve for the solid phase is the same, whether we are talking about pure solvent or the solution. But the triple point (the intersection of the liquid and solid vapor-pressure curves) shifts to a lower temperature in going from pure solvent to the solution. The normal freezing point, which is nearly equal to the triple-point temperature (Figure 11.26, page 380), shifts in the same way. This shift of the freezing point, called the **freezing-point depression,** is indicated in Figure 12.12.

Freezing-point depression, ΔT_f, like boiling-point elevation, is proportional to the molal concentration, c_m (for dilute solutions):

$$\Delta T_f = K_f c_m$$

Here K_f is the *freezing-point-depression constant* and depends only on the solvent. Table 12.4 gives values of K_f for some solvents. The freezing-point-depression constant for benzene is 5.06°C/m. Thus, a 1 m solution freezes at 5.06°C below the freezing point of pure benzene. Pure benzene freezes at 5.46°C; the freezing point of the solution is 5.46°C − 5.06°C = 0.40°C.

Example 12.10

An aqueous solution is 0.222 m glucose. What are the boiling point and the freezing point of this solution?

Solution

Table 12.4 gives K_b and K_f for water as 0.521°C/m and 1.86°C/m, respectively. Therefore,

(Continued)

$$\Delta T_b = K_b c_m$$
$$= 0.521°C/m \times 0.222\ m = 0.116°C$$
$$\Delta T_f = K_f c_m$$
$$= 1.86°C/m \times 0.222\ m = 0.413°C$$

The boiling point of the solution is 100.000°C + 0.116°C = 100.116°C, and the freezing point is 0.000°C − 0.413°C = −0.413°C. Note that ΔT_b is added and ΔT_f is subtracted.

Exercise 12.14

How many grams of ethylene glycol, CH_2OHCH_2OH, must be added to one gallon of water (3779 g) to give a freezing point of −15°C?

(See Problems 12.49 and 12.50.)

The boiling-point elevation and the freezing-point depression of solutions have a number of practical applications. We mentioned in the chapter opening that ethylene glycol is used in automobile radiators as an antifreeze because it lowers the freezing point of the coolant. The same substance also helps prevent the radiator coolant from boiling away by elevating the boiling point. Sodium chloride is spread on icy roads in the winter time to lower the melting point of ice and snow below the temperature of the surrounding air.■ Salt−ice mixtures are used as freezing mixtures in, for example, domestic ice-cream makers. Melting ice cools the ice-cream mixture to the freezing point of the solution, which is well below the freezing point of pure water.

Colligative properties are also used to obtain molecular weights. Freezing-point depression is commonly employed because it is simple to determine a melting point or a freezing point (Figure 12.13). From the freezing-point lowering one can calculate the molal concentration, and from the molality one can obtain the molecular weight. The next two examples illustrate these calculations.

■ Calcium chloride, $CaCl_2$, is also used to melt ice on roadways. It is obtained as a by-product from industrial production of chemicals.

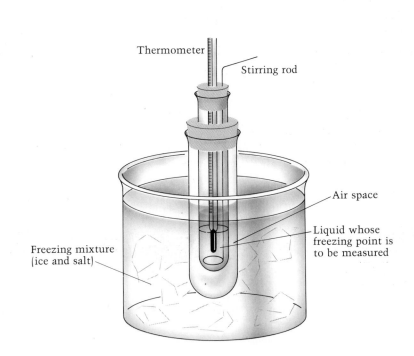

Thermometer

Stirring rod

Air space

Liquid whose freezing point is to be measured

Freezing mixture (ice and salt)

Figure 12.13
Determination of the freezing point of a liquid. The liquid is cooled by means of a freezing mixture. In order to control the rate of decrease of temperature, the liquid is separated by an air space from the freezing mixture.

Example 12.11

A solution is prepared by dissolving 0.131 g of a substance in 25.4 g of water. The molality of the solution is determined by freezing-point depression to be 0.056 m. What is the molecular weight of the substance?

Solution

The molality of an aqueous solution is

$$\text{Molality} = \frac{\text{moles of substance}}{\text{kg H}_2\text{O}}$$

Hence, for the given solution,

$$0.056 \frac{\text{mol}}{\text{kg H}_2\text{O}} = \frac{\text{moles of substance}}{25.4 \times 10^{-3}\,\text{kg H}_2\text{O}}$$

Or, rearranging this equation,

$$\begin{aligned}\text{Moles of substance} &= 0.056 \frac{\text{mol}}{\text{kg H}_2\text{O}} \times 25.4 \times 10^{-3}\,\text{kg H}_2\text{O} \\ &= 1.42 \times 10^{-3}\,\text{mol}\end{aligned}$$

We have retained an extra digit for further computation. The molar mass of the substance equals the mass of the substance divided by the number of moles:

$$\text{Molar mass} = \frac{0.131\,\text{g}}{1.42 \times 10^{-3}\,\text{mol}} = 92\,\text{g/mol}$$

We round the answer to two significant figures. The molecular weight is 92 amu.

Exercise 12.15

A 1.86-g sample of ascorbic acid (vitamin C) was dissolved in 95.0 g of water. The concentration of ascorbic acid, as determined by freezing-point depression, was 0.111 m. What is the molecular weight of ascorbic acid?

(See Problems 12.53 and 12.54.)

Example 12.12

Camphor is a white solid that melts at 179.5°C. It is frequently used to determine the molecular weights of organic compounds, because it has an unusually large freezing-point-depression constant (40°C/m). For that reason, ordinary thermometers may be used in the determination. The organic substance is dissolved in melted camphor, and then the melting point of the solution is determined. A 6.43-mg sample of a compound with the empirical formula CH was dissolved in 78.1 mg of camphor. The solution melted at 158.5°C. What is the molecular formula for the compound?

Solution

The freezing-point lowering is

$$\Delta T_f = (179.5 - 158.5)°C = 21.0°C$$

so the molality of the solution is

$$\frac{\Delta T_f}{K_f} = \frac{21.0°C}{40°C/m}$$
$$= 0.52\,m$$

From this we compute the moles of the compound that are dissolved in 78.1 mg of camphor (see Example 12.11):

$$0.52\,\text{mol/kg} \times 78.1 \times 10^{-6}\,\text{kg} = 4.1 \times 10^{-5}\,\text{mol}$$

The molar mass of the compound is

$$M_m = \frac{6.43 \times 10^{-3}\,\text{g}}{4.1 \times 10^{-5}\,\text{mol}} = 1.6 \times 10^2\,\text{g/mol}$$

The formula weight of CH is 13. Therefore, the number of CH units in the molecule is

$$\frac{1.6 \times 10^2}{13} = 12$$

The molecular formula is $C_{12}H_{12}$.

Exercise 12.16

A 2.05-g sample of white phosphorus was dissolved in 25.0 g of carbon disulfide, CS_2. The boiling-point elevation of the carbon disulfide solution was found to be 1.59°C. What is the molecular weight of the phosphorus in solution? What is the formula of molecular phosphorus?

(See Problems 12.55 and 12.56.)

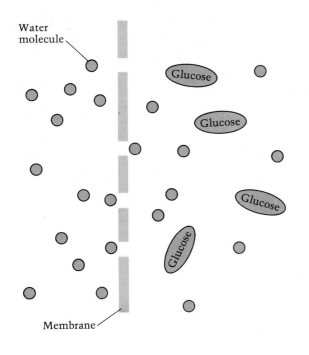

Figure 12.14
A semipermeable membrane separating water and an aqueous solution of glucose. Here the membrane is depicted as similar to a sieve—the pores being too small to allow the glucose molecules to pass. The actual mechanism in particular cases may be more complicated than this. For example, water may dissolve in one side of the membrane itself and diffuse through it to the other side.

12.8 Osmosis

Certain membranes will allow solvent molecules to pass through them but not solute molecules, particularly not those of large molecular weight. Such a membrane is called *semipermeable* and might be an animal bladder, vegetable tissue, or a piece of cellophane. Figure 12.14 depicts the operation of a semipermeable membrane. **Osmosis** is the phenomenon of solvent flow through a semipermeable membrane to equalize the solute concentrations on both sides of the membrane. If two solutions of the same solvent are separated by a semipermeable membrane, solvent molecules will migrate through the membrane from the solution of low concentration to the solution of high concentration. Thus, in osmosis, the migration of solvent is in a direction that tends to equalize the concentrations of the solutions on opposite sides of the membrane.

Figure 12.15 shows an experiment that demonstrates osmosis. A dilute glucose solution is placed in an inverted funnel whose mouth is sealed with a semipermeable membrane. The funnel containing the glucose is then placed in a beaker of pure water. As water flows from the beaker through the membrane into the funnel, the liquid level will rise in the stem of the funnel.

The glucose solution continues to rise up the funnel stem, until the downward pressure exerted by the solution above the membrane eventually stops the upward flow of solvent. This downward pressure, which stops the flow of solvent through the membrane, is called the **osmotic pressure** of the solution. Since osmotic pressure depends on the concentration of solute molecules, but not on the particular characteristics of the substance, it is a colligative property.

The osmotic pressure, π, of a solution is related to the molar concentration of solute, M:

$$\pi = MRT$$

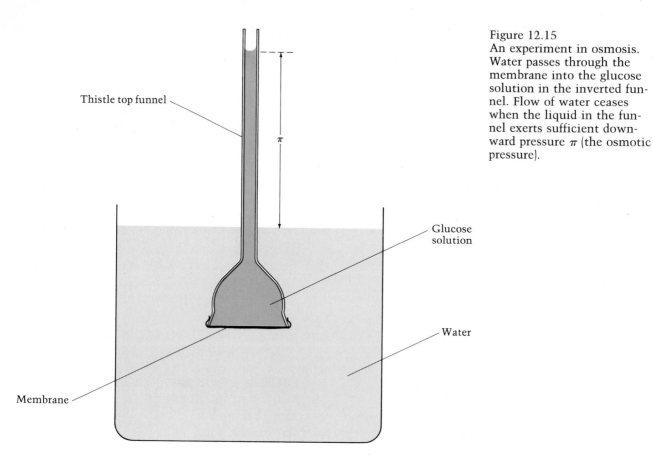

Thistle top funnel

Glucose solution

Water

Membrane

Here R is the gas constant, and T is the absolute temperature. There is a formal similarity between this equation for osmotic pressure and the equation for an ideal gas. Thus, since the molar concentration, M, of a gas equals n/V, we have that $P = (n/V)RT = MRT$.

To see the order of magnitude of the osmotic pressure of a solution, consider the aqueous solution described in Example 12.10. There we found that a solution that is about 0.2 m has a freezing-point depression of about half a degree Celsius. The molarity and molality of a dilute aqueous solution are approximately equal (see Example 12.7). Therefore, this solution will also be about 0.2 M. Hence, the osmotic pressure at 25°C (298 K) is

$$
\begin{aligned}
\pi &= MRT \\
&= 0.2 \; \cancel{\text{mol/L}} \times 0.082 \; \cancel{L} \cdot \text{atm/}(\cancel{K} \cdot \cancel{\text{mol}}) \times 298 \; \cancel{K} \\
&= 5 \text{ atm}
\end{aligned}
$$

If this pressure were to be exerted by a column of water, as in Figure 12.15, the column would have to be more than 150 feet high!

In most osmotic-pressure experiments, much more dilute solutions are employed. Often osmosis is used to determine the molecular weights of macromolecular or polymeric substances. (Polymers are very large molecules generally made up from a simple, repeating unit. A typical molecule of polyethylene, for instance, might have the formula $CH_3(CH_2CH_2)_{2000}CH_3$.)

Although polymer solutions can be fairly concentrated in terms of grams per liter, on a mole per liter basis they may be quite dilute. The freezing-point depression is usually too small to measure, though the osmotic pressure may be appreciable.

Example 12.13

The formula for low-molecular-weight starch is $(C_6H_{10}O_5)_n$, where n averages 2.00×10^2. If 3.19 g of starch are dissolved in 100 mL of water solution, what is the osmotic pressure at 25°C?

Solution

The molecular weight of $(C_6H_{10}O_5)_{200}$ is 32,400. Hence, the number of moles in 3.19 g of starch is

$$3.19 \text{ g starch} \times \frac{1 \text{ mol starch}}{32,400 \text{ g starch}} =$$

$$9.85 \times 10^{-5} \text{ mol starch}$$

The molarity of the solution is

$$\frac{9.85 \times 10^{-5} \text{ mol}}{0.100 \text{ L solution}} = 9.85 \times 10^{-4} \text{ mol/L solution}$$

and the osmotic pressure at 25°C is

$$\pi = MRT$$
$$= 9.85 \times 10^{-4} \text{ mol/L}$$
$$\times 0.0821 \text{ L} \cdot \text{atm/(K} \cdot \text{mol)} \times 298\text{K}$$
$$= 2.41 \times 10^{-2} \text{ atm} = 2.41 \times 10^{-2} \text{ atm} \times \frac{760 \text{ mmHg}}{1 \text{ atm}}$$
$$= 18.3 \text{ mmHg}$$

For comparison, we can calculate the freezing-point depression. Assume that the molality is equal to the molarity (this is in effect the case for very dilute aqueous solutions). Then,

$$\Delta T_f = 1.86°C/m \times 9.85 \times 10^{-4} \text{ } m$$
$$= 1.83 \times 10^{-3}°C$$

which is barely detectable with generally available equipment.

Exercise 12.17

Calculate the osmotic pressure at 20°C of an aqueous solution containing 5.0 g of sucrose, $C_{12}H_{22}O_{11}$, in 100.0 mL of solution.

(See Problems 12.57 and 12.58.)

Osmosis is important to the understanding of many biological processes.■ A cell might be thought of (simplistically) as an aqueous solution enclosed by a semipermeable membrane. The solution surrounding the cell must have an osmotic pressure equal to that within the cell. Otherwise, water will either leave the cell, dehydrating it, or enter the cell and possibly burst the membrane. For intravenous feeding (adding a nutrient solution to the venous blood supply of a patient), it is necessary that the nutrient solution have exactly the osmotic pressure of blood plasma. If it does not, the blood cells may collapse or burst through osmosis.

When the body has portions of organs that are at different osmotic pressures, an active pumping mechanism is required to offset osmosis. For example, cells of the transparent tissue of the exterior eye, the cornea, have a more concentrated optical fluid than does the aqueous humor, a solution just behind the cornea (Figure 12.16). In order to prevent the cornea from taking up additional water from the aqueous humor, cells that pump water are located in the tissue of the cornea adjacent to the aqueous humor. Corneas that are to be stored and used for transplants must be removed from the globe

■ Osmotic pressure appears to be important for the rising of sap in a tree. During the day, water evaporates from the leaves of the tree, so that the aqueous solution in the leaves becomes more concentrated and the osmotic pressure increases. Sap flows upward to dilute the water solution in the leaves.

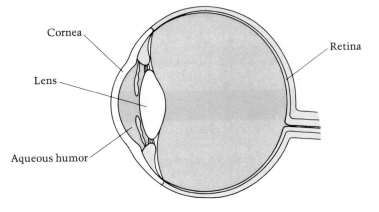

Figure 12.16
Parts of the eye.

Cornea

Retina

Lens

Aqueous humor

of the eye soon after death. This prevents the clouding that occurs when the pumping mechanism fails (as its energy supply is depleted at death).

12.9 Ionic Solutions

Electrolytes are substances that dissolve to form electrically conducting solutions. The theory that ions are formed when electrolytes dissolve was developed in 1884 by Svante Arrhenius (1859–1927) in his doctoral thesis. The theory was not well received by his doctoral committee, but both Wilhelm Ostwald (1853–1932) and Jacobus van't Hoff (1852–1911) accepted and promoted Arrhenius's ideas. Not only did the theory explain the electrical conductivity of electrolyte solutions, but it also explained their anomalous freezing points (and other colligative properties). The values of the freezing-point depression of sodium chloride solutions, for example, are nearly twice those expected from the molality of NaCl. Arrhenius explained this by saying that sodium chloride dissolves in water to form the ions, Na^+ and Cl^-. Each formula unit of NaCl gives two particles. We can write the freezing-point lowering more generally as

$$\Delta T_f = iK_f c_m$$

where i is the number of ions resulting from each formula unit and c_m is the molality computed on the basis of the formula of the ionic compound.

Actually, the freezing points of ionic solutions agree with this equation only when quite dilute, and i values calculated from the freezing-point depression are usually smaller than the number of ions obtained from the formula unit. For example, a 0.029 m aqueous solution of potassium sulfate, K_2SO_4, has a freezing point of $-0.14°C$. Hence,

$$i = \frac{\Delta T_f}{K_f c_m} = \frac{0.14°C}{1.86°C/m \times 0.029\ m} = 2.6$$

We might have expected a value of 3, based on the fact that K_2SO_4 ionizes to give three ions. The discrepancy is believed to be caused by the strong electrical interactions between ions.∎

■ The value of i equal to $\Delta T_f/K_f c_m$ is often called the van't Hoff factor. Thus, the van't Hoff factor for 0.029 m K_2SO_4 is 2.6.

Exercise 12.18

Calculate the freezing point of a 0.10 *m* aqueous solution of $Al_2(SO_4)_3$. Assume the value of *i* based on the formula.

(See Problems 12.59 and 12.60.)

A Checklist for Review

Important Terms

miscible (12.1)
solute (12.1)
solvent (12.1)
hydration (of ions) (12.2)
LeChatelier's principle (12.3)
Henry's law (12.3)
colloid (12.4)
Tyndall effect (12.4)
aerosols (12.4)

emulsion (12.4)
sol (12.4)
hydrophilic colloid (12.4)
hydrophobic colloid (12.4)
coagulation (12.4)
micelles (12.4)
association colloid (12.4)
colligative property (p. 404)
mass percentage of solute (12.5)

molality (12.5)
mole fraction (12.5)
vapor-pressure lowering (12.6)
Raoult's law (12.6)
boiling-point elevation (12.7)
freezing-point depression (12.7)
osmosis (12.8)
osmotic pressure (12.8)

Summary of Facts and Concepts

Solutions are homogeneous mixtures and can be gases, liquids, or solids. Two gases, for example, will mix in all proportions to give a gaseous solution, because gases are completely *miscible* in one another. Often one substance will dissolve in another only to a limited extent. The maximum amount that dissolves at equilibrium is the *solubility* of that substance. Solubility is determined by the natural tendency toward disorder (say, by the mixing of two substances) and by the tendency to give the strongest forces of attraction between species. The dissolving of one molecular substance in another will be limited if intermolecular forces strongly favor the unmixed substances. When two substances have similar types of intermolecular forces, they tend to be soluble ("like dissolves like"). Ionic substances differ greatly in their solubility in water because the solubility depends on the relative values of *lattice energy* and *hydration energy*.

Solubilities of substances usually vary with temperature. Most gases become less soluble in water at higher temperatures, whereas most ionic solids become more soluble. Pressure has a significant effect only on the solubility of a gas. A gas is more soluble in a liquid if the partial pressure of the gas is increased, in agreement with LeChatelier's principle. According to *Henry's law*, the solubility of a gas is directly proportional to the partial pressure of the gas.

A *colloid* is a dispersion of particles of one substance (about 10 Å to 2000 Å in size) throughout another. Colloids can be detected by the *Tyndall effect* (the scattering of light by colloidal-sized particles). They are characterized by the state of the dispersed phase and the state of the continuous phase. An aerosol, for example, consists of liquid droplets or solid particles dispersed in a gas. Colloids in water are also classified as hydrophilic or hydrophobic. A *hydrophilic colloid* consists of a dispersed phase that is attracted to the water. Many of these colloids have macromolecules dissolved in water. In *hydrophobic colloids*, there is little attraction between the dispersed phase and the water. The colloidal particles may be electrically charged and so repel one another, preventing aggregation to larger particles. An ionic solution can neutralize this charge so that the colloid will *coagulate*, that is, aggregate. An *association colloid* consists of molecules with a hydrophobic end and a hydrophilic end dispersed in water. These molecules associate into colloidal-sized groups, or *micelles*.

Colligative properties of solutions depend only on the concentration of solute particles. The concentration may be defined by the molarity, mass percentage of solute, molality, or mole fraction. One example of a colligative property is the *vapor-pressure lowering* of a volatile solvent by the addition of a nonvolatile solute. According to *Raoult's law*, the vapor pressure of the solution depends on the mole fraction of solvent in the solution. Since the addition of a nonvolatile solute lowers the vapor pressure, the boiling point must be raised to bring the vapor pressure back to one atmosphere (*boiling-point elevation*). Such a solution also exhibits a *freezing-point depression*. Boiling-point elevation and freezing-point depression are colligative properties. *Osmosis* is another colligative property. In osmosis, there is a flow of solvent through a semipermeable membrane in order to equalize the concentrations of solutions on the two sides of the membrane. Colligative properties may be used to measure the concentration of solute. In this way, one can determine the molecular weight of the solute.

Operational Skills

1. Given the solubility of a gas at one pressure, find the solubility at another pressure (Example 12.1).

2. Given the mass percent of solute, state how to prepare a given mass of solution (Example 12.2). Given the masses of solute and solvent, find the molality (Example 12.3) and mole fractions (Example 12.4).

3. Given the molality of a solution, calculate the mole fractions of solute and solvent; and given the mole fractions, calculate the molality (Examples 12.5 and 12.6). Given the density, calculate the molality from the molarity, and vice versa (Examples 12.7 and 12.8).

4. Given the mole fractions of solvent in a solution of nonvolatile solute and the vapor pressure of pure solvent, calculate the vapor pressure and vapor-pressure lowering of the solution (Example 12.9).

5. Given the molality of a solution of nonvolatile solute, calculate the boiling-point elevation and freezing-point depression (Example 12.10).

6. Given the masses of solvent and solute and the molality of the solution, find the molecular weight of the solute (Example 12.11). Given the masses of solvent and solute, the freezing-point depression, and K_f, find the molecular weight of the solute (Example 12.12).

7. Given the molarity and the temperature of a solution, calculate the osmotic pressure (Example 12.13).

Review Questions

12.1 Explain why glycerol, $CH_2OHCHOHCH_2OH$, is completely miscible in water, but benzene, C_6H_6, has very limited solubility in water.

12.2 Explain why ionic substances show a wide range of solubilities in water.

12.3 Fewer fish can be kept in an aquarium of fixed size in the summer than in the winter. Explain.

12.4 Explain why a carbonated beverage must be stored in a closed container.

12.5 Pressure has an effect on the solubility of oxygen in water but a negligible effect on the solubility of sugar in water. Why is this?

12.6 One can often see "sunbeams" passing through the less dense portions of clouds. What is the explanation of this?

12.7 Give an example of an aerosol, a foam, an emulsion, a sol, and a gel.

12.8 If electrodes connected to a direct current are dipped into a beaker of colloidal iron(III) hydroxide, a precipitate collects at the negative electrode. Explain.

12.9 Stearic acid can be made to form a monolayer (a layer one molecule thick) on water by placing a drop of solution of stearic acid in benzene onto the surface of the water. The solution spreads over the water and the benzene evaporates. Explain why a monolayer forms.

12.10 Explain why the addition of a nonvolatile solute to a volatile solvent lowers the vapor pressure.

12.11 Explain why the boiling point of a solution containing a nonvolatile solute is higher than the boiling point of a pure solvent.

12.12 It is possible to obtain drinking water from sea water by freezing. Explain the process.

12.13 List several applications of freezing-point depression.

12.14 A green, leafy salad becomes wilted if left too long in a salad dressing containing vinegar and salt. Explain what happens.

Problems

Types of Solutions

12.15 Give an example of a liquid solution prepared by dissolving a gas in a liquid.

12.16 Give an example of a solid solution prepared from two solids.

Solubility

12.17 Would boric acid, $B(OH)_3$, be more soluble in ethanol, C_2H_5OH, or in benzene, C_6H_6?

12.19 Arrange the following substances in order of increasing solubility in hexane, C_6H_{14}: CH_2OHCH_2OH, $C_{10}H_{22}$, H_2O.

12.18 Would naphthalene, $C_{10}H_8$, be more soluble in ethanol, C_2H_5OH, or in benzene, C_6H_6?

12.20 Choose the substance from each of the following pairs that is more soluble in ethanol, C_2H_5OH:
   (a) paradichlorobenzene, $C_6H_4Cl_2$, and parachlorophenol, $C_6H_4(OH)Cl$
   (b) acetic acid, CH_3COOH, and stearic acid, $C_{17}H_{35}COOH$

12.21 Which of the following ions would be expected to have the greater energy of hydration, K^+ or Ca^{2+}?

12.23 Arrange the following alkaline earth-metal iodates in order of increasing solubility in water; explain your reasoning: $Ba(IO_3)_2$, $Ca(IO_3)_2$, $Sr(IO_3)_2$, $Mg(IO_3)_2$.

12.22 Which of the following ions would be expected to have the greater energy of hydration, F^- or Cl^-?

12.24 Explain the trends in solubility of the alkali metal fluorides and permanganates.

Solubility (grams per 100 mL of water)

	F^-	MnO_4^-
Li^+	0.27	71
Na^+	4.2	Very soluble
K^+	92	6.4
Rb^+	131	0.5
Cs^+	367	0.097

12.25 The solubility of carbon dioxide in water is 0.161 g CO_2 in 100 mL of water at 20°C and 1.00 atm. A soft drink is carbonated with carbon dioxide gas at 5.0 atm pressure. What is the solubility of carbon dioxide in water at this pressure?

12.26 Nitrogen, N_2, is soluble in blood and body tissues. At the concentration that will dissolve at higher pressures, nitrogen causes a narcosis or intoxication, which deep-sea divers call the "raptures of the deep." For this reason, the U.S. Navy advises divers using compressed air not to go below 125 feet. The total pressure at this depth is 3.79 atm. If the solubility of nitrogen at 1.00 atm is 1.75×10^{-3} g/100 mL of water, and the mole percent of nitrogen in air is 78.1, what is the solubility of nitrogen in water from air at 3.79 atm?

Colloids

12.27 Give the type of colloid (aerosol, foam, emulsion, sol, gel) for each of the following:
   (a) rain cloud   (b) milk of magnesia
   (c) soapsuds   (d) silt in water

12.28 Give the type of colloid (aerosol, foam, emulsion, sol, gel) for each of the following:
   (a) ocean spray
   (b) beaten egg white
   (c) dust cloud
   (d) salad dressing (oil in vinegar)

12.29 Arsenic(III) sulfide forms a sol with a negative charge. Which of the following ionic substances should be most effective in coagulating the sol?
  (a) KCl   (b) $MgCl_2$
  (c) $Al_2(SO_4)_3$ (d) Na_3PO_4

12.30 Aluminum hydroxide forms a positively charged sol. Which of the following ionic substances should be most effective in coagulating the sol?
  (a) NaCl   (b) $CaCl_2$
  (c) $Fe_2(SO_4)_3$ (d) K_3PO_4

Solution Concentration

12.31 How would you prepare 125 g of an aqueous solution that is 2.50% by mass of potassium iodide, KI?

12.33 Vanillin, $C_8H_8O_3$, occurs naturally in vanilla extract and is used as a flavoring agent. A 37.2-mg sample of vanillin was dissolved in 153.1 mg of diphenyl ether, $(C_6H_5)_2O$. What is the molality of vanillin in the solution?

12.35 Fructose, $C_6H_{12}O_6$, is a sugar occurring in honey and in many fruits. It is the sweetest sugar, nearly twice as sweet as sucrose (cane or beet sugar). How much water should be added to 1.58 g of fructose to give a 0.125 m solution?

12.37 A 100.0-g sample of a brand of rubbing alcohol contains 70.0 g of isopropyl alcohol, C_3H_7OH, and 30.0 g of water. What is the mole fraction of isopropyl alcohol in the solution? What is the mole fraction of water?

12.39 A bleaching solution contains sodium hypochlorite, NaOCl, dissolved in water. If the solution is 0.650 m NaOCl, what is the mole fraction of sodium hypochlorite?

12.41 Concentrated hydrochloric acid contains 1.00 mol HCl dissolved in 3.31 mol H_2O. What is the mole fraction of HCl in concentrated hydrochloric acid? What is the molal concentration of HCl?

12.43 Oxalic acid, $H_2C_2O_4$, occurs as the potassium or calcium salt in many plants, including rhubarb and spinach. An aqueous solution of oxalic acid is 0.585 m $H_2C_2O_4$. The density of the solution is 1.022 g/mL. What is the molar concentration?

12.45 A solution of vinegar is 0.763 M in acetic acid, $HC_2H_3O_2$. The density of the vinegar is 1.004 g/mL. What is the molal concentration of acetic acid?

12.32 What mass of solution containing 2.50% by mass of potassium iodide, KI, contains 235 mg KI?

12.34 Lauryl alcohol, $C_{12}H_{25}OH$, is prepared from coconut oil; it is used to make sodium lauryl sulfate, a synthetic detergent. What is the molality of lauryl alcohol in a solution containing 15.6 g of lauryl alcohol dissolved in 125 g of ethanol, C_2H_5OH?

12.36 Caffeine, $C_8H_{10}N_4O_2$, is a stimulant found in tea and coffee. A sample of the substance was dissolved in 50.0 g of chloroform, $CHCl_3$, to give a 0.0946 m solution. How many grams of caffeine were in the sample?

12.38 An automobile antifreeze solution contains 2.42 kg of ethylene glycol, CH_2OHCH_2OH, and 2.00 kg of water. Find the mole fraction of ethylene glycol in this solution. What is the mole fraction of water?

12.40 An antiseptic solution contains hydrogen peroxide, H_2O_2, in water. If the solution is 0.850 m H_2O_2, what is the mole fraction of hydrogen peroxide?

12.42 Concentrated aqueous ammonia contains 1.00 mol NH_3 dissolved in 2.44 mol H_2O. What is the mole fraction of NH_3 in concentrated aqueous ammonia? What is the molal concentration of NH_3?

12.44 Citric acid, $H_3C_6H_5O_7$, occurs in plants. Lemons contain 5 to 8% of citric acid by mass. The acid is added to beverages and candy. An aqueous solution is 0.710 m in citric acid. The density is 1.049 g/mL. What is the molar concentration?

12.46 A beverage contains tartaric acid, $H_2C_4H_4O_6$, a substance obtained from grapes during wine making. If the beverage is 0.271 M in tartaric acid, what is the molal concentration? The density of the solution is 1.016 g/mL.

Colligative Properties

12.47 Calculate the vapor pressure at 35°C of a solution made by dissolving 20.2 g of sucrose, $C_{12}H_{22}O_{11}$, in 50.0 g of water. The vapor pressure of pure water at 35°C is 42.2 mmHg. What is the vapor-pressure lowering of the solution? (Sucrose is nonvolatile.)

12.48 What is the vapor pressure at 23°C of a solution of 1.20 g of naphthalene, $C_{10}H_8$, in 20.4 g of benzene, C_6H_6? The vapor pressure of pure benzene at 23°C is 86.0 mmHg; the vapor pressure of naphthalene can be neglected. Calculate the vapor-pressure lowering of the solution.

12.49 What is the boiling point of a solution of 1.20 g of glycerol, $C_3H_9O_3$, in 20.0 g of water? What is the freezing point?

12.51 An aqueous solution of a molecular compound freezes at $-0.78°C$. What is the molality of the solution?

12.53 A 0.182-g sample of an unknown substance was dissolved in 2.135 g of benzene. The molality of this solution, determined by freezing-point depression, was found to be 0.698. What is the molecular weight of the unknown substance?

12.55 Safrole is contained in oil of sassafras, which is used to flavor root beer. A 2.39-mg sample of safrole was dissolved in 41.2 mg of diphenyl ether. The solution had a melting point of 23.98°C. Calculate the molecular weight of safrole. The freezing point of pure diphenyl ether is 26.84°C, and the freezing-point-depression constant, K_f, is 8.00°C/m.

12.57 Dextran is a polymeric carbohydrate produced by certain bacteria. It is used as a blood plasma substitute. An aqueous solution contains 0.582 g of dextran in 106 mL of solution at 21°C. It has an osmotic pressure of 1.47 mmHg. What is the average molecular weight of the dextran?

12.59 A 0.0140-g sample of an ionic compound with the formula $Cr(NH_3)_5Cl_3$ was dissolved in water to give 25.0 mL of solution at 25°C. The osmotic pressure was determined to be 119 mmHg. How many ions are obtained from each formula unit when the compound is dissolved in water?

12.50 A solution was prepared by dissolving 8.20 g of sulfur, S_8, in 100.0 g of acetic acid, $HC_2H_3O_2$. Calculate the freezing point and boiling point of the solution.

12.52 Urea, $(NH_2)_2CO$, is dissolved in 100.0 g of water. The solution freezes at $-0.97°C$. How many grams of urea were dissolved to make this solution?

12.54 A solution contains 0.653 g of a compound in 9.75 g of ethanol. The molality of the solution is 0.368. Calculate the molecular weight of the compound.

12.56 Butylated hydroxytoluene (BHT) is used as an antioxidant in processed foods (it prevents fats and oils from becoming rancid). A solution of 2.500 g of BHT in 100.0 g of benzene had a freezing point of 4.880°C. What is the molecular weight of BHT?

12.58 Arginine vasopressin is a pituitary hormone. It helps regulate the amount of water in the blood by reducing the flow of urine from the kidneys. An aqueous solution containing 0.864 g of vasopressin in 100.0 mL of solution had an osmotic pressure at 25°C of 148 mmHg. What is the molecular weight of the hormone?

12.60 In a mountainous location, the boiling point of pure water is found to be 95°C. How many grams of sodium chloride must be added to 1 kg of water to bring the boiling point back to 100°C? Assume that $i = 2$.

Additional Problems

*12.61** A gaseous mixture consists of 80.0 mole percent N_2 and 20.0 mole percent O_2 (the approximate composition of air). Suppose that water is saturated with the gas mixture at 25°C and 1.00 atm total pressure, and then the gas is expelled from the water by heating. What is the composition in mole fractions of the gas mixture that is expelled? The solubilities of N_2 and O_2 at 25°C and 1.00 atm are 0.0175 g/L H_2O and 0.0393 g/L H_2O, respectively.

*12.62** A natural gas mixture consists of 90.0 mole percent CH_4 (methane) and 10.0 mole percent C_2H_6 (ethane). Suppose that water is saturated with the gas mixture at 20°C and 1.00 atm total pressure, and the gas is then expelled from the water by heating. What is the composition in mole fractions of the gas mixture that is expelled? The solubilities of CH_4 and C_2H_6 at 20°C and 1.00 atm are 0.023 g/L H_2O and 0.059 g/L H_2O, respectively.

12.63 An aqueous solution is 8.50% ammonium chloride, NH_4Cl, by mass. The density of the solution is 1.024 g/mL. What are the molality, mole fraction, and molarity of NH_4Cl in the solution?

12.64 An aqueous solution is 22.0% lithium chloride, $LiCl$, by mass. The density of the solution is 1.127 g/mL. What are the molality, mole fraction, and molarity of $LiCl$ in the solution?

12.65 A 58-g sample of a gaseous fuel mixture contains 0.43 mole fraction propane, C_3H_8; the remainder of the mixture is butane, C_4H_{10}. What are the masses of propane and butane in the sample?

12.66 The diving atmosphere used by the U.S. Navy in its undersea Sea-Lab experiments consisted of 0.036 mole fraction O_2 and 0.056 mole fraction N_2, with helium (He) making up the remainder. What are the masses of nitrogen, oxygen, and helium in a 7.84-g sample of this atmosphere?

*12.67 A liquid solution consists of 0.35 mole fraction ethylene dibromide, $C_2H_4Br_2$, and 0.65 mole fraction propylene dibromide, $C_3H_6Br_2$. Both ethylene dibromide and propylene dibromide are volatile liquids with vapor pressures at 85°C of 173 mmHg and 127 mmHg, respectively. Assume that each compound follows Raoult's law in the solution. Calculate the total vapor pressure of the solution.

*12.68 What is the total vapor pressure at 20°C of a liquid solution containing 0.75 mole fraction benzene, C_6H_6, and 0.25 mole fraction toluene, $C_6H_5CH_3$? Assume that Raoult's law holds for each component of the solution. The vapor pressure of pure benzene at 20°C is 75 mmHg; that of toluene at 20°C is 22 mmHg.

12.69 Urea, $(NH_2)_2CO$, has been used to melt ice from sidewalks, since the use of salt is harmful to plants. If the saturated aqueous solution contains 44% urea by mass, what is the freezing point? (The answer will be approximate because the equation in the text applies accurately only to dilute solutions.) Will urea melt ice at $-30°C$?

12.70 Calcium chloride, $CaCl_2$, has been used to melt ice from roadways. Given that the saturated solution is 32% $CaCl_2$ by mass, estimate the freezing point. Will calcium chloride melt ice at $-30°C$?

12.71 The osmotic pressure of blood at 37°C is 7.7 atm. A solution that is given intravenously must have the same osmotic pressure. What should be the molarity of a glucose solution to give an osmotic pressure of 7.7 atm at 37°C?

12.72 Maltose, $C_{12}H_{22}O_{11}$, is a sugar produced by malting (sprouting) grain. A solution of maltose at 25°C has an osmotic pressure of 5.61 atm. What is the molal concentration of maltose?

12.73 Which aqueous solution has the lower freezing point, 0.10 m $CaCl_2$ or 0.10 m glucose?

12.74 Which aqueous solution has the lower boiling point, 0.10 m KCl or 0.10 m $CaCl_2$?

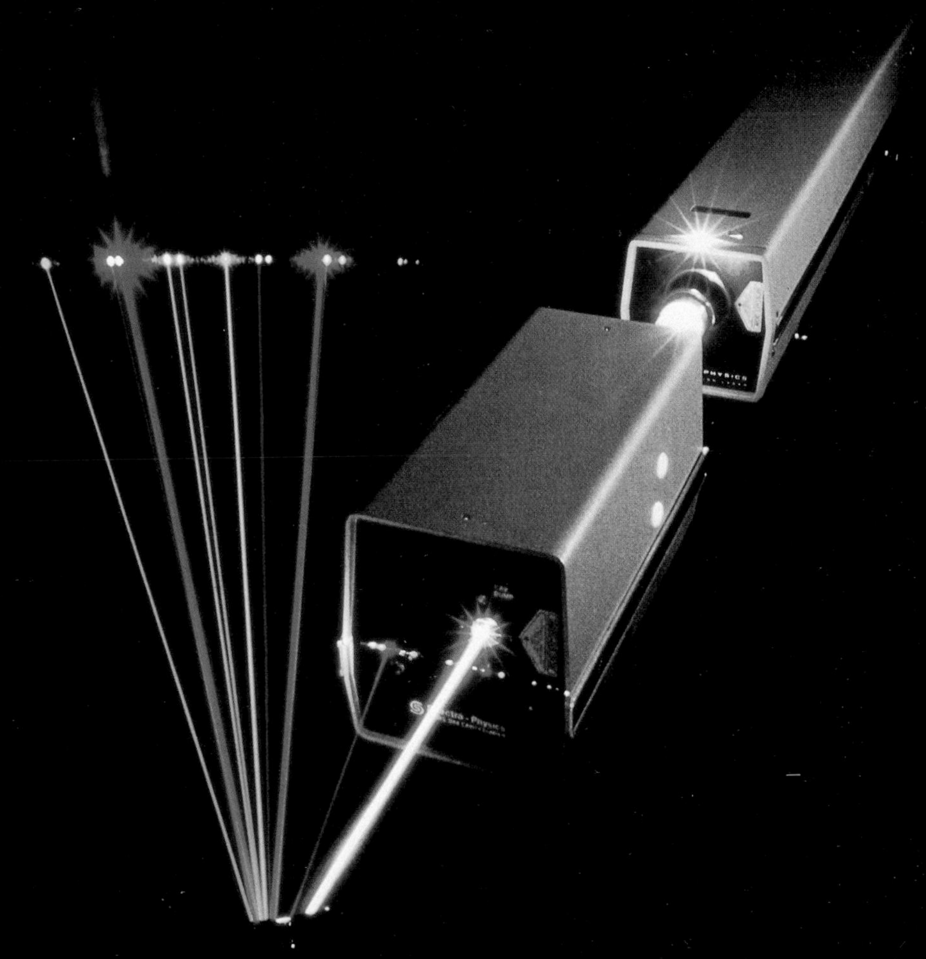

Plate 1. *Top*: A krypton laser emits several discrete wavelengths, which together appear white. This "white" light is spread into its discrete wavelengths after striking a ruled grating. The spectrum, right to left, shows a red, a blue, a yellow-green, and two green lines. The spectrum then repeats; a red, a blue, and a yellow-green line are visible. *Bottom*: White light, entering at the left, strikes a prism, which disperses the light into a continuous spectrum of wavelengths.

VISIBLE SPECTRUM

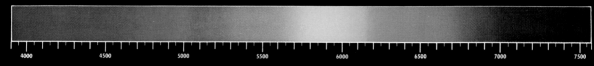

EMISSION (BRIGHT LINE) SPECTRA

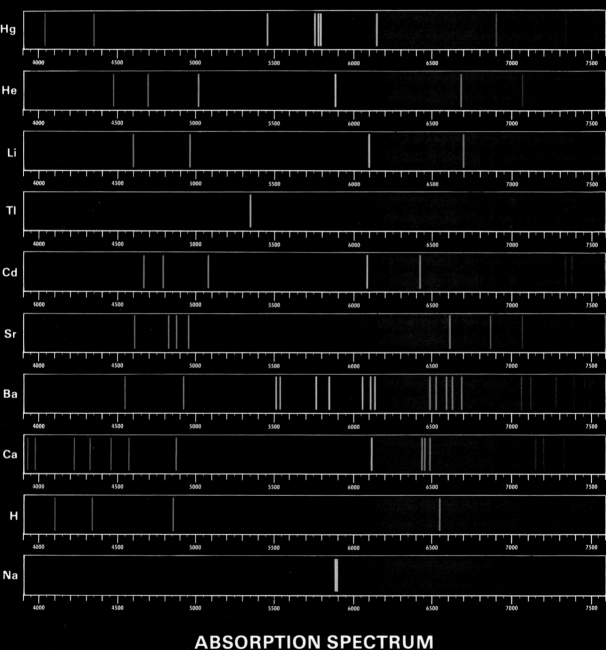

Hg

He

Li

Tl

Cd

Sr

Ba

Ca

H

Na

ABSORPTION SPECTRUM

Plate 2. *Top*: Visible (continuous) spectrum and the emission (line) spectra of various atoms.

Bottom: Absorption spectrum of the sun, showing the black lines resulting from absorption of light by gases in the sun's atmosphere. The black lines, called Fraunhofer lines, correspond to the emission lines of atoms present in the sun's atmosphere.

Plate 3. *Right*: Flame tests of Group IIA elements. The flames are those of strontium (red), calcium (orange), and magnesium (colorless).

Bottom: Study of combustion using laser light. Laser light enters at the right and passes through the flame, where it excites molecular fragments produced during combustion. Each excited molecular fragment then emits its identifying wavelengths, which are detected at the left.

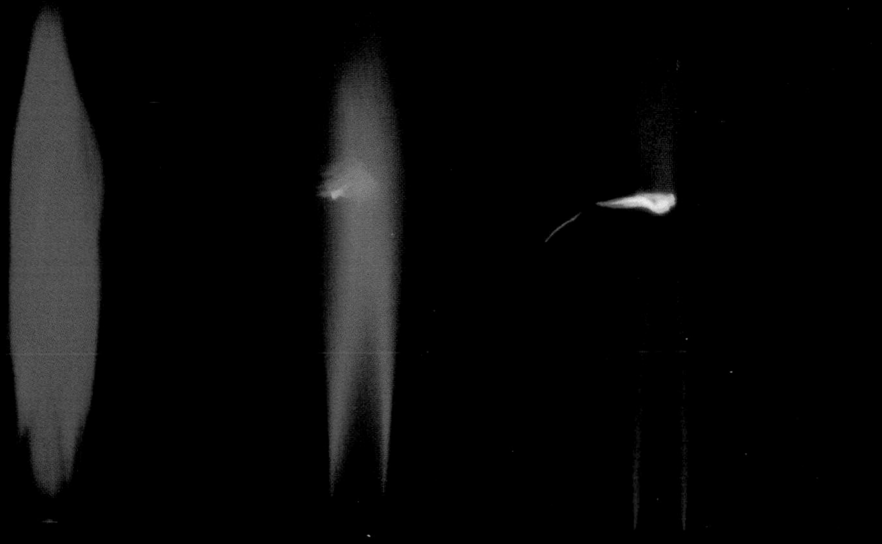

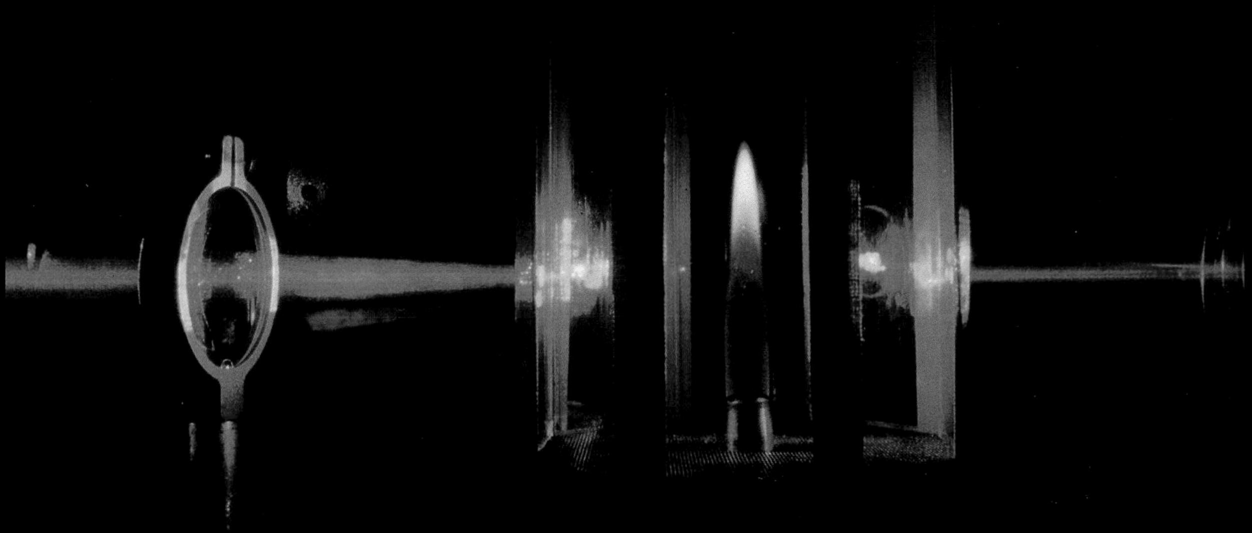

Plate 4. *Top*: Acid-base titration. Sodium hydroxide solution in the flask is titrated by hydrochloric acid from a buret. Phenolphthalein indicator gives a deep pink color with the base. Near the end of the titration, the color is a light pink. A final drop of hydrochloric acid turns the solution in the flask colorless. *Bottom*: Suspensions of metal iodide precipitates. Left to right: mercury (II) iodide, silver iodide, lead (II) iodide.

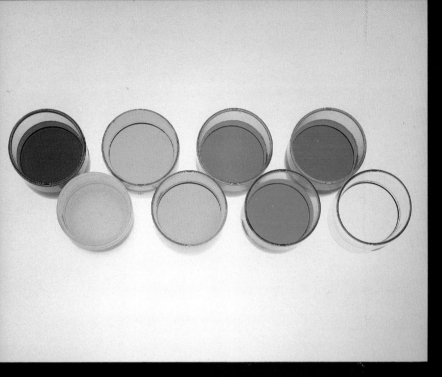

Plate 5. *Top*: Common transition-metal cations in aqueous solution. Left to right: Cr^{3+} (red violet), Mn^{2+} (pale pink), Fe^{2+} (pale green), Fe^{3+} (pale yellow), Co^{2+} (pink), Ni^{2+} (green), Cu^{2+} (blue), Zn^{2+} (colorless). *Bottom*: Change of coordination number of Co^{2+}. Cobalt(II) chloride is dissolved in a solution of water and ethanol. At room temperature, the Co^{2+} ion exists in solution as a tetrahedral complex with blue color. When placed in an ice bath, the Co^{2+} ion exists as an octahedral complex with pink color. The change in color and intensity results from the change from tetrahedral geometry to octahedral geometry.

Plate 6. Minerals. (a) Quartz (amethyst crystal), SiO_2 (b) Orpiment (yellow), As_2O_3, with realgar (red), As_4S_4 (c) Sulfur (yellow), S_8 (d) Realgar (red), As_4S_4 (e) Galena (gray), PbS (f) Malachite, $Cu_2CO_3(OH)_2$ (g) Vanadīnite (red-orange), $Pb_5(VO_4)_3Cl$

Plate 7. Kinetics versus thermodynamics (true equilibrium). *Top*: A bright orange precipitate of tetragonal crystals of HgI_2 is formed from concentrated solutions of $HgCl_2$ and KI (*left*). In dilute solution (*right*), however, the HgI_2 precipitates as rhombic crystals with a lemon-yellow color. Although less stable, the yellow crystals are formed faster in dilute solution. *Bottom*: In less than an hour, the yellow crystals change to the more stable orange crystals.

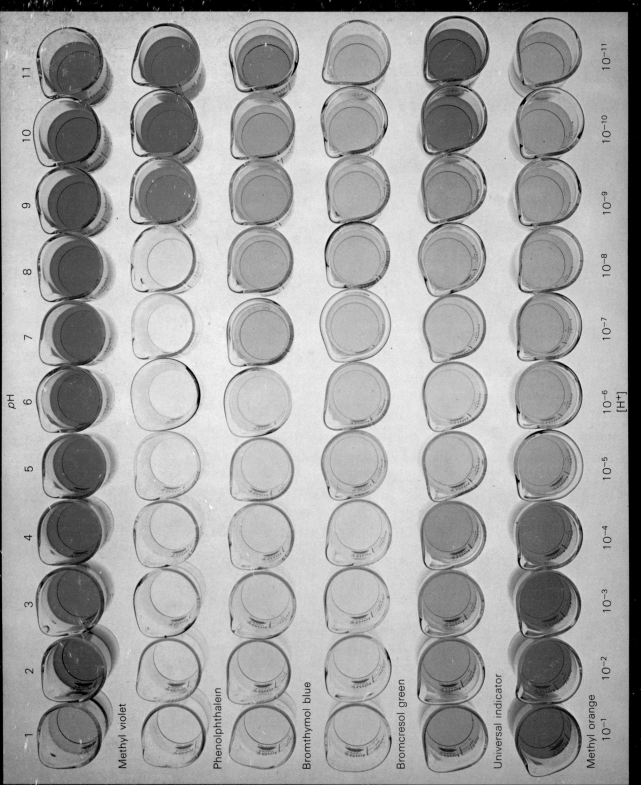

pH   1   2   3   4   5   6   7   8   9   10   11

Methyl violet

Phenolphthalein

Bromthymol blue

Bromcresol green

Universal indicator

Methyl orange

$[H^+]$ 10^{-1} 10^{-2} 10^{-3} 10^{-4} 10^{-5} 10^{-6} 10^{-7} 10^{-8} 10^{-9} 10^{-10} 10^{-11}

Plate 8. Some acid-base indicators in solutions with different H^+ concentrations.

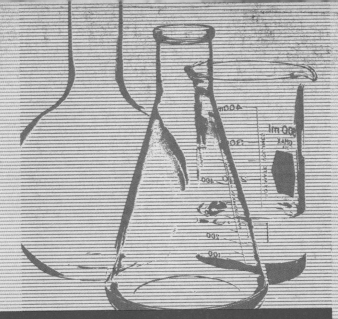

13. The Atmosphere: Oxygen, Nitrogen, and the Noble Gases

The Earth's Atmosphere

13.1 The Present Atmosphere Regions of the Atmosphere/ Heating of the Thermosphere and the Stratosphere

13.2 Evolution of the Atmosphere Lack of Noble Gases in the Atmosphere/ Formation of a Primitive Atmosphere/ Formation of Molecular Oxygen

Oxygen

13.3 Molecular Oxygen Preparation of Oxygen in Small Quantities/ Production of Oxygen in Commercial Quantities/ Structure and Physical Properties of Oxygen/ Uses of Oxygen

13.4 Reactions of Oxygen Reactions with Metals/ Reactions with Nonmetals/ Reactions with Compounds/ Basic and Acidic Oxides

13.5 Ozone Properties of Ozone/ *Aside: Stratospheric Ozone Depletion*

Nitrogen

13.6 Molecular Nitrogen Uses of Nitrogen/ Nitrogen Fixation

13.7 Compounds of Nitrogen Ammonia and the Nitrides / Hydrazine/ Hydroxylamine/ Nitrous Oxide/ Nitric Oxide/ Nitrogen Dioxide/ Dinitrogen Trioxide and Nitrous Acid/ Dinitrogen Pentoxide and Nitric Acid

Noble Gases and Other Constituents of Air

13.8 The Noble Gases Discovery of the Noble Gases/ Compounds of the Noble Gases/ Preparation and Uses of the Noble Gases

13.9 Trace Constituents of Air *Aside: The Earth's Thermal Moderators*

F rom the moon, the earth appears as a beautiful blue ball with swirls of white clouds, quite unlike our view of the moon from the earth. The difference is the earth's atmosphere. In this gaseous envelope, clouds are formed as a result of the physical processes of weather, and blue light is scattered in preference to other wavelengths, giving the earth its characteristic color.

On the surface of the earth, we can easily imagine that we are at the bottom of a limitless sea of air. But though atmospheric gases do extend hundreds of kilometers above us, the pressure falls quickly, from 760 mmHg at sea level to less than 1 mmHg at 50 km in altitude. Ninety per cent of the mass of the atmosphere lies no further than 16 km above sea level. Since the radius of solid earth is 6370 km, we might compare the atmosphere of the earth to the peel of an orange.

Like the peel of an orange, the atmosphere shields the life beneath it. If there were no atmosphere, the surface of the earth would be showered with lethal radiation—cosmic rays and high-energy ultraviolet light. This atmospheric blanket also maintains the moderate thermal environment needed for life. Without it, the surface of the earth would be alternately scorched by the midday sun and frozen during the night.

Not only does the atmosphere shield life, it provides sustenance too. Oxygen in the air is needed by us, and by other forms of life, for respiration. And atmospheric nitrogen and carbon dioxide are incorporated by plants into compounds that ultimately supply us with essential nutrients.

As we investigate the features of the earth's atmosphere in this chapter, we will see that the chemistry we have already learned can be used to understand this gaseous shell. We will also discuss the chemistry of the main constituents of air: oxygen, nitrogen, and the noble gases. We will see that air is a commercial source of chemical substances.

Chapter Overview

This chapter on the earth's atmosphere draws upon basic principles that we studied in previous chapters. We look first at the general features of the atmosphere—its composition, the various regions into which it is divided, and its evolution from an earlier atmosphere. Then, we discuss the chemistry of oxygen, nitrogen, and the noble gases. In the concluding section, we look at some trace constituents of the atmosphere.

The Earth's Atmosphere

As we discussed in the chapter opening, many of the characteristics of the earth—its color viewed from the moon, its moderate thermal conditions, and the presence of life—depend on the fact that it has an atmosphere. We will describe this atmosphere in the next section.

13.1 The Present Atmosphere

The lower atmosphere (to about 80 km in altitude) is kept mixed by turbulence, so that its composition is nearly constant worldwide. It is a mixture of gases, consisting primarily of molecular nitrogen (N_2) and molecular oxygen (O_2), with smaller amounts of argon (Ar) and carbon dioxide (CO_2), and variable amounts of water vapor. It also has many minor or trace constituents. Table 13.1 gives the major components of dry air in the lower atmosphere. The composition is given in percent by volume, which is equivalent to mole percent.■

■ The composition of the atmosphere is often given in terms of the relative volumes of pure gases (at a fixed temperature and pressure) that could be obtained from the mixture. According to Avogadro's law (Section 4.4), the volume of gas (at a fixed temperature and pressure) is proportional to the moles of gas. Therefore, the volume percent of a component gas in a mixture is equivalent to the mole percent (mole fraction × 100%).

Component		Percent by Volume
Nitrogen	N_2	78.08
Oxygen	O_2	20.95
Argon	Ar	0.934
Carbon dioxide	CO_2	0.033

Table 13.1
Major Components of
Dry Air in the Lower
Atmosphere*

*See Table 13.4 for minor components.

Regions of the Atmosphere

The atmosphere is divided into regions in which the temperature either rises
or falls with altitude, as shown in Figure 13.1. Near the surface of the earth,
the atmospheric temperature decreases with altitude—a fact well known to
mountain climbers. This lowest region of the atmosphere is called the
troposphere.

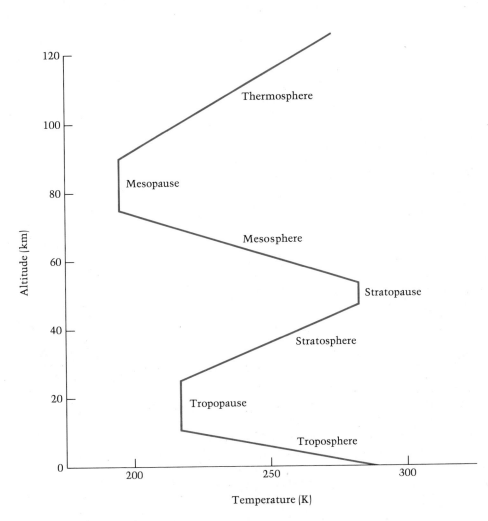

Figure 13.1
Regions of the atmosphere.
The exact temperatures
and altitudes for various regions vary with latitude and
season.

Until the beginning of this century, it was thought that the temperature continued to decrease with increasing altitude until absolute zero was reached at the edge of outer space. Then, in 1902, French meteorologist Léon Philippe Teisserenc de Bort discovered, after years of investigations with balloons, that near an altitude of 10 km the temperature becomes constant at approximately 220 K. At higher altitudes, the temperature increases, reaching a maximum of about 280 K at 50 km. This region of increasing temperature is called the **stratosphere.** The region between the troposphere and the stratosphere in which the temperature remains constant is called the **tropopause.**

Above 50 km, the temperature once again begins to fall, in the region of the atmosphere called the **mesosphere.** The **stratopause** is the transitional region between the stratosphere and the mesosphere.

At about 80 km, the temperature ceases to fall and begins to rise again. The region of the atmosphere showing this temperature rise is called the **thermosphere,** and the transitional region between it and the mesosphere is called the **mesopause.**

Heating of the Thermosphere and the Stratosphere

Why does the atmosphere show these alternate decreases and increases in temperature? To answer this, let us look at what happens as the sun's radiation enters the outer region of the atmosphere, then penetrates into the lower regions.

The major components of the atmosphere, nitrogen and oxygen, are transparent to all wavelengths of the sun's radiation except the most energetic ones. These energetic rays have wavelengths in the ultraviolet portion of the spectrum. Ultraviolet wavelengths stretch from about 400 nm, just beyond the violet end of the visible spectrum down to much smaller wavelengths (20 nm).

When the sun's rays enter the thermosphere, ultraviolet radiation below 200 nm is absorbed (see Figure 13.2). It is absorbed primarily by molecular oxygen, O_2, which is dissociated by the addition of energy from a photon:

$$O_2(g) \xrightarrow{h\nu} O(g) + O(g)$$

The expression above the arrow (Planck's constant times the frequency of radiation) indicates the absorption of frequencies of light with sufficient energy to break the oxygen–oxygen bond.

Example 13.1

Molecular oxygen, O_2, has a bond energy of 494 kJ/mol. What is the longest wavelength of electromagnetic radiation that could possibly dissociate O_2 into atoms, assuming that the dissociation occurs by absorption of a single photon?

Solution

The bond energy of O_2 is the minimum energy needed to break the bond (Section 10.6). Bond energy is usually given for one mole of bonds (in kilojoules per mole). Since we want to find the frequency of radiation, which

(Continued)

is related to the energy of a single photon, we need to first convert the bond energy per mole of O_2 to the bond energy for one bond. For conversion, we can use the fact that there are 6.02×10^{23} bonds per mole:

$$494 \times 10^3 \text{ J/mol} \times \frac{1 \text{ mol}}{6.02 \times 10^{23}} = 8.21 \times 10^{-19} \text{ J}$$

The energy of a photon, E, is related to the frequency of radiation, ν, by the fundamental formula $E = h\nu$ (see Section 5.6). Substituting E (the energy needed to break a single bond) and $h = 6.63 \times 10^{-34}$ J/s and solving for ν gives

$$\nu = \frac{E}{h} = \frac{8.21 \times 10^{-19} \text{ J}}{6.63 \times 10^{-34} \text{ J} \cdot \text{s}}$$
$$= 1.24 \times 10^{15}/\text{s}$$

The wavelength, λ, is

$$\lambda = \frac{c}{\nu} = \frac{3.00 \times 10^8 \text{ m/s}}{1.24 \times 10^{15}/\text{s}}$$
$$= 2.42 \times 10^{-7} \text{ m } (= 242 \text{ nm})$$

Radiation of longer wavelengths has less energy than 494 kJ/mol of photons and cannot dissociate O_2. It should be noted, however, that O_2 does not actually absorb appreciably above 200 nm. Radiation below 200 nm has more than enough energy to break the oxygen–oxygen bond.

Exercise 13.1

What is the longest wavelength of electromagnetic radiation that can be expected to dissociate N_2? (See Table 10.2)

(See Problems 13.29 and 13.30.)

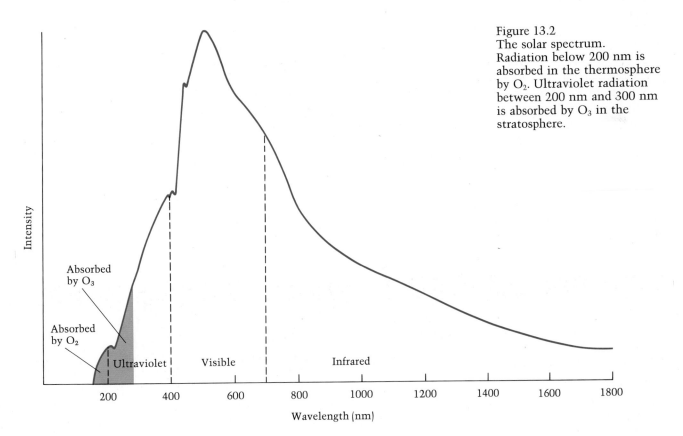

Figure 13.2
The solar spectrum. Radiation below 200 nm is absorbed in the thermosphere by O_2. Ultraviolet radiation between 200 nm and 300 nm is absorbed by O_3 in the stratosphere.

There are two effects of the absorption of high-energy ultraviolet light by O_2. First, it results in a relatively large concentration of oxygen atoms. In fact, at altitudes above 100 km, oxygen atoms become more prevalent than oxygen molecules. Second, the energy imparted to these oxygen atoms shows up as heat. This accounts for the temperature rise seen in the thermosphere. As the sun's radiation passes into lower portions of the thermosphere, less of the energetic ultraviolet radiation is available, and the heating effect diminishes. This results in the temperature minimum at the mesopause.

The heating that we see in the stratosphere (20 to 50 km) is due to ozone, which absorbs ultraviolet radiation between 200 nm and 300 nm (see Figure 13.2). But why is it that most of the ozone occurs in the atmosphere at this altitude? To understand this, we must know how ozone is formed.

Ozone is formed in the stratosphere by the reaction of oxygen atoms, produced there and in higher regions of the atmosphere, with oxygen molecules:

$$O_2(g) \ + \ O(g) \ + \ M(g) \longrightarrow O_3(g) \ + \ M^\star(g)$$

Here M is some molecule, such as N_2 or O_2, which takes away excess energy as an excited molecule, $M^\star$. Unless this energy is carried away, the ozone molecule dissociates immediately back to O_2 and O.

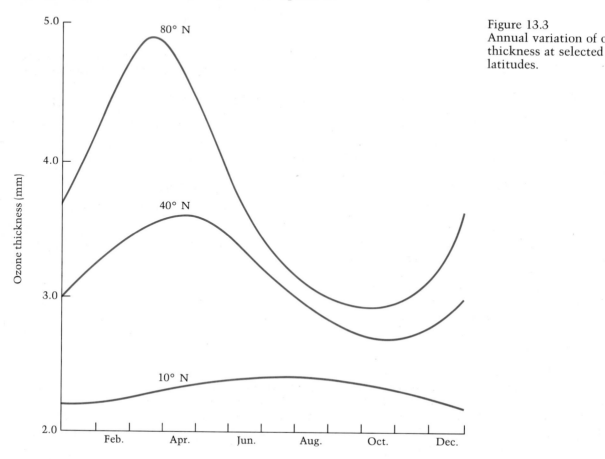

Figure 13.3
Annual variation of ozone thickness at selected latitudes.

Although both oxygen atoms and oxygen molecules are present in the thermosphere, they do not react with one another there because the concentrations of gases are extremely low. The chance of a reactive collision of the three molecules O_2, O, and M in the thermosphere is quite small. Only when the concentrations of molecules become great enough, at the lower altitudes of the stratosphere, do substantial amounts of ozone form. Below the stratosphere, the concentration of oxygen atoms is negligible, since the high-energy ultraviolet radiation needed to dissociate oxygen molecules to atoms has been all absorbed in the upper atmosphere.

By the time the sun's radiation has reached the surface of the earth, all ultraviolet light below 300 nm has been absorbed. This is of vital importance to forms of life that are exposed to sunlight, since high-energy ultraviolet light destroys DNA (deoxyribonucleic acid), the genetic code of life.

Given that ozone in the stratosphere absorbs all of the ultraviolet light between 200 nm and 300 nm, it is remarkable how little ozone is actually present. It is customary to report the amount of ozone in the stratosphere in terms of the thickness with which the ozone would blanket the earth if it were present as a layer of pure gas under standard conditions (0°C, 1 atm). This *ozone thickness* varies with latitude and season (Figure 13.3), but averages only about 3 mm.

Example 13.2

The average amount of ozone throughout the stratosphere is about 1.5 ppm (parts per million) by volume. (Parts per million by volume of a gas mixture is equivalent to mole fraction $\times$ 10^{-6}.) Assume a uniform temperature of -50°C and a uniform pressure of 20 mmHg. Also assume that the ozone extends from 10 km to 50 km in altitude. Calculate the ozone thickness at STP.

Solution

The volume of gas in a shell about the earth is proportional to the thickness of the shell. Let us first calculate the thickness of the stratosphere at STP (proportional to final volume) from its initial thickness (proportional to initial volume). This initial thickness is 50 km $-$ 10 km $=$ 40 km, or 40 $\times$ 10^6 mm.

Stratospheric thickness (at STP)

$$= \text{initial thickness} \times \frac{P_i}{P_f} \times \frac{T_f}{T_i}$$

$$= 40 \times 10^6 \text{ mm} \times \frac{20 \text{ mmHg}}{760 \text{ mmHg}} \times \frac{273 \text{ K}}{223 \text{ K}}$$

$$= 1.9 \times 10^6 \text{ mm}$$

Since ozone makes up only 1.5 ppm or 1.5×10^{-6} mole fraction, we get

Ozone thickness (at STP) $= 1.9 \times 10^6 \text{ mm} \times 1.5 \times 10^{-6}$

$$= 2.9 \text{ mm}$$

Exercise 13.2

The amount of xenon in the troposphere is 0.087 ppm. Calculate the thickness of xenon at STP in the whole of the troposphere. Assume that the troposphere is 10 km high and assume an average temperature of -20°C and an average pressure of 500 mmHg.

(See Problems 13.31 and 13.32.)

13.2 Evolution of the Atmosphere

There are reasons for believing that the composition of the earth's atmosphere has not always been what it is now, but has evolved from a primitive atmosphere. One piece of evidence comes from astronomy.

Lack of Noble Gases in the Atmosphere

According to the prevailing astronomical theory, our solar system was formed out of the dust and gases of the universe by a process in which the mutual gravitational attractions of molecules and particles pulled matter together into clumps that became the sun and planets. If the earth's atmosphere formed when the earth formed, it seems reasonable to expect that this atmosphere would have contained gases known to be abundant in the universe.

Of the first ten most abundant elements in the universe, three are noble gases (helium, neon, and argon). We might expect these noble gases to be present in the atmosphere in large amounts. Except for argon, however, the noble gases are very minor constituents of air. And the argon now present in air is thought to originate from the radioactive decay of potassium-40 in the earth's crust.

What could explain the lack of noble gases in the atmosphere? Since helium is one of the lightest gases, it could well have escaped from the upper atmosphere into outer space. In order for a molecule to escape the gravitational field of earth, it must attain an upward speed of 11 km/s (the *escape velocity*).■ The distribution of molecular speeds depends on molecular mass and temperature.■ The temperature in the upper atmosphere varies from 800 K to 2000 K. At the maximum temperature, the rms speed of helium atoms, calculated from kinetic theory, is 3.5 km/s. Though this is well below the escape velocity of earth, there are sufficient helium molecules at this temperature whose speed is greater than 11 km/s to explain a steady loss. We can therefore account for the lack of helium in the atmosphere.

If we were to account for the lack of neon in the atmosphere by saying that it has escaped from the upper atmosphere into outer space, we should have to assume that the temperature of the earth was at one time much hotter than it is now (see Example 13.3). But if this was true, then much of the original atmosphere—including oxygen and nitrogen—would also have escaped.

■ Any object or particle, whether a space shuttle or a molecule, must reach an upward speed of 11 km/s if it is to overcome the gravitational pull from earth.

■ Maxwell's distribution was discussed in Section 4.8. The rms molecular speed is $(3RT/M_m)^{1/2}$, where M_m is the molar mass and T is the absolute temperature.

Example 13.3

What is the rms molecular speed of helium atoms at 2000 K? What temperature would be required for the rms speed of neon atoms to equal the rms speed of helium atoms at 2000 K?

Solution

The rms molecular speed is given by the formula $(3RT/M_m)^{1/2}$, where M_m is the molar mass (gram molecular weight). Since the molar mass of he-

(Continued)

lium is 4.00 g/mol, or 4.00×10^{-3} kg/mol, and $R = 8.31$ J/(mol·K), the rms speed of helium atoms is

$$\text{Rms speed of He} = \sqrt{\frac{3 \times 8.31 \text{ J/(mol·K)} \times 2000 \text{ K}}{4.00 \times 10^{-3} \text{ kg/mol}}}$$

$$= 3.53 \times 10^3 \text{ m/s}$$

The temperature of neon (molar mass $= 20.2 \times 10^{-3}$ kg/mol), where the rms speed is 3.53×10^3 m/s, is obtained by solving the equation

$$3.53 \times 10^3 \text{ m/s} = \sqrt{\frac{3 \times 8.31 \text{ J/(mol·K)} \times T}{20.2 \times 10^{-3} \text{ kg/mol}}}$$

Squaring both sides and rearranging gives

$$T = (3.53 \times 10^3 \text{ m/s})^2 \times \frac{20.2 \times 10^{-3} \text{ kg/mol}}{3 \times 8.31 \text{ J/(mol·K)}}$$

$$= 1.01 \times 10^4 \text{ K, or } 10,100 \text{ K}$$

Exercise 13.3

The uppermost layer of the atmosphere consists mostly of hydrogen atoms. Calculate the rms speed of hydrogen atoms at 2000 K.

(See Problems 13.33 and 13.34.)

Formation of a Primitive Atmosphere

Whether the original atmosphere was lost or whether the earth formed without an atmosphere, a primitive atmosphere must have formed later. It is assumed to have come from gases that were trapped within the earth's crust. As the earth heated up—either as a result of radioactivity or gravity or after a particular thermal event, such as a near collision with the moon—these gases were expelled in volcanic eruptions and from *fumaroles* (which are vents in the earth in volcanic areas).

There is considerable disagreement about the exact composition of this primitive atmosphere—whether it contained methane and ammonia or whether it contained a more oxidized atmosphere. It is generally agreed, however, that the primitive atmosphere contained little or no molecular oxygen, O_2.

The reactivity of molecular oxygen is one of the main reasons for believing that this gas was absent from the primitive atmosphere. Two pieces of geological evidence support this belief. First, uranium–gold sedimentary deposits in South Africa have been dated at 1.8 to 2.5 billion years old.■ They contain pyrite, which is iron(II) disulfide, FeS_2. But the iron(II) in pyrite is easily oxidized by moist molecular oxygen to iron(III). It does not seem likely, therefore, that these sediments could have been formed while there was an oxidizing atmosphere.

■ Sedimentary deposits are rocks formed from the buildup of layers of sediment as particles settle out of water.

The second piece of evidence concerns the so-called banded iron deposits. These contain iron(II) minerals precipitated from water between 2 and 3 billion years ago. The geologic record shows no such deposits of a more recent age. So presumably about 2 billion years ago, increasing amounts of oxygen in the atmosphere completely oxidized iron(II), after which banded iron deposits were no longer formed.

Example 13.4

After mining operations, pyrite (FeS_2) may be exposed to weathering from moist oxygen in the air. The pyrite is oxidized to iron(III) ion and sulfuric acid. (Acidic water from mines—acid mine drainage—is a serious environmental problem.) Iron(III) is eventually deposited as iron(III) oxide, which leaves a characteristic yellow stain on rocks. Write a balanced equation for the oxidation of pyrite to give iron(III) ion and sulfuric acid.

Solution

Using the half-reaction method (Section 9.8), we first note that FeS_2 results in Fe^{3+} and SO_4^{2-}:

$$FeS_2 \longrightarrow Fe^{3+} + 2SO_4^{2-}$$

This equation is balanced in iron and sulfur. We balance the oxygen atoms by adding H_2O and H^+:

$$FeS_2 + 8H_2O \longrightarrow Fe^{3+} + 2SO_4^{2-} + 16H^+$$

Then we balance the charge with electrons to give the oxidation half-reaction:

$$FeS_2 + 8H_2O \longrightarrow Fe^{3+} + 2SO_4^{2-} + 16H^+ + 15e^-$$

The reduction half-reaction is obtained in a similar way:

$$O_2 + 4H^+ + 4e^- \longrightarrow 2H_2O$$

When these are added to cancel electrons, we get the balanced equation

$$4FeS_2 + 2H_2O + 15O_2 \longrightarrow 4Fe^{3+} + 8SO_4^{2-} + 4H^+$$

The right side is equivalent to $2Fe_2(SO_4)_3 + 2H_2SO_4$.

Exercise 13.4

Arsenic sulfide minerals, such as orpiment, As_2S_3, can be oxidized with moist molecular oxygen to give an acidic solution of arsenate ion, AsO_4^{3-}, and sulfate ion, SO_4^{2-}. The weathering of such minerals can be a source of arsenate ion in fresh water. Write a balanced equation for the weathering of orpiment to given arsenate ion and sulfate ion in acidic solution.

(See Problems 13.35 and 13.36.)

Formation of Molecular Oxygen

The geological evidence seems to indicate that oxygen began to appear in the atmosphere about 2 billion years ago. Some of this oxygen may have come from the **photolysis** (dissociation by light) of water vapor, followed by the loss of hydrogen to outer space:■

$$2H_2O(g) \xrightarrow{h\nu} 2H_2(g) + O_2(g)$$

However, it is believed that most of the oxygen had a biological origin.

When living organisms first appeared, they derived sustenance by breaking down complex molecules, formed perhaps by ultraviolet radiation of the primitive atmosphere. Then organisms evolved that utilized sunlight directly for energy, through the process of photosynthesis. **Photosynthesis** is a series of chemical reactions using sunlight as energy that converts carbon dioxide and water into molecules such as glucose. The reactions can be summarized as follows:

$$6CO_2 + 6H_2O \xrightarrow{h\nu} \underset{\text{glucose}}{C_6H_{12}O_6} + 6O_2 \qquad \textbf{(13.1)}$$

■ Unless the hydrogen is somehow lost, it will recombine with oxygen to give water. Because hydrogen is very light, it easily escapes from the upper portion of the atmosphere. See the discussion preceding Example 13.3.

Molecular oxygen appears as a by-product. Fossil evidence indicates that blue green algae, which are still found today, were among the first oxygen-producing photosynthetic organisms. The energy stored by photosynthesis was recovered by these primitive organisms, which broke up glucose into smaller pieces:

$$C_6H_{12}O_6 \longrightarrow 2CH_3-\overset{\overset{\displaystyle H}{|}}{\underset{\underset{\displaystyle OH}{|}}{C}}-\overset{\overset{\displaystyle O}{\|}}{C}-OH + energy$$

glucose   lactic acid

Eventually, as the partial pressure of oxygen in the atmosphere increased, energy-efficient respiring organisms evolved. These organisms can completely oxidize glucose in a process called *respiration*. The overall equation for respiration is just the reverse of that for photosynthesis:

$$C_6H_{12}O_6 + 6O_2 \longrightarrow 6CO_2 + 6H_2O + energy \qquad \text{(13.2)}$$

Normally, carbon is cycled between photosynthetic organisms, in which carbon dioxide is transformed to organic carbon (reduced forms of carbon), and those organisms in which organic carbon is combined with oxygen to give back carbon dioxide (see Figure 13.4). In other words Reaction 13.1 is

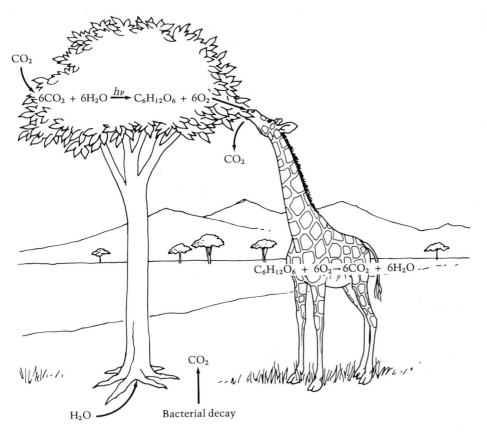

Figure 13.4
The carbon cycle.
Carbon dioxide is incorporated in green plants by photosynthesis, and oxygen is produced as a by-product. When these plants are eaten by animals or decayed by bacteria, the oxygen recombines with the organic material to give carbon dioxide. Of course, part of the glucose is oxidized by the plant itself.

accompanied by Reaction 13.2, so that there is no net build-up of oxygen. Occasionally, however, organic carbon becomes trapped in sediment at the bottom of lakes and seas. This carbon is balanced by a net addition of oxygen to the atmosphere.

The net amount of O_2 produced has been calculated from the estimated total mass of sedimentary rocks (2×10^{24} g).* Sedimentary rock contains an average of 0.4% organic carbon, so that the amount of buried organic carbon is 2×10^{24} g $\times 0.004 = 8 \times 10^{21}$ g. From Reaction 13.1, it follows that for each atom of reduced carbon (from glucose) that is eventually deposited in sediment, one molecule of O_2 is released to the atmosphere. A stoichiometric calculation yields 2×10^{22} g O_2 (see Example 13.5). It is believed that most of this oxygen (about 1.9×10^{22} g) was consumed in the oxidation of inorganic minerals. The rest (1×10^{21} g) has remained in the atmosphere.

Example 13.5

Suppose that 8×10^{21} g of organic carbon is buried in sediment over a long period of time. Assume that the organic carbon originated in the photosynthetic reaction

$$6CO_2 + 6H_2O \xrightarrow{h\nu} C_6H_{12}O_6 + 6O_2$$

How many grams of molecular oxygen would be left in the atmosphere (assuming that none is consumed in subsequent reactions, such as the oxidation of minerals)?

Solution

According to the equation for photosynthesis, for each glucose molecule formed, six molecules of O_2 are pro-

duced. Since each molecule of glucose contains six carbon atoms, we have

$$6C \simeq C_6H_{12}O_6 \simeq 6O_2$$

In other words, for each carbon atom incorporated in glucose, one O_2 molecule is produced. Thus, for each such carbon atom buried in sediment, one molecule of O_2 remains in the atmosphere. We now convert 8×10^{21} g C to grams of O_2:

$$8 \times 10^{21} \text{ g C} \times \frac{1 \text{ mol C}}{12 \text{ g C}} \times \frac{1 \text{ mol O}_2}{1 \text{ mol C}} \times \frac{32 \text{ g O}_2}{1 \text{ mol O}_2}$$
$$= 2 \times 10^{22} \text{ g O}_2$$

Exercise 13.5

The amount of oxygen in the present atmosphere is about 1.2×10^{21} g. If the quantity of fossil fuels burned annually contains 6.5×10^{15} g C, how many years would it take to use up all of the O_2? Assume that the fossil fuels are pure carbon, and that sufficient fossil fuels exist to combine with all of the O_2. How many years would be required to use up all of the O_2 if the fossil fuels were in the form of octane, C_8H_{18}, containing 6.5×10^{15} g C? Note that hydrogen is also present.

(See Problems 13.37 and 13.38.)

Oxygen

Oxygen is the most abundant element on earth, occurring to the extent of 48% in the outer portion (crust, atmosphere, and surface waters). Although there is so much oxygen in the atmosphere, most of the element occurs in

*Bryan Gregor, "Carbon and Atmospheric Oxygen," *Science*, 174 (1971), 316.

combined form, predominantly within the earth's crust as the oxides of silicon and aluminum.

It is not too surprising that oxygen should occur mostly in combined form, since the majority of elements give stable oxides. Fortunately for life on earth, molecular oxygen, O_2, reacts slowly at normal temperatures because the oxygen–oxygen bond is quite strong. Otherwise all organic materials would have burned up, and living organisms would have been impossible. Once burning starts, however, the heat released speeds up the rate of the reaction (reactions usually go faster at higher temperatures). A forest fire is an example of what happens once a reaction of O_2 with organic material begins.

13.3 Molecular Oxygen

The discovery of oxygen is credited separately to the Swedish chemist Karl Wilhelm Scheele and the English chemist Joseph Priestley. Priestley obtained the gas in 1774 by heating mercury(II) oxide (see Figure 13.5).

$$2HgO(s) \xrightarrow{\Delta} 2Hg(l) + O_2(g)$$

(In this equation, the symbol Δ indicates that the reaction proceeds by heating.) Scheele prepared oxygen perhaps as early as 1771, though the published account of his work did not appear until 1777. He produced oxygen by heating various substances, including mercury(II) oxide.

Preparation of Oxygen in Small Quantities

Priestley and Scheele prepared oxygen by thermally decomposing a number of oxygen-containing compounds. Here we will describe how certain oxides, peroxides, and salts of oxyacids may be decomposed to give gaseous oxygen. These methods are useful for preparing small quantities of O_2.

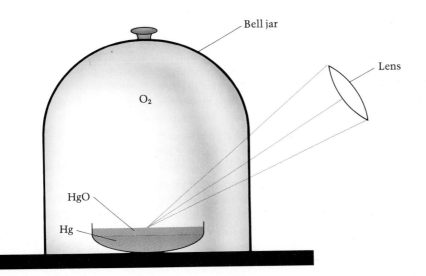

Figure 13.5
Priestley's apparatus for decomposing mercury(II) oxide, HgO, into free oxygen, O_2. Heat was provided by focusing the sun's rays with a lens.

Decomposition of Oxides Oxides are binary compounds of oxygen in the -2 oxidation state. The oxides of some metals are easily decomposed by heating. Mercury(II) oxide, as was mentioned earlier, was used by both Priestley and Scheele to prepare oxygen. Another example of a metal oxide that decomposes when heated is silver oxide, Ag_2O:

$$2Ag_2O(s) \xrightarrow{\Delta} 4Ag(s) + O_2(g)$$

Gold(III) oxide, Au_2O_3, platinum(IV) oxide, PtO_2, and palladium(II) oxide, PdO, also decompose when heated.

Water (hydrogen oxide) is a stable oxide, but does decompose when heated to very high temperatures. It is more easily decomposed by *electrolysis* (the passage of direct electric current through an aqueous solution of electrolytes). An aqueous solution of sulfuric acid is often used in the electrolysis of water. When electrodes (electrical connections) from a battery or other direct-current source are dipped into the solution, oxygen gas forms at the positive electrode and hydrogen forms at the negative electrode (see Figure 14.2 on page 469).

$$2H_2O(l) \xrightarrow{electrolysis} 2H_2(g) + O_2(g)$$

Decomposition of Peroxides Peroxides are compounds with oxygen in the -1 oxidation state. They contain either the peroxide ion, O_2^{2-}, or covalently bonded oxygen in the form $-O-O-$. Many peroxides can be decomposed to give molecular oxygen.

A method for preparing oxygen that was used commercially in the late nineteenth century depended on the thermal decomposition of barium peroxide, BaO_2, an ionic compound. When barium oxide, BaO, is heated in air at about 500°C, it is converted to barium peroxide:

$$2BaO(s) + O_2(g) \xrightarrow{500°C} 2BaO_2(s)$$
barium oxide   barium peroxide

When this peroxide is then heated to above 800°C, it decomposes:

$$2BaO_2(s) \xrightarrow{800°C} 2BaO(s) + O_2(g)$$

The net result is that oxygen is separated from air. Today, oxygen is separated from air by liquefaction followed by fractional distillation, as described later in this section.

Hydrogen peroxide, H_2O_2, is a covalent peroxide.■ Aqueous solutions of hydrogen peroxide decompose in the presence of a catalyst, such as Fe^{3+} or OH^-, to give water and oxygen:

$$2H_2O_2(aq) \xrightarrow{catalyst} 2H_2O(l) + O_2(g)$$

■ Hydrogen peroxide is discussed in Section 14.5.

(A *catalyst* speeds up a reaction, but is not consumed by it. Thus, it does not appear in the equation for the overall reaction.)

When sodium metal burns in air, it forms first the oxide, then the peroxide:

$$4Na(s) + O_2(g) \longrightarrow 2Na_2O(s)$$
$$2Na_2O(s) + O_2(g) \longrightarrow 2Na_2O_2(s)$$
sodium peroxide

Sodium peroxide is a convenient source of oxygen gas. When it is treated with water, it reacts to give hydrogen peroxide, which decomposes immediately in this basic solution to give oxygen:

$$Na_2O_2(s) + 2H_2O(l) \longrightarrow 2Na^+(aq) + 2OH^-(aq) + H_2O_2(aq)$$

$$2H_2O_2(aq) \xrightarrow{OH^-} 2H_2O(l) + O_2(g)$$

Thermal Decomposition of Salts of Oxyacids Certain salts of oxyacids whose central atom is in a high oxidation state decompose when heated. They give oxygen plus the salt of an acid whose central atom is in a lower oxidation state. Potassium nitrate, KNO_3, for example, decomposes when heated to give potassium nitrite, KNO_2, and oxygen:

$$2KNO_3(s) \xrightarrow{\Delta} 2KNO_2(s) + O_2(g)$$

Other alkali metal nitrates undergo a similar reaction.

A common laboratory preparation of oxygen consists in moderate heating of potassium chlorate, $KClO_3$, with manganese dioxide, MnO_2, as a catalyst:

$$2KClO_3(s) \xrightarrow[MnO_2]{\Delta} 2KCl(s) + 3O_2(g)$$

Figure 13.6 shows a laboratory setup for this reaction. In carrying out this reaction, it is very important that the potassium chlorate is not contaminated with substances, such as paper or charcoal, that can be oxidized. Otherwise, an explosion can result.

Exercise 13.6

You have the following chemical reagents available: $Fe_2(SO_4)_3$, Au_2O_3, $NaNO_3$, and H_2O_2. Using equations, describe three different ways you could prepare molecular oxygen from these reagents. Indicate if heat or a particular catalyst is required.

(See Problems 14.39 and 14.40.)

Figure 13.6
A laboratory preparation of oxygen. Oxygen is collected by displacement of water.

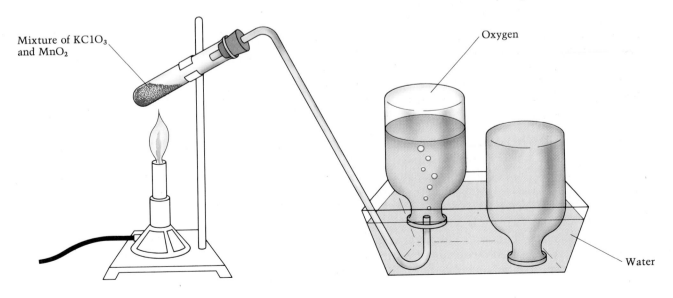

Mixture of $KClO_3$ and MnO_2

Oxygen

Water

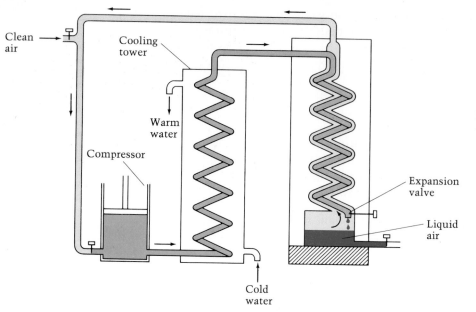

Figure 13.7
Liquid air machine.
Clean air is compressed and
cooled, first in the cooling
tower and then by previously
cooled air. When this cold,
compressed air is allowed to
expand, it cools further and
liquefies.

Production of Oxygen in Commercial Quantities

Today most molecular oxygen used in commercial quantities is separated
from air by using liquid air machines (Figure 13.7). Air is first filtered to
remove dust particles, then compressed and cooled, so that water and carbon
dioxide freeze out. When the cooled compressed air is allowed to expand, it
cools further (to ~ 77 K) and becomes a liquid.

In order to obtain oxygen, the liquid air is subjected to fractional distillation.■ Nitrogen (N_2) and argon are more volatile than oxygen and are
taken off the top of the fractionation column as gases when the liquid air is
warmed, whereas liquid oxygen remains behind at the bottom of the still.

■ Fractional distillation was discussed in Section 2.4.

Structure and Physical Properties of Oxygen

The structure of molecular oxygen is not readily described by Lewis dot
formulas. Neither of the following Lewis formulas is adequate:

$$:\ddot{O}{=}\ddot{O}: \qquad :\ddot{O}{-}\dot{\ddot{O}}:$$

        (a)   (b)

Formula (a) describes oxygen as double-bonded and diamagnetic. Molecular
oxygen, however, is known to be paramagnetic. Formula (b) describes oxygen as paramagnetic, but single-bonded, which is not in agreement with the
known O_2 bond length (1.21 Å). By comparison, the O—O single bond length
in hydrogen peroxide, HOOH, is 1.49 Å. The much smaller value of the O_2
bond length suggests that O_2 is multiple-bonded. Molecular orbital theory
correctly describes the bonding. The paramagnetism, according to this theory, is due to two unpaired electrons in antibonding pi orbitals. Molecular
orbital theory also predicts a double bond (bond order of 2).■

■ The molecular-orbital description of O_2 was given in Section
8.6.

Molecular oxygen is a colorless, odorless gas under standard conditions. The critical temperature is $-118°C$. Thus, oxygen may be liquefied if the gas is first cooled below $-118°C$, then compressed. Both liquid and solid O_2 have a pale blue color. The melting point of the solid is $-218°C$ and the boiling point of the liquid at 1 atm is $-183°C$.

Uses of Oxygen

Over two-thirds of the oxygen produced is used in the making of steel. Iron as it is first obtained from a blast furnace contains 3 to 4% carbon and is soft and brittle. Much of this carbon must be removed by controlled oxidation with O_2 in order to produce a strong steel.

Another important use of oxygen is in the chemical industry. For example, ethylene oxide is prepared from ethylene and oxygen, using a silver metal catalyst:

$$2CH_2{=}CH_2 + O_2 \xrightarrow{\text{Ag}} 2H_2C{-}CH_2$$

ethylene   ethylene oxide

Ethylene oxide is an intermediate in the preparation of a number of organic chemicals, including ethylene glycol, CH_2OHCH_2OH. Ethylene glycol is used as an automobile antifreeze and for the preparation of polyester films and fibers.

Small amounts of oxygen are used in welding and for medical purposes. Also, liquid oxygen (LOX) is commonly used as the oxidizing agent in rocket propulsion.

13.4 Reactions of Oxygen

Molecular oxygen is a very reactive gas, combining with many substances. The products are usually oxides, although peroxides and superoxides sometimes result.

Reactions with Metals

Many metals react readily with oxygen to form oxides. **Oxides** are binary compounds with oxygen in the -2 oxidation state. Oxides of metals in low oxidation states ($+1$ and $+2$) are ionic, whereas those of metals in higher oxidation states have more or less covalent character.

Lithium reacts vigorously with oxygen, as one might expect, and yields the oxide:

$$4Li(s) + O_2(g) \longrightarrow 2Li_2O(s)$$

The other alkali metals (Group IA elements), however, form predominantly peroxides and superoxides. In **peroxides,** oxygen has the oxidation number -1. Metal peroxides are ionic and contain the peroxide ion, O_2^{2-}. In the

superoxides, oxygen has an oxidation number $-\frac{1}{2}$; the compounds contain the superoxide ion, O_2^-.

Molecular-orbital structures of the O_2^- and O_2^{2-} ions are obtained by adding one and two electrons, respectively, into the antibonding pi orbitals of O_2. This decreases the bond order from 2 in O_2 to $1\frac{1}{2}$ for O_2^- and to 1 for O_2^{2-}. As expected, the bond length increases as the bond order decreases: O_2, 1.21 Å; O_2^-, 1.28 Å; and O_2^{2-}, 1.49 Å.

Sodium metal reacts with oxygen to give mainly the peroxide: ∎

$$2Na(s) + O_2(g) \longrightarrow Na_2O_2(s)$$

Potassium, rubidium, and cesium form mainly the superoxides:

$$K(s) + O_2(g) \longrightarrow KO_2(s)$$
potassium
superoxide

The alkaline earth metals (Group IIA elements) also combine vigorously with molecular oxygen. Magnesium is less reactive than the metals below it in Group IIA, but burns in air with a bright white light, which makes it useful for fireworks. Photoflash bulbs contain fine magnesium wire in an atmosphere of oxygen. The product is magnesium oxide:

$$2Mg(s) + O_2(g) \longrightarrow 2MgO(s)$$

Calcium and strontium also burn to give the oxides, but barium forms the peroxide:

$$Ba(s) + O_2(g) \longrightarrow BaO_2(s)$$

Above 800°C, the peroxide decomposes to the oxide, BaO.∎

∎ Solutions of sodium peroxide are used to bleach textiles and wood pulp.

∎ This reaction was described earlier in the chapter as a method of preparing oxygen.

Reactions with Nonmetals

Nonmetals react to form covalent oxides. For example, hydrogen gas burns in oxygen or air to give water:

$$2H_2(g) + O_2(g) \longrightarrow 2H_2O(g)$$

Similarly, carbon (charcoal or graphite) burns in an excess of oxygen to give carbon dioxide:

$$C(s) + O_2(g) \longrightarrow CO_2(g) \qquad \text{(excess } O_2)$$

In a limited supply of oxygen or at higher temperature, carbon monoxide is the usual product:

$$2C(s) + O_2(g) \longrightarrow 2CO(g) \qquad \text{(limited } O_2)$$

Sulfur, S_8, burns in oxygen to give sulfur dioxide, SO_2, a gas with a suffocating odor:

$$S_8(s) + 8O_2(g) \longrightarrow 8SO_2(g)$$

Sulfur forms another oxide, sulfur trioxide, SO_3, but only small amounts are obtained during the burning of sulfur in air. Sulfur trioxide is prepared commercially by the catalytic oxidation of sulfur dioxide. Platinum metal and divanadium pentoxide are often used as catalysts.

$$2SO_2(g) + O_2(g) \xrightarrow{\text{Pt or } V_2O_5} 2SO_3(g)$$

White phosphorus, P_4, combines vigorously with oxygen. It bursts into flame spontaneously in air, giving dense white clouds of solid phosphorus(V) oxide, P_4O_{10}:

$$P_4(s) + 5O_2(g) \longrightarrow P_4O_{10}(s) \quad \text{(excess } O_2)$$

In limited oxygen, white phosphorus forms phosphorus(III) oxide, P_4O_6:

$$P_4(s) + 3O_2(g) \longrightarrow P_4O_6(s) \quad \text{(limited } O_2)$$

Nitrogen, N_2, reacts with oxygen only at higher temperatures. Nitric oxide, NO, is formed during any combustion in air:■

$$N_2(g) + O_2(g) \xrightarrow{\Delta} 2NO(g)$$

■ Nitric oxide is formed this way in automobile exhaust and leads to air pollution. See Section 13.9.

Reactions with Compounds

Compounds in which at least one element is in a reduced state will be oxidized by oxygen, giving the compounds that would be expected to form when the individual elements are burned in oxygen. For example, a hydrocarbon such as octane, C_8H_{18}, burns to give carbon dioxide and water:

$$2C_8H_{18}(l) + 25O_2(g) \longrightarrow 16CO_2(g) + 18H_2O(g)$$

Some other examples are given by the following equations:

$$2H_2S(g) + 3O_2(g) \longrightarrow 2H_2O(g) + 2SO_2(g)$$
$$CS_2(l) + 3O_2(g) \longrightarrow CO_2(g) + 2SO_2(g)$$
$$2ZnS(s) + 3O_2(g) \longrightarrow 2ZnO(s) + 2SO_2(g)$$

Basic and Acidic Oxides

Basic oxides (sometimes called **basic anhydrides**) are ionic oxides that are soluble in water and that react with water to produce basic solutions. The ionic reaction is

$$O^{2-}(aq) + H_2O(l) \longrightarrow 2OH^-(aq)$$

Examples of basic oxides are the oxides of the alkali metals and alkaline earth metals, such as sodium oxide, Na_2O, and calcium oxide, CaO.

$$Na_2O(s) + H_2O(l) \longrightarrow 2Na^+(aq) + 2OH^-(aq)$$
$$CaO(s) + H_2O(l) \longrightarrow Ca^{2+}(aq) + 2OH^-(aq)$$

Many other metal oxides, such as nickel oxide, NiO, though not soluble in water, react with acids. These are also classified as basic oxides.■

Most nonmetal oxides are **acidic oxides,** that is, they react with water to produce acidic solutions. Such oxides are also known as **acid anhydrides.** Carbon dioxide, for example, dissolves in water to give the unstable carbonic acid, H_2CO_3:

■ Oxides of metals in a high oxidation state, however, are often acidic; examples are CrO_3, Mn_2O_7, and WO_3.

$$CO_2(g) + H_2O(l) \rightleftharpoons H_2CO_3(aq)$$
carbonic acid

Sulfur dioxide reacts with water to give sulfurous acid, H_2SO_3, and sulfur trioxide reacts with water to give sulfuric acid, H_2SO_4:

$$SO_2(g) + H_2O(l) \rightleftharpoons H_2SO_3(aq)$$
sulfurous acid

$$SO_3(l) + H_2O(l) \longrightarrow H_2SO_4(aq)$$
sulfuric acid

Some other examples are given by the following equations:

$$P_4O_{10}(s) + 6H_2O(l) \longrightarrow 4H_3PO_4(aq)$$
phosphoric acid

$$N_2O_5(s) + H_2O(l) \longrightarrow 2HNO_3(aq)$$
nitric acid

When the oxides of sulfur and nitrogen react with water vapor in the atmosphere, the resulting build-up of sulfuric and nitric acids produces rainfall with an unusually high acidity. This *acid rain* is a serious problem in industrial areas such as northeastern North America. Acid rain endangers whole populations of organisms (for example, lake trout) by altering the acid-base balance of their environments. It also erodes building materials, such as limestone or marble, through a reaction that produces water-soluble calcium sulfate:

$$CaCO_3(s) + H_2SO_4(aq) \longrightarrow CaSO_4(aq) + CO_2(g) + H_2O(l)$$

Silicon dioxide (silica) is inert in water, but if fused with a basic oxide, it gives a salt. Therefore, it acts like an acidic oxide. For example,

$$CaO + SiO_2 \xrightarrow{\Delta} CaSiO_3$$
calcium silicate

Exercise 13.7

Write equations for the following: (a) burning of cesium metal in air   (b) burning of ethanethiol, C_2H_5SH, in excess oxygen   (c) preparation of sulfuric acid in three steps, starting from sulfur, S_8.

(See Problems 13.41 and 13.42.)

13.5 Ozone

Ozone, O_3, and molecular oxygen, O_2, are *allotropes*, that is, different structural forms of the same element. The electronic structure of ozone may be described by the resonance formulas

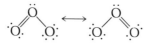

giving the oxygen–oxygen bond in O_3 partial double-bond character. This partial double-bond character is borne out by comparing the oxygen–oxygen bond length in hydrogen peroxide (single bond, 1.49 Å), ozone (1.28 Å), and molecular oxygen (double bond, 1.21 Å). The O—O—O bond angle is 117°, which is close to that predicted from the VSEPR model or from sp^2 hybridization (120°). ■

■ According to the VSEPR model, the three electron groups about the central oxygen atom should have a trigonal planar arrangement (Section 8.1).

Properties of Ozone

Pure ozone is a pale blue gas. When cooled to $-180°C$, a dark blue liquid is obtained that is unstable and can decompose explosively. The solid has a deep violet color and is also unstable.

Ozone is present in air at ground level, usually in concentrations of less than 0.05 parts per million. At these concentrations, the odor of ozone is pleasant and refreshing. The odor of ozone is especially noticeable on bright sunny days. Polluted air and air near electrical equipment may contain much higher concentrations (0.15 ppm or more). At such concentrations, the gas has an irritating odor, produces headache, and is poisonous.

The usual method of preparing ozone consists of passing molecular oxygen through a space subjected to an electrical discharge. Figure 13.8 shows a typical apparatus. The equation for the reaction is

$$3O_2(g) \xrightarrow[\text{discharge}]{\text{electrical}} 2O_3(g)$$

The gaseous mixture that issues from the apparatus contains up to 10% ozone by volume. Ozone is also produced when the ultraviolet light of sterilizing lamps irradiates molecular oxygen in air. The germicidal property of these lamps is due at least in part to ozone.

Ozone is a much more powerful oxidizing agent than molecular oxygen. We can use this fact to detect ozone in air. When ozone is passed through a solution of potassium iodide, it oxidizes iodide ion to iodine, I_2:

$$O_3(g) + 2I^-(aq) + H_2O(l) \longrightarrow 2OH^-(aq) + I_2(aq) + O_2(g)$$

If starch is present in the solution, it reacts with the iodine to give a deep

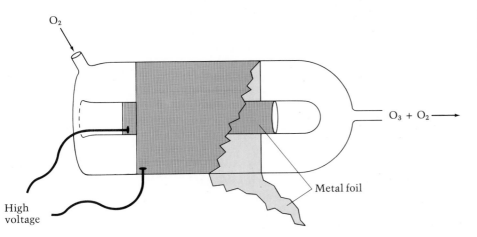

O₂

O₃ + O₂ →

Metal foil

High voltage

Figure 13.8
Ozone generator.
A silent electric discharge passes through the space between the metal foil electrodes.

blue color. Molecular oxygen gives a negative test in neutral or basic solution.■ Other substances, particularly organic ones, react readily with ozone at room temperature. Rubber, for example, is quickly attacked by ozone.

Ozone is used as a bleach and germicide. Some municipalities are investigating the possible replacement of chlorine by ozone for disinfecting water supplies.

■ Acidic solutions of iodide ion are oxidized by molecular oxygen in air.

$$O_2(g) + 4I^-(aq) + 4H^+(aq) \longrightarrow 2I_2(aq) + 2H_2O(l)$$

Aside: Stratospheric Ozone Depletion

Ozone is an unstable allotrope of oxygen and in time will revert to molecular oxygen. The process is normally slow, but becomes rapid in the presence of catalysts. Concern is expressed that the ozone in the stratosphere might be depleted by catalysts arising from the exhaust of supersonic transports (SSTs), from nuclear explosions, and from the chlorofluoromethanes (Freons) used as aerosol spray-can propellants. Ozone in the stratosphere absorbs ultraviolet light that would be detrimental to life. The problem of stratospheric ozone depletion is under continuing investigation.

It is known, for example, that nitric oxide, NO, catalyzes the decomposition of ozone. The series of reactions involved in this decomposition are believed to be

$$NO + O_3 \longrightarrow NO_2 + O_2$$
$$O_3 \xrightarrow{h\nu} O_2 + O$$
$$NO_2 + O \longrightarrow NO + O_2$$

Note that NO is used in the first step, but is recovered in the last one. The net result of these steps is given by the overall equation

$$2O_3 \xrightarrow{h\nu} 3O_2$$

Nitric oxide, therefore, is not consumed and acts as a catalyst.

Whenever air is subjected to high temperature, as in a combustion in air, nitric oxide is formed.

$$N_2 + O_2 \longrightarrow 2NO$$

If this occurs in the stratosphere, there is a possibility of the catalytic decomposition of ozone. Conditions for the formation of NO in the stratosphere exist in the exhaust gases of SSTs and during atmospheric nuclear explosions.

Another, even more efficient, catalyst for the decomposition of ozone is atomic chlorine.■ According to F. Sherwood Rowland and Mario J. Molino at the University of California, Irvine, the chlorofluoromethanes, such as $CClF_3$ and CCl_2F_2, are possible sources of chlorine atoms in the stratosphere. These compounds are some of the most chemically inert substances known and are used as refrigerants, as well as spray-can propellants. They are decomposed by high-energy ultraviolet radiation (185–227 nm), which splits off chlorine atoms:

$$CClF_3 \xrightarrow{h\nu} CF_3 + Cl.$$

It has been shown that the chlorofluoromethanes can diffuse slowly from the troposphere into the stratosphere, where ultraviolet light could decompose them into chlorine atoms.

Although there is no direct evidence for the depletion of ozone in the stratosphere by NO or Cl, there is need for much research. Any significant depletion would be difficult to reverse and would cause an increase of skin cancer and other biological problems.

■ The reactions are
$$Cl + O_3 \longrightarrow ClO + O_2$$
$$ClO + O \longrightarrow Cl + O_2$$
giving the net result
$$O_3 + O \longrightarrow 2O_2$$

Nitrogen

A Scottish botanist and chemist, Daniel Rutherford, is usually given credit for the discovery of nitrogen, although others investigated the same substance. In 1772, Rutherford described the properties of the gas that remains when all of the oxygen is removed from air. He called this residual gas "mephitic air," meaning poisonous air, since it does not support respiration. When this gas was later shown to be an element that was also a constituent of the mineral nitre (KNO_3), it was called nitrogen.

13.6 Molecular Nitrogen

Molecular nitrogen, N_2, is the principal component of air, from which it is obtained by liquefaction and fractional distillation. Small amounts of pure nitrogen can be obtained by heating an aqueous solution of ammonium nitrite: ■

■ Ammonium nitrite is unstable and is not sold commercially. Therefore, in practice, one heats a solution of ammonium chloride and sodium nitrite, which provides NH_4^+ and NO_2^- ions.

$$NH_4^+(aq) + NO_2^-(aq) \xrightarrow{\Delta} N_2(g) + 2H_2O(l)$$

In contrast to oxygen, nitrogen is a rather unreactive gas. Its lack of reactivity is partly due to its relative stability. In addition, the uncatalyzed reactions of molecular nitrogen are very slow because of the strength of the nitrogen–nitrogen triple bond, which must be broken to give the products. Unlike the combustion reactions of oxygen, many reactions of N_2 are endothermic, so are not self-sustaining.

Although molecular nitrogen is often unreactive, it does combine easily with some of the more active metals to form ionic nitrides (binary compounds with nitrogen). For example, when magnesium burns in air, the nitride Mg_3N_2 forms, in addition to the oxide.

Uses of Nitrogen

The major direct use of nitrogen gas stems from its unreactivity and ready availability. The electronics industry uses large amounts of nitrogen as a protective atmosphere or blanket when making components that would be adversely affected by atmospheric oxygen. Similarly, nitrogen is used in the laboratory when reagents would be attacked by oxygen and moisture in the air. Figure 13.9 shows a laboratory glove box, used to carry out reactions in a protective atmosphere. A cylinder of nitrogen can be attached to the box for continuous flushing with N_2.

Nitrogen Fixation

Nitrogen in combined form is an important constituent of all living organisms. Most organisms cannot use nitrogen directly from the atmosphere. **Nitrogen-fixing bacteria,** however, can produce nitrogen compounds in the soil from atmospheric nitrogen. Many of these nitrogen-fixing bacteria live in nodules on the roots of leguminous plants, such as peas, beans, and clover.

Figure 13.9
A laboratory glove box. When a chemical reaction must be run in an inert atmosphere, it may be done in an enclosed box. Nitrogen can be led from a cylinder into the chamber while the operator manipulates equipment inside by placing the hands in the rubber gloves located at the front of the box.

Additional amounts of *fixed nitrogen* (nitrogen in compound form) are produced during storms, when nitrogen and oxygen in the atmosphere react through electrical discharge to give nitric oxide, NO. The nitrogen compounds in the soil are incorporated into plant tissues, some of which are eaten by animals to provide a source of amino acids for protein synthesis.

To complete the cycle of nitrogen through the *biosphere* (that portion of the earth where life is found), bacteria decompose nitrogen-containing organic matter to ammonia. Ammonia is then transformed by other bacteria first to nitrite ion, NO_2^-, then to nitrate ion. Nitrate ion, NO_3^-, is reduced to N_2 by *denitrifying bacteria*. For example,

$$5C_6H_{12}O_6 + 24NO_3^- \longrightarrow 30CO_2 + 18H_2O + 24OH^- + 12N_2$$

(Note the similarity to Reaction 13.2. Here NO_3^- is the oxidizing agent instead of O_2.) Figure 13.10 summarizes the steps in the **nitrogen cycle,** the circulation of the element nitrogen in the biosphere, from nitrogen fixation to the release of free nitrogen by denitrifying bacteria.

Before World War I, the main commercial source of fixed nitrogen (used chiefly for fertilizers and explosives) was various nitrate deposits, such as those in northern Chile (Chile saltpeter, $NaNO_3$). Then the German chemist Fritz Haber discovered the conditions under which nitrogen could be made to react with hydrogen to give ammonia. By 1914, a plant began to operate using this reaction, called the **Haber process,** with a specially prepared catalyst, high temperature, and high pressure. Haber received the Nobel Prize for Chemistry in 1918 for his work on ammonia synthesis. Most chemical

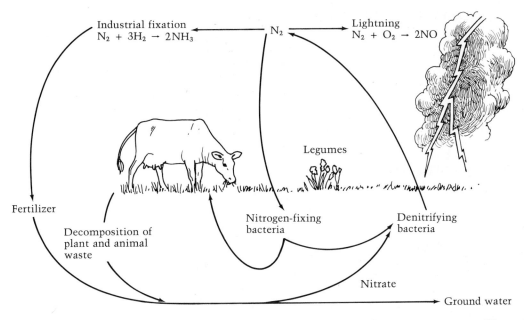

Figure 13.10
The nitrogen cycle. Nitrogen, N_2, is fixed by bacteria, by lightning, and by industrial synthesis of ammonia. Fixed nitrogen is used by plants and enters the food chain of animals. Later, plant and animal waste decomposes. Denitrifying bacteria complete the cycle by producing free nitrogen again.

plants that use the Haber process operate at temperatures of 400° to 600°C and pressures of 200 to 400 atm. The catalyst is metallic iron mixed with various oxides.

$$N_2(g) + 3H_2(g) \xrightarrow[\text{Fe catalyst}]{\text{high temperature and pressure}} 2NH_3(g)$$

Hydrogen for the Haber process is obtained from methane (natural gas and petroleum refining gas), which in effect provides the energy for the fixation of nitrogen. Figure 13.11 is a flow diagram for the industrial synthesis of ammonia.

Exercise 13.8

An early commercial method (no longer used) for the fixation of nitrogen was based on the reaction of nitrogen and oxygen at high temperature. Air was passed through

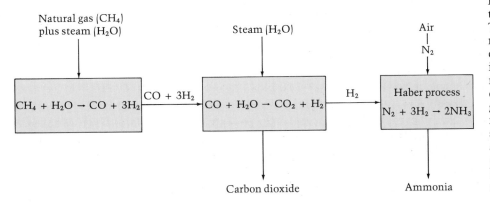

Figure 13.11
Flow diagram of the industrial synthesis of ammonia. The raw materials are natural gas, water, and air. Hydrogen for the Haber process is obtained by reacting natural gas with steam to give carbon monoxide and hydrogen. In the next step, carbon monoxide is reacted with steam to give carbon dioxide and additional hydrogen. The carbon dioxide is removed by dissolving it in water solution.

an electric arc, where it produced nitric oxide, NO. Outside the arc, nitric oxide combined with oxygen in air to give nitrogen dioxide, NO_2:

$$2NO(g) + O_2(g) \longrightarrow 2NO_2(g)$$

This gas was then bubbled with more air through a solution of calcium hydroxide (lime water). The reactions are

$$2NO_2(g) + 2OH^-(aq) \longrightarrow NO_2^-(aq) + NO_3^-(aq) + H_2O(l)$$
$$2NO_2^-(aq) + O_2(g) \longrightarrow 2NO_3^-(aq) \quad \text{(basic solution)}$$

Calcium nitrate was obtained by evaporating the solution. How many metric tons (1.00 metric ton = 1.00×10^6 g) of NO are required to produce 1.00 metric ton of calcium nitrate?

(See Problems 13.43 and 13.44.)

13.7 Compounds of Nitrogen

Nitrogen, a Group VA element, forms compounds in a large range of oxidation states, from -3 to $+5$. Table 13.2 lists some simple compounds, whose properties and uses are described in this section.

Ammonia and the Nitrides (Oxidation State -3)

Ammonia, NH_3, is a colorless gas with a characteristic irritating or pungent odor. It is prepared commercially by the Haber process, discussed in the preceding section. Laboratory amounts can be obtained by the reaction of a strong base, such as NaOH or $Ca(OH)_2$, with an ammonium salt. Ammonia gas evolves on heating:

$$NH_4^+(aq) + OH^-(aq) \xrightarrow{\Delta} NH_3(g) + H_2O(l)$$

Ammonia can also be obtained by reacting an ionic nitride with water as we will discuss.

Table 13.2
Some Compounds of Nitrogen in Various Oxidation States

Oxidation State	Compound	Formula
-3	Ammonia	NH_3
	Lithium nitride	Li_3N
-2	Hydrazine	N_2H_4
-1	Hydroxylamine	NH_2OH
0	Nitrogen	N_2
$+1$	Nitrous oxide	N_2O
$+2$	Nitric oxide	NO
$+3$	Dinitrogen trioxide	N_2O_3
	Nitrous acid	HNO_2
$+4$	Nitrogen dioxide	NO_2
	Dinitrogen tetroxide	N_2O_4
$+5$	Dinitrogen pentoxide	N_2O_5
	Nitric acid	HNO_3

Figure 13.12
Liquid ammonia being applied to the soil as a fertilizer.

Ammonia is an important industrial chemical. It is easily liquefied; the liquid is used as a nitrogen fertilizer (Figure 13.12). Ammonium salts, such as the sulfate and nitrate, are also sold as fertilizers. Large quantities of ammonia are converted to urea, NH_2CONH_2, which is used as a fertilizer and livestock feed supplement. Ammonia and carbon dioxide react in aqueous solution under pressure.

$$2NH_3(aq) + CO_2(aq) \xrightarrow[\text{pressure}]{\Delta} \underset{\text{urea}}{NH_2CONH_2(aq)} + H_2O(l)$$

Ionic nitrides are compounds of nitrogen and active metals in which nitrogen is in the -3 oxidation state. Lithium, as well as magnesium, combines with nitrogen at moderate temperatures to give the nitride:

$$6Li(s) + N_2(g) \longrightarrow 2Li_3N(s)$$

Because the nitride ion is a very strong base, the ionic nitrides react with water, producing ammonia:

$$\underset{\text{nitride ion}}{N^{3-}(aq)} + 3H_2O(l) \longrightarrow NH_3(g) + 3OH^-(aq)$$

Hydrazine (Oxidation State -2)

Hydrazine, N_2H_4, is a colorless liquid with an ammonia-like odor. It is prepared by reacting excess ammonia with sodium hypochlorite in aqueous solution:■

■ Hypochlorite bleaches, such as Clorox, and cleaning compounds containing ammonia must never be mixed, because poisonous fumes of hydrazine are produced.

$$2NH_3(aq) + OCl^-(aq) \longrightarrow N_2H_4(aq) + Cl^-(aq) + H_2O(l)$$

Gelatin is usually added to the reaction mixture and appears to remove transition metal ions that would catalyze the decomposition of hydrazine.

Liquid hydrazine is used as a rocket fuel. When a mixture of hydrazine and hydrogen peroxide is forced through a silver or platinum catalyst, an exothermic reaction occurs, producing a large volume of nitrogen gas:

$$N_2H_4(l) + 2H_2O_2(l) \xrightarrow{\text{Ag or Pt}} N_2(g) + 4H_2O(l)$$

Dinitrogen tetroxide may be used as the oxidizing agent in place of hydrogen peroxide:

$$2N_2H_4(l) + N_2O_4(l) \longrightarrow 3N_2(g) + 4H_2O(g)$$

In this case, the fuel and the oxidizing agent are kept separate and are mixed in the combustion chamber. The hydrazine is often mixed with 1,1-dimethylhydrazine, which ignites spontaneously when it comes in contact with the oxidizing agent.

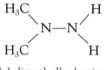

1,1-dimethylhydrazine

Hydroxylamine (Oxidation State −1)

Hydroxylamine, NH_2OH, is a white solid. The molecule has a structure like that of ammonia in which one hydrogen has been replaced by an —OH group:

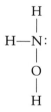

The compound has basic properties similar to those of ammonia and is often isolated as a salt, such as hydroxylammonium chloride, NH_3OHCl. The hydroxylammonium ion is obtained when a cold, weakly acidic solution of an alkali metal nitrite (to give HNO_2) is reduced by the hydrogen sulfite ion:

$$2HSO_3^-(aq) + HNO_2(aq) + H^+(aq) + H_2O(l) \xrightarrow{0°C}$$
$$2HSO_4^-(aq) + NH_3OH^+(aq)$$

Hydroxylamine and its salts are reducing agents. Silver metal, for example, is precipitated from a solution of silver nitrate treated with hydroxylammonium ion:■

$$2Ag^+(aq) + 2NH_3OH^+(aq) \longrightarrow 2Ag(s) + N_2(g) + 4H^+(aq) + 2H_2O(l)$$

■ Hydroxylamine has been used as a photographic developer. A developer reduces light-exposed silver salts more readily than nonexposed salts.

Nitrous Oxide (Oxidation State +1)

Nitrous oxide, N_2O, is a colorless gas with a sweet odor. It may be prepared by careful heating of molten ammonium nitrate. (*Strong heating may cause an explosion.*)

$$NH_4NO_3(s) \xrightarrow{\Delta} N_2O(g) + 2H_2O(g)$$

Nitrous oxide, also called laughing gas, is used as a dental anesthetic. Being soluble in cream, it is also useful as a propellant in whipped-cream dispensers. The gas is dissolved in cream under pressure; when the cream is dispensed, the gas bubbles out, forming a foam.

Nitric Oxide (Oxidation State +2)

Nitric oxide, NO, is a colorless gas whose molecule has one unpaired electron. (The molecular-orbital structure is similar to that of O_2^+, which has the same number of electrons.) The compound can be prepared by direct union of N_2 and O_2 at high temperature. However, large amounts are formed from ammonia as the first step in the commercial preparation of nitric acid. Ammonia is oxidized in the presence of a platinum catalyst to give nitric oxide:

$$4NH_3(g) + 5O_2(g) \xrightarrow{Pt} 4NO(g) + 6H_2O(g)$$

The nitric oxide reacts rapidly with oxygen to give nitrogen dioxide, and this in turn reacts with water to give nitric acid.

Nitrogen Dioxide (Oxidation State +4)

Nitrogen dioxide, NO_2, is a reddish brown gas or liquid (its boiling point is 21°C). It exists in equilibrium with the colorless compound dinitrogen tetroxide, N_2O_4:

$$2NO_2(g) \rightleftharpoons N_2O_4(g)$$

Dinitrogen tetroxide predominates at lower temperatures; above 140°C, the mixture is mostly nitrogen dioxide.

Nitrogen dioxide is formed in large amounts during the commercial production of nitric acid when nitric oxide reacts with oxygen:

$$2NO(g) + O_2(g) \longrightarrow 2NO_2(g)$$

Small quantities can be obtained by reducing nitric acid with copper metal, as described under nitric acid.

Nitrogen dioxide reacts with cold water to give nitrous and nitric acids.∎ In warm water, the nitrous acid decomposes. The overall reaction is

$$3NO_2(g) + H_2O(l) \xrightarrow{\Delta} 2HNO_3(aq) + NO(g)$$

■ Can you write the equation for this oxidation–reduction reaction?

Dinitrogen Trioxide and Nitrous Acid (Oxidation State +3)

Dinitrogen trioxide, N_2O_3, is an unstable compound and exists pure only as a solid or as a liquid as part of the equilibrium shown below. It is formed by cooling equal molar amounts of nitric oxide and nitrogen dioxide to $-21°C$.

$$NO + NO_2 \rightleftharpoons N_2O_3$$

Dinitrogen trioxide is the acidic oxide, or anhydride, of nitrous acid. The acid is prepared by dissolving nitric oxide and nitrogen dioxide in water:

$$\underbrace{NO(g) + NO_2(g)} + H_2O(l) \rightleftharpoons 2HNO_2(aq)$$
$$N_2O_3$$

Nitrous acid is too unstable to isolate, but salts can be prepared from this solution. Sodium nitrite, the most important salt, is prepared commercially by reduction of sodium nitrate with carbon:

$$NaNO_3(s) + C(s) \xrightarrow{\Delta} NaNO_2(s) + CO(g)$$

Dinitrogen Pentoxide and Nitric Acid (Oxidation State +5)

Dinitrogen pentoxide, N_2O_5, is a colorless solid. It is prepared by the dehydration of nitric acid with phosphorus(V) oxide:

$$P_4O_{10}(s) + 4HNO_3(l) \longrightarrow 4HPO_3(s) + 2N_2O_5(g)$$

This reaction is based on the fact that dinitrogen pentoxide is the anhydride of nitric acid; phosphorus(V) oxide removes the water from nitric acid. If dinitrogen pentoxide is added to water, nitric acid is obtained. Dinitrogen pentoxide is a volatile solid; the vapor decomposes at 45°C.

$$2N_2O_5(g) \xrightarrow{\Delta} 4NO_2(g) + O_2(g)$$

Nitric acid, HNO_3, can be prepared by heating a salt, such as sodium nitrate, with a nonvolatile acid, such as sulfuric acid. Nitric acid distills from the mixture:

$$NaNO_3(s) + H_2SO_4(l) \xrightarrow{\Delta} NaHSO_4(s) + HNO_3(g)$$

Industrially, nitric acid is made from ammonia by the **Ostwald process.** The steps in this process, which have been described separately earlier, are

$$4NH_3(g) + 5O_2(g) \xrightarrow{Pt} 4NO(g) + 6H_2O(g)$$
$$2NO(g) + O_2(g) \longrightarrow 2NO_2(g)$$
$$3NO_2(g) + H_2O(l) \longrightarrow 2HNO_3(aq) + NO(g)$$

The nitric oxide produced in the last step is recycled for use in the second step.

Nitric acid is a strong oxidizing agent. Although copper metal does not react with acids to give hydrogen gas (as active metals do), it is oxidized by nitric acid. In dilute acid, nitric oxide is a principal reduction product.

$$3Cu(s) + 8H^+(aq) + 2NO_3^-(aq) \longrightarrow 3Cu^{2+}(aq) + 2NO(g) + 4H_2O(l)$$

In concentrated acid, nitrogen dioxide is obtained.

$$Cu(s) + 4H^+(aq) + 2NO_3^-(aq) \longrightarrow Cu^{2+}(aq) + 2NO_2(g) + 2H_2O(l)$$

Nitric acid is an important industrial chemical. It is used to prepare explosives, nylon, and polyurethane plastics.

Exercise 13.9

You have the following chemical reagents available: H_2SO_4, $NaNO_3$, Cu, $NH_3(aq)$, and water. Write equations in which you prepare N_2O, NO, and NO_2 from these reagents. Several steps are required.

(See Problems 13.45 and 13.46.)

Noble Gases and Other Constituents of Air

Although nitrogen and oxygen are the main components of air, there are many others. One of these is argon. It makes up 96% by volume of the gases remaining after nitrogen and oxygen are removed. Argon is a Group VIIIA element, one of the noble gases, all of which exist as monatomic species.

13.8 The Noble Gases

The noble gases were unknown until argon was discovered in 1894. Other noble gases were discovered soon afterward.

Discovery of the Noble Gases

In 1785, the English chemist and physicist Henry Cavendish reported that air contained, in addition to nitrogen and oxygen, about 1% by volume of a nonreactive gas. To show this, he passed an electrical discharge through a mixture of air and oxygen until all of the nitrogen was converted to nitrogen dioxide, which was dissolved in strong base.■ After the excess oxygen was removed with a reducing agent, a small volume of gas remained. Cavendish did not pursue the matter further.

■ In the electrical discharge, nitrogen combines with oxygen to give nitric oxide. This reacts immediately with more oxygen to give nitrogen dioxide.

Then, in 1892, the English physicist Lord Rayleigh discovered that the density of molecular nitrogen obtained from air (1.2561 g/L at STP) was noticeably greater than the density of nitrogen obtained by decomposition of nitrogen compounds (1.2498 g/L at STP). He concluded that one of these two nitrogen sources was contaminated with another substance.

Soon after this, Rayleigh began collaborating with the Scottish chemist and physicist William Ramsay. Ramsay passed atmospheric nitrogen over hot magnesium to remove the nitrogen as magnesium nitride, Mg_2N_3, and obtained the same nonreactive residual gas that Cavendish had found over a century earlier. He placed this gas in a sealed glass tube and subjected it to a high-voltage electrical discharge, to study the emission of light. Analysis of the emission showed a spectrum of red and green lines. The spectral lines of this nonreactive gas were not those of any known element. In 1894 Ramsay and Rayleigh concluded that they had discovered a new element, which they called *argon* from the Greek word *argos*, meaning "lazy." ■ They also surmised that argon was a member of a new column of elements in the periodic table lying between the halogens and the alkali metals.

■ *Lazy* here refers to argon's lack of chemical reactivity.

Other noble gases were discovered soon after argon. Emission lines of helium (from the Greek *helios*, meaning "sun") had actually been observed in the sun's spectrum before helium was discovered on earth in the uranium ore clevite. Neon, krypton, and xenon were all obtained by fractional distillation of liquid air. Radon was discovered as a gaseous decay product of radium. All known isotopes of radon are radioactive.

Exercise 13.10

In Cavendish's experiments, nitrogen dioxide was dissolved in strong base. Nitrite and nitrate ions were produced. Write a balanced equation for this reaction.
(See Problems 13.47 and 13.48.)

Example 13.6

Assume that nitrogen prepared from air is contaminated with argon in the same ratio as exists in air (see Table 13.1). What are the mole fractions of N_2 and Ar in this mixture? Calculate the average molecular weight of the mixture, then calculate its density at STP, assuming ideal gas behavior. Compare your answer with the value given in the text for the density of nitrogen prepared from air.

Solution

A sample of the mixture containing 0.7808 mol N_2 would contain 0.00934 mol Ar. Hence,

$$\text{Mole fraction } N_2 = \frac{0.7808}{0.7808 + 0.00934} = 0.9882$$

$$\text{Mole fraction Ar} = 1.0000 - 0.9882 = 0.0118$$

The average molecular weight (since that for N_2 is 28.013 amu and that for Ar is 39.948 amu) is

$$28.013 \text{ amu} \times 0.9882 + 39.948 \text{ amu} \times 0.0118$$
$$= 28.15 \text{ amu}$$

The density at STP is

$$d = \frac{PM_m}{RT} = \frac{1 \text{ atm} \times 28.15 \text{ g/mol}}{0.082056 \text{ L} \cdot \text{atm/(mol} \cdot \text{K)} \times 273.15 \text{ K}}$$
$$= 1.256 \text{ g/L}$$

This agrees with the value given in the text to within the precision of the numbers.

Exercise 13.11

Suppose you make up a mixture of synthetic air from nitrogen and oxygen in the proportions given in Table 13.1. What are the mole fractions of N_2 and O_2 in the mixture? What is the average molecular weight? What is the density at STP?
(See Problems 13.49 and 13.50.)

Compounds of the Noble Gases

Because the electronic structures of the noble gases are particularly stable, for a long time it was thought that true compounds of these elements could not be formed. Then Neil Bartlett, at the University of British Columbia, discovered that molecular oxygen reacts with platinum hexafluoride, PtF_6,

Compound	Formula	Description	Table 13.3 Some Compounds of Xenon
Xenon difluoride	XeF_2	Colorless crystals	
Xenon tetrafluoride	XeF_4	Colorless crystals	
Xenon hexafluoride	XeF_6	Colorless crystals	
Xenon trioxide	XeO_3	Colorless crystals, explosive	
Xenon tetroxide	XeO_4	Colorless gas, explosive	

to form the ionic solid $[O_2^+][PtF_6^-]$. Since the ionization energy of xenon (1.17×10^3 kJ/mol) is close to that of molecular oxygen (1.21×10^3 kJ/mol), he reasoned that xenon should react with platinum hexafluoride also. In 1962, Bartlett reported the synthesis of a red compound with the approximate formula $XePtF_6$.■ Later in the same year chemists from Argonne National Laboratory near Chicago reported that xenon reacts directly with fluorine at 400°C to give the tetrafluoride:

■ The product actually has variable composition and can be represented by the formula $Xe(PtF_6)_n$, where n is between 1 and 2.

$$Xe(g) + 2F_2(g) \longrightarrow XeF_4(g)$$

The product is a volatile, colorless solid. Since then a number of noble-gas compounds have been prepared, all involving bonds to the highly electronegative elements fluorine and oxygen. Most are compounds of xenon (see Table 13.3), but a few are compounds of krypton and radon.

Exercise 13.12

Give the Lewis formula for KrF_2. What hybridization is expected for the krypton atom in this molecule? Using the VSEPR model, deduce the molecular geometry of KrF_2. **(See Problems 13.51 and 13.52.)**

Exercise 13.13

Xenon difluoride is a powerful oxidizing agent. In acidic solution it is reduced to xenon. Hydrogen fluoride is the other product. Write a half-reaction for this reduction. Hydrochloric acid is oxidized to Cl_2 by xenon difluoride. Write a balanced equation for this reaction. **(See Problems 13.53 and 13.54.)**

Preparation and Uses of the Noble Gases

Commercially, all of the noble gases except helium and radon are obtained by the fractional distillation of liquid air. The principal sources of helium are certain natural gas wells located in the western United States. Helium has the lowest boiling point (-268.9°C) of any substance and is very important in low-temperature research. The major use of argon is in welding. It is also used as a mixture with nitrogen to fill incandescent light bulbs. Here the gas mixture conducts heat away from the hot tungsten filament. All of the noble gases are used in gas discharge tubes. Neon gives a highly visible red orange emission and has long been used in advertising signs. Radon is used in cancer therapy.

13.9 Trace Constituents of Air

The atmosphere contains many constituents in trace amounts (see Table 13.4). By and large, they are of natural origin. For example, volcanic eruptions spew quantities of particulates, sulfur dioxide, and carbon monoxide into the atmosphere. Forest fires contribute particulates and carbon monoxide, as well as other compounds. Bacterial decompositions give rise to methane, ammonia, and nitric oxide. Furthermore, microorganisms remove many of these same constituents from the atmosphere. Thus, carbon monoxide is removed by certain fungi that inhabit the soil in which grasses grow. So the soil is referred to as a "**sink**" (or absorber) for carbon monoxide.

These trace constituents become atmospheric pollutants when they are produced (say, by people) faster than they can be disposed of by natural sinks, or when because of meteorological conditions they are allowed to concentrate in an area. Carbon monoxide is a common air pollutant in urban areas. It is generated in automobile exhaust and is therefore present in the atmosphere in high concentrations during peak traffic hours. It poisons by attaching itself strongly to the hemoglobin of red blood cells, thereby blocking their oxygen-carrying function. Normally, hemoglobin picks up oxygen in the lungs, and discharges it to oxygen-deficient cells throughout the body.

$$\text{Hemoglobin} + O_2 \rightleftharpoons \text{oxyhemoglobin}$$

Carbon monoxide also combines with hemoglobin, but much more strongly than oxygen does.

$$\text{Hemoglobin} + CO \rightleftharpoons \text{carboxyhemoglobin}$$

Once carboxyhemoglobin forms in a red blood cell, the cell loses its ability to carry oxygen. It recovers slowly if the partial pressure of carbon monoxide in the surrounding air falls to normal trace levels.

The *photochemical smog* of Los Angeles and other cities is another important air pollution problem. The two necessary ingredients of **photochemical smog** are nitrogen oxides (NO and NO_2) and hydrocarbons. Both

Table 13.4
Some Trace Components
of Air in the Lower
Atmosphere[*]

Component	Formula	Amount by Volume (ppm)[†]
Neon	Ne	18.2
Helium	He	5.24
Methane	CH_4	1.5
Krypton	Kr	1.14
Hydrogen	H_2	0.5
Nitrous oxide	N_2O	0.2
Xenon	Xe	0.087
Ozone	O_3	~ 0.01
Carbon monoxide	CO	0.06–0.7
Nitrogen oxides	$NO + NO_2$	0.0005–0.02

[*]Except for helium, the values are taken from Murray J. McEwan and Leon F. Phillips, *Chemistry of the Atmosphere* (London: Edward Arnold Ltd., 1975), p. 7.
[†]To convert ppm to percent by volume, multiply by 10^{-4}.

are present in automobile exhaust: the nitrogen oxides are formed from the O_2 and N_2 of air in the hot exhaust gases; incomplete combustion of gasoline in the engine provides the hydrocarbons. When this mixture is irradiated with sunlight, it produces an irritating haze.

The details of photochemical smog were first studied by Arie J. Haagen-Smit, who was a natural-products chemist from the California Institute of Technology. Haagen-Smit describes how he was first attracted to this problem:

I had just finished the isolation and identification of the odors of the pineapple when I opened the window and smelled what we now call "smog." I passed the smoggy air through the cold traps that had served me faithfully in collecting the vapors of the fruits. Three hundred cubic feet of air (equal to the amount of air we breathe every day) went through the traps, resulting in a glass filled with an evil-smelling water. . . . The dirty glass of water made a deep impression on me. It was a glass of water that anyone would refuse to drink. Yet we inhale it each day, and the lungs are far more efficient in assimilating all the impurities.*

In order to combat photochemical smog, a catalytic converter is attached to the automobile exhaust system to convert the nitrogen oxides to molecular nitrogen and the hydrocarbons to carbon dioxide and water. A catalyst such as platinum or copper(II) oxide is used. The carburetor is set so that the exhaust gases contain relatively high concentrations of CO and H_2 (produced from the reaction of H_2O with CO). These gases enter the first half of the converter, where the nitrogen oxides are reduced:■

■ Leaded gasoline produces volatile lead compounds in exhaust gas that "poison," or inactivate, the catalyst of a catalytic converter. Thus, leaded gasoline cannot be used in an automobile that has a catalytic converter.

$$2NO + 2CO \xrightarrow{\text{catalyst}} N_2 + 2CO_2$$
$$2NO + 2H_2 \xrightarrow{\text{catalyst}} N_2 + 2H_2O$$

Then a stream of air enters the other half of the converter, and there carbon monoxide and hydrocarbons are oxidized to carbon dioxide and water.

Exercise 13.14

An air-monitoring station finds that the air on one occasion has reached 0.50 ppm by volume of sulfur dioxide. Assume an air temperature of 15°C and a barometric pressure of 750 mmHg. How many milligrams of sulfur dioxide would a person breathe from this air in one day? Assume a person breathes in about 8.5×10^3 L of air per day.

(See Problems 13.55 and 13.56.)

Aside: The Earth's Thermal Moderators

The surface of the moon can reach temperatures as high as 100°C during the lunar day, and at night the temperature drops to −150°C. The earth, by contrast, has much more moderate daily changes in temperature.

Within the earth's atmosphere, moisture, carbon dioxide, and small solid particles (particulates) play a dominant role as moderators of the surface temperature.

(*Continued*)

*Arie J. Haagen-Smit, "Theory of Air Conservation," in *Science, Scientists, and Society*, ed. William Beranek, Jr. (Tarrytown-on-Hudson, N.Y.: Bogden and Quigley, 1972), p. 30.

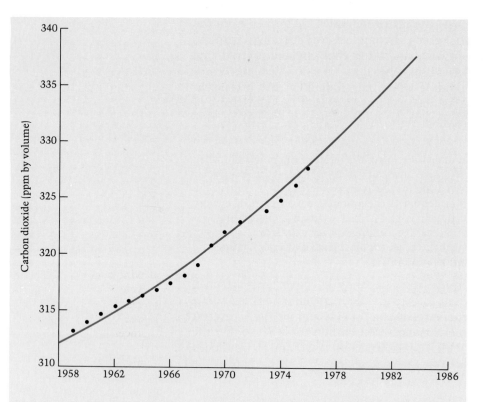

Figure 13.13
Carbon dioxide concentration in the atmosphere, measured at monitoring stations over several decades.

Clouds are very important weather determiners because, besides being a source of rain, they have a significant effect on the thermal balance of the earth. The tops of clouds are quite reflective; as much as 90% of the solar radiation reaching them is reflected back into space.

If too much of the sun's radiation was reflected back into space, there would be a cooling of the earth's surface. But clouds are also absorbers of infrared radiation and by the "greenhouse effect" retain some of the earth's warmth. The greenhouse effect works this way. Solar radiation is absorbed by the surface of the earth, warming it. Heat energy is then radiated from the earth as infrared light. Clouds absorb some of this and reradiate it to earth, producing a heat trap.

Carbon dioxide in the atmosphere contributes to the greenhouse effect because it absorbs strongly in the infrared region, as do clouds, but is transparent to the solar radiation reaching the earth's surface. Recent measurements show a definite increase in the concentration of carbon dioxide in the atmosphere (see Figure 13.13). This increase is believed to be due to the burning of fossil fuels. It is estimated that since 1890, when we began significant burning of fossil fuels, the concentration of carbon dioxide has risen from 290 ppm (parts per million) to over 330 ppm.

Calculations by Syukuro Manabe and Richard Wetherald show that the earth should warm 0.1°C with each 10-ppm increase in carbon dioxide in the atmosphere. Since only a few degrees' change (either increase or decrease) in the average temperature can have a drastic effect on the polar caps, agriculture, and so forth, any increase in atmospheric CO_2 is a matter for great concern. A warming trend was seen during 1900 to 1940, though since then the average temperature of the earth's surface has dropped by 0.2°C. It is possible that there

(Continued)

has been an increase in the reflectivity of the earth (and therefore a cooling) due to increasing concentration of particulates in the atmosphere. But a committee of the U.S. National Academy of Sciences has found (*Energy and Climate,* 1977) that the climatic effects of increasing atmospheric carbon dioxide will outweigh those of particulates or waste heat production. They conclude that the carbon dioxide increase, rather than availability of fossil fuels, may be the limiting factor in the continued use of fossil fuels over the next centuries.

A Checklist for Review

Important Terms

troposphere (13.1)
stratosphere (13.1)
tropopause (13.1)
mesosphere (13.1)
stratopause (13.1)
thermosphere (13.1)
mesopause (13.1)

photolysis (13.2)
photosynthesis (13.2)
oxides (13.4)
peroxides (13.4)
superoxides (13.4)
basic oxides (basic anhydrides) (13.4)
acidic oxides (acid anhydrides) (13.4)

nitrogen-fixing bacteria (13.6)
nitrogen cycle (13.6)
Haber process (13.6)
ionic nitrides (13.7)
Ostwald process (13.7)
sink (13.9)
photochemical smog (13.9)

Summary of Facts and Concepts

The earth's enveloping atmosphere is divided into regions, in which the air temperature alternately rises and falls (*troposphere, stratosphere, mesosphere,* and *thermosphere*). Heating of the thermosphere is due to absorption by O_2 of ultraviolet light of a wavelength below 200 nm. Similar heating of the stratosphere is due to absorption by O_3 of ultraviolet light of a wavelength between 200 and 300 nm.

Our present atmosphere is believed to have evolved from one that contained no molecular oxygen, since sedimentary rocks exist that can be rapidly oxidized by O_2. These rocks could not have been laid down initially in an oxygen-containing atmosphere. *Photosynthesis,* coupled with the burying of a portion of the carbon-containing product in sedimentary rock, appears to account for the molecular oxygen now present.

Molecular oxygen can be made in small quantities by heating oxides of less-active metals (such as HgO and Ag_2O), by decomposing peroxides (BaO_2, H_2O_2, Na_2O_2), and by heating some salts of oxyacids whose central atom is in a high oxidation state (KNO_3, $KClO_3$). Commercial quantities are obtained by the fractional distillation of liquid air. Large amounts of molecular oxygen are used in steelmaking.

Most metals react with oxygen to give *oxides,* although some active metals form *peroxides* (Na_2O_2, BaO_2) or *superoxides* (KO_2). When the nonmetals H_2, C, S_8, or

their compounds burn in excess air, H_2O, CO_2, and SO_2 are formed. Most metal oxides are *basic oxides,* whereas nonmetal ones are *acidic oxides.*

Ozone, O_3, is prepared by passing molecular oxygen through an electrical discharge. This allotrope of oxygen is a powerful oxidizing agent. It is unstable and eventually reverts to molecular oxygen. The process is rapid in the presence of catalysts. Concern exists that the stratosphere might be depleted of its ozone by catalysts (such as NO or Cl atoms) resulting from human activities.

Nitrogen, N_2, is much less reactive than molecular oxygen and is used industrially as a protective blanket. The fixation of molecular nitrogen (conversion to compounds of nitrogen) is accomplished by certain bacteria, or industrially by the *Haber process* for ammonia synthesis. Nitrogen forms compounds in oxidation states from -3 to $+5$. Besides ammonia, one of the most important compounds of nitrogen is nitric acid, HNO_3, produced by the *Ostwald process.*

All of the *noble gases,* except helium and radon, are obtained from air. Helium is found in some natural gas; radon is formed as a radioactive decay product of radium. Although the noble gases are usually unreactive, compounds have been made of xenon, krypton, and radon.

The atmosphere contains trace amounts of many gases, some of which are pollutants produced by the combustion of fuels or by industrial processing.

Operational Skills

The basic problem-solving skills used in this chapter were introduced in Chapters 1 to 12.

Review Questions

13.1 What is the basis for the classification of the atmosphere into regions such as the troposphere and stratosphere?

13.2 What causes the heating that is seen in the thermosphere? in the stratosphere?

13.3 What is the biological importance of ozone in the stratosphere?

13.4 What evidence is there that our present atmosphere evolved from a primitive one formed after the solid earth came into being?

13.5 Helium is still being formed on earth by radioactive decay (giving alpha radiation). Most of this helium has left the earth. Explain why this has happened.

13.6 What evidence is there that the primitive atmosphere did not contain molecular oxygen? What evidence is there that oxygen appeared about 2 billion years ago?

13.7 Explain how early photosynthetic organisms could have produced the oxygen now present in the atmosphere. Why is it that photosynthesis by itself is not an adequate explanation?

13.8 What reaction was used by Priestley in preparing pure oxygen?

13.9 How can oxygen be separated from air using barium oxide? Give the reactions involved.

13.10 What is the most important commercial means of producing oxygen?

13.11 Describe the molecular-orbital configurations of O_2, O_2^-, and O_2^{2-}. State which species are paramagnetic and which are diamagnetic. What is the bond order of each?

13.12 What is the difference between an oxide, a peroxide, and a superoxide? Give an example of each kind of compound.

13.13 Classify each of the following as an acidic oxide or a basic oxide: (a) MgO (b) SO_3 (c) CO_2 (d) FeO (e) Na_2O

13.14 Write an equation for the reaction of each of the following oxides with water: (a) Na_2O (b) CaO (c) P_4O_{10} (d) SO_2

13.15 How is ozone prepared in the laboratory? How is it formed in the stratosphere?

13.16 Farmers sometimes plant corn and clover on the same land in alternate years. Explain what value this could have.

13.17 Describe the steps in the nitrogen cycle.

13.18 What is the Haber process? Why is it important?

13.19 List the different nitrogen oxides. What is the oxidation number of nitrogen in each?

13.20 Match each compound on the right with a usage or characteristic in the left-hand column:
  (a) dental anesthetic NH_3
  (b) anhydride of nitric acid N_2H_4
  (c) rocket propellant N_2O
  (d) fertilizer; gives basic aqueous solution NO_2
  (e) brown gas N_2O_5

13.21 Describe the steps in the Ostwald process for making nitric acid from ammonia.

13.22 Suggest how each of the following might be used to distinguish between a sample of nitrogen and a sample of neon:
  (a) density
  (b) reactivity with magnesium
  (c) emission spectrum
  (d) relative rate of effusion

13.23 Argon (contaminated with other noble gases) can be separated from air by the following chemical reactions. White phosphorus is burned in the air and the product of combustion is dissolved in NaOH solution. Then, the residual gas is passed over hot magnesium. Write the equation for the principal reaction occurring in each step.

13.24 What was the argument used by Bartlett that led him to the first synthesis of a noble-gas compound?

13.25 An incandescent bulb slowly gathers a black coating on the inside wall. Explain this. Why is the bulb filled with nitrogen or argon? What would happen if the bulb were filled with air? What would happen if the bulb were evacuated?

13.26 Why is carbon monoxide poisonous?

13.27 What is photochemical smog? How can the incidence of this type of smog be reduced?

13.28 Why would one expect xenon to only combine with the more electronegative elements (O, F), as illustrated by the compounds listed in Table 13.3?

Problems

The Atmosphere

13.29 What is the longest wavelength of ultraviolet light that can be expected to dissociate the water molecule into H and OH? The energy of the O—H bond is 459 kJ/mol.

13.31 The air in a home is at 21°C and 745 mmHg and contains 50 ppm by volume of carbon monoxide. (At this concentration of CO, definite physiological effects can be observed.) What is the partial pressure of CO in mmHg? If a room is 6.6 m long by 3.7 m wide and 2.4 m high (20 ft × 12 ft × 8 ft), what is the volume of carbon monoxide in this room at STP? Imagine this CO to be an even layer of gas at STP on the floor of the room. What would be the thickness of this layer?

13.33 What is the rms speed of an oxygen molecule, O_2, near the surface of the earth? Assume a temperature of 15°C. What would be the rms speed in the upper atmosphere, where the temperature is 1500 K?

13.35 Iron(II) ion in slightly acidic lake water is oxidized by molecular oxygen to iron(III) ion. Write a balanced equation for the reaction.

13.37 If the quantity of fossil fuels burned since 1860 is equivalent to 1.9×10^{17} g C, how many grams of carbon dioxide were formed? If only 50% of this remains in the atmosphere, what would be the percent of increase of carbon dioxide in the atmosphere? The amount of carbon dioxide in the atmosphere in 1860 is estimated to have been 2.0×10^{18} g.

13.30 Nitrogen, N_2, is ionized to N_2^+ in the upper atmosphere by the absorption of high-energy ultraviolet light. The ionization energy of N_2 is 1495 kJ/mol. What is the longest wavelength of light that can ionize the nitrogen molecule?

13.32 An urban area 16 km by 16 km is covered with smog to a depth of 1500 m. The average concentration of nitrogen dioxide in this air mass is 0.45 ppm by volume. If the average temperature and pressure of this air is 15°C and 735 mmHg, what is the volume at STP of NO_2 in this volume of air? Imagine the NO_2 to be a layer of gas at STP covering this 16-km × 16-km area. What would be the thickness of this layer?

13.34 What is the rms speed of an ozone molecule, O_3, near the surface of the earth? Assume a temperature of 15°C. What would be the rms speed of an ozone molecule in the stratosphere, where the temperature is $-50°C$?

13.36 Hydrogen sulfide, H_2S, in a mineral water was oxidized to free sulfur, S_8, when the water was exposed to molecular oxygen. Write a balanced equation for this oxidation in acidic water.

13.38 If 6.5×10^{15} g C in fossil fuels are burned each year, and if 50% of the carbon dioxide produced is precipitated in the oceans as calcium carbonate, $CaCO_3$, how many grams of calcium carbonate are formed this way per year?

Oxygen

13.39 Describe three different ways you could prepare molecular oxygen from the following reagents: MnO_2, BaO_2, $KClO_3$, Na_2O_2.

13.41 Write equations for the following:
(a) burning of lithium in oxygen
(b) burning of methylamine, CH_3NH_2, in excess oxygen (assume the nitrogen in methylamine ends up as N_2)
(c) the preparation of phosphoric acid, H_3PO_4, from phosphorus, P_4, in two steps

13.40 Describe three different ways you could separate oxygen from air. Two of these ways should be chemical methods. You have sodium and barium metal for use in these separations. Use equations to describe the chemical separations.

13.42 Write equations for the following:
(a) burning of calcium metal in air
(b) burning of phosphine, PH_3, in excess oxygen
(c) preparation of calcium carbonate, $CaCO_3$, from carbon, C, and an aqueous solution of calcium hydroxide, $Ca(OH)_2$, in two steps

Nitrogen

13.43 The cyanamide process has been used to fix atmospheric nitrogen. In the first step, nitrogen was passed over white-hot calcium carbide to form calcium cyanamide:

$$CaC_2(s) + N_2(g) \longrightarrow CaNCN(s) + C(s)$$
$$\text{calcium} \qquad\qquad\qquad \text{calcium}$$
$$\text{carbide} \qquad\qquad\qquad \text{cyanamide}$$

Calcium cyanamide was then treated with steam to yield ammonia:

$$CaNCN(s) + 3H_2O(g) \longrightarrow CaCO_3(s) + 2NH_3(g)$$

How many kilograms of calcium carbide are required to produce 2.00 kg of ammonia?

13.45 You are given the following substances: NH_3, O_2, Pt, and P_4O_{10}. Write equations in which you prepare NO, N_2O_5, and N_2O. Several steps may be required for any substance.

13.44 Atmospheric nitrogen can be fixed by reacting the nitrogen with hot magnesium metal to form magnesium nitride:

$$3Mg(s) + N_2(g) \longrightarrow Mg_3N_2(s)$$

This ionic nitride reacts with water to give ammonia:

$$Mg_3N_2(s) + 6H_2O(l) \longrightarrow 3Mg(OH)_2(s) + 2NH_3(g)$$

How many grams of ammonia can be obtained from 1.00 g of magnesium?

13.46 Give equations for the preparation of N_2O, $NaNO_2$, and HNO_3. You can use C, $NaNO_3$, H_2SO_4, and $(NH_4)_2SO_4$. Several steps may be required for any substance.

Noble Gases

13.47 In his experiments on the composition of air, Cavendish removed residual quantities of molecular oxygen from gas mixtures by reacting O_2 with a solution of potassium sulfide, K_2S. In this solution, sulfide ion reacts with water to give HS^- ion, and molecular oxygen reacts with HS^- ion in basic solution to give the disulfide ion, S_2^{2-} (among other possible oxidation products). Write the balanced equation for the oxidation of HS^- to S_2^{2-} by O_2 in basic solution.

13.49 A sample of argon was discovered to be contaminated with neon. Upon analysis, 1.000 g of the gas was found to contain 0.9507 g of argon and 0.0493 g of neon. Calculate the mole fractions of argon and neon in the mixture. Then calculate the average molecular weight of the sample gas. What is the density of this gas at STP?

13.51 Xenon tetrafluoride, XeF_4, is a colorless solid. Give the Lewis formula for the XeF_4 molecule. What is the hybridization of the xenon atom in this compound? What geometry is predicted by the VSEPR model for this molecule?

13.53 Xenon difluoride, XeF_2, is hydrolyzed (broken up by water) in basic aqueous solution to give xenon, fluoride ion, and molecular oxygen as products. Write a balanced chemical equation for the reaction.

13.48 Rayleigh discovered that the density of molecular nitrogen obtained from air was greater than that obtained by the decomposition of nitrogen compounds. Ammonium salts of oxidizing anions often decompose to N_2 when heated. Write the balanced equation for the decomposition of ammonium dichromate, $(NH_4)_2Cr_2O_7$, to nitrogen, N_2, chromium(III) oxide, Cr_2O_3, and water.

13.50 A sample of krypton contains some xenon. Analysis showed that the mixture was 91.05% krypton by mass, the remainder being xenon. Find the mole fraction of components in the mixture and the average molecular weight and density at STP of the mixture.

13.52 Xenon tetroxide, XeO_4, is a colorless, unstable gas. Give the Lewis formula for the XeO_4 molecule. What is the hybridization of the xenon atom in this compound? What geometry would you expect for this molecule?

13.54 Xenon trioxide, XeO_3, is reduced to xenon in acidic solution by iodide ion. Iodide ion is oxidized to iodine, I_2. Write a balanced equation for the reaction.

Trace Constituents in Air

13.55 The inhaled smoke from a cigarette, mixed with air, has an approximate partial pressure of carbon monoxide of 0.35 mmHg. If a person inhales 3.8×10^2 L of this gas at 40°C and 745 mmHg, how many milligrams of carbon monoxide have been breathed?

13.56 A concentration of 2.5×10^2 ppm by volume of carbon monoxide produces unconsciousness in a person breathing this air after several hours. How many grams of carbon monoxide are there in a room 6.1 m long by 3.7 m wide and 2.4 m high (20 ft × 12 ft × 8 ft) that contains this air at 20°C and 745 mmHg?

Additional Problems

13.57 The carbon–chlorine bond energy is 327 kJ/mol. From this, estimate the maximum wavelength of light that will dissociate the CCl_3F molecule to give a chlorine atom.

13.58 The carbon–oxygen double-bond energy is 799 kJ/mol. From this, estimate the maximum wavelength of light that will dissociate the CO_2 molecule to give an oxygen atom.

13.59 The maximum concentration of ozone in the stratosphere is about 10 ppm by volume. Calculate the number of O_3 molecules in 1.00 cm^3 of stratospheric air (at -50°C and 1.00 atm) that contains 10.0 ppm O_3.

13.60 The concentration of xenon in the troposphere is 0.087 ppm by volume. How many xenon atoms are there in 1.00 cm^3 of tropospheric air at 15°C and 1.00 atm?

*13.61 Calculate the mass percentages of N_2 and O_2 in dry tropospheric air, using the data in Table 13.1.

*13.62 Calculate the mass percentages of Ne and Xe in dry tropospheric air, using the data in Table 13.4.

**13.63 Calculate the mass of O_2 in the entire atmosphere (in grams). To do this, recall that barometric pressure at the earth's surface is due to the weight of the atmosphere above per unit of surface area. Hence, first calculate the total force exerted on the surface of earth due to O_2 in air (assume a barometric pressure of 1.00 atm). The surface area of a sphere with radius r is $4\pi r^2$; the earth's radius is 6370 km. Then, noting that force = mass × g, where g is the constant acceleration due to gravity (9.81 m/s^2), calculate the mass of O_2.

**13.64 Calculate the mass (in grams) of carbon dioxide in the entire atmosphere at the present time, when the concentration is 3.4×10^2 ppm by volume. (See Problem 13.63.) If the concentration of carbon dioxide in the atmosphere was 2.9×10^2 ppm by volume in 1860, what has been the increase in mass (in grams) of CO_2 in the atmosphere since that time?

13.65 Lead(IV) oxide is used in the lead storage battery. The compound decomposes on heating to give lead(II) oxide and oxygen:

$$2PbO_2(s) \xrightarrow{\Delta} 2PbO(s) + O_2(g)$$

What volume of oxygen is evolved from 5.81 g of lead(IV) oxide if the gas is collected over water at 23°C and at a total pressure (oxygen and water vapor) of 736 mmHg? Assume that the gas is saturated with water vapor. The vapor pressure of water at 23°C is 21.1 mmHg.

13.66 Potassium permanganate, $KMnO_4$, can be thermally decomposed to give oxygen:

$$2KMnO_4(s) \xrightarrow{\Delta} K_2MnO_4(s) + MnO_2(s) + O_2(g)$$

This reaction may be used to prepare very pure oxygen in the laboratory. What is the volume of oxygen gas evolved from 1.12 g $KMnO_4$ if the gas is collected over water at 21°C and at a total pressure (oxygen and water vapor) of 748 mmHg? Assume that the gas is saturated with water vapor. The vapor pressure of water at 21°C is 18.7 mmHg.

13.67 Zinc metal reacts with dilute nitric acid to give ammonium ion (as well as molecular nitrogen and nitrogen oxides). Write a balanced equation for the reduction of nitric acid by zinc to give ammonium ion.

13.68 Nitrogen dioxide can be made in the laboratory by heating lead nitrate, $Pb(NO_3)_2$. The other products are lead(II) oxide and oxygen. Write a balanced equation for this reaction.

13.69 Complete and balance the following equations:
(a) $SO_2(g) + NaOH(aq) \longrightarrow$
(b) $Na_2O(s) + CO_2(g) \longrightarrow$
(c) $Na_2O(s) + SO_3(g) \longrightarrow$
(d) $MgO(s) + HCl(g) \longrightarrow$

13.70 Write balanced equations for the following reactions: (a) Nitrous oxide, N_2O, is used as the oxidizing agent and carbon disulfide, CS_2, is the fuel. (b) Sulfur, S_8, is oxidized by nitric acid to give sulfate ion. Nitric acid is reduced to nitrogen dioxide.

13.71 Starting from air, water, and hydrogen (H_2), and any necessary catalysts, show by means of equations how you could prepare nitrous oxide.

13.72 Hydrogen sulfide, H_2S, is separated from natural gas. Starting from hydrogen sulfide, air, water, and ammonia, and any necessary catalysts, show how you could prepare ammonium sulfate.

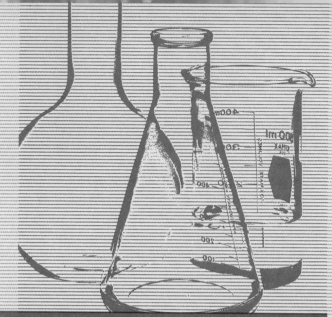

14. Hydrogen and Its Compounds with Oxygen: Water and Hydrogen Peroxide

Hydrogen, Water, and Hydrogen Peroxide

14.1 Discovery of Hydrogen; Composition of Water

14.2 Properties of Hydrogen Physical Properties/ Preparation/ Hydrides/ Uses

14.3 Properties of Water Hydrogen Bonding and the Physical Properties of Water/ Chemical Properties of Water

14.4 Hydrates and Hydrated Ions

14.5 Hydrogen Peroxide

Natural Waters

14.6 Physical Properties of Water and the Environment Effect of Large Heat Capacity of Water/ Effect of Relative Densities of Ice and Water

14.7 Water Pollutants Thermal Pollution/ Biological Oxygen Demand/ Inorganic Ions/ Other Pollutants

14.8 Water Treatment and Purification

14.9 Removing Ions from Water; Desalination

Water is one of the few liquid substances to be found on earth in significant amounts. This liquid is also readily convertible under conditions on earth to the solid and gaseous forms. Perhaps it was these facts that led the ancient Greek philosopher Thales of Miletus (ca. 580 B.C.) to conceive the unitary theory in which all things are composed of water. But water has several unusual properties that set it apart from other substances. For example, its solid phase, ice, is less dense than liquid water, whereas for most substances the solid phase is more dense than the liquid. In addition, water has an unusually large heat capacity. These are some of the properties that are important in determining conditions favorable to life. In fact, it is difficult to think of life without water.

Seventy percent of the earth's surface is covered with this liquid substance. Yet, in spite of its abundance, humankind has always had a problem in finding water of the requisite purity when and where it was wanted. Today, aggravated by an increasing world population, unprecedented industrial usage, and shrinking open lands, the problem of an adequate supply of fresh water looms larger than ever.

In this chapter, we will look at hydrogen and the compounds of hydrogen with oxygen—water and hydrogen peroxide. We will see how the structure of water contributes to its unusual properties.

Chapter Overview

After describing the early experiments that showed water to be a compound of hydrogen and oxygen, we will discuss the *chemistry of hydrogen*. (The chemistry of oxygen was discussed in Chapter 13.) Then, having looked at the elements hydrogen and oxygen, we will examine the physical and chemical *properties of water and hydrogen peroxide*. In the second part of the chapter, we will discuss *natural waters:* the properties of water and their effect on the environment, water pollution, and methods of water purification.

Hydrogen, Water, and Hydrogen Peroxide

A century after Thales, another Greek philosopher, Empedocles, postulated that the world was composed of four elements: fire, air, earth, and water. Remnants of this view persisted until the eighteenth century. Gases, for example, were often called "airs" and were not well differentiated until the work of the British chemist Joseph Black (1728–1799). Another British chemist, Henry Cavendish (1731–1810), was the first person to systematically study hydrogen gas, H_2 (in 1766). This latter study was the essential groundwork for determining the composition of water.

14.1 Discovery of Hydrogen; Composition of Water

Before Cavendish, others (including Robert Boyle) had noticed that a flammable gas was produced when metals reacted with acids. However, carbon monoxide and hydrocarbon gases, which are also flammable, were

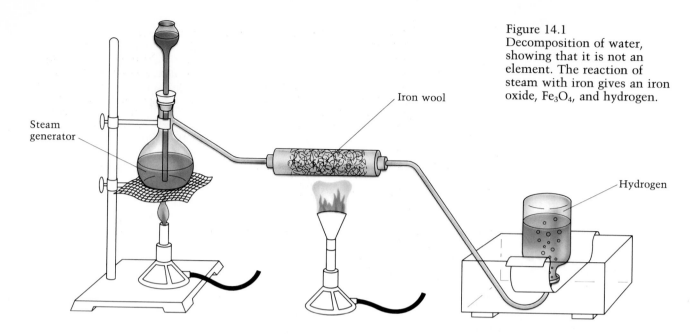

Figure 14.1
Decomposition of water, showing that it is not an element. The reaction of steam with iron gives an iron oxide, Fe_3O_4, and hydrogen.

generally not distinguished from hydrogen. What Cavendish did was to treat the metals zinc, iron, and tin in turn with hydrochloric acid, then with sulfuric acid. In every case, he observed the release of a gas. He collected this gas and showed that it was always flammable and of the same low density. He thus characterized the gas now known as hydrogen.

In 1784, Cavendish showed that hydrogen combines with oxygen during combustion to give water. He also determined the relative volumes of hydrogen and oxygen that react. It was left to Lavoisier, however, to give the modern explanation of Cavendish's work: hydrogen and oxygen are elements that react during combustion to give a compound, water. He strengthened his interpretation of water as a compound by showing that steam could be decomposed by passing it over iron filings to liberate hydrogen (Figure 14.1). The reaction is

$$4H_2O(g) + 3Fe(s) \longrightarrow Fe_3O_4(s) + 4H_2(g)$$

Electrolysis, the decomposition of a substance by an electric current, provides one of the simplest demonstrations of the compound nature of water. In 1800, William Nicholson and Anthony Carlisle passed an electric current from a battery through water that had been made conductive by dissolved substances (for example, by adding small amounts of sulfuric acid). They discovered that bubbles of hydrogen were released at the electrode connected to the negative terminal of the battery, while bubbles of oxygen were released at the other electrode (Figure 14.2). The overall electrolysis can be written ■

■ Electrolysis is discussed in Chapter 21.

$$2H_2O(l) \xrightarrow[\text{energy}]{\text{electrical}} 2H_2(g) + O_2(g)$$

In this experiment, two volumes of hydrogen are produced for each volume of oxygen, in agreement with the formula H_2O.

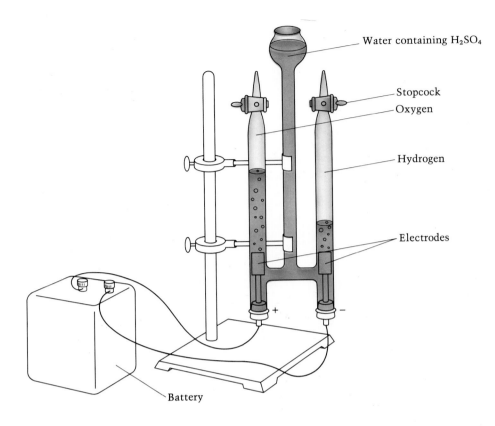

Water containing H_2SO_4

Stopcock

Oxygen

Hydrogen

Electrodes

Battery

Figure 14.2
Electrolysis of water.
Note that the volume of hydrogen released at the negative electrode is twice that of the oxygen released at the other electrode, in agreement with the formula H_2O.

Exercise 14.1

How many moles of H_2 and O_2 are obtained when 2.78 g of water are decomposed by electrolysis? What are the volumes of H_2 and O_2 at 20°C and 782 mmHg?
(See Problems 14.21 and 14.22.)

14.2 Properties of Hydrogen

The element hydrogen occurs on earth principally as a component of water, with smaller amounts in the form of hydrocarbons found in fossil fuels. It occurs in the earth's crust and in the **hydrosphere** (oceans, polar caps, lakes, rivers, and atmospheric moisture) to the extent of less than 1% by mass. Nevertheless, this makes it the ninth most abundant element (by mass) in the accessible portion of earth.

Physical Properties

Hydrogen has an extremely low melting point and boiling point ($-259°C$ and $-253°C$, respectively), which are indicative of the very weak forces of attraction between H_2 molecules. It is thus a gas under usual circumstances. The gas is odorless and colorless and has the lowest density of any substance. Because of its low density, hydrogen was used in lighter-than-air dirigibles,

Figure 14.3
The *Hindenburg* dirigible, which was filled with hydrogen, was destroyed by fire in 1937 at Lakehurst, New Jersey.

although the *Hindenburg* catastrophe (Figure 14.3) was a terrible reminder of its flammability.

Preparation

Hydrogen, H_2, can be prepared by reducing hydrogen ion in aqueous solution with certain metals.■ Table 14.1 lists metals in order of their activity as reducing agents, that is, in order of their ease of losing electrons to give the metal ions in aqueous solution. This list, called the **activity,** or **electromotive, series** of the metals, includes hydrogen for comparison.■ Those metals above H_2 in the series are strong enough reducing agents to release hydrogen gas in acidic solution. For example,

$$Zn(s) + 2H^+(aq) \longrightarrow Zn^{2+}(aq) + H_2(g)$$

The alkali metals, such as sodium, are such strong reducing agents that they react readily with water at room temperature to give hydrogen:

$$2Na(s) + 2H_2O(l) \longrightarrow 2Na^+(aq) + 2OH^-(aq) + H_2(g)$$

This reaction is so rapid and exothermic that the hydrogen may catch fire or explode. Calcium metal also reacts with cold water, but not as vigorously as do the alkali metals.

$$Ca(s) + 2H_2O(l) \longrightarrow Ca^{2+}(aq) + 2OH^-(aq) + H_2(g)$$

■ Oxidation and reduction were discussed in Section 9.7.

■ We will see that these elements are in the order of their voltages for the reduction of metal ion to the metal (see Table 21.1 in Section 21.5).

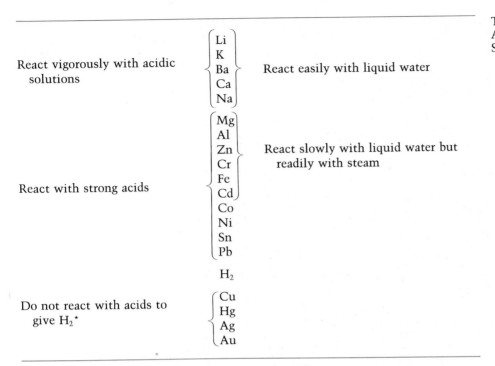

Table 14.1
Activity (Electromotive)
Series of the Metals

*Cu, Hg, and Ag react with HNO_3, but do not produce H_2.

Magnesium reacts slowly with boiling water by a reaction similar to calcium's. The metal reacts rapidly with steam, however, to give magnesium oxide and hydrogen:

$$Mg(s) + H_2O(g) \longrightarrow MgO(s) + H_2(g)$$

Metals from magnesium to cadmium in the activity series react readily in this way with steam to give the metal oxide and hydrogen. We noted the reaction of iron with steam earlier. Table 14.1 summarizes the reactivities of metals with water and steam and with acids.■

Certain metals also react with a strong base, such as sodium hydroxide, to give hydrogen:

$$2Al(s) + 2OH^-(aq) + 6H_2O(l) \longrightarrow 2Al(OH)_4^-(aq) + 3H_2(g)$$
$$\text{aluminate ion}$$

$$Zn(s) + 2OH^-(aq) + 2H_2O(l) \longrightarrow Zn(OH)_4^{2-}(aq) + H_2(g)$$
$$\text{zincate ion}$$

Industrially, hydrogen is prepared from water and hydrocarbons. Until recently the **water–gas reaction** was an important way of preparing hydrogen. In this reaction, steam was passed over red-hot coke to give a gaseous mixture of carbon monoxide and hydrogen:

$$C(s) + H_2O(g) \longrightarrow CO(g) + H_2(g)$$

Now, two general processes are used, both starting with natural gas or petroleum hydrocarbons. In the **steam-reforming process,** hydrocarbons are

■ Nitric acid attacks silver and copper, but in both cases it is the nitrate ion that oxidizes the metal. Hydrogen is not formed.

reacted with steam at high temperature and pressure over a nickel catalyst. For example,

$$CH_4(g) + H_2O(g) \xrightarrow{Ni} CO(g) + 3H_2(g)$$

$$C_3H_8(g) + 3H_2O(g) \xrightarrow{Ni} 3CO(g) + 7H_2(g)$$

A second process involves the partial oxidation of hydrocarbons. Natural gas, for example, is mixed with a limited supply of oxygen and burned at elevated pressures.

$$\underset{\text{natural gas}}{2CH_4(g)} + O_2(g) \longrightarrow 2CO(g) + 4H_2(g)$$

As natural gas and petroleum become more expensive, the water–gas reaction may again become widely used.

The gaseous mixture of carbon monoxide and hydrogen from these reactions (called *synthesis gas*) is used in the preparation of methanol (see later in this section under "Uses"). Hydrogen is obtained from this mixture free of carbon monoxide by means of the **water–gas shift reaction,** which converts the CO to CO_2:■

$$CO(g) + H_2O(g) \xrightarrow{\text{catalyst}} CO_2(g) + H_2(g)$$

Carbon dioxide is removed by dissolving the gas in a basic solution to give carbonate ion.

Small amounts of hydrogen are obtained by electrolysis of water, though the process is economical only when electricity is cheap. In theory, the thermal decomposition of water into its elements should be more efficient (rather than first converting heat to electricity). Although the direct thermal decomposition of water requires too high a temperature (3000°C) to be practical, cyclic processes are being investigated that operate at much lower temperatures. One of these consists of the following steps:

■ Final traces of CO are removed by a process called *catalytic methanation,* in which the H_2 with traces of CO is passed over a nickel catalyst.

$$CO(g) + 3H_2(g) \longrightarrow$$
$$CH_4(g) + H_2O(g)$$

This reaction is discussed at length in Chapter 16.

$$2H_2O(l) + SO_2(g) + I_2(aq) \xrightarrow[\text{temperature}]{\text{room}} H_2SO_4(aq) + 2HI(aq) \qquad \text{Step 1}$$

$$2HI(g) \xrightarrow{300°C} H_2(g) + I_2(g) \qquad \text{Step 2}$$

$$H_2SO_4(g) \xrightarrow{800°C} H_2O(g) + SO_2(g) + \tfrac{1}{2}O_2(g) \qquad \text{Step 3}$$

The essential result of these steps is the decomposition of water:

$$H_2O(l) \xrightarrow[\text{energy}]{\text{thermal}} H_2(g) + \tfrac{1}{2}O_2(g)$$

Exercise 14.2

A 0.389-g sample of a metal reacts with an excess of hydrochloric acid to give 149 mL of hydrogen collected over water at 751 mmHg and 21°C. What is the metal? (Note that the mass of metal per mole of H atoms equals the atomic weight/n for the metal, where n is an integer—the metal ion charge. Thus, by trying values of n and comparing them with the atomic weights of elements, you can identify the metal.)

(See Problems 14.23 and 14.24.)

Hydrides

Hydrogen forms binary compounds, called **hydrides,** with metals and non-metals.■ Most nonmetals form covalent hydrides. Thus, hydrogen reacts readily with halogens, such as chlorine, to give the hydrogen halide:

$$H_2(g) + Cl_2(g) \longrightarrow 2HCl(g)$$

Reaction of hydrogen with oxygen to give water is vigorous, whereas the reaction with sulfur to give hydrogen sulfide, H_2S, is slow except at high temperatures. However, at high temperatures, H_2S begins to decompose into the elements, so it is not normally prepared this way. Reaction of hydrogen with nitrogen to give ammonia, NH_3, occurs at a significant speed only at high temperature and in the presence of a catalyst. Carbon also reacts with hydrogen only at high temperature and in the presence of a catalyst. Methane, CH_4, and other hydrocarbons are produced.

Since the electron affinity of the hydrogen atom is small (73 kJ/mol, compared with 349 kJ/mol for Cl), ionic compounds containing the hydride ion, H^-, are formed only with metals of lowest ionization energy, the alkali metals and the alkaline earth metals.■ Thus, sodium and calcium react with hydrogen gas at moderate temperatures to give the hydrides:

$$2Na(l) + H_2(g) \xrightarrow{\Delta} 2NaH(s)$$
$$Ca(l) + H_2(g) \xrightarrow{\Delta} CaH_2(s)$$

■ The term *hydride* is used as the class name of binary compounds of hydrogen. Thus, NaH and HCl belong to the class of compounds called hydrides. However, the specific name of a compound includes the word *hydride* only if hydrogen is in the formal −1 oxidation state. The *hydride ion* is H^-.

■ The role of ionization energy and electron affinity in ionic bonding is discussed in Section 7.1.

These ionic hydrides conduct electricity when molten, indicating the presence of ions. Hydrogen is liberated at the electrode connected to the positive terminal of the battery, according to the electrode reaction

$$2H^- \longrightarrow H_2(g) + 2e^-$$

Ionic hydrides react with water, giving hydrogen:

$$NaH(s) + H_2O(l) \longrightarrow Na^+(aq) + OH^-(aq) + H_2(g)$$
$$CaH_2(s) + 2H_2O(l) \longrightarrow Ca^{2+}(aq) + 2OH^-(aq) + 2H_2(g)$$

Because of the last reaction, calcium hydride is a convenient, portable source of small quantities of hydrogen.

Exercise 14.3

By means of equations, describe three different methods of preparing H_2. Assume you have any necessary catalyst and other special requirements of the preparation, but use only the following as reactants: H_2O, KH, CH_4.

(See Problems 14.25 and 14.26.)

Exercise 14.4

Give balanced equations for the reactions of H_2 with I_2, with Mg, and with Ag_2O.

(See Problems 14.27 and 14.28.)

Uses

Approximately 40% of the hydrogen produced commercially is used to manufacture ammonia and nitric acid, and about the same amount is used in petroleum refining.■ Hydrogen is also used to synthesize methanol and formaldehyde, which are both starting materials for plastics. Methanol is produced from synthesis gas (a mixture of CO and H_2), using a catalyst at moderately high temperature and pressure:

■ The ammonia synthesis and preparation of nitric acid from ammonia were discussed in Section 13.7.

$$CO(g) + 2H_2(g) \xrightarrow{\text{catalyst}} CH_3OH(g)$$

Methanol vapor is oxidized with a silver catalyst to give formaldehyde:

$$2CH_3OH(g) + O_2(g) \xrightarrow{\text{Ag}} 2HCHO(g) + 2H_2O(g)$$

Hydrogenation is a process in which hydrogen, with a catalyst such as nickel, adds to compounds containing carbon–carbon multiple bonds. For example,

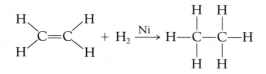

Hydrogenation converts an *unsaturated* organic compound (a carbon compound containing one or more carbon–carbon multiple bonds) to a *saturated* compound (a carbon compound not containing carbon–carbon multiple bonds). Polyunsaturated oils (vegetable oils) when hydrogenated give higher-melting saturated fats (such as Crisco). Hydrogenation is also used in the refining of petroleum to make high-octane gasoline.

Hydrogen can reduce the oxides of metals below chromium in the activity (electromotive) series to give the pure metal and water. Tungsten, for example, is obtained commercially by reduction of tungsten(VI) oxide (obtained from tungsten ores):

$$WO_3(s) + 3H_2(g) \longrightarrow W(s) + 3H_2O(g)$$

(Tungsten is used to make hard steels, light-bulb filaments, and tungsten carbides, WC and W_2C, for cutting tools.)

Recently, hydrogen has been studied as a portable source of energy. It can be burned in internal-combustion engines that have been modified for this purpose. Petroleum may well become too expensive to burn as a fuel, and hydrogen would be an ideal replacement in a number of respects. It is light and would burn in air to produce an exhaust containing only water, with perhaps a small quantity of nitrogen oxides (from N_2 in air). Environmentally, therefore, hydrogen has much to recommend its use. Of course, hydrogen would not be a primary source of energy; coal, solar energy, or nuclear power are possible primary sources. Hydrogen would be obtained from water, perhaps by electrolysis or by thermal decomposition processes that are being investigated.

Exercise 14.5

Give equations for the preparation of (a) CH_3OH and (b) Cu. Start from the following reactants: C_2H_6, H_2O, CuO. Assume that any required catalysts are available. More than one step may be needed for a preparation.

(See Problems 14.29 and 14.30.)

14.3 Properties of Water

Table 14.2 lists the thermal properties of a number of common substances (on a per-gram basis). Note that water has, by comparison, a large heat capacity, a large heat of fusion, and a large heat of vaporization. (Ammonia is similar in having unusually large thermal properties.) In addition, water is unusual in being one of the few substances whose solid phase is less dense than the liquid.

Hydrogen Bonding and the Physical Properties of Water

Ice is less dense than liquid water because each water molecule can form hydrogen bonds to four other water molecules. In order to form the maximum number of hydrogen bonds, four hydrogen atoms are arranged tetrahedrally about a given oxygen atom. The resulting tetrahedral angles give rise to a three-dimensional structure in ice that actually contains open space (see Figure 14.4). When ice melts, hydrogen bonds break, the ice structure partially disintegrates, and the molecules become more compactly arranged.

Even liquid water has significant hydrogen bonding. Only about 15% of the hydrogen bonds are broken when ice melts. We might view the liquid as composed of icelike clusters in which hydrogen bonds are continually breaking and forming, so that clusters disappear and new ones appear. This is sometimes referred to as the "flickering cluster" model of liquid water.

As the temperature rises from 0°C, these clusters tend to break down further, giving an even more compact liquid. Thus, the density rises (see Figure 14.5). However, as is normal in any liquid, when the temperature rises, the molecules begin to move and vibrate faster, and the space occupied by the average molecule increases. For most other liquids this results in a continuous decrease in density with temperature increase. In water, this normal effect is countered by the density increase due to breaking of hydrogen bonds, but at 4°C the normal effect begins to predominate. Water shows a maximum density at 4°C and becomes less dense at higher temperatures.

The unusually large heat capacity of water is also explained by hydrogen bonding. To increase the temperature of liquid water, it is necessary to supply energy to break hydrogen bonds, in addition to the energy normally

Table 14.2
Thermal Properties of
Some Common Substances

Substance	Formula	Melting Point (°C)	Boiling Point (°C)	Heat Capacity of Liquid (J/g · °C)	Heat of Fusion (J/g)	Heat of Vaporization (J/g)
Water	H_2O	0	100	4.18	333	2257
Ammonia	NH_3	−78	−33	4.48	341	1368
Ethanol	C_2H_5OH	−117	78	2.24	104	854
Benzene	C_6H_6	6	80	1.63	127	395
Carbon tetrachloride	CCl_4	−23	77	0.83	17	194
Mercury	Hg	−39	357	0.14	12	295

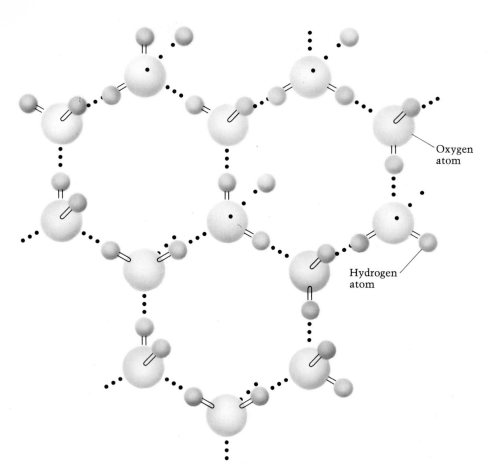

Oxygen
atom

Hydrogen
atom

Figure 14.4
The structure of a portion of ice. Oxygen atoms are represented by large spheres, hydrogen atoms by small spheres. Each oxygen atom is tetrahedrally surrounded by four hydrogen atoms. Two are close, giving the H_2O molecule. Two are further away, held by hydrogen bonding (represented by three dots). The distribution of hydrogen atoms in these two positions is random.

required to increase molecular agitation. The heat of fusion and heat of vaporization are high for the same reason; hydrogen bonds must be broken to melt or to vaporize the substance.

Exercise 14.6

How many joules of heat must be added to 5.0 g of ice at 0°C to change it to water at 20°C?

(See Problems 14.31 and 14.32.)

Chemical Properties of Water

Water is thermally very stable, decomposing into its elements to the extent of only 11% at 2700°C.

$$2H_2O(g) \xrightarrow{\Delta} 2H_2(g) + O_2(g)$$

It is a relatively reactive substance, reacting with many of the elements, both metals and nonmetals. Reactions with the alkali metals, the alkaline earth metals, iron, and carbon were described in the previous sections. Silicon reacts as carbon does at red heat, giving hydrogen and silicon dioxide. Fluorine reacts vigorously to give oxygen,

$$2F_2(g) + 2H_2O(l) \longrightarrow 4HF(aq) + O_2(g)$$

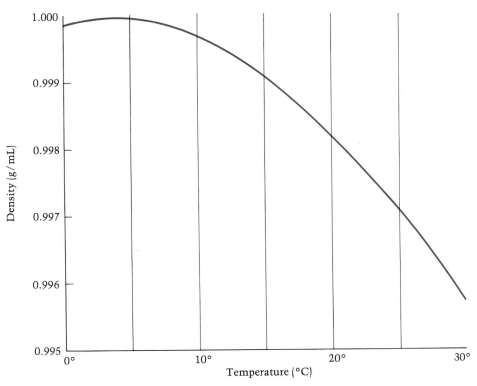

Figure 14.5
The variation of the density of water with temperature.

whereas chlorine *disproportionates*, that is, it is both oxidized and reduced:

$$Cl_2(g) + H_2O(l) \rightleftharpoons H^+(aq) + Cl^-(aq) + HOCl(aq)$$

Bromine similarly disproportionates, though to only a slight extent, and iodine is nearly unreactive in pure water. In basic solution these disproportionations occur easily.■

Reactions of oxides and hydrides with water were described earlier (Sections 13.4 and 14.2, respectively). The ionic dissociation of water

$$H_2O(l) \rightleftharpoons H^+(aq) + OH^-(aq)$$

and related acid–base equilibria are most important reactions of water, and will be discussed in Chapter 17.

■ In basic solution, the disproportionation of Cl_2 is

$$Cl_2(g) + 2OH^-(aq) \longrightarrow$$
$$Cl^-(aq) + OCl^-(aq) + H_2O(l)$$

This reaction is used commercially to produce sodium hypochlorite bleaches, such as Clorox.

Exercise 14.7

Write equations for the reactions of water with the following elements: (a) Si (b) K (c) Br_2

(See Problems 14.33 and 14.34.)

14.4 Hydrates and Hydrated Ions

Interest in water centers on its solvent properties, in addition to its chemical reactions. Hydrogen-bond formation is important in determining the solvent properties of water. Thus, substances like methanol, CH_3OH, dissolve in water through hydrogen-bond formation. But hydrogen bonding is not the only reason for water's solvent properties. Another factor is that water has a large dipole moment, which attracts ions and polar molecules. Moreover, the lone pairs of electrons on water can be donated to metal ions to form *complex ions*, which are ions formed when electron pairs on molecules or anions bond to a metal atom.

A **hydrated ion** is a cation that gathers a sphere of water molecules around itself. In many cases, we know little about the exact structures of hydrated ions, but we do know that a certain number of water molecules are associated with specific cations. In water solutions, the iron(III) ion is associated with six water molecules to give the complex ion $Fe(H_2O)_6^{3+}$ (see Figure 14.6). Oxygen atoms from H_2O molecules are bonded to the metal ion.

Hydrates are salts that crystallize from water solution and that contain weakly bound water molecules. Iron(III) chloride crystallizes with six water molecules. The formula is written $FeCl_3 \cdot 6H_2O$. Since the water is associated with the iron(III) ion, a more informative formula is $[Fe(H_2O)_6]Cl_3$. Copper(II) sulfate pentahydrate, $CuSO_4 \cdot 5H_2O$, crystallizes from aqueous solutions of copper sulfate. Four of the water molecules are associated with the copper(II) ion; the other water molecule is near the sulfate, perhaps hydrogen bonded to an oxygen. Hence, the formula could be written $[Cu(H_2O)_4]SO_4 \cdot H_2O$. When a hydrate is heated, the loosely held water is driven off as water vapor. Heating the blue crystals of $CuSO_4 \cdot 5H_2O$ yields white crystals of the *anhydrous* (without water) copper sulfate, $CuSO_4$. Sometimes a hydrate loses water of hydration when exposed to air at normal temperatures. This process is called *efflorescence*. Substances that remove moisture from air at room temperature are called *hygroscopic*. For example, anhydrous magnesium perchlorate is hygroscopic and forms the hydrate $Mg(ClO_4)_2 \cdot 6H_2O$ on exposure to moist air.■

In the salt hydrates, water molecules are weakly bonded to the ions. In another type of hydrate, water molecules form icelike "cages" that trap other molecules. This is an example of a **clathrate**, a solid solution in which one substance is trapped in the crystal of another. The hydrates formed with the noble gases are examples of clathrates. Methane also forms a clathrate with water. An icelike slush of methane–water clathrate occasionally forms in natural-gas pipelines at temperatures that are well above the freezing point of water. It is thought that general anesthesia with inert substances like xenon occurs through the formation of clathrates in the body and in cell fluids that interfere with nervous communication.

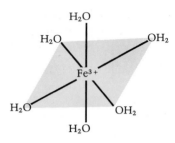

Figure 14.6
The structure of $Fe(H_2O)_6^{3+}$. The water molecules are octahedrally arranged about the Fe^{3+} ion. Oxygen atoms of H_2O molecules are linked by coordinate covalent bonds to the metal atom.

■ This property of magnesium perchlorate, $Mg(ClO_4)_2$, makes it useful as a drying agent, a substance used to remove water from gases and liquids.

Exercise 14.8

A hydrate of sodium sulfate is heated to obtain the anhydrous salt. If 5.26 g of the hydrate yield 2.32 g of the anhydrous salt, what is the formula of the hydrate?
(See Problems 14.35 and 14.36.)

14.5 Hydrogen Peroxide

Hydrogen peroxide, H_2O_2, is another compound of hydrogen and oxygen. It has the structural formula

Pure hydrogen peroxide is a syrupy liquid with a pale blue color. If very pure, it is stable, but in the presence of small quantities of impurities, the liquid may decompose explosively. The decomposition is exothermic.

$$2H_2O_2(l) \longrightarrow 2H_2O(l) + O_2(g); \Delta H = -196.0 \text{ kJ}$$

Platinum metal and ions such as Fe^{2+}, Fe^{3+}, and I^- act as catalysts for this reaction. Pure hydrogen peroxide boils at 152°C at 760 mmHg and freezes at -0.89°C.

In the nineteenth century, hydrogen peroxide was prepared from barium peroxide, BaO_2, by reacting it with sulfuric acid.■ Barium sulfate precipitates, leaving an aqueous solution of hydrogen peroxide:

■ The preparation of barium peroxide was discussed in Section 13.4.

$$BaO_2(s) + H_2SO_4(aq) \longrightarrow BaSO_4(s) + H_2O_2(aq)$$

Today most hydrogen peroxide is produced by the reduction of compounds called anthraquinones to the anthraquinol, followed by reaction with oxygen. For example, 2-ethyl-9,10-anthraquinone in benzene solution is reduced with hydrogen, using a palladium catalyst:

$$\underset{\text{2-ethyl-9,10-anthraquinone}}{C_{16}H_{12}O_2} + H_2 \xrightarrow{\text{Pd}} \underset{\text{2-ethyl-9,10-anthraquinol}}{C_{16}H_{12}(OH)_2}$$

The 2-ethyl-9,10-anthraquinol reacts with oxygen to give the original anthraquinone and hydrogen peroxide:

$$C_{16}H_{12}(OH)_2 + O_2 \longrightarrow C_{16}H_{12}O_2 + H_2O_2$$

The hydrogen peroxide is extracted from the benzene solution with water to give a solution that is 18% H_2O_2. More concentrated solutions are obtained by distillation under vacuum, where their boiling points are well below those at 760 mmHg. (Low temperatures are required to prevent explosive decomposition.) Note that the overall result of these two reactions is to produce H_2O_2 from H_2 and O_2.

Hydrogen peroxide is a strong oxidizing agent, giving water as the reduction product. For example, iodide ion in acidic solution is oxidized to iodine:

$$H_2O_2(aq) + 2H^+(aq) + 2I^-(aq) \longrightarrow 2H_2O(l) + I_2(aq)$$

Hydrogen peroxide behaves as a reducing agent in the presence of strong oxidizing agents, yielding oxygen and water. For example, potassium permanganate solutions (which have a purple color) are reduced by hydrogen peroxide to manganese(II) ion (colorless to pale pink):

$$5H_2O_2(aq) + 2MnO_4^-(aq) + 6H^+(aq) \longrightarrow 8H_2O(l) + 5O_2(g) + 2Mn^{2+}(aq)$$

Dilute solutions of hydrogen peroxide (3 to 6% by mass) are used as an antiseptic and as a bleach. Large quantities of hydrogen peroxide are used commercially to bleach cotton and wool. Concentrated solutions have been used as a rocket propellant, in the following reaction:

$$2H_2O_2(l) \xrightarrow{\text{catalyst}} 2H_2O(g) + O_2(g)$$

This reaction is exothermic and generates large volumes of gases.

Exercise 14.9

Hydrogen peroxide has been used to clean oil paintings. It oxidizes lead sulfide, PbS, a black compound, to lead sulfate, $PbSO_4$, which is white. Write a balanced equation for the reaction.

(See Problems 14.37, 14.38, 14.39, and 14.40.)

Exercise 14.10

A solution of hydrogen peroxide was titrated with 0.135 *M* $KMnO_4$. If 35.4 mL of the $KMnO_4$ solution are needed to titrate 25.0 mL of acidic hydrogen peroxide, what is the molar concentration of H_2O_2 in the original solution?

(See Problems 14.41 and 14.42.)

Exercise 14.11

Hydrogen peroxide is used as a rocket propellant. The reaction is

$$2H_2O_2(l) \xrightarrow{\text{catalyst}} 2H_2O(g) + O_2(g)$$

Calculate the heat released per gram of hydrogen peroxide. The enthalpies of formation of $H_2O_2(l)$ and $H_2O(g)$ are -187.8 kJ/mol and -241.8 kJ/mol, respectively.

(See Problems 14.43 and 14.44.)

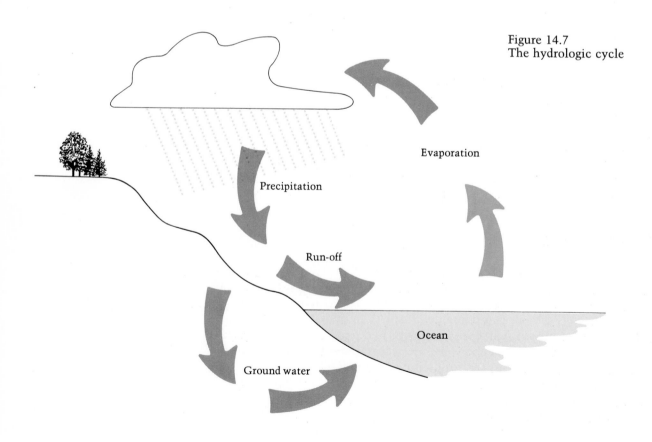

Figure 14.7
The hydrologic cycle

Natural Waters

The total amount of water on earth is estimated to be 1.4×10^{21} kg. This is many times more than the annual human consumption. Of this total, though, freshwater lakes, rivers, and underground water comprise only 0.6%. Fortunately, energy from the sun transports water continuously from the oceans to land. Each year 5×10^{17} kg of water precipitate as rain and snow. One-fourth of this precipitation occurs over land area to replenish streams and ground water. Eventually, the water finds its way back to the oceans. This natural cycle of water from the oceans to fresh water sources and its return to the oceans is called the **hydrologic cycle** (see Figure 14.7).

14.6 Physical Properties of Water and the Environment

Evaporation of the surface waters uses over 30% of the solar energy reaching the earth's surface, or about 1×10^{24} J/yr. (For comparison, the total U.S. consumption of energy is about 7×10^{19} J.) Some of this energy is released in thunderstorms and hurricanes. The transport of energy by evaporation of water and its subsequent condensation is a major factor in the weather on earth. Although the evaporation and condensation of water play a dominant role in our weather, other properties of water are important in making the weather and other environmental conditions congenial to life.

Effect of Large Heat Capacity of Water

The exceptionally large heat capacity of bodies of water has an important moderating effect on the surrounding temperature by warming cold air masses in winter and cooling warm air masses in summer. Worldwide, the oceans are most important, but even inland lakes have a pronounced effect. For example, the Great Lakes give Detroit a more moderate winter compared with cities somewhat further south, but which have no nearby lakes.

Effect of Relative Densities of Ice and Water

Because ice is less dense than water, it forms on top of the liquid when freezing occurs. This has far-reaching effects, both for weather and for aquatic animals. When ice forms on a body of water, it insulates the under-lying water from the cold air and limits further freezing. Fish depend on this for winter survival. Consider what would happen to a lake if ice were more dense than water. The ice would freeze from the bottom of the lake upward. Without the insulating effect at the surface, the lake could well freeze solid, killing the fish. Spring thaw would be prolonged by the fact that it would take much longer for the ice at the bottom of a lake to melt due to the insulating effect of the surface water.

14.7 Water Pollutants

Natural waters are never pure in a chemical sense. Ocean water is principally a solution of sodium chloride (3.5% by mass), but it contains over seventy elements, most in only trace amounts.■ (See Table 14.3 for the principal dissolved ions.) Fresh waters contain a variety of dissolved ions from minerals, including Ca^{2+}, Mg^{2+}, Fe^{3+}, CO_3^{2-} and SO_4^{2-}. In addition, natural waters contain microorganisms. Natural water is said to be polluted when it has been changed by human activity, say by the addition of dissolved substances, so that it no longer supports its normal aquatic life or can no longer be used for purposes such as drinking or industrial processing.

■ The waters of the oceans are commercial sources of sodium chloride, bromine, and magnesium. Other substances, including manganese, are being obtained from the ocean floor.

Thermal Pollution

The addition of heat to natural waters, or **thermal pollution,** might not be thought of as the addition of a pollutant. Yet many biological organisms are seriously affected by temperature changes.■ Rapid changes in water temperature can be particulary harmful, but even gradual elevation of the water temperature will alter the biological community. Many fish species, for example, have a narrow temperature range in which they can thrive.

In a world of expanding energy consumption, thermal pollution may become a serious problem. Eighty percent of the water used by industry is for cooling, most of this by electric power companies. The amount of waste heat from a power facility depends on its efficiency at converting heat to electrical energy. If there is to be increased reliance on nuclear power, thermal pollution may be one of several difficulties to be surmounted. Cooling towers (Figure 14.8) can alleviate the problem of thermal pollution, but they require energy to operate, and they take up valuable land space.

■ The rates of chemical reactions, including the metabolism of a fish, roughly double with each 10°C increase in temperature. This means that the oxygen consumption of a fish increases with water temperature. At the same time, oxygen is less soluble in warmer waters, so that the fish has greater difficulty getting enough oxygen.

Biological Oxygen Demand

Normally, surface waters maintain a certain level of dissolved oxygen by contact with the atmosphere. However, large amounts of decaying organic

Ions	Molality
Cl^-	0.5545
Na^+	0.4756
SO_4^{2-}	0.0286
Mg^{2+}	0.0542
Ca^{2+}	0.0103
K^+	0.0101
HCO_3^-	0.0024
Br^-	0.0008
H_3BO_3	0.0004
Sr^{2+}	0.0001

Table 14.3
Principal Dissolved
Ions in Sea Water

Warm air

Cool air

Air vents

Hot water →
Cold water ←

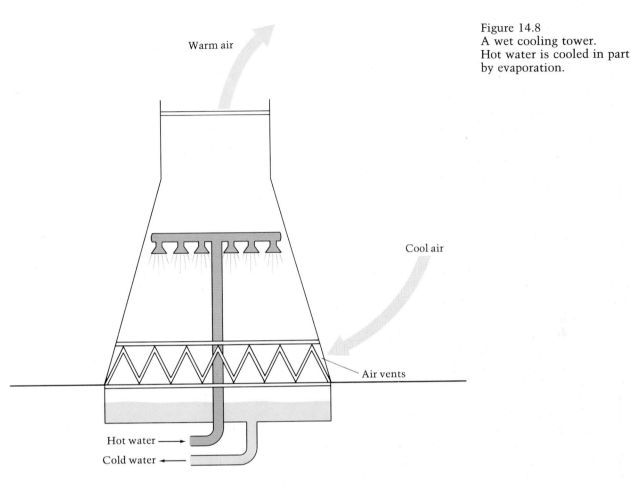

Figure 14.8
A wet cooling tower.
Hot water is cooled in part
by evaporation.

matter from domestic sewage or industrial waste can rapidly deplete water of its dissolved oxygen, since the decay is an oxidation process that consumes oxygen. Such water can no longer support oxygen-consuming organisms. Only *anaerobic* microorganisms survive—that is, organisms that live in oxygen-deficient surroundings. Under these conditions, the waste products of metabolism include many malodorous compounds (such as hydrogen sulfide, H_2S), and the water is then readily identified as polluted.■

Biological oxygen demand (BOD) is an important measure of water quality. Basically, it determines the amount of organic reducing substances (organic wastes, and so forth) in water. The effluent from a sewage treatment plant, for example, is tested for its biological oxygen demand to determine the efficiency of the treatment. To obtain the BOD, the water is first diluted with distilled water containing oxygen, and the amount of oxygen per liter of water is immediately determined. After five days, the amount of dissolved oxygen in the water is again determined. The difference in values is a measure of the oxygen consumed by microorganisms in metabolizing the organic matter. Values are reported as milligrams of O_2 consumed per liter of water. Table 14.4 shows the range of BOD values for different qualities of water.

■ The waste products of anaerobic metabolism are small molecules with elements in low oxidation states, including NH_3, PH_3, and CH_4. The final products of aerobic metabolism are substances in high oxidation states, such as NO_3^-, PO_4^{3-}, and CO_2 or CO_3^{2-}.

Type of Water	BOD Values (mg O_2/L)
Unpolluted stream water	<1.5
Untreated municipal sewage	100–400
Municipal sewage after primary treatment	50–250
Municipal sewage after secondary treatment	10–40
Food-processing wastes	100–10,000

Table 14.4
Representative BOD
Values for Polluted Water

Inorganic Ions

Organic wastes not only deplete water of oxygen, but produce inorganic ions such as nitrate, NO_3^-, and phosphate, PO_4^{3-}, as a result of bacterial degradation. These ions also find their way into surface waters from agricultural run-off (fertilizers) and from household detergents. Being plant nutrients, these ions overfertilize the growth of algae in lakes. Large mats of algae die and sink to the lake bottom, where they decay, using up the oxygen. Many fish species are unable to live in water of diminished oxygen content and are replaced by more tolerant species. Eventually, anaerobic conditions prevail, hastening the end of the lake. As in the normal geological transformation of a lake, it becomes a marsh, then a meadow. But the time scale is enormously speeded up.

Other Pollutants

Various chemical substances, including pesticides and heavy metal compounds, have at times been found as water pollutants. Mercury compounds are toxic to all organisms. In 1970, mercury was detected in fish taken from the St. Clair River (between Michigan and Ontario) at levels exceeding the U.S. Food and Drug Administration limit of 0.5 parts per million (ppm). The source of the mercury was traced to industrial plants involved in the electrolysis of aqueous sodium chloride, where mercury is used as an electrode.■ At the time it was not realized that metallic mercury could be incorporated into the food chain. It has since been discovered that certain anaerobic bacteria in the bottom mud of rivers can convert mercury to methyl mercury, $(CH_3)_2Hg$, which then enters the water and subsequently the food chain. Mercury discharge into surface waters has been now largely curbed.

■ The use of mercury in the electrolysis of aqueous sodium chloride is discussed in Section 23.2.

14.8 Water Treatment and Purification

One of the principal factors limiting human habitation is an adequate supply of drinking water. Even more water is needed for industry than for domestic use. For these reasons, early settlements were almost always located near streams and lakes. And, as the population grew, natural waters increasingly became the disposal sites for domestic and industrial wastes. There are few places in the world today where one can drink from streams without worry that they are polluted.

Purification treatment is considered necessary for most municipal water supplies. For certain uses, water must be further specially purified. We can classify water impurities into four types: suspended materials, colloidal materials, biological organisms, and dissolved substances.

Suspended materials are easily removed by means of sedimentation tanks. Silt and other colloidal materials are usually separated by coagulation and filtration. To accomplish this, aluminum sulfate, $Al_2(SO_4)_3 \cdot 18H_2O$, and lime, CaO, are added to the water. The aluminum ion, Al^{3+}, tends to coagulate colloidal material. ■ Lime gives a basic solution,

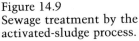
■ Coagulation of colloids by ions is discussed in Section 12.4.

$$CaO(s) + H_2O(l) \longrightarrow Ca^{2+}(aq) + 2OH^-(aq)$$

which with aluminum ion gives a gelatinous precipitate of aluminum hydroxide:

$$Al^{3+}(aq) + 3OH^-(aq) \longrightarrow Al(OH)_3(s)$$

The settling precipitate takes with it coagulated materials as well as some bacteria. The water is filtered through sand to remove the aluminum hydroxide. A disinfectant such as chlorine or ozone is then added.

The treatment of waste water is similar in many respects to that used to obtain drinking water. Any organic matter present must be removed from waste water before it is released into a river or lake. Otherwise, the organic matter can seriously deplete the water of its dissolved oxygen. Primary treatment of waste water consists of sedimentation. Secondary treatment involves biological degradation of organic wastes. The **activated-sludge method** is the most popular secondary treatment. With this method, waste water is fed into tanks that are aerated (Figure 14.9). Aerobic bacteria feed on the waste and degrade it to inorganic ions. The water is then filtered and a disinfectant is added.

Occasionally, a tertiary (third) stage is added to the process. This stage is used to remove dissolved substances and must often be tailored specifically for the substance to be removed. Activated charcoal can be used to remove offensive odors (such as H_2S) and organic compounds that are not removed

Figure 14.9
Sewage treatment by the activated-sludge process.

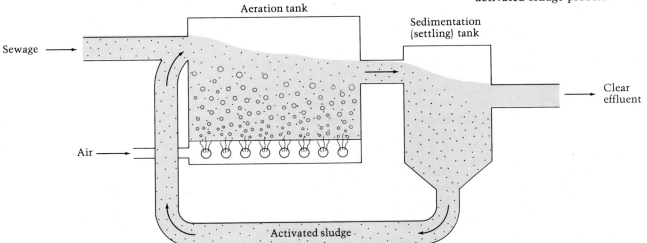

by biological degradation. Lime, CaO, is sometimes added to precipitate phosphates:

$$3Ca^{2+}(aq) + 2PO_4^{3-}(aq) \longrightarrow Ca_3(PO_4)_2(s)$$

With proper treatment, waste water can be returned to natural sources in a condition fit to drink.

14.9 Removing Ions from Water; Desalination

Once it is used, water invariably increases in concentration of inorganic ions. Ideally, waste water would have these ionic concentrations reduced to low levels before being discharged into streams and lakes. Certain ions can be removed by precipitation. Finding economical ways to remove or reduce the concentration of all ions is a problem of current research interest. The process of removing ions from brackish (slightly salty) water or sea water, thus producing water that is fit to drink is called **desalination.**

Ion exchange is used in the home and by industry to remove certain ions from water. Calcium ions in hard water, for example, react with soap to produce a curdy precipitate. Soap (the sodium salt of stearic acid and similar acids) is not only less efficient in water containing calcium ions, but its precipitate—calcium stearate (and calcium salts of similar acids)—adheres to clothes, the skin, and bathtubs. Calcium salts also precipitate out in boilers as "boiler scale." Water softening is accomplished by removing calcium ions. **Zeolites** are natural and synthetic minerals that contain sodium ions within cavities of the mineral. When hard water is flushed through the mineral, the sodium ions are exchanged for calcium ions in the water:

$$Na_2Z(s) + Ca^{2+}(aq) \longrightarrow CaZ(s) + 2Na^+(aq) \qquad (Z = zeolite)$$

Zeolite is recharged after use by flushing it with brine ($NaCl$ solution), which reverses the reaction:

$$CaZ(s) + 2Na^+(aq) \longrightarrow Na_2Z(s) + Ca^{2+}(aq)$$

This process does not remove ions, but merely exchanges troublesome calcium ions for sodium ions. However, it is possible to use ion exchange resins (macromolecular substances) to completely remove cations and anions from water. A **cation exchange resin** removes cations and replaces them with hydrogen ions:

$$HR(s) + Na^+(aq) \longrightarrow NaR(s) + H^+(aq) \qquad (R = exchange\ resin)$$

Similarly, an **anion exchange resin** removes anions and replaces them with hydroxide ions:

$$ROH(s) + Cl^-(aq) \longrightarrow RCl(s) + OH^-(aq)$$

Deionized water is prepared by allowing water to flow first through a column packed with cation exchange resin, then through a column packed with anion exchange resin. The hydrogen ions and hydroxide ions produced then react to give water:

$$H^+(aq) + OH^-(aq) \longrightarrow H_2O(l)$$

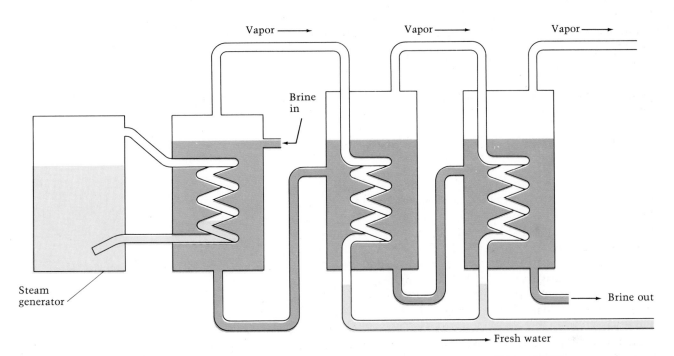

Figure 14.10
Desalination of sea water by multiple-stage distillation. The heat obtained from condensing vapor in one stage is used to heat water for the next stage.

Ion exchange is often used to prepare deionized water for laboratory use, but is too expensive to consider for general treatment of waste water.

Distillation has been used to prepare large quantities of potable water from sea water. In order to make the process economical, the heat released from condensation of water vapor must be used to heat water for distillation. Figure 14.10 shows a multiple-stage distillation apparatus that employs this idea.

Reverse osmosis has been employed with some success in the desalination of ocean water. In osmosis, water will normally flow through a membrane from the dilute solution to the concentrated solution.∎ Osmosis can be stopped and even reversed, however, by the application of sufficient pressure to the concentrated solution. (Recall that the osmotic pressure, π, is just sufficient to stop osmosis.) Waste water or sea water can in effect be forced through a membrane, which will retain ions and other substances while passing pure water.

∎ Osmosis was discussed in Section 12.8.

Exercise 14.12

Suppose sea water to be an aqueous solution of sodium chloride (3.50 mass % NaCl). Calculate the freezing point of this solution, assuming an i value of 2 in the freezing-point formula (see Section 12.9).

(See Problems 14.51 and 14.52.)

Exercise 14.13

What is the osmotic pressure at 25°C of the solution described in Exercise 14.12? The density of this solution is 1.023 g/mL.

(See Problems 14.53 and 14.54.)

Exercise 14.14

The minimum energy (in joules) required to separate impurities from water equals the osmotic pressure (in pascals) times the volume (in cubic meters) of the water solution. What is the minimum energy required to obtain pure water from 1.00 L of the solution described in the previous exercises?

(See Problems 14.55 and 14.56.)

A Checklist for Review

Important Terms

hydrosphere (14.2)
activity (electromotive) series (14.2)
water–gas reaction (14.2)
steam-reforming process (14.2)
water–gas shift reaction (14.2)
hydrides (14.2)
hydrogenation (14.2)

hydrated ion (14.4)
hydrates (14.4)
clathrate (14.4)
hydrologic cycle (p. 481)
thermal pollution (14.7)
biological oxygen demand (BOD) (14.7)
activated-sludge method (14.8)

desalination (14.9)
ion exchange (14.9)
zeolites (14.9)
cation exchange resin (14.9)
anion exchange resin (14.9)
reverse osmosis (14.9)

Summary of Facts and Concepts

Henry Cavendish first characterized hydrogen gas in 1766, and Lavoisier interpreted the combustion of hydrogen in air as a reaction of two elements (hydrogen and oxygen) to give the compound water. Hydrogen may be prepared by reacting metals above hydrogen in the activity series with an acid. Very active metals, such as sodium, will even react with water to give hydrogen. Certain metals, such as aluminum, react with a strong base to give hydrogen. The gas is obtained commercially by the *steam-reforming process* (reaction of hydrocarbons with steam) or by partial oxidation of hydrocarbons. Some H_2 is prepared commercially by the electrolysis of water. Ionic *hydrides*, such as CaH_2, are a convenient source of small amounts of H_2.

Hydrogen is used to manufacture ammonia, methanol (CH_3OH), and formaldehyde (CH_2O). Methanol and formaldehyde are used to make plastics. Hydrogen is also used in the *hydrogenation* of polyunsaturated oils to give saturated fats, and in the reduction of certain metal oxides, such as tungsten(VI) oxide, to the metal.

Water, the most important compound of hydrogen, has several unusual properties. For example, it has a relatively large heat capacity. Also the solid phase is less dense than the liquid; usually, the solid phase of a substance is more dense than the liquid. These unusual characteristics are due to hydrogen bonding. Water is thermally stable, but does react with many substances, including the more active metals and some nonmetals (C, Si, F_2, Cl_2, Br_2). Metal ions form *hydrated* species in water solution and often retain water molecules even in crystalline salts, giving *hydrates*. The water in hydrates, however, need not be associated with the cations.

Hydrogen peroxide, H_2O_2, is another compound of hydrogen and oxygen. It is unstable and decomposes in the presence of catalysts to give water and oxygen. Hydrogen peroxide is prepared in a cyclic process in which the net reaction produces H_2O_2 from H_2 and O_2. The compound behaves as both an oxidizing agent and a reducing agent.

The presence of water on earth is a major factor in determining the weather and the conditions for life. Thus, bodies of water moderate the weather because of their large heat capacity. The fact that ice is less dense than liquid water is important to aquatic life. Water becomes polluted when changed adversely by human activity. Examples of water pollutants are heat (*thermal pollution*), organic wastes (measured by *biological oxygen demand*), and inorganic ions such as NO_3^- and PO_4^{3-}. Municipal water supplies are treated to remove pollutants by filtration, coagulation, and by disinfecting with Cl_2 or O_3. Organic material in waste water is broken down with aerobic bacteria in the *activated-sludge process*. Ions can be removed from water by *ion exchange*. Distillation and *reverse osmosis* are also used to obtain pure water.

Operational Skills

The basic problem-solving skills used in this chapter were introduced in Chapters 1 to 12.

Review Questions

14.1 You are given unlabeled samples of the gases H_2, O_2, CO, and CO_2. Devise a simple scheme that can distinguish these gases.

14.2 Using the reaction involving steam and iron given in the text, carefully describe an experiment that could be used to obtain the simplest formula of water.

14.3 What are the principal natural sources of hydrogen?

14.4 Would you expect silver metal to react with sulfuric acid to give H_2? Explain.

14.5 Tin dissolves in an alkali metal hydroxide solution to form the stannite ion, $Sn(OH)_4^{2-}$, and hydrogen. Write the balanced equation.

14.6 What are the principal methods for the commercial preparation of hydrogen?

14.7 What is the evidence that NaH is an ionic hydride?

14.8 Why might you expect the cost of nitrogen fertilizers to depend on the cost of natural gas and petroleum?

14.9 Explain why water has a larger heat capacity than does benzene, C_6H_6.

14.10 Why does ammonia have unusually large thermal properties compared with most other substances?

14.11 Explain why water becomes more dense from 0°C to 4°C. Why does the density become less from 4°C to 100°C?

14.12 Anhydrous magnesium perchlorate is used as a drying agent, that is, as a substance that picks up moisture from other substances, such as liquids. Explain how this drying action works.

14.13 The reaction of barium peroxide with sulfuric acid to give barium sulfate and hydrogen peroxide can be looked at as an acid–base reaction. Explain.

14.14 Calculate the amount of heat required to evaporate 5×10^{17} kg of water (the total annual amount of rain and snow). Compare with the figure given in the text.

14.15 Explain how the relative densities of water and ice affect the environment on earth.

14.16 How would the thermal pollution of a stream affect the dissolved oxygen? The oxygen requirements of fish increase with temperature. What would you expect to be the effect on fish as the temperature of the water increases?

14.17 What happens to a lake to which large quantities of nitrate and phosphate ions are added from polluted waters?

14.18 What are the four types of water impurities? Describe ways used to remove each type of impurity.

14.19 Why is it necessary to remove organic wastes from water before the water is returned to a natural source?

14.20 Describe some methods of desalination.

Problems

Composition of Water

14.21 When heated, copper(II) oxide combines with hydrogen gas to form copper metal and water. In an experiment, a stream of hydrogen passes over 0.538 g of heated copper(II) oxide until it is completely reduced to copper, weighing 0.430 g. At the beginning of the experiment, the drying agent weighed 12.105 g; at the end it weighed 12.227 g. What is the mass of the water formed? What is the mass of oxygen that is in this water, according to the given data? Show that these data are consistent with the known formula of water.

14.22 Steam reacts with magnesium metal to give magnesium oxide and hydrogen. In one experiment, when steam was passed through a tube containing magnesium, the mass of the tube contents increased by 0.285 g. The volume of H_2 collected from this experiment was 415 mL at 20°C and 785 mmHg. How many moles of H and O are in the water that has reacted with the magnesium, according to these data? Do these amounts agree with the known formula of water?

Hydrogen

14.23 When treated with excess strong acid, a 0.458-g sample of a metal gives 452 mL of hydrogen collected over water at 791 mmHg and 23°C. What is the metal?

14.25 You are given Zn, HCl(aq), NaOH, and H_2O. Write three different equations for the preparation of H_2 using only these reagents. You can use heat, but not electricity.

14.27 Write an equation for the reaction of H_2 with each of the following: F_2, Fe_3O_4, Ca.

14.29 Show by equations how you could prepare (a) HCHO and (b) W. Start from the following substances: C, O_2, H_2O, WO_3. Assume that the necessary catalysts are available.

14.24 A 0.298-g sample of metal was reacted with an excess of sulfuric acid. The volume of hydrogen released was 133 mL, collected over water at 745 mmHg and 19°C. What is the metal?

14.26 You are given Mg, HCl(aq), $C_2H_6(g)$, and H_2O. Write three different equations for the preparation of H_2 using only these reagents. You can use heat, but not electricity.

14.28 Write an equation for the reaction of H_2 with each of the following: K, CuO, Br_2.

14.30 Starting from air, water, and natural gas, describe (with chemical equations where necessary) how you would prepare (a) NH_3 and (b) HCHO. Assume that the necessary catalysts are available.

Water; Hydrates

14.31 How much heat (in joules) would be obtained when 5.0 g of steam at 100°C are changed to water at 80°C?

14.33 By means of equations, show how you could prepare hydrogen chloride gas from Si, Cl_2, and H_2O.

14.35 A 4.78-g sample of a hydrate of sodium carbonate gives 1.77 g of the anhydrous salt when heated. What is the formula of the hydrate?

14.32 How many grams of steam at 100°C must be condensed to water at 100°C to provide the heat necessary to raise 8.0 g of aluminum from 20°C to 25°C? The specific heat of aluminum is 0.903 J/(g · °C).

14.34 By means of equations, show how you could prepare aqueous sodium hypochlorite, NaOCl(aq), from Na, Cl_2, and H_2O.

14.36 Paper impregnated with cobalt(II) chloride hydrates is used to indicate moist air (pink) or dry air (blue). The pink hydrate $CoCl_2 \cdot 6H_2O$ loses moisture in dry air to give a blue hydrate of cobalt(II) chloride. If 6.83 g $CoCl_2 \cdot 6H_2O$ give 5.80 g of the blue hydrate, what is the formula of this latter compound?

Hydrogen Peroxide

14.37 Hydrogen sulfide is oxidized to free sulfur by acidic hydrogen peroxide solution. Write the balanced equation for the reaction.

14.39 Chromium(III) hydroxide precipitates from solution of chromium(III) salts that are made basic. Hydrogen peroxide dissolves this precipitate in basic solution by oxidizing it to chromate ion. Write the balanced equation for the reaction.

14.41 Cerium(IV) ion is reduced to cerium(III) ion by hydrogen peroxide. A 25.0-mL volume of hydrogen peroxide solution is titrated by 26.1 mL of 0.218 M Ce^{4+} in acidic solution. What is the molar concentration of H_2O_2? If the density of the solution is 1.00 g/mL, what is the mass percentage of H_2O_2 in the solution?

14.38 Arsenic(III) oxide, $As_2O_3(s)$, is oxidized to arsenate ion, AsO_4^{3-}, by hydrogen peroxide in acid solution. Write the balanced equation for the reaction.

14.40 A black precipitate of arsenic(III) sulfide is oxidized in a basic solution of hydrogen peroxide to arsenate ion (AsO_4^{3-}) and sulfate ion. Write the balanced equation for the reaction.

14.42 Dichromate ion is reduced to chromium(III) ion by hydrogen peroxide. A 25.0-mL sample of hydrogen peroxide solution is titrated by 30.5 mL of 0.185 M $K_2Cr_2O_7$ in acidic solution. What is the molar concentration of H_2O_2? If the density of the solution is 1.00 g/mL, what is the mass percentage of H_2O_2 in the solution?

14.43 Hydrazine, N_2H_4, and hydrogen peroxide are used together as a rocket propellant. The reaction is

$$N_2H_4(l) + 2H_2O_2(l) \longrightarrow N_2(g) + 4H_2O(g)$$

Calculate the heat released per gram of total reactants (N_2H_4 plus H_2O_2). The enthalpies of formation of $H_2O_2(l)$, $N_2H_4(l)$, and $H_2O(g)$ are -187.8 kJ/mol, 50.6 kJ/mol, and -241.8 kJ/mol, respectively.

14.44 A rocket propellant consists of methanol as the fuel and hydrogen peroxide as the oxidizer. The reaction is

$$CH_3OH(l) + 3H_2O_2(l) \longrightarrow CO_2(g) + 5H_2O(g)$$

Calculate the heat released per gram of total reactants (CH_3OH plus H_2O_2). The enthalpies of formation of $H_2O_2(l)$, $CH_3OH(l)$, $CO_2(g)$, and $H_2O(g)$ are -187.8 kJ/mol, -238.6 kJ/mol, -393.5 kJ/mol, and -241.8 kJ/mol, respectively.

Natural Water

14.45 Bromine, Br_2, is obtained from sea water. (See Table 14.3.) How much sea water (in liters) must be processed to obtain each kilogram of bromine? Assume 100% recovery. The density of sea water is 1.024 g/mL.

14.46 Magnesium metal is obtained from sea water. (See Table 14.3.) How much sea water (in liters) must be processed to obtain each kilogram of magnesium? Assume 100% recovery. The density of sea water is 1.024 g/mL.

***14.47** What would be the five-day biological oxygen demand (BOD) of the effluent from a paper mill that contains 5.9×10^4 kg of starch in 12×10^6 L of water? Assume that 75% of the starch is oxidized in five days to CO_2 and H_2O. The empirical formula of starch is $C_6H_{12}O_5$.

***14.48** A chemical test is sometimes used to determine the amount of reducing material in water. The result, when expressed in terms of milligrams of O_2 equivalent to the amount of oxidizing agent required to effect the oxidation of reducing matter in a liter of water, is called the chemical oxygen demand (COD). To determine the COD, the water is titrated with potassium dichromate in acid solution. Dichromate ion is reduced to chromium(III) ion. (a) Write a balanced equation for the oxidation of glucose, $C_6H_{12}O_6$, to CO_2 and H_2O by dichromate ion. Write a similar equation in which O_2 is the oxidizing agent. (b) One liter of effluent water contains 1.15 g of glucose. Calculate the amount of potassium dichromate needed to oxidize the glucose to carbon dioxide and water. What is the COD of the water?

14.49 Sodium carbonate is often used to soften hard water. Suppose a sample of water contains 1.0 g of calcium sulfate per liter of water solution. How many grams of sodium carbonate decahydrate (washing soda), $Na_2CO_3 \cdot 10H_2O$, should be added per liter of this water to precipitate the calcium ion as calcium carbonate?

14.50 Some waste water contains 5.0 ppm of phosphorus as phosphate ion. How many grams of calcium oxide must be added to one liter of this water to precipitate the phosphate ion as calcium phosphate?

14.51 Some brine water contains 45 g NaCl per liter of solution. Assume an ideal solution (that is, $i = 2$) and calculate the freezing point. The density of the solution is 1.03 g/mL.

14.52 Some brackish well water contains 1.2 g NaCl per liter. What is the freezing point of this water? Assume an ideal solution ($i = 2$). The density of the water solution is 1.00 g/mL.

14.53 What is the osmotic pressure at 25°C of the solution in Problem 14.51?

14.54 What is the osmotic pressure at 25°C of the solution in Problem 14.52?

14.55 What is the minimum energy (in joules) required to produce pure water from one liter of the water described in Problem 14.51?

14.56 What is the minimum energy (in joules) required to produce pure water from one liter of the water described in Problem 14.52?

Additional Problems

14.57 Complete and balance the following:
(a) $Fe(s) + H_2O(g) \longrightarrow$
(b) $C_3H_8(g) + H_2O(g) \longrightarrow$
(c) $NaH(s) + H_2O(l) \longrightarrow$
(d) $F_2(g) + H_2O(l) \longrightarrow$

14.58 Complete and balance the following:
(a) $Zn(s) + HCl(aq) \longrightarrow$
(b) $Zn(s) + KOH(aq) \longrightarrow$
(c) $Zn(s) + H_2O(g) \longrightarrow$
(d) $BaH_2(s) + H_2O(l) \longrightarrow$

14.59 What mass of calcium hydride is required to give 1.00 L H_2 at 738 mmHg and 22°C?

14.60 What mass of aluminum metal is required to react with strong base to give 1.00 L H_2 at 775 mmHg and 20°C?

14.61 Give equations for the preparation of the following (use H_2 in each preparation):
(a) NH_3 (b) CH_3CH_3 (c) W

14.62 Give an equation for the preparation of each of the following, using H_2:
(a) NaH   (b) CH_3OH (c) Cu

14.63 How much heat is released when 1.00 g of sodium reacts with water? Enthalpies of formation are (in kJ/mol): $H_2O(l)$, -285.8; $Na^+(aq)$, -239.7; $OH^-(aq)$, -229.9. If this heat is all absorbed by 25.0 g of water initially at 20°C, what is the final temperature of the water?

14.64 How much heat is released when 1.00 g of calcium reacts with water? Enthalpies of formation are (in kJ/mol): $H_2O(l)$, -285.8; $Ca^{2+}(aq)$, -543.0; $OH^-(aq)$, -229.9. If this heat is all absorbed by 65.0 g of water initially at 20.0°C, what is the final temperature of the water?

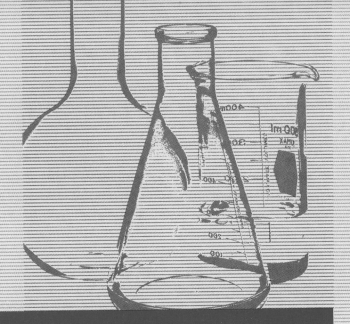

15. Rates of Reaction

Reaction Rates

15.1 Definition of Reaction Rate

15.2 The Experimental Determination of Rate

15.3 Dependence of Rate on Concentration
Reaction Order/ Determining the Rate Law

15.4 Change of Concentration with Time
Concentration–Time Equations/ Half-Life of a
Reaction/ Graphical Plotting

**15.5 Temperature and Rate; Collision and
Transition-State Theories** Collision Theory/
Transition-State Theory/ Potential Energy Diagrams
for Reactions

15.6 Arrhenius Equation

Reaction Mechanisms

15.7 Elementary Reactions Molecularity/ Rate
Equation for an Elementary Reaction

15.8 The Rate Law and the Mechanism Rate-
Determining Step/ Mechanisms with an Initial Fast Step

15.9 Catalysis

C hemical reactions require varying lengths of time for completion, depending on the characteristics of the reactants and products and the conditions under which the reaction is run. Many reactions are over in a fraction of a second, whereas others can take much longer. If we add barium ion to an aqueous solution of sulfate ion, a precipitate of barium sulfate forms almost immediately. On the other hand, the reactions that occur in a cement mixture as it hardens to concrete require years for completion.

The study of the rate, or speed, of a reaction has important applications. In the manufacture of ammonia from nitrogen and hydrogen, we may wish to know what conditions will help the reaction to proceed in a commercially feasible length of time. Or, we may wish to know if the nitric oxide, NO, in the exhaust gases of supersonic transports will destroy the ozone in the stratosphere faster than the ozone is produced. Answering these questions requires knowledge of the rates of reactions.

A further reason for studying reaction rates is to understand how chemical reactions occur. By noting how the rate of a reaction is affected by changing conditions, we can sometimes learn the details of what is happening at the molecular level.

A reaction whose rate has been extensively studied under various conditions is the decomposition of dinitrogen pentoxide, N_2O_5. When this substance is heated in the gas phase, it decomposes to nitrogen dioxide and oxygen:

$$2N_2O_5(g) \longrightarrow 4NO_2(g) + O_2(g)$$

We will look at this reaction in some detail. Questions that we will address are: How is the rate of a reaction like this measured? What are the conditions that affect the rate of a reaction? How do we express the relationship of rate of a reaction to the variables that affect rate? What happens at the molecular level when N_2O_5 decomposes to NO_2 and O_2?

Chapter Overview

After defining what we mean by *reaction rates,* we will briefly describe their experimental determination. Then, we will look at the *variables affecting reaction rates:* reactant and catalyst concentrations and temperature. In the second part of the chapter, we will see how the dependence of reaction rates on concentrations can be explained in terms of the *reaction mechanism,* that is, in terms of the molecular steps in the overall reaction. The final section explains *catalysis* in terms of the reaction mechanism.

Reaction Rates

Chemical kinetics is the study of how reaction rates change under varying conditions and what molecular events occur during the overall reaction. In the first part of this chapter, we will look at reaction rates and the variables that affect them.

What are the variables that affect reaction rates? As we noted in the chapter opening, the rate depends on the characteristics of the reactants in a particular reaction. Some reactions are fast, and others are slow. But the rate of any given reaction is affected by the following factors:

1. *Concentrations of reactants.* Often, the rate of reaction increases when the concentration of a reactant is increased. A piece of steel wool burns with some difficulty in air (20% O_2), but bursts into a dazzling white flame in pure oxygen. The rate of burning has increased by increasing the concentration of O_2. In some reactions, however, the rate is unaffected by the concentration of a particular reactant, so long as it is present at some concentration.

2. *Concentration of catalyst, if one is present.* A **catalyst** is a substance that affects the rate of reaction (speeding it up) without being consumed in the overall reaction. Because it is not consumed by the reaction, it does not appear in the balanced chemical equation (although its presence may be indicated by writing its formula over the arrow). A pure solution of hydrogen peroxide, H_2O_2, is stable, but if hydrobromic acid, HBr(aq), is added, H_2O_2 decomposes rapidly into H_2O and O_2:

$$2H_2O_2(aq) \xrightarrow{\text{HBr}(aq)} 2H_2O(l) + O_2(g)$$

Here, HBr acts as a catalyst to speed decomposition, and after the reaction, all of it can be recovered.

3. *Temperature at which the reaction occurs.* Usually, reactions are speeded up when the temperature increases. It takes less time to boil an egg at sea level than it does on a mountain top, where water boils at a lower temperature. Reactions during cooking go faster at higher temperature.

4. *Surface area of a solid reactant or catalyst.* If a reaction involves a solid with a gas or liquid, the surface area of the solid affects the reaction rate. Because the reaction occurs at the surface of the solid, the rate increases with increasing surface area. A wood fire burns faster if the logs are chopped into smaller pieces. Similarly, the surface area of a solid catalyst is important to determining the rate of reaction. The greater the surface area, the faster the reaction.

In the first sections of this chapter, we will look primarily at reactions in the gas phase and in solution, where factors 1 to 3 are important. Thus, we will look at the effect of *concentrations* and *temperature* on reaction rates. Before we can explore these factors, we need a precise definition of reaction rate.

15.1 Definition of Reaction Rate

The rate of a reaction is the amount of product formed or the amount of reactant used up per unit of time. So that our rate calculations do not depend on the total quantity of reaction mixture used, we express the rate for a unit volume of the mixture. Therefore, the **reaction rate** is the increase in molar concentration of product per unit time, or the decrease in molar concentration of reactant per unit time. The usual unit of reaction rate is moles per liter per second, mol/(L · s).

Consider the gas-phase reaction we discussed in the chapter opening:

$$2N_2O_5(g) \longrightarrow 4NO_2(g) + O_2(g)$$

The rate for this reaction could be found by observing the increase in molar concentration of O_2 produced. We will denote the molar concentration of a substance by enclosing the formula of the substance in square brackets. Thus, $[O_2]$ is the molar concentration of O_2. In a given time interval, Δt, the

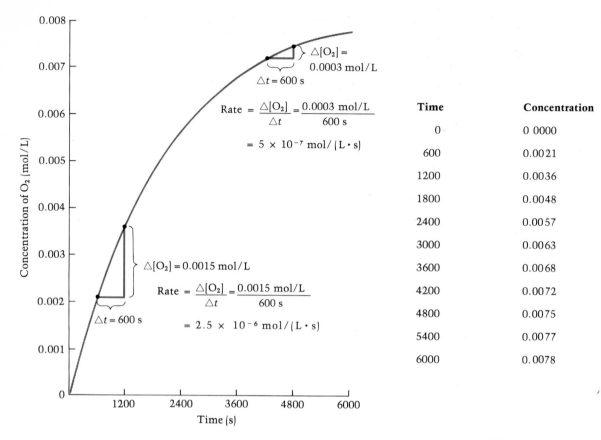

Time	Concentration
0	0 0000
600	0.0021
1200	0.0036
1800	0.0048
2400	0.0057
3000	0.0063
3600	0.0068
4200	0.0072
4800	0.0075
5400	0.0077
6000	0.0078

molar concentration of oxygen, $[O_2]$, in the reaction vessel increases by the amount $\Delta[O_2]$. Recall that the symbol Δ means "change in" and is obtained by subtracting the initial value from the final value. The reaction rate is given by

$$\text{Rate of formation of oxygen} = \frac{\Delta[O_2]}{\Delta t}$$

This equation gives the *average* rate over the time interval Δt. If the time interval is very short, the equation gives the *instantaneous* rate, that is, the rate at a particular instant of time.■

To understand the difference between average rate and instantaneous rate, it may help to think of the speed of an automobile. Speed can be defined as the rate of change of distance, D; that is, speed equals $\Delta D / \Delta t$, where D is the distance traveled. If an automobile travels 84 miles in 2.0 hours, the average speed over this time interval is 84 mi/2.0 hr = 42 mi/hr. However, at any instant during this interval, the speedometer, which registers instantaneous speed, may read more or less than 42 mi/hr. At some moment on the highway, it may read 55 mi/hr, whereas on a congested city street it may read only 20 mi/hr. The quantity $\Delta D / \Delta t$ becomes more nearly an instantaneous speed as the time interval, Δt, is made smaller.

Figure 15.1 shows the increase in concentration of O_2 during the decom-

Figure 15.1
Increase in concentration of O_2 during the decomposition of N_2O_5. The average rate is calculated for two different time intervals. When the time changes from 600 s to 1200 s, the average rate is 2.5×10^{-6} mol/(L · s). Later, when the time changes from 4200 s to 4800 s, the average rate has slowed to 5×10^{-7} mol/(L · s).

■ In calculus, the rate at a given moment (the instantaneous rate) is given by the derivative $d[O_2]/dt$.

position of N_2O_5. It shows the calculation of average rates at two positions on the curve. For example, when the time changes from 600 s to 1200 s ($\Delta t = 600$ s), the O_2 concentration increases by 0.0015 mol/L ($= \Delta[O_2]$). Therefore, the average rate $= \Delta[O_2]/\Delta t = (0.0015 \text{ mol/L})/600 \text{ s} = 2.5 \times 10^{-6}$ mol/(L · s). Later, during the time interval from 4200 s to 4800 s, the average rate is 5×10^{-7} mol/(L · s). Note that the rate decreases as the reaction proceeds.

Because the amounts of products and reactants are related by stoichiometry, any substance in the reaction may be used to express the rate of reaction. In the case of the decomposition of N_2O_5 to NO_2 and O_2, we gave the rate in terms of the rate of formation of oxygen, $\Delta[O_2]/\Delta t$. However, we can also express it in terms of the rate of decomposition of N_2O_5:

$$\text{Rate of decomposition of } N_2O_5 = -\frac{\Delta[N_2O_5]}{\Delta t}$$

Note the negative sign. It always occurs in a rate expression for a reactant in order to indicate a decrease in concentration and to give a positive value for the rate. Thus, since $[N_2O_5]$ decreases, $\Delta[N_2O_5]$ is negative, and $-\Delta[N_2O_5]/\Delta t$ is positive.

The rate of decomposition of N_2O_5 and the rate of formation of oxygen are easily related. Since two moles of N_2O_5 decompose for each mole of oxygen formed, the rate of decomposition of N_2O_5 is twice as fast as the rate of formation of oxygen. To equate the rates, we must divide the rate of decomposition of N_2O_5 by 2 (its coefficient in the balanced chemical equation). We get

$$\frac{\Delta[O_2]}{\Delta t} = -\frac{1}{2}\frac{\Delta[N_2O_5]}{\Delta t}$$

Example 15.1

Consider the reaction of nitrogen dioxide with fluorine to give nitryl fluoride, NO_2F:

$$2NO_2(g) + F_2(g) \longrightarrow 2NO_2F(g)$$

How is the rate of formation of NO_2F related to the rate of reaction of fluorine?

Solution

We have

$$\text{Rate of formation of } NO_2F = \frac{\Delta[NO_2F]}{\Delta t}$$

and

$$\text{Rate of reaction of } F_2 = -\frac{\Delta[F_2]}{\Delta t}$$

We divide each rate by the coefficient of the corresponding substance in the chemical equation, then equate them:

$$\frac{1}{2}\frac{\Delta[NO_2F]}{\Delta t} = -\frac{\Delta[F_2]}{\Delta t}$$

Exercise 15.1

For the reaction given in Example 15.1, how is the rate of formation of NO_2F related to the rate of reaction of NO_2?

(See Problems 15.19 and 15.20.)

15.2 The Experimental Determination of Rate

To obtain the rate of a reaction, we must determine the concentration of reactants or products during the course of the reaction. One way to do this for a slow reaction is to withdraw samples from the reaction vessel at various times and analyze them. The rate of the reaction of ethyl acetate with water in acidic solution was one of the first to be determined this way.■

$$\underset{\text{ethyl acetate}}{CH_3CH_2O\overset{\displaystyle O}{\overset{\|}{C}}CH_3} + H_2O \xrightarrow{H^+} \underset{\text{ethanol}}{CH_3CH_2OH} + \underset{\text{acetic acid}}{HO\overset{\displaystyle O}{\overset{\|}{C}}CH_3}$$

■ Ethyl acetate is a liquid with a fruity odor and belongs to a class of organic (carbon-containing) compounds called esters. Unlike a salt, such as sodium acetate, an ester is a covalent compound.

This reaction is slow, and the acetic acid that is produced is easily measured by titration.

More convenient are techniques that can continuously follow the progress of a reaction by observing the change in some physical property of the system. These physical methods are often adaptable to fast reactions as well as to slow ones. If a gas reaction, for example, involves a change in the number of molecules, the pressure of the system will change when the volume and temperature are held constant. This pressure can be related to the reaction rate. The decomposition of phosgene, $COCl_2$, has been studied this way.

$$COCl_2(g) \longrightarrow CO(g) + Cl_2(g)$$

If one of the reactants or products has a color, this may be used to follow the rate. Thus, in the reaction

$$ClO^-(aq) + I^-(aq) \longrightarrow IO^-(aq) + Cl^-(aq)$$

hypoiodite ion, IO^-, absorbs at the blue end of the spectrum near 400 nm.

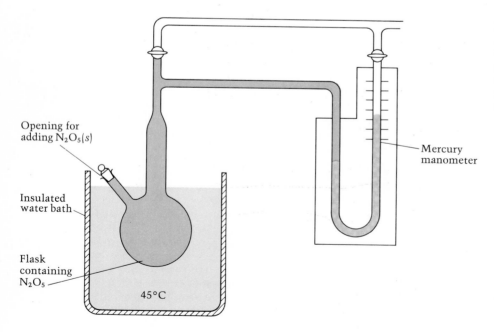

Opening for adding $N_2O_5(s)$

Insulated water bath

Flask containing N_2O_5

45°C

Mercury manometer

Figure 15.2
An experiment to follow the change of pressure during the decomposition of N_2O_5 at 45°C. Pressure values can be related to the concentration of N_2O_5 in the flask.

Time (seconds)	[N$_2$O$_5$] (mol/L)	Rate (mol/(L · s))	Table 15.1 Kinetic Data for the Decomposition of Dinitrogen Pentoxide at 45°C
0	1.65×10^{-2}		
		6.8×10^{-6}	
600	1.24×10^{-2}		
		5.2×10^{-6}	
1200	0.93×10^{-2}		
		3.7×10^{-6}	
1800	0.71×10^{-2}		
		3.0×10^{-6}	
2400	0.53×10^{-2}		
		2.3×10^{-6}	
3000	0.39×10^{-2}		
		1.7×10^{-6}	
3600	0.29×10^{-2}		

The intensity of this absorption is proportional to $[IO^-]$ and can be used to determine the reaction rate.

A reaction that has been well studied is the decomposition of dinitrogen pentoxide in the gas phase, as mentioned in the chapter opening:

$$2N_2O_5(g) \longrightarrow 4NO_2(g) + O_2(g)$$

Dinitrogen pentoxide crystals are sealed in a vessel equipped with a manometer (a pressure-measuring device) (Figure 15.2). The vessel is then plunged into a water bath at 45°C, at which temperature the solid vaporizes and the gas decomposes into NO_2 and O_2. Manometer readings are taken at regular time intervals, and the pressure values are converted to partial pressures or concentrations of N_2O_5. Typical data from such an experiment are given in the first two columns of Table 15.1.

From the data given in Table 15.1, we can calculate the average rate of reaction during each time interval. At time = 600 s, the concentration of N_2O_5 is 1.24×10^{-2} mol/L; at 1200 s the concentration is 0.93×10^{-2} mol/L. In the time interval from 600 s to 1200 s, the concentration of N_2O_5 has changed by

$$\Delta[N_2O_5] = (0.93 - 1.24) \times 10^{-2} \text{ mol/L}$$
$$= -0.31 \times 10^{-2} \text{ mol/L}$$

Note that we calculate $\Delta[N_2O_5]$ by subtracting the initial concentration from the final concentration. (The negative sign indicates a decrease in concentration.) The rate of the reaction during this time interval is obtained by dividing $-\Delta[N_2O_5]$ by the time interval, $\Delta t = (1200 - 600)$ s $= 600$ s:

$$\text{Rate} = -\frac{\Delta[N_2O_5]}{\Delta t}$$
$$= -\frac{-0.31 \times 10^{-2} \text{ mol/L}}{600 \text{ s}}$$
$$= 5.2 \times 10^{-6} \text{ mol/(L · s)}$$

The rate of the reaction during each similar time interval has been calculated in this way and is given in the third column of Table 15.1. Note how the rate slows as the reaction proceeds.

Exercise 15.2

Calculate the rate of production of oxygen from the reaction

$$2N_2O_5(g) \longrightarrow 4NO_2(g) + O_2(g)$$

during the time interval from $t = 600$ s to $t = 1200$ s. Use the following data:

Time	$[O_2]$
600 s	0.26×10^{-2} mol/L
1200 s	0.41×10^{-2} mol/L

(See Problems 15.21 and 15.22.)

15.3 Dependence of Rate on Concentration

Experimentally, it has been found that a reaction rate depends on the concentrations of certain reactants (and the concentration of catalyst, if there is one). Consider the reaction of nitrogen dioxide with fluorine to give nitryl fluoride, NO_2F:

$$2NO_2(g) + F_2(g) \longrightarrow 2NO_2F(g)$$

The rate of this reaction is observed to be proportional to the concentration of nitrogen dioxide. If the concentration of nitrogen dioxide is doubled, the rate doubles. The rate is also proportional to the concentration of fluorine, so that doubling the concentration of fluorine doubles the rate.

A **rate law** relates the rate of a reaction to the concentrations of reactants (and catalyst) raised to various powers. The following equation is the rate law for the above reaction:

$$\text{Rate} = k[NO_2][F_2]$$

Here k, called the **rate constant,** is a proportionality constant in the relationship between rate and concentrations. It has a fixed value at a given temperature, but will vary with temperature. Note that in this rate law both reactant concentrations have an exponent of 1.

As a more general example, consider the reaction of substances A and B to give D and E, according to the balanced equation

$$a\text{A} + b\text{B} \xrightarrow{\text{C}} d\text{D} + e\text{E} \qquad \text{C} = \text{catalyst}$$

We could write the rate law in the form

$$\text{Rate} = k[\text{A}]^q[\text{B}]^r[\text{C}]^s$$

The exponents q, r, and s are frequently, but not always, integers. *They must be determined experimentally*, and cannot be obtained simply by looking at the balanced equation. For example, note that the exponents in the equation Rate $= k[NO_2][F_2]$ have no relationship to the coefficients in the balanced equation $2NO_2 + F_2 \longrightarrow 2NO_2F$.

Once we know the rate law for a reaction and have found the value of the rate constant, we can calculate the rate of a reaction for any values of reactant concentrations. As we will see later, knowledge of the rate law is also useful in understanding how the reaction occurs at the molecular level.

Reaction Order

We can classify a reaction by its orders. The **reaction order** with respect to a given reactant species equals the exponent of the concentration of that species in the rate law as determined experimentally. For the reaction of NO_2 with F_2 to give NO_2F, the reaction is first order with respect to the NO_2 because the exponent of $[NO_2]$ in the rate law is 1. Similarly, the reaction is first order with respect to F_2.

The *overall order* of a reaction equals the sum of the exponents in the rate law. In this example, the overall order is 2; that is, the reaction is second order overall.

Reactions display a variety of reaction orders. Some examples are listed below.

1. Cyclopropane, C_3H_6, has the molecular structure

When heated, the carbon ring (the triangle) opens up to give propylene, $CH_2{=}CHCH_3$. Since the compounds are isomers (different compounds with the same molecular formula), the reaction is called an *isomerization*.

$$C_3H_6(g) \longrightarrow CH_2{=}CHCH_3(g)$$
cyclopropane   propylene

It has the rate law

$$\text{Rate} = k[C_3H_6]$$

The reaction is first order in cyclopropane and first order overall.

2. Nitric oxide, NO, reacts with hydrogen according to the equation

$$2NO(g) + 2H_2(g) \longrightarrow N_2(g) + 2H_2O(g)$$

The experimentally determined rate law is

$$\text{Rate} = k[NO]^2[H_2]$$

Thus, the reaction is second order in NO, first order in H_2, and third order overall.

3. Acetone, CH_3COCH_3, reacts with iodine in acidic solution:

$$CH_3COCH_3(aq) + I_2(aq) \xrightarrow{\text{H}^+} CH_3COCH_2I(aq) + HI(aq)$$

The experimentally determined rate law is

$$\text{Rate} = k[CH_3COCH_3][H^+]$$

The reaction is first order in acetone. It is zero order in iodine, that is, the rate law contains the factor $[I_2]^0 = 1$. Therefore, the rate does not depend on the concentration of I_2, as long as some concentration of I_2 is present. Note that the reaction is first order in the catalyst, H^+. Thus, the overall order is 2.

Although reaction orders frequently have whole number values (particularly 1 or 2), they can be fractional. Negative orders are also possible.

Exercise 15.3

What are the reaction orders with respect to each reactant species for the following reaction?

$$NO_2(g) + CO(g) \longrightarrow NO(g) + CO_2(g)$$

The rate law is

$$Rate = k[NO_2]^2$$

What is the overall order?

(See Problems 15.23 and 15.24.)

Determining the Rate Law

The experimental determination of the rate law for a reaction requires that we find the order of the reaction with respect to each reactant and any catalyst. The *initial-rate method* is a simple way to obtain reaction orders. It consists of doing a series of experiments in which the initial, or starting, concentrations of reactants are varied. Then the initial rates are compared, from which the reaction orders can be deduced.

To see how this method works, again consider the reaction mentioned in the chapter opening:

$$2N_2O_5(g) \longrightarrow 4NO_2(g) + O_2(g)$$

We observe this reaction in two experiments. In Experiment 2, the initial concentration of N_2O_5 is twice that of Experiment 1. We then note the initial rate of disappearance of N_2O_5 in each case. The initial concentrations and corresponding initial rates for two experiments are as follows:■

■ In Figure 15.1, the initial rate corresponds to the slope of the tangent to the curve at $t = 0$.

	Initial N_2O_5 Concentration	Initial Rate
Experiment 1	1.0×10^{-2} mol/L	4.8×10^{-6} mol/(L · s)
Experiment 2	2.0×10^{-2} mol/L	9.6×10^{-6} mol/(L · s)

The rate law for this reaction will have the concentration of reactant raised to a power m:

$$Rate = k[N_2O_5]^m$$

The value of m (the reaction order) must be determined from the experimental data. Note that if the N_2O_5 concentration is doubled, we get a new rate, Rate′, given by the following equation:

$$Rate' = k(2[N_2O_5])^m = 2^m k[N_2O_5]^m$$

This rate is 2^m times the original rate.

We can now see how the rate is affected when the concentration is doubled for various choices of m. Suppose $m = 2$. We get $2^m = 2^2 = 4$. That is, if the initial concentration is doubled, the rate is multiplied by 4. We summarize the results for various choices of m:

m	If the initial concentration is doubled, the rate is multiplied by:
-1	$\frac{1}{2}$
0	1
1	2
2	4

Let us divide the initial rate from Experiment 2 by the initial rate from Experiment 1:

$$\frac{9.6 \times 10^{-6}\, \text{mol/(L} \cdot \text{s)}}{4.8 \times 10^{-6}\, \text{mol/(L} \cdot \text{s)}} = 2$$

In other words, when the N_2O_5 concentration is doubled, the rate is doubled. We see that this corresponds to the result for $m = 1$. The rate law must have the form

$$\text{Rate} = k[N_2O_5]$$

We can determine the value of the rate constant, k, by substituting values of the rate and N_2O_5 concentration from any one of the experiments into the rate law. Using values from Experiment 2, we get

$$9.6 \times 10^{-6}\, \text{mol/(L} \cdot \text{s)} = k \times 2.0 \times 10^{-2}\, \text{mol/L}$$

Hence,

$$k = \frac{9.6 \times 10^{-6}/\text{s}}{2.0 \times 10^{-2}} = 4.8 \times 10^{-4}/\text{s}$$

Example 15.2

Iodide ion is oxidized to hypoiodite ion, IO^-, by hypochlorite ion, ClO^-, in basic solution. The equation is

$$I^-(aq) + ClO^-(aq) \xrightarrow{OH^-} IO^-(aq) + Cl^-(aq)$$

In order to use the initial-rate method to obtain the rate law for this reaction, the following experiments were run:

	Initial Concentrations (mol/L)			Initial Rate (mol/(L · s))
	I^-	ClO^-	OH^-	
Experiment 1	0.010	0.010	0.010	6.1×10^{-4}
Experiment 2	0.020	0.010	0.010	12.2×10^{-4}
Experiment 3	0.010	0.020	0.010	12.2×10^{-4}
Experiment 4	0.010	0.010	0.020	3.0×10^{-4}

Obtain the reaction orders with respect to I^-, ClO^-, and OH^-. Then find the rate law and the rate constant.

Solution

By comparing Experiment 1 and Experiment 2, we can see that when the iodide-ion concentration is doubled, with the concentrations of other species kept constant, the rate is doubled. Thus, the reaction is first order in I^-. Similarly, in a comparison of Experiment 1 and Experiment 3 (in which the hypochlorite ion concentration is doubled, with other concentrations fixed), the rate is seen to double. The reaction is first order in ClO^-. Finally, comparing Experiment 1 and Experiment 4, we see that when the hydroxide ion concentration is doubled (but other concentrations are con-

(Continued)

504

stant), the rate is halved. That is, $2^m = \frac{1}{2}$. Hence, $m = -1$. The rate is inversely proportional to the hydroxide ion concentration. The rate law is

$$\text{Rate} = k\frac{[I^-][ClO^-]}{[OH^-]}$$

We can calculate the rate constant by substituting

values from any one of the experiments. Using Experiment 1, we get

$$6.1 \times 10^{-4}\,\frac{\text{mol}}{\text{L} \cdot \text{s}} = k \times \frac{0.010\,\text{mol/L} \times 0.010\,\text{mol/L}}{0.010\,\text{mol/L}}$$

Therefore, $k = \dfrac{6.1 \times 10^{-4}/\text{s}}{0.010} = 6.1 \times 10^{-2}/\text{s}$

Exercise 15.4

The initial-rate method was applied to the decomposition of nitrogen dioxide.

$$2NO_2(g) \longrightarrow 2NO(g) + O_2(g)$$

It gave the following results:

	Initial NO$_2$ Concentration	Initial Rate of Formation of O$_2$
Experiment 1	0.010 mol/L	7.1×10^{-5} mol/(L · s)
Experiment 2	0.020 mol/L	28×10^{-5} mol/(L · s)

Find the rate law and the value of the rate constant.

(See Problems 15.25, 15.26, 15.27, and 15.28.)

15.4 Change of Concentration with Time

A rate law tells us how the rate of a reaction depends on reactant concentrations at a particular moment. But often we would like to have a mathematical relationship showing how a reactant concentration changes over a period of time. Such an equation would be directly comparable to the experimental data, which are usually obtained as concentrations at various times. In addition to summarizing the experimental data, this equation would predict concentrations for all times. Using it, we could answer questions such as: How long does it take for this reaction to be 50% complete? to be 90% complete?

Moreover, as we will see, knowing exactly how the concentrations change with time for different rate laws suggests ways for graphical plotting of the experimental data. Graphical plotting provides an alternative to the initial-rate method for determining the rate law.

A rate law can be transformed to a mathematical relationship between concentration and time using calculus. Since we will work only with the final equations, we need not go into the derivation here. We will look in some depth at first-order reactions and briefly at second-order reactions.

Concentration–Time Equations

Let us first look at first-order rate laws. The decomposition of dinitrogen pentoxide has the following rate law:

$$\text{Rate} = -\frac{\Delta[N_2O_5]}{\Delta t} = k[N_2O_5]$$

Using calculus, one can show that such a first-order rate law leads to the following relationship between N_2O_5 concentration and time:

$$\log \frac{[N_2O_5]_0}{[N_2O_5]_t} = \frac{kt}{2.303}$$

Here $[N_2O_5]_t$ is the concentration at time t, and $[N_2O_5]_0$ is the initial concentration of N_2O_5 (that is, the concentration at $t = 0$). The symbol "log" denotes the logarithm to the base 10. ■

This equation allows us to calculate the concentration of N_2O_5 at any time, once we are given the initial concentration and the rate constant. Also, we can find the time it takes for the N_2O_5 concentration to decrease to a particular value.

More generally, let A be a substance that reacts to give products according to the equation

$$a\,A \longrightarrow \text{products}$$

where a is the stoichiometric coefficient of reactant A. Suppose that this reaction has a first-order rate law

$$\text{Rate} = -\frac{\Delta[A]}{\Delta t} = k[A]$$

Using calculus, we get the following equation: ■

$$\log \frac{[A]_0}{[A]_t} = \frac{kt}{2.303}$$

Here $[A]_t$ is the concentration of reactant A at time t, and $[A]_0$ is the initial concentration. The next example illustrates how we work with this equation.

■ See Appendix A for a discussion of logarithms.

■ The derivation using calculus is as follows. The rate law

$$\frac{-d[A]}{dt} = k[A]$$

is rearranged to give

$$\frac{-d[A]}{[A]} = k\,dt$$

Integrating this from time $= 0$ to time $= t$,

$$-\int_{[A]_0}^{[A]_t} \frac{d[A]}{[A]} = k \int_0^t dt$$

gives

$$-\{\ln[A]_t - \ln[A]_0\} = k(t - 0)$$

This is easily identified with the equation in the text, if we note that the natural logarithm, denoted ln, equals 2.303 times the logarithm to the base 10.

Example 15.3

The decomposition of N_2O_5 to NO_2 and O_2 is first order, with a rate constant of 4.80×10^{-4}/s at 45°C. (a) If the initial concentration is 1.65×10^{-2} mol/L, what is the concentration after 825 s? (b) How long would it take for the concentration of N_2O_5 to decrease to 1.00×10^{-2} mol/L from its initial value, given in (a)?

Solution

(a) We substitute into the equation

$$\log \frac{[N_2O_5]_0}{[N_2O_5]_t} = \frac{kt}{2.303}$$

to get

$$\log \frac{1.65 \times 10^{-2} \text{ mol/L}}{[N_2O_5]_t}$$

$$= \frac{4.80 \times 10^{-4}/\text{s} \times 825 \text{ s}}{2.303} = 0.172$$

In order to solve for $[N_2O_5]_t$, we take the antilogarithm

of both sides. This removes the log from the left and gives antilog 0.172, or $10^{0.172}$, on the right, which equals 1.49:

$$\frac{1.65 \times 10^{-2} \text{ mol/L}}{[N_2O_5]_t} = 1.49$$

Hence,

$$[N_2O_5]_t = \frac{1.65 \times 10^{-2} \text{ mol/L}}{1.49} = 0.0111 \text{ mol/L}$$

(b) We substitute into the equation relating $[N_2O_5]$ to the time:

$$\log \frac{1.65 \times 10^{-2} \text{ mol/L}}{1.00 \times 10^{-2} \text{ mol/L}} = \frac{4.80 \times 10^{-4}/\text{s} \times t}{2.303}$$

The left side equals log 1.65 = 0.217. The right side equals 2.08×10^{-4}/s $\times t$. Hence,

$$0.217 = 2.08 \times 10^{-4}/\text{s} \times t$$

Or, $\quad t = \dfrac{0.217}{2.08 \times 10^{-4}/\text{s}} = 1.04 \times 10^3 \text{ s } (17.4 \text{ min})$

Exercise 15.5

(a) What would be the concentration of dinitrogen pentoxide in the experiment described in Example 15.3 after 6.00×10^2 s? Compare with the value in Table 15.1. (b) How long would it take for the concentration of N_2O_5 to decrease to 10.0% of its initial value?

(See Problems 15.29 and 15.30.)

Consider the reaction

$$a\,A \longrightarrow \text{products}$$

which has the second-order rate law

$$\text{Rate} = -\frac{\Delta[A]}{\Delta t} = k[A]^2$$

An example is the decomposition of nitrogen dioxide at moderately high temperatures (300°C to 400°C):

$$2NO_2(g) \longrightarrow 2NO(g) + O_2(g)$$

Using calculus, we can obtain the following relationship between the concentration of A and the time:

$$\frac{1}{[A]_t} = kt + \frac{1}{[A]_0}$$

Using this equation, we can calculate the concentration of NO_2 at any time during its decomposition if we know the rate constant and the initial concentration. At 330°C, the rate constant for the decomposition of NO_2 is 0.775 L/(mol · s). Suppose the initial concentration is 0.0030 mol/L. What is the concentration of NO_2 after 645 s? By substituting into the previous equation, we get

$$\frac{1}{[NO_2]_t} = 0.775 \text{ L/(mol · s)} \times 645 \text{ s} + \frac{1}{0.0030 \text{ mol/L}} = 8.3 \times 10^2 \text{ L/mol}$$

If we take the inverse of both sides of this equation, we find that $[NO_2]_t = 0.0012$ mol/L. Thus, after 645 s, the concentration of NO_2 decreased from 0.0030 mol/L to 0.0012 mol/L.

Half-Life of a Reaction

As a reaction proceeds, the concentration of a reactant decreases, since it is being consumed. The **half-life,** $t_{1/2}$, of a reaction is the time it takes for the reactant concentration to decrease by one-half of its initial value.■

For a first-order reaction, such as the decomposition of dinitrogen pentoxide, the half-life is independent of the initial concentration. To see this, let us substitute into the equation

$$\log \frac{[N_2O_5]_0}{[N_2O_5]_t} = \frac{kt}{2.303}$$

In one half-life, the N_2O_5 concentration decreases by one-half, from its initial value, $[N_2O_5]_0$, to $[N_2O_5]_t = \frac{1}{2}[N_2O_5]_0$. After substituting, the equation

■ The concept of half-life is also used to characterize a radioactive nucleus, whose radioactive decay is a first-order process. This is discussed in Chapter 22.

becomes

$$\log \frac{[N_2O_5]_0}{\frac{1}{2}[N_2O_5]_0} = \frac{kt_{1/2}}{2.303}$$

The expression on the left equals log 2 = 0.301. Hence,

$$0.301 = \frac{kt_{1/2}}{2.303}$$

Solving for the half-life, $t_{1/2}$, we get

$$t_{1/2} = \frac{0.301 \times 2.303}{k} = \frac{0.693}{k}$$

Since the rate constant for the decomposition of N_2O_5 at 45°C is 4.80×10^{-4}/s, the half-life is

$$t_{1/2} = \frac{0.693}{4.80 \times 10^{-4}/s} = 1.44 \times 10^3 \text{ s}$$

Thus, the half-life is 1.44×10^3 s, or 24.0 min.

We see that the half-life of N_2O_5 does not depend on the initial concentration of N_2O_5. This means that the half-life is the same at any time during the reaction. If the initial concentration is 0.0165 mol/L, after one half-life (24.0 min) the concentration decreases to $\frac{1}{2} \times$ 0.0165 mol/L = 0.0083 mol/L. After another half-life (another 24.0 min), the N_2O_5 concentration decreases to $\frac{1}{2} \times$ 0.0083 mol/L = 0.0041 mol/L. Every time one half-life passes, the N_2O_5 concentration decreases by one-half again (see Figure 15.3).

The foregoing result for the half-life for the first order decomposition of N_2O_5 is perfectly general. That is, for the general first-order rate law

$$\text{Rate} = -\frac{\Delta[A]}{\Delta t} = k[A]$$

the half-life is related to the rate constant, but it is independent of the concentration of A.

$$t_{1/2} = \frac{0.693}{k}$$

Exercise 15.6

The isomerization of cyclopropane, C_3H_6, to propylene, $CH_2{=}CHCH_3$, is first order in cyclopropane and first order overall. At 1000°C, the rate constant is 9.2/s. What is the half-life of cyclopropane at 1000°C? How long would it take for the concentration of cyclopropane to decrease to 50% of its initial value? to 25% of its initial value?
(See Problems 15.31 and 15.32.)

It can be shown by reasoning similar to that given previously that the half-life of a second-order rate law, Rate = $k[A]^2$, is $1/(k[A]_0)$. In this case, the half-life depends on initial concentration and becomes larger as time goes on. Consider the decomposition of NO_2 at 330°C. It takes 430 s for the concentration to decrease by one-half from 0.0030 mol/L to 0.0015 mol/L. However, it takes 860 s (twice as long) for the concentration to decrease by one-half again. The fact that the half-life changes with time is evidence that the reaction is not first order.

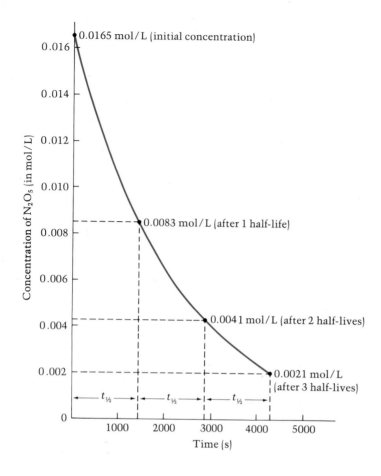

Figure 15.3
A graph illustrating that the half-life of a first-order reaction is independent of initial concentration. In one half-life (1440 s, or 24.0 min), the concentration decreases by one-half, from 0.0165 mol/L to $\frac{1}{2}$ × 0.0165 mol/L = 0.0083 mol/L. After each succeeding half-life, the concentration decreases by one-half again, from 0.0083 mol/L to $\frac{1}{2}$ × 0.0083 mol/L = 0.0041 mol/L, then to $\frac{1}{2}$ × 0.0041 mol/L = 0.0021 mol/L.

Graphical Plotting

We saw earlier that the order of a reaction can be determined by comparing initial rates for several experiments using different initial concentrations (initial-rate method). It is also possible to determine the order of a reaction by graphical plotting of the data for a particular experiment. The experimental data are plotted in several different ways, first assuming a first-order reaction, then a second-order reaction, and so forth. If, for example, the reaction turns out to be first order, the experimental data will best fit this type of plot. To illustrate, we will look at how the plotting should be done for first-order and second-order reactions.

We have seen that the first-order rate law, $-\Delta[A]/\Delta t = k[A]$, gives the following relationship between concentration of A and time:

$$\log \frac{[A]_0}{[A]_t} = \frac{kt}{2.303}$$

This equation can be rewritten in a slightly different form, which we can identify with the equation of a straight line. Using the property of logarithms that $\log (A/B) = \log A - \log B$, we get

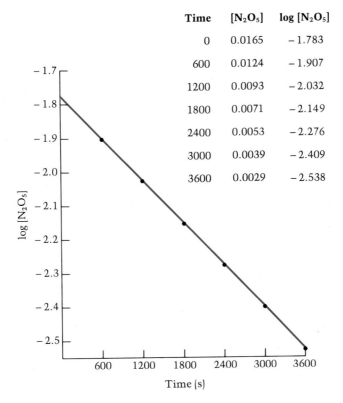

Time	$[N_2O_5]$	$\log [N_2O_5]$
0	0.0165	-1.783
600	0.0124	-1.907
1200	0.0093	-2.032
1800	0.0071	-2.149
2400	0.0053	-2.276
3000	0.0039	-2.409
3600	0.0029	-2.538

Figure 15.4
A plot of $\log [N_2O_5]$ versus time. A straight line can be drawn through the experimental points (black dots). The fact that the straight line fits the experimental data so well confirms that the rate law is first order.

$$\log [A]_t = \left(\frac{-k}{2.303}\right)t + \log [A]_0$$

A straight line has the mathematical form $y = ax + b$, when y is plotted on the vertical axis against x on the horizontal axis. ■ Let us now make the following identifications:

$$\underbrace{\log [A]_t}_{y} = \underbrace{\left(\frac{-k}{2.303}\right)t}_{ax} + \underbrace{\log [A]_0}_{b}$$

■ See Appendix A for a discussion of the mathematics of a straight line.

This means that if we plot $\log [A]_t$ on the vertical axis against the time, t, on the horizontal axis, we will get a straight line. Figure 15.4 shows a plot of $\log [N_2O_5]$ at various times during the decomposition reaction. The fact that the points lie on a straight line is confirmation that the rate law is first order.

The second-order rate law, $-\Delta[A]/\Delta t = k[A]^2$, gives the following relationship between concentration of A and time:

$$\underbrace{\frac{1}{[A]_t}}_{y} = \underbrace{kt}_{ax} + \underbrace{\frac{1}{[A]_0}}_{b}$$

In this case, we get a straight line if we plot $1/[A]_t$ on the vertical axis against the time, t, on the horizontal axis.

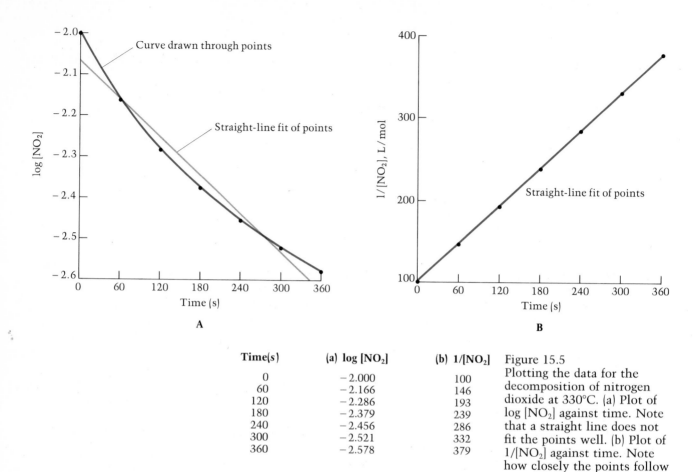

Time(s)	(a) log [NO₂]	(b) 1/[NO₂]
0	-2.000	100
60	-2.166	146
120	-2.286	193
180	-2.379	239
240	-2.456	286
300	-2.521	332
360	-2.578	379

Figure 15.5
Plotting the data for the decomposition of nitrogen dioxide at 330°C. (a) Plot of log [NO₂] against time. Note that a straight line does not fit the points well. (b) Plot of 1/[NO₂] against time. Note how closely the points follow the straight line, indicating that the decomposition is second order.

As an illustration of the determination of reaction order by graphical plotting, consider the following data for the decomposition of NO_2 at 330°C:

$$2NO_2(g) \longrightarrow 2NO(g) + O_2(g)$$

The concentrations of NO_2 for various times are

Time, s	[NO₂], mol/L
0	1.00×10^{-2}
60	0.683×10^{-2}
120	0.518×10^{-2}
180	0.418×10^{-2}
240	0.350×10^{-2}
300	0.301×10^{-2}
360	0.264×10^{-2}

In (a) of Figure 15.5, we have plotted log [NO₂] against t, and in (b) we have plotted 1/[NO₂] against t. Only in (b) do the points closely follow a straight line, indicating that the rate law is second order, that is,

$$\text{Rate} = -\frac{\Delta[NO_2]}{\Delta t} = k[NO_2]^2$$

Table 15.2 summarizes the relationships discussed in this section for first-order and second-order reactions.

Order	Rate Law	Concentration–Time Equation	Half-Life	Graphical Plot	Table 15.2 Relationships for First-Order and Second-Order Reactions
1	Rate $= k[A]$	$\log \dfrac{[A]_0}{[A]_t} = \dfrac{kt}{2.303}$	$0.693/k$	$\log [A]$ vs t	
2	Rate $= k[A]^2$	$\dfrac{1}{[A]_t} = kt + \dfrac{1}{[A]_0}$	$1/(k[A]_0)$	$\dfrac{1}{[A]}$ vs t	

15.5 Temperature and Rate; Collision and Transition-State Theories

We have seen that the rate of reaction depends on the concentrations of reactants and of a catalyst, if any. This dependence on concentrations is summarized by the rate law for the reaction. The rate also depends on temperature. This shows up in the rate law through the rate constant, which is found to vary with temperature. Consider the reaction of nitric oxide, NO, with chlorine, Cl_2, to give nitrosyl chloride, NOCl, and chlorine atoms:

$$NO(g) + Cl_2(g) \longrightarrow NOCl(g) + Cl(g)$$

The rate constant, k, for this reaction is 4.9×10^{-6} L/(mol · s) at 25°C and 1.5×10^{-5} L/(mol · s) at 35°C. Thus, in this case the rate constant, and therefore the rate, is more than tripled for a 10°C rise in temperature. ■ How do we explain this strong dependence of reaction rate on temperature? To understand it, we will need to look at a simple theoretical explanation of reaction rates.

■ The change in rate constant with temperature varies considerably from one reaction to the next. In many cases, the rate of reaction approximately doubles for a 10°C rise.

Collision Theory

Why the rate constant depends on temperature can be explained by collision theory. **Collision theory** of reaction rates assumes that in order for reaction to occur reactant molecules must collide. However, collision of two reactant molecules is not enough for their reaction. For the molecules to react, they must collide with an energy greater than a minimum value, called the **activation energy**, E_a. The value of E_a will depend on the particular reaction. Moreover, the two molecules must collide with the proper orientation.

In collision theory, the rate constant for a reaction is given as a product of three factors: (1) Z, the collision frequency, (2) f, the fraction of collisions having energy greater than the activation energy, and (3) p, the fraction of collisions that occur with the reactant molecules properly oriented. Thus,

$$k = pfZ$$

We will discuss each of these factors in turn.

To have a specific reaction to relate the concepts to, we will consider the previous reaction of NO with Cl_2 in the gas phase. This reaction is believed to occur in a single step. An NO molecule collides with a Cl_2 molecule. If the collision has sufficient energy and if the molecules are properly oriented, they react to produce NOCl and Cl.

Collision frequency, Z, the frequency with which the reactant molecules collide, will depend on temperature. As we will see, however, this dependence of collision frequency on temperature does not explain why reaction rates usually change greatly with small temperature increases. We can easily explain why the collision frequency depends on temperature. As the temperature rises, the gas molecules move faster and therefore collide more frequently. Thus, collision frequency is proportional to the root-mean-square (rms) molecular speed, which in turn is proportional to the square root of the absolute temperature, according to the kinetic theory of gases. ■

From kinetic theory, one can show that at 25°C, a 10°C rise in temperature increases the collision frequency by about 2%. If we were to assume that each collision of reactant molecules resulted in reaction, we would conclude that the rate would increase with temperature at the same rate as the collision frequency, that is, by 2% for a 10°C rise in temperature. This clearly does not explain the tripling of the rate (a 200% increase) that we see in the reaction of NO with Cl_2 when the temperature is raised from 25°C to 35°C.

We see that the collision frequency varies only slowly with temperature. However, f, the fraction of molecular collisions having energy greater than the activation energy, changes rapidly in most reactions with even small temperature changes. It can be shown that f is related to the activation energy, E_a, this way: ■

$$\log f = \frac{-E_a}{2.303\,RT}$$

Here, R is the gas constant and equals 8.31 J/(mol · K). For the reaction of NO with Cl_2, the activation energy is 8.5×10^4 J/mol. By substituting these values and the absolute temperature, T, into this equation, we can obtain log f, hence f. At 25°C (298 K), the fraction of collisions with sufficient energy for reaction is 1.2×10^{-15}. Thus, the fraction of collisions of reactant molecules that actually result in reaction is extremely small. But, the frequency of collisions is very large, so that the reaction rate, which depends on the product of these quantities, is not small. If the temperature is raised by 10°C to 35°C, the fraction of collisions of NO and Cl_2 molecules with sufficient energy for reaction is 3.8×10^{-15}, or over three times larger than the value at 25°C. In other words, the tripling of the reaction rate when the temperature rises 10°C is explained by the temperature dependence of the factor f.

From the previous equation relating f to E_a, we see that log f decreases with increasing values of E_a. Since the rate constant depends on f, this means that reactions with large activation energies have small rate constants, and reactions with small activation energies have large rate constants.

We earlier noted that the reaction rate also depends on p, the proper orientation of the reactant molecules when they collide. This factor is independent of temperature changes. We can see why it is important that the reactant molecules be properly oriented by looking in some detail at the reaction of NO with Cl_2. Figure 15.6 shows two possible molecular collisions. In (a), the NO and Cl_2 molecules collide properly oriented for reaction. The NO molecule approaches with its N atom toward the Cl_2 mole-

■ According to the kinetic theory of gases, the rms molecular speed equals $\sqrt{3RT/M_m}$ (see Section 4.8).

■ The fraction of molecular collisions having energy greater than E_a is

$$f = e^{-E_a/RT}$$

where e = 2.718 · · · . The natural logarithm of this is

$$\ln f = -E_a/RT$$

Using the relation ln f = 2.303 log f gives the equation in the text.

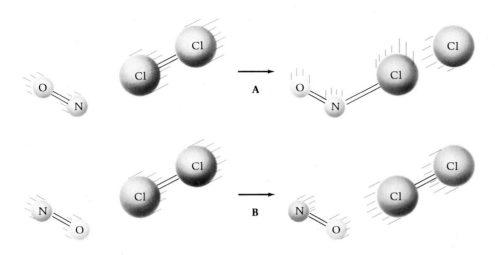

Figure 15.6
Importance of molecular orientation in the reaction of NO and Cl_2. (a) NO approaches with its N atom toward Cl_2, and an N—Cl bond forms. Also, the angle of approach is close to that in the product NOCl. (b) NO approaches with its O atom toward Cl_2. No N—Cl bond can form, so NO and Cl_2 collide, then fly apart.

cule. In addition, the angle of approach is about that expected for the formation of bonds in the product molecule NOCl. In (b), an NO molecule approaches with its O atom toward the Cl_2 molecule. Because this orientation does not allow the formation of a bond between the N atom and a Cl atom, it is ineffective for reaction. The NO and Cl_2 molecules come together, then fly apart. All orientations except those close to that shown in (a) will be ineffective.

Transition-State Theory

Collision theory is a simple theory, but it is limited in that it does not explain the role of activation energy. **Transition-state theory** looks at the collision of two reactant molecules in greater detail. It explains the reaction in terms of the formation of an **activated complex** (transition state), an unstable grouping of reactant molecules that can break up to form products. We can represent the formation of the activated complex this way:

$$O{=}N + Cl{-}Cl \longrightarrow O{=}N\text{----}Cl\text{----}Cl$$

When the molecules come together with proper orientation, an N—Cl bond begins to form. At the same time, the kinetic energy of the collision is absorbed by the activated complex as a vibrational motion of the atoms. This energy becomes concentrated in the bonds denoted by the dashed lines and can flow between them. If at some moment, sufficient energy becomes concentrated in one of the bonds of the activated complex, that bond breaks or falls apart. Depending upon whether the N----Cl or Cl----Cl bond breaks, the activated complex either reverts to the reactants or gives the products:

$$O{=}N + Cl_2 \longleftarrow O{=}N\text{----}Cl\text{----}Cl \longrightarrow O{=}N{-}Cl + Cl$$

reactants   activated complex   products

We can now see why the energy of a collision must be sufficiently large (greater than the activation energy). There must be enough energy to break the bond that results in products.

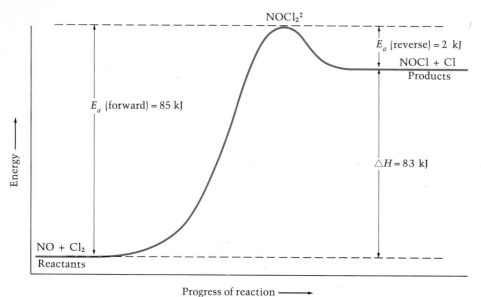

Figure 15.7
Potential energy curve (not to scale) for the reaction

$$NO + Cl_2 \longrightarrow NOCl + Cl$$

In order for NO and Cl_2 to react, at least 85 kJ of energy must be supplied by the collision of reactant molecules. Once the activated complex forms, it may break up to products, releasing 2 kJ of energy. The difference, $(85 - 2)$ kJ $= 83$ kJ, is the heat energy absorbed, ΔH.

Potential Energy Diagrams for Reactions

It is instructive to consider a potential energy diagram for the reaction of NO with Cl_2. We can represent this reaction by the equation

$$NO + Cl_2 \longrightarrow NOCl_2^{\ddagger} \longrightarrow NOCl + Cl$$

Here $NOCl_2^{\ddagger}$ denotes the activated complex. Figure 15.7 shows the change of potential energy (indicated by the solid curve) that occurs during the progress of the reaction. The potential energy curve starts at the left with the potential energy of the reactants, $NO + Cl_2$. Moving along the curve toward the right, the potential energy increases to a maximum, corresponding to the activated complex. Further to the right, the potential energy decreases to that of the products, $NOCl + Cl$.

At the start, the NO and Cl_2 molecules have a certain quantity of kinetic energy. The total energy of these molecules equals their potential energy plus their kinetic energy and will remain constant throughout the reaction (according to the law of conservation of energy). As the reaction progresses (going from left to right in the diagram), the reactants come together. The potential energy increases because the outer electrons of the two molecules repel. Thus, the kinetic energy decreases, and the molecules slow down. Only if the reactant molecules have sufficient kinetic energy will it be possible for the potential energy to increase to the value for the activated complex. This kinetic energy must be equal to or greater than the difference in energy between the activated complex and the reactant molecules (85 kJ/mol). The energy difference will be the activation energy for the forward reaction.

At the maximum in the potential energy curve, the reactant molecules have come together as the activated complex. When the activated complex breaks up into products, the products go to lower potential energy (by 2 kJ/mol) and gain in kinetic energy. Note that the products are at higher

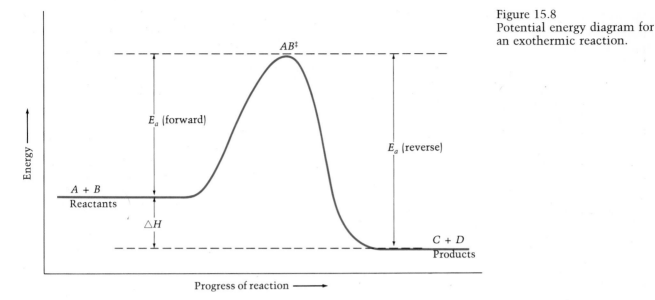

Figure 15.8
Potential energy diagram for
an exothermic reaction.

energy compared to the reactants. The difference in energy equals the heat
of reaction, ΔH. Since the energy increases, ΔH is positive, and the reaction
is endothermic.■

Now, let us look at the reverse reaction

$$NOCl + Cl \longrightarrow NO + Cl_2$$

The activation energy for this reverse reaction is 2 kJ/mol, which is the
difference in energy of the initial molecules, $NOCl + Cl$, and the activated
complex. Since this is a smaller quantity than that for the forward reaction,
the rate constant for the reverse reaction is larger.

Figure 15.8 shows the potential energy diagram for an exothermic reaction. In this case, the energy of the reactants is higher than that of the
products, so heat energy is released when the reaction goes in the forward
direction.

■ The difference in activation
energies equals the difference in
energy, ΔE. In general, $\Delta H = \Delta E$
$+ P\Delta V$ for a reaction at constant
P (see Aside at the end of
Section 10.1). The term $P\Delta V$ is
usually small, and, for a reaction
in which the moles of gases do
not change, $P\Delta V$ is essentially
zero. Therefore, $\Delta H \simeq \Delta E$.

15.6 Arrhenius Equation

Rate constants for most chemical reactions follow an equation of the form

$$\log k = \log A - \frac{E_a}{2.303\ RT}$$

It is known as the **Arrhenius equation,** after the Swedish chemist Svante
Arrhenius, who formulated it.■ Here A, sometimes called the **frequency
factor,** is a constant. Let us make the following identification of symbols:

$$\log k = \log A + \left(\frac{-E_a}{2.303\ R}\right)\left(\frac{1}{T}\right)$$
$$\downarrow \qquad \downarrow \qquad\quad \downarrow \qquad \downarrow$$
$$y \;\;=\;\; b \;\;+\;\;\quad a \qquad x$$

■ Arrhenius published this
equation in 1889, and in that
paper suggested that molecules
must be given enough energy to
become "activated" before they
could react. Collision and
transition-state theories, which
enlarged on this concept, were
developed later (1920s and
1930s, respectively).

This shows that if we plot $\log k$ against $1/T$, we should get a straight line. The slope of this line is $-E_a/(2.303\ R)$, from which we can obtain the activation energy, E_a. The intercept is $\log A$.

We can put the Arrhenius equation into a form that is very useful for computation. Let us write the Arrhenius equation for two different absolute temperatures, T_1 and T_2. We write k_1 for the rate constant at temperature T_1 and k_2 for the rate constant at temperature T_2:

$$\log k_2 = \log A - \frac{E_a}{2.303\ RT_2}$$

$$\log k_1 = \log A - \frac{E_a}{2.303\ RT_1}$$

We eliminate A by subtracting these equations:

$$\log k_2 - \log k_1 = -\frac{E_a}{2.303\ RT_2} + \frac{E_a}{2.303\ RT_1}$$

or,

$$\log \frac{k_2}{k_1} = \frac{E_a}{2.303\ R}\left(\frac{1}{T_1} - \frac{1}{T_2}\right)$$

The next example illustrates the use of this equation.

Example 15.4

The rate constant for the formation of hydrogen iodide from the elements

$$H_2(g) + I_2(g) \longrightarrow 2HI(g)$$

is 2.7×10^{-4} L/(mol · s) at 600 K and 3.5×10^{-3} L/(mol · s) at 650 K. Find the activation energy, E_a. Then calculate the rate constant at 700 K.

Solution

If we substitute the data given in the problem into the last equation, we can solve for E_a.

$$\log \frac{3.5 \times 10^{-3}}{2.7 \times 10^{-4}}$$

$$= \frac{E_a}{2.303 \times 8.31\ \text{J/(mol·K)}}\left(\frac{1}{600\ \text{K}} - \frac{1}{650\ \text{K}}\right)$$

$$\log 1.30 \times 10^1$$

$$= 1.11 = \frac{E_a}{2.303 \times 8.31\ \text{J/mol}} \times (1.28 \times 10^{-4})$$

Hence,

$$E_a = \frac{1.11 \times 2.303 \times 8.31\ \text{J/mol}}{1.28 \times 10^{-4}} = 1.66 \times 10^5\ \text{J/mol}$$

To find the rate constant at 700 K, we write the equation obtained previously,

$$\log \frac{k_2}{k_1} = \frac{E_a}{2.303\ R}\left(\frac{1}{T_1} - \frac{1}{T_2}\right)$$

and substitute $E_a = 1.66 \times 10^5$ J/mol and

$$k_1 = 2.7 \times 10^{-4}\ \text{L/(mol · s)}\ (T_1 = 600\ \text{K})$$
$$k_2 = \text{unknown as yet}\ (T_2 = 700\ \text{K})$$

We get

$$\log \frac{k_2}{2.7 \times 10^{-4}\ \text{L/(mol · s)}}$$

$$= \frac{1.66 \times 10^5\ \text{J/mol}}{2.303 \times 8.31\ \text{J/(mol · K)}} \times \left(\frac{1}{600\ \text{K}} - \frac{1}{700\ \text{K}}\right)$$

$$= 2.07$$

Taking antilogarithms, we get

$$\frac{k_2}{2.7 \times 10^{-4}\ \text{L/(mol · s)}} = 10^{2.07} = 1.2 \times 10^2$$

Hence,

$$k_2 = (1.2 \times 10^2) \times (2.7 \times 10^{-4})\ \text{L/(mol · s)}$$
$$= 3.2 \times 10^{-2}\ \text{L/(mol · s)}$$

Exercise 15.7

Acetaldehyde, CH_3CHO, decomposes when heated:

$$CH_3CHO(g) \longrightarrow CH_4(g) + CO(g)$$

The rate constant for the decomposition is $0.105/(M \cdot s)$ at 759 K and $2.14/(M \cdot s)$ at 836 K. What is the activation energy for this decomposition? What is the rate constant at 865 K?

(See Problems 15.33, 15.34, 15.35, and 15.36.)

Reaction Mechanisms

A balanced chemical equation is a description of the overall result of a chemical reaction. However, what actually happens at the molecular level may be more involved than is represented by this single equation. The reaction may take place in several steps. In the next sections, we will examine some reactions and see how the rate law can give us information about these steps, or *elementary reactions*.

15.7 Elementary Reactions

Consider the reaction of nitrogen dioxide with carbon monoxide:

$$NO_2(g) + CO(g) \longrightarrow NO(g) + CO_2(g) \quad \text{(net chemical equation)}$$

At temperatures below 500 K, this gas-phase reaction is believed to take place in two steps:

$$NO_2 + NO_2 \longrightarrow NO_3 + NO \quad \text{(elementary reaction)}$$
$$NO_3 + CO \longrightarrow NO_2 + CO_2 \quad \text{(elementary reaction)}$$

Each step, called an **elementary reaction,** represents the result of a collision occurring at the molecular level. The set of elementary reactions whose overall effect is given by the net chemical equation is called the **reaction mechanism.**

According to the reaction mechanism just given, two NO_2 molecules collide and react to give the product molecule NO and the reaction intermediate NO_3. A **reaction intermediate** is a species produced during a reaction that does not appear in the net equation because it reacts in a subsequent step in the mechanism. Often it has a fleeting existence and cannot be isolated from the reaction mixture. The NO_3 molecule is known only from its visible light spectrum. It reacts immediately with CO to give the product molecules NO_2 and CO_2.

The net chemical equation, which represents the overall result of these two elementary reactions in the mechanism, is obtained by adding these steps together and canceling species that occur on both sides:

$$NO_2 + NO_2 \longrightarrow NO_3 + NO \quad \text{(elementary reaction)}$$
$$\underline{NO_3 + CO \longrightarrow NO_2 + CO_2} \quad \text{(elementary reaction)}$$
$$NO_2 + \cancel{NO_2} + \cancel{NO_3} + CO \longrightarrow \cancel{NO_3} + NO + \cancel{NO_2} + CO_2$$
$$\text{(net chemical equation)}$$

Exercise 15.8

The iodide ion catalyzes the decomposition of aqueous hydrogen peroxide, H_2O_2. This decomposition is believed to occur in two steps:

$$H_2O_2 + I^- \longrightarrow H_2O + IO^- \qquad \text{(elementary reaction)}$$
$$H_2O_2 + IO^- \longrightarrow H_2O + O_2 + I^- \qquad \text{(elementary reaction)}$$

What is the net chemical equation representing this decomposition? Note that IO^- is a reaction intermediate. The iodide ion is not an intermediate; it was added to the reaction mixture.

(See Problems 15.37 and 15.38.)

Molecularity

Elementary reactions are classified according to their molecularity. The **molecularity** of an elementary reaction is the number of molecules on the reactant side. A **unimolecular** reaction is one that involves one reactant molecule; a **bimolecular** reaction is one that involves two reactant molecules. Bimolecular reactions are the most common. Unimolecular reactions are best illustrated by decomposition of some previously excited species. Some gas-phase reactions are thought to occur in a **termolecular** step, an elementary reaction of three molecules. Higher molecularities are not encountered, presumably because the chance of the correct four molecules coming together at once is extremely small.

As an example of a unimolecular reaction, consider the elementary process in which an energetically excited ozone molecule (symbolized by O_3^*) spontaneously decomposes:

$$O_3^* \longrightarrow O_2 + O$$

Normally, a molecule in a sample of ozone gas is in a lower energy level. But such a molecule may be excited to a higher level if it collides with another molecule or if it absorbs a photon. The energy of this excited molecule will be distributed among its three nuclei as vibrational energy. After a period of time, this energy will become redistributed. If by chance most of the energy finds its way to one oxygen atom, that atom will fly off. In other words, the excited ozone molecule decomposes into an oxygen molecule and an oxygen atom.

Each of the steps in the mechanism of the reaction of NO_2 with CO, given earlier in this section, are bimolecular. Consider the first step, which involves the reaction of two NO_2 molecules:

$$NO_2 + NO_2 \longrightarrow NO_3 + NO$$

When these two NO_2 molecules come together, they form an activated complex of the six atoms, $(NO_2)_2$, which immediately breaks into two new molecules, NO_3 and NO.

The overall reaction of two atoms, say two Br atoms, to form a diatomic molecule (Br_2) is normally a termolecular process. When two bromine atoms collide, they form an excited bromine molecule, Br_2^*. This excited molecule will immediately fly apart, reforming the atoms, unless another atom or molecule is present just at the moment of molecule formation to take away the excess energy. Suppose an argon atom in its ground-energy level and the two bromine atoms all collide at the same moment.

$$Br + Br + Ar \longrightarrow Br_2 + Ar^\star$$

Energy that would have been left with the bromine molecule is now picked up by the argon atom (giving the excited atom $Ar^\star$). The bromine molecule is stabilized by being left in a lower energy level.

Exercise 15.9

The following is an elementary reaction that occurs in the decomposition of ozone in the stratosphere by nitric oxide:

$$O_3 + NO \longrightarrow O_2 + NO_2$$

What is the molecularity of this reaction; that is, is the reaction unimolecular, bimolecular, or termolecular?

(See Problems 15.39 and 15.40.)

Rate Equation for an Elementary Reaction

There is no necessarily simple relationship between the overall reaction and the rate law that we observe for it. As we stressed before, the rate law must be obtained experimentally. However, if we are dealing with an elementary reaction, the rate does have a simple, predictable form. The rate is proportional to the product of the concentration of each reactant molecule.

In order to understand this, let us look at the different possibilities. Consider, for example, the unimolecular reaction

$$A \longrightarrow B + C$$

For each A molecule there is a definite probability or chance that it will decompose into B and C molecules. The more A molecules there are in a given volume, the more A molecules that can decompose in that volume per unit time. In other words, the rate of reaction is proportional to the concentration of A:

$$\text{Rate} = k[A]$$

Now consider a bimolecular reaction, such as

$$A + B \longrightarrow C + D$$

For the reaction to occur, the reactant molecules, A and B, must collide. Reaction will not occur with every collision. Nevertheless, the rate of formation of product will be proportional to the frequency of molecular collisions because a definite fraction of those collisions will produce reaction. Within a given volume, the frequency of collisions will be proportional to the number of A molecules, n_A, times the number of B molecules, n_B. Furthermore, the concentration of A is proportional to n_A, and the concentration of B is proportional to n_B. Therefore, the rate of reaction is proportional to $[A][B]$:

$$\text{Rate} = k[A][B]$$

A termolecular reaction has a rate equation that is obtained by similar reasoning. For the elementary reaction

$$A + B + C \longrightarrow D + E$$

the rate is proportional to the concentrations of A, B, and C:

$$\text{Rate} = k[A][B][C]$$

Any reaction that we observe is likely to consist of several elementary steps, and the rate law that we find will be the combined result of these steps. In the next section, we look at the relationship between a reaction mechanism and the observed rate law.

Exercise 15.10

Write the rate equation, showing the dependence of rate on concentrations, for the elementary reaction

$$NO_2 + NO_2 \longrightarrow N_2O_4$$

(See Problems 15.41 and 15.42.)

15.8 The Rate Law and the Mechanism

The mechanism of a reaction cannot be observed directly. A mechanism is devised to explain the experimental observations. It is like the explanation provided by a detective to explain a crime in terms of the clues found. Other explanations may be possible, and further clues may make one of these other explanations seem more plausible than the currently accepted one. So it is with reaction mechanisms. They are accepted provisionally, with the understanding that further experiments may lead us to accept another mechanism as the more likely explanation.■

An important clue in understanding the mechanism of a reaction is the rate law. The reason for its importance is that once we assume a mechanism, we can predict the rate law. If this prediction does not agree with the experimental rate law, the assumed mechanism must be wrong. Take, for example, the overall equation

$$2NO_2(g) + F_2(g) \longrightarrow 2NO_2F(g)$$

If we follow the rate of disappearance of F_2, we observe that it is directly proportional to the concentration of NO_2 and F_2:

$$\text{Rate} = k[NO_2][F_2] \quad \text{(experimental rate law)}$$

This rate law is a summary of the experimental data. Let us assume that the reaction occurs in a single elementary reaction:

$$NO_2 + NO_2 + F_2 \longrightarrow NO_2F + NO_2F \quad \text{(elementary reaction)}$$

This then is our assumed mechanism. Because this is an elementary reaction, we can immediately write the rate law predicted by it:

$$\text{Rate} = k[NO_2]^2[F_2] \quad \text{(predicted rate law)}$$

However, this does not agree with experiment, and our assumed mechanism

■ We see the scientific method in operation here, which recalls the discussion in Chapter 1. Experiments have been made from which we determine the rate law. Then a mechanism is devised to explain the rate law. This mechanism in turn suggests more experiments. These may confirm the explanation or they may disagree with it. If they disagree, a new mechanism must be devised that explains all of the experimental evidence.

must be discarded. We conclude that the reaction occurs in more than one step.

Rate-Determining Step

The reaction of NO_2 with F_2 is believed to occur in the following steps (elementary reactions):

$$NO_2 + F_2 \xrightarrow{k_1} NO_2F + F \qquad \text{(slow step)}$$
$$F + NO_2 \xrightarrow{k_2} NO_2F \qquad \text{(fast step)}$$

Rate constants have been written over the arrows. The second step is assumed to be much faster than the first, so that as soon as NO_2 and F_2 react, the F atom formed reacts with an NO_2 molecule to give another NO_2F molecule. Therefore, the rate of disappearance of F_2 is determined completely by the slow step, which for this reason is called the **rate-determining step.**

To better understand the significance of the rate-determining step, think of an assembly line for making widgets. Suppose one person on the line has a much lengthier, intricate task to perform than the others on the line. Widgets can be produced only as fast as this difficult task can be completed. This person's task is rate determining.

The rate equation for the rate-determining step of the mechanism under discussion is

$$\text{Rate} = k_1[NO_2][F_2]$$

The mechanism agrees with the experimental rate law if we equate k_1 to k. This agreement is not absolute evidence that the mechanism is correct. However, one can perform experiments to see whether or not fluorine atoms do react very quickly with nitrogen dioxide. Such experiments show that they do.

Exercise 15.11

The iodide ion–catalyzed decomposition of hydrogen peroxide, H_2O_2, is believed to follow the mechanism

$$H_2O_2 + I^- \xrightarrow{k_1} H_2O + IO^- \qquad \text{(slow)}$$
$$H_2O_2 + IO^- \xrightarrow{k_2} H_2O + O_2 + I^- \qquad \text{(fast)}$$

What rate law is predicted by this mechanism? Explain.

(See Problems 15.43 and 15.44.)

Mechanisms with an Initial Fast Step

A somewhat more complicated situation occurs when the rate-determining step follows an initial fast step. The decomposition of dinitrogen pentoxide,

$$2N_2O_5(g) \longrightarrow 4NO_2(g) + O_2(g) \qquad \text{(overall equation)}$$

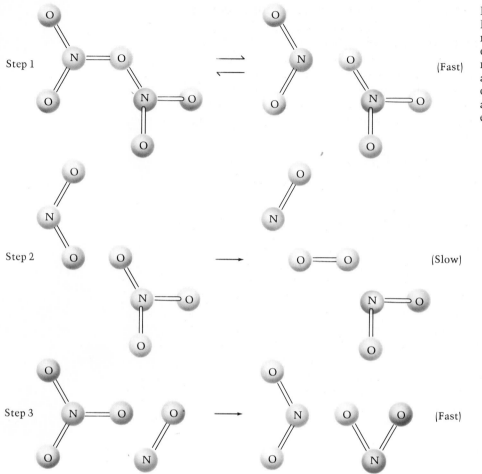

Step 1   (Fast)

Step 2   (Slow)

Step 3   (Fast)

Figure 15.9
Representation of the
mechanism of decomposition
of N_2O_5 using molecular
models. Step 1 occurs twice
as Steps 2 and 3 each occur
once. Note that there is little
atomic rearrangement during
each step.

which we discussed in the chapter opening, is believed to follow this type of
mechanism:

$$N_2O_5 \xrightleftharpoons[k_{-1}]{k_1} NO_2 + NO_3 \qquad \text{(fast)}$$

$$NO_2 + NO_3 \xrightarrow{k_2} NO + NO_2 + O_2 \qquad \text{(slow)}$$

$$NO_3 + NO \xrightarrow{k_3} 2NO_2 \qquad \text{(fast)}$$

The first step occurs twice as the succeeding steps each occur once. The
reasonableness of this mechanism can be better appreciated by looking at
Figure 15.9, which represents the mechanism by molecular models. Let us
show that this mechanism is consistent with the experimentally determined rate law,

$$\text{Rate} = k[N_2O_5]$$

The second step in the mechanism is assumed to be much slower than the
other steps and is therefore rate determining. Hence, the rate law predicted
from this mechanism is

$$\text{Rate} = k_2[NO_2][NO_3]$$

However, this equation cannot be compared directly with experiment because it is written in terms of the reaction intermediate, NO_3. The experimental rate law will be written in terms of substances occurring in the chemical equation and not in terms of reaction intermediates. For purposes of comparison, it is necessary to re-express the rate equation, eliminating $[NO_3]$. To do this, we must look at the first step in the mechanism.

This step is fast and reversible. That is, N_2O_5 dissociates rapidly into NO_2 and NO_3, and these products in turn react to re-form N_2O_5. The rate of the forward reaction (dissociation of N_2O_5) is

$$\text{Forward rate} = k_1[N_2O_5]$$

and the rate of the reverse reaction (formation of N_2O_5 from NO_2 and NO_3) is

$$\text{Reverse rate} = k_{-1}[NO_2][NO_3]$$

When the reaction first begins, there are no NO_2 or NO_3 molecules, and the reverse rate is zero. But as N_2O_5 dissociates, the concentration of N_2O_5 decreases, and the concentrations of NO_2 and NO_3 increase. Therefore, the forward rate decreases and the reverse rate increases. Soon the two rates become equal, so that N_2O_5 molecules form as often as other molecules dissociate. The first step has reached *dynamic equilibrium*. Because these elementary reactions are much faster than the second step, this equilibrium is reached before any significant reaction by the second step occurs. Moreover, this equilibrium is maintained throughout the reaction.■

■ Chemical equilibrium is discussed in detail in the next chapter.

At equilibrium, the forward and reverse rates are equal, so that we can write

$$k_1[N_2O_5] = k_{-1}[NO_2][NO_3]$$

or, rearranging,

$$\frac{k_1}{k_{-1}} = \frac{[NO_2][NO_3]}{[N_2O_5]}$$

The quantity k_1/k_{-1} is called the *equilibrium constant* for the first step. We will designate this equilibrium constant by the symbol K_1. Note that K_1 is the concentrations of products divided by concentrations of reactants in this step.

We can use the equilibrium constant to eliminate the concentration of NO_3 from the rate law that we wrote for the rate-determining step. We have

$$K_1 = \frac{[NO_2][NO_3]}{[N_2O_5]}$$

or

$$[NO_3] = K_1\frac{[N_2O_5]}{[NO_2]}$$

Substituting into the rate law, we get

$$\text{Rate} = k_2[NO_2][NO_3] = k_2[\cancel{NO_2}] \times K_1\frac{[N_2O_5]}{[\cancel{NO_2}]}$$

Or,

$$\text{Rate} = k_2 K_1[N_2O_5]$$

Thus, if we identify $k_2 K_1$ as k, we reproduce the experimental rate law.

Example 15.5

The oxidation of I^- by hypochlorite ion, ClO^-, in basic solution,

$$I^-(aq) + ClO^-(aq) \xrightarrow{OH^-} IO^-(aq) + Cl^-(aq)$$
(overall equation)

is thought to occur by the following mechanism:

$$ClO^- + H_2O \underset{k_{-1}}{\overset{k_1}{\rightleftharpoons}} HClO + OH^- \quad \text{(fast)}$$

$$I^- + HClO \xrightarrow{k_2} HIO + Cl^- \quad \text{(slow)}$$

$$OH^- + HIO \xrightarrow{k_3} H_2O + IO^- \quad \text{(fast)}$$

Verify that this mechanism agrees with the experimentally determined rate law,

$$\text{Rate} = k\frac{[I^-][ClO^-]}{[OH^-]}$$

Solution

According to the rate-determining step (the slow step),

$$\text{Rate} = k_2[I^-][HClO]$$

However, HClO does not appear in the overall equation. Let us try to eliminate [HClO] from the rate law by looking at the first step. This step is fast and reversible

and quickly attains equilibrium. We get the following equilibrium constant, which equals the concentrations of products divided by concentrations of reactants in this step.

$$K_1' = \frac{k_1}{k_{-1}} = \frac{[HClO][OH^-]}{[ClO^-][H_2O]}$$

But since the concentration of water does not vary significantly during the reaction and is essentially constant (because it is the solvent), we can write

$$K_1 = K_1'[H_2O] = \frac{[HClO][OH^-]}{[ClO^-]}$$

Hence, $$[HClO] = \frac{K_1[ClO^-]}{[OH^-]}$$

When we substitute this expression for [HClO] into the rate equation for the rate-determining step, we get

$$\text{Rate} = k_2[I^-][HClO]$$

$$= k_2[I^-] \times K_1\frac{[ClO^-]}{[OH^-]} = k_2K_1\frac{[I^-][ClO^-]}{[OH^-]}$$

We need only identify the observed rate constant k with k_2K_1.

Exercise 15.12

Nitric oxide, NO, reacts with oxygen to produce nitrogen dioxide:

$$2NO(g) + O_2(g) \longrightarrow 2NO_2(g) \quad \text{(overall equation)}$$

If the mechanism is

$$NO + O_2 \underset{k_{-1}}{\overset{k_1}{\rightleftharpoons}} NO_3 \quad \text{(fast)}$$

$$NO_3 + NO \xrightarrow{k_2} NO_2 + NO_2 \quad \text{(slow)}$$

what is the predicted rate law? Remember to express this in terms of substances in the chemical equation.

(See Problems 15.45 and 15.46.)

15.9 Catalysis

Catalysis is the change in speed of a reaction as the result of adding a catalyst. Although a catalyst is not consumed by a reaction, it does take part in the reaction mechanism, entering at one step, but being regenerated in a later step. In the commercial preparation of sulfuric acid, its anhydride sulfur trioxide is obtained by oxidizing sulfur dioxide, SO_2. Sulfuric acid is then obtained from SO_3. The reaction of SO_2 with O_2 to give SO_3 is normally very slow. In an early industrial process, this reaction was carried out in the presence of nitric oxide, NO, because it is then much faster:

$$2SO_2(g) + O_2(g) \xrightarrow{NO} 2SO_3(g)$$

Nitric oxide does not appear in the overall equation, nor does it in any way affect the final composition of the reaction mixture. Thus, it acts as a catalyst.

A possible mechanism for the nitric oxide catalysis is

$$2NO + O_2 \longrightarrow 2NO_2$$
$$NO_2 + SO_2 \longrightarrow NO + SO_3$$

The last step occurs twice each time the first step occurs once. Thus, two molecules of nitric oxide are used up in the first step and are regenerated in the second step.

This catalytic action in a homogeneous phase (that is, within a solution or gas) is called **homogeneous catalysis.** Another example is given by the oxidation of thallium(I) to thallium (III) by cerium(IV):

$$2Ce^{4+}(aq) + Tl^+(aq) \longrightarrow 2Ce^{3+}(aq) + Tl^{3+}(aq)$$

The uncatalyzed reaction is very slow, presumably involving the collision of three positive ions. The reaction can be catalyzed by manganese(II) ion, however. The mechanism is thought to be

$$Ce^{4+} + Mn^{2+} \longrightarrow Ce^{3+} + Mn^{3+}$$
$$Ce^{4+} + Mn^{3+} \longrightarrow Ce^{3+} + Mn^{4+}$$
$$Mn^{4+} + Tl^+ \longrightarrow Tl^{3+} + Mn^{2+}$$

Each step is bimolecular.

Some of the most important industrial reactions involve **heterogeneous catalysis,** that is, the reaction of substances in gas or liquid phase at the surface of a solid catalyst. Such surface or heterogeneous catalysis is thought to occur by chemical adsorption of the reactants onto the surface of the catalyst. *Adsorption* is the attraction of molecules to a surface. In *physical adsorption,* molecules adhere to a surface through weak intermolecular forces. By contrast, in **chemisorption** molecules are attracted to a surface by chemical bonding forces. It may happen that bonds in the molecule are broken during chemisorption.

An example of heterogeneous catalysis involving chemisorption is provided by *catalytic hydrogenation.* This is the addition of H_2 to a compound, such as one with a carbon–carbon double bond, using a catalyst of platinum or nickel metal. Vegetable oils, which contain carbon–carbon double bonds, are changed to solid fats when the bonds are catalytically hydrogenated. In the case of ethylene, C_2H_4, the equation is

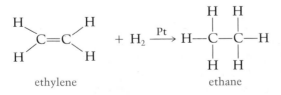

ethylene   ethane

A mechanism for this reaction is represented by the four steps shown in Figure 15.10. In (a), ethylene and hydrogen molecules diffuse to the catalyst

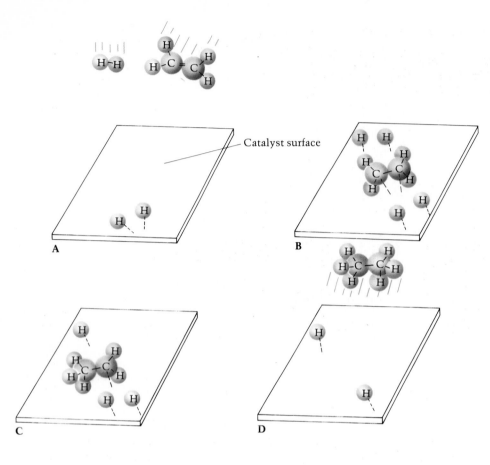

- Catalyst surface

A B C D

surface, where, as shown in (b), they undergo chemisorption. The electrons of ethylene form bonds to the metal, and hydrogen molecules break into H atoms that bond to the metal. In (c), two H atoms migrate to an ethylene molecule bonded to the catalyst, where they react to form ethane. Then, in (d), because ethane does not bond to the metal, it diffuses away. Note that the catalyst surface that was used in (b) is regenerated in (d).

At the beginning of this section, we described the catalytic oxidation of SO_2 to SO_3, the anhydride of sulfuric acid. In an early process, the homogeneous catalyst NO was used. Today, in the *contact process*, a heterogeneous catalyst, Pt or V_2O_5, is used. Surface catalysts are used in the catalytic converters of automobiles (Figure 15.11) to convert substances, such as CO and NO, that would be atmospheric pollutants into harmless substances, such as CO_2 and N_2.

Enzymes are proteins that catalyze biochemical processes. The reactant molecule, referred to as the enzyme *substrate*, attaches itself to a particular place (an *active site*) on the enzyme molecule. In this way, the substrate is held in a configuration that makes subsequent reaction easier to occur. The catalytic activity of enzymes is usually very specific. For example, the enzyme lysozyme, which is found in the tears of the eye, catalyzes the breakdown of the membrane of bacteria. Once the membrane of a bacterial cell has been removed, the cell breaks up as a result of the high osmotic pressure within. Thus, lysozyme in tears protects the surface of the eye from bacterial infection.

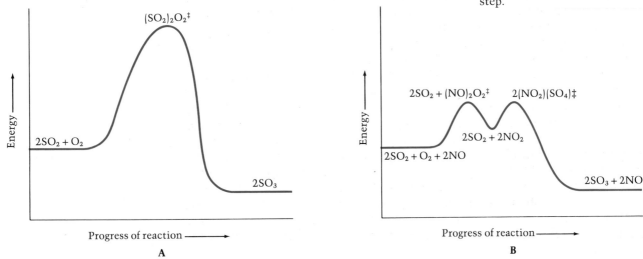

Figure 15.11
Cross-sectional view of an
automobile catalytic con-
verter. Potential pollutants,
such as CO and NO, are
converted to harmless substances, such as CO_2 and N_2.

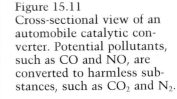

We have seen that a catalyst operates by providing a mechanism that facilitates the reaction. The catalyzed mechanism is faster than the uncatalyzed mechanism. We can look at this in terms of a potential energy diagram. The activation energy for the uncatalyzed oxidation of sulfur dioxide is very large. When a catalyst is added, however, a new mechanism is provided, in which the activation energy is lower than that for the uncatalyzed reaction in the rate-determining step. When the activation energy is less, the reaction occurs faster. Figure 15.12b shows the potential energy diagram for the NO-catalyzed reaction of SO_2 and O_2. Since this occurs in two steps, there are two maximums, one for each activated complex.

Figure 15.12
Potential energy diagram for
the reaction of SO_2 and O_2.
(a) Because the direct reaction of $2SO_2 + O_2$ has a large
activation energy, it is very
slow. (b) The NO catalyst
provides a new mechanism
in which the activation
energy of each step is lower
than in (a). The first step in
the catalytic mechanism,

$$2NO + O_2 \longrightarrow 2NO_2$$

corresponds to the first
maximum and minimum.
Products from this step react
in the second step to give
sulfur trioxide: $NO_2 + SO_2 \longrightarrow NO + SO_3$ (occurs twice
each time the first step
occurs once). Note that NO
is used in the first step and
regenerated in the second
step.

A — graph with Energy (vertical axis) vs Progress of reaction (horizontal axis). Curve starts at $2SO_2 + O_2$, rises to a maximum labeled $(SO_2)_2O_2^{\ddagger}$, then falls to $2SO_3$.

B — graph with Energy (vertical axis) vs Progress of reaction (horizontal axis). Curve starts at $2SO_2 + O_2 + 2NO$, rises to first maximum labeled $2SO_2 + (NO)_2O_2^{\ddagger}$, dips to $2SO_2 + 2NO_2$, rises to second maximum labeled $2(NO_2)(SO_4)^{\ddagger}$, then falls to $2SO_3 + 2NO$.

A Checklist for Review

Important Terms

catalyst (15.1)
reaction rate (15.1)
rate law (15.3)
rate constant (15.3)
reaction order (15.3)
half-life (15.4)
collision theory (15.5)
activation energy (15.5)

transition-state theory (15.5)
activated complex (15.5)
Arrhenius equation (15.6)
frequency factor (15.6)
elementary reaction (15.7)
reaction mechanism (15.7)
reaction intermediate (15.7)
molecularity (15.7)

unimolecular (15.7)
bimolecular (15.7)
termolecular (15.7)
rate-determining step (15.8)
homogeneous catalysis (15.9)
heterogeneous catalysis (15.9)
chemisorption (15.9)

Summary of Facts and Concepts

The *reaction rate* is defined as the increase in moles of product per liter per second (or as the decrease in moles of reactant per liter per second). Rates of reaction are determined by following the change of concentration of a reactant or product, either by chemical analysis or by observing a physical property. It is found that reaction rates are proportional to concentrations of reactants raised to various powers (usually 1 or 2, but they can be fractional or negative). The *rate law* mathematically expresses the relationship between rates and concentrations for a chemical reaction. Although the rate law tells us how the rate of a reaction depends on concentrations at a given moment, it is possible to transform a rate law to show how concentrations change with time. The *half-life* of a reactant is the time it takes for the reactant concentration to decrease to one-half of its original concentration.

Reaction rates can often double or triple with a 10°C rise in temperature. The effect of temperature on the rate can be explained by *collision theory*. According to this theory, two molecules react after colliding only if the energy of collision is greater than the *activation energy* and if the molecules are properly oriented. It is the rapid increase in the fraction of collisions having energy greater than the activation energy that explains the large temperature dependence of reaction rates. *Transition-state theory* explains reaction rates in terms of the formation of an *activated complex* of the colliding mole-

cules. The *Arrhenius equation* is a mathematical relation showing the dependence of a rate constant on temperature.

A chemical equation describes the overall result of a chemical reaction that may take place in one or more steps. These steps are called *elementary reactions*, and the set of elementary reactions that describes what is happening at the molecular level in an overall reaction is called the *reaction mechanism*. In some cases, the overall reaction involves a *reaction intermediate*, a species produced in one step but used up in a subsequent one. The rate of the overall reaction depends on the rate of the slowest step (the *rate-determining step*). This rate is proportional to the product of the concentrations of each reactant molecule in that step. If this step involves a reaction intermediate, its concentration may be eliminated by using the relationship of the concentrations in the preceding fast, *equilibrium* step.

A *catalyst* is a substance that speeds up a chemical reaction without being consumed by it. The catalyst is used up in one step of the reaction mechanism, but is regenerated in a later step. Catalysis is classified as *homogeneous catalysis* if the substances react within one phase or *heterogeneous catalysis* if the substances in a gas or liquid phase react at the surface of a solid catalyst. Many industrial reactions involve heterogeneous catalysis. Catalytic activity operates by providing a mechanism for the reaction that has lower activation energy.

Operational Skills

1. Given the balanced equation for a reaction, relate the different possible ways of defining the rate of the reaction (Example 15.1).

2. Given initial concentrations and initial-rate data (in which the concentration of each species is changed, one at a time, holding the others constant), find the rate law for a reaction (Example 15.2).

3. Given the rate constant and initial reactant concentration for a first-order reaction, calculate the reactant concentration after a definite time or calculate the time it takes for the concentration to decrease to a prescribed value (Example 15.3).

4. Given the values of the rate constant for two temperatures, find the activation energy and calculate the rate constant at a third temperature (Example 15.4).

5. Given a reaction mechanism (in which either the first step is rate determining, or the first step is fast and reversible followed by the rate-determining step), derive the rate law (Example 15.5).

Review Questions

15.1 Define the rate of reaction of HBr in the following reaction:

$$4HBr(g) + O_2(g) \longrightarrow 2Br_2(g) + 2H_2O(g)$$

15.2 Give at least two physical properties that might be used to determine the rate of a reaction.

15.3 A rate of reaction depends on the reactant concentrations, catalyst concentration, and temperature. Explain by means of an example how the rate law deals with each of these variables.

15.4 The rate of a reaction is quadrupled when the concentration of one reactant is doubled. What is the order of the reaction with respect to this reactant?

15.5 A rate law is one-half order with respect to a reactant. What is the effect on the rate if the concentration of this reactant is doubled?

15.6 The reaction

$$2NO(g) + O_2(g) \longrightarrow NO_2(g)$$

is found to be second order with respect to NO and first order with respect to O_2. Write the rate law.

15.7 The reaction $A(g) \longrightarrow B(g) + C(g)$ is known to be first order in $A(g)$. It takes 25 s for the concentration of $A(g)$ to decrease by one-half of its initial value. How long will it take for the concentration of $A(g)$ to decrease to one-fourth of its initial value? to one-eighth of its initial value?

15.8 What two factors are important in order that a collision between two reactant molecules will result in reaction?

15.9 Sketch a potential energy diagram for the exothermic, elementary reaction,

$$A + B \longrightarrow C + D$$

and on it denote the activation energies for the forward and reverse reactions. Also, indicate the reactants, products, and activated complex.

15.10 Draw a structural formula for the activated complex in the following reaction:

$$NO_2 + NO_3 \longrightarrow NO + NO_2 + O_2$$

Refer to Figure 15.9, center, for details of this reaction. Use dashed lines for bonds about to form or break. Use a single line for all other bonds.

15.11 By means of an example, explain what is meant by the term *reaction intermediate*.

15.12 Why is it generally impossible to predict the rate law for a reaction on the basis of the chemical equation only?

15.13 The rate law for the reaction

$$2NO_2Cl(g) \longrightarrow 2NO_2(g) + Cl_2(g) \quad \text{(overall equation)}$$

is first order in nitryl chloride, NO_2Cl:

$$\text{Rate} = k[NO_2Cl]$$

Explain why the mechanism for this reaction cannot be the single elementary reaction

$$2NO_2Cl \longrightarrow 2NO_2 + Cl_2 \quad \text{(elementary reaction)}$$

15.14 There is often one step in a reaction mechanism that is rate determining. What is the characteristic of such a step that makes it rate determining? Explain.

15.15 The dissociation of N_2O_4 into NO_2,

$$N_2O_4(g) \longrightarrow 2NO_2(g)$$

is believed to occur in one step. Express the equilibrium constant for this reaction in terms of the rate constant for the forward reaction k_f, and the rate constant for the reverse reaction k_r.

15.16 How does a catalyst speed up a reaction? How can a catalyst be involved in a reaction without being consumed by it?

15.17 Compare physical adsorption and chemisorption (chemical adsorption).

15.18 Describe the steps in the catalytic hydrogenation of ethylene.

Problems

Reaction Rates

15.19 Relate the rate of decomposition of NO_2 to the rate of formation of O_2 for the following reaction:

$$2NO_2(g) \longrightarrow 2NO(g) + O_2(g)$$

15.21 Azomethane, CH_3NNCH_3, decomposes according to the following equation:

$$CH_3NNCH_3(g) \longrightarrow C_2H_6(g) + N_2(g)$$

The initial concentration of azomethane was 1.50×10^{-2} mol/L. After 10.0 min, this concentration decreased to 1.29×10^{-2} mol/L. Obtain the rate of reaction during this time interval. Express the answer in units of mol/(L · s).

15.20 For the reaction of hydrogen with iodine

$$H_2(g) + I_2(g) \longrightarrow 2HI(g)$$

relate the rate of disappearance of iodine vapor to the rate of formation of hydrogen iodide.

15.22 Nitrogen dioxide, NO_2, decomposes upon heating according to the equation

$$2NO_2(g) \longrightarrow 2NO(g) + O_2(g)$$

At the beginning of an experiment, the concentration of nitrogen dioxide in a reaction vessel was 0.1103 mol/L. After 60.0 s, the concentration decreased to 0.1076 mol/L. What is the rate of decomposition of NO_2 during this time interval, in mol/(L · s)?

Rate Laws

15.23 Propylene dibromide, $C_3H_6Br_2$, reacts with potassium iodide in methanol solution to give propylene, C_3H_6:

$$C_3H_6Br_2 + 3KI \longrightarrow C_3H_6 + 2KBr + KI_3$$

The observed rate law is

$$Rate = k[C_3H_6Br_2][KI]$$

What are the reaction orders with respect to each reactant? What is the overall order?

15.25 In experiments on the decomposition of azomethane,

$$CH_3NNCH_3(g) \longrightarrow C_2H_6(g) + N_2(g)$$

the following data were obtained:

	Initial Concentration of Azomethane	Initial Rate
Experiment 1	1.13×10^{-2} M	2.8×10^{-6} M/s
Experiment 2	2.26×10^{-2} M	5.6×10^{-6} M/s

What is the rate law? What is the value of the rate constant?

15.24 For the reaction of nitric oxide, NO, with chlorine, Cl_2,

$$2NO(g) + Cl_2(g) \longrightarrow 2NOCl(g)$$

the observed rate law is

$$Rate = k[NO]^2[Cl_2]$$

What is the reaction order with respect to nitric oxide? with respect to Cl_2? What is the overall order?

15.26 Ethylene oxide, C_2H_4O, decomposes when heated to give methane and carbon monoxide:

$$C_2H_4O(g) \longrightarrow CH_4(g) + CO(g)$$

The following kinetic data were observed for the reaction at 688 K:

	Initial Concentration of Ethylene Oxide	Initial Rate
Experiment 1	0.00272 M	5.57×10^{-7} M/s
Experiment 2	0.00544 M	1.11×10^{-6} M/s

Find the rate law and the value of the rate constant for this reaction.

6.835   10

15.27 Nitric oxide, NO, reacts with hydrogen to give nitrous oxide, N_2O, and water:

$$2NO(g) + H_2(g) \longrightarrow N_2O(g) + H_2O(g)$$

In a series of experiments, the following initial rates of disappearance of NO were obtained:

	Initial Concentrations		Initial Rate
	NO	H_2	
Experiment 1	$6.4 \times 10^{-3}\ M$	$2.2 \times 10^{-3}\ M$	$2.6 \times 10^{-5}\ M/s$
Experiment 2	$12.8 \times 10^{-3}\ M$	$2.2 \times 10^{-3}\ M$	$1.0 \times 10^{-4}\ M/s$
Experiment 3	$6.4 \times 10^{-3}\ M$	$4.5 \times 10^{-3}\ M$	$5.1 \times 10^{-5}\ M/s$

Find the rate law and the value of the rate constant.

15.28 In a kinetic study of the reaction

$$2NO(g) + O_2(g) \longrightarrow 2NO_2(g)$$

the following data were obtained for the initial rates of disappearance of NO:

	Initial Concentrations		Initial Rate
	NO	O_2	
Experiment 1	$0.0125\ M$	$0.0253\ M$	$0.0281\ M/s$
Experiment 2	$0.0250\ M$	$0.0253\ M$	$0.112\ M/s$
Experiment 3	$0.0125\ M$	$0.0506\ M$	$0.0561\ M/s$

Obtain the rate law. What is the value of the rate constant?

Change of Concentration with Time; Half-Life

15.29 Sulfuryl chloride, SO_2Cl_2, decomposes when heated:

$$SO_2Cl_2(g) \longrightarrow SO_2(g) + Cl_2(g)$$

In an experiment, the initial concentration of sulfuryl chloride was 0.0248 mol/L. If the rate constant is $2.2 \times 10^{-5}/s$, what is the concentration of SO_2Cl_2 after 4.5 hr? The reaction is first order.

15.31 If the rate constant for the first-order reaction

$$SO_2Cl_2(g) \longrightarrow SO_2(g) + Cl_2(g)$$

is $2.2 \times 10^{-5}/s$, what is the half-life? How long would it take for the concentration of SO_2Cl_2 to decrease to 25% of its initial value? to 12.5% of its initial value?

15.30 Cyclopropane, C_3H_6, is converted to its isomer propylene, $CH_2{=}CHCH_3$, when heated. The rate law is first order in cyclopropane, and the rate constant is $6.0 \times 10^{-4}/s$ at 500°C. If the initial concentration of cyclopropane is 0.0226 mol/L what is the concentration after 955 s?

15.32 Dinitrogen pentoxide, N_2O_5, decomposes when heated in carbon tetrachloride solvent:

$$N_2O_5 \longrightarrow 2NO_2 + O_2(g)$$

If the rate constant is $6.2 \times 10^{-4}/min$, what is the half-life? (The rate law is first order in N_2O_5.) How long would it take for the concentration of N_2O_5 to decrease to 25% of its initial value? to 12.5% of its initial value?

Arrhenius Equation

15.33 In a series of experiments on the decomposition of dinitrogen pentoxide, N_2O_5, rate constants were determined at two different temperatures. At 35°C, the rate constant was found to equal $1.4 \times 10^{-4}/s$; at 45°C, the rate constant was $5.0 \times 10^{-4}/s$. What is the activation energy for this reaction? What is the value of the rate constant at 55°C?

15.35 The rate of a particular reaction triples when the temperature is increased from 25°C to 35°C. Calculate the activation energy for this reaction.

15.34 The reaction

$$2NOCl(g) \longrightarrow 2NO(g) + Cl_2(g)$$

has rate constant values of $9.3 \times 10^{-6}/s$ at 350 K and $6.9 \times 10^{-4}/s$ at 400 K. Calculate the activation energy for this reaction. What is the rate constant at 450 K?

15.36 The rate of a particular reaction quadruples when the temperature is increased from 25°C to 35°C. Calculate the activation energy for this reaction.

Reaction Mechanisms

15.37 Nitric oxide, NO, is believed to react with chlorine according to the following mechanism:

$$NO + Cl_2 \rightleftharpoons NOCl_2 \quad \text{(elementary reaction)}$$
$$NOCl_2 + NO \longrightarrow 2NOCl \quad \text{(elementary reaction)}$$

Identify any reaction intermediate. What is the overall equation?

15.39 Identify the molecularity of each of the following elementary reactions:

(a) $NO + O_3 \longrightarrow NO_2 + O_2$
(b) $NOCl_2 + NO \longrightarrow 2NOCl$
(c) $O_3 \longrightarrow O_2 + O$
(d) $H + H + N_2 \longrightarrow H_2 + N_2^*$

15.41 Write a rate equation, showing the dependence of rate on reactant concentrations, for each of the following elementary reactions:

(a) $O_3 \longrightarrow O_2 + O$
(b) $NOCl_2 + NO \longrightarrow 2NOCl$

15.43 The isomerization of cyclopropane, C_3H_6, is believed to occur by the following mechanism:

$$C_3H_6 + C_3H_6 \xrightarrow{k_1} C_3H_6 + C_3H_6^* \quad \text{(Step 1)}$$
$$C_3H_6^* \xrightarrow{k_2} CH_2{=}CHCH_3 \quad \text{(Step 2)}$$

Here $C_3H_6^*$ is an excited cyclopropane molecule. At low pressure Step 1 is much slower than Step 2. Derive the rate law for this mechanism at low pressure. Explain.

15.45 The reaction

$$H_2(g) + I_2(g) \longrightarrow 2HI(g)$$

may occur by the following mechanism:

$$I_2 \xrightarrow{K_1} 2I \quad \text{(fast, equilibrium)}$$
$$I + I + H_2 \xrightarrow{k_2} 2HI \quad \text{(slow)}$$

Here K_1 is the equilibrium constant for the first step. What rate law is predicted by the mechanism?

15.38 The decomposition of ozone occurs in two steps:

$$O_3 \rightleftharpoons O_2 + O \quad \text{(elementary reaction)}$$
$$O_3 + O \longrightarrow 2O_2 \quad \text{(elementary reaction)}$$

Identify any reaction intermediate. What is the overall reaction?

15.40 What is the molecularity of each of the following elementary reactions?

(a) $O + O_2 + N_2 \longrightarrow O_3 + N_2^*$
(b) $NO_2Cl + Cl \longrightarrow NO_2 + Cl_2$
(c) $Cl + H_2 \longrightarrow HCl + H$
(d) $CS_2 \longrightarrow CS + S$

15.42 Write a rate equation, showing the dependence of rate on reactant concentrations, for each of the following elementary reactions:

(a) $CS_2 \longrightarrow CS + S$
(b) $CH_3Br + OH^- \longrightarrow CH_3OH + Br^-$

15.44 The thermal decomposition of nitryl chloride, NO_2Cl,

$$2NO_2Cl(g) \longrightarrow 2NO_2(g) + Cl_2(g)$$

is thought to occur by the following mechanism:

$$NO_2Cl \xrightarrow{k_1} NO_2 + Cl \quad \text{(slow step)}$$
$$NO_2Cl + Cl \xrightarrow{k_2} NO_2 + Cl_2 \quad \text{(fast step)}$$

What rate law is predicted by this mechanism?

15.46 The oxidation of nitric oxide by oxygen,

$$2NO(g) + O_2(g) \longrightarrow 2NO_2(g)$$

may have the following mechanism:

$$NO + O_2 \rightleftharpoons NO_3 \quad \text{(fast, equilibrium)}$$
$$NO_3 + NO \longrightarrow 2NO_2 \quad \text{(slow)}$$

What is the rate law derived from this mechanism?

Catalysis

15.47 Consider the following mechanism, and indicate the species acting as a catalyst:

$$Cl_2 \rightleftharpoons 2Cl$$
$$N_2O + Cl \longrightarrow N_2 + ClO$$
$$ClO + ClO \longrightarrow Cl_2 + O_2$$

Explain why you believe this species is a catalyst. What is the overall reaction? (*Note:* The second step occurs twice each time the first and third steps occur once.) What substance is introduced into the reaction mixture to give the catalytic activity?

15.48 Consider the following mechanism for a reaction in aqueous solution and indicate the species acting as a catalyst:

$$NH_2NO_2 + OH^- \rightleftharpoons H_2O + NHNO_2^-$$
$$NHNO_2^- \longrightarrow N_2O(g) + OH^-$$

Explain why you believe this species is a catalyst. What is the overall reaction? What substance might be added to the reaction mixture to give the catalytic activity?

533

Additional Problems

15.49 A study of the decomposition of azomethane,

$$CH_3NNCH_3(g) \longrightarrow C_2H_6(g) + N_2(g)$$

gave the following concentrations of azomethane at various times:

Time	$[CH_3NNCH_3]$
0 min	1.50×10^{-2} M
10 min	1.29×10^{-2} M
20 min	1.10×10^{-2} M
30 min	0.95×10^{-2} M

Obtain the average rate in units of M/s for each time interval.

15.50 Nitrogen dioxide decomposes when heated:

$$2NO_2(g) \longrightarrow 2NO(g) + O_2(g)$$

During an experiment, the concentration of NO_2 varied with time in the following way:

Time	$[NO_2]$
0.0 min	0.1103 M
1.0 min	0.1076 M
2.0 min	0.1050 M
3.0 min	0.1026 M

Obtain the average rate in units of M/s for each time interval.

*15.51** We can write the rate law for the decomposition of azomethane as

$$Rate = k[CH_3NNCH_3]^n \quad or \quad k = \frac{rate}{[CH_3NNCH_3]^n}$$

where n is the order of the reaction. Note that if we divide the rates at various times by the concentrations raised to the correct power n, we should get the same number (the rate constant, k). Verify that the decomposition of azomethane is first order by dividing each average rate in a time interval (obtained in Problem 15.49) by the average concentration in that interval. Note that each calculation gives nearly the same value. Take the average of these calculated values to obtain the rate constant.

*15.52** Use the technique described in Problem 15.51 to verify that the decomposition of nitrogen dioxide is second order. That is, divide each average rate in a time interval (obtained in Problem 15.50) by the square of the average concentration in that interval. Note that each calculation gives nearly the same value. Take the average of these calculated values to obtain the rate constant.

15.53 Methyl acetate reacts in acidic solution:

$$CH_3COOCH_3 + H_2O \xrightarrow{H^+} CH_3OH + CH_3COOH$$

methyl acetate   methanol   acetic acid

The rate law is first order in methyl acetate in acidic solution and the rate constant at 25°C is 1.26×10^{-4}/s. How long will it take for 85% of the methyl acetate to react?

15.54 Benzene diazonium chloride, C_6H_5NNCl, decomposes by a first-order rate law.

$$C_6H_5NNCl \longrightarrow C_6H_5Cl + N_2(g)$$

If the rate constant at 20°C is 4.3×10^{-5}/s, how long will it take for 85% of the compound to decompose?

15.55 What is the half-life of methyl acetate at 25°C in the acidic solution described in Problem 15.53?

15.56 What is the half-life of benzene diazonium chloride at 20°C? See Problem 15.54 for data.

15.57 A compound decomposes by a first-order reaction. If the concentration of the compound is 0.0250 M after 65 s when the initial concentration was 0.0350 M, what is the concentration of the compound after 98 s?

15.58 A compound decomposes by a first-order reaction. The concentration of compound decreases from 0.1180 M to 0.0950 M in 5.2 min. What fraction of the compound remains after 6.7 min?

*15.59** Plot the data given in Problem 15.49 to verify that the decomposition of azomethane is first order. Determine the rate constant from the slope of the straight-line plot of log $[CH_3NNCH_3]$ versus time.

*15.60** The decomposition of aqueous hydrogen peroxide in a given concentration of catalyst gave the following data:

Time	$[H_2O_2]$
0.0 min	0.1000 M
5.0 min	0.0804 M
10.0 min	0.0648 M
15.0 min	0.0519 M

Verify that the reaction is first order. Determine the rate constant (in units of /s) from the slope of the straight-line plot of log $[H_2O_2]$ versus time.

15.61 The decomposition of nitrogen dioxide,

$$2NO_2(g) \longrightarrow 2NO(g) + O_2(g)$$

has a rate constant of 0.498 M/s at 319°C and a rate constant of 1.81 M/s at 354°C. What are the values of the activation energy and the frequency factor for this reaction? What is the rate constant at 383°C?

15.62 A second-order reaction has a rate constant of $8.7 \times 10^{-4}/(M \cdot s)$ at 30°C. At 40°C, the rate constant is $1.5 \times 10^{-3}/(M \cdot s)$. What are the activation energy and frequency factor for this reaction? Predict the value of the rate constant at 45°C.

15.63 At high temperature, the reaction

$$NO_2(g) + CO(g) \longrightarrow NO(g) + CO_2(g)$$

is thought to occur in a single step. What should be the rate law in that case?

15.64 Methyl chloride, CH_3Cl, reacts in basic solution to give methanol:

$$CH_3Cl + OH^- \longrightarrow CH_3OH + Cl^-$$

This reaction is believed to occur in a single step. If so, what should be the rate law?

15.65 Nitryl bromide, NO_2Br, decomposes into nitrogen dioxide and bromine:

$$2NO_2Br(g) \longrightarrow 2NO_2(g) + Br_2(g)$$

A proposed mechanism is

$$NO_2Br \longrightarrow NO_2 + Br \qquad \text{(slow)}$$
$$NO_2Br + Br \longrightarrow NO_2 + Br_2 \qquad \text{(fast)}$$

Write the rate law predicted by this mechanism.

15.66 Tertiary butyl chloride reacts in basic solution according to the equation

$$(CH_3)_3CCl + OH^- \longrightarrow (CH_3)_3COH + Cl^-$$

The accepted mechanism for this reaction is

$$(CH_3)_3CCl \longrightarrow (CH_3)_3C^+ + Cl^- \qquad \text{(slow)}$$
$$(CH_3)_3C^+ + OH^- \longrightarrow (CH_3)_3COH \qquad \text{(fast)}$$

What should be the rate law for this reaction?

15.67 Urea, $(NH_2)_2CO$, can be prepared by heating ammonium cyanate, NH_4CNO:

$$NH_4CNO \longrightarrow (NH_2)_2CO$$

This reaction may occur by the following mechanism:

$$NH_4^+ + CNO^- \rightleftharpoons NH_3 + HCNO \text{ (fast, equilibrium)}$$
$$NH_3 + HCNO \longrightarrow (NH_2)_2CO \qquad \text{(slow)}$$

What is the rate law predicted by this mechanism?

15.68 Acetone reacts with iodine in acidic aqueous solution to give monoiodoacetone:

$$CH_3COCH_3 + I_2 \xrightarrow{H^+} CH_3COCH_2I + HI$$

acetone   monoiodoacetone

A possible mechanism for this reaction is

Write the rate law that you derive from this mechanism.

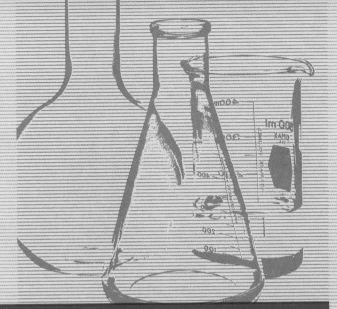

16. Chemical Equilibrium; Gaseous Reactions

Describing Chemical Equilibrium

16.1 Chemical Equilibrium—A Dynamic Equilibrium

16.2 The Equilibrium Constant Definition of the Equilibrium Constant K_c/ Obtaining Equilibrium Constants for Reactions/ The Equilibrium Constant K_p

16.3 Heterogeneous Equilibria

Using an Equilibrium Constant

16.4 Qualitatively Interpreting an Equilibrium Constant

16.5 Predicting the Direction of Reaction

16.6 Calculating Equilibrium Concentrations

Changing the Reaction Conditions and the Application of LeChatelier's Principle

16.7 Adding a Catalyst

16.8 Removing or Adding Reactants or Products

16.9 Changing the Pressure and Temperature Effect of Pressure Change/ Effect of Temperature Change/ Choosing the Optimum Conditions for Reaction

Natural gas, which is primarily methane, CH_4, offers a number of advantages as a fuel. It is easy to use and burns clean. Moreover, it is conveniently transported by pipeline. Unfortunately, supplies of natural gas are limited. Perhaps 80% of U.S. supplies of natural gas will be exhausted in 30 to 40 years. Coal, on the other hand, is more plentiful. It has been estimated that there is sufficient coal to last 300 to 400 years. The availability of this solid fuel has stimulated research into methods of converting it to liquid and gaseous fuels.

One method for obtaining a gaseous fuel that has recently received much attention begins with the conversion of coal with steam to a gaseous mixture of carbon monoxide, CO, and hydrogen, H_2. Under certain conditions this mixture will react to produce methane, which can be used in place of natural gas:

$$CO(g) + 3H_2(g) \longrightarrow CH_4(g) + H_2O(g)$$
$$\text{methane}$$

This reaction, which requires a catalyst to make it occur at a reasonable rate, is called *catalytic methanation*. It was discovered in 1902 by Paul Sabatier, a Nobel Prize–winning French chemist. The reaction is used industrially today to remove traces of carbon monoxide from hydrogen. Present research is directed toward discovering the optimum conditions for converting high concentrations of carbon monoxide to methane, CH_4.

Interestingly, catalytic methanation is the reverse of a reaction presently used to prepare synthesis gas (carbon monoxide–hydrogen mixtures). In this reaction, called *steam reforming*, methane (as natural gas) reacts with steam:

$$CH_4(g) + H_2O(g) \longrightarrow CO(g) + 3H_2(g)$$

The product, synthesis gas, is used to prepare a number of industrial chemicals, such as methanol, CH_3OH. It is also the source of hydrogen for the industrial preparation of ammonia, NH_3, from its elements (Haber process).

The processes of catalytic methanation and steam reforming show us the reversibility of chemical reactions. Because of this reversibility, reactions in general do not go completely to products. Rather, they result in a mixture of reactants and products. Suppose we start with a mixture of CO and H_2. This reacts by catalytic methanation to give CH_4 and H_2O vapor. These products in turn react by steam reforming to give the original reactants, CO and H_2. Depending on the conditions under which the reaction is run, the final reaction mixture may contain a higher concentration of products or a higher concentration of reactants. An important question here is "What conditions favor the production of methane, and what conditions favor CO and H_2?"

A reaction mixture that ceases to change and consists of reactants and products in definite concentrations is said to have reached *chemical equilibrium*. In this chapter, we will see how to determine the composition of a reaction mixture at equilibrium and how to alter this composition by changing the conditions for the reaction.

Chapter Overview

As we described very briefly in Chapter 9, chemical reactions in general do not go to completion, but rather reach a *dynamic equilibrium*. When this equilibrium is reached, forward and reverse reactions occur at the same rate, so that there is no net change. In the first part of this chapter, we will discuss how we can describe this chemical equilibrium in terms of the *equilibrium constant*. Once we have determined the equilibrium constant, we can find the *equilibrium composition* of a reaction mixture for different starting amounts of reactants. In the last sections of the chapter, we will discuss how *changing the reaction conditions*, such as the pressure or temperature, affects the equilibrium composition.

The concepts and calculations in this chapter will be illustrated by gaseous reactions. We will treat reactions in aqueous solution in later chapters.

Describing Chemical Equilibrium

Many chemical reactions are like the catalytic methanation reaction discussed in the chapter opening. Under the proper conditions, such reactions can be made to go in either one direction or the other. Let us look more closely at this reversibility and see how we may characterize it quantitatively.

16.1 Chemical Equilibrium—A Dynamic Equilibrium

When substances react, they eventually form a mixture of reactants and products in *dynamic equilibrium.* ■ This dynamic equilibrium consists of a forward reaction, in which substances react to give products, and a reverse reaction, in which products react to give the original reactants. Both forward and reverse reactions occur at the same rate, or speed.

Consider the catalytic methanation reaction discussed in the chapter opening. It consists of forward and reverse reactions, which we represent by the chemical equation

$$CO(g) + 3H_2(g) \rightleftharpoons CH_4(g) + H_2O(g)$$

Suppose we put 1.000 mol CO and 3.000 mol H_2 into a 10.00-L vessel at 1200 K (927°C). The rate of the reaction of CO and H_2 depends on the concentrations of CO and H_2. At first these concentrations are large, but as the substances react their concentrations decrease (Figure 16.1). Thus, the rate of the forward reaction is large at first but steadily decreases. On the other hand, the concentrations of CH_4 and H_2O, which are zero at first,

■ The concept of dynamic equilibrium in chemical reactions was briefly mentioned in Sections 9.1 and 15.8. Discussions on vapor pressure (Section 11.2) and solubility (Section 12.2) give detailed explanations of the role of dynamic equilibria in those processes.

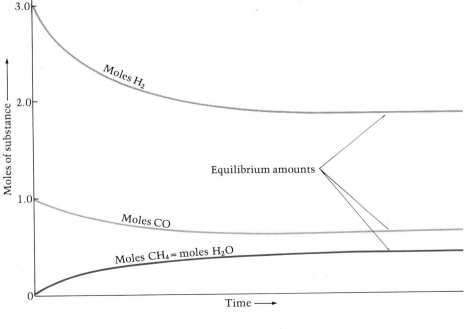

Figure 16.1
Variation in moles of substances during catalytic methanation in a 10.00-L vessel. Note how the amounts of substances become constant at equilibrium.

increase with time, as Figure 16.1 shows. Therefore, the rate of the reverse reaction starts at zero and steadily increases. The forward rate decreases and the reverse rate increases until eventually they become equal. When that happens, CO and H_2 molecules are formed as fast as they react. The concentrations of reactants and products no longer change, and the reaction mixture has reached *equilibrium*. Figure 16.1 shows how the amounts of substances in the reaction mixture become constant when equilibrium is reached.

Chemical equilibrium is the state reached by a reaction mixture when the rates of forward and reverse reactions have become equal, so that *net* change no longer occurs. If we observe the reaction mixture, we see nothing happening, although the forward and reverse reactions are continuing. It is the occurrence of the forward and reverse reactions that makes the equilibrium a dynamic process.

Suppose we place known amounts of reactants in a vessel and let the mixture come to equilibrium. To obtain the composition of the equilibrium mixture, it is only necessary to determine the amount of one of the substances. The amounts of the others can be calculated from the amounts of substances originally placed in the vessel and the equation that represents the reaction. For example, suppose we place 1.000 mol CO and 3.000 mol H_2 in a reaction vessel at 1200 K. After equilibrium is reached, we chill the reaction mixture quickly to condense the water vapor to liquid, which we then weigh and find to be 0.387 mol. The other substances are calculated by the stoichiometry of the reaction, as shown in the following example.

Example 16.1

Carbon monoxide and hydrogen react according to the following equation:

$$CO(g) + 3H_2(g) \rightleftharpoons CH_4(g) + H_2O(g)$$

When 1.000 mol CO and 3.000 mol H_2 are placed in a 10.00-L vessel at 927°C (1200 K) and allowed to come to equilibrium, the mixture is found to contain 0.387 mol H_2O. What is the molar composition of the equilibrium mixture; that is, how many moles of each substance are present?

Solution

In many of the equilibrium problems in this and the following chapters, we *start* with given amounts (or concentrations) of substances, which *change* during the reaction to give *equilibrium* values. It is very useful to set up a table for these starting, change, and equilibrium values so that we can easily see what we have

to calculate. Using the information given in the problem, we set up the table shown below.

In setting up this table, we filled in the values given in the problem. We are given the starting amounts of all substances in the equation, but we are not told explicitly what changes occur in these amounts during the reaction. Therefore, we let x be the molar change. That is, each product increases by x moles multiplied by the coefficients of the substances in the balanced equation. Reactants decrease by x moles multiplied by the corresponding coefficients. The decrease is indicated by a negative sign. Equilibrium values are equal to starting values plus the changes. For example, the starting amount of H_2 is 3.000 mol. Since this changes by $-3x$ mol, the equilibrium amount is $(3.000 - 3x)$ mol. Similarly, the starting amount of H_2O is 0, the change is x mol, and the equilibrium amount is $(0 + x)$ mol $= x$ mol. The only equilibrium amount given in the prob-

Amount (moles)	CO(*g*)	+	3H₂(*g*)	⇌	CH₄(*g*)	+	H₂O(*g*)
Starting	1.000		3.000		0		0
Change	$-x$		$-3x$		$+x$		$+x$
Equilibrium	$1.000 - x$		$3.000 - 3x$		x		$x = 0.387$

(Continued)

lem statement is that for H_2O (= 0.387 mol). This tells us that x = 0.387. Equilibrium amounts for other substances can be calculated from the expressions given in the table, using this value of x. Thus,

Equilibrium amount CO = (1.000 − x) mol
    = (1.000 − 0.387) mol = 0.613 mol

Equilibrium amount H_2 = (3.000 − 3x) mol
    = (3.000 − 3 × 0.387) mol = 1.839 mol

Equilibrium amount CH_4 = x mol = 0.387 mol

Therefore, the amounts of substances in the equilibrium mixture are 0.613 mol CO, 1.839 mol H_2, 0.387 mol CH_4, and 0.387 mol H_2O.

Exercise 16.1

Synthesis gas (a mixture of CO and H_2) is increased in concentration of hydrogen by passing it with steam over a catalyst. This is the so-called water-gas-shift reaction. Some of the CO is converted to CO_2, which can be removed:

$$CO(g) + H_2O(g) \rightleftharpoons CO_2(g) + H_2(g)$$

Suppose we start with a gaseous mixture containing 1.00 mol CO and 1 mol H_2O. When equilibrium is reached at 1000°C, the mixture contains 0.59 mol H_2. What is the molar composition of the equilibrium mixture?

(See Problems 16.13, 16.14, 16.15, and 16.16.)

16.2 The Equilibrium Constant

In the previous section, we found that when 1.000 mol CO and 3.000 mol H_2 react in a 10.00-L vessel by catalytic methanation at 1200 K, they give an equilibrium mixture containing 0.613 mol CO, 1.839 mol H_2, 0.387 mol CH_4, and 0.387 mol H_2O. Let us call this Experiment I. Now consider a similar experiment, Experiment II, in which we start with an additional mole of carbon monoxide. That is, we place 2.000 mol CO and 3.000 mol H_2 in a 10.00-L vessel at 1200 K. At equilibrium, we find that the vessel contains 1.522 mol CO, 1.566 mol H_2, 0.478 mol CH_4, and 0.478 mol H_2O. What we find from the results of Experiments I and II is that the equilibrium composition depends on the amounts of starting substances. Nevertheless, what we will see is that all of the equilibrium compositions for a reaction at a given temperature are related by a quantity called the *equilibrium constant.*

Definition of the Equilibrium Constant K_c

Consider the general reaction

$$a A + b B \rightleftharpoons c C + d D$$

where A, B, C, and D denote reactants and products and a, b, c, and d are coefficients in the balanced chemical equation. The **equilibrium-constant expression** for a reaction is obtained by multiplying the concentrations of products together, dividing by the concentrations of reactants, and raising each concentration term to a power equal to the coefficient in the chemical equation. The **equilibrium constant** K_c is the value obtained for the equilibrium-constant expression when equilibrium concentrations are substituted. For the general reaction, we have ∎

$$K_c = \frac{[C]^c[D]^d}{[A]^a[B]^b}$$

Here, the molar concentration of a substance is denoted by writing the formula of that substance in square brackets. The subscript c on the equilibrium constant denotes that it is defined in terms of molar concentrations. According to the **law of mass action,** the value of K_c is a constant for a particular reaction at a given temperature, whatever equilibrium concentrations are substituted.

As the following example illustrates, the equilibrium-constant expression is defined in terms of the balanced chemical equation. If the equation is rewritten with different coefficients, the equilibrium-constant expression is changed.

∎ In Section 16.3, we will see that we ignore concentrations of pure liquids and solids when we write the equilibrium-constant expression. The concentrations of these substances are constant and are incorporated into the value of K_c.

Example 16.2

(a) Write the equilibrium-constant expression, K_c, for catalytic methanation:

$$CO(g) + 3H_2(g) \rightleftharpoons CH_4(g) + H_2O(g)$$

(b) Write the equilibrium-constant expression, K_c, for the reverse of the reaction in (a), that is

$$CH_4(g) + H_2O(g) \rightleftharpoons CO(g) + 3H_2(g)$$

(c) Write the equilibrium-constant expression, K_c, for the synthesis of ammonia:

$$N_2(g) + 3H_2(g) \rightleftharpoons 2NH_3(g)$$

(d) Write the equilibrium-constant expression, K_c, when the equation for the reaction in (c) is written

$$\tfrac{1}{2}N_2(g) + \tfrac{3}{2}H_2(g) \rightleftharpoons NH_3(g)$$

Solution

(a) The expression for the equilibrium constant is:

$$K_c = \frac{[CH_4][H_2O]}{[CO][H_2]^3}$$

Note that concentrations of products are on the top and concentrations of reactants are on the bottom. Also, note that each concentration term is raised to a power equal to the coefficient of the substance in the chemical equation. (b) When the equation is written in reverse order, the expression for K_c is inverted:

$$K_c = \frac{[CO][H_2]^3}{[CH_4][H_2O]}$$

(c) The equilibrium constant for $N_2 + 3H_2 \rightleftharpoons 2NH_3$ is

$$K_c = \frac{[NH_3]^2}{[N_2][H_2]^3}$$

(d) If the coefficients in the equation in (c) are multiplied by $\tfrac{1}{2}$ to give $\tfrac{1}{2}N_2 + \tfrac{3}{2}H_2 \rightleftharpoons NH_3$, the equilibrium-constant expression becomes

$$K_c = \frac{[NH_3]}{[N_2]^{1/2}[H_2]^{3/2}}$$

which is the square root of the previous expression.

Exercise 16.2

Write the equilibrium-constant expression, K_c, for the equation

$$2NO_2(g) + 7H_2(g) \rightleftharpoons 2NH_3(g) + 4H_2O(g)$$

Write the equilibrium-constant expression, K_c, if this reaction is written

$$NO_2(g) + \tfrac{7}{2}H_2(g) \rightleftharpoons NH_3(g) + 2H_2O(g)$$

(See Problems 16.17, 16.18, 16.19, and 16.20.)

Obtaining Equilibrium Constants for Reactions

At the beginning of this section, we gave data from the results of two experiments, Experiments I and II, involving catalytic methanation. By substituting the molar concentrations from these two experiments into the equilibrium-constant expression for the reaction, we can show that we get the same value, as we expect from the law of mass action. This value equals K_c for methanation at 1200 K. Thus, besides verifying the validity of the law of mass action in this case, we also will see how an equilibrium constant can be obtained from experimental data.

Experiment I The equilibrium composition is 0.613 mol CO, 1.839 mol H_2, 0.387 mol CH_4, and 0.387 mol H_2O. Since the volume of the reaction vessel is 10.00 L, the concentration of CO is

$$[CO] = \frac{0.613 \text{ mol}}{10.00 \text{ L}} = 0.0613 \ M$$

Similarly, the other equilibrium concentrations are: $[H_2] = 0.1839 \ M$, $[CH_4] = 0.0387 \ M$, and $[H_2O] = 0.0387 \ M$.

Let us substitute these values into the equilibrium expression for catalytic methanation. (The expression was obtained in Example 16.2.) Although until now we have consistently carried units along with numbers in calculations, it is the usual practice to write equilibrium constants without units. We will follow that practice here. Substitution of concentrations into the equilibrium-constant expression gives

$$K_c = \frac{[CH_4][H_2O]}{[CO][H_2]^3} = \frac{(0.0387)(0.0387)}{(0.0613)(0.1839)^3} = 3.93$$

Experiment II The equilibrium composition is 1.522 mol CO, 1.566 mol H_2, 0.478 mol CH_4, and 0.478 mol H_2O. Therefore, the concentrations, which are obtained by dividing by 10.00 L, are $[CO] = 0.1522 \ M$, $[H_2] = 0.1566 \ M$, $[CH_4] = 0.0478 \ M$, and $[H_2O] = 0.0478 \ M$. Substituting into the equilibrium-constant expression gives

$$K_c = \frac{[CH_4][H_2O]}{[CO][H_2]^3} = \frac{(0.0478)(0.0478)}{(0.1522)(0.1566)^3} = 3.91$$

Within the precision of the data, these values (3.93 and 3.91) of the equilibrium expression for different starting mixtures of these gases at 1200 K are essentially the same. Other similar experiments (see Figure 16.2) give about the same value for K_c. These experiments verify the law of mass action for this reaction. We can take the equilibrium constant for catalytic methanation at 1200 K to be 3.92, the average of these values.

The following exercise provides additional practice with evaluating an equilibrium constant from the experimentally determined equilibrium composition. Example 16.3 gives the detailed solution of a more complicated problem of this type.

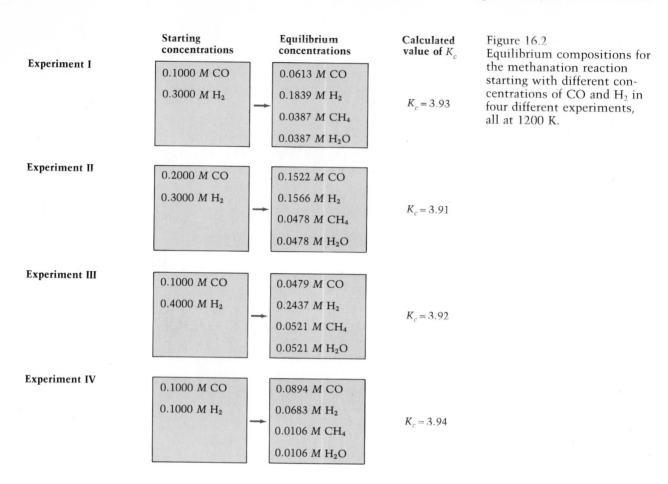

	Starting concentrations	Equilibrium concentrations	Calculated value of K_c
Experiment I	0.1000 M CO 0.3000 M H$_2$	0.0613 M CO 0.1839 M H$_2$ 0.0387 M CH$_4$ 0.0387 M H$_2$O	$K_c = 3.93$
Experiment II	0.2000 M CO 0.3000 M H$_2$	0.1522 M CO 0.1566 M H$_2$ 0.0478 M CH$_4$ 0.0478 M H$_2$O	$K_c = 3.91$
Experiment III	0.1000 M CO 0.4000 M H$_2$	0.0479 M CO 0.2437 M H$_2$ 0.0521 M CH$_4$ 0.0521 M H$_2$O	$K_c = 3.92$
Experiment IV	0.1000 M CO 0.1000 M H$_2$	0.0894 M CO 0.0683 M H$_2$ 0.0106 M CH$_4$ 0.0106 M H$_2$O	$K_c = 3.94$

Figure 16.2
Equilibrium compositions for the methanation reaction starting with different concentrations of CO and H$_2$ in four different experiments, all at 1200 K.

Exercise 16.3

When 1.00 mol each of carbon monoxide and water reach equilibrium at 1000°C in a 10.0-L vessel, the equilibrium mixture contains 0.57 mol CO, 0.57 mol H$_2$O, 0.43 mol CO$_2$, and 0.43 mol H$_2$. Write the chemical equation for the equilibrium. What is the value of K_c? ■

(See Problems 16.21 and 16.22.)

■ The reaction described here is used industrially to adjust the ratio of H$_2$ to CO in synthesis gas (mixture of CO and H$_2$). In this process CO reacts with H$_2$O to give H$_2$, so that the ratio of H$_2$ to CO is increased.

Example 16.3

Hydrogen iodide, HI, decomposes at moderate temperatures according to the equation

$$2HI(g) \rightleftharpoons H_2(g) + I_2(g)$$

The amount of I$_2$ in the reaction mixture can be determined from the intensity of the violet color of I$_2$; the more intense the color, the more I$_2$ in the reaction vessel. When 4.00 mol HI were placed in a 5.00-L vessel

at 458°C, the equilibrium mixture was found to contain 0.442 mol I$_2$. What is the value of K_c for the decomposition of HI at this temperature?

Solution

Note that this problem involves starting, change, and equilibrium amounts of substances. Therefore, we set

(*Continued*)

up a table as we did in Example 16.1. Since we want the concentrations of substances to evaluate K_c, we will set this table up in terms of amounts per liter, or molar concentrations. We first calculate the concentrations of substances whose amounts were given in the problem. We divide the amounts by the volume of the reaction vessel (5.00 L):

$$\text{Starting concentration of HI} = \frac{4.00 \text{ mol}}{5.00 \text{ L}} = 0.800 \ M$$

$$\text{Equilibrium concentration of } I_2 = \frac{0.442 \text{ mol}}{5.00 \text{ L}} = 0.0884 \ M$$

From these values, we set up the following table:

Concentration (M)	$2HI(g)$	$\rightleftharpoons$	$H_2(g)$	$+$	$I_2(g)$
Starting	0.800		0		0
Change	$-2x$		x		x
Equilibrium	$0.800 - 2x$		x		$x = 0.0884$

The equilibrium concentrations of substances can be evaluated from the expressions given in the last line of this table. We know that the equilibrium concentration of I_2 is 0.0884 M, and that this equals x. Therefore

$$[HI] = (0.800 - 2x) \ M = (0.800 - 2 \times 0.0884) \ M$$
$$= 0.623 \ M$$
$$[H_2] = x = 0.0884 \ M$$

Now we substitute into the equilibrium-constant expression for the reaction. From the chemical equation, we write

$$K_c = \frac{[H_2][I_2]}{[HI]^2}$$

Substituting, we get

$$K_c = \frac{(0.0884)(0.0884)}{(0.623)^2} = 0.0201$$

Exercise 16.4

Hydrogen sulfide is a colorless gas with a foul odor. It dissociates on heating:

$$2H_2S(g) \rightleftharpoons 2H_2(g) + S_2(g)$$

When 0.100 mol H_2S was introduced into a 10.0-L vessel and heated to 1132°C, it gave an equilibrium mixture containing 0.0285 mol H_2. What is the value of K_c at this temperature?

(See Problems 16.23, 16.24, 16.25, and 16.26.)

The Equilibrium Constant K_p

The concentration of a gas is proportional to its partial pressure at a fixed temperature (Figure 16.3). We can see this by looking at the ideal gas law, $PV = nRT$. This can be solved for n/V, which is the molar concentration of the gas. We get $n/V = P/RT$. In other words, the molar concentration of a gas equals its partial pressure divided by RT, which is constant at a given temperature.

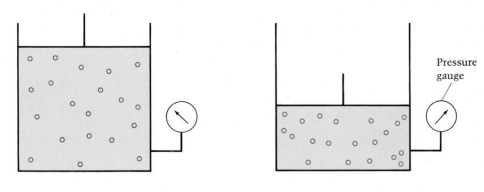

Figure 16.3
The concentration of a gas at a given temperature is proportional to the pressure.

Pressure gauge

Equilibrium constants for gaseous reactions are often written in terms of partial pressures (in atmospheres). This equilibrium constant, denoted K_p, is defined exactly like K_c, except that partial pressures are used in place of concentrations. For example, for catalytic methanation,

$$CO(g) + 3H_2(g) \rightleftharpoons CH_4(g) + H_2O(g)$$

the equilibrium expression in terms of partial pressures becomes

$$K_p = \frac{P_{CH_4}P_{H_2O}}{P_{CO}P_{H_2}{}^3}$$

In general, the value of K_p is different from K_c. From the relationship $n/V = P/RT$, one can show that

$$K_p = K_c(RT)^{\Delta n}$$

where Δn is the sum of the coefficients of gaseous products in the chemical equation minus the sum of the coefficients of gaseous reactants. For the methanation reaction ($K_c = 3.92$), in which 2 mol gaseous products ($CH_4 + H_2O$) are obtained from 4 mol gaseous reactants ($CO + 3H_2$), Δn equals $2 - 4 = -2$. Since the usual unit of partial pressures in K_p is atmospheres, the value of R is 0.0821 L · atm/(K · mol). Hence,

$$K_p = 3.92 \times (0.0821 \times 1200)^{-2} = 4.04 \times 10^{-4}$$

Exercise 16.5

Phosphorus pentachloride dissociates on heating:

$$PCl_5(g) \rightleftharpoons PCl_3(g) + Cl_2(g)$$

If K_c equals 3.26×10^{-2} at 191°C, what is K_p at this temperature?

(See Problems 16.27, 16.28, 16.29, and 16.30.)

16.3 Heterogeneous Equilibria

A **homogeneous equilibrium** is an equilibrium that involves reactants and products in a single phase. Catalytic methanation is an example of a homogeneous equilibrium, because it involves only gaseous reactants and products. On the the other hand, a **heterogeneous equilibrium** is an equilibrium involving reactants and products in more than one phase. For example, the reaction of iron metal filings with steam to produce iron oxide, Fe_3O_4, and hydrogen involves solid phases, Fe and Fe_3O_4, in addition to a gaseous phase:

$$3Fe(s) + 4H_2O(g) \rightleftharpoons Fe_3O_4(s) + 4H_2(g)$$

In writing the equilibrium-constant expression for a heterogeneous equilibrium, we omit concentration terms for pure solids and liquids. For the previous reaction of iron with steam, we would write

$$K_c = \frac{[H_2]^4}{[H_2O]^4}$$

Concentrations of Fe and Fe_3O_4 are omitted because, whereas the concentration of a gas can have various values, the concentration of a pure solid or a pure liquid is a constant at a given temperature and depends on the density. For example, the density of iron is 7.86 g/cm^3, which is equivalent to 141 mol/L or 141 M. In any equilibrium involving pure iron, we have that [Fe] = 141 M.

To see why such constant concentrations are omitted in writing K_c, let us write the equilibrium-constant expression for the reaction of iron with steam, but let us include [Fe] and [Fe_3O_4]. We will call this K_c'.

$$K_c' = \frac{[Fe_3O_4][H_2]^4}{[Fe]^3[H_2O]^4}$$

We can rearrange this equation, putting all of the constant factors on the left side:

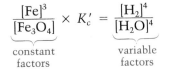

$$\underbrace{\frac{[Fe]^3}{[Fe_3O_4]} \times K_c'}_{\substack{\text{constant} \\ \text{factors}}} = \underbrace{\frac{[H_2]^4}{[H_2O]^4}}_{\substack{\text{variable} \\ \text{factors}}}$$

Note that all terms on the left-hand side are constant, but those on the right (the concentrations of H_2 and H_2O) are variable. The left-hand side equals K_c. Thus, in effect, concentrations of pure solids or pure liquids are incorporated in the value of K_c. ∎

The fact that the concentrations of Fe and Fe_3O_4 do not occur in the equilibrium-constant expression means that the equilibrium is not affected by the amounts of these substances, as long as some of each is present. However, chemical equilibrium cannot exist if either Fe or Fe_3O_4 is absent.

■ In precise work, equilibrium constants are defined in terms of *activities*, which are essentially "effective concentrations." For pure solids and liquids, however, the activities are given the value 1. Therefore, they do not appear in the equilibrium-constant expressions.

Example 16.4

(a) Quicklime (calcium oxide, CaO) is prepared by heating a source of calcium carbonate, $CaCO_3$, such as limestone or seashells:

$$CaCO_3(s) \rightleftharpoons CaO(s) + CO_2(g)$$

Write the expression for K_c. (b) We can write the equilibrium-constant expression for a physical equilibrium, such as vaporization, as well as for chemical

equilibrium. Write the expression for K_c for the vaporization of water:

$$H_2O(l) \rightleftharpoons H_2O(g)$$

Solution

In writing the equilibrium expressions, we ignore pure liquid and solid phases. (a) $K_c = [CO_2]$ (b) $K_c = [H_2O(g)]$

Exercise 16.6

The Mond process for purifying nickel involves the formation of nickel carbonyl, $Ni(CO)_4$, from nickel metal and carbon monoxide:

$$Ni(s) + 4CO(g) \rightleftharpoons Ni(CO)_4(g)$$

Write the expression for K_c for this reaction. ∎

(See Problems 16.31 and 16.32.)

■ In the *Mond process*, carbon monoxide is passed over impure nickel. Volatile nickel carbonyl is formed, leaving the impurities behind. Nickel carbonyl is then led into a heated chamber, where it decomposes and deposits pure nickel.

Using an Equilibrium Constant

In the preceding sections, we described how a chemical reaction reaches equilibrium and how we can characterize this equilibrium by the equilibrium constant. Now we want to see the ways in which an equilibrium constant can be used to answer important questions. We will look at the following uses:

1. *Qualitatively interpreting the equilibrium constant.* By merely looking at the magnitude of K_c, we can tell whether a particular equilibrium favors products or reactants.

2. *Predicting the direction of reaction.* Consider a reaction mixture that is not at equilibrium. By substituting the concentrations of substances that exist in a reaction mixture into an expression similar to the equilibrium constant and comparing with K_c, we can predict whether reaction will be toward products or toward reactants (as defined by the way we write the chemical equation).

3. *Calculating equilibrium concentrations.* Once we know the value of K_c for a reaction, we can determine the composition at equilibrium for any set of starting concentrations.

Let us see what meaning we can attach to the value of K_c.

16.4 Qualitatively Interpreting an Equilibrium Constant

If the value of the equilibrium constant is large, we know immediately that the products are favored at equilibrium. Consider the synthesis of ammonia from its elements:

$$N_2(g) + 3H_2(g) \rightleftharpoons 2NH_3(g)$$

At 25°C, the equilibrium constant K_c equals 4.1×10^8. This means that the numerator (product concentrations) is 4.1×10^8 times larger than the denominator (reactant concentrations). In other words, at this temperature the reaction favors the formation of ammonia at equilibrium.

We can verify this by calculating one possible equilibrium composition for this reaction. Suppose that the equilibrium mixture is 0.010 M in N_2 and 0.010 M in H_2. From these concentrations we can calculate the concentration of ammonia necessary to give equilibrium. Let us substitute the concentrations of N_2 and H_2 and the value of K_c into the equilibrium expression

$$K_c = \frac{[NH_3]^2}{[N_2][H_2]^3}$$

We get

$$\frac{[NH_3]^2}{(0.010)(0.010)^3} = 4.1 \times 10^8$$

Now we can solve for the concentration of ammonia:

$$[NH_3]^2 = 4.1 \times 10^8 \times (0.010)(0.010)^3 = 4.1$$

Figure 16.4
Methane (natural gas) reacts with oxygen to give mostly products, carbon dioxide and water ($CH_4 + 2O_2 \rightleftharpoons CO_2 + 2H_2O$). The equilibrium constant K_c is 10^{140}.

After we take the square root of both sides of this equation, we find that $[NH_3] = 2.0\ M$. Since the concentrations of N_2 and H_2 are each $0.010\ M$, the amount of ammonia formed at equilibrium is 200 times that of any one reactant. Figure 16.4 shows another reaction, with an enormously large equilibrium constant.

If the value of the equilibrium constant is small, the reactants are favored at equilibrium. As an example, consider the reaction of nitrogen and oxygen to give nitric oxide, NO:

$$N_2(g) + O_2(g) \rightleftharpoons 2NO(g)$$

The equilibrium constant K_c equals 4.6×10^{-31} at 25°C. If we assume that the concentrations of N_2 and O_2 are $1.0\ M$, we find that the concentration of NO is $6.8 \times 10^{-16}\ M$. In this case, the equilibrium constant is very small, and the concentration of product is not detectable. Reaction occurs to only a very limited extent.■

When the equilibrium constant is neither large nor small (between about 0.1 and 10) neither reactants nor products are strongly favored. The equilibrium mixture contains significant amounts of all substances in the reaction. For example, in the case of the methanation reaction, the equilibrium constant K_c equals 3.92 at 1200 K. We found that if we start with 1.000 mol CO and 3.000 mol H_2 in a 10.00-L vessel, the equilibrium composition is 0.613 mol CO, 1.839 mol H_2, 0.387 mol CH_4, and 0.387 mol H_2O. Neither reactants nor products are predominant.

■ The equilibrium constant does become large enough at higher temperatures (about 2000°C) to give appreciable amounts of nitric oxide.

In summary, if K_c for a reaction,

$$\underbrace{a\text{A} + b\text{B}}_{\text{reactants}} \rightleftharpoons \underbrace{c\text{C} + d\text{D}}_{\text{products}}$$

is large, the equilibrium mixture is largely products. On the other hand, if K_c is small, the equilibrium mixture is largely reactants. When K_c is between 0.1 and 10, the equilibrium mixture contains appreciable amounts of both reactants and products.

Exercise 16.7

The equilibrium constant K_c for the reaction

$$2\text{NO}(g) + \text{O}_2(g) \rightleftharpoons 2\text{NO}_2(g)$$

equals 4.0×10^{13} at 25°C. Does the equilibrium mixture contain predominantly reactants or products? If $[\text{NO}] = [\text{O}_2] = 0.50\ M$ at equilibrium, what is the equilibrium concentration of NO_2?

(See Problems 16.33, 16.34, 16.35, and 16.36.)

16.5 Predicting the Direction of Reaction

Suppose a gaseous mixture from an industrial plant has the following composition at 1200 K: 0.0200 M CO, 0.0200 M H$_2$, 0.00100 M CH$_4$, and 0.00100 M H$_2$O. If the mixture reacts at 1200 K by catalytic methanation,

$$\text{CO}(g) + 3\text{H}_2(g) \rightleftharpoons \text{CH}_4(g) + \text{H}_2\text{O}(g)$$

would the reaction go toward the right or toward the left? That is, would the mixture form more CH$_4$ and H$_2$O in going toward equilibrium, or would it form more CO and H$_2$?

To answer this question, we substitute the concentrations of substances into the *reaction quotient* and compare its value to K_c. The **reaction quotient, Q**, has the same form as the equilibrium-constant expression, but the concentrations that we substitute are those of a mixture that is not necessarily at equilibrium. For catalytic methanation, the reaction quotient is

$$Q = \frac{[\text{CH}_4][\text{H}_2\text{O}]}{[\text{CO}][\text{H}_2]^3}$$

If we substitute the concentrations of the gaseous mixture described earlier, we get

$$Q = \frac{(0.00100)(0.00100)}{(0.0200)(0.0200)^3} = 6.25$$

Recall that the equilibrium constant K_c for catalytic methanation is 3.92 at 1200 K. For the reaction mixture to go to equilibrium, the value of Q must decrease from 6.25 to 3.92. This will happen if the reaction goes to the left. In that case, the numerator of Q ($[\text{CH}_4][\text{H}_2\text{O}]$) will decrease, and the denominator ($[\text{CO}][\text{H}_2]^3$) will increase. Thus, the gaseous mixture will give more CO and H$_2$.

Consider the problem more generally. We are given a reaction mixture that is not yet at equilibrium. We would like to know in what direction the reaction will go as it approaches equilibrium. To answer this, we substitute the concentrations of substances from the mixture into the reaction quotient, Q. Then, we compare Q to the equilibrium constant K_c:

If $Q > K_c$, the reaction will go to the left.
If $Q < K_c$, the reaction will go to the right.
If $Q = K_c$, the reaction mixture is at equilibrium.

Example 16.5

A 50.0-L reaction vessel contains 1.00 mol N_2, 3.00 mol H_2, and 0.500 mol NH_3. Will more ammonia, NH_3, be formed when the mixture goes to equilibrium at 350°C? The equation is

$$N_2(g) + 3H_2(g) \rightleftharpoons 2NH_3(g)$$

K_c is 69.5 at 350°C.

Solution

The composition of the gas has been given in terms of moles. We convert these to molar concentrations by dividing by the volume (50.0 L). This gives 0.0200 M N_2, 0.0600 M H_2, and 0.0100 M NH_3. Substituting these concentrations into the reaction quotient gives

$$Q = \frac{[NH_3]^2}{[N_2][H_2]^3} = \frac{(0.0100)^2}{(0.0200)(0.0600)^3} = 23.1$$

Since $Q = 23.1$ is less than $K_c = 69.5$, the reaction will go to the right when it approaches equilibrium. Therefore, more ammonia will form.

Exercise 16.8

A 10.0-L vessel contains 0.0015 mol CO_2 and 0.10 mol CO. If a small amount of carbon is added to this vessel and the temperature raised to 1000°C, will more CO form? The reaction is

$$CO_2(g) + C(s) \rightleftharpoons 2CO(g)$$

The value of K_c for this reaction is 1.17 at 1000°C. Assume that the volume of gas in the vessel is 10.0 L.

(See Problems 16.37 and 16.38.)

16.6 Calculating Equilibrium Concentrations

Once we have determined the equilibrium constant for a reaction, we can use it to calculate the concentrations of substances in an equilibrium mixture. The next example illustrates a simple type of equilibrium problem.

Example 16.6

A gaseous mixture contains 0.30 mol CO, 0.10 mol H_2, and 0.020 mol H_2O, plus an unknown amount of CH_4, in each liter. This mixture is in equilibrium at 1200 K.

$$CO(g) + 3H_2(g) \rightleftharpoons CH_4(g) + H_2O(g)$$

What is the concentration of CH_4 in this mixture? The equilibrium constant, K_c, equals 3.92.

Solution

The equilibrium equation is

$$K_c = \frac{[CH_4][H_2O]}{[CO][H_2]^3}$$

Substituting the known concentrations and the value of

(*Continued*)

K_c into this equation gives

$$3.92 = \frac{[CH_4](0.020)}{(0.30)(0.10)^3}$$

We can now solve for $[CH_4]$:

$$[CH_4] = \frac{(3.92)(0.30)(0.10)^3}{(0.020)} = 0.059$$

The concentration of CH_4 in the mixture is 0.059 mol/L.

Exercise 16.9

Phosphorus pentachloride gives an equilibrium mixture of PCl_5, PCl_3, and Cl_2 when heated:

$$PCl_5(g) \rightleftharpoons PCl_3(g) + Cl_2(g)$$

A 1.00-L vessel contains an unknown amount of PCl_5 and 0.020 mol each of PCl_3 and Cl_2 at equilibrium at 250°C. How many moles of PCl_5 are in the vessel if K_c for this reaction is 0.0415 at 250°C?

(See Problems 16.39 and 16.40.)

Usually, we begin a reaction with known starting quantities of substances and want to calculate what the quantities will be at equilibrium. The next example illustrates the steps used to solve this type of problem.

Example 16.7

The reaction

$$CO(g) + H_2O(g) \rightleftharpoons CO_2(g) + H_2(g)$$

is used to increase the ratio of hydrogen in synthesis gas (mixtures of CO and H_2). Suppose we start with 1.00 mol each of carbon monoxide and water in a 50.0-L vessel. How many moles of each substance are in the equilibrium mixture at 1000°C? The equilibrium constant K_c at this temperature is 0.58.

Solution

Step 1 Note that we *start* with known quantities of substances, which *change* to *equilibrium* values. This suggests that we begin by setting up a table in which we list starting, change, and equilibrium quantities of substances. Since we will relate these to K_c, we should list molar amounts per liter, that is, molar concentrations. The *starting concentrations* of CO and H_2O are

$$[CO] = [H_2O] = \frac{1.00 \text{ mol}}{50.0 \text{ L}} = 0.0200 \text{ mol/L}$$

Concentrations of the products, CO_2 and H_2, are 0. The *changes in concentrations* when the mixture goes to equilibrium are not given. However, we can write them all in terms of a single unknown. If we let x be the moles of CO_2 formed per liter, then the mole of H_2 formed per liter is also x. Similarly, x moles each of CO and H_2O are consumed. We write the changes for CO and H_2O as $-x$. We obtain the *equilibrium concentrations* by adding the change in concentrations to the starting concentrations, as shown in the table below.

Step 2 We substitute the equilibrium concentrations into the equilibrium equation,

$$K_c = \frac{[CO_2][H_2]}{[CO][H_2O]}$$

and we get

$$0.58 = \frac{(x)(x)}{(0.0200 - x)(0.0200 - x)}$$

or

$$0.58 = \frac{x^2}{(0.0200 - x)^2}$$

Concentrations (M)	CO(g)	+	H$_2$O(g)	$\rightleftharpoons$	CO$_2$(g)	+	H$_2$(g)
Starting	0.0200		0.0200		0		0
Change	$-x$		$-x$		$+x$		$+x$
Equilibrium	$0.0200 - x$		$0.0200 - x$		x		x

(Continued)

Step 3 We now solve this equilibrium equation for the value of x. Note that the right-hand side is a perfect square. If we take the square root of both sides, we get

$$\pm 0.76 = \frac{x}{0.0200 - x}$$

We have written $\pm$ to indicate that we should consider both positive and negative values, since both are mathematically possible. However, we can dismiss the negative value as physically impossible (x can only be positive, since it represents the concentration of CO_2 formed). Rearranging the equation gives

$$x = \frac{0.0200 \times 0.76}{1.76} = 0.0086$$

If we substitute for x in the last line of the table, the equilibrium concentrations are 0.0114 M CO, 0.0114 M H_2O, 0.0086 M CO_2, and 0.0086 M H_2. To find the moles of each substance in the 50.0-L vessel, we multiply the concentrations by the volume of the vessel. For example, the amount of CO is

$$0.0114 \text{ mol/L} \times 50.0 \text{ L} = 0.570 \text{ mol}$$

We find that the equilibrium composition of the reaction mixture is 0.570 mol CO, 0.570 mol H_2O, 0.43 mol CO_2, and 0.43 mol H_2.

Exercise 16.10

What is the equilibrium composition of a reaction mixture if we start with 0.500 mol each of H_2 and I_2 in a 1.0-L vessel? The reaction is

$$H_2(g) + I_2(g) \rightleftharpoons 2HI(g) \qquad K_c = 49.7 \text{ at } 458°C$$

(See Problems 16.41 and 16.42.)

In the previous example, if we had not started with the same number of moles of reactants, we would not have gotten an equation with a perfect square. In that case we would have had to solve a quadratic equation. The next example illustrates how to solve such an equation. ∎

■ A quadratic equation of the form

$$ax^2 + bx + c = 0$$

has the solution

$$x = \frac{-b \pm \sqrt{b^2 - 4ac}}{2a}$$

which is called the *quadratic formula.*

Example 16.8

Hydrogen and iodine react according to the equation

$$H_2(g) + I_2(g) \rightleftharpoons 2HI(g)$$

Suppose 1.00 mol H_2 and 2.00 mol I_2 are placed in a 1.00-L vessel. How many moles of substances are in the gaseous mixture when it comes to equilibrium at 458°C? The equilibrium constant K_c at this temperature is 49.7.

Solution

As in the previous example, we note that starting concentrations change to equilibrium concentrations. We follow the same three steps: (1) set up a table for starting, change, and equilibrium concentrations (use algebraic expressions for unknown quantities); (2) substitute the expressions for the equilibrium concentrations into the equilibrium equation; (3) solve the equilibrium equation for the unknown, then work out explicit values for the equilibrium concentrations.

Step 1 The table listing concentrations of substances is as follows:

Concentrations (M)	$H_2(g)$	+	$I_2(g)$	$\rightleftharpoons$	2HI(g)
Starting	1.00		2.00		0
Change	$-x$		$-x$		$2x$
Equilibrium	$1.00 - x$		$2.00 - x$		$2x$

Note that the changes in concentrations equal x multiplied by the coefficient of that substance in the balanced chemical equation. The change is negative for a reactant, and positive for a product. Equilibrium concentrations equal starting concentrations plus the changes in concentrations.

Step 2 Substituting into the equilibrium equation,

$$K_c = \frac{[HI]^2}{[H_2][I_2]}$$

(*Continued*)

we get

$$49.7 = \frac{(2x)^2}{(1.00 - x)(2.00 - x)}$$

sign. We get

$$x = 2.33 \quad \text{and} \quad x = 0.93$$

Step 3 Since the right-hand side is not a perfect square, we must use the quadratic formula to solve for x. The previous equation rearranges to give

$$(1.00 - x)(2.00 - x) = (2x)^2/49.7 = 0.0805\,x^2$$

or

$$0.920x^2 - 3.00x + 2.00 = 0$$

Hence,

$$x = \frac{3.00 \pm \sqrt{9.00 - 7.36}}{1.84} = 1.63 \pm 0.70$$

There are two mathematical solutions to a quadratic equation. We obtain one by taking the upper (positive) sign in $\pm$, and the other by taking the lower (negative)

However $x = 2.33$ gives a negative value to $1.00 - x$ (the equilibrium concentration of H_2), which is physically impossible. Thus, we are left with $x = 0.93$. We substitute this value of x into the last line of the table in Step 1 to get the equilibrium concentrations, then multiply these by the volume of the vessel (1.00 L) to get the amounts of substances. (The last line of the table can be written as shown below.) The equilibrium composition is 0.07 mol H_2, 1.06 mol I_2, and 1.87 mol HI.

Concentrations (M)	$H_2(g)$	$+$	$I_2(g)$	$\rightleftharpoons$	$2HI(g)$
Equilibrium	$1.00 - x = 0.07$		$2.00 - x = 1.06$		$2x = 1.87$

Exercise 16.11

Phosphorus pentachloride, PCl_5, decomposes when heated:

$$PCl_5(g) \rightleftharpoons PCl_3(g) + Cl_2(g)$$

If the initial concentration of PCl_5 is 1.00 mol/L, what is the equilibrium composition of the gaseous mixture at 160°C? The equilibrium constant K_c at 160°C is 0.0211.

(See Problems 16.43 and 16.44.)

Changing the Reaction Conditions and the Application of LeChatelier's Principle

Getting the maximum amount of product from a reaction depends on the proper selection of reaction conditions. By changing these conditions, we can increase or decrease the yield of product. There are three ways in which we can alter the equilibrium composition of a gaseous reaction mixture and possibly increase the yield of product. We might change the yield by

1. Removing or adding reactants or products to the reaction vessel.
2. Changing the partial pressure of gaseous reactants and products.
3. Changing the temperature.

Another way of affecting the yield of a reaction is by adding a catalyst. A catalyst cannot alter the equilibrium composition, but it can change the rate at which a product can be formed.

16.7 Adding a Catalyst

A *catalyst* is a substance that affects the rate of a reaction but is not consumed by it. The significance of a catalyst can be seen in the reaction of sulfur dioxide with oxygen to give sulfur trioxide:

$$2SO_2(g) + O_2(g) \rightleftharpoons 2SO_3(g)$$

The equilibrium constant, K_c, for this reaction is 1.7×10^{26}, which indicates that for all practical purposes the reaction should go almost completely to products. Yet, when sulfur is burned in air or oxygen, it forms predominantly SO_2, with very little SO_3. Oxidation of SO_2 to SO_3 is simply too slow to give a significant amount of product. However, the rate of the reaction is appreciable in the presence of a platinum or divanadium pentoxide catalyst. The oxidation of SO_2 in the presence of a catalyst is the main step in the *contact process* for the industrial production of sulfuric acid, H_2SO_4. The acid is prepared by reacting SO_3 with water.■

■ Sulfur dioxide from the combustion of coal and from other sources appears to be a major cause of the marked increase in acidity of rain in the eastern United States in the last few decades. This *acid rain* has been shown to contain sulfuric and nitric acids. The SO_2 is oxidized in moist, polluted air to H_2SO_4.

It is important to understand that *a catalyst has no effect on the equilibrium composition of a reaction mixture. A catalyst merely speeds up the attainment of equilibrium.* For example, suppose we mix 2.00 mol SO_2 and 1.00 mol O_2 in a 100.0-L vessel. In the absence of a catalyst, these substances appear unreactive. Much later, if we analyze the mixture, we find essentially the same amounts of SO_2 and O_2. But if a catalyst is added, the rates of both forward and reverse reactions are very much increased. As a result, the reaction mixture comes to equilibrium in a short time. The amounts of SO_2, O_2, and SO_3 can be calculated from the equilibrium constant. We find that the mixture is mostly SO_3 (2.00 mol), with only 1.7×10^{-8} mol SO_2 and 8.4×10^{-9} mol O_2.

A catalyst is beneficial for a reaction, such as $2SO_2 + O_2 \rightleftharpoons 2SO_3$, that is normally slow but has a large equilibrium constant. However, if the reaction has an exceedingly small equilibrium constant, a catalyst is of little help. The reaction

$$2N_2(g) + 2O_2(g) \rightleftharpoons 2NO(g)$$

has been considered for the industrial production of nitric acid (NO reacts with O_2 and H_2O to give nitric acid). At 25°C, however, the equilibrium constant, K_c, equals 4.6×10^{-31}. An equilibrium mixture would contain an extremely small concentration of NO. We cannot expect a catalyst to give a significant yield at this temperature, since a catalyst merely speeds up the attainment of equilibrium. The equilibrium constant does increase slowly with temperature, so that at 2000°C, air (which is a mixture of N_2 and O_2) forms about 0.4% NO at equilibrium. An industrial plant was set up in Norway in 1905 to prepare nitrate fertilizers using this reaction. The plant was eventually made obsolete by the *Ostwald process* for making nitric acid, in which NO is prepared by the oxidation of ammonia. This latter reaction is more economical than the direct reaction of N_2 and O_2, in part because the equilibrium constant is larger at moderate temperatures.

The Ostwald process is an interesting example of how a catalyst can affect the product obtained from a mixture by making one reaction much faster than other possible reactions. Ammonia will burn in oxygen to give nitrogen:

$$4NH_3(g) + 3O_2(g) \rightleftharpoons 2N_2(g) + 6H_2O(g)$$

However, if a hot platinum wire is inserted into a mixture of NH_3 and O_2,

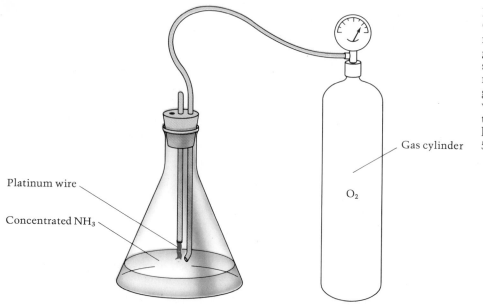

Figure 16.5
Catalytic oxidation of ammonia (Ostwald process). Oxygen is led from a tank to the
surface of concentrated ammonia. A platinum wire
glows from the heat released
when NH_3 reacts with O_2 on
the platinum, which cata-
lyzes the reaction $4NH_3$ +
$5O_2 \rightleftharpoons 4NO + 6H_2O$.

the reaction is different. This time the product is nitric oxide, NO:

$$4NH_3(g) + 5O_2(g) \xrightarrow{\text{Pt}} 4NO(g) + 6H_2O(g)$$

Figure 16.5 shows a demonstration of this reaction, which is the main step
in the Ostwald process for nitric acid. True equilibrium actually favors the
oxidation of NH_3 to N_2 rather than NO. We can see this by noting that the
equilibrium constant, K_c, for the reaction $2NO(g) \rightleftharpoons N_2(g) + O_2(g)$ is 2.2
$\times 10^{30}$. Therefore, any NO that forms should eventually dissociate to the
elements. However, this dissociation of NO is extremely slow at room
temperature. The catalyst in this case speeds up the formation of NO, while
having no effect on its rate of dissociation.■

■ Color Plate 7 shows another
example in which the relative
rates of reaction are important
in determining the product.

Table 16.1 lists the catalysts used in some important gaseous reactions.
Note that carbon monoxide and hydrogen react to give various products
depending on the catalyst. Mixtures of CO and H_2 can be made from
many organic, or carbon-containing, substances by reaction with steam. For
example,

$$CH_4(g) + H_2O(g) \rightleftharpoons CO(g) + 3H_2(g)$$
natural gas

$$C(s) + H_2O(g) \rightleftharpoons CO(g) + H_2(g)$$
coal

In the presence of ZnO and Cr_2O_3, carbon monoxide and hydrogen give
methanol, CH_3OH. Methanol is the starting material for formaldehyde,
CH_2O, which is used to make plastics. Carbon monoxide and hydrogen also
react in catalytic methanation to give methane for substitute natural gas, as
we described in the chapter opening. The catalyst is nickel. In the *Fischer-
Tropsch process*, developed in Germany in 1940, an iron–cobalt catalyst is

Reaction	Catalyst	Commercial Application
$N_2 + 3H_2 \rightleftharpoons 2NH_3$	Fe with K_2O and Al_2O_3	Haber process for ammonia
$4NH_3 + 5O_2 \rightleftharpoons 4NO + 6H_2O$	Pt–Rh mixture	Ostwald process for nitric acid (NO reacts with O_2 and H_2O to give HNO_3)
$2SO_2 + O_2 \rightleftharpoons 2SO_3$	Pt, V_2O_5	Contact process for sulfuric acid (SO_3 reacts with H_2O to give H_2SO_4)
$C_2H_4 + H_2 \rightleftharpoons C_2H_6$	Ni, Pt, Pd	Hydrogenation of compounds with $C{=}C$ (such as unsaturated vegetable oils)
$CO + H_2O \rightleftharpoons CO_2 + H_2$	ZnO–CuO mixture	Water-gas-shift reaction (to produce H_2 or synthesis gas with higher concentration of H_2)
$CO + 3H_2 \rightleftharpoons CH_4 + H_2O$	Ni	Catalytic methanation (preparation of substitute natural gas)
$8CO + 17H_2 \rightleftharpoons C_8H_{18} + 8H_2O$	Fe–Co mixtures	Fischer-Tropsch process (preparation of synthetic gasoline)
$CO + 2H_2 \rightleftharpoons CH_3OH$	ZnO–Cr_2O_3 mixture	Industrial synthesis of methanol
$CH_4 + 2S_2 \rightleftharpoons CS_2 + 2H_2S$	Al_2O_3	Industrial synthesis of carbon disulfide
$2CO + O_2 \rightleftharpoons 2CO_2$ $C_7H_{16} + 11O_2 \rightleftharpoons 7CO_2 + 8H_2O$	Pt, Pd, V_2O_5	Automobile catalytic converter (to reduce pollutants in exhaust gases)

Table 16.1
Catalysts Used in Some
Gaseous Reactions

used to convert mixtures of CO and H_2 to liquid hydrocarbons for synthetic gasoline.■ In each case, the catalyst determines the product by allowing one reaction to attain equilibrium much faster than all others.

Study of the catalysts and reaction conditions for the various reactions of CO and H_2 has become an active research area. These reactions make it possible to convert a variety of organic materials, including coal and organic wastes, to gaseous and liquid fuels and all of the industrial chemicals presently obtained from petroleum.

■ The Fischer-Tropsch process supplied needed gasoline to Germany during World War II. Immediately after the war other countries set up plants to manufacture synthetic gasoline, but these could not compete with cheap petroleum manufacture. With increasing prices of petroleum, synthetic gasoline may again become economical. A commercial plant for producing liquid fuels from coal is presently operating in South Africa.

16.8 Removing or Adding Reactants or Products

One way to increase the yield of a desired product is by removing or adding a substance to the reaction mixture. Consider the methanation reaction,

$$CO(g) + 3H_2(g) \rightleftharpoons CH_4(g) + H_2O(g)$$

If we place 1.000 mol CO and 3.000 mol H_2 in a 10.00-L reaction vessel, the equilibrium composition at 1200 K is 0.613 mol CO, 1.839 mol H_2, 0.387 mol CH_4, and 0.387 mol H_2O. Can we alter this composition by removing or adding one of the substances to improve the yield of methane?

We can apply **LeChatelier's principle** to this question.■ The principle states that *if a system in chemical equilibrium is altered by the change of some condition, chemical reaction occurs to shift the equilibrium composition in a way that attempts to reduce that change of condition.* Suppose

■ LeChatelier's principle was introduced in Section 12.3, where it was used to determine the effect of pressure on solubility.

we move a substance from or add one to the equilibrium mixture, thus altering the concentration of the substance. Chemical reaction then occurs to partially restore the initial concentration of the removed or added substance.

For example, suppose that water vapor is removed from the reaction vessel containing the equilibrium mixture for methanation. LeChatelier's principle predicts that net chemical change will occur to partially reinstate the original concentration of water vapor. This means that the methanation reaction momentarily goes in the forward direction,

$$CO(g) + 3H_2(g) \longrightarrow CH_4(g) + H_2O(g)$$

until equilibrium is re-established. Going in the forward direction, the concentrations of both water vapor and methane increase.

A practical way to remove water vapor in this reaction might be to cool the reaction mixture quickly to condense the water. Liquid water could be removed, and the gases reheated until equilibrium was again established. The concentration of water vapor would build up again as the concentration of methane increased. Table 16.2 lists the amounts of each substance at each stage of this process. Note how the yield of methane has been improved.

It is often useful to add an excess of a cheap reactant in order to force the reaction toward more products. In this way, the more expensive reactant is made to react to a greater extent than it would otherwise.

Consider the ammonia synthesis

$$N_2(g) + 3H_2(g) \rightleftharpoons 2NH_3(g)$$

If we wish to convert as much hydrogen to ammonia as possible, we might increase the concentration of nitrogen. To understand the effect of this, first suppose that a mixture of nitrogen, hydrogen, and ammonia is at equilibrium. If nitrogen is now added to this mixture, the equilibrium is disturbed. According to LeChatelier's principle, the reaction will now go in the direction that will use up some of the added nitrogen:

$$N_2(g) + 3H_2(g) \longrightarrow 2NH_3(g)$$

Consequently, adding more nitrogen than is required by the stoichiometry will have the effect of converting a greater quantity of hydrogen to ammonia.

We can look at these situations in terms of the reaction quotient, Q. Consider the methanation reaction, in which

$$Q = \frac{[CH_4][H_2O]}{[CO][H_2]^3}$$

Stage of Process	Mol CO	Mol H_2	Mol CH_4	Mol H_2O
Original reaction mixture	0.613	1.839	0.387	0.387
After removing water	0.613	1.839	0.387	0
When equilibrium is re-established	0.491	1.473	0.509	0.122

Table 16.2
The Effect of Removing Water Vapor from a Methanation Mixture

If the reaction mixture is at equilibrium, $Q = K_c$. Suppose we remove some H_2O from this equilibrium mixture. Now $Q < K_c$, and from what we said in Section 16.5, the reaction will proceed in the forward direction to restore equilibrium.

We can now summarize the conclusions from this section:

When more reactant is added to or some product is removed from an equilibrium mixture, net reaction occurs left to right (that is, in the forward direction) to give a new equilibrium, and more products are produced.

When more product is added to or some reactant is removed from an equilibrium mixture, net reaction occurs right to left (that is, in the reverse direction) to give a new equilibrium, and more reactants are produced.

Example 16.9

Predict the direction of reaction if H_2 is removed from a mixture in which the following equilibrium is established:

$$H_2(g) + I_2(g) \rightleftharpoons 2HI(g)$$

Solution

When H_2 is removed from the reaction mixture, the reaction goes in the reverse direction (more HI dissociates to H_2 and I_2) to partially restore the H_2 that was removed:

$$H_2(g) + I_2(g) \longleftarrow 2HI(g)$$

Exercise 16.12

Consider each of the following equilibria that are disturbed in the manner indicated. Predict the direction of reaction.
  (a) The equilibrium

$$CaCO_3(s) \rightleftharpoons CaO(s) + CO_2(g)$$

is disturbed by increasing the pressure (that is, concentration) of carbon dioxide.
  (b) The equilibrium

$$2Fe(s) + 3H_2O(g) \rightleftharpoons Fe_2O_3(s) + 3H_2(g)$$

is disturbed by increasing the concentration of hydrogen.

(See Problems 16.47 and 16.48.)

It should be pointed out that a reaction whose equilibrium constant is extremely small will remain almost completely as reactants and cannot be shifted to products by adding an excess of one reactant. For example, the reaction

$$CO_2(g) + 2H_2O(g) \rightleftharpoons CH_4(g) + 2O_2(g)$$

has an equilibrium constant K_c equal to 10^{-140}. The value is so small that the equilibrium mixture practically consists only of carbon dioxide and water. Adding more carbon dioxide to the reaction vessel has no appreciable effect. The reaction is essentially irreversible.

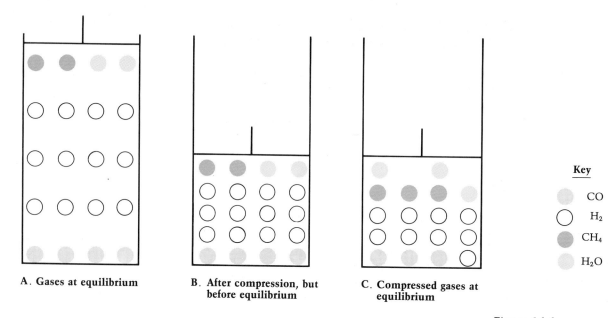

A. Gases at equilibrium   B. After compression, but   C. Compressed gases at
                              before equilibrium   equilibrium

Key

CO

H_2

CH_4

H_2O

16.9 Changing the Pressure and Temperature

The optimum conditions for catalytic methanation involve moderately elevated temperatures and normal to moderately high pressures:

$$CO(g) + 3H_2(g) \xrightarrow[\text{1–100 atm}]{\text{230–450°C}} CH_4(g) + H_2O(g)$$

Let us see if we can gain insight into why these might be the optimum conditions for the reaction.

Effect of Pressure Change

A pressure change *obtained by changing the volume* can affect the yield of product in a gaseous reaction, if the reaction involves a change of total moles of gas. The methanation reaction, $CO + 3H_2 \rightleftharpoons CH_4 + H_2O$, is an example of a change of moles of gas. When the reaction goes in the forward direction, four moles of reactant gas ($CO + 3H_2$) become two moles of product gas ($CH_4 + H_2O$).

To see the effect of such a pressure change, consider what would happen if an equilibrium mixture from the methanation reaction is compressed to one-half of its original volume at a fixed temperature (see Figure 16.6). The total pressure would be doubled ($PV = $ constant at a fixed temperature, according to Boyle's law, so that halving V requires that P double). Because the partial pressures and therefore the concentrations of reactants and products have changed, the mixture is no longer at equilibrium. The direction in which the reaction goes to again attain equilibrium can be predicted by LeChatelier's principle. Reaction should go in the forward direction, because

Figure 16.6
Effect on the equilibrium composition of the methanation reaction of increasing the pressure by halving the volume. (a) The relative numbers of CO, H_2, CH_4, and H_2O molecules in the original equilibrium mixture are shown. (b) The gases are compressed to one-half their original volume, and the mixture is no longer at equilibrium. (c) Equilibrium is re-established when the reaction goes in the forward direction, $CO + 3H_2 \longrightarrow CH_4 + H_2O$. In this way, the total number of molecules is reduced, which reduces the initial pressure increase.

then the moles of gas decrease, and the pressure (which is proportional to moles of gas) decreases. In this way, the initial pressure increase is partially reduced.

We find the same result by looking at the reaction quotient, Q. Let [CO], [H_2], [CH_4], and [H_2O] be the molar concentrations at equilibrium for the methanation reaction. If the volume of an equilibrium mixture is halved, the partial pressures and therefore the concentrations are doubled. The reaction quotient at that moment is obtained by replacing each equilibrium concentration by double its value:

$$Q = \frac{(2[CH_4])(2[H_2O])}{(2[CO])(2[H_2])^3} = \frac{K_c}{4}$$

Since $Q < K_c$, the reaction proceeds in the forward direction.

We can see the quantitative effect of the pressure change by solving the equilibrium problem. Recall that if we put 1.000 mol CO and 3.000 mol H_2 into a 10.00-L vessel, the equilibrium composition at 1200 K, where K_c equals 3.92, is 0.613 mol CO, 1.839 mol H_2, 0.387 mol CH_4, and 0.381 mol H_2O. Suppose the volume of the reaction gases is halved so that the initial concentrations are doubled. The temperature remains at 1200 K, so K_c is still 3.92. But if we solve for the new equilibrium composition (following the method of Example 16.8), we find 0.495 mol CO, 1.485 mol H_2, 0.505 mol CH_4, and 0.505 mol H_2O. Note that the amount of CH_4 has increased from 0.387 mol to 0.505 mol. We conclude that high pressure of reaction gases favors high yields of methane.

In order to decide the direction of reaction when the pressure of the reaction mixture is increased, say, by decreasing the volume, we ignore liquids and solids. These are not much affected by pressure changes because they are nearly incompressible. Consider the reaction

$$C(s) + CO_2(g) \rightleftharpoons 2CO(g)$$

The moles of gas decrease when the reaction goes in the reverse direction (2 mol CO go to 1 mol CO_2). Therefore, if we increase the pressure of the reaction mixture by decreasing its volume, the reaction will go in the reverse direction. The moles of gas will decrease, and the initial increase of pressure will be partially reduced, as we expect by LeChatelier's principle.

It is important to note that an increase or decrease in pressure of a gaseous reaction must result in changes of partial pressures of substances in the chemical equation if it is to have an effect on the equilibrium composition. Only changes in partial pressures, or concentrations, of these substances can affect the reaction quotient. Consider the effect of increasing the pressure in the methanation reaction by adding helium gas. Although the total pressure increases, the partial pressures of CO, H_2, CH_4, and H_2O do not change. Thus, the equilibrium composition is not affected. However, changing the pressure by changing the volume of this system changes the partial pressures of all gases, so that the equilibrium composition is affected.

We can summarize these conclusions. If the pressure is increased by decreasing the volume of a reaction mixture, reaction shifts in the direction of fewer moles of gas. The following example illustrates applications of this statement.

Example 16.10

Look at each of the following equations and decide whether an increase of pressure obtained by decreasing the volume will increase, decrease, or have no effect on the amounts of products:

(a) $CO(g) + Cl_2(g) \rightleftharpoons COCl_2(g)$
(b) $2H_2S(g) \rightleftharpoons 2H_2(g) + S_2(g)$
(c) $C(graphite) + S_2(g) \rightleftharpoons CS_2(g)$

Solution

(a) Reaction decreases the number of molecules of gas (from two to one). According to LeChatelier's principle, an increase of pressure will increase the amount of product. (b) Reaction increases the number of molecules of gas (from two to three); hence, an increase of pressure will decrease the amounts of products. (c) Reaction does not change the number of molecules of gas. (We ignore the change in volume due to consumption of solid carbon because the change in volume of solid is insignificant. Look only at gas volumes when deciding the effect of pressure change on equilibrium composition.) Pressure change has no effect.

Exercise 16.13

Can you increase the amount of product in the following equations by increasing the pressure? Explain.

(a) $CO_2(g) + H_2(g) \rightleftharpoons CO(g) + H_2O(g)$
(b) $4CuO(s) \rightleftharpoons 2Cu_2O(s) + O_2(g)$
(c) $2SO_2(g) + O_2(g) \rightleftharpoons 2SO_3(g)$

(See Problems 16.49 and 16.50.)

Effect of Temperature Change

Temperature has a profound effect on most reactions. In the first place, reaction rates usually increase with an increase in temperature, meaning that equilibrium is reached sooner. Many gaseous reactions are sluggish or have imperceptible rates at room temperature, but speed up enough at higher temperature to become commercially feasible processes.

Second, equilibrium constants vary with temperature. Table 16.3 gives values of K_c for methanation at various temperatures. Note that K_c equals 4.9×10^{27} at 298 K. Thus, an equilibrium mixture at room temperature is mostly methane and water.

Whether we should raise or lower the temperature of a reaction mixture to increase the equilibrium amount of product can be shown by LeChatelier's principle. Again, consider the methanation reaction,

$$CO(g) + 3H_2(g) \rightleftharpoons CH_4(g) + H_2O(g); \Delta H = -206.2 \text{ kJ}$$

Temperature (K)	K_c
298	4.9×10^{27}
800	1.38×10^5
1000	2.54×10^2
1200	3.92

Table 16.3
Equilibrium Constant for Methanation at Different Temperatures

The value of ΔH shows this reaction to be quite exothermic. Thus, as products are formed, considerable heat is released. As the temperature is raised, according to LeChatelier's principle, the reaction shifts to form more reactants, thereby absorbing heat and attempting to counter the increase in temperature:

$$CO(g) + 3H_2(g) \longleftarrow CH_4(g) + H_2O(g) + heat$$

Thus, we would predict the equilibrium constant to be larger for lower temperatures, in agreement with the values of K_c given earlier. ■

The conclusions from LeChatelier's principle regarding temperature effects on an equilibrium can be summarized this way. For an endothermic reaction (ΔH positive), the amounts of products are increased at equilibrium by an increase in temperature (K_c is larger at higher T). For an exothermic reaction (ΔH negative), the amounts of products are increased at equilibrium by a decrease in temperature (K_c is larger at lower T).

■ The quantitative effect of temperature on the equilibrium constant is discussed in Chapter 20.

Example 16.11

Carbon monoxide is formed when carbon dioxide reacts with solid carbon (graphite):

$$CO_2(g) + C(graphite) \rightleftharpoons 2CO(g); \Delta H = 172.5 \text{ kJ}$$

Is a high or low temperature more favorable to the formation of carbon monoxide?

Solution

The reaction absorbs heat in the forward direction:

$$Heat + CO_2(g) + C(graphite) \longrightarrow 2CO(g)$$

As the temperature is raised, reaction occurs in the forward direction, using heat and thereby attempting to lower the temperature. Thus, high temperature is more favorable to the formation of carbon monoxide. This is why combustions of carbon and organic materials can produce significant amounts of carbon monoxide.

Exercise 16.14

Consider the possibility of converting carbon dioxide to carbon monoxide by the endothermic reaction

$$CO_2(g) + H_2(g) \rightleftharpoons CO(g) + H_2O(g)$$

Is a high or low temperature more favorable to the production of carbon monoxide? Explain.

(See Problems 16.51 and 16.52.)

Choosing the Optimum Conditions for Reaction

We are now in a position to understand the optimum conditions for the methanation reaction. Because the reaction is exothermic, low temperatures should favor high yields of methane, that is, the equilibrium constant is large for low temperature. However, gaseous reactions are often very slow at room temperature. In practice, the methanation reaction is run at moderately elevated temperatures (230–450°C) in the presence of a nickel catalyst, where the rate of reaction is sufficiently fast but the equilibrium constant is not too small. Because the methanation reaction involves a decrease in moles of gas, the yield of methane should increase as the pressure increases. However, the equilibrium constant is large at the usual operating temperatures, so that very high pressures are not needed to obtain economical yields of methane. Pressures of 1–100 atm are usual for this reaction.

As another example, consider the Haber process for the synthesis of ammonia:

$$N_2(g) + 3H_2(g) \underset{\text{Fe catalyst}}{\rightleftharpoons} 2NH_3(g); \Delta H = -91.8 \text{ kJ}$$

Because the reaction is exothermic, the equilibrium constant is larger for lower temperatures. But the reaction proceeds too slowly at room temperature to be practical, even in the presence of the best available catalysts.■ The optimum choice of temperature, found experimentally to be about 450°C, is a compromise between an increased rate of reaction at higher temperature and an increased yield of ammonia at lower temperature. Because the formation of ammonia decreases the moles of gases, the yield of product is improved by high pressures. Since the equilibrium constant K_c is only 0.159 at 450°C, higher pressures (up to 1000 atm) are required for an economical yield of ammonia.

■ This technological problem has stimulated a great deal of basic research into understanding how certain bacteria "fix" nitrogen at atmospheric pressure to make NH_3. An enzyme called nitrogenase is responsible for N_2 fixation in these bacteria. This enzyme contains Fe and Mo, which may play a role in the catalysis.

Exercise 16.15

Consider the reaction

$$2CO_2(g) \rightleftharpoons 2CO(g) + O_2(g); \Delta H = 566 \text{ kJ}$$

Discuss the temperature and pressure conditions that would give the best yield of carbon monoxide.

(See Problems 16.53 and 16.54.)

A Checklist for Review

Important Terms

chemical equilibrium (16.1)
equilibrium-constant expression (16.2)
equilibrium constant (16.2)

law of mass action (16.2)
homogeneous equilibrium (16.3)
heterogeneous equilibrium (16.3)

reaction quotient (16.5)
LeChatelier's principle (16.8)

Summary of Facts and Concepts

Chemical equilibrium can be characterized by the *equilibrium constant* K_c. The expression for K_c has the concentration of products in the numerator and concentration of reactants in the denominator. Pure liquids and solids are ignored in writing the equilibrium-constant expression. When K_c is very large, the equilibrium mixture is mostly products, and when K_c is very small, the equilibrium mixture is mostly reactants. The reaction quotient, Q, takes the form of the equilibrium-constant expression. If we substitute the concentrations of substances in a reaction mixture into Q, we can predict the direction the reaction must go to attain equilibrium. We can use K_c to calculate the composition of the reaction mixture at equilibrium, starting from various initial compositions.

The *choice of conditions*, including *catalysts*, can be very important to the success of a reaction. Removing a product from the reaction mixture, for example, will shift the equilibrium composition to give more product. Changing the pressure and temperature can also affect the product yield. *LeChatelier's principle* is useful in predicting the effect of such changes.

Operational Skills

1. Given the starting amounts of reactants and the amount of one substance at equilibrium, find the equilibrium composition (Example 16.1).

2. Given the chemical equation, write the equilibrium-constant expression (Examples 16.2 and 16.4).

3. Given the equilibrium composition, find K_c (Example 16.3).

4. Given the concentrations of substances in a reaction mixture, predict the direction of reaction (Example 16.5).

5. Given K_c and all concentrations of substances but one in an equilibrium mixture, calculate the concentration of this one substance (Example 16.6).

6. Given the starting composition and K_c of a reaction mixture, calculate the equilibrium composition (Examples 16.7 and 16.8).

7. Given a reaction, use LeChatelier's principle to decide the effect of adding or removing a substance (Example 16.9), changing the pressure (Example 16.10), or changing the temperature (Example 16.11).

Review Questions

16.1 Consider the reaction $N_2O_4(g) \rightleftharpoons 2NO_2(g)$. Draw a graph illustrating the changes of concentrations of N_2O_4 and NO_2 as equilibrium is approached when starting with pure N_2O_4. Describe how the rates of forward and reverse reactions change as the mixture approaches dynamic equilibrium. Why is this called a *dynamic* equilibrium?

16.2 When 1.0 mol each of $H_2(g)$ and $I_2(g)$ are mixed at a certain high temperature, they react to give a final mixture consisting of 0.5 mol each of $H_2(g)$ and $I_2(g)$ and 1.0 mol $HI(g)$. Why is it that you obtain the same final mixture if you bring 2.0 mol $HI(g)$ to the same temperature?

16.3 Explain why the equilibrium constant for a gaseous reaction can be written in terms of partial pressures instead of concentrations.

16.4 Which of the following reactions involve homogeneous equilibria and which involve heterogeneous equilibria? Explain the difference.

(a) $2NO(g) + O_2(g) \rightleftharpoons 2NO_2(g)$
(b) $2Cu(NO_3)_2(s) \rightleftharpoons 2CuO(s) + 4NO_2(g) + O_2(g)$
(c) $2N_2O(g) \rightleftharpoons 2N_2(g) + O_2(g)$
(d) $2NH_3(g) + 3CuO(s) \rightleftharpoons$
$\qquad\qquad\qquad 3H_2O(g) + N_2(g) + 3Cu(s)$

16.5 Explain why pure liquids and solids can be ignored when writing the equilibrium expression.

16.6 What qualitative information can one get from the magnitude of the equilibrium constant?

16.7 What is the reaction quotient? How is it useful?

16.8 List the possible ways in which one can alter the equilibrium composition of a reaction mixture.

16.9 Two moles of H_2 are mixed with 1 mol O_2 at 25°C. No observable reaction takes place, although K_c for the reaction to form water is very large at this temperature. When a piece of platinum is added, however, the gases react rapidly. Explain the role of platinum in this reaction. How does it affect the equilibrium composition of the reaction mixture?

16.10 How is it possible for a catalyst to give products from a reaction mixture that are different from what are obtained when no catalyst or a different catalyst is used? Give examples.

16.11 When a stream of hydrogen gas is passed over magnetic iron oxide, Fe_3O_4, metallic iron and water are formed. On the other hand, when a stream of water vapor is passed over metallic iron, Fe_3O_4 and hydrogen are formed. Explain by means of LeChatelier's principle why the reaction goes in one direction in one case, but in the reverse direction in the other.

16.12 List four ways in which the yield of ammonia in the reaction

$$N_2(g) + 3H_2(g) \rightleftharpoons 2NH_3(g); \ \Delta H < 0$$

can be improved for a given amount of H_2. Explain the principle behind each way.

Problems

Reaction Stoichiometry

16.13 A 1.000-mol sample of phosphorus pentachloride, PCl_5, dissociates at 160°C and 1 atm to give 0.135 mol of phosphorus trichloride, PCl_3, at equilibrium:

$$PCl_5(g) \rightleftharpoons PCl_3(g) + Cl_2(g)$$

What is the composition of the final reaction mixture?

16.15 Methanol, CH_3OH, formerly known as wood alcohol, is manufactured commercially by the following reaction:

$$CO(g) + 2H_2(g) \rightleftharpoons CH_3OH(g)$$

A 1.000-L vessel was filled with 0.1000 mol CO and 0.2000 mol H_2. When this mixture came to equilibrium at 500 K, the vessel contained 0.0791 mol CO. How many moles of each substance were in the vessel?

16.14 Nitric oxide, NO, reacts with bromine, Br_2, to give nitrosyl bromide, NOBr:

$$2NO(g) + Br_2(g) \rightleftharpoons 2NOBr(g)$$

A sample of 0.0984 mol NO with 0.0492 mol Br_2 gives an equilibrium mixture containing 0.0584 mol NOBr. What is the composition of the equilibrium mixture?

16.16 In the contact process, sulfuric acid is manufactured by first oxidizing SO_2 to SO_3, which is then reacted with water. The reaction of SO_2 with O_2 is

$$2SO_2(g) + O_2(g) \rightleftharpoons 2SO_3(g)$$

A 1.000-L flask was filled with 0.0200 mol SO_2 and 0.0100 mol O_2. At equilibrium at 900 K, the flask contained 0.0148 mol SO_3. How many moles of each substance were in the flask at equilibrium?

The Equilibrium Constant and Its Evaluation

16.17 Write equilibrium-constant expressions, K_c, for the following reactions:
  (a) $H_2(g) + Br_2(g) \rightleftharpoons 2HBr(g)$
  (b) $CS_2(g) + 4H_2(g) \rightleftharpoons CH_4(g) + 2H_2S(g)$
  (c) $4HCl(g) + O_2(g) \rightleftharpoons 2H_2O(g) + 2Cl_2(g)$
  (d) $CO(g) + 2H_2(g) \rightleftharpoons CH_3OH(g)$

16.19 The equilibrium constant K_c for the equation

$$2HI(g) \rightleftharpoons H_2(g) + I_2(g)$$

at 425°C is 1.84. What is the value of K_c for the following equation?

$$H_2(g) + I_2(g) \rightleftharpoons 2HI(g)$$

16.21 A 6.00-L vessel contained 0.0222 mol of phosphorus trichloride, 0.0189 mol of phosphorus pentachloride, and 0.1044 mol of chlorine at 230°C in an equilibrium mixture. Calculate the value of K_c for the reaction

$$PCl_3(g) + Cl_2(g) \rightleftharpoons PCl_5(g)$$

16.23 Obtain the value of K_c for the following reaction at 500 K:

$$CO(g) + 2H_2(g) \rightleftharpoons CH_3OH(g)$$

Use the data given in Problem 16.15.

16.18 Write equilibrium-constant expressions, K_c, for the following reactions:
  (a) $N_2O_4(g) \rightleftharpoons 2NO_2(g)$
  (b) $2NO(g) + Br_2(g) \rightleftharpoons 2NOBr(g)$
  (c) $2SO_2(g) + O_2(g) \rightleftharpoons 2SO_3(g)$
  (d) $4NH_3(g) + 5O_2(g) \rightleftharpoons 4NO(g) + 6H_2O(g)$

16.20 The equilibrium constant K_c for the equation

$$CS_2(g) + 4H_2(g) \rightleftharpoons CH_4(g) + 2H_2S(g)$$

at 900°C is 27.8. What is the value of K_c for the following equation?

$$\tfrac{1}{2}CS_2(g) + 2H_2(g) \rightleftharpoons \tfrac{1}{2}CH_4(g) + H_2S(g)$$

16.22 A reaction vessel at 491°C contained 0.0812 M H_2, 0.0344 M I_2, and 0.357 M HI. Assuming that the substances are at equilibrium, what is the value of K_c at 491°C for the reaction of hydrogen and iodine to give hydrogen iodide? The equation is

$$H_2(g) + I_2(g) \rightleftharpoons 2HI(g)$$

16.24 Obtain the value of K_c for the following reaction at 900 K:

$$2SO_2(g) + O_2(g) \rightleftharpoons 2SO_3(g)$$

Use the data given in Problem 16.16.

16.25 At 77°C, 2.00 mol of nitrosyl bromide, NOBr, placed in a 1.00-L flask dissociates to the extent of 9.4%; that is, for each mole of NOBr before reaction, (1.000 − 0.094) mol NOBr remains after dissociation. Calculate the value of K_c for the dissociation reaction

$$2NOBr(g) \rightleftharpoons 2NO(g) + Br_2(g)$$

16.27 The value of K_c for the following reaction at 900°C is 0.28:

$$CS_2(g) + 4H_2(g) \rightleftharpoons CH_4(g) + 2H_2S(g)$$

What is K_p at this temperature?

16.29 The reaction

$$SO_2(g) + \tfrac{1}{2}O_2(g) \rightleftharpoons SO_3(g)$$

has K_p equal to 6.55 at 627°C. What is the value of K_c at this temperature?

16.31 Write the expression for the equilibrium constant K_c for the following equations:
   (a) $C(s) + CO_2(g) \rightleftharpoons 2CO(g)$
   (b) $FeO(s) + CO(g) \rightleftharpoons Fe(s) + CO_2(g)$
   (c) $Na_2CO_3(s) + SO_2(g) + \tfrac{1}{2}O_2(g) \rightleftharpoons Na_2SO_4(s) + CO_2(g)$

16.26 A 2.00-mol sample of nitrogen dioxide was placed in an 80.0-L vessel. At 200°C, the nitrogen dioxide was 6.0% decomposed according to the equation

$$2NO_2(g) \rightleftharpoons 2NO(g) + O_2(g)$$

Calculate the value of K_c for this reaction at 200°C. (See Problem 16.25.)

16.28 The equilibrium constant K_c equals 10.5 for the following reaction at 227°C:

$$CO(g) + 2H_2(g) \rightleftharpoons CH_3OH(g)$$

What is the value of K_p at this temperature?

16.30 Fluorine, F_2, dissociates into atoms on heating:

$$\tfrac{1}{2}F_2(g) \rightleftharpoons F(g)$$

The value of K_p at 842°C is 7.55×10^{-2}. What is the value of K_c at this temperature?

16.32 For each of the following equations, give the expression for the equilibrium constant K_c:
   (a) $NH_4Cl(s) \rightleftharpoons NH_3(g) + HCl(g)$
   (b) $C(s) + 2N_2O(g) \rightleftharpoons CO_2(g) + 2N_2(g)$
   (c) $2NaHCO_3(s) \rightleftharpoons Na_2CO_3(s) + H_2O(g) + CO_2(g)$

Using the Equilibrium Constant

16.33 On the basis of the value of K_c, decide whether or not you expect nearly complete reaction at equilibrium for each of the following:
   (a) $2H_2(g) + O_2(g) \rightleftharpoons 2H_2O(g)$; $K_c = 3 \times 10^{81}$
   (b) $2HF(g) \rightleftharpoons H_2(g) + F_2(g)$; $K_c = 1 \times 10^{-95}$

16.35 Hydrogen fluoride decomposes according to the equation

$$2HF(g) \rightleftharpoons H_2(g) + F_2(g)$$

The value of K_c at room temperature is 1.0×10^{-95}. From the magnitude of K_c, do you think the decomposition occurs to any great extent at room temperature? If an equilibrium mixture in a 1.0-L vessel contains 1.0 mol HF, what is the amount of H_2 formed? Does this result agree with what you expect from the magnitude of K_c?

16.37 Methanol, CH_3OH, is manufactured industrially by the reaction

$$CO(g) + 2H_2(g) \rightleftharpoons CH_3OH(g)$$

A gaseous mixture at 500 K is 0.020 M CH_3OH, 0.10 M CO, and 0.10 M H_2. What will be the direction of reaction if this mixture goes to equilibrium? The equilibrium constant K_c equals 10.5 at 500 K.

16.34 Would either of the following reactions go almost completely to product at equilibrium?
   (a) $N_2(g) + 2O_2(g) \rightleftharpoons 2NO_2(g)$; $K_c = 3 \times 10^{-17}$
   (b) $2SO_2(g) + O_2(g) \rightleftharpoons 2SO_3(g)$; $K_c = 8 \times 10^{25}$

16.36 Suppose sulfur dioxide reacts with oxygen at room temperature:

$$2SO_2(g) + O_2(g) \rightleftharpoons 2SO_3(g)$$

The equilibrium constant K_c equals 8.0×10^{35} at this temperature. From the magnitude of K_c, do you think this reaction occurs to any great extent when equilibrium is reached at room temperature? If an equilibrium mixture is 1.0 M SO_3, what are the concentrations of SO_2 and O_2? Assume the concentrations of SO_2 and O_2 are equal. Does this result agree with what you expect from the magnitude of K_c?

16.38 Sulfur trioxide is obtained commercially from sulfur dioxide:

$$2SO_2(g) + O_2(g) \rightleftharpoons 2SO_3(g)$$

Sulfur trioxide is used to manufacture sulfuric acid. The equilibrium constant K_c for this reaction is 4.17×10^{-2} at 727°C. What is the direction of reaction when a mixture that is 0.20 M SO_2, 0.10 M O_2, and 0.40 M SO_3 approaches equilibrium?

16.39 Phosgene, $COCl_2$, a starting material for the manufacture of polyurethane plastics, is prepared from CO and Cl_2:

$$CO(g) + Cl_2(g) \rightleftharpoons COCl_2(g)$$

An equilibrium mixture at 395°C contains 0.10 mol CO and 0.20 mol Cl_2 per liter, as well as $COCl_2$. If K_c at 395°C is 22.5, what is the concentration of $COCl_2$?

16.41 Iodine and bromine react to give iodine monobromide, IBr:

$$I_2(g) + Br_2(g) \rightleftharpoons 2IBr(g)$$

What is the equilibrium composition of a mixture at 150°C that initially contained 0.0010 mol each of iodine and bromine in a 5.0-L vessel? The equilibrium constant K_c for this reaction at 150°C is 1.2×10^2.

16.43 The equilibrium constant K_c for the reaction

$$PCl_3(g) + Cl_2(g) \rightleftharpoons PCl_5(g)$$

equals 49 at 230°C. If 0.500 mol each of phosphorus trichloride and chlorine are added to a 5.0-L reaction vessel, what is the equilibrium composition of the mixture at 230°C?

**16.45 Suppose 1.000 mol CO and 3.000 mol H_2 are put in a 10.00-L vessel at 1200 K. The equilibrium constant K_c for the reaction

$$CO(g) + 3H_2(g) \rightleftharpoons CH_4(g) + H_2O(g)$$

equals 3.92. Find the equilibrium composition of the reaction mixture. Note that you get a quartic equation (one in x^4), which is a perfect square. Thus, the problem can be reduced to a quadratic.

16.40 Nitric oxide, NO, is formed in automobile exhaust by the reaction of N_2 and O_2 (from air):

$$N_2(g) + O_2(g) \rightleftharpoons 2NO(g)$$

The equilibrium constant K_c is 0.0025 at 2127°C. If an equilibrium mixture at this temperature contains 0.016 mol N_2 and 0.036 mol O_2 per liter, what is the concentration of NO?

16.42 Initially a mixture contains 1.00 mol each of N_2 and O_2 in a 10.0-L vessel. Find the composition of the mixture when equilibrium is reached at 3900°C. The reaction is

$$N_2(g) + O_2(g) \rightleftharpoons 2NO(g)$$

and $K_c = 0.0123$ at 3900°C.

16.44 Calculate the composition of the gaseous mixture obtained when 1.00 mol of carbon dioxide is exposed to hot carbon at 800°C in a 1.00-L vessel. The equilibrium constant K_c at 800°C is 14.0 for the reaction

$$CO_2(g) + C(s) \rightleftharpoons 2CO(g)$$

**16.46 The equilibrium constant K_c for the reaction

$$N_2(g) + 3H_2(g) \rightleftharpoons 2NH_3(g)$$

at 450°C is 0.153. Calculate the equilibrium composition when 1.00 mol N_2 is mixed with 3.00 mol H_2 in a 2.00-L vessel. Note that you get a quartic equation (one in x^4), which is a perfect square. The resulting quadratic equation can then be solved.

LeChatelier's Principle

16.47 (a) Predict the direction of reaction when chlorine gas is added to an equilibrium mixture of PCl_3, PCl_5, and Cl_2. The reaction is

$$PCl_3(g) + Cl_2(g) \rightleftharpoons PCl_5(g)$$

(b) What is the direction of reaction if chlorine gas is removed from an equilibrium mixture of these gases?

16.49 What would you expect to be the effect of an increase of pressure on each of the following reactions? Would the pressure change cause reaction to go to the right or left?

(a) $CH_4(g) + 2S_2(g) \rightleftharpoons CS_2(g) + 2H_2S(g)$
(b) $H_2(g) + Br_2(g) \rightleftharpoons 2HBr(g)$
(c) $CO_2(g) + C(s) \rightleftharpoons 2CO(g)$

16.48 Consider the equilibrium

$$FeO(s) + CO(g) \rightleftharpoons Fe(s) + CO_2(g)$$

If carbon dioxide is removed from the equilibrium mixture (say, by passing the gases through water to absorb CO_2), what is the direction of net reaction as the new equilibrium is achieved?

16.50 Indicate if either an increase or a decrease of pressure obtained by changing the volume would increase the amount of product in the following reactions:

(a) $CO(g) + 2H_2(g) \rightleftharpoons CH_3OH(g)$
(b) $2SO_2(g) + O_2(g) \rightleftharpoons 2SO_3(g)$
(c) $N_2O_4(g) \rightleftharpoons 2NO_2(g)$

16.51 Methanol is prepared industrially from synthesis gas (CO and H_2):

$$CO(g) + 2H_2(g) \rightleftharpoons CH_3OH(g); \quad \Delta H = -21.7 \text{ kcal}$$

Would the fraction of methanol obtained at equilibrium be increased by raising the temperature? Explain.

16.53 What would you expect to be the general temperature and pressure conditions for an optimum yield of nitric oxide, NO, by the oxidation of ammonia?

$$4NH_3(g) + 5O_2(g) \rightleftharpoons 4NO(g) + 6H_2O(g); \quad \Delta H < 0$$

16.52 A possible way of preparing hydrogen is by the decomposition of water:

$$2H_2O(g) \rightleftharpoons 2H_2(g) + O_2(g); \quad \Delta H = 484 \text{ kJ}$$

Would you expect the decomposition to be favorable at high or low temperature? Explain.

16.54 Predict the general temperature and pressure conditions for the optimum conversion of ethylene (C_2H_4) to ethane (C_2H_6):

$$C_2H_4(g) + H_2(g) \rightleftharpoons C_2H_6(g); \quad \Delta H < 0$$

Additional Problems

16.55 A mixture of carbon monoxide, hydrogen, and methanol, CH_3OH, is at equilibrium according to the equation

$$CO(g) + 2H_2(g) \rightleftharpoons CH_3OH(g)$$

At 250°C, the mixture is 0.096 M CO, 0.191 M H_2, and 0.015 M CH_3OH. What is K_c for this reaction at 250°C?

16.56 An equilibrium mixture of SO_3, SO_2, and O_2 at 727°C is 0.0160 M SO_3, 0.0056 M SO_2, and 0.0021 M O_2. What is the value of K_c for the following reaction?

$$SO_2(g) + \tfrac{1}{2}O_2(g) \rightleftharpoons SO_3(g)$$

**16.57 At 850°C and 1.000 atm pressure, a gaseous mixture of carbon monoxide and carbon dioxide in equilibrium with solid carbon is 90.55% CO by mass:

$$C(s) + CO_2(g) \rightleftharpoons 2CO(g)$$

Calculate K_c for this reaction at 850°C. *(Hint: Calculate the moles of gas in 100.0 g of mixture; then calculate the volume of gas and the concentrations of CO and CO_2.)*

**16.58 An equilibrium mixture of dinitrogen tetroxide, N_2O_4, and nitrogen dioxide, NO_2, is 65.8% NO_2 by mass at 1.00 atm pressure and 25°C. Calculate K_c at 25°C for the reaction

$$N_2O_4(g) \rightleftharpoons 2NO_2(g)$$

(See hint for Problem 16.57.)

16.59 A 2.00-L vessel contains 1.00 mol N_2, 1.00 mol H_2, and 2.00 mol NH_3. What is the direction of reaction (forward or reverse) needed to attain equilibrium at 400°C? The equilibrium constant K_c for the reaction

$$N_2(g) + 3H_2(g) \rightleftharpoons 2NH_3(g)$$

is 0.51 at 400°C.

16.60 A vessel originally contained 0.200 mol of iodine monobromide (IBr), 0.0010 mol I_2, and 0.0010 mol Br_2.

The equilibrium constant K_c for the reaction

$$I_2(g) + Br_2(g) \rightleftharpoons 2IBr(g)$$

is 1.2×10^2 at 150°C. What is the direction (forward or reverse) needed to attain equilibrium at 150°C?

16.61 A gaseous mixture containing 1.00 mol each of CO, H_2O, CO_2, and H_2 is exposed to a zinc oxide–copper oxide catalyst at 1000°C. The reaction is

$$CO(g) + H_2O(g) \rightleftharpoons CO_2(g) + H_2(g)$$

and the equilibrium constant K_c is 0.58 at 1000°C. What is the direction of reaction (forward or reverse) as the mixture attains equilibrium?

16.62 A 2.0-L reaction flask initially contains 0.10 mol CO, 0.20 mol H_2, and 0.50 mol CH_3OH (methanol). If this mixture is brought in contact with a zinc oxide–chromium(III) oxide catalyst, the equilibrium

$$CO(g) + 2H_2(g) \rightleftharpoons CH_3OH(g)$$

is attained. The equilibrium constant K_c for this reaction at 300°C is 1.1×10^{-2}. What is the direction of reaction (forward or reverse) as the mixture attains equilibrium?

16.63 Hydrogen bromide dissociates when heated according to the equation

$$2HBr(g) \rightleftharpoons H_2(g) + Br_2(g)$$

The equilibrium constant K_c equals 1.6×10^{-2} at 200°C. What are the moles of substances in the equilibrium mixture at 200°C if we start with 0.010 mol HBr in a 1.0-L vessel?

16.64 Iodine monobromide, IBr, occurs as brownish black crystals that vaporize with decomposition:

$$2IBr(g) \rightleftharpoons I_2(g) + Br_2(g)$$

The equilibrium constant K_c at 100°C is 0.026. If 0.010 mol IBr are placed in a 1.0-L vessel at 100°C, what are the moles of substances at equilibrium in the vapor?

16.65 Phosgene, $COCl_2$, is a toxic gas used in the manufacture of urethane plastics. The gas dissociates at high temperature:

$$COCl_2(g) \rightleftharpoons CO(g) + Cl_2(g)$$

At 400°C, the equilibrium constant K_c is 8.05×10^{-4}. Find the percentage of phosgene that dissociates at this temperature when 1.00 mol of phosgene is placed in a 25.0-L vessel.

16.66 Dinitrogen tetroxide, N_2O_4, is a colorless gas (boiling point, 21°C), which dissociates to give nitrogen dioxide, NO_2, a reddish brown gas:

$$N_2O_4(g) \rightleftharpoons 2NO_2(g)$$

The equilibrium constant K_c at 25°C is 0.125. What percentage of dinitrogen tetroxide is dissociated when 0.0300 mol N_2O_4 is placed in a 1.00-L flask at 25°C?

**16.67 Suppose one starts with a mixture of 1.00 mol CO and 4.00 mol H_2 in a 10.00-L vessel. Find the moles of substances present at equilibrium at 1200 K for the reaction

$$CO(g) + 3H_2(g) \rightleftharpoons CH_4(g) + H_2O(g); \quad K_c = 3.92$$

You will get an equation of the form

$$f(x) = 3.92$$

where $f(x)$ is an expression in the unknown x (the amount of CH_4). Solve this equation numerically by guessing values of x, then computing values of $f(x)$. Try to find values of x so that the values of $f(x)$ bracket 3.92. Then choose values of x to get a smaller bracket around 3.92. Continue until x is known to 2 significant figures.

**16.68 What are the moles of substances present at equilibrium at 450°C if 1.00 mol N_2 and 4.00 mol H_2 in a 10.0-L vessel react according to the following equation:

$$N_2(g) + 3H_2(g) \rightleftharpoons 2NH_3(g)$$

The equilibrium constant K_c is 0.153 at 450°C. Use the numerical procedure described in Problem 16.67.

16.69 The amount of nitrogen dioxide formed by dissociation of dinitrogen tetroxide,

$$N_2O_4(g) \rightleftharpoons 2NO_2(g)$$

increases as the temperature rises. Is the dissociation of N_2O_4 endothermic or exothermic?

16.70 The equilibrium constant K_c for the synthesis of methanol, CH_3OH,

$$CO(g) + 2H_2(g) \rightleftharpoons CH_3OH(g)$$

is 4.3 at 250°C and 1.8 at 275°C. Is this reaction endothermic or exothermic?

*16.71 For the reaction

$$N_2(g) + 3H_2(g) \rightleftharpoons 2NH_3(g)$$

show that

$$K_c = K_p(RT)^2$$

Do not use the formula $K_p = K_c(RT)^{\Delta n}$ given in the text. Start from the fact that $P_i = [i]RT$, where P_i is the partial pressure of substance i, and $[i]$ is its molar concentration. Substitute into K_c.

*16.72 For the reaction

$$COCl_2(g) \rightleftharpoons CO(g) + Cl_2(g)$$

show that

$$K_c = K_p/(RT)$$

Do not use the formula $K_p = K_c(RT)^{\Delta n}$ given in the text. See Problem 16.71.

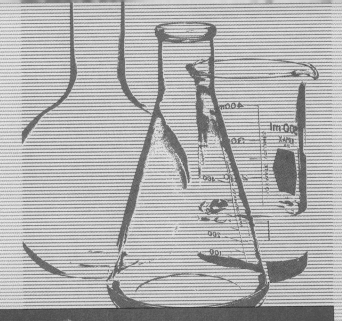

17. Acid–Base Concepts

17.1 Arrhenius Concept of Acids and Bases
17.2 Self-Ionization of Water
17.3 The pH of a Solution
17.4 Brønsted-Lowry Concept of Acids and Bases
17.5 Relative Strengths of Acids and Bases
17.6 Molecular Structure and Acid Strength
17.7 Acid–Base Properties of Salt Solutions;
Hydrolysis
17.8 Lewis Concept of Acids and Bases

A cids and bases were first recognized by simple properties—taste, for example. Acids have a sour taste, whereas bases are bitter. Also, acids and bases change the color of certain dyes called indicators, such as litmus and phenolphthalein. Acids change litmus from blue to red and phenolphthalein from red to colorless. Bases change litmus from red to blue and phenolphthalein from colorless to red. As we can see from these color changes, acids and bases neutralize, or reverse, the action of one another. During neutralization, acids and bases react with each other to produce ionic substances called salts. Acids react with active metals, such as magnesium and zinc, to release hydrogen.

Arrhenius gave the first successful concept of acids and bases. He defined acids and bases in terms of the effect these substances have on water. According to Arrhenius, acids are substances that increase the concentration of H^+ ion in aqueous solution, and bases increase the concentration of OH^- ion in aqueous solution. But many reactions that have characteristics of acid–base reactions in aqueous solution occur in other solvents or without a solvent. Broader acid–base concepts were needed. For example, hydrochloric acid reacts with aqueous ammonia, which is a base because it increases the concentration of OH^- ion in aqueous solution. The reaction can be written

$$HCl(aq) + NH_3(aq) \longrightarrow NH_4Cl(aq)$$

The product is a solution of NH_4Cl, that is, a solution of NH_4^+ and Cl^- ions. A very similar reaction occurs between hydrogen chloride and ammonia dissolved in benzene, C_6H_6. The product is again NH_4Cl, which in this case precipitates from the solution:

$$HCl(benzene) + NH_3(benzene) \longrightarrow NH_4Cl(s)$$

Hydrogen chloride and ammonia even react in the gas phase. If open bottles of concentrated hydrochloric acid and concentrated ammonia are placed next to each other, dense white fumes of NH_4Cl form where HCl gas and NH_3 gas come into contact (Figure 17.1).

$$HCl(g) + NH_3(g) \longrightarrow NH_4Cl(s)$$

In this chapter, we will discuss the Arrhenius, the Brønsted-Lowry, and the Lewis concepts of acids and bases. The Brønsted-Lowry and Lewis concepts apply to nonaqueous and aqueous solutions and enlarge on the Arrhenius concept in other ways as well. No one of these concepts is more correct; each can give us insight into certain applications.

Chapter Overview

We first introduced acids and bases in Section 9.5, and we will extend the discussion in this chapter. After briefly reviewing the definitions given earlier in Chapter 9 (*Arrhenius concept*), we will look at the quantitative measure of acidity, the hydrogen-ion concentration and its expression in terms of *pH*.

We will then look at the *Brønsted-Lowry concept*, which enlarges on the Arrhenius view by looking at an acid–base reaction as a transfer of a proton. This concept is especially useful for discussing the *relative strengths of acids and bases* and the *acid–base characteristics of salt solutions*. The *Lewis concept* views acid–base reactions in terms of the acceptance or donation of an electron pair by a species (molecule or ion) and includes the Brønsted-Lowry concept as a special case.

17.1 Arrhenius Concept of Acids and Bases

Lavoisier was one of the first chemists to try to explain what makes a substance acidic. In 1777, he proposed that oxygen was an essential element

Figure 17.1
Reaction of HCl(g) and NH₃(g) to form NH₄Cl(s). Gases from the concentrated solutions diffuse from their watch glasses (shallow dishes) and react to give a smoke of ammonium chloride.

in acids. (*Oxygen*, which he named, means "acid-former" in Greek.) But, in 1808, Humphrey Davy showed that hydrogen chloride, which dissolves in water to give hydrochloric acid, contains only hydrogen and chlorine. Although some chemists argued that chlorine was a compound of oxygen, chlorine was eventually proved an element. Chemists then saw that hydrogen, not oxygen, was the essential constituent of acids. The cause of acidity and basicity was finally explained in 1884 by Svante Arrhenius.

A modern statement of the **Arrhenius concept** of acids and bases is as follows: *An acid is a substance that when dissolved in water increases the concentration of hydrogen ion,* $H^+(aq)$. *A base is a substance that when dissolved in water increases the concentration of hydroxide ion,* $OH^-(aq)$. The hydrogen ion, $H^+(aq)$, is not a bare proton, but a proton bonded to water molecules. For this reason, the formula for this ion is sometimes written $H_3O^+(aq)$. However, the ion may be a species such as $H_9O_4^+(aq)$ (see Figure 17.2). Water dissociates to only a small extent into H^+ and OH^- ions:

$$H_2O(aq) \rightleftharpoons H^+(aq) + OH^-(aq)$$

Thus, addition of acids or bases alters the equilibrium concentrations of these ions in water.

The definitions of acid and base just given are somewhat broader than the traditional definitions we gave in Section 9.5. There, we excluded salts from the definitions of acids and bases. A salt is any ionic compound but an oxide or a hydroxide. In terms of that restricted view, an acid reacts with a base to

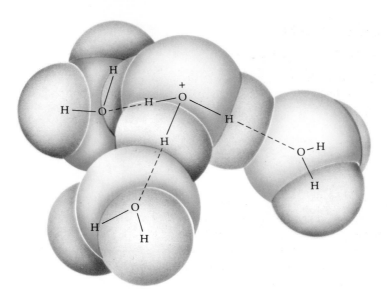

Figure 17.2
The $H^+(aq)$ ion is possibly
$H_9O_4^+$, represented here by a
model.

give a salt. The reaction is called a *neutralization.* For example,

$$HCl(aq) + NaOH(aq) \longrightarrow NaCl(aq) + H_2O(l)$$

acid   base   salt

$$HCl(aq) + NH_3(aq) \longrightarrow NH_4Cl(aq)$$

acid   base   salt

However, many salts do alter the concentrations of H^+ and OH^- ions in aqueous solution, as we will explain later in this chapter. Ammonium chloride, NH_4Cl, increases the H^+ concentration of aqueous solutions. Sodium carbonate, Na_2CO_3, gives solutions with OH^- concentrations about equal to those of NH_3 solutions. Thus, using the broader definitions given here for the Arrhenius concept, we would call NH_4Cl an acid (or acidic substance) and Na_2CO_3 a base (or basic substance).

In Arrhenius's theory, a *strong acid* is a strong electrolyte. It completely dissociates in aqueous solution to give $H^+(aq)$ and an anion. An example is perchloric acid, $HClO_4$:

$$HClO_4(aq) \longrightarrow H^+(aq) + ClO_4^-(aq)$$

Other examples are H_2SO_4, HI, HBr, HCl, and HNO_3. A *strong base* is also a strong electrolyte. It completely dissolves in aqueous solution to give OH^- and a cation. Sodium hydroxide is an example of a strong base:

$$NaOH(s) \longrightarrow Na^+(aq) + OH^-(aq)$$

The principal strong bases are the hydroxides of Group IA elements and Group IIA elements (except Be).■

Evidence for the Arrhenius theory comes from the heat of reaction, ΔH, for neutralization. According to Arrhenius, the neutralization of any strong acid by any strong base is essentially the reaction of $H^+(aq)$ with $OH^-(aq)$ and should therefore always give the same ΔH. For example, if we write the neutralization of $HClO_4$ with NaOH in ionic form, we have

$$[H^+(aq) + \cancel{ClO_4^-(aq)}] + [\cancel{Na^+(aq)} + OH^-(aq)] \longrightarrow$$
$$[\cancel{Na^+(aq)} + \cancel{ClO_4^-(aq)}] + H_2O(l)$$

■ See Table 9.3 for a list of strong acids and bases.

After canceling, we get the net ionic equation

$$H^+(aq) + OH^-(aq) \longrightarrow H_2O(l)$$

Experimentally, it is found that all neutralizations involving strong acids and bases have the same ΔH, -55.90 kJ per mole of H^+. This indicates that the same reaction occurs in each neutralization, as Arrhenius's theory predicted.

17.2 Self-Ionization of Water

Although pure water is often considered a nonelectrolyte (nonconductor of electricity), precise measurements do show a very small conduction. This conduction results from **self-ionization** (or **autoionization**), a reaction in which two like molecules react to give ions. In the case of water, a proton from one H_2O molecule is transferred to another H_2O molecule, leaving behind an OH^- ion and forming a hydronium ion, $H_3O^+(aq)$:

$$H_2O(l) + H_2O(l) \rightleftharpoons H_3O^+(aq) + OH^-(aq)$$

If we write the $H_3O^+(aq)$ ion as $H^+(aq)$, then this self-ionization is simply

$$H_2O(l) \rightleftharpoons H^+(aq) + OH^-(aq)$$

We can see the slight extent to which self-ionization occurs by the small value of the equilibrium constant, K_c. An equilibrium constant, which characterizes a chemical equilibrium, is the constant value (at a given temperature) of the equilibrium-constant expression. This expression is obtained by multiplying the concentrations of products and dividing by the concentrations of reactants, raising each concentration term to a power equal to the coefficient of that substance in the chemical equation. Thus,

$$K_c = \frac{[H^+][OH^-]}{[H_2O]}$$

The value of K_c at room temperature is 1.8×10^{-16}. Because the concentration of ions formed is very small, the concentration of H_2O remains essentially constant, about 56 M at 25°C. If we rearrange this equation, placing $[H_2O]$ with K_c, then the ion product $[H^+][OH^-]$ equals a constant:

$$\underbrace{[H_2O]K_c}_{\text{constant}} = [H^+][OH^-]$$

We call the value of this constant the **ion-product constant for water,** written K_w. ■ At 25°C, the value of K_w is 1.00×10^{-14}.

$$K_w = [H^+][OH^-] = 1.00 \times 10^{-14} \text{ at } 25°C$$

Like any equilibrium constant, K_w varies with temperature. For example, at body temperature, 37°C, K_w equals 2.5×10^{-14}.

Using K_w, we can calculate the concentrations of H^+ and OH^- ions in pure water. Since these ions are produced in equal numbers in pure water, their concentrations are equal. Let $x = [H^+] = [OH^-]$. Then, substituting into the equation for the ion-product constant

$$K_w = [H^+][OH^-]$$

■ In effect, we omit the solvent from the equilibrium expression. This is similar to the omission of pure liquids and solids from the equilibrium expression for heterogeneous equilibria (see Section 16.3).

we get, at 25°C,

$$1.00 \times 10^{-14} = x^2$$

Hence, x equals 1.00×10^{-7}. Thus, the concentrations of H^+ and OH^- are each 1.00×10^{-7} M.

In acidic or basic solutions, the concentrations of hydrogen ion and hydroxide ion will differ from those in pure water. The next example shows how to calculate the concentrations of these ions in a solution of strong acid or base. Solutions of weak acids and bases will be discussed in the next chapter.

Example 17.1

What is the concentration of OH^- in 0.0100 M $Ca(OH)_2$? What is the concentration of H^+ at 25°C?

Solution

Since calcium hydroxide is a strong base, it goes into the solution completely as ions:

$$Ca(OH)_2(s) \longrightarrow Ca^{2+}(aq) + 2OH^-(aq)$$

A liter of 0.0100 M $Ca(OH)_2$ is made up by dissolving 0.0100 mol $Ca(OH)_2$ in sufficient water. In solution, this forms 0.0100 mol Ca^{2+} and 2×0.0100 mol OH^-. We can ignore the amount of OH^- contributed by the self-ionization of water, since it is so small. Thus, the hydroxide ion concentration is 0.0200 M.

We calculate the hydrogen-ion concentration from the equation

$$K_w = [H^+][OH^-]$$

Substituting values, we get

$$1.00 \times 10^{-14} = [H^+] \times 0.0200$$

or,

$$[H^+] = \frac{1.00 \times 10^{-14}}{0.0200} = 5.00 \times 10^{-13}$$

The hydrogen-ion concentration is 5.00×10^{-13} M.

Exercise 17.1

A solution of barium hydroxide at 25°C is 0.125 M $Ba(OH)_2$. What are the concentrations of hydrogen ion and hydroxide ion?

(See Problems 17.17, 17.18, 17.19, and 17.20.)

By dissolving substances in water, we can alter the concentrations of H^+ and OH^- ions. In a *neutral* solution, the concentrations of H^+ and OH^- remain equal, as they are in pure water. In an *acidic* solution, the concentration of H^+ is greater than that of OH^-. In a *basic* solution, the concentration of OH^- is greater than that of H^+. At 25°C, we observe the following conditions:

> In an acidic solution, $[H^+] > 1.00 \times 10^{-7}$ M.
> In a neutral solution, $[H^+] = 1.00 \times 10^{-7}$ M.
> In a basic solution, $[H^+] < 1.00 \times 10^{-7}$ M.

Exercise 17.2

A solution has a hydroxide-ion concentration of 1.0×10^{-5} M at 25°C. Is the solution acidic, neutral, or basic?

(See Problems 17.21 and 17.22.)

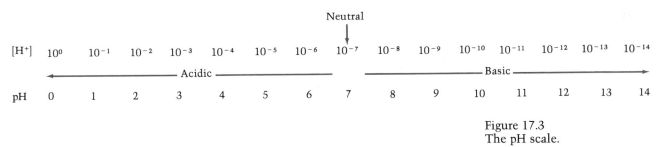

Figure 17.3
The pH scale.

17.3 The pH of a Solution

We see that whether an aqueous solution is acidic, neutral, or basic depends on the hydrogen-ion concentration. We can quantitatively describe the acidity by giving the hydrogen-ion concentration. But because these concentration values may be very small, it is often more convenient to give the acidity in terms of the **pH**, which is defined as the negative of the logarithm of the molar hydrogen ion-concentration: ■

$$pH = -\log [H^+]$$

For a solution in which the hydrogen-ion concentration is $1.0 \times 10^{-3}\,M$, the pH is

$$pH = -\log (1.0 \times 10^{-3}) = 3.00$$

Note that the number of places after the decimal point in the pH equals the number of significant figures reported in the hydrogen-ion concentration. ■

A neutral solution, whose hydrogen-ion concentration at 25°C is $1.00 \times 10^{-7}\,M$, has a pH of 7.000. For an acidic solution, the hydrogen-ion concentration is greater than $1.00 \times 10^{-7}\,M$, and therefore the pH is less than 7.000. Similarly, a basic solution has a pH greater than 7.000. Figure 17.3 shows a diagram of the pH scale.

■ The Danish biochemist S. P. L. Sørensen devised the pH scale while working on the brewing of beer.

■ See Appendix A for a discussion of logarithms.

Example 17.2

A sample of orange juice has a hydrogen-ion concentration of $2.9 \times 10^{-4}\,M$. What is the pH? Is the solution acidic?

Solution

We have that

$$pH = -\log [H^+]$$
$$= -\log (2.9 \times 10^{-4})$$

Using an electronic calculator with a *log* key, we simply enter the H^+ concentration, press *log*, and change the sign of the result. To obtain the logarithm from a table, we use the fact that the logarithm of a product is the sum of the logarithms of the factors. Hence,

$$pH = -(\log 2.9 + \log 10^{-4})$$

Tables of logarithms are normally given only for numbers between 1 and 10. We can therefore look up the logarithm of 2.9, where we find a value of 0.46. The logarithm of 10^{-4} is simply -4. Hence,

$$pH = -(0.46 - 4) = 3.54$$

The solution has pH 3.54, which is less than 7.00 and therefore acidic.

Exercise 17.3

What is the pH of a sample of gastric juice (digestive juice in the stomach) whose hydrogen-ion concentration is 0.045 M?

(See Problems 17.25 and 17.26.)

We can find the pH of a solution of known hydroxide-ion concentration by first solving for the hydrogen-ion concentration, as shown in Example 17.1. However, we can also find the pH simply from the *pOH*, a measure of hydroxide-ion concentration similar to the pH:

$$pOH = -\log [OH^-]$$

Then, since $K_w = [H^+][OH^-] = 1.00 \times 10^{-14}$ at 25°C, we can show that ■

$$pH + pOH = 14.000$$

For example, suppose we wish to find the pH of an ammonia solution whose hydroxide-ion concentration is 1.9×10^{-3} *M*. We first calculate the pOH:

$$pOH = -\log (1.9 \times 10^{-3}) = 2.72$$

Thus, the pH is

$$pH = 14.00 - pOH$$
$$= 14.00 - 2.72 = 11.28$$

■ Taking the logarithm of both sides of the equation
$$[H^+][OH^-] = 1.00 \times 10^{-14}$$
we get
$$\log [H^+] + \log [OH^-]$$
$$= -14.000$$
or $(-\log [H^+]) + (-\log [OH^-])$
$$= 14.000$$
Hence, $pH + pOH = 14.000$

Exercise 17.4

A saturated solution of calcium hydroxide has a hydroxide-ion concentration of 0.025 *M*. What is the pH of the solution?

(See Problems 17.27 and 17.28.)

The usefulness of the pH scale depends on the ease with which we can interconvert pH and hydrogen-ion concentrations. In the next example, we will find the hydrogen-ion concentration, given the pH value.

Example 17.3

The pH of normal arterial blood is 7.40. What is the hydrogen-ion concentration?

Solution

To solve this problem using a table of logarithms, we rearrange the equation for pH to give

$$\log [H^+] = -pH$$

In the present problem, we have

$$\log [H^+] = -7.40$$
$$= -0.40 - 7$$

Here, we separated -7.40 into the decimal part and the whole number. We now add 1 to the decimal part and subtract 1 from the whole number:

$$\log [H^+] = (-0.40 + 1) + (-7 - 1)$$
$$= 0.60 - 8$$

This separates the value of the logarithm into a positive decimal part, which we can look up in a table of logarithms, and a negative whole number. From the table, we find that the antilogarithm of 0.60 is 4.0; that is, $10^{0.60} = 4.0$. Thus,

$$[H^+] = 10^{0.60} \times 10^{-8} = 4.0 \times 10^{-8}$$

We can use an electronic calculator with a *log* key to solve this problem also. If the calculator has an *inverse* key, enter -7.40, press *inverse*, then *log*. The result will be antilog (-7.40), which equals $[H^+]$. If the calculator has a y^x key, then enter 10, press y^x, enter -7.40, and press *equals*. The result is $10^{-7.40} = $ antilog (-7.40).

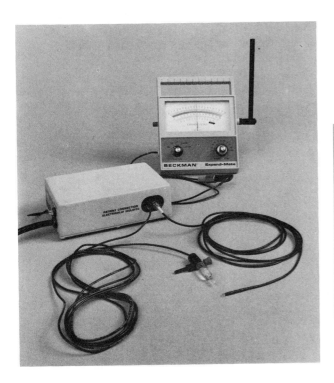

Figure 17.4
Two pH meters. (Left) Electrodes, at the ends of wires, are dipped into the solution whose pH is to be determined. (Right) A digital pH meter.

Exercise 17.5

A brand of beer has a pH of 4.46. What is the hydrogen-ion concentration of the beer?

(See Problems 17.29 and 17.30.)

Exercise 17.6

A 0.010 M solution of ammonia, NH_3, has a pH of 10.6 at 25°C. What is the concentration of hydroxide ion? (One way to solve this problem is to first find the pOH, then calculate the hydroxide-ion concentration.)

(See Problems 17.31 and 17.32.)

 The pH of a solution can be accurately measured by a pH meter (Figure 17.4). This instrument consists of specially designed electrodes that are dipped into the solution. A voltage determined by the pH is generated between the electrodes and is read on a meter calibrated directly in pH. Table 17.1 shows the pH values of some common solutions.
 Although less precise, acid–base indicators are often used to measure pH, since they usually change color within a small pH range. Figure 17.5 shows the color changes for various acid–base indicators (see also Color Plate 8). Phenophthalein, mentioned in the chapter opening, is colorless in acidic and slightly basic solutions, but begins to turn pink at pH 8.0. The color change is complete at pH 9.7. Methyl orange is red in very acidic solutions, but begins to change color at pH 3.1 to orange and at pH 4.4 becomes yellow. By impregnating paper strips with several indicator dyes, one can obtain a "pH paper" that changes to a definite color for each pH value.

■ Litmus contains several substances and changes color over a broader range, from about pH 5 to pH 8.

Solution	pH
Gastric juice	1.0–3.0
Lemon juice	2.2–2.4
Vinegar	2.4–3.4
Carbonated water (soda water)	3.9
Beer	4.0–4.5
Milk	6.4
Blood	7.4
Sea water	7.0–8.3
Baking soda solution (0.1 M)	8.4
Milk of magnesia	10.5
Household ammonia	11.9

Table 17.1
The pH of Some Common Solutions

17.4 Brønsted-Lowry Concept of Acids and Bases

In 1923, Danish chemist Johannes N. Brønsted (1879–1947) and, independently, British chemist Thomas M. Lowry (1874–1936) pointed out that Arrhenius acid–base reactions can be seen as proton-transfer reactions. According to the **Brønsted-Lowry concept,** the acid in a proton-transfer reaction is the species (molecule or ion) donating the proton, and the base is the species accepting the proton:

$$\text{Brønsted-Lowry acid = proton donor}$$
$$\text{Brønsted-Lowry base = proton acceptor}$$

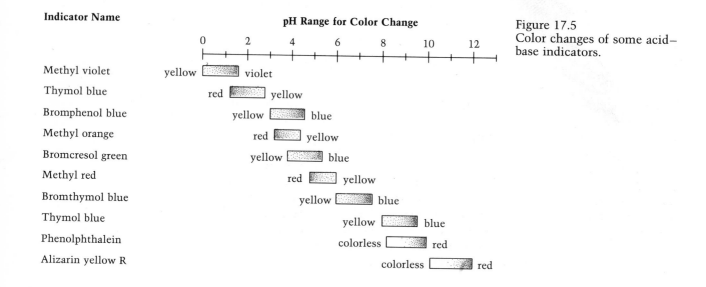

Figure 17.5
Color changes of some acid–base indicators.

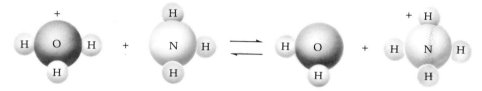

Figure 17.6
A representation of the reaction $H_3O^+ + NH_3 \rightleftharpoons H_2O + NH_4^+$. Note the transfer of a proton, H^+, from H_3O^+ to NH_3.

Consider, for example, the reaction of hydrochloric acid with ammonia, which was mentioned in the chapter opening. Writing it as an ionic equation, we have

$$[H^+(aq) + \cancel{Cl^-(aq)}] + NH_3(aq) \longrightarrow [NH_4^+(aq) + \cancel{Cl^-(aq)}]$$

After canceling Cl^-, we get the net ionic equation. To see the proton transfer, we write $H^+(aq)$ as $H_3O^+(aq)$ for a net ionic equation of

$$H_3O^+(aq) + NH_3(aq) \longrightarrow H_2O(l) + NH_4^+(aq)$$

In this reaction, a proton, H^+, is transferred from the H_3O^+ ion to the NH_3 molecule, giving H_2O and NH_4^+ (see Figure 17.6). Here H_3O^+ is the proton donor, or acid, and NH_3 is the proton acceptor, or base. Note that in the Brønsted-Lowry concept, acids (and bases) can be ions as well as molecular substances.

The Brønsted-Lowry concept can also be applied to the reaction of HCl and NH_3 dissolved in benzene, C_6H_6, which was mentioned in the chapter opening. In benzene, HCl and NH_3 are not ionized. The equation is

$$HCl(benzene) + NH_3(benzene) \longrightarrow NH_4Cl(s)$$
$$\quad\quad\; \text{acid} \quad\quad\quad\quad \text{base}$$

Here HCl is the proton donor, or acid, and NH_3 is the proton acceptor, or base.

In any Brønsted-Lowry acid–base equilibrium, both forward and reverse reactions involve proton transfers. Consider the reaction of NH_3 with H_2O:

$$NH_3(aq) + H_2O(l) \rightleftharpoons NH_4^+(aq) + OH^-(aq)$$
$$\;\text{base} \quad\quad\; \text{acid} \quad\quad\quad \text{acid} \quad\quad\quad \text{base}$$

In the forward reaction, NH_3 accepts a proton from H_2O. Thus, NH_3 is a base and H_2O is an acid. In the reverse reaction, NH_4^+ donates a proton to OH^-. The NH_4^+ ion is the acid and OH^- is the base.

Note that NH_3 and NH_4^+ differ by a proton. That is, NH_3 becomes the NH_4^+ ion by gaining a proton, whereas the NH_4^+ ion becomes the NH_3 molecule by losing a proton. The pair of species NH_4^+ and NH_3 is a conjugate acid–base pair. A **conjugate acid–base pair** consists of two species in an acid–base equilibrium, one acid and one base, that differ by the gain or loss of a proton. The acid in such a pair is called the **conjugate acid** of the base, whereas the base is the **conjugate base** of the acid. Thus, NH_4^+ is the conjugate acid of NH_3, and NH_3 is the conjugate base of NH_4^+.

Example 17.4

In the following equations, label all of the species as acid or base, according to the Brønsted-Lowry concept. Show the conjugate acid–base pairs:

(a) $HCO_3^-(aq) + HF(aq) \rightleftharpoons H_2CO_3(aq) + F^-(aq)$
(b) $HCO_3^-(aq) + OH^-(aq) \rightleftharpoons CO_3^{2-}(aq) + H_2O(l)$

Solution

(a) Examine the equation to find the proton donor on each side. On the left, HF is the proton donor; on the right, H_2CO_3 is the proton donor. The proton acceptors are HCO_3^- and F^-. Once the proton donors and acceptors are identified, the acids and bases can be labeled:

$$HCO_3^-(aq) + HF(aq) \rightleftharpoons H_2CO_3(aq) + F^-(aq)$$
$$\text{base} \qquad \text{acid} \qquad \text{acid} \qquad \text{base}$$

In this reaction, H_2CO_3 and HCO_3^- are a conjugate acid–base pair, as are HF and F^-.

(b) We have

$$HCO_3^-(aq) + OH^-(aq) \rightleftharpoons CO_3^{2-}(aq) + H_2O(l)$$
$$\text{acid} \qquad \text{base} \qquad \text{base} \qquad \text{acid}$$

Here, HCO_3^- and CO_3^{2-} are a conjugate acid–base pair, as are H_2O and OH^-. Note that although HCO_3^- functions as an acid in this reaction, it functions as a base in (a).

Exercise 17.7

For the reaction

$$H_2CO_3(aq) + CN^-(aq) \rightleftharpoons HCN(aq) + HCO_3^-(aq)$$

label the Brønsted-Lowry acids and bases. For the base on the left, what is the conjugate acid?

(See Problems 17.37 and 17.38.)

The Brønsted-Lowry concept defines a species as an acid or base according to its function in the acid–base, or proton-transfer, reaction. As we saw in Example 17.4, some species can act either as an acid or a base. An **amphiprotic species** can act either as an acid or a base by losing or gaining a proton, depending on the other reactant.■ Thus, HCO_3^- acts as an acid in the presence of OH^-, but as a base in the presence of HF. Anions with ionizable hydrogens, such as HCO_3^-, and certain solvents, such as water, are amphiprotic.

The amphiprotic characteristic of water is important in the acid–base properties of aqueous solutions. Consider, for example, the reactions of water with the base NH_3 and with the acid $HC_2H_3O_2$ (acetic acid):

$$NH_3(aq) + H_2O(l) \rightleftharpoons NH_4^+(aq) + OH^-(aq)$$
$$\text{base} \qquad \text{acid} \qquad \text{acid} \qquad \text{base}$$
$$HC_2H_3O_2(aq) + H_2O(aq) \rightleftharpoons C_2H_3O_2^-(aq) + H_3O^+(aq)$$
$$\text{acid} \qquad \text{base} \qquad \text{base} \qquad \text{acid}$$

In the first case, water reacts as an acid with the base NH_3. In the second case, water reacts as a base with the acid $HC_2H_3O_2$.

We have now seen several ways in which the Brønsted-Lowry concept of acids and bases has greater scope than the Arrhenius concept. In the Brønsted-Lowry concept

1. Acids and bases can be ions as well as molecular substances.
2. Acid–base reactions are not restricted to aqueous solution.
3. Some species can act either as acids or bases, depending on the other reactant.

■ *Amphoteric* is a general term referring to a species that can act as an acid or base. The species need not be amphiprotic, however. Thus, aluminum oxide is an amphoteric oxide, because it reacts with acids and bases. It is not amphiprotic since it has no protons.

In the next sections, we will look at two areas in which the Brønsted-Lowry concept has proved particularly useful. One is the concept of relative acid–base strength. The other is the explanation of the acid–base properties of salt solutions.

17.5 Relative Strengths of Acids and Bases

It is instructive to consider a Brønsted-Lowry acid–base equilibrium as a competition for protons. From this point of view, we can order acids and bases by their relative strengths. The stronger acids are those that lose their protons more easily than others. Similarly, the stronger bases are those that hold on to protons more strongly than others.

Recall that an Arrhenius acid is strong if it completely ionizes in water. In the reaction of hydrogen chloride with water, for example, water acts as a base, accepting the proton from HCl:

$$\underset{\text{acid}}{HCl(aq)} + \underset{\text{base}}{H_2O(l)} \rightleftharpoons \underset{\text{base}}{Cl^-(aq)} + \underset{\text{acid}}{H_3O^+(aq)}$$

The reverse reaction occurs only to an extremely small extent. Because the reaction goes almost completely to the right, we say that HCl is a strong acid. However, even though the equilibrium composition is almost completely products, we can consider the reverse reaction. In it, the Cl^- ion acts as the base, accepting a proton from the acid H_3O^+. We have labeled the Brønsted-Lowry acids and bases in the preceding equilibrium.

Let us look at this equilibrium in terms of the relative strengths of the two acids in the equation, HCl and H_3O^+. Because HCl is a strong Arrhenius acid, it must lose its proton readily, more readily than H_3O^+ does. We can say that HCl is a stronger acid than H_3O^+; of the two, H_3O^+ is the weaker acid.

$$\underset{\substack{\text{stronger} \\ \text{acid}}}{HCl(aq)} + H_2O(l) \rightleftharpoons Cl^-(aq) + \underset{\substack{\text{weaker} \\ \text{acid}}}{H_3O^+(aq)}$$

It is important to understand that we use the terms *stronger* and *weaker* only in a relative sense. All strong acids ionize completely in water to give a solution of H_3O^+ ion, and this ion is a fairly strong Brønsted-Lowry acid. However, it is weaker than HCl.■

An acid–base equilibrium always lies in the direction of the weaker acid. We can see that this is the case for the above reaction. By experimentally comparing equilibria between different pairs of acids, we can arrive at a relative order for acid strengths. The first column of Table 17.2 lists acids by their strength, with the strongest at the top of the table. Note that the arrow along the left side of the table points toward the direction of the weaker acid, in the direction the equilibrium lies. For example, in the equilibrium between HCl and H_3O^+, the arrow points from HCl toward H_3O^+, the direction in which the reaction occurs.

This reaction can also be viewed in terms of the bases, H_2O and Cl^-. A stronger base picks up a proton more readily than does a weaker one. Water has greater base strength than the Cl^- ion; that is, H_2O picks up protons

■ The strongest acid that can exist in H_2O is H_3O^+.

Table 17.2
Relative Strengths of Acids
and Bases

	Acid	Base	
Strongest acids	$HClO_4$	ClO_4^-	Weakest bases
	H_2SO_4	HSO_4^-	
	HI	I^-	
	HBr	Br^-	
	HCl	Cl^-	
	HNO_3	NO_3^-	
	H_3O^+	H_2O	
	HSO_4^-	SO_4^{2-}	
	H_2SO_3	HSO_3^-	
	H_3PO_4	$H_2PO_4^-$	
	HNO_2	NO_2^-	
	HF	F^-	
	$HC_2H_3O_2$	$C_2H_3O_2^-$	
	$Al(H_2O)_6^{3+}$	$Al(H_2O)_5OH^{2+}$	
	H_2CO_3	HCO_3^-	
	H_2S	HS^-	
	HClO	ClO^-	
	HBrO	BrO^-	
	NH_4^+	NH_3	
	HCN	CN^-	
	HCO_3^-	CO_3^{2-}	
	H_2O_2	HO_2^-	
	HS^-	S^{2-}	
	H_2O	OH^-	
Weakest acids	NH_3	NH_2^-	Strongest bases
	OH^-	O^{2-}	

more readily than Cl^- does. In fact, Cl^- has little attraction for protons. (If it had more, HCl would not lose its proton so readily.) Because H_2O molecules compete more successfully for protons than Cl^- ions do, the reaction goes almost completely to the right. (Note that the arrow on the right in Table 17.2 points from H_2O to Cl^-.) That is, the equilibrium concentrations of product species (Cl^- and H_3O^+) are greater than the equilibrium concentrations of reactant species.

$$HCl(aq) + H_2O(l) \longrightarrow Cl^-(aq) + H_3O^+(aq)$$
$$\text{stronger base} \qquad \text{weaker base}$$

The equilibrium lies in the direction of the weaker base.

By comparing equilibria between different pairs of bases, we can arrive at a relative order for base strengths, just as we did for acids. We can see a definite relationship between acid and base strengths. When we say an acid loses its proton readily, we can also say that its conjugate base does not hold the proton very tightly. Thus, *the strongest acids have the weakest conjugate bases, and the strongest bases have the weakest conjugate acids*. This means that a list of conjugate bases of the acids in Table 17.2 will be in order of increasing base strength. The weakest bases will be at the top of the table, and the strongest at the bottom.

We can use Table 17.2 to predict whether reactants or products of a given equation are favored at equilibrium. The equilibrium for an acid–base reaction always favors the weaker acid and weaker base; that is, the normal direction of reaction is from the stronger acid and base to the weaker acid and base. For the reaction we have been discussing, we have

$$HCl(aq) + H_2O(l) \longrightarrow Cl^-(aq) + H_3O^+(aq)$$

stronger   stronger   weaker   weaker
acid   base   acid   base

This equilibrium follows the direction of the arrows at the left and right margins in Table 17.2.

Example 17.5

For the reaction

$$SO_4^{2-}(aq) + HCN(aq) \rightleftharpoons HSO_4^-(aq) + CN^-(aq)$$

use the relative strengths of acids and bases (Table 17.2) to decide which species (reactants or products) are favored at equilibrium.

Solution

If we compare the relative strengths of the two acids HCN and HSO_4^-, we see that HCN is weaker. Or, com-

paring the bases SO_4^{2-} and CN^-, we see that SO_4^{2-} is weaker. Hence, the reaction would normally go from right to left (that is, the equilibrium concentrations of SO_4^{2-} and HCN are greater):

$$SO_4^{2-}(aq) + HCN(aq) \longleftarrow HSO_4^-(aq) + CN^-(aq)$$

weaker   weaker   stronger   stronger
base   acid   acid   base

Thus, the reactants are favored at equilibrium.

Exercise 17.8

Decide the direction of reaction for the following from the relative strengths of acids and bases:

$$H_2S(aq) + C_2H_3O_2^-(aq) \rightleftharpoons HC_2H_3O_2(aq) + HS^-(aq)$$

(See Problems 17.39 and 17.40.)

17.6 Molecular Structure and Acid Strength

The strength of an acid depends on how easily the proton, H^+, is lost or removed from an H—X bond in the acid species. Understanding the factors that determine this ease of proton loss will help us to predict the relative strengths of similar acids.

Two factors are important in determining relative acid strengths. One is the polarity of the bond to which the H atom is attached. The H atom should have a positive partial charge:

$\delta+$ $\delta-$
H—X

The more polarized the bond is in this direction, the easier the proton is removed, and the greater the acid strength. The second factor determining acid strength is the strength of the bond, that is, how tightly the proton is held. This in turn depends on the size of atom X. The larger atom X, the weaker the bond, and the greater the acid strength.

For example, consider a series of binary acids, HX, formed from a given column of elements of the periodic table. The acids would be compounds of these elements with hydrogen, such as the binary acids of the Group VIIA elements: HF, HCl, HBr, and HI. As we go down the column of elements, each time adding a shell of electrons to the atom, the radius increases rapidly. For this reason, the size of the atom X becomes the dominant factor in determining acid strength. *In going down a column of elements of the periodic table, the size of atom X increases, the H—X bond strength decreases, and the strength of the binary acid increases.* We thus predict the following order of acid strength:

$$HF < HCl < HBr < HI$$

This is the same order shown in Table 17.2 for these acids.

As we go across a row of elements of the periodic table, the atomic radius decreases slowly. For this reason, the relative strengths of the binary acids of these elements are less dependent on the sizes of atoms X. Now the polarity of the H—X bond becomes the dominant factor in determining acid strength. *Going across a row of elements of the periodic table, the electronegativity increases, the H—X bond polarity increases, and the acid strength increases.* For example, the binary acids of the last three elements of the second period are NH_3, H_2O, and HF. The acid strengths are

$$NH_3 < H_2O < HF$$

This, again, is the order shown in Table 17.2. Hydrogen fluoride, HF, is a weak acid, H_2O is a very weak acid (and a very weak base), and NH_3 is an extremely weak acid. As an acid, NH_3 loses a proton, to form NH_2^-. In aqueous solution, however, NH_3 acts as a base by accepting a proton to give NH_4^+.

Let us examine what determines the bond strength of oxyacids. An oxyacid has the structure

$$H—O—Y—$$

The acidic H atom is always attached to an O atom, which in turn is attached to an atom Y. Other groups, such as O atoms or O—H groups, are attached to Y. Bond polarity appears to be the dominant factor determining relative strengths of the oxyacids. This in turn depends on the electronegativity of atom Y. If the electronegativity of atom Y is large, the H—O bond is relatively polar, and the acid strength large. *For a series of oxyacids of the same structure, differing only in atom Y, the acid strength increases with the electronegativity of Y.* Consider, for example, the acids HClO, HBrO, and HIO.∎ The structures are

$$H—\ddot{O}—\ddot{C}l\!: \qquad H—\ddot{O}—\ddot{B}r\!: \qquad H—\ddot{O}—\ddot{I}\!:$$

The electronegativity of Group VIIA elements decreases going down the column of elements, so that the electronegativity decreases from Cl to Br to I. Therefore, the order of acid strengths is

$$HClO > HBrO > HIO$$

For a series of oxyacids of the same element Y, the H—O bond polarity, and therefore the acid strength, increases with the oxidation state of Y. The

■ The formulas of these acids may be written HXO or HOX, depending on the convention used. Formulas of oxyacids are generally written with the acidic H atoms first, followed by the characteristic element (X), then O atoms. However, the formulas of molecules composed of three atoms are often written in the order in which the atoms are bonded, which in this case is HOX.

series of oxyacids of Cl provides an example. Acid strength is in the following order (oxidation states of Cl are in color):

$$\overset{+7}{HClO_4} > \overset{+5}{HClO_3} > \overset{+3}{HClO_2} > \overset{+1}{HClO}$$

Perchloric acid, $HClO_4$, is the strongest acid listed in Table 17.2. Its structure is

The Cl atom strongly attracts electrons because of its large positive oxidation state. This strongly polarizes the H—O bond, so that the proton is easily lost.

Before we leave the subject of molecular structure and acid strength, let us look at the relative acid strengths of a polyprotic acid and its corresponding acid anions. For example, H_2SO_4 ionizes by losing a proton to give HSO_4^-, which in turn ionizes to give SO_4^{2-}. Since HSO_4^- can lose a proton, it is acidic. However, because of the negative charge of the ion, which tends to attract protons, its acid strength is reduced from that of the uncharged species. That is, the acid strengths are in the order

$$H_2SO_4 > HSO_4^-$$

This shows us that *the acid strength of a polyprotic acid and its anions decrease with increasing negative charge* (see Table 17.2).

Exercise 17.9

Which one of each of the following pairs is the stronger acid? (a) NH_3, PH_3 (b) HI, H_2Te (c) HSO_3^-, H_2SO_3 (d) H_3AsO_4, H_3AsO_3 (e) HSO_4^-, $HSeO_4^-$
(See Problems 17.43 and 17.44.)

17.7 Acid–Base Properties of Salt Solutions; Hydrolysis

One of the successes of the Brønsted-Lowry concept is its explanation of the acid–base properties of salt solutions. Consider the neutralization of hydrochloric acid by aqueous ammonia, discussed in the chapter opening. When we add equal molar amounts of HCl and NH_3, we get a solution of the salt, ammonium chloride, NH_4Cl. But even though we have added equal molar amounts of an acid and a base, we do not get a neutral solution. Solutions of NH_4Cl are acidic; 0.10 M NH_4Cl has a pH of 5.1.

A solution of NH_4Cl is a solution of NH_4^+ ion and Cl^- ion. The acidity of such a solution results from the reaction of NH_4^+ ion with water:

$$NH_4^+(aq) + H_2O(aq) \rightleftharpoons NH_3(aq) + H_3O^+(aq)$$

This is an example of **hydrolysis,** which is the reaction of an ion with water to produce either H_3O^+ or OH^-. It is a type of Brønsted-Lowry acid–base reaction. As we can see from Table 17.2, NH_4^+ is a much weaker acid than H_3O^+. Thus, the hydrolysis equilibrium is toward the left; that is, it favors NH_4^+ and H_2O. Nevertheless, a very small fraction (about 0.01%) of NH_4^+ ions do react with water, and this is sufficient to account for the acidity of NH_4Cl solutions. Salts of organic amines, such as $CH_3NH_3^+Cl^-$, also give acidic solutions because of similar hydrolysis of the cation.

We can ignore the Cl^- ions in these solutions as far as the acidity is concerned because the Cl^- ion is unreactive with water. The hydrolysis reaction would be

$$Cl^-(aq) + H_2O(l) \rightleftharpoons HCl(aq) + OH^-(aq)$$

But the product HCl is a strong acid and is completely ionized in aqueous solution. This simply means that the Cl^- ion is so weak a base that its hydrolysis occurs to an insignificant extent. That is

$$Cl^-(aq) + H_2O(l) \longrightarrow \text{no reaction}$$

As another example, consider a solution of sodium carbonate. Such a solution is basic; $0.10\ M\ Na_2CO_3$ has a pH of 11.6. To explain the basicity, we first note that like most salts, Na_2CO_3 exists in aqueous solution as the ions, in this case, Na^+ and CO_3^{2-}. There is no tendency for Na^+ ions to hydrolyze; that is, Na^+ ions do not react with H_2O. In general, the ions of Groups IA and IIA metals (except Be), which give strong bases, are unreactive with water:

$$Na^+(aq) + H_2O(l) \longrightarrow \text{no reaction}$$

However, carbonate ions, CO_3^{2-}, do hydrolyze:

$$CO_3^{2-}(aq) + H_2O(l) \rightleftharpoons HCO_3^{2-}(aq) + OH^-(aq)$$

It is this hydrolysis that is responsible for the basicity of Na_2CO_3 solutions.

Metal ions, except those of Groups IA and IIA metals (excluding Be), usually hydrolyze by acting as Brønsted-Lowry acids. In these cases, water molecules associated with the metal ion are acidic. Consider the aluminum ion, Al^{3+}, which exists in aqueous solution as the hydrated species $Al(H_2O)_6^{3+}$. The H_2O molecules are bonded to the Al atom through lone pairs of electrons on the O atoms. Because the electrons are drawn away from the O atoms by the Al atom the O atoms in turn tend to draw electrons from the O—H bonds, making them highly polar. As a result, the H_2O molecules in $Al(H_2O)_6^{3+}$ are acidic:

$$Al(H_2O)_6^{3+}(aq) + H_2O(l) \rightleftharpoons Al(H_2O)_5OH^{2+}(aq) + H_3O^+(aq)$$

Other common metal ions that hydrolyze are $Zn^{2+}(aq)$, $Fe^{3+}(aq)$, and $Cu^{2+}(aq)$.

From this discussion, we can summarize the following rules for the hydrolysis of ions:

1. Metal ions of Groups IA and IIA elements (except Be) and anions that are conjugate bases of strong acids do not hydrolyze (the strong acids are listed in Table 9.3, p. 284).

2. Ammonium ion, NH_4^+, and hydrated metal ions, except those of Groups IA and IIA elements (excluding Be), hydrolyze to give H_3O^+.

3. Anions of weak acids hydrolyze to give OH^-.

To decide whether an aqueous solution of a salt is acidic, neutral, or basic, we determine from the preceding rules whether the ions from the salt hydrolyze. If neither ion hydrolyzes, a solution of the salt will be neutral. If one ion hydrolyzes, the solution will be acidic or basic depending on whether the hydrolysis gives H_3O^+ or OH^- ions. When both ions hydrolyze, no simple answer can be given. The solution will be acidic, neutral, or basic depending on the relative acid–base strengths of the two ions. The next example illustrates how we use the preceding rules to decide whether a solution is acidic, neutral, or basic.

Example 17.6

Decide whether aqueous solutions of the following salts will be acidic, basic, or neutral: (a) NaCl (b) KCN (c) $Zn(NO_3)_2$

Solution

(a) The Na^+ and Cl^- ions do not hydrolyze (Rule 1); Na^+ is an ion of a Group IA element; Cl^- is an anion conjugate to a strong acid. Therefore, a solution of NaCl is neutral. (b) The K^+ ion is an ion of a Group IA element and does not hydrolyze (Rule 1). However, CN^- is an anion conjugate to a weak acid and hydrolyzes to give OH^- (Rule 3). A solution of KCN is basic. (c) The Zn^{2+} ion hydrolyzes in water to give H_3O^+ (Rule 2). The NO_3^- ion is an anion conjugate to a strong acid, so does not hydrolyze (Rule 1). A solution of $Zn(NO_3)_2$ is acidic.

Exercise 17.10

Consider solutions of the following salts: (a) NH_4NO_3, (b) KNO_3, (c) $Al(NO_3)_3$. Which solution is acidic? Which is basic? Which is neutral?

(See Problems 17.45 and 17.46.)

17.8 Lewis Concept of Acids and Bases

Certain reactions have characteristics of acid–base reactions, but do not fit the Brønsted-Lowry concept. An example is the reaction of the basic oxide Na_2O with the acidic oxide SO_3 to give the salt Na_2SO_4:

$$Na_2O(s) + SO_3(g) \longrightarrow Na_2SO_4(s)$$

G. N. Lewis, who proposed the electron-pair theory of covalent bonding, realized that the concept of acids and bases could be generalized to include reactions of acidic and basic oxides, and many other reactions, as well as proton-transfer reactions. According to the **Lewis concept**

A Lewis acid is an electron-pair acceptor.

A Lewis base is an electron-pair donor.

The Lewis and the Brønsted-Lowry concepts are simply different ways of looking at certain chemical reactions. Such different views are often helpful in devising new reactions.

Consider again the neutralization of NH_3 by HCl in aqueous solution, mentioned in the chapter opening. The reaction consists of a transfer of a proton from H_3O^+ to NH_3. The transfer of the proton to NH_3 can be written as follows:

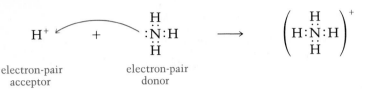

Here, the arrow shows the proton accepting an electron pair from NH_3. Since the proton is an electron-pair acceptor, it is a Lewis acid. Ammonia, NH_3, which has a lone pair of electrons, is an electron-pair donor and therefore a Lewis base.

Now, let us look at the reaction of Na_2O with SO_3. It involves the reaction of the oxide ion, O^{2-}, from the ionic solid, Na_2O, with SO_3:

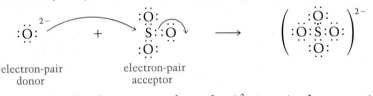

Here, SO_3 accepts the electron pair from the O^{2-} ion. At the same time, an electron pair from the S=O bond moves to the O atom. Thus, O^{2-} is the Lewis base and SO_3 is the Lewis acid.

The formation of *complex ions* can also be looked at as Lewis acid–base reactions. Complex ions are formed when a metal ion bonds to electron pairs from molecules such as H_2O or NH_3 or from anions such as $:C{\equiv}N:^-$. An example of a complex ion is $Al(H_2O)_6^{3+}$. The formation of a hydrated metal ion, such as $Al(H_2O)_6^{3+}$, involves a Lewis acid–base reaction:

$$Al^{3+} + 6(:\ddot{O}-H) \longrightarrow Al(:\ddot{O}-H)_6^{3+}$$
$$\qquad\qquad |\qquad\qquad\qquad\qquad |$$
$$\qquad\qquad H\qquad\qquad\qquad\qquad H$$

Lewis acid   Lewis base

Example 17.7

In the following reactions, identify the Lewis acid and Lewis base:
(a) $Ag^+ + 2NH_3 \rightleftharpoons Ag(NH_3)_2^+$
(b) $B(OH)_3 + H_2O \rightleftharpoons B(OH)_4^- + H^+$

Solution

We write out the equations using Lewis electron-dot formulas, then identify the electron-pair acceptor, or Lewis acid, and electron-pair donor, or Lewis base.

(a) The silver ion, Ag^+, forms a complex ion with two NH_3 molecules:

$$Ag^+ + 2:NH_3 \rightleftharpoons Ag(:NH_3)_2^+$$

Lewis   Lewis
acid   base

(b) The reaction is

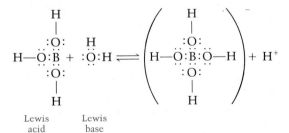

Lewis   Lewis
acid   base

Note that the acidity of boric acid, $B(OH)_3$, does not result from proton transfer, but rather from the gain of an OH^- ion from water, leaving an H^+ ion behind.

Exercise 17.11

Identify the Lewis acid and Lewis base in each of the following reactions. Write out the chemical equations using electron-dot formulas.

(a) BF_3 + $:NH_3 \longrightarrow F_3B:NH_3$ (b) O^{2-} + $CO_2 \longrightarrow CO_3^{2-}$

(See Problems 17.49 and 17.50.)

A Checklist for Review

Important Terms

Arrhenius concept (17.1)
self-ionization (autoionization) (17.2)
ion-product constant for water (K_w) (17.2)
pH (17.3)

Brønsted-Lowry concept (17.4)
conjugate acid–base pair (17.4)
conjugate acid (17.4)
conjugate base (17.4)

amphiprotic species (17.4)
hydrolysis (17.7)
Lewis concept (17.8)

Summary of Facts and Concepts

Acids, according to the *Arrhenius concept,* are substances that increase the hydrogen-ion concentration of an aqueous solution. Bases increase the hydroxide-ion concentration. The concentrations of the hydrogen ion and hydroxide ion in aqueous solution are related by the *ion-product constant for water* (K_w). Thus, we can describe the acidity or basicity of a solution by the hydrogen-ion concentration. We often use the pH, which is the negative logarithm of the hydrogen-ion concentration, as a measure of acidity.

The *Brønsted-Lowry concept* enlarges on the Arrhenius concept by looking at acid–base reactions as proton-transfer reactions. In this view, an acid is a proton donor and a base is a proton acceptor. It is instructive to look at acid–base reactions as a competition for protons. Equi-

librium for such reactions is determined by the *relative strengths of the acids and bases.* Equilibrium favors the weaker acid and weaker base. Acid strength depends on the polarity and strength of the bond involving the acidic H atom. We can therefore relate *molecular structure and acid strength.* A number of rules can be stated that allow us to predict the relative strengths of similar acids. The Brønsted-Lowry concept is also successful at explaining the *acid–base properties of salt solutions.*

The *Lewis concept* is even more general than the Brønsted-Lowry concept. A Lewis acid is an electron-pair acceptor and a Lewis base is an electron-pair donor. Reactions of acidic and basic oxides and formation of complex ions, as well as proton-transfer reactions, can be described in terms of the Lewis concept.

Operational Skills

1. Given the concentration of hydroxide ion (or concentration of strong base), calculate the hydrogen-ion concentration (Example 17.1).

2. Given the hydrogen-ion concentration (or concentration of strong acid), calculate the pH (Example 17.2); given the pH, calculate the hydrogen-ion concentration (Example 17.3).

3. Given a proton-transfer reaction, label the Brønsted-Lowry acids and bases, and name the conjugate acid–base pairs (Example 17.4).

4. Given a Brønsted-Lowry acid–base reaction and the relative strengths of acids (or bases), decide whether reactants or products are favored at equilibrium (Example 17.5).

5. Decide whether an aqueous solution of a given salt will be acidic, basic, or neutral (Example 17.6).

6. Given a reaction involving the donation of an electron pair, identify the Lewis acid and Lewis base (Example 17.7).

Review Questions

17.1 Define an acid and a base according to the Arrhenius concept. Give an example of each.

17.2 Which of the following are strong acids? Which are weak acids?

    (a) $HC_2H_3O_2$ (b) $HClO$
    (c) HCl (d) HNO_3
    (e) HNO_2 (f) HCN

17.3 Describe any thermochemical (heat of reaction) evidence for the Arrhenius concept.

17.4 What is meant by the self-ionization of water? Write the expression for K_w. What is its value at 25°C?

17.5 What is meant by the pH of a solution? Describe two ways of measuring pH.

17.6 What is the pH of a neutral solution at 37°C, where K_w equals 2.5×10^{-14}?

17.7 What is the sum of the pH and the pOH for a solution at 37°C, where K_w equals 2.5×10^{-14}?

17.8 Define an acid and a base according to the Brønsted-Lowry concept. Give an acid–base equation and identify each species as an acid or a base.

17.9 What is meant by a conjugate acid of a base?

17.10 Write an equation in which $H_2PO_3^-$ acts as a Brønsted-Lowry acid and another in which it acts as a Brønsted-Lowry base.

17.11 Describe three ways in which the Brønsted-Lowry concept enlarges on the Arrhenius concept.

17.12 Explain why the equilibrium for an acid–base reaction favors the weaker acid.

17.13 Formic acid, $HCHO_2$, is a stronger acid than acetic acid, $HC_2H_3O_2$. Which of the following is the stronger base, CHO_2^- or $C_2H_3O_2^-$?

17.14 Give two important factors that determine the strength of an acid. How does an increase in each factor affect the acid strength?

17.15 Explain why an aqueous solution of a salt may not be neutral.

17.16 Define an acid and a base according to the Lewis concept. Give a chemical equation to illustrate.

Problems

Self-Ionization of Water

17.17 A solution of hydrochloric acid is 0.020 M HCl. What is the hydrogen-ion concentration? What is the hydroxide-ion concentration at 25°C?

17.19 What are the hydrogen-ion and hydroxide-ion concentrations of a solution at 25°C that is 0.0050 M strontium hydroxide, $Sr(OH)_2$?

17.21 A shampoo solution at 25°C has a hydroxide-ion concentration of 1.5×10^{-9} M. Is the solution acidic, neutral, or basic?

17.18 A solution is 0.050 M HNO_3 (nitric acid). What is the hydrogen-ion concentration? What is the hydroxide-ion concentration at 25°C?

17.20 A saturated solution of magnesium hydroxide is 3.2×10^{-4} M $Mg(OH)_2$. What are the hydrogen-ion and hydroxide-ion concentrations in the solution at 25°C?

17.22 An antiseptic solution at 25°C has a hydroxide-ion concentration of 8.4×10^{-5} M. Is the solution acidic, neutral, or basic?

pH

17.23 Which of the following pH values indicate an acidic solution at 25°C? Which are basic and which are neutral?

    (a) 11.2   (b) 7.0   (c) 1.2   (d) 6.1

17.24 For each of the following, state whether the solution at 25°C is acidic, neutral, or basic: (a) A 0.1 M solution of trisodium phosphate, Na_3PO_4, has a pH of 12.0. (b) A 0.1 M solution of calcium chloride, $CaCl_2$, has a pH of 7.0. (c) A 0.2 M solution of copper(II) sulfate, $CuSO_4$, has a pH of 4.0. (d) A sample of rainwater has a pH of 6.4.

17.25 A sample of vinegar has a hydrogen-ion concentration of 7.5×10^{-3} M. What is the pH of the vinegar?

17.26 Some lemon juice has a hydrogen-ion concentration of 5.0×10^{-3} M. What is the pH of the lemon juice?

17.27 A solution of washing soda (sodium carbonate, Na_2CO_3) has a hydroxide-ion concentration of 0.0040 M. What is the pH at 25°C?

17.28 A solution of lye (sodium hydroxide, NaOH) has a hydroxide-ion concentration of 0.050 M. What is the pH at 25°C?

17.29 The pH of a cup of coffee (at 25°C) was found to be 5.12. What is the hydrogen-ion concentration?

17.30 A wine was tested for acidity, and its pH was found to be 3.85 at 25°C. What is the hydrogen-ion concentration?

17.31 A detergent solution has a pH of 11.63 at 25°C. What is the hydroxide-ion concentration?

17.32 Morphine is a narcotic that is used to relieve pain. A solution of morphine has a pH of 9.61 at 25°C. What is the hydroxide-ion concentration?

17.33 A 1.00-L aqueous solution contained 5.80 g of sodium hydroxide, NaOH. What was the pH of the solution at 25°C?

17.34 A 1.00-L aqueous solution contained 6.78 g of barium hydroxide, $Ba(OH)_2$. What was the pH of the solution at 25°C?

17.35 A sample of rainwater gave a yellow color with methyl red and a yellow color with bromthymol blue. What was the approximate pH of the water? Is the rainwater acidic, neutral, or basic? (See Figure 17.5.)

17.36 A drop of thymol blue gave a yellow color with a solution of aspirin. A sample of the same aspirin solution gave a yellow color with bromphenol blue. What was the pH of the solution? Is the solution acidic, neutral, or basic? (See Figure 17.5.)

Brønsted-Lowry Concept

17.37 For the following reactions, label all of the species as Brønsted-Lowry acids or bases. Indicate the species that are conjugates of one another:
(a) $H_2C_2O_4 + ClO^- \rightleftharpoons HC_2O_4^- + HClO$
(b) $HPO_4^{2-} + NH_4^+ \rightleftharpoons NH_3 + H_2PO_4^-$
(c) $SO_4^{2-} + H_2O \rightleftharpoons HSO_4^- + OH^-$
(d) $Fe(H_2O)_6^{2+} + H_2O \rightleftharpoons H_3O^+ + Fe(H_2O)_5OH^+$

17.38 For the following reactions, label all of the species as Brønsted-Lowry acids or bases. Indicate the species that are conjugates of one another:
(a) $HSO_4^- + HCO_3^- \rightleftharpoons SO_4^{2-} + H_2CO_3$
(b) $CN^- + H_2O \rightleftharpoons HCN + OH^-$
(c) $HCO_3^- + H_2O \rightleftharpoons OH^- + H_2CO_3$
(d) $HSO_4^- + H_2O \rightleftharpoons H_3O^+ + SO_4^{2-}$

17.39 Use Table 17.2 to decide whether species on the left or the right side are favored at equilibrium:
(a) $NH_4^+ + H_2PO_4^- \rightleftharpoons NH_3 + H_3PO_4$
(b) $HCN + HS^- \rightleftharpoons CN^- + H_2S$
(c) $HCO_3^- + OH^- \rightleftharpoons CO_3^{2-} + H_2O$
(d) $Al(H_2O)_6^{3+} + OH^- \rightleftharpoons Al(H_2O)_5OH^{2+} + H_2O$

17.40 Use Table 17.2 to decide whether species on the left or the right side are favored at equilibrium:
(a) $NH_4^+ + CO_3^{2-} \rightleftharpoons NH_3 + HCO_3^-$
(b) $HCO_3^- + H_2S \rightleftharpoons H_2CO_3 + HS^-$
(c) $CN^- + H_2O \rightleftharpoons HCN + OH^-$
(d) $CN^- + H_2CO_3 \rightleftharpoons HCN + HCO_3^-$

17.41 In the following reaction involving trichloroacetic acid, $HC_2Cl_3O_2$, with formate ion, CHO_2^-, the equilibrium favors the formation of trichloroacetate ion, $C_2Cl_3O_2^-$, and formic acid, $HCHO_2$:

$$HC_2Cl_3O_2 + CHO_2^- \longrightarrow C_2Cl_3O_2^- + HCHO_2$$

Which is the stronger acid, trichloroacetic acid or formic acid? Explain.

17.42 In the following reaction involving tetrafluoroboric acid, HBF_4, with the acetate ion, $C_2H_3O_2^-$, the equilibrium favors the formation of tetrafluoroborate ion, BF_4^-, and acetic acid:

$$HBF_4 + C_2H_3O_2^- \longrightarrow BF_4^- + HC_2H_3O_2$$

Which is the weaker base, BF_4^- or acetate ion?

17.43 For each of the following pairs, give the stronger acid:
(a) H_2S, HS^- (b) H_2SO_3, H_2CO_3
(c) HBr, H_2Se (d) HIO_4, HIO_3
(e) H_2S, H_2O

17.44 Order each of the following pairs by acid strength, giving the weaker acid first:
(a) HNO_3, HNO_2 (b) HCO_3^-, H_2CO_3
(c) H_2S, H_2Te (d) HCl, H_2S
(e) H_3PO_4, H_3AsO_4

Acid–Base Properties of Salts

17.45 For each of the following salts, indicate whether the aqueous solution will be acidic, basic, or neutral:
(a) $FeCl_3$ (b) K_2CO_3 (c) $LiCN$ (d) NH_4NO_3

17.46 Label each of the following salt solutions according to whether it is acidic, basic, or neutral:
(a) $Na_2S(aq)$ (b) $CuCl_2(aq)$
(c) $NH_4ClO_4(aq)$ (d) $KNO_3(aq)$

17.47 Write equations for the hydrolysis of each of the following ions:
(a) S^{2-} (b) CN^- (c) $Zn(H_2O)_6^{2+}$ (d) $N_2H_5^+$

17.48 Write equations for the hydrolysis of each of the following ions:
(a) CO_3^{2-} (b) $Cu(H_2O)_6^{2+}$ (c) PO_4^{3-} (d) HCO_2^-

Lewis Acids

17.49 In the following reactions, identify each of the reactants as a Lewis acid or Lewis base:
(a) $Cr^{3+} + 6H_2O \longrightarrow Cr(H_2O)_6^{3+}$
(b) $BF_3 + (C_2H_5)_2\ddot{O}: \longrightarrow F_3B:\ddot{O}(C_2H_5)_2$

17.50 In the following reactions, label each of the reactants as a Lewis acid or Lewis base:
(a) $BeF_2 + 2F^- \longrightarrow BeF_4^{2-}$
(b) $SnCl_4 + 2Cl^- \longrightarrow SnCl_6^{2-}$

Additional Problems

17.51 Identify each of the following as an Arrhenius acid or base. Give the chemical equation for the reaction of the substance with water, showing the origin of the acidity or basicity.
(a) BaO (b) H_2S (c) CH_3NH_2 (d) SO_2

17.52 Which of the following substances is an acid by the Arrhenius concept? Which are bases? Show this acid or base character using chemical equations.
(a) P_4O_{10} (b) K_2O (c) N_2H_4 (d) H_2Se

17.53 Benzoic acid, C_6H_5COOH, is added to fruit juice as a preservative. A 0.10-mol sample of benzoic acid is reacted with 1.0 L of 0.10 M sodium hydroxide solution. Is the resulting solution acidic, neutral, or basic?

17.54 Butylamine, $C_4H_9NH_2$, is a liquid with an ammonia-like odor. A 0.10-mol sample of butylamine is reacted with 1.0 L of 0.10 M hydrochloric acid. Is the resulting solution acidic, neutral, or basic?

17.55 What is the H^+ concentration in a neutral solution at 0°C, where K_w is 1.1×10^{-15}?

17.56 What is the OH^- concentration in a neutral solution at 60°C, where K_w is 9.6×10^{-14}?

17.57 What is the pH of each of the following aqueous solutions at 25°C if its concentration of H^+ or OH^- ion is as given? Is the solution acidic, neutral, or basic?
(a) 1.3×10^{-4} M H^+ (b) 5.4×10^{-8} M OH^-
(c) 2.5×10^{-9} M H^+ (d) 8.1×10^{-3} M OH^-

17.58 What is the pH of each of the aqueous solutions below at 25°C if its H^+ or OH^- concentration is as given? Is the solution acidic, neutral, or basic?
(a) 7.5×10^{-2} M H^+ (b) 3.3×10^{-10} M OH^-
(c) 4.8×10^{-10} M H^+ (d) 6.5×10^{-2} M OH^-

17.59 A cookbook suggests that the odor from cooking fish can be prevented by first sprinkling the fresh fish with lemon juice. The odor is due to organic amines, such as trimethylamine, $(CH_3)_3N$. These amines are converted to salts by reaction with an acid, such as lemon juice. Write the net ionic equation for the reaction of trimethylamine with a strong acid.

17.60 One mole of hydrazine, N_2H_4, reacts with two moles of hydrochloric acid to give a salt. Write a net ionic equation for the reaction.

17.61 Write equations for the hydrolysis reactions, if any, for the following ions:
(a) SO_3^{2-} (b) Ca^{2+} (c) $CH_3NH_3^+$ (d) $Fe(H_2O)_6^{2+}$

17.62 Write equations for the hydrolysis reactions, if any, for the following ions:
(a) Br^- (b) F^- (c) $Ni(H_2O)_6^{2+}$ (d) Cs^+

17.63 Iodine, I_2, is more soluble in aqueous potassium iodide, $KI(aq)$, than it is in pure water. The iodide ion reacts with iodine to give the triiodide ion, I_3^-. Interpret this reaction in terms of the Lewis acid–base concept.

17.64 Consider the reaction of the basic oxide CaO with the acidic oxide CO_2:

$$CaO + CO_2 \longrightarrow CaCO_3$$

This is essentially the reaction of the oxide ion, O^{2-}, with carbon dioxide to give the carbonate ion:

$$O^{2-} + CO_2 \longrightarrow CO_3^{2-}$$

Interpret this reaction in terms of the Lewis acid–base concept.

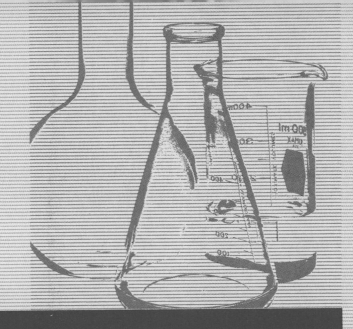

18. Acid–Base Equilibria

Solutions of a Weak Acid or Base or Salt

18.1 Acid Ionization Equilibria Experimental
Determination of K_a/ Calculations with K_a

18.2 Polyprotic Acids

18.3 Base Ionization Equilibria

18.4 Hydrolysis

*Solutions of a Weak Acid or Base
with Another Solute*

18.5 Common-Ion Effect

18.6 Buffers

18.7 Acid–Base Titration Curves Titration of a
Strong Acid by a Strong Base/ Titration of a Weak Acid
by a Strong Base/ Titration of a Weak Base by a Strong
Acid

M any well-known substances are weak acids or bases. The following are weak acids: aspirin (acetylsalicylic acid, a headache remedy), phenobarbital (a sedative), saccharin (a sweetener), and niacin (nicotinic acid, a B vitamin). Because these are weak acids, their reactions with water do not go to completion. To discuss such acid–base reactions, we need to look at the equilibria involved and be able to calculate the concentrations of species in a reaction mixture.

Consider, for example, how we could answer the following questions: What is the pH of 0.10 M niacin (nicotinic acid)? What is the pH of the solution obtained by dissolving one 5.00-grain tablet of aspirin (acetylsalicylic acid) in 0.500 L of water? If these were solutions of strong acids, the calculations would be simple. The acids are monoprotic; that is, they release one H^+ ion for each molecule of acid. Thus, 0.10 M acid would yield 0.10 M H^+ ion, and the pH would be 1.00. However, since niacin is a weak acid, the H^+ ion concentration is less than 0.10 M, and the pH is greater than 1.00. To find the answer, we will need the equilibrium constant for the reaction involved, and we will need to solve an equilibrium problem.

A similar process is involved in finding the pH of 0.10 M sodium nicotinate (sodium salt of niacin). In this case, we will need to look at the hydrolysis equilibrium of the nicotinate ion. Let us see how we answer questions of this sort.

Chapter Overview

In this chapter, we will solve the principal kinds of problems in acid–base equilibria. The essential problem-solving procedure was discussed in Chapter 16, and acid–base reactions were dealt with qualitatively in Chapter 17. We will begin this chapter by looking at solutions containing a single solute, that is, *solutions of a weak acid or a weak base or a salt that hydrolyzes.* The acid–base reaction will involve the reaction of the solute with water, and we will solve for the concentrations of species at equilibrium. In the second part of the chapter, we will examine solutions containing two solutes, that is, *solutions of a weak acid or weak base plus another solute,* either a strong acid or base or a salt with a conjugate ion.

Solutions of a Weak Acid or Base or Salt

The simplest acid–base equilibria we will be concerned with are those in which a single solute—either a weak acid or weak base or salt that hydrolyzes—reacts with water. We will look first at solutions of weak acids.

18.1 Acid Ionization Equilibria

An Arrhenius acid reacts with water to produce the hydrogen ion and the conjugate base ion. The process is called *acid ionization* or *acid dissociation.* For example, consider an aqueous solution of the weak acid acetic acid, $HC_2H_3O_2$, the sour constituent of vinegar. When the acid is added to water, it reacts according to the equation

$$HC_2H_3O_2(aq) + H_2O(l) \rightleftharpoons H_3O^+(aq) + C_2H_3O_2^-(aq)$$

The reaction involves a transfer of a proton from acetic acid to water. This reaction can be written more simply as

$$HC_2H_3O_2(aq) \rightleftharpoons H^+(aq) + C_2H_3O_2^-(aq)$$

In this case, the transfer of the proton to water is not made explicit. However, the equation does succinctly express the production of ions in solution. Since acetic acid is a weak electrolyte, the acid *ionizes* or *dissociates* to a small extent in water (about 1% or less, depending on concentration of acid).

If the acid is strong, it completely ionizes in solution, and the concentrations of ions are determined by the stoichiometry of the reaction from the initial concentration of acid. However, if the acid is weak, as in the above example, the concentrations of ions in solution are determined from the **acid ionization** (or **dissociation**) **constant,** which is the equilibrium constant for the ionization of a weak acid.

To find the acid ionization constant, let us write HA for the general formula of a weak, monoprotic acid. The acid ionization equilibrium in aqueous solution is

$$HA(aq) + H_2O(l) \rightleftharpoons H_3O^+(aq) + A^-(aq)$$

The corresponding equilibrium constant is

$$K_c = \frac{[H_3O^+][A^-]}{[HA][H_2O]}$$

Assuming that this is a dilute solution and that the reaction occurs to only a small extent, the concentration of water will be nearly constant. Rearranging this equation gives

$$K_a = [H_2O]K_c = \frac{[H_3O^+][A^-]}{[HA]}$$

Thus, K_a, the acid ionization constant, equals the constant $[H_2O]K_c$.

Finally, if we write the acid ionization equilibrium in the form

$$HA(aq) \rightleftharpoons H^+(aq) + A^-(aq)$$

the acid ionization constant is

$$K_a = \frac{[H^+][A^-]}{[HA]}$$

This is essentially the result we got for acetic acid, with $[H^+]$ written for $[H_3O^+]$.

Experimental Determination of K_a

The ionization constant for a weak acid is usually determined experimentally by one of two methods. In one, the electrical conductivity or some colligative property of a solution of the acid is measured to obtain its degree of ionization. The **degree of ionization** of a weak electrolyte is the fraction of molecules that react with water to give ions, as compared to the value expected for complete ionization. This may also be expressed as a percentage, giving the *percent ionization*. In the other method, the pH of a solution of weak acid is determined. From the pH one finds the concentration of H^+ ion and then the concentrations of other ions. The next example shows how to calculate K_a from the pH of a solution of nicotinic acid. It also shows how to calculate degree of ionization and percent ionization.

Example 18.1

Nicotinic acid (niacin) is a monoprotic acid with the formula $HC_6H_4NO_2$. A solution that is 0.012 M in nicotinic acid has a pH of 3.39 at 25°C. What is the acid ionization constant, K_a, for this acid at 25°C? What is the degree of ionization of nicotinic acid in this solution?

Solution

It is important to realize that when we say the solution is 0.012 M nicotinic acid, this refers to *how the solution is prepared*. The solution is made up by adding 0.012 mol of nicotinic acid to give a liter of solution. However, once the solution is prepared, some of the nicotinic acid molecules ionize. Hence, the actual concentration of nicotinic acid in solution is somewhat less than 0.012 M. To solve for K_a, we follow the three steps outlined in Chapter 16 for solving equilibrium problems.

Step 1 Let us abbreviate the formula for nicotinic acid as HNic. Then 1 L of solution contains 0.012 mol HNic before ionization. If x mol HNic ionize, x mol each of H^+ and Nic^- are formed, leaving $(0.012 - x)$ mol HNic in solution. We can summarize the situation as follows:

Concentration (M)	HNic(aq) $\rightleftharpoons$ H$^+$(aq) + Nic$^-$(aq)		
Starting	0.012	0	0
Change	$-x$	$+x$	$+x$
Equilibrium	$0.012 - x$	x	x

Thus, the molar concentrations of HNic, H^+, and Nic^- at equilibrium are $(0.012 - x)$, x, and x, respectively.

Step 2 The equilibrium equation is

$$K_a = \frac{[H^+][Nic^-]}{[HNic]}$$

If we substitute the expressions for the equilibrium concentrations, we get

$$K_a = \frac{x^2}{(0.012 - x)}$$

Step 3 The value of x equals the numerical value of the molar hydrogen-ion concentration, and this can be obtained from the pH of the solution:

$$x = [H^+] = \text{antilog} (-pH)$$
$$= \text{antilog} (-3.39) = 4.1 \times 10^{-4} = 0.00041$$

We can substitute this value of x into the equation obtained in Step 2. Note first, however, that

$$0.012 - x = 0.012 - 0.00041 = 0.01159$$
$$\approx 0.012 \text{ (to two significant figures)}$$

This means that the concentration of undissolved acid is equal to the original concentration of the acid within the precision of the data. (We will make use of this type of observation in later problem solving.) Therefore,

$$K_a = \frac{x^2}{(0.012 - x)} \approx \frac{x^2}{0.012}$$
$$\approx \frac{(0.00041)^2}{0.012} = 1.4 \times 10^{-5}$$

To obtain the degree of ionization, note that x mol out of 0.012 mol of nicotinic acid ionize. Hence,

$$\text{Degree of ionization} = \frac{x}{0.012} = \frac{0.00041}{0.012} = 0.034$$

The percent ionization is obtained by multiplying this by 100; we get 3.4%.

Exercise 18.1

Lactic acid, $HC_3H_5O_3$, is found in sour milk, where it is produced by the action of lactobacilli on lactose, or milk sugar. ■ A 0.025 M solution of lactic acid has a pH of 2.75. What is the ionization constant, K_a, for this acid? What is the degree of ionization?

(See Problems 18.17 and 18.18.)

■ Lactic acid has the structural formula

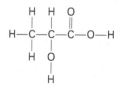

Table 18.1 lists acid ionization constants for various weak acids. The weakest acids have the smallest values.

Table 18.1
Acid Ionization Constants
at 25°C

Substance	Formula	K_a
Acetic acid	$HC_2H_3O_2$	1.7×10^{-5}
Benzoic acid	$HC_7H_5O_2$	6.3×10^{-5}
Boric acid	H_3BO_3	5.9×10^{-10}
Carbonic acid*	H_2CO_3	4.3×10^{-7}
	HCO_3^-	4.8×10^{-11}
Cyanic acid	$HCNO$	3.5×10^{-4}
Formic acid	$HCHO_2$	1.7×10^{-4}
Hydrocyanic acid	HCN	4.9×10^{-10}
Hydrofluoric acid	HF	6.8×10^{-4}
Hydrogen sulfate ion	HSO_4^-	1.1×10^{-2}
Hydrogen sulfide*	H_2S	8.9×10^{-8}
	HS^-	1.2×10^{-13}
Hypochlorous acid	$HClO$	3.5×10^{-8}
Nitrous acid	HNO_2	4.5×10^{-4}
Oxalic acid*	$H_2C_2O_4$	5.6×10^{-2}
	$HC_2O_4^-$	5.1×10^{-5}
Phosphoric acid*	H_3PO_4	6.9×10^{-3}
	$H_2PO_4^-$	6.2×10^{-8}
	HPO_4^{2-}	4.8×10^{-13}
Phosphorous acid*	H_2PHO_3	1.6×10^{-2}
	$HPHO_3^-$	7×10^{-7}
Propionic acid	$HC_3H_5O_2$	1.3×10^{-5}
Pyruvic acid	$HC_3H_3O_3$	1.4×10^{-4}
Sulfurous acid*	H_2SO_3	1.3×10^{-2}
	HSO_3^-	6.3×10^{-8}

*The ionization constants for polyprotic acids are for successive ionizations. Thus, for H_3PO_4, the equilibrium is

$$H_3PO_4 \rightleftharpoons H^+ + H_2PO_4^-$$

For $H_2PO_4^-$, the equilibrium is

$$H_2PO_4^- \rightleftharpoons H^+ + HPO_4^{2-}$$

Calculations with K_a

Once we know the value of K_a for an acid HA, we can calculate the concentrations of species HA, A$^-$, and H$^+$ for solutions of different molarities. This is illustrated in the next example, where we answer the first question posed in the chapter opening.

Example 18.2

What are the concentrations at 25°C of nicotinic acid, hydrogen ion, and nicotinate ion in a solution of 0.10 M nicotinic acid, $HC_6H_4NO_2$? What is the pH of the solution? What is the degree of ionization of nicotinic acid?

The acid ionization constant, K_a, was determined in the previous example to be 1.4×10^{-5}.

(*Continued*)

Solution

Step 1 Let us assume that we have one liter of solution. Then the nicotinic acid in this volume of solution ionizes to give x mol H^+ and x mol Nic^-, leaving $(0.10 - x)$ mol of nicotinic acid. This is summarized in the following table:

Concentration (M)	$HNic(aq) \rightleftharpoons H^+(aq) + Nic^-(aq)$		
Starting	0.10	0	0
Change	$-x$	$+x$	$+x$
Equilibrium	$0.10 - x$	x	x

The equilibrium concentrations of HNic, H^+, and Nic^- are $(0.10 - x)$, x, and x, respectively.

Step 2 We now substitute these concentrations and the value of K_a into the equilibrium equation for acid ionization:

$$\frac{[H^+][Nic^-]}{[HNic]} = K_a$$

We get

$$\frac{x^2}{(0.10 - x)} = 1.4 \times 10^{-5}$$

Step 3 Now we solve this equation for the value of x. This is actually a quadratic equation, but it can be simplified so that the value of x is easily found. Since the acid ionization constant is small, the value of x is small. Let us assume that x is much smaller than 0.10, so that

$$0.10 - x \simeq 0.10$$

We will need to check that this assumption is valid after we obtain a value for x. The equilibrium equation

becomes

$$\frac{x^2}{0.10} \simeq 1.4 \times 10^{-5}$$

Or, $x^2 \simeq 1.4 \times 10^{-5} \times 0.10 = 1.4 \times 10^{-6}$

Hence, $x \simeq 1.2 \times 10^{-3} = 0.0012$

At this point, we should check to make sure our assumption that $0.10 - x \simeq 0.10$ is valid. We substitute the value obtained for x into $0.10 - x$:

$0.10 - x = 0.10 - 0.0012$

$\qquad\qquad = 0.10$ (to two significant figures)

Thus, the assumption is indeed valid.

Now we can substitute the value of x into the last line of the table we wrote in Step 1:

Concentration (M)	$HNic(aq) \rightleftharpoons H^+(aq) + Nic^-(aq)$		
Equilibrium	$0.10 - x$ $\simeq 0.10$	x $\simeq 0.0012$	x $\simeq 0.0012$

Thus, the concentrations of nicotinic acid, hydrogen ion, and nicotinate ion are $0.10\ M$, $0.0012\ M$, and $0.0012\ M$, respectively.

The pH of the solution is

$$pH = -\log [H^+] = -\log 0.0012 = 2.92$$

The degree of ionization equals the amount per liter of nicotinic acid that ionizes ($x = 0.0012$) divided by the total amount per liter of nicotinic acid initially present (0.10). Thus, the degree of ionization is $0.0012/0.10 = 0.012$.

Exercise 18.2

What are the concentrations of hydrogen ion and acetate ion in a solution of $0.10\ M$ acetic acid, $HC_2H_3O_2$? What is the pH of the solution? What is the degree of ionization? See Table 18.1 for the value of K_a.

(See Problems 18.19, 18.20, 18.21, and 18.22.)

In Example 18.2, the degree of ionization of nicotinic acid (0.012) is relatively small in $0.10\ M$ solution. That is, only 1.2% of the molecules ionize, or dissociate. It is the small value of the degree of ionization that allows us to neglect x in the term $0.10 - x$ and thereby simplify the calculation.

The degree of ionization of a weak acid depends on both K_a and the concentration of the acid solution. For a given concentration, the larger the K_a, the greater the degree of ionization. For a given K_a, however, the more dilute the solution, the greater the degree of ionization. We can understand this in terms of LeChatelier's principle. Consider the equilibrium

$$HNic(aq) + H_2O(l) \rightleftharpoons H_3O^+(aq) + Nic^-(aq)$$

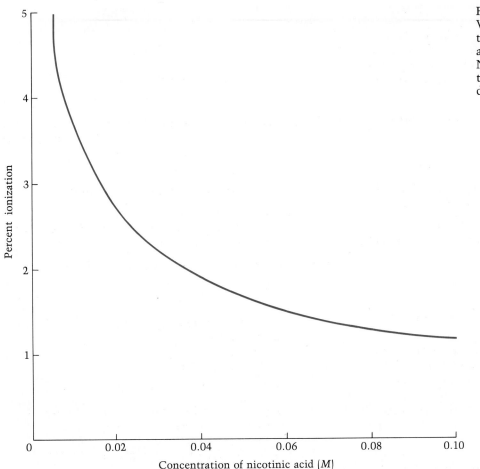

Figure 18.1
Variation of percent ionization of a weak acid (nicotinic acid) with concentration. Note that the percent ionization is greatest for the most dilute solutions.

When water is added to a given solution of nicotinic acid, the equilibrium shifts to the right. That is, more nicotinic acid molecules ionize, and the degree of ionization increases. Figure 18.1 shows how the percent ionization (degree of ionization × 100) varies with the concentration of solution.

How can we know when we can use the simplifying assumption of Example 18.2, where we neglected x in the denominator of the equilibrium equation? *As a rough rule, this simplifying assumption will be valid if* K_a *divided by the concentration is about* 10^{-3} *or less.* ■ Thus, for a K_a equal to 10^{-5} and a concentration of 0.1 M, the assumption will be valid. But it will not be if the concentration of the same acid is 0.001 M.

If the simplifying assumption of Example 18.2 is not valid, we have two ways to proceed. We can solve the equilibrium equation exactly using the *quadratic formula*. On the other hand, we can still use the simplifying assumption if we follow it with the *method of successive approximations*. In this case we take the approximate value of x found by using the simplifying assumption and successively improve on it. Both of these methods are illustrated in the next example, which answers the second question posed in the chapter opening.

■ The degree of ionization of a weak acid is approximately $\sqrt{K_a/C}$, where C is the concentration of acid. Thus, $K_a/C = 1 \times 10^{-3}$ corresponds to a degree of ionization of 0.03 (3%). For this degree of ionization, neglecting x in the denominator of the equilibrium equation gives approximately two significant figures.

Example 18.3

What is the pH at 25°C of the solution obtained by dissolving a 5.00-grain tablet of aspirin (acetylsalicylic acid) in 0.500 L of water? The tablet contains 5.00 grains, or 0.324 g, of acetylsalicylic acid, $HC_9H_7O_4$. The acid is monoprotic, and K_a equals 3.3×10^{-4} at 25°C.

Solution

The molar mass of $HC_9H_7O_4$ is 180.2 g. From this we find that an aspirin tablet contains 0.00180 mol of the acid. Hence, the concentration of the solution is 0.00180 mol/0.500 L, or 0.0036 M. (We will retain only two significant figures, since this is the number of significant figures in K_a.)

Step 1 We will abbreviate the formula for acetylsalicylic acid as HAcs and let x be the amount of H^+ ion formed in one liter of solution. The amount of acetylsalicylate ion formed is also x mol, and the amount of un-ionized acetylsalicylic acid is $(0.0036 - x)$ mol. This is summarized in the following table:

Concentration (M)	HAcs(aq) $\rightleftharpoons$ H$^+$(aq) + Acs$^-$(aq)		
Starting	0.0036	0	0
Change	$-x$	$+x$	$+x$
Equilibrium	$0.0036 - x$	x	x

Step 2 If we substitute the equilibrium concentrations and the value of K_a into the equilibrium equation, we get

$$\frac{[H^+][Acs^-]}{[HAcs]} = K_a$$

$$\frac{x^2}{0.0036 - x} = 3.3 \times 10^{-4}$$

Step 3 Note that K_a divided by the concentration is $3.3 \times 10^{-4}/0.0036 = 9.2 \times 10^{-2}$. Thus, we can expect that x cannot be neglected compared with 0.0036. We can use the quadratic formula or we can use the method of successive approximations. Each approach is now described.

Quadratic Formula Let us rearrange the preceding equation to put it in the form $ax^2 + bx + c = 0$. We get

$$x^2 = (0.0036 - x) \times 3.3 \times 10^{-4}$$
$$= 1.2 \times 10^{-6} - 3.3 \times 10^{-4} x$$

Or, $\qquad x^2 + 3.3 \times 10^{-4} x - 1.2 \times 10^{-6} = 0$

We now substitute into the quadratic formula:

$$x = \frac{-b \pm \sqrt{b^2 - 4ac}}{2a}$$

$$= \frac{-3.3 \times 10^{-4} \pm \sqrt{(3.3 \times 10^{-4})^2 + 4(1.2 \times 10^{-6})}}{2}$$

$$= \frac{-3.3 \times 10^{-4} \pm 2.2 \times 10^{-3}}{2}$$

If we take the lower sign in $\pm$, we will get a negative value for x. But since x equals the concentration of H^+ ion, it must be positive. Therefore, we ignore the negative solution. Taking the upper sign, we get

$$x = [H^+] = 9.4 \times 10^{-4}$$

Now we can calculate the pH:

$$pH = -\log [H^+] = -\log 9.4 \times 10^{-4} = 3.03$$

Method of Successive Approximations We wish to solve the following equation for the value of x:

$$\frac{x^2}{0.0036 - x} = 3.3 \times 10^{-4}$$

Or, $\qquad x^2 = (0.0036 - x) \times 3.3 \times 10^{-4}$

The method of successive approximations consists in choosing an approximate value of x, putting it into the right side of this equation and solving for an improved value of x on the left. This value of x then becomes the new approximation to x. The method is repeated until the value put into the equation on the right equals the value obtained on the left (to the number of significant figures desired, usually two). To get the first approximation to x, we adopt the simplifying assumption used in Example 18.2. We put $x = 0$ on the right. Hence,

$$x^2 = (0.0036 - 0) \times 3.3 \times 10^{-4} = 1.2 \times 10^{-6}$$
$$x = 0.0011$$

We now use this value of x for the next approximation:

$$x^2 = (0.0036 - 0.0011) \times 3.3 \times 10^{-4} = 8.3 \times 10^{-7}$$
$$x = 0.00091$$

For the next approximation, we get

$$x^2 = (0.0036 - 0.00091) \times 3.3 \times 10^{-4} = 8.9 \times 10^{-7}$$
$$x = 0.00094$$

Then, in the next approximation,

$$x^2 = (0.0036 - 0.00094) \times 3.3 \times 10^{-4} = 8.8 \times 10^{-7}$$
$$x = 0.00094$$

(Continued)

Note that the approximations to x improve at each stage, with the changes becoming smaller. The successive approximations are 0.0011, 0.00091, 0.00094, and 0.00094. Because the value of x obtained in the last approximation equals the value put into the right side, we can assume that x equals 0.00094, to two significant figures. Thus, the H^+ ion concentration is 0.00094, and the pH is $-\log 0.00094$, or 3.03.

Exercise 18.3

What is the pH of an aqueous solution that is 0.0030 M pyruvic acid, $HC_3H_3O_3$? ■
(See Problems 18.23, 18.24, 18.25, and 18.26.)

■ Pyruvic acid is an important biochemical intermediate formed during the breakdown of glucose by a cell. In muscle tissue, glucose is broken down to pyruvic acid, which is eventually oxidized to carbon dioxide and water. The structure of pyruvic acid is

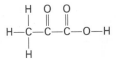

18.2 Polyprotic Acids

In the preceding section, we dealt only with acids releasing one H^+ ion or proton. Some acids, however, have two or more such protons (these are called *polyprotic acids*). Sulfuric acid, for example, loses two protons in aqueous solution. The first one is lost completely (sulfuric acid is a strong acid).

$$H_2SO_4(aq) \longrightarrow H^+(aq) + HSO_4^-(aq)$$

The hydrogen sulfate ion, HSO_4^-, that is left may then lose the second proton. In this case, an equilibrium exists.

$$HSO_4^-(aq) \rightleftharpoons H^+(aq) + SO_4^{2-}(aq)$$

For a weak diprotic acid like carbonic acid, H_2CO_3, there are two simultaneous equilibria to consider:

$$H_2CO_3(aq) \rightleftharpoons H^+(aq) + HCO_3^-(aq)$$
$$HCO_3^-(aq) \rightleftharpoons H^+(aq) + CO_3^{2-}(aq)$$

Carbonic acid is in equilibrium with the hydrogen carbonate ion, HCO_3^-, which is in turn in equilibrium with the carbonate ion, CO_3^{2-}. Each equilibrium has an associated acid ionization constant. For the loss of the first proton, we have

$$K_{a1} = \frac{[H^+][HCO_3^-]}{[H_2CO_3]} = 4.3 \times 10^{-7}$$

and for the loss of the second proton we have

$$K_{a2} = \frac{[H^+][CO_3^{2-}]}{[HCO_3^-]} = 4.8 \times 10^{-11}$$

Note that K_{a2} for carbonic acid is much smaller than K_{a1} (by a factor of about 1×10^{-4}). This indicates that carbonic acid loses the first proton more easily than it does the second one, because the first proton separates from an ion of single negative charge, but the second proton separates from an ion of double negative charge. This double negative charge strongly attracts the proton back to it. In general, the second ionization constant, K_{a2}, of a polyprotic acid is much smaller than the first ionization constant, K_{a1}. In the case of a triprotic acid, the third ionization constant, K_{a3}, is much smaller than the second one, K_{a2}. See the values for phosphoric acid, H_3PO_4, in Table 18.1.

Calculation of the concentrations of various species in a solution of a polyprotic acid might appear to be complicated because several equilibria occur at once. However, reasonable assumptions can be made that simplify the calculation, as we show in the next example.

Example 18.4

Ascorbic acid (vitamin C) is a diprotic acid, $H_2C_6H_6O_6$. What is the pH of a 0.10 M solution? What is the concentration of ascorbate ion, $C_6H_6O_6{}^{2-}$? The acid ionization constants are $K_{a1} = 7.9 \times 10^{-5}$ and $K_{a2} = 1.6 \times 10^{-12}$.

Solution

Calculation of pH Let us abbreviate the formula for ascorbic acid as H_2Asc. Hydrogen ions are produced by the two successive acid ionizations:

$$H_2Asc(aq) \rightleftharpoons H^+(aq) + HAsc^-(aq);$$
$$K_{a1} = 7.9 \times 10^{-5}$$

$$HAsc^-(aq) \rightleftharpoons H^+(aq) + Asc^{2-}(aq);$$
$$K_{a2} = 1.6 \times 10^{-12}$$

To be exact, we should account for both reactions to obtain the pH. However, since K_{a2} is much smaller than K_{a1}, the amount of hydrogen ion produced in the second reaction may be neglected in comparison with that produced in the first. Therefore, we can find the pH by considering only the first reaction.

If we let x be the amount of H^+ formed, we get the following results:

Concentration (M)	$H_2Asc(aq)$	$\rightleftharpoons$ $H^+(aq)$	$+$ $HAsc^-(aq)$
Starting	0.10	0	0
Change	$-x$	$+x$	$+x$
Equilibrium	$0.10 - x$	x	x

We now substitute into the equilibrium equation for the first ionization:

$$\frac{[H^+][HAsc^-]}{[H_2Asc]} = K_{a1}$$

$$\frac{x^2}{0.10 - x} = 7.9 \times 10^{-5}$$

Assuming x to be much smaller than 0.10, we get

$$\frac{x^2}{0.10} \simeq 7.9 \times 10^{-5}$$

Or,

$$x^2 \simeq 7.9 \times 10^{-5} \times 0.10$$
$$x \simeq 2.8 \times 10^{-3} = 0.0028$$

(Note that $0.10 - x = 0.10 - 0.0028 = 0.10$, correct to two significant figures. Thus, the assumption that $0.10 - x \simeq 0.10$ is correct.)

Since the hydrogen-ion concentration is 0.0028 M,

$$pH = -\log [H^+] = -\log (0.0028) = 2.55$$

Ascorbate-Ion Concentration Ascorbate ion, Asc^{2-}, is produced only in the second reaction. We assume the starting concentrations of H^+ and $HAsc^-$ for this reaction to be those from the first equilibrium. Let us write y for the concentration of ascorbate ion produced. The amounts of each species in one liter of solution are as follows:

Concentration (M)	$HAsc^-(aq)$	$\rightleftharpoons$ $H^+(aq)$	$+$ $Asc^{2-}(aq)$
Starting	0.0028	0.0028	0
Change	$-y$	$+y$	$+y$
Equilibrium	$0.0028 - y$	$0.0028 + y$	y

We now substitute into the equilibrium equation for the second ionization:

$$\frac{[H^+][Asc^{2-}]}{[HAsc^-]} = K_{a2}$$

$$\frac{(0.0028 + y)y}{0.0028 - y} = 1.6 \times 10^{-12}$$

This equation can be simplified if we assume that y is much smaller than 0.0028. (Again, this assumes that the reaction occurs to only a small extent, as we expect from the magnitude of the equilibrium constant.) That is,

$$0.0028 + y \simeq 0.0028$$
$$0.0028 - y \simeq 0.0028$$

Then the equilibrium equation reads

$$\frac{(0.0028)y}{0.0028} \simeq 1.6 \times 10^{-12}$$

Hence,

$$y \simeq 1.6 \times 10^{-12}$$

(Note that 1.6×10^{-12} is indeed much smaller than 0.0028, as we assumed.) We see that the concentration of ascorbate ion equals K_{a2}, or 1.6×10^{-12} M.

Exercise 18.4

Sulfurous acid, H_2SO_3, is a diprotic acid with $K_{a1} = 1.3 \times 10^{-2}$ and $K_{a2} = 6.3 \times 10^{-8}$. The acid is formed when sulfur dioxide (a gas with a suffocating odor) dissolves in water. What is the pH of a 0.25 M solution of sulfurous acid? What is the concentration of sulfite ion, SO_3^{2-}, in the solution? Note that K_{a1} is large.

(See Problems 18.27 and 18.28.)

What we see from Example 18.4 is that the concentration of H^+ and HA^- in a solution of a diprotic acid H_2A can be calculated from the first ionization constant, K_{a1}. The concentration of the ion A^{2-}, on the other hand, equals the second ionization constant, K_{a2}.

18.3 Base Ionization Equilibria

Equilibria involving weak bases are treated similarly to those for weak acids. Ammonia, for example, ionizes in water as follows:

$$NH_3(aq) + H_2O(l) \rightleftharpoons NH_4^+(aq) + OH^-(aq)$$

The corresponding equilibrium constant is

$$K_c = \frac{[NH_4^+][OH^-]}{[NH_3][H_2O]}$$

Because the concentration of H_2O is nearly constant, we can rearrange this as we did for acid ionization:

$$K_b = [H_2O]K_c = \frac{[NH_4^+][OH^-]}{[NH_3]}$$

In general, a weak base B with the *base ionization* of

$$B(aq) + H_2O(l) \rightleftharpoons HB^+(aq) + OH^-(aq)$$

has a **base ionization** (or **dissociation**) **constant,** K_b, equal to

$$K_b = \frac{[HB^+][OH^-]}{[B]}$$

Table 18.2 lists ionization constants for some weak bases.

Substance	Formula	K_b
Ammonia	NH_3	1.8×10^{-5}
Aniline	$C_6H_5NH_2$	4.2×10^{-10}
Dimethylamine	$(CH_3)_2NH$	5.1×10^{-4}
Ethylamine	$C_2H_5NH_2$	4.7×10^{-4}
Hydrazine	N_2H_4	1.7×10^{-6}
Hydroxylamine	NH_2OH	1.1×10^{-8}
Methylamine	CH_3NH_2	4.4×10^{-4}
Pyridine	C_5H_5N	1.4×10^{-9}
Urea	NH_2CONH_2	1.5×10^{-14}

Table 18.2
Base Ionization Constants at 25°C

Exercise 18.5

Quinine is an alkaloid, or naturally occurring base, used to treat malaria. A 0.0015 M solution of quinine has a pH of 9.84. The basicity of alkaloids is due to a nitrogen atom that picks up protons from water in the same manner as ammonia.■ What is K_b? (See Example 18.1.)

■ Some other well-known alkaloids and their plant sources are nicotine (tobacco), cocaine (coca bush), mescaline (species of cacti), and caffeine (coffee and tea).

(See Problems 18.29 and 18.30.)

Example 18.5

Morphine, $C_{17}H_{19}NO_3$, is administered medically to relieve pain. It is a naturally occurring base, or *alkaloid*. What is the pH of a 0.0075 M solution of morphine at 25°C? The base ionization constant, K_b, is 1.6×10^{-6} at 25°C.

Solution

Let us abbreviate the formula for morphine as Mor. Morphine ionizes by picking up a proton from water (as does ammonia):

$$Mor(aq) + H_2O(l) \rightleftharpoons HMor^+(aq) + OH^-(aq)$$

Step 1 The morphine in one liter of solution ionizes to give x mol HMor$^+$ and x mol OH$^-$. This is summarized in the table given below.

Step 2 Substituting into the equilibrium equation

$$\frac{[HMor^+][OH^-]}{[Mor]} = K_b$$

gives

$$\frac{x^2}{0.0075 - x} = 1.6 \times 10^{-6}$$

Step 3 If we assume that x is small enough to neglect compared with 0.0075, we have

$$0.0075 - x \simeq 0.0075$$

and we can write the equilibrium equation as

$$\frac{x^2}{0.0075} \simeq 1.6 \times 10^{-6}$$

Hence, $\quad x^2 \simeq 1.6 \times 10^{-6} \times 0.0075 = 1.2 \times 10^{-8}$

$$x \simeq 1.1 \times 10^{-4}$$

(If we evaluate $0.0075 - x$, we get 0.0074. Thus, x may not be quite small enough to ignore. However, if we use the method of successive approximations, we find that $x \simeq 1.1 \times 10^{-4}$, as just given.)

The hydroxide-ion concentration, we see, is 1.1×10^{-4} M. Let us now calculate the hydrogen-ion concentration. The product of the hydrogen-ion concentration and the hydroxide-ion concentration equals 1.0×10^{-14} at 25°C:

$$[H^+][OH^-] = 1.0 \times 10^{-14}$$

Substituting into this equation gives

$$[H^+] \times 1.1 \times 10^{-4} = 1.0 \times 10^{-14}$$

Hence, $\quad [H^+] = \dfrac{1.0 \times 10^{-14}}{1.1 \times 10^{-4}} = 9.1 \times 10^{-11}$

The pH of the solution is

$$pH = -\log[H^+] = -\log(9.1 \times 10^{-11}) = 10.04$$

Note that we could also calculate the pH from the formula pH + pOH = 14.00.

Concentration (M)	Mor(aq) + H$_2$O(l) $\rightleftharpoons$ HMor$^+$(aq) + OH$^-$(aq)		
Starting	0.0075	0	0
Change	$-x$	$+x$	$+x$
Equilibrium	$0.0075 - x$	x	x

Exercise 18.6

What is the hydrogen-ion concentration of a 0.20 M solution of ammonia in water? See Table 18.2 for K_b.

(See Problems 18.31 and 18.32.)

18.4 Hydrolysis

Hydrolysis is the reaction of the ions from a salt with water.■ Since such ions may produce H^+ or OH^- ions, they may give acidic or basic solutions. Consider the hydrolysis of sodium cyanide, NaCN. The salt dissolves in water to give Na^+ and CN^- ions. Sodium ion, Na^+, is unreactive with water, but CN^- ion hydrolyzes as follows:

■ Hydrolysis was qualitatively described in Section 17.7.

$$CN^-(aq) + H_2O(l) \rightleftharpoons HCN(aq) + OH^-(aq)$$

Anions, such as CN^-, that are conjugate bases of weak acids (HCN is the conjugate acid of CN^-), hydrolyze to a small extent to give OH^- ion. The solution is therefore basic.

The equilibrium constant for the hydrolysis, K_h', is

$$K_h' = \frac{[HCN][OH^-]}{[CN^-][H_2O]}$$

Or, since the concentration of H_2O is nearly constant, we can write the simpler **hydrolysis constant, K_h**:

$$K_h = \frac{[HCN][OH^-]}{[CN^-]}$$

The hydrolysis constant for an ion like CN^- is simply a base ionization constant, K_b. Compare the base ionization reaction of NH_3 with the hydrolysis reaction of CN^-:

$$NH_3(aq) + H_2O(l) \rightleftharpoons NH_4^+(aq) + OH^-(aq)$$

$$CN^-(aq) + H_2O(l) \rightleftharpoons HCN(aq) + OH^-(aq)$$

In both reactions, the base reacts with water to give the conjugate acid and OH^- ion. The base ionization constant for NH_3 is

$$K_b = \frac{[NH_4^+][OH^-]}{[NH_3]}$$

which is similar in form to K_h for the CN^- ion. We are, therefore, justified in calling the hydrolysis constant for the CN^- ion its base ionization constant.

Because the hydrolysis constant for CN^- is simply K_b, we can determine the concentrations of species in an aqueous solution of NaCN in the same way that we determined the concentrations of species in a solution of a weak base (see Example 18.5). The only practical difference is that the base ionization constants, K_b, for anions are usually not tabulated. However, as we will show, the K_b for a base is related to the K_a for the conjugate acid. This means that we can obtain K_b (the hydrolysis constant) of CN^- from the K_a for HCN, which is the conjugate acid of CN^-.

To see this relationship between K_a and K_b for conjugate acid–base pairs, let us look at the acid–base pair HCN and CN⁻. The base ionization constant for CN⁻ is

$$K_b = \frac{[\text{HCN}][\text{OH}^-]}{[\text{CN}^-]}$$

Let us multiply the right side by $[\text{H}^+]/[\text{H}^+]$, which equals 1:

$$K_b = \frac{[\text{HCN}][\text{OH}^-]}{[\text{CN}^-]} \times \frac{[\text{H}^+]}{[\text{H}^+]}$$

Note that the terms in color equal the inverse of the acid ionization constant, K_a, of HCN:

$$K_a = \frac{[\text{H}^+][\text{CN}^-]}{[\text{HCN}]}$$

The other terms equal $[\text{H}^+][\text{OH}^-]$, which is the ion-product constant of water, K_w. ∎ Thus, K_b equals K_w/K_a.

The relationship we just obtained is a very general one. It can be written

$$K_a K_b = K_w$$

This equation says that the product of the acid and base ionization constants in aqueous solution for conjugate acid–base pairs equals the ion-product constant for water.

■ The ion-product constant of water was discussed in Section 17.2.

Example 18.6

Use Tables 18.1 and 18.2 to obtain the following at 25°C: (a) K_b for CN⁻ (b) K_a for NH₄⁺

Solution

(a) The conjugate acid of CN⁻ is HCN, whose K_a is 4.9×10^{-10} (Table 18.1). Hence,

$$K_b = \frac{K_w}{K_a} = \frac{1.0 \times 10^{-14}}{4.9 \times 10^{-10}} = 2.0 \times 10^{-5}$$

Note that K_b is approximately equal to the K_b for ammonia (which equals 1.8×10^{-5}). Thus, the base

strength of CN⁻ is comparable to that of NH₃. (b) The conjugate base of NH₄⁺ is NH₃, whose K_b is 1.8×10^{-5} (Table 18.2). Hence,

$$K_a = \frac{K_w}{K_b} = \frac{1.0 \times 10^{-14}}{1.8 \times 10^{-5}} = 5.6 \times 10^{-10}$$

Note that NH₄⁺ is a relatively weak acid. Acetic acid, for example, has a K_a equal to 1.7×10^{-5}.

Exercise 18.7

Calculate the following, using Tables 18.1 and 18.2: (a) K_b for F⁻ (b) K_a for $C_6H_5NH_3^+$ (conjugate acid of aniline, $C_6H_5NH_2$).

(See Problems 18.33 and 18.34.)

If we have a solution of a salt in which only one of the ions hydrolyzes, the calculation of concentrations of species present follows that for solutions of weak acids or bases. The only difference is that we must first obtain the K_a or K_b for the ion that hydrolyzes. The next example illustrates the reasoning and calculations.

Example 18.7

(a) What is the pH of 0.10 M sodium nicotinate at 25°C (a problem posed in the chapter opening)? The K_a for nicotinic acid was determined in Example 18.1 to be 1.4×10^{-5} at 25°C. (b) What is the concentration of NH_3 in 0.15 M NH_4Cl at 25°C? What is the pH of the solution at this temperature? See Example 18.6 for the K_a of NH_4^+.

Solution

(a) Sodium nicotinate gives Na^+ and nicotinate ions in solution. Only the nicotinate ion hydrolyzes. Let us write HNic for nicotinic acid and Nic^- for the nicotinate ion. The hydrolysis of nicotinate ion is

$$Nic^-(aq) + H_2O(l) \rightleftharpoons HNic(aq) + OH^-(aq)$$

Thus, nicotinate ion acts as a base, and we can calculate the concentration of species in solution as in Example 18.5. First, however, we need K_b for the nicotinate ion. This is related to K_a for nicotinic acid by the equation $K_a K_b = K_w$. Hence, substituting into the following equation, we get

$$K_b = \frac{K_w}{K_a} = \frac{1.0 \times 10^{-14}}{1.4 \times 10^{-5}} = 7.1 \times 10^{-10}$$

Now we can proceed with the equilibrium calculation.

Step 1 In one liter of solution, 0.10 mol of nicotinate ion yields x mol of nicotinic acid and x mol of hydroxide ion, as in the table given below.

Step 2 Substituting into the equilibrium equation

$$\frac{[HNic][OH^-]}{[Nic^-]} = K_b$$

gives

$$\frac{x^2}{0.10 - x} = 7.1 \times 10^{-10}$$

Step 3 The base ionization is quite small, so we can assume that x can be neglected compared with 0.10. This gives

$$\frac{x^2}{0.10} \approx 7.1 \times 10^{-10}$$

and

$$x^2 \approx 7.1 \times 10^{-10} \times 0.10$$

Hence,

$$x = 8.4 \times 10^{-6}$$

Note that x is indeed very small compared with 0.10, so our assumption that $0.10 - x = 0.10$ to two significant figures is correct. The concentration of OH^- is 8.4×10^{-6} M. Hence,

$$pH = 14.00 - pOH = 14.00 - (-\log [OH^-])$$
$$= 14.00 + \log (8.4 \times 10^{-6}) = 8.92$$

As expected, the solution has a pH greater than 7.00. (b) We will only sketch the solution to this problem. A solution of NH_4Cl contains NH_4^+ and Cl^- ions. Only NH_4^+ hydrolyzes:

$$NH_4^+(aq) + H_2O(l) \rightleftharpoons NH_3(aq) + H_3O^+(aq)$$

As a result, the solution is acidic. To obtain the concentrations of species in solution, we require the hydrolysis constant, which is K_a for NH_4^+. We have already determined K_a (in Example 18.6) and found it to be 5.6×10^{-10}. If we let $x = [NH_3] = [H^+]$, the equilibrium equation

$$\frac{[H^+][NH_3]}{[NH_4^+]} = K_a$$

gives

$$\frac{x^2}{0.15 - x} = 5.6 \times 10^{-10}$$

Solving this, we find that $x = 9.2 \times 10^{-6}$. Hence, $[NH_3] = 9.2 \times 10^{-6}$ M, and pH = 5.04.

Concentration (M)	$Nic^-(aq)$ + $H_2O(l)$ $\rightleftharpoons$	$HNic(aq)$	+ $OH^-(aq)$
Starting	0.10	0	0
Change	$-x$	$+x$	$+x$
Equilibrium	$0.10 - x$	x	x

Exercise 18.8

Benzoic acid, $HC_7H_5O_2$, and its salts are used as food preservatives. What is the concentration of benzoic acid in an aqueous solution of 0.015 M sodium benzoate? What is the pH of the solution? The K_a for benzoic acid is 6.2×10^{-5}.

(See Problems 18.35, 18.36, 18.37, and 18.38.)

In the case of a salt of a weak acid and a weak base, both ions of the salt hydrolyze. The situation is somewhat complicated, and we will not consider it in quantitative detail. However, we can judge whether the solution is acidic, neutral, or basic by comparing hydrolysis constants for the two ions. Consider the case of ammonium formate, NH_4CHO_2. The hydrolysis reactions for the ions NH_4^+ and CHO_2^- are

$$NH_4^+(aq) + H_2O(l) \rightleftharpoons NH_3(aq) + H_3O^+(aq)$$
$$CHO_2^-(aq) + H_2O(l) \rightleftharpoons HCHO_2(aq) + OH^-(aq)$$

One ion produces H_3O^+ and the other produces OH^-. If the hydrolysis reactions were to proceed to equal extents, we would expect the solution of the salt to be neutral. However, we can easily show that the hydrolysis of NH_4^+ proceeds to a greater extent than does the hydrolysis of CHO_2^-. Let us calculate the hydrolysis constants and compare them. The hydrolysis constant of NH_4^+ equals its K_a, which we calculated earlier (Example 18.6) and found to be 5.6×10^{-10}. The hydrolysis constant of CHO_2^- equals its K_b, which we can calculate from K_a for formic acid, $HCHO_2$:

$$K_b(CHO_2^-) = \frac{K_w}{K_a(HCHO_2)} = \frac{1.0 \times 10^{-14}}{1.7 \times 10^{-4}} = 5.9 \times 10^{-11}$$

Since the hydrolysis constant of NH_4^+ (5.6×10^{-10}) is somewhat greater than that for CHO_2^- (5.9×10^{-11}), we conclude that the hydrolysis of NH_4^+ goes to a greater extent. Therefore, the solution is acidic.

In general, we can decide whether a solution of a salt of a weak acid and weak base is acidic, neutral, or basic by comparing the hydrolysis constants of the two ions from the salt. The hydrolysis constant of the cation will be its K_a, and that for the anion its K_b.

The solution will be acidic if K_a (cation) $> K_b$ (anion).

The solution will be neutral if K_a (cation) $= K_b$ (anion).

The solution will be basic if K_a (cation) $< K_b$ (anion).

Exercise 18.9

Is a solution of ammonium cyanide, NH_4CN, acidic, neutral, or basic?

(See Problems 18.39 and 18.40.)

Solutions of a Weak Acid or Base with Another Solute

In the preceding sections, we looked at solutions that contained either a weak acid, a weak base, or a salt of a weak acid or base. In the remaining sections of this chapter, we will look at the effect of adding another solute to a solution of a weak acid or base. The solutes we will look at are those that significantly affect acid or base ionization, that is, strong acids or bases and salts that contain an ion in common with the weak acid or base. These solutes affect the equilibrium through the *common-ion effect*, which we discuss in the next section.

18.5 Common-Ion Effect

If we add to an ionic equilibrium any solute that provides an ion common to the equilibrium, the equilibrium will shift. This shift is called the **common-ion effect.**

A strong acid provides an ion (H_3O^+) common to the acid ionization equilibrium. Consider, for example, the acid ionization of acetic acid, $HC_2H_3O_2$:

$$HC_2H_3O_2(aq) + H_2O(l) \rightleftharpoons C_2H_3O_2^-(aq) + H_3O^+(aq)$$

What would be the effect on this equilibrium if $HCl(aq)$ is added to a solution of $HC_2H_3O_2$? Since $HCl(aq)$ is a strong acid, it provides H_3O^+ ion, which is present on the right side of the equation for acetic acid ionization. According to LeChatelier's principle, the equilibrium composition should shift to the left.■ Thus, the degree of ionization of acetic acid will be decreased by the addition of a strong acid. The repression of ionization of acetic acid by $HCl(aq)$ is an example of the common-ion effect. The next example illustrates this repression of ionization quantitatively.

■ LeChatelier's principle was applied in Section 16.8 to the problem of adding substances to an equilibrium mixture.

Example 18.8

(a) Calculate the degree of ionization of acetic acid, $HC_2H_3O_2$, in a 0.10 M aqueous solution at 25°C. The K_a at this temperature is 1.7×10^{-5}. (b) Calculate the degree of ionization of $HC_2H_3O_2$ in a 0.10 M solution at 25°C to which sufficient HCl is added to make it 0.010 M HCl. Compare the answers in (a) and (b).

Solution

(a) The calculation is similar to that in Example 18.2, so we will only sketch the solution. We put $x = [H^+] = [C_2H_3O_2^-]$. Then $[HC_2H_3O_2] = 0.10 - x$. Substituting into the equilibrium equation

$$\frac{[H^+][C_2H_3O_2^-]}{[HC_2H_3O_2]} = K_a$$

we get

$$\frac{x^2}{(0.10 - x)} = 1.7 \times 10^{-5}$$

Setting $0.10 - x \simeq 0.10$, the equation becomes $x^2/0.10 = 1.7 \times 10^{-5}$. Solving this, we get $x = 1.3 \times 10^{-3}$. The degree of ionization, or fraction of molecules that have reacted to give ions, is $x/0.10 = 0.013$.

(b) **Step 1** Starting concentrations are $[HC_2H_3O_2] = 0.10$, $[H^+] = 0.010$ (from HCl), and $[C_2H_3O_2^-] = 0$. The acetic acid ionizes to give an additional x mol/L of H^+ and x mol/L of $C_2H_3O_2^-$. The table is

Concentration (M)	$HC_2H_3O_2(aq) \rightleftharpoons$	$H^+(aq)$ +	$C_2H_3O_2^-(aq)$
Starting	0.10	0.010	0
Change	$-x$	$+x$	$+x$
Equilibrium	$0.10 - x$	$0.010 + x$	x

Step 2 We substitute into the equilibrium equation

$$\frac{[H^+][C_2H_3O_2^-]}{[HC_2H_3O_2]} = K_a$$

giving

$$\frac{(0.010 + x)x}{0.10 - x} = 1.7 \times 10^{-5}$$

Step 3 To solve this equation, let us assume that x is small compared to 0.10. Then

$$0.010 + x \simeq 0.010$$
$$0.10 - x \simeq 0.10$$

The equation becomes

$$\frac{0.010x}{0.10} \simeq 1.7 \times 10^{-5}$$

Solving for x, we get

$$x = 1.7 \times 10^{-5} \times \frac{0.10}{0.010} = 1.7 \times 10^{-4}$$

The degree of ionization of $HC_2H_3O_2$ is $x/0.10 = 0.0017$. This is much smaller than the value for 0.10 M $HC_2H_3O_2$ (0.013) because the addition of HCl represses the ionization of $HC_2H_3O_2$.

Exercise 18.10

What is the concentration of formate ion, CHO_2^-, in a solution at 25°C that is 0.10 M $HCHO_2$ and 0.20 M HCl? What is the degree of ionization of formic acid, $HCHO_2$?

(See Problems 18.41 and 18.42.)

As we will see in the following section, solutions that contain a weak acid or weak base and a corresponding salt are especially important. Therefore, we will need to be able to calculate the concentrations of species present in such solutions. A solution of acetic acid and sodium acetate is an example. Note that the acetate ion will repress the ionization of acetic acid by the common-ion effect, just as the addition of H^+ ion did. Therefore, the pH of an acetic acid solution will be raised by adding sodium acetate. (We can also look at this as the addition of a base, $NaC_2H_3O_2$, to an acid, $HC_2H_3O_2$, which of course raises the pH of the acid solution.) The next example shows how we can calculate the concentrations of species in such a solution.

Example 18.9

A solution is prepared to be 0.10 M acetic acid, $HC_2H_3O_2$, and 0.20 M sodium acetate, $NaC_2H_3O_2$. What is the pH of this solution at 25°C? The K_a of acetic acid at 25°C is 1.7×10^{-5} M.

Solution

Step 1 We consider the equilibrium

$$HC_2H_3O_2(aq) \rightleftharpoons H^+(aq) + C_2H_3O_2^-(aq)$$

Initially, 1 L of solution contains 0.10 mol of acetic acid. Since sodium acetate is a strong electrolyte, 1 L of solution will contain 0.20 mol of acetate ion. When the acetic acid ionizes, it gives x mol of hydrogen ion and x mol of acetate ion. This is summarized in the following table:

Concentration (M)	$HC_2H_3O_2(aq) \rightleftharpoons$	$H^+(aq) +$	$C_2H_3O_2^-(aq)$
Starting	0.10	0	0.20
Change	$-x$	$+x$	$+x$
Equilibrium	$0.10 - x$	x	$0.20 + x$

Step 2 The equilibrium equation is

$$\frac{[H^+][C_2H_3O_2^-]}{[HC_2H_3O_2]} = K_a$$

Substituting into this equation gives

$$\frac{x(0.20 + x)}{0.10 - x} = 1.7 \times 10^{-5}$$

Step 3 To solve the equation, let us assume that x is small compared with 0.10 and 0.20. Then

$$0.20 + x \approx 0.20$$
$$0.10 - x \approx 0.10$$

The equilibrium equation becomes

$$\frac{x(0.20)}{0.10} \approx 1.7 \times 10^{-5}$$

Hence $\quad x \approx 1.7 \times 10^{-5} \times \dfrac{0.10}{0.20} = 8.5 \times 10^{-6}$

(Note that x is indeed much smaller than 0.10 or 0.20.) Thus, the hydrogen-ion concentration is 8.5×10^{-6} M, and

$$pH = -\log [H^+] = -\log (8.5 \times 10^{-6}) = 5.07$$

Exercise 18.11

One liter of solution was prepared by dissolving 0.025 mol of formic acid, $HCHO_2$, and 0.018 mol of sodium formate, $NaCHO_2$, in water. What was the pH of the solution? K_a for formic acid is 1.7×10^{-4}.

(See Problems 18.43 and 18.44.)

3.63

The previous examples considered the common-ion effect in solutions of weak acids. We also encounter the common-ion effect in solutions of weak bases. An important case is that of a solution containing a weak base and one of its salts. This problem is very similar to that considered in the preceding example and is shown in detail in the next example.

Example 18.10

What is the pH at 25°C of a solution that is 0.10 M NH_3 and 0.20 M NH_4Cl? The K_b for NH_3 is 1.8×10^{-5} at 25°C.

Solution

Step 1 The base ionization is

$$NH_3(aq) + H_2O(l) \rightleftharpoons NH_4^+(aq) + OH^-(aq)$$

Addition of an ammonium salt to a solution of NH_3 provides the common ion NH_4^+ and represses the ionization of the base. We set up the following table:

Concentration (M)	$NH_3(aq)$ +	$H_2O(l)$ $\rightleftharpoons$	$NH_4^+(aq)$ +	$OH^-(aq)$
Starting	0.10		0.20	0
Change	$-x$		$+x$	$+x$
Equilibrium	$0.10 - x$		$0.20 + x$	x

Step 2 Substituting into the equilibrium equation

$$\frac{[NH_4^+][OH^-]}{[NH_3]} = K_b$$

gives

$$\frac{(0.20 + x)x}{0.10 - x} = 1.8 \times 10^{-5}$$

Step 3 Let us assume that $0.20 + x \simeq 0.20$ and $0.10 - x \simeq 0.10$. The previous equation becomes

$$\frac{0.20x}{0.10} \simeq 1.8 \times 10^{-5}$$

Hence, $x = 1.8 \times 10^{-5} \times \dfrac{0.10}{0.20} = 9.0 \times 10^{-6}$

Note that x is indeed very small compared with 0.10 and 0.20. Since $x = [OH^-]$,

$$pH = 14.00 - pOH = 14.00 + \log [OH^-]$$
$$= 14.00 + \log (9.0 \times 10^{-6}) = 14.00 - 5.05 = 8.95$$

Exercise 18.12

A solution is 0.15 M CH_3NH_2 (methylamine) and 0.10 M CH_3NH_3Cl (methyl-ammonium chloride, a salt of methylamine). What is the pH of this solution at 25°C? See Table 18.2 for the K_b of methylamine.

(See Problems 18.45 and 18.46.)

18.6 Buffers

A **buffer** is a solution characterized by the ability to withstand changes in pH when limited amounts of acid or base are added to it. If 0.01 mol of hydro-chloric acid is added to 1 L of pure water, the pH changes from 7.0 to 2.0—a pH change of 5.0 units. By contrast, the addition of this amount of hydro-chloric acid to 1 L of buffered solution might change the pH by only 0.1 unit. Biological fluids, such as blood, are usually buffer solutions because the control of pH is vital to their proper functioning. The oxygen-carrying func-tion of blood depends on its being maintained at a pH of 7.4. If the pH were to change by several tenths of a unit, the capacity of the blood to carry oxygen would be lost.

Most buffers contain a weak acid and its conjugate base, or a weak base and its conjugate acid. Blood, for example, contains H_2CO_3 and HCO_3^-, as well as other conjugate acid–base pairs. A buffer frequently used in the laboratory contains the conjugate acid–base pair $H_2PO_4^-$ and HPO_4^{2-}. Buffers also have commercial application.

For example, the label on a package of artificial fruit-juice mix says that it contains "citric acid to provide tartness and sodium citrate to regulate tartness." A solution of citric acid and its base conjugate, citrate ion (provided by sodium citrate), functions as an acid–base buffer, which is what "to regulate tartness" means. The pH of the buffer will be in the acid range.

Suppose a buffer contains approximately equal molar amounts of a weak acid HA and its conjugate base, A^-. If a strong acid is added to the buffer, it supplies hydrogen ions that react with the base A^-:

$$H^+(aq) + A^-(aq) \longrightarrow HA(aq)$$

On the other hand, if a strong base is added to the buffer, it supplies hydroxide ions. These ions react with the acid HA:

$$OH^-(aq) + HA(aq) \longrightarrow H_2O(l) + A^-(aq)$$

Thus, a buffer solution resists changes in pH by the ability to combine with both H^+ and OH^- ions.

The solution described in Example 18.9, which is 0.10 M in acetic acid and 0.20 M in sodium acetate, is a buffer solution. It contains the acid $HC_2H_3O_2$ and its conjugate base $C_2H_3O_2^-$. In that example, we found the pH to be 5.07. The pH of this solution will remain near this value if strong acids or bases are added, as long as the quantities are not too large. To see that this solution does indeed act as a buffer, we will calculate the pH of the solution after adding some strong acid to it. We will see that the pH changes by only a few hundredths of a unit. By comparison, pure water will change by several units in pH if the same quantity of strong acid is added to it.

Example 18.11

Calculate the pH of 75 mL of the buffer solution described in Example 18.9 (0.10 M $HC_2H_3O_2$ and 0.20 M $NaC_2H_3O_2$) to which 9.5 mL of 0.10 M hydrochloric acid is added. Compare the pH change with what would occur if this amount of acid were added to pure water.

Solution

When hydrogen ion (from hydrochloric acid) is added to the buffer, it reacts with acetate ion:

$$H^+(aq) + C_2H_3O_2^-(aq) \longrightarrow HC_2H_3O_2(aq)$$

Since acetic acid is a weak acid, we can assume as a first approximation that the reaction goes to completion. This part of the problem is simply one in stoichiometry. Then we assume that the acetic acid ionizes slightly. This part of the problem involves an acid ionization equilibrium.

Stoichiometric Calculation We must first calculate the amounts of hydrogen ion, acetate ion, and acetic acid present in the solution before reaction. The molar amount of hydrogen ion will equal the molar amount of hydrochloric acid added, which we obtain by converting the volume of hydrochloric acid, HCl, to moles of HCl. To do this, we note for 0.10 M HCl that

$$1 \text{ L HCl} \simeq 0.10 \text{ mol HCl}$$

Hence, to convert 9.5 mL $(= 9.5 \times 10^{-3}$ L) of hydrochloric acid, we have

$$9.5 \times 10^{-3} \text{ L HCl} \times \frac{0.10 \text{ mol HCl}}{1 \text{ L HCl}} \simeq 0.00095 \text{ mol HCl}$$

Since hydrochloric acid is a strong acid, it exists in solution as the ions H^+ and Cl^-. Thus, the amount of H^+ added is 0.00095 mol.

(Continued)

The amounts of acetate ion and acetic acid are found in a similar way. The buffer in Example 18.9 contains 0.20 mol of acetate ion and 0.10 mol of acetic acid in 1 L of solution. The amounts in 75 mL (= 0.075 L) of solution are obtained by converting to moles:

$$0.075 \ \text{L soln} \times \frac{0.20 \ \text{mol} \ C_2H_3O_2^-}{1 \ \text{L soln}}$$

$$\simeq 0.015 \ \text{mol} \ C_2H_3O_2^-$$

$$0.075 \ \text{L soln} \times \frac{0.10 \ \text{mol} \ HC_2H_3O_2}{1 \ \text{L soln}}$$

$$\simeq 0.0075 \ \text{mol} \ HC_2H_3O_2$$

We now assume that all of the hydrogen ion added (0.00095 mol) reacts with acetate ion. Therefore, 0.00095 mol of acetic acid are produced and 0.00095 mol of acetate ion are used up. Hence, after reaction we have

Moles of acetate ion = (0.015 − 0.00095) mol $C_2H_3O_2^-$

$$= 0.014 \ \text{mol} \ C_2H_3O_2^-$$

Moles of acetic acid = (0.0075 + 0.00095) mol $HC_2H_3O_2$

$$= 0.0085 \ \text{mol} \ HC_2H_3O_2$$

Equilibrium Calculation We first calculate the concentrations of $HC_2H_3O_2$ and $C_2H_3O_2^-$ present in the solution before we consider the acid ionization equilibrium. Note that the total volume of solution (buffer plus hydrochloric acid) is 75 mL + 9.5 mL or 84 mL (0.084 L). Hence, the starting concentrations are

$$[HC_2H_3O_2] = \frac{0.0085 \ \text{mol}}{0.084 \ \text{L}} = 0.10 \ M$$

$$[C_2H_3O_2^-] = \frac{0.014 \ \text{mol}}{0.084 \ \text{L}} = 0.17 \ M$$

From this, we construct the following table:

Concentration (M)	$HC_2H_3O_2(aq) \rightleftharpoons$	$H^+(aq) +$	$C_2H_3O_2^-(aq)$
Starting	0.10	0	0.17
Change	$-x$	$+x$	$+x$
Equilibrium	$0.10 - x$	x	$0.17 + x$

The equilibrium equation is

$$\frac{[H^+][C_2H_3O_2^-]}{[HC_2H_3O_2]} = K_a$$

Substituting, we get

$$\frac{x(0.17 + x)}{0.10 - x} = 1.7 \times 10^{-5}$$

If we assume that x is small enough that $0.17 + x \simeq 0.17$ and $0.10 - x \simeq 0.10$, this equation becomes

$$\frac{x(0.17)}{0.10} = 1.7 \times 10^{-5}$$

or $\quad x = 1.7 \times 10^{-5} \times \dfrac{0.10}{0.17} = 1.0 \times 10^{-5}$

Note that x is indeed small, so the assumptions we made earlier are correct. Thus, the H^+ ion concentration is 1.0×10^{-5} M. The pH is

$$pH = -\log [H^+] = -\log (1.0 \times 10^{-5}) = 5.00$$

Since the pH of the buffer was 5.07 (see Example 18.8), the pH has changed by 5.07 − 5.00 = 0.07 units.

Adding HCl to Pure Water If 9.5 mL of 0.10 M hydrochloric acid were added to 75 mL of pure water, the hydrogen-ion concentration would change to

$$[H^+] = \frac{\text{amount of } H^+ \text{ added}}{\text{total volume of solution}}$$

$$= \frac{0.00095 \ \text{mol} \ H^+}{0.084 \ \text{L solution}} = 0.011 \ M$$

(The total volume is 75 mL of water plus 9.5 mL HCl, assuming no change of volume on mixing.) The pH is

$$pH = -\log [H^+] = -\log 0.011 = 1.96$$

Since the pH of pure water is 7.00, the change of pH is 7.00 − 1.96 = 5.04 units, compared with 0.07 units for the buffered solution.

Exercise 18.13

What is the pH of the solution described in Exercise 18.11 if 50.0 mL of 0.10 M sodium hydroxide are added to one liter of solution?

(See Problems 18.47 and 18.48.)

Two important characteristics of a buffer are the pH and the *buffer capacity*, which is the amount of acid or base the buffer can react with before giving a significant pH change. Buffer capacity depends on the amount of acid and conjugate base in the solution. Figure 18.2 illustrates the change of

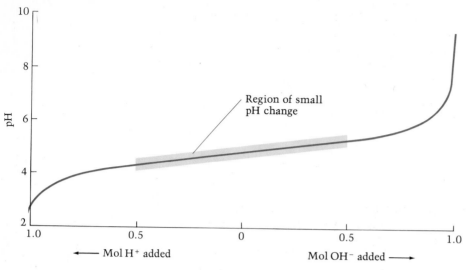

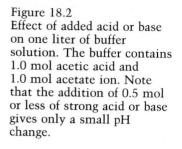

Figure 18.2
Effect of added acid or base on one liter of buffer solution. The buffer contains 1.0 mol acetic acid and 1.0 mol acetate ion. Note that the addition of 0.5 mol or less of strong acid or base gives only a small pH change.

pH of a buffer solution containing 1.0 mol acetic acid and 1.0 mol acetate ion. This buffer changes less than 0.5 pH units as long as no more than 0.5 mol of H^+ or OH^- ion is added. Note that this is one-half or less than the amounts of acid and conjugate base in the solution. The ratio of amounts of acid and conjugate base is also important. Unless this ratio is approximately 1 (between 1:10 and 10:1), the buffer capacity will be too low to be useful.

The other important characteristic of a buffer is its pH. How do we prepare a buffer of given pH? We can show that the buffer must be prepared from a conjugate acid–base pair in which the acid ionization constant is approximately equal to the H^+ ion concentration. To illustrate, consider a buffer made up of a weak acid HA and its conjugate base A^-. The acid ionization equilibrium is

$$HA(aq) \rightleftharpoons H^+(aq) + A^-(aq)$$

and the acid ionization constant is

$$K_a = \frac{[H^+][A^-]}{[HA]}$$

By rearranging this, we can get an equation for the H^+ ion concentration:

$$[H^+] = K_a \times \frac{[HA]}{[A^-]}$$

This equation expresses the H^+ ion concentration in terms of the K_a for the acid and the ratio of concentrations of HA and A^-. Since this equation was derived from the equilibrium constant, the concentrations of HA and A^- should be equilibrium values. But because the presence of A^- represses the ionization of HA, these concentrations do not differ significantly from the values used to prepare the buffer.

We can use the preceding equation to derive an equation for the pH of a buffer. Let us take the negative logarithm of both sides of the equation. That is,

$$-\log [H^+] = -\log \left(K_a \times \frac{[HA]}{[A^-]} \right) = -\log K_a - \log \frac{[HA]}{[A^-]}$$

The left side equals the pH. We can also simplify the right side. The pK_a of a weak acid is defined in a manner similar to pH and pOH, that is,■

$$pK_a = -\log K_a$$

■ Acid and base ionization constants are often listed as pK_a and pK_b $(= -\log K_b)$.

The previous equation can be written

$$pH = pK_a - \log \frac{[HA]}{[A^-]} = pK_a + \log \frac{[A^-]}{[HA]}$$

More generally, we can write

$$pH = pK_a + \log \frac{[base]}{[acid]}$$

This is known as the **Henderson-Hasselbalch equation.** By substituting the value of pK_a for the conjugate acid and the ratio [base]/[acid], we obtain the pH of the buffer.

The question we asked earlier was how to prepare a buffer of a given pH, for example, pH 4.90. We can see that we need to find a conjugate acid–base pair in which the pK_a of the acid is close to the desired pH. Thus, the K_a of acetic acid is 1.7×10^{-5} and its pK_a is $-\log (1.7 \times 10^{-5}) = 4.77$. We can get a pH somewhat higher by increasing the ratio [base]/[acid].

Consider the calculation of pH of a buffer containing 0.10 M NH_3 and 0.20 M NH_4Cl. The conjugate acid is NH_4^+, whose K_a we can calculate from the K_b for NH_3 $(= 1.8 \times 10^{-5})$. The K_a for NH_4^+ is 5.6×10^{-10}, and the pK_a is $-\log (5.6 \times 10^{-10}) = 9.25$. Hence,■

■ We did the equilibrium calculation earlier, in Example 18.10.

$$pH = 9.25 + \log \frac{0.10}{0.20} = 8.95$$

The solution is basic, as we should have guessed, since the buffer contains the weak base NH_3 and the very weak acid NH_4^+.

18.7 Acid–Base Titration Curves

An acid–base titration is a procedure for determining the amount of acid (or base) in a solution by determining the volume of base (or acid) of known concentration that will completely react with it.■ An **acid–base titration curve** is a plot of the pH of a solution of acid (or base) against the volume of added base (or acid). Such curves are used to gain insight into the titration process. We can use the titration curve to choose an indicator that will show when the titration is complete.

■ The technique of titration was discussed in Section 3.12.

Titration of a Strong Acid by a Strong Base

Figure 18.3 shows a curve for the titration of 25.0 mL of 0.100 M HCl by 0.100 M NaOH. Note that the pH changes slowly at first until the molar amount of base added nearly equals that of the acid, that is, until the titration is near the equivalence point. The **equivalence point** occurs when a stoichiometric amount of reactant has been added during a titration. At the equivalence point, the pH of this solution of NaOH and HCl is 7.0 because it contains a salt, NaCl, that does not hydrolyze. However, the pH changes

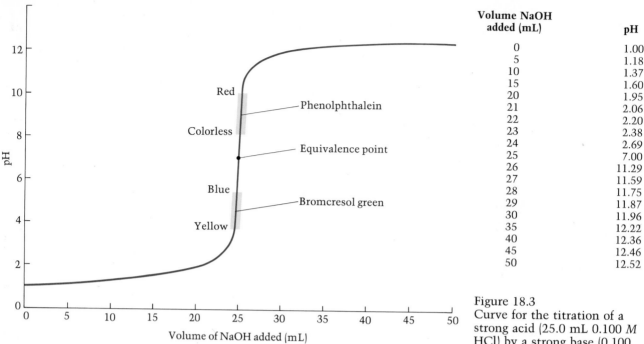

Volume NaOH added (mL)	pH
0	1.00
5	1.18
10	1.37
15	1.60
20	1.95
21	2.06
22	2.20
23	2.38
24	2.69
25	7.00
26	11.29
27	11.59
28	11.75
29	11.87
30	11.96
35	12.22
40	12.36
45	12.46
50	12.52

Figure 18.3
Curve for the titration of a strong acid (25.0 mL 0.100 M HCl) by a strong base (0.100 M NaOH). The portions of the curve where the indicators bromcresol green and phenolphthalein change color are shown. Note that both indicators change color where the pH changes rapidly (nearly vertical part of the curve).

rapidly near the equivalence point, from a pH of about 3 to a pH of about 11. To detect the equivalence point, we add an indicator that changes color within the pH range 3 to 11. Phenolphthalein can be used, since it changes from colorless to red in the pH range 8.2 to 10.0. (Figure 17.5 shows the pH ranges for the color changes of indicators.) Even though this color change occurs on the basic side, only a fraction of a drop of base is required to change the pH several units when the titration is near the equivalence point. The indicator bromcresol green, whose color changes in the pH range 3.8 to 5.4, would also work.

The following example shows how to calculate a point on the titration curve of a strong acid and a strong base.

Example 18.12

Calculate the pH of a solution in which 10.0 mL of 0.100 M NaOH are added to 25.0 mL of 0.100 M HCl.

Solution

Because the reactants are strong electrolytes, the problem is essentially one in stoichiometry. The reaction is

$$H^+(aq) + OH^-(aq) \longrightarrow H_2O(l)$$

We get the amounts of reactants by multiplying the volume (in liters) of each solution by the molar concentrations:

Mol H^+ = 0.0250 L × 0.100 mol/L = 0.00250 mol

Mol OH^- = 0.0100 L × 0.100 mol/L = 0.00100 mol

All of the OH^- reacts, leaving an excess of H^+:

Excess H^+ = (0.00250 − 0.00100) mol

= 0.00150 mol H^+

We obtain the concentration of H^+ by dividing this amount of H^+ by the total volume of solution (= 0.0250 L + 0.0100 L = 0.0350 L).

$$[H^+] = \frac{0.00150 \text{ mol}}{0.0350 \text{ L}} = 0.0429$$

Hence,   pH = $-\log [H^+]$ = $-\log (0.0429)$ = 1.368

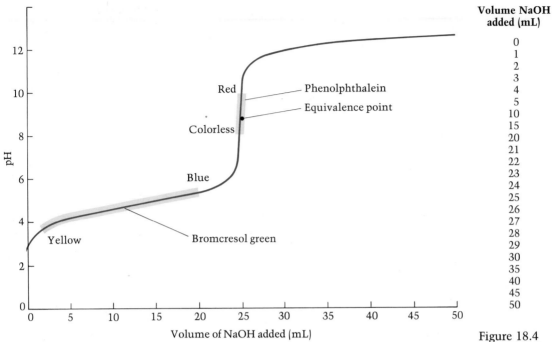

Volume NaOH added (mL)	pH
0	2.92
1	3.47
2	3.79
3	3.98
4	4.13
5	4.25
10	4.67
15	5.03
20	5.45
21	5.57
22	5.72
23	5.91
24	6.23
25	8.78
26	11.30
27	11.60
28	11.78
29	11.90
30	12.00
35	12.30
40	12.48
45	12.60
50	12.70

Figure 18.4
Curve for the titration of a weak acid (25.0 mL 0.100 M nicotinic acid) by a strong base (0.100 M NaOH). Note that bromcresol green changes color during the early part of the titration, well before the equivalence point. Phenolphthalein changes color where the pH changes rapidly (near the equivalence point). Thus, phenolphthalein could be used to indicate the end of the titration, whereas bromcresol green could not.

Exercise 18.14

What is the pH of a solution in which 15 mL of 0.10 M NaOH has been added to 25 mL of 0.10 M HCl?

(See Problems 18.55 and 18.56.)

Titration of a Weak Acid by a Strong Base

The titration of a weak acid by a strong base gives a somewhat different curve. Figure 18.4 shows the curve for the titration of 25.0 mL of 0.100 M nicotinic acid, $HC_6H_4NO_2$, by 0.100 M NaOH. The titration starts at a higher pH than the titration of HCl because nicotinic acid is a weak acid. As before, the pH changes slowly at first, then rapidly near the equivalence point. The pH range in which the rapid change is seen occurs from about pH 7 to pH 11. Note that the pH range is shorter than that for the titration of a strong acid by a strong base. This means that the choice of an indicator is more critical. Phenolphthalein would work, since it changes color in the range 8.2 to 10.0. Bromcresol green would not work. This indicator changes color in the range 3.8 to 5.4, which occurs before the titration curve rises steeply.

Note also that the equivalence point for the titration curve of nicotinic acid occurs on the basic side. This happens because at the equivalence point, the solution is that of the salt, sodium nicotinate, which is basic from the hydrolysis of the nicotinate ion. The optimum choice of indicator would be one that changes color over a range that includes the pH of the equivalence point.■

The following example shows how to calculate the pH at the equivalence point in a titration.

■ Titration can be used to analyze for the amount of niacin (nicotinic acid) in a solution extracted from a vitamin tablet or sample of food.

Example 18.13

Calculate the pH of the solution at the equivalence point when 25 mL of 0.10 M nicotinic acid are titrated by 0.10 M sodium hydroxide. The K_a for nicotinic acid equals 1.4×10^{-5}.

Solution

At the equivalence point, equal molar amounts of nicotinic acid and sodium hydroxide react to give a solution of sodium nicotinate. We first calculate the concentration of nicotinate ion. Then, we find the pH of this solution. (This is a hydrolysis problem.)

Concentration of Nicotinate Ion We assume that the reaction of the base with the acid is complete. In this case, 25 mL of 0.10 M sodium hydroxide are needed to react with 25 mL of 0.10 M nicotinic acid. The molar amount of nicotinate ion formed equals the initial molar amount of nicotinic acid:

$$25 \times 10^{-3} \text{ L soln} \times \frac{0.10 \text{ mol nicotinate ion}}{1 \text{ L soln}}$$

$$= 2.5 \times 10^{-3} \text{ mol nicotinate ion}$$

The total volume of solution is 50 mL (25 mL NaOH solution plus 25 mL of nicotinic acid solution, assuming no change of volume on mixing). Dividing the molar amount of nicotinate ion by the volume of solution in liters gives the molar concentration of nicotinate ion:

$$\text{Molar concentration} = \frac{2.5 \times 10^{-3} \text{ mol}}{50 \times 10^{-3} \text{ L}} = 0.050 \ M$$

Hydrolysis of Nicotinate Ion This portion of the calculation follows the method given in Example 18.7. We find that the K_b of nicotinate ion is 7.1×10^{-10} and the concentration of hydroxide ion is $6.0 \times 10^{-6} \ M$. The pH is 8.78. Try working through the calculation to verify these numbers.

Exercise 18.15

What is the pH at the equivalence point when 25 mL of 0.10 M HF are titrated by 0.15 M NaOH?

(See Problems 18.57 and 18.58.)

Figure 18.5
Curve for the titration of a weak base (25.0 mL 0.100 M NH$_3$) by a strong acid (0.100 M HCl). Methyl red changes color near the equivalence point, so it can be used to indicate the end of the titration.

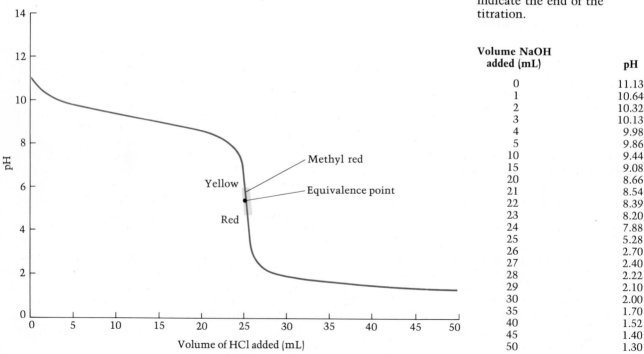

Volume NaOH added (mL)	pH
0	11.13
1	10.64
2	10.32
3	10.13
4	9.98
5	9.86
10	9.44
15	9.08
20	8.66
21	8.54
22	8.39
23	8.20
24	7.88
25	5.28
26	2.70
27	2.40
28	2.22
29	2.10
30	2.00
35	1.70
40	1.52
45	1.40
50	1.30

Titration of a Weak Base by a Strong Acid

If we titrate a weak base with a strong acid, we get a titration curve similar to that for a weak acid by a strong base. Figure 18.5 shows the pH changes during the titration of 25.0 mL 0.100 M NH_3 with 0.100 M HCl. In this case, the pH declines slowly at first, then falls abruptly from about pH 7 to pH 3. Methyl red, which changes color from yellow at pH 6.0 to red at pH 4.8, is a possible indicator for this titration. Note that phenolphthalein could not be used to find the equivalence point.

Exercise 18.16

What is the pH at the equivalence point when 35 mL of 0.20 M ammonia are titrated by 0.12 M hydrochloric acid? The K_b for ammonia is 1.8×10^{-5}.

(See Problems 18.59 and 18.60.)

A Checklist for Review

Important Terms

acid ionization (dissociation) constant (18.1)   hydrolysis constant (18.4)   Henderson-Hasselbalch equation (18.6)
degree of ionization (18.1)   common-ion effect (18.5)   acid–base titration curve (18.7)
base ionization (dissociation) constant (18.3)   buffer (18.6)   equivalence point (18.7)

Summary of Facts and Concepts

When an acid dissolves in water, it ionizes (or dissociates) to give hydrogen ions and the conjugate base ions. The equilibrium constant for this *acid ionization* is $K_a = [H^+][A^-]/[HA]$, where HA is the general formula for the acid. The constant K_a can be determined from the pH of an acid solution of known concentration. Once obtained, the acid ionization constant can be used to find the concentrations of species in any solution of the acid. In the case of a *diprotic* acid, H_2A, the concentration of H^+ and HA are calculated from K_{a1}, and the concentration of A^{2-} equals K_{a2}. Similar considerations apply to *base ionizations*.

Solutions of salts may be acidic, neutral, or basic because of hydrolysis of the ions. The *hydrolysis constant*, or equilibrium constant for hydrolysis, equals K_a for a cation or K_b for an anion. Calculation of the pH of a solution of a salt in which one ion hydrolyzes is fundamentally the same as the calculation of the pH of a solution of an acid or base. However, the K_a or K_b of an ion is usually obtained from the conjugate base or acid by using the equation $K_a K_b = K_w$.

A *buffer* is a solution that can withstand changes in pH when small amounts of acid or base are added to it. A buffer contains either a weak acid and its conjugate base or a weak base and its conjugate acid. The concentrations of acid and base conjugates are approximately equal. Different pH ranges are possible for acid–base conjugates depending on their ionization constants.

An *acid–base titration curve* is a plot of the pH of the solution against the volume of reactant added. During the titration of a strong acid by a strong base, the pH changes slowly at first. Then, as the amount of base nears the stoichiometric value, that is, nears the *equivalence point*, the pH rises abruptly, changing by several units. The pH at the equivalence point is 7.0. A similar curve is obtained when a weak acid is titrated with a strong base. However, the pH changes less at the equivalence point. Moreover, the pH at the equivalence point will be greater than 7.0 because of hydrolysis of the salt produced. An indicator must be chosen that changes color within the pH range near the equivalence point where the pH changes rapidly.

Operational Skills

1. Given the molarity and pH of a solution of a weak acid, calculate the acid ionization constant, K_a (Example 18.1). Given K_a, calculate the hydrogen-ion concentration and pH of a solution of a weak acid of known molarity (Examples 18.2 and 18.3).

2. Given K_{a1}, K_{a2}, and the molarity of a diprotic acid solution, calculate the pH and the concentrations of H^+, HA^-, and A^{2-} (Example 18.4).

3. Given the molarity and pH of a solution of a weak base, calculate the base ionization constant, K_b (similar to Example 18.1). Given K_b, calculate the hydrogen-ion concentration and pH of a solution of a weak base of known molarity (Example 18.5).

4. Calculate the K_a for a cation or the K_b for an anion from the ionization constant of the conjugate base or acid (Example 18.6).

5. Given the concentration of a solution of a salt in which one ion hydrolyzes and given the ionization constant of the conjugate acid or base of this ion, calculate the H^+ ion concentration (Example 18.7).

6. Given the concentrations of weak acid and strong acid in a solution, calculate the degree of ionization and concentration of the anion of the weak acid (Example 18.8).

7. Given the K_a and the concentrations of weak acid and its salt in a solution, calculate the pH (Example 18.9). Given the K_b and the concentrations of weak base and its salt in a solution, calculate the pH (Example 18.10).

8. Calculate the pH of a given volume of buffer solution (given the concentrations of conjugate acid and base in the buffer) to which a specified amount of strong acid or base is added (Example 18.11).

9. Calculate the pH during the titration of a strong acid and strong base, given the volumes and concentrations of the acid and base (Example 18.12).

10. Calculate the pH at the equivalence point for the titration of a specified quantity of weak acid by a strong base, given the K_a (Example 18.13). Be able to do the same type of calculation for the titration of a weak base by a strong acid (similar to Example 18.13).

Review Questions

18.1 Write an equation for the ionization of hydrogen cyanide, HCN, in aqueous solution. What is the equilibrium expression, K_a, for this acid ionization?

18.2 Which one of the following is the weakest acid: $HClO_4$, HCN, $HC_2H_3O_2$? See Table 9.3, p. 284, and Table 18.1.

18.3 Briefly describe two methods for determining the K_a for a weak acid.

18.4 Describe how the degree of ionization of a weak acid changes as the concentration increases.

18.5 Consider a solution of 0.010 M HF ($K_a = 6.8 \times 10^{-4}$). In solving for the concentrations of species in this solution, could you use the simplifying assumption in which you neglect x in the denominator of the equilibrium equation? Explain.

18.6 Phosphorous acid, H_2PHO_3, is a diprotic acid. Write equations for the acid ionizations. Write the expressions for K_{a1} and K_{a2}.

18.7 What is the concentration of oxalate ion, $C_2O_4^{2-}$, in 0.10 M oxalic acid, $H_2C_2O_4$? K_{a1} is 5.6×10^{-2} and K_{a2} is 5.1×10^{-5}.

18.8 Write the equation for the ionization of aniline, $C_6H_5NH_2$, in aqueous solution. Write the expression for K_b.

18.9 Which of the following is the strongest base: NH_3, $C_6H_5NH_2$, CH_3NH_2? See Table 18.2.

18.10 Do you expect a solution of anilinium chloride (aniline hydrochloride), $C_6H_5NH_3Cl$, to be acidic or basic? (Anilinium chloride is the salt of aniline and hydrochloric acid.) Write the equation for the reaction involved. What is the equilibrium expression? How would you obtain the value for the equilibrium constant from Table 18.2?

18.11 What is meant by the common-ion effect? Give an example.

18.12 The pH of 0.10 M CH_3NH_2 (methylamine) is 11.8. If the chloride salt of methylamine, CH_3NH_3Cl, is added to this solution, will the pH increase or decrease? Explain this using Le Chatelier's principle and the common-ion effect.

18.13 Define a buffer. Give an example of one.

18.14 What is meant by the capacity of a buffer? Describe a buffer with low capacity and the same buffer with greater capacity.

18.15 Describe the pH changes that occur during the titration of a weak base by a strong acid. What is meant by the term *equivalence point*?

18.16 If the pH is 8.0 at the equivalence point for the titration of a certain weak acid with sodium hydroxide, what indicator might you use? See Figure 17.5. Explain your choice.

Problems

Note: Values of K_a and K_b that are not given in the problems can be obtained from Tables 18.1 and 18.2.

Acid Ionization

18.17 Calculate the acid ionization constant, K_a, for propionic acid, $HC_3H_5O_2$. The pH of a 0.012 M aqueous solution is 3.40.

18.19 Boric acid, $B(OH)_3$, is used as a mild antiseptic. What is the pH of a 0.025 M aqueous solution of boric acid? What is the degree of ionization of boric acid in this solution? The hydrogen ion arises principally from the reaction

$$B(OH)_3(aq) + H_2O(l) \rightleftharpoons B(OH)_4^-(aq) + H^+(aq)$$

The equilibrium constant for this reaction is 5.9×10^{-10}.

18.21 $C_6H_4NH_2COOH$, *para*-aminobenzoic acid (PABA), is used in some sunscreen agents. Calculate the concentrations of hydrogen ion and *para*-aminobenzoate ion, $C_6H_4NH_2COO^-$, in a 0.050 M solution of the acid. The value of K_a is 2.2×10^{-5}.

18.23 Hydrofluoric acid, HF, unlike hydrochloric acid, is a weak electrolyte. What is the pH of a 0.40 M aqueous solution of HF?

18.25 Sulfanilic acid, $HC_6H_6NO_3S$, is related to the sulfa drug sulfanilamide. What is the pH of a 0.085 M solution of sulfanilic acid? The K_a is 5.9×10^{-4}.

18.27 Oxalic acid, $H_2C_2O_4$, is found in a wide variety of plants (for example, rhubarb, spinach), usually as one of its salts. What is the pH of a 0.50 M aqueous solution of oxalic acid? What is the concentration of oxalate ion, $C_2O_4^{2-}$? Note that K_{a1} is large.

18.18 A solution of 0.020 M nitrous acid, HNO_2, has a pH of 2.53. What is the value of K_a?

18.20 Formic acid, $HCHO_2$, is used to make methyl formate (a fumigant for dried fruit) and ethyl formate (an artificial rum flavor). What is the pH of a 0.12 M solution of formic acid? What is the degree of ionization of $HCHO_2$ in this solution? See Table 18.1 for K_a.

18.22 Barbituric acid, $HC_4H_3N_2O_3$, is used to prepare various barbiturate drugs (used as sedatives). Calculate the concentrations of hydrogen ion and barbiturate ion in a 0.20 M solution of the acid. The value of K_a is 9.8×10^{-5}.

18.24 Chloroacetic acid, $CH_2ClCOOH$, has a greater acid strength than acetic acid because the electronegative chlorine atom pulls electrons away from the O—H bond. Calculate the pH of a 0.15 M solution of chloroacetic acid. K_a is 1.3×10^{-3}.

18.26 Gout is a form of arthritis caused by the precipitation of salts of uric acid in the joints. Uric acid, $C_5H_4N_4O_3$, has a K_a value of 1.3×10^{-4}. What is the pH of a 0.036 M solution of uric acid?

18.28 Hydrogen sulfide, H_2S, is a poisonous gas (with the odor of rotten eggs) that dissolves in water to give an acidic solution (hydrosulfuric acid). What are the concentrations of H^+, HS^-, and S^{2-} in a saturated solution (0.10 M)?

Base Ionization

18.29 Ethanolamine, $HOC_2H_4NH_2$, is a viscous liquid with an ammonia-like odor; it is used to remove hydrogen sulfide from natural gas. A 0.15 M aqueous solution of ethanolamine has a pH of 11.34. What is K_b for ethanolamine?

18.31 What is the concentration of hydroxide ion in a 0.080 M aqueous solution of methylamine, CH_3NH_2? What is the pH?

18.30 Trimethylamine, $(CH_3)_3N$, is a gas with a fishy, ammonia-like odor. An aqueous solution that is 0.25 M trimethylamine has a pH of 11.63. What is K_b for trimethylamine?

18.32 What is the concentration of hydroxide ion in a 0.15 M aqueous solution of hydroxylamine, NH_2OH? What is the pH?

Hydrolysis

18.33 Obtain (a) the K_b for NO_2^-, (b) the K_a for $C_5H_5NH^+$ (pyridinium ion).

18.35 What is the pH of a 0.025 M aqueous solution of sodium propionate, $NaC_3H_5O_2$? What is the concentration of propionic acid in the solution?

18.37 Calculate the concentration of pyridine, C_5H_5N, in a solution that is 0.15 M pyridinium bromide, C_5H_5NHBr. What is the pH of the solution?

18.39 Decide whether solutions of the following salts are acidic, neutral, or basic:
    (a) ammonium acetate   (b) anilinium acetate

18.34 Obtain (a) the K_b for ClO^- (b) the K_a for NH_3OH^+ (hydroxylammonium ion).

18.36 Find the concentrations of H^+, CN^-, and HCN in a 0.010 M aqueous solution of sodium cyanide, NaCN.

18.38 What is the pH of a 0.35 M solution of methylammonium chloride, CH_3NH_3Cl? What is the concentration of methylamine in the solution?

18.40 Decide whether solutions of the following salts are acidic, neutral, or basic:
    (a) ammonium cyanate   (b) pyridinium cyanate

Common-Ion Effect

18.41 Calculate the degree of ionization of
    (a) 0.80 M HF
    (b) the same solution which is also 0.10 M HCl

18.43 What is the pH of a solution that is 0.10 M KNO_2 and 0.15 M HNO_2 (nitrous acid)?

18.45 What is the pH of a solution that is 0.10 M N_2H_4 (hydrazine) and 0.15 M N_2H_5Cl (hydrazinium chloride)? (There is a small effect from the self-ionization of water, which may be ignored.)

18.42 Calculate the degree of ionization of
    (a) 0.20 M $HCHO_2$ (formic acid)
    (b) the same solution which is also 0.10 M HCl

18.44 What is the pH of a solution that is 0.20 M KCNO and 0.10 M HCNO (cyanic acid)?

18.46 What is the pH of a solution that is 0.15 M $C_2H_5NH_2$ (ethylamine) and 0.10 M $C_2H_5NH_3Br$ (ethylammonium bromide)?

Buffers

18.47 What is the pH of a buffer solution that is 0.10 M in NH_3 and 0.10 M in NH_4^+? What is the pH if 12 mL of 0.20 M hydrochloric acid are added to 125 mL of buffer?

18.49 What is the pH of a buffer solution that is 0.10 M chloroacetic acid and 0.15 M sodium chloroacetate? Solve using the Henderson-Hasselbalch equation ($K_a = 1.4 \times 10^{-3}$).

18.51 What is the pH of a buffer solution that is 0.15 M pyridine and 0.10 M pyridinium bromide? Solve using the Henderson-Hasselbalch equation.

*__18.53__ How many moles of sodium acetate must be added to 2.0 L of 0.10 M acetic acid to give a solution that has a pH equal to 5.00? Ignore the volume change due to the addition of sodium acetate.

18.48 A buffer is prepared by mixing 525 mL of 0.50 M formic acid, $HCHO_2$, and 475 mL of 0.50 M sodium formate, $NaCHO_2$. Calculate the pH. What would be the pH of 85 mL of the buffer to which 8.5 mL of 0.15 M hydrochloric acid have been added?

18.50 What is the pH of a buffer solution that is 0.20 M propionic acid and 0.10 M sodium propionate? Solve using the Henderson-Hasselbalch equation.

18.52 What is the pH of a buffer solution that is 0.20 M methylamine and 0.15 M methylammonium chloride? Solve using the Henderson-Hasselbalch equation.

*__18.54__ How many moles of hydrofluoric acid, HF, must be added to 500.0 mL of 0.30 M sodium fluoride to give a buffer of pH 3.50? Ignore the volume change due to the addition of hydrofluoric acid.

Titration Curves

18.55 What is the pH of a solution in which 20 mL of 0.10 M NaOH are added to 25 mL of 0.10 M HCl?

18.56 What is the pH of a solution in which 35 mL of 0.10 M NaOH are added to 25 mL of 0.10 M HCl?

18.57 A 1.24-g sample of benzoic acid was dissolved in water to give 50.0 mL of solution. This solution was titrated with 0.180 M NaOH. What was the pH of the solution when the equivalence point was reached?

18.59 Find the pH of the solution obtained when 32 mL of 0.087 M ethylamine are titrated to the equivalence point with 0.15 M HCl.

**18.61 Calculate the pH of a solution obtained by mixing 500.0 mL of 0.10 M NH$_3$ with 200.0 mL of 0.15 M HCl.

18.58 A 0.400-g sample of propionic acid was dissolved in water to give 50.0 mL of solution. This solution was titrated with 0.150 M NaOH. What was the pH of the solution when the equivalence point was reached?

18.60 What is the pH at the equivalence point when 22 mL of 0.20 M hydroxylamine are titrated with 0.15 M HCl?

**18.62 Calculate the pH of a solution obtained by mixing 35.0 mL of 0.15 M acetic acid with 25.0 mL of 0.10 M sodium acetate.

Additional Problems

18.63 Salicylic acid, $C_6H_4OHCOOH$, is used in the manufacture of acetylsalicylic acid (aspirin) and methyl salicylate (wintergreen flavor). A saturated solution of salicylic acid contains 2.2 g of the acid per liter of solution and has a pH of 2.43. What is the value of K_a?

18.64 Cyanoacetic acid, $CH_2CNCOOH$, is used in the manufacture of barbiturate drugs. An aqueous solution containing 5.0 g in a liter of solution has a pH of 1.89. What is the value of K_a?

18.65 A 0.050 M aqueous solution of sodium hydrogen sulfate, $NaHSO_4$, has a pH of 1.73. Calculate K_{a2} for sulfuric acid. Since sulfuric acid is a strong electrolyte, you can ignore hydrolysis of the HSO_4^- ion.

18.66 A 0.10 M aqueous solution of sodium dihydrogen phosphate, NaH_2PO_4, has a pH of 4.10. Calculate K_{a2} for phosphoric acid. You can ignore hydrolysis of the $H_2PO_4^-$ ion.

*18.67 Compare the acid ionization constant of HCO_3^- with its base ionization (hydrolysis) constant. Which is more extensive for HCO_3^-, acid ionization or hydrolysis? What is the pH of a 0.10 M solution of sodium hydrogen carbonate, $NaHCO_3$? Use the more extensive equilibrium only.

*18.68 Compare the acid ionization constant of HPO_4^{2-} with its base ionization (hydrolysis) constant. Which is more extensive for HPO_4^{2-}, acid ionization or hydrolysis? What is the pH of a 0.10 M solution of disodium hydrogen phosphate, Na_2HPO_4? Use the more extensive equilibrium only.

18.69 Calculate the base ionization constants for CN^- and CO_3^{2-}. Which ion is the stronger base?

18.70 Calculate the base ionization constant for PO_4^{3-} and SO_4^{2-}. Which ion is the stronger base?

*18.71 The pH of a white vinegar solution is 2.45. This vinegar is an aqueous solution of acetic acid with a density of 1.09 g/mL. What is the mass percent of acetic acid in the solution?

*18.72 The pH of a household cleaning solution is 11.87. This cleanser is an aqueous solution of ammonia with a density of 1.00 g/mL. What is the mass percent of ammonia in the solution?

18.73 Calculate the pH of a 0.15 M aqueous solution of aluminum chloride, $AlCl_3$. The acid ionization of hydrated aluminum ion is

$$Al(H_2O)_6^{3+}(aq) + H_2O(l) \rightleftharpoons$$
$$Al(H_2O)_5OH^{2+}(aq) + H_3O^+(aq)$$

and the K_a is 1.4×10^{-5}.

18.74 Calculate the pH of a 0.15 M aqueous solution of zinc chloride, $ZnCl_2$. The acid ionization of hydrated zinc ion is

$$Zn(H_2O)_6^{2+}(aq) + H_2O(l) \rightleftharpoons$$
$$Zn(H_2O)_5OH^+(aq) + H_3O^+(aq)$$

and the K_a is 2.5×10^{-10}.

18.75 An artificial fruit beverage contains 11.0 g of tartaric acid, $H_2C_4H_4O_6$, and 20.0 g of its salt, potassium hydrogen tartrate, per liter. What is the pH of the beverage? $K_{a1} = 1.0 \times 10^{-3}$.

18.76 A buffer is made by dissolving 13.0 g of sodium dihydrogen phosphate, NaH_2PO_4, and 15.0 g of disodium hydrogen phosphate, Na_2HPO_4, in a liter of solution. What is the pH of the buffer?

18.77 Blood contains several acid–base systems that tend to keep its pH constant at about 7.4. One of the most important buffer systems involves carbonic acid and hydrogen carbonate ion. What must be the ratio of $[HCO_3^-]/[H_2CO_3]$ in the blood if the pH is 7.40?

18.78 Codeine, $C_{18}H_{21}NO_3$, is an alkaloid ($K_b = 6.2 \times 10^{-9}$) used as a painkiller and cough suppressant. A solution of codeine is acidified with hydrochloric acid to pH 4.50. What is the ratio of concentrations of the cation $HCod^+$ to that of the free base Cod? (Here Cod is used as an abbreviated formula for codeine.)

18.79 Calculate the pH of a solution that is obtained by mixing 456 mL of 0.10 M hydrochloric acid with 285 mL of 0.15 M sodium hydroxide. Assume the combined volume is the sum of the two original volumes.

18.80 Calculate the pH of a solution that is made up from 2.0 g of potassium hydroxide dissolved in 115 mL of 0.19 M perchloric acid. Assume the change in volume due to adding potassium hydroxide is negligible.

18.81 Find the pH of the solution obtained when 25 mL of 0.065 M benzylamine, $C_7H_7NH_2$, are titrated to the equivalence point by 0.050 M hydrochloric acid. K_b for benzylamine is 4.7×10^{-10}.

18.82 What is the pH of the solution obtained by titrating 1.24 g of sodium hydrogen sulfate, $NaHSO_4$, dissolved in 50.0 mL of water with 0.180 M sodium hydroxide until the equivalence point is reached? Assume any volume change due to adding the sodium hydrogen sulfate or to the mixing of solutions is negligible.

**18.83 Consider a salt of a weak acid and a weak base, such as ammonium formate. In solution, the principal equilibrium involves the exchange of a proton between the acid and base:

$$NH_4^+(aq) + CHO_2^-(aq) \rightleftharpoons NH_3(aq) + HCHO_2(aq)$$

Show that the equilibrium constant, K_c, for this type of reaction is

$$K_c = \frac{K_w}{K_a K_b}$$

where K_a is the acid ionization constant for formic acid and K_b is the base ionization constant for ammonia. What is the ratio of concentrations $[NH_3]/[NH_4^+]$ in 0.10 M ammonium formate?

**18.84 When a salt of a weak acid and a weak base, such as ammonium formate, is dissolved in water, the principal equilibrium is

$$NH_4^+(aq) + CHO_2^-(aq) \rightleftharpoons NH_3(aq) + HCHO_2(aq)$$

Note that $[NH_3]/[NH_4^+] = [HCHO_2]/[CHO_2^-]$ in this solution. (Why?) In Problem 18.83 it was shown that the equilibrium constant for this reaction equals $K_w/K_a K_b$ (K_a is the acid ionization constant for formic acid and K_b is the base ionization constant for ammonia). Hence, show that

$$[H^+] = \sqrt{\frac{K_w K_a}{K_b}}$$

What is the pH of 0.10 M ammonium formate?

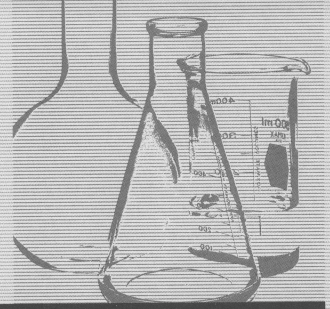

19. Solubility and Complex-Ion Equilibria

Solubility Equilibria

19.1 The Solubility Product Constant

19.2 Solubility and the Common-Ion Effect

19.3 Precipitation Calculations Criterion for Precipitation/ Completeness of Precipitation/ Fractional Precipitation

19.4 Effect of pH on Solubility Qualitative Effect of pH/ Separation of Metal Ions by Sulfide Precipitation

Complex-Ion Equilibria

19.5 Complex-Ion Formation Aside: Stepwise Formation Constants

19.6 Complex Ions and Solubility

An Application of Solubility Equilibria

19.7 Qualitative Analysis of Metal Ions

M any natural processes depend on the precipitation or the dissolving of a slightly soluble salt. For example, caves form in limestone (calcium carbonate) over thousands of years as ground water seeps through cracks, dissolving out cavities in the rock. Kidney stones form when salts such as calcium phosphate or calcium oxalate slowly precipitate in these organs. Poisoning by oxalic acid is also explained by the precipitation of the calcium salt. If oxalic acid is accidentally ingested, the oxalate-ion concentration in the blood may increase sufficiently to precipitate calcium oxalate. Calcium ion, which is needed for proper muscle control, is then removed from the blood, and muscle tissues go into spasm.

To understand such phenomena quantitatively, we must be able to solve problems in solubility equilibria. Calcium oxalate kidney stones form when the concentrations of calcium ion and oxalate ion are sufficiently great. What is the relationship between the concentrations of ions and the solubility of a salt? What is the minimum concentration of oxalate ion that gives a precipitate of the calcium salt from a 0.0025 M solution of Ca^{2+} (the approximate concentration of calcium ion in blood plasma)? What is the effect of pH on the solubility of this salt? We will look at questions such as these in this chapter.

Chapter Overview

Solubility and precipitation were introduced in Section 9.4. In this chapter, we will discuss the quantitative aspects of solubility and complex-ion equilibria.

In the first part of the chapter, we will see how solubility and the equilibrium constant for the solubility process are related. After describing simple problems in solubility and precipitation, we will look at how acidity and complex-ion formation affect solubility. In the concluding section, we will use a scheme for the qualitative analysis of metals to illustrate the principles introduced in this chapter.

Solubility Equilibria

To deal quantitatively with an equilibrium, we require the equilibrium constant. In the next section, we will look at the equilibria of slightly soluble, or nearly insoluble, ionic compounds and show how we can determine their equilibrium constants. Once we find values for various ionic compounds, we can use them to answer solubility or precipitation questions.

19.1 The Solubility Product Constant

When an ionic compound is dissolved in water, it usually goes into solution as the ions. If an excess of the ionic compound is mixed with water, an equilibrium occurs between the solid compound and the ions in the saturated solution. For the salt calcium oxalate, CaC_2O_4, we have the following equilibrium:

$$CaC_2O_4(s) \rightleftharpoons Ca^{2+}(aq) + C_2O_4^{2-}(aq)$$

The equilibrium constant for this solubility process can be written

$$K_c{}' = \frac{[Ca^{2+}][C_2O_4^{2-}]}{[CaC_2O_4]}$$

However, since the concentration of the solid remains constant, we normally combine its concentration with $K_c{}'$ to give the equilibrium constant K_{sp}, which is called the *solubility product constant* of CaC_2O_4:∎

$$K_{sp} = K_c{}'[CaC_2O_4] = [Ca^{2+}][C_2O_4^{2-}]$$

■ This is an example of a heterogeneous equilibrium. See Section 16.3.

In general, the **solubility product constant**, K_{sp}, is the equilibrium constant for a solubility equilibrium of a slightly soluble (or nearly insoluble) ionic compound. It equals the product of the equilibrium concentrations of the ions in the compound, each concentration raised to a power equal to the number of such ions in the formula of the compound. As is true of any equilibrium constant, K_{sp} depends on the temperature, but at a given temperature, it has a constant value for various concentrations of the ions.∎

■ An ionic equilibrium is affected to a small extent by the presence of ions not directly involved in the equilibrium. We will ignore this effect here.

Lead iodide, PbI_2, is another example of a slightly soluble salt. The equilibrium in water is

$$PbI_2(s) \rightleftharpoons Pb^{2+}(aq) + 2I^-(aq)$$

and the equilibrium constant or solubility product constant is

$$K_{sp} = [Pb^{2+}][I^-]^2$$

Example 19.1

Write the solubility product expressions for the following salts: (a) AgCl (b) Hg_2Cl_2 (c) $Pb_3(AsO_4)_2$

Solution

The equilibria and solubility product expressions are

(a) $AgCl(s) \rightleftharpoons Ag^+(aq) + Cl^-(aq)$

$$K_{sp} = [Ag^+][Cl^-]$$

(b) $Hg_2Cl_2(s) \rightleftharpoons Hg_2^{2+}(aq) + 2Cl^-(aq)$

$$K_{sp} = [Hg_2^{2+}][Cl^-]^2$$

Note that the mercury(I) ion is Hg_2^{2+}. (Mercury also has salts with the mercury(II) ion, Hg^{2+}.)

(c) $Pb_3(AsO_4)_2(s) \rightleftharpoons 3Pb^{2+}(aq) + 2AsO_4^{3-}(aq)$

$$K_{sp} = [Pb^{2+}]^3[AsO_4^{3-}]^2$$

Exercise 19.1

Give solubility product expressions for the following: (a) barium sulfate (b) iron(III) hydroxide (c) calcium phosphate

(See Problems 19.11 and 19.12.)

If we can determine the solubility of a slightly soluble ionic compound by experiment, we can calculate its K_{sp}. The next two examples show how to do this calculation.∎

■ Values of K_{sp} can be determined by methods of electrochemistry, discussed in Chapter 21.

Example 19.2

A liter of a solution saturated at 25°C with calcium oxalate, CaC_2O_4, is evaporated to dryness, giving a 0.0061-g residue of CaC_2O_4. Calculate the solubility product constant for this salt at 25°C.

Solution

The solubility of calcium oxalate is 0.0061 g/L *of solution*. We can also express this as a molar solubility, that is, as the number of moles of salt that dissolve per liter of solution. We convert grams per liter to moles per liter (the molecular weight of CaC_2O_4 is 128 amu):

Molar solubility of CaC_2O_4

$$= 0.0061 \text{ g } CaC_2O_4/L \times \frac{1 \text{ mol } CaC_2O_4}{128 \text{ g } CaC_2O_4}$$

$$= 4.8 \times 10^{-5} \text{ mol } CaC_2O_4/L$$

Let us look at the equilibrium problem.

Step 1 Suppose we mix solid CaC_2O_4 in a liter of solution. Of this solid, 4.8×10^{-5} mol will dissolve to form 4.8×10^{-5} mol of each ion. The results are summarized in the table below. (Since the concentration of the solid does not appear in K_{sp}, we do not include it in the following or in subsequent concentration tables.)

Step 2 We now substitute into the equilibrium equation:

$$K_{sp} = [Ca^{2+}][C_2O_4^{2-}] = (4.8 \times 10^{-5})(4.8 \times 10^{-5})$$
$$= 2.3 \times 10^{-9}$$

Concentration (M)	$CaC_2O_4(s) \rightleftharpoons$	$Ca^{2+}(aq)$	$+$	$C_2O_4^{2-}(aq)$
Starting		0		0
Change		$+4.8 \times 10^{-5}$		$+4.8 \times 10^{-5}$
Equilibrium		4.8×10^{-5}		4.8×10^{-5}

Exercise 19.2

Silver ion may be recovered from used photographic fixing solution by precipitating it as silver chloride. The solubility of silver chloride is 1.9×10^{-3} g/L. Calculate K_{sp}.

(See Problems 19.13 and 19.14.)

The next example is similar, except that the salt produces unequal numbers of cations and anions.

Example 19.3

By experiment, it is found that 1.2×10^{-3} mol of lead iodide, PbI_2, will dissolve in one liter of aqueous solution at 25°C. What is the solubility product constant at this temperature?

Solution

Step 1 Suppose that solid lead iodide is mixed into one liter of solution. We find that 1.2×10^{-3} mol dissolve to form 1.2×10^{-3} mol Pb^{2+} and $2 \times (1.2 \times 10^{-3})$ mol I^-, as the table below summarizes:

Step 2 We substitute into the equilibrium equation.

$$K_{sp} = [Pb^{2+}][I^-]^2 = (1.2 \times 10^{-3}) \times (2 \times 1.2 \times 10^{-3})^2$$
$$= 6.9 \times 10^{-9}$$

Concentration (M)	$PbI_2(s) \rightleftharpoons$	$Pb^{2+}(aq)$	$+$	$2I^-(aq)$
Starting		0		0
Change		$+1.2 \times 10^{-3}$		$+2 \times (1.2 \times 10^{-3})$
Equilibrium		1.2×10^{-3}		$2 \times (1.2 \times 10^{-3})$

Exercise 19.3

Lead arsenate, $Pb_3(AsO_4)_2$, has been used as an insecticide. It is only slightly soluble in water. If the solubility is 3.0×10^{-5} g/L, what is the solubility product constant? Assume that the solubility equilibrium is the only important one.

(See Problems 19.15 and 19.16.)

Table 19.1 gives a list of solubility product constants for various ionic compounds. If the solubility product constant is known, the solubility of the compound can be calculated. The problem is the reverse of that for finding K_{sp} from the solubility.

Example 19.4

The mineral fluorite is calcium fluoride, CaF_2. Calculate the solubility (in grams per liter) of calcium fluoride in water from the known solubility product constant (3.4×10^{-11}).

Solution

Step 1 Let x be the molar solubility of CaF_2. When solid CaF_2 is mixed into a liter of solution, x mol dissolve, forming x mol Ca^{2+} and $2x$ mol F^-.

Concentration (M)	$CaF_2(s) \rightleftharpoons Ca^{2+}(aq)$	$+ \ 2F^-(aq)$
Starting	0	0
Change	$+x$	$+2x$
Equilibrium	x	$2x$

Step 2 We substitute into the equilibrium equation:

$$[Ca^{2+}][F^-]^2 = K_{sp}$$

$$(x) \times (2x)^2 = 3.4 \times 10^{-11}$$
$$4x^3 = 3.4 \times 10^{-11}$$

Step 3 We now solve for x:

$$x = \sqrt[3]{\frac{3.4 \times 10^{-11}}{4}} = 2.0 \times 10^{-4}$$

Thus, the molar solubility is 2.0×10^{-4} mol CaF_2 per liter. To get the solubility in grams per liter, we convert, using the fact that the molar mass of CaF_2 is 78.1 g/mol:

$$Solubility = 2.0 \times 10^{-4} \ \text{mol } CaF_2/L \times \frac{78.1 \text{ g } CaF_2}{1 \text{ mol } CaF_2}$$

$$= 1.6 \times 10^{-2} \text{ g } CaF_2/L$$

Exercise 19.4

Anhydrite is a calcium sulfate mineral deposited when sea water evaporates. What is the solubility of calcium sulfate, in grams per liter? Table 19.1 gives the solubility product for calcium sulfate.

(See Problems 19.19, 19.20, 19.21, and 19.22.)

These examples illustrate the relationship between the solubility of an ionic compound in pure water and its solubility product constant. In the next sections, we will see how the solubility product constant can be used to calculate the solubility in the presence of other ions. It is also useful in deciding whether or not to expect precipitation under given conditions.

19.2 Solubility and the Common-Ion Effect

The importance of the solubility product constant becomes apparent when we consider the solubility of one salt in the solution of another having the same cation or anion. For example, suppose we wish to know the solubility

Table 19.1
Solubility Product
Constants, K_{sp}

Substance	Formula	K_{sp}
Aluminum hydroxide	$Al(OH)_3$	4.6×10^{-33}
Barium chromate	$BaCrO_4$	1.2×10^{-10}
Barium fluoride	BaF_2	1.0×10^{-6}
Barium sulfate	$BaSO_4$	1.1×10^{-10}
Cadmium oxalate	CdC_2O_4	1.5×10^{-8}
Cadmium sulfide	CdS	8×10^{-27}
Calcium carbonate	$CaCO_3$	3.8×10^{-9}
Calcium fluoride	CaF_2	3.4×10^{-11}
Calcium oxalate	CaC_2O_4	2.3×10^{-9}
Calcium phosphate	$Ca_3(PO_4)_2$	1×10^{-26}
Calcium sulfate	$CaSO_4$	2.4×10^{-5}
Cobalt sulfide	CoS	4×10^{-21}
Copper(II) hydroxide	$Cu(OH)_2$	2.6×10^{-19}
Copper(II) sulfide	CuS	6×10^{-36}
Iron(II) hydroxide	$Fe(OH)_2$	8×10^{-16}
Iron(II) sulfide	FeS	6×10^{-18}
Iron(III) hydroxide	$Fe(OH)_3$	2.5×10^{-39}
Lead arsenate	$Pb_3(AsO_4)_2$	4×10^{-36}
Lead chloride	$PbCl_2$	1.6×10^{-5}
Lead chromate	$PbCrO_4$	1.8×10^{-14}
Lead iodide	PbI_2	6.5×10^{-9}
Lead sulfate	$PbSO_4$	1.7×10^{-8}
Lead sulfide	PbS	2.5×10^{-27}
Magnesium arsenate	$Mg_3(AsO_4)_2$	2×10^{-20}
Magnesium hydroxide	$Mg(OH)_2$	1.8×10^{-11}
Magnesium oxalate	MgC_2O_4	8.5×10^{-5}
Manganese(II) sulfide	MnS	2.5×10^{-10}
Mercury(I) chloride	Hg_2Cl_2	1.3×10^{-18}
Mercury(II) sulfide	HgS	1.6×10^{-52}
Nickel hydroxide	$Ni(OH)_2$	2.0×10^{-15}
Nickel sulfide	NiS	3×10^{-19}
Silver acetate	$AgC_2H_3O_2$	2.0×10^{-3}
Silver bromide	$AgBr$	5.0×10^{-13}
Silver chloride	$AgCl$	1.8×10^{-10}
Silver chromate	Ag_2CrO_4	1.1×10^{-12}
Silver iodide	AgI	8.3×10^{-17}
Silver sulfide	Ag_2S	6×10^{-50}
Strontium carbonate	$SrCO_3$	9.3×10^{-10}
Strontium chromate	$SrCrO_4$	3.5×10^{-5}
Strontium sulfate	$SrSO_4$	2.5×10^{-7}
Zinc hydroxide	$Zn(OH)_2$	2.1×10^{-16}
Zinc sulfide	ZnS	1.1×10^{-21}

of calcium oxalate in a solution of calcium chloride. Each salt contributes the same cation (Ca^{2+}). The effect of the calcium ion provided by the calcium chloride is to make calcium oxalate less soluble than it would be in pure water.

We can explain this decrease in solubility in terms of LeChatelier's principle. Suppose we first mix crystals of calcium oxalate in a quantity of pure water to establish the equilibrium

$$CaC_2O_4(s) \rightleftharpoons Ca^{2+}(aq) + C_2O_4^{2-}(aq)$$

Now imagine that we add some calcium chloride. Since calcium chloride is a soluble salt, it dissolves to give an increase in calcium-ion concentration. We can regard this increase as a stress on the original equilibrium. According to LeChatelier's principle, the ions will react to remove some of the added calcium ion:

$$CaC_2O_4(s) \longleftarrow Ca^{2+}(aq) + C_2O_4^{2-}(aq)$$

In other words, some calcium oxalate precipitates from the solution. The solution now contains less calcium oxalate. We conclude that calcium oxalate is less soluble in a solution of calcium chloride than it is in pure water.

A solution of soluble calcium chloride has an ion *in common* with the slightly soluble calcium oxalate. The decrease in solubility of calcium oxalate in a solution of calcium chloride is an example of the *common-ion effect.*■ In general, any ionic equilibrium is affected by a substance producing an ion involved in the equilibrium, as we would predict from LeChatelier's principle. Figure 19.1 illustrates the principle of the common-ion effect for potassium perchlorate, $KClO_4$, which is only moderately soluble in water at 25°C. When the very soluble KCl is added to a saturated solution of $KClO_4$, the concentration of common ion, K^+, increases, and $KClO_4$ precipitates. The equilibrium is

$$KClO_4(s) \rightleftharpoons K^+(aq) + ClO_4^-(aq)$$

The next example shows how we can calculate the solubility of a slightly soluble salt in a solution containing a substance with a common ion.

■ The common-ion effect was first discussed in Section 18.5.

Figure 19.1
Demonstration of the common-ion effect. (*Left*) The beaker contains a saturated solution of potassium perchlorate. (*Right*) When an aqueous solution of potassium chloride is added to the beaker, a milky precipitate of potassium perchlorate forms.

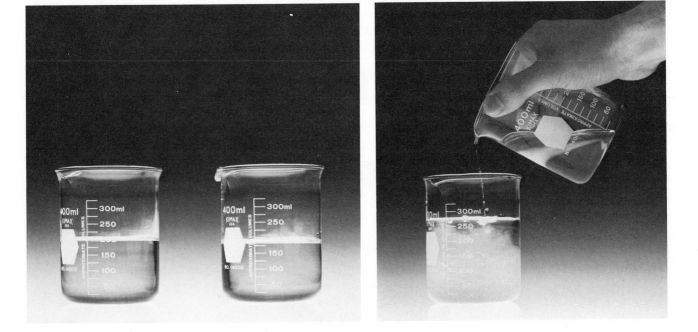

Example 19.5

What is the molar solubility of calcium oxalate in 0.15 M calcium chloride? Compare this molar solubility with that found earlier (see Example 19.2) for pure water ($4.8 \times 10^{-5}\ M$). The solubility product constant for calcium oxalate is 2.3×10^{-9}.

Solution

As we noted earlier, the two salts have a common ion (Ca^{2+}). One of the salts is soluble ($CaCl_2$) and provides Ca^{2+} ion that suppresses the solubility of the slightly soluble salt by the common-ion effect.

Step 1   Suppose solid CaC_2O_4 is mixed into one liter of 0.15 M $CaCl_2$ and that the molar solubility of the CaC_2O_4 is $x\ M$. Thus, at the start (before CaC_2O_4 dissolves) there is 0.15 mol Ca^{2+} in the solution. Of the solid CaC_2O_4, x mol dissolve to give x mol of additional Ca^{2+} and x mol $C_2O_4^{2-}$. The following table summarizes these results:

Concen-tration (M)	$CaC_2O_4(s) \rightleftharpoons Ca^{2+}(aq)\ +$	$C_2O_4^{2-}(aq)$
Starting	0.15	0
Change	$+x$	$+x$
Equilibrium	$0.15 + x$	x

Step 2   We substitute into the equilibrium equation:

$$[Ca^{2+}][C_2O_4^{2-}] = K_{sp}$$
$$(0.15 + x)x = 2.3 \times 10^{-9}$$

Step 3   Let us rearrange this equation in the form

$$x = \frac{2.3 \times 10^{-9}}{0.15 + x}$$

Since calcium oxalate is only slightly soluble, we might expect x to be negligible in comparison with 0.15. In that case,

$$0.15 + x \simeq 0.15$$

and the previous equation becomes

$$x \simeq \frac{2.3 \times 10^{-9}}{0.15} = 1.5 \times 10^{-8}$$

Note that x is indeed much smaller than 0.15, so our assumption was correct. Therefore, the molar solubility of calcium oxalate in 0.15 M $CaCl_2$ is $1.5 \times 10^{-8}\ M$. In pure water, the molar solubility is $4.8 \times 10^{-5}\ M$, which is over 3000 times greater.

Exercise 19.5

(a) Calculate the molar solubility of barium fluoride, BaF_2, in water at 25°C. The solubility product constant for BaF_2 at this temperature is 1.0×10^{-6}.
(b) What is the molar solubility of barium fluoride in 0.15 M NaF at 25°C? Compare the solubility in this case with that of BaF_2 in pure water.

(See Problems 19.23, 19.24, 19.25, and 19.26.)

In the preceding example, the slightly soluble salt did not contribute significantly to the common-ion concentration. When it does contribute, the method of successive approximations can be used. The next example illustrates this.∎

■ The method of successive approximations was introduced in Example 18.3 (p. 600) in discussing acid ionization equilibria.

Example 19.6

What is the molar solubility of lead chloride, $PbCl_2$, in 0.080 M sodium chloride solution? The solubility product constant for lead chloride is 1.6×10^{-5}.

Solution

The common ion in this case is Cl^-. We expect the slightly soluble salt lead chloride to be less soluble in 0.080 M NaCl than in pure water.

Step 1   We let the molar solubility of $PbCl_2$ be $x\ M$. Thus, x mol $PbCl_2$ dissolve to give x mol Pb^{2+} and $2x$ mol of additional Cl^-.

Concentration (M)	$PbCl_2(s) \rightleftharpoons Pb^{2+}(aq)\ +$	$2Cl^-(aq)$
Starting	0	0.080
Change	$+x$	$+2x$
Equilibrium	x	$0.080 + 2x$

(Continued)

Step 2 We substitute into the equilibrium equation:

$$[Pb^{2+}][Cl^-]^2 = K_{sp}$$
$$x(0.080 + 2x)^2 = 1.6 \times 10^{-5}$$

Step 3 The preceding equation rearranges to

$$x = \frac{1.6 \times 10^{-5}}{(0.080 + 2x)^2}$$

If we assume that 2x is small compared to 0.080 (that is, we assume that the concentration of Cl^- from $PbCl_2$ is small compared to that from NaCl), then the preceding equation becomes

$$x \simeq \frac{1.6 \times 10^{-5}}{(0.080)^2} = 0.0025$$

But note that 2x is *not* small compared to 0.080. To get an improved value of x, we substitute this approximate value into the right side of the equation given at the beginning of Step 3:

$$x \simeq \frac{1.6 \times 10^{-5}}{(0.080 + 2 \times 0.0025)^2} = 0.0022$$

If we substitute this new value of x into the right side of the same equation, we again get 0.0022:

$$x \simeq \frac{1.6 \times 10^{-5}}{(0.080 + 2 \times 0.0022)^2} = 0.0022$$

Therefore, x equals 0.0022 to two significant figures, and the molar solubility of $PbCl_2$ in 0.080 M NaCl is 0.0022 M.

Exercise 19.6

Calculate the solubility (in grams per liter) of silver acetate, $AgC_2H_3O_2$, in 0.10 M sodium acetate. The solubility product constant for silver acetate is 2.0×10^{-3}. You can solve the equilibrium equation by successive approximations or by using the quadratic formula.

(See Problems 19.27 and 19.28.)

19.3 Precipitation Calculations

In the chapter opening, we mentioned that calcium oxalate precipitates to form kidney stones. The same salt would precipitate in the body if oxalic acid (a poison) were to be accidentally ingested, since Ca^{2+} ion is present in the blood. To understand processes such as these, we must understand the conditions under which precipitation occurs. Precipitation is merely another way of looking at a solubility equilibrium. Rather than ask how much of a substance will dissolve in a solution, we ask, Will precipitation occur for given starting ion concentrations?

Criterion for Precipitation

The question just asked can be stated more generally: Given the concentrations of substances in a reaction mixture, will the reaction go in the forward or reverse direction? To answer this, we evaluate the *reaction quotient, Q,* and compare it with the equilibrium constant, K_c. ∎ The reaction quotient has the same form as the equilibrium-constant expression, but the concentrations of substances are not necessarily equilibrium values. Rather, they are concentrations at the start of a reaction. To predict the direction of reaction, we compare Q with K_c.

∎ The reaction quotient was discussed in Section 16.5.

If $Q < K_c$, the reaction should go in the forward direction.
If $Q = K_c$, the reaction mixture is at equilibrium.
If $Q > K_c$, the reaction should go in the reverse direction.

Let us apply this test to a precipitation reaction. Suppose lead nitrate, $Pb(NO_3)_2$, and sodium chloride, NaCl, are added to water to give a solution that is $0.050\ M\ Pb^{2+}$ and $0.10\ M\ Cl^-$. Will lead chloride, $PbCl_2$, precipitate? To answer this, we first write the solubility equilibrium:

$$PbCl_2(s) \rightleftharpoons Pb^{2+}(aq) + 2Cl^-(aq)$$

The reaction quotient has the form of the equilibrium-constant expression, which in this case is the K_{sp} expression, though the concentrations are starting values.

$$Q = [Pb^{2+}][Cl^-]^2$$

Here Q for a solubility reaction is often called the **ion product** (rather than reaction quotient), since it is a product of ion concentrations, each concentration raised to a power equal to the number of ions in the formula of the ionic compound.

To evaluate the ion product, Q, we substitute the concentrations of Pb^{2+} and Cl^- ions that are in the solution at the start of the reaction. We have $0.050\ M\ Pb^{2+}$ and $0.10\ M\ Cl^-$. Substituting,

$$Q = (0.050)(0.10)^2 = 5.0 \times 10^{-4}$$

Since K_{sp} for $PbCl_2$ is 1.6×10^{-5} (Table 19.1), we see that Q is greater than K_{sp}. Therefore, the reaction goes in the reverse direction. That is,

$$PbCl_2(s) \longleftarrow Pb^{2+}(aq) + 2Cl^-(aq)$$

In other words, Pb^{2+} and Cl^- react to precipitate $PbCl_2$. As precipitation occurs, the ion concentrations, and hence the ion product, decrease. Precipitation ceases when the ion product equals K_{sp}. The reaction mixture is then at equilibrium.

We can summarize our conclusions in terms of the following criterion for precipitation. Precipitation is expected to occur if the ion product for a solubility reaction is greater than K_{sp}.■ If the ion product is less than K_{sp}, precipitation will not occur (the solution is unsaturated with respect to the ionic compound). If the ion product equals K_{sp}, the reaction is at equilibrium (the solution is saturated with the ionic compound.) The next example is a simple application of this criterion for precipitation.

■ Precipitation may not occur even though the ion product has been exceeded. In such a case, the solution is supersaturated. Usually after a time a small crystal forms, and then precipitation occurs rapidly.

Example 19.7

The concentration of calcium ion in the blood plasma is 0.0025 M. If the concentration of oxalate ion is $1.0 \times 10^{-7}\ M$, do you expect calcium oxalate to precipitate? The K_{sp} for calcium oxalate is 2.3×10^{-9}.

Solution

The ion product for calcium oxalate is

Ion product $= [Ca^{2+}][C_2O_4^{2-}]$
$= (0.0025) \times (1.0 \times 10^{-7})$
$= 2.5 \times 10^{-10}$

Since this is smaller than the solubility product constant, we do not expect precipitation to occur.

Exercise 19.7

Anhydrite is a mineral whose chemical composition is $CaSO_4$ (calcium sulfate). An inland lake has Ca^{2+} and SO_4^{2-} concentrations of 0.0052 M and 0.0041 M, respectively. If these concentrations were to be doubled by evaporation, would you expect calcium sulfate to precipitate?

(See Problems 19.29 and 19.30.)

The following example is a typical industrial or laboratory problem in precipitation. We are given the volumes and concentrations of two solutions and asked whether a precipitate will form when the solutions are mixed.

Example 19.8

Sulfate ion, SO_4^{2-}, in solution is often determined quantitatively by precipitating it as barium sulfate, $BaSO_4$. The sulfate ion may have been formed from a sulfur compound. Analysis for the amount of sulfate ion then indicates the percentage of sulfur in the compound. Is a precipitate expected to form at equilibrium when 50.0 mL of 0.0010 M $BaCl_2$ are added to 50.0 mL of 0.00010 M Na_2SO_4? The solubility product constant for barium sulfate is 1.1×10^{-10}. Assume that the total volume of solution after mixing equals the sum of the volumes of the separate solutions.

Solution

Let us first calculate the concentrations of Ba^{2+} and SO_4^{2-}, assuming no precipitate of barium sulfate has formed. The molar amount of Ba^{2+} present in 50.0 mL ($= 0.0500$ L) of 0.0010 M $BaCl_2$ is

$$\text{Amount of } Ba^{2+} = \frac{0.0010 \text{ mol } Ba^{2+}}{1 \text{ L soln}} \times 0.050 \text{ L soln}$$

$$= 5.0 \times 10^{-5} \text{ mol } Ba^{2+}$$

The molar concentration of Ba^{2+} in the total solution equals the molar amount of Ba^{2+} divided by the total volume (0.0500 L $BaCl_2$ + 0.0500 L Na_2SO_4 = 0.1000 L).

$$[Ba^{2+}] = \frac{5.0 \times 10^{-5} \text{ mol}}{0.1000 \text{ L soln}} = 5.0 \times 10^{-4} \text{ } M$$

Similarly, we find

$$[SO_4^{2-}] = 5.0 \times 10^{-5} \text{ } M$$

(Try the calculations for SO_4^{2-}.) The ion product is

$$Q = [Ba^{2+}][SO_4^{2-}] = (5.0 \times 10^{-4}) \times (5.0 \times 10^{-5})$$

$$= 2.5 \times 10^{-8}$$

Since the ion product is greater than the solubility product constant (1.1×10^{-10}), we expect barium sulfate to precipitate.

Exercise 19.8

A solution of 0.00016 M lead nitrate, $Pb(NO_3)_2$, was poured into 456 mL of 0.00023 M sodium sulfate, Na_2SO_4. Would a precipitate of lead sulfate, $PbSO_4$, be expected to form if 255 mL of the lead nitrate solution were added?

(See Problems 19.31 and 19.32.)

Completeness of Precipitation

For various reasons, we may want to know how completely an ion precipitates from a solution. We may want to know if more than 99% of the Mg^{2+} ion in sea water can be removed by precipitating the ion as $Mg(OH)_2$ with a specified concentration of OH^-. Or, we might want to know if a precipitation is sufficiently complete for the quantitative determination of an ion.

As indicated in Example 19.8, the amount of SO_4^{2-} ion in solution can be determined by precipitating it as $BaSO_4$. The precipitate is filtered from the solution, dried, then weighed. If this determination of SO_4^{2-} ion is to be useful, most of the ion must be precipitated and very little (less than 0.1%) should remain in solution. The next example shows how to calculate the concentration of an ion remaining after precipitation and the percentage of the ion that was not precipitated.

Example 19.9

In Example 19.8, solutions of $BaCl_2$ and Na_2SO_4 were mixed and a precipitate of $BaSO_4$ formed. Before precipitation, we have a solution containing 0.00050 M Ba^{2+} and 0.000050 M SO_4^{2-}. Calculate the molar concentration of SO_4^{2-} ion left after $BaSO_4$ precipitates. What is the percentage of SO_4^{2-} not precipitated?

Solution

Before we try to set up the calculation, let us think about the reaction (a good thing to do in any case). Because the solubility product constant for $BaSO_4$ is very small (1.1×10^{-10}, according to Table 19.1), the Ba^{2+} and SO_4^{2-} ions will react until almost all of the ion that is the limiting reactant is consumed. Since the reaction is $Ba^{2+} + SO_4^{2-} \longrightarrow BaSO_4$, equal molar amounts of the ions react. The solution initially contains 0.00050 mol Ba^{2+} and 0.000050 mol SO_4^{2-} per liter. Therefore, the SO_4^{2-} ion is the limiting reactant. If the reaction were to go to completion, we would get a precipitate of $BaSO_4$ in a solution of unconsumed Ba^{2+} ion. However, we would expect $BaSO_4$ to dissolve to a slight extent in the solution of the common ion, Ba^{2+}, and this would produce some SO_4^{2-} ion.

This suggests that we solve the problem in two parts. First, we assume the reaction goes to completion. We do a *stoichiometry calculation* to find the concentration of Ba^{2+}. Then, to find the SO_4^{2-} concentration we do an *equilibrium calculation* of the solubility of $BaSO_4$ in a solution of the common ion Ba^{2+} (as in Example 19.5).

Stoichiometry Calculation The precipitation equation, with initial and final amounts of substance, assuming the reaction were to go to completion, is shown in the table below.

Equilibrium Calculation The concentration table for the solubility equation is

Concentration (M)	$BaSO_4(s) \rightleftharpoons Ba^{2+}(aq) + SO_4^{2-}(aq)$	
Starting	0.00045	0
Change	$+x$	$+x$
Equilibrium	$0.00045 + x$	x

Since K_{sp} for $BaSO_4$ is 1.1×10^{-10}, the equilibrium equation is

$$[Ba^{2+}][SO_4^{2-}] = K_{sp}$$
$$(0.00045 + x)x = 1.1 \times 10^{-10}$$

If we assume that x can be neglected in comparison to 0.00045, this equation is easily solved for x:

$$x \simeq \frac{1.1 \times 10^{-10}}{0.00045} = 2.4 \times 10^{-7}$$

Note that x can be neglected compared to 0.00045. Thus, the SO_4^{2-} concentration is 2.4×10^{-7} M. Since the concentration of the ion before precipitation of $BaSO_4$ was 5.0×10^{-5} M, the percentage of SO_4^{2-} remaining in solution is

$$\frac{2.4 \times 10^{-7}}{5.0 \times 10^{-5}} \times 100\% = 0.48\%$$

This is somewhat higher than the 0.1% required for a precise analytical procedure. However, we could reduce the concentration of SO_4^{2-} remaining by increasing the concentration of Ba^{2+}.

Moles of Substance	$Ba^{2+}(aq)$	$+ SO_4^{2-}(aq) \rightarrow$	$BaSO_4(s)$
Initial moles	0.00050	0.000050	0
Final moles	$(0.00050 - 0.000050)$ $= 0.00045$	0	0.000050

Exercise 19.9

Lead chromate, $PbCrO_4$, is a yellow pigment used in paints. Suppose 0.50 L of $1.0 \times 10^{-5}\ M\ Pb(C_2H_3O_2)_2$ and 0.50 L of $1.0 \times 10^{-3}\ M\ K_2CrO_4$ are mixed. Calculate the equilibrium concentration of Pb^{2+} ion remaining in solution after lead chromate precipitates. What is the percentage of Pb^{2+} remaining in solution after precipitation of $PbCrO_4$?

(See Problems 19.35 and 19.36.)

Fractional Precipitation

Fractional precipitation is the technique of separating two or more ions from a solution by adding a reactant that first precipitates one ion, then another, and so forth. For example, suppose a solution is $0.10\ M\ Ba^{2+}$ and $0.10\ M\ Sr^{2+}$. As we will show in the following paragraphs, when we slowly add a concentrated solution of potassium chromate, K_2CrO_4, to the solution of Ba^{2+} and Sr^{2+} ions, barium chromate precipitates first. After most of the Ba^{2+} ion has precipitated, strontium chromate begins to come out of solution. It is therefore possible to separate Ba^{2+} and Sr^{2+} ions from a solution by fractional precipitations using K_2CrO_4.

To understand why Ba^{2+} and Sr^{2+} ions can be separated this way, let us calculate the concentration of CrO_4^{2-} necessary to just begin the precipitation of $BaCrO_4$, and that necessary to just begin the precipitation of $SrCrO_4$. We will ignore any volume change in the solution of Ba^{2+} and Sr^{2+} ions resulting from the addition of the concentrated K_2CrO_4 solution. To calculate the CrO_4^{2-} concentration when $BaCrO_4$ begins to precipitate, we substitute the initial Ba^{2+} concentration into the solubility-product equation. The K_{sp} for $BaCrO_4$ is 1.2×10^{-10}.

$$[Ba^{2+}][CrO_4^{2-}] = K_{sp}\ (\text{for } BaCrO_4)$$
$$(0.10)[CrO_4^{2-}] = 1.2 \times 10^{-10}$$
$$[CrO_4^{2-}] = \frac{1.2 \times 10^{-10}}{0.10} = 1.2 \times 10^{-9}$$

In the same way, we can calculate the CrO_4^{2-} concentration when $SrCrO_4$ begins to precipitate. The K_{sp} for $SrCrO_4$ is 3.5×10^{-5}.

$$[Sr^{2+}][CrO_4^{2-}] = K_{sp}\ (\text{for } SrCrO_4)$$
$$(0.10)[CrO_4^{2-}] = 3.5 \times 10^{-5}$$
$$[CrO_4^{2-}] = \frac{3.5 \times 10^{-5}}{0.10} = 3.5 \times 10^{-4}$$

Note that $BaCrO_4$ precipitates first because the CrO_4^{2-} concentration necessary to form the $BaCrO_4$ precipitate is smaller.

From these results, we see that as the solution of K_2CrO_4 is slowly added to the solution of Ba^{2+} and Sr^{2+}, barium chromate begins to precipitate when the CrO_4^{2-} concentration reaches $1.2 \times 10^{-9}\ M$. Barium chromate continues to precipitate as K_2CrO_4 is added. When the concentration of CrO_4^{2-} reaches $3.5 \times 10^{-4}\ M$, strontium chromate begins to precipitate.

What is the percentage of Ba^{2+} ion remaining just as $SrCrO_4$ begins to precipitate? Let us first calculate the concentration of Ba^{2+} at this point. We write the solubility-product equation and substitute $[CrO_4^{2-}] = 3.5 \times 10^{-4}$, which is the concentration of chromate ion when $SrCrO_4$ begins to precipitate:

$$[Ba^{2+}][CrO_4^{2-}] = K_{sp} \text{ (for } BaCrO_4)$$
$$[Ba^{2+}](3.5 \times 10^{-4}) = 1.2 \times 10^{-10}$$
$$[Ba^{2+}] = \frac{1.2 \times 10^{-10}}{3.5 \times 10^{-4}} = 3.4 \times 10^{-7}$$

To calculate the percentage of Ba^{2+} ion remaining, we divide this concentration of Ba^{2+} by the initial concentration $(0.10\ M)$, and multiply by 100%:

$$\frac{3.4 \times 10^{-7}}{0.10} \times 100\% = 0.00034\%$$

We see that the percentage of Ba^{2+} ion remaining in solution is quite low, so that most of the Ba^{2+} ion has precipitated by the time $SrCrO_4$ begins to precipitate. We conclude that Ba^{2+} and Sr^{2+} can indeed be separated by fractional precipitation.

19.4 Effect of pH on Solubility

In discussing solubility in the previous sections, we assumed that the only equilibrium of interest is the one between the solid ionic compound and its ions in solution. Sometimes, however, it is necessary to account for other reactions the ions might undergo. For example, if the anion is the conjugate base of a weak acid, it will react with H^+ ion. We should expect the solubility to be affected by pH. We will look into this possibility in this section.

Qualitative Effect of pH

Consider the equilibrium between solid calcium oxalate, CaC_2O_4, and its ions in aqueous solution:

$$CaC_2O_4(s) \rightleftharpoons Ca^{2+}(aq) + C_2O_4^{2-}(aq)$$

Since the oxalate ion is conjugate to a weak acid (hydrogen oxalate ion, $HC_2O_4^-$), we would expect it to react with any H^+ ion that is added, say from a strong acid:

$$C_2O_4^{2-}(aq) + H^+(aq) \rightleftharpoons HC_2O_4^-(aq)$$

According to LeChatelier's principle, as $C_2O_4^{2-}$ ion is removed by reaction with H^+ ion, more calcium oxalate dissolves to replenish some of the $C_2O_4^{2-}$ ion:

$$CaC_2O_4(s) \longrightarrow Ca^{2+}(aq) + C_2O_4^{2-}(aq)$$

Therefore, we expect calcium oxalate to be more soluble in acidic solution (low pH) than it is in pure water.

Example 19.10

Consider two slightly soluble salts, calcium carbonate and calcium sulfate. Which of these would have its solubility more affected by the addition of strong acid? Would the solubility of the salt increase or decrease?

Solution

Calcium carbonate gives the solubility equilibrium

$$CaCO_3(s) \rightleftharpoons Ca^{2+}(aq) + CO_3^{2-}(aq)$$

If a strong acid is added, the hydrogen ion reacts with carbonate ion, since it is conjugate to a weak acid (HCO_3^-):

$$H^+(aq) + CO_3^{2-}(aq) \rightleftharpoons HCO_3^-(aq)$$

As carbonate ion is removed, calcium carbonate dissolves. Moreover, the hydrogen carbonate ion itself is removed in further reaction:

$$H^+(aq) + HCO_3^-(aq) \longrightarrow H_2CO_3(aq)$$
$$\longrightarrow H_2O(l) + CO_2(g)$$

Bubbles of carbon dioxide gas appear as more calcium carbonate dissolves.

In the case of calcium sulfate, the corresponding equilibria are

$$CaSO_4(s) \rightleftharpoons Ca^{2+}(aq) + SO_4^{2-}(aq)$$
$$H^+(aq) + SO_4^{2-}(aq) \rightleftharpoons HSO_4^-(aq)$$

Again, we see that the anion of the insoluble salt is removed by reaction with hydrogen ion. We would expect calcium sulfate to become more soluble in strong acid (except in sulfuric acid, since that would supply sulfate ion and therefore precipitate calcium sulfate by the common-ion effect). However, HSO_4^- is a much stronger acid than HCO_3^-, as we can see by comparing acid ionization constants. (The values K_{a2} for H_2CO_3 and K_a for HSO_4^- in Table 18.1 are 4.8×10^{-11} and 1.1×10^{-2}, respectively.) Thus, calcium carbonate becomes much more soluble in acidic solution, whereas the solubility of calcium sulfate is less affected.

Exercise 19.10

Which of the following salts would have its solubility more affected by changes in pH: silver chloride or silver cyanide?

(See Problems 19.39 and 19.40.)

Separation of Metal Ions by Sulfide Precipitation

Many metal sulfides are insoluble in water but dissolve in acidic solution. The change in solubility with pH, or hydrogen-ion concentration, can be used to separate a mixture of metal ions. ■ As an example, consider an aqueous solution containing 0.10 M zinc ion, Zn^{2+}, and 0.10 M lead ion, Pb^{2+}. We would like to separate these ions from one another. Let us first look at the possibility of precipitating the metal sulfides by dissolving H_2S gas in the aqueous solution.

When H_2S dissolves in water, it ionizes as a diprotic acid.

$$H_2S(aq) \rightleftharpoons H^+(aq) + HS^-(aq)$$
$$HS^-(aq) \rightleftharpoons H^+(aq) + S^{2-}(aq)$$

Sulfide ion, S^{2-}, produced in this way may react with metal ions to precipitate the metal sulfides. For example,

$$Zn^{2+}(aq) + S^{2-}(aq) \rightleftharpoons ZnS(s)$$

■ A separation scheme based on this variation of solubility is discussed in Section 19.7.

To predict whether we will get a precipitate of ZnS, we calculate the ion product of ZnS and compare it with its K_{sp}. We can similarly predict whether PbS will precipitate.

We know the metal-ion concentrations, but to calculate the ion products for ZnS and PbS, we also need the S^{2-} ion concentration. Recall that the concentration of the anion from the second ionization of a diprotic acid equals K_{a2}.■ Thus, the second ionization of H_2S gives the anion S^{2-}, whose concentration should equal the K_{a2} for H_2S. From Table 18.1, we see that K_{a2} for H_2S is 1.2×10^{-13}. Hence, $[S^{2-}] = 1.2 \times 10^{-13}$.

■ This result was obtained in Section 18.2.

The ion product of ZnS is

$$[Zn^{2+}][S^{2-}] = (0.10)(1.2 \times 10^{-13}) = 1.2 \times 10^{-14}$$

Since this is much larger than K_{sp} for ZnS (1.1×10^{-21}), we conclude that zinc sulfide should precipitate. The ion product of PbS also equals 1.2×10^{-14}. This is larger than the K_{sp} for PbS (2.5×10^{-27}), and therefore lead sulfide is also expected to precipitate. Because ZnS and PbS precipitate together, we have not obtained a separation of the metal ions.

Now consider the effect of adding a strong acid to the solution of the Zn^{2+} and Pb^{2+} ions before saturating it with H_2S gas. The H^+ ion from the strong acid will repress the ionization of H_2S, giving a lower concentration of S^{2-} ion. By adjusting the H^+ ion concentration, we can control the S^{2-} concentration and therefore the ion products for ZnS and PbS. As we will show in the next example, we can adjust the H^+ ion concentration so that only PbS precipitates when the solution is saturated with H_2S. The precipitate of PbS can be filtered from the solution of Zn^{2+} ion. In this way, we can separate a mixture of the metal ions.

In order to obtain the H^+ ion concentration necessary to keep some ions from precipitating while others form the sulfides, let us look at the net, or overall, equation for the acid ionizations of H_2S. We obtain this net equation by taking the sum of the acid ionizations:

$$H_2S(aq) \rightleftharpoons H^+(aq) + HS^-(aq)$$
$$\underline{HS^-(aq) \rightleftharpoons H^+(aq) + S^{2-}(aq)}$$
$$H_2S(aq) \rightleftharpoons 2H^+(aq) + S^{2-}(aq)$$

The equilibrium constant for an equation obtained by taking the sum of two or more equations is obtained by multiplying the equilibrium constants of the equations that were summed. Thus, the equilibrium constant for the equation obtained by taking the sum of the acid ionizations of H_2S is $K_{a1}K_{a2}$, where K_{a1} and K_{a2} are the equilibrium constants for the successive ionizations of H_2S. We can verify this by substituting the expressions for K_{a1} and K_{a2} into $K_{a1}K_{a2}$:

$$K_{a1}K_{a2} = \frac{[H^+][HS^-]}{[H_2S]} \times \frac{[H^+][S^{2-}]}{[HS^-]} = \frac{[H^+]^2[S^{2-}]}{[H_2S]}$$

From Table 18.1, we see that K_{a1} equals 8.9×10^{-8} and K_{a2} equals 1.2×10^{-13}. Hence, $K_{a1}K_{a2}$ equals $(8.9 \times 10^{-8}) \times (1.2 \times 10^{-13})$, or 1.1×10^{-20}.

A saturated solution of hydrogen sulfide is 0.10 *M* H_2S, and this is not significantly altered by changes in H^+ ion concentration. Substituting into the equilibrium equation

$$\frac{[H^+]^2[S^{2-}]}{[H_2S]} = K_{a1}K_{a2}$$

gives

$$\frac{[H^+]^2[S^{2-}]}{0.10} = 1.1 \times 10^{-20}$$

Note that as $[H^+]$ increases, $[S^{2-}]$ decreases. Thus, by adjusting the H^+ ion concentration, we can obtain any desired S^{2-} concentration.

Example 19.11

Consider a solution of 0.10 M Zn^{2+} and 0.10 M Pb^{2+} saturated with H_2S. What range of H^+ ion concentration will give a precipitate of one of the metal sulfides, leaving the other metal ion in solution? The concentration of H_2S in a saturated solution is 0.10 M.

Solution

We first calculate the minimum S^{2-} ion concentration that must be present before each metal sulfide can precipitate. Once we have the S^{2-} ion concentrations necessary to precipitate ZnS and PbS, we can calculate the corresponding H^+ ion concentrations.

Step 1 We find the minimum S^{2-} ion concentration necessary to give a precipitate of ZnS by substituting the Zn^{2+} ion concentration into the equilibrium equation

$$[Zn^{2+}][S^{2-}] = K_{sp}$$

to give

$$(0.10)[S^{2-}] = 1.1 \times 10^{-21}$$

Hence, the minimum S^{2-} ion concentration needed to give a precipitate of ZnS is

$$[S^{2-}] = \frac{1.1 \times 10^{-21}}{0.10} = 1.1 \times 10^{-20}$$

Similarly, the minimum S^{2-} ion concentration needed to precipitate PbS is

$$[S^{2-}] = \frac{K_{sp} \text{ (for PbS)}}{Pb^{2+}} = \frac{2.5 \times 10^{-27}}{0.10} = 2.5 \times 10^{-26}$$

Step 2 If we substitute these minimum S^{2-} ion concentrations into the equilibrium equation for the double ionization of H_2S,

$$\frac{[H^+]^2[S^{2-}]}{[H_2S]} = 1.1 \times 10^{-20}$$

we can obtain the corresponding H^+ ion concentrations. For ZnS, we get

$$[H^+]^2 = 1.1 \times 10^{-20} \times \frac{[H_2S]}{[S^{2-}]}$$

$$= 1.1 \times 10^{-20} \times \frac{0.10}{1.1 \times 10^{-20}} = 0.10$$

Or,

$$[H^+] = 0.32$$

At any concentration of H^+ ion equal to or less than 0.32 M, the S^{2-} ion concentration will be equal to or greater than that necessary to give a precipitate of ZnS. If the H^+ ion concentration is greater than 0.32 M, the Zn^{2+} ion will remain in solution.

For PbS, we get

$$[H^+]^2 = 1.1 \times 10^{-20} \times \frac{[H_2S]}{[S^{2-}]}$$

$$= 1.1 \times 10^{-20} \times \frac{0.10}{2.5 \times 10^{-26}} = 4.4 \times 10^4$$

$$[H^+] = 2.1 \times 10^2$$

At any H^+ ion concentration equal to or less than 2.1×10^2 M, PbS will precipitate. (It is not possible to get an H^+ ion concentration this large, so we can expect PbS to precipitate in any solution.)

When saturated with H_2S, any solution of 0.10 M Zn^{2+} and 0.10 M Pb^{2+} in which the H^+ ion concentration is greater than 0.32 M will give a precipitate of PbS but not of ZnS.

Exercise 19.11

Find the range of pH that will allow only one of the metal ions in a solution that is 0.050 M Cu^{2+}, 0.050 M Fe^{2+}, and saturated with H_2S to precipitate as a sulfide.
(See Problems 19.41 and 19.42.)

Complex-Ion Equilibria

Many metal ions, especially those of the transition elements, form coordinate covalent bonds with molecules or anions having lone pairs of electrons. ■ For example, the silver ion, Ag^+, can react with NH_3 to form the $Ag(NH_3)_2^+$ ion. The lone pair of electrons of the N atom of NH_3 forms a coordinate covalent bond to the silver ion:

■ The nature of this bonding is discussed in Chapter 25.

$$Ag^+ + 2(:NH_3) \longrightarrow (H_3N:Ag:NH_3)^+$$

Note that the molecule or anion having the lone pair of electrons acts as a Lewis base.

Complex ions are ions formed from a metal ion with Lewis bases attached to it by coordinate covalent bonds. A *complex* is a compound containing complex ions. **Ligands** are the Lewis bases that bond to a metal ion to form a complex ion. Thus, $Ag(NH_3)_2^+$ is a complex ion formed from the Ag^+ ion and NH_3 molecules. The NH_3 molecules are the ligands.

19.5 Complex-Ion Formation

The aqueous silver ion forms a complex ion with ammonia by reacting with NH_3 in steps:

$$Ag^+(aq) + NH_3(aq) \rightleftharpoons Ag(NH_3)^+(aq)$$
$$Ag(NH_3)^+(aq) + NH_3(aq) \rightleftharpoons Ag(NH_3)_2^+(aq)$$

If we add these equations, we get the overall equation for the formation of the complex ion $Ag(NH_3)_2^+$:

$$Ag^+(aq) + 2NH_3(aq) \rightleftharpoons Ag(NH_3)_2^+(aq)$$

The **formation constant,** or **stability constant,** K_f, of a complex ion is the equilibrium constant for the formation of the complex ion from the aqueous metal ion and the ligands. Thus, the formation constant of $Ag(NH_3)_2^+$ is

$$K_f = \frac{[Ag(NH_3)_2^+]}{[Ag^+][NH_3]^2}$$

The value of K_f for $Ag(NH_3)_2^+$ is 1.7×10^7. Its large value means that the complex ion is quite stable. If a large amount of NH_3 is added to a solution of Ag^+, we expect most of the Ag^+ ion to react to form the complex ion. Table 19.2 lists values for the formation constants of some complex ions.

The **dissociation constant,** K_d, for a complex ion is the reciprocal, or inverse, value of K_f. The equation for the dissociation of $Ag(NH_3)_2^+$ is

$$Ag(NH_3)_2^+(aq) \rightleftharpoons Ag^+(aq) + 2NH_3(aq)$$

and its equilibrium constant is

$$K_d = \frac{1}{K_f} = \frac{[Ag^+][NH_3]^2}{[Ag(NH_3)_2^+]}$$

The next example shows how to calculate the concentration of aqueous metal ion in equilibrium with a complex ion.

Table 19.2
Formation Constants of
Complex Ions at 25°C

Complex Ion	K_f
$Ag(CN)_2^-$	5.6×10^{18}
$Ag(NH_3)_2^+$	1.7×10^7
$Ag(S_2O_3)_2^{3-}$	2.9×10^{13}
$Cd(NH_3)_4^{2+}$	1.0×10^7
$Cu(CN)_2^-$	1.0×10^{16}
$Cu(NH_3)_4^{2+}$	4.8×10^{12}
$Fe(CN)_6^{4-}$	1.0×10^{35}
$Fe(CN)_6^{3-}$	9.1×10^{41}
$Ni(CN)_4^{2-}$	1.0×10^{31}
$Ni(NH_3)_6^{2+}$	5.6×10^8
$Zn(NH_3)_4^{2+}$	2.9×10^9
$Zn(OH)_4^{2-}$	2.8×10^{15}

Example 19.12

What is the concentration of $Ag^+(aq)$ ion in 0.010 M $AgNO_3$ that is also 1.0 M NH_3? The K_f for $Ag(NH_3)_2^+$ ion is 1.7×10^7.

Solution

Since the formation constant for $Ag(NH_3)_2^+$ is large, the silver exists primarily as $Ag(NH_3)_2^+$. This suggests that we first assume that $Ag^+(aq)$ reacts completely to form $Ag(NH_3)_2^+$, then that this complex ion subsequently dissociates to give a small amount of $Ag^+(aq)$.

In one liter of solution, we initially have 0.010 mol $Ag^+(aq)$ from $AgNO_3$. This reacts to give 0.010 mol $Ag(NH_3)_2^+$, leaving (1.0 − 0.010) mol NH_3, which equals 1.0 mol NH_3 to two significant figures. We now look at the equilibrium for the dissociation of $Ag(NH_3)_2^+$.

Step 1 One liter of solution contains 0.010 mol $Ag(NH_3)_2^+$ and 1.0 mol NH_3. The complex ion dissociates slightly, so that one liter of solution contains x mol Ag^+. This is summarized in the table below.

Step 2 The formation constant is

$$K_f = \frac{[Ag(NH_3)_2^+]}{[Ag^+][NH_3]^2}$$

From this we can write the dissociation constant, which corresponds to the reaction as written in the table:

$$\frac{[Ag^+][NH_3]^2}{[Ag(NH_3)_2^+]} = K_d = \frac{1}{K_f}$$

Substituting into this equation gives

$$\frac{x(1.0 + 2x)^2}{(0.010 - x)} = \frac{1}{1.7 \times 10^7}$$

Step 3 The right-hand side of this equation equals 5.9×10^{-8}. If we assume x to be small in comparison with 0.010, we get

$$\frac{x}{0.010} \simeq 5.9 \times 10^{-8}$$

and

$$x \simeq 5.9 \times 10^{-8} \times 0.010 = 5.9 \times 10^{-10}$$

The silver-ion concentration is thus 5.9×10^{-10} M, over ten million times smaller than its value in 0.010 M $AgNO_3$ that does not contain ammonia.

Concentration (M)	$Ag(NH_3)_2^+(aq) \rightleftharpoons Ag^+(aq) + 2NH_3(aq)$		
Starting	0.010	0	1.0
Change	$-x$	$+x$	$+2x$
Equilibrium	$0.010 - x$	x	$1.0 + 2x$

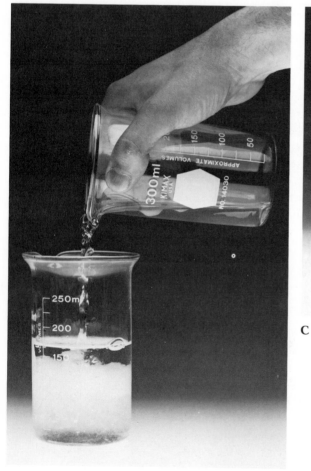

Figure 19.2
Demonstration of the amphoteric behavior of zinc hydroxide. (a) The beaker on the left contains $ZnCl_2$; the one on the right contains NaOH. (b) As some of the solution of NaOH is added to the solution of $ZnCl_2$, a white precipitate of $Zn(OH)_2$ forms. (c) After more NaOH is added, the precipitate dissolves.

A

B

C

Exercise 19.12

What is the concentration of $Cu^{2+}(aq)$ in a solution that was originally 0.015 M $Cu(NO_3)_2$ and 0.100 M NH_3? The Cu^{2+} ion forms the complex ion $Cu(NH_3)_4^{2+}$. Its formation constant is given in Table 19.2.

(See Problems 19.43 and 19.44.)

Amphoteric hydroxides are metal hydroxides that react with bases as well as with acids. Zinc is an example of a metal that forms such a hydroxide. Zinc hydroxide, $Zn(OH)_2$, is an insoluble hydroxide, as are most of the metal hydroxides, except those of Groups IA and IIA elements. It reacts with a strong acid, as one would expect of a base, and the metal hydroxide dissolves:

$$Zn(OH)_2(s) + 2H^+(aq) \longrightarrow Zn^{2+}(aq) + 2H_2O(l)$$

With a base, however, $Zn(OH)_2$ reacts to form the complex ion, $Zn(OH)_4^{2-}$:

$$Zn(OH)_2(s) + 2OH^-(aq) \longrightarrow Zn(OH)_4^{2-}(aq)$$

Thus, the hydroxide also dissolves in strong base.

If a strong base is slowly added to a solution of $ZnCl_2$, a white precipitate of $Zn(OH)_2$ first forms.

$$Zn^{2+}(aq) + 2OH^-(aq) \longrightarrow Zn(OH)_2(s)$$

But as more base is added, the white precipitate dissolves, forming the complex ion $Zn(OH)_4^{2-}$ (see Figure 19.2).

Other common amphoteric hydroxides are those of aluminum, chromium(III), lead(II), tin(II) and tin(IV). The amphoterism of $Al(OH)_3$ is commercially used to separate aluminum oxide from the aluminum ore bauxite. Bauxite contains hydrated Al_2O_3 plus impurities, such as silica sand (SiO_2) and iron oxide (Fe_2O_3). The metal oxides dissolve in hydrochloric acid to give a solution of the ions $Al^{3+}(aq)$ and $Fe^{3+}(aq)$. After filtering off the insoluble impurities, such as SiO_2, sodium hydroxide is added, which precipitates iron(III) hydroxide. Aluminum hydroxide, which is formed from Al_2O_3 with dilute base, reacts with excess OH^- to give the complex ion $Al(OH)_4^-$:

$$Al(OH)_2(s) + 2OH^-(aq) \longrightarrow Al(OH)_4^{2-}$$

The solution of $Al(OH)_4^{2-}$ is acidified to precipitate pure $Al(OH)_3$.

Aside: Stepwise Formation Constants

A complex ion is formed in steps. In ammonia, $Ag^+(aq)$ ion first forms $Ag(NH_3)^+(aq)$, then $Ag(NH_3)_2^+(aq)$ ion. The equilibria and the corresponding equilibrium constants are

$$Ag^+(aq) + NH_3(aq) \rightleftharpoons Ag(NH_3)^+(aq); \; K_{f1} = 2.3 \times 10^3$$

$$Ag(NH_3)^+(aq) + NH_3(aq) \rightleftharpoons Ag(NH_3)_2^+(aq); \; K_{f2} = 7.2 \times 10^3$$

The equilibrium constants K_{f1} and K_{f2} are called *stepwise formation constants.*

The overall equation obtained by taking the sum of the stepwise formation reactions is

(Continued)

$$Ag^+(aq) + 2NH_3(aq) \rightleftharpoons Ag(NH_3)_2^+(aq)$$

The equilibrium constant for this equation is K_f, the *overall formation constant* or, simply, the *formation constant*. Since the overall equation is obtained by taking the sum of the steps in the formation of the final complex ion, the equilibrium constant K_f equals the product of the stepwise formation constants. That is, $K_f = K_{f1}K_{f2} = (2.3 \times 10^3) \times (7.2 \times 10^3) = 1.7 \times 10^7$.

In Example 19.12, we omitted any mention of the $Ag(NH_3)^+$ ion. To the extent that this ion is formed, it should affect the calculation of the concentration of $Ag^+(aq)$ ion. However, we can show that the concentration of $Ag(NH_3)^+$ is small compared with that of $Ag(NH_3)_2^+$. The equilibrium equation for K_{f2} is

$$\frac{[Ag(NH_3)_2^+]}{[Ag(NH_3)^+][NH_3]} = K_{f2} = 7.2 \times 10^3$$

Hence, the ratio of concentration of $Ag(NH_3)^+$ to $Ag(NH_3)_2^+$ is

$$\frac{[Ag(NH_3)^+]}{[Ag(NH_3)_2^+]} = \frac{1}{(7.2 \times 10^3)[NH_3]}$$

If $[NH_3] = 1.0\ M$, as in Example 19.12, this ratio is $1/(7.2 \times 10^3) = 1.4 \times 10^{-4}$. We see that the concentration of the intermediate complex ion $Ag(NH_3)^+$ is much smaller than that of $Ag(NH_3)_2^+$. Thus, we are justified in neglecting the concentration of $Ag(NH_3)^+$. In the calculations we do in this text, we will assume that we can neglect the intermediate complex ions.

19.6 Complex Ions and Solubility

From Example 19.12, we see that the formation of the complex ion $Ag(NH_3)_2^+$ reduces the concentration of $Ag^+(aq)$ in solution. The solution was initially $0.010\ M$ $AgNO_3$, or $0.010\ M$ $Ag^+(aq)$. When this solution is made $1.0\ M$ in NH_3, the $Ag^+(aq)$ ion concentration decreases to $5.9 \times 10^{-10}\ M$. We can see that the ion product for a slightly soluble silver salt could be decreased to below that for K_{sp}. Thus, the slightly soluble salt might not precipitate in a solution containing ammonia, whereas it otherwise would.

Example 19.13

(a) Will silver chloride precipitate from a solution that is $0.010\ M$ $AgNO_3$ and $0.010\ M$ NaCl? (b) Will silver chloride precipitate from this solution if it is also $1.0\ M$ NH_3?

Solution

(a) To see if a precipitate should form, we calculate the ion product and compare it with the K_{sp} for AgCl (1.8×10^{-10}).

$$\text{Ion product} = [Ag^+][Cl^-] = (0.010)(0.010)$$
$$= 1.0 \times 10^{-4}$$

Since this is greater than $K_{sp} = 1.8 \times 10^{-10}$, a precipitate should form. (b) We must first calculate the concentration of $Ag^+(aq)$ in a solution containing $1.0\ M$ NH_3. We did this in Example 19.12 and found that $[Ag^+]$ equals 5.9×10^{-10}. Hence,

$$\text{Ion product} = [Ag^+][Cl^-] = (5.9 \times 10^{-10})(0.010)$$
$$= 5.9 \times 10^{-12}$$

Since the ion product is smaller than $K_{sp} = 1.8 \times 10^{-10}$, no precipitate should form.

Exercise 19.13

Will silver iodide precipitate from a solution that is 0.0045 M $AgNO_3$, 0.15 M NaI, and 0.20 M KCN?

(See Problems 19.45 and 19.46.)

In Example 19.13, we determined whether a precipitate of AgCl is expected to form from a solution made up with given concentrations of Ag^+, Cl^-, and NH_3. The problem involved the $Ag(NH_3)_2^+$ complex in equilibrium. From it, we determined the concentration of free $Ag^+(aq)$, then calculated the ion product of AgCl.

Suppose we want to find the solubility of AgCl in a solution of aqueous ammonia of given concentration. This problem will involve the solubility equilibrium for AgCl, in addition to the complex-ion equilibrium. As silver chloride dissolves to give ions, the Ag^+ ion reacts with NH_3 to give the complex ion $Ag(NH_3)_2^+$. The equilibria are

$$AgCl(s) \rightleftharpoons Ag^+(aq) + Cl^-(aq)$$
$$Ag^+(aq) + 2NH_3(aq) \rightleftharpoons Ag(NH_3)_2^+(aq)$$

When $Ag^+(aq)$ reacts to give the complex ion, more AgCl dissolves to partially replenish the $Ag^+(aq)$ ion, according to LeChatelier's principle. Thus, silver chloride will be more soluble in aqueous ammonia than it is in pure water. The next example shows how we can calculate this solubility.

Example 19.14

Calculate the molar solubility of AgCl in 1.0 M NH_3 at 25°C.

Solution

Let us obtain the overall equation for the process by adding the solubility and complex-ion equilibria:

$$AgCl(s) \rightleftharpoons Ag^+(aq) + Cl^-(aq)$$
$$\underline{Ag^+(aq) + 2NH_3(aq) \rightleftharpoons Ag(NH_3)_2^+(aq)}$$
$$AgCl(s) + 2NH_3(aq) \rightleftharpoons Ag(NH_3)_2^+(aq) + Cl^-(aq)$$

The equilibrium constant for this overall equation is

$$K_c = \frac{[Ag(NH_3)_2^+][Cl^-]}{[NH_3]^2}$$

Since this overall equation is the sum of the solubility and complex-ion equilibria, the equilibrium constant K_c equals the product of the equilibrium constants for the solubility and complex-ion equilibria. We can easily verify this.

$$K_c = K_{sp}K_f = [Ag^+][Cl^-]\frac{[Ag(NH_3)_2^+]}{[Ag^+][NH_3]^2}$$

The value of $K_c = K_{sp}K_f$ can be obtained from the K_{sp} for AgCl in Table 19.1 and the K_f for $Ag(NH_3)_2^+$ in Table 19.2.

$$K_{sp}K_f = (1.8 \times 10^{-10})(1.7 \times 10^7) = 3.1 \times 10^{-3}$$

Let us now solve the equilibrium problem. The concentration table is given below. Substituting into the equilibrium equation

$$\frac{[Ag(NH_3)_2^+][Cl^-]}{[NH_3]^2} = K_{sp}K_f$$

gives

$$\frac{x^2}{(1.0 - 2x)^2} = 3.1 \times 10^{-3}$$

Concentration (M)	AgCl(s) + 2NH$_3$(aq) $\rightleftharpoons$ Ag(NH$_3$)$_2^+$(aq) + Cl$^-$(aq)		
Starting	1.0	0	0
Change	$-2x$	$+x$	$+x$
Equilibrium	$1.0 - 2x$	x	x

(*Continued*)

We can solve this equation by taking the square root of both sides:

$$\frac{x}{1.0 - 2x} = 0.056$$

Rearranging,

$$x = 0.056(1.0 - 2x) = 0.056 - 0.11x$$

Hence,

$$x = \frac{0.056}{1.11} = 0.050$$

Note that the molar solubility of AgCl equals the molar concentration of silver in the solution. Because of the stability of $Ag(NH_3)_2^+$, most of the silver in solution will be in the form of this complex ion. Therefore, since x equals the concentration of $Ag(NH_3)_2^+$, the molar solubility of AgCl equals $0.050\ M$.

Exercise 19.14

What is the molar solubility of AgBr in $1.0\ M\ Na_2S_2O_3$ (sodium thiosulfate)? Silver ion forms the complex ion $Ag(S_2O_3)_2^{3-}$. See Tables 19.1 and 19.2 for data.

(See Problems 19.47 and 19.48.)

An Application of Solubility Equilibria

Qualitative analysis involves the determination of what substances are present in a mixture. One scheme for the qualitative analysis of a solution of metal ions is based on the relative solubilities of the metal sulfides. This scheme is no longer widely used in practical analysis, having been replaced largely by modern instrumental methods. But it is still taught in general chemistry laboratory as a way of imparting some knowledge of inorganic chemistry, while illustrating the concepts of solubility, buffers, and so forth. In the next section, we will look at the main outline of the sulfide scheme.

19.7 Qualitative Analysis of Metal Ions

In the qualitative analysis scheme for metal ions, a cation is usually detected by a characteristic precipitate. For example, silver ion gives a white precipitate with chloride ion. But other ions also give a white precipitate with chloride ion. Therefore, one must first subject a mixture to a procedure that separates individual ions before a precipitation test can be applied to any particular ion.

Figure 19.3 shows a flow diagram illustrating how the metal ions in an aqueous solution are first separated into five analytical groups. The Analytical Group I ions, Ag^+, Hg_2^{2+}, and Pb^{2+}, are separated from a solution of other ions by adding dilute hydrochloric acid. These ions are precipitated as the chlorides, AgCl, Hg_2Cl_2, and $PbCl_2$, which are removed by filtering.

The other metal ions remain in the filtrate, the solution that passes through the filter. Many of these ions can be separated by precipitating them as the metal sulfide with H_2S. This separation takes advantage of the fact that only the least soluble metal sulfides precipitate in acidic solution. After these are removed and the solution is made basic, other metal sulfides precipitate. ∎

■ The solubility of the metal sulfides at different pH was discussed in Section 19.4.

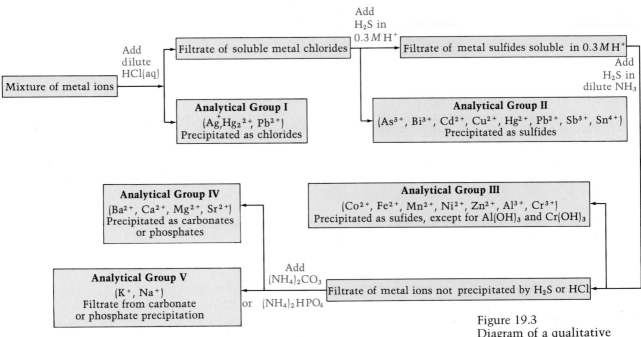

Add
dilute
HCl(aq)

Add
H$_2$S in
0.3 M H$^+$

Add
H$_2$S in
dilute NH$_3$

Mixture of metal ions

Filtrate of soluble metal chlorides

Filtrate of metal sulfides soluble in 0.3 M H$^+$

Analytical Group I
(Ag$^+$, Hg$_2^{2+}$, Pb^{2+})
Precipitated as chlorides

Analytical Group II
(As^{3+}, Bi^{3+}, Cd^{2+}, Cu^{2+}, Hg^{2+}, Pb^{2+}, Sb^{3+}, Sn^{4+})
Precipitated as sulfides

Analytical Group IV
(Ba^{2+}, Ca^{2+}, Mg^{2+}, Sr^{2+})
Precipitated as carbonates
or phosphates

Analytical Group III
(Co^{2+}, Fe^{2+}, Mn^{2+}, Ni^{2+}, Zn^{2+}, Al^{3+}, Cr^{3+})
Precipitated as sufides, except for Al(OH)$_3$ and Cr(OH)$_3$

Analytical Group V
(K$^+$, Na$^+$)
Filtrate from carbonate
or phosphate precipitation

Add
(NH$_4$)$_2$CO$_3$
or (NH$_4$)$_2$HPO$_4$

Filtrate of metal ions not precipitated by H$_2$S or HCl

Figure 19.3
Diagram of a qualitative
analysis scheme for the separation of metal ions into
analytical groups.

Analytical Group II consists of metal ions precipitated as metal sulfides from a solution that is 0.3 M H$^+$ and saturated with H$_2$S.■ The ions are As^{3+}, Bi^{3+}, Cd^{2+}, Cu^{2+}, Hg^{2+}, Pb^{2+}, Sb^{3+}, and Sn^{4+}. Lead(II) ion can appear in Analytical Group II as well as in Group I because PbCl$_2$ is somewhat soluble, and therefore Pb^{2+} may not be completely precipitated by HCl.

Analytical Group III consists of the metal ions in the filtrate from Group II that are precipitated in weakly basic solution with H$_2$S. The ions Co^{2+}, Fe^{2+}, Mn^{2+}, Ni^{2+}, and Zn^{2+} are precipitated as the sulfides. The ions Al^{3+} and Cr^{3+} are precipitated as hydroxides.

The filtrate obtained after the Group III metal ions have precipitated contains the alkali metal and alkaline earth ions. The Analytical Group IV ions, Ba^{2+}, Ca^{2+}, Mg^{2+}, and Sr^{2+}, are precipitated as carbonates or phosphates by adding (NH$_4$)$_2$CO$_3$ or (NH$_4$)$_2$HPO$_4$. The filtrate from this separation contains the Analytical Group V ions, K$^+$ and Na$^+$.

To illustrate the separation scheme, suppose we have a solution containing the metal cations Ag$^+$, Cu^{2+}, Zn^{2+}, Ca^{2+}, and Na$^+$. When we add dilute HCl(aq) to this solution, Ag$^+$ precipitates as AgCl:

$$Ag^+(aq) + Cl^-(aq) \longrightarrow AgCl(s)$$
$$\text{white}$$

We filter off the precipitate of AgCl. The filtrate contains the remaining ions.

If we make the filtrate 0.3 M H$^+$, then saturate it with H$_2$S, the Cu^{2+} ion precipitates as CuS:

$$Cu^{2+}(aq) + S^{2-}(aq) \longrightarrow CuS(s)$$
$$\text{black}$$

■ Hydrogen sulfide is often prepared in the solution by adding
thioacetamide, CH$_3$CSNH$_2$, and
warming.

CH$_3$CSNH$_2$ + H$_2$O

$\longrightarrow$ CH$_3$CONH$_2$ + H$_2$S

Zinc(II) ion also forms a sulfide, but it is soluble in acidic solution. After removing the precipitate of CuS, the solution is made weakly basic with $NH_3(s)$. This increases the S^{2-} concentration sufficiently to precipitate ZnS:

$$Zn^{2+}(aq) + S^{2-}(aq) \longrightarrow ZnS(s)$$
white

The filtrate from this separation contains Ca^{2+} and Na^+. Calcium ion can be precipitated as the phosphate or the carbonate. For example, by adding $(NH_4)_2CO_3$, we get the reaction

$$Ca^{2+}(aq) + CO_3^{2-}(aq) \longrightarrow CaCO_3(s)$$
white

The filtrate from this solution contains Na^+ ion. Now the five ions have been separated.

In the complete analysis, once the ions are separated into analytical groups, these are further separated into the individual ions. We can illustrate this with Analytical Group I, which is precipitated as a mixture of the chlorides, AgCl, Hg_2Cl_2, and $PbCl_2$. Of these chlorides, only $PbCl_2$ is significantly soluble in hot water. Thus, if the Analytical Group I precipitate is mixed with hot water, lead chloride will dissolve and the silver and mercury(I) chlorides can be filtered off. Any lead ion will be in the filtrate, and can be revealed by adding potassium chromate. Lead(II) ion gives a bright yellow precipitate of lead chromate, $PbCrO_4$:

$$Pb^{2+}(aq) + CrO_4^{2-}(aq) \longrightarrow PbCrO_4(s)$$
yellow

If there is precipitate remaining after the extraction with hot water, it consists of either AgCl or Hg_2Cl_2, or both. However, only silver(I) ion forms a stable ammonia complex ion. Thus, silver chloride dissolves in ammonia:

$$AgCl(s) + 2NH_3(aq) \longrightarrow Ag(NH_3)_2^+(aq) + Cl^-(aq)$$

Mercury(I) chloride, however, is simultaneously oxidized and reduced in ammonia solution, giving a precipitate of mercury(II) amido chloride, $HgNH_2Cl$, and mercury metal, which appears black because of its finely divided state.

$$Hg_2Cl_2(s) + 2NH_3(aq) \longrightarrow \underbrace{HgNH_2Cl(s) + Hg(l)}_{\text{black or gray}} + NH_4^+(aq) + Cl^-(aq)$$

The presence of silver ion in the filtrate is revealed by adding hydrochloric acid, which combines with the NH_3, releasing $Ag^+(aq)$ ion. The $Ag^+(aq)$ ion reacts with $Cl^-(aq)$ to give a white precipitate of AgCl:

$$Ag(NH_3)_2^+(aq) + 2H^+(aq) + Cl^-(aq) \longrightarrow AgCl(s) + 2NH_4^+(aq)$$

The other analytical groups are separated in similar fashion. Once a given ion has been separated, its presence is usually confirmed by a particular reactant, perhaps one that gives a distinctive precipitate.

A Checklist for Review

Important Terms

solubility product constant (K_{sp}) (19.1)
ion product (19.3)
fractional precipitation (19.3)
complex ions (p. 642)
ligands (p. 642)

formation (stability) constant (K_f) (19.5)
dissociation constant (of a complex) (K_d) (19.5)
amphoteric hydroxides (19.5)
qualitative analysis (19.7)

Summary of Facts and Concepts

The equilibrium constant for the equilibrium between an ionic solid and its ions in solution is called the *solubility product constant*, K_{sp}. Its value can be determined from the solubility of the solid. Conversely, if the solubility product constant is known, the solubility of the solid can be calculated. The solubility will be decreased by the addition of a soluble salt that supplies a common ion. Qualitatively, this can be seen to follow from LeChatelier's principle. Quantitatively, the *common-ion effect* on solubility can be obtained from the solubility product constant.

Rather than look at the solubility process as the dissolving of a solid in a solution, we can look at it as the *precipitation* of the solid from the solution. We can decide whether precipitation will occur by computing the *ion product*. Precipitation occurs when the ion product is greater than K_{sp}.

Solubility will be affected by the pH, if the compound supplies an anion conjugate to a weak acid. As the pH decreases (H^+ ion concentration increases), the anion concentration decreases because the anion forms the weak acid. Therefore, the ion product decreases and the solubility increases.

The concentration of a metal ion in solution is decreased by *complex-ion formation*. The equilibrium constant for the equilibrium between the aqueous metal ion, the ligand, and the complex ion is called the *formation constant* (or stability constant), K_f. Since complex-ion formation reduces the concentration of aqueous metal ion, an ionic compound of the metal is more soluble in a solution of the ligand.

The sulfide scheme of *qualitative analysis* separates a mixture of metal ions by using precipitation reactions. Variation of solubility with pH and with complex-ion formation is used to aid in the separation.

Operational Skills

1. Write the solubility product expression for a given ionic compound (Example 19.1).

2. Given the solubility of a slightly soluble ionic compound, calculate K_{sp} (Examples 19.2 and 19.3). Given K_{sp}, calculate the solubility of an ionic compound (Example 19.4).

3. Given the solubility product constant, calculate the molar solubility of a slightly soluble ionic compound in a solution containing a common ion (Examples 19.5 and 19.6).

4. Given the concentrations of ions originally in solution, determine if a precipitate is expected to form (Example 19.7). Determine if a precipitate is expected to form when two solutions of known volume and molarity are mixed (Example 19.8). For both problems, you will need the solubility product constant.

5. Calculate the concentration and percentage of an ion remaining after the corresponding ionic compound precipitates from a solution of known concentrations of ions (Example 19.9). K_{sp} is required.

6. Decide whether the solubility of a salt will be greatly increased by decreasing the pH (Example 19.10).

7. Given the metal-ion concentrations in solution, calculate the range of pH required to separate a mixture of two metal ions by precipitating one as the metal sulfide (Example 19.11). K_{sp} is required.

8. Calculate the concentration of aqueous metal ion in equilibrium with the complex ion, given the original metal-ion and ligand concentrations (Example 19.12). The formation constant, K_f, of the complex ion is required.

9. Predict whether an ionic compound will precipitate from a solution of known concentrations of cation, anion, and ligand that complexes with the cation (Example 19.13). K_f and K_{sp} are required.

10. Calculate the molar solubility of a slightly soluble ionic compound in a solution of known concentration of a ligand that complexes with the cation (Example 19.14). K_{sp} and K_f are required.

Review Questions

19.1 Suppose the molar solubility of nickel hydroxide, $Ni(OH)_2$ is x M. Show that K_{sp} for nickel hydroxide equals $4x^3$.

19.2 In your own words, explain why calcium sulfate is less soluble in sodium sulfate solution than it is in pure water.

19.3 What must be the concentration of silver ion in a solution that is in equilibrium with solid silver chloride and which is 0.10 M in Cl^-?

19.4 Discuss briefly how you could predict whether a precipitate will form if solutions of lead nitrate and potassium iodide are mixed. What information do you need to know?

19.5 Explain why barium fluoride dissolves in dilute hydrochloric acid, but is insoluble in water.

19.6 Explain how metal ions such as Pb^{2+} and Zn^{2+} are separated by precipitation with hydrogen sulfide.

19.7 Lead chloride at first precipitates when sodium chloride is added to a solution of lead nitrate. Later, when the solution is made more concentrated in chloride ion, the precipitate dissolves. Explain what is happening. What equilibria are involved? Note that lead ion forms the complex ion $PbCl_4^{2-}$.

19.8 A precipitate forms when a small amount of sodium hydroxide is added to a solution of aluminum sulfate. This precipitate dissolves when more sodium hydroxide is added. Explain what is happening.

19.9 Describe how you would separate the metal ions in a solution containing silver ion, copper(II) ion, and nickel ion using the sulfide scheme of qualitative analysis.

19.10 A solution containing calcium ion and magnesium ion is buffered with ammonia–ammonium chloride. When carbonate ion is added to the solution, calcium carbonate precipitates but magnesium carbonate does not. Explain.

Problems

Solubility and K_{sp}

19.11 Write solubility product expressions for the following: (a) $PbSO_4$ (b) $PbBr_2$ (c) $Ca_3(PO_4)_2$

19.13 The solubility of silver bromate, $AgBrO_3$, in water is 0.0072 g/L. Calculate K_{sp}.

19.15 Calculate the solubility product constant for copper iodate, $Cu(IO_3)_2$. The solubility of copper iodate in water is 0.13 g/100 mL.

19.17 The pH of a saturated solution of magnesium hydroxide ("milk of magnesia") was found to be 10.52. From this, find K_{sp} for magnesium hydroxide.

19.19 Celestite (strontium sulfate) is an important mineral of strontium. Calculate the solubility of strontium sulfate, $SrSO_4$, from the solubility product constant (see Table 19.1).

19.21 What is the concentration of copper(II) ion in a saturated solution of copper(II) hydroxide, $Cu(OH)_2$? See Table 19.1.

19.12 Write solubility product expressions for the following: (a) $MgCO_3$ (b) $Mg(OH)_2$ (c) $Mg_3(AsO_4)_2$

19.14 The solubility of magnesium oxalate, MgC_2O_4, in water is 0.0093 mol/L. Calculate K_{sp}.

19.16 The solubility of silver chromate, Ag_2CrO_4, in water is 0.028 g/L. Calculate K_{sp}.

19.18 A solution saturated in calcium hydroxide ("lime water") has a pH of 12.35. What is K_{sp} for calcium hydroxide?

19.20 Barite (barium sulfate, $BaSO_4$) is a common barium mineral. From the solubility product constant (Table 19.1), find the solubility of barium sulfate in grams per liter of water.

19.22 What is the concentration of silver ion in a saturated solution of silver chromate, Ag_2CrO_4? See Table 19.1.

Common-Ion Effect

19.23 What is the solubility (in grams per liter) of strontium sulfate, $SrSO_4$, in 0.10 M sodium sulfate, Na_2SO_4? See Table 19.1.

19.25 The solubility of magnesium fluoride, MgF_2, in water is 0.0076 g/L. What is the solubility (in grams per liter) of magnesium fluoride in 0.015 M sodium fluoride, NaF?

19.27 What is the solubility (in grams per liter) of magnesium oxalate, MgC_2O_4, in 0.020 M sodium oxalate, $Na_2C_2O_4$? See Table 19.1 for K_{sp}.

19.24 What is the solubility (in grams per liter) of lead chromate, $PbCrO_4$, in 0.15 M potassium chromate, K_2CrO_4? See Table 19.1.

19.26 The solubility of silver sulfate, Ag_2SO_4, in water has been determined to be 8.0 g/L. What is the solubility in 0.15 M sodium sulfate, Na_2SO_4?

19.28 Calculate the molar solubility of strontium sulfate, $SrSO_4$, in 0.0015 M sodium sulfate, Na_2SO_4. See Table 19.1 for K_{sp}.

Precipitation

19.29 Lead chromate, $PbCrO_4$, is used as a yellow paint pigment ("chrome yellow"). If a solution is prepared that is 5.0×10^{-4} M in lead ion, Pb^{2+}, and 5.0×10^{-5} M in chromate ion, CrO_4^{2-}, would you expect some of the lead chromate to precipitate? See Table 19.1.

19.31 The following solutions are mixed: 1.0 L of 0.00010 M NaOH and 1.0 L of 0.0020 M $MgSO_4$. Is a precipitate expected? Explain.

19.33 How many moles of calcium chloride, $CaCl_2$, can be added to 1.5 L of 0.020 M potassium sulfate, K_2SO_4, before a precipitate is expected? Assume that the volume of the solution is not changed significantly by the addition of calcium chloride.

19.35 What is the concentration and the percentage of Ag^+ ion remaining after Ag_2CrO_4 precipitates when 25.0 mL of 0.10 M $AgNO_3$ is added to 25.0 mL of 0.10 M K_2CrO_4?

**19.37 What is the I^- concentration just as AgCl begins to precipitate when 1.0 M $AgNO_3$ is slowly added to a solution containing 0.015 M Cl^- and 0.015 M I^-?

19.30 Lead sulfate, $PbSO_4$, is used as a white paint pigment. If a solution is prepared that is 5.0×10^{-4} M in lead ion, Pb^{2+}, and 1.0×10^{-5} M in sulfate ion, SO_4^{2-}, would you expect some of the lead sulfate to precipitate? See Table 19.1.

19.32 A 45-mL sample of 0.015 M calcium chloride, $CaCl_2$, is added to 55 mL of 0.010 M sodium sulfate, Na_2SO_4. Is a precipitate expected? Explain.

19.34 Magnesium sulfate, $MgSO_4$, is added to 456 mL of 0.040 M sodium hydroxide, NaOH, until a precipitate just forms. How many grams of magnesium sulfate were added? Assume that the volume of the solution is not changed significantly by the addition of magnesium sulfate.

19.36 What is the concentration and the percentage of Ca^{2+} ion remaining after $CaCO_3$ precipitates when 25.0 mL of 0.10 M $CaCl_2$ is added to 25.0 mL of 0.10 M Na_2CO_3?

**19.38 What is the Cl^- concentration just as Ag_2CrO_4 begins to precipitate when 1.0 M $AgNO_3$ is slowly added to a solution containing 0.015 M Cl^- and 0.015 M CrO_4^{2-}?

Effect of pH on Solubility

19.39 Which of the following salts would you expect to dissolve readily in acidic solution: barium sulfate or barium sulfite? Explain.

19.41 A solution is 0.10 M Co^{2+} and 0.10 M Hg^{2+}. Calculate the range of pH in which only one of the metal sulfides will precipitate.

19.40 Which of the following salts would you expect to dissolve readily in acidic solution: calcium phosphate, $Ca_3(PO_4)_2$, or mercury(I) chloride, Hg_2Cl_2? Explain.

19.42 A solution is 0.10 M Ni^{2+} and 0.10 M Cd^{2+}. Calculate the range of pH in which only one of the metal sulfides will precipitate.

Complex Ions

19.43 Sufficient sodium cyanide, NaCN, was added to 0.015 M silver nitrate, $AgNO_3$, to give a solution that was initially 0.100 M in cyanide ion, CN^-. What is the concentration of silver ion, Ag^+, in this solution after $Ag(CN)_2^-$ was formed? The formation constant, K_f, for the complex ion $Ag(CN)_2^-$ is 5.6×10^{18}.

19.45 Predict whether cadmium oxalate, CdC_2O_4, will precipitate from a solution that is 0.0020 M $Cd(NO_3)_2$, 0.010 M $Na_2C_2O_4$, and 0.10 M NH_3. Note that cadmium ion forms the $Cd(NH_3)_4^{2+}$ complex ion.

19.47 What is the molar solubility of CdC_2O_4 in 0.10 M NH_3?

19.44 The formation constant, K_f, for the complex ion $Zn(OH)_4^{2-}$ is 2.8×10^{15}. In a solution that is initially 0.20 M in $Zn(OH)_4^{2-}$, what is the concentration of zinc ion, Zn^{2+}?

19.46 Predict whether nickel hydroxide, $Ni(OH)_2$, will precipitate from a solution that is 0.0020 M $NiSO_4$, 0.010 M NaOH, and 0.10 M NH_3. Note that nickel ion forms the $Ni(NH_3)_6^{2+}$ complex ion.

19.48 What is the molar solubility of NiS in 0.10 M NH_3?

Additional Problems

19.49 What is the solubility of magnesium hydroxide in a solution buffered at pH 8.80?

19.50 What is the solubility of silver oxide, Ag_2O, in a solution buffered at pH 10.50? The equilibrium is $Ag_2O(s) + H_2O(l) \rightleftharpoons 2Ag^+(aq) + 2OH^-(aq)$; $K_c = 2.0 \times 10^{-8}$.

19.51 What must be the concentration of sulfate ion to precipitate calcium sulfate, $CaSO_4$, from a solution that is 0.0030 M Ca^{2+}?

19.52 What must be the concentration of chromate ion to precipitate strontium chromate, $SrCrO_4$, from a solution that is 0.0025 M Sr^{2+}? K_{sp} for strontium chromate is 5.7×10^{-5}.

19.53 How many grams of sodium chloride can be added to 785 mL of 0.0015 M silver nitrate before a precipitate forms?

19.54 How many grams of sodium sulfate can be added to 435 mL of 0.0028 M barium chloride before a precipitate forms?

*__19.55__ A solution is 0.10 M in sodium sulfate, Na_2SO_4. If 50.0 mL of 0.10 M barium nitrate, $Ba(NO_3)_2$, are added to 50.0 mL of this solution, what fraction of the sulfate ion is not precipitated?

*__19.56__ A solution is 0.10 M in sodium chloride. If 50.0 mL of 0.10 M silver nitrate are added to 50.0 mL of this solution, what fraction of the chloride ion is not precipitated?

**__19.57__ What is the molar solubility of silver acetate in a solution that is buffered at pH 2.60? (*Hint:* First show that the ratio $[C_2H_3O_2^-]/[HC_2H_3O_2]$ is small, so that most of the acetate ion from the dissolution of silver acetate will be converted to acetic acid. Thus, the molar solubility of silver acetate equals $[Ag^+] \approx [HC_2H_3O_2]$. From the equations for K_{sp} and K_a, eliminate $[C_2H_3O_2^-]$ and solve for the molar solubility.)

**__19.58__ What is the molar solubility of calcium fluoride in a solution that is buffered at pH 1.15? See hint in Problem 19.57.

**__19.59__ The solubility of zinc oxalate, ZnC_2O_4, in 0.0150 M ammonia is 3.6×10^{-4} mol/L. What is the oxalate-ion concentration in the saturated solution? If the solubility product constant for zinc oxalate is 1.5×10^{-9}, what must be the zinc-ion concentration in the solution? Now calculate the formation constant for the complex ion $Zn(NH_3)_4^{2+}$.

**__19.60__ The solubility of cadmium oxalate, CdC_2O_4, in 0.0150 M ammonia is 6.1×10^{-3} mol/L. What is the oxalate-ion concentration in the saturated solution? If the solubility product constant for cadmium oxalate is 1.5×10^{-8}, what must be the cadmium-ion concentration in the solution? Now calculate the formation constant for the complex ion $Cd(NH_3)_4^{2+}$.

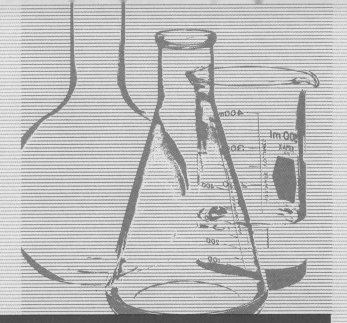

20. Thermodynamics and Equilibrium

20.1 Enthalpy and Heat of Reaction

Spontaneous Processes and Entropy

20.2 Entropy and the Second Law Entropy/ Second Law of Thermodynamics
20.3 Standard Entropies Third Law of Thermodynamics/ Entropy Change for a Reaction

Free Energy and Equilibrium

20.4 Free Energy and Spontaneity Standard Free-Energy Change/ Standard Free Energies of Formation/ $\Delta G°$ as a Criterion for Spontaneity
20.5 Interpretation of Free Energy Maximum Work/ Coupling of Reactions/ Free-Energy Change During Reaction
20.6 Calculation of Equilibrium Constants
20.7 Change of Free Energy with Temperature Spontaneity and Temperature Change/ Calculation of $\Delta G°$ at Various Temperatures

U rea, NH_2CONH_2, is a key industrial chemical. It is used to make synthetic resins for adhesives and melamine plastics. Its largest use, however, is as a nitrogen fertilizer. Urea is produced by reacting ammonia with carbon dioxide:

$$2NH_3(g) + CO_2(g) \rightleftharpoons NH_2CONH_2(aq) + H_2O(l)$$

Suppose we want to determine whether this or some other reaction could be useful for the industrial preparation of urea. Some of our immediate questions would be: Does the reaction naturally go in the direction it is written? Will the reaction mixture contain a sufficient amount of the product at equilibrium? We addressed these questions in the preceding chapters by looking at the equilibrium constant. In the present chapter, we want to discuss these questions from the point of view of thermodynamics.

Thermodynamics is the study of the relationship between heat and other forms of energy involved in a chemical or physical process. With only heat measurements of substances, we can answer the questions just posed. We can predict the natural direction of a chemical reaction. We can also determine the composition of a reaction mixture at equilibrium. Consider then the possibility of reacting NH_3 and CO_2 to give urea. What can thermodynamics tell us about this reaction?

Chapter Overview

In the first section, we will review some concepts, such as enthalpy, H, introduced in Chapter 10. Then we will look at the concept of *entropy*, denoted S, which is a measure of randomness or disorder in a chemical system. *Free energy, G,* which equals $H - TS$, can be used as a criterion of the natural, or *spontaneous*, direction of a chemical reaction. The change of free energy for a reaction can be related to the composition of an equilibrium mixture and will be discussed in the last part of the chapter.

20.1 Enthalpy and Heat of Reaction

Recall that the heat of reaction at constant pressure equals the change of enthalpy, ΔH, of the chemical system. We can calculate $\Delta H°$ for a reaction if we have the enthalpies of formation, $\Delta H_f°$, for all of the substances involved.

Suppose we would like to know how much heat is absorbed or released in the reaction given in the chapter opening. We write the balanced equation with values of $\Delta H_f°$ below every reactant and product. The $\Delta H_f°$ for 1 M urea solution, $NH_2CONH_2(aq)$, is -319.2 kJ/mol. See Table 10.1, p. 329, for other values of $\Delta H_f°$. Each value is multiplied by the corresponding stoichiometric coefficient. Then, to obtain $\Delta H°$, we sum the values for the products and subtract the values for the reactants.■

■ See Example 10.8 for a similar calculation.

$$2NH_3(g) \quad + \quad CO_2(g) \rightleftharpoons NH_2CONH_2(aq) + \quad H_2O(l)$$
$$\Delta H_f°: 2 \times (-45.9) \quad -393.5 \quad\quad -319.2 \quad\quad -285.8 \text{ kJ}$$

Hence,

$$\Delta H° = [(-319.2 - 285.8) - (-2 \times 45.9 - 393.5)] \text{ kJ} = -119.7 \text{ kJ}$$

From the minus sign, we conclude that heat is released. The reaction is exothermic.

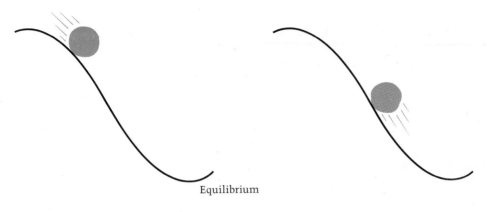

Equilibrium

Figure 20.1 *(left)*
A rock rolling downhill is a spontaneous process. The rock will eventually come to equilibrium at the bottom of the hill.

Figure 20.2 *(right)*
A rock rolling uphill by itself is a nonspontaneous process.

Spontaneous Processes and Entropy

A **spontaneous process** is a physical or chemical change that occurs by itself. It requires no continuing outside agency to make it happen. A rock at the top of a hill rolls down (Figure 20.1). Heat flows from a hot object to a cold one. An iron tool rusts in moist air. Each of these processes occurs spontaneously, or naturally, without requiring an outside force or agency. They continue until equilibrium is reached. If these processes were to go in the opposite direction, they would be *nonspontaneous*. A rock rolling uphill by itself is not a natural process; it is nonspontaneous (see Figure 20.2). The rock could be moved to the top of the hill, but it would require that work be expended. Heat can be made to flow from a cold to a hot object, but a heat pump or refrigerator is needed. Rust can be converted to iron, but the process requires chemical reactions used in the manufacture of iron from its ore (iron oxide).■

■ Each of these nonspontaneous processes requires an expenditure of work. This is discussed in Section 20.5.

20.2 Entropy and the Second Law

When we ask whether a chemical reaction goes in the direction it is written, as we did in the chapter opening, we are asking whether the reaction is spontaneous in this direction. The second law of thermodynamics, which we will discuss in this section, gives us a way to answer the question. The second law is expressed in terms of a quantity called *entropy*.

Entropy

Entropy, denoted by the symbol S, is a thermodynamic quantity that is a measure of the randomness or disorder in a system. As we will discuss later, the SI unit of entropy is joules per kelvin (J/K). Entropy, like enthalpy, is a *state function*. That is, the quantity of entropy in a given amount of substance depends only on variables such as temperature and pressure that determine the state of the substance. If these variables are fixed, the quantity of entropy is fixed. Thus, one mole of ice at 0°C and 1 atm pressure has an entropy that has been determined experimentally to be 41 J/K. One mole of liquid water at 0°C and 1 atm has an entropy of 63 J/K. (We will describe how entropy values are measured in the next section.)

We would expect the entropy to increase when ice melts to the liquid. Ice has an ordered crystalline structure (see Figure 14.4, p. 476). When this ice melts to liquid water, the crystalline structure breaks down, and a less-ordered liquid structure results. In ice, the H_2O molecules occupy regular, fixed positions in a crystal lattice, that is, ice has a relatively low entropy. But in the liquid, the molecules move freely about, giving a disordered structure, one having greater entropy. We can see this in the entropy values for these states; the entropy of ice at 0°C, 1 atm is 41 J/K, that for the liquid under the same conditions is 63 J/K.

We calculate the entropy change, ΔS, for a process similarly to the way we calculate ΔH. What is the entropy change when one mole of ice melts to liquid water at 0°C and 1 atm? Let us write the equation for the melting of ice, placing entropy values below the formulas for each state of water:

$$H_2O(s) \longrightarrow H_2O(l)$$

$$S: \quad 41 \qquad\qquad 63 \quad J/K$$

We calculate the entropy change, ΔS, by subtracting the initial entropy (that of ice) from the final entropy (that of liquid water):

$$\Delta S = (63 - 41) \text{ J/K} = 22 \text{ J/K}$$

Thus, when one mole of ice melts at 0°C, the water increases in entropy by 22 J/K. The entropy increases, as we expect, since the water becomes a more disordered substance when it melts.

Second Law of Thermodynamics

A process occurs naturally as a result of an overall increase in disorder. There is a natural tendency for things to mix and to break down, events that represent increasing disorder. Where we see order or structure being built, it results from the use of greater order elsewhere. Order in one place is used to build order in another. Thus, the building of a house is at the expense of order in the food molecules of workers and order in substances needed to construct building materials. However, in the building process, some of the original order is lost so that the net result is an increase of disorder.

We can put this more precisely in terms of the **second law of thermodynamics.** It states that the total entropy of a system and its surroundings always increases for a spontaneous process. Note that entropy is quite different from energy. Energy can be neither created nor destroyed during chemical change (law of conservation of energy). ■ But entropy is produced or created during a spontaneous, or natural, process.

Entropy accompanies any heat flow. Heat is the transfer of energy into or out of a system as a result of temperature differences. It occurs when molecular kinetic energy is transferred through molecular collisions between a system and its surroundings. Associated with this flow of energy is a flow of entropy, that is, a transfer of disorder between the system and its surroundings. Thus, as a piece of ice absorbs heat, it becomes more disordered and melts. If a quantity of heat, q, is absorbed at absolute temperature T, the quantity of entropy that has flowed into the system is q/T. We can see from q/T that the units for entropy are joules per kelvin, as we noted earlier.

■ The first law of thermodynamics is a particular statement of the law of conservation of energy. It is often written $\Delta E = q - w$. That is, the change of energy of a system equals any heat absorbed less any work done. See the Aside in Section 10.1.

The entropy of a system can change as the system absorbs or releases heat. However, the entropy can also change as it is created or produced in the system during a spontaneous change. This means that the entropy change in a system, ΔS, during a spontaneous change equals q/T plus any entropy that was created. We can express this as

$$\Delta S > \frac{q}{T} \quad \text{(spontaneous process)}$$

Certain processes occur at equilibrium, or more precisely, very close to equilibrium.■ For example, ice at 0°C is in equilibrium with liquid water at 0°C. If heat is slowly absorbed by the system, it remains very near equilibrium, but the ice melts. Under these conditions, no significant amount of entropy is produced. The entropy change results entirely from the absorption of heat. Therefore,

$$\Delta S = \frac{q}{T} \quad \text{(equilibrium process)}$$

■ When a system is at equilibrium, a small change of a condition can make the process go in one direction or the other. Think of a two-pan balance in which the weight on each pan has been adjusted to make the balance beam level, that is, so the beam is at equilibrium. A small weight on either pan will tip the scale one way or the other.

Other phase changes, such as the vaporization of a liquid, can also occur under equilibrium conditions.

We can use the previous equation to obtain the entropy change for a phase change. Consider the melting of ice. The heat absorbed is the heat of fusion, ΔH_{fus}, which is known from experiment to be 6.0 kJ for one mole of ice. We get the entropy change for melting by dividing ΔH_{fus} by the absolute temperature of the phase transition, 273 K (0°C). Because entropy changes are usually expressed in joules per kelvin, we convert ΔH_{fus} to 6.0×10^3 J.

$$\Delta S = \frac{\Delta H_{fus}}{T} = \frac{6.0 \times 10^3 \text{ J}}{273 \text{ K}} = 22 \text{ J/K}$$

Note that this is the ΔS value we obtained earlier for the conversion of ice to water.

Example 20.1

The heat of vaporization, ΔH_{vap}, of carbon tetrachloride, CCl_4, at 25°C is 43 kJ/mol:

$$CCl_4(l) \longrightarrow CCl_4(g); \quad \Delta H_{vap} = 43 \text{ kJ/mol}$$

If one mole of liquid carbon tetrachloride at 25°C has an entropy of 214 J/K, what is the entropy of one mole of the vapor in equilibrium with the liquid at this temperature?

Solution

When the liquid evaporates, it absorbs heat, $\Delta H_{vap} = $ 43 kJ/mol (43×10^3 J/mol) at 25°C, or 298 K. The en-

tropy change, ΔS, is

$$\Delta S = \frac{\Delta H_{vap}}{T} = \frac{43 \times 10^3 \text{ J/mol}}{298 \text{ K}} = 144 \text{ J/(mol} \cdot \text{K)}$$

In other words, one mole of carbon tetrachloride increases in entropy by 144 J/K when it vaporizes. The entropy of one mole of the vapor equals the entropy of one mole of liquid (214 J/K) plus 144 J/K.

$$\text{Entropy of vapor} = (214 + 144) \text{ J/(mol} \cdot \text{K)}$$
$$= 358 \text{ J/(mol} \cdot \text{K)}$$

Exercise 20.1

Liquid ethanol, $C_2H_5OH(l)$, at 25°C has an entropy of 161 J/(mol · K). If the heat of vaporization, ΔH_{vap}, at 25°C is 42.3 kJ/mol, what is the entropy of the vapor in equilibrium with the liquid at 25°C?

(See Problems 20.17, 20.18, 20.19, and 20.20.)

Suppose we propose a chemical reaction for the preparation of a substance. We will also assume that the reaction occurs at constant temperature and pressure. For urea, we propose the reaction

$$2NH_3(g) + CO_2(g) \rightleftharpoons NH_2CONH_2(aq) + H_2O(l)$$

Is this a spontaneous reaction, that is, does it go left to right as written? We can use the second law to answer this question if we know both ΔH and ΔS for the reaction, as we will now show.

Recall that the heat of reaction, q, at constant pressure equals the enthalpy change, ΔH. The second law for a spontaneous reaction at constant temperature and pressure becomes

$$\Delta S > \frac{q}{T} = \frac{\Delta H}{T} \qquad \text{(spontaneous reaction, constant } T \text{ and } P\text{)}$$

Thus, ΔS is greater than $\Delta H/T$ for a spontaneous reaction at constant temperature and pressure. Therefore, if we subtract ΔS from $\Delta H/T$, we will get a negative quantity for such a process. That is,

$$\frac{\Delta H}{T} - \Delta S < 0 \qquad \text{(spontaneous reaction, constant } T \text{ and } P\text{)}$$

Multiplying each term of this inequality by the positive quantity T, we get

$$\Delta H - T\Delta S < 0 \qquad \text{(spontaneous reaction, constant } T \text{ and } P\text{)}$$

This inequality is important. If we had a table of entropies of substances, we could calculate ΔS for the proposed reaction. From Table 10.1, we could also calculate ΔH. If $\Delta H - T\Delta S$ is negative for the reaction, we would predict that it is spontaneous left to right, as written. However, if $\Delta H - T\Delta S$ is positive, we would predict that the reaction is nonspontaneous in the direction written, but spontaneous in the opposite direction. If $\Delta H - T\Delta S$ is zero, the reaction is at equilibrium.

In the next section, we will see how to obtain the entropies of substances and then the entropy change of a reaction.

20.3 Standard Entropies

To experimentally determine the entropy of a substance, we first measure the heat absorbed by the substance by warming it at various temperatures. That is, we find the heat capacity at different temperatures.■ We then calcu- ■ Heat capacity was discussed late the entropy as we will describe. This determination of the entropy is in Section 10.2. based on the third law of thermodynamics.

Third Law of Thermodynamics

The **third law of thermodynamics** states that a substance that is perfectly crystalline at 0 K has an entropy of zero. This seems reasonable. A perfectly crystalline substance at 0 K should have perfect order. When the temperature is raised, however, the substance becomes more disordered as it absorbs heat.

We can determine the entropy of a substance at a temperature other than 0 K, say, at 298 K (25°C), by slowly heating the substance from 0 K to 298 K. Recall that the entropy change, ΔS, that occurs when heat is absorbed at a temperature T is q/T. Suppose we heat the substance from 0.0 K to 2.0 K, and the heat absorbed is 0.19 J. We find the entropy change by dividing the heat absorbed by the average absolute temperature [$\frac{1}{2}(0.0 + 2.0)K = 1.0$ K]. Hence, ΔS equals 0.19 J/1.0 K = 0.19 J/K. This gives us the entropy of the substance at 2.0 K. Now we heat the substance from 2.0 K to 4.0 K, and this time 0.88 J of heat are absorbed. The average temperature is $\frac{1}{2}(2.0 + 4.0)K =$ 3.0 K, and the entropy change is 0.88 J/3.0 K = 0.29 J/K. The entropy of the substance at 4.0 K is (0.19 + 0.29) J/K = 0.48 J/K. Proceeding this way, we can eventually get the entropy at 298 K.■

Figure 20.3 shows how the entropy of a substance changes with temperature. Note that the entropy increases gradually as the temperature increases. But when there is a phase change, for example from solid to liquid, the entropy increases sharply. The entropy change for the phase transition is calculated from the enthalpy of the phase transition, as described earlier (see Example 20.1). **Standard entropies,** also called absolute entropies, $S°$, are entropy values for the *standard state* of substances (indicated by the superscript degree sign). The standard states of substances are 1 atm pressure for pure substances. Ions in solution are 1 M concentration. Table 20.1 gives entropies of various substances at 25°C and 1 atm. Note that the elements

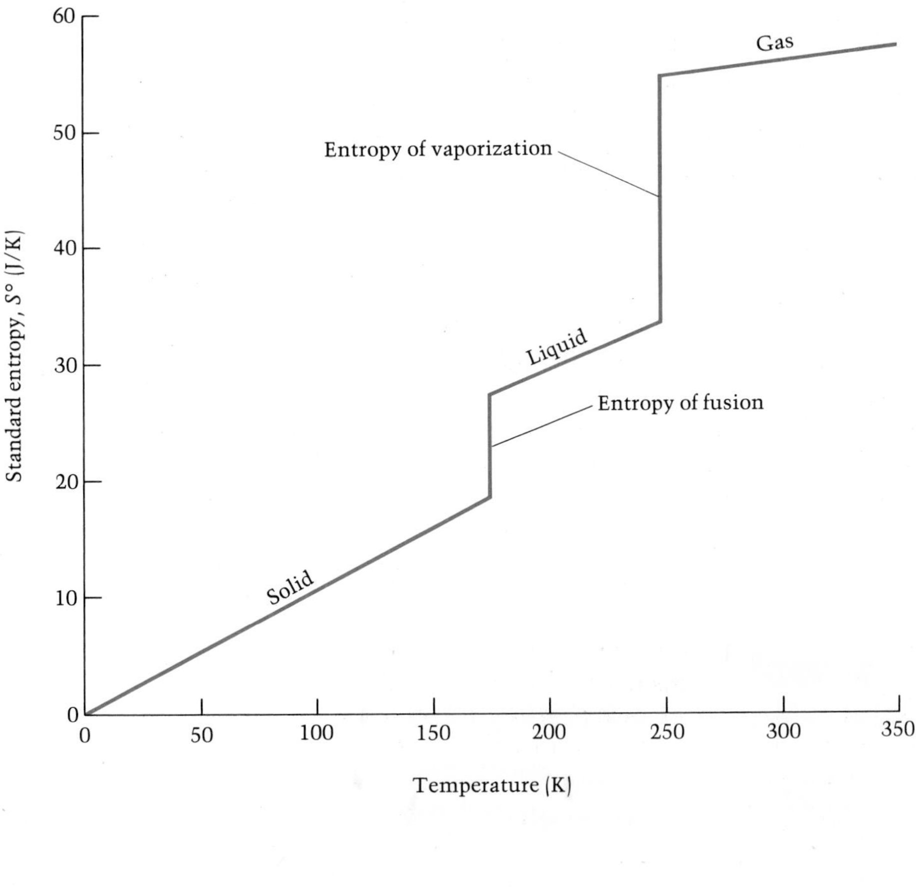

■ The process just described is essentially the numerical evaluation of an integral, which we can write as follows. The heat absorbed for temperature change dT is $C_p(T)dT$, and the entropy change is $C_p(T)dT/T$. The standard entropy at temperature T is

$$\int_0^T \frac{C_p(T)dT}{T}$$

Figure 20.3
Standard entropy of methyl chloride, CH_3Cl, at various temperatures (approximate schematic graph). The entropy rises gradually as the temperature increases, but jumps sharply at each phase transition.

Table 20.1
Standard Entropies (at 25°C)

Formula	$S°$, J/(mol·K)	Formula	$S°$, J/(mol·K)	Formula	$S°$, J/(mol·K)
Hydrogen		*Carbon (continued)*		*Sulfur*	
$H^+(aq)$	0	$HCN(g)$	201.7	$S_2(g)$	228.1
$H_2(g)$	130.6	$CCl_4(g)$	309.7	$S(rhombic)$	31.9
		$CCl_4(l)$	214.4	$S(monoclinic)$	32.6
Sodium		$CH_3CHO(g)$	266	$SO_2(g)$	248.1
$Na^+(aq)$	60.2	$C_2H_5OH(l)$	161	$H_2S(g)$	205.6
$Na(s)$	51.4				
$NaCl(s)$	72.1	*Silicon*		*Fluorine*	
$NaHCO_3(s)$	102	$Si(s)$	18.0	$F^-(aq)$	-9.6
$Na_2CO_3(s)$	139	$SiO_2(s)$	41.5	$F_2(g)$	202.7
		$SiF_4(g)$	285	$HF(g)$	173.7
Calcium					
$Ca^{2+}(aq)$	-55.2	*Lead*		*Chlorine*	
$Ca(s)$	41.6	$Pb(s)$	64.8	$Cl^-(aq)$	55.1
$CaO(s)$	38.2	$PbO(s)$	66.3	$Cl_2(g)$	223.0
$CaCO_3(s)$	92.9	$PbS(s)$	91.3	$HCl(g)$	186.8
Carbon		*Nitrogen*		*Bromine*	
$C(graphite)$	5.7	$N_2(g)$	191.5	$Br^-(aq)$	80.7
$C(diamond)$	2.4	$NH_3(g)$	193	$Br_2(l)$	152.2
$CO(g)$	197.5				
$CO_2(g)$	213.7	*Oxygen*		*Iodine*	
$HCO_3^-(aq)$	95.0	$O_2(g)$	205.0	$I^-(aq)$	109.4
$CH_4(g)$	186.1	$O_3(g)$	238.8	$I_2(s)$	116.1
$C_2H_4(g)$	219.2	$OH^-(aq)$	-10.5		
$C_2H_6(g)$	229.5	$H_2O(g)$	188.7	*Silver*	
$C_6H_6(l)$	172.8	$H_2O(l)$	69.9	$Ag^+(aq)$	73.9
$HCHO(g)$	219			$Ag(s)$	42.7
$CH_3OH(l)$	127			$AgF(s)$	84
$CS_2(g)$	237.8			$AgCl(s)$	96.1
$CS_2(l)$	151.0			$AgBr(s)$	107.1
				$AgI(s)$	114

have nonzero values, in contrast to standard enthalpies of formation, $\Delta H_f°$, which by convention are zero.■ Also, the symbol $S°$ is chosen for standard entropies, rather than $\Delta S°$, to emphasize their origin from the third law.

■ Entropies of substances must be positive. Those for ions, however, can be negative because they are derived by arbitrarily putting $S°$ for $H^+(aq)$ equal to zero.

Entropy Change for a Reaction

Once we determine the standard entropies of all substances in a reaction, we can calculate the change of entropy, $\Delta S°$, for the reaction. A sample calculation is described in Example 20.3. Even without knowing values for the entropies of substances, one can sometimes predict the sign of $\Delta S°$ for a reaction. The entropy will usually increase in the following situations:

1. A reaction in which a molecule is broken into two or more smaller molecules.
2. A reaction in which there is an increase in moles of gas. (This may result from a molecule breaking up, in which case items 1 and 2 are related.)

3. A process in which a solid changes to a liquid or gas or a liquid changes to a gas.

The next example illustrates how we can apply these rules (especially rule 2) to find the sign of $\Delta S°$ for certain reactions involving gases.

Example 20.2

(a) Is $\Delta S°$ positive or negative for the following reaction?

$$C_6H_{12}O_6(s) \longrightarrow 2C_2H_5OH(l) + 2CO_2(g)$$

$$\underset{\text{glucose}}{\qquad} \qquad \underset{\text{ethanol}}{\qquad}$$

Explain. The equation represents the essential change that takes place during the fermentation of glucose (grape sugar) to ethanol (ethyl alcohol). (b) Do you expect the entropy to increase or decrease in the preparation of urea from NH_3 and CO_2,

$$2NH_3(g) + CO_2(g) \rightleftharpoons NH_2CONH_2(aq) + H_2O(l)$$

as described in the chapter opening? Explain. (c) What is the sign of $\Delta S°$ for the following reaction?

$$CO(g) + H_2O(g) \longrightarrow CO_2(g) + H_2(g)$$

Solution

(a) A molecule (glucose) breaks into smaller molecules (C_2H_5OH) and (O_2). Moreover, this results in a gas being released. We predict that $\Delta S°$ for this reaction is positive. That is, the entropy increases. (b) In this reaction, the moles of gas decrease (by 3 moles), which would decrease the entropy. We predict that the entropy should decrease. That is, $\Delta S°$ is negative. (c) Since there is no change in the number of moles of gas, we cannot predict the sign of $\Delta S°$.

Exercise 20.2

Predict the sign of $\Delta S°$ for the following reactions:
(a) $CaCO_3(s) \longrightarrow CaO(s) + CO_2(g)$
(b) $CS_2(l) \longrightarrow CS_2(g)$
(c) $2Hg(l) + O_2(g) \longrightarrow 2HgO(s)$
(d) $2Na_2O_2(s) + 2H_2O(l) \longrightarrow 4NaOH(aq) + O_2(g)$

(See Problems 20.21 and 20.22.)

It is useful to be able to predict the sign of $\Delta S°$. We gain some understanding of the reaction, and the prediction can be used for qualitative work. For quantitative work, however, we need to find the value of $\Delta S°$.

Example 20.3

Calculate $\Delta S°$ at 25°C for the reaction in which urea is formed from NH_3 and CO_2:

$$2NH_3(g) + CO_2(g) \rightleftharpoons NH_2CONH_2(aq) + H_2O(l)$$

The standard entropy of $NH_2CONH_2(aq)$ is 174 J/(mol · K). See Table 20.1 for other values.

Solution

The calculation is similar to that for obtaining $\Delta H°$ from $\Delta H_f°$ values. We put the standard entropy values multiplied by stoichiometric coefficients below the for-

mulas in the balanced equation:

$$2NH_3(g) + CO_2(g) \rightleftharpoons NH_2CONH_2(aq) + H_2O(l)$$
$$S°: \ 2 \times 193 \qquad 214 \qquad\qquad 174 \qquad\qquad 70$$

We can calculate the entropy change by subtracting the entropy of the reactants from the entropy of the products:

$$\Delta S° = [(174 + 70) - (2 \times 193 + 214)] \ \text{J/K} = -356 \ \text{J/K}$$

Note that the sign of $\Delta S°$ agrees with the solution of Example 20.2(b).

Exercise 20.3

Calculate the change of entropy, $\Delta S°$, for the reaction in Example 20.2(a). The standard entropy of glucose, $C_6H_{12}O_6(s)$, is 212 J/(mol · K). See Table 20.1 for other values.

(See Problems 20.23, 20.24, 20.25, and 20.26.)

Free Energy and Equilibrium

At the end of Section 20.2, we saw that the quantity $\Delta H - T\Delta S$ can serve as a criterion of spontaneity of a reaction at constant temperature and pressure. If the value of this quantity is negative, the reaction will be spontaneous. If it is positive, the reaction will be nonspontaneous. If it equals zero, the reaction is at equilibrium.

As an application of this criterion, consider the reaction described in the chapter opening, in which urea is prepared from NH_3 and CO_2. The heat of reaction, $\Delta H°$, was calculated from enthalpies of formation in Section 20.1, where we obtained -119.7 kJ. Then, in Example 20.3, we calculated the entropy of reaction, $\Delta S°$, and found a value of -356 J/K, or -0.356 kJ/K. Let us substitute these values, and $T = 298$ K (25°C), into the expression $\Delta H° - T\Delta S°$:

$$\Delta H° - T\Delta S° = (-119.7 \text{ kJ}) - (298 \text{ K})(-0.356 \text{ kJ/K})$$
$$= -13.6 \text{ kJ}$$

We see that $\Delta H° - T\Delta S°$ is a negative quantity and so conclude that the reaction is spontaneous under standard conditions.

20.4 Free Energy and Spontaneity

It is very convenient to define a new thermodynamic quantity in terms of H and S that will be directly useful as a criterion of spontaneity. For this purpose, the American physicist J. Willard Gibbs (1839–1903) introduced the **free energy,** G, which is defined as follows.

$$G = H - TS$$

This thermodynamic quantity, we will see, gives a direct criterion of spontaneity of reaction.

As a reaction proceeds at a given temperature and pressure, reactants form products and the enthalpy, H, and entropy, S, change. These changes in H and S, denoted ΔH and ΔS, result in a change in free energy, ΔG, given by the equation

$$\Delta G = \Delta H - T\Delta S$$

Note that the change in free energy, ΔG, equals the quantity $\Delta H - T\Delta S$ that we just saw serves as a criterion of spontaneity of a reaction. Thus, if we can show that ΔG for a reaction at a given temperature and pressure is negative, we can predict that the reaction will be spontaneous.

Standard Free-Energy Change

Recall that for purposes of tabulating thermodynamic data, certain *standard states* are chosen, which are indicated by a superscript degree sign on the symbol of the quantity. The standard states are: for pure liquids and solids, 1 atm pressure; for gases, 1 atm partial pressure; for solutions, 1 M concentration. The temperature is usually 25°C (298 K).

The standard free-energy change, $\Delta G°$, is the free-energy change that occurs when reactants in their standard states are converted to products in their standard states. The next example illustrates the calculation of the standard free-energy change, $\Delta G°$, from $\Delta H°$ and $\Delta S°$. ■

■ Note that both $\Delta H°$ and $\Delta S°$, and therefore $\Delta G°$, can be obtained from heat measurements. In Section 20.6, we will relate $\Delta G°$ to equilibrium constants, which we can therefore obtain from heat measurements.

Example 20.4

What is the standard free-energy change, $\Delta G°$, for the following reaction at 25°C?

$$N_2(g) + 3H_2(g) \longrightarrow 2NH_3(g)$$

Use values of $\Delta H_f°$ and $S°$ from Tables 10.1 and 20.1.

Solution

Let us write the balanced equation with values of $\Delta H_f°$ and $S°$ multiplied by stoichiometric coefficients below each formula:

	$N_2(g)$ +	$3H_2(g)$	$\longrightarrow$	$2NH_3(g)$
$\Delta H_f°$:	0	0		$2 \times (-45.9)$ kJ
$S°$:	191.5	3×130.6		2×193 J/K

We calculate $\Delta H°$ for the reaction by taking the enthalpy of products and subtracting the enthalpy of reactants:

$$\Delta H° = [2 \times (-45.9) - 0] \text{ kJ} = -91.8 \text{ kJ}$$

Similarly, for $\Delta S°$ we get

$$\Delta S° = [2 \times 193 - (191.5 + 3 \times 130.6)] \text{ J/K} = -197 \text{ J/K}$$

We now substitute into the equation for $\Delta G°$ in terms of $\Delta H°$ and $\Delta S°$. Note that we substitute $\Delta S°$ in units of kJ/K.

$$\Delta G° = \Delta H° - T\Delta S°$$
$$= -91.8 \text{ kJ} - (298 \text{ K})(-0.197 \text{ kJ/K}) = -33.1 \text{ kJ}$$

Exercise 20.4

Calculate $\Delta G°$ for the following reaction at 25°C. Use data in Tables 10.1 and 20.1.

$$CH_4(g) + 2O_2(g) \longrightarrow CO_2(g) + 2H_2O(g)$$

(See Problems 20.27 and 20.28.)

Standard Free Energies of Formation

The **standard free energy of formation,** $\Delta G_f°$, of a substance is defined similarly to the standard enthalpy of formation. That is, $\Delta G_f°$ is the free-energy change when one mole of substance is formed from its elements in their stablest states at 1 atm and at a specified temperature (usually 25°C). For example, the standard free energy of formation of $NH_3(g)$ is the free-energy change for the following reaction:

$$\tfrac{1}{2}N_2(g) + \tfrac{3}{2}H_2(g) \longrightarrow NH_3(g)$$

The reactants, N_2 and H_2, each at 1 atm, are converted to the product, NH_3, at 1 atm pressure. In Example 20.4, we found the $\Delta G°$ for the formation of

2 mol NH_3 from its elements to be -33.1 kJ. Hence, $\Delta G_f^\circ = -33.1$ kJ/2 mol $= -16.6$ kJ/mol.

As in the case of standard enthalpies of formation, the standard free energies of formation of elements in their stablest states are assigned the value zero. By tabulating ΔG_f° for substances, we can easily calculate ΔG° for any reaction involving those substances. Table 20.2 lists standard free energies of formation. Note that the value of ΔG_f° given for $NH_3(g)$ is -16 kJ/mol, which essentially equals the value we have just calculated.

Example 20.5

Calculate ΔG° for the combustion of one mole of ethanol, C_2H_5OH:

$$C_2H_5OH(l) + 3O_2(g) \longrightarrow 2CO_2(g) + 3H_2O(g)$$

Use standard free energies of formation given in Table 20.2.

Solution

We write the balanced equation with values of ΔG_f° multiplied by stoichiometric coefficients below each formula:

$$C_2H_5OH(l) + 3O_2(g) \longrightarrow 2CO_2(g) + 3H_2O(g)$$
$$\Delta G_f^\circ: \quad -174.8 \qquad 0 \qquad 2(-394.4) \quad 3(-228.6) \text{ kJ}$$

Summing values for products, then subtracting values for reactants, we get

$$\Delta G^\circ = [2(-394.4) + 3(-228.6) - (-174.8)] \text{ kJ}$$
$$= -1299.8 \text{ kJ}$$

Exercise 20.5

Calculate ΔG° for the following reaction, using values of ΔG_f°:

$$CaCO_3(s) \longrightarrow CaO(s) + CO_2(g)$$

(See Problems 20.31 and 20.32.)

ΔG° as a Criterion for Spontaneity

We have already seen that the quantity $\Delta G = \Delta H - T\Delta S$ can be used as a criterion for the spontaneity of a reaction. The change of free energy, ΔG, should be calculated for the conditions at which the reaction occurs. If the reactants are at standard conditions and these give products at standard conditions, then the free-energy change we need to look at is ΔG°. The calculation is simple, as shown in Example 20.5. For other conditions, we should look at the appropriate ΔG value. This would be a more complicated calculation, which we will not illustrate here. Nevertheless, the standard free-energy change, ΔG°, is still a useful guide to the spontaneity of reaction in these cases. The following rules are useful in judging the spontaneity of a reaction:

1. If ΔG° is a large negative number (more negative than -10 kJ), the reaction is spontaneous as written, and reactants transform almost entirely to products when equilibrium is reached.
2. If ΔG° is a large positive number (larger than 10 kJ), the reaction is nonspontaneous as written, and reactants do not give significant amounts of products at equilibrium.

Table 20.2
Standard Free Energies of
Formation (at 25°C)

Formula	ΔG_f°, kJ/mol	Formula	ΔG_f°, kJ/mol	Formula	ΔG_f°, kJ/mol
Hydrogen		*Carbon (continued)*		*Sulfur*	
$H^+(aq)$	0	$CS_2(l)$	63.6	$S_2(g)$	80.1
$H_2(g)$	0	$HCN(g)$	125	$S(rhombic)$	0
		$CCl_4(g)$	−53.7	$S(monoclinic)$	0.10
Sodium		$CCl_4(l)$	−68.6	$SO_2(g)$	−300.2
$Na^+(aq)$	−261.9	$CH_3CHO(g)$	−133.7	$H_2S(g)$	−33
$Na(s)$	0	$C_2H_5OH(l)$	−174.8		
$NaCl(s)$	−348.0			*Fluorine*	
$NaHCO_3(s)$	−851.9	*Silicon*		$F^-(aq)$	−276.5
$Na_2CO_3(s)$	−1048.1	$Si(s)$	0	$F_2(g)$	0
		$SiO_2(s)$	−856.5	$HF(g)$	−275
Calcium		$SiF_4(g)$	−1506		
$Ca^{2+}(aq)$	−553.0			*Chlorine*	
$Ca(s)$	0	*Lead*		$Cl^-(aq)$	−131.2
$CaO(s)$	−603.5	$Pb(s)$	0	$Cl_2(g)$	0
$CaCO_3(s)$	−1128.8	$PbO(s)$	−189	$HCl(g)$	−95.3
		$PbS(s)$	−96.7		
Carbon				*Bromine*	
$C(graphite)$	0	*Nitrogen*		$Br^-(aq)$	−102.8
$C(diamond)$	2.9	$N_2(g)$	0	$Br_2(l)$	0
$CO(g)$	−137.2	$NH_3(g)$	−16		
$CO_2(g)$	−394.4			*Iodine*	
$HCO_3^-(aq)$	−587.1	*Oxygen*		$I^-(aq)$	−51.7
$CH_4(g)$	−50.8	$O_2(g)$	0	$I_2(s)$	0
$C_2H_4(g)$	68.4	$O_3(g)$	163		
$C_2H_6(g)$	−32.9	$OH^-(aq)$	−157.3	*Silver*	
$C_6H_6(l)$	124.5	$H_2O(g)$	−228.6	$Ag^+(aq)$	77.1
$HCHO(g)$	−110	$H_2O(l)$	−237.2	$Ag(s)$	0
$CH_3OH(l)$	−166.2			$AgF(s)$	−185
$CS_2(g)$	66.9			$AgCl(s)$	−109.7
				$AgBr(s)$	−95.9
				$AgI(s)$	−66.3

3. If ΔG° has a small negative or positive value (less than about 10 kJ), the reaction gives an equilibrium mixture with significant amounts of both reactants and products.

Example 20.6

Calculate ΔH° and ΔG° for the reaction

$$2KClO_3(s) \longrightarrow 2KCl(s) + 3O_2(g)$$

Interpret the signs obtained for ΔH° and ΔG°. Values of ΔH_f° (in kJ/mol) are: $KClO_3(s)$, −391.2; $KCl(s)$, −436.7. Similarly, values of ΔG_f° (in kJ/mol) are: $KClO_3(s)$, −289.9; $KCl(s)$, −408.8.

Solution

The problem is set up as follows:

	$2KClO_3(s)$	$\longrightarrow$	$2KCl(s)$	$+$	$3O_2(g)$	
ΔH_f°:	2 × (−391.2)		2 × (−436.7)		0	kJ
ΔG_f°:	2 × (−289.9)		2 × (−408.8)		0	kJ

(Continued)

Then,

$$\Delta H° = 2 \times (-436.7) - 2 \times (-391.2) \text{ kJ}$$
$$= -91.0 \text{ kJ}$$
$$\Delta G° = 2 \times (-408.8) - 2 \times (-289.9) \text{ kJ}$$
$$= -237.8 \text{ kJ}$$

The reaction is exothermic, liberating 91.0 kJ of heat. The large negative value for $\Delta G°$ indicates that the equilibrium composition is predominantly potassium chloride and oxygen.

Exercise 20.6

Which of the following reactions are spontaneous in the direction written? See Table 20.2 for data.
(a) $C(\text{graphite}) + 2H_2(g) \longrightarrow CH_4(g)$
(b) $2H_2(g) + O_2(g) \longrightarrow 2H_2O(l)$
(c) $4HCN(g) + 5O_2(g) \longrightarrow 2H_2O(l) + 4CO_2(g) + 2N_2(g)$
(d) $Ag^+(aq) + I^-(aq) \longrightarrow AgI(s)$

(See Problems 20.33, 20.34, 20.35, and 20.36.)

20.5 Interpretation of Free Energy

We have seen that the free-energy change serves as a criterion of spontaneity of a chemical reaction. This gives us some idea of what free energy is and how we interpret it. In this section, we will look more closely at the meaning of free energy.

Maximum Work

Theoretically, spontaneous reactions can be used to obtain useful work. By useful work, we mean energy that can be used directly to move objects of normal size. We use the combustion of gasoline to move an automobile, and a reaction in a battery to generate electricity to drive a motor. Similarly, biochemical reactions in muscle tissue occur in a way to contract muscle fibers and lift a weight.

Often reactions are not carried out in a way to do useful work. The reactants are simply poured together in a reaction vessel, and products are separated from the mixture. As the reaction occurs, the free energy of the system decreases and entropy is produced. No useful work is obtained.

In principle, if a reaction is carried out to obtain the maximum useful work, no entropy is produced. It can be shown that the maximum useful work, w_{max}, that can be done by a spontaneous reaction is $-\Delta G$.

$$w_{max} = -\Delta G$$

The term *free energy* comes from this result. *The negative of the free-energy change is the maximum energy that is available, or free, to do useful work.* As a reaction occurs in a way to give the maximum useful work, the free energy decreases, and a corresponding quantity of useful work is obtained.∎

The concept of maximum work from a chemical reaction is an idealization. In any real situation, less than this maximum work will be obtained, and some entropy will be produced. When this work is eventually expended, it will appear in the environment as additional entropy.

■ It is possible to obtain the maximum work, and therefore ΔG, for some reactions from electrochemical cells (batteries), as we will show in Chapter 21.

Coupling of Reactions

One kind of useful work is that needed to effect a nonspontaneous chemical change. Consider the direct decomposition of iron(III) oxide, essentially rust, to iron:

$$2Fe_2O_3(s) \longrightarrow 4Fe(s) + 3O_2(g); \Delta G° = +1487 \text{ kJ}$$

The reaction is nonspontaneous, since $\Delta G°$ is a large positive quantity. This is in agreement with common knowledge. Iron tends to rust in air. We do not expect a rusty wrench to turn spontaneously into shiny iron and oxygen. But this does not mean we cannot change iron(III) oxide to iron. It only means that we will have to do work on the iron(III) oxide to reduce it. We must find a way to couple this reaction with one having a more negative $\Delta G°$.■

Consider the reaction

$$2CO(g) + O_2(g) \longrightarrow 2CO_2(g); \Delta G° = -514.4 \text{ kJ}$$

For three moles of O_2, the $\Delta G°$ is -1543 kJ, which is more negative than that for the direct decomposition of two moles of Fe_2O_3 to its elements. Let us add the two reactions:

$$2Fe_2O_3(s) \longrightarrow 4Fe(s) + 3O_2(g); \Delta G° = 1487 \text{ kJ}$$
$$6CO(g) + 3O_2(g) \longrightarrow 6CO_2(g); \Delta G° = -1543 \text{ kJ}$$
$$2Fe_2O_3(s) + 6CO(g) \longrightarrow 4Fe(s) + 6CO_2(g); \Delta G° = -56 \text{ kJ}$$

■ In addition to coupling, free energy can be supplied to chemical reactions by electrical work. The process is called electrolysis (see Chapter 21).

Thus, iron(III) oxide can be reduced spontaneously to free iron with carbon monoxide. In fact, this is the reaction that occurs in a blast furnace, where iron ore (mainly Fe_2O_3) is reduced to iron.

The concept of coupling two chemical reactions, one that is spontaneous with one that is nonspontaneous, to give a spontaneous change is a very useful one in biochemistry.■ Adenosine triphosphate, or ATP, is a large molecule containing phosphate groups. It plays a central role in the transfer of energy in living systems. ATP can react with water in the presence of an enzyme (biochemical catalyst) to give adenosine diphosphate, ADP, and a phosphate ion:

■ Coupling of reactions will be discussed in more detail in Chapter 27.

$$ATP + H_2O \longrightarrow ADP + \text{phosphate ion}; \Delta G° = -31 \text{ kJ}$$

ATP is first synthesized in a living organism using energy from food. The spontaneous reaction of ATP to give ADP is then coupled to various nonspontaneous reactions to accomplish the necessary reactions of the cell.

Free-Energy Change During Reaction

We have seen that the free-energy change is related to the work done during a chemical reaction. Consider the combustion of gasoline in O_2. The reaction is spontaneous, so the free-energy change is negative. That is, the free energy of the system changes to a lower value as the reactants are converted to products. Figure 20.4 shows the free-energy change during this reaction.

If the gasoline is burned in a gasoline stove, the decrease in free energy shows up as an increase in entropy of the system and surroundings. How-

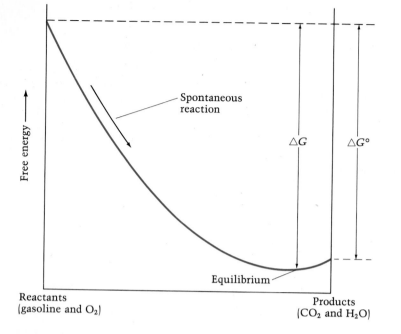

Figure 20.4
Free-energy change during a spontaneous reaction (combustion of gasoline). Note that the free energy decreases as the reaction proceeds. At equilibrium the free energy is a minimum, and the equilibrium mixture is mostly products.

ever, if the gasoline is burned in an automobile engine, some of this decrease in free energy shows up as work done. Theoretically, all of the free-energy decrease can be used to do work. This gives the maximum work. In practice, less work is obtained, and the difference appears as an increase of entropy. Ultimately, the work is itself used up, that is, changed to entropy.

Let us look at Figure 20.4 again. At the start of the reaction, the system contains only reactants, gasoline and O_2. The free energy has a relatively high value. This decreases as the reaction proceeds. The decrease in free energy appears either as an increase in entropy or as work done. Eventually, the free energy of the system reaches its minimum value. Then the net reaction stops; it has come to equilibrium.

We see that a reaction occurs only if the free energy can decrease. Therefore, the total change in free energy in going from the reactants to the equilibrium mixture must be negative. This free-energy change is labeled ΔG in Figure 20.4. It should be compared with the standard free-energy change, $\Delta G°$. This is the free-energy change when reactants are converted to products (reactants and products in their standard states). Note that ΔG and $\Delta G°$ are not identical. We use $\Delta G°$ as a criterion of spontaneity because it is simple to calculate. Moreover, as we will see in the next section, $\Delta G°$ is directly related to the equilibrium constant.

Before we leave this section, let us consider a reaction in which $\Delta G°$ is positive. We would predict that the reaction is nonspontaneous. Figure 20.5 shows the free-energy change as the reaction proceeds. Note that there is a small decrease in free energy as the system goes to equilibrium. Some reaction occurs in order to give the equilibrium mixture. But this mixture is primarily reactants, since the reaction does not go very far before coming to equilibrium. To change reactants completely to products is a nonspontaneous reaction (shown by the arrow along the curve in Figure 20.5).

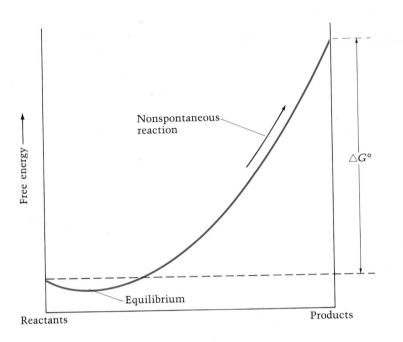

Figure 20.5
Free-energy change during a nonspontaneous reaction. The free energy decreases until equilibrium, at which the minimum value is reached. However, there is very little reaction because the equilibrium mixture is mostly reactants. To change reactants to mostly products would be a nonspontaneous reaction.

20.6 Calculation of Equilibrium Constants

One of the most important equations of chemical thermodynamics relates the equilibrium constant for a reaction to the standard free-energy change, $\Delta G°$. The relationship is

$$\Delta G° = -2.303 \, RT \log K_{th}$$

Here K_{th} is the **thermodynamic equilibrium constant.**

The thermodynamic equilibrium constant has the form of K_c. However, the concentrations of gases are expressed in partial pressures in atmospheres, whereas the concentrations of solutes in liquid solution are in molarities. For a reaction involving only solutes in liquid solution, K_{th} is identical to K_c. For reactions involving only gases, K_{th} equals K_p.

Example 20.7

Write expressions for the thermodynamic equilibrium constants for the following:
(a) The reaction given in the chapter opening,

$$2NH_3(g) + CO_2(g) \rightleftharpoons NH_2CONH_2(aq) + H_2O(l)$$

(b) The solubility process,

$$AgCl(s) \rightleftharpoons Ag^+(aq) + Cl^-(aq)$$

Solution

(a) Note that H_2O is the solvent, and its concentration is essentially constant. We do not include it in the equi-

librium constant.

$$K_{th} = \frac{[NH_2CONH_2]}{P_{CO_2}P_{NH_3}^2}$$

(b) Note that the concentration of $AgCl(s)$ is constant, so it is not included in K_{th}.

$$K_{th} = [Ag^+][Cl^-]$$

This equilibrium constant is identical to K_{sp}.

Exercise 20.7

Give the expression for K_{th} for the following reactions:
(a) $CaCO_3(s) \rightleftharpoons CaO(s) + CO_2(g)$
(b) $PbI_2(s) \rightleftharpoons Pb^{2+}(aq) + 2I^-(aq)$
(c) $H^+(aq) + HCO_3^-(aq) \rightleftharpoons H_2O(l) + CO_2(g)$

(See Problems 20.39 and 20.40.)

The next example illustrates the calculation of a thermodynamic equilibrium constant from the standard free-energy change, $\Delta G°$.

Example 20.8

What is the value of the equilibrium constant K_{th} at 25°C (298 K) for the reaction

$$2NH_3(g) + CO_2(g) \rightleftharpoons NH_2CONH_2(aq) + H_2O(l)$$

The standard free-energy change, $\Delta G°$, at 25°C equals -13.6 kJ. (We calculated this value of $\Delta G° = \Delta H° - T\Delta S°$ just before Section 20.4.)

Solution

We rearrange the equation $\Delta G° = -2.303 \, RT \log K_{th}$ to give

$$\log K_{th} = \frac{\Delta G°}{-2.303 \, RT}$$

Note that $\Delta G°$ and R must be in compatible units. We normally express $\Delta G°$ in joules and put R equal to 8.31

J/(K · mol). Substituting numerical values into this equation, we get

$$\log K_{th} = \frac{-13.6 \times 10^3}{-2.303 \times 8.31 \times 298} = 2.38$$

Hence, K_{th} = antilog 2.38 = $10^{2.38}$ = 2.4×10^2

Note that although the value of K_{th} indicates that products predominate at equilibrium, K_{th} is only moderately large. We would expect that the composition could be easily shifted toward reactants if we could remove either NH_3 or CO_2 (according to LeChatelier's principle). This is what happens when urea is used as a fertilizer. As NH_3 is used up, more NH_3 is produced by the decomposition of urea.

Exercise 20.8

Use data from Table 20.2 to obtain the equilibrium constant K_p for the reaction

$$CaCO_3(s) \rightleftharpoons CaO(s) + CO_2(g)$$

Note that values of $\Delta G_f°$ are needed for $CaCO_3$ and CaO even though the substances do not appear in $K_p = P_{CO_2}$.

(See Problems 20.41, 20.42, 20.43, and 20.44.)

The method of calculating K_{th} from $\Delta G°$ also works for net ionic equations. In the next example, we obtain a solubility-product constant from thermodynamic data.

Example 20.9

Calculate the equilibrium constant K_{sp} at 25°C for the reaction

$$AgCl(s) \rightleftharpoons Ag^+(aq) + Cl^-(aq)$$

Use $\Delta G_f°$ values from Table 20.2.

Solution

Note that K_{sp} equals K_{th}. We first calculate $\Delta G°$. Writing $\Delta G_f°$ values below the formulas in the equation gives

(*Continued*)

$$AgCl(s) \rightleftharpoons Ag^+(aq) + Cl^-(aq)$$

ΔG_f°: -109.7 77.1 -131.2 kJ

Hence, $\Delta G^\circ = [(-131.2 + 77.1) - (-109.7)]$ kJ

         $= 55.6$ kJ (or 55.6×10^3 J)

We now substitute numerical values into the equation relating log K_{th} and ΔG°:

$$\log K_{th} = \frac{\Delta G^\circ}{-2.303\ RT}$$

$$= \frac{55.6 \times 10^3}{-2.303 \times 8.31 \times 298} = -9.75$$

Therefore, $K_{th} = $ antilog $(-9.75) = 10^{-9.75}$

         $= 1.8 \times 10^{-10}$

Exercise 20.9

Calculate the solubility-product constant for $Mg(OH)_2$ at 25°C. The ΔG_f° values (in kJ/mol) are: $Mg^{2+}(aq)$, -456.0; $OH^-(aq)$, -157.3; $Mg(OH)_2(s)$, -933.9.

                                    (See Problems 20.45 and 20.46.)

We can understand the use of ΔG° as a criterion of spontaneity in terms of its relationship to the equilibrium constant, $\Delta G^\circ = -2.303\ RT \log K_{th}$. If the equilibrium constant is greater than 1, $\log K_{th}$ is positive and ΔG° is negative. Similarly, if the equilibrium constant is less than 1, $\log K_{th}$ is negative and ΔG° is positive. This agrees with the first two rules listed at the end of Section 20.4. We can get the third rule if we substitute $\Delta G^\circ = \pm 10 \times 10^3$ J into $\Delta G^\circ = -2.303\ RT \log K_{th}$ and solve for K_{th}. We find that K_{th} is between 0.018 and 57. In this range, the equilibrium mixture will contain significant amounts of reactants as well as products.

20.7 Change of Free Energy with Temperature

In the previous sections, we obtained the free-energy change and equilibrium constant for a reaction at 25°C, the temperature at which the thermodynamic data were given. How do we find ΔG° or K_{th} at another temperature? Precise calculations are possible, but rather involved. Instead, we will look at a simple method that gives approximate results.

In this method, we assume that ΔH° and ΔS° are essentially constant with respect to temperature. We get the value of ΔG° at any temperature by substituting values of ΔH° and ΔS° obtained from tables at 25°C into the following equation.■

■ This approximation is most accurate for temperatures not too different from that for which the ΔH° and ΔS° values are obtained. Much different temperatures give greater error.

$$\Delta G^\circ = \Delta H^\circ - T\Delta S^\circ$$

Spontaneity and Temperature Change

Each of the four possible choices of signs for ΔH° and ΔS°, listed in Table 20.3, give different temperature behaviors for ΔG°. Consider the case in which ΔH° is negative and ΔS° is positive. An example is the reaction

$$C_6H_{12}O_6(s) \longrightarrow 2C_2H_5OH(l) + 2CO_2(g)$$

     glucose   ethanol

This represents the overall change of glucose (grape sugar) to ethanol (ethyl alcohol). The signs of ΔH° and ΔS° are easily explained. The formation of

$$\Delta G^0 = \Delta H^0 - T\Delta S^0$$

$\Delta H°$	$\Delta S°$	$\Delta G°$	Description*	Example
−	+	−	Spontaneous at all T	$C_6H_{12}O_6(s) \longrightarrow 2C_2H_5OH(l) + 2CO_2(g)$
+	−	+	Nonspontaneous at all T	$3O_2(g) \longrightarrow 2O_3(g)$
−	−	+ or −	Spontaneous at low T; nonspontaneous at high T	$2NH_3(g) + CO_2(g) \longrightarrow NH_2CONH_2(aq) + H_2O(l)$
+	+	+ or −	Nonspontaneous at low T; spontaneous at high T	$Ba(OH)_2 \cdot 8H_2O(s) + 2NH_4SCN(s) \longrightarrow$ $Ba(SCN)_2(aq) + 2NH_3(g) + 10H_2O(l)$

*The terms *low* and *high* temperature are relative. For a particular reaction, high temperature could mean room temperature.

Table 20.3
Effect of Temperature on the Spontaneity of Reactions

more stable bonds, such as occur in CO_2, releases energy as heat. Thus, the reaction is exothermic and $\Delta H°$ is negative. As explained in Example 20.2, the breaking up of a molecule $(C_6H_{12}O_6)$ into smaller ones and the formation of a gas are expected to increase the entropy, so that $\Delta S°$ is positive.

When $\Delta H°$ is negative and $\Delta S°$ is positive, both terms in $\Delta G°$ (that is, $\Delta H°$ and $-T\Delta S°$) are negative. Therefore, $\Delta G°$ is always negative and the reaction is spontaneous whatever the temperature.

If the signs of $\Delta H°$ and $\Delta S°$ are reversed, that is, if $\Delta H°$ is positive (endothermic) and $\Delta S°$ is negative, $\Delta G°$ is always positive. Thus, the reaction is nonspontaneous at all temperatures. An example is the reaction in which oxygen gas is converted to ozone:

$$3O_2(g) \longrightarrow 2O_3(g)$$

To accomplish this conversion, oxygen is passed through a tube in which an electrical discharge occurs. The electrical discharge supplies the necessary free energy for this otherwise nonspontaneous change.

The reaction described in the chapter opening,

$$2NH_3(g) + CO_2(g) \longrightarrow NH_2CONH_2(aq) + H_2O(l)$$

is one in which both $\Delta H°$ and $\Delta S°$ are negative. In this case, the sign of $\Delta G°$ depends on the relative magnitudes of the terms $\Delta H°$ and $-T\Delta S°$, which have opposite signs. At some temperature, these terms will just cancel and $\Delta G°$ will equal zero. Below this temperature, $\Delta G°$ will be negative. Above it, $\Delta G°$ will be positive. Therefore, this reaction will be spontaneous at low temperatures, but will become nonspontaneous at sufficiently high temperatures. This particular reaction is spontaneous at 25°C, but becomes nonspontaneous at about 60°C.

A reaction in which both $\Delta H°$ and $\Delta S°$ are positive is the one described in the opening to Chapter 10:

$$Ba(OH)_2 \cdot 8H_2O(s) + 2NH_4SCN(s) \longrightarrow Ba(SCN)_2(aq) + 2NH_3(g) + 10H_2O(l)$$

The reaction is endothermic, and because crystalline solids change to a solution and a gas, the entropy increases. Again, the sign of $\Delta G°$ depends on the relative magnitude of the terms $\Delta H°$ and $-T\Delta S°$. The reaction is spontaneous at room temperature, but would be nonspontaneous at a sufficiently low temperature.■ Table 20.3 summarizes this discussion.

■ The reaction mixture spontaneously cools enough to freeze water. See Figure 10.1, p. 313.

Calculation of $\Delta G°$ at Various Temperatures

As an application of the method of calculating $\Delta G°$ at various temperatures, assuming $\Delta H°$ and $\Delta S°$ are constant, we will look at the following reaction:

$$CaCO_3(s) \rightleftharpoons CaO(s) + CO_2(g)$$

At 25°C, $\Delta G°$ equals $+130.9$ kJ, and the equilibrium partial pressure of CO_2 calculated from this is 1.1×10^{-23} atm. The very small value of this partial pressure shows that $CaCO_3$ is quite stable at room temperature. In the next example, we will see how $\Delta G°$ and K_p for this reaction change at higher temperature.

Example 20.10

(a) What is $\Delta G°$ at 1000°C for the following reaction?

$$CaCO_3(s) \longrightarrow CaO(s) + CO_2(g)$$

Is this reaction spontaneous at 1000°C and 1 atm? (b) What is the value of K_p at 1000°C for this reaction? What is the partial pressure of CO_2?

Solution

(a) From Tables 10.1 and 20.1, we have

	$CaCO_3(s) \longrightarrow$	$CaO(s)$	$+ CO_2(g)$
$\Delta H_f°$:	-1206.9	-635.1	-393.5 kJ
$S°$:	92.9	38.2	213.7 J/K

We calculate $\Delta H°$ and $\Delta S°$ from these values:

$\Delta H° = [(-635.1 - 393.5) - (-1206.9)]$ kJ $= 178.3$ kJ

$\Delta S° = [(38.2 + 213.7) - (92.9)]$ J/K $= 159.0$ J/K

Now we substitute $\Delta H°$, $\Delta S°$ ($= 0.159$ kJ/K), and T ($= 1273$ K) into the equation for $\Delta G°$:

$\Delta G° = \Delta H° - T\Delta S° = 178.3$ kJ $- (1273$ K$)(0.1590$ kJ/K$)$

$= -24.1$ kJ

Since $\Delta G°$ is negative, the reaction should be spontaneous at 1000°C and 1 atm. (b) We substitute the value of $\Delta G°$ at 1273 K, which equals -24.1×10^3 J, into the equation relating $\log K_{th}$ and $\Delta G°$:

$$\log K_{th} = \frac{\Delta G°}{-RT\,2.303} = \frac{-24.1 \times 10^3}{-8.31 \times 1273 \times 2.303}$$

$$= 0.989$$

$$K_{th} = K_p = \text{antilog } 0.989 = 9.75$$

Since $K_p = P_{CO_2}$, the partial pressure of CO_2 is 9.75 atm.

Exercise 20.10

The thermodynamic equilibrium constant for the vaporization of water,

$$H_2O(l) \rightleftharpoons H_2O(g)$$

is $K_p = P_{H_2O}$. Use thermodynamic data to calculate the vapor pressure of water at 45°C. Compare with the value given in Table 4.4, p. 110.

(See Problems 20.51 and 20.52.)

We can easily use the method described in Example 20.10 to find the temperature at which a reaction such as the decomposition of $CaCO_3$ changes from being nonspontaneous to spontaneous under standard conditions (1 atm for reactants and products). At this temperature $\Delta G°$ equals zero.

$$\Delta G° = 0 = \Delta H° - T\Delta S°$$

Solving for T gives

$$T = \frac{\Delta H°}{\Delta S°}$$

For the decomposition of $CaCO_3$, using values obtain in Example 20.10, we get

$$T = \frac{178.3 \text{ kJ}}{0.1590 \text{ kJ/K}} = 1121 \text{ K } (848°C)$$

Thus, $CaCO_3$ should be stable to thermal decomposition to CaO and CO_2 at 1 atm until heated to 848°C.

Exercise 20.11

To what temperature must magnesium carbonate be heated to decompose it to MgO and CO_2 at 1 atm? Is this higher or lower than that for $CaCO_3$? Values of ΔH_f° (in kJ/mol) are: $MgO(s)$, -601.2; $MgCO_3(s)$, -1112. Values of $S°$ (in J/K) are: $MgO(s)$, 26.9; $MgCO_3(s)$, 65.9. Data for CO_2 are in Tables 10.1 and 20.1.

(See Problems 20.53 and 20.54.)

A Checklist for Review

Important Terms

spontaneous process (20.2)
entropy (20.2)
second law of thermodynamics (20.2)

third law of thermodynamics (20.3)
standard entropies (20.3)
free energy (20.4)

standard free energy of formation (20.4)
thermodynamic equilibrium constant (20.6)

Summary of Facts and Concepts

Entropy (S) is a thermodynamic measure of randomness or disorder in a system. According to the *second law of thermodynamics,* the total entropy of a system and its surroundings increases for a *spontaneous process,* one that occurs of its own accord. For a spontaneous reaction at constant T and P, one can show that the entropy change, ΔS, is greater than $\Delta H/T$. For an equilibrium process, such as a phase transition, $\Delta S = \Delta H/T$. The standard entropy, $S°$, of a substance is determined by measuring the heat absorbed in warming it at increasing temperatures at 1 atm. The method depends on the *third law of thermodynamics,* which states that perfectly crystalline substances at 0 K have zero entropy.

Free energy, G, is defined as $H - TS$. The change in free energy, G, for a reaction at constant T and P equals $\Delta H - T\Delta S$, which is negative for a spontaneous change. *Standard free-energy changes,* $\Delta G°$, can be calculated from ΔH_f° and $S°$ values or from *standard free energies of formation,* ΔG_f°. The ΔG_f° for a substance is the standard free-energy change for the formation of the substance

from the elements in their stablest states. The standard free-energy change, $\Delta G°$, serves as a *criterion of spontaneity* of a reaction. A negative value means the reaction is spontaneous. If it equals zero, the reaction is at equilibrium. A positive value means the reaction is nonspontaneous.

The free-energy change, ΔG, is related to the *maximum work* that can be done by a reaction. This maximum work equals $-\Delta G$. A nonspontaneous reaction can be made to go by coupling it with a reaction that has a negative ΔG, so that the ΔG for the overall result is negative. The reaction with the negative ΔG does work on the nonspontaneous reaction.

The *thermodynamic equilibrium constant,* K_{th}, can be calculated from the standard free-energy change by the relationship $\Delta G° = -2.303 \, RT \log K_{th}$. The $\Delta G°$ at various temperatures can be obtained using the approximation that $\Delta H°$ and $\Delta S°$ are constant. Therefore, $\Delta G°$, which equals $\Delta H° - T\Delta S°$, can be easily calculated for any T.

Operational Skills

1. Given the heat of phase transition and the temperature of the transition, calculate the entropy change of the system, ΔS (Example 20.1).

2. Predict the sign of $\Delta S°$ for a reaction to which the rules listed in the text can be clearly applied (Example 20.2).

3. Given the standard entropies of reactants and products, calculate the entropy of reaction, $\Delta S°$ (Example 20.3).

4. Given enthalpies of formation and standard entropies of reactants and products, calculate the standard free-energy change, $\Delta G°$, for a reaction (Example 20.4).

5. Given the free energies of formation of reactants and products, calculate the standard free-energy change, $\Delta G°$, for a reaction (Example 20.5).

6. Use the standard free-energy change to decide the spontaneity of a reaction (Example 20.6).

7. For any balanced chemical equation, write the expression for the thermodynamic equilibrium constant (Example 20.7).

8. Given the standard free-energy change for a reaction, calculate the thermodynamic equilibrium constant (Examples 20.8 and 20.9).

9. Given $\Delta H°$ and $\Delta S°$ at 25°C, calculate $\Delta G°$ for a reaction at a temperature other than 25°C (Example 20.10).

Review Questions

20.1 What is a spontaneous process? Give three examples of spontaneous processes. Give three examples of nonspontaneous processes.

20.2 Which contains greater entropy, a quantity of frozen benzene or the same quantity of liquid benzene at the same temperature? Explain in terms of the degree of order of the substance.

20.3 Give a statement of the second law of thermodynamics.

20.4 The entropy change, ΔS, for a phase transition equals $\Delta H/T$, where ΔH is the enthalpy change. Why is it that the entropy change for a system in which a chemical reaction occurs spontaneously does not equal $\Delta H/T$?

20.5 Describe how the standard entropy of hydrogen gas at 25°C can be obtained from heat measurements.

20.6 Describe what you would look for in a reaction involving gases in order to predict the sign of $\Delta S°$. Explain.

20.7 Define the free energy, G. How is ΔG related to ΔH and ΔS?

20.8 What is meant by the standard free-energy change, $\Delta G°$, for a reaction? What is meant by the standard free energy of formation, $\Delta G_f°$, of a substance?

20.9 Describe how $\Delta G°$ can be used to decide whether a chemical equation is spontaneous in the direction written.

20.10 What is the useful work obtained in the ideal situation in which a chemical reaction with free-energy change, ΔG, is run so that it produces no entropy?

20.11 Give an example of a chemical reaction used to obtain useful work.

20.12 How is the concept of coupling of reactions useful in explaining how a nonspontaneous change could be made to occur?

20.13 Explain how the free energy changes as a spontaneous reaction occurs. Show by means of a diagram how G changes with the extent of reaction.

20.14 Explain how an equilibrium constant can be obtained from thermal data alone (that is, from measurements of heat only).

20.15 Discuss the different sign combinations of ΔH and ΔS that are possible for a process carried out at constant temperature and pressure. For each combination, state whether the process must be spontaneous or not, or whether both situations are possible. Explain.

20.16 Consider a reaction in which $\Delta H°$ and $\Delta S°$ are positive. Suppose the reaction is nonspontaneous at room temperature. How would you estimate the temperature at which the reaction becomes spontaneous?

Problems

Entropy Changes

20.17 Acetic acid, CH_3COOH, freezes at 16.6°C. If the heat of fusion, ΔH_{fus}, is 69.0 J/g, what is the change of ~~entropy,~~ ΔS, when one mole of liquid acetic acid freezes ~~to the~~

20.19 The enthalpy change when liquid methanol, CH_3OH, vaporizes at 25°C is 37.4 kJ/mol. What is the entropy change when one mole of vapor in equilibrium with the liquid condenses to liquid at 25°C? If the entropy of this vapor at 25°C is 252 J/(mol · K), what is the entropy of the liquid at this temperature?

20.21 Without doing any calculations, decide what the sign of $\Delta S°$ will be for the following reactions:

    (a) $2LiOH(aq) + CO_2(g) \longrightarrow Li_2CO_3(aq) + H_2O(l)$
    (b) $(NH_4)_2Cr_2O_7(s) \longrightarrow$
                    $N_2(g) + 4H_2O(g) + Cr_2O_3(s)$
    (c) $2N_2O_5(g) \longrightarrow 4NO_2(g) + O_2(g)$
    (d) $O_2(g) + 2F_2(g) \longrightarrow 2OF_2(g)$

20.23 Calculate $\Delta S°$ for the following reaction using standard entropy values, Table 20.1.

$$CS_2(l) + 3O_2(g) \longrightarrow CO_2(g) + 2SO_2(g)$$

20.25 Calculate $\Delta S°$ for the reaction

$$CH_4(g) + 2O_2(g) \longrightarrow CO_2(g) + 2H_2O(l)$$

See Table 20.1 for values of standard entropies. Does the entropy of the chemical system increase or decrease as you expect? Explain.

20.18 Acetone, CH_3COCH_3, boils at 56°C. The heat of vaporization of acetone at this temperature is 1.83 kJ/mol. What is the entropy change when one mole of liquid acetone vaporizes at 56°C?

20.20 The heat of vaporization of carbon disulfide, CS_2, at 25°C is 29 kJ/mol. What is the entropy change when one mole of carbon disulfide vapor in equilibrium with liquid condenses to liquid at 25°C? If the entropy of this vapor at 25°C is 248 J/(mol · K), what is the entropy of the liquid at this temperature?

20.22 For each of the following reactions decide whether there is an increase or decrease of entropy. Why do you think so? (No calculations are needed.)

    (a) $N_2(g) + 3H_2(g) \longrightarrow 2NH_3(g)$
    (b) $NH_4Cl(s) \longrightarrow NH_3(g) + HCl(g)$
    (c) $CO(g) + 2H_2(g) \longrightarrow CH_3OH(l)$
    (d) $Li_3N(s) + 3H_2O(l) \longrightarrow 3LiOH(aq) + NH_3(g)$

20.24 Calculate $\Delta S°$ for the following reaction using standard entropy values, Table 20.1.

$$CS_2(g) + 4H_2(g) \longrightarrow CH_4(g) + 2H_2S(g)$$

20.26 What is the change of entropy, $\Delta S°$, for the reaction

$$CaCO_3(s) + 2H^+(aq) \longrightarrow Ca^{2+}(aq) + H_2O(l) + CO_2(g)$$

See Table 20.1 for values of standard entropies. Does the entropy of the chemical system increase or decrease as you expect? Explain.

Free-Energy Change and Spontaneity

20.27 Using enthalpies of formation (Table 10.1), calculate $\Delta H°$ for the reaction in Problem 20.23. Combine this with the value of $\Delta S°$ found in that problem to find $\Delta G°$.

20.29 Write the chemical equation whose standard free-energy change is the free energy of formation of $CO(g)$.

20.31 Calculate the standard free-energy change, $\Delta G°$, for the following reaction, using free energies of formation, Table 20.2.

$$CH_4(g) + 2O_2(g) \longrightarrow CO_2(g) + 2H_2O(l)$$

20.28 Calculate $\Delta G°$ for the reaction given in Problem 20.24. Use standard enthalpies of formation (Table 10.1) and the standard entropy change found in Problem 20.24.

20.30 Write the chemical equation whose standard free-energy change is the free energy of formation of $HCl(g)$.

20.32 What is the standard free-energy change for the reaction

$$CaCO_3(s) + 2H^+(aq) \longrightarrow Ca^{2+}(aq) + H_2O(l) + CO_2(g)$$

Use $\Delta G_f°$ values in Table 20.2.

20.33 On the basis of $\Delta G°$ for each of the following, decide whether the reaction is spontaneous or nonspontaneous as written. Or, if you expect an equilibrium mixture with significant amounts of both reactants and products, state so.

(a) $SO_2(g) + 2H_2S(g) \longrightarrow 3S(s) + 2H_2O(g)$;
$$\Delta G° = -91 \text{ kJ}$$

(b) $2H_2O_2(aq) \longrightarrow O_2(g) + 2H_2O(l)$;
$$\Delta G° = -211 \text{ kJ}$$

(c) $HCOOH(l) \longrightarrow CO_2(g) + H_2(g)$;
$$\Delta G° = 119 \text{ kJ}$$

(d) $I_2(s) + Br_2(l) \longrightarrow 2IBr(g)$; $\Delta G° = 7.5 \text{ kJ}$

(e) $NH_4Cl(s) \longrightarrow NH_3(g) + HCl(g)$; $\Delta G° = 92 \text{ kJ}$

20.35 Calculate $\Delta H°$ and $\Delta G°$ for the following reaction using Tables 10.1 and 20.2.

$$H_2(g) + S(\text{rhombic}) \longrightarrow H_2S(g)$$

Interpret the signs of $\Delta H°$ and $\Delta G°$.

20.34 For each of the following, state whether the reaction is spontaneous or nonspontaneous as written, or is easily reversible (that is, is a mixture with significant amounts of reactants and products)

(a) $HCN(g) + 2H_2(g) \longrightarrow CH_3NH_2(g)$;
$$\Delta G° = -92 \text{ kJ}$$

(b) $N_2(g) + O_2(g) \longrightarrow 2NO(g)$; $\Delta G° = 173 \text{ kJ}$

(c) $2NO(g) + 3H_2O(g) \longrightarrow 2NH_3(g) + \frac{5}{2}O_2(g)$;
$$\Delta G° = 479 \text{ kJ}$$

(d) $H_2(g) + Cl_2(g) \longrightarrow 2HCl(g)$; $\Delta G° = -191 \text{ kJ}$

(e) $H_2(g) + I_2(s) \longrightarrow 2HI(g)$; $\Delta G° = 2.6 \text{ kJ}$

20.36 Calculate $\Delta H°$ and $\Delta G°$ for the following reaction using Tables 10.1 and 20.2.

$$H_2(g) + F_2(g) \longrightarrow 2HF(g)$$

Interpret the signs of $\Delta H°$ and $\Delta G°$.

Maximum Work

20.37 Consider the reaction of two moles of $H_2(g)$ at 25°C and 1 atm with one mole of $O_2(g)$ at the same temperature and pressure to produce liquid water at these same conditions. If this reaction is run in a controlled way to generate work, what is the maximum useful work that can be obtained? How much entropy is produced in this case?

20.38 Consider the reaction of one mole of $H_2(g)$ at 25°C and 1 atm with one mole of $Cl_2(g)$ at the same temperature and pressure to produce gaseous HCl at these same conditions. If this reaction is run in a controlled way to generate work, what is the maximum useful work obtained? How much entropy is produced in this case?

Calculation of Equilibrium Constants

20.39 Give the expression for the thermodynamic equilibrium constant for each of the following reactions:

(a) $CO(g) + H_2O(g) \rightleftharpoons CO_2(g) + H_2(g)$

(b) $Mg(OH)_2(s) \rightleftharpoons Mg^{2+}(aq) + 2OH^-(aq)$

(c) $2Li(s) + 2H_2O(l) \rightleftharpoons 2Li^+(aq)$
$$+ 2OH^-(aq) + H_2(g)$$

20.41 What is the standard free-energy change, $\Delta G°$, for the following reaction? Obtain necessary information from Table 20.2.

$$H_2(g) + Cl_2(g) \longrightarrow 2HCl(g)$$

What is the value of the thermodynamic equilibrium constant, K_{th}?

20.43 Calculate the standard free-energy change and the equilibrium constant K_p for the following reaction at 25°C:

$$CO(g) + H_2O(g) \rightleftharpoons CO_2(g) + H_2(g)$$

See Table 20.2 for data.

20.40 Write the expression for the thermodynamic equilibrium constant for each of the following reactions:

(a) $CO(g) + 2H_2(g) \rightleftharpoons CH_3OH(g)$

(b) $2Ag^+(aq) + CrO_4^{2-}(aq) \rightleftharpoons Ag_2CrO_4(s)$

(c) $CaCO_3(s) + 2H^+(aq) \rightleftharpoons Ca^{2+}(aq)$
$$+ H_2O(l) + CO_2(g)$$

20.42 What is the standard free-energy change, $\Delta G°$, for the following reaction? (See Table 20.2.)

$$C(\text{graphite}) + O_2(g) \longrightarrow CO_2(g)$$

Calculate the value of the thermodynamic equilibrium constant, K_{th}.

20.44 Calculate the standard free-energy change and the equilibrium constant K_p for the following reaction at 25°C:

$$CO(g) + 2H_2(g) \rightleftharpoons CH_3OH(g)$$

See Table 20.2 for data.

20.45 Obtain the equilibrium constant K_c from the free-energy change for the following reaction:

$$Mg(s) + Cu^{2+}(aq) \longrightarrow Mg^{2+}(aq) + Cu(s)$$

Standard free energies of formation are (in kJ/mol): $Cu^{2+}(aq)$, 65; $Mg^{2+}(aq)$, −456.

20.46 Calculate the equilibrium constant K_c from the free-energy change for the following reaction:

$$Zn(s) + Cu^{2+}(aq) \longrightarrow Zn^{2+}(aq) + Cu(s)$$

Standard free energies of formation are (in kJ/mol): $Zn^{2+}(aq)$, −147.2; $Cu^{2+}(aq)$, 65.0.

Free Energy and Temperature Change

20.47 What is the sign of $\Delta S°$ for the reaction

$$2N_2O_5(s) \longrightarrow 4NO_2(g) + O_2(g)$$

The reaction is endothermic and spontaneous at 25°C. Explain the spontaneity of the reaction in terms of enthalpy and entropy changes.

20.48 The combustion of acetylene, C_2H_2, is a spontaneous reaction given by the equation

$$2C_2H_2(g) + 5O_2(g) \longrightarrow 4CO_2(g) + 2H_2O(l)$$

As expected for a combustion, the reaction is exothermic. What is the sign of $\Delta H°$? What do you expect for the sign of $\Delta S°$? Explain the spontaneity of the reaction in terms of the enthalpy and entropy changes.

20.49 Estimate the value of ΔH for the following reaction from bond energies (Table 10.2):

$$H_2(g) + Cl_2(g) \longrightarrow 2HCl(g)$$

Is the reaction exothermic or endothermic? Note that the reaction involves the breaking of symmetrical molecules (H_2 and Cl_2) and the formation of a less symmetrical product (HCl). From this, would you expect ΔS to be positive or negative? Comment on the spontaneity of the reaction in terms of the enthalpy and entropy changes.

20.50 Compare the energies of the bonds broken and formed (see Table 10.2) for the reaction

$$HCN(g) + 2H_2(g) \longrightarrow CH_3NH_2(g)$$

From this, conclude whether the reaction is exothermic or endothermic. What is the sign of ΔS? Explain. The reaction is spontaneous at 25°C. Explain this in terms of the enthalpy and entropy changes.

20.51 Use data given in Table 10.1 and 20.1 to obtain the value of K_p at 1000°C for the reaction

$$C(graphite) + CO_2(g) \rightleftharpoons 2CO(g)$$

Carbon monoxide is known to form during combustion of carbon at high temperatures. Do the data agree with this? Explain.

20.52 Use data given in Tables 10.1 and 20.1 to obtain the value of K_p at 2000°C for the reaction

$$N_2(g) + O_2(g) \rightleftharpoons 2NO(g)$$

Nitric oxide is known to form in hot flames in air, which is a mixture of N_2 and O_2. It is present in auto exhaust from this reaction. Are the data in agreement with this result? Explain.

20.53 Sodium carbonate, Na_2CO_3, can be prepared by heating sodium hydrogen carbonate, $NaHCO_3$.

$$2NaHCO_3(s) \longrightarrow Na_2CO_3(s) + H_2O(g) + CO_2(g)$$

Estimate the temperature at which $NaHCO_3$ decomposes to products at 1 atm. Use data from Tables 10.1 and 20.1.

20.54 Oxygen was first prepared by heating mercury(II) oxide, HgO.

$$2HgO(s) \longrightarrow 2Hg(g) + O_2(g)$$

Estimate the temperature at which HgO decomposes to O_2 at 1 atm. Values of $\Delta H_f°$ (kJ/mol) are: HgO(s), −90.8; Hg(g), 61.3. Values of $S°$ (J/(mol·K)) are: HgO(s), 70.3; Hg(g), 174.9. See also Tables 10.1 and 20.1.

Additional Problems

20.55 Nitrogen dioxide reacts with water according to the equation

$$3NO_2(g) + H_2O(l) \longrightarrow 2HNO_3(l) + NO(g)$$

The reaction is used commercially to produce nitric acid. Predict the sign of $\Delta S°$ for this reaction.

20.56 Ethanol burns in air or oxygen according to the equation

$$C_2H_5OH(l) + 3O_2(g) \longrightarrow 2CO_2(g) + 3H_2O(g)$$

Predict the sign of ΔS° for this reaction.

20.57 Acetic acid in vinegar results from the bacterial oxidation of ethanol.

$$C_2H_5OH(l) + \tfrac{1}{2}O_2(g) \longrightarrow CH_3COOH(l)$$

What is ΔS° for this reaction? Use standard entropy values. (See Appendix C for data.)

20.58 Methanol is produced commercially from carbon monoxide and hydrogen.

$$CO(g) + H_2(g) \longrightarrow CH_3OH(l)$$

What is ΔS° for this reaction? Use standard entropy values.

20.59 Is the following reaction spontaneous as written? Explain. Do whatever calculation is needed to answer the question.

$$SO_2(g) + H_2(g) \longrightarrow H_2S(g) + O_2(g)$$

20.60 Is the following reaction spontaneous as written? Explain. Do whatever calculation is needed to answer the question.

$$CH_4(g) + N_2(g) \longrightarrow HCN(g) + NH_3(g)$$

20.61 The reaction

$$CO_2(g) + H_2(g) \longrightarrow CO(g) + H_2O(g)$$

is nonspontaneous at room temperature, but becomes spontaneous at a much higher temperature. What can you conclude from this about the signs of ΔH° and ΔS°, assuming that the enthalpy and entropy changes are not greatly affected by the temperature change? Explain your reasoning.

20.62 The reaction

$$N_2(g) + 3H_2(g) \longrightarrow 2NH_3(g)$$

is spontaneous at room temperature, but becomes non-spontaneous at a much higher temperature. From this fact alone, obtain the signs of ΔH° and ΔS°, assuming that ΔH° and ΔS° do not change much with temperature. Explain your reasoning.

20.63 Calculate ΔG° at 25°C for the reaction

$$CaF_2(s) \rightleftharpoons Ca^{2+}(aq) + 2F^-(aq)$$

The value of ΔG_f° at 25°C for $CaF_2(s)$ is -1162 kJ/mol. See Table 20.2 for other values. What is the value of the solubility-product constant, K_{sp}, for this reaction at 25°C?

20.64 Calculate ΔG° at 25°C for the reaction

$$BaSO_4(s) \rightleftharpoons Ba^{2+}(aq) + SO_4^-(aq)$$

The values of ΔG_f° at 25°C are (in kJ/mol): $BaSO_4(s)$, -1353; $Ba^{2+}(aq)$, -561; $SO_4^{2-}(aq)$, -742. What is the value of the solubility-product constant, K_{sp}, for this reaction at 25°C?

20.65 What is the standard free-energy change at 25°C for the vaporization of carbon disulfide? Calculate the vapor pressure of carbon disulfide at 25°C.

20.66 What is the standard free-energy change at 25°C for the vaporization of carbon tetrachloride, CCl_4? Calculate the vapor pressure of carbon tetrachloride at 25°C.

20.67 Consider the decomposition of phosgene, $COCl_2$:

$$COCl_2(g) \longrightarrow CO(g) + Cl_2(g)$$

Calculate ΔH° and ΔS° at 25°C for this reaction. The standard enthalpy of formation of $COCl_2(g)$ is -220 kJ/mol, and the standard entropy is 284 J/(mol · K). See Tables 10.1 and 20.1 for other data. What is ΔG° at 25°C? Assume that ΔH° and ΔS° are constant with respect to a change of temperature. Now calculate ΔG° at 800°C. Compare the two values of ΔG°. Briefly discuss the spontaneity of the reaction at 25°C and at 800°C.

20.68 Consider the following reaction:

$$CS_2(g) + 4H_2(g) \rightleftharpoons CH_4(g) + 2H_2S(g)$$

Calculate ΔH°, ΔS°, and ΔG° at 25°C for this reaction. Assume ΔH° and ΔS° are constant with respect to a change of temperature. Now calculate ΔG° at 650°C. Compare the two values of ΔG°. Briefly discuss the spontaneity of the reaction at 25°C and at 650°C.

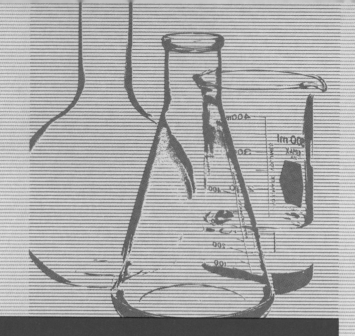

21. Electrochemistry

21.1 Electrochemical Cells

Voltaic Cells

21.2 Some Commercial Voltaic Cells
21.3 Notation for a Voltaic Cell
21.4 Electromotive Force
21.5 Electrode Potentials Cell emf/ Strength of an Oxidizing or Reducing Agent
21.6 Equilibrium Constants from emf's
21.7 Dependence of emf on Concentration Nernst Equation/ Electrode Potentials for Nonstandard Conditions/ Determination of pH

Electrolytic Cells

21.8 Aqueous Electrolysis
21.9 Stoichiometry of Electrolysis

The first battery was invented by Alessandro Volta about 1800. He assembled a pile of metal disks that were alternately zinc and silver. Each disk was separated from adjacent disks by paper soaked in salt water. With a tall pile of metal disks, he could detect a weak electrical shock when he touched the two ends of the pile. Later, Volta showed that any two different metals could be used to make such a battery or voltaic pile.

A battery cell that became popular during the nineteenth century was constructed in 1836 by the English chemist John Frederick Daniell. It used zinc and copper. The basic principle was that of Volta's battery pile, but the solutions surrounding each metal were kept separate by a porous pot. Each metal with its solution was a half-cell; a zinc half-cell and a copper half-cell made up one voltaic cell. This construction became the standard form of such cells, which use the spontaneous chemical reaction

$$Zn(s) + Cu^{2+}(aq) \longrightarrow Zn^{2+}(aq) + Cu(s)$$

to generate electrical energy. In this chapter, we will look at the general principles involved in setting up a chemical reaction as a battery. We will answer such questions as, "What voltage can we expect from a particular battery?" and "How can we relate the battery voltage to the equilibrium constant for the reaction?"

Chapter Overview

Electrochemistry concerns the study of electrical effects in chemical reactions. Such electrical effects are usually carried out in cells, called electrochemical cells. A battery or *voltaic cell* is one kind of electrochemical cell. It uses the work available from a spontaneous reaction to produce an electric current. After looking at some commercial voltaic cells, we will look at the maximum voltage that a cell can develop, relating it to the standard free-energy change and the equilibrium constant for the reaction. An *electrolytic cell* is another kind of electrochemical cell, in which an electric current is used to force a reaction to go in the nonspontaneous direction.

21.1 Electrochemical Cells

In discussing oxidation–reduction reactions in Chapter 9, we described the separation of such reactions into half-reactions. The separation into half-reactions was done only on paper, as a way to balance chemical equations. It is possible in some cases, however, to physically separate a reaction into half-reactions. In that event, the useful energy or work available from a spontaneous reaction may be used to drive electrons from the oxidation half-reaction through an external circuit to the reduction half-reaction. This is what happens in the operation of a battery.

A battery is a kind of electrochemical cell. An **electrochemical cell** is a system consisting of electrodes that dip into an electrolyte and in which a chemical reaction either uses or generates an electric current. A **voltaic,** or **galvanic, cell** is an electrochemical cell in which a spontaneous reaction generates an electric current. An **electrolytic cell** is an electrochemical cell in which an electric current is used to drive a nonspontaneous reaction. The nonspontaneous reaction driven by the electric current is called **electrolysis.**

When a metal dips into a solution of the metal ion, an equilibrium occurs between the metal and the metal ion. Ions enter the solution from the metal, leaving electrons behind, and metal ions in the solution combine with elec-

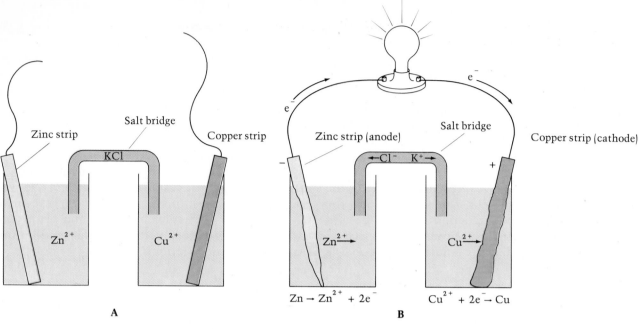

A

B

Figure 21.1
A voltaic cell. (a) A zinc and a copper electrode, without an external circuit; there is no cell reaction. (b) When the two electrodes are connected by an external circuit (light bulb), chemical reaction occurs. As the cell discharges, the anode decreases in mass and the cathode increases.

trons from the metal. The equilibrium of a metal M with a singly charged ion, for example, can be represented by the half-reaction

$$M(s) \rightleftharpoons M^+(aq) + e^-$$

The equilibrium causes an electrical charge, positive or negative, to form on the metal. This charge prevents the half-reaction from proceeding to any significant extent unless the charge is continuously drained away.

A voltaic cell consists of two **half-cells** that are electrically connected. Each half-cell is the portion of the total cell in which a half-reaction takes place. Figure 21.1a is a diagram of a voltaic cell. In its essential features, it is like the Daniell cell, which we described in the chapter opening. The left half-cell consists of a strip of zinc metal in a solution of zinc sulfate, giving the equilibrium

$$Zn(s) \rightleftharpoons Zn^{2+}(aq) + 2e^-$$

The right half-cell is a copper metal strip that dips into copper(II) sulfate; the equilibrium is

$$Cu(s) \rightleftharpoons Cu^{2+}(aq) + 2e^-$$

These half-cells in Figure 21.1a are connected electrically by a **salt bridge** that allows the flow of charged ions but prevents diffusional mixing of the different solutions. If the solutions were to mix, direct chemical reaction would take place, destroying the half-cells. The salt bridge shown contains aqueous potassium chloride (to provide for ion flow) in a gel (to prevent mixing of the separate solutions).■

Since zinc tends to lose electrons more readily than copper, the zinc electrode takes on a negative charge relative to the copper electrode. If the two electrodes are now connected through an external metallic wire, as in

■ The porous pot in the original Daniell cell has the same essential purpose as the salt bridge.

Figure 21.1b, electrons flow from the zinc through the external circuit to the copper. The following half-reactions occur:

$$Zn(s) \longrightarrow Zn^{2+}(aq) + 2e^-$$
$$Cu^{2+}(aq) + 2e^- \longrightarrow Cu(s)$$

The overall voltaic cell reaction is the sum of these two half-reactions.

$$Zn(s) + Cu^{2+}(aq) \longrightarrow Zn^{2+}(aq) + Cu(s)$$

Note that reduction occurs at the copper electrode and oxidation occurs at the zinc electrode.

In a voltaic cell, the electrical work can be used in the external circuit to light a bulb, drive a motor, and so forth. On the other hand, if the external circuit is replaced by a source of electricity that opposes that of the voltaic cell, the electrode reactions can be reversed. Now the external source pushes the electrons in the opposite direction and supplies energy or work to the cell so that the reverse, nonspontaneous reaction occurs.

For the zinc–copper cell, the half-reactions are reversed to give

$$Zn^{2+}(aq) + 2e^- \longrightarrow Zn(s)$$
$$Cu(s) \longrightarrow Cu^{2+}(aq) + 2e^-$$

and the overall reaction becomes

$$Zn^{2+}(aq) + Cu(s) \longrightarrow Zn(s) + Cu^{2+}(aq)$$

Now oxidation occurs at the copper electrode, and reduction occurs at the zinc electrode. The cell operates as an electrolytic cell (Figure 21.2), in which energy from an external source drives a nonspontaneous reaction.

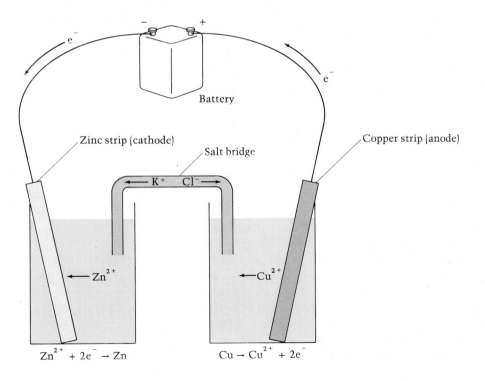

Figure 21.2
An electrolytic cell.
A source of electricity (a battery) reverses the half-reactions of Figure 21.1. The zinc strip is now the cathode and the copper strip is the anode.

Whether a cell operates as a voltaic or an electrolytic cell, the electrode at which reduction occurs is called the **cathode** and the electrode at which oxidation occurs is called the **anode.**

Once we know which electrode is the anode and which is the cathode, we can easily decide the direction of electron flow in the external circuit attached to the electrochemical cell. Electrons are given up by the anode reaction and thus flow from the anode, whereas they are used up by the cathode reaction and so flow into the cathode. Look again at Figures 21.1 and 21.2, and note the direction of electron flow in the external circuit in both cases.

Note also the labeling of electrodes as anode and cathode and the direction of migration of ions. Whether the cell is operating as a voltaic cell or as an electrolytic cell, cations move toward the cathode (or away from the anode) and anions move toward the anode (or away from the cathode).

Example 21.1

Consider the following electrochemical cells. For each one, sketch the cell. Label the anode and cathode, showing the corresponding electrode reactions. Indicate the direction of electron flow in the external circuit and the direction of migration of the cations in the half-cells. (a) A voltaic cell is constructed from a half-cell in which a cadmium electrode dips into a solution of cadmium nitrate and another half-cell in which a silver electrode dips into a solution of silver nitrate. The two half-cells are connected through a salt bridge.

Silver ion is reduced during operation of the voltaic cell. (b) An aqueous solution of copper(II) nitrate is electrolyzed by dipping silver electrodes (attached to a battery) into the solution. The electrode reactions are

$$Ag(s) \longrightarrow Ag^+(aq) + e^-$$
$$Cu^{2+}(aq) + 2e^- \longrightarrow Cu(s)$$

Solution

Sketches for (a) and (b) are given in Figure 21.3.

Exercise 21.1

Consider the following cells:
(a) A voltaic cell consists of a silver–silver ion half-cell and a nickel–nickel(II) ion half-cell. Silver ion is reduced during operation of the cell.
(b) The cell in (a) is operated as an electrolytic cell by connecting the electrodes to a battery.

Sketch each cell, labeling the anode and cathode and indicating the corresponding electrode reactions. Show the direction of electron flow in the external circuit and the direction of cation movement in the half-cells.

(See Problems 21.19, 21.20, 21.21, and 21.22.)

Voltaic Cells

One area of electrochemistry deals with voltaic cells, in which a spontaneous chemical reaction is set up to generate an electric current. We will look first at some commercial uses of voltaic cells, then discuss the basic principles behind them.

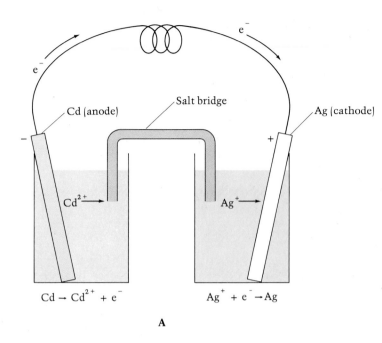

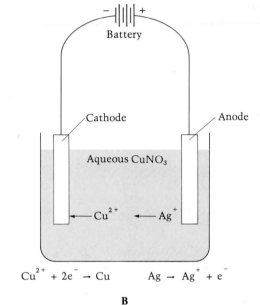

A

B

Figure 21.3
(a) The voltaic cell described in Example 21.1a. (b) The electrolysis described in Example 21.1b.

21.2 Some Commercial Voltaic Cells

A common use of voltaic cells is as convenient, portable sources of energy. Flashlights and radios, for example, are often powered by the **zinc–carbon** or **Leclanché dry cell** (Figure 21.4). This cell has a zinc can as the anode; a graphite rod in the center, immediately surrounded by a paste of manganese dioxide and carbon black, is the cathode. Around this is another paste, this one containing ammonium and zinc chlorides. The electrode reactions are complicated, but are approximately these:

$$Zn(s) \longrightarrow Zn^{2+}(aq) + 2e^- \qquad \text{(anode reaction)}$$
$$2NH_4^+(aq) + 2MnO_2(s) + 2e^- \longrightarrow$$
$$\qquad\qquad Mn_2O_3(s) + H_2O(l) + 2NH_3(aq) \qquad \text{(cathode reaction)}$$

The electrical pressure difference or voltage of this dry cell is initially about 1.5 volts (V), but decreases as current is drawn off. The voltage also deteriorates rapidly in cold weather. An alkaline dry cell is similar to the Leclanché cell, but has potassium hydroxide in place of ammonium chloride. This cell performs better under current drain and in cold weather.

Once a dry cell is completely discharged (has come to equilibrium), the cell cannot be reversed, or recharged, and is discarded. Some types of cell are rechargeable after use, however. The best known of these is the **lead storage cell.** It consists of electrodes of lead alloy grids; one electrode is packed with a spongy lead to form the anode, and the other is packed with lead dioxide to form the cathode (see Figure 21.5). They are bathed in an aqueous solution of sulfuric acid, H_2SO_4. The half-cell reactions during discharge are

$$Pb(s) + HSO_4^-(aq) \longrightarrow PbSO_4(s) + H^+(aq) + 2e^- \qquad \text{(anode reaction)}$$

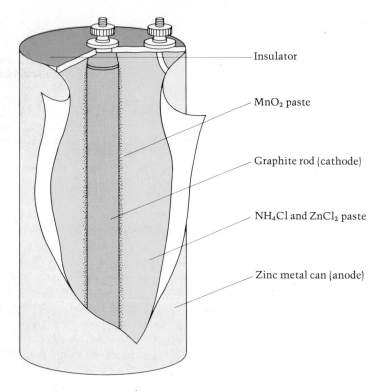

Figure 21.4
A Leclanché dry cell.

Insulator

MnO_2 paste

Graphite rod (cathode)

NH_4Cl and $ZnCl_2$ paste

Zinc metal can (anode)

$$PbO_2(s) \ + \ 3H^+(aq) \ + \ HSO_4^-(aq) \ + \ 2e^- \longrightarrow$$
$$PbSO_4(s) \ + \ 2H_2O(l) \qquad \text{(cathode reaction)}$$

White lead sulfate coats each electrode during discharge, and sulfuric acid is consumed. When the cell is recharged, the half-cell reactions are reversed. (An automobile's alternator causes essentially the same reversal of the half-cell reactions in a car battery after the motor is started, thereby recharging the battery.) Each cell delivers about 2 V, and a battery consisting of six cells in series gives about 12 V.

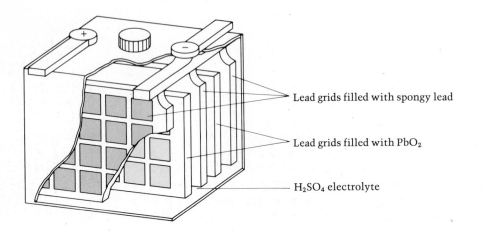

Figure 21.5
A lead storage battery.

Lead grids filled with spongy lead

Lead grids filled with PbO_2

H_2SO_4 electrolyte

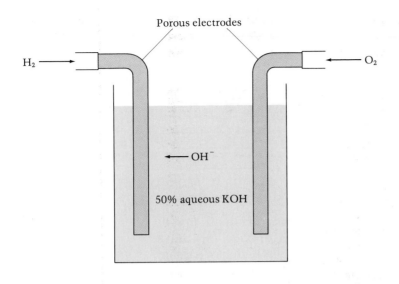

Porous electrodes

$H_2 \longrightarrow$

$\longleftarrow O_2$

$\longleftarrow OH^-$

50% aqueous KOH

Figure 21.6
A hydrogen–oxygen fuel cell.

A **fuel cell** is essentially a battery, but differs from the ones we have discussed so far by operating with a continuous supply of energetic reactants or fuel. Figure 21.6 shows a fuel cell that uses hydrogen and oxygen. ■ At one electrode oxygen passes through a porous material that catalyzes the following reaction:

$$O_2(g) + 2H_2O(l) + 4e^- \longrightarrow 4OH^-(aq) \quad \text{(cathode reaction)}$$

At the other electrode, hydrogen reacts:

$$2H_2(g) + 4OH^-(aq) \longrightarrow 4H_2O(l) + 4e^- \quad \text{(anode reaction)}$$

The sum of these half-cell reactions is

$$2H_2(g) + O_2(g) \longrightarrow 2H_2O(l)$$

and is the net reaction in the fuel cell. Such cells were used in Apollo spacecraft to supply electrical energy. Other types of cells employing hydrocarbon fuels have been constructed. ■

Another use of voltaic cells, besides their use as portable sources of energy, is to control the corrosion of underground pipelines and tanks. Such pipelines and tanks are usually made of steel, an alloy of iron, and their corrosion or rusting is an electrochemical process.

Consider the process of rusting that occurs when a drop of water is in contact with iron. The edge of the water drop exposed to the air becomes one pole of a voltaic cell. (See Figure 21.7.) At this edge, molecular oxygen from air is reduced to hydroxide ion in solution:

$$O_2(g) + 2H_2O(l) + 4e^- \longrightarrow 4OH^-(aq)$$

The electrons for this reduction are supplied by the oxidation of metallic iron at the center of the drop, which acts as the other pole of the voltaic cell:

$$Fe(s) \longrightarrow Fe^{2+}(aq) + 2e^-$$

These electrons flow from the center of the drop through the metallic iron to the edge of the drop. The metallic iron functions as the external circuit between the cell poles. ■

■ The first fuel cell was invented in 1839 by William Groves. He obtained the cell by electrolyzing water and collecting the hydrogen and oxygen in small inverted test tubes. After removing the battery current from the electrolysis cell, he found that a current would flow from the cell. In effect, the electrolysis charged the cell, after which it operated as a fuel cell.

■ Natural gas (methane) can be used in a fuel cell. The fuel cell converts the chemical energy of combustion directly to electrical energy. Theoretically, this is more efficient than using the heat of combustion to drive an engine, which then generates electricity.

■ Any salts that are dissolved in the water drop accelerate the rusting by giving an electrically conducting solution. This explains the corrosive effect of road salt.

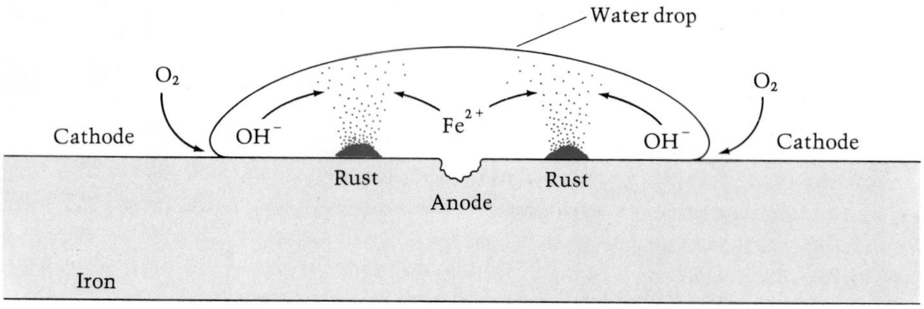

Figure 21.7
The electrochemical process involved in the rusting of iron. Iron is oxidized to iron(II) ion at the anode, and oxygen is reduced to hydroxide ion at the cathode.

Ions move within the water drop, completing the electrical circuit. Iron(II) ions move outward from the center of the drop, and hydroxide ions move inward from the edge. The two ions meet in a doughnut-shaped region, where they react to precipitate iron(II) hydroxide:

$$Fe^{2+}(aq) + 2OH^-(aq) \longrightarrow Fe(OH)_2(s)$$

This precipitate is quickly oxidized by oxygen to rust (approximated by the formula $Fe_2O_3 \cdot H_2O$):

$$4Fe(OH)_2(s) + O_2(g) \longrightarrow 2Fe_2O_3 \cdot H_2O(s) + 2H_2O(l)$$

If a buried steel pipeline (Figure 21.8) is connected to an active metal (that is, a highly electropositive substance), such as magnesium, then a voltaic cell is formed, with the active metal as the anode and iron as the cathode.

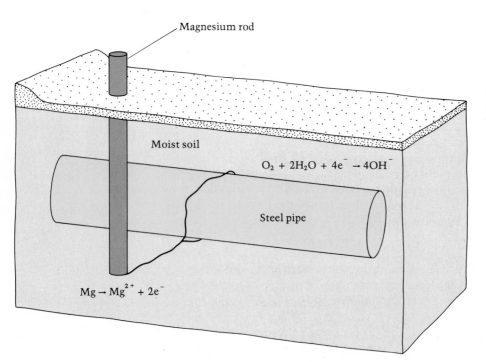

Figure 21.8
Cathodic protection of a buried steel pipe. Iron in the steel becomes the cathode in an iron–magnesium voltaic cell. Magnesium is then oxidized in preference to iron.

Wet soil forms the electrolyte and the electrode reactions are

$$Mg(s) \longrightarrow Mg^{2+}(aq) + 2e^- \qquad \text{(anode reaction)}$$
$$O_2(g) + 2H_2O(l) + 4e^- \longrightarrow 4OH^-(aq) \quad \text{(cathode reaction)}$$

As the cathode, the iron-containing steel pipe is protected from oxidation. Of course, the magnesium rod is eventually consumed and must be replaced, but this is cheaper than digging up the pipeline. This use of an active metal to protect iron from corrosion is called **cathodic protection.**

21.3 Notation for a Voltaic Cell

It is convenient to have a shorthand way of designating particular voltaic cells. The cell described earlier, consisting of a zinc metal–zinc ion half-cell and a copper metal–copper ion half-cell, is written

$$Zn(s)|Zn^{2+}(aq)||Cu^{2+}(aq)|Cu(s)$$

In this notation, the anode or oxidation half-cell is always written on the left, and the cathode or reduction half-cell is written on the right. The two electrodes are electrically connected by means of a salt bridge, denoted by two vertical bars.

$$\underset{\text{anode}}{Zn(s)|Zn^{2+}(aq)} \quad \underset{\text{salt bridge}}{||} \quad \underset{\text{cathode}}{Cu^{2+}(aq)|Cu(s)}$$

The cell terminals are at the extreme ends in this cell notation, and a single vertical bar indicates a phase boundary, say between a solid terminal and the electrode solution. For the anode of the same cell, we have

$$\underset{\text{anode terminal}}{Zn(s)} \qquad \underset{\text{phase boundary}}{|} \qquad \underset{\text{anode electrolyte}}{Zn^{2+}(aq)}$$

When the half-reaction involves a gas, an inert material such as platinum serves as a terminal and as an electrode surface on which the half-reaction occurs. The platinum catalyzes the half-reaction, but otherwise is not involved in it. Figure 21.9 shows a hydrogen electrode; hydrogen bubbles over a platinum electrode that is immersed in an acidic solution. The cathode half-reaction is

$$2H^+(aq) + 2e^- \rightleftharpoons H_2(g)$$

The notation for the hydrogen electrode is

$$H^+(aq)|H_2(g)|Pt$$

Here are several additional examples of this notation for electrodes (written as cathodes). A comma separates ions present in the same solution.

Cathode	Cathode reaction
$Cl^-(aq)|Cl_2(g)|Pt$	$Cl_2(aq) + 2e^- \rightleftharpoons 2Cl^-(aq)$
$Fe^{3+}(aq), Fe^{2+}(aq)|Pt$	$Fe^{3+}(aq) + e^- \rightleftharpoons Fe^{2+}(aq)$
$Cd^{2+}(aq)|Cd(s)$	$Cd^{2+}(aq) + 2e^- \rightleftharpoons Cd(s)$

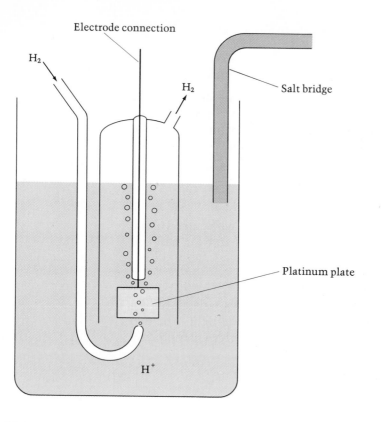

Electrode connection

H_2

H_2

Salt bridge

Platinum plate

H^+

Figure 21.9
A hydrogen electrode.

Exercise 21.2

Write the notation for a cell in which the electrode reactions are

$$Cl_2(aq) + 2e^- \longrightarrow 2Cl^-(aq)$$
$$Zn(s) \longrightarrow Zn^{2+}(aq) + 2e^-$$

(See Problems 21.27 and 21.28.)

The overall cell reaction can be written from the cell notation by first writing the appropriate half-cell reactions, then summing these in such a way that the electrons cancel. Example 21.2 shows how this is done.

Example 21.2

(a) Write the cell reaction for the voltaic cell

$$Tl(s)|Tl^+(aq)||Sn^{2+}(aq)|Sn(s)$$

(b) Write the cell reaction for the voltaic cell

$$Zn(s)|Zn^{2+}(aq)||Fe^{3+}(aq), Fe^{2+}(aq)|Pt$$

Solution

(a) The half-cell reactions are

$$Tl(s) \longrightarrow Tl^+(aq) + e^-$$
$$Sn^{2+}(aq) + 2e^- \longrightarrow Sn(s)$$

Multiplying the anode reaction by 2, then summing the half-cell reactions gives

$$2Tl(s) + Sn^{2+}(aq) \longrightarrow 2Tl^+(aq) + Sn(s)$$

(b) The half-cell reactions are

$$Zn(s) \longrightarrow Zn^{2+}(aq) + 2e^-$$
$$Fe^{3+}(aq) + e^- \longrightarrow Fe^{2+}(aq)$$

and the cell reaction is

$$Zn(s) + 2Fe^{3+}(aq) \longrightarrow Zn^{2+}(aq) + 2Fe^{2+}(aq)$$

Exercise 21.3

Give the overall cell reaction for the voltaic cell

$$Cd(s)|Cd^{2+}(aq)||H^+(aq)|H_2(g)|Pt$$

(See Problems 21.29 and 21.30.)

To fully specify a voltaic cell, it is necessary to give the concentrations of solutions or ions and the pressure of gases. In the cell notation, these are written within parentheses for each species. For example,

$$Zn(s)|Zn^{2+}(1.0\ M)||H^+(1.0\ M)|H_2(1.0\ atm)|Pt$$

21.4 Electromotive Force

It requires an expenditure of work to move a charged particle in a wire or other form of electrical conductor. The amount of work required to move an electrical charge from one point to another is proportional to the difference in electrical potential (electrical pressure) of the two points. In SI units **potential difference** is measured in **volts (V)**. The electrical work expended in moving a charge through a potential difference is

Electrical work = charge × potential difference

Corresponding SI units for the terms in this equation are

Joules = coulombs × volts

The **faraday constant, F,** is the magnitude of charge on one mole of electrons and equals 9.65×10^4 C (96,500 coulombs). In moving one faraday of charge from one electrode to another, the work done, w, by a voltaic cell is the product of the faraday constant, F, times the potential difference between the electrodes:

$$w = F \times \text{potential difference}$$

Under normal operation of a voltaic cell, the potential difference (voltage) across the electrodes is less than the maximum possible voltage of the cell. One reason for this is that it takes energy or work to drive a current through the cell itself. The decrease in cell voltage as current is drawn is a reflection of this energy expenditure. Thus, the cell voltage has its maximum value only when no current flows.

The maximum potential difference between the electrodes of a voltaic cell is referred to as the **electromotive force (emf)** of the cell, denoted E_{cell}. It is measured by balancing the cell voltage against that of a **potentiometer,** a device that has an adjustable voltage (Figure 21.10). When the two voltage sources are in balance, no current flows from the voltaic cell, and the emf is the balancing voltage on the potentiometer. We also can find which electrode of a cell is the anode and which is the cathode from this measurement. A voltaic cell can only be balanced when its electrodes are connected to like electrodes of a potentiometer.

We can now write an expression for the maximum work obtainable from a voltaic cell. Let n be the number of electrons transferred in the overall cell

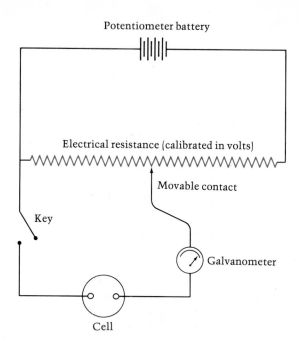

Potentiometer battery

Electrical resistance (calibrated in volts)

Movable contact

Key

Galvanometer

Cell

Figure 21.10
Measurement of cell emf by
a potentiometer. The contact
can be moved along the electrical resistance to vary the
potential that is directed
across the terminals of the
cell from the potentiometer
battery. When no current
flows from the cell (as determined by the lack of deflection in the galvanometer
needle), the voltage from the
potentiometer just balances
that from the cell. The galvanometer is observed while
the circuit is closed briefly
by tapping the key.

reaction. The maximum electrical work obtained from a voltaic cell for molar amounts of reactants (according to the cell reaction *as written*) is

$$w_{max} = nFE_{cell}$$

Here E_{cell} is the cell emf, and F is the faraday constant, 9.65×10^4 C.

Example 21.3

The emf of a particular voltaic cell with the cell reaction

$$Hg_2^{2+}(aq) + H_2(g) \rightleftharpoons 2Hg(l) + 2H^+(aq)$$

is 0.650 V. Calculate the maximum electrical work that can be obtained from this cell when 0.500 g H_2 is consumed.

Solution

To get n, it is necessary to obtain the half-reactions. These may be found by using the method described in Section 9.8 for balancing oxidation–reduction equations. The half-reactions are

$$Hg_2^{2+}(aq) + 2e^- \rightleftharpoons 2Hg(l)$$
$$H_2(g) \rightleftharpoons 2H^+(aq) + 2e^-$$

Thus, n equals 2, and the maximum work for the reaction as written is

$$w_{max} = nFE_{cell} = 2 \times 9.65 \times 10^4 \text{ C} \times 0.650 \text{ V}$$
$$= 1.25 \times 10^5 \text{ V} \cdot \text{C}$$
$$= 1.25 \times 10^5 \text{ J}$$

(Remember that a joule is equal to a volt-coulomb.) For 0.500 g H_2, the maximum work is

$$0.500 \text{ g } H_2 \times \frac{1 \text{ mol } H_2}{2.02 \text{ g } H_2} \times \frac{1.25 \times 10^5 \text{ J}}{1 \text{ mol } H_2} = 3.09 \times 10^4 \text{ J}$$

Note that the conversion factor 1.25×10^5 J/l mol H_2 is determined by the chemical equation as written. In this equation, 1 mol H_2 reacts and the maximum work produced is 1.25×10^5 J.

Exercise 21.4

What is the maximum electrical work that can be obtained from 6.54 g of zinc metal that react in a Daniell cell, described in the chapter opening, whose emf is 1.10 V?

The overall cell reaction is

$$Zn(s) + Cu^{2+}(aq) \longrightarrow Zn^{2+}(aq) + Cu(s)$$

(See Problems 21.33 and 21.34.)

21.5 Electrode Potentials

Electrons flow in an external circuit from low potential to high potential. The anode of a voltaic cell is at low potential and by convention is said to have negative **polarity.** The cathode of a voltaic cell is at higher potential and has positive polarity. (See the labeling of polarity in Figure 21.1.)

■ Positive charge flows from high potential to low potential ("downhill"). Negative charge flows in the opposite direction.

Since the emf of a voltaic cell is the difference in potential of the electrodes, we can find the cell emf once we know the potentials of individual electrodes. Thus,

$$E_{cell} = E_{cathode} - E_{anode}$$

where E_{cell} is the cell emf, $E_{cathode}$ is the cathode potential, and E_{anode} is the anode potential. The anode potential is less than the cathode potential, so that the difference $E_{cathode} - E_{anode}$ is a positive number.

It is not possible to measure electrode potentials directly; only potential differences are obtained experimentally. The problem is similar to obtaining the enthalpies of compounds. In that case one chooses an arbitrary reference (all elements in their stablest state have an enthalpy equal to zero). We do a similar thing here. We arbitrarily choose one electrode to have a potential equal to zero. With such a convention, other electrode potentials have fixed values relative to the one chosen to be zero.

A standard electrode refers to an electrode in which the molarities of ions and the pressure of gases (in atmospheres) are equal to 1 at 25°C. The **standard electrode potential** is designated by a superscript degree sign: $E°$. By convention, the standard hydrogen electrode is taken to have an electrode potential of zero.

$$H^+(1\ M)|H_2(1\ atm)|Pt;\ E° = 0\ V$$

Table 21.1 lists standard electrode potentials for selected electrodes. Values increase from the top of the table to the bottom.

Cell emf

The emf of a voltaic cell constructed from standard electrodes is easily calculated. For example, consider the electrodes

$$Cd^{2+}(aq)|Cd(s);\ E° = -0.40\ V$$
$$Ag^+(aq)|Ag(s);\ E° = 0.80\ V$$

Since the cadmium electrode has the lower potential, it is the anode. The cell notation, with the anode on the left, is

$$Cd(s)|Cd^{2+}(aq)||Ag^+(aq)|Ag(s)$$

Cathode	Cathode Reaction	Standard Potential, $E°$ (volts)
$Li^+ \vert Li$	$Li^+ + e^- \rightleftharpoons Li$	-3.04
$K^+ \vert K$	$K^+ + e^- \rightleftharpoons K$	-2.92
$Ca^{2+} \vert Ca$	$Ca^{2+} + 2e^- \rightleftharpoons Ca$	-2.76
$Na^+ \vert Na$	$Na^+ + e^- \rightleftharpoons Na$	-2.71
$Mg^{2+} \vert Mg$	$Mg^{2+} + 2e^- \rightleftharpoons Mg$	-2.38
$Al^{3+} \vert Al$	$Al^{3+} + 3e^- \rightleftharpoons Al$	-1.66
$Zn^{2+} \vert Zn$	$Zn^{2+} + 2e^- \rightleftharpoons Zn$	-0.76
$Cr^{3+} \vert Cr$	$Cr^{3+} + 3e^- \rightleftharpoons Cr$	-0.74
$Fe^{2+} \vert Fe$	$Fe^{2+} + 2e^- \rightleftharpoons Fe$	-0.41
$Cd^{2+} \vert Cd$	$Cd^{2+} + 2e^- \rightleftharpoons Cd$	-0.40
$Ni^{2+} \vert Ni$	$Ni^{2+} + 2e^- \rightleftharpoons Ni$	-0.23
$Sn^{2+} \vert Sn$	$Sn^{2+} + 2e^- \rightleftharpoons Sn$	-0.14
$Pb^{2+} \vert Pb$	$Pb^{2+} + 2e^- \rightleftharpoons Pb$	-0.13
$Fe^{3+} \vert Fe$	$Fe^{3+} + 3e^- \rightleftharpoons Fe$	-0.04
$H^+ \vert H_2 \vert Pt$	$2H^+ + 2e^- \rightleftharpoons H_2$	0.00
$Sn^{4+}, Sn^{2+} \vert Pt$	$Sn^{4+} + 2e^- \rightleftharpoons Sn^{2+}$	0.15
$Cu^{2+}, Cu^+ \vert Pt$	$Cu^{2+} + e^- \rightleftharpoons Cu^+$	0.16
$ClO_4^-, ClO_3^-, OH^- \vert Pt$	$ClO_4^- + H_2O + 2e^- \rightleftharpoons ClO_3^- + 2OH^-$	0.17
$Cl^- \vert AgCl \vert Ag$	$AgCl(s) + e^- \rightleftharpoons Ag(s) + Cl^-$	0.22
$Cu^{2+} \vert Cu$	$Cu^{2+} + 2e^- \rightleftharpoons Cu$	0.34
$ClO_3^-, ClO_2^-, OH^- \vert Pt$	$ClO_3^- + H_2O + 2e^- \rightleftharpoons ClO_2^- + 2OH^-$	0.35
$IO^-, I^-, OH^- \vert Pt$	$IO^- + H_2O + 2e^- \rightleftharpoons I^- + 2OH^-$	0.49
$Cu^+ \vert Cu$	$Cu^+ + e^- \rightleftharpoons Cu$	0.52
$I^- \vert I_2 \vert Pt$	$I_2(s) + 2e^- \rightleftharpoons 2I^-$	0.54
$ClO_2, ClO^-, OH^- \vert Pt$	$ClO_2^- + H_2O + 2e^- \rightleftharpoons ClO^- + 2OH^-$	0.59
$Fe^{3+}, Fe^{2+} \vert Pt$	$Fe^{3+} + e^- \rightleftharpoons Fe^{2+}$	0.77
$Hg_2^{2+} \vert Hg$	$Hg_2^{2+} + 2e^- \rightleftharpoons 2Hg$	0.80
$Ag^+ \vert Ag$	$Ag^+ + e^- \rightleftharpoons Ag$	0.80
$Hg^{2+} \vert Hg$	$Hg^{2+} + 2e^- \rightleftharpoons Hg$	0.85
$ClO^-, Cl^-, OH^- \vert Pt$	$ClO^- + H_2O + 2e^- \rightleftharpoons Cl^- + 2OH^-$	0.90
$Hg^{2+}, Hg_2^{2+} \vert Pt$	$2Hg^{2+} + 2e^- \rightleftharpoons Hg_2^{2+}$	0.90
$NO_3^-, H^+ \vert NO \vert Pt$	$NO_3^- + 4H^+ + 3e^- \rightleftharpoons NO + 2H_2O$	0.96
$Br^-, Br_2 \vert Pt$	$Br_2(l) + 2e^- \rightleftharpoons 2Br^-$	1.07
$H^+ \vert O_2 \vert Pt$	$O_2 + 4H^+ + 4e^- \rightleftharpoons 2H_2O$	1.23
$Cr_2O_7^{2-}, Cr^{3+}, H^+ \vert Pt$	$Cr_2O_7^{2-} + 14H^+ + 6e^- \rightleftharpoons 2Cr^{3+} + 7H_2O$	1.33
$Cl^- \vert Cl_2 \vert Pt$	$Cl_2 + 2e^- \rightleftharpoons 2Cl^-$	1.36
$Ce^{4+}, Ce^{3+} \vert Pt$	$Ce^{4+} + e^- \rightleftharpoons Ce^{3+}$	1.44
$MnO_4^-, Mn^{2+}, H^+ \vert Pt$	$MnO_4^- + 8H^+ + 5e^- \rightleftharpoons Mn^{2+} + 4H_2O$	1.49
$H_2O_2, H^+ \vert Pt$	$H_2O_2 + 2H^+ + 2e^- \rightleftharpoons 2H_2O$	1.78
$Co^{3+}, Co^{2+} \vert Pt$	$Co^{3+} + e^- \rightleftharpoons Co^{2+}$	1.82
$H^+ \vert O_3, O_2 \vert Pt$	$O_3 + 2H^+ + 2e^- \rightleftharpoons O_2 + H_2O$	2.07
$F^- \vert F_2 \vert Pt$	$F_2 + 2e^- \rightleftharpoons 2F^-$	2.87

and the emf is

$$E°_{cell} = E°_{cathode} - E°_{anode}$$
$$= 0.80 \text{ V} - (-0.40 \text{ V}) = 1.20 \text{ V}$$

Here the emf is the **standard emf** and is designated by a superscript degree sign.

Table 21.1
Standard Electrode
(Reduction) Potentials in
Aqueous Solution at 25°C

If for some reason we were to write the preceding cell as

$$Ag(s)|Ag^+(aq)||Cd^{2+}(aq)|Cd(s)$$

the calculation of emf would give a negative value. From the cell notation, $Ag|Ag^+$ is written as the anode and $Cd^{2+}|Cd$ is written as the cathode. Then, the emf would be

$$E^{\circ}_{cell} = -0.40 \text{ V} - 0.80 \text{ V} = -1.20 \text{ V}$$

A negative emf merely indicates that the cell reaction as written from the cell notation is spontaneous from right to left. The actual anode is on the right, and the cathode is on the left.

Example 21.4

Calculate the standard emf of the cell

$$Zn(s)|Zn^{2+}(aq)||Cl^-(aq)|Cl_2(g)|Pt$$

from standard electrode potentials.

Solution

The standard electrode potentials are

$Zn^{2+}(aq)|Zn(s); \ E^{\circ} = -0.76 \text{ V}$
$Cl^-(aq)|Cl_2(g)|Pt; \ E^{\circ} = 1.36 \text{ V}$

From these values we get

$$E^{\circ}_{cell} = 1.36 \text{ V} - (-0.76 \text{ V})$$
$$= 2.12 \text{ V}$$

Exercise 21.5

Using standard electrode potentials, calculate E°_{cell} for the following zinc–copper cell:

$$Zn(s)|Zn^{2+}(aq)||Cu^{2+}(aq)|Cu(s)$$

(See Problems 21.35 and 21.36.)

We can calculate standard emf's for a given cell using a different arrangement of the work from that given in Example 21.4. We write the oxidation half-reaction and under it the reduction half-reaction. Next to the reduction half-reaction, we write the *reduction potential*, that is, the electrode potential, E°. Then, next to the oxidation half-reaction, we write the *oxidation potential*, that is, the electrode potential with its sign reversed. Note that Table 21.1 lists electrode potentials, or reduction potentials. If a half-reaction in this table is *reversed*, it becomes an oxidation half-reaction, and the oxidation potential is obtained by *reversing* the sign of E°. When the two half-reactions and their corresponding potentials are added, we get the overall cell reaction and the cell emf. The arrangement of work for the solution of Example 21.4 is as follows:

$Zn(s) \longrightarrow Zn^{2+}(aq) + 2e^-$		0.76 V	(oxidation)
$2e^- + Cl_2(g) \longrightarrow 2Cl^-(aq)$		1.36 V	(reduction)
$Zn(s) + Cl_2(g) \longrightarrow Zn^{2+}(aq) + 2Cl^-(aq)$		2.12 V	

Strength of an Oxidizing or Reducing Agent

Apart from calculating emf's, the electrode potentials in Table 21.1 are useful in deciding whether an oxidation–reduction reaction is spontaneous and in judging the strength of a particular oxidizing or reducing agent under standard conditions (where the concentrations of ions are 1 molar, pressures of gases are 1 atmosphere, and the temperature is 25°C). Thus, since electrode potentials are written as reduction potentials by convention, those reduction half-reactions with larger (that is, more positive) electrode potentials in Table 21.1 have a greater tendency to go as written (left to right). The reactants in these half-reactions (such as MnO_4^-, O_3 and F_2) will be the stronger oxidizing agents. On the other hand, those half-reactions with lower (that is, more negative) electrode potentials have a greater tendency to go right to left. The substances on the right (such as Li, K, and Na) will be the stronger reducing agents.

To determine the direction of spontaneity of a reaction, we need only note the relative strengths of the oxidizing or reducing agents on the left and right sides of the reaction. The stronger oxidizing agent (and stronger reducing agent) will be on the reactant side of the equation.

Example 21.5

Consider the reaction

$$Zn^{2+}(aq) + 2Fe^{2+}(aq) \longrightarrow Zn(s) + 2Fe^{3+}(aq)$$

Does the reaction go spontaneously in the direction indicated, under standard conditions?

Solution

In this reaction, Zn^{2+} is the oxidizing agent on the left;

Fe^{3+} is the oxidizing agent on the right. The corresponding electrode potentials are

$$Zn^{2+}(aq) + 2e^- \longrightarrow Zn(s); E° = -0.76 \text{ V}$$
$$Fe^{3+}(aq) + e^- \longrightarrow Fe^{2+}(aq); E° = 0.77 \text{ V}$$

Thus, the electrode potential is greater for the second half-reaction, so that iron(III) is the stronger oxidizing agent. The reaction is nonspontaneous as written.

Exercise 21.6

Does the following reaction occur spontaneously in the direction indicated, under standard conditions?

$$Cu^{2+}(aq) + 2I^-(aq) \longrightarrow Cu(s) + I_2(s)$$

(See Problems 21.37, 21.38, 21.39, and 21.40.)

If we choose to look at reducing agents in Example 21.5, we see that zinc metal is a stronger reducing agent than iron(II) ion. Again we conclude that the reaction is nonspontaneous as written and should go from right to left.

We reach the same conclusion from the sign of $E°$. For the reaction as written, $E° = -1.53$ V; the negative sign means that the reaction is spontaneous in the opposite direction. This can be expressed in a more general manner:

If $E° > 0$, then the reaction is spontaneous.

If $E° < 0$, then the reaction is nonspontaneous.

21.6 Equilibrium Constants from emf's

Some of the most important results of electrochemistry are the relationships among cell emf, free-energy change, and equilibrium constant. Recall that the free-energy change, ΔG, for a reaction equals the negative of the maximum useful work obtained from the reaction (Section 20.5):

$$\Delta G = -w_{max}$$

For a voltaic cell this work is the electrical work, nFE_{cell}, so that when the reactants and products are in their standard states, we have

$$\Delta G° = -nFE°_{cell}$$

With this equation, emf measurements become an important source of thermodynamic information. Alternatively, thermodynamic data can be used to calculate cell emf's. These calculations are shown in the following examples.

Example 21.6

Calculate the standard free-energy change for the reaction

$$Zn(s) + 2H^+(aq) \longrightarrow Zn^{2+}(aq) + H_2(g)$$

using electrode potential data.

Solution

The half-cell reactions and corresponding standard potentials from Table 21.1 are

$$Zn(s) \longrightarrow Zn^{2+}(aq) + 2e^-; \text{ anode}$$
$$2H^+(aq) + 2e^- \longrightarrow H_2(g); \text{ cathode}$$

Looking at Table 21.1, we see that the electrode potentials for $Zn^{2+}|Zn$ and $H^+|H_2$ are 0.76 V and 0 V, respectively. Hence, the standard cell emf, $E°_{cell}$, is $E°_{cathode} - E°_{anode} = 0 V - (-0.76 V) = 0.76 V$. Note that $n = 2$. Thus,

$$\Delta G° = -nFE°_{cell} = -2 \times 9.65 \times 10^4 \text{ C} \times 0.76 \text{ V}$$
$$= -1.5 \times 10^5 \text{ J}$$

Exercise 21.7

What is $\Delta G°$ for the reaction

$$Sn^{2+}(aq) + 2Hg^{2+}(aq) \longrightarrow Sn^{4+}(aq) + Hg_2^{2+}(aq)$$

For data, see Table 21.1.

(See Problems 21.41 and 21.42.)

Example 21.7

Suppose the reaction of zinc metal and chlorine gas is utilized in a fuel cell in which zinc ions and chloride ions are formed in aqueous solution.

$$Zn(s) + Cl_2(g) \xrightarrow{H_2O} Zn^{2+}(aq) + 2Cl^-(aq)$$

Calculate the standard emf for this reaction from standard free energies of formation (see Appendix C).

Solution

We write the equation with $\Delta G_f°$'s beneath:

$$Zn(s) + Cl_2(g) \longrightarrow Zn^{2+}(aq) + 2Cl^-(aq)$$
$$\Delta G_f°: \quad 0 \qquad 0 \qquad \quad -147 \quad 2 \times (-131) \text{ kJ}$$

Hence,

$$\Delta G° = [-147 + 2 \times (-131)] \text{ kJ}$$
$$= -409 \text{ kJ} = -4.09 \times 10^5 \text{ J}$$

Then, since $n = 2$,

$$\Delta G° = -nFE°_{cell}$$
$$-4.09 \times 10^5 \text{ J} = -2 \times 9.65 \times 10^4 \text{ C} \times E°_{cell}$$

Solving for $E°_{cell}$, we get

$$E°_{cell} = 2.12 \text{ V}$$

Exercise 21.8

Use standard free energies of formation (Appendix C) to obtain the standard emf of a cell with the reaction

$$Mg(s) + Cu^{2+}(aq) \longrightarrow Mg^{2+}(aq) + Cu(s)$$

(See Problems 21.43 and 21.44.)

Since the standard free-energy change is related to the equilibrium constant K_{th} (see Section 20.6), cell emf's give us yet another route for finding equilibrium constants of chemical reactions. This is illustrated by the next example.

Example 21.8

Calculate the equilibrium constant at 25°C for the reaction

$$Zn(s) + Cu^{2+}(aq) \rightleftharpoons Zn^{2+}(aq) + Cu(s)$$

The corresponding standard emf is 1.10 V.

Solution

We can write

$$\Delta G° = -nFE°_{cell} = -2.303RT \log K_{th}$$

Substituting for the middle and right-hand expressions,

we get

$$-2 \times 9.65 \times 10^4 \text{ C/mol} \times 1.10 \text{ V} =$$
$$-2.303 \times 8.31 \text{ J/(mol} \cdot \text{K)} \times 298 \text{ K} \times \log K_{th}$$

Solving for $\log K_{th}$, we find

$$\log K_{th} = 37.2$$

Thus
$$K_{th} = 2 \times 10^{37}$$

Note that the antilog of 37.2 gives 1.6×10^{37}, which when rounded to the correct number of significant figures (1) is 2×10^{37}.

Exercise 21.9

Calculate the equilibrium constant K_{th} for the reaction

$$Fe(s) + Sn^{4+}(aq) \rightleftharpoons Fe^{2+}(aq) + Sn^{2+}(aq)$$

The standard emf is 0.56 V.

(See Problems 21.45 and 21.46.)

21.7 Dependence of emf on Concentration

The emf of a cell depends on the concentrations of ions and on gas pressures. For that reason, cell emf's provide a way to measure ion concentrations. The pH meter, for example, depends on the variation of cell emf with hydrogen-ion concentration.

Nernst Equation

The emf of a cell, E_{cell}, is related to the standard emf, $E°_{cell}$, and the ion concentrations by the **Nernst equation:**

$$E_{cell} = E°_{cell} - \frac{2.303RT}{nF} \log Q$$

Here R is the gas constant, equal to 8.31 J/(mol · K), and Q is the reaction quotient.■ The reaction quotient has the form of the equilibrium constant, except that the concentrations are those that exist in the voltaic cell. Concentrations of ions are expressed in molarities; those of gases are expressed as partial pressures in atmospheres. If we substitute 298 K (25°C) for the temperature in the Nernst equation and put in values for R and F, we get

■ This equation was named after the German chemist Walther Nernst (1864–1941), who discovered it. Nernst also formulated the third law of thermodynamics, for which he received the Nobel Prize in 1920.

$$E_{cell} = E°_{cell} - \frac{0.0592}{n} \log Q \text{ (values in volts at 25°C)}$$

We can show from the Nernst equation that the cell emf, E_{cell}, decreases as the cell reaction proceeds. As the reaction occurs in the voltaic cell, the concentrations of products increase and the concentrations of reactants decrease. Therefore, Q and $\log Q$ increase. The second term in the Nernst equation, $(0.0592/n) \log Q$, increases, so that the difference $E°_{cell} - (0.0592/n) \log Q$ decreases. Thus, the cell emf, E_{cell}, becomes smaller. Eventually the cell emf goes to zero, and the cell reaction comes to equilibrium.

As an example of the computation of the reaction quotient, consider the following voltaic cell:

$$Cd(s)|Cd^{2+}(0.0100 \ M)||H^+(1.00 \ M)|H_2(1.00 \ atm)|Pt$$

The cell reaction is

$$Cd(s) + 2H^+(aq) \rightleftharpoons Cd^{2+}(aq) + H_2(g)$$

and the expression for the equilibrium constant is

$$\frac{[Cd^{2+}]P_{H_2}}{[H^+]^2}$$

Note that the hydrogen-gas concentration is given here in terms of the pressure (in atmospheres). The expression for the reaction quotient has the same form as K_{th}, except that the values for the ion concentrations and hydrogen-gas pressures are those that exist in the cell. Hence,

$$Q = \frac{[Cd^{2+}]P_{H_2}}{[H^+]^2} = \frac{0.0100 \times 1.00}{(1.00)^2} = 0.0100$$

The next example illustrates a complete calculation of emf from the ion concentrations in a voltaic cell.

Example 21.9

What is the emf at 25°C of the following cell?

$$Zn(s)|Zn^{2+}(0.00100 \ M)||Cu^{2+}(10.0 \ M)|Cu(s)$$

The standard emf of this cell is 1.10 V.

Solution

The cell reaction is

$$Zn(s) + Cu^{2+}(aq) \rightleftharpoons Zn^{2+}(aq) + Cu(s)$$

Hence, $n = 2$, and the reaction quotient is

$$Q = \frac{[Zn^{2+}]}{[Cu^{2+}]} = \frac{0.00100}{10.0} = 1.00 \times 10^{-4}$$

Since the standard emf is 1.10 V, the Nernst equation becomes

$$E_{cell} = E°_{cell} - \frac{0.0592}{n} \log Q$$

(Continued)

$$= 1.10 - \frac{0.0592}{2} \log (1.0 \times 10^{-4})$$

$$= 1.10 - (-0.12) = 1.22$$

Thus, the cell emf is 1.22 V. This result is qualitatively what we would expect. Because the concentration of product (Zn^{2+}) is less than the standard value ($1\ M$), whereas the concentration of reactant (Cu^{2+}) is greater, the spontaneity of the reaction as measured by E_{cell} is greater.

Exercise 21.10

What is the emf at 25°C of the following cell?

$$Zn(s)|Zn^{2+}(0.200\ M)\|Ag^+(0.00200\ M)|Ag(s)$$

(See Problems 21.47 and 21.48.)

Electrode Potentials for Nonstandard Conditions

In addition to using the Nernst equation to obtain cell emf's, we can use it to calculate the potential of an electrode when the concentration of an ion is other than 1 molar. In other words, we can use this equation to calculate the potential of an electrode under nonstandard conditions.

Consider this problem: What is the potential of the zinc electrode $Zn^{2+}(0.100\ M)|Zn(s)$ at 25°C? Since the standard potential of the hydrogen electrode is defined to be zero, the emf of the cell

$$Pt|H_2(1\ atm)|H^+(1\ M)\|Zn^{2+}(0.100\ M)|Zn(s)$$

equals

$$E_{cell} = E(Zn^{2+}|Zn) - E°(H^+|H_2)$$
$$= E(Zn^{2+}|Zn)$$

That is, the emf of this cell equals the zinc electrode potential. Also, the standard emf of this cell equals the standard potential of the zinc electrode.

$$E_{cell}° = E°(Zn^{2+}|Zn)$$

Thus, the Nernst equation for the cell at 25°C becomes

$$E(Zn^{2+}|Zn) = E°(Zn^{2+}|Zn) - \frac{0.0592}{n} \log Q$$

Values for n and Q follow from the overall cell reaction.

$$Zn^{2+}(aq) + H_2(g) \rightleftharpoons Zn(s) + 2H^+(aq)$$

We see that the number of electrons transferred is 2; hence, $n = 2$. Also, the reaction quotient, Q, is

$$Q = \frac{[H^+]^2}{[Zn^{2+}]P_{H_2}} = \frac{1^2}{[Zn^{2+}] \times 1} = \frac{1}{[Zn^{2+}]}$$

Or, since $[H^+]$ and P_{H_2} have numerical values of 1, Q equals $1/[Zn^{2+}]$.

In effect, the Nernst equation can be applied to the zinc half-reaction

$$Zn^{2+}(aq) + 2e^- \rightleftharpoons Zn(s)$$

The value of n follows from the half-reaction, and the expression required to find Q is obtained by writing a reaction quotient for the half-reaction (ignoring e^-). We get $E(Zn^{2+}|Zn) = -0.76 + (0.0592/2) \log 0.100 = -0.79V$.

Example 21.10

What is the potential of the hydrogen electrode $H^+(0.100\ M)|H_2(1\ atm)|Pt$ at 25°C (298 K)?

Solution

The half-reaction is

$$2H^+(aq) + 2e^- \rightleftharpoons H_2(g)$$

For this equation, $n = 2$ and

$$Q = \frac{P_{H_2}}{[H^+]^2} = \frac{1}{(0.100)^2}$$

Since $E° = 0$ for the standard hydrogen electrode, the Nernst equation for the nonstandard hydrogen electrode is

$$E = E° - \frac{0.0592}{n} \log Q$$

$$= 0 - \frac{0.0592}{2} \log \frac{1}{(0.100)^2}$$

$$= -0.0592\ V$$

Exercise 21.11

What is the potential of the copper electrode $Cu^{2+}(0.0350\ M)|Cu(s)$ at 25°C?

(See Problems 21.49 and 21.50.)

Determination of pH

To measure the pH of any test solution, we can dip a hydrogen electrode into the solution and connect this to a standard hydrogen electrode. The cell is

$$Pt|H_2(1\ atm)|\text{test solution}\|H^+(1\ M)|H_2(1\ atm)|Pt$$

The emf of this cell will equal the potential developed by the test solution half-cell, whose half-reaction is

$$\tfrac{1}{2}H_2(1\ atm) \rightleftharpoons H^+(\text{test solution}) + e^-$$

Then, according to the Nernst equation for this electrode at 25°C, we have

$$E_{cell} = -0.0592 \log [H^+]$$

This equation shows that there is a direct relationship between the cell emf, E_{cell}, that we measure and the hydrogen-ion concentration of the test solution.

If we wish, we can rewrite the preceding equation in terms of the pH:

$$pH = -\log [H^+]$$

We get

$$E_{cell} = 0.0592\ pH$$

Or, at 25°C,

$$pH = \frac{E_{cell}}{0.0592}$$

The standard hydrogen electrode, which requires hydrogen gas at standard pressure, is too cumbersome for routine laboratory use. Also, the platinum foil is easily fouled by the presence of other substances in solution. The hydrogen electrode is often replaced by a **glass electrode.** This compact electrode (see Figure 21.11) consists of a silver wire coated with silver chloride immersed in a solution of dilute hydrochloric acid. (This silver–silver chloride electrode serves as an internal reference.) The electrode solution is separated from the test solution by a thin glass membrane, which develops a potential across it depending upon the hydrogen-ion concentrations on its inner and outer surfaces. As the other electrode, a mercury–mercury(I) chloride (calomel) electrode is often used. The emf of the cell depends linearly on the pH. In a popular arrangement, the emf is measured with a voltmeter that reads pH directly (see Figure 17.4).

The glass electrode is an example of an **ion-selective electrode.** Many electrodes have been developed recently that are sensitive to a particular ion such as K^+, NH_4^+, Ca^{2+}, and Mg^{2+}. They can be used to monitor solutions of that ion. It is even possible to measure the concentration of a non-electrolyte. Thus, to measure urea, NH_2CONH_2, in solution, one uses an electrode selective to NH_4^+ that is coated with a gel containing the enzyme urease. (An enzyme is a biochemical catalyst.) The gel is held in place around the electrode by means of a nylon net. Urease catalyzes the decomposition of urea to ammonium ion, whose concentration is measured.

$$NH_2CONH_2(aq) + 2H_2O(l) + H^+(aq) \xrightarrow{urease} 2NH_4^+(aq) + HCO_3^-(aq)$$

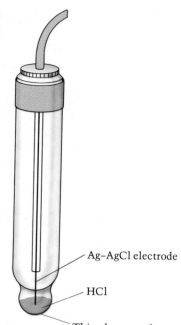

Figure 21.11
A glass electrode for measuring hydrogen-ion concentrations.

Exercise 21.12

What is the nickel(II)-ion concentration in the cell

$$Zn(s)|Zn^{2+}(1.00\ M)\|Ni^{2+}(aq)|Ni(s)$$

if the emf is 0.34 V at 25°C?

(See Problems 21.51 and 21.52.)

Electrolytic Cells

Sodium metal burns in an atmosphere of chlorine gas to give sodium chloride:

$$Na(s) + \tfrac{1}{2}Cl_2(g) \longrightarrow NaCl(s)$$

This reaction is clearly spontaneous in the direction written and for all practical purposes goes to completion. Yet, if sodium chloride is melted and electrodes that are connected to a battery are dipped into the molten salt, a chemical reaction occurs (providing the voltage is sufficiently high). At the electrode connected to the negative pole of the battery globules of sodium metal form, and at the other electrode chlorine gas evolves (see Figure 21.12). The overall reaction is the *electrolysis* of sodium chloride. This is essentially the reverse of the reaction just given and is nonspontaneous, being forced in this direction by the battery voltage.

$$NaCl(l) \longrightarrow Na(l) + \tfrac{1}{2}Cl_2(g)$$

Electrolysis is an important industrial method for obtaining active elements, such as sodium and chlorine, from their compounds. Some of these

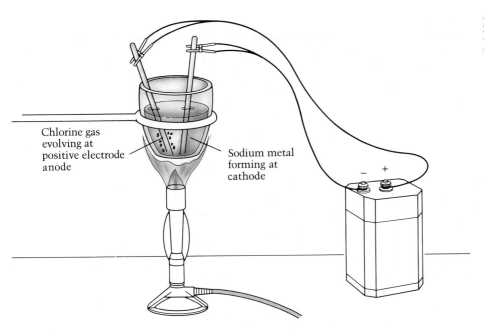

Figure 21.12
Electrolysis of molten
sodium chloride.

Chlorine gas
evolving at
positive electrode
anode

Sodium metal
forming at
cathode

electrolysis reactions are carried out in aqueous solution. We will look at
electrolysis in aqueous solution in the next section.

21.8 Aqueous Electrolysis

Consider the electrolysis of an aqueous solution of sodium chloride, rather
than molten sodium chloride. Now the electrolysis is complicated by the
fact that water may be reduced or oxidized at the electrodes. For example, at
an inert cathode, possible half-reactions are

$$Na^+(aq) + e^- \longrightarrow Na(s)$$
$$2H_2O(l) + 2e^- \longrightarrow H_2(g) + 2OH^-(aq)$$

and for the anode possible half-reactions are

$$2Cl^-(aq) \longrightarrow Cl_2(g) + 2e^-$$
$$2H_2O(l) \longrightarrow O_2(g) + 4H^+(aq) + 4e^-$$

What one actually observes at the cathode are bubbles of hydrogen, indicating that water is reduced in preference to sodium ions. What happens at
the anode depends on the concentration of the sodium chloride. In dilute
solution, oxygen evolves; with a concentrated solution, chlorine evolves.

What factor determines which species react at the electrodes? The electrolysis reaction that is observed is the one with the smallest *decomposition
voltage*. The decomposition voltage is the potential difference needed to give
electrolysis by a given pair of electrode reactions. In a dilute solution of
sodium chloride, the electrode reactions with the smallest decomposition
voltage correspond to the reduction and oxidation of water, and we get the

electrolysis of water. Adding the electrode reactions

$$4H_2O(l) + 4e^- \longrightarrow 2H_2(g) + 4OH^-(aq)$$
$$\underline{2H_2O(l) \qquad\qquad \longrightarrow O_2(g) + 4H^+(aq) + 4e^-}$$
$$6H_2O(l) \qquad\qquad \longrightarrow 2H_2(g) + O_2(g) + \underbrace{4H^+(aq) + 4OH^-(aq)}$$
$$= 4H_2O(l)$$

and canceling $4H_2O$ on both sides gives the overall electrolysis reaction

$$2H_2O(l) \longrightarrow 2H_2(g) + O_2(g)$$

The decomposition voltage depends in part on the difference in electrode potentials for the half-reactions. For an external voltage to push a reaction in the nonspontaneous direction, it must counter the emf generated by the natural tendency of the half-reactions to reverse—to go in the spontaneous direction. This emf will be given by the difference in electrode potentials. For the electrolysis of water, the reverse half-reactions are

$$H_2(g) + 2OH^-(aq) \longrightarrow 2H_2O(l) + 2e^-$$
$$O_2(g) + 4H^+(aq) + 4e^- \longrightarrow 2H_2O(l)$$

The electrode potentials (at pH 7.00) are -0.41 V and 0.82 V, respectively. Therefore, the difference of potential is 0.82 V $- (-0.41$ V$) = 1.23$ V.

To achieve the electrolysis of water, however, requires a decomposition voltage that is greater than this emf by a quantity called the *overvoltage*. This overvoltage arises in part from the work needed to push electrons across the phase boundaries. It depends on the electrolytic current as well as the electrode material and is not easily predicted. Overvoltage usually is small for the plating out of a metal, but may be several tenths of a volt for the discharge of a gas. Thus, the decomposition voltage for the electrolysis of water is several tenths of a volt greater than 1.23 V.

Let us examine the half-reactions for the electrolysis of aqueous sodium chloride to see if we can understand the experimental results. If we ignore the effect of overvoltage, we expect the cathode half-reaction to be the reduction of the more easily reduced species and the anode half-reaction to be the oxidation of the more easily oxidized species. This will correspond to the smallest electrode potential difference (the decomposition voltage, if overvoltage is ignored).

Of the possible cathode half-reactions in 1 *M* NaCl, namely,

$$Na^+(aq) + e^- \longrightarrow Na(s); \ E^\circ = -2.71 \text{ V } (1 \ M \ Na^+)$$
$$2H_2O(l) + 2e^- \longrightarrow H_2(g) + 2OH^-(aq); \ E = -0.41 \text{ V } (pH = 7.00)$$

the reduction of water is easier than that of sodium ion. (The reduction of water has the higher—that is, the less negative—electrode potential.) Because of overvoltage, the discharge of hydrogen gas is more difficult than we would expect from the electrode potential. But the overvoltage is not likely to be so large that sodium ion is reduced in preference to water.■

The possible anode half-reactions in 1 *M* NaCl are

$$2H_2O(l) \longrightarrow O_2(g) + 4H^+(aq) + 4e^- \ (pH = 7.00)$$
$$2Cl^-(aq) \longrightarrow Cl_2(g) + 2e^- \ (1 \ M \ Cl^-)$$

■ Sodium is deposited if the electrode is made of mercury. The sodium dissolves in the mercury, forming sodium amalgam.

Electrode, or reduction, potentials for these are 0.82 V and 1.36 V, respectively. Thus, water is more easily oxidized than chloride ion, since the more easily oxidized species corresponds to the smaller electrode potential. We would expect oxygen to be evolved.

This conclusion might be upset if we included the effect of overvoltage. However, of these two possible half-reactions, only the oxidation of chloride ion depends on the concentration of the sodium chloride solution. By varying the concentration of chloride ions, the potential for the oxidation of Cl^- can be made very small or very large. The half-reaction is pushed to the right by high chloride-ion concentration, according to LeChatelier's principle.■ Thus, at sufficiently low chloride-ion concentration, oxygen is certainly evolved in preference to chlorine. But at high concentrations, chlorine gas should be evolved. This conclusion agrees with what is actually observed.

■ The quantitative effect of chloride-ion concentration is obtained from the Nernst equation for the electrode.

The overall electrolysis reaction in concentrated solution is obtained by adding the electrode reactions.

$$2H_2O(l) + 2e^- \longrightarrow H_2(g) + 2OH^-(aq)$$
$$\underline{2Cl^-(aq) \longrightarrow Cl_2(g) + 2e^-}$$
$$2H_2O(l) + 2Cl^-(aq) \longrightarrow H_2(g) + Cl_2(g) + 2OH^-(aq)$$

If the electrolyte solution is evaporated, sodium hydroxide is recovered. This is the major industrial method for the preparation of sodium hydroxide (known commercially as caustic soda). Chlorine gas is also an important industrial product.

Example 21.11

What do you expect to be the half-reactions in the electrolysis of aqueous copper(II) sulfate? Note that sulfate ion is not easily oxidized.

Solution

One possible cathode reaction would be the reduction of copper(II) ion.

$$Cu^{2+}(aq) + 2e^- \longrightarrow Cu(s); \ E° = 0.34 \text{ V}$$

The other is the reduction of water, which must be considered in any aqueous electrolysis.

$$2H_2O(l) + e^- \longrightarrow H_2(g) + 2OH^-(aq);$$
$$E = -0.41 \text{ V (pH} = 7.00)$$

Since the electrode potential for the copper ion is higher, it is easier to reduce. This conclusion is not changed when we include the effect of overvoltage. Overvoltage is most noticeable in the case of gas evolution. Thus, the reduction of water (with the evolution of H_2) becomes more difficult than we expect from the electrode potential. In either case, we find that the cathode half-reaction is

$$Cu^{2+}(aq) + 2e^- \longrightarrow Cu(s)$$

At the anode, we might expect water or sulfate ion to be oxidized. But since sulfate ion is difficult to oxidize, the anode half-reaction should be

$$2H_2O(l) \longrightarrow O_2(g) + 4H^+(aq) + 4e^-$$

Exercise 21.13

Give the half-reactions when aqueous silver nitrate is electrolyzed. Nitrate ion is not oxidized during the electrolysis.

(See Problems 21.53 and 21.54.)

21.9 Stoichiometry of Electrolysis

During the years 1831 to 1832, the British chemist and physicist Michael Faraday showed that one can relate the amounts of substances released at the electrodes during electrolysis to the total charge that has flowed in the electrical circuit.■ If we look at the electrode reactions, we see that the relationship is a stoichiometric one.

When molten sodium chloride is electrolyzed, sodium ions migrate to the cathode, where they react with electrons:

$$Na^+ + e^- \longrightarrow Na(l)$$

Similarly, chloride ions migrate to the anode and release electrons:

$$Cl^- \longrightarrow \tfrac{1}{2}Cl_2(g) + e^-$$

Therefore, when one mole of electrons reacts with sodium ions, one faraday of charge (the magnitude of charge on one mole of electrons) passes through the circuit. One mole of sodium metal is deposited at one electrode, and one-half mole of chlorine gas evolves at the other.

What is new in this type of stoichiometric problem is the measurement of numbers of electrons. We do not weigh them out as we do substances. Rather, we measure the quantity of electrical charge that has passed through the circuit. We use the fact that one faraday (9.65×10^4 C) is equivalent to the charge on one mole of electrons.

If we know the current in a circuit and the length of time that it has been flowing, we can calculate the electrical charge.

$$\text{Electrical charge} = \text{electric current} \times \text{time lapse}$$

In the International System (SI), the base unit of current is the **ampere** (**A**). The coulomb (C), which is the SI unit of electrical charge, is equivalent to an ampere-second. So a current of 0.50 amperes flowing for 84 seconds gives a charge of 0.50 A $\times$ 84 s $=$ 42 A·s or 42 C.

Given the current and the time of electrolysis, we can calculate the amount of substance produced at an electrode. Conversely, from the amount of substance produced at an electrode and the length of time of electrolysis, we can determine the current. The next two examples illustrate these calculations.

■ Faraday's results were summarized in two laws: (1) The quantity of a substance liberated at an electrode is directly proportional to the quantity of electrical charge that has flowed in the circuit. (2) For a given quantity of electrical charge, the amount of any metal that is deposited is proportional to its equivalent weight (atomic weight divided by the charge on the metal ion). These laws follow directly from the stoichiometry of electrolysis.

Example 21.12

When an aqueous solution of copper(II) sulfate, $CuSO_4$, is electrolyzed, copper metal is deposited.

$$Cu^{2+}(aq) + 2e^- \longrightarrow Cu(s)$$

(The other electrode reaction gives oxygen: $2H_2O \longrightarrow O_2 + 4H^+ + 4e^-$.) If a constant current was passed for 5.00 hr and 404 mg of copper metal were deposited, what must have been the current?

Solution

From the electrode equation for copper, we can write

$$1 \text{ mol Cu} \simeq 2 \text{ mol } e^-$$

Since 1 mol e^- is equivalent to 9.65×10^4 C (1 faraday), the charge equivalent to 404 mg of copper is

$$0.404 \text{ g Cu} \times \frac{1 \text{ mol Cu}}{63.6 \text{ g Cu}} \times \frac{2 \text{ mol } e^-}{1 \text{ mol Cu}} \times \frac{9.65 \times 10^4 \text{ C}}{1 \text{ mol } e^-}$$

$$\simeq 1.23 \times 10^3 \text{ C}$$

The time lapse, 5.00 hr, equals 1.80×10^4 s. Thus,

$$\text{Current} = \frac{\text{charge}}{\text{time}} = \frac{1.23 \times 10^3 \text{ C}}{1.80 \times 10^4 \text{ s}} = 6.83 \times 10^{-2} \text{ A}$$

Exercise 21.14

A constant electric current deposits 365 mg of silver in 216 min from an aqueous silver nitrate solution. What is the current?

(See Problems 21.55 and 21.56.)

Example 21.13

When an aqueous solution of potassium iodide is electrolyzed using platinum electrodes, the half-reactions are

$$2I^-(aq) \longrightarrow I_2(aq) + 2e^-$$
$$2H_2O(l) + 2e^- \longrightarrow H_2(g) + 2OH^-(aq)$$

How many grams of iodine are produced when a current of 8.52 mA flows through the cell for 10.0 min?

Solution

When the current flows for 6.00×10^2 s (10.0 min), the

amount of charge is

$$8.52 \times 10^{-3}\,A \times 6.00 \times 10^2\,s = 5.11\,C$$

Note that two moles of electrons are equivalent to one mole of I_2. Hence,

$$5.11\ \mathcal{C} \times \frac{1\ \text{mole}\ e^-}{9.65 \times 10^4\ \mathcal{C}} \times \frac{1\ \text{mol}\ I_2}{2\ \text{mole}\ e^-} \times \frac{254\ g\ I_2}{1\ \text{mol}\ I_2}$$
$$\simeq 6.73 \times 10^{-3}\ g\ I_2$$

Exercise 21.15

How many grams of oxygen are liberated by the electrolysis of water after 185 s with a current of 0.0565 A?

(See Problems 21.57 and 21.58.)

A Checklist for Review

Important Terms

electrochemical cell (21.1)
voltaic (galvanic) cell (21.1)
electrolytic cell (21.1)
electrolysis (21.1)
half-cells (21.1)
salt bridge (21.1)
cathode (21.1)
anode (21.1)

Leclanché (zinc–carbon) dry cell (21.2)
lead storage cell (21.2)
fuel cell (21.2)
cathodic protection (21.2)
potential difference (21.4)
volt (V) (21.4)
faraday constant (F) (21.4)
electromotive force (emf) (21.4)

potentiometer (21.4)
polarity (21.5)
standard electrode potential (21.5)
standard emf (21.5)
Nernst equation (21.7)
glass electrode (21.7)
ion-selective electrode (21.7)
ampere (A) (21.9)

Summary of Facts and Concepts

Electrochemical cells are of two types: voltaic and electrolytic. A *voltaic cell* uses a spontaneous chemical reaction to generate an electric current. It does this by physically separating the reaction into its oxidation and reduction half-reactions. These half-reactions take place in *half-cells*. The half-cell in which reduction occurs is called the *cathode;* the half-cell in which oxidation occurs is called the *anode*. Electrons flow in the external circuit from the anode to the cathode.

Voltaic cells are used commercially as portable energy

sources (batteries). In addition, the basic principle of the voltaic cell is employed in the *cathodic protection* of buried pipelines and tanks.

The *electromotive force (emf)* is the maximum voltage of a voltaic cell. It can be directly related to the maximum work that can be done by the cell. A *standard electrode potential* or reduction potential refers to the potential of an electrode in which molar concentrations and gas pressures (in atmospheres) have unit values. A table of standard electrode potentials is useful for calculating the *standard emf* of a cell (by subtracting the anode potential from the cathode potential). Standard electrode potentials are also useful for establishing the direction of spontaneity of a redox reaction.

The standard free-energy change, standard emf, and equilibrium constant are all related. Knowing one quantity, we can calculate the others. Electrochemical measurements can, therefore, give us equilibrium or thermodynamic information.

An electrode potential depends on concentrations of the electrode substances, according to the *Nernst equation*. Because of this relationship, cell emf's can be used to measure ion concentrations. This is the basic idea used in a pH meter, a device that measures the hydrogen-ion concentration.

Electrolytic cells are the other type of electrochemical cell. They use an external voltage source to push a reaction in a nonspontaneous direction. The electrolysis of an aqueous solution often involves the oxidation or reduction of water at the electrodes. Electrolysis of concentrated sodium chloride solution, for example, gives hydrogen at the cathode. The amounts of substances released at an electrode are related to the amount of charge passed through the cell. This relationship is stoichiometric and follows from the electrode reactions.

Operational Skills

1. Given a verbal description of a voltaic or electrolytic cell, sketch the cell, labeling the anode and cathode, and give the directions of electron flow and ion migration (Example 21.1).

2. Given the notation for a voltaic cell, write the overall cell reaction (Example 21.2). Alternatively, given the cell reaction, write the cell notation.

3. Given the emf and overall reaction for a voltaic cell, calculate the maximum work that can be obtained from a given amount of reactant (Example 21.3).

4. Given standard electrode potentials, calculate the standard emf of a voltaic cell (Example 21.4). Given standard electrode potentials, decide the direction of spontaneity for an oxidation–reduction reaction under standard conditions (Example 21.5).

5. Given standard electrode potentials, calculate the standard free-energy change for an oxidation–reduction reaction (Example 21.6). Given a table of standard free

energies of formation, calculate the standard emf of a voltaic cell (Example 21.7). Given standard potentials (or standard emf), calculate the equilibrium constant of an oxidation–reduction reaction (Example 21.8).

6. Given standard electrode potentials and the concentrations of substances in a voltaic cell, calculate the cell emf (Example 21.9). Given the standard electrode potential, calculate the electrode potential for nonstandard conditions (Example 21.10).

7. Using values of electrode potentials, decide which electrode reactions actually occur in the electrolysis of an aqueous solution (Example 21.11).

8. Given the amount of product obtained by electrolysis, calculate the amount of charge that flowed (Example 21.12). Given the amount of charge that flowed, calculate the amount of product obtained by electrolysis (Example 21.13).

Review Questions

21.1 Describe the difference between a voltaic cell and an electrolytic cell.

21.2 Define *cathode* and *anode*. Describe the migration of cations and anions in a cell.

21.3 Describe the zinc–carbon or Leclanché dry cell and the lead storage battery.

21.4 What is a fuel cell? Describe an example.

21.5 Explain the electrochemistry of rusting.

21.6 Iron may be protected by coating with tin (tin cans) or with zinc (galvanized iron). Galvanized iron does not corrode as long as zinc is present. By contrast, when a tin can is scratched, the exposed iron underneath corrodes rapidly. Explain the difference between zinc and tin as protective coatings against iron corrosion.

21.7 What is the SI unit of electrical potential?

21.8 Define the *faraday*.

21.9 A voltmeter measures voltage by drawing current through a coil that is suspended on a spring between the poles of a permanent magnet. When current flows through the coil, it acts as an electromagnet and is then deflected by the permanent magnet, depending on the voltage across the voltmeter. This deflection is registered on a meter that is calibrated in volts. Contrast the operation of a voltmeter with that of the potentiometer. Why is the potentiometer preferred for measurements of cell emf?

21.10 How are standard electrode potentials defined?

21.11 Express the SI unit of energy as the product of two electrical units.

21.12 Give the mathematical relationships between each possible pair of the three quantities: $\Delta G°$, $E°_{cell}$, and K_{th}.

21.13 Explain, using the Nernst equation, how the corrosion of iron would be affected by the pH of a water drop.

21.14 The electrolysis of water is often done by passing a current through a dilute solution of sulfuric acid. What is the function of the sulfuric acid?

21.15 Describe a method for the preparation of sodium metal from sodium chloride.

21.16 Potassium was discovered by the British chemist Humphry Davy when he electrolyzed molten potassium hydroxide. What would be the anode reaction?

21.17 Briefly explain why different products are obtained from the electrolysis of molten NaCl and electrolysis of dilute aqueous solution of NaCl.

21.18 Write the Nernst equation for the electrode reaction $2Cl^-(aq) \longrightarrow Cl_2(g) + 2e^-$. With this equation, explain why the electrolysis of concentrated sodium chloride solution might be expected to release chlorine gas rather than oxygen gas at the anode.

Problems

Electrochemical Cells

21.19 A voltaic cell is constructed from the following half-cells: a zinc electrode in zinc sulfate solution and a nickel electrode in nickel sulfate solution. The half-reactions are

$$Zn(s) \longrightarrow Zn^{2+}(aq) + 2e^-$$
$$Ni^{2+}(aq) + 2e^- \longrightarrow Ni(s)$$

Sketch the cell, labeling the anode and cathode (with the electrode reactions), and show the direction of electron flow and the movement of the cations.

21.21 Water may be electrolyzed if an electric current is passed through a dilute aqueous solution of sulfuric acid. The electrode reactions are

$$2H_2O(l) + 2e^- \longrightarrow H_2(g) + 2OH^-(aq)$$
$$2H_2O(l) \longrightarrow O_2(g) + 4H^+(aq) + 4e^-$$

Sketch the electrolytic cell, showing the flow of electrons in the external circuit. Label the anode and cathode, giving the corresponding electrode reactions.

21.20 Half-cells were made from a nickel rod dipping in a nickel sulfate solution and a copper rod dipping in a copper sulfate solution. The half-reactions that occurred in a voltaic cell using these half-cells were

$$Cu^{2+}(aq) + 2e^- \longrightarrow Cu(s)$$
$$Ni(s) \longrightarrow Ni^{2+}(aq) + 2e^-$$

Sketch the cell and label the anode and cathode, showing the corresponding electrode reactions. Give the direction of electron flow and the movement of cations.

21.22 A battery was connected across electrodes that dipped in a solution of copper(II) sulfate. The electrode reactions were

$$Cu^{2+}(aq) + 2e^- \longrightarrow Cu(s)$$
$$2H_2O(l) \longrightarrow O_2(g) + 4H^+(aq) + 4e^-$$

Sketch the electrolytic cell, showing the direction of electron flow in the external circuit. Label the anode and cathode, and indicate the corresponding electrode reactions.

21.23 Cadmium reacts spontaneously with silver ion.

$$Cd(s) + 2Ag^+(aq) \longrightarrow Cd^{2+}(aq) + 2Ag(s)$$

Describe a voltaic cell using this reaction. What are the half-reactions?

21.25 Nickel–cadmium cells are rechargeable batteries used in calculators and similar devices. When the cells are discharging, the half-reactions are

$$Cd(s) + 2OH^-(aq) \longrightarrow Cd(OH)_2(s) + 2e^-$$
$$NiO_2(s) + 2H_2O(l) + 2e^- \longrightarrow Ni(OH)_2(s) + 2OH^-(aq)$$

Identify the anode reaction and the cathode reaction. What are the half-reactions when the battery is being charged? Identify each as the anode or cathode reaction. What is the overall reaction as a voltaic cell?

21.24 Magnesium reacts spontaneously with copper(II) ion.

$$Mg(s) + Cu^{2+}(aq) \longrightarrow Mg^{2+}(aq) + Cu(s)$$

Obtain half-reactions for this, then describe a voltaic cell using these half-reactions.

21.26 A mercury battery, used for hearing aids and electric watches, delivers a constant voltage (1.35 V) for long periods. The half-reactions are

$$HgO(s) + H_2O(l) + 2e^- \longrightarrow Hg(l) + 2OH^-(aq)$$
$$Zn(s) + 2OH^-(aq) \longrightarrow Zn(OH)_2(s) + 2e^-$$

Which half-reaction occurs at the anode and which occurs at the cathode? What is the overall cell reaction?

Voltaic Cell Notation

21.27 Give the notation for a voltaic cell constructed from a hydrogen electrode (cathode) in 1.0 M HCl and a nickel electrode (anode) in 1.0 M $NiSO_4$ solution. The electrodes are connected by a salt bridge.

21.29 Write the overall cell reaction for the following voltaic cell:

$$Ni(s)|Ni^{2+}(aq)\|Ag^+(aq)|Ag(s)$$

21.31 Consider the voltaic cell

$$Cd(s)|Cd^{2+}(aq)\|Ni^{2+}(aq)|Ni(s)$$

Write the half-cell reactions and the overall cell reaction. Make a sketch of this cell and label it. Include labels showing the anode, cathode, and direction of electron flow.

21.28 A voltaic cell has a zinc rod in zinc sulfate solution for the anode half-cell, and chlorine gas (Cl_2) bubbling over a platinum electrode immersed in sodium chloride solution for the cathode half-cell. The half-cells are connected by a salt bridge. Give the notation for this cell.

21.30 Write the overall cell reaction for the following voltaic cell:

$$Pt|H_2(g)|H^+(aq)\|Ce^{4+}(aq), Ce^{3+}(aq)|Pt$$

21.32 Consider the voltaic cell

$$Zn(s)|Zn^{2+}(aq)\|Ag^+(aq)|Ag(s)$$

Write the half-cell reactions and the overall cell reaction. Make a sketch of this cell and label it. Include labels showing the anode, cathode, and direction of electron flow.

Electrode Potentials and Cell emf's

21.33 A voltaic cell whose cell reaction is

$$2Fe^{3+}(aq) + Sn^{2+}(aq) \longrightarrow 2Fe^{2+}(aq) + Sn^{4+}(aq)$$

has an emf of 0.62 V. What is the maximum electrical work that can be obtained from this cell per mole of iron(III) ion?

21.35 Calculate the standard emf of the following cells from electrode potentials:

(a) $Fe(s)|Fe^{3+}(aq)\|Br^-(aq), Br_2(aq)|Pt$
(b) $Pt|I_2(s)|I^-(aq)\|Cl^-(aq)|Cl_2(g)|Pt$

21.34 A particular voltaic cell operates on the reaction

$$Zn(s) + Cl_2(g) \longrightarrow Zn^{2+}(aq) + 2Cl^-(aq)$$

giving an emf of 0.653 V. Calculate the maximum electrical work generated when 10.0 g of zinc metal are consumed.

21.36 Consider a voltaic cell constructed from the following electrodes:

$$Fe^{3+}, Fe^{2+}|Pt \quad \text{and} \quad Ag^+|Ag$$

Obtain the standard emf of this voltaic cell, using standard electrode potentials. Give the notation for this cell (in the conventional order).

21.37 Consider the following reactions. Are they spontaneous in the direction written, under standard conditions?

(a) $Sn^{4+}(aq) + 2Fe^{2+}(aq) \longrightarrow Sn^{2+}(aq) + 2Fe^{3+}(aq)$

(b) $4MnO_4^-(aq) + 12H^+(aq) \longrightarrow$
$$4Mn^{2+}(aq) + 5O_2(g) + 6H_2O(l)$$

21.39 What would you expect to happen if chlorine gas, Cl_2, at 1 atmosphere pressure is bubbled into a solution containing $1.0\ M\ F^-$ and $1.0\ M\ Br^-$? Write a balanced equation for the reaction that occurs.

21.38 Answer the following questions by referring to standard electrode potentials.
(a) Will dichromate ion, $Cr_2O_7^{2-}$, oxidize iron(II) ion in acidic solution under standard conditions?
(b) Will copper metal reduce $1.0\ M\ Ni^{2+}(aq)$ to metallic nickel?

21.40 Dichromate ion, $Cr_2O_7^{2-}$, is added to an acidic solution containing Br^- and Mn^{2+}. Write a balanced equation for any reaction that occurs. Assume standard conditions.

Relationships Among $E°_{cell}$, $\Delta G°$, and K_{th}

21.41 What is $\Delta G°$ for the following reaction?

$$2I^-(aq) + Cl_2(g) \longrightarrow I_2(s) + 2Cl^-(aq)$$

Use data given in Table 21.1.

21.43 Calculate the standard emf of the lead storage cell whose overall reaction is

$PbO_2(s) + 2HSO_4^-(aq) + 2H^+(aq) + Pb(s) \longrightarrow$
$$2PbSO_4(s) + 2H_2O(l)$$

Free energies of formation are (in kJ/mol): $PbO_2(s)$, -219; $PbSO_4(s)$, -811; $HSO_4^-(aq)$, -753. Other values may be found in Table 20.2.

21.45 Copper(I) ion can act as both an oxidizing agent and a reducing agent. Hence, it can react with itself.

$$2Cu^+(aq) \longrightarrow Cu(s) + Cu^{2+}(aq)$$

Calculate the equilibrium constant at 25°C for this reaction, using appropriate values of electrode potentials.

21.42 Using electrode potentials, calculate the standard free-energy change for the following reaction:

$$Na(s) + \tfrac{1}{2}Cl_2(g) \xrightarrow{aq} Na^+(aq) + Cl^-(aq)$$

21.44 Calculate the standard emf of the cell corresponding to the oxidation of oxalic acid, $H_2C_2O_4$, by permanganate ion, MnO_4^-.

$5H_2C_2O_4(aq) + 2MnO_4^-(aq) + 6H^+(aq) \longrightarrow$
$$10CO_2(g) + 2Mn^{2+}(aq) + 8H_2O(l)$$

Free energies of formation are (in kJ/mol): $H_2C_2O_4(aq)$, -698; $Mn^{2+}(aq)$, -223; $MnO_4^-(aq)$, -425. Other values are given in Table 20.2.

21.46 Use electrode potentials to calculate the equilibrium constant at 25°C for the reaction

$$2ClO_3^-(aq) \rightleftharpoons ClO_4^-(aq) + ClO_2^-(aq)$$

Nernst Equation

21.47 Calculate the emf of the following cell at 25°C:

$$Cr(s)|Cr^{3+}(1.0 \times 10^{-2}\ M)||Ni^{2+}(2.0\ M)|Ni(s)$$

21.49 What is the potential of the following electrode at 25°C?

$$Zn^{2+}(2.0 \times 10^{-3}\ M)|Zn(s)$$

21.51 The voltaic cell

$$Cd(s)|Cd^{2+}(aq)||Ni^{2+}(1.0\ M)|Ni(s)$$

has an electromotive force of 0.240 V at 25°C. What is the concentration of cadmium ion? $(E° = 0.170\ V.)$

21.48 What is the emf of the following cell at 25°C?

$$Ni(s)|Ni^{2+}(1.0\ M)||Sn^{2+}(1.0 \times 10^{-4}\ M)|Sn(s)$$

21.50 What is the potential of the following electrode at 25°C?

$$Ag^+(1.0 \times 10^{-7}\ M)|Ag(s)$$

21.52 The emf of the following cell at 25°C is 0.131 V:

$$Pt|H_2(1.00\ atm)|\text{test solution}||H^+(1.00\ M)|$$
$$H_2(1.00\ atm)|Pt$$

What is the pH of the test solution?

Electrolysis

21.53 Describe what you expect to happen when the following solutions are electrolyzed: (a) aqueous Na_2SO_4; (b) aqueous KBr. That is, what are the electrode reactions? What is the overall reaction? Note that sulfate ion is not easily oxidized.

21.55 In the commercial preparation of aluminum, aluminum oxide, Al_2O_3, is electrolyzed at 1000°C. (The mineral cryolite is added as a solvent.) Assume the cathode reaction is

$$Al^{3+} + 3e^- \longrightarrow Al$$

How many coulombs of electricity are required to give 5.12 kg of aluminum?

21.57 When molten lithium chloride, LiCl, is electrolyzed, lithium metal is liberated at the cathode. How many grams of lithium are liberated when 5.00×10^3 C of charge pass through the cell?

21.54 Describe what you expect to happen when the following solutions are electrolyzed: (a) aqueous $CuCl_2$; (b) aqueous $AgNO_3$. That is, what are the electrode reactions? What is the overall reaction? Note that nitrate ion is not oxidized.

21.56 Chlorine, Cl_2, is produced commercially by the electrolysis of aqueous sodium chloride. The anode reaction is

$$2Cl^-(aq) \longrightarrow Cl_2(g) + 2e^-$$

How long will it take to produce 1.18 kg of chlorine, if the current is 5.00×10^2 A?

21.58 How many grams of cadmium are deposited from an aqueous solution of cadmium sulfate, $CdSO_4$, when an electric current of 1.51 A flows through the solution for 156 min?

Additional Problems

21.59 Give the notation for a voltaic cell that uses the reaction

$$Mg(s) + Cl_2(g) \longrightarrow Mg^{2+}(aq) + 2Cl^-(aq)$$

What is the half-cell reaction for the anode? for the cathode? What is the standard emf of the cell?

21.60 Give the notation for a voltaic cell whose overall cell reaction is

$$Mg(s) + 2Ag^+(aq) \longrightarrow Mg^{2+}(aq) + 2Ag(s)$$

What are the half-cell reactions? Label them as anode or cathode reactions. What is the standard emf of this cell?

21.61 Use electrode potentials to answer the following questions: (a) Is the oxidation of nickel by iron(III) ion a spontaneous reaction under standard conditions?

$$Ni(s) + 2Fe^{3+}(aq) \longrightarrow Ni^{2+}(aq) + 2Fe^{2+}(aq)$$

(b) Will iron(III) ion oxidize tin(II) ion to tin(IV) ion under standard conditions?

$$2Fe^{3+}(aq) + Sn^{2+}(aq) \longrightarrow 2Fe^{2+}(aq) + Sn^{4+}(aq)$$

21.62 Use electrode potentials to answer the following questions, assuming standard conditions: (a) Do you expect permanganate ion (MnO_4^-) to oxidize chloride ion to chlorine gas in acidic solution? (b) Will dichromate ion $(Cr_2O_7^{2-})$ oxidize chloride ion to chlorine gas in acidic solution?

*21.63 Determine the emf of the following cell:

$$Pb|PbSO_4(s), SO_4^{2-}(1.0\ M)\|H^+(1.0\ M)|H_2(1.0\ atm)|Pt$$

The anode is essentially a lead electrode, $Pb|Pb^{2+}(aq)$. However, the anode solution is saturated with lead sulfate, so that the lead(II)-ion concentration is determined by the solubility product of $PbSO_4 (= 1.7 \times 10^{-8})$.

*21.64 Determine the emf of the following cell:

$$Pt|H_2(1.0\ atm)|H^+(1.0\ M)\|Cl^-(1.0\ M), AgCl(s)|Ag$$

The cathode is essentially a silver electrode, $Ag^+(aq)|Ag$. However, the cathode solution is saturated with silver chloride, so that the silver-ion concentration is determined by the solubility product of $AgCl (= 1.8 \times 10^{-10})$.

**21.65 An electrode is prepared by dipping a silver strip into a solution saturated with silver thiocyanate, AgSCN, and containing 0.10 M SCN$^-$. The emf of the voltaic cell constructed by connecting this, as the cathode, to the standard hydrogen half-cell as the anode is 0.45 V. What is the solubility product of silver thiocyanate?

**21.66 An electrode is prepared from liquid mercury in contact with a saturated solution of mercury(I) chloride, Hg_2Cl_2, containing 0.10 M Cl$^-$. The emf of the voltaic cell constructed by connecting this, as the anode, to the standard hydrogen half-cell as the cathode is 0.28 V. What is the solubility product of mercury(I) chloride?

*21.67 (a) Calculate the equilibrium constant for the following reaction at 25°C:

$$Sn(s) + Pb^{2+}(aq) \rightleftharpoons Sn^{2+}(aq) + Pb(s)$$

The standard emf of the corresponding voltaic cell is 0.010 V. (b) If an excess of tin metal is added to 1.0 M Pb^{2+}, what is the concentration of Pb^{2+} at equilibrium?

*21.68 (a) Calculate the equilibrium constant for the following reaction at 25°C:

$$Ag^+(aq) + Fe^{2+}(aq) \rightleftharpoons Ag(s) + Fe^{3+}(aq)$$

The standard emf of the corresponding voltaic cell is 0.030 V. (b) If equal volumes of 1.0 M solutions of Ag^+ and Fe^{2+} are mixed, what is the equilibrium concentration of Fe^{2+}?

21.69 How many faradays are required for each of the following processes? How many coulombs are required?

(a) Reduction of 1.0 mol Na^+ to Na
(b) Reduction of 1.0 mol Cu^{2+} to Cu
(c) Oxidation of 1.0 g H_2O to O_2
(d) Oxidation of 1.0 g Cl^- to Cl_2

21.70 How many faradays are required for each of the following processes? How many coulombs are required?

(a) Reduction of 1.0 mol Fe^{3+} to Fe^{2+}
(b) Reduction of 1.0 mol Fe^{3+} to Fe
(c) Oxidation of 1.0 g Sn^{2+} to Sn^{4+}
(d) Reduction of 1.0 g Au^{3+} to Au

*21.71 In an analytical determination of arsenic, a solution containing arsenious acid, H_3AsO_3, potassium iodide, and a small amount of starch is electrolyzed. The electrolysis produces free iodine from iodide ion, and the iodine immediately oxidizes the arsenious acid to hydrogen arsenate ion, $HAsO_4^{2-}$:

$$I_2(aq) + H_3AsO_3(aq) + H_2O(l) \longrightarrow$$
$$2I^-(aq) + HAsO_4^{2-}(aq) + 4H^+(aq)$$

When the oxidation of arsenic is complete, the free iodine combines with the starch to give a deep blue color. If, during a particular run, it takes 65.3 s for a current of 10.5 mA to give an endpoint (indicated by the appearance of the blue color of starch–iodine complex), how many grams of arsenic are present in the solution?

*21.72 The amount of lactic acid, $HC_3H_5O_3$, produced in a sample of muscle tissue was analyzed by reaction with hydroxide ion. Hydroxide ion was produced in the sample mixture by electrolysis. The cathode reaction was

$$2H_2O(l) + 2e^- \longrightarrow H_2(g) + 2OH^-(aq)$$

Hydroxide ion reacts with lactic acid as soon as it is produced. The endpoint of the reaction is detected with an acid–base indicator. It required 115 s for a current of 15.6 mA to reach the endpoint. How many grams of lactic acid (a monoprotic acid) were present in the sample?

22. *Nuclear Chemistry*

Radioactivity and Nuclear Bombardment Reactions

22.1 Radioactivity Nuclear Equations/ Nuclear Stability/ Types of Radioactive Decay/ Radioactive Decay Series

22.2 Nuclear Bombardment Reactions Transmutation/ Transuranium Elements

22.3 Radiations and Matter: Detection and Biological Effects Radiation Counters/ Biological Effects and Radiation Dosage

22.4 Rate of Radioactive Decay Rate of Radioactive Decay and Half-Life/ Radioactive Dating

22.5 Applications of Radioactive Isotopes Chemical Analysis/ Medical Therapy and Diagnosis

Energy of Nuclear Reactions

22.6 Mass–Energy Calculations Mass–Energy Equivalence/ Nuclear Binding Energy

22.7 Nuclear Fission and Nuclear Fusion Nuclear Fission; Nuclear Reactors/ Breeder Reactors/ Nuclear Fusion

Technetium is an unusual element. Although a *d*-transition element (under manganese in Group VIIB) with small atomic number ($Z = 43$), it has no stable isotopes. The nucleus of every technetium isotope is radioactive and decays, or disintegrates, to give an isotope of another element. Many of the technetium isotopes decay by emitting an electron from the nucleus.

Because of its nuclear instability, technetium is not found naturally on earth. Nevertheless, it is produced commercially in kilogram quantities from other elements by nuclear reactions, processes in which nuclei are transformed into different nuclei. Technetium (from the Greek *tekhnetos*, meaning "artificial") was the first new element produced in the laboratory from another element. It was discovered in 1937 by Carlo Perrier and Emilio Segrè when the element molybdenum was bombarded with deuterons (nuclei of hydrogen, each having one proton and one neutron). Later technetium was found to be a product of the fission, or splitting, of uranium nuclei. Thus, technetium is produced in nuclear fission reactors used to generate electricity.

Technetium is one of the principal isotopes used in medical diagnostics based on radioactivity. A compound of technetium is injected into a vein where it is attracted to certain body organs. The energy emitted by technetium nuclei is detected by special equipment and gives an image of these body organs. Figure 22.1 shows the image of a person's skeleton obtained from technetium that was administered. The technetium is eliminated by the body after several hours.

In this chapter, we will look at nuclear processes, such as those that we have described for technetium. We will answer such questions as, How do we describe the radioactive decay of technetium? How do we describe the transformation of a molybdenum nucleus to technetium? How is technetium produced from uranium by nuclear fission, or splitting? What are some practical applications of nuclear processes?

Chapter Overview

We will begin the chapter by recalling what we learned in Chapter 5 about *radioactivity*, or spontaneous nuclear disintegration. After describing a nuclear process by means of a nuclear equation, we will look at various *nuclear bombardment reactions*, in which a nucleus is bombarded, or struck, by a particle such as a neutron or other nucleus. In this way new isotopes are produced, many of which are radioactive. By detecting the rate at which radiation from a radioactive isotope is produced, we can determine the *rate of radioactive decay* of this isotope.

A nucleus decays to another nucleus, one having lower energy. The *energy of nuclear reactions* is useful to know, particularly when looking at nuclear fission (the splitting of one nucleus into two) and nuclear fusion (the reaction of two nuclei to give one).

Radioactivity and Nuclear Bombardment Reactions

In chemical reactions, only the outer electrons of the atoms are disturbed. The nuclei of the atoms are not affected. In nuclear reactions, however, nuclear changes occur. The compound in which the atoms are found is immaterial.

We will look at two types of nuclear reactions. One type is **radioactive decay,** in which a nucleus spontaneously disintegrates, giving off radiation.

Natural radioactive elements emit, or give off, three kinds of radiation, called alpha, beta, and gamma radiation. *Alpha rays* consist of helium-4 nuclei (with two protons and two neutrons). *Beta rays* consist of electrons. *Gamma rays* are electromagnetic radiation, like light, with very short wavelengths (about 0.01 Å or 1×10^{-12} m).

The second type of nuclear reaction is a **nuclear bombardment reaction.** In this reaction, a nucleus is bombarded, or struck, by another nucleus or by a nuclear particle. If there is sufficient energy in this collision, the nuclear particles of the reactants rearrange to give a product nucleus or nuclei. In the next section, we will look at radioactive decay.

22.1 Radioactivity

The phenomenon of radioactivity was discovered by Antoine Henri Becquerel in 1896. He discovered that photographic plates developed bright spots when exposed to uranium minerals and concluded that the mineral was giving off some sort of radiation. It was later shown that uranium nuclei emit alpha particles and thereby decay, or disintegrate, to thorium nuclei.

A sample of uranium-238 decays, or disintegrates, spontaneously over a period of billions of years. After about 30 billion years, the sample would be nearly gone. On the other hand, strontium-90, formed by nuclear reactions that occur in nuclear weapons testing and nuclear power reactors, decays more rapidly. A sample of strontium-90 would be nearly gone after a couple of hundred years. In either case, it is impossible to know when a particular nucleus will decay, although as we will see in Section 22.4, precise information can be given about the rate of decay of any radioactive sample.

We can write an equation for the nuclear reaction corresponding to the decay of uranium-238 much as we would write an equation for a chemical reaction. Let us see how to do that.

Nuclear Equations

A nucleus is composed of **nucleons,** or nuclear particles, of which there are two kinds: protons and neutrons. Each kind of nucleus, called a **nuclide,** is characterized by the number of protons and the number of neutrons it contains. Two atoms whose nuclei have the same number of protons but different numbers of neutrons are called *isotopes*. The **nuclide symbol** indicates the atomic number, or nuclear charge (number of protons in the nucleus), as a left subscript and the mass number (number of nucleons, or protons plus neutrons, in the nucleus) as a left superscript to the symbol of the element. Thus, the most abundant isotope of uranium, which has 92 protons and 146 neutrons in its nucleus, has the nuclide symbol

$$\text{mass number} \rightarrow \ ^{238}_{\ 92}\text{U} \ \leftarrow \text{atomic number}$$

Note that the mass number is $92 + 146 = 238$. ∎

The radioactive decay of $^{238}_{92}$U by alpha-particle emission (loss of a $^4_2$He nucleus) is written

$$^{238}_{92}\text{U} \longrightarrow \ ^{234}_{90}\text{Th} + \ ^4_2\text{He}$$

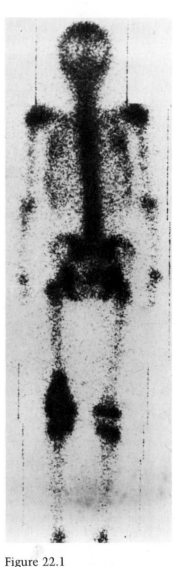

Figure 22.1
Image of person's skeleton obtained from an excited form of technetium-99.

■ Symbols for a nucleus (nuclide symbols) were introduced in Section 5.3. See Example 5.2 in that section.

This is an example of a **nuclear equation,** which symbolically represents a nuclear reaction. Normally, only the nuclei are represented. It is not necessary to indicate the chemical compound or electron charges on any ions involved, since the chemical environment has no effect on nuclear processes.

Reactant and product nuclei are represented in nuclear equations by their nuclide symbols. Other particles are given the following symbols, in which the subscript equals the charge and the superscript equals the number of nucleons (mass number):

Proton	$_1^1\text{H}$ or	$_1^1\text{p}$
Neutron	$_0^1\text{n}$	
Electron	$_{-1}^0\text{e}$ or	$_{-1}^0\beta$
Positron	$_1^0\text{e}$ or	$_1^0\beta$
Gamma photon	$_0^0\gamma$	

A **positron** is a particle similar to an electron, with the same mass but a positive charge. **Gamma photons** are particles of electromagnetic radiation of short wavelength (0.01 Å or 10^{-12} m) and high energy per photon.

Example 22.1

Write the nuclear equation for the radioactive decay of radium-226 by alpha decay to give radon-222. A radium-226 nucleus emits one alpha particle, leaving behind a radon-222 nucleus.

Solution

Looking at a list of elements (inside back cover) we find that the atomic number of radium is 88. Hence, the nuclide symbol is $_{88}^{226}\text{Ra}$. Similarly, the nuclide symbols for radon-222 and the alpha particle are $_{86}^{222}\text{Rn}$ and $_2^4\text{He}$. Therefore, the equation is

$$_{88}^{226}\text{Ra} \longrightarrow {}_{86}^{222}\text{Rn} + {}_2^4\text{He}$$

Exercise 22.1

Potassium-40 is a naturally occurring radioactive isotope. It decays to calcium-40 by beta emission. When a potassium-40 nucleus decays by beta emission, it emits one beta particle and gives a calcium-40 nucleus. Write the nuclear equation for this decay.

(See Problems 22.19 and 22.20.)

The total charge is conserved, or remains constant, during a nuclear reaction. This means that the sum of the subscripts for the products must equal the sum of the subscripts for the reactants. For the equation in Example 22.1, the subscript for the reactant $_{88}^{226}\text{Ra}$ is 88. For the products, the sum of the subscripts is $86 + 2 = 88$.

Similarly, the total number of nucleons is conserved, or remains constant, during a nuclear reaction. This means that the sum of the superscripts (the mass numbers) for the reactants equals the sum of the superscripts for the products. For the equation in Example 22.1 the superscript for the reactant nucleus is 226. For the products, the sum of the superscripts is $222 + 4 = 226$.

Note that if all reactants and products but one are known in a nuclear equation, the identity of that one nucleus or particle can be easily obtained. This is illustrated in the next example.

Example 22.2

Technetium-99 is the longest-lived radioactive isotope of technetium. Each nucleus decays by emitting one beta particle. What is the product nucleus?

Solution

Let us write the nuclear equation for the decay of technetium-99. Looking at a list of elements, we see that technetium has atomic number 43. Thus, the nuclide symbol is $_{43}^{99}\text{Tc}$. A beta particle is an electron, whose symbol is $_{-1}^{0}\text{e}$. For the unknown product nucleus, we write $_{Z}^{A}\text{X}$, where A is the mass number and Z is the atomic number. The nuclear equation is

$$_{43}^{99}\text{Tc} \longrightarrow {}_{Z}^{A}\text{X} + {}_{-1}^{0}\text{e}$$

From the superscripts, we can write

$$99 = A + 0, \text{ or } A = 99$$

Similarly, from the subscripts, we get

$$43 = Z - 1, \text{ or } Z = 43 + 1 = 44$$

Hence, $A = 99$ and $Z = 44$, so the product is $_{44}^{99}\text{X}$. Since element 44 is ruthenium, symbol Ru, we write the product nucleus as $_{44}^{99}\text{Ru}$.

Exercise 22.2

Plutonium-239 is used in nuclear weapons. It decays by alpha emission, with each nucleus emitting one alpha particle. What is the other product of this decay?

(See Problems 22.21 and 22.22.)

Nuclear Stability

At first glance, the existence of several protons in the small space of a nucleus is puzzling. Why wouldn't the protons be stongly repelled by their like electrical charges? The existence of stable nuclei with more than one proton is due to the nuclear force. The **nuclear force** is a strong force of attraction between nucleons that acts only at very short distances (about 10^{-15} m). Beyond nuclear distances, these nuclear forces become negligible. Therefore, two protons that are much further apart than 10^{-15} m will repel one another by their like electrical charges. However, inside the nucleus, two protons will be close enough that the nuclear force between them is effective. This force in a nucleus can more than compensate for the repulsion of electrical charges and thereby give a stable nucleus.

The protons and neutrons in a nucleus appear to have energy levels much as the electrons in an atom have energy levels. According to the **shell model of the nucleus,** the protons and neutrons each have a shell structure, analogous to the shell structure that exists for electrons in an atom. Recall that, in an atom, filled shells of electrons are associated with the special stability of the noble gases. The total numbers of electrons for these stable atoms are 2 (for He), 10 (for Ne), 18 (for Ar), and so forth. Experimentally, it was noted that nuclei with certain numbers of protons or neutrons appeared to be very stable. These numbers, called **magic numbers,** associated with specially stable nuclei, were later explained by the shell model. According to this theory, magic numbers are the numbers of nuclear particles in completed shells of protons or neutrons. Because nuclear forces differ from electrical forces, these numbers are not the same as those for electrons in atoms. For protons, the magic numbers are 2, 8, 20, 28, 50, and 82. Neutrons have these same magic numbers, as well as the magic number 126. For protons, calculations show that 114 should also be a magic number.

Some of the evidence for these magic numbers, and therefore for the shell model, is as follows. Many radioactive nuclei decay by emitting alpha particles, or $_2^4$He nuclei. There appears to be special stability in the $_2^4$He nucleus. It contains 2 protons and 2 neutrons, that is, it contains a magic number of protons (2) and a magic number of neutrons (also 2).

Another piece of evidence is seen in the final products obtained in natural radioactive decay. For example, uranium-238 decays to thorium-234, which in turn decays to protactinium-234, and so forth. Each product is radioactive and decays to another nucleus until the final product, $_{82}^{206}$Pb, is reached. This nucleus is stable. Note that it contains 82 protons, which is a magic number. Other radioactive decay series end at $_{82}^{207}$Pb and $_{82}^{208}$Pb, both of which have a magic number of protons. Note that $_{82}^{208}$Pb also has a magic number of neutrons $(208 - 82 = 126)$.

Evidence also points to the special stability of pairs of protons and pairs of neutrons, analogous to the stability of pairs of electrons in molecules. Table 22.1 lists the number of stable isotopes having an even number of protons and an even number of neutrons (157). By comparison, there are only 5 stable isotopes having an odd number of protons and an odd number of neutrons.

Finally, we note that if we plot each stable nuclide on a graph with the number of protons (Z) on the horizontal axis and the number of neutrons (N) on the vertical axis, these nuclides fall in an area or band, called the **band of stability**. Figure 22.2 shows the band of stability; the rest of the figure is explained later in this section. For nuclides up to $Z = 20$, the ratio of neutrons to protons is about 1.0 to 1.1. As Z increases, however, the neutron-to-proton ratio increases to about 1.5. This increase in neutron-to-proton ratio with Z is believed to result from the increasing repulsions of protons from their electrical charges. More neutrons are required to give attractive nuclear forces to offset these replusions.

It appears that when the number of protons becomes very large the proton–proton repulsions become so great that stable nuclides are impossible. Thus, no stable nuclides are known with atomic numbers greater than 83. On the other hand, all elements with Z equal to 83 or less have one or

	Number of Stable Isotopes			
	157	52	50	5
Number of Protons	Even	Even	Odd	Odd
Number of Neutrons	Even	Odd	Even	Odd

Table 22.1
Number of Stable Isotopes with Even and Odd Numbers of Protons and Neutrons

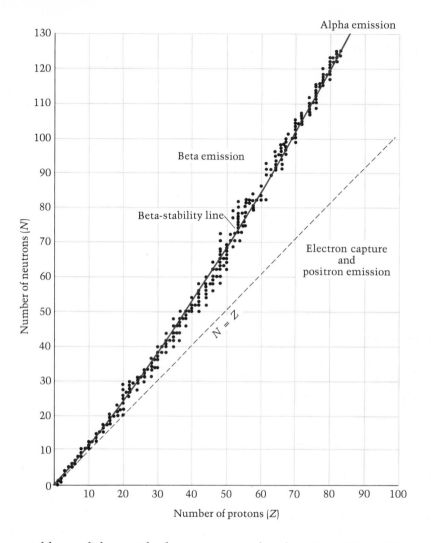

Figure 22.2
Band of stability. The stable nuclides, indicated by black dots, cluster in a band. This band follows the beta-stability line, which separate nuclides that undergo beta decay from those that undergo positron emission and electron capture. Nuclides of $Z > 83$ often decay by alpha emission.

more stable nuclides, with the exception of technetium ($Z = 43$), as we noted in the chapter opening, and promethium ($Z = 61$).

Example 22.3

One of the nuclides in each of the following pairs is radioactive; the other is stable. Which is radioactive and which is stable? Explain.
(a) $^{208}_{84}\text{Po}$, $^{209}_{83}\text{Bi}$ (b) $^{39}_{19}\text{K}$, $^{40}_{19}\text{K}$ (c) $^{71}_{31}\text{Ga}$, $^{76}_{31}\text{Ga}$

Solution

Note that the problem states that one nucleus is radioactive and the other stable. We must decide which is more likely to be radioactive or which is more likely to be stable. (a) Since polonium has an atomic number greater than 83, $^{208}_{84}\text{Po}$ is radioactive. Bismuth-209 has 126 neutrons (a magic number), so $^{209}_{83}\text{Bi}$ is expected to be stable. (b) Of these two isotopes, $^{39}_{19}\text{K}$ has a magic number of neutrons (20), so $^{39}_{19}\text{K}$ is expected to be stable. The isotope $^{40}_{19}\text{K}$ has an odd number of protons (19) and an odd number of neutrons (21). Because stable odd–odd nuclei are rare, we might expect $^{40}_{19}\text{K}$ to be radioactive. (c) Of the two isotopes, $^{76}_{31}\text{Ga}$ lies further from the center of the band of stability, so it is more likely to be radioactive. For this reason, we expect $^{76}_{31}\text{Ga}$ to be radioactive and $^{71}_{31}\text{Ga}$ to be stable.

Exercise 22.3

Of the following nuclides, two are radioactive. Which are radioactive, and which is stable? Explain. (a) $^{118}_{50}Sn$ (b) $^{76}_{33}As$ (c) $^{227}_{89}Ac$.

(See Problems 22.23 and 22.24.)

Types of Radioactive Decay

There are five common types of radioactive decay. These are listed in Table 22.2 and discussed here.

1. Alpha emission (abbreviated α): emission of a ^4_2He nucleus, or alpha particle, from a nucleus. An example is the radioactive decay of radium-226, written as

$$^{226}_{88}Ra \longrightarrow {}^{222}_{86}Rn + {}^4_2He$$

The product nucleus has an atomic number that is two less, and a mass number that is four less, than that of the original nucleus.

2. Beta emission (abbreviated β or β^-): emission of an electron from a nucleus. Beta emission is equivalent to the conversion of a neutron to a proton.■

$$^1_0n \longrightarrow {}^1_1p + {}^0_{-1}e$$

■ Outside the nucleus, the neutron is unstable and decays in a few minutes to a proton and an electron.

An example of beta emission is the radioactive decay of carbon-14:

$$^{14}_6C \longrightarrow {}^{14}_7N + {}^0_{-1}e$$

The product nucleus has an atomic number that is one more than that of the original nucleus. The mass number remains the same.

3. Positron emission (abbreviated β^+): emission of a positron from a nucleus. A positron, denoted in nuclear equations as ^0_1e, is a particle identical to an electron in mass but having a positive charge instead of a negative one. Positron emission is equivalent to the conversion of a proton to a neutron.■

$$^1_1p \longrightarrow {}^1_0n + {}^0_1e$$

■ Positrons are annihilated as soon as they encounter electrons. When a positron and an electron collide, both particles vanish with the emission of two gamma photons that carry away the energy.

$$^0_1e + {}^0_{-1}e \longrightarrow 2{}^0_0\gamma$$

An example of positron emission is

$$^{95}_{43}Tc \longrightarrow {}^{95}_{42}Mo + {}^0_1e$$

Table 22.2
Types of Radioactive Decay

Type of Decay	Radiation	Equivalent Process	Nuclear Change		Usual Nuclear Condition
			Atomic Number	Mass Number	
Alpha emission (α)	^4_2He	—	-2	-4	$Z > 83$
Beta emission (β)	$^0_{-1}e$	$^1_0n \longrightarrow {}^1_1p + {}^0_{-1}e$	$+1$	0	N/Z too large
Positron emission (β^+)	^0_1e	$^1_1p \longrightarrow {}^1_0n + {}^0_1e$	-1	0	N/Z too small
Electron capture (EC)	x rays	$^1_1p + {}^0_{-1}e \longrightarrow {}^1_0n$	-1	0	N/Z too small
Gamma emission (γ)	$^0_0\gamma$	—	0	0	Excited nucleus

The product nucleus has an atomic number of one less than that of the original nucleus. The mass number remains the same.

4. Electron capture (abbreviated EC): change of a nucleus by capturing, or picking up, an electron from an inner orbital of an atom. In effect, a proton is changed to a neutron, as in positron emission.

$$_1^1p + {}_{-1}^0e \longrightarrow {}_0^1n$$

An example is given by potassium-40, which has a natural abundance of 0.0012%. It can decay by electron capture, as well as by beta and positron emissions. The equation for electron capture in potassium-40 is■

$$_{19}^{40}K + {}_{-1}^0e \longrightarrow {}_{18}^{40}Ar$$

The product nucleus has an atomic number of one less than that of the original nucleus. The mass number remains the same. When another orbital electron fills the vacancy in the inner-shell orbital created by electron capture, an x-ray photon is emitted.

5. Gamma emission (abbreviated γ): emission from a nucleus of a gamma photon, corresponding to radiation with a wavelength of about 10^{-12} m. In many cases, radioactive decay results in a product nucleus that is in an excited state. As in the case of atoms, the excited state is unstable and goes to a lower-energy state with the emission of electromagnetic radiation. For nuclei, this radiation is in the gamma-ray region of the spectrum.

Often, gamma emission occurs very quickly after radioactive decay. In some cases, however, an excited state has significant lifetime before it emits a gamma photon. A **metastable nucleus** is a nucleus in an excited state with a lifetime of at least one nanosecond (10^{-9} s). In time, the metastable nucleus decays by gamma emission. An example is metastable technetium-99, denoted $_{43}^{99m}Tc$, which is used in medical diagnosis, as discussed in Section 22.5.

$$_{43}^{99m}Tc \longrightarrow {}_{43}^{99}Tc + {}_0^0\gamma$$

The product nucleus is simply a lower-energy state of the original nucleus, so there is no change of atomic number or mass number.

In Figure 22.2, a curve called the **beta-stability line** separates the radioactive nuclei that decay by beta emission from those that decay by positron emission or electron capture. Nuclei in which the neutron-to-proton ratio (N/Z) is above this line tend to decay by beta emission. Beta emission reduces the N/Z ratio, since a neutron changes to a proton. Nuclei in which the N/Z ratio is below this line tend to decay by either positron emission or electron capture. Both increase the N/Z ratio. Lighter nuclei ($Z < 20$) are more likely to decay by positron emission rather than by electron capture. Heavier nuclei ($Z > 80$) are more likely to decay by electron capture.

As examples of radioactive isotopes, consider carbon-14 and phosphorus-30. Looking at Figure 22.2, we see that carbon-14 lies above the beta-stability line. Thus, we expect it to decay by beta emission, which is what is observed. On the other hand, phosphorus-30 lies below the beta-stability line. We expect it to decay by either positron emission or electron capture. Because its atomic number (15) is less than 20, positron emission is more likely. Positron emission is, in fact, observed.

■ Most of the argon in the atmosphere is believed to have resulted from the radioactive decay of $_{19}^{40}K$.

Heavier nuclei, especially those with Z greater than 83, often decay by alpha emission. Although uranium-238 lies only slightly above the beta-stability line, it decays by alpha emission, as we have noted earlier. The isotopes $^{226}_{88}Ra$ and $^{232}_{90}Th$ are other examples of alpha emitters. However, the decay product of uranium-238, thorium-234, which lies above the beta stability line, decays by beta emission. Californium-245, which lies below the beta-stability line, decays by both alpha emission and electron capture.

$$^{245}_{98}Cf \longrightarrow {}^{241}_{96}Cm + {}^{4}_{2}He$$
$$^{245}_{98}Cf + {}^{0}_{-1}e \longrightarrow {}^{245}_{97}Bk$$

Example 22.4

Using Figure 22.2, predict the possible type of radioactive decay for the following radioactive nuclides: (a) $^{208}_{84}Po$ (b) $^{47}_{20}Ca$ (c) $^{25}_{13}Al$

Solution

(a) Since the atomic number is greater than 83, alpha emission is likely. However, if we look in Figure 22.2 at the position of the nuclide with 84 protons and 208 − 84 = 124 neutrons, we find that it lies below the beta-stability line. Thus, positron emission or electron capture are also possible. (Electron capture is more likely,

since $Z > 80$.) Polonium-208 is observed to decay primarily by alpha emission, but electron capture is also seen. (b) The nucleus of calcium-47 has 20 protons and 47 − 20 = 27 neutrons. This places the nuclide above the beta-stability line. Thus, it is expected to decay by beta emission, which is observed. (c) Aluminum-25 ($^{25}_{13}Al$) has 13 protons and 12 neutrons, placing it below the beta-stability line. We expect it to decay either by positron emission (likely, because $Z < 20$) or by electron capture. Positron emission is actually observed.

Exercise 22.4

Predict the type of decay expected for the following radioactive nuclides: (a) $^{13}_{7}N$ (b) $^{26}_{11}Na$ (c) $^{210}_{86}Rn$. Refer to Figure 22.2.

(See Problems 22.25 and 22.26.)

Radioactive Decay Series

All nuclides with atomic number greater than $Z = 83$ are radioactive, as we have noted. Most of these nuclides decay by alpha emission. Alpha particles, or $^{4}_{2}He$ nuclei, are especially stable and are formed in the radioactive nucleus at the moment of decay. By emitting an alpha particle, the nucleus reduces its atomic number, becoming more stable. However, if the nucleus has a very large Z, the product nucleus will also be radioactive. Natural radioactive elements, such as uranium-238, give a **radioactive decay series,** in which one radioactive nucleus decays to a second, which then decays to a third, and so forth. Eventually, a stable nucleus is reached. For the natural radioactive decay series, this stable nucleus is an isotope of lead.

There are three radioactive decay series found naturally. One of these begins with uranium-238. Figure 22.3 shows the sequence of nuclear decay processes. In the first step, uranium-238 decays by alpha emission to thorium-234:

$$^{238}_{92}U \longrightarrow {}^{234}_{90}Th + {}^{4}_{2}He$$

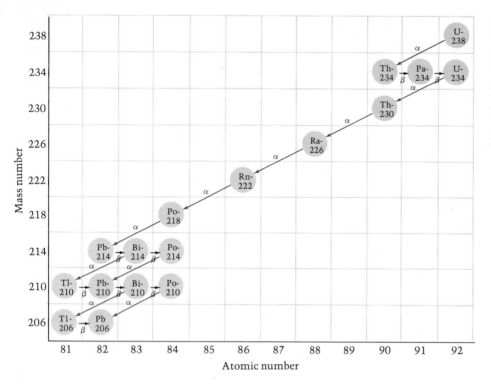

Figure 22.3
Uranium-238 radioactive decay series. Each nuclide occupies a position on the
graph according to its atomic
number and mass number.
Alpha decay is shown by a
diagonal line. Beta decay is
shown by a short, horizontal
line.

Note that this step is represented in Figure 22.3 by a long diagonal line, labeled α. Each alpha decay reduces the atomic number by 2 and the mass number by 4. The atomic number and mass number of each nuclide is given by its horizontal and vertical position on the graph. Thorium-234 in turn decays by beta emission to protactinium-234, which decays by beta emission to uranium-234:

$$^{234}_{90}\text{Th} \longrightarrow {}^{234}_{91}\text{Pa} + {}^{0}_{-1}\text{e}$$
$$^{234}_{91}\text{Pa} \longrightarrow {}^{234}_{92}\text{U} + {}^{0}_{-1}\text{e}$$

Note that a beta emission is represented in Figure 22.3 by a short horizontal line. Each beta decay increases the atomic number by one, but has no effect on the mass number. After the decay of protactinium-234 and the formation of uranium-234, there are a number of alpha-decay steps. The final product of the series is lead-206.

Natural uranium is 99.28% $^{238}_{92}\text{U}$, which decays as we have described. However, the natural element also contains 0.72% $^{235}_{92}\text{U}$. This isotope starts a second radioactive decay series. This series consists of a sequence of alpha and beta decays, ending with lead-207. The third naturally occurring radioactive decay series begins with thorium-232 and ends with lead-208.■

■ Thorium is thought to be three times more abundant than uranium and may become a major source of nuclear power (see Section 22.7).

22.2 Nuclear Bombardment Reactions

The nuclear reactions discussed in the previous section are radioactive decay reactions, in which a nucleus spontaneously decays to another nucleus and emits a particle, such as an alpha or beta particle. In 1919, Rutherford discovered that it was possible to change the nucleus of one element into the nucleus of another by processes we could control in the laboratory. **Transmutation** is the change of one element to another by bombarding the nucleus of the element with nuclear particles or nuclei.

Transmutation

In his experiments, Rutherford used a radioactive element as a source of alpha particles and allowed these particles to collide with nitrogen nuclei. He discovered that protons were ejected in the process. The equation for the nuclear reaction is

$$^{14}_{7}\text{N} + ^{4}_{2}\text{He} \longrightarrow ^{17}_{8}\text{O} + ^{1}_{1}\text{H}$$

These experiments were repeated on other light nuclei, most of which were transmuted to other elements with the ejection of a proton. Two significant results were obtained from these experiments. First, it strengthened the view that all nuclei contain protons. Second, it showed for the first time that it was possible to change one element into another under laboratory control.

When beryllium was bombarded with alpha particles, a penetrating radiation was given off that was not deflected by electric or magnetic fields. Therefore, the radiation did not consist of charged particles. The British physicist James Chadwick (1891–1974) suggested in 1932 that the radiation from beryllium consisted of neutral particles, each with a mass approximately that of a proton. The particles were called neutrons. The reaction that resulted in the discovery of the neutron is

$$^{9}_{4}\text{Be} + ^{4}_{2}\text{He} \longrightarrow ^{12}_{6}\text{C} + ^{1}_{0}\text{n}$$

In 1933, a nuclear bombardment reaction was used to produce the first artificial, radioactive isotope. Irene and Frederic Joliot-Curie found that aluminum bombarded with alpha particles produced phosphorus-30, which decayed by emitting positrons. The reactions are

$$^{27}_{13}\text{Al} + ^{4}_{2}\text{He} \longrightarrow ^{30}_{15}\text{P} + ^{1}_{0}\text{n}$$
$$^{30}_{15}\text{P} \longrightarrow ^{30}_{14}\text{Si} + ^{0}_{1}\text{e}$$

Phosphorus-30 was the first radioactive nucleus produced in the laboratory. Since then over a thousand radioactive isotopes have been made.■

Nuclear bombardment reactions are often referred to by an abbreviated notation. For example, the reaction

$$^{14}_{7}\text{N} + ^{4}_{2}\text{He} \longrightarrow ^{17}_{8}\text{O} + ^{1}_{1}\text{H}$$

is abbreviated $^{14}_{7}\text{N}(\alpha, \text{p})^{17}_{8}\text{O}$. In this notation, one first writes the nuclide symbol for the original nucleus, then in parentheses the symbol for the projectile particle (particle in), followed by a comma and the symbol for the ejected particle. After the last parenthesis, the nuclide symbol for the product nucleus is given. The following symbols are used for particles:

■ Some of the uses of radioactive isotopes are discussed in Section 22.5.

Neutron	n
Proton	p
Deuteron, $^{2}_{1}\text{H}$	d
Alpha, $^{4}_{2}\text{He}$	α

Example 22.5

(a) Write the abbreviated notation for the following bombardment reaction, in which neutrons were first discovered:

$$^{9}_{4}Be + ^{4}_{2}He \longrightarrow ^{12}_{6}C + ^{1}_{0}n$$

(b) Write the nuclear equation for the bombardment reaction denoted $^{27}_{13}Al(p, d)^{26}_{13}Al$.

Solution

(a) The notation is $^{9}_{4}Be(\alpha, n)^{12}_{6}C$. (b) The nuclear equation is

$$^{27}_{13}Al + ^{1}_{1}H \longrightarrow ^{26}_{13}Al + ^{2}_{1}H$$

Exercise 22.5

(a) Write the abbreviated notation for the reaction

$$^{40}_{20}Ca + ^{2}_{1}H \longrightarrow ^{41}_{20}Ca + ^{1}_{1}H$$

(b) Write the nuclear equation for the bombardment reaction $^{12}_{6}C(d, p)^{13}_{6}C$.
(See Problems 22.29, 22.30, 22.31, and 22.32.)

Elements of large atomic number merely scatter, or deflect, alpha particles from natural sources, rather than give a transmutation reaction. These elements have nuclei of large positive charge, and the alpha particle must be traveling very fast in order to penetrate the nucleus and react. Alpha particles from natural sources do not have sufficient kinetic energy. In order to shoot charged particles into heavy nuclei, it is necessary to accelerate, or speed up, the charged particles.

Particle accelerators are devices used to accelerate electrons, protons, and alpha particles and other ions to very high speeds. Essentially, a charged particle in the electric field between charged plates is accelerated toward the plate with opposite charge to that of the particle. It is customary to measure the kinetic energies of these particles in units of electron volts. An **electron volt** (eV) is the energy imparted to an electron (whose charge is 1.602×10^{-19} C) that is accelerated by one volt potential difference.

$$1 \text{ eV} = (1.602 \times 10^{-19} \text{ C}) \times (1 \text{ V}) = 1.602 \times 10^{-19} \text{ J}$$

Typically, particle accelerators give charged particles energies of millions of electron volts (MeV). In order to keep the accelerated particles from colliding with molecules of gas, the apparatus is enclosed and evacuated to low pressures, about 10^{-6} mmHg, or less.

Figure 22.4 shows a diagram of a **cyclotron**, a type of particle accelerator with circular shape. The cyclotron consists of two hollow, semicircular metal electrodes, called *dees* (because the shape of a dee resembles the letter *D*). Ions introduced at the center of the cyclotron are accelerated in the space between the two dees. Magnet poles (not shown in the figure) above and below the dees keep the ions moving in an enlarging spiral path. The dees are connected to a high-frequency electric current that changes the polarity of the dees so that each time the ion moves into the space between them it is accelerated. Thus, the ion is continually accelerated until it finally leaves the cyclotron at high speed. Outside the cyclotron, the ions are directed toward a target element in order to study nuclear reactions or to prepare isotopes.

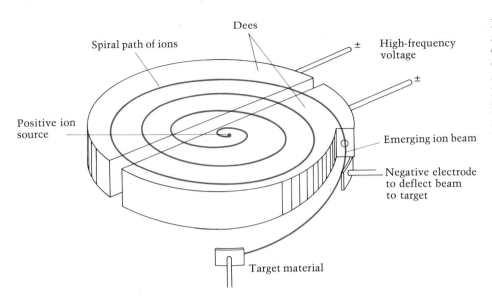

Figure 22.4
A cyclotron. Ions are introduced at the center of the cyclotron and are accelerated in an enlarging spiral by magnet poles above and below the dees (magnet not shown). Eventually the ions exit to hit a target material.

Technetium, whose discovery was mentioned in the chapter opening, was first prepared by directing **deuterons,** or nuclei of hydrogen-2, from a cyclotron at a molybdenum target.■ The nuclear reaction is $^{96}_{42}\text{Mo}(d, n)^{97}_{43}\text{Tc}$, or

$$^{96}_{42}\text{Mo} + {}^{2}_{1}\text{H} \longrightarrow {}^{97}_{43}\text{Tc} + {}^{1}_{0}\text{n}$$

■ The hydrogen-2 atom is often called deuterium and given the symbol D. It is a stable isotope with a natural abundance of 0.015%. Deuterium was discovered in 1931 by Harold Urey and coworkers and was prepared in pure form by G. N. Lewis.

Transuranium Elements

The **transuranium elements** are those elements with atomic numbers greater than that of uranium ($Z = 92$), the naturally occurring element of greatest Z. In 1940, E. M. McMillan and P. H. Abelson, at the University of California at Berkeley, discovered the first transuranium element. They produced an isotope of element 93, which they named neptunium, by bombarding uranium-238 with neutrons. This gave uranium-239, by the capture of a neutron, and this nucleus decayed in a few days by beta emission to neptunium-239.

$$^{238}_{92}\text{U} + {}^{1}_{0}\text{n} \longrightarrow {}^{239}_{92}\text{U}$$
$$^{239}_{92}\text{U} \longrightarrow {}^{239}_{93}\text{Np} + {}^{0}_{-1}\text{e}$$

The next transuranium element to be discovered was plutonium ($Z = 94$). Deuterons, the positively charged nuclei of hydrogen-2, were accelerated by a cyclotron and directed at a uranium target to give neptunium-238, which decayed to plutonium-238.

$$^{238}_{92}\text{U} + {}^{2}_{1}\text{H} \longrightarrow {}^{238}_{93}\text{Np} + 2{}^{1}_{0}\text{n}$$
$$^{238}_{93}\text{Np} \longrightarrow {}^{238}_{94}\text{Pu} + {}^{0}_{-1}\text{e}$$

Plutonium-238 is used to power batteries for heart pacemakers. Another isotope of plutonium, plutonium-239, is now produced in large quantity in nuclear reactors, as described in Section 22.7. Plutonium-239 is used for nuclear weapons.

The discovery of the next two transuranium elements, americium ($Z =$

95) and curium ($Z = 96$), depended upon an understanding of the correct positions in the periodic table of the elements beyond actinium ($Z = 89$). It had been thought that these elements should be placed after actinium under the d-transition elements. Thus, uranium was placed in Group VIIB under tungsten. However, Glenn T. Seaborg at the University of California, Berkeley, postulated a second series of elements to be placed at the bottom of the periodic table, under the lanthanides, as shown in modern tables (see inside front cover). These elements, the actinides, would be expected to have chemical properties similar to the lanthanides. Once understanding this, Seaborg and others soon identified curium and americium. Later, elements with Z up to 107 were discovered. ■ Then, in late 1982, element 109 was believed to be produced by West German physicists Gottfried Munzenberg and Peter Armbruster by bombarding bismuth-209 with iron-58 ions from a particle accelerator. The postulated nuclear reaction is

$$^{209}_{83}\text{Bi} + ^{58}_{26}\text{Fe} \longrightarrow ^{267}_{109}\text{X}$$

Experiments are now in progress to make elements with atomic numbers near 114 and mass numbers near 298. These elements are expected to be relatively stable. ■ The experiments could give important information on the structure of nuclei.

■ Both a Russian group, headed by Georgi N. Flerov, and an American group, headed by Albert Ghiorse, claim priority for the discovery of elements 104, 105, and 106. The privilege of naming an element is usually given to the discoverer, but because of the dispute over these elements, the International Union of Pure and Applied Chemistry (IUPAC) has recommended a system of naming further elements based on the atomic number, using Latin prefixes for each digit. Element 104 would be called unnilquadium (*un* = one, *nil* = zero, and *quad* = four).

■ The number 114 is a magic number of protons; 184 is a magic number of neutrons. Thus, $^{298}_{114}\text{X}$ should be relatively stable.

Example 22.6

Plutonium-239 was bombarded by alpha particles. Each $^{239}_{94}\text{Pu}$ nucleus was struck by one alpha particle and emitted one neutron. What was the product nucleus?

Solution

We can write the nuclear equation as follows:

$$^{239}_{94}\text{Pu} + ^{4}_{2}\text{He} \longrightarrow ^{A}_{Z}\text{X} + ^{1}_{0}\text{n}$$

To balance this equation in charge (subscripts) and mass number (superscripts), we write the equations

$$239 + 4 = A + 1 \quad \text{(from superscripts)}$$
$$94 \;+ 2 = Z + 0 \quad \text{(from subscripts)}$$

Hence,

$$A = 239 + 4 - 1 = 242$$
$$Z = 94 \;+ 2 = 96$$

The product is $^{242}_{96}\text{Cm}$.

Exercise 22.6

Carbon-14 is produced in the upper atmosphere by bombarding a particular nucleus by neutrons. A proton is ejected for each nucleus that reacts. What is the identity of the nucleus that produces carbon-14 by this reaction?

(See Problems 22.35, 22.36, 22.37, and 22.38.)

22.3 Radiations and Matter: Detection and Biological Effects

Radiations from nuclear processes affect matter in part by dissipating energy in it. An alpha, beta, or gamma particle traveling through matter dissipates energy by ionizing atoms or molecules, producing positive ions and electrons. In some cases, these radiations may also excite electrons in matter. When these electrons undergo transitions back to their ground states, light

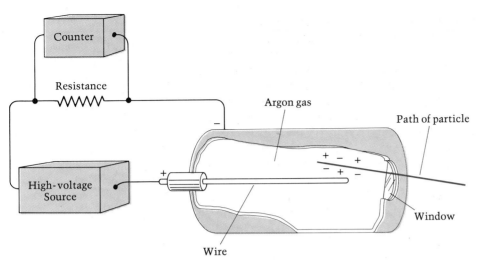

Figure 22.5
A Geiger counter. Particles of radiation enter the thin window and pass into the gas. Energy from the particles ionizes gas molecules, giving positive ions and electrons, which are accelerated to the electrodes. The electrons, which move faster, strike the wire anode, and create a pulse of current. These pulses are counted.

is emitted. The ions, free electrons, and light produced in matter can be used to detect nuclear radiations. Because nuclear radiations can ionize molecules and break chemical bonds, it adversely affects biological organisms. We will first look at the detection of nuclear radiations, then briefly discuss biological effects and radiation dosage in humans.

Radiation Counters

Two types of devices are used to count particles emitted from radioactive nuclei and other nuclear processes, *ionization counters* and *scintillation counters*. Ionization counters depend on the production of ions in matter. Scintillation counters detect the production of scintillations, or flashes of light.

A **Geiger counter** (Figure 22.5), a kind of ionization counter, consists of a metal tube filled with gas, such as argon. The tube is fitted with a thin glass or plastic window, through which radiation enters. A wire runs down the center of the tube and is insulated from the tube. The tube and wire are connected to a high-voltage source so that the tube becomes the negative electrode, and the wire the positive electrode. Normally, the gas in the tube is an insulator and no current flows through it. However, if an alpha particle, $_2^4\text{He}^{2+}$, passes through the window of the tube and into the gas, atoms are ionized. Free electrons are quickly accelerated to the wire. As they are accelerated to the wire, additional atoms may be ionized from collisions with these electrons and more electrons set free. An avalanche of electrons is created, and this gives a pulse of current that is detected by electronic equipment. The amplified pulse activates a digital counter or gives an audible "click."

Alpha, beta, and gamma particles can be detected directly by a Geiger counter. ■ To detect neutrons, boron trifluoride is added to the gas in the tube. Neutrons react with boron-10 nuclei to produce alpha particles, which can then be detected.

$$_0^1\text{n} + {}_5^{10}\text{B} \longrightarrow {}_3^7\text{Li} + {}_2^4\text{He}$$

■ Gamma particles, however, are better detected by scintillation counters. These are filled with solids or liquids, which are more likely to stop gamma rays. Low-energy alpha particles may be absorbed by the window and go undetected.

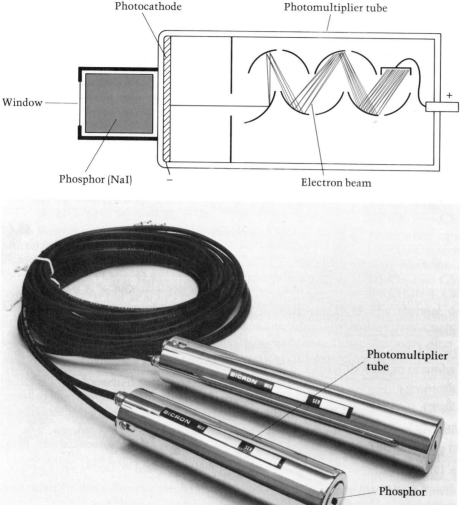

Figure 22.6
A scintillation counter probe. (*Above*) Radiation passes through the window into a phosphor (here, NaI). Flashes of light from the phosphor fall on the photocathode, which ejects electrons by the photoelectric effect. The electrons are accelerated to special electrodes that eject many more electrons, multiplying the original signal. (*Below*) Photograph of a scintillation probe.

A **scintillation counter** (Figure 22.6) consists of a phosphor and a device to detect the light emitted by it. A *phosphor* is a substance that emits flashes of light when struck by radiation. Rutherford had used zinc sulfide as a phosphor to detect alpha particles. A sodium iodide crystal containing thallium(I) iodide is used as a phosphor for gamma radiation. (Excited technetium-99 emits gamma rays and is used for medical diagnostics, as mentioned in the chapter opening. The gamma rays are detected by a scintillation counter.)

The flashes of light from the phosphor are detected by a *photomultiplier*. A photon of light from the phosphor hits a photoelectric sensitive surface (the photocathode).■ This emits an electron, which is accelerated to another electrode, from which several electrons are emitted. These electrons are accelerated to the next electrode, from which more electrons are emitted.

■ The photoelectric effect was discussed in Section 5.6.

The result is that a single electron may produce a million electrons and therefore a detectable pulse of electrical current.

A radiation counter can be used to measure the rate of nuclear disintegrations in a radioactive material. The **activity of a radioactive source,** or number of nuclear disintegrations per unit time, is often measured in units of curies. A **curie** (Ci) equals 3.700×10^{10} disintegrations per second. ■ Thus, a sample of technetium having an activity of 1.0×10^{-2} Ci is decaying at the rate of $(1.0 \times 10^{-2}) \times (3.7 \times 10^{10}) = 3.7 \times 10^{8}$ nuclei per second.

■ The curie was originally defined as the number of disintegrations per second from 1.0 g of radium.

Biological Effects and Radiation Dosage

Although the quantity of energy dissipated in a biological organism from a radiation dosage might be small, the effects can be quite adverse. This is because important chemical bonds may be broken. DNA in the chromosomes of the cell is especially affected, which in turn interferes with cell division. Cells that divide the fastest, such as those in the blood-forming tissue in bone marrow, are most affected by nuclear radiations.

To monitor the effect of nuclear radiations on biological tissue, it is necessary to have a measure of radiation dosage. The **rad** (from *radiation absorbed dose*) is the dosage of radiation that deposits 1×10^{-2} J of energy per kilogram of tissue. However, the biological effect of radiation depends not only on the energy deposited in the tissue but also on the type of radiation. For example, neutrons are more destructive than gamma rays of the same radiation dosage measured in rads. To relate the various kinds of radiations in terms of biological destruction, the unit of radiation dosage called the rem is defined. A **rem** equals the rad times a factor for the type of radiation, called the *relative biological effectiveness* (RBE):

$$\text{rems} = \text{rads} \times \text{RBE}$$

Beta and gamma radiations have an RBE of about 1, whereas neutron radiation has an RBE of about 5 and alpha radiation an RBE of about 10. ■

The effects of radiation on a person depend not only on the dosage but also on the length of time in which the dose was received. A series of small doses has less overall effect than the same dosage given all at once. A single dose of about 500 rems is fatal to most people, and survival from a much smaller dose can be uncertain or leave the person chronically ill. Detectable effects are seen with doses as low as 30 rems. Continuous exposure to such low levels of radiation may result in cancer or leukemia. At even lower levels, the answer to whether the radiation dose is safe rests with understanding the possible genetic effects of the radiation. Because radiation can cause chromosome damage, inheritable defects are possible.

■ Sources of alpha radiation outside the body are relatively harmless, since the radiation is absorbed by the skin. Internal sources, however, are very destructive.

Safe limits for radiations are much debated. Although there may not be a strictly safe limit, it is important to have in mind the magnitude of the radiations humans may be subjected to. A natural background radiation, which we all receive, results from cosmic rays (radiations from space) and natural radioactivity. This averages about 0.1 rem per year, but varies considerably with location. Radium and its decay products in the soil are an important source of this radiation background. Another source is potassium-40, a radioactive isotope with natural abundance of 0.0012%. In addition to natural background radiation, we may receive radiation from other fairly common sources. The most important of these are x rays used in medical diagnosis. The average person receives a radiation dose from this source that is about equal to that of the natural background. Very small radiation sources include consumer items such as television sets and luminous watches.

The background radiation to which we are all subjected has increased slightly since nuclear technology began. Fallout from atmospheric testing of nuclear weapons increased this background by several percent, but this has decreased since atmospheric testing was banned. The radiation contributed by nuclear power plants is only a fraction of a percent of the natural background.

22.4 Rate of Radioactive Decay

Although technetium-99 is radioactive and decays by emitting beta particles (electrons), it is impossible to say when a particular nucleus will disintegrate. A sample of technetium-99 will continue to give off beta rays for millions of years. Thus, a particular nucleus might disintegrate the next instant, or several million years later. The rate of this radioactive decay cannot be changed by varying the temperature, or pressure, or chemical environment of the technetium nucleus. Radioactivity, whether from technetium or some other nucleus, is not affected by those variables that affect the rate of a chemical reaction. In this section, we will look at how we can quantitatively express the rate of radioactive decay.

Rate of Radioactive Decay and Half-Life

The rate of radioactive decay, that is, the number of nuclei disintegrating per unit time, is found to be proportional to the number of radioactive nuclei in the sample. We can express this mathematically as

$$\text{Rate} = kN_t$$

Here, N_t is the number of radioactive nuclei at time t, and k is the **radioactive decay constant,** or rate constant for radioactive decay. This rate constant is a characteristic of the radioactive nuclide, each nuclide having a different value.

The preceding rate equation has the same form as the rate law for a first-order chemical reaction. Indeed, radioactive decay is a first-order rate process, and the mathematical relationships that we derived in Chapter 15 for first-order reactions apply here. We will use this fact in the calculations in this section.

We can obtain the decay constant for a radioactive nucleus by counting the nuclear disintegrations over a period of time. The original definition of the curie (3.7×10^{10} disintegrations per second) was the activity or decay rate of 1.0 g of radium-226. We can use this with the equation just given to obtain the decay constant of radium-226. Radium-226 has a molar mass of 226 g. A 1.0-g sample of radium-226 contains the following number of nuclei:

$$1.0 \text{ g Ra-226} \times \frac{1 \text{ mol Ra-226}}{226 \text{ g Ra-226}} \times \frac{6.02 \times 10^{23} \text{ Ra-226 nuclei}}{1 \text{ mol Ra-226}}$$

$$= 2.7 \times 10^{21} \text{ Ra-226 nuclei}$$

This equals the value of N_t. If we solve rate $= kN_t$ for k, we get

$$k = \frac{\text{rate}}{N_t}$$

Substituting into this gives

$$k = \frac{3.7 \times 10^{10} \text{ nuclei/s}}{2.7 \times 10^{21} \text{ nuclei}} = 1.4 \times 10^{-11}/\text{s}$$

Example 22.7

A 1.0-mg sample of technetium-99 has an activity of 1.7×10^{-5} Ci (Ci = curies), decaying by beta emission. What is the decay constant for $^{99}_{43}\text{Tc}$?

Solution

Since an activity of 1.0 Ci is 3.7×10^{10} nuclei/s, the rate of decay in this sample is

$$\text{Rate} = 1.7 \times 10^{-5} \text{ Ci} \times \frac{3.7 \times 10^{10} \text{ nuclei/s}}{1.0 \text{ Ci}}$$

$$= 6.3 \times 10^{5} \text{ nuclei/s}$$

The number of nuclei in this sample of 1.0×10^{-3} g $^{99}_{43}\text{Tc}$ can be obtained by noting that the molar mass in grams is approximately equal to the mass number. (We can assume that all digits in the whole number are significant). Hence,

$$1.0 \times 10^{-3} \text{ g Tc-99} \times \frac{1 \text{ mol Tc-99}}{99 \text{ g Tc-99}}$$

$$\times \frac{6.02 \times 10^{23} \text{ Tc-99 nuclei}}{1 \text{ mol Tc-99}} = 6.1 \times 10^{18} \text{ Tc-99 nuclei}$$

We now solve the rate equation for k and substitute. We get

$$k = \frac{\text{rate}}{N_t} = \frac{6.3 \times 10^{5} \text{ nuclei/s}}{6.1 \times 10^{18} \text{ nuclei}} = 1.0 \times 10^{-13}/\text{s}$$

Exercise 22.7

The nucleus $^{99m}_{43}\text{Tc}$ is a metastable nucleus of technetium-99, used in medical diagnostic work. Technetium-99m decays by emitting gamma rays. If a 2.5-microgram sample has an activity of 13 Ci, what is the decay constant (in units of /s)?

(See Problems 22.39, 22.40, 22.41, and 22.42.)

Recall that the half-life of a chemical reaction, A $\longrightarrow$ products, is the time it takes for one-half of A to react. We can similarly define the **half-life** of a radioactive nucleus as the time it takes for one-half of the nuclei in a sample to decay. In Chapter 15, we learned that the half-life for a first-order chemical reaction is independent of the concentration of reactant. Similarly, the half-life of a radioactive nucleus is independent of the amount of sample.

In some cases, we could find the half-life of a radioactive sample by observing how long it takes for one-half of the sample to decay. For example, we find that 1.000 g of iodine-131, an isotope used in treating thyroid cancer, decays to 0.500 g in 8.07 days.■ Thus, its half-life is 8.07 days. In another 8.07 days, this sample would decay to one-half of 0.500 g, or 0.250 g, and so forth. Figure 22.7 shows this decay pattern.

Although we might be able to obtain the half-life of a radioactive nucleus by direct observation in some cases, this is impossible for many nuclei be-

■ Iodine is attracted to the thyroid gland, which incorporates the element into the growth hormone thyroxine. Radiation from iodine-131 would kill cancer cells in the thyroid gland.

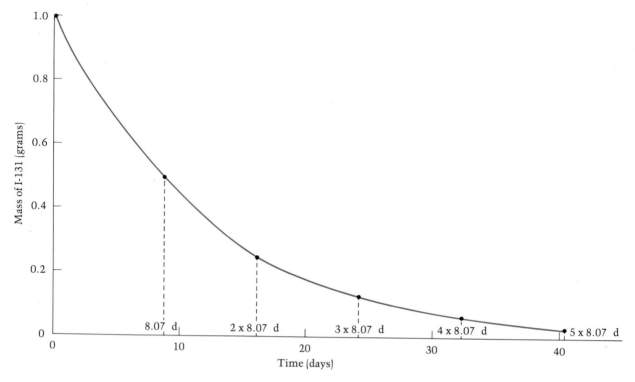

Figure 22.7
Radioactive decay of a
1.000-g sample of iodine-131.
Note how the sample decays
by one-half in each half-life
of 8.07 days.

cause they decay too quickly or too slowly. Uranium-238, for example, has a half-life of 4.51 billion years, which is too long to be directly observed. The usual method of determining the half-life is by measuring decay rates and relating them to half-lives.

We can relate the half-life for radioactive decay, $t_{1/2}$, to the decay constant, k, in the same way that we relate the half-life and rate constant for a first-order chemical reaction.

$$t_{1/2} = \frac{0.693}{k}$$

The next example illustrates the calculation.

Example 22.8

The decay constant for the beta decay of $_{43}^{99}$Tc was obtained in Example 22.7. There we found that k equals 1.0×10^{-13}/s. What is the half-life of this isotope in years?

Solution

We substitute the value of k into the equation

$$t_{1/2} = \frac{0.693}{k} = \frac{0.693}{1.0 \times 10^{-13}/s} = 6.9 \times 10^{12} \text{ s}$$

We convert this half-life in seconds to years:

$$6.9 \times 10^{12} \text{ s} \times \frac{1 \text{ min}}{60 \text{ s}} \times \frac{1 \text{ h}}{60 \text{ min}} \times \frac{1 \text{ d}}{24 \text{ h}} \times \frac{1 \text{ y}}{365 \text{ d}}$$

$$= 2.2 \times 10^5 \text{ y}$$

Exercise 22.8

Cobalt-60, used in cancer therapy, decays by beta and gamma emission. If the decay constant is 4.18×10^{-9}/s, what is the half-life in years?

(See Problems 22.43 and 22.44.)

Tables of radioactive nuclei often list the half-life. If we want the decay constant or the activity of a sample, we can calculate these from the half-life. This is illustrated in the next example.

Example 22.9

Tritium, $^3_1$H, is a radioactive nucleus of hydrogen. It is used in luminous watch dials. Tritium decays by beta emission with a half-life of 12.3 years. What is the decay constant (in /s)? What is the activity in curies of a sample containing 2.5 μg of tritium? The atomic mass of $^3_1$H is 3.02 amu.

Solution

Let us convert the half-life to its value in seconds, then calculate k. After that we can use the rate equation to find the rate of decay of the sample (in nuclei/s) and finally the activity. The conversion of the half-life to seconds gives

$$12.3 \text{ y} \times \frac{365 \text{ d}}{1 \text{ y}} \times \frac{24 \text{ h}}{1 \text{ d}} \times \frac{60 \text{ min}}{1 \text{ h}} \times \frac{60 \text{ s}}{1 \text{ min}}$$

$$= 3.88 \times 10^8 \text{ s}$$

Since $t_{1/2} = 0.693/k$, we solve this for k and substitute the half-life in seconds:

$$k = \frac{0.693}{t_{1/2}} = \frac{0.693}{3.88 \times 10^8 \text{ s}} = 1.79 \times 10^{-9}/\text{s}$$

Before substituting into the rate equation, we need to know the number of tritium nuclei in a sample containing 2.5×10^{-6} g of tritium. We get

$$2.5 \times 10^{-6} \text{ g H-3} \times \frac{1 \text{ mol H-3}}{3.02 \text{ g H-3}}$$

$$\times \frac{6.02 \times 10^{23} \text{ H-3 nuclei}}{1 \text{ mol H-3}} = 5.0 \times 10^{17} \text{ H-3 nuclei}$$

Now we substitute into the rate equation:

$$\text{Rate} = kN_t = 1.79 \times 10^{-9}/\text{s} \times 5.0 \times 10^{17} \text{ nuclei}$$
$$= 9.0 \times 10^8 \text{ nuclei/s}$$

The activity of the sample is obtained by dividing the rate in disintegrations of nuclei per second by 3.70×10^{10} disintegrations of nuclei per second per curie:

$$\text{Activity} = \frac{9.0 \times 10^8 \text{ nuclei/s}}{3.70 \times 10^{10} \text{ nuclei/(s} \cdot \text{Ci)}} = 0.024 \text{ Ci}$$

Exercise 22.9

Strontium-90, $^{90}_{38}$Sr, is a radioactive decay product of nuclear fallout from nuclear weapons testing. Because of its chemical similarity to calcium, it is incorporated into the bones if present in food. The half-life of strontium-90 is 28.1 y. What is the decay constant of this isotope? What is the activity of a sample containing 5.2 ng (5.2×10^{-9} g) of strontium-90?

(See Problems 22.45, 22.46, 22.47, and 22.48.)

Once we know the decay constant for a radioactive isotope, we can calculate the fraction of the radioactive nuclei that remain after a given period of time. Recall from Chapter 15 that the reactant concentration at time t, $[A]_t$, is given by the equation

$$\log \frac{[A]_0}{[A]_t} = \frac{kt}{2.303}$$

where $[A]_0$ is the concentration of A at $t = 0$. A similar equation holds for radioactive decay:

$$\log \frac{N_0}{N_t} = \frac{kt}{2.303}$$

Here N_0 is the number of nuclei in the original sample ($t = 0$). After a period of time, t, the number of nuclei decay to the number N_t. The next example illustrates this.

Example 22.10

Phosphorus-32 is a radioactive isotope with a half-life of 14.3 d. A biochemist has a vial containing a compound of phosphorus-32. If the compound is used in an experiment 5.5 d after the compound was prepared, what fraction of the radioactive isotope originally present remains? Suppose the sample in the vial originally contained 0.28 g of phosphorus-32. How many grams remain after 5.5 d?

Solution

If N_0 is the original number of P-32 nuclei in the vial, and N_t is the number after 5.5 d, the fraction remaining is N_t/N_0. We can obtain this fraction from the equation

$$\log \frac{N_0}{N_t} = \frac{kt}{2.303}$$

We substitute $k = 0.693/t_{1/2}$.

$$\log \frac{N_0}{N_t} = \frac{0.693t}{2.303t_{1/2}}$$

Since $t = 5.5$ d and $t_{1/2} = 14.3$ d, we obtain

$$\log \frac{N_0}{N_t} = \frac{0.693 \times 5.5\ \text{d}}{2.303 \times 14.3\ \text{d}} = 0.116$$

(We have retained an additional digit for further calculation.) Hence,

$$\frac{N_0}{N_t} = 10^{0.116} = 1.31$$

The fraction of $_{15}^{32}$P nuclei remaining is N_t/N_0.

$$\text{Fraction nuclei remaining} = \frac{N_t}{N_0} = \frac{1}{1.31} = 0.77$$

We round the answer to two significant figures. Thus, 77% of the nuclei remain. The mass of $_{15}^{32}$P in the vial after 5.5 d is

$$0.28\ \text{g} \times 0.77 = 0.22\ \text{g}$$

Exercise 22.10

A nuclear power plant emits into the atmosphere a very small amount of krypton-85, a radioactive isotope with a half-life of 10.76 y. What fraction of this krypton-85 remains after 25.0 y?

(See Problems 22.51 and 22.52.)

Radioactive Dating

Fixing the dates of relics and stone implements or pieces of charcoal from ancient campsites is an application based on radioactive decay rates. Because the rate of radioactive decay of a nuclide is constant, this rate can serve as a clock for dating very old rocks and human implements. Dating wood and similar carbon-containing objects that are several thousand to fifty thousand years old can be done with radioactive carbon, carbon-14, which has a half-life of 5,730 y.

Carbon-14 is present in the atmosphere as a result of cosmic-ray bombardment of earth. Cosmic rays are radiations from space, consisting of protons and neutrons, as well as other particles. Nitrogen nuclei in the upper atmosphere are bombarded with neutrons from cosmic rays and produce carbon-14 nuclei.

$$^{14}_{7}\text{N} + ^{1}_{0}\text{n} \longrightarrow ^{14}_{6}\text{C} + ^{1}_{1}\text{H}$$

Carbon dioxide containing carbon-14 mixes with the lower atmosphere. Because of the constant production of $^{14}_{6}\text{C}$ and its radioactive decay, a small, constant fractional abundance of carbon-14 is maintained in the atmosphere.

Living plants, which continuously use atmospheric carbon dioxide, also maintain a constant abundance of carbon-14. Similarly, living animals, by feeding on plants, have a constant fractional abundance of carbon-14. But once an organism dies, it is no longer in chemical equilibrium with atmospheric CO_2. The ratio of carbon-14 to carbon-12 begins to decrease by radioactive decay. Thus, this ratio of carbon isotopes becomes a clock measuring the time since the death of the organism.

If we assume that the ratio of carbon isotopes in the lower atmosphere has remained at the present level for the last 50,000 years (presently one out of 10^{12} carbon atoms is carbon-14), then we can deduce the age of any dead organic object by measuring the level of beta emissions that arise from the radioactive decay of carbon-14. ∎

$$^{14}_{6}\text{C} \longrightarrow ^{14}_{7}\text{N} + ^{0}_{-1}\text{e}$$

In this way, bits of campfire charcoal, parchment, jaw bones, and so on, have been dated.

■ Analyses of tree rings have shown that this assumption is not quite valid. Before 1000 B.C., the levels of carbon-14 were somewhat higher than they are today. Moreover, recent human activities (burning of fossil fuels and atmospheric nuclear testing) have changed the fraction of carbon-14 in atmospheric CO_2.

Example 22.11

A piece of charcoal from a tree killed by the volcanic eruption that formed the crater in Crater Lake (in Oregon) gave 7.0 disintegrations of carbon-14 nuclei per minute per gram of total carbon. Present-day carbon (in living matter) gives 15.3 disintegrations per minute per gram of total carbon. Determine the date of the volcanic eruption.

Solution

We substitute $k = 0.693/t_{1/2}$ into the equation for the number of nuclei in a sample after time t:

$$\log \frac{N_0}{N_t} = \frac{kt}{2.303} = \frac{0.693t}{2.303t_{1/2}}$$

Hence,

$$t = \frac{2.303t_{1/2}}{0.693} \log \frac{N_0}{N_t}$$

To get N_0/N_t, we assume that the ratio of $^{14}_{6}\text{C}$ to $^{12}_{6}\text{C}$ in the atmosphere has remained constant. Then we can say that 1.00 gram of total carbon from the living tree gave 15.3 disintegrations per minute. The ratio of the number of $^{14}_{6}\text{C}$ nuclei originally present to the number that existed at the time of dating equals the ratio of rates of disintegration. That is,

$$\frac{N_0}{N_t} = \frac{15.3}{7.0} = 2.2$$

Therefore, substituting this value of N_0/N_t and $t_{1/2} = 5,730$ y into the previous equation gives

$$t = \frac{2.303t_{1/2}}{0.693} \log \frac{N_0}{N_t} = \frac{2.303 \times 5,730\,\text{y}}{0.693} \log 2.2$$
$$= 6.5 \times 10^3 \text{ y}$$

Thus, the date of the eruption was about 4500 B.C.

Exercise 22.11

A jawbone from the archeological site at Folsom, New Mexico, was dated by its radioactive carbon. The activity of the carbon from the jawbone was 4.5 disintegrations per minute per gram of total carbon. What was the age of the jawbone? Carbon from living material gives 15.3 disintegrations per minute per gram of carbon.

(See Problems 22.57 and 22.58.)

For the age of rocks and meteorites, other similar methods of dating have been used. One method depends on the radioactivity of naturally occurring potassium-40, which decays by positron emission and electron capture (as well as by beta emission).

$$^{40}_{19}K \longrightarrow \, ^{40}_{18}Ar + \, ^{0}_{1}e$$

$$^{40}_{19}K + \, ^{0}_{-1}e \longrightarrow \, ^{40}_{18}Ar$$

Potassium occurs in many rocks. Once such a rock forms by solidification of molten material, the argon from the decay of potassium-40 is trapped. To obtain the age of a rock, one first determines the number of $^{40}_{19}K$ atoms and the number of $^{40}_{18}Ar$ atoms in a sample. The number of $^{40}_{19}K$ atoms equals N_t. The number originally present, N_0, equals N_t plus the number of argon atoms, since each argon atom resulted from the decay of a $^{40}_{19}K$ nucleus. One then calculates the age of the rock from the ratio N_0/N_t.

The oldest rocks on earth have been dated at 3.8×10^9 y. Since the rocks at the earth's surface have been subjected to extensive weathering, even older rocks may have existed. This age, 3.8×10^9 y, therefore represents the minimum possible age of the earth, the time since the solid crust first formed. Ages of meteorites, which are assumed to have solidified at the same time as other solid objects in the solar system, including earth, have been determined to be 4.4×10^9 y to 4.6×10^9 y. It is now believed from this and other evidence that the age of the earth is 4.6×10^9 y.

22.5 Applications of Radioactive Isotopes

We have already described two applications of nuclear chemistry. One was the preparation of elements not available naturally. We noted that the discovery of the transuranium elements clarified the position of the heavy elements in the periodic table. In the section just completed, we discussed the use of radioactivity in dating objects. We will discuss practical uses of nuclear energy in the last section of the chapter. Here we will look at the applications of radioactive isotopes to chemical analysis and to medicine.

Chemical Analysis

A **radioactive tracer** is a radioactive isotope added to a chemical, biological, or physical system to study the system. The advantage of a radioactive tracer is that it behaves chemically just as a nonradioactive isotope does, but it can be detected in exceedingly small amounts by measuring the radiations emitted.

As an illustration of the use of radioactive tracers, consider the problem of establishing that chemical equilibrium is a dynamic process. Let us look at the equilibrium of solid lead iodide and its saturated solution, containing $Pb^{2+}(aq)$ and $I^-(aq)$. The equilibrium is

$$PbI_2(s) \rightleftharpoons Pb^{2+}(aq) + 2I^-(aq)$$

In two separate beakers, we prepare saturated solutions of PbI_2 in contact with the solid. One beaker contains only natural iodine atoms with nonradioactive isotopes. The other beaker contains radioactive iodide ion, $^{131}I^-$. Some of the solution, but no solid, containing the radioactive iodide ion is now added to the other beaker containing nonradioactive iodide ion. Since both solutions are saturated, the amount of solid in this beaker remains constant. Yet, after a time the solid lead iodide, which was originally non-radioactive, becomes radioactive. This is evidence for a dynamic equi-librium, in which radioactive iodide ions in the solution substitute for non-radioactive iodide ions in the solid.

With only naturally occurring iodine available, it would have been impos-sible to detect the dynamic equilibrium. By using ^{131}I as a radioactive tracer, we can easily follow the substitution of radioactive iodine into the solid by measuring its radioactivity.

A series of experiments using tracers was carried out in the late 1950s by Melvin Calvin at the University of California, Berkeley, in order to discover the mechanism of photosynthesis in plants.■ The overall process of photo-synthesis involves the reaction of CO_2 and H_2O to give glucose, $C_6H_{12}O_6$, and O_2. Energy for photosynthesis comes from the sun.

■ Melvin Calvin received the Nobel prize in chemistry in 1961 for his work on photosynthesis.

$$6CO_2(g) \; + \; 6H_2O(l) \; \xrightarrow{\text{sunlight}} \; C_6H_{12}O_6(aq) \; + \; 6O_2(g)$$

This equation represents only the net result of photosynthesis. As Calvin was able to show, the actual process consists of many separate steps. In several experiments, algae (single-celled plants) were exposed to carbon di-oxide containing much more radioactive carbon-14 than occurs naturally. Then the algae were extracted with a solution of alcohol and water. The various compounds in this solution were separated by chromatography and identified.■ Those compounds that contained radioactive carbon were pro-duced in the different steps of photosynthesis. Eventually, Calvin was able to use tracers to show the main steps in photosynthesis.

■ Chromatography was dis-cussed in Section 2.4.

Another example of the use of radioactive tracers in chemistry is provided by the technique of isotope dilution. **Isotope dilution** is a technique designed to determine the quantity of a substance in a mixture or the total volume of solution by adding a known amount of an isotope to it. After removing a portion of the mixture, the fraction by which the isotope has been di-luted provides a way of determining the quantity of substance or volume of solution.

As an example, suppose we wish to obtain the volume of water in a tank, without being able to drain the tank. We add 100 mL of water containing a radioactive isotope. After allowing this to mix completely with the water in the tank, we withdraw 100 mL of solution from the tank. We find that the activity of this solution in curies is 1/1000 of the original solution. Since the isotope has been diluted by a factor of 1000, the volume of the tank is 1000×100 mL $= 100,000$ mL (100 L).

A typical chemical example of isotope dilution is the determination of the amount of vitamin B_{12} in a sample of food. Although part of the vitamin in food can be obtained in pure form, not all of the vitamin can be separated. Therefore, we cannot precisely determine the quantity of vitamin B_{12} in a sample of food by separating the pure vitamin and weighing it. But we can determine the amount of vitamin B_{12} by isotope dilution. Suppose we add 2.0×10^{-7} g of vitamin B_{12} containing radioactive cobalt-60 to 125 g of food and mix well. We then separate 5.4×10^{-7} g of pure vitamin B_{12} from the food and find that the activity in curies of this quantity of the vitamin contains 5.6% of the activity added from the radioactive cobalt. The mass of vitamin B_{12} in the food, including the amount added $(2.0 \times 10^{-7}$ g), is

$$5.4 \times 10^{-7} \text{ g} \times \frac{100}{5.6} = 9.6 \times 10^{-6} \text{ g}$$

Subtracting the amount added in the analysis gives

$$9.6 \times 10^{-6} \text{ g} - 2.0 \times 10^{-7} \text{ g} = 9.4 \times 10^{-6} \text{ g}$$

Neutron activation analysis is another, somewhat different, method of analysis based on the conversion of stable isotopes to radioactive isotopes by bombarding a sample with neutrons. Human hair contains trace amounts of many elements. By determining the exact amounts and the position of the elements in the hair shaft, we could identify who the hair comes from. Consider the analysis of human hair for arsenic, for example. When the natural isotope $^{75}_{33}\text{As}$ is bombarded with neutrons, a metastable nucleus $^{76m}_{33}\text{As}$ is obtained.

$$^{75}_{33}\text{As} + {}^{1}_{0}\text{n} \longrightarrow {}^{76m}_{33}\text{As}$$

A metastable nucleus is in an excited state. It decays, or undergoes a transition, to a lower state by emitting gamma rays. The frequencies, or energies, of the gamma rays emitted are characteristic of the element and serve to identify it. Also, the intensity of the gamma rays emitted are proportional to the amount of the element present. The method is very sensitive, identifying as little as 10^{-9} g of arsenic. ∎

Medical Therapy and Diagnosis

It is safe to say that the use of radioactive isotopes has had a profound effect on the practice of medicine. Radioisotopes were first used in medicine in the treatment of cancer. This treatment is based on the fact that rapidly dividing cells, such as those in cancer, are more adversely affected by radiation from radioactive substances than are those cells that divide more slowly. Radium-226 and its decay product radon-222 were used for cancer therapy, or treatment, a few years after the discovery of radioactivity. Today, gamma radiation from cobalt-60 is more commonly used.

Cancer therapy, however, is only one of the ways in which radioactive isotopes are used in medicine. The greatest advances in the use of radioactive isotopes have been in the diagnosis of disease. Radioactive isotopes are used for diagnosis in two ways. They are used to develop images of internal body organs, so that their functioning can be examined. And, they are used as tracers in the analysis of minute amounts of substances, such as a growth hormone in blood, in order to deduce possible disease conditions.

Technetium-99m is the radioactive isotope most often used to develop pictures or images of internal body organs. It has a half-life of 6.02 h, decaying by gamma emission to technetium-99 in its nuclear ground state. The image is prepared by scanning the body for gamma rays with a scintillation detector. Figure 22.1, described in the chapter opening, shows the image of a person's skeleton obtained with technetium-99m. The technetium is soon

■ Neutron activation analysis has been used to authenticate oil paintings by giving an exact analysis of pigments used. Since pigment compositions have changed, it is possible to detect fraudulent paintings done with more modern pigments. The analysis can be done without affecting the painting.

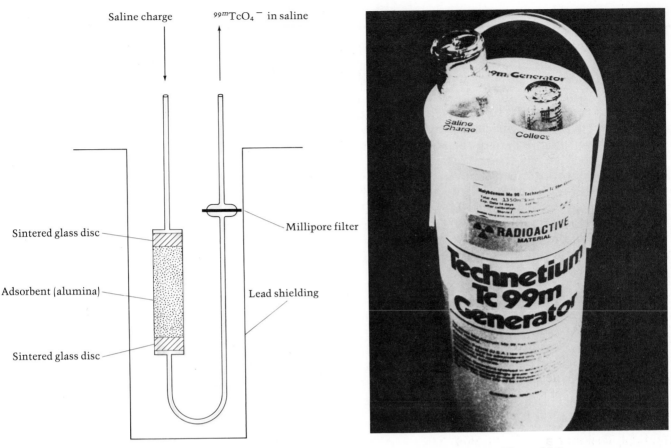

Figure 22.8
A technetium-99m generator. (*Left*) A schematic view of the generator. Molybdenum-99, in the form of MoO_4^{2-} ion adsorbed on alumina, decays to technetium-99m. The technetium is leached from the generator with a salt solution (saline charge) as TcO_4^-. (*Right*) Photograph of a technetium-99m generator.

excreted by the body, and the gamma radioactivity decays to negligible levels within hours.

In a hospital, the technetium isotope is produced in a special container, or generator, shown in Figure 22.8. The generator contains radioactive molybdate ion, MoO_4^{2-}, adsorbed on alumina granules. Radioactive molybdenum-99 is produced at a nuclear reactor facility by bombarding the natural, nonradioactive, isotope molybdenum-98 with neutrons.

$$\ce{^{98}_{42}Mo + ^{1}_{0}n -> ^{99}_{42}Mo}$$

This radioactive molybdenum, adsorbed on alumina, is placed in the generator and sent to the hospital. Pertechnetate ion is obtained when the molybdenum-99 nucleus in MoO_4^{2-} decays. The nuclear equation is

$$\ce{^{99}_{42}Mo -> ^{99m}_{43}Tc + ^{0}_{-1}e}$$

Each day pertechnetate ion, TcO_4^-, is leached from the generator with a salt solution whose osmotic pressure is the same as that of blood. Pertechnetate ion can be used to obtain images of the brain. However, other compounds of technetium are prepared to obtain images of other organs. Certain complex compounds of technetium bind to damaged heart tissues. They can be used to diagnose heart attacks. An active area of research is the synthesis of compounds of radioactive isotopes that may make it possible to see the functioning of the various organs of the body.■

Radioimmunoassay is a newly developed technique for analyzing blood and other body fluids for very small quantities of biologically active substances. The technique depends on the reversible binding of the substance to an antibody. Antibodies are produced in animals as protection against foreign substances. They protect by binding to the substance, and countering its biological activity. Consider, for example, the analysis for insulin in a sample of blood from a patient. Before the analysis, a solution of insulin-binding antibodies has been prepared from laboratory animals. This solution is combined with insulin containing a radioactive isotope, in which the antibodies bind with radioactive insulin. Now the sample containing an unknown amount of insulin is added to the antibody–radioactive insulin mixture. The nonradioactive insulin replaces some of the radioactive insulin bound to the antibody. As a result, the antibody loses some of its radioactivity. The loss in radioactivity is a measure of the amount of insulin in the blood sample.

■ Over a hundred radioactive isotopes have been used in medicine. Some examples are iodine-131, used to measure thyroid gland activity; phosphorus-32, used to locate tumors; and iron-59, used to measure the rate of formation of red blood cells.

Energy of Nuclear Reactions

Nuclear reactions, like chemical reactions, involve changes of energy. However, the changes of energy in nuclear reactions are enormous by comparison with chemical reactions. The energy released by certain nuclear reactions is used in nuclear power reactors.

22.6 Mass–Energy Calculations

When nuclei decay, they form products of lower energy. The change of energy is related to changes of mass, according to the mass–energy equivalence relation derived by Albert Einstein in 1905. Using this relation, we can obtain the energies of nuclear reactions from mass changes.

Mass–Energy Equivalence

One of the conclusions from Einstein's theory of special relativity is that the mass of a particle changes with its speed: the greater the speed, the greater the mass. Or, since kinetic energy depends on speed, we can say that the greater the kinetic energy of a particle, the greater its mass. This result, according to Einstein, is even more general. Energy and mass are equivalent and are related by the equation

$$E = mc^2$$

Here c is the speed of light, 3.00×10^8 m/s.

The meaning of this equation is that to any mass there is an associated energy, or to any energy there is an associated mass. Thus, if any system loses energy, it must also lose mass. For example, when carbon burns in oxygen, it releases heat energy:

$$C(graphite) + O_2(g) \longrightarrow CO_2(g), \quad \Delta H = -393.5 \text{ kJ}$$

Since this chemical system loses energy, it should also lose mass. In principle, we could obtain ΔH for the reaction by measuring the change in mass and relating this by Einstein's equation to the change in energy.■ However, weight measurements are of no practical value in determining heats of reaction. The changes in mass are simply too small to detect.

Calculation of the mass change for a typical chemical reaction, the burning of carbon in oxygen, will show just how small this quantity is. When the energy changes by an amount ΔE, the mass changes by an amount Δm. We can write Einstein's equation in the form

$$\Delta E = (\Delta m)c^2$$

The change in energy when one mole of carbon reacts is -3.935×10^5 J, or -3.935×10^5 kg · m²/s². Hence,

$$\Delta m = \frac{\Delta E}{c^2} = \frac{-3.935 \times 10^5 \text{ kg} \cdot \text{m}^2/\text{s}^2}{(3.00 \times 10^8 \text{ m/s})^2} = -4.37 \times 10^{-12} \text{ kg}$$

For comparison, a good analytical balance can detect a mass as small as 1×10^{-7} kg, but this is ten thousand times greater than the mass change caused by the release of heat during the combustion of one mole of carbon.

By contrast, the mass changes in nuclear reactions are approximately a

■ Einstein's equation gives ΔE. The enthalpy change, ΔH, equals $\Delta E + P\Delta V$ (see the Aside at the end of Section 10.1). For the reaction given in the text, $P\Delta V$ is essentially zero. In general, however, the $P\Delta V$ term would have to be added to ΔE to get ΔH.

Table 22.3
Masses of Some Nuclei and other Atomic Particles*

Symbol	Z	A	Mass (amu)	Symbol	Z	A	Mass (amu)
e⁻	−1	0	0.000549	Co	27	59	58.9184
n	0	1	1.00867	Ni	28	58	57.9199
H or p	1	1	1.00728	Pb	82	206	205.9295
	1	2	2.01345		82	207	206.9309
	1	3	3.01550		82	208	207.9316
He	2	3	3.01493	Po	84	210	209.9368
	2	4	4.00150		84	218	217.9628
Li	3	6	6.01347	Rn	86	222	221.9703
	3	7	7.01435	Ra	88	226	225.9771
Be	4	9	9.00999	Th	90	230	229.9837
B	5	10	10.0102		90	234	233.9942
	5	11	11.0066	Pa	91	234	233.9931
C	6	12	11.9967	U	92	233	232.9890
	6	13	13.0001		92	234	233.9904
O	8	16	15.9905		92	235	234.9934
Cr	24	52	51.9273		92	238	238.0003
Fe	26	56	55.9206	Pu	94	239	239.0006

*The mass of an atom is obtained by adding the masses of the electrons to the nuclear mass given in this table. For example, the mass of the $^{12}_{6}$C atom is 11.9967 + 6(0.000549) = 12.0000. (From R. C. Weast, ed., *CRC Handbook of Chemistry and Physics*, 59th ed. [© CRC Press, Inc., Boca Raton, Florida, 1978]. With permission of CRC Press, Inc.)

million times larger per mole of reactant than those in chemical reactions. Consider the alpha decay of uranium-238 to thorium-234. The nuclear equation is

$$\underset{238.0003}{\overset{238}{_{92}}\text{U}} \longrightarrow \underset{233.9942}{\overset{234}{_{90}}\text{Th}} + \underset{4.00150 \text{ amu}}{\overset{4}{_{2}}\text{He}}$$

Here, we have written the nuclear mass (in amu) beneath each nuclide symbol. (Table 22.3 lists masses of some nuclei and other atomic particles.) The change in mass for this nuclear reaction, starting with molar amounts, is

$$\Delta m = (233.9942 + 4.00150 - 238.0003) \text{ g} = -0.0046 \text{ g}$$

As in calculating ΔH and similar quantities, we subtract the value for the reactant from the sum of the values for the products. The minus sign indicates a loss of mass. This loss of mass is clearly large enough to detect.

From a table of nuclear masses, such as Table 22.3, we can use Einstein's equation to calculate the energy change for a nuclear reaction. This is illustrated in the next example. Recall from Section 22.2 that 1 eV = 1.602 × 10^{-19} J. Therefore, 1 MeV equals 1.602 × 10^{-13} J.■

■ From Einstein's equation, one can show that 1 amu is equivalent to 931 MeV. This number can be used to obtain energies of nuclear reactions. Thus, for the reaction given in Example 22.12, the mass change is −0.0196 amu. Multiplying this by 931 MeV gives the energy change, −18.2 MeV, which agrees with the answer given in that example, within the precision of the data.

Example 22.12

(a) Calculate the energy change in joules (3 significant figures) for the following nuclear reaction per mol of $_{1}^{2}$H:

$$_{1}^{2}\text{H} + {_{2}^{3}}\text{He} \longrightarrow {_{2}^{4}}\text{He} + {_{1}^{1}}\text{H}$$

Nuclear masses are given in Table 22.3. (b) What is this energy change in MeV for one $_{1}^{2}$H nucleus?

Solution

We write the nuclear masses below each nuclide symbol, then calculate Δm. Once we have Δm, we can obtain ΔE.

$$\underset{2.01345}{\overset{2}{_{1}}\text{H}} + \underset{3.01493}{\overset{3}{_{2}}\text{He}} \longrightarrow \underset{4.00150}{\overset{4}{_{2}}\text{He}} + \underset{1.00728 \text{ amu}}{\overset{1}{_{1}}\text{H}}$$

Hence,

$$\Delta m = (4.00150 + 1.00728 - 2.01345 - 3.01493) \text{ amu}$$
$$= -0.01960 \text{ amu}$$

(a) To obtain the energy change for molar amounts, we note that the molar mass of a nucleus in grams is numerically equal to the mass of a single nucleus in amu. Therefore, the mass change for molar amounts in this nuclear reaction is −0.0196 g, or −1.96 × 10^{-5} kg. The energy change is

$$\Delta E = (\Delta m)c^2 = (-1.96 \times 10^{-5} \text{ kg})(3.00 \times 10^8 \text{ m/s})^2$$
$$= -1.76 \times 10^{12} \text{ kg} \cdot \text{m}^2/\text{s}^2, \text{ or } -1.76 \times 10^{12} \text{ J}$$

(b) The mass change for the reaction of one $_{1}^{2}$H nucleus is −0.0196 amu. Let us change this to grams. Recall that 1 amu equals 1/12 the mass of a $_{6}^{12}$C atom, whose mass is 12 g/6.02 × 10^{23}. Thus, 1 amu = 1 g/6.02 × 10^{23}. Hence, the mass change in grams is

$$\Delta m = -0.0196 \text{ amu} \times \frac{1 \text{ g}}{1 \text{ amu} \times 6.02 \times 10^{23}}$$
$$= -3.26 \times 10^{-26} \text{ g (or } -3.26 \times 10^{-29} \text{ kg)}$$

Then,

$$\Delta E = (\Delta m)c^2 = (-3.26 \times 10^{-29} \text{ kg})(3.00 \times 10^8 \text{ m/s})^2$$
$$= -2.93 \times 10^{-12} \text{ J}$$

We now convert this to MeV:

$$\Delta E = -2.93 \times 10^{-12} \text{ J} \times \frac{1 \text{ MeV}}{1.602 \times 10^{-13} \text{ J}} = -18.3 \text{ MeV}$$

Exercise 22.12

(a) Calculate the energy change in joules when 1.00 g $_{90}^{234}$Th decays to $_{91}^{234}$Pa by beta emission. (b) What is the energy change in MeV when one $_{90}^{234}$Th nucleus decays? Use Table 22.3 for these calculations.

(See Problems 22.61 and 22.62.)

Nuclear Binding Energy

The equivalence of mass and energy explains the otherwise puzzling fact that the mass of a nucleus is always less than the sum of the masses of its constituent nucleons. For example, the helium-4 nucleus consists of two protons and two neutrons, giving the following sum:

$$\begin{aligned}
\text{Mass of 2 protons} &= 2 \times 1.00728 \text{ amu} = &2.01456 \text{ amu} \\
\text{Mass of 2 neutrons} &= 2 \times 1.00867 \text{ amu} = &\underline{2.01734 \text{ amu}} \\
\text{Total mass of nucleons} &= &4.03190 \text{ amu}
\end{aligned}$$

Since the mass of the helium-4 nucleus is 4.00150 amu (see Table 22.3), the mass difference is

$$\Delta m = (4.00150 - 4.03190) \text{ amu} = -0.03040 \text{ amu}$$

This mass difference is explained as follows. When the nucleons come together to form a nucleus, energy is released. (The nucleus has lower energy and is therefore more stable than the separate nucleons). There must be an equivalent decrease in mass, according to Einstein's equation.

The **binding energy** of a nucleus is the energy needed to break a nucleus into its individual protons and neutrons. Thus, the energy of the helium-4 nucleus is the energy change for the reaction

$$^4_2\text{He} \longrightarrow 2^1_1\text{p} + 2^1_0\text{n}$$

The **mass defect** of a nucleus equals the total nucleon mass minus the nuclear mass. In the case of helium-4, the mass defect is 4.03190 amu − 4.00150 amu = 0.03040 amu (this is the positive value of the mass difference we calculated earlier). Both the binding energy and the corresponding mass defect are reflections of the stability of the nucleus.

To compare the stabilities of various nuclei, it is useful to compare binding energies per nucleon. Figure 22.9 shows values of this quantity (in MeV

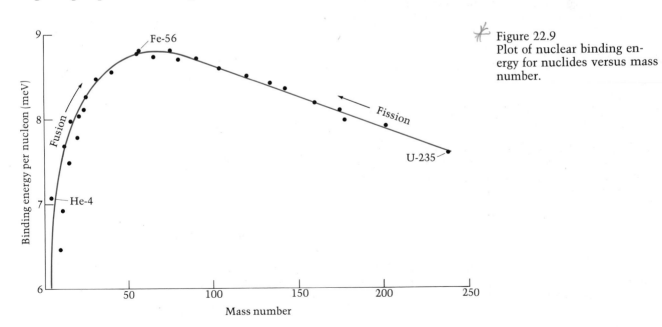

Figure 22.9
Plot of nuclear binding energy for nuclides versus mass number.

per nucleon) plotted against the mass number for various nuclides. Most of the points lie near the smooth curve drawn on the graph.

Note that nuclides near mass number 50 have the largest binding energies per nucleon. This means that a group of nucleons would tend to form those nuclides, because they would thus form nuclei of the lowest energy. For this reason, heavy nuclei might be expected to split to give lighter nuclei, and light nuclei might be expected to react, or combine, to form heavier nuclei.

Nuclear fission is a nuclear reaction in which a heavy nucleus splits into lighter nuclei and energy is released. For example, californium-252 decays both by alpha emission (97%) and by spontaneous fission (3%). During spontaneous fission, the nucleus splits into two stabler, lighter nuclei plus several neutrons. There are many possible ways in which the nucleus can split. One way is represented by the following equation:

$$^{252}_{98}Cf \longrightarrow {}^{142}_{56}Ba + {}^{106}_{42}Mo + 4{}^{1}_{0}n$$

In some cases, a nucleus can be induced to undergo fission by bombarding it with neutrons. An example is the nuclear fission of uranium-235. When a neutron strikes the $^{235}_{92}U$ nucleus, the nucleus splits into roughly equal parts, giving off several neutrons. Three possible splittings are shown in the following equations:

$$^{1}_{0}n + {}^{235}_{92}U \longrightarrow \begin{cases} {}^{142}_{54}Xe + {}^{90}_{38}Sr + 4{}^{1}_{0}n \\ {}^{139}_{56}Ba + {}^{94}_{36}Kr + 3{}^{1}_{0}n \\ {}^{144}_{55}Cs + {}^{90}_{37}Rb + 2{}^{1}_{0}n \end{cases}$$

Nuclear fission is discussed further in Section 22.7.

Nuclear fusion is a nuclear reaction in which light nuclei combine to give a more stable, heavier nucleus plus possibly several neutrons, and energy is released. An example of nuclear fusion is

$$^{2}_{1}H + {}^{3}_{1}H \longrightarrow {}^{4}_{2}He + {}^{1}_{0}n$$

Even though a nuclear reaction is energetically favorable, the reaction may be imperceptibly slow unless the correct conditions are present. This aspect of nuclear fusion reactions is discussed in Section 22.7.

22.7 Nuclear Fission and Nuclear Fusion

We have seen that the stablest nuclei are those of intermediate size (with mass numbers around 50). Nuclear fission and nuclear fusion are reactions in which nuclei attain sizes closer to this intermediate range. In doing so, these nuclei release tremendous amounts of energy. Nuclear fission of uranium-235 is employed in nuclear power plants to generate electricity. Nuclear fusion may supply energy in the future.

Nuclear Fission; Nuclear Reactors

Nuclear fission was discovered as a result of experiments to produce transuranium elements. Soon after the neutron was discovered in 1932, experimenters realized that this particle, being electrically neutral, should easily penetrate heavy nuclei. Many experimenters began using neutrons in bombardment reactions, hoping to produce isotopes that would decay to new elements. In late 1938, Otto Hahn and Fritz Strassmann in Berlin identified barium in uranium samples that had been bombarded with neutrons. Soon afterward, the presence of barium was explained as a result of the fission of the uranium-235 nucleus. When this nucleus is struck by a neutron, it splits into two nuclei. Fissions of uranium nuclei produce approximately 30 different elements of intermediate mass, including barium.

When the uranium-235 nucleus splits, approximately 2 to 3 neutrons are released. If the neutrons from each nuclear fission are absorbed by other uranium-235 nuclei, these nuclei will split and release even more neutrons. Thus, a **chain reaction** can occur, in which the number of nuclei that fission quickly multiply as a result of the absorption of neutrons released from previous nuclear fissions. Figure 22.10 shows how such a chain reaction occurs. The chain reaction of nuclear fissions is the basis of nuclear power and nuclear weapons.

Figure 22.10
Representation of a chain reaction of nuclear fissions. The number of fissions produced can multiply quickly.

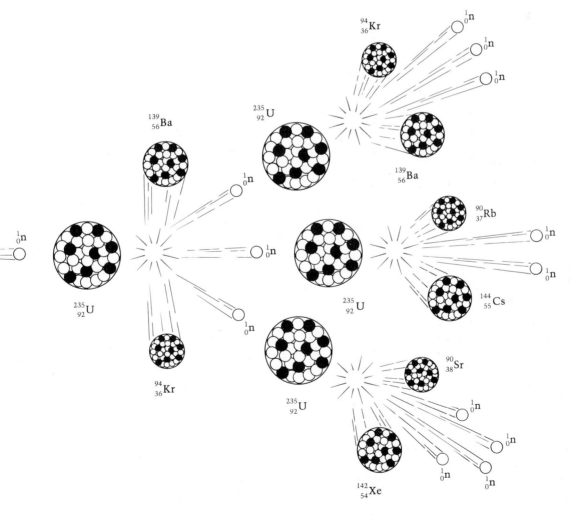

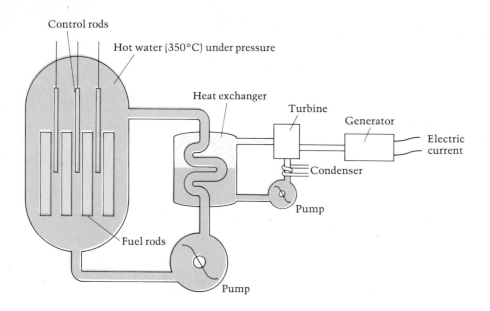

Control rods

Hot water (350°C) under pressure

Heat exchanger

Turbine

Generator

Electric current

Condenser

Pump

Fuel rods

Pump

Figure 22.11
Light water nuclear reactor (pressurized-water design). The fuel rods heat the water that is circulated to a heat exchanger. Steam is produced in the heat exchanger and passes to the turbine that drives a generator.

In order to sustain a chain reaction in a sample of fissionable material, a nucleus that splits must give an average of one neutron that results in the fission of another nucleus, and so on. If the sample is too small, many of the neutrons leave the sample before they have a chance to be absorbed. There is thus a **critical mass** for a particular fissionable material, which is the smallest mass in which a chain reaction can be sustained. If the mass is much larger than this (a *supercritical* mass), the number of nuclei that split multiply rapidly. An atomic bomb is detonated with a small explosive device that pushes together two or more masses of fissionable material to get a supercritical mass. A rapid chain reaction results in the splitting of most of the fissionable nuclei, and the release of an enormous amount of energy.

A **nuclear fission reactor** is a device that permits a controlled chain reaction of nuclear fissions. In power plants, a nuclear reactor is used to produce heat, which in turn is used to drive an electric generator. A nuclear reactor consists of fuel rods alternating with control rods contained within a vessel. The **fuel rods** contain fissionable material. In the light water (ordinary water) reactors commonly used in the United States (see Figure 22.11), these fuel rods contain uranium dioxide pellets in a zirconium alloy tube. Natural uranium contains only 0.72% uranium-235, which is the isotope that undergoes fission. The uranium used for fuel in these reactors is "enriched" so that it contains about 3% of the uranium-235 isotope. **Control rods,** composed of boron or cadmium, absorb neutrons, and can therefore slow the chain reaction. By varying the depth of the control rods within the fuel-rod assembly, one can increase or decrease the absorption of neutrons. If necessary, these rods can be dropped all the way into the fuel-rod assembly to stop the chain reaction.

A **moderator,** which is a substance that slows down neutrons, is required if uranium-235 is the fuel and this isotope is present as a small fraction of the total fuel. The neutrons that are released by the splitting of uranium-235 nuclei are absorbed more readily by uranium-238 than by other uranium-235 nuclei. However, when the neutrons are slowed down by a moderator, they are more readily absorbed by uranium-235, so it is possible to sustain a chain reaction with low fractional abundance of this isotope. Commonly used moderators are heavy water ($_1^2H_2O$), light water (ordinary water), and graphite.

In the light water reactor, ordinary water acts both as a moderator and a coolant. Figure 22.11 shows a pressurized-water design of this type of reactor. Water in the reactor is maintained at about 350°C under high pressure (150 atm) so it does not boil. The hot water is circulated to a heat exchanger, where the heat is used to produce steam to run a turbine and generate electricity.

After a period of time, fission products that absorb neutrons accumulate in the fuel rods. This interferes with the chain reaction, so eventually the fuel rods must be replaced. Originally, the intention was to send these fuel rods to *reprocessing plants*, where fuel material could be chemically separated from the radioactive wastes. Opposition to constructing these plants has been intense, however. Plutonium-239 would be one of the fuel materials separated from the spent fuel rods. This isotope is produced during the operation of the reactor when uranium-238 is bombarded with neutrons. It is fissionable and can be used to construct atomic bombs. For this reason, many people believe that the availability of this element in large quantities will increase the chance that many countries and terrorist groups would be able to divert enough plutonium to produce atomic bombs.

Whether or not the spent fuel rods are reprocessed, a pressing problem facing the nuclear power industry is how to safely dispose of radioactive wastes. One of many proposals is to encase the waste in a ceramic material and store the solids deep in the earth, perhaps in salt mines.

Breeder Reactors

Only about one percent of the energy available in natural uranium is actually released in a conventional nuclear reactor. This energy is released primarily from the splitting of uranium-235, which makes up only 0.72% of natural uranium. An additional small quantity of energy is obtained from the splitting of plutonium-239, produced when uranium-238 absorbs neutrons. In principle, if more of the uranium-238, the most abundant isotope of uranium, could be converted to plutonium, more energy could be obtained from a given quantity of uranium. This is the basic motivation for developing breeder reactors.

A **breeder reactor** is a reactor designed to produce more fissionable material than it consumes. The proposed breeder reactor at Clinch River, Tennessee, would use a fuel of plutonium-239 and uranium-235 oxides. This would be surrounded by uranium-238. No moderator would be used, but the high concentration of nuclear fuel would allow a chain reaction to occur. Because no moderator is used, many more of the neutrons would be absorbed by uranium-238 and so more of this isotope would be converted to plutonium-239. The net result of the breeder reactor would be to produce energy and to convert uranium-238 to plutonium-239, which can be used for additional fuel.

Critics of this design say that because reprocessing is needed to obtain the plutonium fuel, plutonium could be diverted to the manufacture of atomic bombs. For this reason, breeder reactors have been proposed that use a molten salt bath containing dissolved nuclear fuels. This molten salt would circulate from the reactor to a reprocessing plant, then back to the reactor. This enclosed reprocessing operation is designed to reduce the risk of theft of bomb materials.

So far, we have discussed the possibility of using breeder reactors only to extend supplies of uranium. However, it is possible to convert thorium-232, the most abundant isotope of thorium, to fissionable uranium-233 in a breeder reactor. Reserves of thorium are more plentiful than those of uranium.

Nuclear Fusion

As we noted in Section 22.6, energy can be obtained by combining light nuclei into a heavier nucleus by nuclear fusion. Such fusion reactions have been observed in the laboratory by means of bombardment using particle accelerators. Deuterons ($^2_1$H nuclei), for example, can be accelerated toward

Figure 22.12
Tokamak nuclear fusion test reactor. This reactor at Princeton University's Plasma Physics Laboratory will test the feasibility of nuclear fusion power. A doughnut-shaped plasma is contained by a magnetic field. Initially, the plasma will be heated by a combination of methods, including electric heating and injection of a heated neutral atom beam into the plasma. Once heated to about 100 million °C, the plasma would be maintained at this temperature by nuclear fusions.

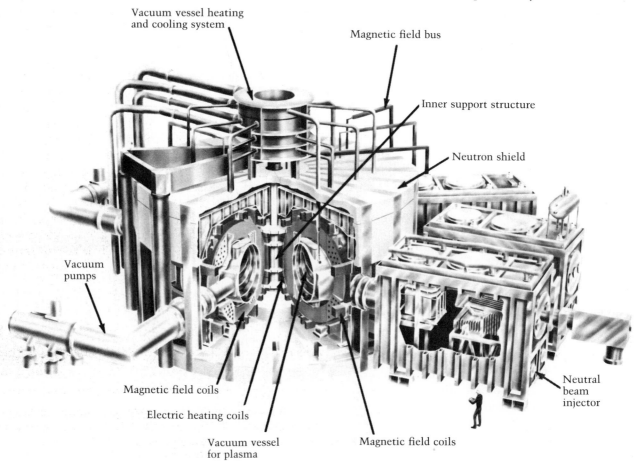

Vacuum vessel heating and cooling system

Magnetic field bus

Inner support structure

Neutron shield

Vacuum pumps

Magnetic field coils

Electric heating coils

Vacuum vessel for plasma

Magnetic field coils

Neutral beam injector

targets containing deuterium ($_1^2$H atoms) or tritium ($_1^3$H atoms). The reactions are

$$_1^2H + {_1^2}H \longrightarrow {_2^3}He + {_0^1}n$$
$$_1^2H + {_1^3}H \longrightarrow {_2^4}He + {_0^1}n$$

To get the nuclei to react, the bombarding nucleus must have enough kinetic energy to overcome the repulsion of electrical charges of the nuclei. The first reaction uses only deuterium, which is present in ordinary water.■ It is therefore very attractive as a source of energy. But, as we will discuss, the second reaction is more likely to be used first.

■ Natural hydrogen contains 0.015% deuterium.

Energy cannot be obtained in a practical way using particle accelerators. Another way to give nuclei sufficient kinetic energy to react is by heating the nuclear materials to a sufficiently high temperature. The reaction of deuterium, $_1^2$H, and tritium, $_1^3$H, turns out to require the lowest temperature of any fusion reaction and for this reason will likely be the first developed as an energy source. For practical purposes, this temperature will need to be about 100 million °C.■ At this temperature, all atoms will have been stripped of their electrons, so that a plasma results. A **plasma** is an electrically neutral gas of ions and electrons. At 100 million °C, the plasma is essentially separate nuclei and electrons. Thus, the development of nuclear fusion requires the study of the properties of plasmas at high temperatures.

■ Some nuclear fusion occurs at a much lower temperature, but not enough to be practical. Nuclear fusion occurs at the center of the sun, where the temperature is estimated to be 15 million °C. However, the sun is so massive that the total energy released is enormous.

It is now believed that the energy of stars, including our sun, where extremely high temperatures exist, derives from nuclear fusion. The hydrogen bomb also employs nuclear fusion for its destructive power. High temperature is first attained by a fission bomb. This then ignites fusion reactions in surrounding material of deuterium and tritium.

The main problem in developing controlled nuclear fusion is to be able to heat a plasma to high temperature and maintain those temperatures. If a plasma touches any material whatever, heat is quickly conducted away, and the plasma temperature quickly falls.■ A *tokamak nuclear fusion reactor* uses a doughnut-shaped magnetic field to hold the plasma away from any material (see Figure 22.12). A *laser fusion reactor* employs a bank of lasers aimed at a single point. Pellets containing deuterium and tritium would drop into the reactor where they would be heated to 100 million °C by bursts of laser light.

■ Even though the plasma is nearly 100 million °C above the melting point of any material, the total quantity of heat that could be transferred from the plasma is very small because its concentration is extremely low. Therefore, if the plasma were to touch the walls of the reactor, the plasma would cool, but the walls would not melt.

A Checklist for Review

Important Terms

radioactive decay (22.1)
nuclear bombardment
   reaction (22.1)
nucleons (22.1)
nuclide (22.1)
nuclide symbol (22.1)
nuclear equation (22.1)
positron (22.1)

gamma photons (22.1)
nuclear force (22.1)
shell model of the nucleus (22.1)
magic numbers (22.1)
band of stability (22.1)
alpha emission (22.1)
beta emission (22.1)
positron emission (22.1)

electron capture (22.1)
gamma emission (22.1)
metastable nucleus (22.1)
beta-stability line (22.1)
radioactive decay series (22.1)
transmutation (22.2)
particle accelerators (22.2)
electron volt (22.2)

cyclotron (22.2)

deuterons (22.2)

transuranium elements (22.2)

Geiger counter (22.3)

scintillation counter (22.3)

activity of a radioactive
    source (22.3)

curie (22.3)

rad (22.3)

rem (22.3)

radioactive decay constant (22.4)

half-life (22.4)

radioactive tracer (22.5)

isotope dilution (22.5)

neutron activation
    analysis (22.5)

binding energy (22.6)

mass defect (22.6)

nuclear fission (22.6)

nuclear fusion (22.6)

chain reaction (22.7)

critical mass (22.7)

nuclear fission reactor (22.7)

fuel rods (22.7)

control rods (22.7)

moderator (22.7)

breeder reactor (22.7)

plasma (22.7)

Summary of Facts and Concepts

Nuclear reactions are of two types, *radioactive decay* and *nuclear bombardment reactions*. Such reactions are represented by nuclear equations, each nucleus being denoted by a *nuclide* symbol. These equations must be balanced in charge (subscripts) and in nucleons (superscripts).

According to the *nuclear shell model*, the nucleons are arranged in shells. *Magic numbers* are the numbers of nucleons in a completed shell of protons or neutrons. Nuclei with magic numbers of protons or neutrons are especially stable. Pairs of protons and pairs of neutrons are also especially stable. Stable nuclei, when placed on a plot of N versus Z, fall in a *band of stability*. Those radioactive nuclides that fall above the *beta-stability line* in this plot usually decay by beta emission. Those radioactive nuclides that fall below this line usually decay by positron emission or electron capture. However, nuclides with $Z > 83$ usually decay by alpha emission. Uranium-238 forms a *radioactive decay series*. In such series, one element decays to another, which decays to another, and so forth, until a stable isotope is reached (lead-206 in the case of the uranium-238 series).

Transmutation of elements has been carried out in the laboratory by bombarding nuclei with various atomic particles. Alpha particles from natural sources can be used as reactants with light nuclei. For heavier nuclei, positive ions such as alpha particles must be first accelerated in a *particle accelerator*. Many of the *transuranium elements* have been obtained by bombardment of elements with accelerated particles. For example, plutonium was first made by bombarding uranium-238 with deuterons ($_1^2$H nuclei) from a *cyclotron*, a type of particle accelerator.

Particles of radiation from nuclear processes can be counted by Geiger counters or scintillation counters. In a *Geiger counter*, the particle ionizes a gas, which then conducts a pulse of electricity between two electrodes. In a *scintillation counter*, the particle hits a phosphor, and this emits a flash of light that is detected by a photomultiplier tube. The activity of a radioactive source, or

number of nuclear disintegrations per unit time, is measured in units of *curies* (3.700×10^{10} disintegrations per second).

Radiation affects biological organisms by breaking chemical bonds. The *rad* is the measure of radiation dosage that deposits 1×10^{-2} J of energy per kilogram of tissue. A *rem* equals the number of rads times a factor to account for the relative biological effectiveness of the radiation.

Radioactive decay is a first-order rate process. The rate is characterized by the *decay constant, k*, or by the *half-life, $t_{1/2}$*. These quantities, k and $t_{1/2}$, are related. Knowing one or the other, we can calculate how long it will take for a given radioactive sample to decay by a certain factor. Methods of *radioactive dating* depend on determining the fraction of a radioactive isotope that has decayed and from this the time that has elapsed.

Radioactive isotopes are used as *radioactive tracers* in chemical analysis and medicine. *Isotope dilution* is one application of radioactive tracers in which the dilution of the tracer can be related to the original quantity of nonradioactive isotope. *Neutron activation analysis* is a method of analysis that depends on the conversion of elements to radioactive isotopes by neutron bombardment.

According to Einstein's *mass–energy equivalence*, mass is related to energy by the equation $E = mc^2$. A nucleus will have less mass than the sum of the masses of the separate nucleons. The positive value of this mass difference is called the *mass defect* and is equivalent to the *binding energy* of the nucleus. Nuclides having mass numbers near 50 have the largest binding energies per nucleon. It follows that heavy nuclei should tend to split, a process called *nuclear fission*, and light nuclei should tend to combine, a process called *nuclear fusion*. Tremendous amounts of energy are released in both processes. Nuclear fission is used in conventional nuclear power reactors. Nuclear fusion reactors are in the experimental stage.

Operational Skills

1. Given a word description of a radioactive decay process, write the nuclear equation (Example 22.1). Given all but one of the reactants and products in a nuclear reaction, find that one nuclide (Examples 22.2 and 22.6).

2. Given a number of nuclides, determine which are most likely radioactive and which are most likely stable (Example 22.3).

3. Predict the type of radioactive decay that is most likely for given nuclides (Example 22.4).

4. Given an equation for a nuclear bombardment reaction, write the abbreviated notation, or vice versa (Example 22.5).

5. Given the activity (disintegrations per second) of a radioactive isotope, obtain the decay constant (Example 22.7).

6. Given the decay constant of a radioactive isotope, obtain the half-life (Example 22.8), or vice versa (Example 22.9). With the decay constant and mass of a radioactive isotope, calculate the activity of the sample (Example 22.9).

7. Given the half-life of a radioactive isotope, calculate the fraction remaining after a specified time (Example 22.10).

8. Given the disintegrations of carbon-14 nuclei per gram of carbon in a dead organic object, calculate the age of the object, that is, the time since its death (Example 22.11).

9. Given nuclear masses, calculate the energy change for a nuclear reaction (Example 22.12). Obtain the answer in joules per mole or MeV per particle.

Review Questions

22.1 What are the two types of nuclear reaction? Give an example of a nuclear equation for each type.

22.2 What are *magic numbers?* Give several examples of nuclei with magic numbers of protons.

22.3 List items to look for in a nucleus in order to predict whether it is stable.

22.4 What are the five common types of radioactive decay? What is the usual condition that leads to each type of decay?

22.5 What are the isotopes that begin each of the naturally occurring radioactive decay series?

22.6 Give equations for (a) the first transmutation of an element obtained in the laboratory by nuclear bombardment, (b) the reaction that produced the first artificial, radioactive isotope.

22.7 What is a particle accelerator, and how does one operate? Why are they required for certain nuclear reactions?

22.8 In what major way has the discovery of transuranium elements affected the form of modern periodic tables?

22.9 Describe how a Geiger counter works. How does a scintillation counter work?

22.10 Define the units *curie, rad,* and *rem.*

22.11 The half-life of cesium-137 is 30.2 y. How long will it take for a sample of cesium-137 to decay to 1/8 its original mass?

22.12 What is the age of a rock that contains equal numbers of $^{40}_{19}K$ and $^{40}_{18}Ar$ nuclei? The half-life of $^{40}_{19}K$ is 1.28×10^9 y.

22.13 What is a radioactive tracer? Give an example of the use of such a tracer in chemistry.

22.14 Isotope dilution has been used to obtain the volume of blood supply in a living animal. Explain how this could be done.

22.15 Briefly describe neutron activation analysis.

22.16 The deuteron, ^2_1H, has a mass that is smaller than the sum of the masses of its constituents, the proton plus the neutron. Explain why this is so.

22.17 Certain stars obtain their energy from nuclear reactions such as

$$^{12}_6C + {}^{12}_6C \longrightarrow {}^{23}_{11}Na + {}^1_1H$$

Explain in a sentence or two why this reaction might be expected to release energy.

22.18 Briefly describe how the following operate: (a) nuclear fission reactor (b) breeder reactor (c) tokamak fusion reactor.

Problems

Radioactivity

22.19 Write the nuclear equation for the decay of phosphorus-32 to sulfur-32 by beta emission. A phosphorus-32 nucleus emits a beta particle and gives a sulfur-32 nucleus.

22.21 Francium-212 decays by emitting alpha particles, each nucleus emitting one alpha particle. Write the nuclear equation for the decay. What element results from this decay?

22.23 Choose the nuclide from each of the following pairs that is radioactive (one is known to be radioactive, the other stable). Explain your choice.
(a) $^{122}_{51}Sb$, $^{136}_{54}Xe$ (b) $^{204}_{82}Pb$, $^{204}_{85}At$ (c) $^{87}_{37}Rb$, $^{80}_{37}Rb$

22.25 Predict the type of radioactive decay processes that are likely for the following nuclides. See Figure 22.2.
(a) $^{228}_{92}U$ (b) $^{8}_{5}B$ (c) $^{68}_{29}Cu$

22.27 Four radioactive decay series are known, three naturally occurring, and one beginning with the synthetic isotope $^{241}_{94}Pu$. To which of these decay series does the isotope $^{219}_{86}Rn$ belong? To which series does $^{220}_{86}Rn$ belong? Note that each isotope in these series decays by either α or β emission. How do these decay processes affect the mass number?

22.20 Write the nuclear equation for the decay of fluorine-18 to oxygen-18 by positron emission. A fluorine-18 nucleus emits a positron and gives an oxygen-18 nucleus.

22.22 Sodium-24 decays by emitting beta particles, each nucleus emitting one beta particle. Write the nuclear equation for the decay. What element results from this decay?

22.24 Choose the nuclide from each of the following pairs that is radioactive (one is known to be radioactive, the other stable). Explain your choice.
(a) $^{102}_{47}Ag$, $^{109}_{47}Ag$ (b) $^{25}_{12}Mg$, $^{24}_{10}Ne$ (c) $^{203}_{81}Tl$, $^{223}_{90}Th$

22.26 Predict the type of radioactive decay processes that are likely for the following nuclides. See Figure 22.2.
(a) $^{60}_{30}Zn$ (b) $^{6}_{2}He$ (c) $^{241}_{93}Np$

22.28 Four radioactive decay series are known, three naturally occurring, and one beginning with the synthetic isotope $^{241}_{94}Pu$. To which of these decay series does the isotope $^{227}_{89}Ac$ belong? To which series does $^{225}_{89}Ac$ belong? Note that each isotope in these series decays by either α or β emission. How do these decay processes affect the mass number?

Nuclear Bombardment Reactions

22.29 Write the abbreviated notations for the following bombardment reactions:
(a) $^{27}_{13}Al + {}^{2}_{1}H \longrightarrow {}^{25}_{12}Mg + {}^{4}_{2}He$
(b) $^{63}_{29}Cu + {}^{1}_{1}H \longrightarrow {}^{63}_{30}Zn + {}^{1}_{0}n$

22.31 Write out the nuclear equations for the following bombardment reactions:
(a) $^{63}Cu(\alpha, n)^{66}Ga$ (b) $^{33}S(n, p)^{33}P$

22.33 A proton is accelerated to 12.5 MeV per particle. What is this energy in kJ/mol?

22.35 Fill in the missing spaces in the following:
(a) $^{6}_{3}Li + {}^{1}_{0}n \longrightarrow ? + {}^{3}_{1}H$
(b) $^{232}_{90}Th(?, n)^{235}_{92}U$

22.37 Curium was first synthesized by bombarding an element with alpha particles. The products were curium-242 and a neutron. What was the target element?

22.30 Write the abbreviated notations for the following bombardment reactions:
(a) $^{10}_{5}B + {}^{4}_{2}He \longrightarrow {}^{13}_{6}C + {}^{1}_{1}H$
(b) $^{45}_{21}Sc + {}^{1}_{0}n \longrightarrow {}^{42}_{19}K + {}^{4}_{2}He$

22.32 Write out the nuclear equations for the following bombardment reactions:
(a) $^{23}Na(d, p)^{24}Na$ (b) $^{60}Ni(\alpha, n)^{63}Zn$

22.34 An alpha particle is accelerated to 24.2 MeV per particle. What is this energy in kJ/mol?

22.36 Fill in the missing spaces in the following:
(a) $^{27}_{13}Al + {}^{3}_{1}H \longrightarrow {}^{27}_{12}Mg + ?$
(b) $^{12}C({}^{3}H, ?)^{14}C$

22.38 Californium was first synthesized by bombarding an element with alpha particles. The products were californium-245 and a neutron. What was the target element?

Rate of Radioactive Decay

22.39 Tritium, or hydrogen-3, is prepared by bombarding lithium-6 with neutrons. A 0.45-mg sample of tritium decays at the rate of 1.61×10^{11} disintegrations per second. What is the decay constant (in /s) of tritium, whose atomic mass is 3.02 amu?

22.40 The first isotope of plutonium discovered was plutonium-238. It is used to power batteries for heart pacemakers. A sample of plutonium-238 weighing 3.5×10^{-6} g decays at the rate of 2.3×10^6 disintegrations per second. What is the decay constant of plutonium-238 in reciprocal seconds (/s)?

22.41 Sulfur-35 is a radioactive isotope used in chemical and medical research. A 0.58-mg sample of sulfur-35 has an activity of 24.6 Ci. What is the decay constant of sulfur-35 (in /s)?

22.42 Sodium-24 is used in medicine to study the circulatory system. A sample weighing 3.2×10^{-6} g has an activity of 27.9 Ci. What is the decay constant of sodium-24 (in /s)?

22.43 Rubidium-87 is a radioactive isotope occurring in natural rubidium. The decay constant is 4.6×10^{-19}/s. What is the half-life in years?

22.44 Neptunium-237 was the first isotope of a transuranium element to be discovered. The decay constant is 1.03×10^{-14}/s. What is the half-life in years?

22.45 Carbon-14 has been used to study the mechanisms of reactions involving organic compounds. The half-life of carbon-14 is 5.73×10^3 y. What is the decay constant (in /s)?

22.46 Promethium-147 has been used in luminous paint for dials. The half-life of this isotope is 2.5 y. What is the decay constant (in /s)?

22.47 Gold-198 has a half-life of 2.69 d. What is the activity in curies of a 0.43-mg sample?

22.48 Cesium-134 has a half-life of 2.05 y. What is the activity in curies of a 0.75-mg sample?

22.49 A sample of a phosphorus compound contains phosphorus-32. If this sample of radioactive isotope is decaying at the rate of 6.0×10^{12} disintegrations per second, how many grams of ^{32}P are in the sample? The half-life of ^{32}P is 14.3 d.

22.50 A sample of sodium thiosulfate, $Na_2S_2O_3$, contains sulfur-35. Determine the mass of ^{35}S in the sample from the decay rate, which was determined to be 7.7×10^{11} disintegrations per second. The half-life of ^{35}S is 88 d.

22.51 A sample of sodium-24 was administered to a patient to test for faulty blood circulation by comparing the radioactivity reaching various parts of the body. What fraction of the sodium-24 nuclei would remain undecayed after 24.0 h? The half-life is 15.0 h. If a sample contains 5.0 μg of ^{24}Na, how many micrograms remain after 24.0 h?

22.52 A solution of sodium iodide containing iodine-131 was given to a patient to test for malfunctioning of the thyroid gland. What fraction of the iodine-131 nuclei would remain undecayed after 7.0 d? If a sample contains 5.0 μg of ^{131}I, how many micrograms remain after 7.0 d? The half-life of I-131 is 8.07 d.

22.53 If 28.0% of a sample of silver-112 decays in 1.52 h, what is the half-life of this isotope (in hours)?

22.54 If 18.0% of a sample of zinc-65 decays in 69.9 d, what is the half-life of this isotope (in days)?

*22.55 A sample of iron-59 initially registers 125 counts per second on a radiation counter. After 10.0 d, the sample registers 107 counts per second. What is the half-life (in days) of iron-59?

*22.56 A sample of copper-64 gives a reading of 88 counts per second on a radiation counter. After 9.5 h, the sample gives a reading of 53 counts per second. What is the half-life (in hours) of copper-64?

22.57 Carbon from a cypress beam obtained from the tomb of Sneferu, a king of ancient Egypt, gave 8.1 disintegrations of ^{14}C per minute per gram of carbon. How old is the cypress beam? Carbon from living material gives 15.3 disintegrations of ^{14}C per minute per gram of carbon.

22.58 Carbon from the Dead Sea Scrolls, very old manuscripts found in Israel, gave 12.1 disintegrations of ^{14}C per minute per gram of carbon. How old are the manuscripts? Carbon from living material gives 15.3 disintegrations of ^{14}C per minute per gram of carbon.

Mass–Energy Equivalence

22.59 Find the change of mass (in grams) resulting from the release of heat when 1 mol H_2 reacts with 1 mol Cl_2.

$$H_2(g) + Cl_2(g) \longrightarrow 2HCl(g), \Delta H = -185 \text{ kJ}$$

22.60 Find the change of mass (in grams) resulting from the release of heat when 1 mol of SO_2 is formed from the elements.

$$S(s) + O_2(g) \longrightarrow SO_2(g), \Delta H = -297 \text{ kJ}$$

22.61 Calculate the energy change for the following nuclear reaction (in joules per mole of ^2_1H):

$$^2_1\text{H} + {}^3_1\text{H} \longrightarrow {}^4_2\text{He} + {}^1_0\text{n}$$

Give the energy change in MeV per ^2_1H nucleus. See Table 22.3.

22.63 Obtain the mass defect (in amu) and binding energy (in MeV) for the ^6_3Li nucleus. What is the binding energy (in MeV) per nucleon? See Table 22.3.

22.62 Calculate the change in energy, in joules per mole of ^1_1H, for the following nuclear reaction:

$$^1_1\text{H} + {}^1_1\text{H} \longrightarrow {}^2_1\text{H} + {}^0_1\text{e}$$

Give the energy change in MeV per ^1_1H nucleus. See Table 22.3

22.64 Obtain the mass defect (in amu) and binding energy (in MeV) for the $^{56}_{26}\text{Fe}$ nucleus. What is the binding energy (in MeV) per nucleon? See Table 22.3.

Additional Problems

22.65 Sodium-23 is the only stable isotope of sodium. Predict how sodium-20 and how sodium-26 will decay.

22.66 Aluminum-27 is the only stable isotope of aluminum. Predict how aluminum-24 and how aluminum-30 will decay.

22.67 A uranium-235 nucleus decays by a series of alpha and beta emissions until it reaches lead-207. How many alpha emissions and how many beta emissions occur in this series of decays?

22.68 A thorium-232 nucleus decays by a series of alpha and beta emissions until it reaches lead-208. How many alpha emissions and how many beta emissions occur in this series of decays?

22.69 A bismuth-209 nucleus reacts with an alpha particle to produce an astatine nucleus and two neutrons. Write the complete nuclear equation for this reaction.

22.70 A bismuth-209 nucleus reacts with a deuteron to produce a polonium nucleus and a neutron. Write the complete nuclear equation for this reaction.

22.71 Complete the following equation by filling in the blank:

$$^{238}_{92}\text{U} + {}^{12}_{6}\text{C} \longrightarrow \underline{\hspace{1cm}} + 4{}^1_0\text{n}$$

22.72 Complete the following equation by filling in the blank:

$$^{246}_{96}\text{Cm} + {}^{12}_{6}\text{C} \longrightarrow \underline{\hspace{1cm}} + 4{}^1_0\text{n}$$

*22.73** Tritium, or hydrogen-3, is formed in the upper atmosphere by cosmic rays, similar to the formation of carbon-14. Tritium has been used to determine the age of wines. A wine that has been aged in a bottle has a tritium content only 78% of that in a similar wine of the same mass that has just been bottled. How long has the aged wine been in the bottle? The half-life of tritium is 12.3 y.

*22.74** The naturally occurring isotope rubidium-87 decays by beta emission to strontium-87. This decay is the basis of a method for determining the ages of rocks. A sample of rock contains 102.1 μg ^{87}Rb and 5.3 μg ^{87}Sr. What is the age of the rock? The half-life of rubidium-87 is 4.8×10^{10} y.

*22.75** When a positron and an electron collide, they are annihilated and two gamma photons of equal energy are emitted. Calculate the wavelength corresponding to this gamma emission.

*22.76** When technetium-99*m* decays to technetium-99, a gamma photon corresponding to an energy of 0.143 MeV is emitted. What is the wavelength of this gamma emission? What is the difference in mass between Tc-99*m* and Tc-99?

22.77 Calculate the energy released when 1.00 kg of uranium-235 undergoes the following fission process:

$$^1_0\text{n} + {}^{235}_{92}\text{U} \longrightarrow {}^{136}_{53}\text{I} + {}^{96}_{39}\text{Y} + 4{}^1_0\text{n}$$

The masses of $^{136}_{53}\text{I}$ and $^{96}_{39}\text{Y}$ nuclei are 135.8401 amu and 95.8629 amu, respectively. Other masses are given in Table 22.3. Compare this energy with the heat released when 1.00 kg C(graphite) burns to $CO_2(g)$.

22.78 Calculate the energy released when 1.00 kg of hydrogen-1 undergoes fusion to helium-4, according to the following reaction:

$$4{}^1_1\text{H} \longrightarrow {}^4_2\text{He} + 2{}^0_1\text{e}$$

This reaction is one of the principal sources of energy from the sun. See Table 22.3 for data. Compare the energy released by 1.00 kg of ^1_1H in this reaction to the heat released when 1.00 kg of C(graphite) burns to $CO_2(g)$.

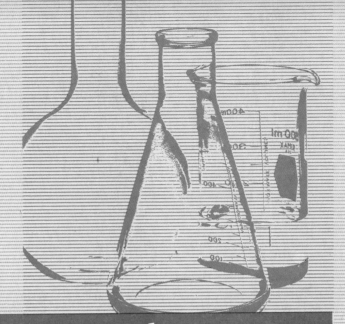

23. The Main-Group Elements: Groups IA to IIIA

23.1 General Observations About the Main-Group Elements Metallic–Nonmetallic Character/ Oxidation States/ Differences in Behavior of the Second-Row Elements

23.2 Group IA: The Alkali Metals Properties of the Elements/ Preparation of the Elements/ Uses of the Elements and Their Compounds

23.3 Group IIA: The Alkaline Earth Metals Properties of the Elements/ Preparation of the Elements/ Uses of the Elements and Their Compounds

23.4 Group IIIA: Boron and the Aluminum Family Metals Properties of the Elements/ Preparation of the Elements/ Uses of the Elements and Their Compounds

23.5 Metallurgy Preliminary Treatment/ Reduction/ Refining of a Metal

The main-group or representative elements —those elements in the A columns of the periodic table—have valence-shell configurations $ns^a np^b$. Within a particular column or family of elements, the valence-shell configurations are similar. As a result, the elements in a column often show similar physical and chemical properties. The alkali metals, for example, are very much alike both physically and chemically. So, too, are the alkaline earth metals.

Perhaps even more noteworthy are the systematic trends that can be seen in reading down a given column or across a given period. Ionization energies, electron affinities, electronegativities, and other properties change from one element to the next in a fairly definite way. Indeed, this is what allowed Mendeleev to predict the properties of germanium fifteen years before the element was discovered.

In this chapter and the next, we will see the similarities within a family of elements and the trends in a given column or period.

Chapter Overview

In this and the following chapters, we will look at the properties of some important elements, applying the basic principles described in earlier chapters. Before looking in detail at the main-group elements, we will make some general observations about the behavior of these elements. Except for boron, the elements of Groups IA to IIIA described in this chapter are active metals. In the last section, we will examine the steps in the production of a metal from its natural source.

23.1 General Observations About the Main-Group Elements

We pointed out in the chapter opening that we can observe trends in the behavior of the elements across a row or down a column of the periodic table. As you study the chemical and physical properties of the main-group elements, you should have these trends firmly in mind. Let us first look at the variation of metallic–nonmetallic character of these elements.

Metallic–Nonmetallic Character

Table 23.1 compares properties of the metallic and nonmetallic elements. A metal has a characteristic shine or luster and is a conductor of heat and electricity. Most metallic elements are malleable, ductile solids. The nonmetallic elements are nonlustrous. Most are diatomic gases (O_2, N_2, Cl_2) or hard, brittle solids (C, S_8, I_2) at 20°C, 1 atm.

The metallic elements have low ionization energies and low electronegativities compared with the nonmetallic elements. As a result, the metals often form cations in compounds or in aqueous solution (Na^+, Ca^{2+}, Al^{3+}). Nonmetals, on the other hand, form monatomic anions (O^{2-}, Cl^-) and oxyanions (NO_3^-, SO_4^{2-}).

Oxides of the metallic elements are usually basic. The reason for the basicity of the oxides of the most active metals is simple. These oxides are ionic and contain the oxide ion, O^{2-}. Oxide ions react with water to give a

Metals	Nonmetals	Table 23.1 Comparison of Metallic and Nonmetallic Elements
Lustrous	Nonlustrous	
Solids at 20°C (except Hg, which is liquid)	Solids or gases at 20°C (except Br$_2$, which is liquid)	
Solids are malleable and ductile	Solids are usually hard and brittle	
Conductors of heat and electricity	Nonconductors of heat and electricity (except graphite, an allotrope of C)	
Low ionization energies	Moderate to high ionization energies	
Low electronegativities	Moderate to high electronegativities	
Form cations	Form monatomic anions or oxyanions	
Oxides are basic (unless metal is in a high oxidation state)	Oxides are acidic	

basic solution, and they react with acids to give water:

$$O^{2-}(aq) + H_2O(l) \longrightarrow 2OH^-(aq)$$
$$O^{2-}(aq) + 2H^+(aq) \longrightarrow H_2O(l)$$

As the bonding in a metal oxide becomes less ionic (as it does in higher oxidation states), the oxide becomes less basic. ■Aluminum oxide, Al$_2$O$_3$, is an example of a metal oxide that is amphoteric. An **amphoteric** substance has both acidic and basic properties. Aluminum oxide dissolves in acids to produce the cation, as expected for a metal oxide:

$$Al_2O_3(s) + 6H^+(aq) \longrightarrow 2Al^{3+}(aq) + 3H_2O(l)$$

But the oxide also dissolves in strong base:

$$Al_2O_3(s) + 3H_2O(l) + 2OH^-(aq) \longrightarrow 2Al(OH)_4^-(aq)$$

In this case, the aluminate anion, Al(OH)$_4^-$, is formed. Note that Al—O bonds, which have considerable covalent character, are retained in going from Al$_2$O$_3$ to the ion Al(OH)$_4^-$.

Nonmetal oxides are acidic. In this case, covalent bonds between oxygen and the element are retained in reacting with water or base by forming oxyanions. For example,

$$SO_2(g) + H_2O(l) \rightleftharpoons H^+(aq) + HSO_3^-(aq)$$
$$SO_2(g) + 2OH^-(aq) \longrightarrow H_2O(l) + SO_3^{2-}(aq)$$

Figure 23.1 shows a periodic table in which only the main-group elements appear. Those elements to the left of the heavy "staircase" line are largely metallic in behavior; those to the right are largely nonmetallic. Elements along the line are metalloids, or semimetals. These elements have characteristics of both metals and nonmetals. Though nonmetallic in much of their chemical behavior, these elements have allotropes that have a luster and conduct electricity. They are called **semiconductors** because the pure elements are only slightly conducting at room temperature, but become moderately good conductors at higher temperatures. By contrast, metals become less conducting as the temperature increases.

■ Some transition-metal oxides in high oxidation states are acidic. For example, chromium trioxide, CrO$_3$, is an acidic oxide. The corresponding acid is chromic acid, H$_2$CrO$_4$, which forms chromate salts.

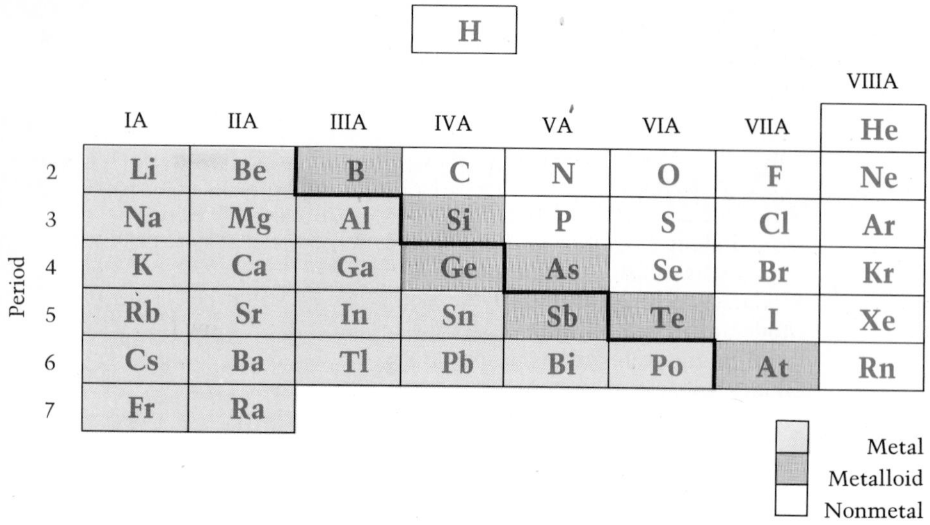

Figure 23.1
The main-group elements. Elements to the left of the heavy line are largely metallic in behavior, while those to the right are largely nonmetallic.

The metallic characteristics of the elements in the periodic table decrease in going across a period from left to right. Figure 23.1 illustrates this in a broad way. In any period, the elements on the left are metals and those on the right are nonmetals. Table 23.2, which lists the oxides of the main-group elements, more clearly shows the gradual change from metallic to nonmetallic characteristics in a given period. Sodium oxide, Na_2O, on the far left, is a strongly basic oxide (characteristic of a metal). Proceeding to the right, we have magnesium oxide, MgO (strongly basic), aluminum oxide, Al_2O_3 (amphoteric), silicon dioxide, SiO_2 (weakly acid), phosphorus(V) oxide, P_4O_{10} (moderately acidic), sulfur trioxide, SO_3 (strongly acidic), and dichlorine heptoxide, Cl_2O_7 (strongly acidic).

The metallic characteristics of the elements in the periodic table become more important as one goes down any column (group). This trend is most pronounced in Groups IIIA to VA. For example, in Group IVA, carbon (in the

Table 23.2
Acid–Base Behavior of the Oxides of the Main-Group Elements*

Period	Group						
	IA	IIA	IIIA	IVA	VA	VIA	VIIA
2	Li_2O (s.b.)	BeO (amph.)	B_2O_3 (w.a.)	CO_2 (w.a.)	N_2O_5 (s.a.) N_2O_3 (w.a.)	—	—
3	Na_2O (s.b.)	MgO (s.b.)	Al_2O_3 (amph.)	SiO_2 (w.a.)	P_4O_{10} (a.) P_4O_6 (w.a.)	SO_3 (s.a.) SO_2 (w.a.)	Cl_2O_7 (s.a.) Cl_2O (w.a.)
4	K_2O (s.b.)	CaO (s.b.)	Ga_2O_3 (amph.)	GeO_2 (w.a.)	As_2O_5 (w.a.) As_4O_6 (amph.)	SeO_3 (s.a.) SeO_2 (w.a.)	Br_2O (w.a.)
5	Rb_2O (s.b.)	SrO (s.b.)	In_2O_3 (w.b.)	SnO_2 (amph.) SnO (amph.)	Sb_2O_5 (w.a.) Sb_4O_6 (amph.)	TeO_3 (w.a.) TeO_2 (amph.)	I_2O_5 (a.)
6	Cs_2O (s.b.)	BaO (s.b.)	Tl_2O_3 (w.b.) Tl_2O (s.b.)	PbO_2 (amph.) PbO (amph.)	Bi_2O_5 (w.a.) Bi_2O_3 (w.b.)	PoO_2 (amph.) PoO (w.b.)	—

*The acid–base behavior is indicated as follows: s.b. = strongly basic; w.b. = weakly basic; amph. = amphoteric; w.a. = weakly acidic; a. = moderately acidic; s.a. = strongly acidic.

second period) is a nonmetal, germanium (in the fourth period) is a metalloid, and lead (in the sixth period) is a metal. Note the changes in acid–base behavior of the oxides in each column of elements.

Example 23.1

Choose the more metallic element in each of the following pairs: (a) Li or Be, (b) Be or Mg, (c) Al or K. Explain your answers in terms of their positions in the periodic table.

Solution

(a) Li and Be are in the same period, with Li to the left of Be. Therefore, Li is the more metallic.

(b) Be and Mg are in the same column, with Mg below Be. Therefore, Mg is the more metallic.

(c) Al is in the same period as Na, which is to its left and therefore is more metallic. K is below Na and more metallic than it. Hence, K is more metallic than Al.

Exercise 23.1

For each of the following pairs, select the one that is more nonmetallic: (a) Ca or Ga, (b) Ga or B, (c) Be or Cs. Explain your answers.

(See Problems 23.41 and 23.42.)

Oxidation States

Table 23.3 shows the oxidation states displayed by compounds of the main-group elements. From this, we can draw a number of conclusions:

1. The elements of Groups IA to IIIA have a common oxidation state equal to the group number (the number of valence electrons). This is the only oxidation state found in compounds of Groups IA and IIA metals, and the only important one in compounds of Group IIIA elements, except for thallium. In the case of thallium, the +1 oxidation state is most common.

2. The more electronegative nonmetals have a most common oxidation state equal to the group number minus 8. The absolute value of this difference equals the number of electrons gained or shared in bonding. Thus, for oxygen (Group VIA), the most common oxidation state is $6 - 8 = -2$. For fluorine (Group VIIA), the oxidation state $7 - 8 = -1$ is the only one exhibited by compounds.

3. Except for oxygen and fluorine, the nonmetals and metalloids of Groups IVA to VIIA have a variety of oxidation states in compounds, extending from the most positive (equal to the group number) to the most negative (equal to the group number minus 8). With the exception of nitrogen, the oxidation states for any one of these elements are either all even or all odd. For example, chlorine (Group VIIA) has oxidation states of +7, +5, +3, +1, and −1 in its compounds.

4. For Periods 2 and 3 of Groups IIIA to VIA, the most common positive oxidation state equals the group number (except for oxygen, which has no common positive oxidation states). But, as one progresses down a column of these elements, the oxidation state equal to the group number minus 2 assumes greater importance. In the heavier metals (Periods 5 and 6), this

Table 23.3
Oxidation States in Compounds of the Main-Group Elements*

	Group						
	IA	IIA	IIIA	IVA	VA	VIA	VIIA
Period 2	Li +1	Be +2	B +3	C +4 +2 -4	N +5 +4 +3 +2 +1 -3	O -1 -2	F -1
Period 3	Na +1	Mg +2	Al +3	Si +4 -4	P +5 +3 -3	S +6 +4 +2 -2	Cl +7 +5 +3 +1 -1
Period 4	K +1	Ca +2	Ga +3	Ge +4 +2	As +5 +3 -3	Se +6 +4 -2	Br +7 +5 +1 -1
Period 5	Rb +1	Sr +2	In +3 +1	Sn +4 +2	Sb +5 +3 -3	Te +6 +4 -2	I +7 +5 +1 -1
Period 6	Cs +1	Ba +2	Tl +3 +1	Pb +4 +2	Bi +5 +3	Po +4 +2	At +5 -1

*The most common oxidation state is shown in color. Some uncommon oxidation states are not shown.

oxidation state is an important one, being the most common in the Period 6 metals (Tl, Pb, Bi, Po). It appears as though the *s* electrons of the valence shell are less likely to be involved in bonding in these elements. ■ Consider the Group VA elements. The most common positive oxidation state of nitrogen is +5, the group number. But in arsenic, the $+3 (= 5 - 2)$ oxidation state becomes important, and in antimony and bismuth this is the most common oxidation state.

Elements that display multiple oxidation states can have several oxides. As you can see from Table 23.2, the oxide in which the element is in the lower oxidation state is more basic than the oxide with the element in a higher oxidation state. For example, arsenic(V) oxide is weakly acidic, but arsenic(III) oxide is amphoteric. The reason for this general behavior is that, as the oxidation state increases, the bonding becomes more covalent. As we saw earlier, the more covalent the oxide, the more acidic it is. Greater covalent character of the higher oxidation state can be seen in other com-

■ This noninvolvement of the *ns* electrons in the heavier metals is sometimes called the "inert pair" effect. The explanation depends on the fact that bond strengths decrease with increasing atomic size. To form more bonds in the higher oxidation state requires promotional energy ($ns^2np^2 \longrightarrow ns^1np^3$ in Group IVA elements), but less energy is available from bond formation in the heavier elements.

pounds also. For example, chlorides, bromides, and iodides show a noticeable increase in covalent character with increase in oxidation state. Thus, tin(II) chloride is a white, crystalline solid melting at 246°C, but tin(IV) chloride is a colorless liquid that freezes at -33°C. That tin(IV) chloride has such a low freezing point is an indication of its molecular nature and the covalent character of the Sn—Cl bond.

Example 23.2

Lead forms two chlorides, $PbCl_2$ and $PbCl_4$. Which compound do you expect to be less stable? In which compound is the bonding more covalent?

Solution

Since lead is a metal of the sixth period, the oxidation number corresponding to the group number is expected to be less stable than the oxidation number equal to the

group number minus 2 $(4 - 2 = +2)$. We would expect $PbCl_2$ to be more stable than $PbCl_4$. Moreover, we expect the bonding in $PbCl_4$ to be more covalent than that in $PbCl_2$. (Lead(IV) chloride is a yellow, oily liquid that explodes near its boiling point of 105°C. Lead(II) chloride is a white, crystalline solid that is slightly soluble in water, dissolving to give Pb^{2+} ions. Its melting point is 501°C.)

Exercise 23.2

Thallium forms two chlorides, $TlCl$ and $TlCl_3$. One chloride is a crystalline substance that melts at 25°C and decomposes on heating. The other is a crystalline substance melting at 430°C. Identify the formulas of the chlorides with the properties given.

(See Problems 23.43 and 23.44.)

Differences in Behavior of the Second-Row Elements

The chemical and physical properties of a second-row element are often rather different from those of the other elements in the same group. In part, this difference in behavior is due to the relatively small atom in these elements. Thus, the second-row elements of Groups IA to IIIA would give relatively small, polarizing cations. These elements, therefore, display greater covalent character in their compounds. Beryllium compounds are often covalent, whereas compounds of other Group IIA elements (Mg, Ca, Sr, Ba) are primarily ionic. Similarly, all boron compounds are covalent, but aluminum and other Group IIIA elements (Ga, In, Tl) have many ionic compounds.

The relatively small atom in a second-row nonmetal gives rise to relatively high electronegativity (electron-withdrawing power). For example, nitrogen has an electronegativity of 3.0, but other Group VA elements have electronegativities between 1.9 and 2.1. Similarly, oxygen has an electronegativity of 3.5; other Group VIA elements have electronegativities between 2.0 and 2.5.

Another reason for the difference in behavior of a second-row element and the other elements of the same group has to do with the fact that bonding in the second-row elements involves only s and p orbitals, whereas the other elements may use d orbitals. This places a limit on the types of compound formed by the second-row elements. For example, although nitrogen forms only the trihalides (such as NCl_3), phosphorus has both trihalides (PCl_3) and pentahalides (PCl_5), which it forms by using $3d$ orbitals.

The second-row nonmetals, unlike other elements of the same group, often display strong multiple bonding, in which π orbitals are formed by the overlap of p orbitals. For example, carbon has many compounds with multiple bonds; silicon does not. ■ The structures of the oxides CO_2 and SiO_2 are quite different. Carbon dioxide is molecular (O=C=O), but silicon dioxide is macromolecular and has silicon–oxygen single bonds. Elementary nitrogen, N_2, has triply bonded atoms (N≡N); phosphorus consists of P_4 molecules, with phosphorus–phosphorus single bonds. The explanation for this difference is that π bonds form between two sidewise approaching p orbitals only if they can get close enough for good overlap. The overlap will be largest for the small atoms of the second-row elements. It should be noted, however, that the elements beyond the second row (such as phosphorus and sulfur) do show some multiple bonding involving d orbitals.

■ Only recently has a compound with a Si=Si bond been prepared.

Although the elements of the second row show differences from other elements in the same group, the first three members of the second row exhibit many similarities to those elements located diagonally below them in the periodic table:

$$\begin{array}{cccc} Li & Be & B & C \\ Na & Mg & Al & Si \end{array}$$

The elements Li and Mg, for example, are said to exhibit a **diagonal relationship.** Lithium shows a strong resemblance to the other alkali metals, yet it also has many similarities to magnesium. For example, when lithium is heated in nitrogen, it forms a nitride:

$$6Li(s) + N_2(g) \longrightarrow 2Li_3N(s)$$

No other alkali metal combines directly with nitrogen, but magnesium forms a nitride when it burns in air:

$$3Mg(s) + N_2(g) \longrightarrow Mg_3N_2(s)$$

Table 23.4 compares some properties of lithium with those of the other alkali metals and magnesium. Note the similarity of lithium and magnesium. As we will see, the diagonal relationships between beryllium and aluminum and between boron and silicon are even more striking.

Table 23.4
Comparison of Some Properties of Lithium with Those of Other Alkali Metals and Magnesium

Other Alkali Metals	Lithium	Magnesium
Burn in air to form peroxides or superoxides, but no nitrides	Burns in air to form normal oxide and nitride	Burns in air to form normal oxide and nitride
Carbonates decompose when strongly heated	Carbonate decomposes when moderately heated	Carbonate decomposes when moderately heated
Fluorides, carbonates, and phosphates are soluble	Fluoride, carbonate, and phosphate are slightly soluble	Fluoride, carbonate, and phosphate are insoluble

23.2 Group IA: The Alkali Metals

Group IA elements, also known as the alkali metals, are the most reactive metallic elements, readily losing the ns^1 valence electron to form compounds in the $+1$ oxidation state. The majority of these compounds are ionic and water-soluble.

Because of their reactivity, Group IA elements always occur in nature as compounds, never as free metals. Sodium and potassium are abundant in the rocks of the earth's crust. Most of these rocks are composed of insoluble aluminosilicate minerals—substances containing silicon, aluminum, and oxygen with positive ions such as Na^+ and K^+. ■ (A **mineral** is a naturally occurring solid substance or solid solution with definite crystalline form.) Normal weathering of these insoluble minerals releases the alkali metal ions as soluble salts, which eventually find their way to the oceans. Sea water is principally a water solution of the salts of the alkali and alkaline earth (Group IIA) metals, with sodium chloride as the main constituent.

■ The structure of silicates and aluminosilicates is discussed in Section 24.1.

When sea water evaporates, the dissolved salts crystallize out, often as pure substances. The deposits of soluble minerals of sodium and potassium found in many parts of the world, including halite or rock salt ($NaCl$), sylvite (KCl), and carnallite ($KCl \cdot MgCl_2 \cdot 6H_2O$), probably originated when enclosed bodies of sea water slowly evaporated. These minerals are now important commercial sources of sodium chloride and potassium chloride. Sodium chloride is also obtained commercially from sea water by solar evaporation.

The other alkali metals (except francium) are obtained mainly from aluminosilicate minerals. Lithium is present in spodumene, $LiAl(SiO_3)_2$, an aluminosilicate mineral that occurs in several areas of the world. Cesium is available from the relatively rare aluminosilicate mineral pollucite, $CsAl(SiO_3)_2 \cdot H_2O$. And rubidium occurs as an impurity in many minerals, including pollucite; however, there are no known minerals in which rubidium is a major constituent.

Francium is a very rare, radioactive element. It was discovered in 1939 by Marguerite Perey, a French chemist. She found that about 1% of actinium-227 decays by alpha emission to francium-223, the remainder decaying by beta emission to thorium-227.

$$\ce{^{227}_{89}Ac} \longrightarrow \ce{^{223}_{87}Fr} + \ce{^{4}_{2}He}$$

Francium-223, the longest-lived isotope of the element, has a half-life of 21 minutes. It has been estimated that there is less than 25 grams of francium on earth. The element has no commercial uses, and very little of its chemistry is known.

Properties of the Elements

All Group IA elements are silvery white, metallic solids, although cesium, which melts at 28°C, would be liquid on a warm day. The metals are soft; sodium can be easily cut with a knife. Table 23.5 lists melting points and other physical properties of the alkali metals. Note that lithium, sodium, and potassium have densities less than 1.00 g/mL and would float on water. (But the three metals react vigorously with water!)

Property	Lithium	Sodium	Potassium	Rubidium	Cesium
Electron configuration	$[He]2s^1$	$[Ne]3s^1$	$[Ar]4s^1$	$[Kr]5s^1$	$[Xe]6s^1$
Melting point, °C	181	97.8	63.6	38.9	28.4
Boiling point, °C	1347	883	774	688	678
Density, g/cm³	0.53	0.97	0.86	1.53	1.88
Ionization energy (first), kJ/mol	520	496	419	403	376
(second), kJ/mol	7298	4562	3051	2632	2420
Electronegativity (Pauling scale)	1.0	0.9	0.8	0.8	0.7
Standard potential (volts), $M^+ + e^- \longrightarrow M$	−3.04	−2.71	−2.92	−2.92	−2.92
Atomic radius, Å	1.52	1.86	2.31	2.44	2.62
Ionic radius, Å	0.60	0.95	1.33	1.48	1.69

Table 23.5
Properties of Group IA Elements

The softness of the metals and their low melting points are indications of weak metal bonding. This weak bonding is due to the relatively large size of the alkali metal atoms, compared with those of the elements that follow in the same period, and to the fact that there is only one valence electron. Note how the melting points decrease from lithium to cesium. This is expected because the atomic radii increase, decreasing the strength of the metal bonding.

The alkali metals, as we noted earlier, are all chemically very reactive. They are strong reducing agents, as we can see from their standard potentials (see Table 23.5 for physical data). And from their low electronegativities, we expect them to form ionic compounds. This is also apparent from their low first ionization energies. Second ionization energies are quite high, however, so that only +1 ions are formed in compounds.

All of the alkali metals react with water. Lithium reacts readily. Sodium and potassium react more vigorously, and because the reactions are exothermic, the evolved hydrogen may catch fire. The reactions of rubidium and cesium are even more violent.

As we noted when discussing oxygen, the alkali metals burn in O_2.■ Lithium forms lithium oxide, Li_2O, and sodium forms mainly sodium peroxide, Na_2O_2. The other metals form the superoxides, such as KO_2, a yellow solid. The alkali metals also react easily with the halogens (F_2, Cl_2, Br_2, I_2) to give the halides and at elevated temperatures with hydrogen to give ionic hydrides (such as Li^+H^-). Lithium, as we discussed earlier, is the only alkali metal that combines directly with nitrogen.

■ The reactions of the alkali metals with O_2 were discussed in Section 13.4.

Because of their reactivity with air and moisture, the alkali metals must be stored under an inert liquid such as kerosene or in an inert atmosphere. They must not be picked up with the bare fingers, since they will react with any moisture, producing a burn (from the alkali metal hydroxide and heat of reaction).

Preparation of the Elements

Sodium and lithium metals are most easily prepared by electrolysis of their fused salts. Sodium was first isolated in 1807 by Humphry Davy, who

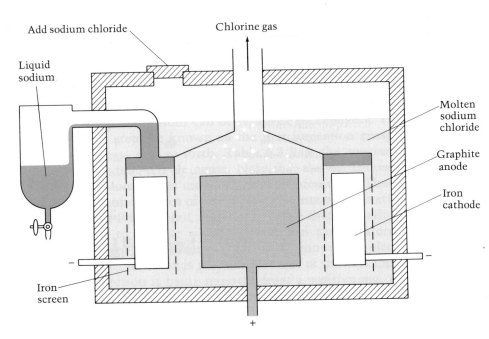

Add sodium chloride

Chlorine gas

Liquid sodium

Molten sodium chloride

Graphite anode

Iron cathode

Iron screen

−

+

−

Figure 23.2
Downs cell for the electrolysis of molten sodium chloride. Liquid sodium metal forms at the cathodes. It rises to the top of the molten sodium chloride and collects in a tank.

electrolyzed molten sodium hydroxide.■ Until recently, this was a major industrial method of preparation. Today most sodium is produced by the electrolysis of molten sodium chloride–calcium chloride mixtures; calcium chloride is added to lower the melting point to about 500°C. Figure 23.2 shows a **Downs cell,** used to electrolyze sodium chloride. Lithium is obtained from the electrolysis of molten lithium chloride–potassium chloride mixtures.

Potassium is more difficult to prepare by electrolysis because it attacks graphite electrodes. It is easier to reduce potassium chloride chemically. In the commercial process, molten potassium chloride reacts with sodium metal at 870°C:

$$Na(l) + KCl(l) \longrightarrow NaCl(l) + K(g)$$

The reaction goes in the direction written because potassium vapor leaves the reaction chamber and is condensed.

The other alkali metals, rubidium and cesium, are also prepared by chemical reduction of their salts. For example, when molten cesium chloride is heated at 700°C to 800°C with calcium metal at low pressure, cesium vapor distills over:

$$2CsCl(l) + Ca(l) \longrightarrow CaCl_2(l) + 2Cs(g)$$

■ The electrode reactions are

$$Na^+ + e^- \longrightarrow Na$$
$$4OH^- \longrightarrow 2H_2O + O_2 + 4e^-$$

Uses of the Elements and Their Compounds

Sodium is the most important of the alkali metals. The largest single use of the metal has been in the production of tetraethyllead, $(C_2H_5)_4Pb$, a gasoline additive that controls engine knocking. Sodium–lead alloy reacts with ethyl chloride, C_2H_5Cl, to give tetraethyllead:

$$4NaPb + 4C_2H_5Cl \longrightarrow (C_2H_5)_4Pb + 4NaCl + 3Pb$$

Table 23.6
Uses of Alkali Metal
Compounds

Compound	Use
Li_2CO_3	Aluminum production (added to the molten electrolyte)
	Preparation of LiOH
LiOH	Manufacture of lithium soaps for lubricating greases
LiH	Reducing agent in organic syntheses
$LiNH_2$	Preparation of antihistamines and other pharmaceuticals
NaCl	Source of sodium and sodium compounds
	Condiment and food preservative
	Soap manufacture (precipitates soap from reaction mixture)
NaOH	Pulp and paper industry
	Extraction of aluminum oxide from ore
	Manufacture of viscose rayon
	Petroleum refining
	Manufacture of soap
Na_2CO_3	Manufacture of glass
	Used in detergents and water softeners
Na_2O_2	Textile bleach
$NaNH_2$	Preparation of indigo dye for denim (blue jeans)
KCl	Fertilizer
	Source of other potassium compounds
KOH	Manufacture of soft soap
	Manufacture of other potassium compounds
K_2CO_3	Manufacture of glass
KNO_3	Fertilizers
	Explosives and fireworks

This use of sodium is declining as leaded gasolines are being phased out. Lead in gasoline interferes with the action of the catalytic converters of new automobiles. Moreover, leaded gasolines have been criticized as a source of environmental lead pollution.

Large quantities of sodium are used in the preparation of certain sodium compounds, such as sodium peroxide, Na_2O_2, and sodium amide, $NaNH_2$. Because sodium is a strong reducing agent, the metal is useful in the production of other metals (titanium, for example) and in the preparation of organic compounds. Small amounts of the metal are used in sodium vapor lamps, those bright yellow lamps along roadways. Sodium is also being increasingly used as a heat-transfer agent, for example in nuclear reactors.

Of the other alkali metals, only lithium has much commercial use. Lithium is used in metal processing as a *scavenger*, a substance added to molten metals to remove impurities such as oxygen and sulfur. The metal is also used in several alloys. It is added to lead alloys to harden them for use in bearings and is mixed with magnesium to give a strong, light alloy for aerospace applications. Lithium is also used in organic syntheses and in the preparation of some lithium compounds, such as lithium hydride, LiH, and lithium amide, $LiNH_2$. Since lithium has low density and large negative standard potential, it is a concentrated source of electrical energy and is used as a battery anode.

Small amounts of potassium are prepared, primarily to make potassium superoxide, KO_2, for use in rebreathing gas masks. These gas masks consist of a closed system in which air is circulated through a canister of KO_2. Oxygen is released when moisture in the breath attacks the superoxide:

$$4KO_2(s) + 2H_2O(l) \longrightarrow 4KOH(s) + 3O_2(g)$$

The potassium hydroxide produced in this reaction removes carbon dioxide from the exhaled air:

$$KOH(s) + CO_2(g) \longrightarrow KHCO_3(s)$$

Potassium also forms a liquid alloy with sodium that can be used as a heat-transfer agent.

Cesium and rubidium are used to make photoelectric cells. They are also used in the manufacture of vacuum tubes, as *getters*, which are placed in vacuum tubes to remove traces of oxygen.

Table 23.6 lists some of the most important compounds of the alkali metals. Lithium carbonate is a slightly soluble salt obtained from the processing of lithium ores. It is used to make lithium hydroxide. Calcium hydroxide (lime) reacts with solutions of lithium carbonate to precipitate calcium carbonate, leaving a solution of LiOH:

$$[Ca^{2+}(aq) + 2OH^-(aq)] + [2Li^+(aq) + CO_3^{2-}(aq)] \longrightarrow$$
$$CaCO_3(s) + 2[Li^+(aq) + OH^-(aq)]$$

Lithium hydroxide is used to manufacture lithium soaps for lubricating greases.

Sodium chloride is the source of sodium and most sodium compounds. One of the most important sodium compounds is sodium hydroxide, known commercially as caustic soda. It is prepared by the electrolysis of aqueous sodium chloride. In the *mercury cell* (Figure 23.3), mercury metal is used as

Figure 23.3
A mercury cell for the electrolysis of sodium chloride solution (brine). Sodium is deposited at the mercury cathode and chlorine gas bubbles up from the graphite anodes. Sodium amalgam (sodium–mercury alloy) is circulated to the amalgam decomposer, where the amalgam and graphite particles form the electrodes of many small voltaic cells. The net result is the reaction of sodium and water to give $NaOH(aq)$ and H_2.

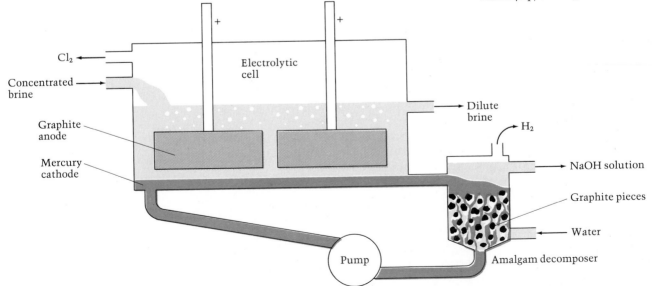

Cl₂ ←
Concentrated brine →
Electrolytic cell
Dilute brine →
H₂ →
Graphite anode
Mercury cathode
NaOH solution
Graphite pieces
Water ←
Pump
Amalgam decomposer

the cathode, where sodium ions are reduced to form a sodium–mercury alloy, called sodium amalgam:

$$Na^+ + e^- \xrightarrow{\ Hg\ } Na(amalgam)$$

The sodium amalgam is circulated from the electrolytic cell to the amalgam decomposer. Here the amalgam and graphite particles form the electrodes of many small voltaic cells in which sodium reacts with water to give sodium hydroxide solution and hydrogen gas. Loss of mercury from these cells has been a source of mercury pollution of waterways. Because of environmental regulations, filters and coolers have been added to reduce the mercury losses. In another type of cell, graphite electrodes are used, and the anode and cathode are separated by an asbestos diaphragm or specially designed membrane. The half-cell reactions are

$$2Cl^-(aq) \longrightarrow Cl_2(g) + 2e^- \qquad \text{(anode)}$$
$$2H_2O(l) + 2e^- \longrightarrow H_2(g) + 2OH^-(aq) \qquad \text{(cathode)}$$

Sodium ion migrates to the cathode, giving sodium hydroxide. Chlorine is also an important product of the electrolysis.

Sodium hydroxide is a strong base and has many important applications in chemical processing. Large quantities are used to make paper, to separate aluminum oxide from its ore, and to refine petroleum.■

Sodium carbonate is another important compound of sodium. The anhydrous compound, Na_2CO_3, is called soda ash. Large quantities of soda ash are consumed in making glass. The decahydrate, $Na_2CO_3 \cdot 10H_2O$, called washing soda, is used as a water softener and is added to detergent preparations. Most of the soda ash used in the United States is obtained from the mineral trona ($Na_2CO_3 \cdot NaHCO_3 \cdot H_2O$), which is mined from deposits in southwestern Wyoming. Sodium carbonate is also prepared from sodium chloride by the **Solvay process.** In this process, carbon dioxide and ammonia are bubbled into a cold solution (0°C) of sodium chloride. At this temperature, the sodium hydrogen carbonate (also called baking soda) precipitates:

$$CO_2(g) + NH_3(g) + [Na^+(aq) + Cl^-(aq)] + H_2O(l) \longrightarrow$$
$$NaHCO_3(s) + [NH_4^+(aq) + Cl^-(aq)]$$

The sodium hydrogen carbonate is filtered off the solution of ammonium chloride and washed. When heated to about 300°C, it decomposes to sodium carbonate:

$$2NaHCO_3(s) \xrightarrow{\ \Delta\ } Na_2CO_3(s) + CO_2(g) + H_2O(g)$$

The carbon dioxide gas from this step is recycled to produce more $NaHCO_3$. Ammonia is recovered from the ammonium chloride solution by heating it with calcium hydroxide, and this ammonia is also recycled:

$$2[NH_4^+(aq) + Cl^-(aq)] + [Ca^{2+}(aq) + 2OH^-(aq)] \xrightarrow{\ \Delta\ }$$
$$2NH_3(g) + 2H_2O(l) + [Ca^{2+}(aq) + 2Cl^-(aq)]$$

Potassium chloride is the most important compound of potassium. Over 90% of KCl is used as an agricultural fertilizer, since potassium ion is an important plant nutrient.■ The remainder is used to prepare potassium and potassium compounds. Potassium hydroxide, from which many potassium compounds are produced, is obtained by the electrolysis of aqueous KCl.

■ Pure aluminum oxide is obtained from bauxite, an aluminum ore, by the *Bayer process,* which is discussed in Section 23.5.

■ Potassium ion is also required in human nutrition, as is sodium ion. Potassium ion is found predominantly in the fluid within cells, and sodium ion is found in the fluids outside cells.

Example 23.3

Show by equations how potassium nitrate, KNO_3, could be prepared from potassium chloride, KCl, and nitric acid (in two steps).

Solution

Potassium hydroxide is produced commercially from potassium chloride by electrolysis of aqueous solutions:

$$2[K^+(aq) + Cl^-(aq)] + 2H_2O(l) \xrightarrow{\text{electrolysis}}$$
$$2[K^+(aq) + OH^-(aq)] + H_2(g) + Cl_2(g)$$

If the aqueous potassium hydroxide from this electrolysis is neutralized with nitric acid, the salt potassium nitrate is obtained:

$$[K^+(aq) + OH^-(aq)] + [H^+(aq) + NO_3^-(aq)] \longrightarrow$$
$$[K^+(aq) + NO_3^-(aq)] + H_2O(l)$$

Exercise 23.3

Starting from potassium chloride, show by means of equations how potassium superoxide is prepared (in two steps).

(See Problems 23.53 and 23.54.)

23.3 Group IIA: The Alkaline Earth Metals

Group IIA elements, or alkaline earth metals, are chemically reactive, though less so than the alkali metals. During reaction, the alkaline earth elements use the ns^2 valence electrons to form compounds in the +2 oxidation state. In the case of calcium, strontium, barium, and radium, these compounds are nearly always ionic and contain the +2 metal ion. However, bonding in magnesium often shows some covalent character, and in beryllium it is predominantly covalent.

Like the alkali metals, Group IIA elements always occur in nature as compounds. Magnesium and calcium are very abundant in the rocks of the earth's crust. This outer portion of the earth was originally in the form of silicates (compounds of silicon and oxygen with cations) and aluminosilicates. Magnesium and calcium, with sodium and potassium, are present in these rocks as cations. Weathering of the rocks produces soluble compounds of these cations, which eventually reach the seas. Calcium ion in sea water is used by shellfish to form their outer shell of calcium carbonate. Shells from dead animals have accumulated over geological time periods to form limestone deposits. Magnesium ion in sea water has reacted with these calcium carbonate sediments to form dolomite, $CaCO_3 \cdot MgCO_3$. Most of the magnesium ion has remained in the oceans, however.

The chief commercial sources of magnesium are sea water, underground brines, and the minerals dolomite and magnesite, $MgCO_3$. Calcium compounds are obtained from seashells and limestone. Gypsum, $CaSO_4 \cdot 2H_2O$, is also an important mineral. Other alkaline earth elements are much less common than magnesium and calcium. The principal ore of beryllium is the aluminosilicate mineral beryl, $Be_3Al_2(SiO_3)_6$. (An **ore** is a rock or mineral from which a metal can be economically produced.) Gem-quality forms of beryl are aquamarine (light blue) and emerald (dark green). Strontium is found in celestite, $SrSO_4$, and strontianite, $SrCO_3$. Barium is found in barite, $BaSO_4$, and witherite, $BaCO_3$. Radium occurs in small amounts in uranium ores, as the radioactive isotope of mass number 226, which has a half-life of 1620 years, decaying by alpha emission.

$$^{226}_{88}Ra \longrightarrow {}^{222}_{86}Rn + {}^{4}_{2}He$$

All other isotopes of radium are also radioactive.

Properties of the Elements

Beryllium is a gray metal almost as hard as iron and hard enough to scratch glass. The other alkaline earth elements are silvery metals and much softer than beryllium, but still harder than the alkali metals. Table 23.7 lists some of their properties. Note that the melting points of Group IIA metals are well above those of Group IA metals (see Table 23.5). The higher melting points and greater hardness of Group IIA metals are due to the increased strength of bonding from two valence electrons.

The first two ionization energies of Group IIA elements are relatively low (see Table 23.7), and as expected the elements give $+2$ cations. Thus, the most active elements (Ca, Sr, Ba, Ra) react vigorously with water, producing the metal ions:

$$Ca(s) + 2H_2O(l) \longrightarrow [Ca^{2+}(aq) + 2OH^-(aq)] + H_2(g)$$

Magnesium reacts slowly with water at normal temperatures, but rapidly with steam. Beryllium is rather unreactive with water. Beryllium is also unlike the other Group IIA elements in forming complex anions, such as $Be(OH)_4^{2-}$. This is an indication of its partial nonmetallic character. Thus, beryllium not only reacts with acids, but dissolves in strong base (as does aluminum, with which beryllium has a diagonal relationship):

$$Be(s) + 2[Na^+(aq) + OH^-(aq)] + 2H_2O(l) \longrightarrow$$
$$[2Na^+(aq) + Be(OH)_4^{2-}(aq)] + H_2(g)$$
$$\text{sodium beryllate}$$

Table 23.7
Properties of Group IIA
Elements

Property	Beryllium	Magnesium	Calcium	Strontium	Barium
Electron configuration	$[He]2s^2$	$[Ne]3s^2$	$[Ar]4s^2$	$[Kr]5s^2$	$[Xe]6s^2$
Melting point, °C	1278	649	839	769	725
Boiling point, °C	2970	1090	1484	1384	1640
Density, g/cm³	1.85	1.74	1.54	2.6	3.51
Ionization energy (first), kJ/mol	899	738	590	549	503
(second), kJ/mol	1757	1451	1145	1064	965
(third), kJ/mol	14848	7733	4912	4210	3430
Electronegativity (Pauling scale)	1.5	1.2	1.0	1.0	0.9
Standard potential (volts), $M^{2+} + 2e^- \longrightarrow M$	−1.70	−2.38	−2.76	−2.89	−2.90
Atomic radius, Å	1.11	1.60	1.97	2.15	2.17
Ionic radius, Å	0.30	0.65	0.99	1.13	1.35

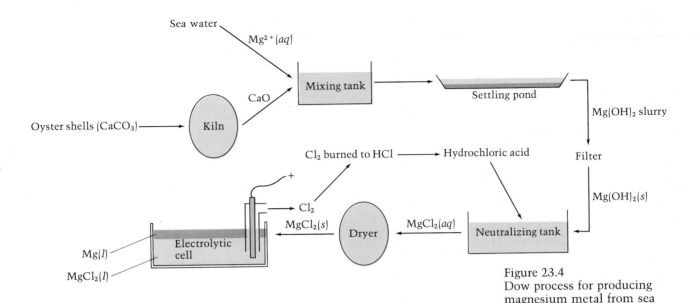

Figure 23.4
Dow process for producing magnesium metal from sea water.

Magnesium metal is unreactive with basic solutions. Of course, Group IIA metals below magnesium would react with the water in such solutions.

The alkaline earth metals burn in oxygen to form the oxides, or in the case of barium, the peroxide. Barium peroxide forms at low temperatures, but begins to decompose to the oxide at 700°C. Calcium, strontium, and barium react exothermically with hydrogen to form the ionic hydrides. For example,

$$Ca(s) + H_2(g) \longrightarrow CaH_2(s); \Delta H = -183 \text{ kJ}$$

Magnesium reacts with hydrogen only at high pressure and in the presence of a catalyst (MgI_2). All Group IIA elements react directly with the halogens to form halides and with nitrogen on heating to give the nitrides.

Preparation of the Elements

The alkaline earth metals are prepared by electrolysis of the molten halides (usually the chlorides) or by chemical reduction of either the halides or oxides. Magnesium, which is commercially the most important alkaline earth metal, is produced by the electrolysis of fused magnesium chloride. Sea water provides an inexhaustible source of magnesium ion, and seashells, which are mostly calcium carbonate, are a source of needed base to isolate the magnesium ion. The process for doing this, developed by the Dow Chemical Company, is shown in Figure 23.4. When oyster shells are heated, the calcium carbonate decomposes to the oxide:■

$$CaCO_3(s) \xrightarrow{\Delta} CaO(s) + CO_2(g)$$

Addition of calcium oxide (a basic oxide) to sea water precipitates magnesium hydroxide:

$$Mg^{2+}(aq) + CaO(s) + H_2O(l) \longrightarrow Mg(OH)_2(s) + Ca^{2+}(aq)$$

The magnesium hydroxide is filtered off, then treated with hydrochloric acid

■ Dolomite, $CaCO_3 \cdot MgCO_3$, is sometimes used in place of calcium carbonate. When heated, it yields MgO as well as CaO. The magnesium oxide provides magnesium ion, in addition to that from sea water.

to convert it to the chloride:

$$Mg(OH)_2(s) + 2[H^+(aq) + Cl^-(aq)] \longrightarrow [Mg^{2+}(aq) + 2Cl^-(aq)] + 2H_2O(l)$$

The dry salt, obtained by evaporation of the solution, is melted and electro-lyzed at 700°C:

$$MgCl_2(l) \xrightarrow{\text{electrolysis}} Mg(l) + Cl_2(g)$$

The by-product chlorine can be sold or burned with methane (natural gas) to provide hydrochloric acid for the process:

$$2CH_4(g) + O_2(g) + 4Cl_2(g) \longrightarrow 8HCl(g) + 2CO(g)$$

Magnesium is also obtained from magnesite or dolomite by decomposing it to MgO, then reducing the oxide with ferrosilicon, an alloy of silicon and iron.

Other alkaline earth metals are manufactured in smaller amounts. Beryllium can be obtained by electrolysis of beryllium chloride, $BeCl_2$, to which sodium chloride is added to increase the conductivity of the molten salt. However, most beryllium is prepared by chemical reduction of the fluoride with magnesium:

$$BeF_2(l) + Mg(l) \xrightarrow[950°C]{\Delta} MgF_2(l) + Be(s)$$

Calcium is prepared by electrolysis of molten calcium chloride and by reduction of calcium oxide by aluminum in a vacuum, where the calcium produced distills off. The reaction can be written ∎

$$3CaO(s) + 2Al(l) \xrightarrow[1200°C]{\Delta} 3Ca(g) + Al_2O_3(s)$$

Barium is also produced by reduction of the oxide by aluminum, and although very little strontium is used commercially, it can be produced by a similar process.

∎ Calcium oxide (a basic oxide) reacts with aluminum oxide (an amphoteric oxide) to give tri-calcium aluminate:

$$3CaO + Al_2O_3 \longrightarrow Ca_3Al_2O_6$$

Therefore, the overall reaction of CaO with Al can be written

$$6CaO + 2Al \longrightarrow 3Ca + Ca_3Al_2O_6$$

Uses of the Elements and Their Compounds

Commercially, the most important Group IIA metal is magnesium. It is used in large quantities to make aluminum alloy, to which it imparts hardness as well as corrosion resistance. The alloy appears to gain corrosion resistance from each metal. Magnesium is resistant to bases (but dissolves in acids); aluminum is somewhat resistant to acids (but dissolves easily in bases). Similarly, magnesium alloys often contain small quantities of aluminum. These alloys are especially useful in the aircraft and missile industry, where lightness is required. They are also used to make such things as truck bodies, auto parts, and ladders. Smaller quantities of magnesium are used as a reducing agent to prepare other metals, such as uranium and beryllium (see the previous section, Preparation of the Elements). Photographic flash lamps use a fine magnesium wire in oxygen and give off a brilliant white light when the metal burns.

Beryllium is an expensive metal, but it has some special characteristics that make it useful. When added in small amounts to copper, it gives an alloy as hard as steel. The alloy is used to make nonsparking electrical contacts. Since beryllium is transparent to x rays, the metal is used to make windows for x-ray tubes.■ Beryllium is also a moderator of neutrons (that is, it slows them down) and is used in nuclear reactors and weapons. Special precautions must be taken when working with beryllium and its alloys as the metallic dust is very toxic. (Beryllium compounds are also very toxic.)

■ X rays are scattered by the electrons in atoms. For this reason, atoms of high atomic number (which have many electrons) scatter x rays more effectively than atoms of low atomic number. Therefore, elements of low atomic number, such as Be, are transparent to x rays; elements such as lead are opaque.

Calcium is used as a scavenger (agent to remove impurities in materials) in the production of certain metals and in preparing various alloys. When added to lead, for example, it produces a hard metal for storage battery grids (electrodes). Calcium is also used as a reducing agent in preparing some of the less common metals, such as thorium:

$$ThO_2(s) + 2Ca(l) \xrightarrow[1000°C]{\Delta} Th(s) + 2CaO(s)$$

Barium is used in small amounts in the manufacture of television and vacuum tubes to remove traces of air. Strontium has few commercial uses.

Some of the most important compounds of Group IIA elements are formed by calcium. Calcium carbonate minerals and seashells are the chief commercial sources of these compounds. When heated to 900°C, the carbonate decomposes, releasing carbon dioxide:

$$CaCO_3(s) \xrightarrow[900°C]{\Delta} CaO(s) + CO_2(g)$$

The product is calcium oxide, known commercially as lime or quicklime. It is one of the most important industrial chemicals, second only to sulfuric acid in tons produced. (See Table 23.8 for a list of alkaline earth compounds

Table 23.8
Uses of Alkaline Earth Compounds

Compound	Use
MgO	Refractory bricks (for furnaces)
	Animal feeds
$Mg(OH)_2$	Source of magnesium for the metal and compounds
	Milk of magnesia (antacid and laxative)
$MgSO_4 \cdot 7H_2O$	Fertilizer
	Medicinal uses (laxative and analgesic)
	Mordant (used in dyeing fabrics)
CaO and $Ca(OH)_2$	Manufacture of steel
	Neutralizer for chemical processing
	Water treatment
	Mortar
	Stack-gas scrubber (to remove H_2S and SO_2)
$CaCO_3$	Paper coating and filler
	Antacids, dentifrices
$CaSO_4$	Plaster, wallboard
	Portland cement
$Ca(H_2PO_4)_2$	Soluble phosphate fertilizer
$BaSO_4$	Oil-well drilling mud
	Gastrointestinal x-ray photography
	Paint pigment (lithopone)

	Hydroxides	Carbonates	Sulfates	
Be	Insoluble $K_{sp} = 2.0 \times 10^{-18}$	—*	Soluble	Table 23.9 Solubilities of Some Group IIA Compounds at 25°C
Mg	Insoluble $K_{sp} = 5.5 \times 10^{-10}$	Slightly soluble $K_{sp} = 1.0 \times 10^{-5}$	Soluble	
Ca	Slightly soluble $K_{sp} = 5.5 \times 10^{-6}$	Insoluble $K_{sp} = 3.8 \times 10^{-9}$	Slightly soluble $K_{sp} = 2.4 \times 10^{-5}$	
Sr	Soluble	Insoluble $K_{sp} = 9.3 \times 10^{-10}$	Insoluble $K_{sp} = 2.5 \times 10^{-7}$	
Ba	Soluble	Insoluble $K_{sp} = 4.9 \times 10^{-9}$	Insoluble $K_{sp} = 1.1 \times 10^{-10}$	

*Only an insoluble hydroxide carbonate, $Be(OH)_2 \cdot BeCO_3$, is known.

and their uses.) Most of the calcium oxide is consumed in steel making. Added to molten iron containing silicates or silicon dioxide, it combines to give a slag (a glassy waste material) that floats to the top of the metal. This is essentially an acid–base reaction.

$$CaO(s) \ + \ SiO_2(s) \longrightarrow CaSiO_3(l)$$

base   acid   calcium silicate
oxide   oxide   slag

Calcium oxide reacts exothermically with water to produce the hydroxide, known commercially as slaked lime:

$$CaO(s) \ + \ H_2O(l) \longrightarrow Ca(OH)_2(s); \ \Delta H = -65.7 \text{ kJ}$$

Since the heat released is sufficient to ignite paper, care must be taken in storing the oxide.

Table 23.9 gives the solubilities of some Group IIA compounds. Aqueous solutions of calcium hydroxide react with carbon dioxide to give a white precipitate of calcium carbonate:

$$[Ca^{2+}(aq) \ + \ 2OH^-(aq)] \ + \ CO_2(g) \longrightarrow CaCO_3(s) \ + \ H_2O(l)$$

The reaction is the basis of a test for carbon dioxide. It is also used to prepare a pure form of finely divided calcium carbonate as a filler in making paper, and for toothpowders, antacids, and other purposes. Mortar, used in brick-laying, is made by mixing slaked lime with sand and water. The mortar hardens as the mixture dries and calcium hydroxide crystallizes. Later, it slowly sets to a harder solid as the hydroxide reacts with carbon dioxide in the air to form calcium carbonate.

Interestingly, large amounts of calcium oxide (or hydroxide) are used to remove calcium ion from hard water. Water that has passed through lime-stone deposits usually contains soluble calcium hydrogen carbonate, $Ca(HCO_3)_2$. Although calcium carbonate is insoluble in pure water, it does dissolve in water containing acidic substances, such as carbon dioxide (con-tained in ground water exposed to air). (See Figure 23.5.) We can write the

Figure 23.5
A cave formed in limestone by acidic ground water. The iciclelike formations (stalagmites and stalactites) in the cave are caused by redeposition of calcium carbonate.

reaction as

$$CaCO_3(s) + CO_2(g) + H_2O(l) \longrightarrow [Ca^{2+}(aq) + 2HCO_3^-(aq)]$$
limestone $\qquad\qquad\qquad\qquad$ calcium hydrogen carbonate

When hard water containing $Ca(HCO_3)_2$ is treated with the correct amount of calcium hydroxide, nearly all the calcium ion is precipitated as calcium carbonate:

$$[Ca^{2+}(aq) + 2HCO_3^-(aq)] + [Ca^{2+}(aq) + 2OH^-(aq)] \longrightarrow$$
$$2CaCO_3(s) + 2H_2O(l)$$

Magnesium carbonate, like calcium carbonate, decomposes when heated:

$$MgCO_3(s) \xrightarrow{\Delta} MgO(s) + CO_2(g)$$

If the carbonate is heated strongly (above 1400°C), the magnesium oxide that results is chemically rather inert. In this form, it is used to make refractory bricks (to line high-temperature furnaces). When the oxide is prepared at lower temperatures (about 700°C), however, it is obtained as a powder that dissolves easily in acids. It is used in animal feeds to provide magnesium ion as a nutrient.

The most important compound of barium is the sulfate, which occurs as the mineral barite. Barium sulfate is a very insoluble compound. It is used in oil drilling in the form of a mud, which makes a seal around the drilling bit. Barium, like all elements with high atomic number, is opaque to x rays. For this reason, barium sulfate is given orally or as an enema to obtain diagnostic x-ray photographs of the stomach and intestines. Soluble barium compounds cannot be used because they are toxic, but suspensions of barium sulfate provide negligible amounts of barium ion.

Beryllium is unlike the other alkaline earth elements in several ways. As we saw earlier, it forms the beryllate ion, $Be(OH)_4^{2-}$. Because of this, beryllium hydroxide, $Be(OH)_2$, is amphoteric, in contrast to the other Group IIA hydroxides. It dissolves in acids to form $Be^{2+}(aq)$, but also dissolves easily in strong base to form $Be(OH)_4^{2-}$. For example,

$$2[Na^+(aq) + OH^-(aq)] + Be(OH)_2(s) \longrightarrow [2Na^+(aq) + Be(OH)_4^{2-}(aq)]$$

In this reaction, beryllium hydroxide is acting as a Lewis acid, accepting electron pairs from OH^- ions to give an octet of electrons about the Be atom. As we will see later, a similar type of reaction occurs with aluminum hydroxide, which is also amphoteric, thus showing the diagonal relationship between Be and Al.

The tendency of Be^{2+} to accept electrons is due to its small size. As a result, bonding in its compounds is highly covalent. The melting point of beryllium chloride (450°C), for example, is low compared with that of the other chlorides of Group IIA elements ($MgCl_2$ melts at 714°C). Also, the molten salt has a low electrical conductivity, so that electrolysis of $BeCl_2$ requires the addition of a salt such as NaCl. In the solid phase, beryllium chloride has a polymeric structure (that is, a covalent substance with an infinitely repeating unit), in which each chlorine atom "bridges" two beryllium atoms:

Some aluminum halides form molecular substances with a somewhat similar structure.

Example 23.4

By means of chemical tests, distinguish between $MgCl_2$ and $BaCl_2$.

Solution

From Table 23.9, we note that $MgSO_4$ is soluble in water, but $BaSO_4$ is insoluble. Therefore, if we add sodium sulfate to an aqueous solution of each compound, $BaCl_2$ will give a precipitate of the sulfate and $MgCl_2$ will not.

$$Na_2SO_4(aq) + BaCl_2(aq) \longrightarrow 2NaCl(aq) + BaSO_4(s)$$

Exercise 23.4

Use chemical tests to distinguish $BeCl_2$ from $MgCl_2$.

(See Problems 23.57, 23.58, 23.59, and 23.60.)

Example 23.5

Show how you could prepare CaH_2 from limestone $(CaCO_3)$, aluminum, and hydrogen in three steps.

$$3CaO(s) + 2Al(l) \xrightarrow{\Delta} 3Ca(g) + Al_2O_3(s)$$
$$Ca(s) + H_2(g) \longrightarrow CaH_2(s)$$

Solution

The steps are

$$CaCO_3(s) \xrightarrow{\Delta} CaO(s) + CO_2(g)$$

Exercise 23.5

By means of equations, show how you could prepare magnesium sulfate in three steps from limestone, sea water, and sulfuric acid.

(See Problems 23.61 and 23.62.)

23.4 Group IIIA: Boron and the Aluminum Family Metals

As we noted earlier, the metallic character of the elements increases in moving down a column of the periodic table. This behavior is quite evident in the Group IIIA elements. The first element, boron, is a semiconducting solid, and its compounds display distinctly nonmetallic properties. For example, the oxide B_2O_3 is acidic (see Table 23.2). The rest of the elements in this group are metals, but the oxides and hydroxides change from amphoteric for aluminum and gallium to basic for indium and thallium. Bonding to boron is covalent, with the atom sharing its $2s^2 2p^1$ valence electrons to give the +3 oxidation state. Although aluminum has many covalent compounds, it also has definitely ionic ones, such as AlF_3. The cation $Al^{3+}(aq)$ is present in aqueous solutions of aluminum salts. Gallium, indium, and thallium also have ionic compounds and give cations in aqueous solution. The +3 oxidation state is the only important one for aluminum, but some +1 compounds are known for gallium and indium, and in the case of thallium, both +1 and +3 states are important.

Aluminum is the third most abundant element in the earth's crust (after oxygen and silicon). It occurs primarily in aluminosilicate minerals found in the original rocks of this outer portion of the earth. Weathering of these rocks has formed aluminum-containing clays, which are an essential part of most soils. Further weathering of clays has given deposits of bauxite, an aluminum ore containing $AlO(OH)$ and $Al(OH)_3$ in various proportions. Corundum is a hard mineral of aluminum oxide, Al_2O_3. The pure oxide is colorless, but the presence of impurities can give various colors to it. Sapphire (usually blue) and ruby (deep red) are gem-quality corundum.

Boron, although not an abundant element, is well known because it occurs concentrated in accessible minerals, such as borax $(Na_2B_4O_7 \cdot 10H_2O)$, kernite $(Na_2B_4O_7 \cdot 4H_2O)$, and colemanite $(Ca_2B_6O_{11} \cdot 5H_2O)$. These minerals were formed by the evaporation of hot springs and salt lakes. The major world supply of boron comes from southern California. The other elements of

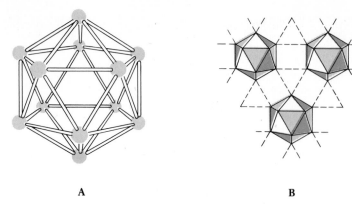

A   B

Figure 23.6
(a) The repeating unit in several allotropes of boron. It consists of 12 boron atoms arranged at the corners of a regular icosahedron. (b) The structure of alpha-rhombohedral boron. Icosahedra of boron atoms are linked by chemical bonds to form a solid.

Group IIIA, gallium, indium, and thallium, are relatively rare and are obtained as by-products from the processing of other metals. For example, gallium is obtained from bauxite during aluminum production.

Properties of the Elements

Boron has a number of allotropic forms. Three of these are well characterized, and all consist of repeating units of twelve boron atoms arranged in regular icosahedra (20-sided figures, as shown in Figure 23.6). These icosahedra of boron atoms are covalently bonded to one another to form covalent network solids; the allotropes differ in the exact arrangement of these icosahedra. The allotropes are semiconductors, and the most stable one is a hard, black solid.■ It has a high melting point (2300°C), as expected for a covalent network solid. Table 23.10 lists properties of boron and other Group IIIA elements. Except at very high temperatures, crystalline boron is inert to chemical attack.

■ The three well-characterized allotropes of boron are alpha-rhombohedral, beta-rhombohedral, and tetragonal boron. The alpha-rhombohedral allotrope is red; the others are black solids. Beta-rhombohedral boron is the most thermodynamically stable.

Table 23.10
Properties of Group IIIA Elements

Property	Boron	Aluminum	Gallium	Indium	Thallium
Electron configuration	$[He]2s^22p^1$	$[Ne]3s^23p^1$	$[Ar]3d^{10}4s^24p^1$	$[Kr]4d^{10}5s^25p^1$	$[Xe]4f^{14}5d^{10}6s^26p^1$
Melting point, °C	2300	660	30	157	304
Boiling point, °C	2550	2467	2403	2080	1457
Density, g/cm^3	2.34	2.70	5.90	7.30	11.85
Ionization energy (first), kJ/mol	801	578	579	558	589
(second), kJ/mol	2427	1817	1979	1821	1971
(third), kJ/mol	3660	2745	2963	2704	2878
(fourth), kJ/mol	25025	11577	6200	5200	4890
Electronegativity (Pauling scale)	2.0	1.5	1.6	1.7	1.8
Standard potential (volts), $M^{3+} + 3e^- \longrightarrow M$	—	−1.66	−0.56	−0.34	0.72
Atomic radius, Å	0.88	1.43	1.22	1.62	1.71
Ionic radius, Å	0.20	0.50	0.62	0.81	0.95

The other Group IIIA elements are metals. Aluminum has a moderately high melting point (660°C); the others melt at lower temperatures. However, there is no simple pattern as we go down the column of elements. Gallium melts easily when held in the hand, but indium and thallium have higher melting points. Except for thallium, the metals form an adherent oxide coating that gives them some protection against further oxidation. The elements form the +3 metal oxides, such as Al_2O_3. Thallium also forms the +1 metal oxide, Tl_2O. Similarly, aluminum, gallium, and indium combine directly with the halogens to give the metal(III) halides. Thallium forms both thallium(I) halides and thallium(III) halides (except for thallium(III) iodide, which is unknown)■

As expected for active metals, Group IIIA elements react with acids to release hydrogen:

$$2Al(s) + 6[H^+(aq) + Cl^-(aq)] \longrightarrow 2[Al^{3+}(aq) + 3Cl^-(aq)] + 3H_2(g)$$

(Nitric acid, however, renders aluminum unreactive, or *passive*, because a thick, adhering oxide layer forms.) Aluminum and gallium also dissolve in strong base to form the hydroxo anions and hydrogen:

$$2Al(s) + 2[Na^+(aq) + OH^-(aq)] + 6H_2O(l) \longrightarrow$$
$$2[Na^+(aq) + Al(OH)_4^-(aq)] + 3H_2(g)$$
sodium aluminate

$$2Ga(s) + 2[Na^+(aq) + OH^-(aq)] + 6H_2O(l) \longrightarrow$$
$$2[Na^+(aq) + Ga(OH)_4^-(aq)] + 3H_2(g)$$
sodium gallate(III)

■ The iodide ion presumably reduces thallium(III) to thallium(I). A compound TlI_3 is known, but it consists of $Tl^+I_3^-$ and not $Tl^{3+}(I^-)_3$. It is therefore thallium(I) triiodide.

Preparation of the Elements

Aluminum is the only element in Group IIIA produced in large quantity. It is made from aluminum oxide, Al_2O_3, which is obtained from bauxite. In the **Hall-Héroult process,** the oxide is dissolved in a fused, or melted, fluoride mixture (primarily cryolite, Na_3AlF_6, and aluminum fluoride, AlF_3) at about 950°C■ The fused mixture is then electrolyzed (Figure 23.7). Although the electrolysis is complicated and not completely understood, the final result can be represented by the following simplified electrode reactions:

$$4Al^{3+} + 12e^- \longrightarrow 4Al(l) \qquad \text{(cathode)}$$
$$\underline{6O^{2-} + 3C(s) \longrightarrow 3CO_2(g) + 12e^-} \qquad \text{(anode)}$$
$$2[\underbrace{2Al^{3+} + 3O^{2-}}_{Al_2O_3}] + 3C(s) \xrightarrow{\text{electrolysis}} 4Al(l) + 3CO_2(g) \qquad \text{(overall)}$$

■ Cryolite, Na_3AlF_6, is an ore of aluminum found in Greenland. For use in aluminum production, it is now usually prepared from aluminum hydroxide obtained from bauxite:

$$12HF + 2Al(OH)_3 + 6NaOH \longrightarrow$$
$$2Na_3AlF_6 + 12H_2O$$

The carbon anodes are made from carbonized petroleum and must be continuously replaced.

One difficulty of the Hall-Héroult process is controlling environmental fluoride pollution. This difficulty is by-passed in a new process developed by Alcoa (Aluminum Corporation of America). In the Alcoa process, aluminum oxide and carbon are heated in a stream of chlorine gas to obtain aluminum

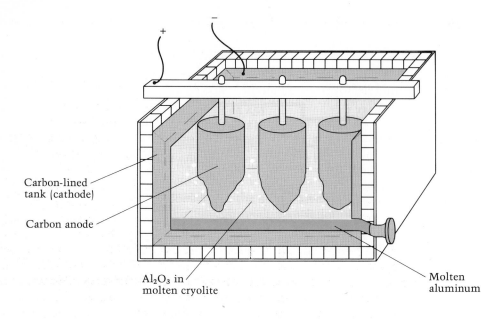

Figure 23.7
Hall-Héroult cell for the production of aluminum by the electrolysis of Al_2O_3 in molten cryolite.

chloride, which sublimes:

$$2Al_2O_3(s) + 3C(s) + 6Cl_2(g) \xrightarrow{\Delta} 4AlCl_3(g) + 3CO_2(g)$$

Solid aluminum chloride is recovered, then fused and electrolyzed to give aluminum and chlorine gas. The chlorine is recycled to obtain more of the chloride from the oxide.

Boron may be prepared by reducing boric oxide, B_2O_3, with magnesium:

$$B_2O_3(l) + 3Mg(l) \xrightarrow{\Delta} 2B(s) + 3MgO(s)$$

After the magnesium oxide has been leached out with hydrochloric acid, the product is a brown, amorphous powder of rather impure boron. Pure boron can be prepared by reduction of boron trichloride vapor with hydrogen on a hot tungsten wire:

$$2BCl_3(g) + 3H_2(g) \xrightarrow[1500°C]{W} 2B(s) + 6HCl(g)$$

Lustrous, black crystals are obtained.

Uses of the Elements and Their Compounds

Commercially, the most important Group IIIA element is aluminum, which is made in large quantities for alloys. Although the pure metal is soft and corrodes easily, the addition of small quantities of other metals, such as copper, magnesium, and manganese, gives hard, corrosion-resistant alloys. Some aluminum is used to produce other metals. Chromium is obtained by the **Goldschmidt process,** in which the metal oxide is reduced in an exothermic reaction with powdered aluminum:

$$Cr_2O_3(s) + 2Al(l) \longrightarrow Al_2O_3(l) + 2Cr(l); \Delta H = -536 \text{ kJ}$$

Table 23.11
Uses of Boron and
Aluminum Compounds

Compound	Use
$Na_2B_4O_7 \cdot 10H_2O$ (borax)	Source of boron and its compounds
	Detergent formulations
$B(OH)_3$	Preparation of boric oxide
	Antiseptic and preservative
B_2O_3	Fiber glass and borosilicate glass
	Enamels and glazes
Al_2O_3	Source of aluminum and its compounds
	Abrasive
	Refractory bricks and furnace linings
	Synthetic sapphires and rubies
$Al_2(SO_4)_3 \cdot 18H_2O$	Making of paper
	Water purification
	Mordant
$AlCl_3$	Preparation of aluminum
	Catalyst in organic reactions
$AlCl_3 \cdot 6H_2O$	Antiperspirant

A similar reaction produces iron for welding from a mixture of powdered aluminum and iron(III) oxide. Once the mixture (called *thermite*) is ignited, the reaction is self-sustaining and gives a spectacular incandescent shower. The reaction is also the basis of incendiary bombs.

The other Group IIIA elements have few commercial uses. Boron has been used to make filaments to reinforce plastic and metal parts for special applications. Gallium and indium have been used to make semiconductors for solid-state electronics.

Compounds of both boron and aluminum are important (see Table 23.11). Boron resembles silicon in forming a large number of compounds and polymeric materials containing bonds between the element and oxygen.■ Boron and oxygen atoms are often arranged in rings and more complicated structures involving trigonal planar BO_3 units and tetrahedral BO_4 units. Figure 23.8 shows the structure of the anion in the mineral borax, also known as sodium tetraborate decahydrate. Note the alternating trigonal planar and tetrahedral boron atoms in the structure. The chemical formula of borax is usually written $Na_2B_4O_7 \cdot 10H_2O$, but the compound consists of an anion whose formula is better written as $B_4O_5(OH)_4^{2-}$. Thus, borax is $Na_2[B_4O_5(OH)_4] \cdot 8H_2O$.

■ This characteristic is indicative of the diagonal relationship between boron and silicon.

Borax is used in soaps and detergents. But much of the borax produced is converted to boric acid, then to boric oxide. Boric acid, $B(OH)_3$, is prepared by adding sulfuric acid to borax solutions:

$$[2Na^+(aq) + B_4O_5(OH)_4^{2-}(aq)] + [2H^+(aq) + SO_4^{2-}(aq)] + 3H_2O(l) \longrightarrow$$
$$[2Na^+(aq) + SO_4^{2-}(aq)] + 4B(OH)_3(s)$$

borax   boric acid

The boric acid crystallizes from the solution. Although the formula is often written H_3BO_3, the acid is monoprotic and functions as a Lewis acid by

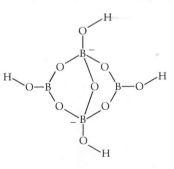

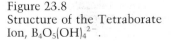

Figure 23.8
Structure of the Tetraborate Ion, $B_4O_5(OH)_4^{2-}$.

accepting an electron pair of OH^- (from water):

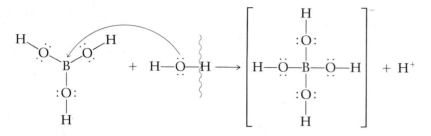

Boric acid is toxic to bacteria and fungi and is used as an antiseptic and as a preservative for wood and leather.

Most of the boric acid produced is converted to boric oxide, by heating:

$$2B(OH)_3(s) \xrightarrow{\Delta} B_2O_3(s) + 3H_2O(g)$$

Large amounts of boric oxide are used to make fiber glass and borosilicate glass, which has a very small coefficient of thermal expansion (Pyrex is a brand name). Boric oxide is also used in porcelain enamels and ceramic glazes.

Boron forms a series of hydrides called **boranes,** which are of interest because of their unique electronic structure. The simplest member is diborane, B_2H_6, a gas that is spontaneously flammable in moist air. Its structure may be written

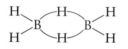

The end B—H bonds are the normal covalent sort. Each interior B—H—B bond, however, consists of a single electron pair involving three nuclei. To get a more detailed picture of the bonding, let us assume that the bonding orbitals on the boron atoms are sp^3 hybrids. Two of the sp^3 hybrid orbitals on each boron atom are involved in bonding to the end hydrogen atoms. The other sp^3 hybrid orbitals are used in bonding to the "bridge" hydrogens through the formation of **three-center bonds.** Each such bond, containing two electrons, is obtained by overlapping of an sp^3 hybrid orbital from each boron with the $1s$ of hydrogen, giving a three-center molecular orbital (see Figure 23.9). Note that the plane containing the end hydrogens must be perpendicular to the plane containing the bridge hydrogens.

The most important compound of aluminum is alumina, or aluminum oxide, Al_2O_3. It is prepared by heating aluminum hydroxide, which is obtained from bauxite:

$$2Al(OH)_3(s) \xrightarrow{\Delta} Al_2O_3(s) + 3H_2O(g)$$

Prepared at low temperature (550°C), it is a porous or powdery, white solid. It is used as a carrier for catalysts and for other purposes, though most alumina is used in the production of aluminum metal. When aluminum oxide is fused at high temperature (2045°C), it forms corundum and can be used as an abrasive or refractory. When it is fused with certain metal impurities, synthetic sapphires and rubies are obtained. Thus, synthetic rubies

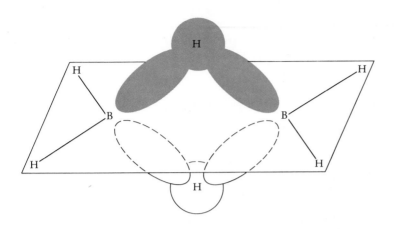

Figure 23.9
Bonding in diborane, B_2H_6.
The three-center B—H—B
bonds (one shown in color)
are perpendicular to the
plane of the rest of the
molecule.

contain about 2.5% chromium oxide, Cr_2O_3. They are used primarily in watches and instruments as "jewel" bearings.

Aluminum sulfate octadecahydrate, $Al_2(SO_4)_3 \cdot 18H_2O$, is the commonest soluble salt of aluminum. It is prepared by dissolving bauxite in sulfuric acid. The salt is acidic in water solution. In water, aluminum ion forms a strong hydration complex, $Al(H_2O)_6^{3+}$, and this ion in turn hydrolyzes:

$$Al(H_2O)_6^{3+}(aq) + H_2O(l) \rightleftharpoons Al(H_2O)_5OH^{2+}(aq) + H_3O^+(aq)$$

Aluminum sulfate is used in large amounts by the paper industry during the *sizing* of paper. In this process, substances such as clay are added to the cellulose fibers to give a less-porous material with a smooth finish. Colloidal clay is coagulated onto the paper fibers by aluminum ion. ■ Aluminum sulfate is also used to treat waste water in paper pulping, as well as to purify municipal water. When a base is added to an aqueous solution of an aluminum salt, a gelatinous precipitate of aluminum hydroxide forms:

$$Al^{3+}(aq) + 3OH^-(aq) \longrightarrow Al(OH)_3(s)$$

- Coagulation of colloids by multiply charged ions was discussed in Section 12.4.

The aluminum ion coagulates colloidal suspensions, which are then adsorbed on the aluminum hydroxide precipitate and can be filtered off. Aluminum hydroxide is often precipitated onto the fibers in fabrics where it adsorbs certain dyes. It is said to act as a *mordant* (fixative).

Aluminum hydroxide dissolves in acid to form the Al^{3+} ion, and in excess base to form the aluminate ion, $Al(OH)_4^-$:

$$Al(OH)_3(s) + 3H^+(aq) \longrightarrow Al^{3+}(aq) + 3H_2O(l)$$
$$Al(OH)_3(s) + OH^-(aq) \longrightarrow Al(OH)_4^-(aq)$$

Thus, aluminum hydroxide is amphoteric.

Aluminum chloride hexahydrate, $AlCl_3 \cdot 6H_2O$, can be prepared by dissolving alumina in hydrochloric acid. It is used as an antiperspirant and disinfectant. The anhydrous salt, $AlCl_3$, cannot be prepared by heating the hydrate, because it decomposes to give HCl. Anhydrous aluminum chloride is obtained by reacting scrap aluminum with chlorine gas:

$$2Al(s) + 3Cl_2(g) \longrightarrow 2AlCl_3(s)$$

It is also produced by the Alcoa process for aluminum, which was mentioned earlier. The solid is believed to be ionic, but when the substance is heated at near its melting point, the molecular substance Al_2Cl_6 is formed. Aluminum chloride sublimes at its melting point (180°C), and Al_2Cl_6 molecules are present in the vapor. The structure of Al_2Cl_6 has chlorine "bridges" as in solid $BeCl_2$:

At higher temperatures, the Al_2Cl_6 molecules in the vapor dissociate to the trigonal planar $AlCl_3$ molecules. Anhydrous aluminum chloride is used as a catalyst, for example, in the preparation of ethylbenzene. Ethylbenzene is used to prepare styrene, the starting material for polystyrene plastic and foam (Styrofoam).

Example 23.6

A solution is either $NaCl(aq)$, $MgCl_2(aq)$, or $AlCl_3(aq)$. Devise a chemical test to differentiate these solutions. You may use only one other substance.

Solution

If $NaOH(s)$ is slowly added to $AlCl_3(aq)$, a white precipitate of $Al(OH)_3$ forms. This later dissolves as more NaOH is added to give the soluble aluminate, $Na[Al(OH)_4]$. If NaOH is added to a solution of $MgCl_2$, a white precipitate of $Mg(OH)_2$ forms. This does not dissolve when more base is added. Sodium chloride solution does not give a precipitate with NaOH.

Exercise 23.6

The following white solids have been placed in separate unlabeled beakers so that you do not know which is which: KOH; $Al_2(SO_4)_3 \cdot 18H_2O$; $BaCl_2 \cdot 2H_2O$. Devise a series of chemical tests, using only water in addition to the substances in these beakers, to establish the identity of each solid.

(See Problems 23.73 and 23.74.)

23.5 Metallurgy

With the exception of boron, the elements we have studied in this chapter are metals. Many of them have commercial importance. In fact, the world production of aluminum exceeds that of all metals but iron. **Metallurgy** is the scientific study of the production of metals and metal products from natural sources. Natural sources are ores or, as in the case of magnesium, a natural water. In the present section, we wish to look at the basic steps in metallurgy, which are

1. **Preliminary treatment.** Usually, the natural source must first be concentrated in the metallic element. The result may be a relatively pure compound of the metal. Often, this compound must then be converted by chemical reaction to one that can be reduced to the metal.
2. **Reduction.** A compound of the metal is reduced to the free element by electrolysis or chemical reduction.
3. **Refining.** The metal obtained from the reduction may need to be purified through the process of **metal refining.**

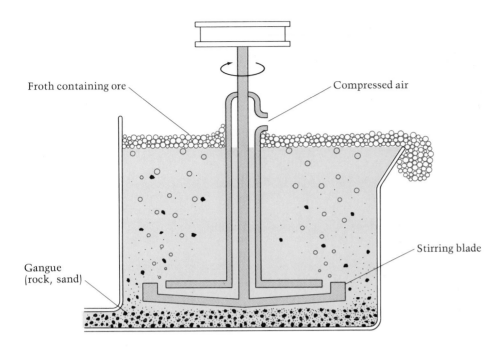

Figure 23.10
Flotation process for the concentrating of certain ores. The ore attaches to bubbles of air and is carried off in the froth, whereas the gangue settles to the bottom of the tank.

Froth containing ore

Compressed air

Stirring blade

Gangue
(rock, sand)

Preliminary Treatment

Most metals are obtained from ores that contain varying amounts of impurities along with the mineral of the metal. These impurities in an ore are called **gangue**. Ores may be concentrated in the metallic element by separating off the gangue, using physical or chemical methods. Panning for gold is a simple physical separation. It depends on the difference in density of gold and sand. Water easily flushes sand and dirt aside, leaving grains of gold at the bottom of the pan. **Flotation** is a physical separation process that depends on differences in wettability of the mineral and the gangue. The ore is crushed to a fine powder, then mixed into a tank of water (usually containing wetting agents) where a fast stream of air forms a froth (see Figure 23.10). Particles that are less easily wetted by water adhere to the bubbles of air and are swept out of the tank. Sulfide ores, such as those of copper, lead, and zinc, are concentrated this way. The gangue is wetted by the water and sinks to the bottom of the tank. The sulfide mineral is carried off in the froth.

The **Bayer process** for obtaining purified aluminum oxide from bauxite is an example of a chemical method of concentrating an ore. It depends on the fact that aluminum hydroxide is amphoteric. Bauxite is an ore containing aluminum hydroxide, $Al(OH)_3$, or the oxide hydroxide, $AlO(OH)$. When bauxite is mixed with hot, aqueous sodium hydroxide solution, the aluminum minerals dissolve to give the aluminate ion, $Al(OH)_4^-$:

$$Al(OH)_3(s) + OH^-(aq) \longrightarrow Al(OH)_4^-(aq)$$
$$AlO(OH)(s) + OH^-(aq) + H_2O(l) \longrightarrow Al(OH)_4^-(aq)$$

Silicate and iron oxide impurities remain undissolved and are filtered off. As the hot solution of sodium aluminate cools, aluminum hydroxide precipitates. In practice, the solution is seeded with aluminum hydroxide to start the precipitation of $Al(OH)_3$.

$$Al(OH)_4{}^-(aq) \longrightarrow Al(OH)_3(s) + OH^-(aq)$$

Aluminum hydroxide can also be precipitated by acidifying the solution.■
The precipitate is filtered off and heated to convert it to a white powder of
anhydrous aluminum oxide:

$$2Al(OH)_3(s) \xrightarrow{\Delta} Al_2O_3(s) + 3H_2O(g)$$

■ Gallium can be separated
from bauxite at this stage of the
process. Gallate ion is more stable than aluminate ion and
tends to stay in solution,
whereas aluminate ion reverts
to the hydroxide on cooling or
with mild acidification with CO_2.
Gallium hydroxide precipitates
only after further acidification.

Once an ore is concentrated, it may be necessary to convert the mineral
to a compound suitable for reduction. Sulfide ores, for example, are usually
heated in air to obtain the oxide. The process is called **roasting.** Thus, zinc
ore (ZnS) is converted to zinc oxide by roasting in air:

$$2ZnS(s) + 3O_2(g) \longrightarrow 2ZnO(s) + 2SO_2(g)$$

The roasting is exothermic and once started does not require additional
heating.

Reduction

The free metal is obtained by reduction of a compound, using either electrolysis or a chemical reducing agent. In the commercial production of zinc, zinc
oxide is reduced by heating it with some form of carbon, such as anthracite
coal or coke.

$$ZnO(s) + C(s) \xrightarrow{\Delta} Zn(g) + CO(g)$$

Zinc metal vapor leaves the reaction chamber and condenses to the solid.
Hydrogen and active metals, such as sodium, magnesium, and aluminum,
are also used as reducing agents when carbon is unsuitable. Tungsten, for
example, combines chemically with carbon. To produce the pure metal,
hydrogen is used instead:

$$WO_3(s) + 3H_2(g) \xrightarrow{\Delta} W(s) + 3H_2O(g)$$

Several examples of the production of a metal using an active metal as
reducing agent were given in earlier sections.

Refining of a Metal

Often the metal obtained from a reduction process contains impurities and
must be purified or refined before it can be used. Zinc, for example, is
contaminated by lead, cadmium, and iron. It is refined by fractional distillation. Some metals, such as copper, nickel, and aluminum, are refined
electrolytically. In the **Hoopes process** for the refining of aluminum, impure
aluminum forms the anode and pure aluminum forms the cathode. A
Hoopes cell contains three liquid layers (Figure 23.11). The bottom layer is
molten, impure aluminum; the middle is a fused salt layer containing aluminum fluoride; and the top layer is pure aluminum. At the anode (bottom
layer), aluminum passes into solution as aluminum ions, Al^{3+}, and at the
cathode (top layer), these ions are reduced to the pure metal. In operation,
impure, molten metal is added to the bottom of the cell and pure aluminum
is drawn off the top.

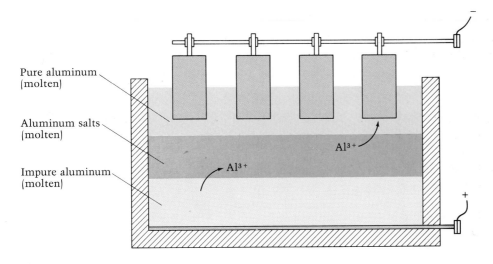

Figure 23.11
Hoopes cell for the electrolytic refining of aluminum.

Pure aluminum (molten)

Aluminum salts (molten)

Impure aluminum (molten)

Al³⁺

Al³⁺

Exercise 23.7

(a) Write equations for the metallurgy of magnesium that illustrate the concentration of magnesium ion from sea water and the conversion of the resulting compound to one that can be electrolytically reduced.
(b) Write the equation for the electrolytic reduction of this compound to give metallic magnesium.

(See Problems 23.75 and 23.76.)

A Checklist for Review

Important Terms

amphoteric (23.1)
semiconductors (23.1)
diagonal relationship (23.1)
mineral (23.2)
Downs cell (23.2)
Solvay process (23.2)

ore (23.3)
Hall-Héroult process (23.4)
Goldschmidt process (23.4)
boranes (23.4)
three-center bonds (23.4)
metallurgy (23.5)

metal refining (23.5)
gangue (23.5)
flotation (23.5)
Bayer process (23.5)
roasting (23.5)
Hoopes process (23.5)

Summary of Facts and Concepts

Before examining the properties of the main-group elements, we can make several *general observations*. First, the metallic characteristics of the elements decrease in going across a period and increase in going down a column. Also, the oxidation states of these elements vary predictably. Finally, the elements of the second row differ somewhat from the other elements in the same column; in fact, lithium, beryllium, and boron are similar in many ways to the elements located diagonally to them (*diagonal relationships*).

Group IA elements (alkali metals) occur as ions in the aluminosilicate rocks of the earth's crust. Commercial sources of sodium are halite (NaCl) and sea water; sources of potassium are sylvite (KCl) and carnallite (KCl·MgCl₂·6H₂O). The alkali metals are soft and chemically very reactive. All react vigorously with water, for example. Sodium and lithium are prepared by electrolysis of the molten chlorides. Other Group IA elements are prepared by reduction of the molten chlorides with sodium. Sodium is the most important alkali metal and is used to produce tetraethyllead, a gasoline anti-knock additive. Important compounds of sodium are sodium hydroxide and sodium carbonate. Sodium hydroxide, obtained by electrolysis of aqueous NaCl, is used to make paper and to extract aluminum oxide from bauxite. Sodium carbonate is used to make glass. Potassium chloride is important as a fertilizer.

Group IIA elements (alkaline earth metals) also occur as ions in aluminosilicate rocks. The main commercial source of magnesium ion is sea water, and calcium compounds are prepared from seashells and limestone. The most active alkaline earth metals are calcium and the metals below it; they react easily with water. Magnesium is less reactive with water but dissolves readily in acids. Beryllium reacts with both acids and bases. Magnesium is prepared by electrolysis of the chlorides; the other Group IIA elements are usually prepared by chemical reduction. Magnesium is an important metal used in lightweight alloys. Calcium oxide, obtained by heating calcium carbonate, is produced in large quantities for the manufacture of steel, for water treatment, and for various chemical processes.

Of the Group IIIA elements, aluminum is most abundant (the third most abundant element on earth). Boron is much less abundant, but it occurs in accessible minerals, such as borax ($Na_2B_4O_7 \cdot 10H_2O$). Elemental boron is a semiconductor, but is nonmetallic in its chemical properties. Aluminum is a reactive metal, dissolving in both acids and bases. The metal is prepared in large quantity for alloys by the electrolytic reduction of aluminum oxide (*Hall-Héroult process*), obtained from the ore bauxite. Boron compounds are used to make detergents (borax) and glass (boric oxide). *Boranes* are a class of compounds, including diborane (B_2H_6), that are of interest because of their unique bonding; they contain *three-center bonds* (B—H—B). The most important aluminum compound is the sulfate, $Al_2(SO_4)_3 \cdot 18H_2O$; it is used by the paper industry and for water purification. Aluminum chloride is an ionic compound that vaporizes as Al_2Cl_6 molecules, which have a bridge chlorine structure.

The last section discusses the basic steps in *metallurgy*. These steps are *preliminary treatment* of the ore or other natural source (concentration and conversion to a reducible compound), *reduction* to the metal, and finally *metal refining*, or purification. In the production of aluminum, pure aluminum oxide is prepared from bauxite by treating it with hot sodium hydroxide solution (*Bayer process*). The aluminum minerals dissolve, leaving impurities to be filtered off. Aluminum hydroxide precipitates from the cool solution and is converted to the oxide by heating. The oxide is reduced to the metal (*Hall-Héroult process*) and refined electrolytically (*Hoopes process*). In a Hoopes cell, impure aluminum forms the anode, with pure aluminum as the cathode. During electrolysis, aluminum leaves the anode as the ion and deposits as the pure metal at the cathode.

Operational Skills

The problem-solving skills used in this chapter were discussed in previous chapters.

Review Questions

23.1 Give four characteristics of a metal.

23.2 Describe how the metallic characteristics of the elements vary in going through the periodic table.

23.3 Consider the Group VIA elements. For each element, state the rule that applies for predicting the most common oxidation state.

23.4 Without looking at Table 23.2, choose the oxide from the following pairs that you would expect to be more acidic. Explain your choices. (a) As_4O_6, As_2O_5 (b) Tl_2O, Tl_2O_3 (c) Sb_2O_5, Sb_4O_6 (d) SO_2, SO_3 (e) SnO, SnO_2

23.5 What are some reasons for the difference in behavior of a second-row element from the remaining elements in the same column?

23.6 Explain what is meant by a diagonal relationship between elements. Illustrate for some pair of elements.

23.7 Give a commercial source (ore or other natural source) of each of the following elements: Na, K, Mg, Ca, B, Al.

23.8 Francium was discovered as a radioactive-decay product of actinium-227. Write the equation for the formation of francium from actinium-227 by alpha emission.

23.9 Write balanced equations for the reactions of lithium with oxygen, with water, and with nitrogen. Do the same for sodium. Compare. (If no reaction occurs, indicate this by writing NR.)

23.10 Ethanol, C_2H_5OH, reacts with sodium metal because the hydrogen atom attached to oxygen is slightly acidic. Write a balanced equation for the reaction of sodium with ethanol to give the salt sodium ethylate, $NaOC_2H_5$.

23.11 (a) Write electrode reactions for the electrolysis of fused lithium chloride. (b) Do the same for fused lithium hydroxide. The hydroxide ion is oxidized to oxygen and water.

23.12 Write the equation for the preparation of rubidium by the reduction of rubidium chloride with calcium. Be sure to show the phase for each substance. How is the reaction forced in the direction of the product (Rb)?

23.13 What is the principal use of sodium metal? Of potassium?

23.14 How is sodium hydroxide manufactured? What is another important product in this process?

23.15 How is lithium hydroxide produced from lithium carbonate?

23.16 One of the alkaline earth elements has only radioactive isotopes. What is the name of this element?

23.17 Write equations for the reactions of calcium metal with (a) water, (b) oxygen, (c) hydrogen, (d) nitrogen.

23.18 Starting from sea water, limestone, and hydrochloric acid, write equations for the preparation of magnesium metal.

23.19 Strontium can be produced by chemical reduction of strontium oxide. Write a chemical equation showing how this can be done.

23.20 Calcium carbonate is used in some antacid preparations to neutralize HCl in the stomach. Write the equation for the antacid reaction.

23.21 Write the equation that represents the setting of mortar.

23.22 Barium sulfate is given orally to obtain x-ray photographs of the stomach. Barium carbonate is a water-insoluble compound. Why could not the carbonate be used in place of the sulfate?

23.23 Write equations in which (a) $Be(OH)_2$ reacts with $HCl(aq)$ and (b) $Be(OH)_2$ reacts with KOH. Compare the behavior of $Be(OH)_2$ in these reactions with that of $Mg(OH)_2$ with the same substances.

23.24 What is the principal chemical constituent in ruby?

23.25 Aluminum reacts with both acids and bases. Write balanced equations illustrating this.

23.26 Nitric acid may be shipped in aluminum drums. Explain how this is possible even though aluminum is an active metal.

23.27 Aluminum can be produced by electrolysis. Write the electrode reactions for two different methods.

23.28 Describe a method for preparing pure boron from boron trichloride.

23.29 Aluminum oxide has a large negative enthalpy of formation. How is this useful for welding with thermite?

23.30 How is boric acid obtained from borax? Write the chemical equation for the reaction.

23.31 Explain the acidity of boric acid. How does its acidity differ from that of an acid like HCN?

23.32 Describe the bonding in tetraborane, B_4H_{10}:

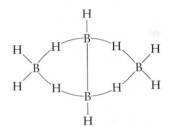

23.33 Aluminum nitrate dissolves in water to give acidic solutions. Explain why this is so.

23.34 Compare the structures of solid $BeCl_2$ and the molecule Al_2Cl_6. How are they similar? How are they different?

23.35 What are the basic steps in the production of a pure metal from a natural source? Illustrate each step with the production of aluminum.

23.36 How does flotation separate an ore mineral from impurities?

23.37 What is the purpose of the roasting of zinc ore?

23.38 Carbon is often used as a cheap reducing agent, but is sometimes unsuitable. Give an example where another reducing agent is preferred.

Problems

General Observations About the Main-Group Elements

23.39 Decide whether each of the following main-group elements is a metal, a metalloid, or a nonmetal. List all of the evidence that points to this conclusion. (a) Element X is a nonlustrous, yellow solid that burns in air producing an oxide XO_2. The oxide dissolves in water to give a solution that turns blue litmus red. When NaOH is added to this solution, the ion XO_3^{2-} is formed. (b) Element X is a lustrous solid that conducts electricity. The oxide reacts with HCl(aq), forming a solution of the salt XCl_3. It reacts with NaOH(aq) to form a solution of the salt $NaX(OH)_4$.

23.41 Identify the more metallic element in each of the following pairs: (a) Na or Mg, (b) Na or Be, (c) Ga or Cs. Explain your answers.

23.43 Lead forms the oxides PbO and PbO_2. One of these oxides is a yellow solid that melts at 888°C, and the other one is a dark brown powder that decomposes at 290°C. Identify the formula of each oxide with its properties. Explain.

23.45 After studying Sections 23.2 to 23.4, make a list of ways in which the second-row elements Li, Be, and B are similar to other members of the same group. Then make a list of ways in which they differ.

23.40 For each of the following main-group elements, decide whether it is a metal, a metalloid, or a nonmetal. List all of the evidence that points to this conclusion. (a) Element X is a lustrous, black solid that is a poor conductor of electricity at room temperature, but becomes a moderately good conductor at elevated temperatures. The oxide of the element forms an acid H_3XO_3. (b) Element X is a soft, lustrous solid that reacts with HCl(aq), forming a solution of the salt XCl_2. The element is a good conductor of electricity at room temperature.

23.42 Identify the more metallic element in each of the following pairs: (a) K or Ga, (b) K or Ca, (c) Rb or Al. Explain your answers.

23.44 Antimony has two fluorides. One is a colorless, oily liquid that freezes at 7°C. The other is a colorless solid that melts at 292°C. What is the likely formula of each of these compounds? Explain.

23.46 After studying Sections 23.3 and 23.4, make a list of the properties that are similar for beryllium and aluminum.

The Group IA Elements

23.47 Francium was discovered as a minor decay product of actinium-227. The major decay reaction is by beta emission to thorium. Write the nuclear equation for the decay of actinium-227 by beta emission.

23.49 Titanium metal is produced from its ores ilmenite ($FeTiO_3$) or rutile (TiO_2) by chlorination to give titanium tetrachloride. For example,

$$3TiO_2(s) + 4C(s) + 6Cl_2(g) \longrightarrow$$
$$3TiCl_4(g) + 2CO(g) + 2CO_2(g)$$

The titanium tetrachloride vapor is reduced by passing it into liquid sodium or magnesium. Write the balanced equation for the reduction of $TiCl_4$ with sodium.

23.51 Potassium hydroxide may be prepared by the aqueous reaction of potassium sulfate with barium hydroxide (a strong, soluble base). Write the balanced equation for this reaction, including phases for the species. Why does the reaction go in this direction?

23.48 Francium-223 is a radioactive alkali metal that decays by beta emission. Write the nuclear equation for this decay process.

23.50 Liquid ammonia has many chemical properties similar to water. Sodium metal, for example, reacts with liquid ammonia, producing sodium amide ($NaNH_2$) and hydrogen. (The reaction is slow, however, unless a catalyst such as Fe^{3+} is present.) Write the balanced equation for this reaction.

23.52 Pure sodium hydrogen carbonate (baking soda) is prepared by dissolving carbon dioxide gas in a saturated solution of sodium carbonate. Sodium hydrogen carbonate precipitates from the solution. Write the balanced equation, indicating phases for each species.

23.53 Caustic soda, NaOH, has been manufactured from sodium carbonate in a manner similar to lithium hydroxide. Write balanced equations (in three steps) for the preparation of NaOH starting from lime ($Ca(OH)_2$), salt (NaCl), carbon dioxide, and ammonia.

23.54 Sodium cyanide, NaCN, is used in electroplating. It is made from hydrogen cyanide, HCN, which is prepared from ammonia, air, and natural gas (CH_4):

$$2NH_3(g) + 3O_2(g) + 2CH_4(g) \xrightarrow[1030°C]{Pt} 2HCN(g) + 6H_2O(g)$$

Write balanced equations for the preparation of NaCN from salt (NaCl), ammonia, air, and natural gas.

The Group IIA Elements

23.55 Radium is found in uranium mineral. It results from the alpha decay of thorium-230, which is the decay product of uranium-234. Write the nuclear equation representing the decay of thorium-230 to radium.

23.56 A short-lived isotope of radium decays by alpha emission to radon-219. Write the nuclear equation for this decay process.

23.57 Describe a chemical test to distinguish between beryllium and magnesium metals.

23.58 Barium hydroxide and sodium hydroxide are strong bases. What simple test could you use to tell a solution of one from the other?

23.59 Devise a chemical method for separating a solution of $BeCl_2$, $MgCl_2$, and $BaCl_2$ into three solutions or compounds each containing only one of the elements Be, Mg, and Ba.

23.60 Describe a chemical method for separating a solution of NaCl, $MgCl_2$, and $SrCl_2$ into three solutions or compounds each containing only one of the elements Na, Mg, and Sr.

23.61 How could you prepare beryllium metal from beryllium fluoride, magnesium hydroxide, and hydrochloric acid in three steps? Write the balanced equations.

23.62 How could you prepare calcium metal from gypsum ($CaSO_4 \cdot 2H_2O$), soda ash, and aluminum? Write the balanced equations for the preparation of calcium from gypsum (in three steps).

23.63 When magnesium metal burns in air and the resulting ash is treated with water, the odor of ammonia can be detected. Write the balanced equation for the formation of the compound during combustion that subsequently gives the ammonia. Write the equation representing the reaction of the compound with water.

23.64 Finely divided magnesium burns in air. Neither water nor carbon dioxide can be used to extinguish magnesium fires because in both cases magnesium reacts to produce the oxide. Water is reduced to hydrogen, and carbon dioxide is reduced to carbon. Write balanced equations for these reactions.

The Group IIIA Elements

23.65 Write the balanced equation for the reaction that occurs when thermite (Al and Fe_2O_3) is ignited.

23.66 Manganese metal can be obtained from Mn_3O_4 by the Goldschmidt process. Write the balanced equation for this.

23.67 Hans Christian Oersted, a Danish physicist and chemist, first prepared aluminum by the reduction of aluminum chloride with potassium (actually potassium amalgam, which is a potassium–mercury alloy). He obtained the aluminum chloride by reaction of alumina (Al_2O_3) and charcoal with chlorine. Write the balanced equation for this preparation of aluminum, starting with alumina.

23.68 Boron was discovered by the French chemists Joseph Louis Gay-Lussac and Louis Jacques Thenard. They isolated boron by heating boric acid with potassium. During this heating, boric acid is first transformed to boric oxide, which is then reduced to boron. Write balanced equations for the reactions.

23.69 What are the reactions that occur during the neutralization of boric acid by aqueous ammonia, $NH_3(aq)$? Write the equations in terms of electron-dot formulas.

23.70 Hydrofluoric acid, HF, reacts with boron trifluoride to give fluoroboric acid, HBF_4. Write the equation for this reaction in terms of electron-dot formulas.

23.71 Baking powder contains sodium (or potassium) hydrogen carbonate with an acidic substance. When water is added, carbon dioxide is released. One kind of baking powder contains $NaHCO_3$ and sodium aluminum sulfate, $NaAl(SO_4)_2 \cdot 12H_2O$. Write the net ionic equation for the reaction that occurs in water solution.

23.73 The following solid substances are in separate, but unlabeled, test tubes: $AlCl_3 \cdot 6H_2O$, $BaCl_2 \cdot 2H_2O$, $BeSO_4 \cdot 4H_2O$, and KOH. Describe how they can be identified by chemical tests, using only these substances and water.

23.72 When aluminum sulfate is dissolved in water, it produces an acidic solution. Suppose the pH of this solution is raised by the dropwise addition of aqueous sodium hydroxide. (a) Describe what you would observe as the pH continues to rise. (b) Write balanced equations for any reactions that occur.

23.74 Unlabeled test tubes contain solid $AlCl_3 \cdot 6H_2O$ in one, $Ba(OH)_2 \cdot 8H_2O$ in another, and $MgSO_4 \cdot 7H_2O$ in another. How could you find out what is in each test tube, using chemical tests that involve only the compounds in these test tubes? Water is available as a solvent.

Metallurgy

23.75 Galena, PbS, is an important lead ore. (a) Write the balanced equation for the roasting of galena. Assume that the mineral is converted to lead(II) oxide. (b) Write a chemical equation for the reduction of lead(II) oxide by carbon to give metallic lead. (c) Describe what happens at each electrode in the electrolytic refining of lead. Electrodes, one of impure lead and the other of pure lead, are suspended in a bath of fluorosilicic acid, H_2SiF_6.

23.76 Nickel ore, NiS, is roasted to nickel(II) oxide, then converted with hydrogen to the metal. Nickel is refined by the Mond process, in which nickel combines with carbon monoxide to form volatile nickel carbonyl, $Ni(CO)_4$. When nickel carbonyl is heated to 200°C, it decomposes to free nickel and carbon monoxide gas. Write balanced equations for these reactions. Label each reaction according to whether it involves preliminary treatment, reduction, or refining.

Additional Problems

23.77 Aluminum chloride vaporizes as Al_2Cl_6 molecules. At still higher temperatures, these molecules dissociate to $AlCl_3$ molecules. What is the geometry of Cl atoms about aluminum in the $AlCl_3$ molecule? What is the geometry of Cl atoms about aluminum in the Al_2Cl_6 molecule?

23.78 Beryllium chloride has a polymeric bridge Cl structure in the solid phase, but consists of $BeCl_2$ molecules in the gas phase. What is the geometry of Cl atoms about beryllium in solid $BeCl_2$? What is the geometry of Cl atoms about beryllium in the $BeCl_2$ molecule?

23.79 Natural sources of potassium contain 0.00118% of the radioactive isotope $^{40}_{19}K$. This isotope has a half-life of 1.28×10^9 years. What is the value of the radioactive decay constant for this isotope? What fraction of a sample of this isotope would remain after 100 million years $(1.00 \times 10^8$ years)?

23.80 Natural rubidium contains 27.85% of the radioactive isotope $^{87}_{37}Rb$, which decays by beta emission to $^{87}_{38}Sr$. This decay of rubidium to strontium has been used to date mineralogical samples. It is assumed that when a rubidium mineral first crystallized, the ratio of strontium atoms to rubidium atoms was zero. What would be

the value of this ratio 5×10^9 years (approximately the age of the earth) after such a mineral crystallized? The half-life of $^{87}_{37}Rb$ is 5×10^{11} years.

23.81 Calculate the enthalpy change, $\Delta H°$, for the production of one mole of Fe by the following reaction at 25°C (the value of $\Delta H°$ changes rather slowly with temperature, so that the value at 25°C is an approximate value for higher temperatures).

$$\tfrac{1}{2}Fe_2O_3(s) + Al(s) \longrightarrow Fe(s) + \tfrac{1}{2}Al_2O_3(s)$$

Is the reaction exothermic or endothermic? See Appendix C for data.

23.82 Calculate the enthalpy change, $\Delta H°$, for the production of one mole of Ca by the following reaction at 25°C:

$$CaO(s) + \tfrac{2}{3}Al(s) \longrightarrow Ca(s) + \tfrac{1}{3}Al_2O_3(s)$$

Is the reaction exothermic or endothermic? See Appendix C for data.

23.83 Estimate the temperature at which strontium carbonate decomposes to strontium oxide and CO_2 at 1 atm.

$$SrCO_3(s) \longrightarrow SrO(s) + CO_2(g)$$

To do this, calculate $\Delta H°$ and $\Delta S°$ at 25°C. Assume that these values do not change appreciably with temperature. Then calculate the temperature at which $\Delta G° = 0$. See Appendix C for data.

23.84 Estimate the temperature at which barium carbonate decomposes to barium oxide and CO_2 at 1 atm.

$$BaCO_3(s) \longrightarrow BaO(s) + CO_2(g)$$

See Problem 23.83.

23.85 Calculate $E°$ for the disproportionation of $In^+(aq)$:

$$3In^+(aq) \rightleftharpoons 2In(s) + In^{3+}(aq)$$

(*Disproportionation* is a reaction in which a species undergoes both oxidation and reduction.) Use the following standard potentials:

$$In^+(aq) + e^- \rightleftharpoons In(s) \qquad E° = -0.21 \text{ V}$$
$$In^{3+}(aq) + 2e^- \rightleftharpoons In^+(aq) \qquad E° = -0.40 \text{ V}$$

From $E°$, calculate $\Delta G°$ for the reaction (in kilojoules). Does the disproportionation occur spontaneously?

23.86 Calculate $E°$ for the disproportionation of $Tl^+(aq)$

$$3Tl^+(aq) \rightleftharpoons 2Tl(s) + Tl^{3+}(aq)$$

from the following standard potentials:

$$Tl^+(aq) + e^- \rightleftharpoons Tl(s) \qquad E° = -0.34 \text{ V}$$
$$Tl^{3+}(aq) + 2e^- \rightleftharpoons Tl^+(aq) \qquad E° = 1.25 \text{ V}$$

From $E°$, calculate $\Delta G°$ for the reaction (in kilojoules). Does the disproportionation occur spontaneously? See Problem 23.85.

**23.87 Metallic sodium has a body-centered cubic lattice. Assume that sodium atoms in the metal consist of spheres packed as close as allowed by this type of lattice. From the known density of the metal (see Table 23.5),

calculate the radius of a sodium atom in angstroms. (Note that spheres packed body-centered cubic would be touching one another along the diagonal running from opposite corners of a unit cell.)

**23.88 Metallic lithium has a body-centered cubic lattice. Assume that lithium atoms in the metal consist of spheres packed as close as allowed by this type of lattice. From the known density of the metal (see Table 23.5), calculate the radius of a lithium atom in angstroms. (See Problem 23.87.)

23.89 Sea water contains 1272 g of magnesium ion per metric ton (1 metric ton = 1 megagram). What is the minimum amount of slaked lime, $Ca(OH)_2$, that must be added to 1.00 metric ton of sea water to precipitate the magnesium ion?

23.90 Bauxite contains approximately 52% Al_2O_3. How many grams of aluminum can be obtained from 1.00 metric ton (1.00×10^6 g) of bauxite containing 52.0% Al_2O_3?

23.91 Lithium hydroxide has been used in spaceships to absorb carbon dioxide exhaled by astronauts. Assuming that the product is lithium carbonate, what mass of lithium hydroxide is needed to absorb the carbon dioxide from 1.00 L of air containing 30.0 mmHg partial pressure of CO_2 at 25°C?

23.92 Potassium chlorate, $KClO_3$, is used in fireworks and explosives. It can be prepared by bubbling chlorine into hot potassium hydroxide:

$$6KOH(aq) + 3Cl_2(g) \longrightarrow$$
$$KClO_3(s) + 5KCl(aq) + 3H_2O(l)$$

How many grams of $KClO_3$ can be obtained from 156 L Cl_2 whose pressure is 784 mmHg at 25°C?

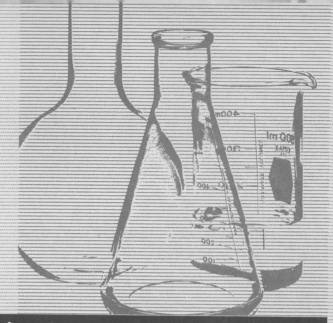

24. The Main-Group Elements: Groups IVA to VIIA

24.1 Group IVA: The Carbon Family Properties of the Elements/ Preparation and Uses of the Elements/ *Aside: Zone Melting and Zone Refining*/ Important Compounds

24.2 Group VA: The Phosphorus Family Properties of the Elements/ Preparation and Uses of the Elements/ Important Compounds

24.3 Group VIA: The Sulfur Family Properties of the Elements/ Preparation and Uses of the Elements/ Important Compounds

24.4 Group VIIA: The Halogens Properties of the Elements/ Preparation and Uses of the Elements/ Important Compounds

n studying the elements of Groups IA to IIIA, we saw the emergence of the two major trends in metallic and nonmetallic character. First, in going across a period from left to right, the elements become less metallic and more nonmetallic. This trend is especially pronounced in the second period. Thus, lithium, in Group IA, is an active metal that reacts in most cases to give ionic compounds; the oxide is basic. Beryllium, in Group IIA, is a metal that shows a tendency toward covalent bonding; the oxide is amphoteric. Boron, in Group IIIA, is a metalloid with definite nonmetallic characteristics. The element has a covalent network structure, and the oxide B_2O_3 is acidic.

The second major trend we saw in the elements of Groups IA to IIIA was an increase in metallic character in going down a column. Thus, the first element in Group IIIA, boron, is a metalloid, but the other elements are metals. The oxides progress from acidic (B_2O_3) to amphoteric (Al_2O_3, Ga_2O_3) to weakly basic (In_2O_3, Tl_2O_3).

As we study the elements of Groups IVA to VIIA, we see these major trends with even greater clarity. And where the elements of Groups IA to IIIA were metals (except for boron), the elements of these groups are primarily nonmetals. Only those elements at the bottom of the first columns are metals, and even they show some nonmetallic characteristics.

Chapter Overview

In this chapter, we continue to look at the main-group elements. Two of these, nitrogen (Group VA) and oxygen (Group VIA) were discussed earlier. Each has properties that are rather different from those of the remaining elements in its group, as we might expect from the discussion of the second-period elements in Section 23.1. As you read this chapter, you will want to pay particular attention to the trends in metallic–nonmetallic character, especially how these trends are revealed by the acid–base behavior of the oxides and by any aqueous ions formed. Also note the stability of the different oxidation states for the elements in a group. (You may wish to reread Section 23.1 before continuing.)

24.1 Group IVA: The Carbon Family

The elements of Group IVA show in a more striking fashion than the earlier groups the normal trend to greater metallic character in progressing down a column. The first element in the group, carbon, is distinctly nonmetallic. Silicon and germanium are metalloids; tin and lead are metals. All Group IVA elements have compounds in the +4 oxidation state, though these compounds are essentially covalent. Carbon and silicon also have compounds in the −4 oxidation state. In some of the metal carbides, such as Al_4C_3, the negative ion C^{4-} is present. Most of the compounds in the −4 state are covalent. The tetravalent state (that is, the +4 or −4 oxidation state) is expected from the ns^2np^2 valence configuration of the atoms, and covalent bonding is usually tetrahedral, involving sp^3 hybridization. However, all of these elements but carbon can also form six bonds by utilizing sp^3d^2 hybridization. Some examples are the complex ions SiF_6^{2-}, $GeCl_6^{2-}$, $Sn(OH)_6^{2-}$, and $Pb(OH)_6^{2-}$. The divalent (+2) state is important for tin and lead, and +2 cations of these elements are found in compounds and aqueous solutions.

Silicon is the most abundant element of Group IVA and occurs in silica (SiO_2) and silicate minerals. The next most abundant element is carbon, which is found in all living matter, as well as in the fossil fuels. The largest sources of carbon, however, are the carbonate minerals limestone and dolomite. Carbon also occurs as CO_2 in the atmosphere and as graphite and diamond, which are allotropic forms of the free element. The other Group IVA elements are much less abundant. Germanium is rather rare and is found in several minerals, such as argyrodite, a mixed silver–germanium sulfide ore ($4Ag_2S \cdot GeS_2$). The most important commercial source of this element is flue dust from zinc ore roasting and from coal combustion. The principal tin ore is cassiterite (SnO_2). Lead is obtained mainly from the mineral galena (PbS).

Properties of the Elements

The crystalline allotropes of carbon—graphite and diamond—are covalent network solids. Graphite is a soft, black solid with a slippery feel. It has a luster and conducts electricity. The solid has a layer structure. ■ Within a layer, each carbon atom is covalently bonded to three others so that a hexagonal or "chicken-wire" pattern is formed. The bonding involves delocalized π electrons and these are responsible for the conductivity of the solid. Two layers are attracted to each other by London forces, so that the layers easily slide over one another. This accounts for the softness and slippery feel of graphite. Amorphous forms of carbon, such as charcoal and carbon black, also have a layer structure like that of graphite. The difference lies in the stacking of layers. In graphite, the stacking is ordered; in amorphous carbon, the stacking is more or less disordered.

■ The structures of graphite and diamond were discussed in detail in Section 11.5.

Diamond is a colorless, transparent solid. ■ It is very hard (the hardest substance known) and an electrical insulator (nonconductor). The difference in properties from those of graphite is explained by the difference in structure. Diamond has a three-dimensional network structure. Each carbon atom is covalently bonded to four other atoms. The bonding is tetrahedral, involving sp^3 hybridization of the atoms. Diamond requires the breaking of many strong C—C bonds in order to move one portion of the crystal relative to another. Once these strong directional bonds are broken, they are unlikely to reform. Thus, diamond is both hard and brittle. As expected for a covalent network solid, the melting point of diamond is high (about 3550°C; see Table 24.1).

■ Impurities and crystal imperfections can give diamond various colors. A trace of boron, for example, gives a blue diamond.

The graphite-type structure is unique to carbon, but the diamond-type structure is not. ■ Silicon, germanium, and one allotrope of tin have a crystal structure like that of diamond. In contrast to diamond, however, these substances are semiconductors. They are slightly conducting at room temperature, but become good conductors at higher temperatures. Silicon is a hard, lustrous gray solid, melting at 1410°C. Germanium has a silvery gray metallic luster, but unlike most metals it shatters as easily as glass.

■ The $p\pi$ bonding in a graphite-type structure would be expected to become weaker as the atomic size increases and the orbital overlap decreases.

Tin exists in two allotropic forms. The common form, white tin, is metallic and malleable. It is a metallic conductor of electricity; that is, it conducts moderately well at room temperature, but becomes less conducting at elevated temperatures. White tin undergoes a transition at 13°C to gray tin, a

Property	Carbon	Silicon	Germanium	Tin	Lead
Electron configuration	$[He]2s^22p^2$	$[Ne]3s^23p^2$	$[Ar]3d^{10}4s^24p^2$	$[Kr]4d^{10}5s^25p^2$	$[Xe]4f^{14}5d^{10}6s^26p^2$
Melting point, °C	3550 (diamond)	1410	937	232	328
Boiling point, °C	4827	2355	2830	2260	1740
Density, g/cm^3	3.51 (diamond)	2.33	5.35	7.28 (white Sn)	11.3
Ionization energy (first), kJ/mol	1086	786	762	709	716
Electron affinity, kJ/mol	−122	−134	−120	−121	−110
Electronegativity (Pauling scale)	2.5	1.8	1.8	1.8	1.8
Standard potential (volts), $M^{2+} + 2e^- \rightleftharpoons M$	—	—	—	−0.14	−0.13
Covalent radius, Å	0.77	1.11	1.22	1.40	1.47
Ionic radius (for M^{2+}), Å	—	—	—	1.06	1.33

Table 24.1
Properties of Group IVA
Elements

brittle semiconductor. It is this allotrope that has the diamondlike structure.

$$\text{Gray tin} \xrightleftharpoons{13°C} \text{white tin}$$

The transition between these allotropes is normally slow, and white tin can be cooled below 13°C without changing to gray tin. However, a tin pipe, for example, kept for some time at below 13°C, will begin to crumble to gray tin. The crystallization begins at isolated points that form centers for further accelerated crystallization. Crumbly spots start to grow all over the pipe, an effect called tin disease or tin pest.

Lead occurs only in a metallic form with a cubic closest-packed structure. When the metal is freshly cut, it has a bright silvery luster with bluish cast. It soon acquires a dull gray appearance as an adherent oxide coating forms.

At room temperature, Group IVA elements are relatively unreactive, particularly if they are in massive crystalline form. They do all burn in air once that temperature has been raised. Carbon burns in an excess of oxygen with a flameless glow to give carbon dioxide. In an inadequate supply of oxygen, it burns to a mixture of CO and CO_2. Silicon, germanium, and tin all react with oxygen to yield the dioxides. Tin, for example, burns in air with a white flame, giving tin(IV) oxide, SnO_2. Lead, however, usually gives lead(II) oxide, PbO, unless the temperature is kept below 500°C. In this case, it forms the red oxide Pb_3O_4, with lead in +2 and +4 oxidation states.

Group IVA elements are relatively unreactive with water, although carbon and silicon do react at red heat with steam:

$$C(s) + H_2O(g) \xrightarrow{\Delta} CO(g) + H_2(g)$$
$$Si(s) + 2H_2O(g) \xrightarrow{\Delta} SiO_2(s) + 2H_2(g)$$

Group IVA elements, except for the metals tin and lead, are also unreactive with dilute acids. Even the metallic elements react only very slowly at room temperature, but upon heating they react more vigorously with acids to release hydrogen gas:

$$Sn(s) + 2[H^+(aq) + Cl^-(aq)] \longrightarrow [Sn^{2+}(aq) + 2Cl^-(aq)] + H_2(g)$$

$$Pb(s) + 2[H^+(aq) + Cl^-(aq)] \longrightarrow [Pb^{2+}(aq) + 2Cl^-(aq)] + H_2(g)$$

Tin and lead are also attacked by hot solutions of alkali metal hydroxides, giving hydrogen and the hydroxo ions, $Sn(OH)_3^-$ or $Pb(OH)_3^-$. ■ For example,

$$Sn(s) + [Na^+(aq) + OH^-(aq)] + 2H_2O(l) \longrightarrow$$
$$[Na^+(aq) + Sn(OH)_3^-(aq)] + H_2(g)$$

Except for diamond and silicon, Group IVA elements can be oxidized by concentrated nitric acid. Germanium and tin give the dioxides, for example,

$$3Ge(s) + 4HNO_3(aq) \longrightarrow 3GeO_2(s) + 4NO(g) + 2H_2O(l)$$

whereas lead gives Pb^{2+}:

$$3Pb(s) + 8HNO_3(aq) \longrightarrow 3[Pb^{2+}(aq) + 2NO_3^-(aq)] + 2NO(g) + 4H_2O(l)$$

Group IVA elements Si, Ge, Sn, and Pb combine with the halogens at elevated temperatures. The tetrahalide is the usual product, except for lead, where the lead(II) halide is usually formed. Here again is displayed the stability of the +2 oxidation state of lead.

■ Hydroxo ions such as $Sn(OH)_3^-$ result from the acidic character of the aqueous metal ion. We can write $Sn^{2+}(aq)$ as $Sn(H_2O)_6^{2+}(aq)$ to indicate explicitly those water molecules directly bonded to the metal atom. In basic solution, we have

$$Sn(H_2O)_6^{2+}(aq) + OH^-(aq) \rightleftharpoons Sn(H_2O)_5(OH)^+(aq) + H_2O(l)$$

Transfer of three protons gives the anion $Sn(H_2O)_3(OH)_3^-$, also written $Sn(OH)_3^-(aq)$.

Preparation and Uses of the Elements

Although carbon occurs in the free state as diamond and graphite and in amorphous form as coal, commercial quantities of the various allotropes of carbon are also manufactured. Carbon in the form of carbon black, for example, is a major chemical product. It is prepared by burning natural gas (CH_4) or petroleum hydrocarbons in a limited supply of air so that heat "cracks" the hydrocarbon.

$$CH_4(g) \xrightarrow{\Delta} C(s) + 2H_2$$
carbon black

Carbon black is used in the manufacture of rubber tires (to increase wear) and as a pigment in black printing inks. Coke is an amorphous carbon obtained by heating coal or petroleum in the absence of air. Petroleum coke is used wherever a pure form of carbon is needed.

Although natural graphite is essential for certain uses, such as the making of pencil "leads," synthetic graphite must be manufactured to fill the demands of the electrochemical industry for electrodes. Graphite is prepared by heating petroleum coke in an electric furnace to 3500°C, using sand and iron as catalysts. Amorphous carbon and graphite fibers, used for reinforcing high-temperature materials, are made by heating textile fibers to high temperature.

The positive free energy of formation of diamond at 1 atm shows that it is thermodynamically unstable and should change to graphite.

$$C(graphite) \longrightarrow C(diamond); \Delta G_f^\circ = +2.9 \text{ kJ/mol}$$

Fortunately, the change of diamond to graphite is imperceptibly slow at normal temperatures. If diamond is heated in an inert atmosphere to 1000°C, however, the change is rapid. Of course, the possibility of changing graphite to diamond is much more interesting. Since graphite is less dense than

Figure 24.1
Synthetic diamonds. These diamonds, manufactured by General Electric, are used in saw blades for cutting stone and concrete.

diamond, we might expect it to change to diamond at sufficiently high pressure.■ High temperature would be needed also so that the change would occur in a reasonable period of time. Although people have tried to synthesize diamond ever since these conditions for the transformation were understood, the first successful experiments were performed only in 1955 by scientists at the General Electric Company. The process, which produces cheap industrial diamonds for grinding and cutting tools, requires temperatures of about 3000°C and pressures of 100,000 atm (1×10^{10} Pa or 10 GPa), in addition to a transition-metal catalyst. Gem-quality diamonds have been made, but the cost is too high to compete with natural diamonds. Some synthetic diamonds are shown in Figure 24.1.

The production of other Group IVA elements involves at some stage the reduction of an oxide. Silicon, for example, is prepared by reducing quartz sand (SiO_2) with coke (C) in an electric furnace at 3000°C.

$$SiO_2(l) + 2C(s) \xrightarrow{\Delta} Si(l) + 2CO(g)$$

When used for metallurgical purposes, silicon is prepared as an alloy with iron, called ferrosilicon. This alloy is obtained by reducing iron(III) oxide along with sand.

Ultrapure silicon is required for solid-state semiconductor devices, such as transistors, solar cells, and microcomputer chips. To prepare it, impure silicon (from the reduction of sand) is heated with chlorine to give silicon tetrachloride, $SiCl_4$, a low-boiling liquid (b.p. 58°C) purified by fractional distillation.

$$Si(s) + 2Cl_2(g) \xrightarrow{\Delta} SiCl_4(l)$$

■ High pressure would be expected to shift the composition of the equilibrium

Graphite $\rightleftharpoons$ diamond

toward the solid that has the smaller volume, that is, to the more dense diamond phase, according to LeChatelier's principle.

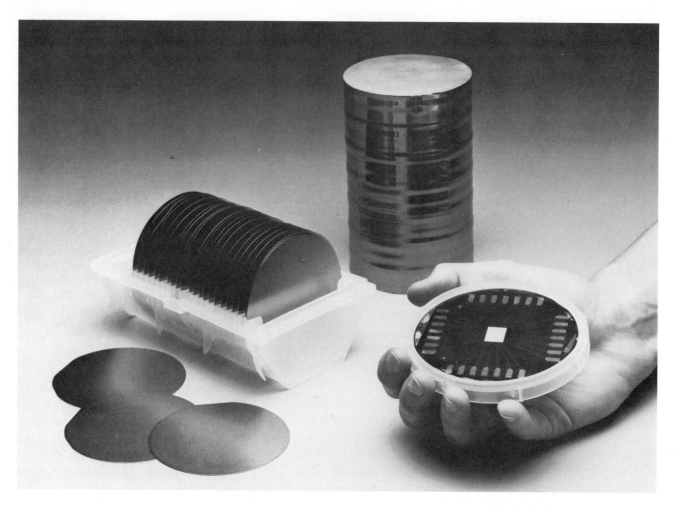

Figure 24.2
Ultrapure silicon rods and
wafers cut from rods. The
lower right shows an electrochemical display device
consisting of silver pads deposited on a silicon wafer.

Then, a mixture of silicon tetrachloride vapor and hydrogen is passed through a hot tube, where very pure silicon crystallizes on the surface of a pure silicon rod.

$$\text{SiCl}_4(g) \ + \ 2\text{H}_2(g) \ \xrightarrow{\Delta} \ \text{Si}(s) \ + \ 4\text{HCl}(g)$$

Silicon prepared this way has only $10^{-8}\%$ impurities (Figure 24.2).

Germanium is prepared from its ores or from flue dusts obtained in the production of zinc. The material is treated with concentrated hydrochloric acid to produce germanium tetrachloride, GeCl_4, which is distilled from the mixture. After purifying by fractional distillation, the germanium tetrachloride is reacted with water (hydrolyzed) to give the dioxide:

$$\text{GeCl}_4(l) \ + \ 2\text{H}_2\text{O}(l) \ \longrightarrow \ \text{GeO}_2(s) \ + \ 4\text{HCl}(aq)$$

Germanium dioxide is reduced to the free element by heating with hydrogen or carbon. The element is purified for semiconductor applications by **zone refining,** a process in which a solid rod of the substance is melted in a small, moving band or zone that carries the impurities out of the rod.■ (See Figure

■ For details, see the Aside on zone melting and refining.

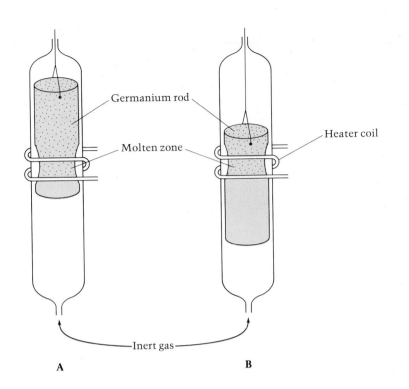

Figure 24.3
Zone refining. In the method shown here, a germanium rod is slowly lowered through a heater coil, which melts the rod. The impurities concentrate in the molten zone that moves upward. Surface tension holds the two parts of the rod together at the molten zone. After cooling, the previously molten zone at the end of the rod can be cut off, removing all the impurities.

Germanium rod

Molten zone

Heater coil

Inert gas

A

B

24.3.) In addition to its use in semiconductor devices, germanium is used to make infrared-transmitting optics.

Tin is obtained from purified tin(IV) oxide, SnO_2, by reducing it with carbon in a furnace:

$$SnO_2(s) + 2C(s) \xrightarrow{\Delta} Sn(l) + 2CO(g)$$

The metal is used in tin-plating of steel, as a protective coating. Some tin is also used for alloys such as bronze and solder.

Lead is produced by roasting galena, a lead(II) sulfide ore. In roasting, the ore is burned in air, where the sulfide is converted to the oxide:

$$2PbS(s) + 3O_2(g) \longrightarrow 2PbO(s) + 2SO_2(g)$$

Some free metal is also produced during the roasting:

$$PbS(s) + 2PbO(s) \longrightarrow 3Pb(s) + SO_2(g)$$

The fused mass from the roasting is broken up and fed into blast furnaces. Here the lead(II) oxide is reduced with carbon monoxide, produced by the partial oxidation of coke:

$$PbO(s) + CO(g) \longrightarrow Pb(l) + CO_2(g)$$

Large quantities of the metal are used to make electrodes for storage batteries. It is also used in alloys for cable coverings and for solder (lead with tin). Important compounds made from lead are tetraethyllead, $(C_2H_5)_4Pb$ (gasoline additive), litharge, PbO, and red lead, Pb_3O_4.

Aside: *Zone Melting and Zone Refining*

Zone melting is a technique in which a rod of some material is melted in a narrow band or zone that travels the length of the rod. One of the more important uses of zone melting is in the purification of certain substances. The process is called zone refining. Consider the zone refining of germanium. In this process, a germanium rod is melted at one end by moving either the germanium rod or the heater. Because a solution containing dissolved solute (impurities in the germanium solvent) normally has a lower melting point than pure solvent, impurities concentrate in the melted zone, which moves from one end of the rod to the other. The process can be repeated until most of the impurities have been removed.

Zone melting is also used to make single crystals of silicon for semiconductor applications. When a silicon rod is first produced, it consists of a tightly packed mass of small crystals, unsuitable for semiconductor devices. To form this into a single crystal, the rod is placed on top of a wafer of single-crystal silicon. This part of the rod is melted, and, as the melted zone is moved upward, the single crystal of silicon grows until it becomes the entire rod. See Figure 24.2.

Important Compounds

Carbon is unique in that it forms stable compounds containing long chains of atoms of the element covalently bonded to one another. **Catenation** is the ability of an atom to bond covalently to like atoms. Because the strengths of carbon–carbon bonds (C—C, C=C, C≡C) are comparable to those of carbon bonds with hydrogen and with oxygen, a wide variety of carbon compounds have been prepared. Over a million such compounds (called organic compounds) are known.

Other Group IVA elements also display limited catenation in the hydrides (binary compounds with hydrogen); Table 24.2 shows these. Silicon, for example, forms hydrides called **silanes,** which can contain chains of up to six single-bonded Si atoms:

hexasilane

Element	Hydride with Longest Chain of the Element
Carbon	Unlimited length
Silicon	Si_6H_{14}
Germanium	Ge_9H_{20}
Tin	Sn_2H_6
Lead	PbH_4

Table 24.2
Hydrides of the Group IVA Elements

Compound	Use	Table 24.3 Uses of Group IVA Compounds
CO	Fuel; reducing agent	
	Synthesis of methanol, CH_3OH	
CO_2	Refrigerant	
	Carbonation of beverages	
SiO_2	Source of silicon and its compounds	
	Abrasives	
	Glass	
SnO_2	Source of tin and its compounds	
PbO	Lead storage batteries	
	Ceramic glazes and glass	
PbO_2	Cathode in lead storage batteries	
Pb_3O_4	Pigment for painting structural steel	

Although multiple bonds are common in organic compounds, none were known in the other Group IVA elements until recently when a compound containing Si=Si bonded to organic groups was discovered.

Carbon Oxides The monoxides and dioxides for all Group IVA elements are known. Carbon dioxide is the usual product of combustion of organic compounds and elementary carbon when there is an excess of oxygen. However, an equilibrium exists among carbon, CO, and CO_2 that favors carbon monoxide at temperatures above about 700°C:

$$CO_2(g) + C(s) \rightleftharpoons 2CO(g)$$

For this reason, CO is always a product of the combustion of carbon unless excess oxygen is present. Commercially, carbon monoxide is prepared from natural gas (CH_4) or petroleum hydrocarbons. For example,

$$CH_4(g) + H_2O(g) \xrightarrow{\Delta} CO(g) + 3H_2(g)$$

The resulting mixture is used to prepare methanol, CH_3OH. (See Table 24.3 for a list of uses of Group IVA compounds.) Many other organic compounds could be made from carbon monoxide if supplies of petroleum, the present source of most organic compounds, become scarce. In that case, carbon monoxide could be made from any source of carbon by the water–gas reaction.

$$C(s) + H_2O(g) \xrightarrow{\Delta} CO(g) + H_2(g)$$

Although carbon monoxide is not soluble or reactive with water under normal conditions, it does react with hot, aqueous sodium hydroxide at high pressure to give sodium formate:

$$CO(g) + OH^-(aq) \xrightarrow{\Delta} \left(H-C \begin{smallmatrix} O \\ \\ O \end{smallmatrix} \right)^-$$

formate ion

Formic acid, HCOOH, can be dehydrated with concentrated sulfuric acid to give carbon monoxide.■ This is a convenient laboratory preparation for small quantities of CO. Thus, in a formal way, CO is the anhydride of formic acid, HCOOH.

■ Concentrated sulfuric acid has a strong attraction for water.

Carbon dioxide is the anhydride of unstable carbonic acid, H_2CO_3. The gas dissolves in water to give the equilibrium

$$CO_2(aq) + H_2O(l) \rightleftharpoons H_2CO_3(aq)$$

Although carbonic acid is unstable and cannot be isolated from an aqueous solution, many of its salts, the carbonates and hydrogen carbonates, are well known. Carbon dioxide is easily liquefied by pressure (about 5 atm), and the liquid is used in large quantities to refrigerate foods. Cooling is caused by vaporization of the liquid. Large quantities of the gas are also used to carbonate beverages.

Silicon Dioxide and Silicates The principal oxide of silicon is silica or silicon dioxide, SiO_2. It has a quite different structure from carbon dioxide, which is a linear molecule. Silica is a covalent network solid in which each Si atom is covalently bonded to four O atoms that are in turn bonded to other Si atoms (see Figure 24.4). Quartz is one of several different crystalline forms of silica and is a constituent of many rocks. Weathering of these rocks releases quartz particles, which are a major component of most sands. Quartz melts at 1610°C to give a viscous liquid that cools to form a glass (an amorphous solid), called silica glass.

Silica reacts with hot, aqueous alkali metal hydroxide or fused alkali metal carbonates to form soluble silicates. **Silicates** are compounds of silicon and oxygen, with various metals. They may be thought of as derived from silicic acid, H_4SiO_4 or $Si(OH)_4$, although this substance has never been isolated from solution. Solutions of sodium silicate appear to contain the ion

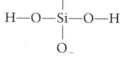

Figure 24.4
Structure of silica (SiO_2). Silica consists of SiO_4 tetrahedra linked by common oxygen atoms. A portion of a plane of atoms within silica is shown. Alternate Si atoms have bonds to O atoms pointing upward; the other Si atoms have bonds to O atoms pointing downward. These O atoms, in turn, have bonds to layers of Si atoms above and below this plane. Thus, silica has a three-dimensional network structure.

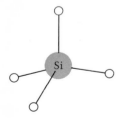

SiO₄ tetrahedron

Portion of a plane within silica

Si
O

An SiO₄ tetrahedron

but undoubtedly they contain ions with two or more silicon atoms as well. Such ions can form from the simple silicate ions by condensation reactions. In a **condensation reaction,** two molecules or ions are joined by the elimination of a small molecule such as H_2O:

$$
\begin{array}{ccc}
& O^- & & O^- \\
& | & & | \\
H-O-Si-O-H & + & H-O-Si-O-H \longrightarrow \\
& | & & | \\
& O_- & & O_-
\end{array}
$$

$$
\begin{array}{cc}
O^- & O^- \\
| & | \\
H-O-Si-O-Si-O-H + H_2O \\
| & | \\
O_- & O_-
\end{array}
$$

Ordinary glass, which is obtained by melting silica with lime (CaO) and sodium carbonate, is a polymeric silicate with a network of silicon–oxygen bonds. Silicon–oxygen chains and networks are a dominant feature of silicon chemistry.

An enormous variety of silicate minerals exists (Figure 24.5). The structure of each can be understood in terms of SiO_4 tetrahedra as basic units. A few minerals contain SiO_4^{4-} as a discrete ion. Zircon, $ZrSiO_4$, is an example. It has a very high melting point (2550°C), as would be expected from the high charge on the ions. Bonding between ions certainly has considerable covalent character.

Usually, silicate minerals consist of more complicated structures in which SiO_4 tetrahedra are linked through a common oxygen, for example:

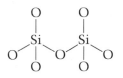

In this way, it is possible to form chain or ring anions or polymeric chains of indefinite length (as, for example, in the fibrous asbestos minerals). Silicate anion chains are linked to one another through cations. In the micas, there is a two-dimensional crosslinking of chains to form silicate sheets. These anionic sheets are held together through bonds to cations. This type of mineral cleaves easily along the silicate sheets. The structures of different silicate minerals are shown in Figure 24.6.

If SiO_4 tetrahedra are linked in a three-dimensional network, silica results. The **aluminosilicate minerals,** such as the feldspars, are three-dimensional networks in which some of the SiO_4 tetrahedra are replaced by AlO_4 tetrahedra. In replacing Si by Al, additional positive charge is needed and is usually supplied by cations of Groups IA and IIA elements.

Tin and Lead Compounds Tin(II) oxide, a black solid, is obtained by heating the white precipitate of $Sn(OH)_2$ formed by reacting Sn^{2+} ions with hydroxide ions:

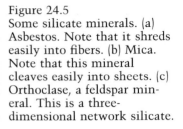

Figure 24.5
Some silicate minerals. (a) Asbestos. Note that it shreds easily into fibers. (b) Mica. Note that this mineral cleaves easily into sheets. (c) Orthoclase, a feldspar mineral. This is a three-dimensional network silicate.

 Represents the SiO_4 tetrahedron.

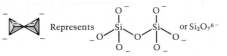 Represents 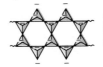 or $Si_2O_7{}^{6-}$

 Here SiO_4 tetrahedra are linked into a ring to give the $Si_6O_{18}{}^{12-}$ ion. Beryl, $Be_3Al_2Si_6O_{18}$, is the best-known mineral with this ion. Emerald is a green variety of beryl.

An infinite chain of SiO_4 tetrahedra, giving the empirical formula $SiO_3{}^{2-}$. The mineral diopside, $CaMg(SiO_3)_2$, is an example with this structure.

An infinite double chain of SiO_4 tetrahedra, giving the empirical formula $Si_4O_{11}{}^{6-}$. The asbestos mineral tremolite, $Ca_2Mg_5(OH)_4(Si_4O_{11})_2$, is an example.

Figure 24.6
Structure of some silicate minerals. A number of silicate minerals consist of finite silicate anions, such as $SiO_4{}^{4-}$ and $Si_2O_7{}^{6-}$. Beryl contains the cyclic anion $Si_6O_{18}{}^{12-}$. Other silicate minerals have anions with very long chains or double chains, as in the asbestos mineral tremolite. The chains are linked to one another by cations. In asbestos, this linkage between chains is weak, and the mineral frays easily into fibers.

$$Sn^{2+}(aq) + 2OH^-(aq) \longrightarrow Sn(OH)_2(s)$$

$$Sn(OH)_2(s) \xrightarrow{\Delta} SnO(s) + H_2O(g)$$

The oxide and hydroxide are amphoteric. They dissolve in acids to form Sn^{2+} and in base to form the stannite ion, $Sn(OH)_3{}^-$. For example,

$$Sn(OH)_2(s) + [Na^+(aq) + OH^-(aq)] \longrightarrow Na^+(aq) + Sn(OH)_3{}^-(aq)$$
$$\text{stannite ion}$$

Tin(IV) oxide, SnO_2, is also amphoteric. It dissolves in sulfuric acid to give a solution of tin(IV) sulfate, $Sn(SO_4)_2$, and in aqueous sodium hydroxide to give a solution of stannate ion, $Sn(OH)_6{}^{2-}$:

$$SnO_2(s) + 2[Na^+(aq) + OH^-(aq)] + 2H_2O(l) \longrightarrow 2Na^+(aq) + Sn(OH)_6{}^{2-}(aq)$$
$$\text{stannate ion}$$

The stannite ion is a strong reducing agent and is oxidized to the stannate ion, for example by H_2O_2.

Lead(II) oxide, PbO, is prepared commercially by exposing molten lead to air. This oxide, which is a reddish yellow solid at normal temperatures, is called litharge in commercial trade. Its largest use is in making lead storage batteries (a paste of litharge and sulfuric acid is packed into the electrode grids). It is also used in glasses and enamels. Lead(II) oxide and lead(II) hydroxide are amphoteric, as are the corresponding tin compounds. For example, normally insoluble lead(II) hydroxide dissolves in acids to give Pb^{2+} and in bases to give the plumbite ion, $Pb(OH)_3{}^-$. Lead(IV) oxide, PbO_2, can be prepared by oxidizing plumbite ion with hypochlorite ion, OCl^-:

$$Pb(OH)_3{}^-(aq) + OCl^-(aq) \longrightarrow PbO_2(s) + OH^-(aq) + H_2O(l) + Cl^-(aq)$$

It is a dark brown solid and a strong oxidizing agent in acid solution. It is formed in the cathodes of lead storage batteries during charging.■ Lead(IV) oxide reacts with strong base to form the plumbate ion, $Pb(OH)_6{}^{2-}$. The oxide PbO_2 is thermally unstable and decomposes on heating to PbO and Pb_3O_4. Trilead tetroxide, Pb_3O_4, also called red lead, is a bright red powder used as a pigment in a protective paint for steel. It contains both lead(II) and lead(IV).

■ The lead storage cell was described in Section 21.2.

Example 24.1

Sodium stannate, $Na_2Sn(OH)_6$, is a white, crystalline solid, solutions of which are used in the electrolytic plating of tin. Show by means of balanced equations how you could prepare sodium stannate from tin metal in two steps.

Solution

First prepare a solution of sodium stannite:

$$Sn(s) + NaOH(aq) + 2H_2O(l) \longrightarrow$$
$$NaSn(OH)_3(aq) + H_2(g)$$

Then oxidize the stannite ion to stannate ion, say with H_2O_2. Hydrogen peroxide is reduced to H_2O. Note that the reaction must be in basic solution, since the stannite ion would decompose to Sn^{2+} in acid solution. The balanced equation is

$$Sn(OH)_3^-(aq) + H_2O_2(aq) + OH^-(aq) \longrightarrow$$
$$Sn(OH)_6^{2-}(aq)$$

or $NaSn(OH)_3(aq) + H_2O_2(aq) + NaOH(aq) \longrightarrow$
$$Na_2Sn(OH)_6(aq)$$

Exercise 24.1

Use balanced equations to describe how lead(IV) oxide can be prepared from lead(II) oxide (in two steps).

(See Problems 24.45 and 24.46.)

Exercise 24.2

Solutions of sodium stannite, $NaSn(OH)_3$, are unstable and decompose to Sn and $Sn(OH)_6^{2-}$. Write the balanced equation for this reaction.

(See Problems 24.47 and 24.48.)

24.2 Group VA: The Phosphorus Family

Like the carbon family, Group VA elements show the distinct trend from nonmetallic to metallic. The first members, nitrogen and phosphorus, are definitely nonmetals; arsenic and antimony are metalloids; bismuth is a metal. Nitrogen, which we considered earlier, shows only slight resemblance to the other members. We see this, for example, in the formulas of the elements and compounds. Elementary nitrogen is N_2; white phosphorus is P_4. Similarly, the common +5 oxyacid of nitrogen is HNO_3; that for phosphorus is H_3PO_4.

Except for bismuth, Group VA elements have stable compounds in the +5 oxidation state. In the case of nitrogen, these compounds are oxidizing agents. Thus, nitric acid, HNO_3, is reduced to NO_2 (oxidation state +4), NO (+2), N_2 (0), and NH_3 (−3). The +5 state of phosphorus is quite stable, however, and phosphoric acid, H_3PO_4, is nonoxidizing. For the remaining elements, the +3 state becomes progressively more stable.

Phosphorus is the most abundant Group VA element and occurs in phosphate minerals, such as fluorapatite, $Ca_5(PO_3)_3F$—also written $3Ca_3(PO_4)_2 \cdot CaF_2$ to emphasize the presence of calcium phosphate. The other elements, except for nitrogen, are much less abundant and occur as oxide and sulfide ores. Bismuth also occurs as the free element.

Property	Nitrogen	Phosphorus	Arsenic	Antimony	Bismuth
Electron configuration	$[He]2s^22p^3$	$[Ne]3s^23p^3$	$[Ar]3d^{10}4s^24p^3$	$[Kr]4d^{10}5s^25p^3$	$[Xe]4f^{14}5d^{10}6s^26p^3$
Melting point, °C	−210	44 (white P)	613 (sublimes)	631	271
Boiling point, °C	−196	280		1750	1560
Density, g/cm³	1.25×10^{-3}	1.82	5.73	6.68	9.80
Ionization energy (first), kJ/mol	1402	1012	1521	834	703
Electron affinity, kJ/mol	≥0	−72	−77	−101	−110
Electronegativity (Pauling scale)	3.0	2.1	2.0	1.9	1.9
Covalent radius, Å	0.70	1.11	1.21	1.41	1.46
Ionic radius, Å	1.32 (N^{3-})	1.85 (P^{3-})	0.72 (As^{3+})	0.90 (Sb^{3+})	1.17 (Bi^{3+})

Table 24.4
Properties of Group VA
Elements

Properties of the Elements

Phosphorus has two common allotropes, white phosphorus and red phosphorus. White phosphorus is a waxy, white solid. It is soluble in nonpolar solvents, such as carbon disulfide. As indicated by its melting point (44°C; see Table 24.4), white phosphorus is a molecular solid, P_4. The phosphorus atoms are arranged at the corners of a regular tetrahedron so that each atom is single-bonded to the other three (Figure 24.7a). The P—P—P bond angle is 60° and thus is much smaller than the normal bond angle for p bonding (90°). This gives a weaker P—P bond than otherwise (because there is smaller overlap of the p orbitals) and accounts for the great reactivity seen in this phosphorus allotrope. Red phosphorus is rather unreactive. It is a polymeric substance whose structure may be that shown in Figure 24.7b.

Arsenic is normally a brittle, gray solid of metallic luster. Gray arsenic sublimes at 615°C. If the vapor is rapidly cooled, a yellow nonmetallic solid crystallizes. Yellow arsenic is believed to be a molecular solid, As_4, analogous to white phosphorus. It is unstable at room temperature and reverts to gray arsenic. Antimony is a silvery, lustrous solid. A yellow nonmetallic form is known, but is stable only at very low temperatures. Bismuth is a white metal with pinkish tinge.

Figure 24.7
(a) Structure of the P_4 molecule. (b) Possible (hypothesized) structure of red phosphorus. The P_4 units are linked to give long chains.

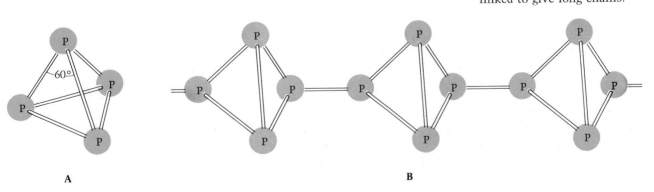

A   B

White phosphorus, as we noted earlier, is chemically very reactive. It ignites spontaneously in air, forming a dense white cloud of phosphorus oxides. In an excess of oxygen, the element burns to phosphorus(V) oxide, P_4O_{10}:

$$P_4(s) + 5O_2(g) \longrightarrow P_4O_{10}(s)$$

In a deficient supply of oxygen, phosphorus(III) oxide, P_4O_6, forms. Arsenic, antimony, and bismuth burn if heated in air. Arsenic forms arsenic(III) oxide, As_4O_6. Antimony forms antimony(III) oxide, Sb_4O_6, and diantimony tetroxide, Sb_2O_4, with antimony in $+3$ and $+5$ oxidation states. Bismuth forms bismuth(III) oxide, Bi_2O_3.

Phosphorus and the heavier elements (As, Sb, and Bi) react directly with the halogens. Phosphorus gives the pentahalides (PF_5, PCl_5, PBr_5, but not PI_5), as well as the trihalides (PF_3, PCl_3, PBr_3, and PI_3). The other elements give primarily the trihalides, although SbF_5, $SbCl_5$, and AsF_5 can form.

Preparation and Uses of the Elements

White phosphorus, a major industrial chemical, is prepared by heating phosphate rock (fluorapatite, $3Ca_3(PO_4)_2 \cdot CaF_2$) with coke (C) and sand (SiO_2) in an electric furnace. The reaction can be written

$$2Ca_3(PO_4)_2(s) + 6SiO_2(s) + 10C(s) \xrightarrow{>1500°C} 6CaSiO_3(l) + 10CO(g) + P_4(g)$$
$$\text{calcium silicate}$$

The gases from the furnace are cooled to condense phosphorus vapor to the liquid, which is stored under water until pumped into tank cars. Slag, consisting of calcium silicate and calcium fluoride (from fluorapatite), is periodically drained from the furnace. Most of the white phosphorus produced is used to manufacture phosphoric acid, H_3PO_4. For this, phosphorus is burned in excess air, and the oxide mist sprayed with water. Some white phosphorus is converted to red phosphorus, for use in making matches, by heating at 240°C in an inert atmosphere.

Arsenic may be obtained from various ores, such as the sulfide, As_4S_6, from which it is prepared by roasting in air, followed by reduction of the oxide with coke:

$$As_4S_6(s) + 9O_2(g) \longrightarrow As_4O_6(s) + 6SO_2(g)$$
$$As_4O_6(s) + 6C(s) \xrightarrow{\Delta} As_4(g) + 6CO(g)$$

Arsenic(III) oxide, present in flue gases from the roasting of copper ores, is also used as a source of arsenic. Antimony is obtained from stibnite, Sb_4S_6, by roasting to the oxide, followed by reduction with coke. Bismuth is obtained as a by-product in the electrolytic refining of copper. It is present in the mud that collects near the anode.■

■ Electrolytic refining was discussed in Section 23.5.

Arsenic, antimony, and bismuth are used in alloys. Arsenic is used to harden lead for lead shot, antimony is alloyed with lead for storage battery plates, and bismuth is mixed with tin and lead in low-melting alloys for automatic fire sprinklers and other uses.

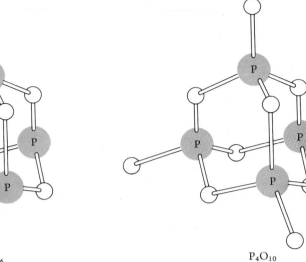

P_4O_6

P_4O_{10}

Figure 24.8
Structures of the phosphorus oxides. The phosphorus atoms in P_4O_6 have tetrahedral positions as do the atoms in P_4. However, the phosphorus atoms are not directly bonded to one another. Rather, an oxygen atom forms a P—O—P bridge between each pair of phosphorus atoms. The P_4O_{10} molecule is similar, except that an additional oxygen atom is bonded to each phosphorus (in the position where the lone pair would be in P_4O_6).

Important Compounds

Phosphorus Oxides and Oxyacids The phosphorus oxides P_4O_6 and P_4O_{10} have related structures (Figure 24.8). Phosphorus(III) oxide, P_4O_6, has a tetrahedron of phosphorus atoms with oxygen atoms between each pair of phosphorus atoms to give P—O—P bonds. Phosphorus(V) oxide, P_4O_{10}, is similar, but has an additional oxygen atom bonded to each phosphorus atom. These phosphorus–oxygen bonds are much shorter than the other P—O bonds (1.39 Å versus 1.62 Å); hence, they have considerable double-bond character.

Phosphorus(III) oxide is a low-melting solid (m.p., 23°C) and is the anhydride of phosphor*ous* acid, H_3PO_3, whose structure is ■

■ Note the suffix *-ous* indicating the lower oxidation state (+3) of phosphorus.

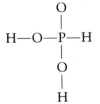

Note that one of the hydrogen atoms is directly attached to phosphorus. It is not acidic, so that phosphorous acid is diprotic and forms salts with anions $H_2PO_3^-$ and HPO_3^{2-}.

Phosphorus(V) oxide is a white solid that sublimes at 360°C. It combines vigorously with water, making it useful in the laboratory as a drying agent. The substance is produced in large quantities by burning white phosphorus in excess air. Most of the P_4O_{10} produced is not isolated, but is reacted immediately with excess water to obtain orthophosphoric acid, H_3PO_4:

$$P_4O_{10}(s) + 6H_2O(l) \longrightarrow 4H_3PO_4(aq)$$

Orthophosphoric acid (often called simply phosphoric acid) is a colorless solid, melting at 42°C when pure. It is usually sold as an aqueous solution. Orthophosphoric acid is triprotic and has the structure

$$H-O-\underset{\underset{H}{\overset{\displaystyle O}{|}}}{\overset{\overset{\displaystyle O}{\|}}{P}}-O-H$$

The possible sodium salts of phosphoric acid are sodium dihydrogen phosphate (NaH_2PO_4), disodium hydrogen phosphate (Na_2HPO_4), and trisodium phosphate (Na_3PO_4). Phosphoric acid produced from phosphorus, as described in the previous paragraph, is relatively pure and is used primarily by the detergent and food and beverage industries. An impure acid, produced in large quantities for the manufacture of fertilizers, is obtained by treating phosphate rock (fluorapatite) with sulfuric acid. We can write this reaction as■

$$Ca_3(PO_4)_2(s) + 3H_2SO_4(aq) \longrightarrow 3CaSO_4(s) + 2H_3PO_4(aq)$$

When phosphate rock is treated with orthophosphoric acid, it dissolves to give a solution of calcium dihydrogen phosphate, $Ca(H_2PO_4)_2$:

$$Ca_3(PO_4)_2(s) + 4H_3PO_4(aq) \longrightarrow 3Ca(H_2PO_4)_2(aq)$$

By this process, insoluble phosphate rock is converted to a soluble phosphate fertilizer. In the trade, this fertilizer is called triple superphosphate. Uses of some phosphorus compounds are listed in Table 24.5.

Phosphorus(V) oxide is not only the anhydride of orthophosphoric acid, but is also the anhydride of two series of acids obtained by condensation reactions from orthophosphoric acid.■ For example,

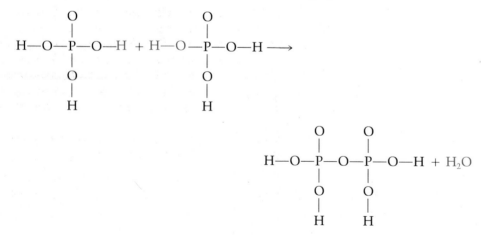

One series consists of the linear **polyphosphoric acids**, with the general formula $H_{n+2}P_nO_{3n+1}$, which are formed from chains of P—O bonds, such as

■ Since phosphate rock contains CaF_2, hydrofluoric acid, HF, can also form. When this reacts with silica (SiO_2) in the phosphate rock, it forms hexafluorosilicic acid, H_2SiF_6. This by-product is used to make AlF_3 and synthetic cryolite for the aluminum industry.

■ Condensation reactions were described in Section 24.1.

Table 24.5
Uses of Some Phosphorus
Compounds

Compound	Use
$Ca(H_2PO_4)_2 \cdot H_2O$	Phosphate fertilizer
	Baking powder
$CaHPO_4 \cdot 2H_2O$	Animal feed additive
	Toothpowder
H_3PO_4	Manufacture of phosphate fertilizers
PCl_3	Manufacture of $POCl_3$
	Manufacture of pesticides
$POCl_3$	Manufacture of plasticizers (substances that keep plastics pliable)
	Manufacture of flame retardants
P_4S_{10}	Manufacture of lubricant additives and pesticides
$Na_5P_3O_{10}$	Detergent builder

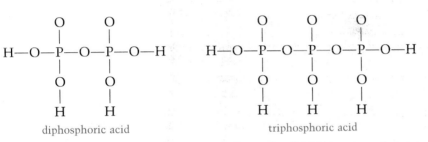

diphosphoric acid   triphosphoric acid

In the other series, with the general formula $(HPO_3)_n$, are the **meta-phosphoric acids.** Figure 24.9 shows the structure of a cyclic, or ring, metaphosphoric acid. If a linear polyphosphoric acid chain is very long, the formula becomes $(HPO_3)_n$, with n very large. This is called a poly-metaphosphoric acid.

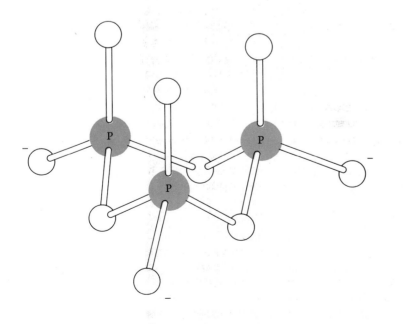

Figure 24.9
The structure of tri-metaphosphate ion, $P_3O_9^{3-}$, a cyclic metaphosphate.

The polyphosphates and metaphosphates are used as *detergent builders*, which complex with metal ions that would otherwise precipitate with dirt onto clothes. They also buffer the pH above 7 where a detergent operates most effectively. Sodium triphosphate ($Na_5P_3O_{10}$), one of the most common detergent builders, is manufactured by adding sufficient sodium carbonate to phosphoric acid to give a solution of the salts NaH_2PO_4 and Na_2HPO_4. When this solution is sprayed into a hot kiln, the phosphate ions condense to sodium triphosphate. The use of phosphate builders in detergents has been criticized on the ground that they contribute to the overfertilization of algae in lakes. Such lakes become oxygen-deficient from decomposing algae, and the fish die.

Arsenic, Antimony, and Bismuth Compounds Compounds of the elements As, Sb, and Bi are produced in small quantities. Arsenic(III) oxide is a white powder and slightly soluble in water, producing a solution of the very weak acid H_3AsO_3 (arsenious acid). Metal arsenites, with the ion AsO_3^{3-}, are known. Arsenic acid, H_3AsO_4, is obtained when arsenic(III) oxide is oxidized with nitric acid. Arsenates, such as Na_3AsO_4, are well-known compounds. Arsenic(V) oxide, As_2O_5, whose structure is unknown, is obtained by heating arsenic acid. Arsenic compounds are poisonous to many animals and have been used as insecticides.

Antimony(III) oxide, Sb_4O_6, is amphoteric and forms salts of the cation Sb^{3+}, such as $Sb_2(SO_4)_3$, as well as antimonites, salts of the anion SbO_2^-. Antimony(V) oxide, Sb_2O_5, whose structure is unknown, is prepared by oxidation of Sb_4O_6 with nitric acid. It is weakly acidic and reacts with strong base, such as KOH, to form the antimonate anion, $Sb(OH)_6^-$. Sodium antimonate, $Na[Sb(OH)_6]$, is one of the few insoluble sodium salts.

Bismuth(III) oxide, Bi_2O_3, which is an ionic oxide, is weakly basic and forms salts, such as $Bi(NO_3)_3$. A white, gelatinous precipitate of $Bi(OH)_3$ is obtained when NaOH is added to a solution of bismuth salt.

Example 24.2

With balanced equations, describe how you might prepare sodium antimonate, $Na[Sb(OH)_6]$, from antimony(III) oxide.

Solution

We can oxidize Sb_4O_6 to Sb_2O_5 with nitric acid. Antimony(III) oxide is amphoteric and first dissolves in nitric acid to form Sb^{3+}:

$$Sb_4O_6(s) + 12H^+(aq) \longrightarrow 4Sb^{3+}(aq) + 6H_2O(l)$$

The Sb^{3+} ion is oxidized to Sb_2O_5 by nitrate ion. If we assume the NO_3^- is reduced to NO, we get

$$6Sb^{3+}(aq) + 4NO_3^-(aq) + 7H_2O(l) \longrightarrow$$
$$3Sb_2O_5(s) + 14H^+(aq) + 4NO(g)$$

The +5 oxide can be dissolved in a strong base, such as KOH,

$$Sb_2O_5(s) + 2OH^-(aq) + 5H_2O(l) \longrightarrow 2Sb(OH)_6^-(aq)$$

and sodium antimonate is precipitated from the solution by adding a sodium salt:

$$Na^+(aq) + Sb(OH)_6^-(aq) \longrightarrow Na[Sb(OH)_6](s)$$

Exercise 24.3

With balanced equations, describe how you can prepare trisodium phosphate, Na_3PO_4, from white phosphorus.

(See Problems 24.55 and 24.56.)

Exercise 24.4

Arsenic compounds can be reduced to arsine, AsH_3, by zinc in acidic solution. This forms the basis of the Marsh test for arsenic. Arsine is a gas. When passed through a heated tube, it decomposes, leaving a black mirror of arsenic. Write the balanced equation for the reduction of arsenious acid by zinc in acidic solution.

(See Problems 24.57, 24.58, 24.59, and 24.60.)

24.3 Group VIA: The Sulfur Family

Group VIA elements, like those of Groups IVA and VA, show the trend from nonmetallic to metallic. Oxygen and sulfur are strictly nonmetals. Although the chemistries of selenium and tellurium are predominantly those of non-metals, they do have semiconducting allotropes, as expected of metalloids. Polonium is metallic.

Oxygen, which we studied earlier, has rather different properties from the other members of Group VIA.■ It is a very electronegative element and bonding involves only *s* and *p* orbitals. For the other members of the group, *d* orbitals become a factor in bonding. Oxygen has compounds mainly in the −2 oxidation state. The other Group VIA elements have compounds in this state also, but the +4 and +6 states are common. Selenium and tellurium in the +6 state are strong oxidizing agents, showing the greater stability of the +4 state in the heavier Group VIA elements.

■ Oxygen was discussed in Chapter 13.

Sulfur is an abundant element. It occurs in sulfate minerals, such as gypsum ($CaSO_4 \cdot 2H_2O$), and in sulfide minerals, which are important metal ores. Sulfur is present in coal and petroleum as organic sulfur compounds and in natural gas as hydrogen sulfide. Free sulfur occurs in some volcanic areas, perhaps formed by the reaction of hydrogen sulfide and sulfur dioxide, which are present in volcanic gases.

$$16H_2S(g) + 8SO_2(g) \longrightarrow 16H_2O(l) + 3S_8(s)$$

Commercial deposits of the free element also occur in salt domes, which are massive columns of salt embedded in rock a hundred meters or more below the earth's surface.■ None of the other Group VIA elements is abundant. Selenium and tellurium occur mixed with sulfide ores, and polonium-210 occurs in thorium and uranium ores. Polonium-210 has a half-life of 138 days, decaying by alpha emission.

■ Such sulfur deposits are found in the United States along the shore of the Gulf of Mexico.

Properties of the Elements

The stable form of sulfur, called rhombic sulfur, is a yellow, crystalline solid with a lattice of crown-shaped S_8 molecules (Figure 24.10). The catenation evident here also shows up in some sulfur compounds, such as the metal polysulfides, which have linear chain ions like S_6^{2-}. Rhombic sulfur melts at 113°C (see Table 24.6) to give a straw-colored liquid. Upon continued heating, this changes to a dark reddish-brown, viscous liquid. The original melt consists of S_8 molecules, but these open up, and the fragments join to give long spiral chains of sulfur atoms. The viscosity increases as compact

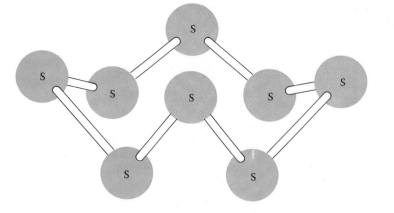

Figure 24.10
Structure of the S_8 molecule.

S_8 molecules are replaced by long molecular chains that can intertwine.■ At temperatures about 180°C, the chains begin to break apart, and the viscosity decreases. Sulfur boils at 445°C, giving a vapor of S_8, S_6, S_4, and S_2 molecules.

Selenium also exists in several allotropic modifications. An amorphous, red selenium, Se_8, precipitates when solutions of selenious acid, H_2SeO_3, are reduced. When this is heated below its melting point, it changes to a crystalline, red selenium. The stable, gray form crystallizes from molten selenium. Gray selenium is a **photoconductor;** that is, it is normally a poor conductor of electricity, but becomes a good conductor when light falls on it. Tellurium exists only as a silvery solid with metallic luster. Polonium is a metal, the only one with a simple cubic lattice.

Sulfur reacts with nearly all elements. It burns in air with a characteristic blue flame, giving off sulfur dioxide gas, recognizable by its sharp, choking odor. Although sulfur also has the trioxide, SO_3, very little of it forms even when sulfur burns in excess air. Like sulfur, the elements selenium and tellurium burn in air to form the dioxides, which in these cases are solids.

■ If this viscous liquid is cooled quickly, it gives a rubbery form of sulfur. The rubberlike properties are due to the long spiral molecular chains that can be stretched along their length.

Table 24.6
Properties of Group VIA Elements

Property	Oxygen	Sulfur	Selenium	Tellurium	Polonium
Electron configuration	$[He]2s^2 2p^4$	$[Ne]3s^2 3p^4$	$[Ar]3d^{10}4s^2 4p^4$	$[Kr]4d^{10}5s^2 5p^4$	$[Xe]4f^{14}5d^{10}6s^2 6p^4$
Melting point, °C	-218	113	217 (gray Se)	452	254
Boiling point, °C	-183	445	685	1390	962
Density, g/cm^3	1.43×10^{-3}	2.07	4.81 (gray Se)	6.25	9.32
Ionization energy (first), kJ/mol	1314	1000	941	869	812
Electron affinity, kJ/mol	-141	-200	-195	-190	-180
Electronegativity (Pauling scale)	3.5	2.5	2.4	2.1	2.0
Standard potential (volts), $X + 2e^- \rightleftharpoons X^{2-}$	—	-0.51	-0.78	-0.92	—
Covalent radius, Å	0.66	1.04	1.17	1.37	1.46
Ionic radius (for X^{2-}), Å	1.26	1.70	1.84	2.07	—

Sulfur, selenium, and tellurium also react directly with the halogens. For example, they react vigorously with fluorine to give the hexafluorides.

$$S(s) + 3F_2(g) \longrightarrow SF_6(g)$$

(We will henceforth follow convention and write the formula of sulfur with the symbol S, rather than the strictly correct S_8, in order to simplify the coefficients in equations.) Group VIA elements react with most metals. Thus, sulfur gives sulfides (containing S^{2-}), and in some cases both sulfides and disulfides (with S_2^{2-}):■

■ Many of the metal sulfides (FeS and CuS, for example) are nonstoichiometric compounds, that is, solid substances that deviate from their idealized formulas and can exhibit variable composition.

$$Hg(l) + S(s) \xrightarrow[\text{temperature}]{\text{room}} HgS(s)$$

$$Fe(s) + S(s) \xrightarrow{\Delta} FeS(s)$$
$$\text{iron(II) sulfide}$$

$$Fe(s) + 2S(s) \xrightarrow{\Delta} FeS_2(s)$$
$$\text{iron(II) disulfide}$$

Sulfur reacts with hot, concentrated nitric acid to give H_2SO_4, an acid in the +6 oxidation state:

$$S(s) + 6HNO_3(aq) \longrightarrow H_2SO_4(aq) + 6NO_2(g) + 2H_2O(l)$$

Selenium and tellurium are oxidized only to the +4 acids. For example,

$$Se(s) + 4HNO_3(aq) \longrightarrow H_2SeO_3(aq) + 4NO_2(g) + H_2O(l)$$

Preparation and Uses of the Elements

Free sulfur is mined by the **Frasch process.** In this process, underground deposits of solid sulfur are melted in place with superheated water (Figure 24.11). Molten sulfur is forced upward as a froth, using air under pressure. The sulfur obtained this way is 99.6% pure. It is used primarily in the manufacture of sulfuric acid. Selenium is obtained from the flue dusts from the roasting of sulfide ores and from the anode mud formed by the electrolytic refining of copper. When these materials are leached with various oxidizing agents, H_2SeO_3 and H_2SeO_4 are obtained. These are reduced to elemental selenium with sulfur dioxide. For example,

$$H_2SeO_3(aq) + 2SO_2(g) + H_2O(l) \longrightarrow Se(s) + 2H_2SO_4(aq)$$
$$\text{red selenium}$$

Selenium has been used to make photoelectric cells and rectifiers for electronic equipment. These uses have been largely displaced by cheaper silicon semiconductor devices. Red selenium is also used to color red glass and enamels. Tellurium is useful in alloys.

Important Compounds

Hydrogen Sulfide, Selenide, and Telluride Hydrogen sulfide, H_2S, is a colorless gas with the strong odor of rotten eggs. It is quite poisonous.■ Hydrogen

■ Hydrogen sulfide is more poisonous than hydrogen cyanide, HCN, which is used in one method of capital punishment.

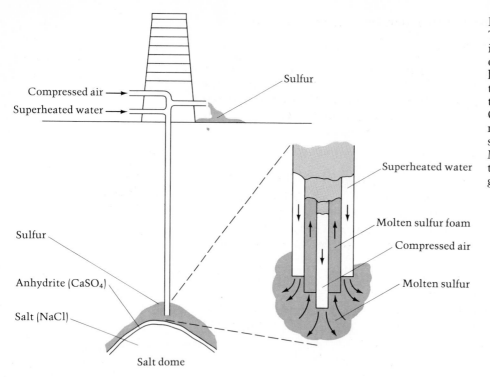

Figure 24.11
The Frasch process for mining sulfur. The well consists of concentric pipes. Superheated water passing down the outer jacket exits into the sulfur deposit, melting it. Compressed air from the inner pipe pushes the molten sulfur up the middle jacket. Molten sulfur flows from the top of the well onto the ground to cool.

sulfide is a very weak, diprotic acid ($K_1 = 8.9 \times 10^{-8}$; $K_2 = 1.2 \times 10^{-13}$). It forms hydrogen sulfide salts (such as $NaHS$) and sulfide salts (Na_2S). In acid solution, hydrogen sulfide is a reducing agent and with mild oxidizing agents goes to sulfur:

$$2Fe^{3+}(aq) + H_2S(aq) \longrightarrow 2Fe^{2+}(aq) + 2H^+(aq) + S(s)$$

Stronger oxidizing agents give sulfate ion.

Hydrogen sulfide is used in qualitative analysis laboratories to separate metal ions. The separation is based on the different solubilities of the metal sulfides formed from the metal ions with H_2S. The gas can be prepared by the reaction of an acid on a metal sulfide:

$$2[H^+(aq) + Cl^-(aq)] + ZnS(s) \longrightarrow [Zn^{2+}(aq) + 2Cl^-(aq)] + H_2S(g)$$

An aqueous solution of H_2S is conveniently prepared in the laboratory by warming a solution of thioacetamide, CH_3CSNH_2:

$$CH_3CSNH_2(aq) + 2H_2O(l) \xrightarrow{\Delta} NH_4^+(aq) + CH_3COO^-(aq) + H_2S(aq)$$
thioacetamide   ammonium acetate

Hydrogen selenide, H_2Se, and hydrogen telluride, H_2Te, are also poisonous gases. They are weak acids, but stronger than H_2S. Both are strong reducing agents and give the elements as products.

Sulfur Oxides and Oxyacids Sulfur dioxide, SO_2, obtained by burning sulfur or sulfides, is a colorless gas with suffocating odor. The gas dissolves in water to give an acidic solution of sulfurous acid, H_2SO_3. These solutions contain only a small fraction of H_2SO_3 and are primarily $SO_2(aq)$:

$$SO_2(aq) + H_2O(l) \rightleftharpoons H_2SO_3(aq)$$

Sulfurous acid, which is stable only in aqueous solution, is diprotic, the first dissociation being moderately strong ($K_1 = 1.3 \times 10^{-2}$), the second rather weak ($K_2 = 6.3 \times 10^{-8}$). Two series of salts are known, the hydrogen sulfites (also called bisulfites) and the sulfites. Acidified solutions of these salts release SO_2:

$$2H^+(aq) + SO_3{}^{2-}(aq) \longrightarrow H_2O(l) + SO_2(g)$$

Sulfur dioxide solutions and sulfites are reducing agents, being oxidized to sulfates. Calcium hydrogen sulfite, made from SO_2 and $Ca(OH)_2$, is used to manufacture paper pulp, by dissolving the natural cement (lignin) that holds the cellulose fibers together in wood. Sulfur dioxide gas is used to preserve dried fruit by inhibiting the growth of fungi. (Table 24.7 lists uses of sulfur compounds.)

Sulfur trioxide, SO_3, is formed in only small amounts when sulfur burns, although thermodynamically it should be favored. The reaction to form SO_3 is too slow. In the **contact process** for manufacturing sulfuric acid, sulfur trioxide (the anhydride of H_2SO_4) is prepared from SO_2 and O_2, using a catalyst such as platinum or divanadium pentoxide, V_2O_5:■

$$2SO_2(g) + O_2(g) \xrightarrow[V_2O_5]{\Delta} 2SO_3(g)$$

Sulfur trioxide exists in two solid forms and as a volatile liquid at room temperature. The liquid contains SO_3 and S_3O_9 molecules (Figure 24.12). The substance reacts with water to give sulfuric acid:

$$SO_3(g) + H_2O(l) \longrightarrow H_2SO_4(aq)$$

The SO_3 mists that result from the contact process are difficult to dissolve completely in water. Therefore, in the industrial preparation of sulfuric acid, the trioxide is first dissolved in concentrated H_2SO_4. The major species in this solution is $H_2S_2O_7$ (pyrosulfuric acid). The solution is diluted with water to give concentrated sulfuric acid.

Sulfuric acid is diprotic; the first dissociation in water is strong, the second moderately strong ($K_2 = 1.1 \times 10^{-2}$). Both sulfate and hydrogen sulfate

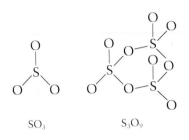

Figure 24.12
The structure of SO_3 and S_3O_9. These molecules are in equilibrium in liquid sulfur trioxide.

■ Sulfur dioxide for the contact process is obtained from several sources. Traditionally, sulfur mined by the Frasch process has been used. However, anti-pollution laws have made it necessary to recover SO_2 from plants that roast sulfide ores. Hydrogen sulfide from natural gas is another source of SO_2 (by burning the H_2S).

Table 24.7
Uses of Some Sulfur Compounds

Compound	Use
CS_2	Manufacture of rayon and cellophane
	Manufacture of CCl_4
SO_2	Manufacture of H_2SO_4
	Food preservative
	Textile bleach
H_2SO_4	Manufacture of phosphate fertilizers
	Petroleum refining
	Manufacture of various chemicals
$Na_2S_2O_3$	Photographic fixer

(bisulfate) salts are known. The sulfate anion, which has a tetrahedral geometry, might be written

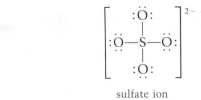

sulfate ion

though the S—O bond length (1.51 Å), when compared with the sum of the sulfur and oxygen single-bonded covalent radii (1.70 Å), indicates substantial double-bond character.■ Partial double bonding can be achieved by overlapping of $2p$ orbitals on the oxygen atoms with $3d$ orbitals on the sulfur atom. One may represent this by resonance structures of the following sort:

■ Covalent radii are given in Table 24.6.

$$\left[\begin{array}{c} :O: \\ \parallel \\ :\ddot{O}{-}S{-}\ddot{O}: \\ \parallel \\ :O: \end{array} \right]^{2-}$$

Concentrated sulfuric acid is a viscous liquid and a powerful dehydrating agent. When it is poured over sucrose (cane sugar, $C_{12}H_{22}O_{11}$), the elements of water are exothermically extracted; a charred mass of carbon remains (Figure 24.13). The concentrated acid is also an oxidizing agent. Copper is not dissolved by most acids ($E°$ for $Cu^{2+}|Cu$ is 0.34 V), but it is dissolved by concentrated sulfuric acid, which is reduced to sulfur dioxide.

$$Cu(s) + 2H_2SO_4(l) \longrightarrow Cu^{2+}(aq) + SO_4^{2-}(aq) + SO_2(g) + 2H_2O(l)$$

More sulfuric acid is manufactured than any other chemical. Most of the acid is used to make soluble phosphate and ammonium sulfate fertilizers. Sulfuric acid is also used in petroleum refining and in the manufacture of many chemicals.

Thiosulfate ion, $S_2O_3^{2-}$, is related structurally to the sulfate ion, with sulfur replacing an oxygen atom. (The prefix *thio-* means that sulfur has replaced oxygen in the compound or ion whose name follows the prefix.)

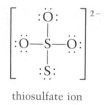

thiosulfate ion

Note that it has sulfur in the -2 oxidation state, as well as $+6$.■ Sodium thiosulfate is prepared by heating a slurry of sulfur in sodium sulfite solution:

$$[2Na^+(aq) + SO_3^{2-}(aq)] + S(s) \xrightarrow{\Delta} [2Na^+(aq) + S_2O_3^{2-}(aq)]$$

■ To find the oxidation numbers, assign the electrons in each S—O bond to the more electronegative O atom, and give one electron in the S—S bond to each S atom. The oxidation number is the charge left on each atom after this assignment of electrons.

Thiosulfate ion decomposes in acidic solution to give sulfur dioxide and a precipitate of sulfur, essentially reversing the reaction just given.

Sodium thiosulfate pentahydrate, $Na_2S_2O_3 \cdot 5H_2O$, is known as photographer's "hypo." Photographic film has a layer of silver halide embedded in gelatin. When exposed to light, the silver halide decomposes to very small grains of metallic silver, forming a "latent" image. A *developer* enlarges these grains and brings out the image, by reducing nearby crystals of silver halide. Then a *fixer* of sodium thiosulfate solution dissolves unexposed silver halide from the film by forming the complex ion, $Ag(S_2O_3)_2^{3-}$.

Figure 24.13
The dehydrating action of concentrated sulfuric acid is shown in this simple demonstration. When the concentrated acid is poured over sucrose (table sugar, $C_{12}H_{22}O_{11}$), it leaves a charred mass. The concentrated acid has extracted the hydrogen and oxygen as water, leaving carbon.

Selenium and Tellurium Oxides and Oxyacids Selenium dioxide is a colorless, volatile solid, prepared by burning selenium in air. The vapor phase contains SeO_2 molecules. The solid dissolves in water to give a solution of selenious acid, from which crystals of selenious acid may be isolated. Tellurium dioxide, prepared by burning tellurium in air, is a colorless solid. It is barely soluble in water, and no aqueous acid is known. Tellurium dioxide does dissolve readily in solutions of alkali metal hydroxides, giving tellurite salts, such as Na_2TeO_3. The dioxide also dissolves in acids, showing that it is amphoteric.

Both selenium trioxide and tellurium trioxide are prepared by heating the corresponding acids. For example,

$$H_2SeO_4(s) \longrightarrow H_2O(g) + SeO_3(s)$$
selenic acid

rioxide is a colorless solid. It dissolves in water, forming the acid.
trioxide is an orange solid. Selenic acid is prepared by strongly
elenious acid, for example, by heating it with 30% hydrogen
peroxide:

$$H_2SeO_3(aq) + H_2O_2(aq) \longrightarrow [H^+(aq) + HSeO_4^-(aq)] + H_2O(l)$$

It is a strong, diprotic acid. Both SeO_3 and H_2SeO_4 are strong oxidizing
agents. Telluric acid may be prepared similarly to selenic acid, by oxidizing
TeO_2 with $H_2O_2(aq)$. Surprisingly, the acid has the formula H_6TeO_6 or
$Te(OH)_6$, rather than the formula we would expect by analogy with sulfuric
and selenic acids. Telluric acid is a weak, diprotic acid and a strong oxidizing
agent. The weak acidity is an indication of the greater metallic character of
tellurium.

Example 24.3

Prepare sulfuric acid from $H_2S(g)$ and air (plus any required catalysts).

Solution

Hydrogen sulfide burns in air to give H_2O and SO_2:

$$2H_2S(g) + 3O_2(g) \longrightarrow 2H_2O(g) + 2SO_2(g)$$

Then, using heat and a catalyst, we obtain

$$2SO_2(g) + O_2(g) \xrightarrow[\text{catalyst}]{\Delta} 2SO_3(g)$$

Finally, we dissolve SO_3 in water:

$$SO_3(g) + H_2O(l) \longrightarrow H^+(aq) + HSO_4^-(aq)$$

Exercise 24.5

Prepare sodium thiosulfate from sulfur, air, and sodium carbonate solution.

(See Problems 24.63 and 24.64.)

Exercise 24.6

Thiosulfate ion is used to titrate iodine by reducing it to I^-. Thiosulfate ion is
oxidized to tetrathionate ion, $S_4O_6^{2-}$, which has the structure

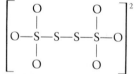

Write the balanced equation for the reaction.

(See Problems 24.65 and 24.66.)

24.4 Group VIIA: The Halogens

For the most part, Group VIIA elements, the halogens, have very similar
properties or at least ones that change smoothly in progressing down the
column. All are reactive nonmetals, except perhaps for astatine, whose
chemistry is not well known. Fluorine, as a second-period element, does
exhibit some differences from the other elements of Group VIIA, although
these are not as pronounced as we have seen for the second-period elements

in Groups IIIA to VIA. The solubilities of the fluorides in water, for example, are often quite different from those of the other halides. Calcium chloride, bromide, and iodide are very soluble in water. Calcium fluoride, however, is insoluble. Silver chloride, bromide, and iodide are insoluble, but silver fluoride is soluble.

All of the halogens form stable compounds in the -1 oxidation state. For fluorine compounds, this is the only oxidation state. Chlorine, bromine, and iodine also have compounds in which the halogen is in one of the positive oxidation states—$+1$, $+3$, $+5$, or $+7$.

Fluorine and chlorine are abundant elements. Fluorine is widely available in fluorapatite, $3Ca_3(PO_4)_2 \cdot CaF_2$, and in fluorite, CaF_2. Chlorine is abundant in the oceans and in salt deposits as NaCl. Bromine is much less abundant, but occurs as Br^- in recoverable concentrations in certain brines and ocean water. Iodine occurs in very low concentration as I^- in sea water, but this is assimilated by kelp (seaweed), which is a commercial source of the element. Iodine also occurs as sodium iodide in certain oil-well brines and as sodium iodate, $NaIO_3$, in Chilean nitrate deposits.

Properties of the Elements

All of the halogens exist as diatomic molecular substances, F_2, Cl_2, Br_2, and I_2. Fluorine and chlorine are gases; fluorine is pale yellow, chlorine is greenish yellow. Bromine is a volatile, reddish brown liquid with reddish brown vapor. Iodine is a shiny, black solid that sublimes readily, giving a violet vapor. Melting and boiling points and other properties are listed in Table 24.8. The increasing melting points and boiling points from F_2 to I_2 are explained by the argument that the London forces between molecules increase as the halogen molecules increase in size.

The halogens are all oxidizing agents, the oxidizing power decreasing from F_2 to I_2. We can see this readily by considering the oxidation of water by the halogens. The emf of these reactions can be calculated from the standard

Table 24.8
Properties of Group VIIA Elements

Property	Fluorine	Chlorine	Bromine	Iodine	Astatine
Electron configuration	[He]$2s^22p^5$	[Ne]$3s^23p^5$	[Ar]$3d^{10}4s^24p^5$	[Kr]$4d^{10}5s^25p^5$	[Xe]$4f^{14}5d^{10}6s^26p^5$
Melting point, °C	-220	-101	-7	114	—
Boiling point, °C	-188	-35	59	184	—
Density, g/cm³	1.69×10^{-3}	3.21×10^{-3}	3.12	4.93	—
Ionization energy (first), kJ/mol	1681	1251	1140	1008	—
Electron affinity, kJ/mol	-328	-349	-325	-295	-270
Electronegativity (Pauling scale)	4.0	3.0	2.8	2.5	2.2
Standard potential (volts), $X_2 + 2e^- \rightleftharpoons 2X^-$	2.87	1.36	1.06	0.54	0.3
Covalent radius, Å	0.64	0.99	1.14	1.33	1.45
Ionic radius (for X^-), Å	1.19	1.67	1.82	2.06	—

potentials for the electrodes $X_2|X^-$, given in Table 24.8, and that for the following half-reaction in neutral water:■

$$4H^+(aq) + O_2(g) + 4e^- \longrightarrow 2H_2O(l); \; E = 0.82 \text{ V}$$

For the fluorine reaction, we get

$$2F_2(g) + 2H_2O(l) \longrightarrow 4H^+(aq) + 4F^-(aq) + O_2(g); \; E^\circ_{cell} = 2.05 \text{ V}$$

Thus, fluorine reacts vigorously with water. For the reaction of water with Cl_2, Br_2, and I_2, we get 0.54 V, 0.24 V, and -0.28 V, respectively. The emf values show that chlorine and bromine should react to some extent with water to give oxygen, whereas the reverse reaction occurs for iodine: I^- is oxidized by molecular oxygen to I_2. In the case of chlorine and bromine, however, the following reactions are faster and compete more favorably, particularly in basic solution:

$$2Cl_2(g) + 2H_2O(l) \rightleftharpoons 2HClO(aq) + 2[H^+(aq) + Cl^-(aq)]$$
$$2Br_2(aq) + 2H_2O(l) \rightleftharpoons 2HBrO(aq) + 2[H^+(aq) + Br^-(aq)]$$

The relative oxidizing powers of Cl_2, Br_2 and I_2 can be seen by observing which of the halogens react with the halide ion of another. When chlorine water, $Cl_2(aq)$, is added to dilute solutions of bromide or iodide ion, the elements are released:

$$Cl_2(aq) + 2Br^-(aq) \longrightarrow 2Cl^-(aq) + Br_2(aq)$$
$$Cl_2(aq) + 2I^-(aq) \longrightarrow 2Cl^-(aq) + I_2(aq)$$

Although bromine is not strong enough as an oxidizing agent to oxidize Cl^-, it will easily oxidize I^-.

$$Br_2(aq) + 2I^-(aq) \longrightarrow 2Br^-(aq) + I_2(aq)$$

Iodine, of course, is not strong enough to oxidize either Cl^- or Br^-. The first two reactions have been used as tests of Br^- and I^-. When methylene chloride, CH_2Cl_2, is poured into a test tube of Br^- or I^- to which chlorine water has been added, it forms a colored layer of the halogen in CH_2Cl_2 at the bottom of the tube. Bromine gives an orange layer; iodine gives a violet layer.

The halogens react directly with most other elements. They react with themselves to form **interhalogens**, binary compounds of one halogen with another, such as ClF, BrF, IBr, ClF_3, ClF_5, and IF_7.

■ In neutral water solution, $[H^+] = 1.0 \times 10^{-7} M$. The concentrations of all other species have standard values.

Preparation and Uses of the Elements

Fluorine is such a strong oxidizing agent that it has only been prepared by electrolysis. The cell electrolyte is potassium fluoride dissolved in liquid hydrogen fluoride. Electrolysis produces hydrogen at the cathode and fluorine at the anode. Fluorine is produced in commercial quantities for the manufacture of uranium nuclear fuel rods. Uranium metal is reacted with excess fluorine to produce uranium hexafluoride, UF_6, a volatile, white solid. The vapor of this compound is separated by diffusion to give mixtures containing more of the fissionable uranium-235 isotope than is present in the natural source. Nuclear fuel rods contain 3 to 4% uranium-235, compared with 0.72% in natural uranium.

Chlorine, Cl_2, is a major industrial chemical. It is prepared by the electrolysis of aqueous sodium chloride.■ The major use of chlorine is in the manufacture of various chlorinated hydrocarbons, such as vinyl chloride, $CH_2=CHCl$ (for plastics), carbon tetrachloride, CCl_4 (for fluorocarbons), and methyl chloride, CH_3Cl (for silicones and tetramethyllead). Large quantities of chlorine are also used to disinfect water supplies and to bleach paper pulp and textiles.

■ The electrolysis of NaCl(aq) was described in Section 23.2.

Bromine, Br_2, can be obtained from sea water or brine by oxidation of the bromide ion in solution with chlorine. Most of the bromine produced in the United States is obtained from brine wells in Arkansas and Michigan. These brines contain such high concentrations of bromide ion that the bromine can be distilled from the oxidized solution. In the commercial process, hot brine is led into the top of a tower, while steam and chlorine are fed in at the bottom. Bromine and water vapor leaving from the top of the tower are condensed, giving a distillate of separate layers, bromine on the bottom, water on top. The bromine layer is drawn off and purified by further distillation.

Most bromine is used to produce ethylene dibromide, $C_2H_4Br_2$, as an additive to leaded gasolines. This use will decline as leaded gasolines are phased out. Bromine is also used to manufacture other bromine compounds, including the alkali metal bromides for sedatives and for the production of silver bromide for photographic film.

Iodine, I_2, is produced from natural brines by oxidizing I^- with chlorine:

$$2I^-(aq) + Cl_2(g) \longrightarrow I_2(s) + 2Cl^-(aq)$$

The precipitated iodine is filtered from the solution. The element is also produced from sodium iodate, an impurity in Chilean saltpeter, $NaNO_3$, by reducing iodate ion with sodium hydrogen sulfite, $NaHSO_3$. Iodine is used to make silver iodide for photographic film.

Important Compounds

The most important inorganic compounds of the halogens are the *hydrogen halides*, the *halogen oxyacids*, and their salts. Table 24.9 lists uses of some halogen compounds.

Hydrogen Halides Each of the halogens forms a binary compound with hydrogen: HF, HCl, HBr, and HI. All are colorless gases with sharp, penetrating odors. Hydrogen fluoride, whose boiling point (20°C) is near room temperature, is easily liquefied. Its boiling point is high in comparison with those of the other hydrogen halides (HCl, -85°C; HBr, -67°C; HI, -35°C), whose boiling points increase as expected for molecular substances having little or no hydrogen bonding. The much higher boiling point of hydrogen fluoride is good evidence for strong hydrogen bonding in this substance.

The hydrogen halides dissolve in water to give acidic solutions, called **hydrohalic acids.** For example, hydrogen chloride gas dissolves in water to give hydrochloric acid, HCl(aq). Hydrochloric, hydrobromic, and hydriodic

Compound	Use
AgBr, AgI	Photographic film
CCl$_4$	Manufacture of fluorocarbons
C$_2$H$_4$Br$_2$	Lead scavenger in leaded gasolines
C$_2$H$_4$Cl$_2$	Manufacture of vinyl chloride (plastics)
	Lead scavenger in leaded gasolines
C$_2$H$_5$Cl	Manufacture of tetraethyllead for gasoline
HCl	Metal treating
	Food processing
NaClO	Household bleach
	Manufacture of hydrazine for rocket fuel
NaClO$_3$	Paper pulp bleaching (with ClO$_2$)

Table 24.9
Uses of Some Halogen
Compounds

acids are strong acids, the molecular species HX ionizing completely in solution.

$$HCl(g) \xrightarrow{H_2O} H^+(aq) + Cl^-(aq)$$

Hydrofluoric acid, by contrast, is a weak acid. An equilibrium exists in aqueous solution between molecular HF and the ions H$^+$ and F$^-$:

$$HF(aq) \rightleftharpoons H^+(aq) + F^-(aq)$$

The hydrogen halides can be formed by direct combination of the elements. Fluorine and hydrogen react violently. The reaction has no commercial importance because fluorine is normally prepared from hydrogen fluoride. However, chlorine is burned in an excess of hydrogen to produce hydrogen chloride for industrial use.

$$H_2(g) + Cl_2(g) \longrightarrow 2HCl(g)$$

Hydrogen bromide and hydrogen iodide are prepared in a similar way. In these cases, the elements are heated with a platinum catalyst, for example:

$$H_2(g) + Br_2(g) \xrightarrow[Pt]{\Delta} 2HBr(g)$$

The hydrogen halides can be prepared by heating the salts with a nonvolatile acid. The commercial method of preparing hydrogen fluoride uses concentrated sulfuric acid:

$$CaF_2(s) + H_2SO_4(l) \xrightarrow{\Delta} CaSO_4(s) + 2HF(g)$$

The reaction is driven to the right by removing gaseous hydrogen fluoride, the other substances being relatively nonvolatile. Hydrogen chloride can be prepared in a similar reaction from sodium chloride and sulfuric acid:

$$NaCl(s) + H_2SO_4(l) \xrightarrow{\Delta} NaHSO_4(s) + HCl(g)$$

On stronger heating, the reaction goes to sodium sulfate:

$$NaCl(s) + NaHSO_4(s) \xrightarrow{\Delta} Na_2SO_4(s) + HCl(g)$$

Neither HBr nor HI can be prepared with H$_2$SO$_4$, since the hot concentrated acid oxidizes Br^- and I^- to the elements. For these halides, phosphoric acid can be used because it is nonoxidizing, as well as nonvolatile:

$$NaBr(s) + H_3PO_4(l) \xrightarrow{\Delta} HBr(g) + NaH_2PO_4(s)$$

Although some hydrogen chloride is prepared commercially from sodium chloride and sulfuric acid, as well as by direct combination of the elements, today most HCl is obtained as a by-product of the chlorination of organic compounds. For example, when methane, CH_4, reacts with chlorine, it produces various chloromethanes, such as methyl chloride (monochloromethane), and HCl:

$$CH_4(g) + Cl_2(g) \longrightarrow CH_3Cl(g) + HCl(g)$$
$$\text{methyl chloride}$$

The quantity of chlorinated hydrocarbons produced for various purposes is so large that by-product HCl satisfies most of the commercial demand.

A major use of hydrogen fluoride is in the preparation of organic fluorine compounds. For example, trichlorofluoromethane, CCl_3F, used as a refrigerant and aerosol propellant, is prepared from carbon tetrachloride and HF, using antimony pentafluoride as catalyst:

$$CCl_4(l) + HF(g) \xrightarrow{SbF_5} CCl_3F(l) + HCl(g)$$

Polytetrafluoroethylene plastic (Teflon) is also an organic fluorine material requiring HF in its manufacture. Another major use of HF is in the preparation of the electrolyte for the production of aluminum.■ The reactions of hydrofluoric acid with silica (SiO_2) and glass are unique among the hydrohalic acids. If we represent glass by the formula $CaSiO_3$ (calcium silicate), we can write the equation for the reaction as

■ The production of aluminum was discussed in Section 23.4.

$$CaSiO_3(s) + 8HF(aq) \longrightarrow H_2SiF_6(aq) + CaF_2(aq) + 3H_2O(l)$$
$$\quad\text{glass} \qquad\qquad\qquad \text{hexafluorosilicic acid}$$

Small quantities of hydrofluoric acid are used to etch glass by this reaction.

Hydrochloric acid is the fourth most important industrial acid (after sulfuric, phosphoric, and nitric acids). It is used to clean metal surfaces of oxides (a process called "pickling") and to extract certain metal ores, such as those of tungsten. Its use in the preparation of magnesium was mentioned earlier.■

■ See Section 23.3.

Halogen Oxyacids The halogens form a variety of oxyacids (Table 24.10). Figure 24.14 shows the structures of the oxyacids of chlorine. The acidic character of these acids increases with the number of oxygen atoms bonded to the halogen atom; that is, acid strength increases from HClO to $HClO_4$. Perchloric acid is the strongest of the common acids. All of the halogen oxyacids are oxidizing agents.

The chemistry of the chlorine oxyacids can be understood in part by referring to the **standard potential diagram** shown in Figure 24.15. Such a diagram is a convenient summary of standard potentials. Species of different oxidation states are connected in the diagram by an arrow, with the standard potential (in volts) for the half-reaction between them written over the

Oxidation State	Fluorine Oxyacids	Chlorine Oxyacids	Bromine Oxyacids	Iodine Oxyacids	General Name
+1	HFO*	HClO†	HBrO†	HIO†	Hypohalous acid
+3	—	HClO₂†	HBrO₂†	—	Halous acid
+5	—	HClO₃†	HBrO₃†	HIO₃	Halic acid
+7	—	HClO₄	HBrO₄†	HIO₄	Perhalic acid
				H₅IO₆	

*The oxidation state of F in HFO is −1.
†These acids are known only in aqueous solution.

Table 24.10
Halogen Oxyacids

arrow. For example,

$$HClO \xrightarrow{+1.63} Cl_2 \quad \text{(acidic solution)}$$

represents the half-reaction for the reduction of HClO to Cl_2 in acid solution. If we write this out in full, we have ■

$$2HClO(aq) + 2H^+(aq) + 2e^- \longrightarrow Cl_2(g) + 2H_2O(l); \quad E° = 1.63 \text{ V}$$

The feasibility of the simultaneous oxidation and reduction of Cl_2 (that is, the **disproportionation** of Cl_2) in basic solution can be seen from a glance at the diagram. The relevant portion is

$$ClO^- \xrightarrow{+0.44} Cl_2 \xrightarrow{+1.36} Cl^- \quad \text{(basic solution)}$$

If Cl_2 is to disproportionate, it must be reduced to Cl^- and oxidized to ClO^-. The emf for this reaction is $+1.36$ V $- (+0.44$ V$) = +0.92$ V. Since this is positive, the reaction is spontaneous. In general, a species disproportionates if the standard potential on the left in the diagram is less than that on the right.

The disproportionation of chlorine in basic solution is used to prepare sodium hypochlorite, NaClO. Chlorine, released by the electrolysis of aqueous sodium chloride, is allowed to mix with the cold sodium hydroxide solution also obtained in the electrolysis. The reaction is

$$Cl_2(g) + 2[Na^+(aq) + OH^-(aq)] \longrightarrow$$
$$[Na^+(aq) + ClO^-(aq)] + [Na^+(aq) + Cl^-(aq)] + H_2O(l)$$

Solutions of sodium hypochlorite are sold as a bleach (Clorox). The corresponding acid, HClO, is unstable except in dilute aqueous solution. Like the other hypohalous acids, it is a weak acid.

Hypochlorite ion is itself unstable, disproportionating into chlorate ion,

■ The method of obtaining the half-reaction was described in Section 9.8.

Figure 24.14
Structures of the chlorine oxyacids.

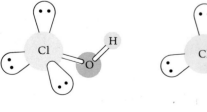

HClO

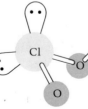

HClO₂

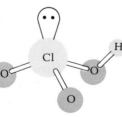

HClO₃

HClO₄

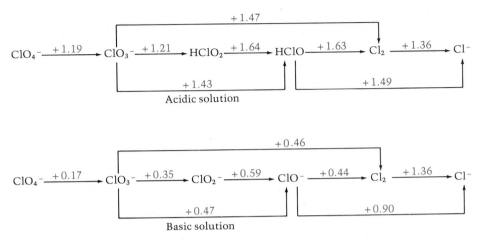

Figure 24.15
Standard potential diagram
for chlorine and its com-
pounds.

ClO_3^-, and chloride ion:

$$3ClO^-(aq) \longrightarrow ClO_3^-(aq) + 2Cl^-(aq); \ E_{cell}^\circ = +0.43 \ V \ (basic)$$

The reaction in cold basic solution is very slow, but it is fast in hot solution. Therefore, when chlorine reacts with hot sodium hydroxide solution, sodium chlorate is a product, instead of NaClO:

$$3Cl_2(g) + 6[Na^+(aq) + OH^-(aq)] \longrightarrow$$
$$[Na^+(aq) + ClO_3^-(aq)] + 5[Na^+(aq) + Cl^-(aq)] + 3H_2O(l)$$

The sodium chlorate is obtained from the solution by partial evaporation. Sodium chloride crystallizes from the hot solution. The filtrate is then evaporated to recover $NaClO_3$. Chloric acid is stable only in aqueous solution.

Sodium or potassium perchlorate are produced commercially by electrolysis of a saturated solution of the chlorate. The anode reaction is

$$ClO_3^-(aq) + H_2O(l) \longrightarrow ClO_4^-(aq) + 2H^+(aq) + 2e^-$$

Hydrogen is evolved at the cathode. Perchloric acid, $HClO_4$, can be prepared by treating a perchlorate salt with sulfuric acid:

$$KClO_4(s) + H_2SO_4(l) \longrightarrow KHSO_4(s) + HClO_4(l)$$

The perchloric acid is distilled from the mixture at reduced pressure (to keep the temperature below 92°C, where perchloric acid decomposes explosively).

Example 24.4

Sodium hexafluorosilicate, Na_2SiF_6, is used for the fluoridation of water supplies. (The fluorine becomes incorporated into tooth enamel, which is then less susceptible to decay.) Describe with balanced equations how this salt can be obtained from calcium fluoride, sand (SiO_2), sulfuric acid, sodium hydroxide, and water.

Solution
The reactions are

$$CaF_2(s) + H_2SO_4(l) \longrightarrow CaSO_4(s) + 2HF(g)$$

$$HF(g) \xrightarrow{H_2O} HF(aq)$$

$$6HF(aq) + SiO_2(s) \longrightarrow H_2SiF_6(aq) + 2H_2O(l)$$

$$H_2SiF_6(aq) + 2NaOH(aq) \longrightarrow Na_2SiF_6(aq) + 2H_2O(l)$$

Exercise 24.7

Chlorine dioxide, ClO_2, is a reddish yellow gas, used to bleach paper pulp. It may be prepared by reducing aqueous sodium chlorate with sulfur dioxide in acid solution. (This gas is safely prepared only in low concentrations; otherwise it may detonate.) Describe with balanced equations how chlorine dioxide could be made from aqueous sodium chloride, sulfur dioxide, and sulfuric acid.

(See Problems 24.73 and 24.74.)

Exercise 24.8

Perbromic acid was discovered in 1968. The sodium salt may be prepared by oxidation of sodium bromate by fluorine in basic solution. Write the balanced equation for this reaction.

(See Problems 24.75, 24.76, 24.77, and 24.78.)

Exercise 24.9

Show by calculating the standard emf that the disproportionation of Cl_2 in acid solution is not spontaneous under standard conditions (see Figure 24.15 for data).

(See Problems 24.79 and 24.80.)

A Checklist for Review

Important Terms

zone refining (24.1)
catenation (24.1)
silanes (24.1)
silicates (24.1)
condensation reaction (24.1)

aluminosilicate minerals (24.1)
polyphosphoric acids (24.2)
metaphosphoric acids (24.2)
photoconductor (24.3)
Frasch process (24.3)

contact process (24.3)
interhalogen (24.4)
hydrohalic acids (24.4)
standard potential diagram (24.4)
disproportionation (24.4)

Summary of Facts and Concepts

Group IVA elements show a definite progression from nonmetallic to metallic in going down the column. Carbon occurs as its nonmetallic allotropes, graphite and diamond. Silicon and germanium have diamondlike structures, but whereas diamond is an electrical insulator, silicon and germanium are semiconductors. Although one allotrope of tin has the diamondlike structure, the common form is metallic. Lead has only a metallic form. The +4 oxidation state is common for carbon, silicon, and germanium, but the +2 state occurs in tin and is the most important oxidation state for lead. We see this in the reaction of the elements. All elements but lead burn in excess oxygen to form the dioxides; lead gives PbO. Germanium and tin are oxidized to the dioxides by concentrated nitric acid, but lead gives Pb^{2+}.

Silicon is the second most abundant element in the earth's crust, where it occurs as silica (SiO_2) and *silicates*. These minerals consist of SiO_4 tetrahedra linked through a common oxygen. Silica is acidic and reacts with basic oxides, such as CaO, to form glass. Tin(II) and lead(II) oxides are amphoteric, forming the cations Sn^{2+} and Pb^{2+} in acid and the anions $Sn(OH)_3^-$ and $Pb(OH)_3^-$ in base.

Phosphorus, in Group VA, has two common allotropes. White phosphorus is very reactive; it has a molecular structure with the formula P_4. Red phosphorus is less reactive. The common forms of arsenic and antimony are metallike; bismuth is definitely metallic. Group VA elements have compounds in +5 and +3 oxidation states, though +3 is the only stable state of bismuth. Phosphorus burns in air, forming P_4O_6 and P_4O_{10}. The other Group VA elements form +3 oxides. Phosphorus is the most important element in this group.

The element and its oxyacid H_3PO_4 are major industrial chemicals, and the soluble phosphates are important agricultural fertilizers. Orthophosphoric acid, H_3PO_4, undergoes condensation reactions to form the poly-phosphoric and metaphosphoric acids. (In a *condensation reaction*, two molecules are joined by eliminating a small molecule, in this case H_2O.) Salts of these condensed phosphates are used in detergent formulations.

Sulfur, selenium, and tellurium, in Group VIA, are similar in their chemical properties. Sulfur is strictly nonmetallic, whereas selenium and tellurium are metalloids. Thus, sulfur is a molecular solid, S_8. Selenium has both a molecular form, Se_8, and a gray, semiconducting allotrope. Tellurium has only a metallike form. The oxidation states $+6$, $+4$, and -2 are all important in these elements, although the $+6$ state is more strongly oxidizing in selenium and tellurium. Free sulfur is mined by the *Frasch process* for use in manufacturing sulfuric

acid by the *contact process*. In this process, SO_2 is catalytically oxidized with O_2 to SO_3, which is then dissolved in concentrated sulfuric acid. When this solution is diluted, it gives the concentrated acid.

Group VIIA elements, or halogens, have similar properties. They are nonmetals with the molecular formula X_2. Fluorine and chlorine are abundant and important elements. Fluorine, F_2, is very reactive; it oxidizes water to O_2. Although less reactive, chlorine gas is nevertheless an active nonmetal. It is obtained by the electrolysis of aqueous NaCl. The gas dissolves readily in cold basic solution, disproportionating to Cl^- and OCl^-. (In *disproportionation*, a species is both the oxidizing and the reducing agent.) In hot solution, the hypochlorite ion, OCl^-, disproportionates to Cl^- and ClO_3^-. The chlorate ion can be electrolytically oxidized to perchlorate ion, ClO_4^-. Thus, a series of oxyacid salts can be made readily.

Operational Skills

The problem-solving skills used in this chapter were discussed in previous chapters.

Review Questions

24.1 Why does silicon form the ion SiF_6^{2-}, but carbon has no similar ion?

24.2 How is carbon black similar to graphite? How does it differ?

24.3 Which Group IVA elements have allotropes with diamondlike structures?

24.4 Why would high pressure be an expected condition for the transformation of graphite to more dense diamond?

24.5 Describe the steps in preparing ultrapure silicon from quartz sand.

24.6 What is meant by the term *catenation?* Give an example of a compound that displays catenation.

24.7 By means of equations, show how an ion with three silicon atoms could form from $Si(OH)_2O_2^{2-}$ ions.

24.8 Give reactions for the dioxides of carbon, silicon, tin, and lead that show their acid–base behavior.

24.9 Write equations for the reactions of silicon, tin, and lead with (a) Br_2, (b) O_2, (c) $HCl(aq)$, and (d) $HNO_3(aq)$. Write NR if no reaction occurs.

24.10 Which of the following ions is the better reducing agent: $Pb(OH)_3^-$ or $Sn(OH)_3^-$? Explain why this is expected.

24.11 Describe the structure of white phosphorus. How does the structure account for its chemical reactivity?

24.12 What are the products when each Group VA element—phosphorus, arsenic, antimony, and bismuth —reacts with O_2?

24.13 Write the equations for the production of antimony from stibnite, Sb_4S_6.

24.14 What are the acids that correspond to the following anhydrides: P_4O_6, P_4O_{10}, As_4O_6, As_2O_5?

24.15 Hypophosphorous acid, H_3PO_2, has the structure

Do you expect this acid to be monoprotic, diprotic, or triprotic? Explain.

24.16 Describe two different methods used to manufacture phosphoric acid, H_3PO_4, starting from $Ca_3(PO_4)_2$.

24.17 By means of an equation, show how triphosphoric acid could be formed from phosphoric acid and diphosphoric acid.

24.18 What is the purpose of a detergent builder? Give an example of a common builder.

24.19 List three natural sources of sulfur or sulfur compounds.

24.20 What is the structure of the stable form of sulfur?

24.21 Describe the changes that occur as sulfur melts and the temperature of the liquid rises. Explain what happens.

24.22 Sulfur hexafluoride is used as a gaseous insulator in electrical transformers. How could you prepare this compound?

24.23 Describe the Frasch process for mining sulfur.

24.24 Each of the following substances often reacts as an oxidizing or reducing agent. Fill in the table, noting whether the substance listed usually acts as an oxidizing or reducing agent; then give the usual product formed.

Substance	Oxidizing or reducing agent?	Usual product
H_2S		
$Na_2SO_3(aq)$		
Hot, conc. H_2SO_4		
$Na_2S_2O_3(aq)$		

24.25 Give equations for two different methods of preparing each of the following: (a) H_2S, (b) SO_2.

24.26 Give the equations for the steps in the contact process for manufacturing sulfuric acid from sulfur.

24.27 Pyrosulfuric acid molecules are in equilibrium with H_2SO_4 in concentrated sulfuric acid. The reaction is a condensation, similar to that in which pyrophosphoric (diphosphoric) acid is formed from phosphoric acid. Write the equation for the equilibrium between H_2SO_4 and $H_2S_2O_7$, using structural formulas for the species involved.

24.28 Describe the preparation of sodium thiosulfate.

24.29 Thiosulfuric acid cannot be prepared. Explain.

24.30 Explain how sodium thiosulfate is used in photography. What would happen to the negative if it was not fixed?

24.31 Fill in the following table, giving the formula of the acid corresponding to the anhydride listed. Then state whether the acid is strong or weak. If the corresponding acid is unknown, give this information in the column under acid.

Anhydride	Acid	Strong or weak?
SO_2		
SeO_2		
TeO_2		
SO_3		
SeO_3		
TeO_3		

24.32 What is the difference in behavior of F_2 and Cl_2 with water? Give equations for the reactions.

24.33 Complete and balance the following equations. Write NR if no reaction occurs.

(a) $I_2(aq) + Cl^-(aq) \longrightarrow$
(b) $Cl_2(aq) + Br^-(aq) \longrightarrow$
(c) $Br_2(aq) + I^-(aq) \longrightarrow$
(d) $Br_2(aq) + Cl^-(aq) \longrightarrow$

24.34 A test tube contains a solution of one of the following salts: NaCl, NaBr, NaI. Describe a single test that can distinguish among these possibilities.

24.35 What is an interhalogen? Give an example.

24.36 Give a natural source for each of the halogens (except astatine). Describe how the element is obtained from that source.

24.37 Hydrogen fluoride has a boiling point near room temperature, but hydrogen chloride boils at $-85°C$. Explain why HF has the higher boiling point, when we might have expected otherwise on the basis of the molecular weights of HF and HCl.

24.38 Hydrogen chloride can be prepared by heating NaCl with concentrated sulfuric acid. Why is it that substituting NaBr for sodium chloride is not a satisfactory way to prepare HBr?

24.39 Phosphate rock is primarily fluorapatite, a calcium fluoride phosphate mineral. In the preparation of phosphate fertilizer, hydrogen fluoride is produced, which at one time was vented into the surrounding atmosphere. Today it is converted to compounds such as hexafluorosilicic acid. Write the equation for the reaction of $HF(aq)$ with silica (SiO_2).

24.40 What is the standard potential for the reduction of ClO_3^- to $HClO_2$? Write the half-reaction.

24.41 How is sodium hypochlorite prepared? Give the balanced equation.

24.42 Do you expect an aqueous solution of sodium hypochlorite to be acidic, neutral, or basic? What about an aqueous solution of sodium perchlorate?

Problems

The Group IVA Elements

24.43 Formic acid, HCOOH, is used to produce certain artificial flavorings. The acid is produced from sodium formate. By means of balanced equations, show how sodium formate can be made, starting from methane, CH_4 (in two steps).

24.45 Show how you could prepare a solution of tin(II) chloride from tin(IV) oxide. Give balanced equations.

24.47 A test for bismuth ion, Bi^{3+}, consists of precipitating it as $Bi(OH)_3$ (white), then reducing this with stannite ion to finely divided bismuth metal (black). Write the balanced equation for the reduction of $Bi(OH)_3$ to Bi by sodium stannite solution.

24.49 The $Sn^{2+}(aq)$ ion can be written more completely as $Sn(H_2O)_6^{2+}$. This ion is acidic by hydrolysis. Write a possible equation for this hydrolysis.

24.51 Solid tin(II) chloride dihydrate, $SnCl_2 \cdot 2H_2O$, consists of molecules of $SnCl_2(H_2O)$, in which H_2O is directly attached to the tin atom by a coordinate bond from oxygen. The other H_2O molecule is not bonded to tin. What is the expected geometry of the $SnCl_2(H_2O)$ molecule?

24.44 One of the most important salts of carbonic acid is sodium carbonate, Na_2CO_3. Use balanced equations to show how sodium carbonate could be made in two steps, starting with carbon.

24.46 Write balanced equations for the preparation of $Pb(OH)_2(s)$ from lead metal.

24.48 Lead(IV) oxide will oxidize hydrochloric acid to chlorine, Cl_2. Write the balanced equation for this reaction.

24.50 Lead(II) nitrate is one of the few soluble lead salts and gives a solution with a pH of about 3 to 4. Write an equation for a possible reaction to explain why the solution is not neutral.

24.52 Ethylene dichloride, $C_2H_4Cl_2$, and ethylene dibromide, $C_2H_4Br_2$, are added to leaded gasolines to produce volatile compounds, such as PbBrCl, in the engine (to keep lead from depositing). What is the expected geometry of the PbBrCl molecule?

The Group VA Elements

24.53 Give the formulas of three compounds for each of the following oxidation states of phosphorus: (a) +3, (b) +5.

24.55 Using balanced equations, describe how arsenic(V) oxide can be prepared from arsenic(III) oxide.

24.57 Phosphorous acid is oxidized to phosphoric acid by hot, concentrated sulfuric acid, which is reduced to SO_2. Write the balanced equation for this reaction.

24.59 Arsenate ion is reduced to arsenite ion in aqueous solution by hydrazinium chloride, $(N_2H_5)Cl$. Note that aqueous solutions of hydrazinium chloride are acidic (why?). Hydrazinium chloride is oxidized to N_2. Write the balanced equation for this reaction.

24.61 Although phosphorus pentabromide exists as PBr_5 molecules in the vapor, the solid is ionic, with the structure $[PBr_4^+]Br^-$. What is the expected geometry of PBr_4^+? Describe the bonding to phosphorous.

24.54 Give the formulas of three compounds for each of the following oxidation states of arsenic: (a) +3, (b) +5.

24.56 Starting from bismuth metal, show how to prepare $Bi(OH)_3$. Show the steps with balanced equations.

24.58 Phosphorous acid is oxidized to phosphoric acid by nitric acid, which is reduced to NO. Write a balanced equation for this reaction.

24.60 Arsenic(III) sulfide is oxidized to arsenate ion by hydrogen peroxide in basic solution. Write the balanced equation for this reaction.

24.62 Antimony(V) oxide dissolves in strong base to give the $Sb(OH)_6^-$ ion. What is the expected geometry of this ion? Describe the bonding to antimony in this ion.

The Group VIA Elements

24.63 With equations, show how sulfur, S_8, could be prepared from only $H_2S(g)$ and air.

24.65 Selenium dioxide can be detected in qualitative analysis by converting it to selenious acid and reducing this with H_2S to selenium. A yellow precipitate of sulfur and selenium forms. Write the balanced equation for the reaction of selenious acid and H_2S.

24.67 Describe the bonding in the following: (a) H_2Se, (b) SeF_4. What are the expected geometries?

24.69 What are the oxidation numbers of sulfur in each of the following: (a) SF_6, (b) SO_3, (c) H_2S, (d) $CaSO_3$?

24.64 With equations, describe how selenic acid could be prepared from selenium, air, H_2O_2, and water.

24.66 Concentrated sulfuric acid oxidizes iodide ion to iodine, I_2. Write the balanced equation for this reaction.

24.68 Describe the bonding in the following: (a) SO_3^{2-}, (b) SeF_6. What are the expected geometries?

24.70 What are the oxidation numbers of sulfur in each of the following: (a) S_8, (b) CaS, (c) $CaSO_4$, (d) SCl_4?

The Group VIIA Elements

24.71 When silica or glass reacts with hydrogen fluoride gas, the product is silicon tetrafluoride, $SiF_4(g)$. Write the balanced equation for the reaction of silica with $HF(g)$.

24.73 Uranium hexafluoride is a volatile solid and is used in the gaseous-diffusion method of separating uranium isotopes. With balanced equations, show how this compound can be prepared from uranium metal, calcium fluoride, sulfuric acid, and potassium fluoride.

24.75 Chlorine may be prepared in the laboratory by oxidizing chloride ion with a sufficiently strong oxidizing agent. In one method, hydrochloric acid is heated with potassium dichromate, $K_2Cr_2O_7$. Write the balanced equation for the reaction. Dichromate ion is reduced to Cr^{3+}.

24.77 Iodine is prepared from sodium iodate with sodium hydrogen sulfite. Write the balanced equation for the reaction.

24.79 By calculating the standard emf, decide whether sodium hypochlorite will oxidize $Fe^{2+}(aq)$ to $Fe^{3+}(aq)$ in acidic solution under standard conditions. See Figure 24.15 and Table 21.1 for data.

24.81 Discuss the bonding in the following molecules or ions. What is the expected geometry? (a) Cl_2O (b) BrO_3^- (c) BrF_3

24.72 A solution of chloric acid may be prepared by reacting a solution of barium chlorate with sulfuric acid. Barium sulfate precipitates. Write the balanced equation for the reaction.

24.74 Silver iodide is a light-sensitive compound, used in photographic film. With balanced equations, show how this compound could be prepared, starting from a solution of sodium iodate, silver nitrate, sodium hydrogen sulfite, and hydrogen.

24.76 Iodic acid can be prepared by oxidizing elemental iodine with concentrated nitric acid, which is reduced to $NO_2(g)$. Write the balanced equation for this reaction.

24.78 If solid potassium chlorate is carefully heated, it disproportionates to give potassium chloride and potassium perchlorate. Write the balanced equation for this reaction.

24.80 Using the data from Figure 24.15, calculate E°_{cell} for the disproportionation of chlorous acid to hypochlorous acid and chlorate ion. Is the reaction spontaneous under standard conditions?

24.82 Discuss the bonding in the following molecules or ions. What is the expected geometry? (a) HFO (b) $SiCl_4$ (c) SiF_6^{2-}

Additional Problems

24.83 Identify each of the following substances from the description:

(a) A yellow solid that burns with a blue flame, giving off a gas with a choking odor

(b) A white solid that reacts with water to give phosphoric acid

(c) A reddish brown liquid that reacts vigorously with sodium metal, to give a white solid

(d) A metal that crumbles to a powder when exposed to temperatures below 13°C for a period of time

24.84 Identify each of the following substances from the description:

(a) A white, waxy solid, normally stored under water because it spontaneously bursts into flames when exposed to air

(b) A viscous liquid that reacts with table sugar, giving a charred mass

(c) An acid that etches glass

(d) A pale green gas that dissolves in aqueous sodium hydroxide to give a solution used as a bleach

24.85 You are given three unlabeled test tubes. One test tube contains a solution of sodium sulfate, one contains a solution of sodium hydrogen sulfate, and the third contains a solution of sodium hydrogen sulfite. Describe how you could identify the solutions.

24.86 You are given three unlabeled test tubes. One contains a solution of sodium fluoride, the second contains a solution of sodium chloride, and the third contains a solution of sodium iodide. Describe how you could find the identity of each solution.

24.87 Arsenious acid is a very weak acid.

$$H_3AsO_3(aq) \rightleftharpoons H^+(aq) + H_2AsO_3^-(aq);$$
$$K_a = 6 \times 10^{-10}$$

What is the pH of 0.050 M sodium dihydrogen arsenite, NaH_2AsO_3?

24.88 Calculate the pH of a 0.015 M solution of arsenic acid. The K_a for the ionization of the first hydrogen is 6.0×10^{-3}. Neglect any further ionizations of the acid.

24.89 Chlorine gas can be prepared by adding dilute $HCl(aq)$ dropwise onto potassium permanganate crystals, $KMnO_4$. The $KMnO_4$ is reduced to $Mn^{2+}(aq)$. What volume (in liters) of 1.50 M $HCl(aq)$ is required to react with 12.0 g $KMnO_4$?

24.90 Iodic acid can be prepared by oxidizing iodine with concentrated nitric acid. What volume (in liters) of 15.8 M HNO_3 is required to produce 15.0 g of iodic acid? Assume that the nitric acid is reduced to NO_2.

24.91 The main ingredient in many phosphate fertilizers is calcium dihydrogen phosphate monohydrate, $Ca(H_2PO_4)_2 \cdot H_2O$. What is the mass percentage of phosphorus in this salt? If a fertilizer is 15.5 mass percent P, and all of this phosphorus is present in the fertilizer as $Ca(H_2PO_4)_2 \cdot H_2O$, what is the mass percentage of this salt in the fertilizer?

24.92 A fertilizer contains phosphorus in two compounds, $Ca(H_2PO_4)_2 \cdot H_2O$ and $CaHPO_4$. The fertilizer contains 30.0 mass percent $Ca(H_2PO_4)_2 \cdot H_2O$ and 10.0 mass percent $CaHPO_4$. What is the mass percentage of phosphorus in the fertilizer?

24.93 Sodium hypochlorite solution is produced by the electrolysis of cold sodium chloride solution. The electrolysis is carried out so as to thoroughly mix the products ($NaOH$ and Cl_2). How long must a cell operate to produce 1.00×10^3 L of 5.25% solution of $NaClO$ if the cell current is 2.50×10^3 A? Assume that the density of the solution is 1.00 g/mL.

24.94 Sodium perchlorate is produced by electrolysis of sodium chlorate. If a current of 2.50×10^3 A passes through an electrolytic cell, how many kilograms of sodium perchlorate are produced per hour?

**24.95 The amount of sodium hypochlorite in a bleach solution can be determined by using a given volume of bleach to oxidize excess iodide ion to iodine, since the reaction goes to completion. The amount of iodine produced is then determined by titration with sodium thiosulfate, which is oxidized to sodium tetrathionate, $Na_2S_4O_6$. Potassium iodide was added in excess to 5.00 mL of bleach (density = 1.00 g/mL). This solution, containing the iodine released in the reaction, was titrated with 0.100 M $Na_2S_2O_3$. If 34.6 mL of sodium thiosulfate were required to reach the end point (detected by disappearance of the blue color of starch–iodine complex), what was the mass percent of $NaClO$ in the bleach?

**24.96 Ascorbic acid (vitamin C), $C_6H_8O_6$, is a reducing agent. It can be determined quantitatively by a titration procedure involving iodine, I_2:

$$C_6H_8O_6 + I_2 \longrightarrow C_6H_6O_6 + 2[H^+ + I^-]$$
ascorbic acid   dehydroascorbic acid

A 30.0-g sample of an orange-flavored beverage mix was placed in a flask to which 10.00 mL of 0.0500 M KIO_3 and excess KI were added. The IO_3^- and I^- ions react in acid solution to give I_2, which then reacts with ascorbic acid. Excess iodine is titrated with sodium thiosulfate (see Problem 24.95). If 29.5 mL of 0.0300 M $Na_2S_2O_3$ were required to titrate the excess I_2, how many grams of ascorbic acid are there in 100.0 g of beverage mix?

24.97 Consider the following standard potentials:

$$Sn(OH)_3^-(aq) + 2e^- \rightleftharpoons Sn(s) + 3OH^-(aq);$$
$$E° = -0.79 \text{ V}$$

$$Sn(OH)_6^{2-}(aq) + 2e^- \rightleftharpoons Sn(OH)_3^-(aq) + 3OH^-(aq);$$
$$E° = -0.96 \text{ V}$$

Would you expect stannite ion to disproportionate in basic solution to give metallic tin and stannate ion? Explain.

24.98 Consider the following standard potentials:

$$H_3PO_3(aq) + 2H^+(aq) + 2e^- \rightleftharpoons H_3PO_2(aq) + H_2O(l);$$
$$E° = -0.50 \text{ V}$$

$$H_3PO_4(aq) + 2H^-(aq) + 2e^- \rightleftharpoons H_3PO_3(aq) + H_2O(l);$$
$$E° = -0.28 \text{ V}$$

Would you expect phosphorous acid to disproportionate into phosphoric acid and hypophosphorous acid, H_3PO_2? Explain.

**24.99 Complete the following standard potential diagram by calculating the standard potential for the change indicated by the question mark.

Hint: The standard potentials are related to the standard free-energy changes for the reaction involving the indicated reduction and the oxidation of H_2 to H^+. Also, the free-energy changes for reactions can be added, as in Hess's law to obtain other values.

**24.100 Complete the following standard potential diagram by calculating the standard potential for the change indicated by the question mark. (See Problem 24.99.)

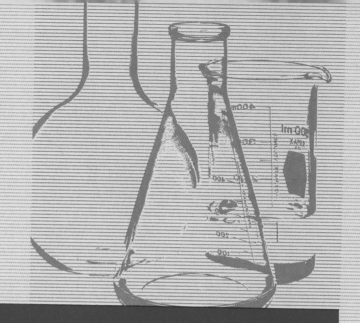

25. *The Transition Elements*

Properties of the Transition Elements

25.1 Periodic Trends in the Transition Elements
Electron Configurations/ Melting Points, Boiling Points, and Hardness/ Atomic Radii/ Ionization Energies/ Oxidation States

25.2 The Chemistry of Selected Transition Metals Chromium/ Iron/ Copper

Complex Ions and Coordination Compounds

25.3 Formation and Structure of Complexes Basic Definitions/ Polydentate Ligands/ Discovery of Complexes; Formula of a Complex/ *Aside: On the Stability of Chelates*

25.4 Naming Coordination Compounds

25.5 Structure and Isomerism in Coordination Compounds Structural Isomerism/ Stereoisomerism

25.6 Valence Bond Theory of Complexes
Octahedral Complexes/ Tetrahedral and Square Planar Complexes

25.7 Crystal Field Theory Effect of an Octahedral Field on the *d* Orbitals/ High-Spin and Low-Spin Complexes/ Tetrahedral and Square Planar Complexes/ Visible Spectra of Transition-Metal Complexes/
Aside: The Cooperative Release of Oxygen from Oxyhemoglobin

In the previous chapters, we studied the main-group elements, the A columns in the periodic table. Between columns IIA and IIIA are ten columns of the transition elements (the B groups). Among these elements are metals with familiar commercial applications: iron tools, copper wire, silver jewelry and coins. Many catalysts for important industrial reactions involve transition elements. Examples are found in petroleum refining and in the synthesis of ammonia from N_2 and H_2.

In addition to their commercial usefulness, many transition elements have biological importance. Iron compounds, for example, are found throughout the plant and animal kingdoms. Iron is present in hemoglobin, the molecule in red blood cells that is responsible for the transport of O_2 from the lungs to other body tissue. Myoglobin, in muscle, is a very similar molecule containing iron. It takes O_2 from hemoglobin, holding it until required by the muscle cells. Cytochromes are iron-containing compounds within each cell and are involved in the oxidation of food molecules. In these cases, the transition element is central to the structure and function of the biological molecule. Hemoglobin and myoglobin are examples of *metal complexes* or *coordination compounds*, in which the metal atom is surrounded by other atoms bonded to it by the electron pairs these atoms donate. In hemoglobin and myoglobin, the O_2 molecule bonds to the iron atom.

Chapter Overview

We will begin the chapter by describing the properties and general periodic behavior of the *transition elements*. We will then discuss the chemical properties of three transition elements: chromium, iron, and copper. These elements illustrate some general characteristics of the transition elements, such as a multiplicity of oxidation states and the formation of complexes. The second part of the chapter covers the structure of *complex ions* and *coordination compounds*. Three features of complexes that we will discuss are (1) isomerism (existence of two or more compounds with the same formula), (2) paramagnetism, and (3) color. Isomerism can be explained in terms of the geometry of the complex. Paramagnetism and color, however, require an understanding of the electronic structure of coordination compounds. Valence bond theory was the earliest theory of electronic structure to be applied to these compounds. Later, *crystal field theory* was developed to explain both paramagnetism and color in these compounds. This theory is essentially a simplified form of molecular orbital theory.

Properties of the Transition Elements

The transition elements are strictly defined as those elements having a partially filled d or f subshell in any common oxidation state (including the 0 oxidation state). For example, copper, which has the configuration $[Ar]3d^{10}4s^1$ in the 0 oxidation state, has the configuration $[Ar]3d^9$ in the common $+2$ oxidation state. Thus, copper(II) has a partially filled d subshell, and therefore copper is a transition element. Figure 25.1 shows the various divisions of the transition elements.

The **d-block transition elements** are those transition elements with an unfilled d subshell in common oxidation states. These elements are fre-

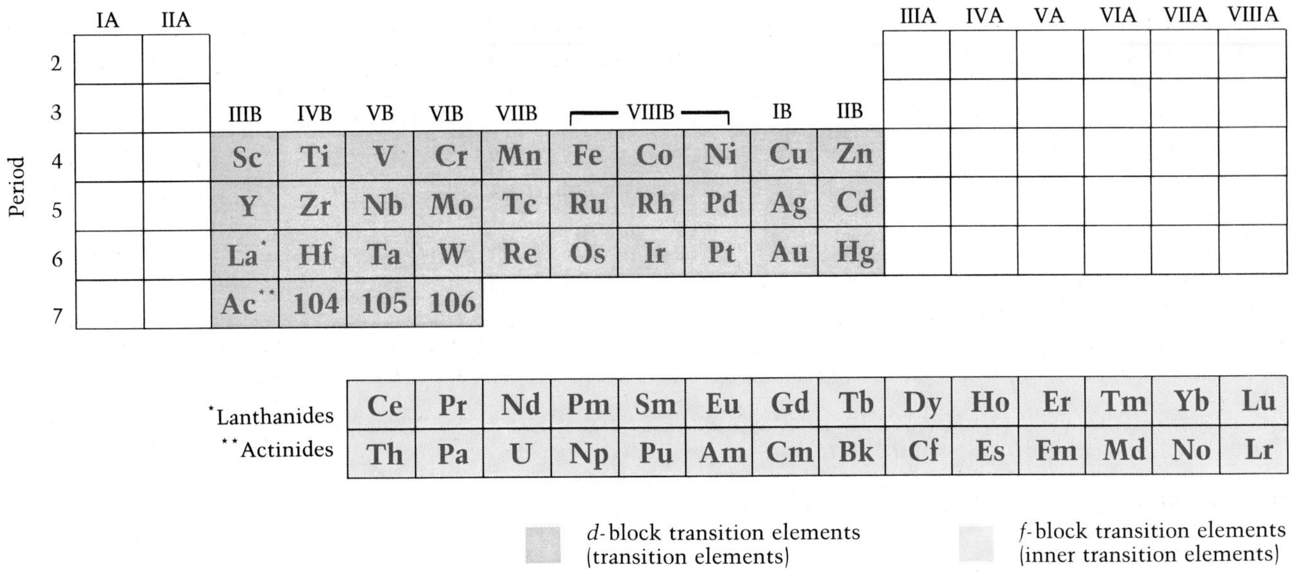

	IA	IIA											IIIA	IVA	VA	VIA	VIIA	VIIIA
Period 2																		
Period 3			IIIB	IVB	VB	VIB	VIIB	┌──── VIIIB ────┐			IB	IIB						
Period 4			Sc	Ti	V	Cr	Mn	Fe	Co	Ni	Cu	Zn						
Period 5			Y	Zr	Nb	Mo	Tc	Ru	Rh	Pd	Ag	Cd						
Period 6			La*	Hf	Ta	W	Re	Os	Ir	Pt	Au	Hg						
Period 7			Ac**	104	105	106												

*Lanthanides	Ce	Pr	Nd	Pm	Sm	Eu	Gd	Tb	Dy	Ho	Er	Tm	Yb	Lu
**Actinides	Th	Pa	U	Np	Pu	Am	Cm	Bk	Cf	Es	Fm	Md	No	Lr

◻ d-block transition elements (transition elements)

◻ f-block transition elements (inner transition elements)

Figure 25.1
Classification of the transition elements.

quently referred to simply as "the transition elements." Although zinc, cadmium, and mercury, in Group IIB, are not d-block transition elements by the strict definition, they are often included with them because of their similar properties. We will accept this broader definition and concentrate our attention in this chapter on these elements in the center of the periodic table.

The elements with a partially filled f subshell in common oxidation states are known as **f-block transition elements** or **inner transition elements.** The f-block transition elements are the two rows of elements at the bottom of the periodic table. The elements in the first row are called the *lanthanides* or *rare earths.*■ The elements in the second row are called *actinides*. All of the actinides are radioactive and from neptunium (Element 93) on are synthetic. They have been produced in nuclear reactors or by using particle accelerators.

■ The lowest energy configuration of the lutetium atom is $[Xe]4f^{14}5d^16s^2$, and in its only common oxidation state ($+3$) the configuration is $[Xe]4f^{14}$. Although both configurations have a filled f subshell, lutetium is usually considered a lanthanide element.

25.1 Periodic Trends in the Transition Elements

The transition elements have a number of characteristics that set them apart from the main-group elements.

1. All of the transition elements are metals and, except for the IIB elements, have high melting points, high boiling points, and are hard solids. Thus, in the fourth-period elements from scandium to copper, the lowest-melting metal is copper (1083°C) and the highest-melting metal is vanadium (1890°C). In the main-group metals, only beryllium melts above 1000°C, and the rest melt at appreciably lower temperatures.

2. With the exception of the IIIB and IIB elements, each transition element has several oxidation states. Chromium, for example, exists in all oxidation states from 0 to $+6$. Of the main-group metals, only the heavier ones display several oxidation states. Because of their multiplicity of oxidation states, the transition elements are often involved in oxidation–reduction reactions.

3. Transition-metal compounds are often colored, and many are paramagnetic. Most of the compounds of the main-group metals are colorless and diamagnetic.

We will look at points 1 and 2 in some detail in this section. In particular, we will look at trends in melting points, boiling points, hardness, and oxidation states. We will also look at trends in covalent radii and ionization energies, which we can relate to chemical properties. We will discuss the color and paramagnetism of transition-metal complexes later in the chapter.

Electron Configurations

Electronic structure is central to any discussion of the transition elements. Table 25.1 lists the electron configurations of the fourth-period transition elements. For the most part, these configurations are those predicted by the building-up principle. Following the building-up principle, the $3d$ subshell begins to fill after calcium (configuration $[Ar]4s^2$). Thus, scandium has the configuration $[Ar]3d^14s^2$, and, as we go across the period, additional electrons go into the $3d$ subshell. We get the configuration $[Ar]3d^24s^2$ for titanium, and $[Ar]3d^34s^2$ for vanadium. Then in chromium the configuration predicted by the building-up principle is $[Ar]3d^44s^2$, but the actual configuration is $[Ar]3d^54s^1$. This is usually explained as due to the special stability of a half-filled d subshell. The configurations for the rest of the elements are those predicted by the building-up principle, until we get to copper. Here, the predicted configuration is $[Ar]3d^94s^2$, but the actual configuration is $[Ar]3d^{10}4s^1$. This is explained as due to the stability of a filled d subshell.

Melting Points, Boiling Points, and Hardness

In Table 25.1, we see that the melting points of the transition metals increase from 1541°C for scandium to 1890°C for vanadium and 1857°C for chromium, then decrease to 1083°C for copper and 420°C for zinc. The same pattern is observed in the fifth-period and sixth-period elements. In going across a row of transition elements, the melting points increase, reaching a

Table 25.1
Properties of the
Fourth-Period Transition
Elements

Property	Scandium	Titanium	Vanadium	Chromium	Manganese
Electron configuration	$[Ar]3d^14s^2$	$[Ar]3d^24s^2$	$[Ar]3d^34s^2$	$[Ar]3d^54s^1$	$[Ar]3d^54s^2$
Melting point, °C	1541	1660	1890	1857	1244
Boiling point, °C	2831	3287	3380	2672	1962
Hardness (Mohs scale)	—	—	—	9.0	5.0
Density, g/cm^3	3.0	4.5	6.0	7.2	7.2
Electronegativity (Pauling scale)	1.3	1.5	1.6	1.6	1.5
Covalent radius, Å	1.44	1.32	1.22	1.18	1.17
Ionic radius (for M^{2+}), Å	—	1.00	0.93	0.87	0.81

maximum at the Group VB or VIB elements, after which the melting points decrease. Thus, tungsten in Group VIB has the highest melting point (3410°C) of any metallic element, and mercury in Group IIB has the lowest (−39°C). Similar trends can be seen in the boiling points and in hardness of the metals (Table 25.1). All of these properties depend on the strengths of metal bonding, which in turn depend roughly on the number of unpaired electrons in the metal atoms. At the beginning of a period of transition elements, there is one unpaired d electron. The number of unpaired d electrons increases in going across a period, until Group VIB, after which the electrons begin pairing.

Atomic Radii

Trends in atomic radii are of concern because chemical properties are determined in part by atomic size. Looking at the covalent radii (one measure of atomic size) in Table 25.1, we see that they decrease quickly from scandium (1.44 Å) to titanium (1.32 Å) and vanadium (1.22 Å). This decrease in atomic size in going across a row is the behavior we see in the main-group elements. It is explained as due to an increase in *effective nuclear charge* that acts on the outer electrons and pulls them in more strongly. The effective nuclear charge is the positive charge "felt" by an electron and equals the nuclear charge minus the shielding or screening of the positive charge by intervening electrons. After vanadium, the covalent radii decrease slowly from 1.18 Å for chromium to 1.15 Å for nickel. Then the covalent radii increase slightly to 1.17 Å for copper and 1.25 Å for zinc.■ The relative constancy in covalent radii for the later elements is partly responsible for the similarity in properties of the Group VIIIB elements iron, cobalt, and nickel.

■ The small increase in covalent radius for copper and zinc has no simple explanation.

Figure 25.2 compares the covalent radii of the transition elements. The atomic radii increase in going from a fourth-period to a fifth-period element within any column. For example, in going down the Group IIIB elements, the covalent radius of scandium is 1.44 Å and the radius of yttrium is 1.62 Å. We expect an increase of radius from scandium to yttrium because of the addition of a shell of electrons. In continuing down the column of IIIB elements, we find a small increase of covalent radius to 1.69 Å in lanthanum. But all of the remaining elements of the sixth period have nearly the same covalent

Table 25.1 (continued)

Property	Iron	Cobalt	Nickel	Copper	Zinc
Electron configuration	$[Ar]3d^64s^2$	$[Ar]3d^74s^2$	$[Ar]3d^84s^2$	$[Ar]3d^{10}4s^1$	$[Ar]3d^{10}4s^2$
Melting point, °C	1535	1495	1453	1083	420
Boiling point, °C	2750	2870	2732	2567	907
Hardness (Mohs scale)	4.5	—	—	3.0	2.5
Density, g/cm³	7.9	8.9	8.9	8.9	7.1
Electronegativity (Pauling scale)	1.8	1.8	1.8	1.9	1.6
Covalent radius, Å	1.17	1.16	1.15	1.17	1.25
Ionic radius (for M^{2+}), Å	0.75	0.79	0.83	0.87	0.88

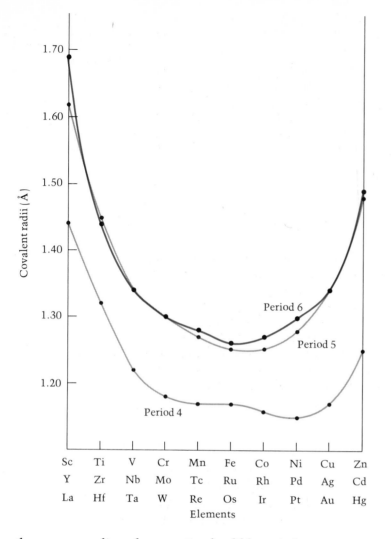

Figure 25.2
Comparison of the covalent radii for different periods of transition elements.

radii as the corresponding elements in the fifth period.

This similarity of radii in the fifth- and sixth-period transition elements can be explained in terms of the *lanthanide contraction*. Between lanthanum and hafnium in the sixth period are 14 lanthanide elements (cerium to lutetium), in which the 4f subshell fills. The covalent radius decreases slowly from cerium (1.65 Å) to lutetium (1.56 Å), but the total decrease is substantial. By the time the 4f subshell is complete, the covalent radii of the transition elements from hafnium on are similar to those of the elements in the preceding row of the periodic table. Thus, the covalent radius of hafnium (1.46 Å) is about the same as that of zirconium (1.45 Å).

The chemical properties of the transition elements parallel the pattern seen in covalent radii. That is, the fourth-period elements have substantially different properties from the elements in the same group in the fifth and sixth periods. However, elements in the same group of the fifth period and the sixth period are very much alike. Hafnium, for example, is so much like zirconium that it remained undiscovered until it was identified in 1923 in zirconium ores from its x-ray spectrum.■ Zirconium (mixed with hafnium as an impurity) was discovered over a hundred years earlier, in 1787.

■ The atomic number of an atom can be obtained from its x-ray spectrum, which therefore gives unequivocal identification of an element. See the Aside at the end of Section 6.2.

Table 25.2
First Ionization Energies
of the Transition Elements
(in kJ/mol)

	IIIB	IVB	VB	VIB	VIIB		VIIIB		IB	IIB
Fourth Period	Sc	Ti	V	Cr	Mn	Fe	Co	Ni	Cu	Zn
	631	658	650	652	717	759	758	737	745	906
Fifth Period	Y	Zr	Nb	Mo	Tc	Ru	Rh	Pd	Ag	Cd
	616	660	664	685	702	711	720	805	731	868
Sixth Period	La	Hf	Ta	W	Re	Os	Ir	Pt	Au	Hg
	538	680	761	770	760	840	880	870	890	1007

Ionization Energies

Looking at the first ionization energies of the fourth-period elements (Table 25.2), we see that although they vary somewhat irregularly, they tend to increase from left to right. The other rows of transition elements behave similarly. Most noteworthy, however, is that all of the sixth-period elements after lanthanum have ionization energies higher than the fourth-period and fifth-period transition elements. This behavior is opposite to what we found in the main-group elements, where the ionization energies decrease regularly in going down a column. The high ionization energies of the sixth-period elements from osmium to mercury is no doubt one determinant of the relative unreactivity of these elements.

Oxidation States

Table 25.3 gives oxidation states in compounds of the fourth-period transition elements. Scandium occurs as the 3 + ion and zinc occurs as the 2 + ion. The other elements, however, exhibit several oxidation states. This multiplicity of oxidation states is due to the varying involvement of the d electrons in bonding.

Table 25.3
Oxidation States of the
Fourth-Period Transition
Elements

IIIB	IVB	VB	VIB	VIIB		VIIIB		IB	IIB
Sc	Ti	V	Cr	Mn	Fe	Co	Ni	Cu	Zn
							+1	+1	
	+2	+2	+2	+2*	+2*	+2*	+2*	+2*	+2*
+3*	+3	+3	+3*	+3	+3*	+3	+3	+3	
	+4*	+4*	+4	+4	+4	+4	+4		
		+5	+5	+5	+5				
			+6	+6	+6				
				+7					

Key: Common oxidation states are in boldface; asterisks denote the stablest states. Additional oxidation states, particularly zero and negative values, may be observed in complexes with CO and with organic compounds.

Most of the transition elements have a doubly filled *ns* orbital. Since the *ns* electrons ionize before the $(n - 1)d$ electrons, we might expect the $+2$ oxidation state to be common. This oxidation state is in fact seen in all of the fourth-period elements except scandium, where the $+3$ ion with an Ar configuration is especially stable.

The maximum oxidation state possible equals the number of *s* and *d* electrons in the valence shell (which equals the group number for the elements up to iron). Thus, we see that titanium in Group IVB has a maximum oxidation state of $+4$. Similarly, vanadium (Group VB), chromium (Group VIB), and manganese (Group VIIB) exhibit maximum oxidation states of $+5$, $+6$, and $+7$, respectively.■

The total number of oxidation states actually observed for the elements increases from scandium ($+3$) to manganese (all states from $+2$ to $+7$). From iron on, however, the maximum oxidation state is not attained, so that the number of observed states decreases. The highest oxidation state seen in iron is $+6$.■ Thereafter, the highest observed oxidation number decreases until for zinc compounds we find only the $+2$ oxidation state. The Zn^{2+} ion has a stable $[Ar]3d^{10}$ configuration.

The stablest oxidation states in the first elements of the fourth period are the maximum values. Later in the period, lower oxidation states become more stable. The $+2$ oxidation state is particularly stable from manganese onward. (The $+3$ oxidation state is also stable in iron.)

In the fifth and sixth periods, there is also a multiplicity of oxidation states, particularly in the middle of the periods. The higher oxidation states of these elements are more stable, however. As an example, we can compare Cr and W in Group VIB. The CrO_4^{2-} ion, with chromium(VI), is a strong oxidizing agent (that is, it is unstable) and is reduced to Cr^{3+} (which is stable). On the other hand, the WO_4^{2-} ion, with tungsten(VI), is not oxidizing. Thus, the $+6$ state in tungsten is very stable. The stability of the higher oxidation states in the heavier elements is also evident in their greater range of observed oxidation numbers. Osmium, in the sixth period of Group VIIIB, has oxidation numbers from $+2$ to $+8$, the $+8$ state being observed in the tetroxide, OsO_4.■ However, iron in the fourth period of the same group, as we said, has $+6$ as its highest observed state.

As in the main-group elements, the transition elements become less characteristically metallic in higher oxidation states. Thus, the halides become more covalent, and the oxides more acidic, as the metal atom goes from a lower to a higher oxidation state. For example, titanium dichloride, $TiCl_2$, is a high-melting solid (m.p. 1035°C), whereas titanium tetrachloride, $TiCl_4$, is a liquid (m.p. -24°C). We conclude that titanium dichloride is ionic and titanium tetrachloride is a covalent, molecular substance. As for the oxides, titanium monoxide, TiO, is weakly basic and titanium dioxide, TiO_2, is amphoteric (acidic and basic).

■ The maximum oxidation state is generally found in compounds of the transition element with very electronegative elements, such as F and O (in oxides and oxyanions).

■ Presumably, Fe(VIII) would oxidize any element to which it would bond.

■ As one might expect, osmium tetroxide is a strong oxidizing agent and can be reduced to lower oxides or to the metal, both of which are black. It is used to stain biological specimens for microscopic examination; the specimen reduces OsO_4 to give a black stain. The substance is hazardous, since its vapors can similarly stain the eyeball.

Exercise 25.1

Compound A is a colorless liquid melting at 20°C. Compound B is a greenish yellow powder melting at 1406°C. One of these compounds is VF_3; the other is VF_5. Which is Compound A? Why?
(See Problems 25.31 and 25.32.)

As in the main-group elements, the transition elements become more metallic going down a column. Thus, the oxides of the fifth- and sixth-period elements are generally more basic than the oxides of the corresponding elements in the fourth period. For example, chromium (VI) oxide, CrO_3, is strongly acidic. Molybdenum(VI) and tungsten(VI) oxides, MoO_3 and WO_3, are weakly acidic.

Example 25.1

Consider the +3 oxidation state of cobalt and rhodium. In one of these elements, the 3+ metal ion, $M^{3+}(aq)$, is stable in water, and 3+ salts can be crystallized from aqueous solution. In the other, the $M^{3+}(aq)$ ion oxidizes water and therefore is unstable in aqueous solution. Which metal ion is unstable in aqueous solution?

Solution

Of two transition elements in the same column and same oxidation state, the one in the fourth period is expected to be the more oxidizing. Thus, we expect $Co^{3+}(aq)$ to be more oxidizing than $Rh^{3+}(aq)$. Since the problem states that one of these ions oxidizes water, this ion must be $Co^{3+}(aq)$, which would therefore be unstable in water.

Exercise 25.2

Consider the compounds $KMnO_4$ and $KReO_4$. Which would you expect to be the stronger oxidizing agent? Explain.

(See Problems 25.33 and 25.34.)

25.2 The Chemistry of Selected Transition Metals

Iron and titanium are the most abundant of the transition elements. In fact, iron is the fourth most abundant element in the earth's crust, after oxygen, silicon, and aluminum. However, many transition elements, particularly some in the sixth period, are rare. Surprisingly, technetium (Element 43) has only radioactive isotopes and does not occur naturally. Promethium (Element 61), a lanthanide, is the only other element with atomic number less than 83 that has no stable isotopes.

The transition elements occur in most cases as oxide or sulfide ores. Table 25.4 lists some of these ores. The least reactive elements, such as gold, silver, copper, and platinum, are found as uncombined metals. Many of the transition metals are used commercially in alloys. The most important of these are the steels, which are alloys of iron with small amounts of carbon and often other transition elements such as chromium, manganese, and nickel. In this section, we will look at the chemistry of three transition elements: chromium (Group VIB), iron (Group VIIIB), and copper (Group IB). We will see some of the similarities as well as the diversity among the transition elements.

Chromium

Chromium metal reacts with acids, releasing hydrogen gas and giving a bright blue solution of the chromium(II) or chromous ion, $Cr(H_2O)_6^{2+}(aq)$:

Transition Metal	Ore
Titanium, Ti	Ilmenite, $FeTiO_3$
	Rutile, TiO_2
Vanadium, V	Carnotite, $K_2(UO_2)_2(VO_4)_2 \cdot 3H_2O$
	Vanadinite, $Pb_5(VO_4)_3Cl$
Chromium, Cr	Chromite, $FeCr_2O_4$
Molybdenum, Mo	Molybdenite, MoS_2
Tungsten, W	Scheelite, $CaWO_4$
	Wolframite, $FeWO_4\text{-}MnWO_4$
Manganese, Mn	Pyrolusite, MnO_2
Iron, Fe	Hematite, Fe_2O_3
	Magnetite, Fe_3O_4
	Siderite, $FeCO_3$
Cobalt, Co	Cobaltite, $CoAsS\text{-}FeAsS$
	Skutterudite, $CoAs_3\text{-}NiAs_3$
Nickel, Ni	Pentlandite, $NiS\text{-}FeS$
Copper, Cu	Native copper, Cu
	Chalcocite, Cu_2S
	Cuprite, Cu_2O
Zinc, Zn	Sphalerite (zinc blende), ZnS

Table 25.4
Some Ores of the Transition Metals

$$Cr(s) + 2[H^+(aq) + Cl^-(aq)] + 6H_2O(l) \longrightarrow$$
$$[Cr(H_2O)_6^{2+}(aq) + 2Cl^-(aq)] + H_2(g)$$

The chromium(II) ion is easily oxidized by O_2 to the chromium(III) ion, $Cr(H_2O)_6^{3+}$, so to prepare chromium(II) salts, this reaction must be carried out in the absence of air. Chromium(II) ion is even oxidized by hydrogen ion, but the reaction is slow.

The most stable oxidation state of chromium is +3. Chromium metal burns in oxygen to given chromium(III) oxide, Cr_2O_3 (also called chromic oxide). This oxide is a hard, green solid and is amphoteric. In acid, the oxide dissolves, forming the violet-colored ion $Cr(H_2O)_6^{3+}(aq)$. Addition of base to a solution of this ion precipitates chromium(III) hydroxide:

$$Cr(H_2O)_6^{3+}(aq) + 3OH^-(aq) \longrightarrow Cr(H_2O)_3(OH)_3(s) + 3H_2O(l)$$

When further base is added, the precipitate dissolves, giving a green solution of the hydroxo ion, whose formula may be $Cr(H_2O)_2(OH)_4^-$.

$$Cr(H_2O)_3(OH)_3(s) + OH^-(aq) \longrightarrow Cr(H_2O)_2(OH)_4^-(aq) + H_2O(l)$$

Chromium(III) oxide also reacts with strong base to give hydroxo ions. The above reactions are sometimes written more simply without showing the water molecules bonded to the metal atom, that is, as

$$Cr^{3+}(aq) + 3OH^-(aq) \longrightarrow Cr(OH)_3(s)$$
$$Cr(OH)_3(s) + OH^-(aq) \longrightarrow Cr(OH)_4^-(aq)$$

Chromium(III) can be oxidized to chromium(VI) species, especially in basic solution. (Transition-metal ions are often more easily oxidized to oxyanions in basic solution.) For example, hydrogen peroxide oxidizes chromium(III) hydroxide to the chromate ion, CrO_4^{2-}:

$$2Cr(OH)_3(s) + 3HO_2^-(aq) + OH^-(aq) \longrightarrow 2CrO_4^{2-}(aq) + 5H_2O(l)$$

(We have written hydrogen peroxide as HO_2^-, since it is a weak acid and exists in basic solution as the anion.) The violet solution of Cr^{3+} turns to the bright yellow of the CrO_4^{2-} ion.

When the yellow solution of a chromate salt is acidified, it turns red orange. The color change is due to the formation of the dichromate ion, $Cr_2O_7^{2-}(aq)$:

$$\underset{\text{yellow}}{2CrO_4^{2-}(aq)} + 2H^+(aq) \rightleftharpoons \underset{\text{orange}}{Cr_2O_7^{2-}(aq)} + H_2O(l)$$

The chromate and dichromate ions are in equilibrium, which is sensitive to pH changes. Both chromate and dichromate salts are well known. The dichromate ion is a strong oxidizing agent in acid solution.

$$Cr_2O_7^{2-}(aq) + 14H^+(aq) + 6e^- \longrightarrow 2Cr^{3+}(aq) + 7H_2O(l); \ E° = 1.33 \text{ V}$$

Chromate ion in basic solution, however, is much less oxidizing.

$$CrO_4^{2-}(aq) + 4H_2O(l) + 3e^- \longrightarrow Cr(OH)_3(s) + 5OH^-(aq); \ E° = -0.12 \text{ V}$$

Chromium(VI) oxide (chromium trioxide, CrO_3) is a red, crystalline compound. It precipitates when concentrated sulfuric acid is added to concentrated solutions of a dichromate salt:■

$$[2K^+(aq) + Cr_2O_7^{2-}(aq)] + 2H_2SO_4(l) \longrightarrow$$
$$[2K^+(aq) + 2HSO_4^-(aq)] + 2CrO_3(s) + H_2O(l)$$

■ Chromium trioxide is a strong oxidizing agent. Vapors of ethanol, C_2H_5OH, spontaneously ignite in the presence of the solid oxide.

Chromium trioxide is an acidic oxide. It dissolves in water, forming acidic solutions of chromate and dichromate ions. When the oxide is heated above its melting point of 196°C, it decomposes by losing oxygen to give chromium(III) oxide:

$$4CrO_3(s) \xrightarrow{\Delta} 2Cr_2O_3(s) + 3O_2(g)$$

The chief commercial source of chromium is the ore chromite, $FeCr_2O_4$, which is an iron(II) chromium(III) oxide. This is reduced with carbon in an electric furnace to obtain an alloy of iron and chromium:

$$FeCr_2O_4(s) + 4C(s) \xrightarrow{\Delta} Fe(l) + 2Cr(l) + 4CO(g)$$

This alloy, called ferrochrome, is used to make steels. Pure chromium is prepared by the exothermic reaction of chromium(III) oxide with aluminum (the Goldschmidt process). Once the mixture is ignited, the heat of reaction produces molten chromium.

$$Cr_2O_3(s) + 2Al(s) \longrightarrow 2Cr(l) + Al_2O_3(l)$$

Chromium(III) oxide and other chromium compounds are obtained from sodium chromate. This salt is prepared by strongly heating chromite ore with sodium carbonate in air:

$$4FeCr_2O_4(s) + 8Na_2CO_3(s) + 7O_2(g) \xrightarrow{1100°C}$$
$$8Na_2CrO_4(s) + 2Fe_2O_3(s) + 8CO_2(g)$$

Sodium chromate is soluble and is leached from the mixture with water. It can be converted to sodium dichromate with acid. Sodium dichromate is reduced to chromium(III) oxide by heating with carbon:

$$Na_2Cr_2O_7(s) + 2C(s) \longrightarrow Cr_2O_3(s) + Na_2CO_3(s) + CO(g)$$

Iron

Iron metal reacts when treated with an acid, giving off hydrogen gas and producing the pale green iron(II) ion, $Fe(H_2O)_6^{2+}$ (also called ferrous ion):

$$Fe(s) + 2[H^+(aq) + Cl^-(aq)] + 6H_2O(l) \longrightarrow$$
$$[Fe(H_2O)_6^{2+}(aq) + 2Cl^-(aq)] + H_2(g)$$

The iron(II) ion is a mild reducing agent and is slowly oxidized in aqueous solution by molecular oxygen:

$$4Fe(H_2O)_6^{2+}(aq) + O_2(aq) + 4H^+(aq) \longrightarrow 4Fe(H_2O)_6^{3+}(aq) + 2H_2O(l)$$

The iron(III) ion $Fe(H_2O)_6^{3+}$ is a very pale purple, but readily forms hydroxo species, such as $Fe(H_2O)_5(OH)^{2+}$, which are yellow. Solutions of iron(III) salts (also called ferric salts) usually have a yellow color.

When solutions of iron(II) salts are made basic in the absence of oxygen, a white precipitate of iron(II) hydroxide appears:

$$Fe^{2+}(aq) + 2OH^-(aq) \longrightarrow Fe(OH)_2(s)$$

In the presence of air, however, this precipitate is quickly oxidized to a red brown precipitate, which we can represent as $Fe(OH)_3$:

$$4Fe(OH)_2(s) + O_2(aq) + 2H_2O(l) \longrightarrow 4Fe(OH)_3(s)$$

The substance $Fe(OH)_3$ is perhaps better represented as the hydrated oxide $Fe_2O_3 \cdot xH_2O$; when it is heated, iron(III) oxide is obtained. ■

Iron(II) oxide is weakly basic and readily dissolves in acids. Iron(III) oxide also dissolves easily in acids, but is amphoteric and dissolves slightly in strong bases. There is some evidence for the hydroxo ion $Fe(OH)_6^{3-}$ in these solutions. The iron(III) ion does show definite acidic properties, and in water it hydrolyzes, forming hydroxo ions such as $Fe(H_2O)_5(OH)^+$:

■ Rust is essentially the hydrated oxide $Fe_2O_3 \cdot xH_2O$.

$$Fe(H_2O)_6^{3+}(aq) + H_2O(l) \longrightarrow Fe(H_2O)_5(OH)^{2+}(aq) + H_3O^+(aq);$$
$$K_a = 6 \times 10^{-3}$$

The ferrate(VI) ion, FeO_4^{2-}, is prepared by oxidizing a suspension of iron(III) oxide with hypochlorite ion in basic solution:

$$Fe_2O_3(s) + 3ClO^-(aq) + 4OH^-(aq) \longrightarrow 2FeO_4^{2-}(aq) + 3Cl^-(aq) + 2H_2O(l)$$

This ion has a purple color. It is stable in basic solution, but decomposes in acidic solution, where it is a powerful oxidizing agent.

Metallic iron is produced commercially by reduction of its ores, hematite (Fe_2O_3), magnetite (Fe_3O_4), and siderite $(FeCO_3)$. The reduction is done at high temperature in a blast furnace (Figure 25.3). A charge of iron ore, coke (C), and limestone is added at the top of the furnace, and a blast of heated

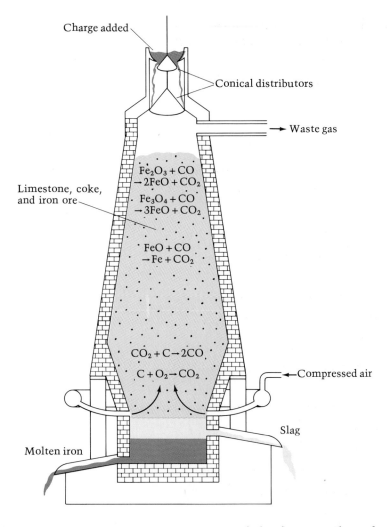

Charge added

Conical distributors

Waste gas

$$Fe_2O_3 + CO \\ \rightarrow 2FeO + CO_2$$

$$Fe_3O_4 + CO \\ \rightarrow 3FeO + CO_2$$

Limestone, coke,
and iron ore

$$FeO + CO \\ \rightarrow Fe + CO_2$$

$$CO_2 + C \rightarrow 2CO$$

$$C + O_2 \rightarrow CO_2$$

Compressed air

Slag

Molten iron

Figure 25.3
Blast furnace for the reduction of iron ore.

air enters at the bottom. Near the bottom of the furnace, the coke burns to carbon dioxide. As the CO_2 rises through the heated coke, it is reduced to carbon monoxide. The carbon monoxide then reduces the iron oxides to metallic iron. The overall reduction of Fe_2O_3 can be written

$$Fe_2O_3(s) + 3CO(g) \longrightarrow 2Fe(l) + 3CO_2(g)$$

Molten iron flows to the bottom of the blast furnace, where it is drawn off. Impurities in the iron ore react with calcium oxide, from the limestone, to produce a glassy material called slag. Molten slag collects in a layer floating on the molten iron and is drawn off periodically.

Steels are alloys containing over 50% iron and up to about 1.5% carbon. The iron obtained from a blast furnace (called pig iron) contains a number of impurities, including 3–4% carbon, that make it brittle. The first step in the production of steel is to remove these impurities and reduce the carbon content of the iron through controlled oxidation. In the **basic oxygen process,** this is accomplished by blowing oxygen into the molten iron until the amount of carbon is reduced to the required level. Carbon in the molten iron reacts to give carbon monoxide. Other metals may be added to the steel to give it desired properties. Stainless steels, for example, contain 12–18% chromium and about 8% nickel.

Copper

Copper, and the other Group IB elements silver and gold, can be found as the free elements. This is a reflection of the stability of the zero oxidation states of these elements. Copper is not attacked by most acids (compare the standard potentials, Table 21.1, for Cu^{2+}/Cu and H^+/H_2). It does react with concentrated sulfuric acid and with nitric acid. In these cases, the anion of the acid acts as the oxidizing agent (rather than H^+, which is the usual oxidizing agent in acids). Sulfuric acid is reduced to SO_2.

$$Cu(s) + 2H_2SO_4(l) \longrightarrow [Cu^{2+}(aq) + SO_4^{2-}(aq)] + SO_2(g) + 2H_2O(l)$$

Dilute nitric acid is reduced to NO, concentrated acid to NO_2.

$$3Cu(s) + 8[H^+(aq) + NO_3^-(aq)] \longrightarrow$$
$$3[Cu^{2+}(aq) + 2NO_3^-(aq)] + 2NO(g) + 4H_2O(l)$$
$$Cu(s) + 4[H^+(aq) + NO_3^-(aq)] \longrightarrow$$
$$[Cu^{2+}(aq) + 2NO_3^-(aq)] + 2NO_2(g) + 2H_2O(l)$$

The oxidation product of copper in the above reactions is the copper(II) ion, $Cu(H_2O)_6^{2+}(aq)$, also called the cupric ion. It has a bright blue color. Hydrated copper(II) salts, such as copper(II) sulfate pentahydrate, $CuSO_4 \cdot 5H_2O$, also have a blue color. Four of the water molecules of copper(II) sulfate pentahydrate are associated with Cu^{2+}, and the fifth one is hydrogen bonded to the sulfate ion as well as to the water molecules on the copper ion. We can write the formula as $[Cu(H_2O)_4]SO_4 \cdot H_2O$ to better represent its structure. When this salt is heated, it loses its water of hydration in stages. The hydrate $CuSO_4 \cdot H_2O$ and the anhydrous salt $CuSO_4$ are colorless substances, which show that it is the complexing of water molecules to the copper ion that is responsible for the blue color.

When solutions of copper(II) ion are made basic, a blue green precipitate of copper(II) hydroxide, $Cu(OH)_2$, forms. By heating the solution, the hydroxide is converted to copper(II) oxide, CuO, which is black. The oxide and hydroxide are amphoteric and dissolve in strong base to form deep blue solutions, perhaps of the anion $Cu(OH)_4^-$.

Although most of the aqueous chemistry of copper involves the +2 oxidation state, there are a number of important binary compounds and complexes of copper(I). When copper is heated in oxygen below 1000°C, it forms the black copper(II) oxide, CuO. But above this temperature, it forms the red copper(I) oxide, Cu_2O. This oxide is found naturally as the mineral cuprite.

In water, the copper(I) ion $Cu(H_2O)_6^+$ disproportionates (that is, it reduces and oxidizes itself):

$$2Cu(H_2O)_6^+(aq) \longrightarrow Cu(s) + Cu(H_2O)_6^{2+}(aq) + 6H_2O(l)$$

For this reason, copper(I) compounds that might be expected to give the $Cu(H_2O)_6^+$ ion are unstable in aqueous solution. However, the above reaction can be shifted to the left if an insoluble copper(I) compound is formed.

Thus, insoluble copper(I) chloride can be prepared by boiling an acidic mixture of copper(II) chloride and copper metal:

$$[Cu^{2+}(aq) + 2Cl^-(aq)] + Cu(s) \longrightarrow 2CuCl(s)$$
$$\text{white precipitate}$$

Similarly, the reduction of Cu^{2+} to Cu^+ is easily accomplished when the concentration of Cu^+ is kept low by the formation of a sufficiently insoluble copper(I) salt. Consider the reduction of Cu^{2+} to copper(I) by the iodide ion to form a white precipitate of copper(I) iodide:

$$2Cu^{2+}(aq) + 4I^-(aq) \longrightarrow 2CuI(s) + I_2(aq)$$

The reaction goes to the right, even though the iodide ion is a very mild reducing agent, because of the formation of very insoluble copper(I) iodide.

The principal commercial use of copper is as an electrical conductor. Common ores of copper are native copper (the free metal), copper oxides, and copper sulfides. Most copper is presently obtained by open pit mining of low-grade rock containing only a few per cent copper as copper sulfides. This ore is concentrated in copper by flotation.■ In this process, a slurry of the crushed ore is agitated with air, and the copper sulfides are carried away in the froth. This concentrated ore is then treated in several steps that result in the production of molten copper(I) sulfide, Cu_2S. This molten material, called *matte*, is reduced to copper by blowing air through it.

■ The general steps in the production of a metal from its ore, including flotation, are discussed in Section 23.5.

$$Cu_2S(l) + O_2(g) \longrightarrow 2Cu(l) + SO_2(g)$$

The metal produced in this step is called *blister copper* and is about 99% pure. For electrical use, the copper must be further purified or refined by electrolysis, using an anode of impure copper and a cathode of pure copper in a copper(II) sulfate bath (Figure 25.4). During the electrolysis, copper(II) ions leave the anode and plate out on the cathode. Less reactive metals, such as gold, silver, and platinum, form a valuable mud that collects on the bottom of the electrolytic cell. Metals more reactive than copper remain as ions in the electrolytic bath.

Example 25.2

Write molecular equations describing the preparation of chromium trioxide from chromium(III) sulfate.

Solution

Chromium(III) ion is oxidized in basic solution to chromate ion. When this solution is acidified, it forms the dichromate ion. If the solution is concentrated, and concentrated sulfuric acid added, chromium trioxide precipitates. The molecular equations are

$$Cr_2(SO_4)_3 + 3H_2O_2 + 10KOH \longrightarrow$$
$$2K_2CrO_4 + 3K_2SO_4 + 8H_2O$$
$$2K_2CrO_4 + 2H_2SO_4 \longrightarrow K_2Cr_2O_7 + 2KHSO_4 + H_2O$$
$$K_2Cr_2O_7 + 2H_2SO_4 \longrightarrow 2CrO_3 + 2KHSO_4 + H_2O$$

Exercise 25.3

Write molecular equations for the preparation of copper(I) iodide from copper metal.
(See Problems 25.35 and 25.36.)

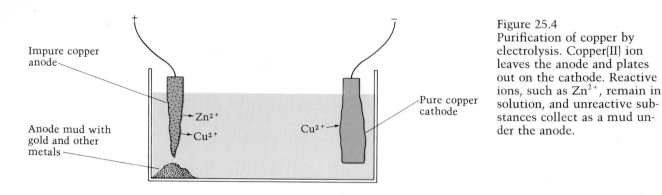

Figure 25.4
Purification of copper by electrolysis. Copper(II) ion leaves the anode and plates out on the cathode. Reactive ions, such as Zn^{2+}, remain in solution, and unreactive substances collect as a mud under the anode.

Exercise 25.4

What is the oxidation number of tungsten in scheelite, $CaWO_4$?

(See Problems 25.37 and 25.38.)

Exercise 25.5

Write a balanced equation for the reaction of dichromate ion and iodide ion in acidic solution.

(See Problems 25.39 and 25.40.)

Complex Ions and Coordination Compounds

As shown in the previous section, ions of the transition elements exist in aqueous solution as *complex ions.* Iron(II) ion, for example, exists in water as $Fe(H_2O)_6^{2+}$. The water molecules in this ion are arranged about the iron atom with their oxygen atoms bonded to the metal by donating electron pairs to it. Replacing the H_2O molecules by six CN^- ions gives the $Fe(CN)_6^{4-}$

Table 25.5
Fourth-Period Transition Elements Essential to Human Nutrition

Element	Some Biochemical Substances	Function
Vanadium		*
Chromium	Glucose tolerance factor	Utilization of glucose
Manganese	Isocitrate dehydrogenase	Cell energetics
Iron	Hemoglobin and myoglobin	Transport and storage of oxygen
	Cytochrome *c*	Cell energetics
	Catalase	Decomposition of hydrogen peroxide
Cobalt	Cobalamin (vitamin B_{12})	Development of red blood cells
Nickel		*
Copper	Ceruloplasmin	Synthesis of hemoglobin
	Cytochrome oxidase	Cell energetics
Zinc	Carbonic anhydrase	Elimination of carbon dioxide
	Carboxypeptidase A (pancreatic juice)	Protein digestion
	Alcohol dehydrogenase	Oxidation of ethanol

*Believed to be essential, but function not identified.

ion. The biological activity of the transition elements and their role in human nutrition (Table 25.5) depend in most cases on the formation of *complexes*, or *coordination compounds*, which exhibit the type of bonding that occurs in $Fe(H_2O)_6^{2+}$ and $Fe(CN)_6^{4-}$. In the chapter opening, for example, we saw that hemoglobin, a complex of iron, is vital to the transport of oxygen by the red blood cells.

25.3 Formation and Structure of Complexes

A metal atom, particularly a transition-metal atom, often functions in chemical reactions as a Lewis acid, accepting electron pairs from molecules or ions.■ Thus, Fe^{2+} and H_2O can bond to one another in a Lewis acid–base reaction:

■ Lewis acid–base reactions are discussed in Section 17.8.

$$Fe^{2+} \quad + \quad :\overset{..}{\underset{H}{O}}—H \quad \longrightarrow \quad \left[Fe:\overset{..}{\underset{H}{O}}—H \right]^{2+}$$

Coordinate covalent bond

Lewis acid   Lewis base

A pair of electrons on the oxygen atom of H_2O forms a coordinate covalent bond to Fe^{2+}. In water, the Fe^{2+} ion ultimately bonds to six H_2O molecules to give the $Fe(H_2O)_6^{2+}$ ion.

The Fe^{2+} ion also undergoes a similar Lewis acid–base reaction with cyanide ions, CN^-. In this case, the Fe^{2+} ion bonds to the electron pair on the carbon atom of CN^-:

$$Fe^{2+} + :C{\equiv}N:^- \longrightarrow (Fe:C{\equiv}N:)^+$$

Finally, a very stable ion $Fe(CN)_6^{4-}$ is obtained with six cyanide ions.■ Note that the charge on the $Fe(CN)_6^{4-}$ ion equals the sum of the charges on the ions from which it is formed: $+2 + 6(-1) = -4$.

In some cases, a neutral species is produced from a metal ion with anions. *Cis*-platin, the anticancer drug discussed in the opening of Chapter 1, has the structure

■ Although the ion $Fe(CN)_6^{4-}$ is a complex of Fe^{2+} and CN^- ions, a solution of $Fe(CN)_6^{4-}$ contains negligible concentration of CN^-. Thus, a substance such as $K_4[Fe(CN)_6]$ is relatively nontoxic, even though the free cyanide ion is a violent poison.

$$\overset{\textstyle :\overset{..}{\underset{..}{Cl}}:}{\underset{\textstyle \overset{..}{\underset{..}{N}H_3}}{H_3N:\overset{..}{\underset{..}{Pt}}:\overset{..}{\underset{..}{Cl}}:}}$$

It consists of Pt^{2+} with two NH_3 molecules (neutral) and two Cl^- ions, giving a neutral species. Iron pentacarbonyl, $Fe(CO)_5$, is an example of a neutral species formed from a neutral iron atom with CO molecules.

Basic Definitions

A **complex ion** is a metal atom or ion with Lewis bases attached to it through coordinate covalent bonds. A **complex** is a compound consisting either of complex ions with other ions of opposite charge, for example, the compound $K_4[Fe(CN)_6]$ of the complex ion $Fe(CN)_6^{4-}$ with four K^+ ions, or a neutral species, such as *cis*-platin. Because the Lewis bases are attached to the metal atom through coordinate covalent bonds, a complex is also called a **coordination compound.**

Complex	Coordination Number
$Ag(NH_3)_2{}^+$	2
$HgI_3{}^-$	3
$PtCl_4{}^{2-}$, $Ni(CO)_4$	4
$Fe(CO)_5$, $Co(CN)_5{}^{3-}$	5
$Co(NH_3)_6{}^{3+}$, $W(CO)_6$	6
$Mo(CN)_7{}^{3-}$	7
$W(CN)_8{}^{4-}$	8

Table 25.6
Examples of Complexes of Various Coordination Numbers

Ligands are the Lewis bases attached to the metal atom in a complex. Since they are electron-pair donors, ligands may be neutral molecules (such as H_2O or NH_3) or anions (such as CN^- or Cl^-) that have at least one atom with a lone pair of electrons. Cations only rarely function as ligands. We might expect this, since an electron pair on a cation is held securely by the positive charge, so it would not be involved in coordinate bonding. ∎

The **coordination number** of a metal atom in a complex is the total number of bonds the metal forms with ligands. In $Fe(H_2O)_6{}^{2+}$, the iron atom bonds to each of the oxygen atoms in the six water molecules. Therefore, the coordination number of iron in this ion is 6, by far the most common coordination number. Coordination number 4 is also well known, and many examples of number 5 have been recently discovered. Table 25.6 gives some examples of complexes for the coordination numbers 2 to 8. The coordination number for an atom depends on several factors, but size of the metal atom is important. Thus, coordination numbers 7 and 8 are seen primarily in fifth- and sixth-period elements. ∎

■ A cation in which the positive charge is far removed from an electron pair that could be donated can function as a ligand. An example is the pyrazinium ion

■ Very high coordination numbers (9 to 12) are known for some complex ions of the lanthanide elements.

Polydentate Ligands

The ligands we have discussed so far bond to the metal atom through one atom of the ligand. Water bonds through the oxygen atom; ammonia bonds through the nitrogen atom. These are called **unidentate ligands** (meaning "one-toothed" ligands). **Bidentate ligands** ("two-toothed" ligands) can bond through two atoms. Ethylenediamine is an example of such a ligand:

ethylenediamine

Nitrogen atoms at the ends of the molecule have lone pairs of electrons that can form coordinate covalent bonds. In forming a complex, the ethylenediamine molecule bends around so that both nitrogen atoms coordinate to the metal atom, M:

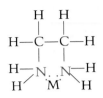

Since ethylenediamine is a common bidentate ligand, it is frequently abbreviated in formulas as "en." Figure 25.5 shows the structure of the stable ion $Co(en)_3^{3+}$. The oxalate ion, $C_2O_4^{2-}$, is another common bidentate ligand:

oxalate ion

   The hemoglobin molecule, in red blood cells, is an example of a complex with a **quadridentate ligand**—one that bonds to the metal atom through four ligand atoms. Hemoglobin consists of a protein *globin* chemically bonded to *heme*, whose structure is shown in Figure 25.6 (*left*). Heme is a planar molecule consisting of iron(II) to which a quadridentate ligand is bonded through its four nitrogen atoms.
   Ethylenediaminetetraacetate ion (EDTA) is a ligand that bonds through six of its atoms.

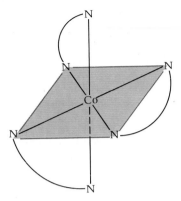

Figure 25.5
Structure of tris(ethylenediamine) cobalt(III) ion, $Co(en)_3^{3+}$. Note that N⏜N represents ethylenediamine.

Figure 25.6
Left: The structure of heme. *Right:* Complex of Fe^{2+} and ethylenediaminetetraacetate ion (EDTA). Notice how the EDTA ion envelops the metal ion.

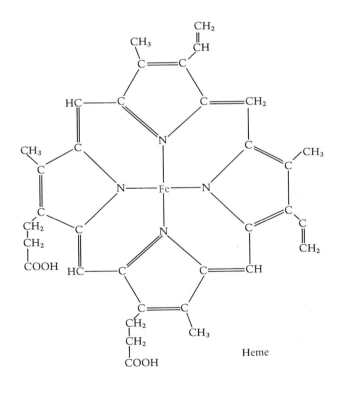

Heme

Complex of Fe^{2+} and EDTA

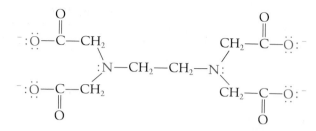

It can completely envelop a metal atom, simultaneously occupying all six positions in an octahedral geometry (Figure 25.6, *right*).

Polydentate ligands ("having many teeth") can bond with two or more atoms to a metal atom. Complexes formed by polydentate ligands are frequently quite stable and are called **chelates.**■ Because of the stability of chelates, polydentate ligands (also called *chelating agents*) are often used to remove metal ions from a chemical system. EDTA, for example, is added to certain canned foods to remove transition-metal ions that can catalyze the deterioration of the food. The same chelating agent has been used to treat lead poisoning because it binds Pb^{2+} ions as the chelate, which can then be excreted by the kidneys.

■ The term *chelate* is derived from the Greek *chele* for "claw," because a polydentate ligand appears to attach itself to the metal atom like crab claws to some object.

Discovery of Complexes; Formula of a Complex

In 1798 B. M. Tassaert found that a solution of cobalt(II) chloride in aqueous ammonia exposed to air (an oxidizing agent) deposits orange yellow crystals. He assigned the formula $CoCl_3 \cdot 6NH_3$ to these crystals. Later, a similar compound of platinum was assigned the formula $PtCl_4 \cdot 6NH_3$. Because these formulas suggested that the substances were somehow composed of two stable compounds, platinum(IV) chloride and ammonia in the latter case, they were called complex compounds or simply complexes.

The basic explanation for the structure of these complexes was given by the Swiss chemist Alfred Werner in 1893. According to Werner, a metal atom exhibits two kinds of valences, a primary valence and a secondary valence. The primary valence is what we now call the oxidation number of the metal. The secondary valence corresponds to what we now call the coordination number, which is often 6. In Werner's view, the substance previously represented by the formula $PtCl_4 \cdot 6NH_3$ is composed of the ion $Pt(NH_3)_6^{4+}$, with six ammonia molecules directly attached to the platinum atom. The charge of this ion is balanced by four Cl^- ions, giving a neutral compound with the structural formula $[Pt(NH_3)_6]Cl_4$.

Table 25.6 lists a series of platinum(IV) complexes studied by Werner. It gives both the old and Werner's modern formulas. (Note that square brackets are used in modern formulas to separate the metal and its associated groups from the rest of the formula.) According to Werner's theory, these complexes should dissolve to give different numbers of ions per formula unit. For example, $[Pt(NH_3)_4Cl_2]Cl_2$ dissolves to give three ions: $Pt(NH_3)_4Cl_2^{2+}$ and two Cl^- ions. Werner was able to show that the electrical conductances of solutions of these complexes were equal to what was expected for the number of ions predicted by his formulas. He also demonstrated that the chloride ions in the platinum complexes were of two kinds, those that could be precip-

Table 25.7
Some Platinum(IV)
Complexes Studied
by Werner

Old Formula	Modern Formula	Number of Ions	Number of Free Cl^- Ions
$PtCl_4 \cdot 6NH_3$	$[Pt(NH_3)_6]Cl_4$	5	4
$PtCl_4 \cdot 4NH_3$	$[Pt(NH_3)_4Cl_2]Cl_2$	3	2
$PtCl_4 \cdot 3NH_3$	$[Pt(NH_3)_3Cl_3]Cl$	2	1
$PtCl_4 \cdot 2NH_3$	$[Pt(NH_3)_2Cl_4]$	0	0

itated from solution as AgCl using silver nitrate and those that could not. He explained that the chloride ions within the platinum complex ion are securely attached to the metal atom, and only those outside the complex ion can be precipitated with silver nitrate. The number of free Cl^- ions (those not attached to platinum) determined this way agreed exactly with his formulas. Over a period of twenty years, Werner carried out many experiments on complexes, all of them in basic agreement with his theory. He received the Nobel Prize for his work in 1913.

Exercise 25.6

Another complex studied by Werner had a composition corresponding to the formula $PtCl_4 \cdot 2KCl$. From electrical-conductance measurements, he determined that each formula unit contained three ions. He also found that silver nitrate did not give a precipitate of AgCl with this complex. Write a formula for this complex that agrees with this information.

(See Problems 25.41 and 25.42.)

Aside: On the Stability of Chelates

The special stability of chelates stems from the additional entropy obtained when they are formed. This leads to a large negative $\Delta G°$, which is equivalent to a large equilibrium constant for the formation of the chelate. To understand how this happens, consider the formation of the chelate $Co(en)_3^{3+}$ from the complex ion $Co(NH_3)_6^{3+}$, with unidentate ligands NH_3:

$$Co(NH_3)_6^{3+} + 3en \rightleftharpoons Co(en)_3^{3+} + 6NH_3$$

Each en molecule replaces two NH_3 molecules. Therefore, the number of particles in the reaction mixture will be increased if the reaction goes to the right. In most cases, an increase in number of particles increases the possibilities for randomness or disorder. Thus, when the reaction goes to the right, there is an increase in entropy; that is, $\Delta S°$ is positive.

At the same time, the reaction involves very little change of energy ($\Delta E°$) or enthalpy ($\Delta H° = \Delta E° + P\Delta V$) because the bonds are similar (all consist of a nitrogen atom coordinated to a cobalt atom). Six nitrogen-metal bonds in $Co(NH_3)_6^{3+}$ are broken, and six new nitrogen-metal bonds of about the same energy are formed in $Co(en)_3^{3+}$. Thus, $\Delta H° \approx 0$.

(Continued)

The spontaneity of a reaction depends on $\Delta G°$, which equals $\Delta H° - T\Delta S°$. But since $\Delta H° \simeq 0$,

$$\Delta G° = \Delta H° - T\Delta S°$$
$$= -T\Delta S°$$

Because the entropy change is positive, $\Delta G°$ is negative, and the reaction is spontaneous from left to right.

The fact that the equilibrium for the above reaction favors the chelate $Co(en)_3^{3+}$ means that the chelate has thermodynamic stability. A similar argument could be made for any equilibrium involving the replacement of unidentate ligands by polydentate ligands. Reaction tends to favor the chelate.

25.4 Naming Coordination Compounds

Thousands of coordination compounds are now known. A systematic method of naming such compounds, or *nomenclature*, can give us basic information about the structure of each coordination compound. What is the metal in the complex? Does the metal atom occur in the cation or anion? What is the oxidation state of the metal? What are the ligands? We can answer these questions by following the rules of nomenclature agreed upon by the International Union of Pure and Applied Chemistry (IUPAC). These rules are essentially an extension of those originally given by Werner.

1. When naming a salt, the name of the cation precedes the name of the anion. (This is a familiar rule.) For example, the coordination compound

$K_4[Fe(CN)_6]$ is named potassium hexacyanoferrate(II)

cation — anion

and

$[Co(NH_3)_6]Cl_3$ is named hexaamminecobalt(III) chloride

cation — anion

2. The name of the complex, whether anion, cation, or neutral species, consists of two parts, written together as one word. Ligands are named first, and the metal atom is named second. For example,

$Fe(CN)_6^{4-}$ is named hexacyanoferrate(II) ion

ligand name — metal name

$Co(NH_3)_6^{3+}$ is named hexaamminecobalt(III) ion

ligand name — metal name

3. The complete ligand name consists of a Greek prefix denoting the number of ligands, followed by the specific name of the ligand. If there are two or more ligands, the ligands are written in alphabetical order (disregarding Greek prefixes).

a. Anionic ligands end in -*o*. Some examples are given in the following table:

Anion Name	Ligand Name
Bromide, Br^-	Bromo
Carbonate, CO_3^{2-}	Carbonato
Chloride, Cl^-	Chloro
Cyanide, CN^-	Cyano
Fluoride, F^-	Fluoro
Hydroxide, OH^-	Hydroxo
Oxalate, $C_2O_4^{2-}$	Oxalato
Oxide, O^{2-}	Oxo
Sulfate, SO_4^{2-}	Sulfato

b. Neutral ligands are usually given the name of the molecule. There are, however, several important exceptions, which are given in the following table:

Molecule	Ligand Name
Ammonia, NH_3	Ammine
Carbon monoxide, CO	Carbonyl
Water, H_2O	Aqua ■

■ *Aquo* is used in earlier literature.

c. The prefixes used to denote the number of ligands are *mono-* (1) (usually omitted), *di-* (2), *tri-* (3), *tetra-* (4), *penta-* (5), *hexa-* (6), and so forth. To see how the ligand name is formed, consider the complex ions

$$Fe(CN)_6^{4-} \text{ or hexacyanoferrate(II) ion}$$

6   CN^- ligands

$$Co(NH_3)_6^{3+} \text{ or hexaamminecobalt(III) ion}$$

6   NH_3 ligands

d. If the name of the ligand also has a Greek prefix, the number of ligands is denoted bis (2), tris (3), tetrakis (4), and so forth. The name of the ligand follows in parentheses. For example, the complex $[Co(en)_3]Cl_3$ is named

tris(ethylenediamine)cobalt(III) chloride

3   ligand name

4. The complete metal name consists of the name of the metal, followed by -*ate* if the complex is an anion, which in turn is followed by the oxidation number of the metal indicated by Roman numerals in parentheses. (An oxidation state of zero is indicated by 0 in parentheses.) If there is a Latin name for the metal, it is used to name the anion (except for mercury). These names are given in the following table:

English Name	Latin Name	Anion Name
Copper	Cuprum	Cuprate
Gold	Aurum	Aurate
Iron	Ferrum	Ferrate
Lead	Plumbum	Plumbate
Silver	Argentum	Argentate
Tin	Stannum	Stannate

Examples are

hexacyanoferrate(II)   hexaamminecobalt(III)

ferrum | oxidation   metal   oxidation
= iron | number 2   name   number 3

indicates
an anion

Example 25.3

Give the IUPAC name of the following coordination compounds: (a) $[Pt(NH_3)_4Cl_2]Cl_2$, (b) $[Pt(NH_3)_2Cl_2]$, (c) $K_2[PtCl_6]$

Solution

(a) The cation is listed first in the formula.

$$[Pt(NH_3)_4Cl_2]Cl_2$$

cation   anions

Since there are two Cl^- anions, the charge on the cation is $+2$: $Pt(NH_3)_4Cl_2^{2+}$. The oxidation number of plat-

inum plus the sum of the charges on the ligands (-2) equals the cation charge $+2$. Therefore, the oxidation number of Pt is $+4$. Hence, the name of the compound is tetraamminedichloroplatinum(IV) chloride. Note that the ligands are listed in alphabetical order (ammine before chloro). (b) This is a neutral complex species. The oxidation number of platinum must balance that of the two chloride ions. The name is diamminedichloroplatinum(II). (c) The complex anion is $PtCl_6^{2-}$. The oxidation number of platinum is $+4$, and the name of the compound is potassium hexachloroplatinate(IV).

Exercise 25.7

Give the IUPAC names of (a) $[Co(NH_3)_5Cl]Cl_2$. (b) $K_2[Co(H_2O)(CN)_5]$, (c) $Fe(H_2O)_5(OH)^{2+}$.

(See Problems 25.49, 25.50, 25.51, and 25.52.)

Example 25.4

Write the structural formulas corresponding to the following IUPAC names: (a) hexaaquairon(II) chloride, (b) potassium tetrafluoroargentate(III), (c) pentachlorotitanate(II) ion

Solution

(a) The complex cation hexaaquairon(II) is Fe^{2+} with six H_2O ligands: $Fe(H_2O)_6^{2+}$. The formula of the compound

is $[Fe(H_2O)_6]Cl_2$. (Remember to enclose the formula of the complex ion in parentheses.) (b) The compound contains the complex anion tetrafluoroargentate(III), that is, Ag^{3+} with four F^- ligands. The formula of the ion is AgF_4^-. Hence, the formula of the compound is $K[AgF_4]$. (c) The ion contains Ti^{2+} and five Cl^- ligands. Thus, the charge on the ion is $2 + 5(-1) = -3$. The formula of the complex ion is $TiCl_5^{3-}$.

Exercise 25.8

Write formulas for the following: (a) potassium hexacyanoferrate(III), (b) tetraamminedichlorocobalt(III) chloride, (c) tetrachloroplatinate(II) ion

(See Problems 25.53 and 25.54.)

25.5 Structure and Isomerism in Coordination Compounds

Although we described the formation of a complex as a Lewis acid–base reaction (Section 25.3), we did not go into any details of structure. We did not look at the geometry of complex ions, nor did we inquire about the precise nature of the bonding. There are three properties of complexes that have proved pivotal in determining these details.

1. *Isomerism*   Isomers are compounds with the same molecular formula (or the same simplest formula in the case of ionic compounds) but with different arrangements of atoms. Because their atoms are differently arranged, isomers have different properties. There are many possibilities for isomerism in coordination compounds. The study of isomerism can lead to information about atomic arrangement in coordination compounds. Werner's research on the isomerism of coordination compounds finally convinced others that his views were essentially correct.

2. *Paramagnetism*   Paramagnetic substances are attracted to a strong magnetic field. Paramagnetism is due to unpaired electrons in a substance. (*Ferromagnetism* in solid iron is also due to unpaired electrons, but in this case the magnetism of many iron atoms is aligned, giving a magnetic effect that is perhaps a million times stronger than that seen in paramagnetic substances.) Many complex compounds are paramagnetic. The magnitude of this paramagnetism can be measured with a *Gouy balance,* in which the force of magnetic attraction is balanced with weights (Figure 25.7). Since

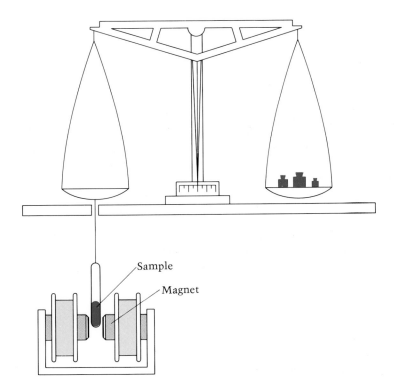

Figure 25.7
Gouy balance for measuring the paramagnetism of a substance.

Sample

Magnet

paramagnetism is related to the electron configuration of the complex, these measurements can give information about the bonding.

3. *Color*   A substance is colored because it absorbs light in the visible region of the spectrum. The absorption of visible light is due to a transition between closely spaced electronic energy levels. As we have seen, many coordination compounds are highly colored. This color is related to the electronic structure of the compounds. ∎

In this section, we will investigate the relationship between structure and isomerism. We will look into explanations of the paramagnetism and color of coordination compounds in the following sections.

∎ Recall that the wavelength of light absorbed by a substance is related to the difference between two energy levels in the substance. See Section 5.7.

Structural Isomerism

There are two major kinds of isomers. **Structural isomers** are isomers that differ in how the atoms are joined together, that is, by the order in which these atoms are bonded to one another. (For example, H—N=C=O and N=C—O—H are structural isomers because the H atom is bonded to N in one case, and to O in the other.) **Stereoisomers,** on the other hand, are alike in that the same atoms are bonded to each other in the same order, but they differ in the precise arrangement of these atoms in space.

   The simplest structural isomers of complexes are ionization isomers. **Ionization isomers** differ in which anion is coordinated to the metal atom. The following cobalt complexes provide an example:

$[Co(NH_3)_5(SO_4)]Br$ red compound

$[Co(NH_3)_5Br]SO_4$ violet compound

These isomers differ in whether the sulfate ion or the bromide ion is attached to the cobalt atom. The red compound gives a precipitate of AgBr when mixed with silver nitrate solution, but no precipitate of $BaSO_4$ with barium chloride solution. This implies that the sulfate ion is strongly attached to the metal. The violet compound, however, gives a precipitate of $BaSO_4$ when treated with barium chloride, but no precipitate of AgBr with silver nitrate.

   Hydrate isomers differ in the placement of water molecules in the complex. A well-studied example is that of the three chromium complexes with composition $CrCl_3 \cdot 6H_2O$:

$[Cr(H_2O)_6]Cl_3$ violet compound

$[Cr(H_2O)_5Cl]Cl_2 \cdot H_2O$ light green compound

$[Cr(H_2O)_4Cl_2]Cl \cdot 2H_2O$ dark green compound

In the violet complex, all of the water molecules are coordinated to the chromium atom. In the others, one or two water molecules are not directly attached to the metal atom, but occur elsewhere in the crystal lattice. The isomers can be differentiated by conductance measurements (which determine the number of ions per formula unit) and by the amount of AgCl precipitated by excess silver nitrate (Cl^- bonded to Cr is not precipitated). Moreover, the water molecules not directly coordinated to a Cr atom are easily lost. Thus, the isomers behave differently when exposed to a dry atmosphere in contact with concentrated sulfuric acid. (Concentrated sulfuric acid is a dehydrating agent.)

In **coordination isomers** both cation and anion are complex, and the ligands are distributed in different ways between the two metal atoms. An example is

$$[Cu(NH_3)_4][PtCl_4] \quad \text{and} \quad [Pt(NH_3)_4][CuCl_4]$$

In the first compound, the NH_3 ligands are associated with copper, and the Cl^- ligands with platinum. In the second compound, the ligands are transposed. It is also possible for ligands to be distributed in different ways between two metal atoms of the same element. An example of this sort of coordination isomerism is

$$[Pt(NH_3)_4][PtCl_4] \quad \text{and} \quad [Pt(NH_3)_3Cl][Pt(NH_3)Cl_3]$$

Linkage isomers have a more subtle structural difference. A ligand such as the nitrite ion, NO_2^-, can coordinate to a metal atom either with the electron pair on the nitrogen atom or with an electron pair on an oxygen atom. A complex with an O-bonded NO_2 ligand has the structure $[Co(NH_3)_5(ONO)]Cl_2$, where the nitrite ligand is written ONO to emphasize its O-bonding to cobalt. The compound is red and slowly changes to the yellow brown isomer $[Co(NH_3)_5(NO_2)]Cl_2$, in which the nitrite group is N-bonded. Ligands such as NO_2^- that can bond through one atom or another are called *ambidentate*. Another ambidentate ligand is the thiocyanate ion, SCN^-, which can bond through the sulfur atom or the nitrogen atom.

Exercise 25.9

Give the name describing the type of structural isomerism displayed by the following:
(a) $[Co(en)_3][Cr(CN)_6]$ and $[Cr(en)_3][Co(CN)_6]$
(b) $[Mn(CO)_5(SCN)]$ and $[Mn(CO)_5(NCS)]$
(c) $[Co(NH_3)_5(NO_3)]SO_4$ and $[Co(NH_3)_5(SO_4)]NO_3$
(d) $[Co(NH_3)_4(H_2O)Cl]Cl_2$ and $[Co(NH_3)_4Cl_2]Cl \cdot H_2O$

(See Problems 25.55 and 25.56.)

Exercise 25.10

A complex has the composition $Co(NH_3)_4(H_2O)Cl_3$. If conductance measurements show that there are three ions per formula unit and if precipitation of AgCl with silver nitrate shows that there are two Cl^- ions not coordinated to cobalt, what is the structural formula of the compound? Write the structural formula of an isomer.

(See Problems 25.57 and 25.58.)

Stereoisomerism

The existence of various types of structural isomers is strong evidence for the view that complexes consist of groups directly bonded to a central metal atom. The existence of stereoisomers not only strengthens this view, but also helps answer the question of how these groups are arranged about the central atom.

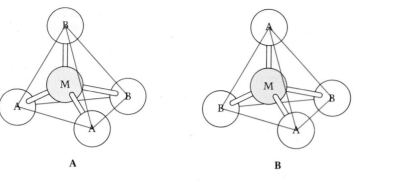

A   B

Figure 25.8
The tetrahedral complex MA_2B_2 has no geometric isomers. Molecule (a) can be rotated to look like (b).

Geometric isomers are isomers in which the atoms are joined to one another in the same way but differ because some atoms occupy different relative positions in space. For an example, consider the complex diamminedichloroplatinum(II), $[Pt(NH_3)_2Cl_2]$. Two compounds of this composition are known. One is an orange yellow compound with solubility at 25°C of 0.252 grams per 100 grams of water. The other compound is pale yellow and is much less soluble (0.037 grams per 100 grams of water at 25°C). ■ How do we explain the occurrence of these two isomers?

There are two symmetrical geometries of complexes of four ligands: tetrahedral and square planar. Let us write MA_2B_2 for a complex with ligands A and B about the metal atom M. As Figure 25.8 shows, a tetrahedral geometry for MA_2B_2 allows only one arrangement of ligands. The square planar geometry for MA_2B_2, however, gives two different arrangements. (These two arrangements for $[Pt(NH_3)_2Cl_2]$ are shown in Figure 25.9.) One, labeled *cis*, has the two A ligands on one side of the square and the two B ligands on the other side. The other arrangement, labeled *trans*, has A and B ligands across the square from one another. The *cis* and *trans* arrangements of MA_2B_2 are examples of geometric isomers.

That there are two isomers of $[Pt(NH_3)_2Cl_2]$ is evidence for the square planar geometry in this complex. But how do we identify the substances and their properties with the *cis* and *trans* arrangements? We can distinguish between these arrangements by predicting their polarity. The *trans* arrangement, being completely symmetrical, is nonpolar. However, the *cis* arrangement, with electronegative Cl atoms on one side of the platinum atom, is polar. This difference between *cis* and *trans* isomers should be reflected in their solubilities in water, since polar substances are more soluble in water

■ The isomers of $[Pt(NH_3)_2Cl_2]$ also differ in their biological properties. The orange yellow compound acts as an anticancer drug, but the other isomer has little anticancer activity.

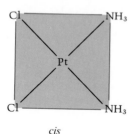

cis

Orange yellow compound
Solubility = 0.252 g/100 g H_2O

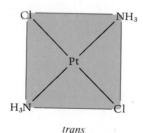

trans

Pale yellow compound
Solubility = 0.037 g/100 g H_2O

Figure 25.9
Geometric isomers of the square planar complex diamminedichloroplatinum(II), $[Pt(NH_3)_2Cl_2]$.

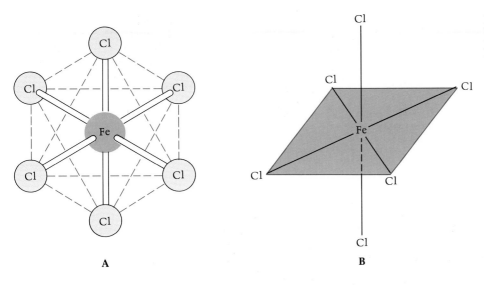

Figure 25.10
The octahedral geometry. Each position in this geometry is equivalent, as (a) shows. The octahedral geometry is often shown simply as in (b).

A   B

(itself a polar substance) than nonpolar substances. Thus, we expect the more soluble isomer to have the *cis* arrangement.

Although the difference in solubility is revealing, the most direct evidence of polarity in a molecule comes from measurements of dipole moment. These show that the less soluble platinum(II) complex has no dipole moment and must be *trans*, and the other isomer does have a dipole moment and must be *cis*. Figure 25.9 shows the two isomers and their properties.

Six-coordinate complexes have only one symmetrical geometry: octahedral (Figure 25.10). Geometric isomers are possible for this geometry also. Consider the complex MA_4B_2 in which two of the A ligands occupy positions just opposite one another. The other four ligands have a square planar arrangement in which *cis–trans* isomers are possible. Tetraamminedichlorocobalt(III) chloride, $[Co(NH_3)_4Cl_2]Cl$, is an example of a complex with such geometric isomers. The *trans* compound is green; the *cis* compound is purple. See Figure 25.11.

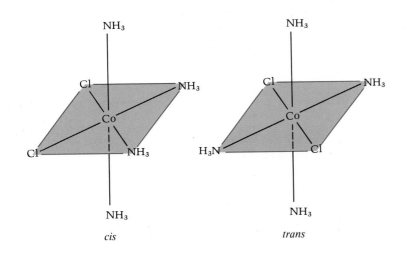

cis   *trans*

Figure 25.11
Geometric isomers of tetraamminedichlorocobalt(III) ion, $Co(NH_3)_4Cl_2{}^+$.

Example 25.5

Are there any geometric isomers of the stable octahedral complex $[Co(NH_3)_3(NO_2)_3]$? If yes, draw them.

Solution

Yes, there are geometric isomers. This is easily seen if we first draw an NH_3 and an NO_2^- ligand opposite one another in an octahedral geometry. Then, the other ligands can have *cis* or *trans* arrangements on the square perpendicular to the axis of the first NH_3 and NO_2^- ligands. See Figure 25.12.

Exercise 25.11

Do any of the following stable octahedral complexes have geometric isomers? If so, draw them.
(a) $Co(NH_3)_5Cl^{2+}$
(b) $Co(NH_3)_4(H_2O)_2^{3+}$
(c) $[Cr(NH_3)_3(SCN)_3]$
(d) $Co(NH_3)_6^{3+}$

(See Problems 25.59 and 25.60.)

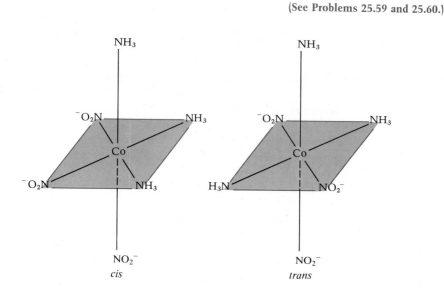

cis   trans

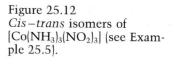

Figure 25.12
Cis–trans isomers of $[Co(NH_3)_3(NO_2)_3]$ (see Example 25.5).

Bidentate ligands increase the potential for isomerism in octahedral complexes. Figure 25.13 shows the isomers of the dichlorobis(ethylenediamine)-cobalt(III) ion, $CoCl_2(en)_2^+$. Isomer (a) is the *trans* isomer (green color). Notice that both (b) and (c) are *cis* isomers (both have a violet color). Yet (b) and (c) are not identical molecules. They are **enantiomers**, or **optical isomers;** that is, they are isomers that are nonsuperimposable mirror images of one another.

To better understand the nature of enantiomers, note that the two *cis* isomers have the same relationship to each other as your left hand and right hand have to each other. The mirror image of a left hand looks like the right hand, and vice versa (Figure 25.14). But neither the right hand nor the mirror image of the left hand can be turned in any way to look exactly like the left hand; that is, the right and left hands cannot be *superimposed* on one another. (Remember that a left-handed glove does not fit a right hand.) Any physical object possessing this handedness characteristic—whose mirror

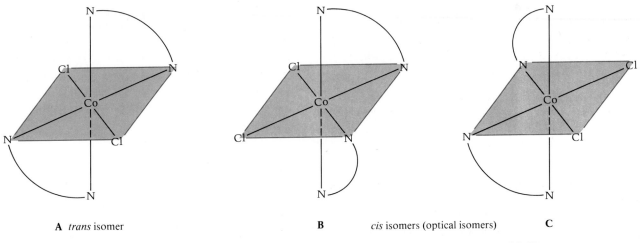

A *trans* isomer   B   *cis* isomers (optical isomers)   C

Figure 25.13
Isomers of dichlorobis-
(ethylenediamine)cobalt(III)
ion, $CoCl_2(en)_2{}^+$. (a) The
trans isomer. (b) and (c) Optical isomers of the *cis* form.
Note that N⌒N represents
ethylenediamine.

image is not identical with itself—is said to be **chiral** (from the Latin *chiro*, hand). Isomer (a) in Figure 25.13 is *achiral*; its mirror image is identical to itself. However, isomer (b) is chiral, and its mirror image, isomer (c), is not superimposable on (b). ∎

Enantiomers (optical isomers) have identical properties in a symmetrical environment. They have identical melting points and the same solubilities and colors. Enantiomers are usually differentiated by the manner in which they affect *plane-polarized* light. Normally, a light beam consists of electromagnetic waves vibrating in all possible planes about the direction of the beam (Figure 25.15). One of these planes may be selected out by passing a beam of light through a *polarizer* (say, through a Polaroid lens). When this plane-polarized light is passed through a solution containing an enantiomer, such as one isomer of *cis*-[CoCl$_2$(en)$_2$]Cl, the plane of the light wave is twisted. One of the enantiomers twists the plane to the right (as the light comes out toward you); the other isomer twists the plane to the left by the same angle. Because of this ability to turn the plane of light waves, enantiomers are said to be **optically active.** (This is also the origin of the term

■ A pencil is *achiral*—there are no left-handed or right-handed pencils.

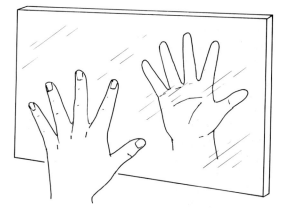

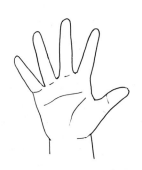

Figure 25.14
The mirror image of the left hand looks like the right hand. But the left hand itself cannot be superimposed on the right hand.

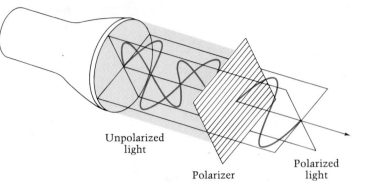

Figure 25.15
Polarization of light. Light
from the source consists of
waves vibrating in various
planes along any axis. The
polarizer filters out all waves
except those vibrating in a
particular plane.

optical isomer.) Figure 25.16 shows a sketch of a *polarimeter,* an instrument that determines the angular change in the plane of a light wave made by an optically active compound.

A compound that rotates the plane of polarized light to the right is called **dextrorotatory** and is labeled *d.* A compound that rotates the plane of polarized light to the left is called **levorotatory** and is labeled *l.* Thus, the dextrorotatory isomer of *cis-*$[CoCl_2(en)_2]^+$ is *d-cis-*dichlorobis(ethylenediamine)cobalt(III).

A chemical reaction normally produces an equal mixture of optically active isomers, with 50% of the *d* isomer and 50% of the *l* isomer. Such a product is called a **racemic mixture.** Because each isomer rotates the plane of polarized light in equal but opposite directions, a racemic mixture has no net effect on polarized light.

In order to show that optical isomers exist, a racemic mixture must be separated into its *d* and *l* isomers; that is, the racemic mixture must be *resolved.* One way to resolve a mixture containing *d* and *l* complex ions is to prepare a salt with an optically active ion of opposite charge. For example, the tartaric acid, $H_2C_4H_4O_6$, prepared from the white substance in wine vats is the optically active isomer *d-*tartaric acid. If the racemic mixture of *cis-*$[CoCl_2(en)_2]Cl$ is treated with *d-*tartaric acid, the *d-*tartrate salts of *d-* and *l-cis-*$[CoCl_2(en)_2]^+$ may be crystallized. These salts will no longer be optical isomers of one another and will have different solubilities.

Figure 25.16
A sketch of a polarimeter.
[From Hart, *Organic Chemistry: A Short Course,* 6th ed. (Boston: Houghton Mifflin Co., 1983), p. 124.]

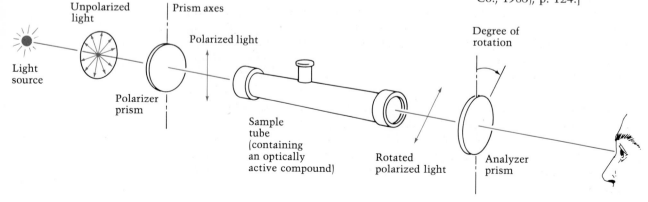

Example 25.6

Are there optical isomers of the complex [Co(en)₃]Cl₃? If yes, draw them.

Solution

Yes, there are optical isomers, because the complex ion

[Co(en)₃]³⁺ has nonsuperimposable mirror images. See Figure 25.17.

Exercise 25.12

Do any of the following have optical isomers? If so, draw them.
(a) *trans*-Co(en)₂(NO₂)₂⁺   (b) *cis*-Co(en)₂(NO₂)₂⁺
(c) Ir(en)₃³⁺   (d) *cis*-[Ir(H₂O)₃Cl₃]

(See Problems 25.61 and 25.62.)

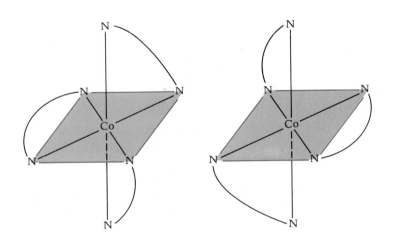

Figure 25.17
Optical isomers of the complex ion Co(en)₃³⁺ (see Example 25.6).

25.6 Valence Bond Theory of Complexes

A complex is formed, as we noted earlier, when electron pairs from ligands are donated to a metal ion. This does not explain how the metal ion can accept electron pairs. Nor does it explain the paramagnetism often observed in complexes. *Valence bond theory* provided the first detailed explanation of the electronic structure of complexes. This explanation is essentially an extension of the view of covalent bonding we described earlier.■ According to this view, a covalent bond is formed by the overlap of two orbitals, one from each bonding atom. In the usual covalent bond formation, each orbital originally holds one electron, and after the overlap, a bond is formed that holds two electrons. In the formation of a coordinate covalent bond in a complex, however, a ligand orbital containing two electrons overlaps an unoccupied orbital on the metal atom. Figure 25.18 diagrams these two bond formations.

■ See Section 8.3.

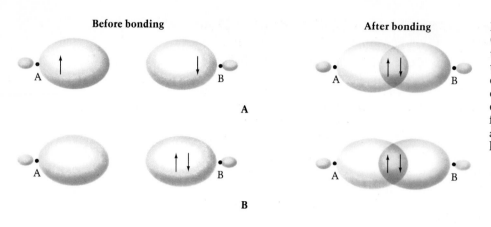

Before bonding   **After bonding**

A

B

Figure 25.18
Covalent bond formation between atoms A and B. (a) In the usual case, each of the overlapping orbitals contains one electron. (b) When a coordinate covalent bond forms, one orbital containing a lone pair of electrons overlaps an empty orbital.

Octahedral Complexes

Among the first complexes studied were those of chromium(III). There are a great number of these, including the violet complex hexaaquachromium(III) chloride, $[Cr(H_2O)_6]Cl_3$, and the yellow complex hexaamminechromium(III) chloride, $[Cr(NH_3)_6]Cl_3$. Almost all chromium(III) complexes have the coordination number 6, and thus are octahedral complexes. All are paramagnetic, displaying a magnetism equivalent to that of three unpaired electrons. Let us see how valence bond theory explains the bonding and paramagnetism of the complex ions of chromium(III), such as $Cr(NH_3)_6^{3+}$.

The electron configuration of the chromium atom is $[Ar]3d^54s^1$. In forming a transition-metal ion, the outer s electrons ($4s^1$, here) are lost first, then the outer d electrons. Thus, Cr^{3+} has the configuration $[Ar]3d^3$. The orbital diagram is

Cr^{3+}: [Ar] (↑)(↑)(↑)( )( )( ) ( ) ( )( )( )( ) ( )( )( )( )( )( )

   $3d$ $4s$ $4p$ $4d$

Notice that the $3d^3$ electrons are placed in separate orbitals with their spins in the same direction, following Hund's rule.

If electron pairs from six $:NH_3$ ligands are to bond to Cr^{3+} forming six equivalent bonds, hybrid orbitals will be required. Hybrid orbitals with an octahedral arrangement can be formed with two d orbitals, the $4s$ orbital, and the three $4p$ orbitals. The d orbitals could be either $3d$ or $4d$; the two available $3d$ orbitals are used because they have lower energy. We will call these d^2sp^3 hybrid orbitals (rather than sp^3d^2) to emphasize that the d orbitals have a principal quantum number of 1 less than that of the s and p orbitals. We can now write the orbital diagram for the metal atom in the complex:

$Cr(NH_3)_6^{3+}$: [Ar] (↑)(↑)(↑)│(↑ ↓)(↑ ↓) (↑ ↓) (↑ ↓)(↑ ↓)(↑ ↓)│ ( )( )( )( )( )( )

   $3d$ │ $4s$ $4p$ │ $4d$

d^2sp^3 bonds to ligands

Electron pairs donated from ligands are shown in color. Note that there are three unpaired electrons in $3d$ orbitals on the chromium atom, thus explaining the paramagnetism of this complex ion. The bonding in other octahedral complexes of chromium(III) is essentially the same.

The bonding in complexes of iron(II) is more diverse. Most of the complexes are octahedral and paramagnetic, and the magnitude of this paramagnetism indicates four unpaired electrons. An example is the pale green hexaaquairon(II) ion, $Fe(H_2O)_6^{2+}$. However, the yellow hexacyanoferrate(II) ion, $Fe(CN)_6^{4-}$, is an example of a diamagnetic iron(II) complex ion. (Iron in oxidation state $+2$ also occurs occasionally in complexes of coordination number 4, with tetrahedral geometry.)

Let us consider the bonding in $Fe(H_2O)_6^{2+}$. The configuration of the iron atom is $[Ar]3d^64s^2$, and the configuration of Fe^{2+} is $[Ar]3d^6$. The orbital diagram of the ion is

Fe^{2+}: [Ar] $(\uparrow\downarrow)(\uparrow \ )(\uparrow \ )(\uparrow \ )(\uparrow \ )$ $(\ \ )$ $(\ )(\ )(\ )$ $(\ )(\ )(\ )(\ )(\ )$
$\qquad\qquad\qquad$ $3d$ $\qquad\qquad\qquad\qquad$ $4s$ $\qquad$ $4p$ $\qquad\qquad\qquad$ $4d$

According to Hund's rule, the first five of the $3d^6$ electrons go into separate orbitals; the sixth electron must then pair up with one of the others. Since the $3d$ orbitals have electrons in them, they cannot be used for bonding to ligand orbitals unless some of the electrons are moved. Suppose we use two of the empty $4d$ orbitals instead. We will call the hybrid orbitals formed from them sp^3d^2 to emphasize that the d orbitals have the same principal quantum number as that of the s and p orbitals. Then, the orbital diagram of the complex ion would be

$Fe(H_2O)_6^{2+}$: [Ar] $(\uparrow\downarrow)(\uparrow \ )(\uparrow \ )(\uparrow \ )(\uparrow \ )$ $\boxed{(\uparrow\downarrow) \ (\uparrow\downarrow)(\uparrow\downarrow)(\uparrow\downarrow) \ (\uparrow\downarrow)(\uparrow\downarrow)}$ $(\ )(\ )(\ )$
$\qquad\qquad\qquad\qquad$ $3d$ $\qquad\qquad\qquad$ $4s$ $\qquad\qquad$ $4p$ $\qquad\qquad\qquad\quad$ $4d$
$\qquad\qquad\qquad\qquad\qquad\qquad\qquad\qquad$ sp^3d^2 bonds to ligands

This bonding picture correctly explains the paramagnetism of the complex ion and shows that it should correspond to that of four unpaired electrons, in agreement with experiment.

Suppose, however, that in forming a complex ion of Fe^{2+}, two of the $3d$ electrons pair up so that two $3d$ orbitals are unoccupied and can be used to form d^2sp^3 hybrid orbitals. The configuration of this excited state of the Fe^{2+} ion is

Fe^{2+}(excited): [Ar] $(\uparrow\downarrow)(\uparrow\downarrow)(\uparrow\downarrow)$ $\boxed{(\ )(\ ) \ (\ )(\ )(\ \ )(\ )}$ $(\ )(\ )(\ )(\ )(\ )$
$\qquad\qquad\qquad\qquad\qquad$ $3d$ $\qquad\qquad\quad$ $4s$ $\qquad\quad$ $4p$ $\qquad\qquad\qquad\qquad$ $4d$
$\qquad\qquad\qquad\qquad\qquad\qquad$ available for d^2sp^3 bonds

Since the $3d$ orbitals are lower in energy than the $4d$ orbitals, they will be preferred for bonding if available. However, pairing of the electrons to make the two $3d$ orbitals available for d^2sp^3 bonding requires energy. We would expect this type of bonding to occur only if the bonding is sufficiently strong to provide the energy needed for electron pairing. This is apparently the case for the hexacyanoferrate(II) ion, $Fe(CN)_6^{4-}$. (The CN^- ion tends to give strong bonding to metal ions, in contrast to the weaker bonding of H_2O.) The orbital diagram for the iron atom in this complex is

$Fe(CN)_6^{4-}$: [Ar] (↑↓)(↑↓)(↑↓) | (↑↓)(↑↓) (↑↓) (↑↓)(↑↓)(↑↓) | ( )( ) ( ) ( ) ( )
                     $3d$ $4s$ $4p$ $4d$

d^2sp^3 bonds to ligands

This diagram correctly describes the ion as diamagnetic (no unpaired electrons).

If we similarly examine any transition-metal ion that has configurations d^4, d^5, d^6, or d^7, we will find two bonding possibilities. In one case, the bonding is sp^3d^2. In the other, some electrons are paired and the bonding is d^2sp^3. The number of unpaired electrons in the latter type of complex is lower than that with sp^3d^2 bonding. For that reason, it is called a **low-spin complex ion.** The complex ion with more unpaired electrons is called a **high-spin complex ion.** Since low-spin complex ions require energy for pairing of electrons, they are expected to occur with ligands that bond relatively strongly. Weakly bonding ligands form high-spin complex ions.

Example 25.7

Cobalt(II) has both high-spin and low-spin octahedral complex ions. Most of the octahedral complex ions are high-spin, such as the pink ion $Co(H_2O)_6^{2+}$. There are also a few low-spin complex ions such as $Co(CN)_6^{4-}$. Describe the bonding in both $Co(H_2O)_6^{2+}$ and $Co(CN)_6^{4-}$ using valence bond theory. How many unpaired electrons are there in each complex ion?

Solution

The electron configuration of cobalt is $[Ar]3d^74s^2$, and that of the cobalt(II) ion is $[Ar]3d^7$. A high-spin complex ion such as $Co(H_2O)_6^{2+}$ would be obtained from sp^3d^2 hybrid orbitals (see below). The complex ion would be paramagnetic, with three unpaired electrons.

Bonding of the cobalt(II) ion using d^2sp^3 hybrid orbitals would require that two unpaired electrons in the $3d$ subshell be moved. One could pair up with another electron in a $3d$ orbital, but the other would have to be promoted to higher unoccupied orbital ($4d$). The orbital diagram for a low-spin complex ion, such as $Co(CN)_6^{4-}$, is given below. The complex is paramagnetic, with one unpaired electron.

$Co(H_2O)_6^{2+}$: [Ar] (↑↓)(↑↓)(↑)(↑)(↑) | (↑↓) (↑↓)(↑↓)(↑↓) (↑↓)(↑↓) | ( )( ) ( )
            $3d$ $4s$ $4p$ $4d$

sp^3d^2 bonds to ligands

$Co(CN)_6^{3-}$: [Ar] (↑↓)(↑↓)(↑↓) | (↑↓)(↑↓) (↑↓) (↑↓)(↑↓)(↑↓) | (↑)( ) ( ) ( ) ( )
            $3d$ $4s$ $4p$ $4d$

d^2sp^3 bonds to ligands

Exercise 25.13

Cobalt(III) forms many stable complex ions, including $Co(NH_3)_6^{3+}$. Most of these are octahedral and diamagnetic. The complex ion CoF_6^{3-}, however, is paramagnetic. Describe the bonding in $Co(NH_3)_6^{3+}$ and CoF_6^{3-} using valence bond theory. How many unpaired electrons are there in each complex ion?

(See Problems 25.63 and 25.64.)

If we examine metal ions with configurations d^8 and d^9, we again find only one spin type of octahedral complex possible. (Recall that the d^3 configuration, as in Cr^{3+}, gives only one type octahedral complex; d^1 and d^2 are similar.) An example of a d^8 complex is the green ion $Ni(H_2O)_6^{2+}$. The nickel atom has the configuration $[Ar]3d^8 4s^2$, and the Ni^{2+} ion has the configuration $[Ar]3d^8$. The orbital diagram for this ion is

Ni^{2+}: [Ar] (↑↓)(↑↓)(↑↓)(↑)(↑) (   ) (   ) (   )(   ) (   ) (   )(   )(   )(   )(   )
                                    $3d$ $4s$ $4p$ $4d$

Since three of the $3d$ orbitals are doubly occupied, two empty orbitals cannot be created by pairing of electrons. However, we could obtain octahedral hybrid orbitals using the $4d$ orbitals. Thus,

$Ni(H_2O)_6^{2+}$: [Ar] (↑↓)(↑↓)(↑↓)(↑)(↑) | (↑↓) (↑↓)(↑↓)(↑↓) (↑↓)(↑↓) | (   )(   )(   )(   )
                                        $3d$ | $4s$ $4p$ | $4d$

$sp^3 d^2$ bonds to ligands

This would predict a paramagnetic complex ion with two unpaired electrons, which is what is observed experimentally.

Tetrahedral and Square Planar Complexes

Although coordination number 6 (octahedral complexes) is very common, coordination number 4 is nearly so. In this case, the geometries are either tetrahedral or square planar. Table 25.8 lists the hybrid orbitals used to describe various geometries. The tetrahedral geometry is described by sp^3 hybrid orbitals. The square planar geometry uses dsp^2 orbitals.

Nickel(II) complex ions with four ligands are common. Most of these, such as $Ni(CN)_4^{2-}$, are diamagnetic. That $Ni(CN)_4^{2-}$ is diamagnetic is evidence for a square planar geometry.■ To see this, we recall that the electron configuration of the Ni^{2+} ion is $[Ar]3d^8$, with three doubly occupied $3d$ orbitals and two singly occupied $3d$ orbitals. To form dsp^2 hybrid orbitals, the two unpaired $3d$ electrons must first be paired, giving an unoccupied $3d$ orbital. The orbital diagram for the complex ion is

■ Care must be used when predicting geometry from the magnetic characteristics of a complex, however. For example, a complex of nickel(II) with the bidentate ligand acetylacetonate (= acac) has the simplest formula $[Ni(acac)_2]$ and is paramagnetic. It appears from this formula that the Ni atom is four-coordinate and because of its paramagnetism has a tetrahedral geometry (see Example 25.8). In fact, the molecular formula is $[Ni(acac)_2]_3$, in which the Ni atom has octahedral geometry. This geometry is expected to be paramagnetic.

$Ni(CN)_4^{2-}$: [Ar] (↑↓)(↑↓)(↑↓)(↑↓) | (↑↓) (↑↓) (↑↓)(↑↓) | (   ) (   ) (   )(   )(   )(   )
                                $3d$ | $4s$ $4p$ | $4d$

dsp^2 bonds to ligands

Example 25.8

Nickel(II) forms some tetrahedral complex ions such as $Ni(NH_3)_4^{2+}$. Discuss the bonding in this complex ion and describe its magnetic characteristics.

Solution

The orbital diagram for Ni^{2+}, whose electron configuration is $[Ar]3d^8$, is given below. A tetrahedral ge-

(Continued)

ometry uses sp^3 hybrid orbitals. Therefore, none of the $3d$ electrons in Ni^{2+} need to be paired up or promoted. The orbital diagram of $Ni(NH_3)_4^{2+}$ is given below. Since there are two unpaired electrons, the tetrahedral complex ions of nickel(II) are paramagnetic, in contrast to the square planar complex ions, which are diamagnetic.

Ni²⁺: [Ar] (↑↓)(↑↓)(↑↓)(↑)(↑) (   ) (   )(   )(   ) (   )(   )(   )(   )(   )
                    3d   4s   4p   4d

Ni(NH₃)₄²⁺: [Ar] (↑↓)(↑↓)(↑↓)(↑)(↑) │ (↑↓) (↑↓)(↑↓)(↑↓) │ (   )(   )(   )(   )(   )(   )
                          3d   4f   4p   4d
                                              sp^3 bonds to ligands

Exercise 25.14

The complex ion $CoCl_4^{2-}$ is paramagnetic, with a magnetism corresponding to three unpaired electrons. If the complex ion indeed has four ligands as indicated by the formula, what geometry is indicated?

(See Problems 25.65 and 25.66.)

25.7 Crystal Field Theory

Although valence bond theory explains the bonding and magnetic properties of complexes in straightforward fashion, it is limited in two important ways. First, the theory cannot simply explain the color of complexes. Second, the theory is difficult to extend quantitatively. Consequently, another theory—crystal field theory—has emerged as the prevailing view of transition-metal complexes.

  Crystal field theory is a simplified way of dealing with the electronic structure of transition-metal complexes. It was so named because it was developed by physicists as a way of explaining the spectra of transition-metal impurities in crystals.■ According to this theory, the ligands in a transition-metal complex are treated as point charges. Thus, a ligand anion becomes simply a point of negative charge. A neutral molecule, with its electron pair that it donates to the metal atom, is replaced by a partial negative charge, representing the negative end of the molecular dipole. In the electrical field of these negative charges, the five d orbitals of the metal atom

■ The colors of many gems, such as ruby, are due to transition-element impurities. Ruby has Cr^{3+} ions distributed in Al_2O_3.

Table 25.8
Hybrid Orbitals for Various Coordination Numbers and Geometries

Coordination Number	Geometry	Hybrid Orbital
2	Linear	sp
4	Tetrahedral	sp^3
	Square planar	dsp^2
6	Octahedral	d^2sp^3 or sp^3d^2

no longer have exactly the same energy. The result, as we will see, explains both the paramagnetism and color observed in certain complexes.

The simplifications used in crystal field theory are drastic. Treating the ligands as point charges is essentially the same as treating the bonding as ionic. However, it turns out that the theory can be extended to include covalent character in the bonding. Simple extension is usually referred to as *ligand field theory*, but after including several levels of refinements, the theory becomes equivalent to molecular orbital theory.

Effect of an Octahedral Field on the *d* Orbitals

All five *d* orbitals of an isolated metal atom have the same energy. But if the atom is brought into the electrical field of several point charges, these *d* orbitals may be affected in different ways, and therefore may have different energies. To understand how this can happen, we must first see what these *d* orbitals look like. We will then be able to picture what happens to them in the crystal field theory of an octahedral complex.

Figure 25.19 shows the shapes of the five *d* orbitals. The orbital labeled d_{z^2} has a dumbbell shape along the *z*-axis, with a collar in the *x–y* plane surrounding this dumbbell. Remember that this shape represents the volume most likely to be occupied by an electron in this orbital. The other four *d* orbitals have "cloverleaf" shapes, each differing from one another only in the orientation of the orbitals in space. Thus, the "cloverleaf" orbital $d_{x^2-y^2}$ has its lobes along the *x*- and *y*-axes. Orbitals d_{xy}, d_{yz}, and d_{xz} have their lobes

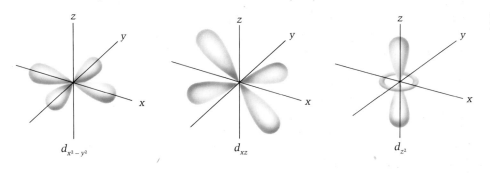

Figure 25.19
The five *d* orbitals.

$d_{x^2-y^2}$ d_{xz} d_{z^2}

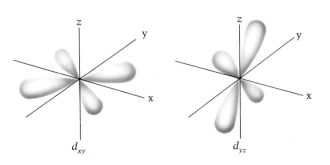

d_{xy} d_{yz}

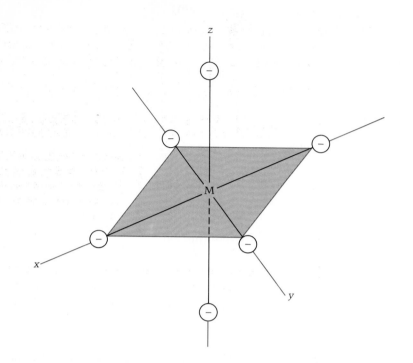

Figure 25.20
Location of the negative charges (ligands) on the *xyz* axes in the crystal field model of an octahedral complex.

directed between the two sets of axes designated in the orbital label. Orbital d_{xy}, for example, has its lobes lying between the x- and y-axes.

A complex ion with six ligands will have those ligands arranged about the metal atom to reduce their mutual repulsions. Normally, an octahedral arrangement will achieve this. Now, imagine that the ligands are replaced by negative charges. If the ligands are anions, they are replaced by the anion charge. If the ligands are neutral molecules, they are replaced by the partial negative charge from the molecular dipole. These six charges are placed at equal distances from the metal atom, one charge on each of the positive and negative sides of the x-, y-, and z-axes (see Figure 25.20).

Fundamentally, the bonding in this model of a complex is due to the attraction of the positive metal ion for the negative charges of the ligands. However, an electron in a d orbital of the metal atom is repelled by the negative charge of the ligands. This repulsion alters the energy of the d orbital depending upon whether it is directed *toward* ligands or is directed *between* ligands. For example, consider the difference in the repulsive effect of ligands on metal-ion electrons in the d_{z^2} and the d_{xy} orbitals. Because the d_{z^2} orbital is directed at the two ligands on the z-axis (one on the $-z$ side and the other on the $+z$ side), an electron in the orbital is rather strongly repelled by them. Thus, the energy of the d_{z^2} orbital becomes greater. Similarly, an electron in the d_{xy} orbital is repelled by the negative charge of the ligands, but since the orbital is not pointed directly at the ligands, this repulsive effect is smaller. Its energy is raised, but less than is that of the d_{z^2} orbital.

If we look at the five d orbitals in an octahedral field (electric field of octahedrally arranged charges), we see that we can divide them into two sets. Orbitals d_{z^2} and $d_{x^2-y^2}$ are both directed toward ligands, and orbitals d_{xy}, d_{yz}, and d_{xz} are directed between ligands. The orbitals in the first set (d_{z^2} and

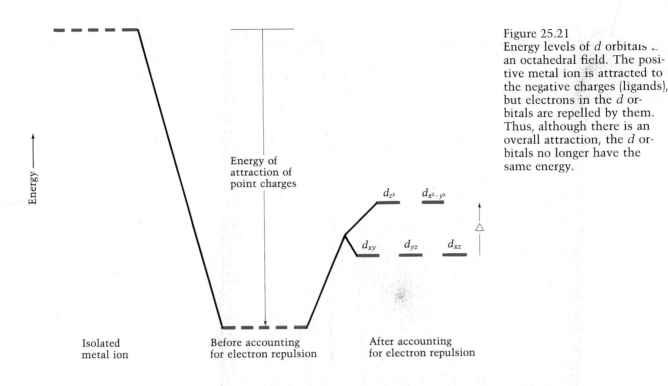

$d_{x^2-y^2}$) have higher energy than those in the second set (d_{xy}, d_{yz}, and d_{xz}). Figure 25.21 shows the energy levels of the d orbitals in an octahedral field. The difference in energy between these two sets of d orbitals is called the **crystal field splitting, Δ.**

High-Spin and Low-Spin Complexes

Once we have the energy levels for the d orbitals in an octahedral complex, we can decide how the d electrons of the metal ion are distributed in them. Knowing this distribution, we can predict the magnetic characteristics of the complex.

Consider the complex ion $Cr(NH_3)_6^{3+}$. According to crystal field theory, this consists of the Cr^{3+} ion surrounded by NH_3 molecules treated as partial negative charges. The effect of these charges is to split the d orbitals of Cr^{3+} into two sets as shown in Figure 25.21. We now ask how the d electrons are distributed among the d orbitals of the Cr^{3+} ion. Since the electron configuration of Cr^{3+} is $[Ar]3d^3$, there are three d electrons to distribute. These electrons are placed in the d orbitals of lower energy, following Hund's rule (Figure 25.22). We see that the complex ion $Cr(NH_3)_6^{3+}$ has three unpaired electrons and is therefore paramagnetic.

As another example, consider the complex ion $Fe(H_2O)_6^{2+}$. What are its magnetic characteristics? Remember, we need only look at the d electrons of the metal ion, Fe^{2+}. The electron configuration of the ion is $[Ar]3d^6$. We distribute six electrons among the d orbitals of the complex ion in such a way as to get the lowest total energy. If we place all six electrons into the lower three d orbitals, we get the distribution shown by the energy-level

Figure 25.22
Occupation of the $3d$ orbitals in an octahedral complex of Cr^{3+}. Note that the electrons occupy different orbitals with the same spin (Hund's rule).

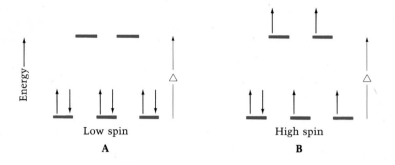

Low spin   High spin

A   B

Figure 25.23
Occupation of the $3d$ orbitals
in complexes of Fe^{2+}. (a) Low
spin. (b) High spin.

diagram in Figure 25.23a. Since all of the electrons are paired, we would predict that this distribution gives a diamagnetic complex. The Fe(H$_2$O)$_6^{2+}$ ion, however, is paramagnetic. Thus, the distribution we gave in Figure 25.23a is not correct. What was our mistake?

The mistake we made above was to ignore the **pairing energy, P,** the energy required to put two electrons into the same orbital. When an orbital is already occupied by an electron, it requires energy to put another electron into that orbital because of their mutual repulsion. Suppose that this pairing energy is greater than the crystal field splitting; that is, suppose $P > \Delta$. In that case, once the first three electrons have singly occupied the three lower energy d orbitals, the fourth electron will go into one of the higher d orbitals. It will cost less energy to do that than to pair up with an electron in one of the lower energy orbitals. Similarly, the fifth electron will go into the last empty d orbital. The sixth electron must pair up, and so it goes into one of the lower energy orbitals. Figure 25.23b shows this electron distribution. In this case, there are four unpaired electrons, and the complex ion will be paramagnetic.

We see that crystal field theory predicts two possibilities, a *low-spin complex* when $P < \Delta$, and a *high-spin complex* when $P > \Delta$. The value of Δ, as we will explain later, can be obtained from the spectrum of a complex, and the value of P can be calculated theoretically. Even in the absence of these numbers, however, we see that the theory predicts that a paramagnetic octahedral complex of Fe^{2+} should have a magnetism equal to that of four unpaired electrons. This is what we find for the Fe(H$_2$O)$_6^{2+}$ ion.

We would expect low-spin diamagnetic Fe^{2+} complexes to occur for ligands that bind strongly to the metal ion, that is, for those giving large Δ. Ligands can be arranged according to the relative strengths of the crystal field splittings they induce in the d orbitals of the metal ion. This order, which remains approximately the same whatever the metal or its oxidation state, is called the **spectrochemical series.** The following short version of the spectrochemical series lists some common ligands:

Weak-bonding ligands   Strong-bonding ligands

I$^-$ < Br$^-$ < Cl$^-$ < F$^-$ < OH$^-$ < H$_2$O < NH$_3$ < en < NO$_2^-$ < CN$^-$ < CO

Increasing $\Delta \longrightarrow$

From this series, we see that the CN$^-$ ion bonds more strongly than H$_2$O, which explains why Fe(CN)$_6^{4-}$ is a low-spin complex ion and Fe(H$_2$O)$_6^{2+}$ is a high-spin complex. We can also see why carbon monoxide might be expected to be poisonous. We know that O$_2$ bonds reversibly to the Fe(II) atom of hemoglobin, and so the bonding is only moderately strong. According to the spectrochemical series, however, carbon monoxide, CO, forms a strong bond. The bonding in this case is irreversible (or practically so) and forms a very stable complex of CO with hemoglobin that cannot function as a transporter of O_2.

Example 25.9

Describe the distribution of d electrons in the complex ions $Co(H_2O)_6^{2+}$ and $Co(CN)_6^{4-}$ using crystal field theory. The hexaaquacobalt(II) ion is a high-spin complex, and the hexacyanocobaltate(II) ion is a low-spin complex. How many unpaired electrons are there in each ion?

Solution

The electron configuration of Co^{2+} is $[Ar]3d^7$. The high-spin and low-spin distributions in the d orbitals are

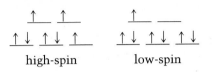

Thus, $Co(H_2O)_6^{2+}$, a high-spin complex ion, has three unpaired electrons, and $Co(CN)_6^{4-}$, a low-spin complex ion, has one unpaired electron.

Exercise 25.15

Describe the distribution of d electrons in $Ni(H_2O)_6^{2+}$ using crystal field theory. How many unpaired electrons are there in this ion?

(See Problems 25.67 and 25.68.)

Tetrahedral and Square Planar Complexes

When a metal ion bonds with tetrahedrally arranged ligands, the d orbitals of the ion split to give two d orbitals at lower energy and three d orbitals at higher energy (just the opposite of what we found for an octahedral field). See Figure 25.24a. On the other hand, in the field of ligands in a square planar arrangement, the d orbitals split as shown in Figure 25.24b.

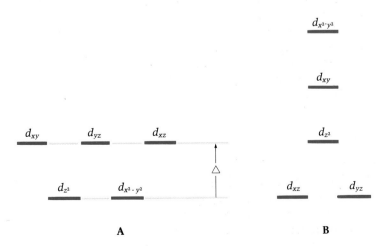

A

B

Figure 25.24
(a) Energy splitting of the d orbitals in a tetrahedral field. The crystal field splitting, Δ, is smaller than in a comparable octahedral complex. (b) Energy splitting of the d orbitals in a square planar field.

The observed splittings in a tetrahedral field are approximately one-half the size of those seen in comparable octahedral complexes. As a result, only high-spin complexes are observed. In the square planar case, only low-spin complexes have been found.

Example 25.10

Describe the d-electron distributions of the complexes $Ni(NH_3)_4^{2+}$ and $Ni(CN)_4^{2-}$ according to crystal field theory. The tetraamminenickel(II) ion is paramagnetic, and the tetracyanonickelate(II) ion is diamagnetic.

Solution

We expect the tetrahedral field to give high-spin complexes and the square planar field to give low-spin complexes. Thus, the geometry of the $Ni(NH_3)_4^{2+}$ ion, which is paramagnetic, is probably tetrahedral. The distribution of d electrons in the Ni^{2+} ion (configuration d^8) is

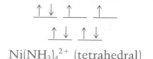

$Ni(NH_3)_4^{2+}$ (tetrahedral)

The geometry of $Ni(CN)_4^{2-}$, which is diamagnetic, is probably square planar; the distribution of d electrons would be

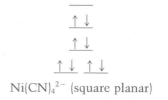

$Ni(CN)_4^{2-}$ (square planar)

Exercise 25.16

Describe the distribution of d electrons in the $CoCl_4^{2-}$ ion. The ion has a tetrahedral geometry. Assume a high-spin complex.

(See Problems 25.69 and 25.70.)

Visible Spectra of Transition-Metal Complexes

Frequently, substances absorb light only in regions outside the visible spectrum, and reflect or pass on (transmit) all of the visible wavelengths. As a result, these substances appear white or colorless (since white light is a mixture of all visible wavelengths). However, some substances absorb cer-

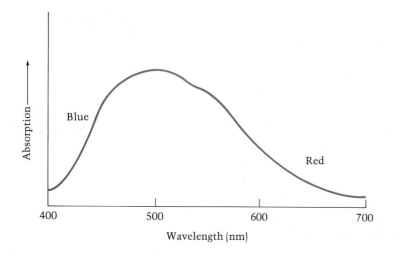

Figure 25.25
Visible spectrum of $Ti(H_2O)_6^{3+}$. Unlike atomic spectra, which show absorption lines, spectra for ions and molecules show broad bands due to changes in nuclear motion accompanying the electronic transition.

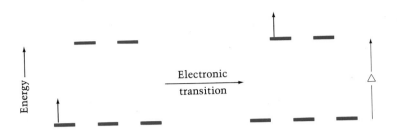

Figure 25.26
The electronic transition responsible for the visible absorption in $Ti(H_2O)_6^{3+}$.

tain wavelengths in the visible spectrum and transmit the remaining ones. Thus they appear colored. Many transition-metal complexes, as we have noted, are colored substances. The color results from electron jumps, or transitions, between the two closely spaced d orbital energy levels that come from the crystal field splitting.

The spectrum of a d^1 configuration complex is particularly simple. Hexaaquatitanium(III) ion, $Ti(H_2O)_6^{3+}$, is an example. Titanium has the configuration $[Ar]3d^24s^2$, and Ti^{3+} has the configuration $[Ar]3d^1$. According to crystal field theory, the d electron occupies one of the lower-energy d orbitals of the octahedral complex. Figure 25.25 shows the visible spectrum of $Ti(H_2O)_6^{3+}$. It results from a transition, or jump, of the d electron from a lower-energy d orbital to a higher-energy d orbital, as shown in Figure 25.26. Notice that the energy change equals the crystal field splitting, Δ. Consequently, the wavelength , λ, of light that is absorbed is related to Δ:■

■ Recall from Section 5.7 that the energy change during a transition equals $h\nu$.

$$\Delta = h\nu = \frac{hc}{\lambda}$$

Or,
$$\lambda = \frac{hc}{\Delta}$$

When white light, which contains all visible wavelengths (from 400 nm to 750 nm), falls on a solution containing $Ti(H_2O)_6^{3+}$, blue green light is absorbed. (The maximum light absorption is observed at 500 nm, which is blue green light. See Table 25.9.) The other wavelengths of visible light, including

Wavelength Absorbed (nm)	Color Absorbed	Approximate Color Observed*
410	Violet	Green yellow
430	Violet blue	Yellow
480	Blue	Orange
500	Blue green	Red
530	Green	Purple
560	Green yellow	Violet
580	Yellow	Violet blue
610	Orange	Blue
680	Red	Blue green
720	Purple red	Green

Table 25.9
Color Observed for Given Absorption of Light by an Object

*The exact color depends on the relative intensities of various wavelengths coming from the object and the response of the eye to those wavelengths.

red and some blue light, pass through the solution, giving it a red purple color.

If the ligands in the Ti^{3+} complex are changed, this will change Δ, and therefore change the color of the complex. For example, replacing H_2O ligands by weaker F^- ligands should give a smaller crystal field splitting and therefore an absorption at longer wavelengths. The absorption of TiF_6^{3-} is at 590 nm, in the yellow, rather than the blue green, and the color observed is violet blue.

From this discussion, we see that the visible spectrum can be related to the crystal field splitting, and values of Δ can therefore be obtained by spectroscopic analysis. However, when there is more than one d electron, several excited states can be formed. Consequently, the spectrum generally consists of several lines, and the analysis is more complicated than that for Ti^{3+}.

Example 25.11

When water ligands in $Ti(H_2O)_6^{3+}$ are replaced by CN^- ligands to give $Ti(CN)_6^{3-}$, the maximum absorption shifts from 500 nm to 450 nm. Is this shift in the expected direction? Explain. What color do you expect to observe for this ion?

Solution

According to the spectrochemical series, CN^- is a more strongly bonding ligand than H_2O. Consequently, Δ should increase, and the wavelength of the absorption $(\lambda = hc/\Delta)$ should decrease. Thus, the shift of the absorption is in the expected direction. Because the absorbed light is between blue and violet blue (see Table 25.8), the observed color is orange yellow (this is the *complementary color* of the color between blue and violet blue).

Exercise 25.17

The $Fe(H_2O)_6^{3+}$ ion has a pale purple color, and the $Fe(CN)_6^{3-}$ ion has a ruby red color. What are the approximate wavelengths of the maximum absorption for each ion? Is the shift of wavelength in the expected direction? Explain.

(See Problems 25.71, 25.72, 25.73, and 25.74.)

Aside: The Cooperative Release of Oxyhemoglobin

Hemoglobin is an iron-containing substance in red blood cells responsible for the transport of O_2 from the lungs to various parts of the body. Myoglobin is a similar substance in muscle tissue. It acts as a reservoir for the storage of O_2 and as a transporter of O_2 within muscle cells. The explanation for the different actions of these two substances involves some fascinating transition-metal chemistry.

Myoglobin consists of heme, a complex of Fe(II) bonded to a quadridentate ligand (Figure 25.6), and globin. Globin, a protein, is attached through a nitrogen atom to one of the octahedral positions of Fe(II). The sixth position is vacant in free myoglobin, but is occupied by O_2 in oxymyoglobin. Hemoglobin is essentially a four-unit structure of myoglobin-like units, that is, a *tetramer* of myoglobin.

(Continued)

Myoglobin and hemoglobin exist in equilibrium with oxygenated forms, oxymyoglobin and oxyhemoglobin, respectively. For example, hemoglobin (Hb) and O_2 are in equilibrium with oxyhemoglobin (HbO_2):

$$Hb + O_2 \rightleftharpoons HbO_2$$

Although hemoglobin is a tetramer of myoglobin, it does not function simply as four independent units of myoglobin. For it to function efficiently as a transporter of O_2 from the lungs, then be able to release that O_2 easily to myoglobin, hemoglobin must be less strongly attached to O_2 in the vicinity of a muscle cell than is myoglobin. In hemoglobin the release of O_2 from one heme group triggers the release of O_2 from another heme group of the same molecule. In other words, there is a *cooperative release* of O_2 from hemoglobin that makes it possible for it to give up its O_2 to myoglobin.

The mechanism that has been postulated for this cooperative release of O_2 depends on a change of iron(II) from a low-spin form to a high-spin form, with a corresponding change in radius of the iron atom. In oxyhemoglobin, iron(II) exists in the low-spin form. When O_2 leaves, the iron atom goes to a high-spin form with two electrons in the higher-energy d orbital. These higher-energy orbitals are antibonding molecular orbitals and somewhat larger in size than the lower-energy d orbitals, which are nonbonding.

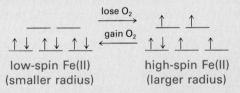

low-spin Fe(II)
(smaller radius)

high-spin Fe(II)
(larger radius)

When an O_2 molecule leaves a heme group, the radius of the iron atom increases, and the atom pops out of the heme plane by about 0.7 Å. In hemoglobin, this change triggers the cooperative release of another O_2 molecule. As the iron atom moves, the attached globin group moves with it. This motion of one globin group causes an adjacent globin group in the tetramer to alter its shape, which in turn makes possible the easy release of an O_2 molecule from its heme unit.

A Checklist for Review

Important Terms

d-block transition elements (p. 842)
f-block (inner) transition element (p. 843)
basic oxygen process (25.2)
complex ion (25.3)
complex (25.3)
coordination compound (25.3)
ligands (25.3)
coordination number (25.3)
unidentate (bidentate, quadridentate, polydentate) ligands (25.3)

chelates (25.3)
structural isomers (25.5)
stereoisomers (25.5)
ionization isomers (25.5)
hydrate isomers (25.5)
coordination isomers (25.5)
linkage isomers (25.5)
geometric isomers (25.5)
enantiomers (optical isomers) (25.5)
chiral (25.5)

optically active (25.5)
dextrorotatory (levorotatory) (25.5)
racemic mixture (25.5)
low-spin (high-spin) complex ion (25.6)
crystal field theory (25.7)
crystal field splitting, Δ (25.7)
pairing energy, P (25.7)
spectrochemical series (25.7)

Summary of Facts and Concepts

Transition elements are defined as those elements having a partially filled *d* subshell in any common oxidation state. They have a number of characteristics, including high melting points and a multiplicity of oxidation states. Compounds of transition elements are frequently colored and many are paramagnetic. These properties are due to the participation of *d* orbitals in bonding. We described the chemical properties of three transition elements: Cr, Fe, and Cu. Chromium metal reacts with acids to give Cr^{2+} ion, which is readily oxidized to Cr^{3+}. Although the +3 oxidation state is stable, Cr^{3+} can be oxidized in basic solution to the chromium(VI) species CrO_4^{2-}. Iron metal reacts with acids to give Fe^{2+}; salts of Fe^{2+} can be oxidized to Fe^{3+}. Copper metal reacts only with acids having strongly oxidizing anions, such as HNO_3; it gives Cu^{2+} ion. Copper(I) ion disproportionates in aqueous solution to Cu and Cu^{2+}, but insoluble compounds of copper(I) can be prepared in water solutions.

Transition metal atoms often function as Lewis acids, reacting with groups called *ligands* by forming coordinate covalent bonds to them. The metal atom with its ligands is a *complex ion* or neutral *complex*. Ligands that bond through more than one atom are called *polydentate*, and the complex that is formed is called a *chelate*. The IUPAC has agreed upon a nomenclature of complexes that gives basic structural information about the species. The presence of isomers in coordination compounds is evidence for particular geometries. For example, $[Pt(NH_3)_2Cl_2]$ has two isomers, which is evidence for a square planar geometry having *cis–trans isomers*. Octahedral complexes with bidentate ligands may have *optical isomers*, that is, isomers that are mirror images of one another.

Valence bond theory gave the earliest theoretical description of the electronic structure of a complex. According to this theory, a complex forms when doubly occupied ligand orbitals overlap unoccupied orbitals of the metal atom. Octahedral complexes with metal-ion configurations d^4, d^5, d^6, or d^7 give rise to two bonding types. In the case of strong-bonding ligands, *d* electrons are paired so that d^2sp^3 hybrid orbitals can be used, and a *low-spin complex* forms. Otherwise, the *d* electrons of the metal ion remain unpaired, and a *high-spin complex* forms using sp^3d^2 hybrid orbitals.

Crystal field theory also predicts high-spin and low-spin complexes in these configurations of the metal ion. Following this theory, the ligands are treated as electrical charge points that affect the energy of the *d* orbitals of the metal ion. In an octahedral complex, two of the *d* orbitals have higher energy than the other three. A high-spin complex forms when the *pairing energy* is greater than the *crystal field splitting*, so that electrons would prefer to occupy a higher-energy *d* orbital rather than pair up with an electron in a lower-energy orbital. When the pairing energy is smaller than the crystal field splitting, the *d* orbitals are occupied in the normal fashion, giving a low-spin complex. Color in transition-metal complexes is explained as due to a jump of an electron from the lower-energy *d* orbitals to the higher-energy. The crystal field splitting can be obtained experimentally from the visible spectrum of a complex.

Operational Skills

1. Given similar compounds of two transition elements in the same column having the same oxidation state, decide which is the stronger oxidizing agent (Example 25.1).

2. Write balanced equations describing the preparation of a compound of Cr, Fe, or Cu (Example 25.2).

3. Given the structural formulas of coordination compounds, write the IUPAC names (Example 25.3); given the IUPAC names of complexes, write the structural formulas (Example 25.4).

4. Given the formula of a complex, decide if geometric isomers are possible and draw them (Example 25.5). Given the structural formula of a complex, decide if enantiomers (optical isomers) are possible and draw them (Example 25.6).

5. Given a transition-metal ion, describe the bonding types (high-spin and low-spin, if both exist) using valence bond theory for octahedral and four-coordinate complexes; give the number of unpaired electrons in the complex (Examples 25.7 and 25.8). Do the same using crystal field theory (Examples 25.9 and 25.10).

6. Given two complexes that differ only in the ligands, predict on the basis of the spectrochemical series which one absorbs at higher wavelength; given the absorption maximums, predict the colors of the complexes (Example 25.11).

Review Questions

25.1 What characteristics of the transition elements set them apart from the main-group elements?

25.2 According to the building-up principle, what is the electron configuration of the ground state of the technetium atom (atomic number 43)?

25.3 The highest melting point for metals in the fifth period occurs for molybdenum. Explain why this is expected.

25.4 Iron, cobalt, and nickel are similar in properties and are sometimes studied together as the "iron triad." For example, each is a fairly active metal and reacts with acids to give hydrogen and the $+2$ ions. In addition to the $+2$ ions, the $+3$ ions of the metals also figure prominently in the chemistries of the elements. Explain why these elements are similar.

25.5 Palladium and platinum are very similar to one another. Thus, they are unreactive toward most acids. However, nickel, which is in the same column of the periodic table, is an active metal. Explain why this difference exists.

25.6 Write balanced equations for the reactions of Cr, Fe, and Cu metals with $HCl(aq)$. If no reaction occurs, write N.R.

25.7 Write balanced equations for the reactions of chromium(III) oxide with a strong acid and with a strong base.

25.8 Explain the acidity of an aqueous solution of iron(III) sulfate, $Fe_2(SO_4)_3$.

25.9 Give the balanced equation for the net result of the reduction of iron(III) oxide in a blast furnace.

25.10 Describe the structure of copper(II) sulfate pentahydrate. What color change occurs when the salt is heated? What causes the color change?

25.11 Copper(I) sulfate dissolves in water to give copper(II) sulfate. What is the other product? Write the balanced equation for the reaction.

25.12 What evidence did Werner obtain to show that the platinum complex $PtCl_4 \cdot 4NH_3$ has the structural formula $[Pt(NH_3)_4Cl_2]Cl_2$?

25.13 Define the terms *complex ion, ligand,* and *coordination number.* Use an example to illustrate the use of these terms.

25.14 Define the term *bidentate ligand.* Give two examples.

25.15 Rust spots on clothes can be removed by dissolving them in oxalic acid. The oxalate ion forms a stable complex with Fe^{3+}. Using an electron dot formula, indicate how an oxalate ion bonds to the metal ion.

25.16 Describe step-by-step how the name potassium hexacyanoferrate(II) leads one to the structural formula $K_4[Fe(CN)_6]$.

25.17 What three properties of coordination compounds have been important in determining the details of their structure and bonding?

25.18 Define each of the following and give an example: (a) ionization isomerism, (b) hydrate isomerism, (c) coordination isomerism, (d) linkage isomerism.

25.19 Define *geometric isomerism* and *optical isomerism* and give an example of each type.

25.20 Compounds A and B are known to be stereoisomers of one another. Compound A has a violet color; compound B has a green color. Are compounds A and B geometric or optical isomers?

25.21 Explain the difference in behavior of d and l isomers with respect to polarized light.

25.22 What is a racemic mixture? Describe one method of resolving a racemic mixture.

25.23 Describe the formation of a coordinate covalent bond between a metal-ion orbital and a ligand orbital.

25.24 (a) Describe the steps in the formation of a high-spin octahedral complex of Fe^{2+} in valence bond terms. (b) Do the same for a low-spin complex.

25.25 Explain why d orbitals of a transition metal atom may have different energies in the octahedral field of six negative charges. Describe how each of the d orbitals is affected by the octahedral field.

25.26 What is meant by the term *crystal field splitting?* How is it determined experimentally?

25.27 What is meant by the term *pairing energy?* How do the relative values of pairing energy and crystal field splitting determine whether a complex will be low spin or high spin?

25.28 (a) Use crystal field theory to describe a high-spin octahedral complex of Fe^{2+}. (b) Do the same for a low-spin complex.

25.29 What is the spectrochemical series? Use the ligands CN^-, H_2O, Cl^-, NH_3 to illustrate the term and arrange them in order, describing the meaning of this order.

25.30 A complex absorbs red light from a single electron transition. What color would this complex be?

Problems

Properties of the Transition Elements

25.31 Chromium forms several oxides. One of these is a dark red, crystalline solid melting at 197°C. It dissolves readily in water giving an acidic solution. Another oxide is a dark green solid melting at 2435°C. It is insoluble in water, but dissolves in acids and bases. If one of these oxides is CrO_3 and the other is Cr_2O_3, which is which? Explain your answer.

25.33 Of the elements chromium and tungsten, only one is known to form a hexachloride. Which element would that be? Explain.

25.32 Iron forms iron(II) chloride, $FeCl_2$, and iron(III) chloride, $FeCl_3$. One of these chlorides is a dark brown solid melting at 306°C; the other is a white crystalline solid with greenish tint and melts at 674°C. Which description fits iron(II) chloride? Why do you think so?

25.34 Manganese and rhenium form oxides in the $+7$ oxidation state. One of these metal(VII) oxides is a bright yellow solid melting at 300°C. The other is a red brown oily liquid that explodes on warming. Which of these describes Mn_2O_7? Why do you think so?

Chemistry of Chromium, Iron, and Copper

25.35 Give balanced equations, using complete formulas, for the preparation of $Fe(OH)_3$ from iron metal.

25.37 Find the oxidation numbers of the transition metal in the following compounds: (a) $FeCO_3$, (b) MnO_2, (c) $CuCl$, (d) CrO_2Cl_2

25.39 Write the balanced equation for the reaction of iron(II) ion with nitrate ion in acidic solution. Nitrate ion is reduced to NO.

25.36 Give balanced equations, with complete formulas, for the preparation of chromium(III) sulfate from chromium metal.

25.38 Find the oxidation numbers of the transition metal in the following compounds: (a) $CoSO_4$, (b) Ta_2O_5, (c) $Cu_2(OH)_3Cl$

25.40 Write the balanced equation for the reaction of sulfurous acid with dichromate ion.

Structural Formulas and Naming of Complexes

25.41 A cobalt complex whose composition corresponded to the formula $Co(NO_2)_2Cl \cdot 4NH_3$ gave an electrical conductance equivalent to two ions per formula unit. Excess silver nitrate solution precipitates one mole AgCl per formula unit. Write a structural formula consistent with these results.

25.43 Give the coordination number of the transition-metal atom in the following complexes:
  (a) $Au(CN)_4^-$ (b) $[Co(NH_3)_4(H_2O)_2]Cl_3$
  (c) $[Au(en)_2]Cl_3$ (d) $Cr(en)_2(C_2O_4)^+$

25.45 Determine the oxidation number of the transition element in the following complexes:
  (a) $K_2[Ni(CN)_4]$ (b) $Mo(en)_3^{3+}$
  (c) $Cr(C_2O_4)_3^{3-}$ (d) $[Co(NH_3)_5(NO_2)]Cl_2$

25.42 A cobalt complex has a composition corresponding to the formula $Co(NO_3)Cl_2 \cdot 4NH_3$. From electrical-conductance measurements, it was determined that there are two ions per formula unit. Silver nitrate solution gave no precipitate. Write a structural formula consistent with this information.

25.44 Give the coordination number of the transition element in the following complexes:
  (a) $[Ni(NH_3)_6](ClO_3)_2$ (b) $[Cu(NH_3)_4]SO_4$
  (c) $[Cr(en)_3]Cl_3$ (d) $K_2[Ni(CN)_4]$

25.46 For each of the following complexes, determine the oxidation state of the transition-metal atom:
  (a) $[CoCl(en)_2(NO_2)]NO_2$ (b) $PtCl_4^{2-}$
  (c) $K_3[Cr(CN)_6]$ (d) $Fe(H_2O)_5(OH)^{2+}$

25.47 Consider the complex ion $Cr(NH_3)_2Cl_2(C_2O_4)^-$.
(a) What is the oxidation state of the metal atom?
(b) Give the formula and name of each ligand in the ion.
(c) What is the coordination number of the metal atom?
(d) What would be the charge on the complex ion if all ligands were chloride ions?

25.49 Write the IUPAC names for the following coordination compounds:
(a) $K_3[FeF_6]$
(b) $Cu(NH_3)_2(H_2O)_2^{2+}$
(c) $(NH_4)_2[Fe(H_2O)F_5]$
(d) $Ag(CN)_2^-$

25.51 Give the IUPAC names for the following:
(a) $Fe(CO)_5$
(b) $Rh(CN)_2(en)_2^+$
(c) $[Cr(NH_3)_4SO_4]Cl$
(d) MnO_4^-

25.53 Write the structural formulas for the following compounds:
(a) potassium hexacyanomanganate(III)
(b) sodium tetracyanozincate(II)
(c) tetraamminedichlorocobalt(III) nitrate
(d) hexaamminechromium(III) tetrachlorocuprate(II)

25.48 Consider the complex ion $Mn(NH_3)_2(H_2O)_3(OH)^{2+}$.
(a) What is the oxidation state of the metal atom?
(b) Give the formula and name of each ligand in the ion.
(c) What is the coordination number of the metal atom?
(d) What would be the charge on the complex ion if all ligands were chloride ions?

25.50 Name the following complexes, using IUPAC rules:
(a) $K_4[Mo(CN)_8]$
(b) CrF_6^{3-}
(c) $V(C_2O_4)_3^{2-}$
(d) $K_2[FeCl_4]$

25.52 Give the IUPAC names for the following:
(a) $W(CO)_8$
(b) $[Co(H_2O)_2(en)_2](SO_4)_3$
(c) $K[Mo(CN)_8]$
(d) CrO_4^{2-}

25.54 Give structural formulas for the following complexes:
(a) diaquadicyanocopper(II)
(b) potassium hexachloroplatinate(IV)
(c) tetraamminenickel(II) perchlorate
(d) tetraammineplatinum(II) tetrachlorocuprate(II)

Isomerism

25.55 Give the type of structural isomerism shown by each of the following:
(a) $Co(CN)_5(NCS)^{3-}$ and $Co(CN)_5(SCN)^{3-}$
(b) $[Co(NH_3)_6][Cr(C_2O_4)_3]$ and $[Cr(NH_3)_6][Co(C_2O_4)_3]$
(c) $[Co(NH_3)_3(H_2O)_2Cl]Br_2$ and $[Co(NH_3)_3(H_2O)ClBr]Br \cdot H_2O$
(d) $[Co(NH_3)_4Cl(NO_2)]Cl$ and $[Co(NH_3)_4Cl_2]NO_2$

25.57 A complex has a composition corresponding to the formula $CoBr_2Cl \cdot 4NH_3$. What is the structural formula if conductance measurements show two ions per formula unit? Silver nitrate solution gives a precipitate of AgCl, but no AgBr. Write the structural formula of an isomer.

25.59 Draw cis—trans structures of any of the following square planar or octahedral complexes that exhibit geometric isomerism. Label the drawings as cis or trans.
(a) $[Pd(NH_3)_2Cl_2]$
(b) $Pd(NH_3)_3Cl^+$
(c) $Pd(NH_3)_4^{2+}$
(d) $Ru(NH_3)_4Br_2^+$

25.61 Determine if there are optical isomers of any of the following. If yes, sketch the isomers.
(a) cis-$Co(NH_3)_2(en)_2^{3+}$
(b) $trans$-$IrCl_2(C_2O_4)_2^{3-}$

25.56 Give the type of structural isomerism shown by each of the following:
(a) $[CoCl(en)_2(NO_2)]NO_2$ and $[Co(en)_2(NO_2)_2]Cl$
(b) $[Co(NH_3)_6][Cr(NO_2)_6]$ and $[Cr(NH_3)_6][Co(NO_2)_6]$
(c) $[Co(en)_2(ONO)_2]Cl$ and $[Co(en)_2(NO_2)_2]Cl$
(d) $[Cr(H_2O)Cl_2(py)_2]Cl$ and $[Cr(H_2O)Cl_3(py)_2] \cdot H_2O$ (py = pyridine, C_5H_5N)

25.58 Studies of a complex gave composition corresponding to the formula $CoBr(C_2O_4) \cdot 4NH_3$. Conductance measurements indicate that there are two ions per formula unit. If calcium nitrate gives no precipitate of calcium oxalate, but silver nitrate precipitates silver bromide, what is the structural formula of the complex? Write the structural formula of an isomer.

25.60 If any of the following octahedral complexes display geometric isomerism, draw the structures and label them as cis or trans.
(a) $Co(NO_2)_4(NH_3)_2^-$
(b) $Co(NH_3)_5(NO_2)^{2+}$
(c) $Pt(NH_3)_3Br_3^+$
(d) $Cr(NH_3)_5Cl^{2+}$

25.62 Sketch mirror images of each of the following. From these sketches determine whether optical isomers exist and note this fact on the drawings.
(a) $Rh(en)_3^{3+}$
(b) cis-$Cr(NH_3)_2(SCN)_4^-$

Valence Bond and Crystal Field Theories

25.63 For each of the following, first draw orbital diagrams for the isolated metal atom and metal ion. Then, using valence bond theory, draw the orbital diagram for the metal atom in the octahedral complex.
   (a) $V(H_2O)_6^{3+}$ (b) $Fe(en)_3^{3+}$ (high spin)
   (c) $Rh(CN)_6^{3-}$ (low spin)

25.65 Each of the following complex ions is either tetrahedral or square planar. On the basis of the number of unpaired electrons (given in parentheses), decide which is the correct geometry. Explain your answers using valence bond theory.
   (a) $Pt(NH_3)_4^{2+}$ (0)   (b) $Co(en)_2^{2+}$ (1)
   (c) $FeCl_4^-$ (5)   (d) $Co(NCS)_4^{2-}$ (3)

25.67 Using crystal field theory, sketch the energy level diagram for the d orbitals in an octahedral field; then fill in the electrons for the metal ion in each of the following complexes. How many unpaired electrons are there in each case?
   (a) $V(CN)_6^{3-}$ (b) $Co(C_2O_4)_3^{4-}$ (high spin)
   (c) $Mn(CN)_6^{3-}$ (low spin)

25.69 Obtain the distribution of d electrons in the complex ions listed in 25.65 using crystal field theory.

25.64 For each of the following, first draw orbital diagrams for the isolated metal atom and metal ion. Then, using valence bond theory, draw the orbital diagram for the metal atom in the octahedral complex.
   (a) $V(H_2O)_6^{2+}$ (b) FeF_6^{3-} (high spin)
   (c) $Co(en)_3^{3+}$ (low spin)

25.66 Each of the following complex ions is either tetrahedral or square planar. On the basis of the number of unpaired electrons (given in parentheses), decide which is the correct geometry. Explain your answers using valence bond theory.
   (a) $Pt(NH_3)_2(NO_2)_2^{2+}$ (2)   (b) $MnCl_4^{2-}$ (5)
   (c) $NiCl_4^{2-}$ (2)   (d) AuF_4^- (0)

25.68 Using crystal field theory, sketch the energy level diagram for the d orbitals in an octahedral field; then fill in the electrons for the metal ion in each of the following complexes. How many unpaired electrons are there in each case?
   (a) $ZrCl_6^{4-}$ (b) $OsCl_6^{2-}$ (low spin)
   (c) $MnCl_6^{4-}$ (high spin)

25.70 Obtain the distribution of d electrons in the complex ions listed in 25.66 using crystal field theory.

Color

25.71 The $Co(SCN)_4^{2-}$ ion has a maximum absorption at 530 nm. What color do you expect for this ion?

25.73 The $Co(NH_3)_6^{3+}$ ion has a yellow color, but when one NH_3 ligand is replaced by H_2O to give $Co(NH_3)_5(H_2O)^{3+}$, the color shifts to red. Is this shift in the expected direction? Explain.

25.75 What is the value of Δ (in kJ/mol) when $\lambda = 500$ nm, corresponding to an electron jump between d orbital levels in a complex with d^1 configuration?

25.72 The $Co(en)_3^{3+}$ ion has a maximum absorption at 470 nm. What color do you expect for this ion?

· **25.74** The $Co(en)_3^{3+}$ ion has a yellow color, but the CoF_6^{3-} ion has a blue color. Is the shift from yellow to blue expected when ethylenediamine ligands are replaced by F^- ligands? Explain.

25.76 What is the value of Δ (in kJ/mol) when $\lambda = 680$ nm, corresponding to an electron jump between d orbital levels in a complex with d^1 configuration?

Additional Problems

25.77 The hexaaquascandium(III) ion, $Sc(H_2O)_6^{3+}$, is colorless. Explain why this might be expected.

25.78 The tetraaquazinc(II) ion, $Zn(H_2O)_4^{2+}$, is colorless. Explain why this might be expected.

25.79 There are only two geometric isomers of the tetraamminedichlorocobalt(III) ion, $Co(NH_3)_4Cl_2^+$. How many geometric isomers would be expected for this ion if it had a regular planar hexagonal geometry? Give drawings for them. Does this rule out a planar hexagonal geometry for $Co(NH_3)_4Cl_2^+$? Explain.

25.80 There are only two geometric isomers of the triamminetrichloroplatinum(IV) ion, $Pt(NH_3)_3Cl_3^+$. How many geometric isomers would be expected for this ion

if it had a regular planar hexagonal geometry? Give drawings for them. Does this rule out a planar hexagonal geometry for $Pt(NH_3)_3Cl_3^+$? Explain.

*25.81 Find the concentrations of $Cu^{2+}(aq)$, $NH_3(aq)$, and $Cu(NH_3)_4^{2+}(aq)$ at equilibrium when 0.10 mol $Cu^{2+}(aq)$ and 0.40 mol $NH_3(aq)$ are made up to 1.00 L of solution. The dissociation constant, K_d, for the complex ion $Cu(NH_3)_4^{2+}$ is 5.0×10^{-14}.

*25.82 Find the concentrations of $Ag^+(aq)$, $NH_3(aq)$, and $Ag(NH_3)_2^+(aq)$ at equilibrium when 0.10 mol $Ag^+(aq)$ and 0.20 mol $NH_3(aq)$ are made up to 1.00 L of solution. The dissociation constant, K_d, for the complex ion $Ag(NH_3)_2^+$ is 5.9×10^{-8}.

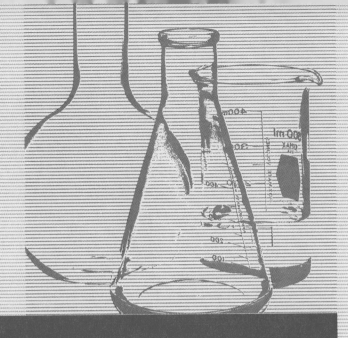

26. Organic Chemistry

Hydrocarbons

26.1 Alkanes and Cycloalkanes Methane, the Simplest Hydrocarbon/ The Alkane Series/ Nomenclature of Alkanes/ Cycloalkanes/ Sources of Alkanes and Cycloalkanes

26.2 Alkenes and Alkynes Alkenes/ Alkynes

26.3 Aromatic Hydrocarbons Derivatives of Benzene/ Sources and Uses of Aromatic Hydrocarbons

26.4 Reactions of Hydrocarbons Oxidation/ Substitution Reactions of Alkanes/ Addition Reactions of Alkenes/ Substitution Reactions of Aromatic Hydrocarbons/ Petroleum Refining

Derivatives of Hydrocarbons

26.5 Organic Compounds Containing Oxygen Alcohols and Ethers/ Aldehydes and Ketones/ Carboxylic Acids and Esters

26.6 Reactions of Oxygen-Containing Organic Compounds Oxidation–Reduction Reactions/ Esterification and Saponification/ Polyesters

26.7 Organic Compounds Containing Nitrogen and Sulfur Amines and Amides/ Thiols and Disulfides

Organic chemistry is the chemistry of compounds containing carbon. The majority of known compounds are organic—over a million such compounds have been described and thousands of new ones are discovered every year. We encounter many organic compounds every day. Some examples are ethanol (grain alcohol), ethylene glycol (automobile antifreeze), and acetone (nail polish remover). What is it about the carbon atom that makes possible such a diversity of substances?

A unique feature of carbon is its ability to bond to other carbon atoms to give chains and rings of various lengths. Several elements have limited ability to form such chains or rings of like atoms, but only carbon does this with more than a few atoms. Petroleum, for example, consists of molecules that have up to 30 or more carbon atoms bonded together, and chains of thousands of carbon atoms exist in molecules of polyethylene plastic.

The tetravalence of carbon, that is, its covalence of four, also makes possible the branching of chains and the fusion of several rings. Moreover, other kinds of atoms, such as oxygen, nitrogen, and sulfur, may be attached to the carbon atoms by single or multiple bonds. Thus, great variety can be found, even among the smaller organic molecules.

In this chapter, we will discuss structural features of organic molecules. We will also discuss some of the important chemical reactions of organic compounds.

Chapter Overview

The simplest organic compounds are the *hydrocarbons* (compounds of only carbon and hydrogen). After looking at the different structural types and the nomenclature (naming) of hydrocarbons, we will describe some important reactions of hydrocarbons. All other organic compounds, for classification purposes, are considered *derivatives of hydrocarbons*, with hydrogen atoms replaced by groups of atoms not containing carbon. Groups of atoms with distinctive chemical properties are called *functional groups*. In the last part of the chapter, we will look at compounds with functional groups containing oxygen, nitrogen, and sulfur.

Hydrocarbons

Organic compounds are compounds of carbon.■ Usually they contain hydrogen atoms and perhaps other atoms, such as oxygen, nitrogen, sulfur, and halogens. The simplest organic compounds are **hydrocarbons,** compounds containing only carbon and hydrogen. All other organic compounds are considered for classification purposes to be derived from hydrocarbons.

26.1 Alkanes and Cycloalkanes

Hydrocarbons are classified into two main types: aliphatic and aromatic (see Figure 26.1). **Aromatic hydrocarbons** contain benzene rings or similar structural features. (Benzene consists of a ring of six carbon atoms with alternating single and double carbon–carbon bonds.) All other hydrocarbons are classified as **aliphatic hydrocarbons.** The simplest hydrocarbon is methane.

■ At one time organic compounds were thought to be made only by living organisms. Urea (NH_2CONH_2), for example, was first obtained from urine, where it occurs from the breakdown of proteins. Then, in 1828, Friedrich Wöhler prepared urea by heating ammonium cyanate, NH_4CNO, a compound obtained from nonliving materials. Although we still use the term *organic,* it no longer distinguishes between compounds of living or nonliving origin.

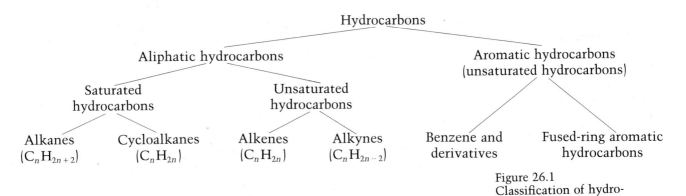

Figure 26.1
Classification of hydro-
carbons.

Methane, the Simplest Hydrocarbon

A carbon atom has four valence electrons and forms four covalent bonds. Therefore, the simplest hydrocarbon consists of one carbon atom to which four hydrogen atoms are bonded. The C—H bonds in this simplest hydro- carbon, called methane, are formed from tetrahedrally directed sp^3 hybrid orbitals on the carbon atom. We can represent the structure of methane by its *molecular formula*, CH_4, which gives the number of different atoms in the molecule, or by its *structural formula*, which shows how these atoms are bonded to one another:

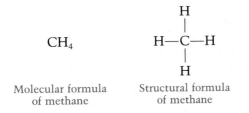

Molecular formula of methane

Structural formula of methane

A structural formula is not meant to convey information about the three- dimensional arrangement of atoms. To do this, we would draw a three- dimensional formula or construct a model of the molecule (see Figure 26.2).

The Alkane Series

Methane is a **saturated hydrocarbon,** that is, a hydrocarbon in which all carbon atoms are bonded to the maximum number of hydrogen atoms. (There are no carbon–carbon double or triple bonds.) One series of saturated hydrocarbons is the alkane series. The **alkanes,** also called *paraffins,* are

Figure 26.2
Three-dimensional models of methane, CH_4. The "space- filling" model on the left rep- resents the atoms to correct scale. Certain details, such as bond angles and length, are obscured, however. The "ball-and-stick" model in the center distorts the sizes of at- oms, but shows bond re- lationships more clearly. A perspective drawing of meth- ane is shown at the right.

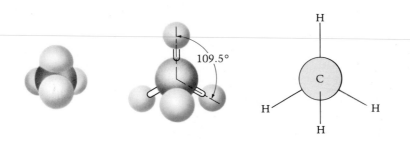

compounds with the general formula C_nH_{2n+2}.■ For $n = 1$, we get the formula CH_4; for $n = 2$, C_2H_6; for $n = 3$, C_3H_8; and so forth. The structural formulas of the first four alkanes are

■ The term *paraffin* comes from the Latin *parum affinus*, meaning "little affinity." The alkanes do not react with (have little affinity for) many reagents.

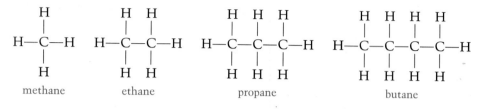

methane   ethane   propane   butane

Structures of organic compounds are often given by *condensed structural formulas*, where

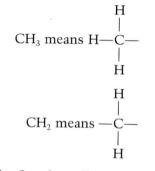

and

Condensed formulas of the first four alkanes are

$$CH_4 \qquad CH_3CH_3 \qquad CH_3CH_2CH_3 \qquad CH_3CH_2CH_2CH_3$$

methane   ethane   propane   butane

Note that the formula of one alkane differs from the preceding one by a —CH_2— group. For example, compare propane with butane:

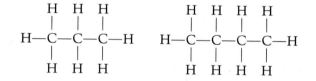

A **homologous series** is a series of compounds in which one compound differs from a preceding one by a —CH_2— group. The alkanes thus constitute a homologous series. Members of a homologous series have similar chemical properties and their physical properties change throughout the series in a regular way. Table 26.1 lists the melting points and boiling points of the first ten *straight-chain* alkanes. (These alkanes have all carbon atoms bonded to one another to give a single chain; hydrogen atoms fill out the four valencies of each carbon atom.) They are also called *normal* alkanes. Note that the melting points and boiling points generally increase in the series from methane to decane. This is explained as a result of increasing intermolecular forces, which tend to increase with molecular weight.

In addition to the straight-chain alkanes, *branched-chain* alkanes are possible. For example, isobutane (or 2-methylpropane) has the structure

Name	Number of Carbons	Formula	Melting Point (°C)	Boiling Point (°C)
Methane	1	CH_4	−183	−162
Ethane	2	CH_3CH_3	−172	−89
Propane	3	$CH_3CH_2CH_3$	−187	−42
Butane	4	$CH_3(CH_2)_2CH_3$	−138	0
Pentane	5	$CH_3(CH_2)_3CH_3$	−130	36
Hexane	6	$CH_3(CH_2)_4CH_3$	−95	69
Heptane	7	$CH_3(CH_2)_5CH_3$	−91	98
Octane	8	$CH_3(CH_2)_6CH_3$	−57	126
Nonane	9	$CH_3(CH_2)_7CH_3$	−54	151
Decane	10	$CH_3(CH_2)_8CH_3$	−30	174

Table 26.1
Physical Properties of Straight-Chain Alkanes*

*Adapted from Whitaker, Fernandez, and Tsokos, *Concepts of General, Organic, and Biological Chemistry* (Boston: Houghton Mifflin Co., 1981), Table 9.1, p. 231.

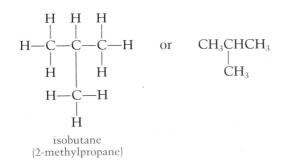

isobutane
(2-methylpropane)

Note that the molecular formula of isobutane is C_4H_{10}, the same as for butane, the straight-chain hydrocarbon. Butane and isobutane are *structural isomers* of one another (compounds with the same molecular formula but different structural formulas).■ Because they have different structures, they have different properties. The number of structural isomers corresponding to the molecular formula C_nH_{2n+2} increases rapidly with n. For example, there are three structural isomers with the molecular formula C_5H_{12} (these compounds are called pentanes; see Figure 26.3), five structural isomers of C_6H_{14} (hexanes), and 75 of $C_{10}H_{22}$ (decanes).

■ Formerly, the names of the straight-chain alkanes were distinguished from branched-chain isomers by the prefix *n-* for *normal.* This designation is still common (butane is called *n*-butane) but is not used in the IUPAC name, which we will discuss next.

Figure 26.3
Isomers of pentane.

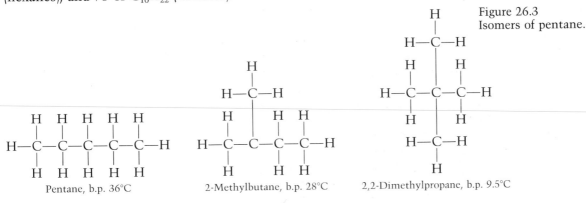

Pentane, b.p. 36°C   2-Methylbutane, b.p. 28°C   2,2-Dimethylpropane, b.p. 9.5°C

Nomenclature of Alkanes

As we noted in the chapter opening, over a million organic compounds are known. A nomenclature for these substances has developed over the years as a way of understanding and classifying their structures. This nomenclature is now formulated in rules agreed upon by the International Union of Pure and Applied Chemistry (IUPAC).

The first four straight-chain alkanes (methane, ethane, propane, and butane) have long-established names. Higher members of the series are named from the Greek words indicating the number of carbon atoms in the molecule, with the suffix *-ane* added. For example, the straight-chain alkane with formula C_5H_{12} is named pentane. Table 26.1 gives the names of the first ten straight-chain alkanes.

The following IUPAC rules are applied in naming the branched-chain alkanes:

1. Determine the longest continuous (not necessarily straight) chain of carbon atoms in the molecule. The base name of the branched-chain alkane is that of the straight-chain alkane (Table 26.1), corresponding to the number of carbon atoms in this longest chain. For example, in

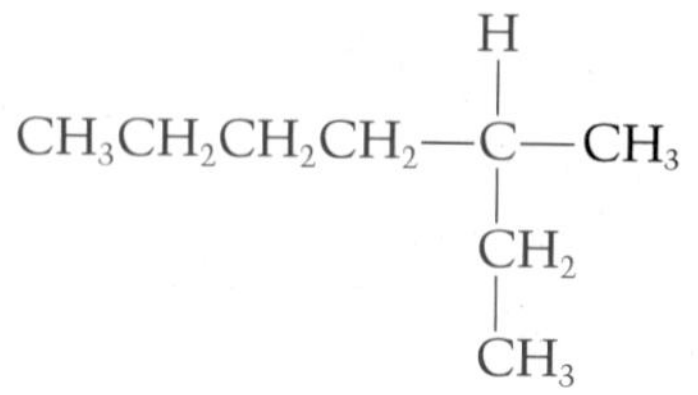

$$CH_3CH_2CH_2CH_2 - \overset{\displaystyle H}{\underset{\displaystyle \underset{CH_3}{\overset{|}{CH_2}}}{\overset{|}{\underset{|}{C}}}} - CH_3$$

the longest continuous carbon chain, shown in color, has seven carbon atoms, giving the base name heptane. The full name for the alkane includes the names of any branched chains. These names are placed in front of the base name, as described below.

2. Any chain branching off the longest chain is named as an alkyl group. An

Original Alkane	Structure of Alkyl Group	Name of Alkyl Group
Methane, CH_4	CH_3-	Methyl
Ethane, CH_3CH_3	CH_3CH_2-	Ethyl
Propane, $CH_3CH_2CH_3$	$CH_3CH_2CH_2-$	Propyl
Propane, $CH_3CH_2CH_3$	$CH_3\underset{|}{CH}CH_3$	Isopropyl
Butane, $CH_3CH_2CH_2CH_3$	$CH_3CH_2CH_2CH_2-$	Butyl
Isobutane, $CH_3\underset{|}{\underset{CH_3}{CH}}CH_3$	$CH_3\underset{|}{\underset{CH_3}{C}}CH_3$	*Tertiary*-butyl (*t*-butyl)

Table 26.2
Important Alkyl Groups

alkyl group is an alkane less one hydrogen atom. (Table 26.2 lists some alkyl groups.) When a hydrogen atom is removed from an end carbon atom of a straight-chain alkane, the alkane is named by changing the -*ane* suffix to -*yl*. Thus, removing a hydrogen atom from methane gives the methyl group, —CH_3. The structure shown in rule 1 has a methyl group as a branch on the heptane chain.

3. The complete name of a branch requires a number that locates that branch on the longest chain. For this purpose, we number each carbon atom on the longest chain in whichever direction gives the smaller numbers for the locations of all branches. The structural formula in rule 1 is numbered as follows:

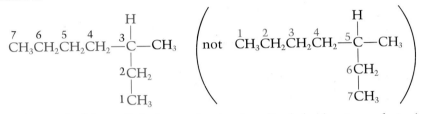

Thus, the methyl branch is located at carbon 3 of the heptane chain (not carbon 5). The complete name of the branch is 3-methyl, and the compound is named 3-methylheptane. Note that the branch name and base name are written as a single word, with a hyphen following the number.

4. When there is more than one alkyl branch of the same kind (say, two methyl groups), this number is indicated by a Greek prefix, such as *di-*, *tri-*, or *tetra-*, used with the name of the alkyl group. The position of each group on the longest chain is given by numbers. For example,

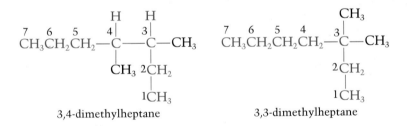

3,4-dimethylheptane   3,3-dimethylheptane

Note that the position numbers are separated by a comma and are followed by a hyphen.

When there are two or more different alkyl branches, the name of each branch, with its position number, precedes the base name. The branch names are placed in alphabetical order. For example,

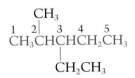

3-ethyl-2-methylpentane

Note the use of hyphens.

Example 26.1

Give IUPAC names for the following compounds:

(a)
$$CH_3CH_2CH_2$$
$$CHCH_2CH_2CH_3$$
$$CH_3CH_2CH_2CH_2$$

(b)
$$CH_3$$
$$CH_3—C—CH_3$$
$$CH_2$$
$$CH_2$$
$$CH_2$$
$$CH_3$$

Solution

(a) The longest continuous chain is numbered as follows:

$$CH_3CH_2CH_2$$
$$\overset{8}{C}H_3\overset{7}{C}H_2\overset{6}{C}H_2\overset{5}{C}H_2 \overset{4}{C}H\overset{3}{C}H_2\overset{2}{C}H_2\overset{1}{C}H_3$$

The name of the compound is 4-propyloctane. If the longest chain had been numbered in the opposite direction, we would have gotten the name 5-propyloctane. But since 5 is larger than 4, this name is unacceptable. (b) The numbering of the longest chain is

$$^1 CH_3$$
$$CH_3—\overset{2}{C}—CH_3$$
$$^3 CH_2$$
$$^4 CH_2$$
$$^5 CH_2$$
$$^6 CH_3$$

Any one of the methyl carbon atoms could be given the number 1, and the other methyl groups branch off carbon atom 2. Hence, the name is 2,2-dimethylhexane.

Exercise 26.1

What are the IUPAC names of the following hydrocarbons?

(a)
$$CH_3$$
$$CH_3CHCHCH_3$$
$$CH_3$$

(b)
$$CH_2CH_2CH_3$$
$$CH_3CHCHCH_2CH_3$$
$$CH_3$$

(See Problems 26.21 and 26.22.)

Example 26.2

Write the condensed structural formula of 4-ethyl-3-methylheptane.

Solution

First write out the carbon skeleton for heptane:

Then attach the alkyl groups:

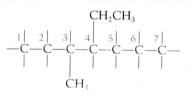

After filling out the structure with H atoms, we have

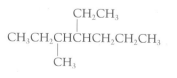

Exercise 26.2

Write the condensed structural formula of 3,3-dimethyloctane.

(See Problems 26.23 and 26.24.)

Cycloalkanes

Besides the straight-chain and branched-chain alkanes that we have discussed, **cycloalkanes** are saturated hydrocarbons in which the carbon atoms form a ring. The general formula of cycloalkanes is C_nH_{2n}. Figure 26.4 gives the names and structural formulas of the first four members of the cycloalkane series. In the condensed structural formulas, a carbon atom and its attached hydrogen atoms are assumed to be at each corner.

Sources of Alkanes and Cycloalkanes

Fossil fuels (natural gas, petroleum, coal) are the principal sources of hydrocarbons. Natural gas is mainly methane with smaller amounts of the other gaseous alkanes (ethane, propane, butane). Petroleum is a mixture of alkanes and cycloalkanes with smaller amounts of aromatic hydrocarbons. These hydrocarbons are separated by distillation into fractions such as gasoline and kerosene (see Table 26.3). These fractions are usually processed further, for example, to obtain a greater quantity of gasoline with the desired fuel characteristics. This processing, called *petroleum refining*, includes various chemical reactions we will discuss in Section 26.4.

26.2 Alkenes and Alkynes

Unsaturated hydrocarbons are hydrocarbons not containing the maximum number of hydrogen atoms for a given carbon-atom framework. Such compounds have carbon–carbon multiple bonds and under the proper conditions

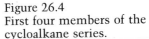

Figure 26.4
First four members of the cycloalkane series.

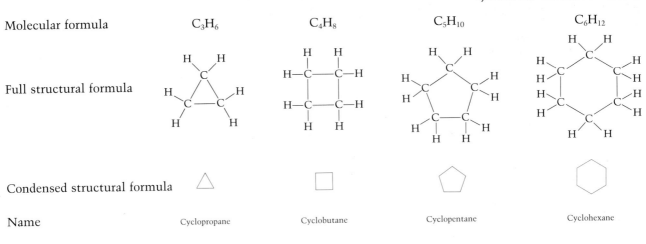

Table 26.3
Fractions from the
Distillation of Petroleum

Boiling Range, °C	Name	Range of Carbon Atoms per Molecule	Use
Below 20	Gases	C_1 to C_4	Heating, cooking, and chemical raw material
20–200	Naphtha; straight run gasoline	C_5 to C_{12}	Fuel; lighter fractions (such as petroleum ether, bp 30–60°C) are also used as laboratory solvents
200–300	Kerosene	C_{12} to C_{15}	Fuel
300–400	Fuel oil	C_{15} to C_{18}	Heating homes, diesel fuel
Over 400		Over C_{18}	Lubricating oil, greases, paraffin waxes, asphalt

From Harold Hart, *Organic Chemistry: A Short Course,* 6th Ed. (Boston: Houghton Mifflin Co., 1983), p. 92.

will add molecular hydrogen to give a saturated compound. For example, ethylene adds hydrogen to give ethane:

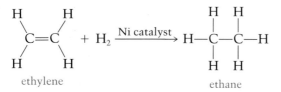

ethylene   ethane

Alkenes

Alkenes are hydrocarbons with the general formula C_nH_{2n} containing a carbon–carbon double bond. (These compounds are also called *olefins.*)■ The simplest alkene, ethylene, has a condensed formula of $CH_2{=}CH_2$. Ethylene is a gas with a sweetish odor. It is obtained from the refining of petroleum and is an important raw material of the chemical industry. Plants also produce it, and exposure of fruit to ethylene speeds ripening. In ethylene and other alkenes, all atoms connected to the two carbon atoms of a double bond lie in a single plane, as Figure 26.5 shows. This is due to the need for maximum overlap of $2p$ orbitals on carbon atoms to form a pi bond.■

We obtain the IUPAC name for an alkene by finding the longest chain containing the double bond. As with the alkanes, this longest chain gives us the stem name, except that the suffix is *-ene* rather than *-ane*. The carbon atoms of the longest chain are numbered from the end nearest the carbon–carbon double bond, and the position of the double bond is given the number of the first carbon atom of that bond. This number is written in front of the stem name of the alkene. Branch chains are named as in the alkanes. The simplest alkene, $CH_2{=}CH_2$, is called ethene, although the common name is ethylene.

■ Olefin means "oil-forming." Many alkenes react with Cl_2 to form oily compounds.

■ Bonding in ethylene was discussed in detail in Section 8.4.

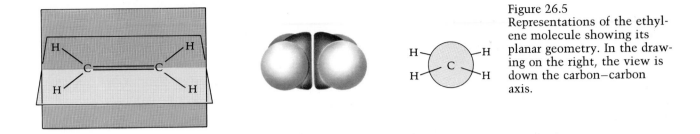

Figure 26.5
Representations of the ethylene molecule showing its
planar geometry. In the drawing on the right, the view is
down the carbon–carbon
axis.

Example 26.3

What are the IUPAC names of the following alkenes?

(a) $CH_2\!\!=\!\!CHCHCH_2CH_3$
$\qquad\qquad\quad |$
$\qquad\qquad CH_3$

(b) $CH_3CH_2CH_2CH_2CHCH\!\!=\!\!CHCH_3$
$\qquad\qquad\qquad\qquad |$
$\qquad\qquad\qquad\quad CH_2$
$\qquad\qquad\qquad\qquad |$
$\qquad\qquad\qquad\quad CH_2$
$\qquad\qquad\qquad\qquad |$
$\qquad\qquad\qquad\quad CH_2$
$\qquad\qquad\qquad\qquad |$
$\qquad\qquad\qquad\quad CH_2$
$\qquad\qquad\qquad\qquad |$
$\qquad\qquad\qquad\quad CH_3$

Solution

(a) The numbering of the carbon chain is

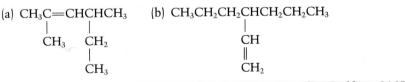

Since the longest chain containing a double bond has
five carbon atoms, this is a pentene. Moreover, it is a

1-pentene, because the double bond is between carbon
atoms 1 and 2. The name of the compound is 3-methyl-
1-pentene. Note the placement of hyphens. If the numbering had been in the opposite direction, we would
have named the compound as a 4-pentene. However,
this is unacceptable, since 4 is greater than 1. (b) The
numbering of the longest chain containing the double
bond is

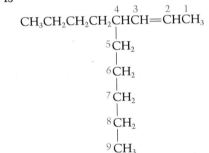

This gives the name 4-butyl-2-nonene. There is a longer
chain (of 10 carbon atoms), but it does not contain the
double bond.

Exercise 26.3

Give IUPAC names for the following compounds:

(a) $CH_3C\!\!=\!\!CHCHCH_3$
$\qquad\quad |\qquad\ \ |$
$\qquad CH_3\ \ CH_2$
$\qquad\qquad\qquad |$
$\qquad\qquad\ \ CH_3$

(b) $CH_3CH_2CH_2CHCH_2CH_2CH_3$
$\qquad\qquad\qquad\ |$
$\qquad\qquad\qquad CH$
$\qquad\qquad\qquad ||$
$\qquad\qquad\qquad CH_2$

(See Problems 26.25 and 26.26.)

Exercise 26.4

Write the condensed structural formula of 2,5-dimethyl-2-heptene.

(See Problems 26.27 and 26.28.)

Rotation about a carbon–carbon double bond cannot occur without breaking the pi bond. Since this requires energies comparable to those in chemical
reactions, rotation normally does not occur. This lack of rotation about the
double bond gives rise to isomers in certain alkenes. For example, there are
two isomers of 2-butene:

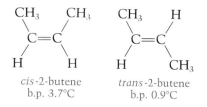

cis-2-butene   *trans*-2-butene
b.p. 3.7°C   b.p. 0.9°C

The different boiling points indicate that these are indeed different compounds.

The type of isomerism that occurs in the butenes is called **geometric,** or
cis–trans, isomerism. The *cis* isomer has identical groups (in the case of the
butenes, alkyl groups) attached to the same side of the double bond, whereas
the *trans* isomer has them on opposite sides. An alkene with the general
formula

has a geometric isomer only if the groups A and B are different and groups
D and E are different. Thus, there is no isomer of propene, CH_2=$CHCH_3$.

Example 26.4

For each of the following alkenes, decide if *cis–trans*
isomers are possible. If yes, draw structural formulas for
the isomers and give their IUPAC names (labeling *cis* or
trans).
(a) CH_3CH_2CH=$C(CH_3)_2$ (b) CH_3CH=$CHCH_2CH_3$

Solution

(a) If we write out the structure, we have

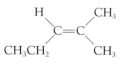

Since two methyl groups are attached to the second
carbon atom of the double bond, geometric isomers are
not possible. (b) Geometric isomers are possible. They
are

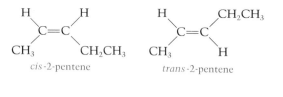

cis-2-pentene   *trans*-2-pentene

Exercise 26.5

Decide whether *cis–trans* isomers are possible for each of the following compounds.
If any have isomers, draw the structural formulas and give the IUPAC names, labeling *cis* or *trans*.   (a) CH_3CH=$CHCH_2CH_3$ (b) CH_3CH_2CH=CH_2

(See Problems 26.29 and 26.30.)

Alkynes

Alkynes are unsaturated hydrocarbons containing a carbon–carbon triple bond. The general formula is C_nH_{2n-2}. The simplest alkyne is acetylene (ethyne):■

■ The *-ene* ending in the common name *acetylene* does not follow IUPAC rules.

$$H—C\equiv C—H$$

Acetylene has a linear molecule. It is a very reactive gas, used to form other chemical compounds and plastics. It burns with oxygen in the oxyacetylene torch to give a very hot flame (about 3000°C). Acetylene is produced commercially from methane:

$$2CH_4 \xrightarrow{1600°C} CH\equiv CH + 3H_2$$

The alkynes are named by IUPAC rules in the same way as the alkenes, except that the stem name is determined from the longest chain containing the carbon–carbon triple bond. The suffix for this stem name is *-yne*.

Exercise 26.6

Give IUPAC names for the following alkynes:

(a) $CH_3C\equiv CH$ (b) $CH\equiv CCHCH_3$
                         |
                      CH_2CH_3

(See Problems 26.31 and 26.32.)

26.3 Aromatic Hydrocarbons

Aromatic hydrocarbons contain benzene rings, or six-membered rings of carbon atoms with alternating single and double carbon–carbon bonds. The electronic structure of these rings may be represented by resonance formulas. For benzene, we have

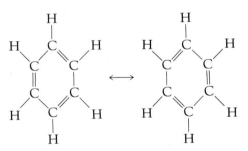

This electronic structure can also be described in molecular-orbital terms (Figure 26.6). In this description, pi molecular orbitals encompass the entire carbon-atom ring. The pi electrons are said to be delocalized. Benzene is given the condensed formula

where the circle represents the delocalization of pi electrons.

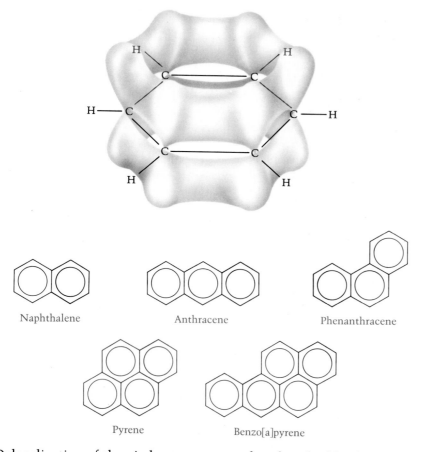

Figure 26.6
Lowest energy pi molecular
orbital of benzene. Note how
the orbital encompasses the
entire ring of carbon atoms.

Figure 26.7
Formulas of some fused-ring
aromatic hydrocarbons.

Naphthalene   Anthracene   Phenanthracene

Pyrene   Benzo[a]pyrene

Delocalization of the pi electrons means that the "double" bonds in benzene do not behave as isolated double bonds. When we discuss the chemical reactions of benzene, we will see that it behaves quite differently from an alkene. Figure 26.7 gives formulas of some *fused-ring*, or *polynuclear*, aromatic hydrocarbons. In the fused-ring hydrocarbons, two or more rings share carbon atoms.

Derivatives of Benzene

Alkyl groups may replace one or more hydrogen atoms of benzene, giving hydrocarbon derivatives of benzene. For example, toluene (methylbenzene) has the structural formula

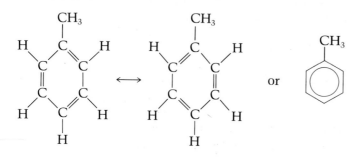

When two groups are on the benzene ring, three isomers are possible. They may be distinguished by the prefixes *ortho-* (*o*), *meta-* (*m*), and *para-* (*p*). For example,

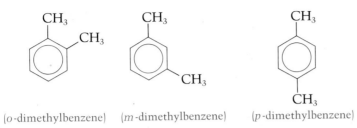

(*o*-dimethylbenzene)   (*m*-dimethylbenzene)   (*p*-dimethylbenzene)

A numbering system is also used, to show the positions of two or more groups:

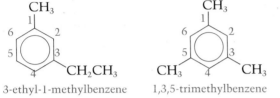

3-ethyl-1-methylbenzene   1,3,5-trimethylbenzene

It is sometimes preferable to name a compound containing a benzene ring by regarding the ring as a group in the same manner as alkyl groups. Pulling a hydrogen atom from benzene leaves the phenyl group, C_6H_5—. For example, the compound diphenylmethane is named using methane as the stem name.

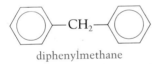

diphenylmethane

Exercise 26.7

Write structural formulas of (a) ethylbenzene, (b) 1,2-diphenylethane.

(See Problems 26.33 and 26.34.)

Sources and Uses of Aromatic Hydrocarbons

The original source of aromatic hydrocarbons was coal tar. Coal tar is a by-product in the preparation of coke (a form of carbon) for the making of steel. Coal is heated in the absence of air to drive off volatile substances, leaving the coke behind. Besides coal gases, these volatile substances are ammoniacal liquor (mostly aqueous ammonia) and coal tar. The coal tar fraction contains many aromatic hydrocarbons, including benzene, toluene, naphthalene, and anthracene. This process is still the principal source of naphthalene and anthracene, but benzene is now chiefly obtained by petroleum refining.

Benzene is used to make styrene, $C_6H_5CH{=}CH_2$, for polystyrene plastic (Styrofoam). It is also used to make cyclohexane, C_6H_{12}, a starting material for nylon. Naphthalene and anthracene are used to make dyes. At one time benzene was used extensively as a solvent; due to its toxicity, it is rarely so used now. Most other aromatic hydrocarbons are also toxic. Some of the fused-ring aromatic hydrocarbons are known cancer-causing substances (carcinogens). For example, benzo[a]pyrene, which is present in coal tar and cigarette smoke, is a carcinogen.

26.4 Reactions of Hydrocarbons

Natural gas and petroleum, which are mixtures of hydrocarbons, are the major sources of organic chemicals. By various reactions they are converted to final products—solvents, plastics, textile fibers, and so forth. The saturated hydrocarbons, forming the bulk of petroleum, are not readily reactive at normal temperatures, except with a few reagents. At higher temperatures with the proper catalysts, however, they can be made to break up or rearrange, giving unsaturated hydrocarbons. These unsaturated hydrocarbons are quite reactive.

Oxidation

All hydrocarbons burn in an excess of O_2 to give carbon dioxide and water.

$$C_2H_6(g) + \tfrac{7}{2}O_2(g) \longrightarrow 2CO_2(g) + 3H_2O(l); \Delta H = -1560 \text{ kJ/mol}$$
ethane

$$C_6H_6(l) + \tfrac{15}{2}O_2(g) \longrightarrow 6CO_2(g) + 3H_2O(l); \Delta H = -3267 \text{ kJ/mol}$$
benzene

The high values of ΔH for these reactions make the hydrocarbons useful as fuels.

Unsaturated hydrocarbons are oxidized under milder conditions than saturated hydrocarbons. For example, when aqueous potassium permanganate, $KMnO_4(aq)$, is added to an alkene or alkyne, the purple color of $KMnO_4$ fades and a precipitate of brown manganese dioxide forms:

$$3C_4H_9CH\text{=}CH_2 + 2MnO_4^-(aq) + 4H_2O \longrightarrow$$
1-hexene

$$
\begin{array}{c}
\quad\quad \text{H} \quad \text{H} \\
\quad\quad | \quad\ \ | \\
3C_4H_9C\text{—}C\text{—H} + 2MnO_2(s) + 2OH^-(aq) \\
\quad\quad | \quad\ \ | \\
\quad\quad \text{O} \quad \text{O} \\
\quad\quad | \quad\ \ | \\
\quad\quad \text{H} \quad \text{H}
\end{array}
$$

Saturated hydrocarbons are unreactive with $KMnO_4(aq)$, so this reagent can be used to distinguish them from unsaturated hydrocarbons (called the *Bayer test* of unsaturation).■

■ A positive Bayer test indicates only that the compound is easily oxidized. Other tests are required to definitely identify a compound as an alkene.

Substitution Reactions of Alkanes

Alkanes react with the halogens F_2, Cl_2, and Br_2. Reaction with Cl_2 requires sunlight (indicated $h\nu$) or heat. For example,

$$
\overset{\displaystyle H}{\underset{\displaystyle H}{H-\overset{|}{\underset{|}{C}}-H}} + Cl-Cl \xrightarrow{h\nu} \overset{\displaystyle H}{\underset{\displaystyle H}{H-\overset{|}{\underset{|}{C}}-Cl}} + H-Cl
$$

This is an example of a substitution reaction. In a **substitution reaction,** a part of the reagent molecule is substituted for an H atom on a hydrocarbon or hydrocarbon group. All of the H atoms of an alkane may undergo substitution, leading to a mixture of products.

$$CH_3Cl + Cl_2 \xrightarrow{h\nu} CH_2Cl_2 + HCl$$

$$CH_2Cl_2 + Cl_2 \xrightarrow{h\nu} CHCl_3 + HCl$$

$$CHCl_3 + Cl_2 \xrightarrow{h\nu} CCl_4 + HCl$$

Fluorine is very reactive with saturated hydrocarbons and usually gives complete substitution. Bromine is less reactive than Cl_2 and often requires elevated temperatures for substitution.

Addition Reactions of Alkenes

Alkenes are more reactive than alkanes because of the presence of the double bond. Many reagents *add* to the double bond. A simple example is the addition of a halogen, such as Br_2, to propene:

$$
CH_3CH=CH_2 + Br_2 \longrightarrow \underset{\underset{Br\quad Br}{|\quad\ |}}{CH_3CH-CH_2}
$$

In an **addition reaction,** parts of a reagent are added to each carbon atom of a carbon–carbon multiple bond, which then becomes a C—C single bond. The addition of Br_2 to an alkene is fast. In fact, it occurs so readily that bromine dissolved in carbon tetrachloride, CCl_4, is a useful reagent to test for unsaturation. When a few drops of the solution are added to an alkene, the red brown color of the Br_2 is immediately lost. There is no reaction with alkanes, and the solution retains the red brown color of the Br_2.

Unsymmetrical reagents, such as HCl or HBr, add to unsymmetrical alkenes to give two products, which are isomers of one another. For example,

$$
CH_3CH=CH_2 + HBr \longrightarrow \overset{3\quad\ 2\quad\ 1}{\underset{\underset{Br\quad H}{|\quad\ |}}{CH_3CH-CH_2}}
$$
2-bromopropane

$$
CH_3CH=CH_2 + HBr \longrightarrow \overset{3\quad\ 2\quad\ 1}{\underset{\underset{H\quad Br}{|\quad\ |}}{CH_3CH-CH_2}}
$$
1-bromopropane

In one case the hydrogen atom of HBr adds to carbon atom 1, giving 2-bromopropane, and in the other case the hydrogen atom of HBr adds to carbon atom 2, giving 1-bromopropane. (The name 1-bromopropane means that a bromine atom is substituted for a hydrogen atom at carbon atom 1.)

The two products are not formed in equal amounts; one is more likely to form. **Markownikoff's rule** states that the major product formed by the addition of an unsymmetrical reagent such as H—Cl, H—Br, or H—OH is the one obtained when the H atom of the reagent adds to the carbon atom of the double bond that already has the more hydrogen atoms attached to it. In the above example, the H atom of HBr should add preferentially to carbon atom 1, which has two hydrogen atoms attached to it. The major product then is 2-bromopropane.

Example 26.5

What is the major product of the following reaction?

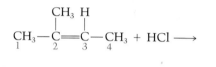

Solution

There is one H atom attached to carbon 3, none to carbon 2. Thus, the major product is

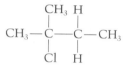

Exercise 26.8

Predict the main product when HBr is added to 1-butene.

(See Problems 26.37 and 26.38.)

Under the appropriate conditions, alkenes may even add to themselves to form chains of hundreds of carbon atoms. This giant molecule (macromolecule) of repeating units is called a *polymer* of the alkene. For example, when propylene (the IUPAC name is propene) is heated under pressure with a catalyst, it forms polypropylene:■

$$\ldots + \overset{\overset{\displaystyle CH_3}{\displaystyle |}}{CH}=CH_2 + \overset{\overset{\displaystyle CH_3}{\displaystyle |}}{CH}=CH_2 + \overset{\overset{\displaystyle CH_3}{\displaystyle |}}{CH}=CH_2 + \ldots \longrightarrow$$

$$-\overset{\overset{\displaystyle CH_3}{\displaystyle |}}{CH}-CH_2-\overset{\overset{\displaystyle CH_3}{\displaystyle |}}{CH}-CH_2-\overset{\overset{\displaystyle CH_3}{\displaystyle |}}{CH}-CH_2-$$

polypropylene

■ Polymerization of an alkene is usually carried out with a substance called an *initiator,* rather than with a true catalyst. An initiator molecule attacks an alkene molecule, making it reactive to addition. After polymerization, the initiator cannot be recovered, because it occupies the end positions on the polymer chains.

Because the polymer is formed by linking molecules with multiple bonds by an addition reaction, it is called an **addition polymer.**

Exercise 26.9

Saran is an addition polymer of vinylidene chloride, $CH_2=CCl_2$. Write the structure of the addition polymer.

(See Problems 26.39 and 26.40.)

The alkynes also undergo addition reactions, usually adding two molecules of the reagent for each C≡C bond. The major product is the isomer predicted by Markownikoff's rule. Thus,

$$CH_3-C\equiv C-H + 2HCl \longrightarrow CH_3-\underset{\underset{Cl}{|}}{\overset{\overset{Cl}{|}}{C}}-\underset{\underset{H}{|}}{\overset{\overset{H}{|}}{C}}-H$$

Substitution Reactions of Aromatic Hydrocarbons

Although benzene, C_6H_6, is an unsaturated hydrocarbon, it usually does not undergo addition reactions. The delocalized π electrons of benzene are more stable than localized π electrons. Therefore, benzene does not react readily with Br_2 in carbon tetrachloride, as an alkene would. The usual reactions of benzene are substitution reactions. In the presence of iron(III) bromide as a catalyst, a bromine atom of Br_2 substitutes for a hydrogen atom on the benzene ring:

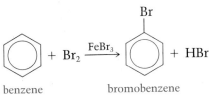

benzene   bromobenzene

Similarly, benzene undergoes substitution with nitric acid in the presence of sulfuric acid to give nitrobenzene:

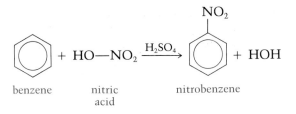

benzene   nitric   nitrobenzene
               acid

Petroleum Refining

Petroleum, as we noted earlier, is a mixture of hydrocarbons, principally alkanes and cycloalkanes. The object of *petroleum refining* is to obtain various hydrocarbon products from this mixture. In part, this is accomplished by fractional distillation of the petroleum, but the demand for certain products, particularly gasoline, is greater than what can be supplied by distillation. For that reason, petroleum refiners resort to various chemical processes to increase the amount of desired products.

One of these chemical processes is *catalytic cracking* (Figure 26.8). Although alkanes are usually stable compounds, when the hydrocarbon vapor is passed over a heated catalyst of alumina (Al_2O_3) and silica (SiO_2), the hydrocarbon molecule breaks up or "cracks" to give hydrocarbons of lower

Figure 26.8
Catalytic cracking unit of a
petroleum refinery.

molecular weight. For example,

$$CH_3CH_2CH_2CH_3 \xrightarrow[\Delta]{Al_2O_3\ +\ SiO_2} CH_4\ +\ CH_2{=}CHCH_3$$

butane ⟶ methane + propylene

$$\xrightarrow[\Delta]{Al_2O_3\ +\ SiO_2} CH_3CH_3 +\ CH_2{=}CH_2$$

ethane + ethylene

Here a four-carbon alkane, butane, has been cracked to give hydrocarbons
with one, two, and three carbon atoms. Thus, large hydrocarbon molecules
are cracked to give smaller ones. Hydrogen may be added to the alkenes to
give stable saturated hydrocarbons, but the smaller alkenes, particularly
ethylene and propylene, are useful for making other products, such as polyethylene or polypropylene.

In order for a gasoline to function properly in an engine, it should not begin
to burn before it is ignited by the spark plug. If it does, it gives engine
"knock." The antiknock characteristics of a gasoline are rated by the *octane
number* scale. This scale is based on heptane, which is given an octane
number of 0, and 2,2,4-trimethylpentane (an octane isomer), which is given
an octane number of 100. The higher the octane number, the better the
antiknock characteristics.

Substances, such as tetraethyllead, may be added to gasoline to raise the octane number, but the octane rating can be improved by changing straight-chain hydrocarbons to branched-chain isomers. This can be accomplished by heating the hydrocarbon vapor with the Lewis acid $AlCl_3$.

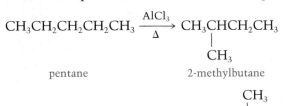

$CH_3CH_2CH_2CH_2CH_3 \xrightarrow[\Delta]{AlCl_3} CH_3CHCH_2CH_3$
    |
    CH_3

pentane   2-methylbutane

The process is called *isomerization*.

Another process that improves the octane rating of a gasoline is *catalytic reforming*. In this process, an alkane or cycloalkane is transformed to an aromatic hydrocarbon.

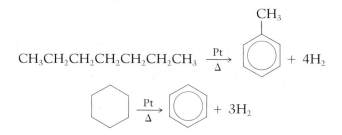

Catalytic reforming is also a major source of benzene, which is used to make styrene, $C_6H_5CH{=}CH_2$, for polystyrene plastic.

Derivatives of Hydrocarbons

Certain groups of atoms in many organic molecules are particularly reactive and have characteristic chemical properties. A **functional group** is a reactive portion of a molecule that undergoes predictable reactions. The $C{=}C$ bond in a compound, for example, reacts readily with reagents Br_2 and HBr in addition reactions. For this reason, the $C{=}C$ bond is a functional group. Many functional groups contain an atom other than carbon and have lone pairs of electrons on it. These lone pairs contribute to the reactivity of the functional group. Others, such as $C{=}O$, have multiple bonds that are reactive. Table 26.4 lists some common functional groups.

In the previous sections of the chapter, we discussed the hydrocarbons and their reactions. All other organic compounds can be considered as derivatives of hydrocarbons. In these compounds, one or more hydrogen atoms of a hydrocarbon have been replaced by atoms other than carbon to give a functional group.

Structure of General Compound* (functional group in color)	Name of Functional Group
R—C̈l: R—B̈r:	Organic halide
R—Ö—H	Alcohol
R—Ö—R′	Ether
:O: ‖ R—C—H	Aldehyde
:O: ‖ R—C—R′	Ketone
:O: ‖ R—C—Ö—H	Carboxylic acid
:O: ‖ R—C—Ö—R′	Ester

R—N̈—H   R—N̈—H   R—N̈—R″
    | | |
   H   R′   R′
	Amine

:O:
‖
R—C—N̈—R′
      |
     H
	Amide

R—S̈—H	Thiol (mercaptan)
R—S̈—S̈—R′	Disulfide

Table 26.4
Some Organic
Functional Groups

*R, R′, and R″ are general hydrocarbon groups.

26.5 Organic Compounds Containing Oxygen

Many of the important functional groups in organic compounds contain oxygen. Examples are alcohols, ethers, aldehydes, ketones, carboxylic acids, and esters. We will look at characteristics of these compounds in this section.

Alcohols and Ethers

Structurally, we may think of an **alcohol** as a compound obtained by substituting a *hydroxyl group* (—OH) for an —H atom on an aliphatic hydrocarbon. Some examples are

$$CH_3OH \qquad CH_3CH_2OH \qquad \overset{\displaystyle OH}{\underset{\displaystyle |}{CH_3CHCH_3}}$$

methanol   ethanol   2-propanol
(methyl alcohol)   (ethyl alcohol)   (isopropyl alcohol)

Alcohols are named by IUPAC rules similar to those used for naming the hydrocarbons.

1. Select the longest carbon chain to which the —OH group is attached. Name this longest chain as a hydrocarbon. Then, change the *-e* of the hydrocarbon name to *-ol*. This becomes the stem name of the alcohol. For example,

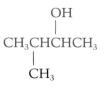

is named as a butanol.

2. The stem name should contain a number designating the position of the carbon atom to which the —OH group is attached if there is more than one nonequivalent position possible. Choose the order of numbering to give the smaller number for this position. For example,

$$\underset{\displaystyle |}{\overset{\displaystyle OH}{\underset{\displaystyle CH_3}{\overset{4 \quad 3 \quad |2 \ 1}{CH_3CHCHCH_3}}}}$$

has a stem name of 2-butanol.

3. The complete name of the alcohol should include the names of any alkyl groups attached to the stem name. These alkyl names are placed in front of the stem name, and their positions denoted by the number assigned to the carbon atoms of the stem hydrocarbon chain. The complete name of the above alcohol is 3-methyl-2-butanol.

Example 26.6

Name the following compound by IUPAC rules:

Solution

Since the compound has an —OH group, it is an alcohol. The numbering of the longest carbon chain to

which the —OH group is attached is shown below:

The complete name is 2-methyl-2-propanol.

Exercise 26.10

Give the IUPAC name of the following compound:

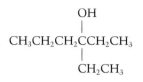

$$\underset{\substack{|\\ \text{CH}_2\text{CH}_3}}{\overset{\substack{\text{OH}\\ |}}{\text{CH}_3\text{CH}_2\text{CH}_2\text{CCH}_2\text{CH}_3}}$$

(See Problems 26.45 and 26.46.)

Alcohols are usually classified by the number of carbon atoms attached to the carbon atom to which the —OH group is bonded. A **primary alcohol** has one such carbon atom, a **secondary alcohol** has two, and a **tertiary alcohol** has three. The following are examples:

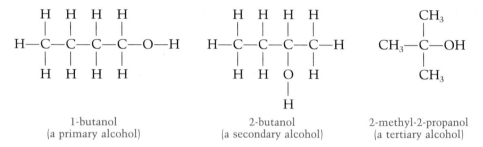

1-butanol
(a primary alcohol)

2-butanol
(a secondary alcohol)

2-methyl-2-propanol
(a tertiary alcohol)

Methanol, ethanol, ethylene glycol, and glycerol are some common alcohols. Methanol, CH_3OH, was at one time separated from the liquid distilled from sawdust, hence the common name wood alcohol. It is a toxic liquid prepared by reacting carbon monoxide with hydrogen at high pressure in the presence of a catalyst. It is used as a solvent and as the starting material for the preparation of formaldehyde. Ethanol is manufactured by the fermentation of glucose (a sugar) or by the addition of water to the double bond of ethylene. The latter reaction is carried out by heating ethylene with water in the presence of sulfuric acid:

$$\text{CH}_2{=}\text{CH}_2 + \text{HOH} \xrightarrow{\text{H}_2\text{SO}_4} \underset{\substack{|\quad\quad|\\ \text{H}\quad \text{OH}}}{\text{CH}_2{-}\text{CH}_2}$$

ethylene   ethanol

Alcoholic beverages contain ethanol. Ethanol is also a solvent and a starting material for many organic compounds. It is mixed with gasoline and sold as gasohol, an automotive fuel. Ethylene glycol (IUPAC name 1,2-ethanediol) and glycerol (1,2,3-propanetriol) are alcohols containing more than one hydroxyl group:

$$\underset{\substack{|\quad\quad|\\ \text{OH}\quad \text{OH}}}{\text{CH}_2{-}\text{CH}_2} \qquad \underset{\substack{|\quad\quad|\quad\quad|\\ \text{OH}\quad \text{OH}\quad \text{OH}}}{\text{CH}_2{-}\text{CH}{-}\text{CH}_2}$$

ethylene glycol   glycerol

Ethylene glycol is a liquid prepared from ethylene and is used as an antifreeze agent. It is also used in the manufacture of polyester plastics and fibers.■ Glycerol is a nontoxic, sweet-tasting liquid obtained from fats dur-

■ Polyesters are discussed in Section 26.6.

ing the making of soap. It is used in foods and candies to keep them soft and moist.

Formally, an alcohol may be thought of as a derivative of water in which one H atom of H_2O has been replaced by a hydrocarbon group, R. Similarly, an **ether** is formed by replacing both H atoms of H_2O by hydrocarbon groups R and R'.

$$H—O—H \qquad R—O—H \qquad R—O—R'$$

water   an alcohol   an ether

Common names for ethers are formed by naming the hydrocarbon groups followed by the word *ether*. Thus, $CH_3OCH_2CH_2CH_3$ is called methyl propyl ether. By IUPAC rules, the ethers are named as derivatives of the longest hydrocarbon chain. For example, $CH_3OCH_2CH_2CH_3$ is 1-methoxypropane; the methoxy group is $CH_3O—$. The best known ether is diethyl ether, $CH_3CH_2OCH_2CH_3$, a volatile liquid used as a solvent and as an anesthetic.

Exercise 26.11

Give the common names of the following compounds: (a) CH_3OCH_3 (b) $CH_3OCH_2CH_3$

(See Problems 26.49 and 26.50.)

Aldehydes and Ketones

Aldehydes and ketones are compounds containing a *carbonyl group:*

carbonyl group

In an **aldehyde,** the carbonyl group has at least one H atom attached to it:

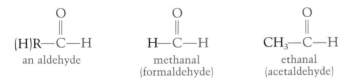

an aldehyde   methanal   ethanal
(formaldehyde)   (acetaldehyde)

Here (H)R indicates a hydrocarbon group or H atom. The aldehyde function is usually abbreviated —CHO, and the structural formula of acetaldehyde is written CH_3CHO.

A **ketone** has the carbonyl group attached to two hydrocarbon groups:

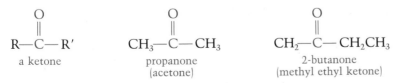

a ketone   propanone   2-butanone
              (acetone)   (methyl ethyl ketone)

The ketone functional group is abbreviated —CO—; thus, acetone is written CH_3COCH_3.

Aldehydes and ketones are named according to IUPAC rules similar to those for alcohols. We first locate the longest carbon chain containing the carbonyl group to get the stem hydrocarbon name. Then, we change the -e ending of the hydrocarbon to *-al* for aldehydes and *-one* for ketones. In the case of aldehydes, the carbon atom of the —CHO group is always the number 1 carbon. In ketones, however, the carbonyl group may occur in various nonequivalent positions on the carbon chain. When that happens, the position of the carbonyl group is indicated by a number before the stem name, just as the position of the hydroxyl group is indicated in alcohols. The carbon chain is numbered to give the smaller number for the position of the carbonyl group.

Example 26.7

Give IUPAC names for the following compounds:

$$
\text{(a)} \quad CH_3CHCH_2CH_2\overset{\displaystyle O}{\overset{\displaystyle \|}{C}}\!-\!H \qquad \text{(b)} \quad \overset{\displaystyle CH_3}{\underset{\displaystyle CH_3}{\overset{\displaystyle |}{CH}}}\!-\!\overset{\displaystyle O}{\overset{\displaystyle \|}{C}}\!-\!CH_2CH_2CH_3
$$
$$
\underset{\displaystyle CH_3}{|}
$$

Solution

(a) Note that an H atom is attached to the carbonyl group. Thus, the compound is an aldehyde. The numbering of the stem carbon chain is

$$
\overset{5}{CH_3}\overset{4}{\underset{\displaystyle |}{CH}}\overset{3}{CH_2}\overset{2}{CH_2}\overset{1}{CHO}
$$
$$
\underset{\displaystyle CH_3}{|}
$$

The IUPAC name is 4-methylpentanal. (b) Note that the two carbon atoms are attached to the carbonyl group (no H atom directly attached). Thus, the compound is a ketone. The numbering of the stem carbon chain is

$$
\overset{1}{CH_3}
$$
$$
\overset{2}{\underset{\displaystyle CH_3}{\overset{\displaystyle |}{CH}}}\overset{}{CO}\overset{3}{CH_2}\overset{4}{CH_2}\overset{5}{CH_2}\overset{6}{CH_3}
$$

The IUPAC name is 2-methyl-3-hexanone.

Exercise 26.12

Name the following compounds by IUPAC rules:

$$
\text{(a)} \quad CH_3CH_2CH_2\!-\!\overset{\displaystyle O}{\overset{\displaystyle \|}{C}}\!-\!CH_3 \qquad \text{(b)} \quad H\!-\!\overset{\displaystyle O}{\overset{\displaystyle \|}{C}}\!-\!CH_2CH_2CH_3
$$

(See Problems 26.51 and 26.52.)

The aldehydes of lower molecular weight have sharp, penetrating odors. Formaldehyde (methanal), HCHO, and acetaldehyde (ethanal), CH_3CHO, are examples. With increasing molecular weight, the aldehydes become more fragrant. Some aldehydes of aromatic hydrocarbons have especially pleasant odors (see Figure 26.9). Formaldehyde is a gas produced by the oxidation of

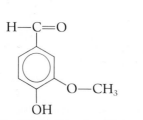

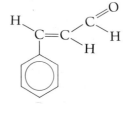

Figure 26.9
Some aldehydes of aromatic hydrocarbons.

Benzaldehyde (oil of almonds)   Vanillin (extract of vanilla)   Cinnamaldehyde (oil of cinnamon)

methanol. The gas is very soluble in water, and a 37% aqueous solution, called Formalin, is marketed as a disinfectant and as a preservative of biological specimens. The main use of formaldehyde is in the manufacture of plastics and resins. Acetone, CH_3COCH_3, is the simplest ketone. It is a liquid with a fragrant odor. The liquid is an important solvent for lacquers, paint removers, and nail polish remover.

Carboxylic Acids and Esters

A **carboxylic acid** has the *carboxyl group:*

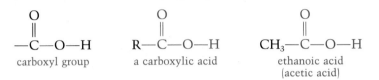

carboxyl group   a carboxylic acid   ethanoic acid (acetic acid)

The formula of the carboxyl group is abbreviated —COOH or —CO$_2$H. These compounds are named by IUPAC rules like those for the aldehydes except that the ending on the stem name is *-oic* followed by the word *acid*. Many carboxylic acids have been known for a long time and are usually referred to by common names (see Table 26.5). The carboxylic acids are weak acids because of the acidity of the H atom on the carboxyl group. Acid ionization constants are about 10^{-5}.

An **ester** is a compound formed from a carboxylic acid, RCOOH, and an alcohol, R'OH. The general structure is

Table 26.5
Common Carboxylic Acids

Carbon Atoms	Formula	Source	Common Name	IUPAC Name
1	HCOOH	Ants (Latin, *formica*)	Formic acid	Methanoic acid
2	CH_3COOH	Vinegar (Latin, *acetum*)	Acetic acid	Ethanoic acid
3	CH_3CH_2COOH	Milk (Greek, *protos pion*, first fat)	Propionic acid	Propanoic acid
4	$CH_3(CH_2)_2COOH$	Butter (Latin, *butyrum*)	Butyric acid	Butanoic acid
5	$CH_3(CH_2)_3COOH$	Valerian root (Latin, *valere*, to be strong)	Valeric acid	Pentanoic acid
6	$CH_3(CH_2)_4COOH$	Goats (Latin, *caper*)	Caproic acid	Hexanoic acid
7	$CH_3(CH_2)_5COOH$	Vine blossom (Greek, *oenanthe*)	Enanthic acid	Heptanoic acid
8	$CH_3(CH_2)_6COOH$	Goats (Latin, *caper*)	Caprylic acid	Octanoic acid
9	$CH_3(CH_2)_7COOH$	Pelargonium (an herb with stork-shaped capsules; Greek, *pelargos*, stork)	Pelargonic acid	Nonanoic acid
10	$CH_3(CH_2)_8COOH$	Goats (Latin, *caper*)	Capric acid	Decanoic acid

From Harold Hart, *Organic Chemistry: A Short Course*, 6th Ed. (Boston: Houghton Mifflin Co., 1983), p. 237.

We will discuss these compounds in the next section dealing with reactions of oxygen-containing organic compounds.

26.6 Reactions of Oxygen-Containing Organic Compounds

Alcohols, aldehydes, ketones, and carboxylic acids are especially reactive compounds. (Ethers are rather unreactive.) In this section, we will discuss the oxidation and reduction of these oxygen-containing organic compounds. We will also discuss the reaction of alcohols with carboxylic acids to form esters.

Oxidation–Reduction Reactions

It is convenient when discussing organic reactions to define *oxidation* as the addition of oxygen atoms to or the removal of hydrogen atoms from an organic compound. *Reduction* is defined as the addition of hydrogen atoms to or the removal of oxygen atoms from an organic compound. Then, a general oxidizing agent is written as (O), and a general reducing agent as (H). For example, the oxidation of ethanol, CH_3CH_2OH, to acetaldehyde, CH_3CHO, is written

$$CH_3CH_2OH \quad + \quad (O) \quad \longrightarrow \quad CH_3CHO \quad + \quad H_2O$$
ethanol   an oxidizing   acetaldehyde
            agent   (ethanal)

Notice that two hydrogen atoms have been taken away from ethanol to give acetaldehyde. Acetaldehyde may be oxidized further; we have

$$CH_3CHO \quad + \quad (O) \quad \longrightarrow \quad CH_3COOH$$
acetaldehyde   an oxidizing   acetic
(ethanal)   agent   acid

In this case, an oxygen atom has been added to acetaldehyde to give acetic acid.

Example 26.8

2-Propanol can be oxidized to acetone (2-propanone):

$$CH_3CHCH_3 + (O) \longrightarrow CH_3CCH_3 + H_2O$$
$\quad\quad |$ $\quad\quad\quad\quad\quad\quad ||$
$\quad\quad OH$ $\quad\quad\quad\quad\quad\quad O$

Write the balanced equation for the oxidation of 2-propanol to acetone by dichromate ion, $Cr_2O_7^{2-}$, in acidic solution. (The half-reaction method of balancing oxidation–reduction equations was discussed in Section 9.8.)

Solution

The oxidation half-reaction is

$$CH_3CHCH_3 \longrightarrow CH_3CCH_3 + 2H^+ + 2e^-$$
$\quad\quad |$ $\quad\quad\quad\quad\quad ||$
$\quad\quad OH$ $\quad\quad\quad\quad\quad O$

Since dichromate ion is reduced to chromium(III) ion, the reduction half-reaction is

(Continued)

$$Cr_2O_7^{2-} + 14H^+ + 6e^- \longrightarrow 2Cr^{3+} + 7H_2O$$

orange   green

(A change from the orange color of $Cr_2O_7^{2-}$ to the green color of Cr^{3+} is seen as the reaction occurs.) The balanced equation for the oxidation–reduction reaction is

$$3CH_3\underset{\underset{OH}{|}}{C}HCH_3 + Cr_2O_7^{2-} + 8H^+ \longrightarrow$$

$$3CH_3\underset{\underset{O}{\|}}{C}CH_3 + 2Cr^{3+} + 7H_2O$$

Exercise 26.13

Write the balanced equation for the oxidation of ethanol to acetaldehyde by permanganate ion in acidic solution. Permanganate is reduced to Mn^{2+} in acidic solution.

(See Problems 26.53 and 26.54.)

Primary and secondary alcohols are easily oxidized. In both, hydrogen atoms are bonded to the carbon atom carrying the hydroxyl group, and these hydrogen atoms are removed in the oxidation to give a carbonyl group. In general, for oxidizing agents such as dichromate or permanganate ions in acidic solution, we get the following reactions:

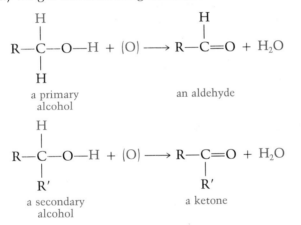

An aldehyde is easily oxidized to a carboxylic acid. Therefore, unless the aldehyde is removed from the reaction vessel, a primary alcohol will give the carboxylic acid on oxidation. Because aldehydes boil at temperatures below those for the corresponding primary alcohols, however, aldehydes are easily distilled from the reaction vessel. Tertiary alcohols are unreactive with these oxidizing agents.

$$R-\underset{\underset{R''}{|}}{\overset{\overset{R'}{|}}{C}}-O-H + (O) \longrightarrow N.R.$$

a tertiary
alcohol

Of course, they can be oxidized to H_2O and CO_2 with very strong oxidizing agents (such as O_2), as can most organic compounds.

As we mentioned, aldehydes are easily oxidized to the corresponding carboxylic acid. Ketones, however, are resistant to oxidation, except under severe conditions where the carbon chain is broken up. With oxidizing agents under moderate conditions, we get the following:■

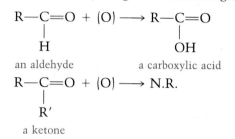

an aldehyde   a carboxylic acid

a ketone

■ The ease of oxidation of an aldehyde is the basis of several methods for their detection. Tollen's test uses silver ion in aqueous ammonia as a mild oxidizing agent; silver(I) is present in the solution as $Ag(NH_3)_2^+$. Silver(I) is reduced to silver metal, which deposits as a reflective coating on the inside of the test tube. Mirrors are made by this reaction.

Note that, as with an alcohol, the presence of an H atom on the carbon atom carrying the functional group determines the reactivity with oxidizing agents.

Aldehydes, ketones, and carboxylic acids can all be reduced to the corresponding alcohols. Lithium aluminum hydride, $LiAlH_4$, is a reducing agent commonly used for these reactions.

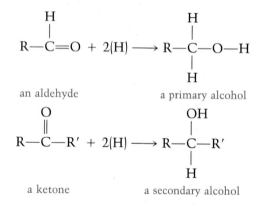

an aldehyde   a primary alcohol

a ketone   a secondary alcohol

Exercise 26.14

Complete the following equations. If no reaction occurs, write N.R. Name any organic products.

(a) $(CH_3)_3COH + (O) \longrightarrow$

(b) $CH_3CHCH_2CH_3 + (O) \longrightarrow$
 |
 CHO

(c) $CH_3CHCH_2CH_3 + (H) \longrightarrow$
 |
 CHO

(d) $CH_3CHCH_2CH_3 + (O) \longrightarrow$
 |
 OH

(See Problems 26.55 and 26.56.)

Esterification and Saponification

Carboxylic acids will react with alcohols when these substances are heated together in the presence of an inorganic acid, such as H_2SO_4. An ester is the organic product of the reaction of a carboxylic acid and an alcohol.

Table 26.6
Odors of Esters

Name	Formula	Odor*
Ethyl formate	$HCOOCH_2CH_3$	Rum
Pentyl acetate	$CH_3COOCH_2CH_2CH_2CH_2CH_3$	Banana
Octyl acetate	$CH_3COOCH_2CH_2CH_2CH_2CH_2CH_2CH_2CH_3$	Orange
Methyl butyrate	$CH_3CH_2CH_2COOCH_3$	Apple
Ethyl butyrate	$CH_3CH_2CH_2COOCH_2CH_3$	Pineapple
Pentyl butyrate	$CH_3CH_2CH_2COOCH_2CH_2CH_2CH_2CH_3$	Apricot
Methyl salicylate	$o\text{-}C_6H_4(OH)COOCH_3$	Wintergreen

*Natural flavors are generally complex mixtures of esters and other constituents.

$$\underset{\substack{\text{a carboxylic}\\\text{acid}}}{R-\overset{\overset{\displaystyle O}{\|}}{C}-O-H} + \underset{\text{an alcohol}}{H-O-R'} \underset{}{\overset{H^+}{\rightleftharpoons}} \underset{\text{an ester}}{R-\overset{\overset{\displaystyle O}{\|}}{C}-O-R'} + H_2O$$

Esters are intriguing substances because many of them have very pleasant odors. They are present in natural flavors and are used to make artificial flavorings. Table 26.6 lists some esters and the odors associated with them. Esters are named by giving the name of the hydrocarbon group corresponding to the alcohol part of the molecule followed by the name of the carboxylic acid, with -oic (or -ic) changed to -ate (similar to the naming of salts). Thus, ethanol and acetic acid give the ester ethyl acetate (Figure 26.10).

The reaction of a carboxylic acid and an alcohol is reversible. Thus, an ester reacts with water in the presence of hydrogen ion to give the carboxylic acid and the alcohol. Such a reaction of a compound with water is called a *hydrolysis*. The hydrolysis will go to completion in the presence of a base, and in this case the products will be the alcohol and the salt of the carboxylic acid.

Saponification is the hydrolysis of an ester in the presence of a base. The term, which comes from the Latin for "soap" (*sapon*), originated from the

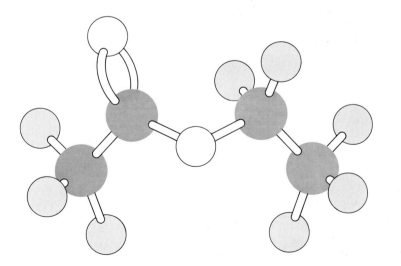

Figure 26.10
Model of the ethyl acetate molecule, $CH_3COOCH_2CH_3$.

soap-making process. In this process, an animal fat or vegetable oil is boiled with a strong base, usually NaOH. An animal fat or vegetable oil is an ester of glycerol with various long-chain carboxylic acids, called *fatty acids.*■ The saponification of a fat that is an ester of glycerol with stearic acid, $CH_3(CH_2)_{16}COOH$ or $C_{17}H_{35}COOH$, is shown below. In general, the three fatty-acid groups in a fat or oil need not be identical.

■ Fatty acids will be discussed further in the next chapter. See Table 27.3 for a list of some important fatty acids.

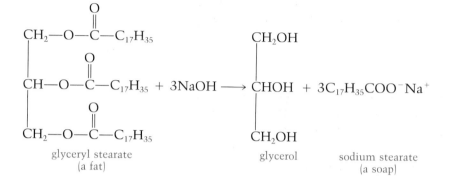

glyceryl stearate (a fat)   glycerol   sodium stearate (a soap)

Exercise 26.15

Write an equation for the preparation of methyl propionate. Note any catalyst used.
(See Problems 26.57 and 26.58.)

Polyesters

A substance with two alcohol groups reacts with one containing two carboxylic acid groups to form a *polyester,* a polymer whose repeating units are joined by ester groups. The reactant molecules join as links of a chain to form a very long molecule. The polyester Dacron, used as a textile fiber, is prepared from ethylene glycol and terphthalic acid:

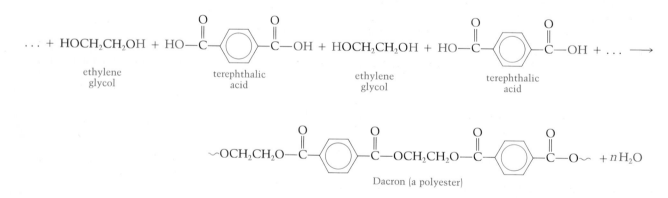

Dacron (a polyester)

Note that the polyester is formed when molecules join by splitting out small molecules, in this case H_2O. This type of reaction is called a *condensation* reaction and the polymer formed is a **condensation polymer.**■ Recall that polyethylene and polypropylene are addition polymers. In general, polymers are either addition polymers or condensation polymers.

■ We discussed condensation reactions of silicic and phosphoric acids in Sections 24.1 and 24.2.

26.7 Organic Compounds Containing Nitrogen and Sulfur

Alcohols and ethers, you may recall, can be considered as derivatives of H_2O, where one or both H atoms are replaced by hydrocarbon groups, R. Thus, the general formula of an alcohol is ROH, and that of an ether ROR′. Important classes of organic compounds are obtained by similarly substituting R groups for the H atoms of ammonia, NH_3, and hydrogen sulfide, H_2S.

Amines and Amides

Most organic bases are **amines,** which are compounds that are structurally derived from ammonia. One, two, or three H atoms of NH_3 can be replaced by hydrocarbon groups to obtain a *primary amine*, a *secondary amine*, or a *tertiary amine*, respectively:

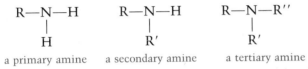

a primary amine   a secondary amine   a tertiary amine

Table 26.7 lists some common amines. Note that the first four are gases; the rest are liquids.

Amines are bases because the nitrogen atom has an unshared electron pair that can accept a proton to form a substituted ammonium ion. For example,

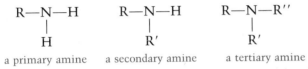

methylamine   methylammonium ion

Like ammonia, amines are weak bases. Table 18.2 gives base ionization constants of some amines.

Amides are compounds derived from the reaction of ammonia, or a primary or secondary amine, with a carboxylic acid. For example, if ammonia is strongly heated with acetic acid, they react to give the amide acetamide:■

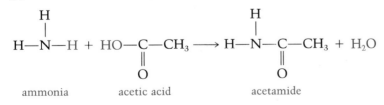

ammonia   acetic acid   acetamide

With methylamine and acetic acid, we get N-methylacetamide:

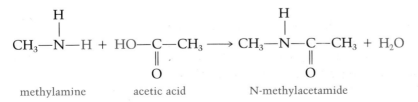

methylamine   acetic acid   N-methylacetamide

■ The condensation to give an amide occurs under milder conditions if the ammonia or amine is reacted with the acid chloride, a derivative of the carboxylic acid obtained by replacing

—COH by —CCl. In this reaction, HCl is a product.

Name	Formula	Boiling Point, °C
Methylamine	CH_3—NH_2	−6.5
Dimethylamine	CH_3—$\overset{\displaystyle}{\underset{\displaystyle H}{N}}$—$CH_3$	7.4
Trimethylamine	CH_3—$\overset{\displaystyle}{\underset{\displaystyle CH_3}{N}}$—$CH_3$	3.5
Ethylamine	CH_3CH_2—NH_2	16.6
Diethylamine	CH_3CH_2—$\overset{\displaystyle}{\underset{\displaystyle H}{N}}$—$CH_2CH_3$	55.5
Triethylamine	CH_3CH_2—$\overset{\displaystyle}{\underset{\displaystyle CH_2CH_3}{N}}$—$CH_2CH_3$	89.5
Piperidine	N—H	106
Morpholine	O N—H	129
Aniline	—NH_2	184

Table 26.7
Some Common Amines

From Robert D. Whitaker et al., *Concepts of General, Organic, and Biological Chemistry* (Boston: Houghton Mifflin Co., 1981), p. 343.

When a compound containing two amine groups reacts with one containing two carboxylic acid groups, a condensation polymer called a *polyamide* is formed (Figure 26.11). Nylon-66 is an example. It is prepared by heating hexamethylene diamine (1,6-diaminohexane) and adipic acid (hexanedioic acid):

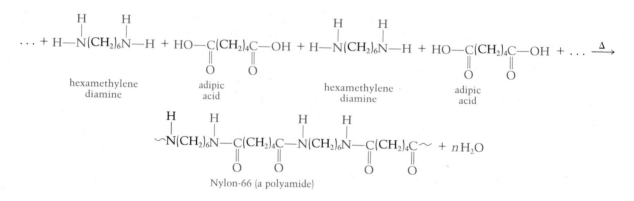

hexamethylene diamine   adipic acid   hexamethylene diamine   adipic acid

Nylon-66 (a polyamide)

Figure 26.11
Nylon is being pulled from a reaction mixture. It is formed by condensing hexamethylene diamine with adipic acid or, as in this experiment, with adipoyl chloride, $ClCO(CH_2)_4COCl$.

Thiols and Disulfides

Thiols (also called **mercaptans**) are sulfur analogs of the alcohols. They have the general formula RSH, compared with ROH for alcohols. These compounds usually have very disagreeable odors; the liquid ejected by skunks contains thiols and other organic sulfur compounds. Ethanethiol (ethyl mercaptan), CH_3CH_2SH, is added to purified natural gas, which is odorless, to give it an odor so that a person can become aware of a gas leak. Thiols can be oxidized to **disulfides**, compounds containing the —S—S— group:

$$2CH_3CH_2SH + H_2O_2 \longrightarrow CH_3CH_2—S—S—CH_2CH_3 + 2H_2O$$
ethanethiol   diethyl disulfide

The disulfides revert to the thiols with a mild reducing agent.

A Checklist for Review

Important Terms

hydrocarbons (p. 895)
aromatic and aliphatic hydrocarbons (26.1)
saturated hydrocarbons (26.1)
alkanes (26.1)
homologous series (26.1)
alkyl group (26.1)
cycloalkanes (26.1)
unsaturated hydrocarbons (26.2)
alkenes (26.2)
geometric (*cis–trans*) isomerism (26.2)

alkynes (26.2)
substitution reaction (26.4)
addition reaction (26.4)
Markownikoff's rule (26.4)
addition polymer (26.4)
functional group (26.5)
alcohol (26.5)
primary, secondary, tertiary alcohols (26.5)
ether (26.5)

aldehyde (26.5)
ketone (26.5)
carboxylic acid (26.5)
ester (26.6)
saponification (26.6)
condensation polymer (26.6)
amines (26.7)
amides (26.7)
thiols (mercaptans) (26.7)
disulfides (26.7)

Summary of Facts and Concepts

Organic compounds are hydrocarbons or derivatives of hydrocarbons. The two main types of hydrocarbons are *aromatic* (containing benzene rings) or *aliphatic*. The simplest aliphatic hydrocarbons are *alkanes*, compounds with the general formula C_nH_{2n+2}. These are *saturated hydrocarbons*, that is, hydrocarbons in which the carbon atoms are bonded to four other atoms (either hydrogens or other carbons). Another *homologous series* of saturated hydrocarbons are the *cycloalkanes*, compounds of the general formula C_nH_{2n} in which the carbon atoms are joined in a ring. The *alkenes* and *alkynes* are unsaturated hydrocarbons, that is, hydrocarbons not containing the maximum number of H atoms but containing double or triple carbon–carbon bonds. An *aromatic hydrocarbon* is one containing benzene rings, that is, six-membered rings of carbon atoms with alternating single and double bonds in resonance formulas.

The alkanes and aromatic hydrocarbons usually undergo *substitution reactions*. Alkenes and alkynes undergo *addition reactions*. *Markownikoff's rule* predicts the major product in the addition of an unsymmetrical reagent to an unsymmetrical alkene.

A *functional group* is a portion of an organic molecule that reacts readily in predictable ways. Important functional groups containing oxygen are *alcohols* (ROH), *aldehydes* (RCHO), *ketones* (RCOR'), and *carboxylic acids* (RCOOH). Alcohols (except tertiary alcohols) and aldehydes are readily oxidized. Primary alcohols are oxidized to aldehydes, which are then oxidized to carboxylic acids. Secondary alcohols are oxidized to ketones. Aldehydes, ketones, and carboxylic acids can be reduced to the corresponding alcohols. Carboxylic acids and alcohols react to produce *esters*. In basic solution, an ester can be *saponified* to give the alcohol and carboxylate salt. *Amines* are organic derivatives of ammonia. They react with carboxylic acids to give *amides*. *Thiols*, RSH, are sulfur analogs of alcohols.

Operational Skills

1. Given the structure of an organic compound, state the IUPAC name (Examples 26.1, 26.3, 26.6, and 26.7). Given the IUPAC name of an organic compound, write the structural formula (Example 26.2).
2. Given a condensed structural formula of an alkene, decide if *cis* and *trans* isomers are possible, then draw the structural formulas (Example 26.4).

3. Predict the major product in the addition of an unsymmetrical reagent to an unsymmetrical alkene (Example 26.5).
4. Write a complete balanced equation for the oxidation or reduction of an organic compound (Example 26.8).

Review Questions

26.1 Give the molecular formula of an alkane with 30 carbon atoms.

26.2 Why would you expect the melting points of the alkanes to increase in the series methane, ethane, propane, and so on?

26.3 Draw structural formulas of the five isomers of C_6H_{14}.

26.4 Draw structural formulas of an alkane, a cyclo-alkane, an alkene, and an aromatic hydrocarbon, each with seven carbon atoms.

26.5 Explain why there are two isomers of 2-butene. Draw their structural formulas and name the isomers.

26.6 Draw structural formulas for the isomers of ethyl-methylbenzene.

26.7 Fill in the following table.

Hydrocarbon	Source	Use
Methane		
Octane		
Ethylene		
Acetylene		
Benzene		
Naphthalene		

26.8 Give condensed structural formulas of all possible substitution products of ethane and Cl_2.

26.9 Define the terms *substitution reaction* and *addition reaction*. Give examples of each.

26.10 What would you expect to be the major product when 2 molecules of HCl add successively to acetylene? Explain.

26.11 Describe three chemical processes used in petroleum refining. What is the purpose of each?

26.12 What is a functional group? Give an example of one and explain how it fits this definition.

26.13 An aldehyde contains the carbonyl group. Ketones, carboxylic acids, and esters also contain the carbonyl group. What distinguishes these latter compounds from an aldehyde?

26.14 Fill in the following table.

Compound	Source	Use
Methanol		
Ethanol		
Ethylene glycol		
Glycerol		
Formaldehyde		

26.15 Identify and name the functional group in each of the following:

(a) CH_3COCH_3 (b) $CH_3OCH_2CH_3$
(c) $CH_3CH=CH_2$ (d) CH_3CH_2COOH
(e) CH_3CH_2CHO (f) $CH_3CH_2CH_2OH$

26.16 What are the products, if any, of oxidation by an inorganic reagent (such as acidic $K_2Cr_2O_7$) of a primary alcohol, a secondary alcohol, a tertiary alcohol, an aldehyde, and a ketone? Write an equation for the oxidation of a particular compound of each type. Use (O) for the oxidizing agent. If no reaction occurs, write N.R.

26.17 Consider the following formulas of two esters:

$$CH_3CH_2-\overset{\overset{O}{\|}}{C}-O-CH_3 \qquad CH_3CH_2-O-\overset{\overset{O}{\|}}{C}-CH_3$$

One of these is ethyl ethanoate (ethyl acetate) and one is methyl propanoate (methyl propionate). Which is which?

26.18 Write the equation for the saponification of ethyl acetate.

26.19 What is the difference between an addition polymer and a condensation polymer? Give an example of each, writing the equation for its formation.

26.20 What is the source of basicity of an amine? Illustrate by writing the equation for the reaction of triethylamine, $(CH_3CH_2)_3N$, with acetic acid.

Problems

Naming Hydrocarbons

26.21 Give the IUPAC name for each of the following hydrocarbons:

(a) $CH_3CHCH_2CHCH_3$
       | |
      CH_3 CH_3

26.22 What are the IUPAC names of the following compounds?

                CH_3
                 |
(a) $CH_3CH_2CHCHCH_3$
               |
              CH_3

(Continued)   (Continued)

$$\text{(b) } CH_3\overset{\overset{\displaystyle CH_3}{|}}{C}CH_2CH_2\overset{\overset{\displaystyle CH_3}{|}}{C}H\overset{\overset{\displaystyle}{}}{C}CH_3$$

(with CH₃ groups below the first and third substituted carbons)

(b) CH₃CCH₂CH₂CHCCH₃ with CH₃ above first C and above last-but-one C, CH₃ below first C and below CHCCH₃

(c) CH₃CH₂CHCH₂CH₂CH₂CH₃
       |
       CH₂CH₂CH₃

(d) CH₃CHCHCH₂CH₂CH₂
       | |
       CH₂CH₃   CH₃
   (with CH₃ above the second carbon)

(b) CH₃CH₂CH₂CHCH₂CH₃
       |
       CH₂CH₃
   (with CH₂CH₃ above)

(c) CH₃CH₂CH₂CHCH₂CH₂CH₃
       |
       CH₃CHCH₃
   (with CH₃CHCH₃ below, and CH₃ below that)

(d) CH₃CHCH₂CH₂CHCH₂CH₃
       | |
       CH₂CH₃   CH₃ / CH₂CH₃

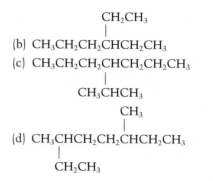

26.23 Write the condensed structural formula for each of the following compounds:
  (a) 2,3-dimethylhexane
  (b) 3-ethylhexane
  (c) 2-methyl-4-isopropylheptane
  (d) 2,2,3,3-tetramethylpentane

26.24 Write the condensed structural formula for each of the following compounds:
  (a) 2,2-dimethylbutane
  (b) 3-isopropylhexane
  (c) 3-ethyl-4-methyloctane
  (d) 3,4,4,5-tetramethylheptane

26.25 Give the IUPAC name of each of the following:
  (a) $CH_2{=}CHCH_2CH_2CH_3$
  (b) $CH_3C{=}CHCH_2CHCH_3$
        with CH₃ below the second carbon and CH₃ below the fifth carbon

26.26 For each of the following, write the IUPAC name:
  (a) CH_3CH_2 and CH_3CH_2 both bonded to $C{=}CHCH_2CH_3$
  (b) $CH_3CH_2CCH_2CH_2CH_3$ with CH_2 double bonded below

26.27 Give the condensed structural formula for each of the following compounds:
  (a) 3-ethyl-2-pentene
  (b) 4-ethyl-2-methyl-2-hexene

26.28 Write condensed structural formulas for the following compounds:
  (a) 2,3-dimethyl-2-pentene
  (b) 2-methyl-4-propyl-3-heptene

26.29 If there are geometric isomers for the following, draw structural formulas showing the isomers. Label the isomers with their IUPAC names, including *cis* and *trans* designations.
  (a) $CH_3CH_2CH{=}CHCH_2CH_3$
  (b) $CH_3C{=}CHCH_2CH_3$
        with CH₂CH₃ below

26.30 One or both of the following have geometric isomers. Draw the structures of any geometric isomers and label the isomers with IUPAC names, including the prefix *cis* or *trans*.
  (a) $CH_3CHCH{=}CHCH_3$
        with CH₃ below
  (b) $CH_3C{=}CHCH_2CH_3$
        with CH₃ below

26.31 Give IUPAC names for the following compounds:
  (a) $CH_3C{\equiv}CCH_3$ (b) $CH{\equiv}CCHCH_3$ with CH₃ below

26.32 Write the IUPAC name of each of the following hydrocarbons:
  (a) $CH_3CHC{\equiv}CH$ with CH₃ below   (b) $CH_3C{\equiv}CCH_2CH_3$

26.33 Write structural formulas for (a) 1,1,1-triphenylethane, (b) *o*-ethylmethylbenzene.

26.34 Write structural formulas for (a) 1,2,3-trimethylbenzene, (b) *p*-diethylbenzene.

Reactions of Hydrocarbons

26.35 Complete and balance the following equations. Note any catalyst used.

(a) $C_3H_6 + O_2 \longrightarrow$

    cyclopropane

(b) $CH_2{=}CH_2 + MnO_4^- + H_2O \longrightarrow$

(c) $CH_2{=}CH_2 + Br_2 \longrightarrow$

(d)

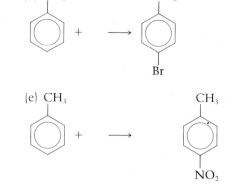

(e)

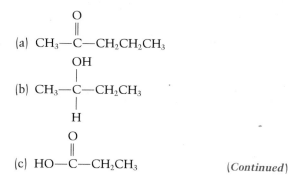

26.36 Complete and balance the following equations. Note any catalyst used.

(a) $C_4H_{10} + O_2 \longrightarrow$

(b)

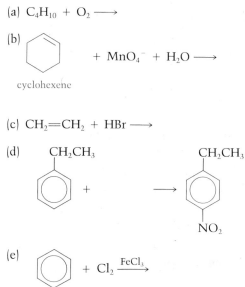

(c) $CH_2{=}CH_2 + HBr \longrightarrow$

26.37 What is the major product when HBr is added to methylpropene?

26.38 Complete the following equation, giving only the main product:

$$CH_2{=}CHCH_3 + H{-}OH \xrightarrow[\text{catalyst}]{H_2SO_4}$$

26.39 Teflon is an addition polymer of 1,1,2,2-tetrafluoroethene. Write the equation for the formation of the polymer.

26.40 Polyvinyl chloride (PVC) is an addition polymer of vinyl chloride, $CH_2{=}CHCl$. Write the equation for the formation of the polymer.

26.41 Write two equations for the cracking of pentane into different products.

26.42 Write two equations for the cracking of hexane into different products.

Naming Oxygen-Containing Organic Compounds

26.43 Circle and name the functional group in each compound.

(a) $CH_3{-}\overset{\displaystyle \overset{O}{\|}}{C}{-}CH_2CH_2CH_3$

(b) $CH_3{-}\underset{\displaystyle \underset{H}{|}}{\overset{\displaystyle \overset{OH}{|}}{C}}{-}CH_2CH_3$

(c) $HO{-}\overset{\displaystyle \overset{O}{\|}}{C}{-}CH_2CH_3$

                                                     (*Continued*)

26.44 Circle and name the functional group in each compound.

(a) $CH_2{=}CHCH_3$

(b) $O{=}\underset{\displaystyle \underset{CH_3CHCH_3}{|}}{C}{-}OH$

(c) $HO{-}CH_2\underset{\displaystyle \underset{CH_3}{|}}{\overset{\displaystyle \overset{CH_3}{|}}{C}}H$

                                                     (*Continued*)

$$\text{O}$$
$$\parallel$$
(d) $H-C-CH_2CH_3$

26.45 Give the IUPAC name for each of the following:

(a) $HOCH_2CH_2CH_2CH_2CH_3$

(b) $CH_3CHCH_2CH_2CH_3$
         |
       OH

(c) $CH_3CH_2CH_2CHCH_2CH_2CH_3$
             |
          $H-C-OH$
              |
              H
             OH
              |

(d) $CH_3CH_2CH_2CHCH_2CH_2CH_3$

(d) $O=C-OH$
        |
   CH_3CHCH_3

26.46 Write the IUPAC name for each of the following:

(a) $HOCH_2CHCH_2CH_3$
              |
           $CH_2CH_2CH_3$

(b) $HOCH_2CH_2CH_2CH_2$
                    |
                 CH_3

(c) $CH_3CHCH_2CH_3$
      |
   $H-C-OH$
         |
       CH_3

(d) $HOCH_2CHCH_2CH_3$
             |
           CH_3

26.47 State whether each of the following alcohols is primary, secondary, or tertiary:

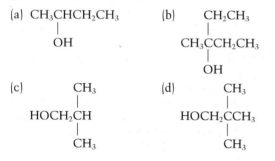

(a) $CH_3CHCH_2CH_3$
      |
    OH

(b) CH_2CH_3
     |
  $CH_3CCH_2CH_3$
     |
    OH

(c) CH_3
   |
 $HOCH_2CH$
      |
     CH_3

(d) CH_3
   |
 $HOCH_2CCH_3$
      |
     CH_3

26.48 Classify each of the following as a primary, secondary, or tertiary alcohol.

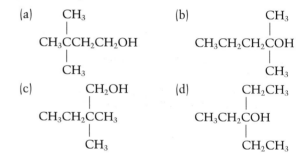

(a) CH_3
   |
 $CH_3CCH_2CH_2OH$
    |
   CH_3

(b) CH_3
   |
 $CH_3CH_2CH_2COH$
      |
     CH_3

(c) CH_2OH
    |
 $CH_3CH_2CCH_3$
     |
    CH_3

(d) CH_2CH_3
     |
 CH_3CH_2COH
     |
    CH_2CH_3

26.49 What are the common names of the following compounds?

(a) $CH_3CH_2OCH_2CH_2CH_3$

(b) CH_3
   |
 $H-COCH_3$
   |
  CH_3

26.50 What are the common names of the following compounds?

(a) CH_3
   |
 CH_3OCCH_3
    |
   CH_3

(b)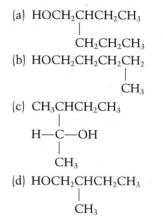

26.51 According to IUPAC rules, what are the names of the following?

(a) $CH_3COCH_2CH_3$

(b) $CH_3CH_2CH_2CHO$

(c) $$\text{O}\qquad\quad CH_3$$
$$\parallel\qquad\quad |$$
$$H-CCH_2CH_2CCH_3$$
             |
          CH_3

(d) $$\text{O}$$
$$\parallel$$
$$CH_3CHCCH_3$$
    |
 CH_2CH_3

26.52 Write IUPAC names for the following compounds:

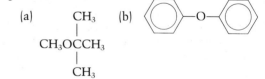

(a) CH_3CHCH_3
    |
  CHO

(b) CH_3CHCH_3
    |
  $COCH_3$

(c) CH_3CH_2
     |
   CH_2C-H
        $\parallel$
        O

(d) CH_3CHCH_3
    |
  $CH_2-C-CH_2CH_3$
         $\parallel$
         O

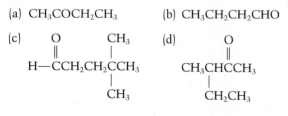

Reactions of Oxygen-Containing Organic Compounds

26.53 Benzaldehyde, C_6H_5CHO, is oxidized by potassium permanganate, $KMnO_4$, in basic solution to benzoic acid, C_6H_5COOH. Permanganate ion is reduced to manganese dioxide. Write a balanced equation for the reaction.

26.54 Cyclohexanol, $C_6H_{11}OH$, is oxidized by chromium trioxide, CrO_3, in acidic solution to cyclohexanone, $C_6H_{10}O$. Chromium trioxide is reduced to Cr^{3+}. Write a balanced equation for the reaction.

26.55 Complete the following equations. If no reaction occurs, write N.R. Name any organic products.

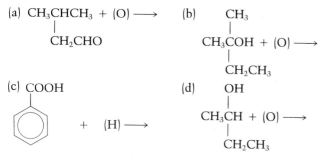

(a) $CH_3CHCH_3 + (O) \longrightarrow$
    |
    CH_2CHO

(b) CH_3
    |
    $CH_3COH + (O) \longrightarrow$
    |
    CH_2CH_3

(c) COOH + (H) $\longrightarrow$

(d)   OH
    |
    $CH_3CH + (O) \longrightarrow$
    |
    CH_2CH_3

26.56 Give the structural formula of and write the IUPAC name for the organic product, if any, in the following. Note any cases where no reaction occurs.

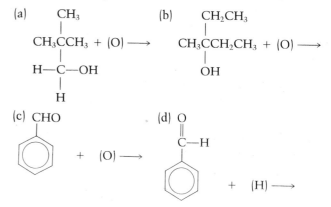

(a) CH_3
    |
    $CH_3CCH_3 + (O) \longrightarrow$
    |
    $H-C-OH$
    |
    H

(b) CH_2CH_3
    |
    $CH_3CCH_2CH_3 + (O) \longrightarrow$
    |
    OH

(c) CHO + (O) $\longrightarrow$

(d) O
    ||
    C-H + (H) $\longrightarrow$

26.57 Write equations for the following. Note any catalyst used.

    (a) Preparation of ethyl butyrate
    (b) Saponification of methyl formate

26.58 Write equations for the following. Note any catalyst used.

    (a) Preparation of isopropyl acetate
    (b) Saponification of a fat that is an ester of glycerol with stearic, palmitic, and oleic acids. (*Note:* The structures of these acids are given in the next chapter in Table 27.3.)

Organic Compounds Containing Nitrogen and Sulfur

26.59 Identify each of the following compounds as a primary, secondary, or tertiary amine, amide, thiol (mercaptan), or disulfide:

(a) NH_2

    (b) $CH_3CH_2NHCH_2CH_3$
    (c) $CH_3CH_2-S-S-CH_3$
    (d) $CH_3CH_2CH_2CH_2SH$

26.60 Identify each of the following compounds as a primary, secondary, or tertiary amine, amide, thiol (mercaptan), or disulfide:

(a) CH_2SH

(b) $O=C-NH_2$

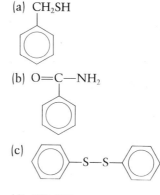

(c)

    (d) $CH_3CH_2CH_2NH_2$

Additional Problems

26.61 Give the IUPAC name of each of the following compounds:

(a) CH₃CHCH₂COOH
      |
     CH₃

(b)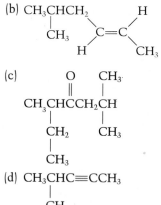

(c)
      O   CH₃
      ‖ |
  CH₃CHCCH₂CH
      | |
     CH₂   CH₃
      |
     CH₃

(d) CH₃CHC≡CCH₃
     |
    CH₃

26.62 Give the IUPAC name of each of the following compounds:

(a)   CHO
       |
   CH₃CCH₃
      |
   CH₂CH₃

(b)   COOH
       |
  CH₃CCH₂CH₃
      |
    CH₃

(c)   CH₃
      |
  CH₃CCOCH₃
      |
   CH₂CH₃

(d)   CH₃
      |
  CH₃CCH₂CH₃
      |
    CH₃

26.63 Write structural formulas for the following compounds:

(a) isopropyl propionate
(b) *t*-butylamine
(c) 2,2-dimethylhexanoic acid
(d) *cis*-3-hexene

26.64 Write structural formulas for the following compounds:

(a) 3-ethyl-1-pentene
(b) 1,1,2,2-tetraphenylethane
(c) 1-phenyl-2-butanone
(d) cyclopentanone

26.65 Describe chemical tests that could distinguish the following:

(a) propionaldehyde (propanal) and acetone (propanone)
(b) CH₂=CH—C≡C—CH=CH₂ and benzene

26.66 Describe chemical tests that could distinguish the following:

(a) acetic acid and acetaldehyde (ethanal)
(b) toluene (methylbenzene) and 2-methylcyclohexene

26.67 Identify each of the compounds below from the description given.

(a) A gas with a sweetish odor that promotes the ripening of green fruit.
(b) An unsaturated compound of the formula C_7H_8 that gives a negative test with bromine in carbon tetrachloride.
(c) A compound with an ammonia-like odor that acts as a base; its molecular formula is CH_5N.
(d) An alcohol used as a starting material for the manufacture of formaldehyde.

26.68 Identify each of the compounds below from the description given.

(a) An acidic compound that also has properties of an aldehyde; its molecular formula is CH_2O_2.
(b) A compound used as a preservative for biological specimens and as a raw material for plastics.
(c) A saturated hydrocarbon boiling at 0°C, which is liquefied and sold in cylinders as a fuel.
(d) A saturated hydrocarbon that is the main constituent of natural gas.

**26.69 A compound containing 85.6% C and 14.4% H with a molecular weight of 56.1 amu reacts with water and sulfuric acid to produce a compound that reacts with acidic potassium dichromate solution to produce a ketone. What is the name of the original hydrocarbon?

**26.70 A compound with fragrant odor reacts with dilute acid to give two organic compounds, A and B. Compound A is identified as an alcohol with a molecular weight of 32.0 amu. Compound B is identified as an acid. It can be reduced to give a compound whose composition is 60.0% C, 13.4% H, and 26.6% O and whose molecular weight is 60.1 amu. What is the name of the original compound?

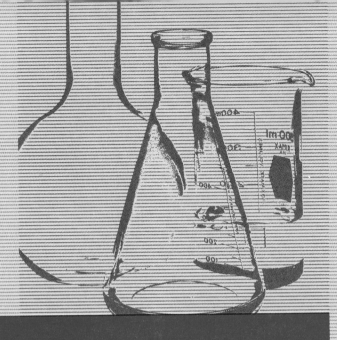

27. Biochemistry

Introduction to Biological Systems

27.1 The Cell: Unit of Biological Structure
27.2 Energy and the Biological System

Biological Molecules

27.3 Biological Polymers
27.4 Proteins Amino Acids/ Protein Primary Structure/ Protein Conformations/ Enzymes
27.5 Carbohydrates Monosaccharides/ Oligosaccharides and Polysaccharides
27.6 Nucleic Acids Nucleotides/ Polynucleotides and Their Conformations/ DNA and the Nature of the Genetic Code/ RNA and the Transmission of the Genetic Code/ Nucleotides and Metabolism
27.7 Lipids Fats and Oils/ Biological Membranes

In 1856, the chemist Louis Pasteur, who was a professor at Lille, France, was hired by one of his students to discover why the student's father was having difficulty manufacturing alcohol from fermenting sugar beets. (In this process, the beet sugar is broken down to alcohol and carbon dioxide in the fermenting liquor.) Examining droplets of the liquor under the microscope, Pasteur discovered that different microorganisms were present in the good and bad samples. He concluded that particular yeasts (previously identified by others as living organisms) were necessary for the desired production of alcohol in the ferments. He postulated that the beet sugar provided food for the yeast and that alcohol was the residue of the food breakdown. Later, Pasteur was asked to investigate the diseases of wines and discovered that good wines resulted only when wild yeasts normally present in the bloom of the grape were present to ferment the grape sugar. Wines of poor quality resulted when other microorganisms contaminated the fermenting grape juice and formed products that caused bad flavors. Pasteur also showed that certain bacteria that ferment milk sugar to lactic acid were responsible for souring of milk, and he made many other contributions to the then-infant sciences of bacteriology and biological chemistry.

Yeasts and bacteria used in the production of foods and beverages provide an everyday example of the ability of organisms to transform matter and energy. These are major concerns of biological chemistry, which we introduce in this chapter.

Chapter Overview

In this chapter, we will introduce molecules that are basic to biological systems and discuss some of their interesting and important chemistry. We will introduce the cell, the unit of biological structure, and briefly note the transformations of matter and energy that take place in it. We will examine the four classes of biological molecules—proteins, carbohydrates, nucleic acids, and lipids—paying particular attention to their structures and to how their structures relate to their functions in the cell. We will highlight the role of proteins as *enzymes*, which catalyze and regulate cellular reactions, and the roles of nucleic acids in the storage and transmission of genetic information.

Introduction to Biological Systems

Living systems contain only about 20 of the most common chemical elements, but they have features not shared by nonliving systems.■ They are able to exchange matter and energy with their surroundings and to respond to changes in those surroundings. They can transform energy and matter into different forms according to their needs. They can grow and reproduce. All these properties are due to the highly organized state of living systems.

The elaborate organization of matter in even the most complex living organisms begins with water and small organic molecules, such as sugars, amines, and various hydrocarbon derivatives. Many of these small molecules are combined into large polymers, including proteins, complex carbohydrates, and nucleic acids. These molecules aggregate, forming fibers and membranes, which in turn are organized into cells. Cells, then tissues and body organs, and finally organisms, represent the culmination of this hierarchy of structural patterns.

■ The 21 elements found in living systems are

H	S	Mn
C	Cl	Fe
N	K	Co
O	Ca	Cu
Na	B	Zn
Mg	Si	Mo
P	V	Se

27.1 The Cell: Unit of Biological Structure

The cell is the smallest structure that shows all the attributes of life—ability to grow, to transform matter and energy, to respond to external stimuli, and to reproduce.■ Most of the plants and animals large enough to see with the naked eye are multicellular organisms. In multicellular organisms, most cells possess specialized functions. There are muscle cells, blood cells, nerve cells, and so forth. The red blood cell, for example, is little more than a bag of hemoglobin. *Hemoglobin* is a protein that captures O_2 molecules when O_2 concentrations (partial pressures) are high (in the lungs) and releases them when O_2 concentrations are low (in other tissues).

Cells are constructed from all four classes of biological molecules: proteins, carbohydrates, nucleic acids, and lipids. In Sections 27.4 to 27.7, we will examine the chemistry and the biological roles of each of these classes of molecules. The organic molecules and complex aggregates of molecules that make up cells are manufactured by the cells themselves. Cells usually employ as raw materials the organic molecules taken in as foods, although green plants and the photosynthetic bacteria and algae are able to construct complex organic molecules from simple inorganic species (CO_2, NH_3 or NO_3^-, SO_4^{2-}, PO_4^{3-}) and water.

Metabolism is the process of building up and breaking down organic molecules in cells. It involves a great variety of organic reactions catalyzed by many different enzymes. **Enzymes** are proteins specialized for catalysis, and much of the protein content of a cell is in its many enzymes.■ A sequence of interconnected enzyme-catalyzed reactions is called a *metabolic pathway*. The reactions in this sequence are subject to *regulation*, that is, to being slowed down or speeded up by controlling the enzyme activity. In this way the cell can exert control over the rates of synthesis and breakdown of biological molecules. We will further consider enzymes in Section 27.4.

■ Most cells range in radius between 0.5 and 20 μm. Most covalently bonded atoms are between 0.05 and 0.15 nm in radius. The lower limit on cell size is imposed because any living cell must have a certain minimum molecular content.

■ Enzymes often increase reaction rates a billionfold or more.

27.2 Energy and the Biological System

A continuing input of energy is necessary to maintain life. The living system represents a highly organized state, thus a very low-entropy state. Recall from the discussion of thermodynamics in Chapter 20 that for any reaction or process to proceed spontaneously it must have a negative free-energy change associated with it. The free-energy change, ΔG, is related to the changes in enthalpy and entropy that occur in the process:

$$\Delta G = \Delta H - T\Delta S$$

If a process yields a more highly organized state, as do most processes in the formation and maintenance of biological systems, the entropy change will be negative and $(-T\Delta S)$ will be a positive number. Unless the ΔH term is sufficiently large and negative to override the entropy term (which is not usually the case), the ΔG will be positive, and the reaction or process will not occur spontaneously unless energy is made available to it.

Say, for example, a cell needs to synthesize a quantity of protein, making low-entropy complex molecules out of high-entropy simpler molecules. This process requires not only raw materials (the simpler molecules), but also a source of energy. This energy can be supplied through **coupling.** If a process requiring free energy, such as protein synthesis, and a process releasing free energy are *coupled,* that is, caused to proceed together, the overall free-energy change for the two coupled processes is the sum of their ΔG's, and the unfavorable reaction can occur.■

What is the source of energy that powers cell metabolism, the mechanical work of muscle contraction, and the electrical work of nerve-impulse transmission? The original source of energy for all purposes is the sun. However, most cells are unable to directly use solar energy. Only photosynthetic cells can absorb radiant energy and transform it into the chemical energy of biological molecules.■ Through photosynthesis, green plants use solar energy to convert simple inorganic substances into the complex organic chemical compounds that all other organisms oxidize for energy. Thus all animals, most bacteria, and nonphotosynthetic plants such as fungi depend on the food-making abilities of photosynthetic organisms. This chapter bypasses the important topic of photosynthesis, focusing instead on how cells use the chemical energy of food (derived either directly or indirectly from plants) for metabolism.

■ Synthesis of one gram of protein would require about 4 kcal.

■ Plant cells containing one of the green pigments called *chlorophylls* trap radiant energy for conversion to chemical energy.

Biological Molecules

Biological molecules, or biomolecules, are classified into four groups: proteins, carbohydrates, nucleic acids, and lipids. Many of these biomolecules are polymers. In this section we will first discuss some features of biological polymers. Then we will look closely at how each of the four types of biomolecules contributes to cell structure and function (metabolism).

27.3 Biological Polymers

Biological polymers are proteins, polysaccharides or complex carbohydrates, and nucleic acids. These very large molecules, often called **macromolecules,** have molecular weights ranging into the millions of atomic mass units and are many times more massive than ordinary organic compounds. However, there is an underlying simplicity in the structures of these biological polymers. They are usually linear chains of small, similar molecules covalently bonded together. We can picture each type of biological polymer as being built up by *condensation reactions* between small units or building blocks.■

The three-dimensional shapes of biological polymers are fundamental to their properties and biological functions. Such large molecules can theoretically take up many different shapes or conformations in space (Figure 27.1). **Conformations** are arrangements of molecules in space that result from rotations of atoms about single bonds. Usually only one such conformation is highly favored energetically and is assumed spontaneously by a molecule in aqueous solution. Loss of the normal conformation is almost always associated with dramatic changes in properties and loss of the biological function of the macromolecule. **Denaturation** is this loss of normal

■ In a condensation reaction, water is lost when a new covalent bond is formed (see Chapter 26 for discussion of condensation polymers).

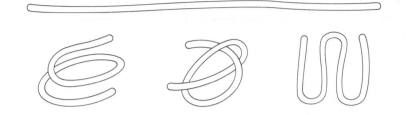

Figure 27.1
Several possible con-
formations of a polymer.

conformation and function. Agents that disrupt normal conformation are thus very damaging to biological molecules and systems. Excessive heat, acid, alkali, nonpolar solvents, strong salt solutions and detergents, and other denaturing agents can destroy the three-dimensional structure of proteins and other biomolecules, leading to loss of function.■

The three-dimensional conformations of polymers depend on *noncovalent bonds,* which anchor different parts of the large molecules together in stable arrangements. These noncovalent bonds include hydrogen bonds (see Section 11.1), such as in

■ Everyday examples of protein denaturation are the cooking of eggs and meats and the killing of bacteria by detergents and disinfectants that denature their proteins.

$$\diagup\!\!\diagdown C=O \cdots H-N \diagup\!\!\diagdown \quad \text{and} \quad \diagup\!\!\diagdown C-O-H \cdots O=C \diagup\!\!\diagdown$$

and ionic bonds (see Section 7.1), such as in

$$\overset{\displaystyle O}{\underset{\displaystyle \|}{-C}}-O^- \ + \ ^+H_3N-$$

Also important to the formation of three-dimensional conformations is **hydrophobic interaction,** which is the tendency of hydrophobic (nonpolar) molecules or parts of molecules to cluster in aqueous solution, minimizing their contact with water molecules. (*Hydrophobic* means "water fearing.") As an illustration, soap molecules in water spontaneously cluster into spherical globules with their hydrocarbon portions buried inside and their charged carboxylate groups on the surface where they can interact with water (see Figure 27.2). These globular aggregates, called **micelles,** are stable and can remain suspended in water indefinitely.■ We will see later how hydrophobic interaction relates to the conformations of proteins and nucleic acids and to the structure and function of lipids in biological membranes.

■ If a soap solution contacts a greasy substance, soap molecules surround the droplets of grease and incorporate them into micelles. This is how soap "dissolves" grease and allows it to be washed away.

Figure 27.2
The formation of micelles in water.

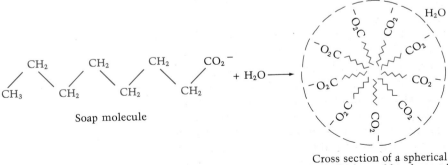

Soap molecule

Cross section of a spherical micelle formed by clustered soap molecules

27.4 Proteins

Proteins are polymers of small building blocks called *amino acids.* Their molecular weights range from 6000 to many millions, so they can be very large molecules. Many proteins also contain non–amino acid components such as metal ions (e.g., Fe^{2+}, Zn^{2+}, Cu^+, Mg^{2+}) or certain complex organic molecules that are usually derived from vitamins. (Vitamins are organic molecules necessary in small quantities for normal cell structure and metabolism.)

Proteins are very important molecules in cells and organisms, playing both structural and functional roles. The proteins of bone and connective tissue, for example, are of major structural importance. Some proteins are enzymes, catalyzing specific metabolic reactions, some transport materials in the bloodstream or across biological membranes, and some (hormones) carry chemical messages to coordinate the body's activities. *Insulin* and *glucagon*, for example, are protein hormones made in the pancreas and secreted to regulate the body's blood-sugar level.

Amino Acids

Amino acids are molecules containing a protonated amino group (NH_3^+) and an ionized carboxyl group (COO^-). The building blocks of protein are alpha-amino acids (α-amino acids). An α-amino acid has the general structure

$$
\begin{array}{ccc}
 & H & O \\
 & | & \| \\
R\!-\!\!\!& C\!-\!\!\!& C\!-\!O^- \\
 _\alpha\!\!\!\!\!\!\nearrow & | & \\
 & NH_3^+ & \\
\end{array}
$$

Here, the carbon atom next to the carboxyl carbon is labeled α and is the one bearing the amino group. Note that the amino acid is shown in doubly ionized form, called the **zwitterion,** since this is the form that predominates in the near-neutral pH of biological systems. The carboxyl group is a fairly strong acid, so its proton dissociates to a large extent. Conversely, the amino group is a fairly strong base, so it holds on to protons quite firmly at neutral pH. If the amino acid were in acidic solution, the $-COO^-$ group would tend to bind hydrogen ions,

$$-COO^- + H^+ \rightleftharpoons -COOH$$

to an extent dictated by its acid dissociation constant (K_a). Similarly, in basic solution the $-NH_3^+$ group would tend to lose protons,

$$-NH_3^+ \rightleftharpoons -NH_2 + H^+$$

depending on the base dissociation constant (K_b). Note that the total charge (net charge, + or −) of the amino-acid molecule will therefore depend on the pH of its solution. ■

■ Various laboratory methods for separating mixtures of compounds are based on differences in net charges. Ion exchange chromatography is an example.

Example 27.1

The structure of the amino acid azaserine (an antibiotic and anticancer agent produced by some microorganisms but not found in proteins) is

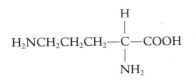

At neutral pH, is the carboxyl group negatively or positively charged? Draw the zwitterion of azaserine.

Solution

Since the carboxyl group is a fairly strong acid, at neutral pH most of these carboxyl groups will be ionized, and the resulting ions are negatively charged.

The zwitterion has both carboxyl and amino groups in ionized form:

$$N{=}N{=}C{-}\overset{\overset{\displaystyle H}{|}}{C}{-}\overset{\overset{\displaystyle O}{\|}}{}{-}O{-}CH_2{-}\overset{\overset{\displaystyle NH_3^{+}}{|}}{CH}{-}COO^{-}$$

Exercise 27.1

A common amino acid in the body is ornithine. It is involved in the excretion of excess nitrogen into the urine. The structural formula of ornithine is

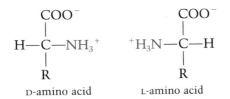

Write the fully ionized form of the molecule. (Note that there are *2* ionizable amino groups.)

(See Problems 27.27 and 27.28.)

An organic compound with four different substituents on any one carbon is *chiral* (Section 25.5), capable of existing as isomers that are non-superimposable mirror images of one another. These isomers are called **enantiomers,** or **D-** and **L-isomers.** Amino acids are chiral (except for glycine, CH_2NH_2COOH), and thus each can exist as the D-isomer or as the L-isomer. All the amino acids of known naturally occurring proteins are L-amino acids, however.

$$\underset{\text{D-amino acid}}{\overset{\displaystyle COO^{-}}{\underset{\displaystyle R}{\overset{\displaystyle |}{H{-}C{-}NH_3^{+}}}}} \qquad \underset{\text{L-amino acid}}{\overset{\displaystyle COO^{-}}{\underset{\displaystyle R}{\overset{\displaystyle |}{{}^{+}H_3N{-}C{-}H}}}}$$

To get the three-dimensional, tetrahedral arrangement about a carbon atom from these flat formulas, proceed as follows. The groups that are attached along the horizontal bonds bend toward you. Those that are attached along the vertical bonds bend away from you. Thus, in the D-amino acid the H atom and NH_3^{+} group are toward you, and the COO^{-} and R groups are away from you. The L-amino acid is similar, except that it is the mirror image of the D-amino acid.

In a protein, amino acids are linked together by **peptide** (or **amide**) **bonds,** which form in a condensation reaction between the carboxyl group of one amino acid and the amino group of a second amino acid:

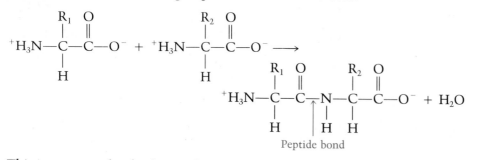

■ A polypeptide is *any* amino acid polymer, which may or may not have an identifiable function. The term *protein* describes polypeptides that have definite functions.

This is an example of a *dipeptide,* a molecule formed by linking together two amino acids. Similarly, a *tripeptide* is formed from three amino acids. A **polypeptide** is a polymer formed by the linking of many amino acids by peptide bonds.■

Example 27.2

Two common amino acids are

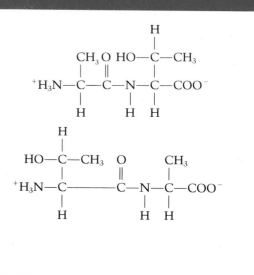

Write the structural formulas of the two dipeptides that they could form.

Solution

The carboxyl group of either of the amino acids could be peptide bonded to the amino group of the other. The structural formulas of these dipeptides are:

Exercise 27.2

Write the structural formulas of all possible tripeptides with the composition of two glycines and one serine. (See the structural formulas immediately following.)

(See Problems 27.29 and 27.30.)

Of the many known amino acids, twenty different kinds are found in most proteins. Each has a different R group, or **side chain**. The side chain determines the protein conformation through bonding and hydrophobic interaction.

Nine of these amino acids have nonpolar, or hydrocarbon, side chains (shown below in color). Beneath the name of each amino acid is its commonly used three-letter abbreviation.

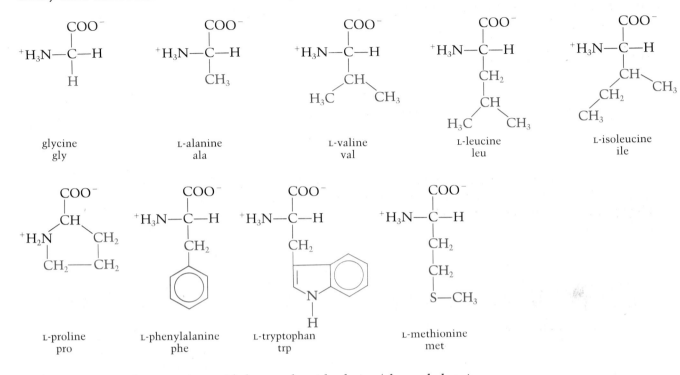

glycine
gly

L-alanine
ala

L-valine
val

L-leucine
leu

L-isoleucine
ile

L-proline
pro

L-phenylalanine
phe

L-tryptophan
trp

L-methionine
met

The remaining eleven amino acids have polar side chains (shown below in color), with groups capable of ionizing or forming hydrogen bonds with other groups.

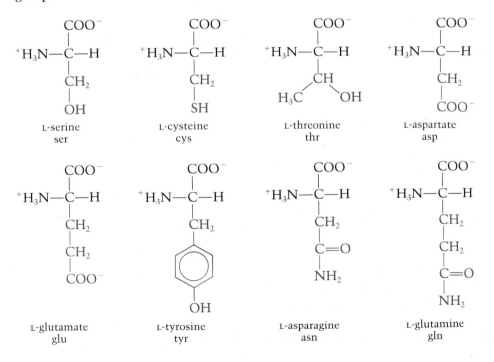

L-serine
ser

L-cysteine
cys

L-threonine
thr

L-aspartate
asp

L-glutamate
glu

L-tyrosine
tyr

L-asparagine
asn

L-glutamine
gln

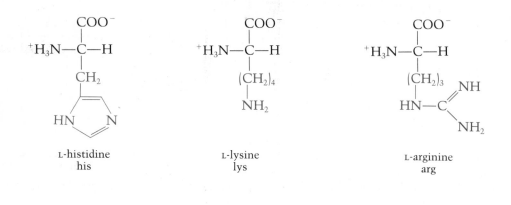

L-histidine
his

L-lysine
lys

L-arginine
arg

Example 27.3

Two amino acids found in most proteins are

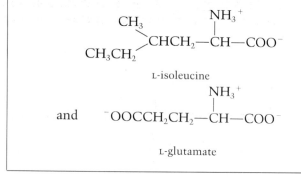

L-isoleucine

and

L-glutamate

By looking at the structural formulas, determine which of these has a polar side chain and which has a nonpolar side chain? Explain.

Solution

By examining the side chains, we can see that L-glutamate has an ionizable group (the carboxyl group). The side chain on L-isoleucine contains only carbon and hydrogen atoms. Thus, L-glutamate is polar and L-isoleucine is nonpolar.

Exercise 27.3

Phenylalanine and tyrosine have very similar structures, yet one has a polar side chain and the other has a nonpolar side chain. Determine which is which and explain why.

(See Problems 27.31 and 27.32.)

Example 27.4

Show by sketching the structural formulas of the molecules (refer to amino-acid formulas given earlier) how two amino-acid side chains might form a hydrogen bond between them.

Solution

Since a hydrogen bond only requires bringing together

an oxygen or nitrogen with an attached hydrogen atom and another group containing an oxygen or nitrogen, there are several pairs of polar amino acids we could choose. One such pair is L-serine and L-aspartate (shown below):

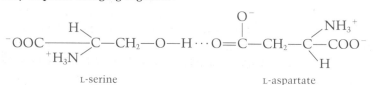

L-serine   L-aspartate

Exercise 27.4

Show how the side chains of L-threonine and L-glutamate could form a hydrogen bond between them.

(See Problems 27.33 and 27.34.)

Protein Primary Structure

Primary structure of a protein refers to the order or sequence of the amino-acid units in the polymer. This order of amino acids is conveniently shown by denoting the amino acids using their three-letter codes, each amino-acid code in the sequence being separated by a dash. In this notation, it is understood that the carboxyl group of the amino acid is on the right, and the amino group is on the left. The dipeptides in Example 27.2 would be written ala-thr (first one given in the solution) and thr-ala. Consider a polypeptide containing five amino-acid units: glycine, alanine, valine, histidine, and serine. There are many ways in which these five units can be ordered to make a polypeptide. Three possible primary structures of the polypeptide are

> glycine–serine–alanine–valine–histidine
>
> histidine–alanine–serine–valine–glycine
>
> serine–alanine–glycine–histidine–valine

Each of the possible sequences would produce a different peptide with different properties. In all there are 120 sequences possible for a polypeptide with five amino acids■ Each sequence has the same percent composition of amino acids, but different primary structures.

In any protein, which may contain as few as 50 or more than 1000 amino-acid units, the amino acids are arranged in one unique sequence, the primary structure of that protein molecule. Figure 27.3 diagrams the primary structure of one form of the hormone insulin.

■ The number of *permutations* (ordered arrangements) of 5 objects taken 5 at a time is 5! (read 5 factorial), which is equal to $5 \cdot 4 \cdot 3 \cdot 2 \cdot 1$, or 120, permutations.

Example 27.5

Write the sequences of all possible tripeptides with the amino-acid composition lysine, alanine, leucine.

Solution

There are six possible sequences:

lys–ala–leu	ala–leu–lys
lys–leu–ala	leu–lys–ala
ala–lys–leu	leu–ala–lys

Exercise 27.5

Write the sequences of all possible tripeptides with the amino-acid composition lysine, lysine, alanine.

(See Problems 27.35 and 27.36.)

Protein Conformations

A protein molecule of a unique amino-acid sequence spontaneously folds and coils into a characteristic three-dimensional conformation in aqueous

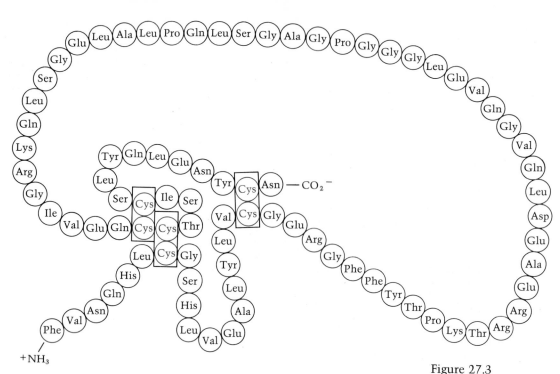

Figure 27.3
Primary structure of human proinsulin. The formula is arranged to show the position of the disulfide linkages, and is not an indication of three-dimensional shape.

solution (Figure 27.4). The side chains of the amino-acid units, with their different chemical properties—nonpolar, polar, negatively or positively charged—determine the characteristic conformation. For an energetically stable conformation, those parts of the chain with nonpolar amino-acid side chains are buried within the structure away from water, because nonpolar groups are hydrophobic. Conversely, most polar groups are most stable on the surface where they can hydrogen bond with water or with other polar side chains. Occasionally, side chains form ionic bonds.

One type of covalent linkage is important in protein conformation. The amino acid cysteine, with its thiol (—SH) side chain, is able to react with a second cysteine in the presence of an oxidizing agent, as follows:

$$\text{cys—SH} + \text{HS—cys} + (0) \longrightarrow \text{cys—S—S—cys} + H_2O$$

The resulting group (—S—S—) is a *disulfide.* When two cysteine side chains in a polypeptide are brought close together by folding of the molecule, then oxidized, they form a **disulfide cross link,** which helps anchor the folded chain into position. ■

Proteins may be classed as fibrous proteins or globular proteins on the basis of their general conformations. **Fibrous proteins** are polypeptides that form long coils or align themselves in parallel to form long water-insoluble fibers. The relatively simple coiled or parallel arrangement of a protein molecule is called its **secondary structure. Globular proteins** are polypeptides in which long coils also fold into more compact, roughly spherical conformations. Thus, in addition to their secondary structure (long coil), they have a **tertiary structure,** or folded structure. Most globular proteins are water soluble because they are relatively small in dimension and have hydrophilic ("water loving") surfaces that bind water molecules.

■ Hair can be permanent waved by the use of disulfide cross links. Hair proteins have many disulfides that can be chemically reduced and reformed again while the hair is rolled on curlers. The new disulfides hold the hair shafts in their new wavy conformation.

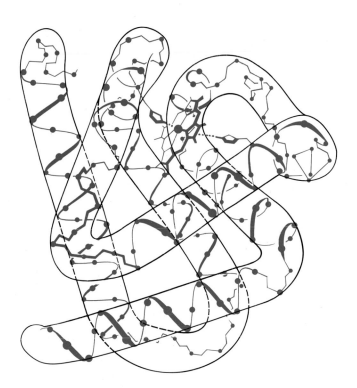

Figure 27.4
Three-dimensional structure of the protein myoglobin. Note the regions of helical structure. This structure was deduced from x-ray diffraction data.

The characteristic coiling or folding pattern of protein molecules produces unique *surface features* such as grooves or indentations, and hydrophobic or charged areas. These surface features are fundamental to protein function. They allow protein molecules to interact with other molecules in specific ways. For example, certain protein molecules characteristically aggregate due to complementary surface features, forming tubules or filaments or other organized structures. Particular surface features are also responsible for specific binding and catalysis by enzyme proteins.

Enzymes

Enzymes, probably the most important of all proteins, are marvelously efficient and specific catalysts. They are usually globular proteins. Enzymes possess characteristic **active sites** at which **substrates** (molecules whose reactions the enzyme will catalyze) bind and where catalysis takes place. Figure 27.5 shows the substrate molecule fitting into the active site as does a key in a lock. Some flexibility is probably possible, however, both in the conformation of the active site and in that of the substrate. ■

The rate of an enzyme-catalyzed reaction does not show a linear dependence on the substrate concentration because the product forms only *after* the substrate has been bound by the enzyme to form an **enzyme–substrate complex.** It is the concentration of the complex on which the rate depends.

■ Many poisons prevent normal enzyme function because they have the right structure to bind to particular enzymes in place of the normal reactants.

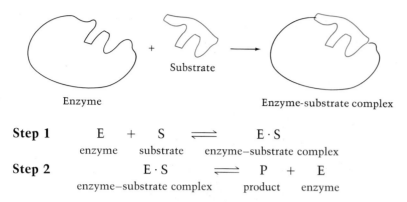

Figure 27.5
Lock-and-key model of the interaction of enzyme and substrate.

Step 1   E + S ⇌   E · S
           enzyme   substrate   enzyme–substrate complex

Step 2   E · S ⇌ P + E
           enzyme–substrate complex   product   enzyme

Many enzymes are capable of increasing reaction rates by a factor of 10^9 or more. Reactions that would be so slow as to be insignificant may proceed extremely rapidly in the presence of the appropriate enzyme. Living organisms would not be possible if they had to depend on uncatalyzed reaction rates for oxidation of foods or synthesis of cellular structures.

In addition to speed, a second feature of enzyme function is **specificity**. Because of the special properties of its active site, an enzyme is said to be *specific* for only certain substrates or certain types of reaction.

27.5 Carbohydrates

Carbohydrates are polyhydroxy aldehydes or ketones, or substances that yield such compounds when they react with water.■ Some carbohydrates, such as cellulose and chitin, are structural elements in plants and animals; others, such as glucose, provide energy and raw materials for cell activities. Beet sugar, grape sugar, and milk sugar, introduced in the chapter opening as important compounds in Pasteur's work, are carbohydrates.

■ Many carbohydrates have empirical formulas $C_m(H_2O)_n$; hence the name carbohydrate.

Carbohydrates may be divided into three categories: **monosaccharides**, or simple sugars; **oligosaccharides**, short polymers of two to ten simple sugar units; and **polysaccharides**, polymers consisting of more than ten simple sugar units.

Monosaccharides

A monosaccharide has three to nine carbon atoms, all but one of which bear a hydroxyl group. The remaining carbon is always a carbonyl carbon, either an aldehyde or ketone, or a derivative of these. The *-ose* suffix is commonly used to designate sugars. Monosaccharides exist as D- and L-isomers, since there is at least one carbon atom in each molecule with four different groups attached. Due to their polar structure, monosaccharides are highly soluble in water.

Only a few of the many monosaccharides are of major biological importance. All are D-sugars, and most are 5- or 6-carbon sugars (pentoses or hexoses)■ D-glucose, D-fructose, D-ribose, and 2-deoxy-D-ribose, are by far the most important sugars. D-glucose is common blood sugar, an important energy source for cell function. D-fructose is the common sugar in fruits

■ Name of Sugar	Number of Carbon Atoms
Triose	3
Tetrose	4
Pentose	5
Hexose	6
Heptose	7
Octose	8
Nonose	9

(such as grapes) and a food source. D-ribose and 2-deoxy-D-ribose are parts of nucleic acids, which we will study later.

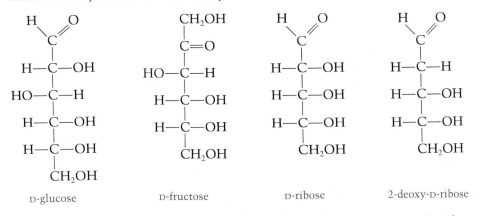

| D-glucose | D-fructose | D-ribose | 2-deoxy-D-ribose |

Although we often draw the simple sugars as straight-chain molecules, they do not exist predominantly in that form. This results from the fact that carbonyl groups react readily with alcohol groups. With aldehydes, alcohols give **hemiacetals,** or compounds in which an —OH group, an —OR group, and an H atom are attached to the same carbon atom.

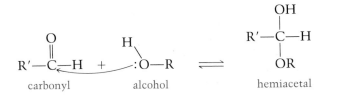

| carbonyl | alcohol | hemiacetal |

A ketone and an alcohol give a *hemiketal*, which is similar to a hemiacetal, except the H atom is replaced by an R″ group.

The straight-chain form of D-glucose, whose formula we gave earlier, is a flexible molecule that can bend around to give a cyclic hemiacetal.

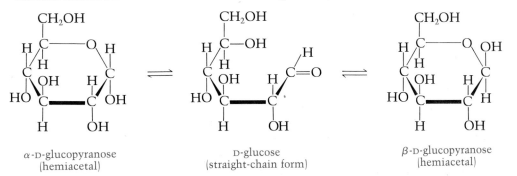

α-D-glucopyranose
(hemiacetal)

D-glucose
(straight-chain form)

β-D-glucopyranose
(hemiacetal)

The straight-chain form, in the center, is shown bent around so that an —OH group and the —CHO group can react to give a cyclic hemiacetal. A new —OH group is obtained (on the right in each hemiacetal), and can point either down or up. Thus, there are two isomers of the hemiacetal, one labeled α (—OH down) and the other labeled β (—OH up). Note that these hemiacetals have six-membered rings. Monosaccharides with six-membered rings are called *pyranoses;* those with five-membered rings are called *furanoses.* In the case of glucose, the hemiacetals are called α-D-glucopyranose and β-D-glucopyranose to indicate that they are six-membered rings.

Example 27.6

Write the reactions showing the equilibria between the straight-chain form of D-ribose and its furanose, hemiacetal forms. The formula of D-ribose was given earlier.

Solution

We write the straight-chain form of D-ribose bent so that the —CHO group can react with that —OH group that gives a five-membered ring.

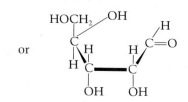

or

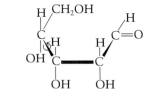

Note that the formula at the left is simply the formula given earlier, but laid on its side and bent around. Thus, all —OH groups that were on the right are now pointing down; those on the left are now pointing up. In the formula at the left, the —CH₂OH, the H, and ⁻OH groups can be rotated to get the formula above. (Remember, we can twist a portion of a molecule about a C—C bond.) In that position, the —OH and —CHO groups can react to give a five-membered ring. The equilibria are as follows:

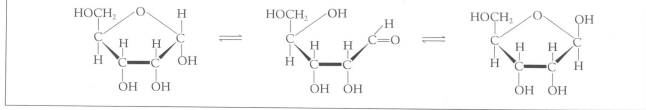

Exercise 27.6

Write the equilibria for the reaction between the straight-chain form of D-fructose and each of its furanose, hemiketal forms. The formula of D-fructose was given earlier.

(See Problems 27.37 and 27.38.)

Oligosaccharides and Polysaccharides

Oligosaccharides and polysaccharides are formed from monosaccharide building blocks. The hemiacetal carbon atom of one monosaccharide is attached to the alcohol oxygen atom of another monosaccharide by a condensation reaction. The most common oligosaccharide is sucrose (a disaccharide), found in sugar beets and sugar cane. In this molecule, glucose and fructose are joined by an O atom from carbon 1 of glucose to carbon 2 of fructose. (The carbon atoms of each monosaccharide are numbered from the end closest to the carbonyl group, or what was the carbonyl group.)

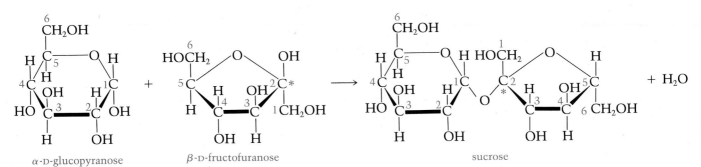

α-D-glucopyranose   β-D-fructofuranose   sucrose   + H₂O

Note that the fructose is flipped upside down and rotated 180° to form the bond (compare the position of the starred carbon atom in β-D-fructofuranose and sucrose). *Maltose* and *lactose* are also common disaccharides with properties similar to sucrose. Maltose is composed of two α-D-glucose units.

Example 27.7

Lactose (milk sugar), introduced in the chapter opening, is composed of β-D-galactopyranose and α-D-glucopyranose, joined by an O atom from carbon 1 of galactose to carbon 4 of glucose. Diagram the reaction of its formation from the monosaccharides. D-galactose has the straight chain form

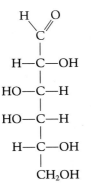

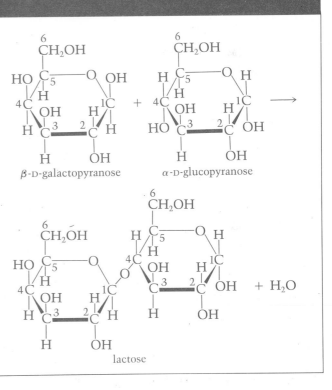

β-D-galactopyranose   α-D-glucopyranose

lactose

Solution

We first write the hemiacetal structures, then link them by a condensation reaction.

Exercise 27.7

Write the structural formula for maltose, which consists of two α-D-glucopyranose units linked by an O atom from carbon 1 of one unit to carbon 4 of the other.
(See Problems 27.39 and 27.40.)

Polysaccharides usually contain only one type of building block, or sometimes two alternating building-block units. Most polysaccharides are linear polymers, one main chain with no branching units.

Polysaccharides are structural molecules and energy-storage polymers. *Cellulose*, a polysaccharide, is a structural carbohydrate that is abundant among plant species. Plant cell walls contain fibers of cellulose that preserve the structural integrity of the cells they surround. Cellulose makes possible the *turgor* (water pressure in cells) that gives nonwoody plants their shapes.

Cellulose is a linear polymer of β-D-glucopyranose units:

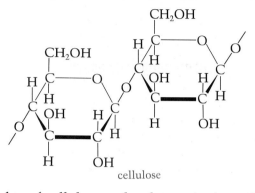

cellulose

Molecular weights of cellulose molecules are in the millions.

Chitin, whose structure is closely related to cellulose, is the skeletal polysaccharide found in the exoskeletons of insects and other invertebrates. Chitin is a linear polymer of a β-D-glucopyranose derivative.

The major energy-storage polysaccharides in plants are starches. One kind of starch, *amylose*, contains long unbranched chains of α-D-glucopyranose units:

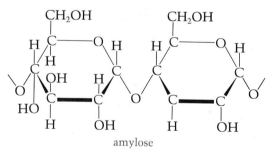

amylose

A second kind of starch, *amylopectin*, is a branched polysaccharide. The structure is like that of amylose, except that a branch occurs about every 18–20 units along the main chain. Starch molecules vary considerably in size, but molecular weights of several million are probably common.

Humans cannot use cellulose for food because our starch-digesting enzyme attacks only α-glucose polymers, not β-glucose polymers such as cellulose.

Animals store sugars in the form of the polysaccharide *glycogen*. Glycogen is very similar in structure to amylopectin except that branches occur more frequently. The molecular weight of glycogen obtained from animal tissues is similarly very high. Plant and animal polysaccharides represent a rich source of chemical energy, and they are stored along with fats (discussed in Section 27.7) within cells to provide an energy reservoir to be drawn from as needed■ Branched polysaccharides are thought to be more quickly added to (or broken down) when the cell's needs dictate, since there are many more "free ends" in such a polymer than in a single linear chain.

■ Runners who do *carbohydrate loading* are trying to increase the amount of glycogen stored in their muscle cells so they will have a larger energy reserve for a race. They deplete their muscle glycogen by a long run, minimize the carbohydrate in their diets for a few days, then load with carbohydrate by eating a starchy food, such as spaghetti, for a day or so before the race.

27.6 Nucleic Acids

Nucleic acids are vital to the life cycles of cells because they are the carriers of species inheritance. ■There are two types of nucleic acids: **DNA,** or **deoxyribonucleic acid;** and **RNA,** or **ribonucleic acid.** Both are polymers of *nucleotide* building blocks. We'll examine their structures in detail in this section.

■ *Genetic engineering* refers to the manipulation of nucleic acids to change the characteristics of organisms, for example, to correct genetic flaws.

Nucleotides

Nucleotides are the building blocks of nucleic acids. They are of two types: **ribonucleotides,** found in RNA and in certain other important molecules; and **deoxyribonucleotides,** found in DNA. The general structure of nucleotides consists of an organic base (described below) linked to carbon 1 of a pentose (5-carbon sugar), which is linked in turn at carbon 5 to a phosphate group (or perhaps a diphosphate or triphosphate group). The base plus the sugar group is called a **nucleoside.** The only difference between the two types of nucleotides is in the sugar. Ribonucleotides contain β-D-ribose; deoxyribonucleotides contain 2-deoxy-β-D-ribose. Both sugar groups have furanose rings.

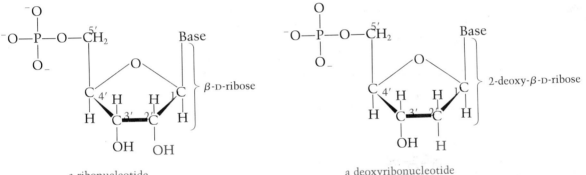

a ribonucleotide   a deoxyribonucleotide

The carbon atoms of ribose are labeled with numbers 1', 2', ... 5'. (Numbers without primes are used for carbon atoms in the organic base.) Note that the phosphate group is shown in ionized form, as this form predominates near neutral pH in cells.

Five organic bases (amines) are most often found in nucleotides. They are *adenine, guanine, cytosine, uracil,* and *thymine.* They link to the sugars through the indicated nitrogen in each of the following structural formulas:

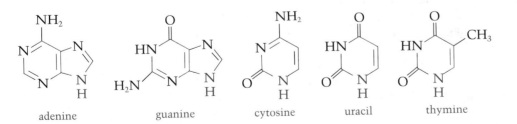

adenine   guanine   cytosine   uracil   thymine

Both DNA and RNA contain nucleotides with adenine, guanine and cytosine bases. However, DNA contains thymine but not uracil, and RNA contains uracil but not thymine.

Organic Base	Nucleoside		5'-Nucleotide	
	With Ribose	With 2'-Deoxyribose	With Ribose (abbreviation)	With 2'-Deoxyribose (abbreviation)
Adenine	Adenosine	Deoxyadenosine	Adenosine-5'-monophosphate (AMP)	Deoxyadenosine-5'-monophosphate (dAMP)
Guanine	Guanosine	Deoxyguanosine	Guanosine-5'-monophosphate (GMP)	Deoxyguanosine-5'-monophosphate (dGMP)
Cytosine	Cytidine	Deoxycytidine	Cytidine-5'-monophosphate (CMP)	Deoxycytidine-5'-monophosphate (dCMP)
Uracil	Uridine	(Deoxyuridine)*	Uridine-5'-monophosphate (UMP)	Deoxyuridine-5'-monophosphate* (dUMP)
Thymine	(Ribothymidine)*	Thymidine	Thymidine-5'-monophosphate* (TMP)	Thymidine-5'-monophosphate (dTMP)

*Uncommon forms.
Source: Robert D. Whitaker et al., *Concepts of General, Organic, and Biological Chemistry* (Boston: Houghton Mifflin Co., 1981), p. 482.

Table 27.1
Naming of Bases, Nucleosides, and Nucleotides

Nucleosides are named from the bases. For example, the nucleoside composed of adenine with β-D-ribose is called adenosine. The nucleoside composed of adenine with 2-deoxy-β-D-ribose is called deoxyadenosine. A nucleotide is named by adding monophosphate (or diphosphate or triphosphate) after the nucleoside name. A number with a prime indicates the position of the phosphate group on the ribose ring. Thus, adenosine-5'-monophosphate is a nucleotide composed of adenine, β-D-ribose, and a phosphate group at the 5' position of β-D-ribose. Table 27.1 lists names and abbreviations for the nucleotides of DNA and RNA.

Example 27.8

Draw the structural formula of the nucleotide containing guanine, 2-deoxy-β-D-ribose, and a phosphate group at the 5' position. Name the compound.

Solution

The structural formula is shown below. The name of the compound is deoxyguanosine monophosphate.

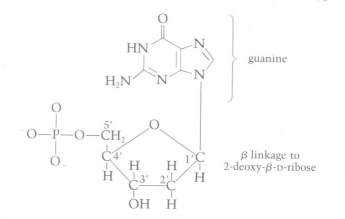

Exercise 27.8

Write the structural formula for cytidine monophosphate.

(See Problems 27.45 and 27.46.)

Polynucleotides and Their Conformations

A **polynucleotide** is a linear polymer of nucleotide units linked from the hydroxyl group at the 3′ carbon of the pentose of one nucleotide, to the phosphate group of the other nucleotide. For example,

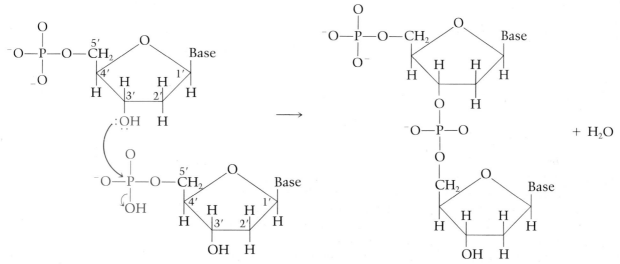

Just as the unique sequence of amino acids in a protein determines the protein's nature, so the sequence of nucleotides determines the particular properties and functions of a polynucleotide.

Nucleic acids are polynucleotides folded or coiled into specific three-dimensional conformations. **Complementary bases** are certain nucleotide bases that form strong hydrogen bonds with one another. Adenine and thymine are complementary bases, as are adenine and uracil, and guanine and cytosine. Hydrogen bonding of complementary bases, called **base pairing,** is the key to nucleic acid structure and function (see Figure 27.6).

Base pairing allows a single polynucleotide strand folded back upon itself to be held in a particular conformation. Below, nucleotides are denoted by the initial letter of their bases.

Loops such as this are a feature of RNA structure. More than one loop may be present. Figure 27.7 shows how base pairing between complementary sequences of nucleotides is important to the structures of two kinds of RNA. (These RNA molecules will be discussed later in the section.)

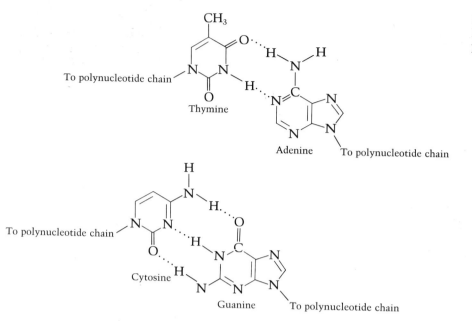

Figure 27.6
Hydrogen-bonded com-
plementary bases.

A DNA molecule, on the other hand, consists of *two* polynucleotide chains with base pairing along their entire lengths. The two polynucleotide chains are coiled about each other in a **double helix** (Figure 27.8). Base pairing is not the only factor in nucleic-acid conformation, but it is the most important one.

Example 27.9

Noting the three complementary base pairs and which bases are found in DNA or RNA, write the DNA sequence complementary to the following sequence:

ATGCTACGGATTCAA

Solution

Since DNA does not contain uracil, the complementary base for adenine is thymine. Thus, the proper sequence of base pairs is

ATGCTACGGATTCAA
TACGATGCCTAAGTT

Exercise 27.9

Write the RNA sequence complementary to the sequence given in Example 27.9 above.

(See Problems 27.51 and 27.52.)

DNA and the Nature of the Genetic Code

Photomicrographs of dividing cells show structures, called chromosomes, as dense, thick rods (Figure 27.9). *Chromosomes* are cell structures that con-

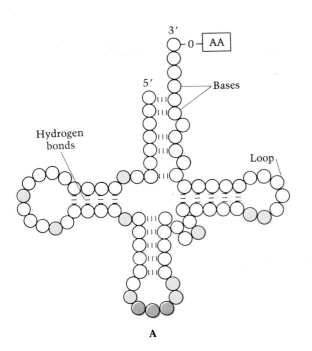

Figure 27.7
Examples of secondary structures in (a) a transfer RNA
(t RNA) and (b) a ribosomal
RNA (r RNA).

A

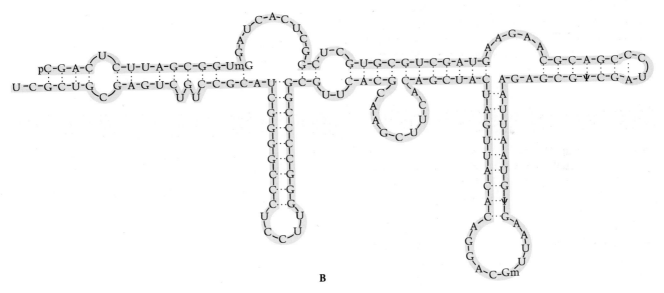

B

tain DNA and proteins; the DNA contains the genetic inheritance of the cell
and organism. Before cell division, the cell synthesizes a new and identical
set of chromosomes, or more particularly a new and identical set of DNA
molecules—the genetic information—to be transmitted to the new cell.
Thus the new cell will have all the necessary instructions for normal structure and function.

What is the nature and role of genetic information? The **genetic code** is the
relationship between the nucleotide sequence in DNA and the amino-acid

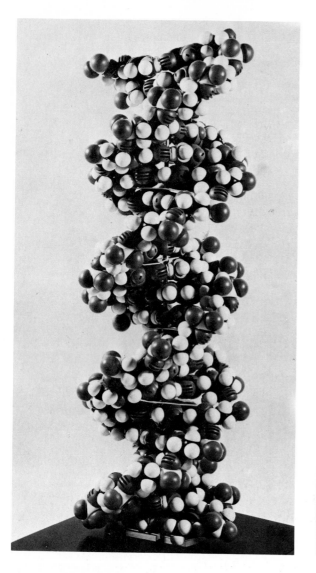

Figure 27.8
Double helix of the DNA molecule. The structure was first deduced by James Watson, an American scientist, and Francis Crick, an English scientist, in the early 1950s.

Figure 27.9
A dividing cell with chromosomes appearing as dense, thick rods.

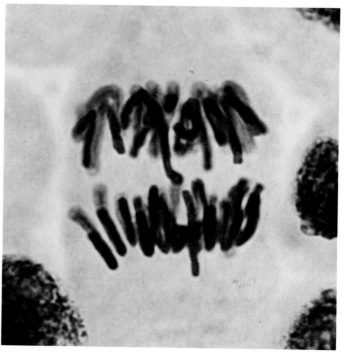

sequence in proteins. Genetic information is coded into the linear sequence of nucleotides in the DNA molecules. This coding then directs the synthesis of specific proteins that make the cell and species what it is. The code structures are called **codons,** each of which is a particular sequence of *three* nucleotides (a triplet codon) usually denoted simply by their bases. An example is adenine–guanine–cytosine (AGC). Each codon specifies an amino acid to be used to construct a specific amino-acid sequence. A **gene** is a *sequence of nucleotides* in a DNA molecule that codes for a given protein. (A single DNA molecule may contain hundreds of genes.) For example, suppose that a gene for a protein is 333 nucleotides long. It then contains 333/3, or 111, codons in a unique sequence. This gene will direct the formation of a polypeptide with 111 amino acids arranged in a unique sequence corresponding exactly to the coded information stored in the gene.

Occasionally, an error is made during the synthesis of new DNA. Such a change in the genetic information, or genetic error, is called a *mutation*. One possible consequence could be the synthesis of a faulty or inactive protein. We know of various genetic diseases that result from a single error leading to a change in amino-acid sequence that has profound effects on the protein's activity and thus the life of the organism.

Example 27.10

A polypeptide of 160 amino-acid units would be coded for by a gene of how many nucleotide units?

Solution

Each codon is a sequence of three nucleotides. A sequence of 160 amino acids requires a sequence of 160 codons, so 160×3 nucleotides per codon = 480 nucleotide units.

Exercise 27.10

How many amino acids would be found in a polypeptide coded for by a gene 1,200 nucleotides in length?

(See Problems 27.57 and 27.58.)

RNA and the Transmission of the Genetic Code

RNA is used to translate the genetic information stored in DNA into protein structure. There are three classes of RNA: ribosomal RNA, messenger RNA, and transfer RNA.

Ribosomes are tiny cellular particles, constructed of numerous proteins plus three or four **ribosomal RNA** molecules.■ Ribosomes are protein "factories," since they provide a surface on which to organize the process of protein synthesis, and they also contain enzymes that catalyze the process.

Messenger RNA molecules are relatively small RNAs that diffuse about the cell and attach themselves to ribosomes, where they serve as patterns for protein synthesis. The first step in protein synthesis, called *transcription*, is synthesis of a messenger RNA molecule that has a sequence of bases complementary to that of a gene. Using our previous example of a 333-nucleotide gene, transcription would involve synthesis of a complementary 333-nucleotide messenger RNA molecule.

When the messenger RNA molecule attaches itself to a ribosome, its codons provide the sequence of amino acids in a protein that will be synthesized. *Translation* is the synthesis of protein using messenger RNA codons. Table 27.2, the complete messenger RNA code-word dictionary, shows the specific amino acids coded for by messenger RNA codons. As we indicated earlier, codons are triplets. There are 64 possible arrangements of the four RNA bases, so there are 64 codons. Three of the codons do not signify amino acids but signify the end of a message, and thus are called **termination codons.** The remaining 61 codons signify particular amino acids, and since there are only 20 amino acids in proteins, there are a number of instances in which 2, 3, 4 or even 6 codons translate to the same amino acid.

■ Bacterial ribosomes are roughly spherical particles about 20 nm in diameter. Animal and plant ribosomes are slightly larger.

Example 27.11

What amino-acid sequence would result if the following messenger-RNA sequence were translated from left to right?

AGAGUCCGAGACUUGACGUGA

Solution

We mark the message off into triplets, beginning at the left, and consult the codon dictionary in Table 27.2 to obtain

AGA	GUC	CGA	GAC	UUG	ACG	UGA
arg	val	arg	asp	leu	thr	end

Exercise 27.11

Give one of the nucleotide sequences that would translate to the peptide lys–pro–ala–phe–trp–glu–his–gly.

(See Problems 27.61 and 27.62.)

Transfer RNA molecules are the smallest RNAs, with molecular weights around 25,000 amu. Transfer RNAs provide the key to the mechanism of translation because it is they that really "read" the coded message. They bond to the amino acids and carry them to the ribosomes, then attach themselves (through base pairing) to messenger-RNA codons.

To picture this process, imagine that we have a ribosome, with a messenger RNA attached in the proper way for translation, and the first codon is in

Table 27.2
Genetic Code Dictionary*

	U		**C**		**A**		**G**	
U	UUU	Phe	UCU	Ser	UAU	Tyr	UGU	Cys
	UUC	Phe	UCC	Ser	UAC	Tyr	UGC	Cys
	UUA	Leu	UCA	Ser	UAA	End	UGA	End
	UUG	Leu	UCG	Ser	UAG	End	UGG	Trp
C	CUU	Leu	CCU	Pro	CAU	His	CGU	Arg
	CUC	Leu	CCC	Pro	CAC	His	CGC	Arg
	CUA	Leu	CCA	Pro	CAA	Gln	CGA	Arg
	CUG	Leu	CCG	Pro	CAG	Gln	CGG	Arg
A	AUU	Ile	ACU	Thr	AAU	Asn	AGU	Ser
	AUC	Ile	ACC	Thr	AAC	Asn	AGC	Ser
	AUA	Ile	ACA	Thr	AAA	Lys	AGA	Arg
	AUG	Met	ACG	Thr	AAG	Lys	AGG	Arg
G	GUU	Val	GCU	Ala	GAU	Asp	GGU	Gly
	GUC	Val	GCC	Ala	GAC	Asp	GGC	Gly
	GUA	Val	GCA	Ala	GAA	Glu	GGA	Gly
	GUG	Val	GCG	Ala	GAG	Glu	GGG	Gly

*The 4 nucleotide bases of RNA—U, C, A, and G—are arranged along the left side of the table and the top. Combining these two, and then adding a third (again U, C, A or G) gives the 64 three-nucleotide codons shown in capital letters in the table. Next to each codon appears the abbreviation for the amino acid it codes for during protein synthesis. In three cases, the word *End* appears because each of these codons serves as a signal to terminate protein synthesis.
From Robert D. Whitaker et al., *Concepts of General, Organic, and Biological Chemistry* (Boston: Houghton Mifflin Co., 1981), p. 697.

position to be read:

AUG GGA CCG ACG UGC GAG CUC...messenger-RNA codons

The first messenger-RNA codon is AUG, codon for methionine. Methionine is carried to the ribosome by a transfer RNA that has a triplet sequence, called an **anticodon,** complementary to the codon AUG. The messenger-RNA codon and the transfer-RNA anticodon with methionine pair up. The next codon in the message is GGA, which specifies glycine. A transfer RNA with the anticodon complementary to GGA carries glycine into position to be peptide bonded to methionine. Each succeeding codon in the message is handled in a similar way until a termination codon appears to signal the end of the polypeptide chain. Then the finished product is released from the ribosome.

Nucleotides and Metabolism

Besides their part in nucleic-acid structure, nucleotides have two other important functions. First, certain nucleotides more complex than those of DNA and RNA serve as coenzymes, the enzyme "helpers" needed in certain reactions.

Second, some nucleotides are used to store chemical energy. Metabolic pathways that promote the oxidation (breakdown) of food—carbohydrates, fats, and proteins—release energy that must be captured in a new chemical form. Most of this energy is initially stored in the form of **ATP** (adenosine triphosphate). ATP is synthesized from ADP and phosphate in reactions coupled to food oxidation. ATP is the energy currency of the living cell. When hydrolyzed back to ADP and phosphate, ATP releases energy that can power a reaction that is coupled to it:

$$ATP + H_2O \longrightarrow ADP + HPO_4^{2+} + energy$$

27.7 Lipids

Lipids are biological substances that are soluble in nonpolar organic solvents, such as chloroform and carbon tetrachloride. They include the familiar food fats and oils, the lipids of biological membranes, steroid hormones, and many more unusual and exotic compounds. In this section we will discuss fats and oils and the membrane lipids. Lipids differ from the other classes of biological molecules in that there are no lipid polymers analogous to polymers of amino acids, sugars, or nucleotides. We will see, however, that lipid structures formed with noncovalent bonds are very important.

Fats and Oils

At normal temperatures, *fats* are solids and *oils* are liquids. Nevertheless, fats and oils have the same basic structure. Both are **triacylglycerols,** which are esters formed from glycerol (a trihydroxy alcohol) and three fatty acids (long chain carboxylic acids):■

■ Esters were discussed in Section 26.6.

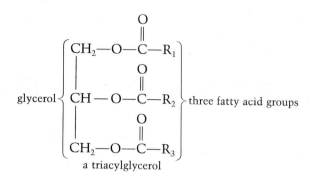

a triacylglycerol

Hydrolysis of a triacylglycerol with strong base (a process called saponification) yields fatty acid salts (soaps) and free glycerol.

Usually, in naturally occurring fats and oils, fatty acids (called "fatty" because of their long nonpolar "tails") are long-chain molecules of 12, 14, 16, 18, 20, 22, or 24 carbons. They may be saturated (containing no double bonds) or unsaturated (containing one or more double bonds). A fat or oil molecule typically is formed from two or three different fatty acids. Table 27.3 summarizes structures of the most common fatty acids. Vegetable oils, which contain unsaturated fatty acids, are liquids at room temperature. This is because fatty acids with double bonds do not pack together well. Conversely, animal fats, composed of the more easily aligned saturated fatty acids, are usually solids at room temperature.

In most organisms, triacylglycerols are a long-term form of energy storage. Excess food taken in is stored as fat, whether it originated as carbohydrate, protein, or fat itself. Fats are a rich source of chemical energy. Stored in

Table 27.3
Some Naturally Occurring
Fatty Acids

Structure	Common Name	m.p., °C
Saturated fatty acids		
$CH_3(CH_2)_{10}CO_2H$	Lauric acid	44.2
$CH_3(CH_2)_{12}CO_2H$	Myristic acid	53.9
$CH_3(CH_2)_{14}CO_2H$	Palmitic acid	63.1
$CH_3(CH_2)_{16}CO_2H$	Stearic acid	69.6
$CH_3(CH_2)_{18}CO_2H$	Arachidic acid	76.5
$CH_3(CH_2)_{22}CO_2H$	Lignoceric acid	86.0
Unsaturated fatty acids		
$CH_3(CH_2)_5CH=CH(CH_2)_7CO_2H$	Palmitoleic acid	−0.5
$CH_3(CH_2)_7CH=CH(CH_2)_7CO_2H$	Oleic acid	13.4
$CH_3(CH_2)_4CH=CHCH_2CH=CH(CH_2)_7CO_2H$	Linoleic acid	−5
$CH_3CH_2CH=CHCH_2CH=CHCH_2CH=CH(CH_2)_7CO_2H$	Linolenic acid	−11
$CH_3(CH_2)_4(CH=CHCH_2)_3CH=CH(CH_2)_3CO_2H$	Arachidonic acid	−49.5

From Robert D. Whitaker et al., *Concepts of General, Organic, and Biological Chemistry* (Boston: Houghton Mifflin Co., 1981), p. 507.

compact, nonhydrated droplets in cells, they are an ideal energy reservoir.■
Fat stored in animal *adipose tissue*, specialized fat-storage tissue, also serves
a structural purpose, providing insulation against cold and padding to protect delicate body organs. The higher-melting saturated fats of animals are
better suited to these purposes than liquid oils would be.

■ Fats yield 9 cal/g compared to
4 cal/g for oxidation of carbohydrates and proteins.

Example 27.12

Write a structural formula for a triacylglycerol containing two 16-carbon saturated fatty acids and one
18-carbon unsaturated fatty acid (see Table 27.3).

Solution

A triacylglycerol contains glycerol in ester linkage with
three fatty acids. We chose two palmitic acids and an
oleic acid to construct this triacylglycerol:

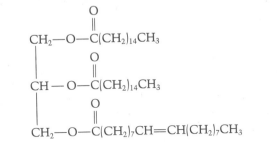

Exercise 27.12

Write a structural formula for a triacylglycerol containing three different unsaturated
fatty acids.

(See Problems 27.67 and 27.68.)

Biological Membranes

Membranes are basic for life processes. The membrane that surrounds a cell
is the boundary between cell and "not cell." The properties of the membrane
largely dictate what can or cannot get into or out of the cell. Most cells also
contain extensive systems of internal membranes, which organize and regulate cell function.

A biological membrane (Figure 27.10) is composed of proteins inserted
into a phospholipid matrix. A *phospholipid* resembles a triacylglycerol, but
only two fatty acids are present; the third glycerol —OH is bonded to a
phosphate group that is bonded in turn to an alcohol. Phospholipids have
both hydrophobic and hydrophilic properties. They are hydrophobic at the
end with the hydrocarbon chains of fatty acids, and they are hydrophilic at
the end with the phosphate group and the alcohol (called the *polar head*).
Because of these properties, phospholipids, like soaps, tend to aggregate
spontaneously in water to form micelles (see Section 27.3) or to form an
extensive sheet two molecules in thickness, called a **phospholipid bilayer.**
The interior of the lipid bilayer is hydrophobic, consisting, as shown in
Figure 27.10, of the hydrocarbon chains of the fatty acids of phospholipids.
It presents an effective barrier to charged or polar substances, which cannot
spontaneously enter or pass through a hydrophobic environment. Usually,
most substances found in living organisms do not readily pass through membranes because they are charged, polar, and/or relatively large in size. The
lipid bilayer does, however, provide a supporting framework for proteins that
catalyze transport of particular substances across the membrane.

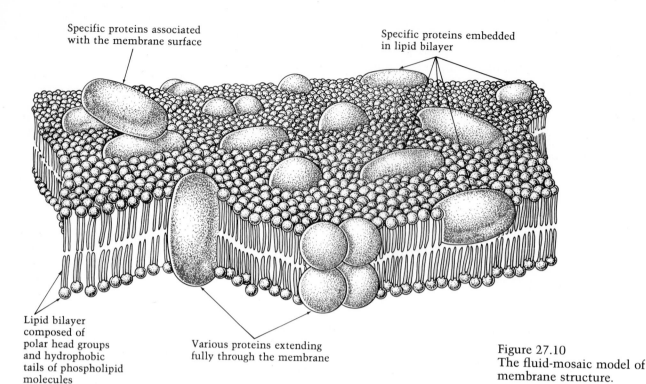

Specific proteins associated with the membrane surface

Specific proteins embedded in lipid bilayer

Lipid bilayer composed of polar head groups and hydrophobic tails of phospholipid molecules

Various proteins extending fully through the membrane

Figure 27.10
The fluid-mosaic model of membrane structure.

A Checklist for Review

Important Terms

metabolism (27.1)
enzymes (27.1)
coupling (27.2)
macromolecules (27.3)
conformations (27.3)
denaturation (27.3)
hydrophobic interaction (27.3)
micelles (27.3)
proteins (27.4)
amino acids (27.4)
zwitterion (27.4)
D- and L-isomers (27.4)
side chain (27.4)
peptide (amide) bonds (27.4)
polypeptide (27.4)
primary structure (27.4)
disulfide cross link (27.4)
fibrous proteins (27.4)

secondary structures (27.4)
globular proteins (27.4)
tertiary structure (27.4)
active sites (27.4)
substrates (27.4)
enzyme–substrate complex (27.4)
specificity (27.4)
carbohydrates (27.5)
monosaccharides (27.5)
oligosaccharides (27.5)
polysaccharides (27.5)
hemiacetals (27.5)
DNA (deoxyribonucleic acid) (27.6)
RNA (ribonucleic acid) (27.6)
nucleotides (27.6)
ribonucleotides (27.6)
deoxyribonucleotides (27.6)
nucleoside (27.6)

polynucleotide (27.6)
nucleic acids (27.6)
complementary bases (27.6)
base pairing (27.6)
double helix (27.6)
genetic code (27.6)
codon (27.6)
gene (27.6)
ribosomes (27.6)
ribosomal RNA (27.6)
messenger RNA (27.6)
termination codons (27.6)
transfer RNA (27.6)
anticodon (27.6)
ATP (27.6)
lipids (27.7)
triacylglycerols (27.7)
phospholipid bilayer (27.7)

Summary of Facts and Concepts

The cell is the smallest organizational unit possessing all the attributes of life. Cells are constructed of *proteins, carbohydrates, nucleic acids,* and *lipids.*

Living organisms transform matter and energy in sequences of *enzyme-*catalyzed reactions called *metabolic pathways,* which together comprise *metabolism.* Life requires input of free energy. Most reactions responsible for the highly organized living system have positive free eneries and must occur *coupled* to reactions releasing free energy.

Macromolecules—proteins, polysaccharides, and nucleic acids—are composed of simple building-block units and form unique *conformations* stabilized by hydrogen bonds, ionic bonds, and *hydrophobic interaction. Denaturation* is loss of normal conformation and associated loss of biological function.

Proteins are polymers of *α-amino acids* linked by *peptide bonds.* The 20 common amino acids of protein have different *side chains;* these may be *polar* or *nonpolar.* The unique *primary structure* of a polypeptide (the amino-acid sequence) is responsible for spontaneous folding into a unique conformation maintained by noncovalent bonds between side chains and *disulfide cross links.* The conformation produces unique *surface features* responsible for the protein's function. An enzyme has an *active site* where a *substrate* binds and catalysis takes place. The size, shape, and polarity of the active site determine *specificity.*

Carbohydrates include *monosaccharides, oligosaccharides,* and *polysaccharides.* A monosaccharide is a water-soluble polyhydroxyl carbonyl compound of 3–9 carbons. Straight-chain monosaccharides exist in equilibrium with cyclic *hemiacetals* having α and β isomers. D-glucose, D-ribose, 2-deoxy-D-ribose, D-fructose, and their derivatives are the most common monosaccharides. Oligosaccharides contain 2–10 monosaccharides; polysaccharides are longer polymers. Monosaccharides and oligosaccharides are water soluble; polysaccharides are not. Structural polysaccharides include cellulose and chitin, and energy-storage polysaccharides include amylose, amylopectin and glycogen.

Nucleic acids are polymers of *ribo-* or *deoxyribonucleotides.* Nucleotides are base-sugar-phosphate compounds. Conformations of nucleic acids are the result of hydrogen bonding, called base pairing, between *complementary bases. DNA* is the genetic material of chromosomes. DNA consists of two complementary polynucleotides coiled into a *double helix;* each has all the information necessary to direct synthesis of a new complementary strand. The *genetic code* is the relationship between nucleotide sequence and protein amino-acid sequence. Sixty-four *codons* (3-nucleotide sequences) are the basis of the code. *Transcription* is the synthesis of a *messenger RNA* molecule, which represents a copy of the information in a DNA *gene.* The messenger RNA binds to a *ribosome,* where *translation* of the message into an amino-acid sequence occurs. *Transfer RNA* molecules attach to amino acids and bind to the messenger RNA by *codon-anticodon* pairing, bringing each amino acid in turn into the position specified by the codon sequence.

Lipids are classified on the basis of their solubility in organic solvents and poor solubility in water. Fats and oils are *triacylglycerols,* which are energy-storage lipids. *Phospholipids* are structural lipids and are the main components of the *phospholipid bilayer* of membranes. Membranes also contain proteins inserted into or bound to the surfaces of the bilayer to transport the water-soluble materials into the cell and to catalyze many other important reactions.

Operational Skills

1. Given the structural formula of an amino acid with a nonpolar side chain, draw the zwitterion (Example 27.1).

2. Given the structural formulas of two α-amino acids, draw the structure of the two dipeptides they could form (Example 27.2).

3. Given the structural formula of an amino acid, decide whether its side chain is polar or nonpolar (Example 27.3).

4. Given the structural formulas of amino acids with polar side chains, show how two amino-acid side chains might form a hydrogen bond between them (Example 27.4).

5. Given the number and names of amino acids composing a peptide, write the different possible sequences in the peptide using abbreviations for the amino acids (Example 27.5).

6. Given the straight-chain structural formula of a 6-carbon monosaccharide, write the reactions for the equilibria existing between that form and the cyclic forms; or given the cyclic form of the sugar, draw the straight-chain form (Example 27.6).

7. Given the structural formulas for the cyclic forms of two 6-carbon monosaccharides, diagram the reaction to give a disaccharide that could form between them, linked from carbon 1 of the monosaccharide to carbon 4 of the other (Example 27.7).

8. Given a pentose, an organic base, and a phosphate group, draw the structural formula of the nucleotide they could form and name it (Example 27.8).

9. Given a DNA sequence denoted by the bases, write a complementary sequence (Example 27.9).

10. Given a polypeptide of a specified number of amino-acid units, determine the number of nucleotides in the gene that codes for the polypeptide. Or, given a gene with a specified number of nucleotides, determine the number of amino acids in the polypeptide coded for by that gene (Example 27.10).

11. Given a messenger-RNA sequence denoted by the bases, write the amino-acid sequence that would result. Or, given the amino-acid sequence, write the messenger-RNA sequence (Example 27.11). See Table 27.2.

12. Given the structures of three fatty acids, draw the structural formula for the triacylglycerol they would give. Or, given the structural formula of a triacylglycerol, identify each fatty-acid unit (Example 27.12). See Table 27.3.

Review Questions

27.1 Most reactions concerned with formation and maintenance of biological systems have positive free-energy changes. How are such reactions caused to proceed?

27.2 Describe the concept that simplifies the study of the structure of biological macromolecules.

27.3 Explain why soap molecules spontaneously form micelles in water.

27.4 Describe the primary structure of protein. What makes one protein different from another protein of the same size? What is the basis of the unique conformation of a protein?

27.5 Distinguish between secondary and tertiary structures of protein.

27.6 Briefly discuss protein denaturation and the factors that can cause it.

27.7 What is an enzyme? What physical features does this molecule have that explain enzyme specificity? Describe the lock-and-key model of catalysis.

27.8 Define the terms *monosaccharide*, *oligosaccharide*, and *polysaccharide*. Name the four most important monosaccharides and their biological functions.

27.9 What is the difference in structure of cellulose and amylose?

27.10 Relate the structure of D-glucose to its solubility in the blood.

27.11 What are the structural forms of D-glucose that are present in the blood?

27.12 Name the complementary base pairs. Describe the DNA double helix.

27.13 How do ribonucleotides and deoxyribonucleotides differ in structure? Do they form polymers in the same way?

27.14 Explain the nature of the genetic code.

27.15 What is the primary chemical role of the genetic code? In other words, how does a cell or species member become what it is?

27.16 Define *codon* and *anticodon*. How do they interact?

27.17 Show mathematically why there are 64 possible triplet codons.

27.18 Outline how the genetic message in a gene is translated to produce a polypeptide. Include the roles of the gene, messenger RNA, ribosomes, and transfer RNA.

27.19 Of what importance are nucleotides in addition to being the building blocks of the nucleic acids?

27.20 Distinguish between a fat and an oil.

27.21 Describe the structure of a triacylglycerol.

27.22 Outline the structure of the biological membrane. Relate the structure of the membrane and its lipid components to its tendency to allow or bar passage to hydrophobic or hydrophilic molecules.

Problems

Energy and Metabolism

27.23 What is the free-energy change associated with a pair of coupled reactions, if the ΔG of one is $+7.32$ kcal/mol and the ΔG of the second is -10.07 kcal/mol?

27.25 A metabolic pathway catalyzes the breakdown of fat into CO_2 and H_2O. Would you expect such a pathway to release free energy or to require the input of free energy? Explain.

27.24 The overall free-energy change for a pair of coupled reactions is -1.5 kcal/mol. What is the ΔG of the energy-requiring reaction if the energy-releasing reaction has $\Delta G = -5.2$ kcal/mol?

27.26 Some 12 enzymes work together to catalyze the formation of glucose (a 6-carbon sugar) from simpler raw materials (3- and 4-carbon compounds) in the animal liver. Would you expect the overall change to release free energy or to require free energy? Explain.

Amino Acids and Primary Structure

27.27 Alanine has the structure

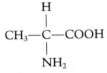

Draw the zwitterion that would exist at neutral pH.

27.29 Write the structural formula of a dipeptide formed from the reaction of L-alanine and L-histidine. How many dipeptides are possible?

27.31 Which of the following amino acids have polar side chains?

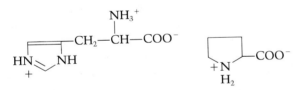

27.33 Draw the following amino acids with hydrogen-bonding between their side chains:

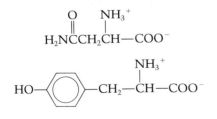

27.28 Valine has the structure

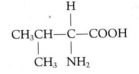

Draw the zwitterion that would exist at neutral pH.

27.30 Write the structural formulas of two tripeptides formed from the reaction of L-tryptophan, L-glutamate, and L-tyrosine. How many tripeptides are possible?

27.32 Which of the following amino acids have nonpolar side chains?

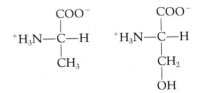

27.34 Which two of these three amino acids have side chains that could undergo hydrophobic interaction?

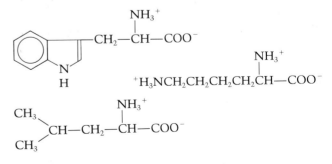

27.35 How many unique sequences of a pentapeptide of five different amino acids are possible?

27.36 Write the structural formula of any *one* of the possible pentapeptides having the composition ala, trp, ser, glu, phe.

Monosaccharides, Oligosaccharides, and Polysaccharides

27.37 Write the reactions for the equilibria existing in a solution of D-galactose:

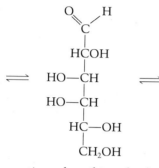

27.38 Write the reactions for the equilibria existing in a solution of D-sorbose:

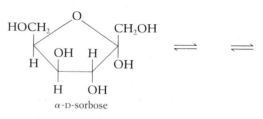

α-D-sorbose

27.39 Write reactions describing the formation of two different disaccharides involving α-D-glucopyranose and β-D-galactopyranose linked between carbon 1 of one molecule and carbon 4 of the other. Is one of your products lactose?

27.40 Write a reaction to show formation of a trisaccharide with two α-D-glucopyranose molecules and one β-D-galactopyranose, linked carbon 1 of one molecule to carbon 4 of another. How many trisaccharides are possible?

27.41 One of the products of the digestion of starch is maltose. Write a balanced equation that illustrates the hydrolysis of an oligosaccharide containing six glucose units to maltose.

27.42 When celluose is digested by the microorganisms in the digestive systems of cattle and other ruminants, it is broken down first to *cellobiose*, a disaccharide of two β-D-glucose units linked carbon 1 to carbon 4. Write a balanced equation to illustrate hydrolysis of a cellulose containing 100 glucose units to cellobiose.

Nucleotides and Polynucleotides

27.43 If adenine, thymine, guanine, and cytosine were all analyzed separately in a sample of DNA, what molar ratios of A:T and G:C would you expect to find?

27.44 If a sample of DNA isolated from a microorganism culture were analyzed and found to contain 1.5 moles of cytosine nucleotides and 0.5 moles of adenosine nucleotides, what would be the amounts of guanine and thymine nucleotides in the sample?

27.45 Write a structural formula for the nucleotide cytidine-5'-monophosphate.

27.46 Write a structural formula for the nucleotide deoxyadenosine-5'-monophosphate.

27.47 Write a structural formula for a dinucleotide consisting of GMP and UMP.

27.48 Write a structural formula for a dinucleotide consisting of dTMP and dCMP.

27.49 How many hydrogen bonds link a guanine–cytosine base pair? An adenine–uracil base pair? Would you expect any difference between the strength of guanine–cytosine bonding and adenine–uracil bonding? Explain.

27.50 How many hydrogen bonds link an adenine–thymine base pair? Would there be any difference in strength between adenine–thymine bonding and adenine–uracil bonding? Between adenine–thymine and cytosine–guanine bonding? Explain.

27.51 Write the DNA sequence complementary to the sequence ACTGACGCAATTGACCGC.

27.52 Write the sequence of an RNA strand that could base pair with the sequence UAGCUUUACGAA-GUGGA.

DNA, RNA, and Protein Synthesis

27.53 If the codon were two nucleotides, how many codons would be possible? Would this be a workable code for the purpose of protein synthesis?

27.55 Nucleic acids can be denatured by heat as proteins can. What bonds are broken when a DNA molecule is denatured? Would DNA of greater percent composition of guanine and cytosine denature more or less readily than DNA of greater percent composition of adenine and thymine?

27.57 A protein containing 400 amino acids would be coded for by a gene of how many nucleotide units?

27.59 If a protein contained 400 amino acids, but consisted of 4 identical polypeptides each 100 amino acids long, how many nucleotide units would comprise the gene for this protein?

27.61 Consulting Table 27.2, write the amino-acid sequence resulting from left-to-right translation of the mRNA sequence

GGAUCCCGCUUUGGGCUGAAAUAG

27.63 List the codons to which the following anticodons would form base pairs:

Anticodon: GAC UGA GGG ACC

Codon:

27.65 Give one of the nucleotide sequences that would translate to

leu–ala–val–glu–asp–cys–met–trp–lys

27.54 If the codon were four nucleotides, how many codons would be possible? Would this be workable as an amino-acid code?

27.56 Would RNA of a certain percent composition of guanine and cytosine denature more or less readily than RNA with a lower percent of guanine and cytosine? Why?

27.58 A gene of 3000 nucleotide units could code for a protein of how many amino acids?

27.60 If a gene containing 360 nucleotides coded for a polypeptide, and 6 of those polypeptides aggregated to form an enzyme protein, how many amino acids would that enzyme consist of?

27.62 Write the amino-acid sequence obtained from left-to-right translation of the mRNA sequence

AUUGGCGCGAGAUCGAAUGAGCCCAGU

See Table 27.2.

27.64 List the anticodons to which the following codons would form base pairs:

Codon: UUG CAC ACU GAA

Anticodon:

27.66 Write one nucleotide sequence that would translate to

tyr–ile–pro–his–leu–his–thr–ser–phe–met

Lipids and Membranes

27.67 Write a reaction for the formation of a triacylglycerol from a mole of glycerol and three moles of stearic acid (see Table 27.3).

27.69 Write a reaction for the saponification (base hydrolysis) of the product of the reaction in Problem 27.67.

27.71 Which of the following molecules or ions would you expect to diffuse slowly across a cell membrane? Which would diffuse more rapidly?

27.68 Write a reaction for the formation of a triacylglycerol from a mole of glycerol, two moles of oleic acid, and one mole of myristic acid (see Table 27.3).

27.70 Write a reaction for the base hydrolysis of the product of the reaction in Problem 27.68.

27.72 In general, would ionic or neutral compounds more readily diffuse across cell membranes?

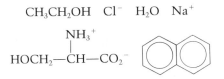

27.73 Directions on the labels of pesticides and herbicides always caution against allowing contact of the product with skin. These products contain complex aromatic hydrocarbons as active ingredients. Explain why contact between your skin and poisons that have hydrocarbon structures is to be particularly avoided.

27.74 Which of the following nutrients would you expect to enter cells more readily (assuming that passive diffusion is the only means of entering the cell)? Explain.

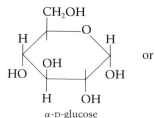

α-D-glucose

or

$$CH_3CH_2CH_2CH_2CH_2CH_2CH_2CH_2CH_2CH_2CH_2CH_2CH_2CO_2H$$

myristic acid

Additional Problems

27.75 A portion of a DNA gene has the nucleotide sequence AGTCGACCGTTAAT. Write a complementary sequence.

27.76 An RNA transcribed from a DNA sequence has the nucleotide sequence UAGCACGGGACUUGG. Write the complementary DNA sequence.

27.77 A peptide contains six amino acids: L-arginine, L-proline, L-glutamate, L-glycine, L-asparagine, and L-glutamine. How many different peptides of this amino-acid composition are possible?

27.78 Using abbreviations for the amino acids, give three possible sequences of the peptide described in Problem 27.77.

27.79 Name each of the three major constituents of the nucleotide adenosine-5′-monophosphate.

27.80 Write the structural formula of deoxyguanosine-5′-monophosphate.

27.81 Which of the following amino acids has a nonpolar side chain?

$$\underset{CH_3SCH_2CH_2CH-CO_2^-}{\overset{\overset{NH_3^+}{|}}{}}\qquad\underset{HSCH_2CH-CO_2^-}{\overset{\overset{NH_3^+}{|}}{}}$$

27.82 Which of the following amino acids has a polar side chain?

$$\underset{^+H_3NCH_2CH_2CH_2CH_2CH-CO_2^-}{\overset{\overset{NH_3^+}{|}}{}}$$

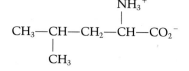

27.83 Write the structural formula of one of the two dipeptides that could be formed from the amino acids in Problem 27.81.

27.84 Write the structural formula of one of the two peptides that could be formed by the amino acids in Problem 27.82.

*27.85** Draw the structure of the ionic form expected for L-valine at high pH.

*27.86** Draw the structure of the ionic form expected for L-valine at low pH. Write an equation for the acid dissociation reaction that occurs when the pH is raised to neutral.

27.87 A common triacylglycerol is

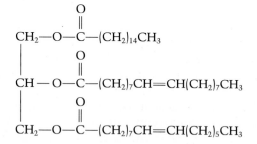

Write structural formulas for each molecule from which the triacylglycerol is constructed.

27.88 Write a balanced equation to describe the saponification of the triacylglycerol shown in Problem 27.87.

Appendix A.
Mathematical Skills

Only a few basic mathematical skills are required for the study of general chemistry. But in order to concentrate your attention on the concepts of chemistry, you will find it necessary to have a firm grasp on these basic mathematical skills. In this appendix, we will review scientific (or exponential) notation, logarithms, simple algebraic operations, quadratic equations, and straight-line graphs.

A.1 Scientific (Exponential) Notation

In chemistry, we frequently encounter very large and very small numbers. Thus, the number of molecules in a liter of air at 20°C and normal barometric pressure is 25,000,000,000,000,000,000,000. Similarly, the distance between two hydrogen atoms in a hydrogen molecule is 0.000,000,000,074 meters. In these forms, such numbers are both inconvenient to write and difficult to read. For this reason, we normally express them in scientific, or exponential, notation. Scientific calculators also use this notation.

In scientific notation, a number is written in the form $A \times 10^n$. A is a number greater than or equal to 1 and less than 10, and the exponent n (the nth power of ten) is a positive or negative integer. For example, 4853 would be written in scientific notation as 4.853×10^3, which is 4.853 multiplied by three factors of 10:

$$4.853 \times 10^3 = 4.853 \times 10 \times 10 \times 10 = 4853$$

The number 0.0568 would be written in scientific notation as 5.68×10^{-2}, which is 5.68 divided by two factors of 10:

$$5.68 \times 10^{-2} = \frac{5.68}{10 \times 10} = 0.0568$$

Any number can be conveniently transformed to scientific notation by moving the decimal point in the number to obtain a number, A, greater than or equal to 1 and less than 10. If the decimal point is moved to the left, we multiply A by 10^n, where n equals the number of places moved. If the decimal point is moved to the right, we multiply A by 10^{-n}. To transform a number written in scientific notation to one in usual form, the process is reversed. If the exponent is positive, the decimal point is shifted right. If the exponent is negative, the decimal point is shifted left.

Example 1

Express the following numbers in scientific notation:
(a) 843.4   (b) 0.00421   (c) 1.54

Solution

Shift the decimal point to get a number between 1 and 10; count the number of positions shifted.
(a) $8\underset{\frown}{4\,3}.4 = 8.434 \times 10^2$
(b) $0.\underset{\frown}{0\,0\,4\,2}1 = 4.21 \times 10^{-3}$
(c) leave as is or write 1.54×10^0

Exercise 1

Express the following numbers in scientific notation:
(a) 4.38 (b) 4380 (c) 0.000483

Example 2

Convert the following numbers in scientific notation to usual form:
(a) 6.39×10^{-4} (b) 3.275×10^{2}

Solution

(a) $0.0006.39 \times 10^{-4} = 0.000639$
(b) $3.27.5 \times 10^{2} = 327.5$

Exercise 2

Convert the following numbers in scientific notation to usual form:
(a) 7.025×10^{3} (b) 8.97×10^{-4}

Addition and Subtraction

Before adding or subtracting two numbers written in scientific notation, it is necessary to express both to the same power of 10. After adding or subtracting, it may be necessary to shift the decimal point to express the result in scientific notation.

Example 3

Carry out the following arithmetic; give the result in scientific notation.

$$(9.42 \times 10^{-2}) + (7.6 \times 10^{-3})$$

Solution

We can shift the decimal point in either number in order to obtain both to the same power of 10. For example, to get both numbers to 10^{-2} we shift the decimal point one place to the left and add 1 to the exponent in the expression 7.6×10^{-3}.

$$7.6 \times 10^{-3} = 0.76 \times 10^{-2}$$

Now we can add the two numbers.

$$(9.42 \times 10^{-2}) + (0.76 \times 10^{-2}) = (9.42 + 0.76) \times 10^{-2}$$
$$= 10.18 \times 10^{-2}$$

Since 10.18 is not between 1 and 10, we shift the decimal point to express the final result in scientific notation.

$$10.18 \times 10^{-2} = 1.018 \times 10^{-1}$$

Exercise 3

Add the following and express the sum in scientific notation:

$$(3.142 \times 10^{-4}) + (2.8 \times 10^{-6})$$

Multiplication and Division

To multiply two numbers in scientific notation, first we multiply the two powers of 10 by adding their exponents. Then we multiply the remaining factors. Division is handled similarly. We first place both powers of 10 in the numerator (note that a power of 10 in the denominator can be placed in the numerator if the sign of the exponent is reversed). After multiplying the two powers of 10, we carry out the indicated division.

Example 4

Do the following arithmetic, and express the answers in scientific notation.

(a) $(6.3 \times 10^2) \times (2.4 \times 10^5)$ (b) $\dfrac{6.4 \times 10^2}{2.0 \times 10^5}$

(b) $\dfrac{6.4 \times 10^2}{2.0 \times 10^5} = \dfrac{6.4}{2.0} \times 10^2 \times 10^{-5} = \dfrac{6.4}{2.0} \times 10^{-3}$

$= 3.2 \times 10^{-3}$

Solution

(a) $(6.3 \times 10^2) \times (2.4 \times 10^5) = (6.3 \times 2.4) \times 10^7$
$= 15.12 \times 10^7$
$= 1.512 \times 10^8$

Exercise 4

Perform the following operations, expressing the answers in scientific notation:

(a) $(5.4 \times 10^{-7}) \times (1.8 \times 10^8)$ (b) $\dfrac{5.4 \times 10^{-7}}{6.0 \times 10^{-5}}$

Powers and Roots

A number $A \times 10^n$ raised to a power p is evaluated by raising A to the power p and multiplying the exponent in the power of 10 by p:

$$(A \times 10^n)^p = A^p \times 10^{n \times p}$$

We extract the rth root of a number $A \times 10^n$ by first moving the decimal point in A so that the exponent in the power of 10 is exactly divisible by r. Suppose this has been done, so that n in the number $A \times 10^n$ is exactly divisible by r. Then

$$\sqrt[r]{A \times 10^n} = \sqrt[r]{A} \times 10^{n/r}$$

Example 5

Simplify the following expressions:
(a) $(5.29 \times 10^2)^3$ (b) $\sqrt{2.31 \times 10^7}$

(b) $\sqrt{2.31 \times 10^7} = \sqrt{23.1 \times 10^6} = \sqrt{23.1} \times 10^{6/2}$
$= 4.81 \times 10^3$

Solution

(a) $(5.29 \times 10^2)^3 = (5.29)^3 \times 10^6 = 148 \times 10^6$
$= 1.48 \times 10^8$

Exercise 5

Obtain the values of the following, and express them in scientific notation:
(a) $(3.56 \times 10^3)^4$ (b) $\sqrt[3]{4.81 \times 10^2}$

Electronic Calculators

Scientific calculators will perform all of the arithmetic operations we have just described (as well as those discussed in the next section). The basic operations of addition, subtraction, multiplication, and division are usually similar and straightforward on most calculators. However, more variation exists in raising a number to a power and extracting a root. If you have the instructions, by all means read them. Otherwise, the following information may help.

Squares and square roots are usually obtained with special keys, perhaps labeled x^2 and $\sqrt{x}$. Thus, to obtain $(5.15)^2$, you enter 5.15 and press x^2. To obtain $\sqrt{5.15}$, you enter 5.15 and press $\sqrt{x}$ (or perhaps INV, for inverse, and x^2). Other powers and roots require a y^x (or a^x) key. The answer to Example 5(a), $(5.29 \times 10^2)^3$, would be obtained by a sequence of steps such as the following. Enter 5.29×10^2, press the y^x key, enter 3, and press the $=$ key.

The same sequence can be used to extract a root. Suppose we want $\sqrt[5]{2.18 \times 10^6}$. This is equivalent to $(2.18 \times 10^6)^{1/5}$ or $(2.18 \times 10^6)^{0.2}$. If the calculator has a $1/x$ key, the sequence would be as follows. Enter 2.18×10^6, press the y^x key, enter 5, press the $1/x$ key, then press the $=$ key. Some calculators have a $\sqrt[x]{y}$ key, so this can be used to extract the xth root of y, using a sequence of steps similar to that for y^x.

A.2 Logarithms

The *logarithm* to the *base* a of a number x, denoted $\log_a x$, is the exponent of the constant a needed to equal the number x. For example, suppose a is 10, and we would like the logarithm of 1000, that is, we would like the value of $\log_{10} 1000$. This is the exponent, y, of 10 such that 10^y equals 1000. The value of y is 3. Thus, $\log_{10} 1000 = 3$.

Common logarithms are logarithms in which the base is 10. The common logarithm of a number x is often denoted simply as $\log x$. It is easy to see how to obtain the common logarithms of 10, 100, 1000, and so forth. But logarithms are defined for all positive numbers, not just the powers of 10. In general, the exponents or values of the logarithm will be decimal numbers. To understand the meaning of a decimal exponent, consider $10^{0.400}$. This is equivalent to $10^{400/1000} = 10^{2/5} = \sqrt[5]{10^2} = 2.51$. Therefore, $\log 2.51 = 0.400$. Any decimal exponent is essentially a fraction, p/r, so by evaluating the expressions $10^{p/r} = \sqrt[r]{10^p}$ one could construct a table of logarithms. In practice, power series or other methods are used. A table of common logarithms is given in Appendix B.

The following are fundamental properties of all logarithms.

$$\log_a 1 = 0 \tag{1}$$

$$\log_a(A \times B) = \log_a A + \log_a B \tag{2}$$

$$\log_a \frac{A}{B} = \log_a A - \log_a B \tag{3}$$

$$\log_a A^p = p \log_a A \tag{4}$$

$$\log_a \sqrt[r]{A} = \frac{1}{r} \log_a A \tag{5}$$

These properties are very useful in working with logarithms.

Obtaining the Logarithm of a Number

Electronic calculators that evaluate logarithms are now available for about $20 or so. Their simplicity of operation makes them well worth the price. To obtain the logarithm of a number, you enter the number and press the LOG key. Obtaining the logarithm from a table is not difficult, however, and by working with the table, you may gain some understanding of logarithms.

Appendix B gives a table of common logarithms of numbers between 1.00 and 9.99. The first two digits of this number are given in the column at the extreme left. (In many tables the decimal point is omitted between these two digits.) The third digit is given in the top row of the table. To obtain the logarithm, we look at the position at which the row containing the first two digits intersects the column containing the third digit.

Suppose we want log 2.35. We find 2.3 in the extreme left column and follow along the row containing these digits. In this row, we read .3711 under 5 at the top of the table. (In many tables the decimal point for numbers in the body of the table is omitted but still understood to be at the left.)

The table in Appendix B lists logarithms of numbers between 1 and 10. To obtain the logarithm of a number outside this range, we use the properties of logarithms. We first express the number in scientific notation, $A \times 10^n$. Using property 2, given earlier, we can write

$$\log(A \times 10^n) = \log A + \log 10^n$$

From property 4 and the fact that $\log 10 = 1$, the last term becomes $n \log 10$ or n. Therefore,

$$\log(A \times 10^n) = \log A + n$$

Example 6

Find the value of log 0.003720.

Solution

We write 0.003720 in scientific notation as 3.720×10^{-3}. Therefore,

$\log 0.003720 = \log(3.720 \times 10^{-3}) = \log 3.720 - 3$

From Appendix B, we find that $\log 3.720 = 0.5705$. Thus,

$$\log 3.720 - 3 = 0.5705 - 3 = -2.4295$$

Exercise 6

Find the values of   (a) log 0.00582   (b) log 689

Antilogarithm

The antilogarithm (abbreviated antilog) is the inverse of the common logarithm. Thus, antilog x is simply 10^x. If an electronic calculator has a 10^x key, you obtain the antilogarithm of a number by entering the number and pressing the 10^x key. (It may be necessary to press an inverse key before pressing a 10^x / LOG key.) If the calculator has a y^x (or a^x) key, you enter 10, press y^x, enter x, then press the = key.

We use the name antilogarithm, rather than simply 10^x, because the values are obtained from a table of logarithms by reversing the process used to obtain a logarithm. Suppose we want to evaluate antilog 0.619. We look in the body of the table in Appendix B for 0.619 (remembering that in some tables the decimal point is understood to be at the left of each entry). When we find the closest entry (.6191), we read the digits at the extreme left (4.1) and the digit at the top of the table (6). Antilog 0.619 equals 4.16.

Note that a table of logarithms normally only gives the antilogarithm of a positive, decimal number. To obtain the antilogarithm of other numbers, we use the property of exponents that $10^{(a+b)} = 10^a \times 10^b$. Suppose we want antilog 2.480, that is, $10^{(2+0.480)}$. We can write this as $10^2 \times 10^{0.480}$. Or,

$$\text{antilog } 2.480 = \text{antilog } 2 \times \text{antilog } 0.480$$

Looking in Appendix B, we find that antilog 0.480 is 3.03. Therefore,

$$\text{antilog } 2.480 = 3.03 \times \text{antilog } 2 = 3.03 \times 10^2$$

The antilogarithm of a negative number is obtained by expressing the negative number as a negative integer plus a positive decimal part. This is illustrated in Example 7.

Example 7

What is the value of antilog (-4.295)?

Solution

The number -4.295 is written first as an integer and a decimal part.

$$\text{antilog } (-4.295) = \text{antilog } (-4 - 0.295)$$

We then add 1 to the negative decimal part and subtract 1 from the negative integer part.

$$\text{antilog } (-4 - 0.295) = \text{antilog } (-5 + 0.705)$$
$$= \text{antilog } (-5) \times \text{antilog } 0.705$$

From Appendix B, we find that antilog 0.705 equals 5.07. Therefore,

$$\text{antilog } (-4.295) = 5.07 \times 10^{-5}$$

Exercise 7

Evaluate   (a) antilog 5.728   (b) antilog (-5.728)

Natural Logarithms

The mathematical constant $e = 2.71828\ldots$, like π, occurs in many scientific and engineering problems. It is frequently seen in the natural exponential function $y = e^x$. The inverse function is called the *natural logarithm*, $x = \ln y$, where $\ln y$ is simplified notation for $\log_e y$.

It is possible to express the natural logarithm in terms of the logarithm to the base 10, or the common logarithm. Let us take the common logarithm of both sides of the equation $y = e^x$. Using property 4, given earlier, we get

$$\log y = \log e^x = x \log e$$

Since x equals $\ln y$, and $\log e$ is 0.4343, we can write this equation as

$$\log y = \ln y \log e = 0.4343 \ln y$$

Finally, solving for ln y

$$\ln y = \frac{1}{0.4343} \log y = 2.303 \log y$$

A.3 Algebraic Operations and Graphing

Often we are given an algebraic formula that we would like to rearrange in order to solve for a particular quantity. As an example, suppose we would like to solve the following equation for V.

$$PV = nRT$$

We can eliminate P from the left-hand side by dividing by P. But to maintain the equality, we must perform the same operation on both sides of the equation:

$$\frac{PV}{P} = \frac{nRT}{P}$$

Or,

$$V = \frac{nRT}{P}$$

Quadratic Formula

A quadratic equation is one involving only powers of x in which the highest power is two. The general form of the equation can be written

$$ax^2 + bx + c = 0$$

where a, b, and c are constants. For given values of these constants, only certain values of x are possible (in general, there will be two values). These values of x are said to be the solutions of the equation.

These solutions are given by the *quadratic formula:*

$$x = \frac{-b \pm \sqrt{b^2 - 4ac}}{2a}$$

In this formula, the symbol $\pm$ means that there are two possible values of x, one obtained by taking the positive sign, the other by taking the negative sign.

Example 8

Obtain the solutions of the following quadratic equation:

$$2.00x^2 - 1.72x - 2.86 = 0$$

Solution

Using the quadratic formula, we substitute $a = 2.00$, $b = -1.72$, and $c = -2.86$. We get

$$x = \frac{1.72 \pm \sqrt{(-1.72)^2 - 4 \times 2.00 \times (-2.86)}}{2 \times 2.00}$$

$$= \frac{1.72 \pm 5.08}{4.00} = -0.84 \text{ and } +1.70$$

Although mathematically there are two solutions, in any real problem one may not be allowed. For example, if the solution is some physical quantity that can have only positive values, a negative solution must be rejected.

Exercise 8

Find the positive solution (or solutions) to the following equation:

$$1.80x^2 + 0.850x - 9.50 = 0.$$

The Straight-Line Graph

A graph is a visual means of representing a mathematical relationship or physical data. Consider the following data in which values of y from some experiment are given for four values of x.

x	y
1	-1
2	1
3	3
4	5

By plotting these x, y points on a graph (Figure A.1), we can see that they fall on a straight line. This suggests (but does not prove) that other points from this type of experiment might fall on the same line. It would be useful to have the mathematical equation for this line.

The general form of a straight line is

$$y = ax + b$$

The constant a is called the *slope* of the straight line. It is obtained by dividing the vertical distance between any two points on the line by the

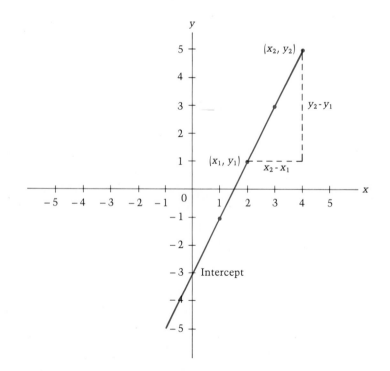

Figure A.1
A straight-line plot of some data.

horizontal distance. If the two points are (x_1, y_1) and (x_2, y_2), the slope is given by the following formula.

$$\text{slope} = \frac{y_2 - y_1}{x_2 - x_1}$$

Suppose we choose the points $(2, 1)$ and $(4, 5)$ from our data. Then

$$\text{slope} = \frac{5 - 1}{4 - 2} = 2$$

Thus, $a = 2$ for the straight line in Figure 1.

The constant b is called the *intercept*. It is the value of y at $x = 0$. From Figure 1, we see that the intercept is -3. Therefore, $b = -3$. Hence, the equation of the straight line is

$$y = 2x - 3$$

Appendix B.
Common Logarithms to Four Places

	0	1	2	3	4	5	6	7	8	9
1.0	.0000	.0043	.0086	.0128	.0170	.0212	.0253	.0294	.0334	.0374
1.1	.0414	.0453	.0492	.0531	.0569	.0607	.0645	.0682	.0719	.0755
1.2	.0792	.0828	.0864	.0899	.0934	.0969	.1004	.1038	.1072	.1106
1.3	.1139	.1173	.1206	.1239	.1271	.1303	.1335	.1367	.1399	.1430
1.4	.1461	.1492	.1523	.1553	.1584	.1614	.1644	.1673	.1703	.1732
1.5	.1761	.1790	.1818	.1847	.1875	.1903	.1931	.1959	.1987	.2014
1.6	.2041	.2068	.2095	.2122	.2148	.2175	.2201	.2227	.2253	.2279
1.7	.2304	.2330	.2355	.2380	.2405	.2430	.2455	.2480	.2504	.2529
1.8	.2553	.2577	.2601	.2625	.2648	.2672	.2695	.2718	.2742	.2765
1.9	.2788	.2810	.2833	.2856	.2878	.2900	.2923	.2945	.2967	.2989
2.0	.3010	.3032	.3054	.3075	.3096	.3118	.3139	.3160	.3181	.3201
2.1	.3222	.3243	.3263	.3284	.3304	.3324	.3345	.3365	.3385	.3404
2.2	.3424	.3444	.3464	.3483	.3502	.3522	.3541	.3560	.3579	.3598
2.3	.3617	.3636	.3655	.3674	.3692	.3711	.3729	.3747	.3766	.3784
2.4	.3802	.3820	.3838	.3856	.3874	.3892	.3909	.3927	.3945	.3962
2.5	.3979	.3997	.4014	.4031	.4048	.4065	.4082	.4099	.4116	.4133
2.6	.4150	.4166	.4183	.4200	.4216	.4232	.4249	.4265	.4281	.4298
2.7	.4314	.4330	.4346	.4362	.4378	.4393	.4409	.4425	.4440	.4456
2.8	.4472	.4487	.4502	.4518	.4533	.4548	.4564	.4579	.4594	.4609
2.9	.4624	.4639	.4654	.4669	.4683	.4698	.4713	.4728	.4742	.4757
3.0	.4771	.4786	.4800	.4814	.4829	.4843	.4857	.4871	.4886	.4900
3.1	.4914	.4928	.4942	.4955	.4969	.4983	.4997	.5011	.5024	.5038
3.2	.5051	.5065	.5079	.5092	.5105	.5119	.5132	.5145	.5159	.5172
3.3	.5185	.5198	.5211	.5224	.5237	.5250	.5263	.5276	.5289	.5302
3.4	.5315	.5328	.5340	.5353	.5366	.5378	.5391	.5403	.5416	.5428
3.5	.5441	.5453	.5465	.5478	.5490	.5502	.5514	.5527	.5539	.5551
3.6	.5563	.5575	.5587	.5599	.5611	.5623	.5635	.5647	.5658	.5670
3.7	.5682	.5694	.5705	.5717	.5729	.5740	.5752	.5763	.5775	.5786
3.8	.5798	.5809	.5821	.5832	.5843	.5855	.5866	.5877	.5888	.5899
3.9	.5911	.5922	.5933	.5944	.5955	.5966	.5977	.5988	.5999	.6010
4.0	.6021	.6031	.6042	.6053	.6064	.6075	.6085	.6096	.6107	.6117
4.1	.6128	.6138	.6149	.6160	.6170	.6180	.6191	.6201	.6212	.6222
4.2	.6232	.6243	.6253	.6263	.6274	.6284	.6294	.6304	.6314	.6325
4.3	.6335	.6345	.6355	.6365	.6375	.6385	.6395	.6405	.6415	.6425
4.4	.6435	.6444	.6454	.6464	.6474	.6484	.6493	.6503	.6513	.6522
4.5	.6532	.6542	.6551	.6561	.6571	.6580	.6590	.6599	.6609	.6618
4.6	.6628	.6637	.6646	.6656	.6665	.6675	.6684	.6693	.6702	.6712
4.7	.6721	.6730	.6739	.6749	.6758	.6767	.6776	.6785	.6794	.6803
4.8	.6812	.6821	.6830	.6839	.6848	.6857	.6866	.6875	.6884	.6893
4.9	.6902	.6911	.6920	.6928	.6937	.6946	.6955	.6964	.6972	.6981
5.0	.6990	.6998	.7007	.7016	.7024	.7033	.7042	.7050	.7059	.7067
5.1	.7076	.7084	.7093	.7101	.7110	.7118	.7126	.7135	.7143	.7152
5.2	.7160	.7168	.7177	.7185	.7193	.7202	.7210	.7218	.7226	.7235
5.3	.7243	.7251	.7259	.7267	.7275	.7284	.7292	.7300	.7308	.7316
5.4	.7324	.7332	.7340	.7348	.7356	.7364	.7372	.7380	.7388	.7396
	0	1	2	3	4	5	6	7	8	9

	0	1	2	3	4	5	6	7	8	9
5.5	.7404	.7412	.7419	.7427	.7435	.7443	.7451	.7459	.7466	.7474
5.6	.7482	.7490	.7497	.7505	.7513	.7520	.7528	.7536	.7543	.7551
5.7	.7559	.7566	.7574	.7582	.7589	.7597	.7604	.7612	.7619	.7627
5.8	.7634	.7642	.7649	.7657	.7664	.7672	.7679	.7686	.7694	.7701
5.9	.7709	.7716	.7723	.7731	.7738	.7745	.7752	.7760	.7767	.7774
6.0	.7782	.7789	.7796	.7803	.7810	.7818	.7825	.7832	.7839	.7846
6.1	.7853	.7860	.7868	.7875	.7882	.7889	.7896	.7903	.7910	.7917
6.2	.7924	.7931	.7938	.7945	.7952	.7959	.7966	.7973	.7980	.7987
6.3	.7993	.8000	.8007	.8014	.8021	.8028	.8035	.8041	.8048	.8055
6.4	.8062	.8069	.8075	.8082	.8089	.8096	.8102	.8109	.8116	.8122
6.5	.8129	.8136	.8142	.8149	.8156	.8162	.8169	.8176	.8182	.8189
6.6	.8195	.8202	.8209	.8215	.8222	.8228	.8235	.8241	.8248	.8254
6.7	.8261	.8267	.8274	.8280	.8287	.8293	.8299	.8306	.8312	.8319
6.8	.8325	.8331	.8338	.8344	.8351	.8357	.8363	.8370	.8376	.8382
6.9	.8388	.8395	.8401	.8407	.8414	.8420	.8426	.8432	.8439	.8445
7.0	.8451	.8457	.8463	.8470	.8476	.8482	.8488	.8494	.8500	.8506
7.1	.8513	.8519	.8525	.8531	.8537	.8543	.8549	.8555	.8561	.8567
7.2	.8573	.8579	.8585	.8591	.8597	.8603	.8609	.8615	.8621	.8627
7.3	.8633	.8639	.8645	.8651	.8657	.8663	.8669	.8675	.8681	.8686
7.4	.8692	.8698	.8704	.8710	.8716	.8722	.8727	.8733	.8739	.8745
7.5	.8751	.8756	.8762	.8768	.8774	.8779	.8785	.8791	.8797	.8802
7.6	.8808	.8814	.8820	.8825	.8831	.8837	.8842	.8848	.8854	.8859
7.7	.8865	.8871	.8876	.8882	.8887	.8893	.8899	.8904	.8910	.8915
7.8	.8921	.8927	.8932	.8938	.8943	.8949	.8954	.8960	.8965	.8971
7.9	.8976	.8982	.8987	.8993	.8998	.9004	.9009	.9015	.9020	.9025
8.0	.9031	.9036	.9042	.9047	.9053	.9058	.9063	.9069	.9074	.9079
8.1	.9085	.9090	.9096	.9101	.9106	.9112	.9117	.9122	.9128	.9133
8.2	.9138	.9143	.9149	.9154	.9159	.9165	.9170	.9175	.9180	.9186
8.3	.9191	.9196	.9201	.9206	.9212	.9217	.9222	.9227	.9232	.9238
8.4	.9243	.9248	.9253	.9258	.9263	.9269	.9274	.9279	.9284	.9289
8.5	.9294	.9299	.9304	.9309	.9315	.9320	.9325	.9330	.9335	.9340
8.6	.9345	.9350	.9355	.9360	.9365	.9370	.9375	.9380	.9385	.9390
8.7	.9395	.9400	.9405	.9410	.9415	.9420	.9425	.9430	.9435	.9440
8.8	.9445	.9450	.9455	.9460	.9465	.9469	.9474	.9479	.9484	.9489
8.9	.9494	.9499	.9504	.9509	.9513	.9518	.9523	.9528	.9533	.9538
9.0	.9542	.9547	.9552	.9557	.9562	.9566	.9571	.9576	.9581	.9586
9.1	.9590	.9595	.9600	.9605	.9609	.9614	.9619	.9624	.9628	.9633
9.2	.9638	.9643	.9647	.9652	.9657	.9661	.9666	.9671	.9675	.9680
9.3	.9685	.9689	.9694	.9699	.9703	.9708	.9713	.9717	.9722	.9727
9.4	.9731	.9736	.9741	.9745	.9750	.9754	.9759	.9763	.9768	.9773
9.5	.9777	.9782	.9786	.9791	.9795	.9800	.9805	.9809	.9814	.9818
9.6	.9823	.9827	.9832	.9836	.9841	.9845	.9850	.9854	.9859	.9863
9.7	.9868	.9872	.9877	.9881	.9886	.9890	.9894	.9899	.9903	.9908
9.8	.9912	.9917	.9921	.9926	.9930	.9934	.9939	.9943	.9948	.9952
9.9	.9956	.9961	.9965	.9969	.9974	.9978	.9983	.9987	.9991	.9996

	0	1	2	3	4	5	6	7	8	9

Appendix C.
Thermodynamic Quantities for Substances at 25°C

Substance	ΔH_f° (kJ/mol)	ΔG_f° (kJ/mol)	S° (J/K · mol)
$e^-(g)$	0	0	20.87
$H^+(g)$	1536.3	1517.1	108.83
$H^+(aq)$	0	0	0
$H(g)$	218.00	203.30	114.60
$H_2(g)$	0	0	130.6
Group IA			
$Li^+(g)$	687.163	649.989	132.91
$Li^+(aq)$	−278.46	−293.8	14
$Li(g)$	161	128	138.67
$Li(s)$	0	0	29.10
$LiF(s)$	−616.9	−588.7	35.66
$LiCl(s)$	−408	−384	59.30
$LiBr(s)$	−351	−342	74.1
$LiI(s)$	−270	−270	85.8
$Na^+(g)$	609.839	574.877	147.85
$Na^+(aq)$	−239.66	−261.87	60.2
$Na(g)$	107.76	77.299	153.61
$Na(s)$	0	0	51.446
$NaF(s)$	−575.4	−545.1	51.21
$NaCl(s)$	−411.1	−384.0	72.12
$NaBr(s)$	−361	−349	86.82
$NaI(s)$	−288	−285	98.5
$NaHCO_3(s)$	−947.7	−851.9	102
$Na_2CO_3(s)$	−1130.8	−1048.1	139
$K^+(g)$	514.197	481.202	154.47
$K^+(aq)$	−251.2	−282.28	103
$K(g)$	89.2	60.7	160.23
$K(s)$	0	0	64.672
$KF(s)$	−568.6	−538.9	66.55
$KCl(s)$	−436.68	−408.8	82.55
$KBr(s)$	−394	−380	95.94
$KI(s)$	−328	−323	106.39
$Rb^+(g)$	495.04		
$Rb^+(aq)$	−246	−282.2	124
$Rb(g)$	85.81	55.86	169.99
$Rb(s)$	0	0	69.5
$RbF(s)$	−549.28		
$RbCl(s)$	−430.58		
$RbBr(s)$	−389.2	−378.1	108.3
$RbI(s)$	−328	−326	118.0
$Cs^+(g)$	458.5	427.1	169.72
$Cs^+(aq)$	−248	−282.0	133
$Cs(g)$	76.7	49.7	175.5

Substance	ΔH_f° (kJ/mol)	ΔG_f° (kJ/mol)	S° (J/K · mol)
Cs(s)	0	0	85.15
CsF(s)	−554.7	−525.4	88
CsCl(s)	−442.8	−414	101.18
CsBr(s)	−395	−383	121
CsI(s)	−337	−333	130
Group IIA			
$Mg^{2+}(g)$	2351		
$Mg^{2+}(aq)$	−461.96	−456.01	−118
$Mg^{+}(g)$	894.1		
Mg(g)	150	115	148.55
Mg(s)	0	0	32.69
$MgCl_2(s)$	−641.6	−592.1	89.630
MgO(s)	−601.2	−569.0	26.9
$Mg_3N_2(s)$	−461	−401	88
$MgCO_3(s)$	−1112	−1028	65.86
$Ca^{2+}(g)$	1934.1		
$Ca^{2+}(aq)$	−542.96	−553.04	−55.2
$Ca^{+}(g)$	788.6		
Ca(g)	192.6	158.9	154.78
Ca(s)	0	0	41.6
$CaF_2(s)$	−1215	−1162	68.87
$CaCl_2(s)$	−795.0	−750.2	114
CaO(s)	−635.1	−603.5	38.2
$CaCO_3(s)$	−1206.9	−1128.8	92.9
$CaSO_4(s)$	−1432.7	−1320.3	107
$Ca_3(PO_4)_2(s)$	−4138	−3899	263
$Sr^{2+}(g)$	1784		
$Sr^{2+}(aq)$	−545.51	−557.3	−39
$Sr^{+}(g)$	719.6		
Sr(g)	164	110	164.54
Sr(s)	0	0	54.4
$SrCl_2(s)$	−828.4	−781.2	117
SrO(s)	−592.0	−562.4	55.5
$SrCO_3(s)$	−1218	−1138	97.1
$SrSO_4(s)$	−1445	−1334	122
$Ba^{2+}(g)$	1649.9		
$Ba^{2+}(aq)$	−538.36	−560.7	13
$Ba^{+}(g)$	684.6		
Ba(g)	175.6	144.8	170.28
Ba(s)	0	0	62.5
$BaCl_2(s)$	−806.06	−810.9	126
BaO(s)	−548.1	−520.4	72.07
$BaCO_3(s)$	−1219	−1139	112
$BaSO_4(s)$	−1465	−1353	132

Substance	ΔH_f° (kJ/mol)	ΔG_f° (kJ/mol)	S° (J/K · mol)
Group IIIA			
B(β-rhombohedral)	0	0	5.87
$B_2O_3(s)$	-1272	-1193	53.8
Al(s)	0	0	28.3
$Al^{3+}(aq)$	-524.7	-481.2	-313
$Al_2O_3(s)$	-1676	-1582	50.94
Group IVA			
C(g)	715.0	669.6	158.0
C(graphite)	0	0	5.686
C(diamond)	1.896	2.866	2.439
CO(g)	-110.5	-137.2	197.5
$CO_2(g)$	-393.5	-394.4	213.7
$CO_2(aq)$	-412.9	-386.2	121
$CO_3^{2-}(aq)$	-676.26	-528.10	-53.1
$HCO_3^-(aq)$	-691.11	-587.06	95.0
$H_2CO_3(aq)$	-698.7	-623.42	191
$CH_4(g)$	-74.87	-50.81	186.1
$C_2H_2(g)$	227	209	200.85
$C_2H_4(g)$	52.47	68.36	219.22
$C_2H_6(g)$	-84.667	-32.89	229.5
$C_6H_6(l)$	49.0	124.5	172.8
$CH_3OH(g)$	-201.2	-161.9	238
$CH_3OH(l)$	-238.6	-166.2	127
HCHO(g)	-116	-110	219
$HCOO^-(aq)$	-410	-335	91.6
HCOOH(l)	-409	-346	129.0
HCOOH(aq)	-410	-356	164
$C_2H_5OH(l)$	-277.63	-174.8	161
$CH_3CHO(g)$	-166	-133.7	266
$CH_3COOH(l)$	-487.0	-392	160
$CN^-(aq)$	151	166	118
HCN(g)	135	125	201.7
HCN(l)	105	121	112.8
HCN(aq)	105	112	129
$CS_2(g)$	117	66.9	237.79
$CS_2(l)$	87.9	63.6	151.0
$CH_3Cl(g)$	-83.7	-60.2	234
$CH_2Cl_2(l)$	-117	-63.2	179
$CHCl_3(l)$	-132	-71.5	203
$CCl_4(g)$	-96.0	-53.7	309.7
$CCl_4(l)$	-139	-68.6	214.4
$COCl_2(g)$	-220	-206	283.74
Si(s)	0	0	18.0
$SiO_2(s)$	-910.9	-856.5	41.5
Sn(gray)	3	4.6	44.8
Sn(white)	0	0	51.5
$SnCl_4(l)$	-545.2	-474.0	259
$Pb^{2+}(aq)$	1.6	-24.3	21
Pb(s)	0	0	64.785

Substance	ΔH_f° (kJ/mol)	ΔG_f° (kJ/mol)	S° (J/K · mol)
$PbO(s)$	−218	−198	68.70
$PbO_2(s)$	−276.6	−219.0	76.6
$PbS(s)$	−98.3	−96.7	91.3
$PbCl_2(s)$	−359	−314	136
$PbSO_4(s)$	−918.39	−811.24	147
Group VA			
$N(g)$	473	456	153.2
$N_2(g)$	0	0	191.5
$NO(g)$	90.29	86.60	210.65
$NO_2(g)$	33	51	239.9
$N_2O_4(g)$	9.1	97.7	304.3
$N_2O_5(g)$	11	118	346
$NH_3(g)$	−45.9	−16	193
$NH_3(aq)$	−80.83	26.7	110
$NO_3^-(aq)$	−206.57	−110.5	146
$HNO_3(l)$	−173.23	−79.914	155.6
$HNO_3(aq)$	−206.57	−110.5	146
$P(g)$	333.9	292.0	163.1
$P(red)$	0	0	22.8
$P_4(white)$	68	48	164
$P_2(g)$	179	127	218
$P_4(g)$	129	72.5	280
$PCl_3(g)$	−271	−258	312
$PCl_5(g)$	−382	−313	353
$P_4O_{10}(s)$	−2942	−2675	229
$PO_4^{3-}(aq)$	−1266	−1013	−218
$HPO_4^{2-}(aq)$	−1281	−1082	−36
$H_2PO_4^-(aq)$	−1285	−1135	89.1
$H_3PO_4(aq)$	−1277		
Group VIA			
$O(g)$	249.2	231.7	160.95
$O_2(g)$	0	0	205.0
$O_3(g)$	143	163	238.82
$OH^-(aq)$	−229.94	−157.30	−10.54
$H_2O(g)$	−241.826	−228.60	188.72
$H_2O(l)$	−285.840	−237.192	69.940
$S(g)$	279	239	168
$S_2(g)$	129	80.1	228.1
$S_8(g)$	101	49.1	430.211
$S(rhombic)$	0	0	31.9
$S(monoclinic)$	0.30	0.096	32.6
$S^{2-}(aq)$	41.8	83.7	22
$HS^-(aq)$	−17.7	12.6	61.1
$H_2S(g)$	−20	−33	205.6
$H_2S(aq)$	−39	−27.4	122
$SO_2(g)$	−296.8	−300.2	248.1
$SO_3(g)$	−396	−371	256.66
$SO_4^{2-}(aq)$	−907.51	−741.99	17

Substance	ΔH_f° (kJ/mol)	ΔG_f° (kJ/mol)	S° (J/K · mol)
$HSO_4^-(aq)$	− 885.75	− 752.87	126.9
$H_2SO_4(l)$	− 813.989	− 690.059	156.90
$H_2SO_4(aq)$	− 907.51	− 741.99	17
Group VIIA			
$F(g)$	78.9	61.8	158.64
$F^-(g)$	− 255.6	− 262.5	145.47
$F^-(aq)$	− 329.1	− 276.5	− 9.6
$F_2(g)$	0	0	202.7
$HF(g)$	− 273	− 275	173.67
$Cl(g)$	121.0	105.0	165.1
$Cl^-(g)$	− 234	− 240	153.25
$Cl^-(aq)$	− 167.46	− 131.17	55.10
$Cl_2(g)$	0	0	223.0
$HCl(g)$	− 92.31	− 95.30	186.79
$HCl(aq)$	− 167.46	− 131.17	55.06
$Br(g)$	111.9	82.40	174.90
$Br^-(g)$	− 218.9		
$Br^-(aq)$	− 120.9	− 102.82	80.71
$Br_2(g)$	30.91	3.13	245.38
$Br_2(l)$	0	0	152.23
$HBr(g)$	− 36	− 53.5	198.59
$I(g)$	106.8	70.21	180.67
$I^-(g)$	− 194.7		
$I^-(aq)$	− 55.94	− 51.67	109.4
$I_2(g)$	62.442	19.38	260.58
$I_2(s)$	0	0	116.14
$HI(g)$	25.9	1.3	206.33
Group IB			
$Cu^+(aq)$	51.9	50.2	− 26
$Cu^{2+}(aq)$	64.39	64.98	− 98.7
$Cu(g)$	341.1	301.4	166.29
$Cu(s)$	0	0	33.1
$Ag^+(aq)$	105.9	77.111	73.93
$Ag(g)$	289.2	250.4	172.892
$Ag(s)$	0	0	42.702
$AgF(s)$	− 203	− 185	84
$AgCl(s)$	− 127.03	− 109.72	96.11
$AgBr(s)$	− 99.50	− 95.939	107.1
$AgI(s, II)$	− 62.38	− 66.32	114
$Ag_2S(s)$	− 31.8	− 40.3	146
Group IIB			
$Zn^{2+}(aq)$	− 152.4	− 147.21	− 106.5
$Zn(g)$	130.5	94.93	160.9
$Zn(s)$	0	0	41.6
$ZnO(s)$	− 348.0	− 318.2	43.9
$ZnS(s$, zinc blende)	− 203	− 198	57.7

Substance	ΔH_f° (kJ/mol)	ΔG_f° (kJ/mol)	S° (J/K · mol)
$Cd^{2+}(aq)$	−72.38	−77.74	−61.1
$Cd(g)$	112.8	78.20	167.64
$Cd(s)$	0	0	51.5
$CdS(s)$	−144	−141	71
$Hg^{2+}(aq)$		164.8	
$Hg_2^{+}(aq)$		153.9	
$Hg(g)$	61.30	31.8	174.87
$Hg(l)$	0	0	76.027
$HgCl_2(s)$	−230	−184	144
$Hg_2Cl_2(s)$	−264.9	−210.66	196
$HgO(s)$	−90.79	−58.50	70.27
Group VIB			
$[Cr(H_2O)_6]^{3+}(aq)$	−1971		
$Cr(s)$	0	0	23.8
$CrO_4^{2-}(aq)$	−863.2	−706.3	38
$Cr_2O_7^{2-}(aq)$	−1461	−1257	214
Group VIIB			
$Mn^{2+}(aq)$	−219	−223	−84
$Mn(s, \alpha)$	0	0	31.8
$MnO_2(s)$	−520.9	−466.1	53.1
$MnO_4^{-}(aq)$	−518.4	−425.1	190
Group VIIIB			
$Fe^{3+}(aq)$	−47.7	−10.5	−293
$Fe^{2+}(aq)$	−87.9	−84.94	113
$Fe(s)$	0	0	27.3
$FeO(s)$	−272.0	−251.4	60.75
$Fe_2O_3(s)$	−825.5	−743.6	87.400
$Fe_3O_4(s)$	−1121	−1018	145.3
$Co^{2+}(aq)$	−67.4	−51.5	−155
$Co(s)$	0	0	30
$Ni^{2+}(aq)$	−64.0	−46.4	−159
$Ni(s)$	0	0	30.1

Answers to Exercises

Chapter 1

1.1 0.20 g, 0.20 g **1.2** (a) 4.9 (b) 2.48 (c) 0.08
(d) 3 **1.3** (a) 1.84 nm (b) 5.67 ps (c) 7.85 mg
(d) 9.7 km (e) 2.34 dm **1.4** (a) 195 K (b) 39.2°C
1.5 3.24 m **1.6** 1.6×10^{-2} mm;
1.6×10^{-2} mm = 16 μm **1.7** 6.76×10^{-29} m^3
1.8 7.87 g/cm^3; the object is made of iron
1.9 38.4 cm^3 **1.10** .964 atm; 9.77×10^4 Pa
1.11 kg/(m · s)

Chapter 2

2.1 7.2×10^{24} hydrogen atoms **2.2** C_3H_8
2.3 Na_2SO_4 **2.4** 4 **2.5** (a) $O_2 + 2PCl_3 \rightarrow 2POCl_3$
(b) $P_4 + 6N_2O \rightarrow P_4O_6 + 6N_2$
(c) $2As_2S_3 + 9O_2 \rightarrow 2As_2O_3 + 6SO_2$
(d) $Ca_3(PO_4)_2 + 4H_3PO_4 \rightarrow 3Ca(H_2PO_4)_2$
2.6 (a) mixture (b) element (c) mixture
(d) compound (e) element **2.7** (a) Sand and salt can
be separated by dissolving the salt in water, then
filtering off the sand. The salt can be recovered by
evaporating the water. (b) Use a magnet to separate
the iron filings. As above, add water to the mixture to
dissolve the salt. Filter off the sand. Evaporate the
water to recover the salt. (c) Benzene and toluene
can be separated by fractional distillation. Benzene,
b.p. 80°C, has a lower boiling point than toluene, b.p.
111°C, and it will distill off first. (d) Gaseous ethane
and ethylene can be separated by gas phase
chromatography. Each substance would have a
different affinity for the stationary phase.

Chapter 3

3.1 (a) 46.0 amu (b) 180 amu (c) 40.0 amu
(d) 58.3 amu **3.2** 1.03×10^{-22} g **3.3** 30.9 g H_2O_2
3.4 0.435 mol NH_4NO_3 **3.5** 1.2×10^{21} HCN
molecules **3.6** % H = 6.71%; % C = 40.0%;
% O = 53.3% **3.7** 40.9% C, 4.57% H, 54.5% O
3.8 $C_7H_6O_2$ **3.9** $C_6H_{12}O_2$
3.10
H_2　　　+　　　Cl_2　　　→　　　2HCl
　　1 molecule H_2 + 1 molecule Cl_2 → 2 molecules HCl
　　1 mol H_2　　+ 1 mol Cl_2　　→ 2 mol HCl
　　2.02 g H_2　　+ 70.9 g Cl_2　　→ 2 × 36.5 g HCl
3.11 178 g Na **3.12** 2.46 kg O_2 **3.13** 81.1 g Hg
3.14 11.0 g ZnS **3.15** theoretical yield = 7.92 g;
percentage yield = 67% **3.16** 0.0464 M NaCl
3.17 0.0075 mol NaCl; 0.44 g NaCl **3.18** 10.1 mL
3.19 400 mL **3.20** 3.38 mL **3.21** 8.4 mL
3.22 48.4 mL $NiSO_4$ **3.23** 160 mL **3.24** 5.07%

Chapter 4

4.1 24.1 L **4.2** 4.47 dm^3 **4.3** 5.54 dm^3
4.4 6.15×10^{21} N_2 molecules **4.5** Use the ideal gas
law: $PV = nRT$. Solve for n: $n = PV/RT = P(V/RT)$.
Everything within parentheses is constant, therefore,
n = constant × P (or $n \propto P$). **4.6** 46.0 atm
4.7 0.513 L CO_2 **4.8** Density of He = 0.164 g/L;
difference in mass between 1 L of air and 1 L of
He = 1.024 g **4.9** 64.2 amu **4.10** P_{O_2} = 0.0769 atm;
P_{CO_2} = 0.0311 atm; P = 0.108 atm **4.11** Partial
pressure of each gas in mmHg: methane, 612;
nitrogen, 102; ethane, 32; propane, 21; butane, 10;

pentane, 5 **4.12** 0.0159 mol O_2; $V = 0.406$ L
4.13 219 m/s **4.14** $T = 52.4$ K, 728 K **4.15** 44.0 amu
4.16 10.0 s

Chapter 5

5.1 The neutral atom has 92 electrons. The U^{+2} ion
has 90 electrons. **5.2** $^{35}_{17}Cl$ **5.3** Carbon-14 is an
isotope of carbon. It differs from carbon-12 in the
mass number because it has two additional neutrons
in its nucleus. **5.4** 35.453 amu **5.5** 767 nm
5.6 6.58×10^{14}/s **5.7** The energy of each photon is
2.0×10^{-19} J, 2.0×10^{-17} J, and 2.0×10^{-15} J,
respectively. The photon with the shortest wavelength
(1.0×10^{-10} m) has the greatest amount of energy per
photon. The photon with the longest wavelength
(1.0×10^{-6} m) has the least amount of energy per
photon. **5.8** 103 nm **5.9** 3.38×10^{-19} J **5.10** (a) The
value of n must be a positive whole number. (b) The
values of l range from 0 to $n - 1$. Here, l has a value
greater than n. (c) The values for m_l range from $-l$ to
$+l$. Here, m_l has a value greater than that of l.
(d) The values for m_s are $+\frac{1}{2}$ or $-\frac{1}{2}$, not 0.

Chapter 6

6.1 (a) Possible orbital diagram. (b) Possible orbital
diagram. (c) Impossible orbital diagram. There are
two electrons in a $2p$ orbital with the same spin.
(d) Possible electron configuration. (e) Impossible
electron configuration. Only two electrons are allowed
in an s subshell. (f) Impossible electron
configuration. Only six electrons are allowed in a p
subshell. **6.2** $1s^2 2s^2 2p^6 3s^2 3p^6 3d^6 4s^2$ **6.3** Building-up
configuration: $[Kr]4d^4 5s^2$. One might expect
$[Kr]4d^5 5s^1$ due to the tendency toward half-filled
orbitals.
6.4 (↑↓) (↑↓) (↑↓)(↑↓)(↑↓) (↑↓) (↑↑)(↑↑)(↑↑)
  $1s$ $2s$ $2p$ $3s$ $3p$
6.5 $4s^2 4p^3$ **6.6** Lead is in Period 6 and Group IVA,
and is a main-group element. **6.7** In order of
increasing radius: Be, Mg, Na **6.8** It is most likely
that 1000 kJ/mol is the ionization energy for iodine
because ionization energies tend to decrease with
atomic number in a column. **6.9** The electron
affinity of barium should be positive because it has a
filled subshell. **6.10** (a) H_2Se (b) $CaSeO_4$

Chapter 7

7.1 $\cdot Mg \cdot + :\ddot{O}\cdot \rightarrow \left[:\ddot{\ddot{O}}:\right]^{2-} + Mg^{2+}$ **7.2** The electron
configuration for Ca^{2+} is [Ar] and its Lewis symbol is
Ca^{2+}. The electron configuration for S^{2-} is $[Ne]3s^2 3p^6$
and its Lewis symbol is $\left[:\ddot{\ddot{S}}:\right]^{2-}$. **7.3** The electron
configurations of Pb and Pb^{2+} are $[Xe]4f^{14} 5d^{10} 6s^2 6p^2$
and $[Xe]4f^{14} 5d^{10} 6s^2$, respectively. **7.4** $[Ar]3d^5$
7.5 K_2CrO_4 **7.6** One would expect S^{2-}, because it has
two additional electrons. **7.7** In order of increasing
ionic radius: Mg^{2+}, Ca^{2+}, Sr^{2+} **7.8** In order of
increasing ionic radius: Ca^{2+}, Cl^-, P^{3-} **7.9** C—O is
the most polar bond. **7.10** $:\ddot{Cl}:$
        $:\ddot{F}:\overset{\displaystyle|}{\underset{\displaystyle|}{C}}:\ddot{F}:$
             $:\ddot{Cl}:$

7.11 $\ddot{O}::C::\ddot{O}$ or $\ddot{O}=C=\ddot{O}$

7.12 (a) $\left[H:\overset{\displaystyle H}{\ddot{O}}:H \right]^+$ (b) $\left[:\ddot{O}:\ddot{Cl}:\ddot{O}: \right]^-$

7.13 $:\ddot{Cl}:Be:\ddot{Cl}:$ **7.14**

         $:\ddot{F} \diagdown \diagup \ddot{F}:$
              S
         $:\ddot{F} \diagup \diagdown \ddot{F}:$

7.15 $\left[:\ddot{O}:\overset{\displaystyle :O:}{N}:\ddot{O}: \right]^- \leftrightarrow \left[\ddot{O}::\overset{\displaystyle :O:}{N}:\ddot{O}: \right]^- \leftrightarrow \left[:\ddot{O}:\overset{\displaystyle :O:}{N}::\ddot{O} \right]^-$

7.16 0.96 Å **7.17** 1.23 Å **7.18** $x_{Cr} = +6$
7.19 $x_{Mn} = +7$ **7.20** (a) calcium oxide (b) lead(II)
chromate (c) sulfur trioxide **7.21** $Tl(NO_3)_3$

Chapter 8

8.1 (a) trigonal pyramidal (b) bent (c) tetrahedral
8.2 T-shaped **8.3** (b), (c) **8.4** (b) **8.5** For the N
atom:
N (ground state) = [He] (↑↓) (↑)(↑)(↑) and
          $2s$ $2p$
N (hybridized) = [He] (↑↓)(↑)(↑)(↑)
              sp^3
Each N—H bond is formed by the overlap of a $1s$
orbital of a hydrogen atom with one of the sp^3 hybrid
orbitals of the nitrogen atom. **8.6** Each P—Cl bond is
formed by the overlap of a phosphorus $sp^3 d$ hybrid
orbital with a singly occupied $3p$ chlorine orbital.
8.7 The C atom is sp hybridized. A carbon–oxygen
bond is double and is composed of a σ bond and a π
bond. The σ bond is formed by the overlap of a hybrid
C orbital with a $2p$ orbital of O that lies along the

axis. The π bond is formed by the sidewise overlap of a $2p$ on C with a $2p$ on O. **8.8** The structural formulas for the isomers are as follows:

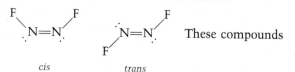

cis	*trans*

These compounds exist as separate isomers. For these to interconvert, one end of the molecule would have to rotate with respect to the other end. This would require breaking the π bond and a considerable expenditure of energy.

8.9 (↿⇂) (↿⇂) (↿⇂) (↿⇂) (↿⇂)(↿⇂);
$\quad\sigma_{1s}\quad\sigma_{1s}^{*}\quad\sigma_{2s}\quad\sigma_{2s}^{*}\quad\pi_{2p}$
$(\sigma_{1s})^{2}(\sigma_{1s}^{*})^{2}(\sigma_{2s})^{2}(\sigma_{2s}^{*})^{2}(\pi_{2p})^{4}$; diamagnetic; bond order = 2.

8.10 (↿⇂) (↿⇂) (↿⇂) (↿⇂) (↿⇂)(↿⇂) (↿⇂);
$\quad\sigma_{1s}\quad\sigma_{1s}^{*}\quad\sigma_{2s}\quad\sigma_{2s}^{*}\quad\pi_{2p}\quad\sigma_{2p}$
$(\sigma_{1s})^{2}(\sigma_{1s}^{*})^{2}(\sigma_{2s})^{2}(\sigma_{2s}^{*})^{2}(\pi_{2p})^{4}(\sigma_{2p})^{2}$; diamagnetic; bond order = 3.

Chapter 9

9.1 (a) $H^{+}(aq) + OH^{-}(aq) \rightarrow H_2O(l)$
(b) $Pb^{2+}(aq) + SO_4^{2-}(aq) \rightarrow PbSO_4(s)$
9.2 (a) Metathesis (b) Oxidation-reduction; S goes from -2 and $+4$ to 0 oxidation states. (c) Oxidation-reduction; Na goes from 0 to $+1$ and H from $+1$ to 0 oxidation states. (d) Metathesis
9.3 $2NaI(aq) + Pb(C_2H_3O_2)_2(aq) \rightarrow$
$\qquad\qquad\qquad PbI_2(s) + 2NaC_2H_3O_2(aq)$;
$2I^{-}(aq) + Pb^{2+}(aq) \rightarrow PbI_2(s)$
9.4 $HCN(aq) + KOH(aq) \rightarrow H_2O(l) + KCN(aq)$;
$HCN(aq) + OH^{-}(aq) \rightarrow H_2O(l) + CN^{-}(aq)$
9.5 $H_2SO_4(aq) + KOH(aq) \rightarrow H_2O(l) + KHSO_4(aq)$
$KHSO_4(aq) + KOH(aq) \rightarrow H_2O(l) + K_2SO_4(aq)$
9.6 Molecular equation: $Na_2SO_3(aq) + 2HCl(aq) \rightarrow$
$\qquad\qquad 2NaCl(aq) + H_2O(l) + SO_2(g)$
Net ionic equation: $SO_3^{2-}(aq) + 2H^{+}(aq) \rightarrow$
$\qquad\qquad\qquad H_2O(l) + SO_2(g)$
9.7 (a) $KClO_4(s) + H_2SO_4(l) \rightarrow HClO_4(g) + KHSO_4(s)$
(b) $AgClO_4(aq) + HCl(aq) \rightarrow HClO_4(aq) + AgCl(s)$
9.8 (a) One possibility is
$Na_2CO_3(aq) + Ca(OH)_2(aq) \rightarrow 2NaOH(aq) + CaCO_3(s)$.
Remove $CaCO_3$ by filtering, then evaporate filtrate.

(b) One possibility is $(NH_4)_2SO_4(aq) + Ca(OH)_2(aq) \xrightarrow{\Delta}$
$\qquad\qquad 2NH_3(g) + 2H_2O(l) + CaSO_4(s)$.
9.9 (a) $2HClO_4(aq) + Ca(OH)_2(aq) \rightarrow$
$\qquad\qquad Ca(ClO_4)_2(aq) + 2H_2O(l)$
(b) $ZnS(s) + 2HCl(aq) \rightarrow ZnCl_2(aq) + H_2S(g)$

(c) One possibility is $Pb(NO_3)_2(aq) + H_2SO_4(aq) \rightarrow$
$\qquad\qquad PbSO_4(s) + 2HNO_3(aq)$
9.10 Skeleton equation: $\overset{+4}{S}O_3^{2-} + \overset{0}{Br_2} \rightarrow \overset{+6}{S}O_4^{2-} + 2\overset{-1}{Br}^{-}$

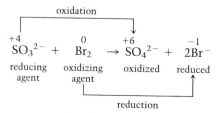

9.11 $4Zn + NO_3^{-} + 10H^{+} \rightarrow 4Zn^{2+} + NH_4^{+} + 3H_2O$
9.12 $I_2 + 10HNO_3 \rightarrow 2HIO_3 + 10NO_2 + 4H_2O$
9.13 22.5 g; 45.0 g **9.14** 89.3%

Chapter 10

10.1 exothermic; $q = -1170$ kJ
10.2 $2N_2H_4(l) + N_2O_4(l) \rightarrow 3N_2(g) + 4H_2O(g)$;
$\qquad\qquad\qquad \Delta H = -1.15 \times 10^3$ kJ
10.3 0.905 J/(g · °C); the metal is aluminum
10.4 2.44 J/(g · °C); 113 J/(mol · °C) **10.5** sp. ht. (Hg) = 0.2505 J/(g · °C) **10.6** -54 kJ **10.7** 180 kJ
10.8 (a) $N_2H_4(l) + \frac{1}{2}N_2O_4(l) \rightarrow \frac{3}{2}N_2(g) + 2H_2O(g)$
(b) $4H_2O(g) + 3N_2(g) \rightarrow N_2O_4(l) + 2N_2H_4(l)$
10.9 $\Delta H = 1661.9$ kJ **10.10** $\Delta H_{vap}^{\circ} = 44.0$ kJ
10.11 1661.9 kJ; same answer **10.12** 62.0 kJ
10.13 -1304 kJ **10.14** $X_S - X_H = 0.59$

Chapter 11

11.1 (a) hydrogen bonding, dipole-dipole forces, and London forces (b) London forces (c) dipole-dipole forces and London forces **11.2** In order of increasing vapor pressure we have: butane (C_4H_{10}), propane (C_3H_8), and ethane (C_2H_6). London forces increase with increasing molecular weight. Therefore, one would expect the lowest vapor pressure to correspond to the highest molecular weight. **11.3** Hydrogen bonding is negligible in methyl chloride but strong in ethanol, which explains the lower vapor pressure of ethanol. **11.4** (a) metallic solid (b) ionic solid (c) covalent network solid (d) molecular solid
11.5 2 atoms **11.6** $N_A = 6.02 \times 10^{23}/\text{mol}$
11.7 5.33 Å **11.8** 282 kJ; 845 g H_2O **11.9** C_2H_5OH, molecular; CH_4, molecular; CH_3Cl, molecular; $MgSO_4$, ionic. In order of increasing melting point: CH_4, CH_3Cl, C_2H_5OH, and $MgSO_4$. **11.10** (a) Methyl

chloride can be liquefied by sufficiently increasing the pressure as long as the temperature is below 144°C. (b) Oxygen can be liquefied by compression as long as the temperature is below −119°C.

Chapter 12

12.1 A dental filling, made up of liquid mercury and solid silver, is a solid solution. **12.2** C_4H_9OH
12.3 Na^+ **12.4** 8.45×10^{-3} g O_2/L **12.5** $AlCl_3$
12.6 7.07 g HCl; 27.9 g H_2O **12.7** 3.09 m $C_6H_5CH_3$
12.8 Mole fraction toluene = 0.194; mole fraction benzene = 0.806 **12.9** Mole fraction methanol = 0.00550; mole fraction ethanol = 0.995
12.10 7.24 m CH_3OH **12.11** 2.97 M $(NH_2)_2CO$
12.12 2.20 m $(NH_2)_2CO$ **12.13** $P = 155$ mmHg
12.14 1.9×10^3 g CH_2OHCH_2OH **12.15** 176 amu
12.16 124 g/mol; P_4 **12.17** $\pi = 3.5$ atm
12.18 −0.93°C

Chapter 13

13.1 128 nm **13.2** 0.62 mm **13.3** 7.03×10^3 m/s
13.4 $As_2S_3 + 6H_2O + 7O_2 \rightarrow 2AsO_4^{3-} + 3SO_4^{2-} + 12H^+$
13.5 6.9×10^4 years; 4.4×10^4 years

13.6 $2Au_2O_3(s) \xrightarrow{\Delta} 4Au(s) + 3O_2(g)$

$2H_2O_2(aq) \xrightarrow{Fe^{3+}} 2H_2O(l) + O_2(g)$

$2NaNO_3(s) \xrightarrow{\Delta} 2NaNO_2(s) + O_2(g)$

13.7 (a) $Cs(s) + O_2(g) \rightarrow CsO_2(s)$
(b) $2C_2H_5SH(l) + 9O_2(g) \rightarrow$
$\qquad\qquad 4CO_2(g) + 6H_2O(l) + 2SO_2(g)$
(c) $S_8(s) + 8O_2(g) \rightarrow 8SO_2(g)$

$2SO_2(g) + O_2(g) \xrightarrow{Pt \text{ or } V_2O_5} 2SO_3(g)$
$SO_3(g) + H_2O(l) \rightarrow H_2SO_4(aq)$ **13.8** 0.366 metric tons
13.9 To prepare N_2O:

$H_2SO_4(l) + NaNO_3(s) \xrightarrow{\Delta} NaHSO_4(s) + HNO_3(g)$
$HNO_3(aq) + NH_3(aq) \rightarrow NH_4NO_3(aq)$

$NH_4NO_3(s) \xrightarrow{\Delta} N_2O(g) + 2H_2O(g)$
To prepare NO and NO_2, use HNO_3 from above.
$3Cu(s) + 8HNO_3$ (dilute) $\rightarrow$
$\qquad\qquad 3Cu(NO_3)_2(aq) + 2NO(g) + 4H_2O(l)$
$Cu(s) + 4HNO_3$ (conc) $\rightarrow$
$\qquad\qquad Cu(NO_3)_2(aq) + 2NO_2(g) + 2H_2O(l)$

13.10 $2NO_2 + 2OH^- \rightarrow NO_2^- + NO_3^- + H_2O$
13.11 0.7884, 0.2116, 28.86 amu, 1.288 g/L

13.12 $:\ddot{F}:\ddot{Kr}:\ddot{F}:$, sp^3d, linear

13.13 $XeF_2 + 2H^+ + 2e^- \rightarrow Xe + 2HF$
$2HCl + \quad XeF_2^{\,\cdot} \rightarrow Cl_2 + Xe + 2HF$
13.14 11 mg

Chapter 14

14.1 0.154 mol H_2; 0.0772 mol O_2; volume H_2 = 3.61 L; volume O_2 = 1.80 L **14.2** zinc

14.3 $KH(s) + H_2O(l) \rightarrow K^+(aq) + OH^-(aq) + H_2(g)$

$CH_4(g) + H_2O(g) \xrightarrow{Ni} CO(g) + 3H_2(g)$

$2H_2O(l) \xrightarrow{electrolysis} 2H_2(g) + O_2(g)$
14.4 $H_2(g) + I_2(g) \rightarrow 2HI(g)$
$Mg(l) + H_2(g) \xrightarrow{\Delta} MgH_2(s)$
$Ag_2O(s) + H_2(g) \rightarrow 2Ag(s) + H_2O(g)$
14.5 (a) $C_2H_6(g) + 2H_2O(g) \xrightarrow{Ni} 2CO(g) + 5H_2(g)$

$CO(g) + 2H_2(g) \xrightarrow{catalyst} CH_3OH(g)$
(b) Use H_2 from (a):
$CuO(s) + H_2(g) \rightarrow Cu(s) + H_2O(g)$
14.6 2.1×10^3 J
14.7 (a) $Si(s) + 2H_2O(g) \rightarrow SiO_2(s) + 2H_2(g)$
(b) $2K(s) + 2H_2O(l) \rightarrow 2K^+(aq) + 2OH^-(aq) + H_2(g)$
(c) $Br_2(aq) + H_2O(l) \rightleftharpoons H^+(aq) + Br^-(aq) + HOBr(aq)$
14.8 $Na_2SO_4 \cdot 10H_2O$
14.9 $PbS + 4H_2O_2 \rightarrow PbSO_4 + 4H_2O$
14.10 0.478 M H_2O_2 **14.11** − 1.588 kJ/g
14.12 −2.31°C **14.13** 30.0 atm **14.14** 3.04×10^3 J

Chapter 15

15.1 $\dfrac{\Delta[NO_2F]}{\Delta t} = -\dfrac{\Delta[NO_2]}{\Delta t}$
15.2 2.5×10^{-6} mol/(L·s) **15.3** Zero order in [CO]; second order in $[NO_2]$; second order overall.
15.4 Rate = $k[NO_2]^2$; $k = 0.71$ L/(mol·s)
15.5 (a) $[N_2O_5]_t = 0.0124$ mol/L (b) $t = 4.80 \times 10^3$ s
15.6 half-life = 0.075 s; decreases to 50% of initial concentration in 0.075 s, to 25% in 0.150 s.
15.7 $E_a = 2.1 \times 10^5$ J; $k = 5.8$ L/(mol·s)
15.8 $2H_2O_2 \rightarrow 2H_2O + O_2$ **15.9** bimolecular
15.10 Rate = $k[NO_2]^2$ **15.11** Rate = $k_1[H_2O_2][I^-]$
15.12 Rate = $k_2K_1[NO]^2[O_2]$

Chapter 16

16.1 0.41 mol CO, 0.41 mol H_2O, 0.59 mol CO_2, and 0.59 mol H_2

16.2 $K_c = \dfrac{[NH_3]^2[H_2O]^4}{[NO_2]^2[H_2]^7}$; $K_c = \dfrac{[NH_3][H_2O]^2}{[NO_2][H_2]^{7/2}}$

16.3 $CO(g) + H_2O(g) \rightleftharpoons H_2(g) + CO_2(g)$; 0.57

16.4 $K_c = 2.3 \times 10^{-4}$ **16.5** $K_p = 1.24$

16.6 $K_c = \dfrac{[Ni(CO)_4]}{[CO]^4}$ **16.7** products; $2.2 \times 10^6\ M$

16.8 $Q = 0.66$; more CO will form **16.9** 0.0096 moles of PCl_5 **16.10** 0.11 mol H_2, 0.11 mol I_2, 0.78 mol HI **16.11** 0.865 mol/L PCl_5, 0.135 mol/L PCl_3, 0.135 mol/L Cl_2 **16.12** (a) in the reverse direction (b) in the reverse direction **16.13** (a) Pressure has no effect. (b) No (c) Yes **16.14** high temperature **16.15** High temperatures and low pressure would give the best yield of product.

Chapter 17

17.1 Hydrogen ion concentration is $4.00 \times 10^{-14}\ M$; hydroxide ion concentration is 0.250 M. **17.2** basic **17.3** 1.35 **17.4** 12.40 **17.5** $3.5 \times 10^{-5}\ M$ **17.6** $4 \times 10^{-4}\ M$

17.7 $H_2CO_3(aq) + CN^-(aq) \rightleftharpoons HCN(aq) + HCO_3^-(aq)$

     acid   base   acid   base

    HCN is the conjugate acid of CN^-

17.8 The reactants are favored. **17.9** (a) PH_3 (b) HI (c) H_2SO_3 (d) H_3AsO_4 (e) $HSeO_4^-$ **17.10** (a) acidic (b) neutral (c) acidic

17.11 (a)

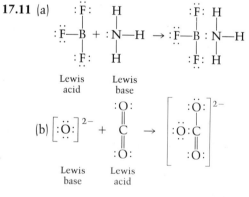

Lewis acid   Lewis base

(b)

Lewis base   Lewis acid

Chapter 18

18.1 1.4×10^{-4}, 0.071

18.2 $[H^+] = [C_2H_3O_2^-] = 1.3 \times 10^{-3}\ M$; pH = 2.89; 0.013 **18.3** 3.24 **18.4** pH = 1.24; $[SO_3^{2-}] \approx 6.3 \times 10^{-8}\ M$ **18.5** 3.2×10^{-6} **18.6** $5.3 \times 10^{-12}\ M$ **18.7** (a) 1.5×10^{-11} (b) 2.4×10^{-5} **18.8** 8.19 **18.9** basic **18.10** $8.5 \times 10^{-5}\ M$, 8.5×10^{-4} **18.11** 3.63 **18.12** 10.82 **18.13** 3.83 **18.14** 1.60 **18.15** 7.97 **18.16** 5.19

Chapter 19

19.1 (a) $K_{sp} = [Ba^{2+}][SO_4^{2-}]$ (b) $K_{sp} = [Fe^{3+}][OH^-]^3$ (c) $K_{sp} = [Ca^{2+}]^3[PO_4^{3-}]^2$ **19.2** 1.8×10^{-10} **19.3** 4.5×10^{-36} **19.4** 0.67 g/L **19.5** (a) 6.3×10^{-3} mol/L (b) 4.4×10^{-5} mol/L **19.6** 2.9 g/L **19.7** precipitation is expected **19.8** no precipitate will form **19.9** $3.6 \times 10^{-11}\ M\ Pb^{2+}$, $3.6 \times 10^{-4}\%$ **19.10** silver cyanide **19.11** pH below 2.5 will precipitate only CuS **19.12** $1.2 \times 10^{-9}\ M$ **19.13** no precipitate will form **19.14** 0.44 mol/L

Chapter 20

20.1 303 J/(mol·K) **20.2** (a) positive (b) positive (c) negative (d) positive **20.3** 537 J/K **20.4** 801 kJ **20.5** 130.9 kJ **20.6** All four reactions are spontaneous in the direction written. **20.7** (a) $K_{th} = P_{CO_2}$

(b) $K_{th} = [Pb^{2+}][I^-]^2$ (c) $K_{th} = \dfrac{P_{CO_2}}{[H^+][HCO_3^-]}$

20.8 1.12×10^{-23} **20.9** 2.325×10^{-29} **20.10** 0.095 atm (72 mmHg) **20.11** $T = 671.4$ K, which is lower than that for $CaCO_3$

Chapter 21

21.1 (a) Anode, $Ni(s) \rightarrow Ni^{2+}(aq) + 2e^-$; cathode, $Ag^+(aq) + e^- \rightarrow Ag(s)$; electrons flow from anode to cathode; Ag^+ flows to Ag, Ni^{2+} flows away from Ni. (b) Anode, $Ag(s) \rightarrow Ag^+(aq) + e^-$; cathode, $Ni^{2+}(aq) + 2e^- \rightarrow Ni(s)$; electrons flow from anode to cathode; Ag^+ flows from anode, Ni^{2+} flows to cathode.

21.2 $Zn(s)|Zn^{2+}(aq)\|Cl^-(aq)|Cl_2(g)|Pt(s)$ **21.3** $Cd(s) + 2H^+(aq) \rightarrow Cd^{2+}(aq) + H_2(g)$ **21.4** 2.12×10^4 J **21.5** 1.10 V **21.6** No **21.7** -1.4×10^5 J/mol **21.8** 2.70 V **21.9** 9×10^{18} **21.10** 1.42 V **21.11** 0.30 V **21.12** $4 \times 10^{-7}\ M$ **21.13** $Ag^+(aq) + e^- \rightarrow Ag(s)$ (cathode) $2H_2O(l) \rightarrow O_2(g) + 4H^+(aq) + 4e^-$ (anode) **21.14** 2.52×10^{-2} A **21.15** 8.66×10^{-4} g

Chapter 22

22.1 $^{40}_{19}K \rightarrow ^{40}_{20}Ca + ^{0}_{-1}e^-$ **22.2** $^{235}_{92}U$ **22.3** (a) is stable; (b) and (c) are radioactive **22.4** (a) positron emission (b) beta emission (c) alpha emission

22.5 (a) $^{40}_{20}Ca(d,p)^{41}_{20}Ca$ (b) $^{12}_{6}C + ^{2}_{1}H \rightarrow ^{13}_{6}C + ^{1}_{1}H$
22.6 $^{14}_{7}N$ **22.7** $3.2 \times 10^{-5}/s$ **22.8** 5.26 years
22.9 $7.82 \times 10^{-10}/s$; 7.4×10^{-7} Ci **22.10** 0.200
22.11 1.0×10^{4} years **22.12** (a) -2.12×10^{8} J;
(b) -0.514 MeV

Chapter 23

23.1 (a) Ga (b) B (c) Be **23.2** TlCl melts at 430°C;
$TlCl_3$ melts at 25°C
23.3 $Na(l) + KCl(l) \rightarrow NaCl(l) + K(g)$
$K(s) + O_2(g) \rightarrow KO_2(s)$
23.4 Add NaOH; hydroxides precipitate, but $Be(OH)_2$
then dissolves to give $Be(OH)_4^{2-}$.

23.5 $CaCO_3(s) \xrightarrow{\Delta} CaO(s) + CO_2(g)$
$CaO(s) + Mg^{2+}(aq) + H_2O(l) \rightarrow Mg(OH)_2(s) + Ca^{2+}(aq)$
$Mg(OH)_2(s) + H_2SO_4(aq) \rightarrow MgSO_4(aq) + 2H_2O(l)$
23.6 Make solutions of the compounds. If any two are
mixed and give no precipitate, one is KOH and the
other is $BaCl_2$. The one that gives a precipitate with
the third solution, which is $Al_2(SO_4)_3$, must be $BaCl_2$.
23.7 (a) $Mg^{2+}(aq) + 2OH^-(aq) \rightarrow Mg(OH)_2(s)$;
$Mg(OH)_2(s) + 2HCl(aq) \rightarrow MgCl_2(aq) + 2H_2O(l)$
(b) $MgCl_2(l) \rightarrow Mg(l) + Cl_2(g)$

Chapter 24

24.1 $PbO(s) + OH^-(aq) + H_2O(l) \rightarrow Pb(OH)_3^-$
$Pb(OH)_3^-(aq) + OCl^-(aq) \rightarrow$
$\qquad PbO_2(s) + OH^-(aq) + H_2O(l) + Cl^-(aq)$
24.2 $2NaSn(OH)_3(aq) \rightarrow Sn(s) + Na_2Sn(OH)_6(aq)$
24.3 $P_4(s) + 5O_2(g) \rightarrow P_4O_{10}(s)$;
$P_4O_{10}(s) + 6H_2O(l) \rightarrow 4H_3PO_4(aq)$;
$H_3PO_4(aq) + 3NaOH(aq) \rightarrow Na_3PO_4(aq) + 3H_2O(l)$

24.4 $H_3AsO_3(aq) + 3Zn(s) + 6H^+(aq) \rightarrow$
$\qquad AsH_3(g) + 3Zn^{2+}(aq) + 3H_2O(l)$
24.5 $S(s) + O_2(g) \rightarrow SO_2(g)$
$SO_2(g) + Na_2CO_3(aq) \rightarrow Na_2SO_3(aq) + CO_2(g)$
$Na_2SO_3(aq) + S(s) \xrightarrow{\Delta} Na_2S_2O_3(aq)$
24.6 $I_2(aq) + 2S_2O_3^{2-}(aq) \rightarrow 2I^-(aq) + S_4O_6^{2-}(aq)$

24.7 $2NaCl(aq) + 2H_2O(l) \xrightarrow{electrolysis}$
$\qquad 2NaOH(aq) + H_2(g) + Cl_2(g)$
$3Cl_2(g) + 6NaOH(aq) \rightarrow$
$\qquad NaClO_3(aq) + 5NaCl(aq) \rightarrow 3H_2O(l)$
$2NaClO_3(aq) + SO_2(g) + H_2SO_4(aq) \rightarrow$
$\qquad 2ClO_2(g) + 2NaHSO_4(aq)$
24.8 $NaBrO_3(aq) + F_2(g) + 2NaOH(aq) \rightarrow$
$\qquad NaBrO_4(aq) + 2NaF(aq) + H_2O(l)$

24.9 $E°_{cell} = -0.11$ V for disproportionation to Cl^- and
ClO_3^-.

Chapter 25

25.1 VF_5 **25.2** $KMnO_4$
25.3 $Cu(s) + 2H_2SO_4(l) \rightarrow$
$\qquad [Cu^{2+}(aq) + SO_4^{2-}(aq)] + SO_2(g) + 2H_2O(l)$
$2Cu^{2+}(aq) + 4I^-(aq) \rightarrow 2CuI(s) + I_2(aq)$
25.4 +6
25.5 $Cr_2O_7^{2-}(aq) + 14H^+(aq) + 6I^-(aq) \rightarrow$
$\qquad 2Cr^{3+}(aq) + 3I_2(aq) + 7H_2O(l)$
25.6 $K_2[PtCl_6]$ **25.7** (a) pentaamminechlorocobalt(III)
chloride (b) potassium aquapentacyanocobaltate(III)
(c) pentaaquahydroxoiron(III) ion
25.8 (a) $K_3[Fe(CN)_6]$ (b) $[Co(NH_3)_4Cl_2]Cl$ (c) $PtCl_4^{2-}$
25.9 (a) coordination isomers (b) linkage isomers
(c) ionization isomers (d) hydrate isomers
25.10 $[Co(NH_3)_4(H_2O)Cl]Cl_2$; $[Co(NH_3)_4Cl_2]Cl \cdot H_2O$ is
a possibility
25.11 Geometric isomers are possible for (b) and (c).
25.12 Only (b) and (c); (b) is similar to Figure 25.13;
(c) is similar to Figure 25.17.
25.13 $Co(NH_3)_6^{3+}$ has d^2sp^3 bonding and 0 unpaired
electrons; $CoCl_6^{3-}$ has sp^3d^2 bonding and 4 unpaired
electrons **25.14** Tetrahedral **25.15**

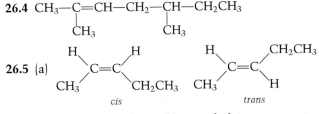

2 unpaired electrons **25.16**

25.17 530 nm, 500 nm, yes

Chapter 26

26.1 (a) 2,3-dimethylbutane (b) 3-ethyl-2-
methylhexane

26.2
$$CH_3CH_2-\overset{\overset{CH_3}{|}}{\underset{\underset{CH_3}{|}}{C}}-CH_2CH_2CH_2CH_2CH_3$$

26.3 (a) 2,4-dimethyl-2-hexene (b) 3-propyl-1-hexene

26.4 $CH_3-\overset{\overset{|}{C}}{\underset{\underset{CH_3}{|}}{}}=CH-CH_2-\overset{\overset{|}{CH}}{\underset{\underset{CH_3}{|}}{}}-CH_2CH_3$

26.5 (a)

(b) none **26.6** (a) propyne (b) 3-methyl-1-pentyne

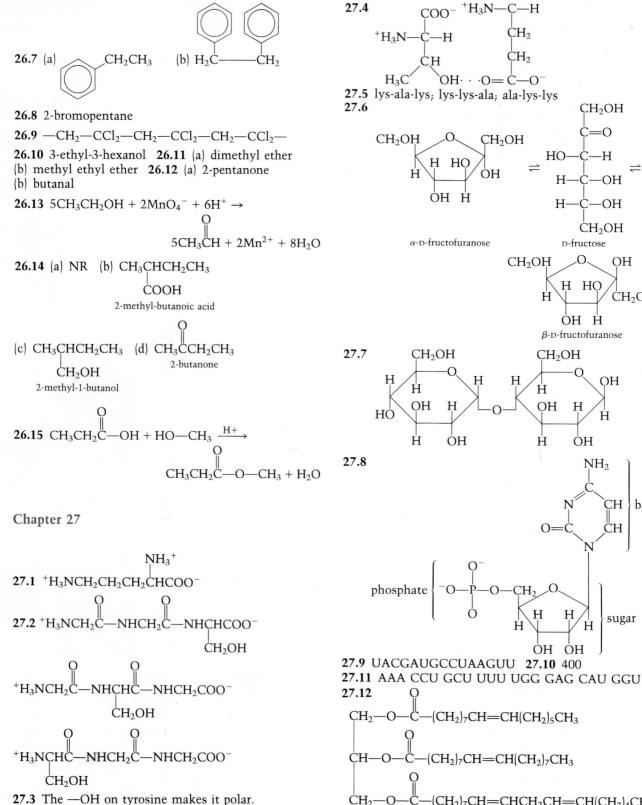

26.7 (a) $C_6H_5CH_2CH_3$ (b) H_2C-CH_2 (cyclopropane)

26.8 2-bromopentane

26.9 $-CH_2-CCl_2-CH_2-CCl_2-CH_2-CCl_2-$

26.10 3-ethyl-3-hexanol **26.11** (a) dimethyl ether (b) methyl ethyl ether **26.12** (a) 2-pentanone (b) butanal

26.13 $5CH_3CH_2OH + 2MnO_4^- + 6H^+ \rightarrow$
$$5CH_3\overset{O}{\overset{\|}{C}}H + 2Mn^{2+} + 8H_2O$$

26.14 (a) NR   (b) $CH_3CHCH_2CH_3$ with $COOH$ — 2-methyl-butanoic acid

(c) $CH_3CHCH_2CH_3$ with CH_2OH — 2-methyl-1-butanol   (d) $CH_3\overset{O}{\overset{\|}{C}}CH_2CH_3$ — 2-butanone

26.15 $CH_3CH_2\overset{O}{\overset{\|}{C}}-OH + HO-CH_3 \xrightarrow{H+}$
$$CH_3CH_2\overset{O}{\overset{\|}{C}}-O-CH_3 + H_2O$$

Chapter 27

27.1 $^+H_3NCH_2CH_2CH_2\overset{NH_3^+}{\overset{|}{C}}HCOO^-$

27.2 $^+H_3NCH_2\overset{O}{\overset{\|}{C}}-NHCH_2\overset{O}{\overset{\|}{C}}-NH\overset{|}{C}HCOO^-$ with CH_2OH

$^+H_3NCH_2\overset{O}{\overset{\|}{C}}-NH\overset{|}{C}H\overset{O}{\overset{\|}{C}}-NHCH_2COO^-$ with CH_2OH

$^+H_3N\overset{|}{C}H\overset{O}{\overset{\|}{C}}-NHCH_2\overset{O}{\overset{\|}{C}}-NHCH_2COO^-$ with CH_2OH

27.3 The —OH on tyrosine makes it polar.

27.4

27.5 lys-ala-lys; lys-lys-ala; ala-lys-lys

27.6

α-D-fructofuranose   D-fructose

β-D-fructofuranose

27.7

27.8

base / phosphate / sugar

27.9 UACGAUGCCUAAGUU **27.10** 400

27.11 AAA CCU GCU UUU UGG GAG CAU GGU

27.12

$CH_2-O-\overset{O}{\overset{\|}{C}}-(CH_2)_7CH=CH(CH_2)_5CH_3$

$CH-O-\overset{O}{\overset{\|}{C}}-(CH_2)_7CH=CH(CH_2)_7CH_3$

$CH_2-O-\overset{O}{\overset{\|}{C}}-(CH_2)_7CH=CHCH_2CH=CH(CH_2)_4CH_3$

Answers to Odd-Numbered Problems

Chapter 1

1.7 4.4 g **1.9** (a) 5 (b) 3 (c) 5 (d) 4 (e) 4 (f) 3
1.11 1.000×10^3 **1.13** (a) 8.73 (b) 57.7 (c) 3.56
(d) 8.76 (e) 5.04×10^4 (f) 8.54×10^2 **1.15** (a) 4.7
(b) 82.5 (c) 111 (d) 2.3×10^3 **1.17** (a) 2.31 pm
(b) 5.43 ns (c) 8.7 μg (d) 9.3 mm **1.19** (a) 0°C
(b) −50°C (c) 20°C (d) −24°C (e) 99°F (f) −107°F
1.21 (a) −273°C (b) −460°F **1.23** 3.10×10^6 yd^2
1.25 6.494 gallons **1.27** 3.25×10^6 g **1.29** 4435 m
1.31 (a) 7.36×10^6 mg (b) 6.55×10^{-1} ms
(c) 5.7×10^{13} nm (d) 2.54 cm **1.33** 3.73×10^{17} m^3,
3.73×10^{20} L **1.35** 7.56 g/cm^3 **1.37** Benzene
1.39 1.3×10^2 g **1.41** 25.1 cm^3 **1.43** 1.58 g/cm^3
1.45 1.02 atm; 103 kPa **1.47** 567.4 kJ **1.49** 9.4 cal
1.51 2.1×10^3 J **1.53** kg·m/s **1.55** 5.4 g
1.57 Chloroform **1.59** 1.59×10^3 J/g **1.61** 34% (by
mass), 79 proof **1.63** 10.1 m

Chapter 2

2.11 (a) Potassium, sulfur, carbon, nitrogen
(b) Sodium, phosphorus, oxygen (c) Copper, sulfur,
oxygen (d) Carbon, chlorine
2.13 4.6×10^{22} C atoms, 5.2×10^{22} H atoms
2.15 (a) N_2H_4 (b) H_2O_2 (c) C_3H_8O **2.17** $Al_2(SO_4)_3$
2.19 1 to 6 **2.21** $Ba_3(PO_4)_2$ **2.23** 12 O atoms
2.25 (a) $Sn + 2NaOH \rightarrow Na_2SnO_2 + H_2$
(b) $8Al + 3Fe_3O_4 \rightarrow 4Al_2O_3 + 9Fe$
(c) $2CH_3OH + 3O_2 \rightarrow 2CO_2 + 4H_2O$
(d) $P_4O_{10} + 6H_2O \rightarrow 4H_3PO_4$
(e) $PCl_5 + 4H_2O \rightarrow H_3PO_4 + 5HCl$

2.27 (a) $SbCl_5 + H_2O \rightarrow SbOCl_3 + 2HCl$
(b) $2MgO + Si \rightarrow 2Mg + SiO_2$
(c) $CaCl_2 + Na_2CO_3 \rightarrow CaCO_3 + 2NaCl$
(d) $2C_6H_6 + 15O_2 \rightarrow 12CO_2 + 6H_2O$
(e) $Al_2S_3 + 6H_2O \rightarrow 2Al(OH)_3 + 3H_2S$
2.29 $Ca_3(PO_4)_2 + 3H_2SO_4 \rightarrow 2H_3PO_4 + 3CaSO_4$
2.31 $2NH_4Cl + Ba(OH)_2 \rightarrow 2NH_3 + BaCl_2 + 2H_2O$
2.33 (a) compound (b) element (c) mixture
(d) mixture (e) compound **2.35** (a) gas (b) solid
(c) gas (d) liquid **2.37** (a) Fractional distillation
(b) Add hot water to the mixture to dissolve NaCl.
Then remove AgCl by filtration and recover NaCl
from the water by evaporation. (c) Use a magnet to
remove the iron. (d) Use thin layer or paper
chromatography to separate the pigments.
2.39 7.0×10^{24} O atoms
2.41 $4NH_3 + 5O_2 \rightarrow 4NO + 6H_2O$ **2.43** Add CS_2 to
the final reaction mixture to dissolve the sulfur,
which can then be filtered away from the FeS and Fe
and recovered by evaporation of the CS_2. Remove Fe
from FeS with a magnet.

Chapter 3

3.15 (a) 32.0 amu (b) 137 amu (c) 138 amu
(d) 366 amu **3.17** 158 g/mol to 3 significant figures
3.19 1.2×10^{-22} g **3.21** 33.6 g
3.23 5.81×10^{-3} mol $CaSO_4$; 1.16×10^{-2} mol H_2O
3.25 3.34×10^{19} CCl_4 molecules **3.27** 86.27%
3.29 34.8% **3.31** 0.656 kg N **3.33** 58.8% C,
27.4% N, 13.8% H **3.35** Both compounds have the
same empirical formula. **3.37** 20.0% C, 6.74% H,

26.6% O   **3.39** K_2MnO_4 **3.41** $C_3H_4O_2$
3.43 $C_4H_{12}N_2$ **3.45** $C_2H_2O_4$
3.47 C_2H_4 + $3O_2$ → $2CO_2$ + $2H_2O$

1 molecule C_2H_4 + 3 molecules O_2 → 2 molecules CO_2 + 2 molecules H_2O
1 mole C_2H_4 + 3 moles O_2 →   2 moles CO_2 + 2 moles H_2O
28.054 g C_2H_4 + 95.9964 g O_2 → 88.018 g CO_2 + 36.0306 g H_2O

3.49 3.81×10^3 g W   **3.51** 6.45×10^6 g NO_2
3.53 15.5 g CS_2 **3.55** 5.33 g CO_2 **3.57** H_2SO_4 is the
limiting reactant; 3.72 g HCl produced; 4.0 g
unreacted NaCl   **3.59** The limiting reactant is
salicylic acid; theoretical yield is 2.61 g; % yield is
84.7%   **3.61** 1.36 M **3.63** 0.106 M
3.65 3.8×10^{-5} mol heme   **3.67** 0.37 g $K_2Cr_2O_7$
3.69 1.25 L   **3.71** 35.8 mL   **3.73** 1.6×10^2 mL
3.75 4.0 mL   **3.77** Dilute 2.51 mL concentrated HNO_3
to a volume of 255 mL   **3.79** 63.7 mL HNO_3
3.81 74.2 mL   **3.83** 2.89%   **3.85** 49.5% C; 5.19% H;
28.8% N; 16.5% O   **3.87** N_2O_3 **3.89** $C_6H_4Cl_2$
3.91 $AuCl_3$ **3.93** $C_{12}H_8Cl_6O$ **3.95** C_4H_4S, C_4H_4S
3.97 616 amu   **3.99** 86.2%   **3.101** 1.63×10^3 g S_8
3.103 58.2%   **3.105** 60.3 g Zn   **3.107** 2.03×10^{-2} M;
2.03×10^{-2} M Ca^{2+}; 4.06×10^{-2} M Cl^-
3.109 329 mL   **3.111** 0.610 M **3.113** 9.66%
3.115 1.14 kg CaC_2 **3.117** 16.2%   **3.119** 1.00×10^2 kg;
76.2%

Chapter 4

4.21 652 Pa   **4.23** 9.97 m   **4.25** 3.41 L   **4.27** 679 L
4.29 3.32×10^{-4} kPa   **4.31** 2.50 mL   **4.33** $-78°C$
4.35 30.6 mL   **4.37** 1 volume NO   **4.39** 15.7 L
4.41 Solve the ideal gas law for V:
$V = \dfrac{nRT}{P} = (nRT)\left(\dfrac{1}{P}\right)$. If the temperature and
number of moles are held constant, then the first
expression in parentheses is a constant:
$V = \text{constant} \times \dfrac{1}{P}$ or $V \propto \dfrac{1}{P}$ **4.43** 35°C

4.45 4.99 atm   **4.47** 0.00224 mol, 0.0985 g CO_2
4.49 6.80 g Zn   **4.51** 585 L   **4.53** 1.18 g/L
4.55 2.38 g/L   **4.57** At the same T and P, the density
of a gas of NH_4Cl would be greater than that of a
mixture of NH_3 and HCl.   **4.59** 45.1 amu
4.61 $P_{O_2} = 3.80 \times 10^{-3}$ atm; $P_{He} = 1.2 \times 10^{-2}$ atm;
$P = 1.6 \times 10^{-2}$ atm   **4.63** $P_{CO_2} = 494$ mmHg;
$P_{H_2} = 190$ mmHg; $P_{HCl} = 41$ mmHg; $P_{HF} = 21$ mmHg;
$P_{SO_2} = 13$ mmHg; $P_{H_2S} = 0.8$ mmHg;   **4.65** 7.59 g
4.67 153 m/s   **4.69** 0.169 km/s   **4.71** 146 amu
4.73 4.6 s   **4.75** 80.7 cm^3 **4.77** 165 mL
4.79 6.5 dm^3 **4.81** 3.01×10^{20} atoms
4.83 1.2×10^3 g   **4.85** 28.979 g/mol   **4.87** $-95°C$
4.89 28.08 g/mol   **4.91** O_2, CO_2; $P_{O_2} = 0.167$ atm;
$P_{CO_2} = 0.333$ atm   **4.93** Van der Waal's: V at
1.00 atm = 22.5 L/mol; V at 10.00 atm = 2.30 L/mol.

Ideal gas law: V at 1.00 atm = 22.4 L/mol; V at
10.00 atm = 2.24 L/mol.

Chapter 5

5.21 9.05×10^{-31} kg   **5.23** 63, 60   **5.25** $^{64}_{30}Zn$
5.27 Cl-35: 18; Cl-37: 20   **5.29** $^{45}_{21}Sc^{3+}$ **5.31** 192 amu
5.33 24.31 amu   **5.35** B-10: 0.20; B-11: 0.80
5.37 1.9×10^2 s   **5.39** 214 m   **5.41** 5.00×10^{14}/s
5.43 9.276×10^{-28} J   **5.45** 3.72×10^{-19} J
5.47 1.60×10^{14}/s   **5.49** 1.22×10^{-7} m   **5.51** 7
5.53 2.50×10^{-19} J   **5.55** $\ell = 0, 1, 2, 3$;
$m_\ell = -3, -2, -1, 0, 1, 2, 3$ **5.57** 3; 7   **5.59** (a) $3p$
(b) $4d$ (c) $4s$ (d) $4f$ **5.61** (a) impossible
(b) impossible (c) possible (d) impossible
(e) possible   **5.63** -1.60×10^{-19} C; 2, 4, 6, 7
5.65 6.51×10^{14}/s, 4.3×10^{-19} J   **5.67** ^9_4Be: 4 protons,
5 neutrons; $^{25}_{12}Mg$: 12 protons, 13 neutrons; $^{59}_{27}Co$: 27
protons, 32 neutrons   **5.69** $^{138}_{56}Ba$ **5.71** 0.482
5.73 31.972 amu; 33.967 amu; 31.971 amu;
33.968 amu; 1, 1, 2, 2.   **5.75** 5.41×10^{14}/s
5.77 205 nm; near ultraviolet   **5.79** 3.10 pm

Chapter 6

6.19 (a) not allowed   (b) allowed; $1s^2 2s^2 2p^2$ (c) not
allowed   (d) allowed; $1s^2 2s^2 2p^4$ **6.21** (a) impossible;
$2s$ subshell can hold no more than 2 electrons
(b) possible   (c) impossible; $3p$ subshell can hold no
more than 6 electrons   (d) possible
6.23 $1s$ $2p$
(↑↓) (↑)( )( )
(↑↓) (↓)( )( )
(↑↓) ( )(↑)( )
(↑↓) ( )(↓)( )
(↑↓) ( )( )(↑)
(↑↓) ( )( )(↓)
6.25 $1s^2 2s^2 2p^6 3s^2 3p^1$ **6.27** $1s^2 2s^2 2p^6 3s^2 3p^6 3d^2 4s^2$
6.29 (a) $[Xe]4f^{14} 5d^9 6s^2$;   (b) $[Xe]4f^{14} 5d^{10} 6s^1$
6.31 $[Ar](↑↓)(↑↓)(↑)(↑)(↑)$ (↑↓)
         $3d$ $4s$
6.33 diamagnetic   **6.35** $5s^2 5p^1$ **6.37** $5d^5 6s^2$ **6.39** 6th
period, group IIIA, main group   **6.41** O → S → P
6.43 Ga → Al → Si   **6.45** small positive: Kr, Cd, Sr;
small negative: P; large negative: S, F   **6.47** As_2O_3 or
As_2O_5 **6.49** $[Rn]5f^{14} 6d^{10} 7s^2 7p^2$, metal, eka-PbO or
eka-PbO_2 **6.51** ~370 kJ/mol
6.53 $1s^2 2s^2 2p^6 3s^2 3p^6 3d^3 4s^2$; Group VB, period 4,
transition element
6.55 $[Kr](↑)(↑)(↑)(↑)( )$ (↑)
         $4d$ $5s$
6.57 (a) Cl_2 (b) Na (c) Sb (d) Ar

Chapter 7

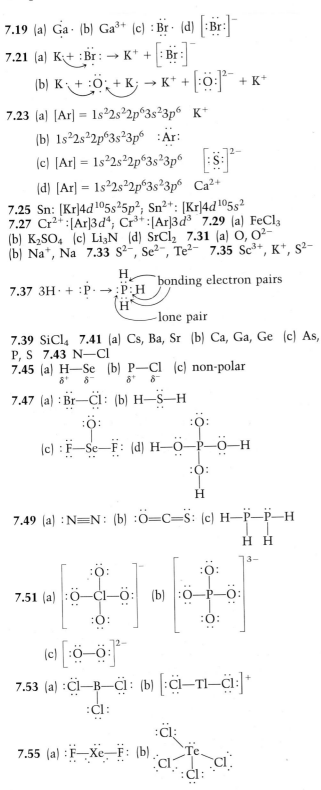

7.19 (a) Ga· (b) Ga^{3+} (c) :Br· (d) [:Br:]$^-$

7.21 (a) K· + :Br: → K$^+$ + [:Br:]$^-$

(b) K· + :O· + K· → K$^+$ + [:O:]$^{2-}$ + K$^+$

7.23 (a) [Ar] = $1s^2 2s^2 2p^6 3s^2 3p^6$ K$^+$

(b) $1s^2 2s^2 2p^6 3s^2 3p^6$:Ar:

(c) [Ar] = $1s^2 2s^2 2p^6 3s^2 3p^6$ [:S:]$^{2-}$

(d) [Ar] = $1s^2 2s^2 2p^6 3s^2 3p^6$ Ca^{2+}

7.25 Sn: [Kr]$4d^{10}5s^25p^2$; Sn^{2+}: [Kr]$4d^{10}5s^2$
7.27 Cr^{2+}:[Ar]$3d^4$; Cr^{3+}:[Ar]$3d^3$ **7.29** (a) FeCl$_3$
(b) K$_2$SO$_4$ (c) Li$_3$N (d) SrCl$_2$ **7.31** (a) O, O^{2-}
(b) Na$^+$, Na **7.33** S^{2-}, Se^{2-}, Te^{2-} **7.35** Sc^{3+}, K$^+$, S^{2-}

7.37 3H· + ·P· → structure showing bonding electron pairs and lone pair

7.39 SiCl$_4$ **7.41** (a) Cs, Ba, Sr (b) Ca, Ga, Ge (c) As, P, S **7.43** N—Cl

7.45 (a) H—Se (b) P—Cl (c) non-polar
δ^+ δ^- δ^+ δ^-

7.47 (a) :Br—Cl: (b) H—S—H

(c) :F—Se—F: (d) H—O—P—O—H (with O and H)

7.49 (a) :N≡N: (b) :O=C=S: (c) H—P—P—H (with H H)

7.51 (a) [:O—Cl—O:]$^-$ (b) [:O—P—O:]$^{3-}$

(c) [:O—O:]$^{2-}$

7.53 (a) :Cl—B—Cl: (b) [:Cl—Tl—Cl:]$^+$

7.55 (a) :F—Xe—F: (b) TeCl structure

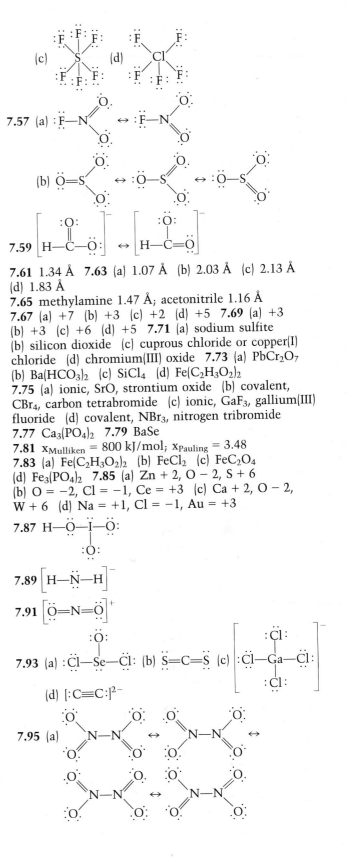

7.57 (a) resonance structures of F—N—O
(b) O=S—O resonance structures

7.59 [H—C—O:]$^-$ ↔ [H—C=O]$^-$ resonance

7.61 1.34 Å **7.63** (a) 1.07 Å (b) 2.03 Å (c) 2.13 Å
(d) 1.83 Å
7.65 methylamine 1.47 Å; acetonitrile 1.16 Å
7.67 (a) +7 (b) +3 (c) +2 (d) +5 **7.69** (a) +3
(b) +3 (c) +6 (d) +5 **7.71** (a) sodium sulfite
(b) silicon dioxide (c) cuprous chloride or copper(I)
chloride (d) chromium(III) oxide **7.73** (a) PbCr$_2$O$_7$
(b) Ba(HCO$_3$)$_2$ (c) SiCl$_4$ (d) Fe(C$_2$H$_3$O$_2$)$_2$
7.75 (a) ionic, SrO, strontium oxide (b) covalent,
CBr$_4$, carbon tetrabromide (c) ionic, GaF$_3$, gallium(III)
fluoride (d) covalent, NBr$_3$, nitrogen tribromide
7.77 Ca$_3$(PO$_4$)$_2$ **7.79** BaSe
7.81 x$_{Mulliken}$ = 800 kJ/mol; x$_{Pauling}$ = 3.48
7.83 (a) Fe(C$_2$H$_3$O$_2$)$_2$ (b) FeCl$_2$ (c) FeC$_2$O$_4$
(d) Fe$_3$(PO$_4$)$_2$ **7.85** (a) Zn + 2, O − 2, S + 6
(b) O = −2, Cl = −1, Ce = +3 (c) Ca + 2, O − 2,
W + 6 (d) Na = +1, Cl = −1, Au = +3

7.87 H—O—I—O: (with O)

7.89 [H—N—H]$^-$

7.91 [O=N=O]$^+$

7.93 (a) :Cl—Se—Cl: (b) S=C=S (c) [:Cl—Ga—Cl:]$^-$ (with Cl)

(d) [:C≡C:]$^{2-}$

7.95 (a) resonance structures of N—N with O groups

(b) $\left[\ddot{\text{O}}\!=\!\text{C}\!=\!\ddot{\text{N}}\right]^{-} \leftrightarrow \left[:\ddot{\text{O}}\!-\!\text{C}\!\equiv\!\text{N}:\right]^{-}$

Chapter 8

8.17 (a) bent (b) linear (c) bent (d) linear
(e) tetrahedral (f) tetrahedral (g) trigonal pyramidal
(h) trigonal pyramidal **8.19** (a) linear (b) bent
(c) trigonal planar (d) tetrahedral (e) trigonal
pyramidal **8.21** (a) distorted tetrahedral (b) trigonal
bipyramidal (c) square pyramidal **8.23** (a) trigonal
bipyramidal (b) linear (c) square planar
8.25 (a) trigonal pyramidal, T-shaped (b) bent
8.27 (b) and (d) **8.29** (a) sp (b) sp^3d^2 (c) sp^3d
(d) sp^3 **8.31** (a) The Hg atom is first promoted to
$[\text{Xe}]4f^{14}5d^{10}6s^16p^1$, then sp hybridized. An Hg—Cl
bond is formed by overlapping an Hg hybrid orbital
with a $3p$ orbital of Cl. (b) The P atom is sp^3
hybridized; each hybrid overlaps a $3p$ orbital of a Cl
atom. **8.33** P is sp^3d^2 hybridized. Each P—Cl bond is
formed by the overlap of a hybrid orbital on P with a
$3p$ orbital on Cl. **8.35** (a) The C atom is sp^2
hybridized and forms a double bond with O. The σ
bond of C=O is formed by the overlap of a C hybrid
orbital with a $2p$ O orbital oriented along the bond
axis. The π bond is formed by the sidewise overlap of
a $2p$ orbital on C with a $2p$ orbital on O. Each of the
remaining sp^2 hybrid orbitals forms a bond by
overlapping with a $1s$ H orbital. (b) Each N atom is
sp hybridized and forms a triple bond (a σ and two π
bonds) to the other N atom. The σ bond of N≡N is
formed by the overlap of hybrid orbitals. Each π bond
is formed by the sidewise overlap of $2p$ orbitals from
the N atoms.

8.37

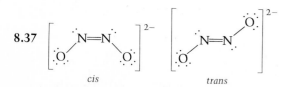

cis   *trans*

8.39 (a) $(\sigma_{1s})^2(\sigma_{1s}^*)^2(\sigma_{2s})^2(\sigma_{2s}^*)(\pi_{2p})^2$; bond order = 1;
stable; paramagnetic. (b) $(\sigma_{1s})^2(\sigma_{1s}^*)^2(\sigma_{2s})^2(\sigma_{2s}^*)(\pi_{2p})^1$;
bond order = $\frac{1}{2}$; stable; paramagnetic.
8.41 $(\sigma_{1s})^2(\sigma_{1s}^*)^2(\sigma_{2s})^2(\sigma_{2s}^*)^2(\pi_{2p})^4(\sigma_{2p})^2$; bond order = 3;
diamagnetic **8.43** (a) and (c) **8.45** The first two C
atoms are sp^2; the third one is sp^3. **8.47** *cis* isomer
has a net dipole moment **8.49** $(\sigma_{1s})^2$; stable
8.51 $(\sigma_{1s})^2(\sigma_{1s}^*)^2(\sigma_{2s})^2(\sigma_{2s}^*)^2(\pi_{2p})^4(\sigma_{2p})^2$ bond order = 3
8.53 The bond order is greater in O_2^+, so the O_2^+
bond length is shorter. However, O_2^- has a smaller
bond order.

Chapter 9

9.19 (a) $\text{HF}(aq) + \text{OH}^-(aq) \rightarrow \text{F}^-(aq) + \text{H}_2\text{O}(l)$
(b) $\text{Ag}^+(aq) + \text{Br}^-(aq) \rightarrow \text{AgBr}(s)$ **9.21** (a) oxidation-
reduction (b) metathesis (c) metathesis
(d) oxidation-reduction **9.23** (a) insoluble
(b) soluble (c) insoluble (d) soluble **9.25** (a) NR
(b) $\text{Na}_2\text{CO}_3(aq) + \text{CaCl}_2(aq) \rightarrow \text{CaCO}_3(s) + 2\text{NaCl}(aq)$;
$\text{Ca}^{2+}(aq) + \text{CO}_3^{2-}(aq) \rightarrow \text{CaCO}_3(s)$
(c) $\text{MgSO}_4(aq) + 2\text{NaOH}(aq) \rightarrow$
$$\text{Mg(OH)}_2(s) + \text{Na}_2\text{SO}_4(aq);$$
$\text{Mg}^{2+}(aq) + 2\text{OH}^-(aq) \rightarrow \text{Mg(OH)}_2(s)$ (d) NR
9.27 (a) $\text{HNO}_3(aq) + \text{NaOH}(aq) \rightarrow$
$$\text{H}_2\text{O}(l) + \text{NaNO}_3(aq);$$
$\text{H}^+(aq) + \text{OH}^-(aq) \rightarrow \text{H}_2\text{O}(l)$
(b) $\text{HC}_2\text{H}_3\text{O}_2(aq) + \text{KOH}(aq) \rightarrow \text{H}_2\text{O}(l) + \text{KC}_2\text{H}_3\text{O}_2(aq)$;
$\text{HC}_2\text{H}_3\text{O}_2(aq) + \text{OH}^-(aq) \rightarrow \text{H}_2\text{O}(l) + \text{C}_2\text{H}_3\text{O}_2^-(aq)$
9.29 (a) $\text{H}_2\text{C}_2\text{O}_4(aq) + \text{NaOH}(aq) \rightarrow$
$$\text{H}_2\text{O}(l) + \text{NaHC}_2\text{O}_4(aq)$$
(b) $\text{NaHC}_2\text{O}_4(aq) + \text{NaOH}(aq) \rightarrow$
$$\text{H}_2\text{O}(l) + \text{Na}_2\text{C}_2\text{O}_4(aq)$$
9.31 $\text{FeS}(s) + 2\text{HCl}(aq) \rightarrow \text{FeCl}_2(aq) + \text{H}_2\text{S}(g)$;
$\text{FeS}(s) + 2\text{H}^+(aq) \rightarrow \text{Fe}^{2+}(aq) + \text{H}_2\text{S}(g)$ **9.33** Add strong
acid; for example,
$\text{CaSO}_3(s) + 2\text{HCl}(aq) \rightarrow \text{CaCl}_2(aq) + \text{H}_2\text{O}(l) + \text{SO}_2(g)$
9.35 $\text{NaI}(s) + \text{H}_3\text{PO}_4(l) \rightarrow \text{NaH}_2\text{PO}_4(s) + \text{HI}(g)$
9.37 $\text{Ba(ClO}_3)_2(aq) + \text{H}_2\text{SO}_4(aq) \rightarrow$
$$\text{BaSO}_4(s) + 2\text{HClO}_3(aq)$$
9.39 $\text{Ba(OH)}_2(aq) + \text{Li}_2\text{SO}_4(aq) \rightarrow$
$$\text{BaSO}_4(s) + 2\text{LiOH}(aq)$$
9.41 $\text{N}_2\text{H}_6\text{Cl}_2(aq) + \text{Ca(OH)}_2(aq) \xrightarrow{\Delta}$
$$\text{N}_2\text{H}_4(g) + 2\text{H}_2\text{O}(l) + \text{CaCl}_2(g)$$
9.43 (a) $2\text{Al(OH)}_3(s) + 3\text{H}_2\text{SO}_4(aq) \rightarrow$
$$6\text{H}_2\text{O}(l) + \text{Al}_2(\text{SO}_4)_3(aq)$$
(b) $\text{CuS}(s) + \text{H}_2\text{SO}_4(aq) \rightarrow \text{CuSO}_4(aq) + \text{H}_2\text{S}(g)$
(c) $3\text{CaCl}_2(aq) + 2\text{Na}_3\text{PO}_4(aq) \rightarrow$
$$\text{Ca}_3(\text{PO}_4)_2(s) + 6\text{NaCl}(aq)$$
9.45 (a) $\text{MgSO}_4(aq) + \text{BaCl}_2(aq) \rightarrow$
$$\text{BaSO}_4(s) + \text{MgCl}_2(aq)$$
(b) $\text{SrI}_2(aq) + \text{Pb(NO}_3)_2(aq) \rightarrow \text{PbI}_2(s) + \text{Sr(NO}_3)_2(aq)$
9.47 (a) oxidizing agent: F_2 (b) oxidizing agent: Hg^{2+};
reducing agent: Al   reducing agent: NO_2^-

9.49

$$\text{CoS} + \text{NO}_3^- \rightarrow \text{S} + \text{NO}$$

with oxidation numbers: -2, $+5$, 0, $+2$; oxidation from -2 to 0; reduction from $+5$ to $+2$; reducing agent, oxidizing agent, oxidized, reduced

9.51 oxidation: $\text{Ni} \rightarrow \text{Ni}^{2+} + 2\text{e}^-$
reduction: $2\text{e}^- + \text{Cu}^{2+} \rightarrow \text{Cu}$

9.53 (a) $8S^{2-} + 32H^+ + 16NO_3^- \rightarrow S_8 + 16NO_2 + 16H_2O$
(b) $3Cu + 2NO_3^- + 8H^+ \rightarrow 3Cu^{2+} + 2NO + 4H_2O$
(c) $5SO_2 + 2MnO_4^- + 2H_2O \rightarrow 5SO_4^{2-} + 2Mn^{2+} + 4H^+$
(d) $3Sn(OH)_3^- + 2Bi(OH)_3 + 3OH^- \rightarrow$
$$3Sn(OH)_6^{2-} + 2Bi$$
9.55 (a) $3AsH_3 + 4KClO_3 \rightarrow 3H_3AsO_4 + 4KCl$
$\qquad\qquad$ oxidizing
$\qquad\qquad$ agent
(b) $2SnCl_2 + O_2 + 8HCl \rightarrow 2H_2SnCl_6 + 2H_2O$
$\qquad\qquad$ oxidizing
$\qquad\qquad$ agent
(c) $MnO_2 + 4HBr \rightarrow Br_2 + MnBr_2 + 2H_2O$
$\quad$ oxidizing
$\quad$ agent
(d) $P_4 + 3NaOH + 3H_2O \rightarrow 3NaH_2PO_2 + PH_3$
$\quad$ oxidizing
$\quad$ agent
9.57 29.16 g; 0.300 g **9.59** 49.00 g; 4.10×10^{-3} eq; 0.201 g
9.61 (a) $LiOH(aq) + HCN(aq) \rightarrow H_2O(l) + LiCN(aq)$;
$OH^-(aq) + HCN(aq) \rightarrow H_2O(l) + CN^-(aq)$
(b) $Li_2CO_3(aq) + 2HNO_3(aq) \rightarrow$
$$H_2O(l) + CO_2(g) + 2LiNO_3(aq);$$
$CO_3^{2-}(aq) + 2H^+(aq) \rightarrow H_2O(l) + CO_2(g)$
(c) $LiCl(aq) + AgNO_3(aq) \rightarrow LiNO_3(aq) + AgCl(s)$
$Cl^-(aq) + Ag^+(aq) \rightarrow AgCl(s)$ (d) NR **9.63** (a) Add $BaCl_2$ to $CuSO_4(aq)$; filter off $BaSO_4$ and evaporate filtrate. (b) Dissolve $CaCO_3$ in $HC_2H_3O_2(aq)$; CO_2 bubbles off; evaporate H_2O to give $Ca(C_2H_3O_2)_2$.
(c) Dissolve Na_2SO_3 in HNO_3. SO_2 bubbles off, leaving $NaNO_3(aq)$. (d) Neutralize $Mg(OH)_2$ with HCl; evaporate H_2O.
9.65 (a) $2MnO_4^- + 3S^{2-} + 4H_2O \rightarrow$
$$2MnO_2 + 3S + 8OH^-$$
(b) $IO_3^- + 3HSO_3^- \rightarrow I^- + 3SO_4^{2-} + 3H^+$
(c) $P_4 + 3OH^- + 3H_2O \rightarrow PH_3 + 3H_2PO_2^-$
(d) $Cl_2 + 2OH^- \rightarrow Cl^- + ClO^- + H_2O$
9.67 $4Fe(OH)_2 + O_2 + 2H_2O \rightarrow 4Fe(OH)_3$
9.69 0.521 N

Chapter 10

10.13 endothermic; $q = 164$ kJ
10.15 $HgO(s) \rightarrow Hg(l) + \frac{1}{2}O_2(g)$, $\Delta H = 90.8$ kJ
10.17 0.385 J/(g·°C) **10.19** 601 J **10.21** 0.86 J/(g·°C), 1.3×10^2 J/(mol·°C) **10.23** 0.383 J/(g·°C)
10.25 20.5 kJ **10.27** 9.256 kJ **10.29** +40.6 kJ
10.31 +411 kJ **10.33** −430.4 kJ **10.35** 709.0 kJ
10.37 −137 kJ **10.39** +43 kJ **10.41** +735 kJ
10.43 −835 kJ **10.45** −75.2 kJ **10.47** −78 kJ
10.49 0.97, 3.1 **10.51** 250°C **10.53** −4.96 kJ, −40.9 kJ **10.55** −23.6 kJ **10.57** −701.8 kJ
10.59 −20 kJ **10.61** +206.2 kJ **10.63** 178 kJ

10.65 323.8 kJ **10.67** 860 kJ **10.69** 0.937
10.71 −349.2 kJ

Chapter 11

11.19 (a) London forces (b) London forces, dipole-dipole, H-bonding (c) London forces, dipole-dipole (d) London forces **11.21** CCl_4, $SiCl_4$, $GeCl_4$
11.23 Acetic acid molecules have the ability to hydrogen-bond to one another to form dimers. Some of these dimers exist in the vapor, giving an average molecular weight for the vapor roughly midway between 60 (for CH_3COOH) and 120 (for the dimer).
11.25 CCl_4 **11.27** Propylamine is capable of hydrogen bonding. Because of the stronger intermolecular forces, propylamine should have a lower vapor pressure.
11.29 (a) metallic (b) covalent network (c) molecular (d) molecular **11.31** 1
11.33 9.26×10^{-23} g; 9.27×10^{-23} g from the molar mass **11.35** 3.61 Å **11.37** 4, face-centered cubic
11.39 19.25 g/cm^3 **11.41** 1.60 Å **11.43** 1.91 kJ
11.45 1.05 g **11.47** 21 g **11.49** 53.3 kJ
11.51 153 mmHg **11.53** CH_4, C_2H_6, CH_3OH, CH_2OHCH_2OH **11.55** (a) low-melting, brittle (b) high melting, hard, brittle (c) malleable, electrically conducting (d) hard, high melting
11.57 SO_2, C_2H_2; to liquefy CH_4, cool below −82°C and compress; to liquefy CO, cool below −140°C and compress. **11.59** (a) solid (b) liquid
11.61 191.0 amu **11.63** 1.28 Å **11.65** 68.0%
11.67 (a) KCl (b) CCl_4 (c) Zn (d) C_2H_5Cl
11.69 Ethylene glycol is capable of hydrogen bonding, but pentane is not. **11.71** 60.7% **11.73** 80°C
11.75 (a) will condense to a liquid at high pressure (b) will condense to a solid when pressure exceeds 5.1 atm (c) will remain as a gas at all pressures

Chapter 12

12.15 One example is $NH_3(g)$ in H_2O **12.17** ethanol
12.19 H_2O, CH_2OHCH_2OH, $C_{10}H_{22}$ **12.21** Ca^{2+}
12.23 $Ba(IO_3)_2$, $Sr(IO_3)_2$, $Ca(IO_3)_2$, $Mg(IO_3)_2$
12.25 0.80 g per 100 mL **12.27** (a) aerosol (b) sol (c) foam (d) sol **12.29** $Al_2(SO_4)_3$ **12.31** 3.12 g KI in 122 g H_2O **12.33** 1.60 molal **12.35** 70.2 g H_2O
12.37 0.412, 0.588 **12.39** 0.0116 **12.41** 16.8 m
12.43 0.568 M **12.45** 0.796 m **12.47** 41.3 mmHg, 0.9 mmHg **12.49** 100.339°C, −1.21°C **12.51** 0.42 m
12.53 122 amu **12.55** 162 amu
12.57 6.85×10^4 amu **12.59** 3 **12.61** $X_{N_2} = .671$, $X_{O_2} = .329$ **12.63** 1.74 m, 1.63 M, $X_{NH_4Cl} = 0.0303$

12.65 21 g C_3H_8, 37 g C_4H_{10} **12.67** 143 mmHg
12.69 $-24°C$, no   **12.71** 0.30 M **12.73** $CaCl_2$

Chapter 13

13.29 261 nm   **13.31** 0.037 mmHg, 2.7 L, 0.11 mm
13.33 4.74×10^2 m/s; 1.08×10^3 m/s
13.35 $4Fe^{2+} + O_2 + 4H^+ \rightarrow 4Fe^{3+} + 2H_2O$
13.37 7.0×10^{17}; 17%

13.39 $2KClO_3(s) \xrightarrow[MnO_2]{\Delta} 2KCl(s) + 3O_2(g)$

$Na_2O_2(s) + H_2O(l) \rightarrow 2NaOH(aq) + H_2O(l) + O_2(g)$

$2BaO_2(s) \xrightarrow{\Delta} 2BaO(s) + O_2(g)$
13.41 (a) $4Li(s) + O_2(g) \rightarrow 2Li_2O(s)$
(b) $4CH_3NH_2(g) + 9O_2(g) \rightarrow$
$$4CO_2(g) + 10H_2O(g) + 2N_2(g)$$
(c) $P_4(s) + 5O_2(g) \rightarrow P_4O_{10}(s)$
$P_4O_{10}(s) + 6H_2O(l) \rightarrow 4H_3PO_4(aq)$ **13.43** 3.76 kg
13.45 To prepare NO:

$4NH_3(g) + 5O_2(g) \xrightarrow{Pt} 4NO(g) + 6H_2O(g)$
To prepare N_2O_5, use this NO:
$2NO(g) + O_2(g) \rightarrow 2NO_2(g)$

$3NO_2(g) + H_2O(l) \xrightarrow{\Delta} 2HNO_3(aq) + NO(g)$
$P_4O_{10}(s) + 4HNO_3(l) \rightarrow 4HPO_3(aq) + 2N_2O_5(g)$
To prepare N_2O, use HNO_3 just prepared:
$NH_3(g) + HNO_3(aq) \rightarrow NH_4NO_3(aq)$

$NH_4NO_3(s) \xrightarrow{\Delta} N_2O(g) + 2H_2O(g)$
13.47 $4HS^- + O_2 \rightarrow 2S_2^{2-} + 2H_2O$ **13.49** 0.9069,
0.0931; 38.11 amu; 1.700 g/L

13.51 :F̈   F̈: sp^3d^2, square planar
       Xe
  :F̈   F̈:

13.53 $2XeF_2 + 4OH^- \rightarrow 2Xe + 4F^- + O_2 + 2H_2O$
13.55 1.9×10^2 mg   **13.57** 366 nm   **13.59** 3.29×10^{14}
O_3 molecules   **13.61** 75.52% N_2, 23.14% O_2
13.63 1.10×10^{21} g O_2 **13.65** 314 mL
13.67 $4Zn + HNO_3 + 9H^+ \rightarrow 4Zn^{2+} + NH_4^+ + 3H_2O$
13.69 (a) $SO_2(g) + 2NaOH(aq) \rightarrow Na_2SO_3(aq) + H_2O(l)$
(b) $Na_2O(s) + CO_2(g) \rightarrow Na_2CO_3(aq)$
(c) $Na_2O(s) + SO_2(g) \rightarrow Na_2SO_3(aq)$
(d) $MgO(s) + 2HCl(g) \rightarrow MgCl_2(aq) + H_2O(l)$
13.71 Prepare NH_3 by the Haber process:
Air $\rightarrow N_2(g)$

$N_2(g) + 3H_2(g) \xrightarrow{Fe} 2NH_3(g)$
Prepare HNO_3 by the Ostwald process:

$4NH_3(g) + 5O_2(g) \xrightarrow{Pt} 4NO(g) + 6H_2O(g)$
$2NO(g) + O_2(g) \rightarrow 2NO_2(g)$

$3NO_2(g) + H_2O(l) \xrightarrow{\Delta} 2HNO_3(aq) + NO(g)$

Then prepare NH_4NO_3 and finally N_2O:
$NH_3(g) + HNO_3(aq) \rightarrow NH_4NO_3(aq)$

$NH_4NO_3(s) \xrightarrow{\Delta} N_2O(g) + 2H_2O(g)$

Chapter 14

14.21 0.122 g H_2O, 0.108 g O   **14.23** Mg
14.25 $Zn(s) + 2HCl(aq) \rightarrow ZnCl_2(aq) + H_2(g)$
$Zn(s) + 2OH^-(aq) + 2H_2O(l) \rightarrow Zn(OH)_4^{2-} + H_2(g)$

$Zn(s) + H_2O(g) \xrightarrow{\Delta} ZnO(s) + H_2(g)$
14.27 $H_2(g) + F_2(g) \rightarrow 2HF(g)$
$Fe_3O_4(s) + 4H_2(g) \rightarrow 3Fe(s) + 4H_2O(g)$

$Ca(l) + H_2(g) \xrightarrow{\Delta} CaH_2(s)$

14.29 (a) $C(s) + H_2O(g) \xrightarrow{\Delta} CO(g) + H_2(g)$

$CO(g) + 2H_2(g) \xrightarrow{catalyst} CH_3OH(g)$

$2CH_3OH(g) + O_2(g) \xrightarrow{Ag} 2HCHO(g) + 2H_2O(g)$

(b) $C(s) + H_2O(g) \xrightarrow{\Delta} CO(g) + H_2(g)$
$WO_3(s) + 3H_2(g) \rightarrow W(s) + 3H_2O(g)$
14.31 1.2×10^4 J

14.33 $Si(s) + 2H_2O(g) \xrightarrow{\Delta} SiO_2(s) + 2H_2(g)$
$H_2(g) + Cl_2(g) \rightarrow 2HCl(g)$
14.35 $Na_2CO_3 \cdot 10H_2O$
14.37 $H_2S + H_2O_2 \rightarrow S + 2H_2O$
14.39 $2Cr(OH)_3 + 3H_2O_2 + 4OH^- \rightarrow 2CrO_4^{2-} + 8H_2O$
14.41 0.114 M, 0.387%   **14.43** 6.417 kJ/g
14.45 2×10^4 L   **14.47** 4.7×10^3 mg O_2/L
14.49 2.1 g $Na_2CO_3 \cdot 10H_2O$ **14.51** $-2.9°C$
14.53 38 atm   **14.55** 3.8×10^3 J
14.57 (a) $3Fe(s) + 4H_2O(g) \rightarrow Fe_3O_4(s) + 4H_2(g)$
(b) $C_3H_8(g) + 3H_2O(g) \rightarrow 3CO(g) + 7H_2(g)$
(c) $NaH(s) + H_2O(l) \rightarrow NaOH(aq) + H_2(g)$
(d) $2F_2(g) + 2H_2O(l) \rightarrow 4HF(aq) + O_2(g)$

14.59 0.844 g   **14.61** (a) $N_2(g) + 3H_2(g) \xrightarrow{\Delta} 2NH_3(g)$

(b) $CH_2{=}CH_2 + H_2 \xrightarrow{Ni} CH_3CH_3$
(c) $WO_3(s) + 3H_2(g) \rightarrow W(s) + 3H_2O(g)$ **14.63** 8.00 kJ
96°C

Chapter 15

15.19 $-\dfrac{1}{2}\dfrac{\Delta[NO_2]}{\Delta t} = \dfrac{\Delta[O_2]}{\Delta t}$
15.21 3.5×10^{-6} mol/(L·s)   **15.23** first order in
$C_3H_6Br_2$, first order in KI, overall second order
15.25 Rate = $k[CH_3NNCH_3]$, $k = 2.5 \times 10^{-4}$/s
15.27 Rate = $k[NO]^2[H_2]$, $k = 2.9 \times 10^2$ L²/(mol²·s)
15.29 1.7×10^{-2} mol/L   **15.31** 3.2×10^4 s, 6.3×10^4 s,
9.4×10^4 s   **15.33** 1.0×10^2 kJ/mol, 2×10^{-3}/s

15.35 84 kJ/mol   **15.37** $NOCl_2$ is a reaction intermediate; $2NO + Cl_2 \rightarrow 2NOCl$
15.39 (a) bimolecular (b) bimolecular (c) unimolecular (d) termolecular
15.41 (a) rate $= k[O_3]$ (b) rate $= k[NOCl_2][NO]$
15.43 rate $= k[C_3H_6]^2$ **15.45** rate $= k[H_2][I_2]$
15.47 The overall reaction is $2N_2O \rightarrow 2N_2 + O_2$; Cl_2 is introduced to give catalytic activity.
15.49 3.5×10^{-6} M/s, 3.2×10^{-6} M/s, 2.5×10^{-6} M/s
15.51 $k_{ave} = 2.5 \times 10^{-4}/s$ **15.53** 1.5×10^4 s
15.55 5.5×10^3 s   **15.57** $0.021\ M$ **15.59** From the slope, $k = 2.6 \times 10^{-4}/s$ **15.61** $E_a = 1.14 \times 10^5$ J, $A = 5.7 \times 10^9$, $k = 4.8\ M/s$ **15.63** rate $= k[NO_2][CO]$
15.65 rate $= k[NO_2Br]$ **15.67** rate $= kK_1[NH_4^+][CNO^-]$

Chapter 16

16.13 0.135 mol PCl_3, 0.135 mol Cl_2, 0.865 mol PCl_5
16.15 0.0791 mol CO, 0.1582 mol H_2, 0.0209 mol CH_3OH **16.17** (a) $K_c = \dfrac{[HBr]^2}{[H_2][Br_2]}$
(b) $K_c = \dfrac{[CH_4][H_2S]^2}{[CS_2][H_2]^4}$ (c) $K_c = \dfrac{[H_2O]^2[Cl_2]^2}{[O_2][HCl]^4}$
(d) $K_c = \dfrac{[CH_3OH]}{[H_2]^2[CO]}$ **16.19** 0.543   **16.21** 48.9
16.23 10.6   **16.25** $1.0 \times 10^{-3}\ M$
16.27 3.0×10^{-5} **16.29** 56.3   **16.31** (a) $\dfrac{[CO]^2}{[CO_2]}$
(b) $\dfrac{[CO_2]}{[CO]}$ (c) $\dfrac{[CO_2]}{[O_2]^{1/2}[SO_2]}$ **16.33** (a) yes (b) no
16.35 no; 3.2×10^{-48} mol; yes   **16.37** to the left
16.39 0.45 M **16.41** 0.0002 mol I_2; 0.0002 mol Br_2; 0.0017 mol IBr   **16.43** 0.18 mol PCl_3, 0.18 mol Cl_2, 0.32 mol PCl_5 **16.45** 0.61 mol CO, 1.84 mol H_2, 0.39 mol CH_4, 0.39 mol H_2O **16.47** (a) forward
(b) reverse   **16.49** (a) no effect (b) no effect (c) left
16.51 no   **16.53** low temperature, low pressure
16.55 4.3   **16.57** 0.153   **16.59** reverse
16.61 reverse   **16.63** 0.008 mol HBr, 0.0010 mol H_2, 0.0010 mol Br_2 **16.65** 13.2%   **16.67** 0.48 mol CO, 2.4 mol H_2, 0.52 mol CH_4, 0.52 mol H_2O
16.69 endothermic

Chapter 17

17.17 0.020 M, $5.0 \times 10^{-13}\ M$ **17.19** $1.0 \times 10^{-12}\ M$, 0.010 M **17.21** acidic   **17.23** (c) and (d)   **17.25** 2.12
17.27 11.60   **17.29** $7.6 \times 10^{-6}\ M$ **17.31** $4.3 \times 10^{-3}\ M$
17.33 13.161   **17.35** 6, acidic

17.37

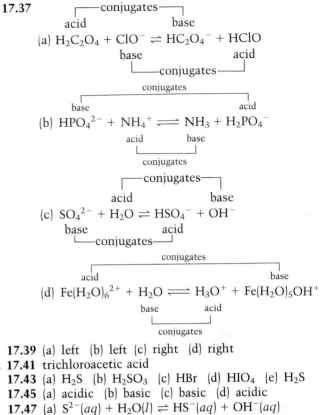

17.39 (a) left (b) left (c) right (d) right
17.41 trichloroacetic acid
17.43 (a) H_2S (b) H_2SO_3 (c) HBr (d) HIO_4 (e) H_2S
17.45 (a) acidic (b) basic (c) basic (d) acidic
17.47 (a) $S^{2-}(aq) + H_2O(l) \rightleftharpoons HS^-(aq) + OH^-(aq)$
(b) $CN^-(aq) + H_2O(l) \rightleftharpoons HCN(aq) + OH^-(aq)$
(c) $Zn(H_2O)_6^{2+}(aq) + H_2O(l) \rightleftharpoons$
$\qquad\qquad Zn(H_2O)_5OH^+(aq) + H_3O^+(aq)$
(d) $N_2H_5^+(aq) + H_2O(l) \rightleftharpoons N_2H_4(aq) + H_3O^+(aq)$
17.49 (a) Cr^{3+}, acid; H_2O, base (b) BF_3, acid; $(C_2H_5)_2O$, base   **17.51** (a) base;
$BaO(s) + H_2O(l) \rightleftharpoons Ba^{2+}(aq) + 2OH^-(aq)$ (b) acid;
$H_2S(aq) + H_2O(l) \rightleftharpoons H_3O^+(aq) + HS^-(aq)$ (c) base;
$CH_3NH_2(aq) + H_2O(l) \rightleftharpoons CH_3NH_3^+(aq) + OH^-(aq)$
(d) acid; $SO_2(aq) + H_2O(l) \rightleftharpoons HSO_3^-(aq) + H^+(aq)$
17.53 basic   **17.55** $3.3 \times 10^{-8}\ M$ **17.57** (a) 3.89, acidic (b) 6.73, acidic (c) 8.60, basic (d) 11.91, basic
17.59 $H_3O^+(aq) + (CH_3)_3N(aq) \rightarrow$
$\qquad\qquad\qquad (CH_3)_3NH^+(aq) + H_2O(l)$
17.61 (a) $SO_3^{2-}(aq) + H_2O(l) \rightarrow HSO_3^-(aq) + OH^-(aq)$
(b) Metal ions of Group IIA elements do not hydrolyze.
(c) $CH_3NH_3^+(aq) + H_2O(l) \rightarrow CH_3NH_2(aq) + H_3O^+(aq)$
(d) $Fe(H_2O)_6^{2+}(aq) + H_2O(l) \rightarrow Fe(H_2O)_5OH^+(aq) + H_3O^+(aq)$
17.63 I_2 acts as a Lewis acid; I^- acts as a Lewis base

Chapter 18

18.17 1.4×10^{-5} **18.19** 5.42, 0.00015
18.21 $1.0 \times 10^{-3}\ M$, $1.0 \times 10^{-3}\ M$ **18.23** 1.79

18.25 2.17   **18.27** 0.85, $5.1 \times 10^{-5}\ M$
18.29 3.2×10^{-5} **18.31** $5.7 \times 10^{-3}\ M$, 11.76
18.33 (a) 2.2×10^{-11} (b) 7.1×10^{-6} **18.35** 8.64,
$4.4 \times 10^{-6}\ M$ **18.37** $1.0 \times 10^{-3}\ M$, 2.99
18.39 (a) neutral (or slightly basic) (b) acidic
18.41 (a) 0.029 (b) 0.0064   **18.43** 3.17   **18.45** 8.05
18.47 9.26, 9.09   **18.49** 3.03   **18.51** 5.32
18.53 0.34 mol   **18.55** 2.0   **18.57** 8.59   **18.59** 5.97
18.61 9.08   **18.63** 1.1×10^{-3} **18.65** 1.1×10^{-2}
18.67 hydrolysis, 9.68   **18.69** 2.0×10^{-5}, 2.1×10^{-4},
CO_3^{2-} **18.71** 4.1%   **18.73** 2.84   **18.75** 3.16
18.77 11   **18.79** 2.4   **18.81** 3.11   **18.83** 0.0018

Chapter 19

19.11 (a) $K_{sp} = [Pb^{2+}][SO_4^{2-}]$ (b) $K_{sp} = [Pb^{2+}][Br^-]^2$
(c) $K_{sp} = [Ca^{2+}]^3[PO_4^{3-}]^2$
19.13 9.3×10^{-10} **19.15** 1.3×10^{-7}
19.17 1.8×10^{-11} **19.19** 0.092 g/L
19.21 $4.0 \times 10^{-7}\ M$ **19.23** 4.6×10^{-4} g/L
19.25 2.0×10^{-6} g/L   **19.27** 0.40 g/L   **19.29** lead
chromate will precipitate   **19.31** no precipitation
19.33 0.0018 mol   **19.35** $6.6 \times 10^{-6}\ M$, 0.013%
19.37 $6.9 \times 10^{-9}\ M$ **19.39** $BaSO_3$ **19.41** pH less than
0.8   **19.43** $5.5 \times 10^{-19}\ M$ **19.45** calcium oxalate will
not precipitate   **19.47** $3.0 \times 10^{-3}\ M$ **19.49** 26 g/L
19.51 $8.0 \times 10^{-3}\ M$ **19.53** 5.5×10^{-6} g
19.55 2.1×10^{-4} **19.57** 0.54 M
19.59 $3.6 \times 10^{-4}\ M$, $4.2 \times 10^{-6}\ M$, 2.5×10^9

Chapter 20

20.17 -14.3 J/(mol·K)   **20.19** -126 J/K·mol;
126 J/K·mol   **20.21** (a) negative (b) positive
(c) positive (d) negative   **20.23** -56.1 J/K
20.25 -242.6 J/K   **20.27** -1075.0 kJ; -1058.3 kJ
20.29 C(graphite) $+ \frac{1}{2}O_2(g) \rightarrow CO(g)$ **20.31** -818.0 kJ
20.33 (a) spontaneous (b) spontaneous
(c) nonspontaneous (d) nonspontaneous
(e) nonspontaneous   **20.35** -20 kJ; -33 kJ;
exothermic and spontaneous   **20.37** 474.4 kJ; no
entropy is produced   **20.39** (a) $K_{th} = \dfrac{P_{CO_2}P_{H_2}}{P_{CO}P_{H_2O}}$
(b) $K_{th} = [Mg^{2+}][OH^-]^2$ (c) $K_{th} = [Li^+]^2[OH^-]^2P_{H_2}$
20.41 -190.6 kJ; 3×10^{33} **20.43** -28.6 kJ; 1.0×10^5
20.45 2×10^{91} **20.47** $\Delta S°$ is positive   **20.49** -184 kJ;
exothermic; ΔS is positive; ΔG is negative, so reaction
is spontaneous   **20.51** 124, yes   **20.53** 383 K
20.55 negative   **20.57** -104 J/K   **20.59** not
spontaneous   **20.61** $\Delta H°$ is positive; $\Delta S°$ is positive

20.63 56 kJ; 2×10^{-10} **20.65** 3.3 kJ/mol; 0.26 atm
20.67 $\Delta H° = 1.10 \times 10^2$ kJ; $\Delta S° = 136$ J/K; $\Delta G°$ at 25°C
is 69 kJ; $\Delta G°$ at 800°C is -36 kJ

Chapter 21

21.19 Anode: $Zn \rightarrow Zn^{2+} + 2e^-$
Cathode: $Ni^{2+} + 2e^- \rightarrow Ni$
21.21 Anode: $2H_2O \rightarrow O_2 + 4H^+ + 4e^-$
Cathode: $2H_2O + 2e^- \rightarrow H_2 + 2OH^-$
21.23 Anode: $Cd(s) \rightarrow Cd^{2+}(aq) + 2e^-$ (oxidation)
Cathode: $Ag^+(aq) + e^- \rightarrow Ag(s)$ (reduction)
21.25 Anode: $Cd(s) + 2OH^-(aq) \rightarrow Cd(OH)_2(s) + 2e^-$
Cathode: $NiO_2(s) + 2H_2O(l) + 2e^- \rightarrow$
$$Ni(OH)_2(s) + 2OH^-(aq)$$
The reactions are reversed when the cell is being
charged.
Anode: $Ni(OH)_2(s) + 2OH^-(aq) \rightarrow$
$$NiO_2(s) + 2H_2O(l) + 2e^-$$
Cathode: $Cd(OH)_2(s) + 2e^- \rightarrow Cd(s) + 2OH^-(aq)$
Overall: $Cd(s) + NiO_2(s) + 2H_2O(l) \rightarrow$
$$Cd(OH)_2 + Ni(OH)_2$$
21.27 $Ni(s)|Ni^{2+}(aq) \| H^+(aq)|H_2(g)|Pt$
21.29 $2Ag^+(aq) + Ni(s) \rightarrow Ni^{2+}(aq) + 2Ag(s)$
21.31 Anode: $Cd(s) \rightarrow Cd^{2+}(aq) + 2e^-$
Cathode: $Ni^{2+}(aq) + 2e^- \rightarrow Ni(s)$
Overall: $Cd(s) + Ni^{2+}(aq) \rightarrow Cd^{2+}(aq) + Ni(s)$
21.33 5.98×10^4 J/mol Fe^{3+} **21.35** (a) 1.11 V
(b) 0.82 V   **21.37** (a) not spontaneous
(b) spontaneous   **21.39** Cl_2 will oxidize Br^-;
$Cl_2(s) + 2Br^-(aq) \rightarrow Br_2(l) + 2Cl^-(aq)$
21.41 -1.6×10^5 J   **21.43** 1.92 V   **21.45** 1.2×10^6
21.47 0.56 V   **21.49** -0.84 V   **21.51** $4 \times 10^{-3}\ M$
21.53 (a) Anode: $2H_2O(l) \rightarrow O_2(g) + 4H^+(aq) + 4e^-$;
cathode: $2H_2O(l) + 2e^- \rightarrow H_2(g) + 2OH^-(aq)$; Overall:
$2H_2O(l) \rightarrow 2H_2(g) + O_2(g)$ (b) Anode:
$2Br^-(aq) \rightarrow Br_2(aq) + 2e^-$; cathode:
$2H_2O(l) + 2e^- \rightarrow H_2(g) + 2OH^-(aq)$; overall:
$2Br^-(aq) + 2H_2O(l) \rightarrow Br_2(aq) + H_2(g) + 2OH^-(aq)$
21.55 5.49×10^7 C   **21.57** 0.360 g   **21.59** 3.74 V
21.61 (a) spontaneous (b) Fe^{3+} will oxidize Sn^{2+} to
Sn^{4+} **21.63** 0.36 V   **21.65** 1×10^{-7} **21.67** (a) 2.2
(b) 0.31 M **21.69** (a) 1.0 F, 9.6×10^4 C (b) 2.0 F,
1.9×10^5 C (c) 0.11 F, 1.1×10^4 C (d) 2.8×10^{-2} F,
2.7×10^3 C   **21.71** 2.66×10^{-4} g

Chapter 22

22.19 $^{32}_{15}P \rightarrow ^{32}_{16}S + ^{\ 0}_{-1}e$ **22.21** astatine;
$^{212}_{87}Fr \rightarrow ^{208}_{85}At + ^4_2He$ **22.23** (a) $^{122}_{51}Sb$ (b) $^{204}_{85}At$
(c) $^{80}_{37}Rb$ **22.25** (a) α-emission (b) positron emission

or electron capture (c) β-emission **22.27** $^{235}_{92}U$ decay series; $^{232}_{90}Th$ decay series. **22.29** (a) $^{27}_{13}Al(d,\alpha)^{25}_{12}Mg$ (b) $^{63}_{29}Cu(p,n)^{63}_{30}Zn$ **22.31** (a) $^{63}_{29}Cu + ^4_2He \rightarrow ^{66}_{31}Ga + ^1_0n$ (b) $^{33}_{16}S + ^1_0n \rightarrow ^{33}_{15}P + ^1_1H$ **22.33** 1.21×10^9 kJ/mol **22.35** (a) ^4_2He (b) α **22.37** plutonium-239 **22.39** 1.8×10^{-9}/s **22.41** 9.1×10^{-8}/s **22.43** 4.8×10^{10} years **22.45** 3.84×10^{-12}/s **22.47** 1.1×10^2 Ci **22.49** 5.7×10^{-4} g **22.51** 0.330, 1.6 μg **22.53** 3.21 h **22.55** 44.6 d **22.57** 5.3×10^3 y **22.59** -2.06×10^{-9} g **22.61** -1.687×10^{12} J; -17.49 MeV **22.63** 0.03438 amu; 32.03 MeV; 5.338 MeV **22.65** Na-20 is expected to decay by electron capture or positron emission. Na-26 is expected to decay by beta emission. **22.67** 7 alpha emissions and 4 beta emissions. **22.69** $^{209}_{83}Bi + ^4_2He \rightarrow ^{211}_{85}At + 2^1_0n$ **22.71** $^{246}_{98}Cf$ **22.73** 4.4 y **22.75** 2.42 pm **22.77** -1.01×10^{11} kJ; -3.28×10^4 kJ

Chapter 23

23.39 (a) nonmetal (b) metal **23.41** (a) Na is more metallic (b) Na is more metallic (c) Cs is more metallic **23.43** Brown solid with lower melting point is lead(IV) oxide, PbO_2 **23.47** $^{227}_{89}Ac \rightarrow ^{227}_{90}Th + ^0_{-1}e$ **23.49** $TiCl_4(g) + 4Na(l) \rightarrow Ti(s) + 4NaCl(s)$ **23.51** $K_2SO_4(aq) + Ba(OH)_2(aq) \rightarrow 2KOH(aq) + BaSO_4(s)$ **23.53** $CO_2(g) + NH_3(g) + NaCl(aq) + H_2O(l) \rightarrow$
$$NaHCO_3(s) + NH_4Cl(aq)$$
$2NaHCO_3(s) \xrightarrow{\Delta} Na_2CO_3(s) + CO_2(g) + H_2O(g)$
$Ca(OH)_2(aq) + Na_2CO_3(aq) \rightarrow$
$$2NaOH(aq) + CaCO_3(s)$$
23.55 $^{230}_{90}Th \rightarrow ^{226}_{88}Ra + ^4_2He$ **23.57** Place a small amount of metal in strong base. If it is Be, it will react with evolution of H_2 gas. Mg gives no reaction. **23.59** Add Na_2SO_4; $BaSO_4$ precipitates and is filtered off. Add NaOH to filtrate; $Mg(OH)_2$ precipitates, leaving $Be(OH)_4^{2-}$. **23.61** $Mg(OH)_2(s) + 2HCl(aq) \rightarrow MgCl_2(aq) + 2H_2O(l)$

$MgCl_2(l) \xrightarrow{\text{electrolysis}} Mg(l) + Cl_2(g)$

$BeF_2(l) + Mg(l) \xrightarrow[950°C]{\Delta} Be(s) + MgF_2(l)$

23.63 $3Mg(s) + N_2(g) \rightarrow Mg_3N_2(s)$
$Mg_3N_2(s) + 6H_2O(l) \rightarrow 3Mg(OH)_2(aq) + 2NH_3(aq)$
23.65 $Fe_2O_3(s) + Al(l) \rightarrow Al_2O_3(l) + Fe(l)$

23.67 $2Al_2O_3(s) + 3C(s) + 6Cl_2(g) \xrightarrow{\Delta}$
$$4AlCl_3(g) + 3CO_2(g)$$
$AlCl_3(s) + 3K(\text{amalgam}) \rightarrow 3KCl(s) + Al(s)$

23.69

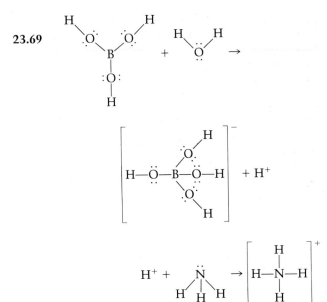

23.71 $Al(H_2O)_6^{3+}(aq) + HCO_3^-(aq) \rightarrow$
$$Al(H_2O)_5OH^{2+}(aq) + H_2O(l) + CO_2(g)$$
23.73 Test portions of solutions of each compound with the others; the results can differentiate the compounds. $AlCl_3$ reacts only with KOH, giving a precipitate that dissolves in excess KOH. $BaCl_2$ reacts only with $BeSO_4$, giving a precipitate. $BeSO_4$ gives a precipitate with $BaCl_2$. It also gives a precipitate with KOH that dissolves in excess KOH.
23.75 (a) $2PbS(s) + 3O_2(g) \rightarrow 2PbO(s) + 2SO_2(g)$

  (b) $PbO(s) + C(s) \xrightarrow{\Delta} Pb(s) + CO(g)$

(c) At the anode the impure lead is oxidized and goes into solution as Pb^{2+}. At the cathode the Pb^{2+} ions are reduced to Pb and deposited on the cathode, leaving the impurities in solution or at the impure anode. **23.77** $AlCl_3$ has a trigonal planar geometry; the geometry of Cl atoms about Al in Al_2Cl_6 is tetrahedral. **23.79** 5.41×10^{-10}/y; 0.947 **23.81** -425 kJ; exothermic **23.83** 1.35×10^3 K **23.85** -37 kJ **23.87** 1.9 Å **23.89** 3878 g **23.91** 7.73×10^{-2} g

Chapter 24

24.43 $CH_4(g) + H_2O(g) \xrightarrow{\Delta} CO(g) + 3H_2(g)$
$CO(g) + NaOH(aq) \xrightarrow{\Delta} HCOONa(aq)$
24.45 $SnO_2(s) + 2C(s) \xrightarrow{\Delta} Sn(l) + 2CO(g)$
$Sn(s) + 2HCl(aq) \rightarrow SnCl_2(aq) + H_2(g)$
24.47 $3NaOH + 3NaSn(OH)_3 + 2Bi(OH)_3 \rightarrow$
$$3Na_2Sn(OH)_6 + 2Bi$$

24.49 $Sn(H_2O)_6^{2+}(aq) + H_2O(l) \rightarrow$
$$Sn(H_2O)_5(OH)^+(aq) + H_3O^+(aq)$$
24.51 trigonal pyramidal **24.53** (a) P_4O_6, H_3PO_3, PF_3
(b) P_4O_{10}, H_3PO_4, PF_5
24.55 $3As_4O_6(s) + 8HNO_3(aq) + 14H_2O(l) \rightarrow$
$$12H_3AsO_4(aq) + 8NO(g)$$

$2H_3AsO_4(s) \xrightarrow{\Delta} As_2O_5(s) + 3H_2O(g)$
24.57 $H_3PO_3 + H_2SO_4 \rightarrow H_3PO_4 + SO_2 + H_2O$
24.59 $2AsO_4^{3-} + N_2H_5^+ \rightarrow$
$$2AsO_3^{3-} + N_2 + 2H_2O + H^+$$
24.61 tetrahedral; sp^3
24.63 $2H_2S(g) + 3O_2(g) \rightarrow 2H_2O(g) + 2SO_2(g)$
$16H_2S(g) + 8SO_2(g) \rightarrow 16H_2O(l) + 3S_8(s)$
24.65 $H_2SeO_3 + 2H_2S \rightarrow Se + S + 3H_2O$
24.67 (a) sp^3; bent (b) sp^3d hybrid; distorted
tetrahedral **24.69** (a) +6 (b) +6 (c) −2 (d) +4
24.71 $SiO_2(s) + 4HF(g) \rightarrow 2H_2O(l) + SiF_4(g)$
24.73 $CaF_2(s) + H_2SO_4(l) \rightarrow CaSO_4(s) + 2HF(g)$

$2HF(l) \xrightarrow[KF]{electrolysis} H_2(g) + F_2(g)$

$U(s) + 3F_2(g) \rightarrow UF_6(s)$
24.75 $8H^+ + 6HCl + K_2Cr_2O_7 \rightarrow$
$$3Cl_2 + 2Cr^{3+} + 7H_2O + 2K^+$$
24.77 $2IO_3^- + 5HSO_3^- \rightarrow I_2 + 5SO_4^{2-} + 3H^+ + H_2O$
or
$2NaIO_3 + 5NaHSO_3 \rightarrow I_2 + 2Na_2SO_4 + 3NaHSO_4 + H_2O$
24.79 E°_{cell} = 0.86 V, ClO^- will oxidize Fe^{2+} to Fe^{3+}
24.81 (a) sp^3; bent (b) sp^3, trigonal pyramidal
(c) sp^3d; T-shaped **24.83** (a) S_8 (b) P_4O_{10} (c) Br_2
(d) Sn **24.85** Test the pH. $Na_2SO_4(aq)$ is neutral; other
solutions are acidic. Add HCl to acidic solutions.
$NaHSO_3(aq)$ will evolve SO_2 (characteristic odor).
24.87 11.0 **24.89** 0.405 L **24.91** 24.6% P, 63.1%
$Ca(H_2PO_4)_2 \cdot H_2O$ **24.93** 15.1 hr **24.95** 2.58%
NaOCl **24.97** disproportionation is expected
24.99 −0.16 V

Chapter 25

25.31 CrO_3 is dark red and low-melting **25.33** W
25.35 $Fe(s) + 2HCl(aq) \rightarrow FeCl_2(aq) + H_2(g)$
$FeCl_2(aq) + 2NaOH(aq) \rightarrow Fe(OH)_2(s) + 2NaCl(aq)$
$4Fe(OH)_2(s) + O_2(g) \rightarrow 2H_2O(l) + 4Fe(OH)_3(s)$
25.37 (a) +2 (b) +4 (c) +1 (d) +6
25.39 $3Fe^{2+} + NO_3^- + 4H^+ \rightarrow 3Fe^{3+} + NO + 2H_2O$
25.41 $[Co(NH_3)_4(NO_2)_2]Cl$ **25.43** (a) 4 (b) 6 (c) 4
(d) 6 **25.45** (a) +2 (b) +3 (c) +3 (d) +3
25.47 (a) +3 (b) *Formula   Name*   (c) 6 (d) −3

Formula	Name
NH_3	ammine
Cl^-	chloro
$C_2O_4^{2-}$	oxalato

25.49 (a) potassium hexafluoroferrate(III)

(b) diamminediaquacopper(II) ion (c) ammonium
aquapentafluoroferrate(III) (d) dicyanoargentate(I) ion
25.51 (a) pentacarbonyliron(0)
(b) dicyanobis(ethylenediamine)rhodium(III) ion
(c) tetraaminesulfatochromium(III) chloride
(d) tetraoxomanganate(VII) ion
25.53 (a) $K_3[Mn(CN)_6]$ (b) $Na_2[Zn(CN)_4]$
(c) $[Co(NH_3)_4Cl_2]NO_3$ (d) $[Cr(NH_3)_6]_2[CuCl_4]_3$
25.55 (a) linkage (b) coordination (c) hydrate
(d) ionization **25.57** $[Co(NH_3)_4Br_2]Cl$; an ionization
isomer —$[Co(NH_3)_4BrCl]Br$

25.59 (a)

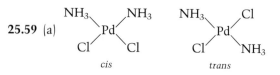

cis   *trans*

(b) no geometric isomerism (c) no geometric
isomerism

(d)

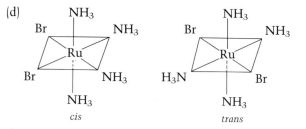

cis   *trans*

25.61 (a) Has optical isomers, similar to those in
Figure 25.13(b) and (c). (b) No

25.63 (a) V:[Ar](↑)(↑)(↑)( )( )   (↑↓)   ( )( )( )
            $3d$ $4s$ $4p$

  V^{3+}:[Ar](↑)(↑)( )( )( )   ( )   ( )( )( )
            $3d$ $4s$ $4p$

$V(H_2O)_6^{3+}$: [Ar](↑)(↑)( ) (↑↓)(↑↓)   (↑↓)   (↑↓)(↑↓)(↑↓)
            $3d$ $4s$ $4p$
        d^2sp^3 bonds to ligands

(b) Fe: [Ar](↑↓)(↑)(↑)(↑)(↑) (↑↓) ( )( )( ) ( )( )( )( )( )
            $3d$ $4s$ $4p$ $4d$
Fe^{3+}: [Ar](↑)(↑)(↑)(↑)(↑) ( ) ( )( )( ) ( )( )( )( )
            $3d$

$Fe(en)_3^{3+}$: [Ar](↑)(↑)(↑)(↑)(↑) (↑↓) (↑↓)(↑↓) (↑↓)(↑↓)( )( )( )
            sp^3d^2 bonds to ligands

(c) Rh: [Kr](↑↓)(↑↓)(↑)(↑)(↑) (↑↓) ( )( )( )
            $4d$ $5s$ $5p$
Rh^{3+}: [Kr](↑↓)(↑)(↑)(↑)(↑) ( ) ( )( )( )

$Rh(CN)_6^{3-}$: [Kr](↑↓)(↑↓)(↑↓) (↑↓)(↑↓) (↑↓) (↑↓)(↑↓)(↑↓)
            d^2sp^3 bonds to ligands

25.65 (a) square planar
(b) square planar geometry (c) tetrahedral
(d) tetrahedral **25.67** (a) 2 unpaired electrons (b) 3
unpaired electrons (c) 2 unpaired electrons
25.69 (a) 8 electrons in square planar field (low spin)
(b) 7 electrons in square planar field (low spin) (c) 5
electrons in tetrahedral field (high spin) (d) 7
electrons in tetrahedral field (high spin) **25.71** purple
25.73 yes, H_2O is a weaker bonding ligand.
25.75 239 kJ/mol **25.77** no d electrons **25.79** 3
25.81 $[Cu^{2+}] = 4.6 \times 10^{-4} \, M;$ $[NH_3] = 1.8 \times 10^{-3} \, M;$
$[Cu(NH_3)_4{}^{2+}] = 0.10 \, M$

Chapter 26

26.21 (a) 2,4-dimethylpentane (b) 2,2,6,6-
tetramethylheptane (c) 4-ethyloctane (d) 3,4-
dimethyloctane

26.23 (a)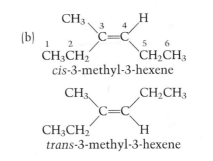

26.25 (a) 1-pentene (b) 2,5-dimethyl-2-hexene
26.27 (a)

26.29 (a)

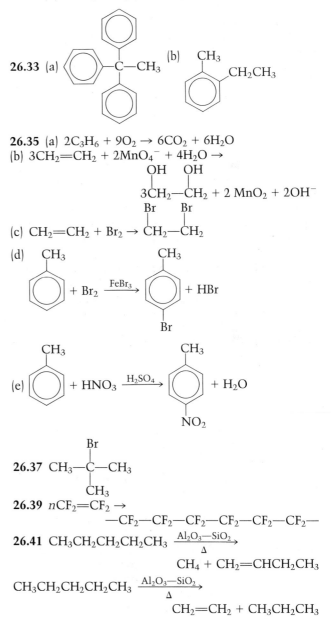

26.31 (a) 2-butyne (b) 3-methyl-1-butyne

26.33 (a) (b)

26.35 (a) $2C_3H_6 + 9O_2 \rightarrow 6CO_2 + 6H_2O$
(b) $3CH_2{=}CH_2 + 2MnO_4^- + 4H_2O \rightarrow$

26.37

26.39 $nCF_2{=}CF_2 \rightarrow$
$-CF_2-CF_2-CF_2-CF_2-CF_2-CF_2-$

26.41 $CH_3CH_2CH_2CH_2CH_3 \xrightarrow[\Delta]{Al_2O_3-SiO_2}$
$CH_4 + CH_2{=}CHCH_2CH_3$

$CH_3CH_2CH_2CH_2CH_3 \xrightarrow[\Delta]{Al_2O_3-SiO_2}$
$CH_2{=}CH_2 + CH_3CH_2CH_3$

26.43 (a)  (b)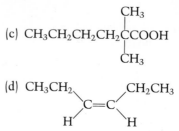

ketone   alcohol

(c) 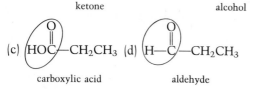 (d)

carboxylic acid   aldehyde

26.45 (a) 1-pentanol (b) 2-pentanol
(c) 2-propyl-1-pentanol (d) 4-heptanol
26.47 (a) secondary alcohol (b) tertiary alcohol
(c) primary alcohol (d) primary alcohol
26.49 (a) ethyl propyl ether (b) methyl isopropyl
ether **26.51** (a) 2-butanone (b) butanal
(c) 4,4-dimethylpentanal (d) 3-methyl-2-pentanone

26.53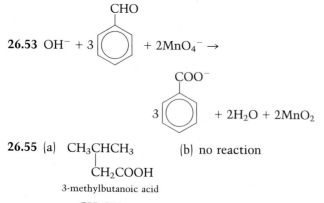

26.55 (a) CH_3CHCH_3 (b) no reaction
           |
           CH_2COOH

3-methylbutanoic acid

(c)   (d) $CH_3\overset{O}{\overset{\|}{C}}CH_2CH_3$

phenylmethanol   2-butanone

26.57 (a) $CH_3CH_2CH_2\overset{O}{\overset{\|}{C}}{-}OH + HO{-}CH_2CH_3$

$CH_3CH_2CH_2\overset{O}{\overset{\|}{C}}{-}O{-}CH_2CH_3 + H_2O$

(b) $H\overset{O}{\overset{\|}{C}}OCH_3 + NaOH \rightarrow H\overset{O}{\overset{\|}{C}}O^-Na^+ + CH_3OH$
26.59 (a) primary amine (b) secondary amine
(c) disulfide (d) thiol **26.61** (a) 3-methylbutanoic
acid (b) *trans*-5-methyl-2-hexene (c) 2,5-dimethyl-4-
heptanone (d) 4-methyl-2-pentyne

26.63 (a) $CH_3CH_2\overset{O}{\overset{\|}{C}}{-}O{-}\underset{\underset{CH_3}{|}}{\overset{\overset{CH_3}{|}}{C}}H$ (b) $CH_3{-}\underset{\underset{CH_3}{|}}{\overset{\overset{CH_3}{|}}{C}}{-}NH_2$

(c) $CH_3CH_2CH_2CH_2\underset{\underset{CH_3}{|}}{\overset{\overset{CH_3}{|}}{C}}COOH$

(d)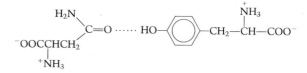

26.65 (a) Addition of dichromate ion in acidic
solution to propionaldehyde will cause the reagent to
change from orange to green as the aldehyde is
oxidized. Under similar conditions, acetone would not
react. (b) Add Br_2 in CCl_4 to each compound. First
compound reacts; the color of Br_2 fades. Benzene does
not react. **26.67** (a) ethylene (b) toluene
(c) methylamine (d) methanol **26.69** 1-butene
or 2-butene

Chapter 27

27.23 -2.75 kcal/mol **27.25** Releases free energy

27.27 $CH_3{-}\underset{\underset{+}{\overset{|}{NH_3}}}{\overset{\overset{H}{|}}{C}}{-}COO^-$ **27.29** 2 dipeptides are possible

27.31 histidine

27.33

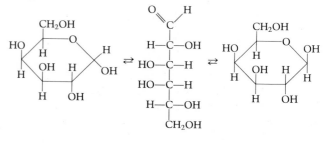

27.35 120

27.37

27.39

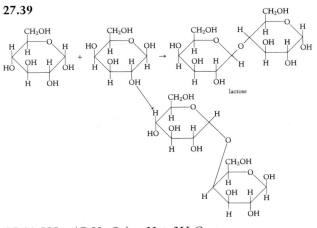

27.41 HO—(C$_6$H$_{10}$O$_5$)$_6$—H + 2H$_2$O →
3HO—(C$_6$H$_{10}$O$_5$)$_2$—H

27.43 A∶T, 1∶1; G∶C, 1∶1

27.45 cytidine triphosphate

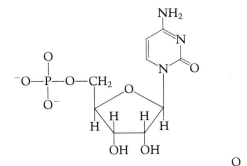

27.47

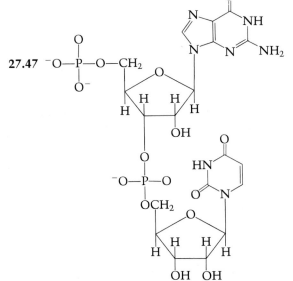

27.49 3; 2; Guanine-cytosine bonding is stronger
27.51 TGACTGCGTTAACTGGCG **27.53** 16; no
27.55 hydrogen bonds; less readily **27.57** 1200
nucleotides

27.59 300 nucleotides **27.61** gly-ser-arg-phe-gly-leu-
lys-end **27.63** CUG, ACU, CCC, UGG **27.65** CUU
GCU GUU GAA GAU UGU AUG UGG AAA
27.67 CH$_2$OH

CHOH + 3CH$_3$(CH$_2$)$_{16}$COOH →

CH$_2$OH

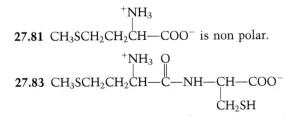

$$CH_2O\overset{O}{\overset{\|}{C}}(CH_2)_{16}CH_3$$

$$CHO\overset{O}{\overset{\|}{C}}(CH_2)_{16}CH_3 \ + 3H_2O$$

$$CH_2O\overset{O}{\overset{\|}{C}}(CH_2)_{16}CH_3$$

27.69
$$CH_2O\overset{O}{\overset{\|}{C}}(CH_2)_{16}CH_3$$
$$CHO\overset{O}{\overset{\|}{C}}(CH_2)_{16}CH_3 \ + 3NaOH →$$
$$CH_2O\overset{O}{\overset{\|}{C}}(CH_2)_{16}CH_3$$

CH$_2$OH

CHOH + 3CH$_3$(CH$_2$)$_{16}$COO$^-$Na$^+$

CH$_2$OH

27.71 will diffuse most readily.

27.73 Poisons with hydrocarbon structures would be
likely to diffuse across the cell membranes of the skin
and their action might thus be speeded up
27.75 TCAGCTGGCAATTA **27.77** 720
27.79 adenine, ribose, phosphate ion

27.81 CH$_3$SCH$_2$CH$_2$$\overset{^+NH_3}{\overset{|}{C}H}$—COO$^-$ is non polar.

27.83 CH$_3$SCH$_2$CH$_2$$\overset{^+NH_3}{\overset{|}{C}H}$—$\overset{O}{\overset{\|}{C}}$—NH—CH—COO$^-$

CH$_2$SH

HSCH$_2$$\overset{^+NH_3}{\overset{|}{C}H}$—$\overset{O}{\overset{\|}{C}}$—NH—CH—COO$^-$

CH$_2$

CH$_2$SCH$_3$

27.85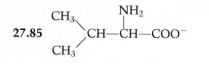

27.87 CH$_2$OH
     |
    CHOH
     |
    CH$_2$OH
     glycerol

CH$_3$(CH$_2$)$_{14}$COOH   palmitic acid

CH$_3$(CH$_2$)$_7$CH=CH(CH$_2$)$_7$COOH   oleic acid

CH$_3$(CH$_2$)$_5$CH=CH(CH$_2$)$_7$COOH   palmitoleic acid

Glossary

Absolute temperature scale   a temperature scale in which the lowest temperature that can be attained theoretically is zero. **(1.4)**

Accuracy   the agreement of a measured value with an accepted value. **(1.3)**

Acid   a substance other than a salt that increases the concentration of hydrogen ion in water solution. (See also *Arrhenius concept, Brønsted–Lowry concept,* and *Lewis concept.*) **(9.5)**

Acid-base indicator   a dye that changes color in acid or base and is used to show the acidity or basicity of a solution. **(9.5)**

Acid-base titration curve   a plot of the pH of a solution of acid (or base) against the volume of added base (or acid). **(18.7)**

Acidic oxides   an oxide that reacts with water to produce acidic solutions, or an oxide that reacts with bases. **(13.4)**

Acid ionization (acid dissociation) constant K_a, the equilibrium constant for the ionization of a weak acid. For the acid HA, $K_a = \dfrac{[H^+][A^-]}{[HA]}$. **(18.1)**

Acid salt   a salt that has one or more acidic hydrogen atoms; such salts can undergo neutralization with bases. **(9.5)**

Actinide element   one of the 14 elements following actinium in the periodic table, in which the $5f$ subshell is filling. **(6.4)**

Activated complex   in transition-state theory, an unstable grouping of reactant molecules that can break up to form products. **(15.5)**

Activated-sludge method   a water purification process in which waste water is degraded by aerobic bacteria. **(14.8)**

Activation energy   the minimum energy that a molecule must possess in order to react upon collision. **(15.5)**

Active site   the region on the surface of an enzyme where the substrate binds. **(27.4)**

Activity of a radioactive source   the number of nuclear disintegrations of a radioactive source per unit time, often measured in curies. **(22.3)**

Activity series (electromotive series)   a list of metals in order of their ease of losing electrons to the hydrogen ion to give metal ions in aqueous solution. **(14.2)**

Actual yield   the actual amount of product obtained from a chemical reaction. (See also *percentage yield* and *theoretical yield.*) **(3.9)**

Addition polymer   a polymer (giant molecule) formed by linking molecules having multiple bonds by an addition reaction. **(26.4)**

Addition reaction   a reaction in which parts of a reagent are added to each atom of a carbon–carbon multiple bond. **(26.4)**

Adenosine triphosphate (ATP)   a nucleotide that stores energy for the cell's needs. **(27.6)**

Aerosol   a colloid consisting of liquid droplets or solid particles dispersed in a gas. **(12.4)**

Alcohol   a compound in which a hydroxyl group (—OH) takes the place of an —H atom on an aliphatic hydrocarbon. **(26.5)**

Aldehyde   a compound containing the carbonyl group with at least one hydrogen atom attached to it. **(26.5)**

Aliphatic hydrocarbon a hydrocarbon that is not aromatic. (**26.1**)

Alkali metal a Group IA element. (**6.6**)

Alkaline earth metal a Group IIA element. (**6.6**)

Alkanes saturated hydrocarbons with the general formula C_nH_{2n+2}. (**26.1**)

Alkenes hydrocarbons with carbon–carbon double bonds, having the general formula C_nH_{2n}. (**26.2**)

Alkyl group an alkane less one hydrogen atom. (**26.1**)

Alkynes hydrocarbons with carbon–carbon triple bonds, having the general formula C_nH_{2n-2}. (**26.2**)

Allotropes forms of an element with different properties. (**6.6**)

Alpha emission the emission of a ^4_2He nucleus, or alpha particle, from an unstable nucleus. (**22.1**)

Aluminosilicate mineral a mineral that contains both SiO_4 and AlO_4 tetrahedra. (**24.1**)

Amide a compound derived from the reaction of ammonia or a primary or secondary amine with a carboxylic acid. (**26.7**)

Amine an organic compound in which one, two, or three hydrogen atoms of ammonia are replaced by hydrocarbon groups. (**26.7**)

Amino acids a compound containing a protonated amino group ($-NH_3^+$) and an ionized carboxyl group ($-COO^-$). (**27.4**)

Amorphous solids solids without well-defined, regular shapes; solids that do not have an ordered structure. (**11.4**)

Ampere (A) the SI base unit of electrical current. (**21.9**)

Amphiprotic species a species that can function either as a Brønsted–Lowry acid or base, that is, it can either lose or gain a proton. (**17.4**)

Amphoteric referring to a substance that has both acidic and basic properties. (**23.1**)

Amphoteric hydroxide a metal hydroxide that reacts with bases as well as with acids. (**19.5**)

Angstrom (Å) a unit of length; $1\ \text{Å} = 10^{-10}$ m. (**1.4**)

Angular momentum (azimuthal) quantum number (l) a quantum number for an atomic orbital, having any integer value from 0 to $n - 1$ and associated with a distinctive orbital shape. (**5.9**)

Anion a negatively charged ion. (**2.1**)

Anion exchange resin a resin (organic polymer) that removes anions from solution and replaces them with hydroxide ions. (**14.9**)

Anode the electrode at which oxidation occurs. (**21.1**)

Antibonding orbital a molecular orbital in which electrons spend little time between nuclei; the energy of the orbital is above that of the orbitals for the separated atoms. (**8.5**)

Anticodon in a transfer RNA molecule, a triplet sequence complementary to the codon on a messenger RNA molecule. (**27.6**)

Aromatic hydrocarbons hydrocarbons that contain benzene rings or similar structural features. (**26.1**)

Arrhenius concept a concept of acids and bases in which (*a*) an acid is a substance that, when dissolved in water, increases the concentration of the hydrogen ion and (*b*) a base is a substance that, when dissolved in water, increases the concentration of hydroxide ion. (**17.1**)

Arrhenius equation the mathematical equation that expresses the dependence of reaction rate on temperature; $\log k = \log A - E_a/(2.303\ RT)$. (**15.6**)

Association colloid a colloid in which the dispersed phase consists of micelles. (**12.4**)

Atmosphere (atm) a unit of pressure; 1 atm = 101.325 kPa (exact). (**1.6**)

Atomic mass unit (amu) a mass unit determined by arbitrarily assigning the carbon-12 isotope a mass of 12 atomic mass units. (**3.1**)

Atomic number (Z) the number of protons in the nucleus of an atom of an element; it identifies the element. (**5.3**)

Atomic orbital the region of space about the nucleus of an atom in which there is a high probability of finding an electron of given quantum state. (**5.9**)

Atomic symbol one- or two-letter "abbreviations" used to designate an element. (**2.1**)

Atomic theory a theory of the structure of matter. This theory postulates the existence of exceedingly small particles (called atoms) from which all matter is composed. (**2.1**)

Atomic weight the average atomic mass for a naturally occurring element expressed in atomic mass units (amu). (**3.1**)

Atoms minute particles of which matter is composed. An atom is the smallest part of an element that can enter into chemical reaction. (**2.1**)

Aufbau principle see building-up principle.

Avogadro's law the principle that states that equal volumes of any two gases at the same temperature and pressure contain the same number of molecules. (**4.4**)

Avogadro's number (N_A) the number of atoms in exactly 12 g of carbon-12, equal to 6.02×10^{23} to three significant figures. (**3.3**)

Axial direction one of the directions in which an electron pair can point in a trigonal bipyramidal arrangement. Another possible direction is equatorial. (**8.1**)

Band of stability in a plot of number of protons (Z) against number of neutrons for nuclides, the region in which stable nuclides lie. **(22.1)**

Barometer a device for measuring the pressure of the atmosphere. **(4.1)**

Base a substance other than a salt that increases the concentration of hydroxide ion in aqueous solution. **(9.5)**

Base ionization constant K_b, the equilibrium constant for the ionization of a weak base. For the base B, $K_b = \dfrac{[HB^+][OH^-]}{[B]}$. **(18.3)**

Base pairing the hydrogen bonding of complementary bases. **(27.6)**

Base units SI units of measurement from which all other units are derived. **(1.4)**

Basic oxide an ionic oxide that reacts with water to produce basic solutions, or a metal oxide that reacts with acids. **(13.4)**

Basic oxygen process the first step in the production of steel from pig iron in which impurities are removed and the carbon content is reduced by blowing oxygen into the molten iron. **(25.2)**

Bayer process a process in which purified aluminum oxide is obtained from bauxite by dissolving the ore in a solution of NaOH. **(23.5)**

Beta emission the emission of a high-speed electron from an unstable nucleus. **(22.1)**

Beta-stability line on a plot of neutrons against atomic number for nuclides, the curve separating the radioactive nuclei that decay by beta emission from those that decay by positron emission or electron capture. **(22.1)**

Bidentate ligand a ligand that bonds to a metal atom through two atoms of the ligand. **(25.3)**

Bimolecular reaction an elementary reaction that involves two reactant molecules. **(15.7)**

Binary compound a compound composed of two elements. **(7.11)**

Binding energy (of a nucleus) the energy required to break a nucleus into its individual protons and neutrons. **(22.6)**

Biological oxygen demand (BOD) a measure of water quality, given in milligrams of O_2 per liter of solution, which determines the amount of organic-reducing substances in a sample of water. **(14.7)**

Body-centered cubic (bcc) refers to a cubic unit cell that has atoms at the corners and an atom at the center. **(11.4)**

Boiling point the temperature at which the vapor pressure of a liquid equals the atmospheric pressure. **(11.8)**

Boiling-point elevation a colligative property of a solution, equal to the boiling point of the solution minus the boiling point of the pure solvent. **(12.7)**

Bond energy (BE) the average enthalpy change for the breaking of a bond in a molecule in the gaseous phase. **(10.6)**

Bonding orbital a molecular orbital in which the electrons occupying the orbital tend to be in the region between two bonded nuclei. **(8.5)**

Bonding pair an electron pair shared between two atoms. **(7.4)**

Bond length (bond distance) the distance between the nuclei of the two atoms in a bond. **(7.9)**

Bond order the number of electron pairs in a bond in a Lewis electron-dot formula **(7.9)**. In molecular orbital theory, the bond order equals $1/2 \, (n_b - n_a)$. **(8.5)**

Boranes binary compounds of boron and hydrogen. **(23.4)**

Boyle's law the principle that states that the volume occupied by any given sample of gas at constant temperature varies inversely with the pressure. **(4.2)**

Breeder reactor a nuclear reactor designed to produce more fissionable material than it consumes. **(22.7)**

Brønsted-Lowry concept a concept of acids and bases in which an acid is a proton donor and a base is a proton acceptor. **(17.4)**

Brownian motion the random motion of microscopic particles suspended in a liquid or a gas that results from molecular motion. **(4.7)**

Buffer a solution characterized by ability to withstand changes in pH when limited amounts of acid or base are added to it. It usually contains a weak acid and its conjugate base, or a weak base and its conjugate acid. **(18.6)**

Building-up principle (Aufbau principle) A scheme used to reproduce the electron configurations of the ground states of atoms by successively filling subshells with electrons in a specific order (the building-up order). **(6.2)**

Calorie (cal) a unit of energy; 1 cal = 4.184 J (exact). **(1.6)**

Calorimeter a device to measure the heat released from or absorbed by a system. **(10.2)**

Carbohydrate a polyhydroxyl aldehyde or polyhydroxyl ketone, or their condensation polymers. **(27.5)**

Carboxylic acid an organic compound containing the carboxyl group, $-\overset{\overset{\textstyle O}{\|}}{C}-O-H$. **(26.5)**

Catalyst a substance that affects the rate of a reaction without being consumed in the overall reaction. **(15.1)**

Catenation the ability of an atom to bond covalently to like atoms. **(24.1)**

Cathode   the electrode at which reduction occurs. **(21.1)**

Cathode rays   the rays that are emitted by the cathode in a gas discharge tube and have been shown to be a beam of electrons. **(5.1)**

Cathodic protection   the use of an active metal (such as magnesium) to protect a less active metal (such as iron) from corrosion. **(21.2)**

Cation   a positively charged ion. **(2.1)**

Cation exchange resin   a resin (organic polymer) that removes cations from solution and replaces them with hydrogen ions. **(14.9)**

Celsius scale   the temperature scale in general scientific use. There are exactly 100 units between the freezing point and normal boiling point of water. **(1.4)**

Chain reaction, nuclear   a self-sustaining series of nuclear fissions caused by the absorption of neutrons released from previous nuclear fissions. **(22.7)**

Change of state (phase change)   the change of a substance from one state to another. **(p. 371)**

Charles's law   the principle that states that the volume occupied by any given sample of gas at a constant pressure is directly proportional to the absolute temperature. **(4.3)**

Chelate   a complex formed by a metal atom and a polydentate ligand. **(25.3)**

Chemical change   the transformation of substances into new ones with different chemical identities; a chemical reaction. **(2.1)**

Chemical equation   a symbolic representation of a chemical reaction. **(2.2)**

Chemical equilibrium   the state reached by a reaction mixture when the rates of forward and reverse reactions have become equal, so that net change no longer occurs. **(16.1)**

Chemical property   a property of a substance involving a chemical change. **(2.3)**

Chemical reaction   a change in the chemical identity of matter resulting from a change of one or more substances into one or more different substances. **(2.1)**

Chemisorption   attraction of a substance through chemical bonding to the surface of another substance. **(15.9)**

Chiral   the quality of possessing handedness. A chiral object has a mirror image that is not identical to the object. **(25.5)**

Chromatography   one of several techniques for separating mixtures based on the difference in rates of movements of substances in a mobile phase relative to a stationary phase. **(2.4)**

Clathrate   a solid solution in which one substance is trapped in the crystal. **(14.4)**

Clausius–Clapeyron equation   an equation relating the vapor pressure of a liquid to the absolute temperature and the heat of vaporization, $\triangle H_{vap}$: $\log \frac{P_2}{P_1} = \frac{\triangle H_{vap}}{2.303\,R}\left(\frac{T_2 - T_1}{T_2\,T_1}\right)$. Here P_1 and P_2 are vapor pressures at absolute temperatures T_1 and T_2. **(11.8)**

Coagulation (of a colloid)   the process in which a colloid is made to aggregate and thus come out of dispersion. **(12.4)**

Codon   a part of the genetic code; a sequence of three nucleotides that signifies a specific amino acid. **(27.6)**

Coefficient   a numeral placed in front of a formula in a chemical equation giving the relative number of the species involved in the chemical reaction. **(2.2)**

Colligative properties   properties of solutions that depend on the concentration of solute particles (molecules or ions) in a given quantity of solvent, but not on the nature of the solute. **(p. 811)**

Collision theory   a theory that assumes that in order for reaction to occur, reactant molecules must collide with a certain minimum quantity of energy (activation energy). **(15.5)**

Colloid   a mixture of two or more substances that appears to be homogeneous but in fact consists of comparatively large particles of one substance dispersed throughout another substance or solution. **(12.4)**

Common ion effect   the shift in an ionic equilibrium caused by the addition of a solute that provides an ion that takes part in the equilibrium. **(18.5)**

Complementary bases   nucleotide bases that form strong hydrogen bonds with one another. **(27.6)**

Complex   a compound consisting of complex ions with other ions of opposite charge. **(25.3)**

Complex ion   a metal atom or ion with Lewis bases attached to it through coordinate covalent bonds. **(25.3)**

Compound   a kind of matter composed of atoms of two or more elements. A compound is a pure substance that can be chemically decomposed into two or more elements. **(2.1)**

Condensation   the change of a gas to either the liquid or solid state. **(11.7)**

Condensation polymer   a giant molecule formed by the joining of many similar molecules with the splitting out of small molecules such as H_2O. **(26.6)**

Condensation reaction   a reaction in which two or more molecules or ions are joined by the elimination of small molecules such as H_2O. **(24.1)**

Conformations   the different arrangements of a molecule in space that result from the rotations of atoms about single bonds. **(27.3)**

Conjugate acid in a conjugate acid–base pair, the species that can donate a proton. (**17.4**)

Conjugate acid–base pair in an acid–base equilibrium, two species that differ by the loss or gain of a proton. (**17.4**)

Conjugate base in a conjugate acid–base pair, the species that can accept a proton. (**17.4**)

Contact process a process for the manufacture of sulfuric acid by the oxidation of sulfur dioxide to sulfur trioxide with a solid catalyst, followed by the reaction of sulfur trioxide and water. (**24.3**)

Continuous spectrum light that has been dispersed into a continuous band of wavelengths. (**5.7**)

Control rods in a nuclear fission reactor, cylinders that contain elements such as boron or cadmium that absorb neutrons and can therefore slow a chain reaction. (**22.7**)

Conversion factor a ratio equivalent to 1 that changes a value in one unit to a value in another unit. (**1.5**)

Coordinate covalent bond a covalent bond in which both electrons are donated by one atom. (**7.4**)

Coordination compound a compound consisting of complex ions with other ions of opposite charge. (**25.3**)

Coordination isomers isomers consisting of complex cations and complex anions that differ in the way the ligands are distributed between the metal atoms. (**25.5**)

Coordination number in a crystal, the number of nearest neighbor atoms of an atom (**11.5**). In a complex, the total number of bonds the metal atom forms with ligands. (**25.3**)

Coupling (of reactions) a combination of a reaction that requires free energy with one that releases it, such that the overall free-energy change for the two reactions is negative and therefore the overall change is spontaneous. (**27.2**)

Covalent bond a chemical bond in which a pair of electrons is shared between two atoms. (**7.4**)

Covalent network solid a solid that consists of atoms held together in large networks or chains by covalent bonds. (**11.3**)

Covalent radius a radius of an atom joined by covalent bonds. (**7.9**)

Critical mass the smallest mass of fissionable material in which a chain reaction can be sustained. (**22.7**)

Critical pressure the minimum pressure that must be applied to liquefy a gas at its critical temperature. (**11.10**)

Critical temperature the temperature above which the liquid state of a substance no longer exists. (**11.10**)

Crystal a solid that contains a regular, repeating arrangement of ions, atoms or molecules. (**2.1**)

Crystal field splitting the difference in energy between two sets of five d orbitals on a central metal ion that has been split by the electric field of ligands. (**25.7**)

Crystal field theory a simple theory for describing the electronic structure of complexes. (**25.7**)

Crystalline solid a solid with well-defined, regular shapes; the basic units of the solid are arranged in a regular array. (**11.4**)

Crystal systems the seven basic shapes possible for unit cells; a classification of crystals. (**11.4**)

Cubic closest-packed structure a crystal structure composed of closest-packed atoms (or other units) with the stacking ABCABCABCA . . . . The structure has a face-centered cubic unit cell. (**11.5**)

Curie (Ci) the unit in which the activity of a radioactive source is measured; $1 \text{ Ci} = 3.700 \times 10^{10}$ nuclear disintegrations/s. (**22.3**)

Cyclotron a type of particle accelerator consisting of two semi-circular "dees" (or metal electrodes) which are charged by a high-frequency alternating current. The particle is continually accelerated in the region between the dees. (**22.2**)

Dalton's law of partial pressures a law that states that the sum of the partial pressures of all the different gases in a mixture is equal to the total pressure of the mixture. (**4.6**)

_d_-Block transition elements an element with an unfilled d subshell in a common oxidation state. (**25.1**)

de Broglie relation the equation relating the wavelength associated with a particle of matter to its mass times speed: $\lambda = h/mv$. (**5.8**)

Degree of ionization the fraction of molecules that react with water to give ions. (**18.1**)

Delocalized bonding bonding in which a bonding pair of electrons is spread over a number of atoms rather than localized between two. (**7.8**)

Denaturation loss of normal conformation of a biological macromolecule usually resulting in loss of the biological function. (**27.3**)

Density the mass per unit volume of a substance or solution. (**1.6**)

Deoxyribonucleic acid (DNA) the genetic constituent of cells composed of two parallel polynucleotide chains with base pairing along their entire lengths. (**27.6**)

Deoxyribonucleotides a nucleotide found in DNA. (**27.6**)

Derived unit a unit obtained by combining base units. (**1.6**)

Desalination the removal of ions from brackish or sea water. (**14.8**)

Deuterons a nucleus of a hydrogen-2 atom. (**22.2**)

Dextrorotatory   rotating the plane of polarized light to the right, pertaining to a solution of a compound. **(25.5)**

Diagonal relationship   the resemblance of each of the first three elements of the second row of the periodic table to the element located diagonally to the right of it in the third row; for example, Li resembles Mg. **(23.1)**

Diamagnetic   not attracted or weakly repelled by a magnetic field. **(16.3)**

Diffusion   the process in which a gas spreads out (through another gas) to occupy a space with uniform partial pressure. **(4.9)**

Dimensional analysis   a method of calculation in which one includes the units for quantities, treating them as algebraic quantities. **(1.5)**

Dipole-dipole force   the attractive force resulting from the tendency of polar molecules to align themselves so that the positive end of one molecule is near the negative end of another. **(11.1)**

Dipole moment   a quantitative measure of the charge separation in a molecule. **(8.2)**

D-isomer   one of a pair of enantiomers. The D-amino acid has the structural formula

$$H - \overset{\displaystyle COO^-}{\underset{\displaystyle R}{\overset{|}{\underset{|}{C}}}} - NH_3^+. \quad \textbf{(27.4)}$$

Disproportionation   the simultaneous oxidation and reduction of a species having an element in an intermediate oxidation state. **(24.4)**

Dissociation constant K_d, the reciprocal, or inverse, of the formation constant (stability constant) for a complex ion. **(19.5)**

Distillation   the process in which a liquid is vaporized and condensed; used to separate substances differing in volatility. **(2.4)**

Disulfide   an organic compound that contains the —S—S— group. **(26.7)**

Disulfide cross link   an —S—S— bonding between different parts of a folded polypeptide chain, helping to hold it in a given conformation. **(27.4)**

Double bond   a covalent bond in which two pairs of electrons are shared by two atoms. **(7.4)**

Double helix   the coiling of the two polynucleotide chains of DNA around each other. **(27.6)**

Downs cell   a commercial electrolytic cell used to electrolyze molten sodium chloride. **(23.2)**

Effective nuclear charge   the nuclear charge acting on an electron, equal to the nuclear charge less any shielding by intervening electrons. **(6.5)**

Efflorescence   the process in which a particular hydrate loses water of hydration when exposed to air at normal temperatures. **(14.4)**

Effusion   the escape of a gas into a vacuum through a very small hole at the same speed it had in the container. **(4.9)**

Electrochemical cell   a system consisting of electrodes that dip into an electrolyte and in which either an electric current produces a chemical change or a chemical change produces an electric current. **(21.1)**

Electrolysis   the nonspontaneous reaction occurring in an electrolytic cell. **(21.1)**

Electrolyte   a substance that dissolves in water to give an electrically conducting solution. **(9.1)**

Electrolytic cell   an electrochemical cell in which an electric current is used to force a reaction to go in the nonspontaneous direction. **(21.1)**

Electromagnetic spectrum   the range of frequencies or wavelengths of electromagnetic radiation. **(5.5)**

Electromotive force (emf)   the maximum potential difference between the electrodes of a voltaic cell. **(21.4)**

Electromotive series   see Activity series.

Electron   a negatively charged particle in an atom, existing in the region about the nucleus. **(5.1)**

Electron affinity   the energy involved when an electron is added to a neutral atom in the gaseous state to form a negative ion. **(6.5)**

Electron capture   the decay of a radioactive nucleus by picking up an electron from an inner orbital of the atom. **(22.1)**

Electron configuration   a particular distribution of electrons among available subshells. For example, the ground state of the nitrogen atom is $1s^2 2s^2 2p^3$, meaning there are two electrons in the $1s$ subshell, two electrons in the $2s$ subshell, and three in the $2p$ subshell. **(6.1)**

Electronegativity   a measure of the ability of an atom in a molecule to draw bonding electrons to itself. **(7.5)**

Electron spin quantum number (m_s)   see Spin quantum number.

Electron volt   a unit of energy equal to 1.602×10^{-19} J; the energy imparted to an electron that is accelerated through 1 volt potential difference. **(22.2)**

Element   a kind of matter composed of only one chemically distinct type of atom. An element is a pure substance that cannot be chemically decomposed into simpler substances. **(2.1)**

Elemental composition   a listing of the mass percentages of all the elements in a compound. **(3.4)**

Elementary reaction   a single molecular event, such as a molecular collision, resulting in a reaction. **(15.7)**

Empirical formula (simplest formula) a chemical formula with the smallest integer subscripts. (**3.6**)

Emulsion a colloid composed of liquid droplets dispersed in another liquid. (**12.4**)

Enantiomers (optical isomers) isomers that are non-superimposable mirror images of one another. (**25.5**)

Endothermic refers to a chemical reaction or physical change in which heat is absorbed. (**10.1**)

Energy the capacity to move matter or do work. (**1.6**)

Energy levels the specific energies that any one atom can have. (**5.7**)

Enthalpy (H) a thermodynamic property, which can be used to obtain the heat absorbed or released in a chemical reaction of physical change at constant pressure. It equals $E + PV$. (**10.1**)

Enthalpy of reaction ($\triangle H$) the difference between the enthalpies of the reactants and products of a chemical reaction at a particular temperature and pressure. It equals the heat of reaction at constant pressure. (**10.1**)

Entropy (S) a thermodynamic quantity that is a measure of the randomness or disorder in a system. (**20.2**)

Enzyme a protein that catalyzes a biochemical reaction. (**27.1**)

Enzyme-substrate complex the result of the binding of a substrate to an enzyme. (**27.4**)

Equatorial direction one of the directions in which an electron pair can point in a trigonal bipyramidal arrangement. Another possible direction is axial. (**8.1**)

Equilibrium constant the value obtained for the equilibrium-constant expression when equilibrium concentrations are substituted. (**16.2**)

Equilibrium-constant expression the expression obtained by multiplying the concentrations of products together, dividing by the concentrations of reactants, and raising each concentration term to a power equal to the coefficient in the chemical equation. (**16.2**)

Equivalence point the point in a titration when the stoichiometric amount of reactant has been added. (**18.7**)

Equivalent the mass that reacts with an equivalent of another reactant. An equivalent of an acid yields 1 mol H^+ in a neutralization. An equivalent of a base yields 1 mol OH^- in a neutralization. An equivalent of an oxidizing or reducing agent is the mass that provides or uses one mole of electrons. (**9.9**)

Ester an organic compound obtained by the reaction of an alcohol with a carboxylic acid. (**26.6**)

Ether a compound with the general formula ROR′, in which R and R′ are hydrocarbon groups. (**26.5**)

Excited state a quantum-mechanical state of an atom or molecule associated with any energy level except the lowest, which corresponds to the ground state. (**6.2**)

Exothermic refers to a chemical reaction or physical change in which heat is evolved, or released. (**10.1**)

Experiment a well-defined controlled procedure carried out so others may duplicate the results. (**1.2**)

Face-centered cubic (fcc) refers to a cubic unit cell that has atoms at the corners and an atom in the center of each face. (**11.4**)

Faraday constant the magnitude of charge on one mole of electrons, equal to 9.65×10^4 C/mol. (**21.4**)

f-Block (inner) transition element an element with an unfilled f subshell in a common oxidation state. (**25.1**)

Fibrous protein a protein formed by coiling or alignment of molecules to form long water-insoluble fibers. (**27.4**)

Filtration the method of separating solid particles suspended in a liquid solution by pouring the mixture through a filter paper or similar porous material. (**2.4**)

First ionization energy (ionization potential) the minimum energy needed to remove the highest-energy (the outermost) electron from a neutral atom in the gaseous state. (**6.5**)

First law of thermodynamics a law relating the total change in energy of a system, $\triangle E$, and the heat and work. At constant pressure, when the only work is pressure-volume work ($P\triangle V$), this relation is $\triangle E = q_p - P\triangle V$, where q_p is the heat at constant pressure. (**Aside to 10.1**)

Flotation a method of separating an ore from the gangue that depends on differences in wettability of the ore and the gangue. (**23.5**)

Formation (stability) constant the equilibrium constant for the formation of a complex ion from the aqueous metal ion and the ligands. (**19.5**)

Formula the symbolic representation of a substance giving the ratios of different kinds of atoms in it. (**2.1**)

Formula weight (FW) the sum of the atomic weights of all the atoms in a given formula. (**3.1**)

Fractional (isotopic) abundance the fraction of the total number of atoms in a naturally occurring element that is composed of a particular isotope. (**5.4**)

Fractional precipitation the technique for separating two or more ions from a solution by adding a reactant that first precipitates one ion, then another, and so forth. (**19.3**)

Frasch process the extraction of free sulfur from underground deposits by means of superheated water and compressed air. (**24.3**)

Free energy (G) a thermodynamic quantity defined by the equation $G = H - TS$. (**20.4**)

Freezing the change of a liquid to a solid. (**11.7**)

Freezing point the temperature at which a liquid freezes. **(11.8)**

Freezing-point depression a colligative property of a solution, equal to the freezing point of the solution minus the freezing point of the pure solvent. **(12.7)**

Frequency the number of wavelengths of a wave that pass a fixed point in one unit of time (usually one second). **(5.5)**

Frequency factor (*A*) a constant in the Arrhenius equation. **(15.6)**

Fuel cell a voltaic cell operating with a continuous supply of energetic reactants or fuel. **(21.2)**

Fuel rods in a nuclear fission reactor, the cylinders that contain the fissionable material. **(22.7)**

Functional group a reactive portion of an organic molecule that undergoes predictable reactions. **(26.5)**

Gamma emission the emission of a photon of very short wavelength (about 10^{-12} m) from a nucleus. **(22.1)**

Gamma photon a particle of electromagnetic radiation of short wavelength (0.01 Å or 10^{-12} m) and high energy per particle. **(22.1)**

Gangue the impurities in an ore. **(23.5)**

Gas a state of matter that is fluid, relatively low in density, and highly compressible. **(2.3)**

Gas constant (*R*) See Molar gas constant.

Geiger counter a kind of ionization counter. It counts particles emitted by radioactive nuclei. **(22.3)**

Gene a sequence of nucleotides in a DNA molecule that codes for a given protein. **(27.6)**

Genetic code the relationship between the nucleotide sequence in DNA and the amino acid sequence in proteins. **(27.6)**

Geometric isomers (*cis-trans* isomers) isomers in which the atoms are joined to one another in the same way but differ because some atoms occupy different relative positions in space **(25.5)**. In compounds with carbon–carbon double bonds, isomers arise from the lack of rotation about the double bond. **(26.2)** Examples:

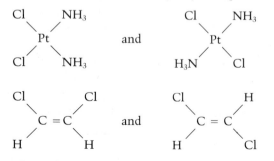

Glass electrode a compact electrode used to determine pH by emf measurements. **(21.7)**

Globular protein a polypeptide that folds into a roughly spherical conformation. **(27.4)**

Goldschmidt process the reduction of a metallic oxide to the metal by aluminum; chromium is obtained in this way. **(23.4)**

Graham's law of effusion a law that states that the rate of effusion of gas molecules from a particular hole is inversely proportional to the square root of the molecular weight of the gas at constant temperature and pressure. **(4.9)**

Ground state a quantum-mechanical state of an atom or molecule associated with the lowest energy level. States associated with higher energy levels are called excited states. **(6.2)**

Group (of the periodic table) vertical column of elements in a modern periodic table. **(6.4)**

Haber process the industrial process for the preparation of ammonia from nitrogen and hydrogen in the presence of a catalyst. **(13.6)**

Half-cell the half of an electrochemical cell in which a half-reaction (oxidation or reduction) takes place. **(21.1)**

Half-life ($t_{1/2}$) the time required for the reactant concentration to decrease by one-half of its initial value **(15.4)**. The time required for the number of radioactive nuclei in a sample to decrease by one-half. **(22.4)**

Half-reaction one of two parts of an oxidation-reduction equation representing either oxidation or reduction. **(9.8)**

Hall-Héroult process the commercial process for the preparation of aluminum by the electrolysis of aluminum oxide in molten cryolite. **(23.4)**

Halogen a Group VIIA element. **(6.6)**

Heat energy that has been transferred between objects as a result of their difference in temperature. **(1.6)**

Heat capacity the quantity of heat needed to raise the temperature of a sample of substance by one degree celsius (or one kelvin). **(10.2)**

Heat of fusion the enthalpy change for the melting of a given amount of a solid. **(11.8)**

Heat of reaction the total amount of heat that is evolved or absorbed by a system at a particular temperature because of chemical reaction. **(10.1)**

Heat of vaporization enthalpy change for the vaporization of a given amount of a liquid. **(11.8)**

Hemiacetal an organic compound with the general formula

$$R' {-} \overset{\displaystyle OH}{\underset{\displaystyle OR}{C}} {-} H.$$ **(27.5)**

Henderson-Hasselbalch equation an equation relating the pH of a buffer for different concentrations of conjugate acid and base: $\mathrm{pH} = \mathrm{p}K_a + \log \dfrac{[\text{base}]}{[\text{acid}]}$. **(18.6)**

Henry's law a law stating that the solubility of a gas in a liquid is directly proportional to the partial pressure of the gas above the solution. **(12.3)**

Hess's law of heat summation a law that states that the total enthalpy change for a reaction equals the sum of the enthalpy changes for all intermediate steps in the reaction (if the overall reaction and intermediate steps are carried out under the same conditions). **(10.4)**

Heterogeneous catalysis the use of a catalyst that exists in a different phase from the reacting species, usually a solid catalyst in a liquid or gaseous solution. **(15.9)**

Heterogeneous equilibrium an equilibrium that involves reactants and products in more than one phase. **(16.3)**

Heteronuclear diatomic molecule a molecule composed of two different atoms. **(8.6)**

Hexagonal closest-packed structure a crystal structure composed of closest-packed atoms (or other units) with the stacking ABABABA . . . . The structure has a hexagonal unit cell. **(11.5)**

High-spin complex a complex with the greater number of unpaired electrons. **(25.6)**

Homogeneous catalysis the use of a catalyst that exists in the same phase as the reacting species. **(15.9)**

Homogeneous equilibrium an equilibrium that involves reactants and products in a single phase. **(16.3)**

Homologous series a series of organic compounds in which one compound differs from a preceding one by a —CH_2— group. **(26.1)**

Homonuclear diatomic molecule a molecule composed of two like atoms. **(8.6)**

Hoopes process a process for the electrolytic purification of aluminum. **(23.5)**

Hund's rule a rule applying to ground-state electron configurations. It states that when electrons fill a subshell, every orbital in the subshell is occupied by a single electron before any orbital is doubly occupied, and all electrons in singly occupied orbitals have their spins in the same direction. **(6.3)**

Hybrid orbitals an orbital used to describe bonding, obtained by taking a combination of atomic orbitals on an isolated atom. **(8.3)**

Hydrate a substance containing weakly bonded water molecules. **(14.4)**

Hydrated ion an ion with a sphere of water molecules around it. **(14.4)**

Hydrate isomers isomers of a complex that differ in the placement of water molecules. **(25.5)**

Hydration the interaction of an ion with water molecules. **(12.2)**

Hydride a binary compound of hydrogen with another element. The hydride ion is H^-, which exists in ionic hydrides. **(14.2)**

Hydrocarbon a compound consisting of only carbon and hydrogen. **(26.1)**

Hydrogenation a reaction in which hydrogen is added, in the presence of a catalyst, to carbon–carbon multiple bonds. **(14.2)**

Hydrogen bonding a weak to moderate attractive force between a hydrogen atom bonded to an electronegative atom (F, O, or N) and a lone pair of electrons of another electronegative atom (F, O, or N). **(11.1)**

Hydrohalic acid an aqueous solution of a hydrogen halide. **(24.4)**

Hydrologic cycle the natural cycle of water from the oceans to freshwater sources and its return to the oceans. **(p. 481)**

Hydrolysis the reaction of an ion with water to produce either a hydrogen ion or a hydroxide ion. **(17.7)**

Hydrolysis constant the equilibrium constant for the reaction of water with either an anion to produce the hydroxide ion and the conjugate acid or a cation to produce the hydrogen ion and the conjugate base. **(18.4)**

Hydronium ion the H_3O^+ ion; the hydrogen ion, $H^+(aq)$. **(9.1)**

Hydrophilic colloid a colloid in which there is a strong attraction between the dispersed phase and the continuous phase (water). Many such colloids consist of macromolecules dispersed in water and are like true solutions. **(12.4)**

Hydrophobic colloid a colloid in which there is a lack of attraction between the dispersed phase and the continuous phase (water). These colloids are unstable and in time the dispersed phase will come out of solution by aggregating into larger particles. **(12.4)**

Hydrophobic interaction the tendency of nonpolar molecules or parts of molecules to cluster together in aqueous solution. **(27.3)**

Hydrosphere the waters of the earth: the oceans, polar caps, lakes, rivers, and atmospheric moisture. **(14.2)**

Hygroscopic refers to a substance that removes moisture from the air at room temperature to form a hydrate. **(14.4)**

Hypothesis a new and untested explanation of the behavior of nature. **(1.2)**

Ideal gas law a mathematical equation relating the volume (V), pressure (P), absolute temperature (T), and moles (*n*) of a gas. The equation is $PV = nRT$; here R is the molar gas constant. (**4.5**)

Inner-transition element a lanthanide or an actinide (**6.4**); an element with unfilled *f* subshells in a common oxidation state. (**25.1**)

Interhalogen a binary compound of one halogen with another. (**24.4**)

Intermolecular force the weak attractive force between molecules. (**11.1**)

International System of Units (SI) a particular choice of metric units adopted by the General Conference of Weights and Measures in 1960. (**1.4**)

Ion an electrically charged atom or group of atoms. (**2.1**)

Ion exchange the process in which a zeolite or ion exchange resin is used to replace ions in water solution by other ions, for example, to exchange Ca^{2+} ions for Na^+ ions. (**14.9**)

Ionic bond the electrostatic attraction between positive and negative ions. (**7.1**)

Ionic equation a chemical equation with the strong electrolytes written as ions. (**9.2**)

Ionic nitride a compound of nitrogen and an active metal, in which nitrogen is in the -3 oxidation state. (**13.7**)

Ionic radius the radius of an ion based on the assumption that ions are spheres; the ionic radius is a measure of the spherical region around the nucleus within which the electrons are most likely to be found. (**7.3**)

Ionic solid a solid consisting of cations and anions held together by the electrical attraction of opposite charges. (**11.3**)

Ionization energy see first ionization energy.

Ionization isomers isomers that differ in which anion is coordinated to which metal atom. (**25.5**)

Ion product the value of the product of ion concentrations, each concentration raised to a power equal to the number of ions in the formula of a slightly soluble ionic compound. (**19.3**)

Ion-product constant for water (K_w) the equilibrium value of $[H^+][OH^-]$. It equals 1.00×10^{-14} at 25°C. (**17.2**)

Ion-selective electrode a specially constructed electrode that is responsive to a particular ion in solution. (**21.7**)

Isoelectronic referring to atoms or molecules having the same electron configuration. (**7.3**)

Isomers compounds of the same molecular formula but with different arrangements of the atoms. (**8.4**)

Isotope dilution a technique designed to determine the quantity of substance in a mixture or the total volume of solution by adding a known amount of an isotope to it. (**22.5**)

Isotopes atoms of the same element having different masses. (**3.1**)

Joule (*J*) the SI unit of energy; $1\ J = 1\ kg \cdot m^2/s^2$; (also $4.184\ J = 1\ cal$). (**1.6**)

Kelvin (K) the SI base unit of temperature; a unit of absolute temperature. (**1.4**)

Ketone a compound containing the carbonyl group to which two hydrocarbon groups are attached. (**26.5**)

Kilogram (kg) the SI base unit of mass (about 2.2 pounds). (**1.4**)

Kinetic energy the energy associated with a body by virtue of its motion. (**1.6**)

Kinetics, chemical the study of how reaction rates change under varying conditions and what molecular events occur during the overall reaction. (**p. 494**)

Kinetic theory of gases a theory of gases based on the assumption that gases are composed of molecules in constant, random motion. (**4.7**)

Lanthanide elements one of the 14 elements following lanthanum in the periodic table, in which the *4f* subshell is filling. (**6.4**)

Lattice energy the energy required to break a solid ionic compound into isolated ions in the gas phase. (**10.7**)

Law (principle) a generalization about some behavior or nature that can be stated briefly, perhaps as a mathematical equation. (**1.2**)

Law of combining volumes a law that states that gases at the same temperature and pressure react with one another in volume ratios of small whole numbers. (**Aside to 3.1**)

Law of conservation of energy a law that states that energy may be converted from one form to another, but the total quantity of energy remains constant. (**1.6**)

Law of conservation of mass a law that states that the mass remains constant during a chemical change (chemical reaction). (**1.1**)

Law of definite proportions (law of constant composition) a law that states that a compound always contains a definite or constant proportion of elements by mass. (**Aside to 3.3**)

Law of mass action a law that states that the value of the equilibrium expression is constant for a particular reaction at a given temperature. (**16.2**)

Law of multiple proportions a law that states that different compounds of the same two elements have mass ratios of the elements that are simple multiples of one another. (**Aside to 3.3**)

Lead storage cell a common voltaic cell (car battery) in which lead is the anode and lead dioxide is the cathode. (**21.2**)

LeChatelier's principle a principle stating that if a system in chemical equilibrium is altered by the change of some condition, chemical change occurs to shift the equilibrium composition in a way that attempts to reduce that change of condition. (**16.8**)

Leclanche (zinc–carbon) dry cell a common voltaic cell (battery) in which zinc is the anode and a carbon rod, surrounded by a manganese dioxide–carbon paste, is the cathode. (**21.2**)

Levorotatory rotating the plane of polarized light to the left, pertaining to a solution of a compound. (**25.5**)

Lewis concept a concept of acids and bases in which an acid is an electron-pair acceptor and a base is an electron-pair donor. (**17.8**)

Lewis electron-dot formula a formula of a molecule or ion that uses dots to represent the valence electrons (bonding pairs and lone pairs). (**7.4**)

Lewis electron-dot symbol a representation of an atom or monatomic ion showing the valence electrons, if present, as dots placed around the letter symbol of the element. (**7.1**)

Ligand a Lewis base that bonds to a metal ion to form a complex ion. (**19.5**)

Limiting reactant (reagent) the reactant that is entirely consumed when a reaction goes to completion. (**3.9**)

Line spectrum the distribution of discrete wavelengths of electromagnetic radiation emitted by an atom; in the visible region the spectrum shows several lines of different color. (**5.7**)

Linkage isomers isomers of a complex in which a ligand bonds through either one of two different atoms. (**25.5**)

Lipid a natural substance that is soluble in hydrocarbons and insoluble in water. (**27.7**)

Liquid the state of matter that has a definite volume but not a definite shape. (**2.3**)

L-isomer one of a pair of enantiomers. The D-amino acid has the structural formula $H-\overset{\displaystyle COO^-}{\underset{\displaystyle R}{\overset{|}{\underset{|}{C}}}}-NH_3^+$. (**27.4**)

Liter (L) a traditional unit of volume; $1\ L = 1\ dm^3$. (**1.6**)

London forces the weak attractive forces between molecules resulting from the small, instantaneous dipoles that occur as a result of the varying positions of the electrons during their motion about their nuclei. (**11.1**)

Lone (nonbonding) pair an electron pair that remains on one atom and is not used in bonding. (**7.4**)

Low-spin complex a complex with the smaller number of unpaired electrons. (**25.6**)

Macromolecule a very large molecule having a molecular weight that may be several million atomic mass units. (**27.3**)

Magic number the number of nuclear particles in a completed shell of protons or neutrons. (**22.1**)

Magnetic quantum number (m_l) an integer from $-l$ to $+l$ designating a specific orbital within a subshell. (**5.9**)

Main group (representative) element an element in an A column of the periodic table, in which an outer s or p subshell is filling. (**6.4**)

Manometer a device that measures the pressure of a gas in a sealed vessel. (**4.1**)

Markownikoff's rule a generalization stating that when an unsymmetrical reagent such as HCl is added to an alkene, the major product formed is the one obtained when the hydrogen atom of the reagent adds to the carbon of the double bond that already has the more hydrogen atoms attached to it. (**26.4**)

Mass the quantity of matter in a material. (**1.1**)

Mass defect for a nucleus, the total nucleon mass minus the nuclear mass. (**22.6**)

Mass-energy equivalence relation the equation $E = mc^2$ relating the energy, E, associated with a given mass, m. (**22.6**)

Mass number the total number of protons and neutrons in the nucleus of an atom. (**5.3**)

Mass percentage parts per hundred in terms of mass. The mass percentage of A is equal to $\dfrac{\text{mass of A in the whole}}{\text{mass of the whole}} \times 100\%$. (**3.4**)

Mass percentage of solute the percentage by mass of solute contained in a solution. (**12.5**)

Mass spectrometer an instrument used to determine the actual masses of atoms from the deflection of atomic ions in electric and magnetic fields. (**5.4**)

Matter anything that occupies space and can be perceived by our senses. (**1.1**)

Maxwell's distribution of molecular speeds a theoretical relationship that predicts the relative number of molecules at various speeds for a given sample of gas at a particular temperature. (**4.8**)

Melting the change of a solid to a liquid. (**11.7**)

Melting point the temperature at which the liquid and solid are in dynamic equilibrium. (**11.8**)

Mesopause transitional region of the atmosphere between the mesosphere and the thermosphere. (**13.1**)

Mesosphere the region of the atmosphere above 50 km, where the temperature begins to fall. (**13.1**)

Messenger RNA a relatively small RNA molecule that

diffuses about a cell, finally attaching itself to a ribosome where it serves as the pattern for the synthesis of a protein molecule. (27.6)

Metabolism the process of building up and breaking down organic molecules in cells. (27.1)

Metal a substance that is lustrous, malleable, and a good conductor of heat and electricity. (6.6)

Metallic solid a solid that consists of atoms held together by metallic bonding. (11.3)

Metalloids (semimetals) elements bordering the staircase line on the periodic table, exhibiting both metallic and nonmetallic properties. (6.6)

Metallurgy the scientific study of the production of metals and metal products from natural sources. (23.5)

Metal refining in metallurgy, the purification of a metal. (23.5)

Metaphosphoric acids acids of the general formula $(HPO_3)_n$. (24.2)

Metastable nucleus a nucleus in an excited state with a lifetime of at least 10^{-9} s (1 nanosecond). (22.1)

Metathesis (double replacement) reaction a reaction in which the positive and negative parts of the two reactants are exchanged to give the products. If the reactants are ionic, these positive and negative parts are the cations and anions. (9.3)

Meter the SI base unit of length (about 39 inches). (1.4)

Micelle a colloidal-sized particle formed in water by the aggregation of molecules each having a hydrophobic end and a hydrophilic end. The hydrophobic ends point inward toward one another, while the hydrophilic ends are on the outside of the micelle facing the water. (27.3)

Millimeters of mercury (mmHg) a unit of pressure; 760 mmHg = 101.325 kPa (exact). (1.6)

Mineral a naturally occurring solid substance or solid solution with definite crystalline form. (23.2)

Miscible soluble in each other, in reference to two fluids. (12.1)

Mixture a material containing two or more substances and having properties that vary depending on the relative amounts of these substances in the mixture. (2.3)

Moderator in a nuclear fission reactor, a substance that can slow down neutrons. (22.7)

Molality (m) the moles of solute dissolved in one kilogram of solvent. (12.5)

Molar concentration; molarity (M) the moles of solute dissolved in one liter (cubic decimeter) of solution. (3.10)

Molar gas constant (R) the constant factor relating the pressure, volume, moles, and absolute temperature in the ideal gas law, $PV = nRT$. (4.5)

Molar gas volume the volume occupied by one mole of any gas at a given temperature and pressure. (4.4)

Molar mass the mass of one mole of substance. It is numerically equal to the formula or molecular weight expressed in grams. However, the molar mass can be expressed in other units. Thus, the molar mass of H_2O is 18.0 g/mol or 0.0180 kg/mol. (3.3)

Mole (mol) the amount of substance that contains as many molecules or formula units as the number of atoms in exactly 12 grams of carbon-12. The amount of substance containing Avogadro's number of molecules or formula units. (3.3)

Molecular equation an equation written as if the substances were molecular, even though they may actually exist in solution as ions. (9.2)

Molecular formula the combination of atomic symbols that gives the kinds of atoms and their number in a molecule. (2.1)

Molecularity the number of molecules on the reactant side of an elementary reaction. (15.7)

Molecular orbital theory a theory of the electronic structure of molecules in terms of molecular orbitals, which may spread over several atoms or the entire molecule. (8.5)

Molecular solid a solid that consists of atoms or molecules held together by intermolecular forces. (11.3)

Molecular weight (MW) the average mass of a molecule expressed in atomic mass units. It equals the sum of the atomic weights of all the atoms in the molecule. (3.1)

Molecule a tightly bonded group of two or more atoms of the same or different elements. (2.1)

Mole fraction the fraction of moles of a component substance in the total moles of a mixture. (4.6 and 12.5)

Monosaccharide a simple sugar containing three to nine carbon atoms, all but one of which bear a hydroxyl group, the remaining one being a carbonyl carbon. (27.5)

Nernst equation an equation relating the emf of a cell, E_{cell}, to its standard emf, $E°_{cell}$, and the reaction quotient, Q. At 25°C, the equation is $E_{cell} = E°_{cell} - (0.0592/n)\log Q$. (21.7)

Net ionic equation an equation showing only those ions that react. (9.2)

Neutralization a reaction of an acid and a base to produce a salt. (9.5)

Neutron a neutral particle with a mass approximately the same as a proton that exists in atomic nuclei. (p. 129)

Neutron activation analysis an analysis of elements based on the conversion of stable isotopes to radioactive isotopes by the bombardment of the sample with neutrons. (22.5)

Nitrogen cycle the circulation of the element nitrogen in the biosphere. (13.6)

Nitrogen fixation a natural or commercial process in which N_2 from air is converted to nitrogen compounds. (13.6)

Nitrogen fixing bacteria bacteria that can produce nitrogen compounds from atmospheric nitrogen. (13.6)

Noble gas (rare or inert gas) a Group VIIIA element. (6.6)

Nonelectrolyte a substance that dissolves in water to give a nonconducting, or very poorly conducting, solution. (9.1)

Nonmetal an element not exhibiting metallic characteristics. These elements are found in the upper right of the periodic table. (6.6)

Normality the number of equivalents of a substance dissolved in one liter of solution. (9.9)

Nuclear bombardment reaction a nuclear reaction in which one nucleus is struck by another. The nuclear particles of the reactant nuclei rearrange to give a product nucleus or nuclei. (22.1)

Nuclear equation a symbolic representation of a nuclear reaction. (22.1)

Nuclear fission a nuclear reaction in which a nucleus splits into lighter nuclei and energy is released. (22.6)

Nuclear fission reactor a device that permits a controlled chain reaction of nuclear fissions. (22.7)

Nuclear force a strong force of attraction between nucleons at very short distances (about 10^{-15} m). (22.1)

Nuclear fusion a nuclear reaction in which nuclei combine to give a more stable heavier nucleus, plus possibly several neutrons, and energy is released. (22.6)

Nucleic acid a polymer of nucleotides. Nucleic acids are found in cells and are carriers of the species inheritance. (27.6)

Nucleon a particle composing an atomic nucleus, either a proton or a neutron. (22.1)

Nucleoside part of a nucleotide, consisting of a base and a sugar. (27.6)

Nucleotide a building block of nucleic acids; composed of an organic base linked to a 5-carbon sugar, which in turn is linked to one phosphate group or to a string of two or three phosphate groups. (27.6)

Nucleus the very small, dense, positively charged core of an atom. (5.1)

Nuclide a specific type or kind of nucleus characterized by the number of protons and the number of neutrons it contains. (22.1)

Nuclide symbol a symbol for a nuclide, in which the atomic number is given as a left subscript and the mass number is given as a left superscript to the symbol of the element. (22.1)

Number of significant figures the number of digits reported for the value of a measured or calculated quantity, indicating the precision of the value. (1.3)

Octet rule the tendency of atoms in molecules to have eight electrons in their valence shells. (7.4)

Oligosaccharide a short polymer containing two to ten simple sugar units. (27.5)

Optically active referring to a solution of a substance, the ability to rotate the plane of polarized light. (25.5)

Orbital diagram a notation showing how the orbitals of a subshell are occupied by electrons. (6.1)

Ore a rock or mineral from which a metal can be economically produced. (23.3)

Osmosis the phenomenon of solvent flow through a semipermeable membrane from a solution of lower concentration to one of higher concentration. (12.8)

Osmotic pressure a colligative property of a solution; the pressure required to stop osmosis. (12.8)

Ostwald process an industrial preparation of nitric acid using the catalytic oxidation of ammonia. (13.7)

Oxidation a half-reaction in which an atom increases in oxidation number and thereby loses electrons. (9.7)

Oxidation number (state) the charge an atom in a covalently bonded species would have if all the bonding electrons belonged to the more electronegative atom in the group; the ionic charge on a monatomic ion. (7.10)

Oxidation-reduction (redox) reaction a reaction in which atoms change oxidation states. (9.3)

Oxide a binary compound with oxygen in the -2 oxidation state. (13.4)

Oxidizing agent a reactant that acts on another substance to increase the oxidation number of one of its atoms. A species that accepts electrons from another species. (9.7)

Oxyacid an acid containing hydrogen and oxygen with another element, called the characteristic or central element. (7.11)

Pairing energy the energy required to put two electrons into the same orbital. (25.7)

Paramagnetic referring to an atom or molecule that has a weak attraction for a magnetic field. The attraction is caused by unpaired electrons. (6.3)

Partial pressure the pressure exerted by a particular gas in a mixture. (4.6)

Particle accelerator a device used to impart high speeds to electrons, protons, alpha particles and other ions. (22.2)

Pascal (Pa) the SI unit of pressure; $1 \text{ Pa} = 1 \text{ kg/}(\text{m·s}^2)$. (1.6)

Pauli exclusion principle a rule that states that no two electrons in an atom can have the same four quantum numbers. **(6.1)**

Peptide (amide) bond the C—N bond resulting from the condensation reaction between the carboxylate group of one amino acid with the amino group of a second amino acid. **(27.4)**

Percentage yield the quantity equal to (actual yield of product/theoretical yield of product) × 100%. **(3.9)**

Period (of a periodic table) a horizontal row of elements in a modern periodic table. **(6.4)**

Periodic law a law that states that when the elements are arranged by atomic number their physical and chemical properties vary periodically. **(6.5)**

Periodic table an arrangement of elements in rows and columns so that there is a regular repetition of properties. **(6.4)**

Peroxide a compound that contains the —O—O— group or the O_2^{2-} ion. Oxygen is in the -1 oxidation state. **(13.4)**

pH a measure of acidity defined as $-\log [H^+]$. **(17.3)**

Phase a homogeneous portion of a material separated from other parts of the material by an observable boundary. **(2.3)**

Phase diagram a graph that summarizes the conditions of temperature and pressure under which the three states of a substance are stable. **(11.10)**

Phospholipid bilayer a part of a biological membrane consisting of two layers of phospholipid molecules. These molecules have hydrophobic and hydrophilic ends, and they aggregate to give this layer structure, similar to the formation of micelles by soap molecules.

Photochemical smog air pollution with a high level of ozone produced by the action of sunlight on O_2 and NO_2 from auto exhaust. **(13.9)**

Photoconductor a material whose conductivity increases when light falls on it. **(24.3)**

Photoelectric effect the ejection of electrons from the surface of a metal when light of the proper frequency shines on it. **(5.6)**

Photolysis the dissociation of a molecule by light. **(13.2)**

Photomultiplier electronic device (tube) which magnifies (amplifies) the effect of the incidence of a photon. **(22.4)**

Photon a particle of light or electromagnetic energy. **(5.6)**

Photosynthesis a series of chemical reactions that converts carbon dioxide and water into glucose, using sunlight as energy. **(13.2)**

Physical change a transformation in the form of a material but not in its composition. **(2.1)**

Physical property a characteristic of a pure substance that does not involve chemical change. **(2.3)**

Pi (π) bond a chemical bond with electron density above and below the bond axis, formed by sidewise overlap of two parallel p orbitals. **(8.4)**

Planck's constant (h) the constant of proportionality relating energy of a photon and the frequency of light; $h = 6.63 \times 10^{-34}$ J·s. **(5.6)**

Plasma an electrically neutral gas of ions and electrons. **(22.7)**

Polar covalent bond a covalent bond in which the bonding electrons are not shared equally, but spend more time near one atom than the other. **(7.5)**

Polarity in an external circuit, the relative potential; negative polarity means low potential, positive polarity means high potential. **(21.5)**

Polyatomic ion a charged group of atoms linked together by covalent bonds. **(7.2)**

Polydentate ligand a ligand bonded to a metal atom through two or more atoms on the ligand. **(25.3)**

Polynucleotide a linear polymer of nucleotide units. **(27.6)**

Polypeptide a polymer formed by the linking of many amino acids by peptide bonds. **(27.4)**

Polyphosphoric acid an acid of the general formula $H_{n+2}P_nO_{3n+1}$. It may be formed by the condensation of phosphoric acid molecules, H_3PO_4. **(24.2)**

Polyprotic acid an acid that can give up two or more moles of H^+ per mole of acid during neutralization. **(9.5)**

Polysaccharide a polymer consisting of more than ten simple sugar units. **(27.5)**

Positron a particle similar to the electron, with the same mass but with a positive charge. **(22.1)**

Positron emission emission of a positron from a nucleus. **(22.1)**

Potential difference the difference in electrical potential between two points; measured in volts. **(21.4)**

Potential energy the energy an object has by virtue of its position. **(1.6)**

Potentiometer a device that has an adjustable voltage; used to measure potential differences. **(21.4)**

Precipitate a solid formed in solution by a reaction. **(9.4)**

Precision the closeness of the set of values obtained from identical measurements of a quantity on the same instrument. **(1.3)**

Prefix (metric) letters placed before a metric unit to indicate a power of ten. **(1.4)**

Pressure the force per unit area. **(1.6)**

Primary alcohol an alcohol in which the hydroxyl group is attached to a carbon atom which is itself bonded to only one other carbon atom. **(26.5)**

Primary structure (of a protein)   the order or sequence of the amino acids in a protein. (**27.4**)

Principal quantum number (n)   the quantum number on which the energy of an electron in an atom principally depends. It can have any positive integer value. (**5.9**)

Product   a substance that results from a chemical reaction. It appears to the right of the arrow in the chemical equation. (**2.2**)

Protein   a polymer of amino acids with high molecular weight. (**27.4**)

Proton   a positively charged particle in atomic nuclei. A hydrogen nucleus, H^+. (**5.1**)

Pseudonoble-gas configuration   an electron arrangement in an ion having a filled d subshell (and a filled f subshell in the case of thallium) in addition to a noble-gas core. (**7.2**)

Pure substance   a material that has constant composition and definite properties. (**2.3**)

Quadridentate ligand   a ligand bonded to a metal atom through four atoms of the ligand. (**25.3**)

Qualitative analysis   the determination of the identity of substances present in a mixture. (**19.7**)

Quantum (wave) mechanics   the branch of physics that mathematically describes the wave properties of submicroscopic particles. (**5.8**)

Racemic mixture   a mixture containing equal amounts of two optical isomers. (**25.5**)

Rad   the dosage of radiation that deposits 1×10^{-2} J of energy per kilogram of tissue (from radiation absorbed dose). (**22.3**)

Radioactive decay   the process in which a nucleus spontaneously disintegrates, giving off radiation. (**22.1**)

Radioactive decay constant   the rate constant for radioactive decay. (**22.4**)

Radioactive decay series   the paths by which a natural radioactive element decays to a stable nuclide. (**22.1**)

Radioactive tracer   a radioactive isotope added to a chemical, biological, or physical system to facilitate study of the system. (**22.5**)

Radioactivity   the spontaneous emission of radiation by an unstable nucleus. (**5.2**)

Raoult's law   the law that states that vapor pressure, P, of the solution of a nonvolatile solute equals the vapor pressure of the pure solvent, P^0, multiplied by the mole fraction of the solvent in the solution, X: $P = P^0 X$. (**12.6**)

Rate constant   the proportionality constant in the relationship between rate and reactant concentrations. (**15.3**)

Rate-determining step   the slowest step in a reaction mechanism. The rate of this step governs the overall rate. (**15.8**)

Rate law   the relationship between the rate of a reaction and the concentrations of reactants (and catalyst). (**15.3**)

Reactant   a starting substance in a chemical reaction. It appears to the right of the arrow in the chemical equation. (**2.2**)

Reaction intermediate   a species that is produced during a reaction but does not appear in the net equation. (**15.7**)

Reaction mechanism   the set of elementary reactions whose overall effect is given by the net chemical equation. (**15.7**)

Reaction order   the experimentally determined exponent of the concentration of a speces in a rate law. (**15.3**)

Reaction quotient   an expression of the same form as the equilibrium-constant expression but whose concentration values are not necessarily those at equilibrium. (**16.5**)

Reaction rate   the increase in molar concentration of product per unit time or the decrease in molar concentration of reactant per unit time. The usual unit of reaction rate is mol/(L · s). (**15.1**)

Reducing agent   a reactant that acts on another substance to decrease the oxidation numbers of one of the atoms. A species that donates electrons to another species. (**9.7**)

Reduction   a half-reaction in which an atom is reduced in oxidation number and thereby gains electrons. (**9.7**)

Rem   a unit of radiation dosage used to relate the various kinds of radiation in terms of biological destruction; rems = rads × RBE, where RBE is a factor for each type of radiation (the relative biological effectiveness). (**22.4**)

Resonance description   a representation of a molecule with delocalized bonding showing all possible electron-dot formulas. (**7.8**)

Reverse osmosis   the process in which a solvent, such as water, is forced through a semipermeable membrane from a concentrated solution to a dilute one. (**14.9**)

Ribonucleic acid (RNA)   a polynucleotide used by a cell to translate genetic information stored in DNA into protein structure. (**27.6**)

Ribonucleotide   a nucleotide found in RNA. (**27.6**)

Ribosomal RNA   the RNA contained in ribosomes. (**27.6**)

Ribosome   a tiny cellular particle in which protein synthesis takes place. (**27.6**)

Roasting   in metallurgy, the heating of a mineral in air to obtain the oxide. (**23.5**)

Root-mean-square (rms) molecular speed   the speed of a molecule having the average kinetic energy; the square root of the average value of the square of the molecular speed: $u = \sqrt{\dfrac{3RT}{M_m}}$. **(4.8)**

Rounding dropping nonsigificant figures in a calculation result and adjusting the last digit reported. **(1.3)**

Salt an ionic substance that contains an anion other than OH^- or O^{2-}. **(9.5)**

Salt bridge a tube of an electrolyte in a gel that connects two half-cells of a voltaic cell, allowing the flow of ions but preventing the diffusional mixing of the different solutions. **(21.1)**

Saponification the hydrolysis of an ester in the presence of a base. **(26.6)**

Saturated hydrocarbon a hydrocarbon in which all carbon atoms are bonded to the maximum number of hydrogen atoms. **(26.1)**

Saturated solution a solution in equilibrium with the solute. It is unable to dissolve additional solute. **(9.4)**

Scientific method the creative process of understanding the physical world that involves hypothesis formation, experimentation, modification of theory, followed by more experimentation, and so forth. **(1.2)**

Scientific notation the representation of numbers in the form $A \times 10^n$, where A is a number with a single nonzero digit to the left of the decimal point, and n is a whole number. **(1.3)**

Scintillation counter a device that detects nuclear radiations from the flashes of light generated in a material by the radiation. **(22.3)**

Second (s) the SI base unit of time. **(1.4)**

Secondary alcohol an alcohol in which the hydroxyl group is attached to a carbon atom which is itself bonded to two carbon atoms. **(26.5)**

Secondary structure (of a protein) a simple coiled or parallel arrangement of a protein molecule. **(27.4)**

Second law of thermodynamics a law that states that the total entropy of a system and its surroundings always increases for a spontaneous process. **(20.2)**

Self-ionization (autoionization) a reaction in which two like molecules react to give ions. **(17.2)**

Semiconductor a substance that is only slightly conducting at room temperature but becomes a moderately good conductor at higher temperatures. **(23.1)**

Shell model of the nucleus the nuclear model in which protons and neutrons exist in levels, or shells, analogous to the shell structure that exists for electrons in an atom. **(22.1)**

Side chain (of an amino acid) the hydrocarbon group of an amino acid that is attached to the carbon atom to which the amino group is bonded. **(27.4)**

Sigma (σ) bond chemical bond with cylindrical shape about the bond axis, formed when two s orbitals overlap or by the overlap along the axis of orbitals with directional character. **(8.4)**

Silane a hydride of silicon containing chains of up to six single-bonded silicon atoms. **(24.1)**

Silicate a compound of silicon and oxygen with various metals. **(24.1)**

Simple cubic unit cell a unit cell in which the atoms are arranged only at the corners of a cube. **(11.4)**

Single bond a covalent bond in which a single pair of electrons is shared by two atoms. **(7.4)**

Sink in environmental chemistry, the means by which a substance is removed from one area of the environment. Thus, certain fungi in soil provide a sink for CO in the atmosphere. **(13.9)**

Sol a colloid in which very small solid particles are dispersed in a liquid. **(12.4)**

Solid the state of matter that has a definite volume and shape. **(2.3)**

Solubility the amount of a substance that dissolves in a given quantity of solvent (such as water) at a given temperature. **(9.4)**

Solubility-product constant (K_{sp}) the equilibrium constant for the dissolution of a slightly soluble (or nearly insoluble) ionic compound. **(19.1)**

Solute in the case of a solution of a gas or solid dissolved in a liquid, the gas or solid is called the solute. In other cases, the solute is the component of a solution in smaller amount. **(12.1)**

Solution a homogeneous mixture. **(2.3)**

Solvay process a commercial method of obtaining soda ash (sodium carbonate) from limestone (calcium carbonate) and salt. **(23.2)**

Solvent in a solution of a gas or solid in a liquid, the liquid is called the solvent. In other cases, the solvent is the component in greater amount. **(12.1)**

Specific heat the quantity of heat required to raise the temperature of one gram of matter one degree Celsius (or one kelvin). **(1.6)**

Specificity (of enzyme action) the property of enzyme action where only certain substrates will bind to an enzyme or only certain types of reactions will be catalyzed. **(27.4)**

Spectator ion an ion that occurs on both sides of the arrow in an equation and does not take part in the reaction. **(9.2)**

Spectrochemical series an arrangement of ligands according to the relative strengths of the crystal field splittings they induce in the d orbitals of a metal ion. **(25.7)**

Spin quantum number (m$_s$) a quantum number designating one of the possible orientations allowed for the spin of a particle. For an electron, m_s is either $+\frac{1}{2}$ or $-\frac{1}{2}$. **(5.9)**

Spontaneous process a physical or chemical change that occurs by itself. **(20.2)**

Standard electrode potential the emf of the cell consisting of the particular electrode under standard conditions, and the standard hydrogen electrode, which is taken to have an electrode potential of zero. **(21.5)**

Standard emf the maximum potential difference between the electrodes of a cell when both electrodes are under standard conditions. **(21.5)**

Standard enthalpy of formation (ΔH_f) the enthalpy change for the formation of one mole of substance from its elements in their stablest states at standard pressure (1 atm) and at a specified temperature (25°C unless otherwise noted). **(10.5)**

Standard entropy (ΔS) the entropy of a substance at 1 atmosphere and usually 25°C (ions in solution are 1 M concentration), obtained by applying the third law of thermodynamics. **(20.3)**

Standard free energy of formation ($\Delta G_f°$) the free-energy change for the formation of one mole of substance from its elements in their stablest states at one atmosphere pressure and (usually) 25°C. **(20.4)**

Standard potential diagram a convenient graphic presentation of the standard potentials of an element. **(24.4)**

Standard temperature and pressure (STP) reference conditions for a gas chosen by convention to be 0°C and 1 atm pressure. **(4.4)**

States of matter the three different forms in which matter exists: solid, liquid, and gas. **(2.3)**

Steam-reforming process the industrial preparation of hydrogen and carbon monoxide mixtures by the reaction of steam and hydrocarbons at high temperatures and pressures over a nickel catalyst. **(14.2)**

Stereoisomers isomers in which the same atoms are bonded to each other in the same order, but which differ in the precise arrangement of these atoms in space. **(25.5)**

Stoichiometry the calculation of the quantities of reactants and products involved in a chemical reaction. **(3.7)**

Stratopause the transitional region of the atmosphere between the stratosphere and the mesosphere. **(13.1)**

Stratosphere the region of the atmosphere just above the troposphere, where the temperature increases. **(13.1)**

Strong acid an acid that ionizes completely in water. **(9.5)**

Strong base a base that is present in aqueous solution as ions only. **(9.5)**

Strong electrolyte an electrolyte that exists in solution almost entirely as ions. **(9.1)**

Structural formula a representation of a molecule showing by means of atomic symbols and lines how the atoms are bonded. **(2.1)**

Structural isomers isomers that differ in how the atoms are joined together. **(25.5)**

Sublimation the change of a solid directly to the vapor. **(11.7)**

Substitution reaction a reaction in which part of a reagent molecule replaces a hydrogen atom on a hydrocarbon or hydrocarbon group. **(26.4)**

Substrate a molecule whose reaction an enzyme will catalyze. **(27.4)**

Superoxide a compound that contains the O_2^{2-} ion, in which the oxygen is in the $-\frac{1}{2}$ oxidation state. **(13.4)**

Supersaturated solution an unstable solution that contains more dissolved solute than a saturated solution. **(9.4)**

Surface tension the energy required to increase the surface area of a liquid by unit amount. **(11.2)**

Surroundings everything in the vicinity of a thermodynamic system. **(10.1)**

System (thermodynamic) substance or mixture of substances under study in which a change occurs. **(10.1)**

Termination codon a codon that signifies the end of a protein synthesis. **(27.6)**

Termolecular reaction an elementary reaction that involves three reactant molecules. **(15.7)**

Tertiary alcohol an alcohol in which the hydroxyl group is attached to a carbon atom that is itself bonded to three carbon atoms. **(26.5)**

Tertiary structure (of a protein) the manner in which a protein coil is folded. Thus, a globular protein consists of a coil that is folded into a compact, roughly spherical conformation. **(27.4)**

Theoretical yield maximum amount of product that can be obtained in a reaction; the amount calculated based on the limiting reactant. **(3.9)**

Theory a generalization of wide scope about the behavior of nature. **(1.2)**

Thermal equilibrium a state in which heat does not flow between a system and its surroundings because they are both at the same temperature. **(10.1)**

Thermal pollution the addition of heat to natural waters. **(14.7)**

Thermodynamic equilibrium constant the equilibrium constant, K_{th}, as defined by the equation, $\Delta G° = -2.303$

$RT \log K_{th}$. K_{th} for a reaction involving only solutes equals K_c; for a reaction involving only gases, K_{th} equals K_p. **(20.6)**

Thermosphere   the region of the atmosphere above 80 km where the temperature ceases to fall and begins to rise. **(13.1)**

Thiols (mercaptans)   the sulfur analogs of the alcohols; organic compounds that contain the —SH group. **(26.7)**

Third law of thermodynamics   the law that states that a substance that is perfectly crystalline at 0 K has an entropy of zero. **(20.3)**

Three-center bond   a bond in which three atoms are held together by two electrons. **(23.4)**

Titration   a procedure for determining the amount of substance A by adding a carefully measured volume of a solution of known concentration of solution B until the reaction is just complete. **(3.11)**

Torr   an alternate name for millimeters of mercury (mmHg). **(1.6)**

Transfer RNA   the smallest type of RNA molecule; each one bonds to an amino acid and carries it to a ribosome, attaching itself through base pairing to a messenger RNA codon. **(27.6)**

Transition-state theory   a theory that assumes that the collision of two reactant molecules gives an unstable grouping or activated complex that can break up to form products. **(15.5)**

Transmutation   the change of one element to another element by bombardment with nuclear particles or nuclei. **(22.2)**

Transuranium element   an element with atomic number greater than 92. **(22.2)**

Triacylglycerol   an ester formed from glycerol and three fatty acids. **(27.7)**

Triple bond   a covalent bond in which three pairs of electrons are shared by two atoms. **(7.4)**

Triple point   the point on a phase diagram at which all three states of a substance are in equilibrium. **(11.10)**

Tropopause   the transitional region of the atmosphere between the troposphere and the stratosphere where the temperature remains constant. **(13.1)**

Troposphere   the lowest region of the atmosphere. **(13.1)**

Tyndall effect   the scattering of light by colloidal-sized particles. **(12.4)**

Uncertainty principle   a relation that states that the product of the uncertainty in position (Δx) and the uncertainty in momentum $(m\Delta v_x)$ of a particle can be no smaller than Planck's constant (h) divided by 4π:

$$(\Delta x)(m\Delta v_x) \geqslant \frac{h}{4\pi}. \quad \textbf{(5.8)}$$

Unidentate ligand   a ligand bonded to the metal atom through one atom of the ligand. **(25.3)**

Unimolecular reaction   an elementary reaction that involves one reactant molecule. **(15.7)**

Unit   a standard of measurement. **(1.3)**

Unit cell   the smallest repeating unit from which a given crystal can be constructed. **(11.4)**

Unsaturated hydrocarbon   a hydrocarbon in which carbon atoms are not bonded to the maximum number of hydrogen atoms. Such compounds contain carbon–carbon multiple bonds. **(26.2)**

Unsaturated solution   a solution that is able to dissolve additional solute. **(9.4)**

Valence bond theory   an explanation of chemical bonding. According to valence bond theory, a bond forms as the result of the overlap of the outer atomic orbitals on the atoms. An electron pair in this bond is shared between the atoms. **(8.3)**

Valence electron   an outer ns or np electron that can be involved in chemical bonding. **(6.4)**

Valence-shell electron-pair repulsion (VSEPR) model   a model that predicts molecular geometries based on the assumption that the repulsion of valence-shell electron pairs is the effect that determines molecular geometry. **(8.1)**

van der Waals equation   a mathematical equation describing the behavior of gases at high pressures and low temperatures. **(4.10)**

van der Waals forces   the weak attractive forces between molecules. **(11.1)**

Vaporization   the change of a solid or liquid to a vapor. The change of a solid to a vapor is specifically referred to as sublimation. **(11.7)**

Vapor pressure   the partial pressure of the vapor of a substance over the liquid measured at equilibrium. **(11.2)**

Vapor pressure lowering   a colligative property of a solution. It equals the vapor pressure of the pure solvent minus the vapor pressure of the solution. **(12.6)**

Viscosity   the resistance of a liquid to flow. **(11.2)**

Volt (V)   a unit of potential difference; 1 V = 1 J/C. **(21.4)**

Voltaic (galvanic) cell   an electrochemical cell that uses the work available from a spontaneous reaction to produce an electric current. **(21.1)**

Water–gas reaction   an industrial process in which steam is passed over red-hot coke to give a gaseous mixture of CO and H_2: $C(s) + H_2O(g) \rightarrow CO(g) + H_2(g)$. **(14.7)**

Water–gas shift reaction   an industrial process in which CO and steam are reacted in the presence of a catalyst to produce CO_2 and H_2: $CO(g) + H_2O(g) \xrightarrow{\text{catalyst}} CO_2(g) + H_2(g)$. **(14.7)**

Wavelength   the distance between any two adjacent similar points of a wave. **(5.5)**

Weak acid   an acid that only partially dissociates into ions in aqueous solution. The acid molecule and the ions are in equilibrium. **(9.5)**

Weak base   a base that is present in solution in an equilibrium involving the molecular species and ions (including OH$^-$). **(9.5)**

Weak electrolyte   an electrolyte that exists in solution as an equilibrium between a molecular substance and a small concentration of ions. **(9.1)**

Zeolites   a natural or synthetic mineral which is used in ion exchange processes. **(14.9)**

Zone refining   in metallurgy, a purification process that depends on the greater solubility of impurities in the liquid phase than in the solid phase. **(24.1)**

Zwitterion   an amino acid in the double ionized form, in which the carboxyl group has lost an H$^+$ to give —COO$^-$ and the amino group has gained an H$^+$ to give —NH$_3^+$. **(27.4)**

Index

A (frequency factor), 515
Abelson, P. H., 729
Absolute entropies, 660–664
  changes for reaction, 662–664
  table, 662
  third law of thermodynamics, 660–662
Absolute temperature, 13
Acceleration, 16, 19
Accelerators, particle, 728–729
Accuracy, defined, 7
Acetaldehyde, 917, 918, 920
Acetate, 612
  common-ion effect, 633
  formula, 203
  salt solubilities, 280
Acetate buffers, 610, 611–614
Acetic acid, 284, 286, 286
  acetaldehyde oxidation, 920
  common-ion effect, 609
  formula, 919
  ionization constant, 597, 598
Acetone, 917, 919
  acetylene solubility, 390
  2-propanol oxidation, 920–921
Acetylene
  bonding in, 255
  Lewis formula, 209
  solubility in acetone, 390
Acetylsalicylic acid, acid ionization constant, 600
Achiral isomers, 871
Acid anhydrides, formation of, 443–444
Acid anions, nomenclature, 227
Acid-base behavior
  chromium compounds, 851
  complex ions, 857
  iron compounds, 852
  metal vs. nonmetal elements, 760, 761

metal vs. nonmetal oxides, 761–762
transition element compounds, 848–849
Acid-base concepts
  Arrhenius concept, 570–573
  Brønsted-Lowry acids and bases, 578–581
  hydrolysis, 585–587
  Lewis acids and bases, 587–588
  molecular structure and acid strength, 583–585
  pH, 575–577, 578
  relative strengths of acids and bases, 581–583
  of salt solutions, 585–587
  water, self ionization, 573–574
Acid-base equilibria
  buffers, 611–615
  common-ion effect, 609–611
  hydrolysis, 605–608
  ionization equilibria
    acid, 595–601
    base, 603–604
  polyprotic acids, 601–603
  titration curves
    strong acids and bases, 615–617
    weak acids and strong bases, 617–618
    weak bases and strong acids, 618
Acid-base indicators, 284, 577, 578
Acid-base pairs, conjugate, hydrolysis of, 605–608
Acid ionization constants, *see* Ionization constants
Acidic oxides, formation of, 443–444
Acidity, main-group oxides, 761, 762
Acid mine drainage, 434
Acid rain, 444, 553n
Acids
  amino acids as, 940. *See also* Amino acids

biological molecule denaturation, 939
buffers and, 613–614
copper reactions with, 854
defined, 283
equivalents and normality, 303–305
gas formation, 286–289
Group IIIA element reactivities, 783
Group IVA element reactivities, 801–802
hydrogen halide preparation, 830–831
hydrohalic, 191, 829–831
hydrolysis, 587
iron reactions with, 852
and metals, hydrogen generation, 467
neutralization, 285–286
nomenclature, 226
preparation of, 288–289
strengths, relative, 581–583
strong, 284
  in Arrhenius's theory, 572
  common-ion effect, 609
  gas formation, 288
  molecular structure, 583–585
  neutralization of, 285, 286
  salt preparation, 291
  salts of, 586
  sulfuric, 823–824, 825
  telluric, 825
  titration of, by strong base, 615–617
  titration of, by weak base, 618, 619
tables, 283
  gas formation, 288
  ionization equilibria, 597
  relative strengths, 582
weak
  buffers, 514, 610, 611–615
  common-ion effect, 609–610
  halogen oxyacids, 833

Acids, weak *(cont.)*
    hydrogen sulfide, selenide and tel-
      luride, 822
    hydrolysis of salts, 608
    ionization equilibria, 594–601
    neutralization of, 286
    salts of, 586
    telluric, 825
    titration of, by strong base, 617–
      618
    see also Acid-base equilibria
    see also Oxyacids
Acid salts, defined, 286
Actinides, 176, 177, 179, 843
Actinium, 730, 767
Activated charcoal, water treatment,
    485
Activated complexes, 513, 518
Activated-sludge treatment, 485
Activation energy
    in collision theory, 511
    potential energy diagrams, 514–515
Active sites, enzymes, 526, 947, 948
Activity series, 470, 471
Actual yield, defined, 17. *See also*
    Yield calculations
Addition polymers, alkene, 910
Addition reactions
    alkenes, 909–910
    alkynes, 910–911
Adenine, 953–954
    base pairing, 955–956
    codon dictionary, 960
Adenosine, 954
Adenosine triphosphate, 669, 961
Adipose tissue, 963
Adsorption, 525
Aerobic metabolism, waste products,
    483
Aerosols, 400, 401
Aerosol spray-can propellants, 446,
    831
Ag-AgCl electrode, 704
Air
    as gaseous solution, 38
    liquid, fractional distillation, 457
    see also Atmosphere
-al, 918
ʟ-Alanine, 942, 943
Alcoa process, 783–784
Alcohol groups, sugars, 949
Alcohols, 914–917
    esterification, 922–923
    esters, 919
    hydroxyl groups, 914–915
    nomenclature, 915–916
    oxidation-reduction reactions, 920–
      921
    solubility in water, 393–394
    structure of, 914
    sulfur analogs, 927
    uses of, 916–917
Aldehydes, 917, 918, 919, 922
    biological molecules, *see* Carbohy-
      drates
    oxidation of, 921–922

oxidation-reduction reactions, 920–
    921
    primary alcohol oxidation to, 921
    reduction of, 922
    structure of, 914
Algae, water pollution and, 484, 818
Aliphatic hydrocarbons, 894, 895. *See*
    also Alkanes; Alkenes;
    Alkynes; Hydrocarbons
Alizarin yellow R, 578
Alkali metals, 188
    analysis of ions, 649
    basic oxides, 443
    crystal structure, 365
    formation of oxides, 441
    and hydride formation, 473
    hydrogen from, 470
    ionization energies, 183
    oxidation number, 223
    perchlorates, 833
    properties of, 766
    reaction with oxygen, 441
    salts, solubilities, 280
    see also Group IA elements; *specific*
      elements
Alkaline dry cells, 687, 688
Alkaline earth hydroxides, solubilities,
    394–395
Alkaline earth metals, 188–189
    formation of oxides, 442
    and hydride formation, 473
    oxidation number, 223
    oxides, 443
    oxygen reactions with, 442
    precipitation of ions, 649
    see also Group IIA elements; *specific*
      elements
Alkaloids, 604
Alkanes, 895–897
    catalytic reforming, 913
    nomenclature of, 898–900
    from petroleum, 911, 912, 913
    sources of, 901, 902
    substitution reactions of, 908–909
Alkane series, 895–897
Alkenes, 895
    addition reactions of, 909–910
    defined, 902
    multiple bonds, 901–902
    nomenclature, 902, 903
    structure, 903, 904
Alkyl groups
    benzene derivatives, 906–907
    defined, 898–899
    table, 898
Alkynes, 895, 905, 910–911
Allotropes
    boron, 782
    carbon, 189, 800
    crystal structure, 367
    defined, 189
    enthalpies of formation for, 328
    Group IVA elements, 800–801
    Group VIA elements, 819
    oxygen, 444
    phosphorus, 813, 814

Alloys
    chromium, 851
    Group IIA elements, 776–777
    Group IIIA elements, 784
    Group IVA elements, 803, 805
    Group VA elements, 814
    silicon, 803
    sodium–potassium, 391, 771
    as solid solutions, 392
    steel, 853–854
    tellurium, 821
α-Amino acids, 940
Alpha particles, 721
    bombardment reactions, 727–729
    and neutron emission, 135
Alpha particle scattering, 134–135
Alpha radiation, 723
    atomic number and, 725
    band of stability, 722
    defined, 718
    detection of, 731
    discovery of, 133–134
    effects of, 730–731
    francium decay as, 767
    mass losses in, 746
    radioactive decay as, 725–726
    radium decay as, 773–774
    relative biological effectiveness, 733
Alternators, 688
Alumina, 786, 911–912
Aluminate ion, 787
Aluminosilicates, 781
    elements in, 767, 773
    structure, 809, 810, 811
Aluminum, 189
    amphoteric hydroxides, 645, 761
    calcium oxide reduction, 776
    diagonal relationships, 766
    electromotive series, 471
    Goldschmidt process, 851
    Group IIA element alloys, 776
    ionization energies, 200
    precipitation of the ion, 649
    preparation of
      Bayer process, 789–790
      by-products, 782
      Hoopes process, 790, 791
    properties of, 782, 783
    radioactive isotope decay, 725
    uses of, 784–785
Aluminum bromide, Lewis formula,
    216
Aluminum chloride, preparation and
    uses of, 787–788
Aluminum family, *see* Group IIIA ele-
    ments
Aluminum fluoride, aluminum produc-
    tion from, 781, 783
Aluminum hydroxide
    amphoteric properties, 645, 761
    Bayer process, 789–790
Aluminum oxide
    aluminum production from, 781, 783
    minerals, 781
    sodium hydroxide and, 772
Aluminum salts, uses of, 787–788

Aluminum sulfate, preparation and uses, 787
Amalgam, sodium, 706, 772
Americium, 729
Amide bonds, 942
Amides, 914, 925–926, 927
Amines, 914, 925
  nucleotides, 953–954, 955, 956
  tables, 914, 926, 954
Amino acids
  bond formation, 942, 943
  charge on, 940
  codons, 958, 959, 960
  isomers, 941
  protein structure
    primary, 945, 946
    secondary, 945–947
  side chains, 942–945
  structural formulas, 943–944
para-Aminobenzoic acid, 621
Amino groups, amino acids, 940, 942
Ammine ligand, 863
Ammonia, 623
  atmospheric, 458
  base ionization constant, 603
  from coal, 907
  complex ions, 642, 643, 646–648
    ligand nomenclature, 863
    stepwise formation, 645–646
  dissociation of, 324
  as fertilizer, 451
  formation of, 287
    equilibrium constant, 549
    free energy changes, 665
  formulas and ions, 287
  formulas and models, 32
  geometry, 238, 239
  Haber process, see Ammonia, synthesis of
  hydrogen bonding in, 350
  hydrogen in, 473
  and hypochlorite bleaches, 451–452
  ionization constant, 603
  ionization of, 274
  neutralization of, 286. See also Neutralization, ammonia-hydrochloric acid
  nitrides and, 450
  Ostwald process, 453, 454, 553–554, 555
  oxidation state, 450
  preparation of, 290–291
  solutions, molarity calculations, 67
  synthesis of
    equilibrium constants, 546–547
    Haber process, 66–68, 448–449
    optimal conditions, 562
  titration with hydrochloric acid, 618, 619. See also Neutralization, ammonia-hydrochloric acid
  as unidentate ligand, 858
  urea formation, 656. See also Urea
  water solutions, equilibrium, 273–274

Ammonia fountain, 72
Ammonium chloride, 224, 587
  hydrolysis of, 585
  in voltaic cells, 687, 688
Ammonium compounds, solubilities, 280
Ammonium formate, 608, 624
Ammonium ion
  formula, 203
  water solutions, equilibrium, 273–274
Ammonium nitrate, 447
Ammonium thiocyanate, 313, 315
Amontons, Guillaume, 102
Amontons' law, 102
Amorphous solids, characteristics of, 358–359
Ampere, 11n, 12, 708
Amphiprotic species, defined, 580
Amphoteric compounds, 644, 645, 762
  aluminum, 787
  and chromium, 850
  copper, 854
  defined, 580, 761
  Group IIIA elements, 781
  Group IVA elements, lead and tin, 811
  iron, 852
  tellurium dioxide, 825
  titanium oxide, 848
Amu, see Atomic mass units
Amylopectin, 952
Amylose, 952
Anaerobic conditions, water pollution, 484
Anaerobic metabolism, waste products, 483
Analytical chemistry
  halogen test, 828
  qualitative analysis of metals, 648–650, 821–822
  radioactive isotopes in, 740–742
Analyzer, mass spectrometer, 137–138
-ane, 898
Anesthesia, 453, 478
Angstrom, defined, 11
Angular arrangement, 238, 239, 245
Angular momentum quantum number, 153–154
Anhydrides, formation of, 443–444
Anhydrite, 629, 635
Anhydrous, defined, 478
Aniline, 926, 603
Animal feeds, magnesium carbonate use in, 777, 779
Anion exchange resin, defined, 486
Anionic detergents, 404
Anionic ligands, nomenclature, 863
Anions
  atomic vs. ionic radii, 205
  coordination compounds, nomenclature, 862, 863
  defined, 32
  in ionic solids, 365
  main-group elements, 200, 201, 202
  metallic vs. nonmetallic elements, 761

  nomenclature, 34, 227
  salts and, 283
Anode reactions, in voltaic cells, 687
Anodes
  defined, 686
  lithium, 770
  notation, 691
  zinc carbon cells, 687
  see also Electrolysis
Anthracene, 906, 907
Anthraquinones, hydrogen peroxide preparation, 479
Antibonding orbitals, 258–261, 887
Anticodons, 961
Antifreeze, 413, 441
Antimony
  oxidation states, 764
  precipitation of, 649
  preparation and uses of, 814
Antiseptics
  boric acid, 786
  hydrogen peroxide as, 479
  see also Disinfectants
Aqua ligand, 863
Aquamarine, 773
Aqueous humor, 417
Aqueous solutions, reactions in
  acids, bases, and salts
    gas formation, 286–289
    neutralization, 285–286
    preparation of, 288–292
    reactions of, 282–288
  electrolysis, 705–707
  electrolytes, 271–276
    chemical equilibrium, 273–274
    hydrogen ions, 273
    strong and weak, 274
  equivalents and normality, 303–305
  ionic equations, 274–276
  ions in solution, 271–277
  metathesis, 277–292
    acids, bases, and salts, preparation of, 288–292
    acids, bases, and salts, reactions of, 282–288
    solubility and precipitation, 277–282
  oxidation-reduction reactions
    equations, balancing, 297–303
    equations, understanding, 294–297
    equivalents and normality, 303–305
    terminology, 292–294
  solubility and precipitation
    precipitation reactions, 279, 280–282
    solubility rules, 278, 279–280
  types of, 276–277
Area
  defined, 16
  in pressure definition, 18
Arginine, 944
Argon
  atmospheric, 432
  discovery of, 456
  electron distribution, 181

Argon *(cont.)*
  potassium and, 724
  p subshell, 170
  uses of, 457
  see also Electron configuration;
    Noble-gas configuration
Argyrodite, 800
Armbruster, Peter, 730
Aromatic hydrocarbons, 895
  benzene derivatives, 906–907
  defined, 894
  ethylbenzene, 788
  sources and uses, 907–908
  structures, 905–906
  substitution reactions, 911
Arrhenius, Svante, 272, 282, 418, 515,
    571
Arrhenius acids and bases, 570–573,
    580
Arrhenius equation, 515–517
Arsenic, 190
  neutron activation analysis, 742
  oxidation states, 764
  precipitation of, 649
  properties, 812, 813, 814
Arsenic disulfide, 434
Asbestos, structure, 809, 810, 811
Ascorbic acid, 602
Asparagine, 943
Aspartate, 943, 944
Association colloids, 403–404
Astatine, 190, 826, 827
-ate, defined, 34, 227, 923
Atmosphere, 426
  carbon-14 in, 738–739
  components of, 427
  defined, 91
  evolution of
    lack of noble gases, 432–433
    molecular oxygen, formation of,
      434–436
    primitive, formation of, 433–434
  nitrogen, compounds of, 450–455
  noble gases, 455–457
  oxygen
    formation of, 434–436
    forms of, 436–437
    molecular, 437–440
    ozone, 444–446
    preparation of, 437–440
    in primitive atmosphere, 433
    reactions of, 441–444
  pollution, 458–459, 553n
  regions of, 427–428
  temperature
    heating of thermosphere and strat-
      osphere, 428–431
    moderators of, 459–461
  trace constituents, 458–459
Atmosphere (unit), defined, 19
Atmospheric pressure, 90n, 373–374
Atomic line spectra, 145–146
Atomic mass units, 135, 138
  chromium, 139
  defined, 52
  in formula weights, 55

and mass number, 136
nuclear and particles, 745
nucleons, 747
table, 54
Atomic number
  alpha decay and, 726
  and alpha particle bombardment, 728
  vs. atomic radii, 181
  defined, 136
  electron affinities vs., 186
  ionization energies vs., 184
  nuclide symbol, 718
  periodic table arrangement by, 176,
    178
  in radioactive decay, 723
  x-ray spectroscopy, 172, 173, 777n,
    846
Atomic orbitals, *see* Orbitals, atomic
Atomic particles, masses, 745
Atomic radii
  vs. atomic number, 181
  Group IA elements, 768
  Group IIA elements, 774
  Group IIIA elements, 782
  vs. ionic radii, 205
  periodic patterns, 180–183
  transition metals, 844–846
Atomic size, ionization energies, 184
Atomic structure
  atomic orbitals, 153–157
    quantum numbers, 153–155
    shapes of, 156–157
    see also Orbitals, atomic
  atomic weights, 138–139
  electron, discovery of
    cathode rays, 130, 131–132
    Millikan's experiments, 132, 133
    Thompson's experiments, 131–132
  hydrogen atom, Bohr theory, 144–
    149
    atomic line spectra, 145–146
    Bohr's postulates, 146–149
  light
    particle nature of, 141–144
    photoelectric effect, 142–144
    wave nature of, 139–141
  mass spectrometry, 137–138
  nuclear model
    alpha particle scattering, 134–135
    alpha rays, 133–134
    radioactivity discovered, 133–134
  nuclear structure, 135–137
  quantum effects and photons, 141–
    144
    photoelectric effect, 142–144
    photon momentum, 144
    Planck's constant, 142
  quantum mechanics, 149–152
    de Broglie waves, 149–150
    wave functions, 151
  quantum numbers, 153–154
  see also Electronic structure of
    atoms
Atomic symbols, 29, 30
Atomic theory
  atoms, 28–29

chemical constitution of matter, 37–
  38
classification of matter, 37–41
compounds, 37–38
elements, 37–38
equations, balancing, 35–37
gaseous state, 39–40
ions, 31–34
and laws of chemistry, 59–60
liquid state, 39–40, 41
mixtures
  matter classified as, 37–38
  separation of, 41–45
molecules, 29–31, 32
nomenclature, 34
physical states of matter, 39–40
reactions, chemical, 34–35
solid state, 39–40, 41
Atomic weights, 52–55
  and boiling points, 377
  defined, 138
  in formula weights, 55
  periodic table arrangement by, 176
  table, 54
Atoms, 28–29
  Boyle's particle view, 91n
  compound composition, 37–38
  defined, 28–29
  electron configuration, ground state,
    169
  in molecular solids, 363
  physical state and, 40, 41
  structure of, 32. *See also* Atomic
    structure; Atomic theory; Elec-
    tronic structure of atoms; Peri-
    odic table; Orbitals, atomic ta-
    bles
    electron configuration, ground
      states, 169
    physical states of matter, 41
  of symbols, 30
ATP, 669, 961
Attractive forces, 32, 418
  dipole-dipole, 348–349
  dipole moment, 243–246
  ion-dipole, 394
  in solids, 358
Aufbau principle, 167–172
Autoionization, of water, 573–574
Automobile exhaust, and smog, 459
Average force, and gas pressure, 114
Average rate of reaction, 496
Avogadro, Amedeo, 54
Avogadro's law, 99–100, 101, 102
Avogadro's number, 56, 370
Axial directions, 240
Azaserine, 941
Azeotrope, defined, 43n
Azimuthal quantum number, 153–154

Background radiation, 733
Bacteria, *see* Biochemistry; Microorga-
  nisms
Baking powder, 817
Baking soda, 772
Balances, 2–3, 4

Balancing equations
  by inspection, 36
  oxidation-reduction, 297–303
Ball and stick model, methane, 895
Balmer, J. J., 146
Balmer series, energy level transition,
  148
Balmer's formula, 146, 147–148
Banded iron deposits, 433
Band of stability, 721
Barbituric acid, 621
Barite, 773, 780
Barium
  compounds
    fractional precipitation of, 637–638
    uses of, 777, 780
  electromotive series, 471
  minerals containing, 773
  oxygen reaction with, 442
  preparation of, 776
  properties of, 774, 775
  uses of, 777
Barium chloride, preparation of, 291
Barium fluoride, common-ion effect,
  632
Barium hydroxide octahydrate, 313,
  315
Barium oxide, 438
Barium peroxide, 442
  hydrogen peroxide preparation, 479
  oxygen preparation from, 438
Barium sulfate, 627
  precipitation reactions, 635, 636
  uses of, 780
Barometer, 89–90
Bartlett, Neil, 456
Base ionization constants, see Ionization constants, base
Base pairing, 955–956
Bases
  amines, 925
  amino acids as, 940
  biological molecule denaturation, 939
  buffers and, 613–614
  chromium oxide reactions, 850
  defined, 283–284
  equivalents and normality, 303–305
  gas formation, 286–289
  Group IIA element reactivity, 783
  Group IVA element reactivity, 802
  hydrolysis, 587
  metals and, hydrogen generation, 471
  neutralization, 285–286
  nitride formation, 450
  nucleotides, 953–954, 955, 956
  preparation of, 290–291
  saponification, 923–924, 962
  strengths, relative, 581–583, 587
  strong, 284–285
    in Arrhenius's theory, 572
    gas formation, 288
    neutralization reactions, 285
    titration of strong acids by, 615–
      617
    titration of weak acids by, 617–
      618

tables, 283
  gas formation, 288
  relative strengths, 582
  strong, 284
  weak, 287
weak, 285, 287
  buffers, 610, 611–615
  common-ion effect, 611
  hydrolysis of salts, 608
  nitrogen-containing, 286, 287
  titration by strong acid, 618, 619
  see also Acid–base behavior; Acid–
    base concepts; Acid–base equi-
    libria
Base units, metric system, 11, 12
Basic anhydrides, formation of, 443
Basic oxides, formation of, 443
Basic oxygen process, 853
Basic solution, balancing half-reactions
  in, 300–301
Batteries
  antimony–lead alloys, 814
  lead storage, 687–688, 811
  see also Voltaic cells
Bauxite, 772n
  aluminum production, 781, 782, 783,
    787
  Bayer process, 772n, 789–790
Bayer process, 772n, 789–790
Bayer test, 908
Becker, H., 135
Becquerel, Antoine Henri, 133, 718
Beet sugar, 936, 948
Bent arrangement, 238, 239, 245
Benzaldehyde, 917
Benzene, 895
  catalytic reformation, 913
  derivatives of, 788, 906–907
  sources and uses of, 907–908
  structure, 905–906
  substitution reactions, 911
Benzo[a]pyrene, formula, 906
Benzoic acid, 607
Benzylamine, 624
Bernoulli, Daniel, 112, 114n
Beryl, 112, 773
Beryllium
  compounds, uses of, 777, 780
  diagonal relationships, 766
  ionization energies, 184, 185
  melting point, 843
  orbital diagram, 174
  ores of, 773
  preparation of, 776
  properties of, 774
  uses of, 776–777, 780
Beryllium difluoride, hybrid orbitals in,
  250
Beryllium fluoride, 237, 238
Beryllium hydroxide, uses of, 780
Beryllium oxide, crystal structures, 366
Berzelius, Jons Jakob, 29
Berzelius' dualistic theory, 196, 197
Beta radiation, 723
  actinium-227, 767
  band of stability, 722

defined, 718
detection of, geiger counters, 731
effects of, 730–731
nuclides decaying with, 725, 726
potassium-40, 740
relative biological effectiveness,
  733
Beta-stability line, 722, 724
Bicarbonate
  buffers, 612
  formula, 203
  in Solvay process, 772
  see also Carbonic acid
Bidentate ligands
  defined, 858–859
  and isomerism, 870
Bimolecular reactions, 518
Binary compounds
  halogen, 828
  naming, 224–226
Binding energy, nuclear, 747–748
Biochemistry
  carbohydrates, 948–952
    monosaccharides, 948–950
    oligosaccharides and polysaccha-
      rides, 950–952
  cell, 937
  energy metabolism, 937–938
  lipids
    biological membranes, 963, 964
    fats and oils, 961–963
  nucleic acids, 953–961
    genetic code, DNA and, 956–959
    genetic code, RNA and, 959–961
    nucleotides, 953–955
    nucleotides, in metabolism, 961
    polynucleotides, 955–956
  polymers, 938–939
  proteins, 940–952
    amino acids, 940–945
    enzymes, 947–948
    structure, conformation, 945–947
    structure, primary, 945, 946
  see also Biological systems
Biological oxygen demand, 482–483,
  484
Biological specimens, osmium tetroxide staining, 848n
Biological systems
  atmospheric oxygen origin in, 434–
    435
  buffers, 611
  cells, 937
  complex ions in, 856, 857
  energy metabolism, 937–938
  enzymes in, 526
  isomerism and, 868
  osmosis in, 417–418
  radiation effects, 733–734
  transition elements in, 842
  see also Biochemistry; Microorga-
    nisms
Bismuth, 190
  as free element, 812
  ionic compounds, 200, 201
  oxidation states, 764

Bismuth *(cont.)*
  preparation and uses of, 814
  properties, 812, 813, 814
Bisulfate, formula, 203
Bisulfites, 203, 823
Black, Joseph, 467
Bleaches
  and ammonia, 451–452
  chlorine use as, 829, 830, 832
  disproportionation reaction, 477
  hydrogen peroxide as, 479
  ozone as, 446
  sulfur compound uses, 823
Blood
  buffers, 611
  osmotic pressure, 417
Blood pressure, 90n
Body-centered cubic arrangement
  of metallic solids, 365
  structure, 360–361
Bohr, Niels, 145, 149
Bohr orbit, 153
Bohr's formula, 153
Bohr's postulates, 146–149
Bohr theory, of hydrogen atom, 144–
    149
Boiling point elevation constants, 412
Boiling points
  aldehydes, 921
  amines, 926
  of common substances, 475
  defined, 43
  Group IA elements, 768
  Group IIA elements, 774
  Group IIIA elements, 782
  Group IVA elements, 801
  Group VA elements, 813
  Group VIA elements, 820
  Group VIIA elements, 827
  hydrogen, 469
  hydrogen halides, 829
  oxygen, 441
  periodic table and, 177, 178
  straight-chain alkanes, 897
  structure and, 377–378
  sulfur, 820
  tables, 374, 475
  transition elements, 843, 844–845
Boltzmann, Ludwig, 112
Bomb calorimeter, 320, 322
Bond angles
  methane, 895
  ozone, 445
Bonding, 29
  beryllium compounds, 780
  in biological molecules
    amino acid side chains and, 942–
      944
    and conformation, 939
    peptide, 942
  boron icosahedra, 782
  complex-ion, structural isomers, 866
  coordinate covalent, *see* Complex-ion
    formation
  covalent, *see* Covalent bonds
  in covalent network solids, 367

delocalized, 218–219, 220, 264–265
dipole moment and, 243–246
directional, *see* Molecular geometry
graphite vs. diamond, 800
Group IIA elements, 773
Group IIIA elements, 781, 782
Group IVA elements, 799
ionic, *see* Ionic bonds
ligands, spectrochemical series, 882
metallic, 211n, 365
metallic vs. nonmetallic elements,
    761
multiple, 253–257. *See also* Multiple
    bonds
oxidation state and, 764–765
peptide, 942
second-row elements, 765–766
three-centered, 786, 787
types of forces, 347
valence bond theory, *see* Valence
    bond theory
and valence orbitals, 261
see also Orbitals; *specific orbitals*
Bonding electrons, and oxidation num-
    bers, 222
Bonding energy, *see* Energy, bonding
Bonding orbitals
  electron configurations, 258–261
  Group VIA elements, 819
  see also Orbitals; *specific orbitals*
Bonding pairs
  formulas for, 208
  valence-shell electron-pair repulsion
    model, 236–243
Bonding theory
  dipole moment and molecular geom-
    etry, 243–246
  molecular orbital theory, 257–265
  multiple bonding, 253–257
  valence bond theory, 246–253
  valence-shell electron-pair repulsion
    model, 236–243
  see also Covalent bonds; Ionic
    bonds; Molecular geometry;
    Multiple bonds; Orbitals
Bond length, 220
  vs. bond order, 221
  in covalent network solids, 367
  electron delocalization and, 222
  molecular oxygen, 440
  ozone, 445
  sulfate ion, 824
Bond order, 221, 260
Bonds, 196
  and bond energy, 333
  nuclear radiation, 731
  peptide, 942
  saturated, 895, 908
  second-row elements, 766
  symbols for, 31, 32
  see also Bonding; Covalent bonds;
    Ionic bonds; Multiple bonds;
    Unsaturated hydrocarbons
Bond strength
  and acid strength, 583, 584
  Group IIA elements, 773–774

ligand, and spin, 876
molecular oxygen, 437
spectrochemical series, 882–883
Boranes, 786
Borate ions, structure, 785–786
Borax, 781, 785
Boric acid, 621
Boric oxide, reduction of, 784
Born-Haber cycle, 336–338
Boron
  allotropes, 782
  diagonal relationships, 766
  ionization energies, 184, 185
  ions, 200
  Lewis formula, 215–216
  orbital diagram, 174
  preparation of, 784
  properties of, 782–783
  p subshell, 170
  reactor control rods, 750
  uses of, 785–786
Boron family, *see* Group IIIA elements
Boron trifluoride
  geometry, 238, 239
  hybrid orbitals in, 250
  Lewis formula, 215–216
  structure, 235
Bothe, W, 135
Boyle, Robert, 91n, 467
Boyle's law, 91–95, 98, 102, 103
Branched-chain alkanes
  nomenclature, 898–900
  structure, 896–897
Branched-chain isomers, 913
Brass, x-ray spectra, 172
Breeder reactors, 751–752
Bridging
  aluminum chloride, 788
  beryllium compounds, 780
  see also Complex-ion formation
Brines, halogens in, 827, 829
Bromine, 190
  alkane substitution reactions, 908–
    909
  compounds
    covalent bonding, oxidation state
      and, 765
    hydrogen bromide, 829–831
    nomenclature, 225
    oxyacids, 832
    solubilities, 280
  enthalpies of formation, 329
  in ocean water, 482
  oxidation number, 223
  periodic table placement, 178
  permanganate test, 294–295, 296
  preparation and uses, 829
  properties, 827–828
  reaction with water, 477
  standard entropies, 662
  standard free energies of formation,
    667
  sulfite reaction with, 297
Bromobenzene, 911
Bromocresol green, 578, 617, 618
Bromophenol blue, 578

Bromthymol blue, 578
Brønsted, Johannes N., 578
Brønsted-Lowry acids and bases, 578–581
    relative strengths, 581–583
    salt solution, properties of, 585–587
Bronze, 189
Brown, Robert, 113
Brownian motion, 113, 114
Buffer capacity, defined, 613–614
Buffers, 611–615, 623
Building-up principle, 167–172
Bunsen, Robert, 141
Burning, see Combustion
Butane, 40n, 896, 897
Butanone, 917
Butene, isomers, 904
Butyl groups, 898. See also Alkyl
        groups
Butyric acid, 62–63, 919

Cadmium
    electromotive series, 471
    precipitation of, 649, 650
    reactor control rods, 750
    and transition elements, 843
    voltaic cells, 691
Cadmium electrodes, 686, 687
Caffeine, 604
Calcium
    common-ion effect, 629–631, 632,
        634
    compounds, uses of, 777–778
    electromotive series, 471
    enthalpies of formation, 329
    and hydride ion formation, 473
    hydrogen from, 470
    oxalates and, 626
    oxygen reaction with, 442
    preparation of, 776
    properties of, 774
    radioactive isotope, 725
    standard entropies, 662
    standard free energies of formation,
        667
    uses of, 777
    x-ray spectra, 172
Calcium carbonate, 626
    formation of, ionic equations, 275,
        276
    in shellfish, 773
    and sodium carbonate, 275
    solubility equilibrium, 639
    uses of, 777, 778–779
    see also Limestone
Calcium dihydrogen phosphate, 816
Calcium fluorite, 629
Calcium hydrogen carbonate, 778–779
Calcium hydroxide, solubility of, 397
Calcium oxalate, solubility product
        constant, 626, 628
Calcium oxide
    uses and preparation of, 777–778
    in water treatment, 486
Calcium silicate, hydrofluoric acid and,
        831

Calcium sulfate, solubility of, 397, 629,
        639
Calculations
    atomic theory, and laws of chemis-
        try, 59–60
    atomic weights, 52–55
    carbon, hydrogen, and oxygen com-
        position, 61–63
    cell electromotive force, 695–697
    Charles's law, 97–98
    chemical formulas, determining,
        60–65
    concentrations, see Molar concentra-
        tion; Molarity; Moles
    decay constants, 736–738
    elemental analysis, 61–63
    enthalpy of reaction from bond ener-
        gies, 334–335
    equations
        balancing, 35–37
        molar interpretation of, 66–67
        oxidation-reduction, 297–303
    equilibria
        buffer, 613
        equilibrium concentrations, 549–
            552
        gaseous reactions, 549–552
    equilibrium constants, thermody-
        namic, 671–673
    equivalents, 303–304
    formulas, mass percentages from,
        60–61
    formulas, molecular, determination
        of, 63–65
    formula weights, 55
    half-life, 736–738
    limiting reactant; yields with, 69–71
    molar concentrations, 72–74
    molar interpretation of chemical
        equation, 66–67
    molecular formulas, determination
        of, 63–65
    moles, 55–60
    normality, 303–305
    oxidation-reduction equations, bal-
        ancing, 297–303
    percentages, mass, 60–61
    percentage composition, carbon, hy-
        drogen, and oxygen, 61–63
    precipitation, 633–638
    significant figures in, 8–9
    solutions, 71–78
    standard free energy changes
        from emfs, 699
        at various temperatures, 675–
            676
    stoichiometry
        of chemical reactions, 66–69
        mole-mass relations in reactions,
            66–71
        solution reactions, 76–78
    see also Conversion factors
Calculators, significant figures, 9
Calibration, and precision vs. accuracy,
        7n
Californium-245, 725

Calories
    in biological systems
        metabolism, 963
        for protein synthesis, 938n
    defined, 21
Calorimetry, 319–323
    calorimeters, 319, 320
    heat capacity and specific heat, 320–
        322
    reaction heat measurement, 322–323
Calvin, Melvin, 741
Cancer
    radiation and, 733
    radiation therapy, 457, 742–744
Cannizaro, Stanislao, 54
Capacitance, polar molecules and, 243–
        244
Capillary rise, 355–356
Capric acid, formula, 919
Caproic acid, formula, 919
Caprylic acid, formula, 919
Carbide, 225
Carbides, metal, 799
Carbohydrate loading, 952n
Carbohydrates
    monosaccharides, 948–950
    oligosaccharides and polysaccharides,
        950–952
    see also Glucose
Carbon, 189
    atomic mass, 52, 53
    atoms, magnetic characteristics,
        175
    elemental analysis, 61–63
    enthalpies of formation, 328, 329
    Hund's rule, 173
    hydrides, 806. See also Hydrocarbons
    hydrogen reactions with, 473
    ionization energies, 185
    isotopes of, 52, 53, 724, 738–740
    magnetic properties, 173
    metal reduction, 790. See also Re-
        duction, metallurgical; Coke
    minerals, 800
    multiple bonding, 204, 474, 525–526,
        766
        hydrogenation of, 474, 525–526
        triple, 209, 255
        see also Multiple bonds
    orbital diagram, 174
    oxides of, see Carbon dioxide; Car-
        bon monoxide; Hydrocarbons,
        oxygen-containing
    oxygen reaction with, 442
    percentage composition, determina-
        tion of, 61–63
    preparation and uses, 802–803
    properties, 800, 801
    p subshells, 170
    standard entropies, 662
    standard free energies of formation,
        667
    see also Biochemistry; Group IVA
        elements; Hydrocarbons
Carbon-12 atomic mass scale, 52
Carbon-14, 724, 738–740

Carbonates
  acids and, 70, 271
  alkali metals, magnesium, and lithium, 766
  of Group IA elements, uses of, 770, 771, 772
  of Group IIA elements
    solubilities of, 778–779
    uses of, 777–778
  ions
    formula, 203
    resonance description, 219
  metal ion precipitation as, 650
  minerals, 800. *See also* Limestone
  salt preparation from, 291
  solubilities, 280
Carbon black, 800, 802
Carbon cycle, 434–435
Carbon dioxide, 800
  atmospheric, and temperature moderation, 459–461
  carbonic acid formation, 287–288, 443
  formation of
    acids and, 70, 288–289
    carbon oxidation, 442
    hydrocarbon oxidation, 443, 908
  geometry, 237, 239
  graphite reduction, 561
  liquid and solid, 346
  phase diagram, 381, 382
  photosynthesis, 434–435
  solid, crystal structure of, 364
  solid and gaseous states of, 39
  solubility of, pressure and, 398–399
  urea formation, 656
  uses and properties, 807–808
  van der Waals constant, 119
  in water, *see* Carbonic acid
Carbon disulfide, 823
  heat of vaporization, 330
  oxygen reaction with, 443
  preparation of, 331
  synthesis of, catalyst, 555
Carbon family, *see* Group IVA elements
Carbonic acid, 612, 808
  acid ionization constant, 601
  limestone dissolution, 287–288, 443
Carbon monoxide, 561
  in air, 458
  atmospheric, parts per million, 458
  catalysts and reaction conditions, 555
  catalytic converters and, 526
  catalytic methanation, *see* Catalytic methanation
  complex ion ligand nomenclature, 863
  formation of, 442
  ligand bonding strength, 883
  in steam-reforming process, 472
  uses and properties, 807–808
Carbon tetrachloride, 105–106
Carbonyl chloride, 214, 560
Carbonyl groups
  aldehydes and ketones, 917–919

sugars, 949
Carbonyl ligand, 863
Carboxyhemoglobin, 458
Carboxyl groups
  in amino acids, 940, 942
  in carboxylic acids, 919
Carboxylic acids, 919
  aldehydes, oxidation to, 921–922
  amide formation, 925–926
  esterification, 922–923
  fatty acids, 924
  long chain, 924
  reduction of, 922
  structure of, 914
  tables, 919
    ester formation, 923
    structure, 914
Carlisle, Anthony, 468
Carlsbad caverns, 271
Carnallite, 767
Carnotite, 850
Carrier, in gas chromatography, 45
Cartesian diver, 91, 92
Cassiterite, 800
Catalysis, metabolic enzymes in, 937
Catalysts, 452
  aluminum chloride as, 788
  catalytic converter, 459
  in contact process, 823
  defined, 111, 438, 495
  enzymes, 937, 947–948
  gaseous reactions, 552–554
  hydrogenation, 474
  in petroleum refining, 911–912
  and reaction rate, 495
  in steam-reforming process, 472
  table, 555
Catalytic converters, 459, 526, 527, 555
Catalytic cracking, 911–912
Catalytic hydrogenation, 525–526
Catalytic methanation, 472
  catalyst, 555
  optimum conditions, 561–562
  pressure changes and, 558–559
  reaction quotient, 548
  temperature changes and, 560–561
  water vapor removal, effect of, 556–557
Catalytic reforming, 913
Catenation
  defined, 806
  sulfur and compounds, 819–820
Cathode rays, 130, 131, 132, 133, 173
Cathode reactions
  notation, 691
  in voltaic cells, 687
Cathodes
  cathode-ray tube, 130, 131
  defined, 686
  notation, 691
Cathodic protection, 690–691
Cation exchange resin, defined, 486
Cationic detergents, 404
Cations, 197
  atomic vs. ionic radii, 205
  coordination compounds, nomenclature, 862

defined, 32
  in ionic solids, 365
  main-group elements, 200, 201, 202
  metallic vs. nonmetallic elements, 761
  nomenclature, 34
  weak bases, 287
Caustic soda, 771
Cavendish, Henry, 455, 467, 468
Celestite, 773
Cell emf, 695–697
Cell reactions, voltaic cells, 691–693
Cells, biological, 937
  carbohydrates, 952
  lipid storage, 963
  membrane, 963, 964
  osmotic pressure, 417
  potassium in, 772n
  radiation and, 733
Cells, voltaic, *see* Voltaic cells
Cellulose, 948, 952
Celsius scale, 13, 14, 96–97
Centi-, defined, 11, 12
Centigrade scale, 13, 14. *See also* Celsius scale
Centimeter, 7n, 15
Centimeter, cubic, 17
Centrifuge, defined, 441n
Ceramics, borates in, 786
Cerium, 525, 846
Cerium selenate, 397
Cesium, 141
  oxygen reaction with, 442
  preparation of, 769
  properties of, 768
  sources of, 767
Cesium chloride, crystal lattice, 365–366
Chadwick, James, 135, 727
Chain reactions, 749–750
Chalcocite, 850
Charcoal, structure, 800
Charge
  and alpha particle bombardment, 728
  amino acids, 940
  crystal field theory, 878–879
  on detergents, 404
  dipole, instantaneous, 348
  dipole-dipole forces, 348–349
  dipole moment, 243–246
  effective nuclear, 845
  on electrons, 129–130, 132, 136n
  and hydrophobic interactions, 402, 939
  ionic substances, 31–33. *See also* Ionic bonds
  in Lewis formulas, 213
  and nomenclature, 34
  nuclear, 168, 169, 718, 845
  during nuclear reactions, 719
  oil drops, 132
  on protons, 129–130, 136n
  and repulsion of electron pairs, 236
  table, 135
  transition element cations, 201
  units of, 132, 708
  see also Electronegativity

Charles, Jacques Alexandre, 95n
Charles's law, 95–96, 98
  calculations, 97–98
  ideal gas law and, 102, 103
  volume, temperature and, 96–97
Chelates, 860, 861–862
Chemical analysis, *see* Analytical
  chemistry
Chemical bonds, *see* Bonding; Bonding
  theory; Bonds
Chemical constitution of matter,
  37–38
Chemical equations
  balancing, 35–37, 297–303
  ionic, 274–276, 572–573
  molar interpretation of, 66–67
  molecular, ionic reactions, 274–275
  neutralization, 572–573
  oxidation number in, 293
  oxidation-reduction
    balancing, 297–303
    understanding, 294–297
  stoichiometry of reaction heats, 323–
    324
Chemical equilibrium, 273–274, 537–
  539
  defined, 538
  radioactive tracers and, 740
  see also Acid-base equilibria; Gase-
    ous reactions, equilibria
    in; Solubility equilibria; Ther-
    modynamics
Chemical formulas
  amines, 926
  complex ions, 860, 861
  cycloalkanes, 901
  elemental analysis, 61–63
  empirical
    defined, 63
    from elemental composition, 63–
      64
    molecular formulas from, 64–65
  ionic compounds, 202–204
  mass percentages from, 60–61
  molar interpretation, 66–67
  molecular, 29–31, 30–31
    cycloalkanes, 901
    defined, 31, 32
    determination of, 63–65
    from empirical formula, 64–65
    and mass percentages, 61
    methane, 895
    straight chain alkanes, 897
  molecular formula determination,
    63–65
  with moles, 56
  nomenclature, 34
  nuclear, 718–720
  percentage composition calculations,
    61–63
  structural, 31, 212–213
    alkanes, 896–900
    alkyl groups, 898–899
    amino acids, 942, 943–944
    benzene derivatives, 905, 906–907
    carbohydrates, 949, 950, 951, 952
    condensed, 896, 901

cycloalkanes, 901
  defined, 31, 32
  ethylene glycol, 57
  hydrogen peroxide, 478
  lactic acid, 596n
  lipids, 962, 963
  methane, 895, 896
  nucleotides, 953, 954, 956
  pyruvic acid, 601n
Chemical impurities, and crystal defects, 362
Chemical industry, oxygen use, 441
Chemical kinetics, *see* Reaction mechanisms; Reaction rates
Chemical names, *see* Nomenclature
Chemical properties
  defined, 38
  of water, 476–477
Chemical reactions, *see* Reactions,
  chemical
Chemical synthesis, vapor phase chromatography in, 45
Chemisorption, 525, 526
Chemistry
  development of, 2–5
  experiment and theory, 5–6, 7
Chilean nitrate, 448, 827, 829
Chiral isomerism
  amino acids, 941
  complex ions, 871
Chitin, 948, 952
Chlorates
  formula, 203
  ions, 832, 833
  solubilities, 280
Chlorides, 225
  copper, 854–855
  covalent bonding, oxidation state
    and, 765
  ions
    common-ion effect, 632–633
    formation of, 32
    and sodium chloride electrolysis,
      707
  metal ion precipitation as, 650
  in ocean water, 482
  solubilities, 280
Chlorine, 28, 34, 190
  alkane substitution reactions, 908–
    909
  atomic
    ionization energy, 185
    and ozone, 446
  atomic vs. ionic radii, 205
  compounds
    hydrogen chloride, 829–831. *See
      also* Hydrogen chloride; Hydro-
      chloric acid
    oxyacids, 830, 831–834
  from electrolysis
    of magnesium chloride, 68–69,
      775, 776
    of sodium chloride, 705, 707, 714
  enthalpies of formation, 329
  oxidation number, 223
  preparation and uses, 829
  properties, 827–828

reaction with water, 477
  sodium chloride formula, 196, 197
  standard entropies, 662
  standard free energies of formation,
    667
  standard potential diagram, 833
Chlorine dioxide, 834
Chlorine tetrafluoride, geometry, 241
Chlorite, formula, 203
Chloroacetic acid, 621
Chlorofluoromethanes, and ozone, 446
Chloromethanes, 831
Chlorophylls, 938n
Chlorous acid, preparation of, 289
Chromate, 851
  formula, 203
  fractional precipitation with, 637
  metal ion precipitation as, 650
Chromatography, 44–45, 940n
Chromite, 850, 851
Chromium, 139
  acids, reaction with, 849–850
  amphoteric hydroxide, 645
  biological substances, 856
  building-up principle exceptions,
    171
  complex ions
    isomers, 866
    magnetic properties, 881–883
    valence bond theory, 874–875
  compounds, preparation of, 851–852
  electromotive series, 471
  Goldschmidt process, 785
  oxidation states, 850–851
  physical properties, 844–848
  precipitation of, 649
  in stainless steel, 854
  x-ray spectra, 172
Chromosomes, 956–957, 958
Churchill, M. R., 220
Cinnamaldehyde, 917
Cis-trans isomers, 868–870
  alkenes, 904
  interconversion, 256–257
Cisplatin, 2, 7, 857, 868–869
Citrate buffers, 612
Citric acid, 612
Classification of matter, 37–41
Clathrates, 478
Clausius, Rudolf, 112
Clausius-Clapeyron equation, 375–
  376
Clays, aluminum-containing, 781
Climate, and atmospheric temperature
  moderators, 459–461
Closest-packed arrangement
  ionic solids, 366
  metallic solids, 365
  molecular solids, 363–364
Coagulation, 402–403
  with aluminum ions, 787
  in water purification, 485
Coal
  metal reduction, 790. *See also* Coke
  sulfur in, 819
Coal tar, as hydrocarbon source, 907
Cobalamin, 856

Cobalt
  biological substances, 856
  complex ions, 860, 861–862
  cis-trans isomers, 870, 871
    ionization isomers, 866
    linkage isomers, 867
    optical isomers, 872, 873
    tris(ethylenediamine), structure of, 859
  coordination numbers, 858
  electromotive series, 471
  ores, 850
  physical properties, 844–848
  precipitation of, 649
  radioactive isotopes, 737, 741–742
  x-ray spectra, 172
Cobaltite, 850
Cocaine, 604
Codeine, 623
Codons, 958, 959, 960–961
Coefficients
  in chemical equations, 35, 36
  in molar interpretation of chemical equations, 67
Coenzymes, nucleotides in, 961
Coke, 802
  coal tar from, 907
  lead reduction, 805
  metal reduction, 790
  silicon dioxide reduction, 803
Colemanite, 781
Colligative properties
  boiling point elevation and freezing point lowering, 411–414
  concentration and, 404–409
  vapor pressure, 409–411
Collisions, elastic, 113–114
Collision theory, 511–513, 515n
Colloids
  aluminum ion coagulation, 787
  association, 403–404
  coagulation, 402–403, 787
  defined, 399–400
  hydrophilic and hydrophobic, 400–402
  Tyndall effect, 400, 401
  types of, 400, 401
Colors
  changes of, reaction rate determinations, 498–499
  complex ions
    chromium, 849–850, 851
    copper, 854
    iron, 852
    isomerism and, 868
    structural isomers, 866
    visible spectra, 884–886
  corundum, impurities and, 781
  diamond, 800
  halogen gases, 828
  transition metal cations, 202
  see also Indicators; Pigments
Column, fractionating, 43–44
Column chromatography, 44–45, 940n
Combining volumes, law of, 53–54, 99

Combustion
  alkali metals, magnesium, and lithium, 766
  as chemical reactions, 34–35
  conservation of mass in, 3–4, 5
  gasoline, free energy changes during, 669–670
  graphite, 322–323, 325–327
  Group IA elements, 768
  Group IIA elements, 775
  Group IVA elements, 801
  Group VA elements, 814
  Group VIA elements, 820
  hydrocarbons, products of, 802
  methane, 323–324, 547
  selenium, 825
Combustion method, carbon and hydrogen percentage composition, 62
Common-ion effect, 629–633
  acid–base equilibria, 609–611
  pH and, 638–639
Complex-ion equilibria
  amphoteric hydroxides, 644, 645
  common-ion effect, 629–633
  complex-ion formation, 642–645
  complex ions, and solubility, 646–648
  stepwise formation constant, 645–646
Complex-ion formation, 642–645
  amphoteric hydroxides, 644, 645. *See also* Amphoteric hydroxide
  beryllium, 774, 780
  chelate stability, 861–862
  chromium, 850–851
  coordination compounds
    naming, 862–864
    stereoisomerism, 866, 867–873
    structural isomerism, 866–867
  copper, 854
  crystal field theory, *see* Crystal field theory
  definition, 857–858
  discovery of, 860–861
  formation constants, 643, 645–646
  formulas, 860–861
  Group IVA elements, 799
  iron, 851–852
  in photography, 825
  polydentate ligands, 858–860
  tables, 858
    formation constants, 643
    formulas, 861
  valence bond theory, 873–878
    octahedral complexes, 874–877
    tetrahedral and square planar complexes, 877–878
    see also Valence bond theory
Compounds, 37–38
  coordination, *see* Coordination compounds
  defined, 29
  mass percentage calculations, 60–61
  noble gases, 191
Compressibility, 39, 40, 41

Computations, *see* Calculations
Concentrated, defined, 72
Concentration
  from colligative properties, 409–414
  defined, 72
  and electromotive force, 700–704
  and reaction rates, 494–495, 495–497
    concentration–time equations, 504–506
    graphical plotting, 508–510, 511
    half-life of reaction, 506–507, 508
    rate law determination, 502–504
    reaction order, 501–502
  units of
    conversion, 407–409
    equivalents and normality, 303–305
    table, 12
    see also Molar concentration; Molarity; Moles
Concentration-time equation, first-order vs. second-order reactions, 511
Condensation, defined, 351–352, 372–373
Condensation polymers, 924
Condensation reactions
  in biological molecules, 938
    amide formation, 925–926, 927
    peptide bonding, 942
  phosphoric acids, 816
  polymers, 924
  silicate ions, 809
Condensed structural formulas
  alkanes, 896
  cycloalkanes, 901
Condenser, distillation, 43, 44
Conductance, isomer differentiation, 866
Conductivity, *see* Electrical conductivity
Conformation, biological molecules
  defined, 938–939
  DNA double helix, 956, 958
  polynucleotides, 955–956
  proteins, 942–944, 945–947
  transfer RNA, 957
Conjugate acid-base pairs, 579
  buffers, 611–615
  hydrolysis of, 605–608
Conservation of energy, law of, 22, 658
Conservation of mass, law of, 3–4, 5, 59–60
Constant composition, law of, 60
Constant of proportionality, 115
Constructive interference, x-ray, 368, 369
Contact process, 526
  catalyst, 555
  sulfuric acid production, 823
Continuous spectrum, 145
Control rods, 750
Conversion factors
  concentration units, 407–409
  defined, 13

Conversion factors *(cont.)*
  Fahrenheit to Celsius, 13
  grams-to-moles, 58
  kilojoules to kilocalories, 323
  limiting reagents, 70
  mole ratio, 67n
  table, 15
  temperature, 13
Cooling towers, 482, 483
Cooperative release, oxygen from hemoglobin, 886–887
Coordinate covalent bonds, 208–209
  boron trifluoride, 216
  water, and metal, 478
  see also Complex-ion formation
Coordination, complex-ion formation, 642–645
Coordination compounds, 842
  defined, 857
  naming, 862–864
  structure and function, 866–873
  see also Complex-ion formation
Coordination isomerism, 867
Coordination numbers
  in body-centered lattice, 365
  in closest-packed arrangement, 364
  defined, 858
  geometry and, 878
  tables, 858, 878
Copper
  in biological molecules, 856, 940
  building-up principle exceptions, 171
  complex ions, 465, 856, 867
  crystal structure, 365
  electrochemical cells, 683–686
  electromotive series, 471
  ores, 850
  oxidation of, 296–297
  physical properties, 844–848
  precipitation of the ion, 649
  preparation and uses, 855, 856
  reactions, 854–855
    with acids, 296–297, 824
    oxidation-reduction, 293–294
  in voltaic cells, 691
  x-ray spectra, 172
Copper electrodes, 686, 687
Copper oxide
  in catalytic converter, 459
  nitric acid oxidation, 301–303
Copper sulfate pentahydrate, 478
Core electrons, 179n, 185
Cornea, 417
Corrosion, electrochemistry of, 689–690
Corundum, 781
Cosmic rays, 733
Coulomb (unit), 132, 708
Coulomb-meters, 243
Coulomb's law, 394
Coupled processes, in biological systems, 937
Coupling of reactions, 669
Covalent bonds, 196, 206–222
  beryllium compounds, 780

bond length and bond order, 220–222
complex-ion formation, 642–645
coordinate, 208–209, 216, 478
delocalized, 218–219, 220
graphite vs. diamond, 800
Group IIA elements, 773
Group IIIA elements, 781
Lewis formulas, 207–208, 211–215
metallic vs. nonmetallic elements, 761
multiple, 209. *See also* Multiple bonds
nomenclature, *see* Nomenclature
octet rule, 209
octet rule exceptions, 215–218
and oxidation numbers, 222, 223n
oxidation state and, 764–765
polar, electronegativity, 209–212
second-row elements, 765–766
structure, skeleton, 212–213
valence bond theory, 873–878
water, and metal, 478
see also Complex-ion formation
Covalent compounds
  electrical conductivity of, 379
  melting and boiling points, 374
  and oxidation numbers, 222
  oxides, 442–443
  see also Hydrocarbons
Covalent network solids, 367
  defined, 357, 358
  hardness of, 378
  melting point of, 376
  physical properties, 379
Covalent radii
  and bond length, 220, 221
  Group IVA elements, 801
  Group VA elements, 813
  Group VIA elements, 820
  Group VIIA elements, 827
  sulfate ion, 824
  transition metals, 844–846
Cracking, petroleum, 911–912
Critical mass, 750
Critical pressure, 381–382
Critical temperature, 381–382, 441
Crooks, William, 130n
Cryolite, 783, 784
Crystal defects, 362–363
Crystal field splitting
  and color, 885
  defined, 881
  spectrochemical series, 882–883
  transition metal complex spectra, 884–886
Crystal field theory
  hemoglobin, oxygen release from, 886–887
  high-spin and low-spin complexes, 881–883
  octahedral fields and *d* orbitals, 879–881
  pairing energy in, 882
  tetrahedral and square planar complexes, 883–884
  visible spectrum, 884–886

Crystal lattice
  defined, 359
  geometries, 360–362
  ionic solids, 365–366
  see also Lattice energies
Crystallization
  and hydrophobic sols, 402
  from supersaturated solutions, 279
  see also Precipitation reactions
Crystals, 196, 394
  characteristics of, 358–359
  defects, 362–363
  defined, 32, 33
  delocalized bonding, 220
  hardness of, 378, 379
  hydrates, 315n
  ionic bonds, 196, 197
  sodium chloride, 32–33
  solubility and precipitation, 277–282
  structure
    covalent, 367
    cubic unit cell, 360–362
    Group VIA elements, 820
    ionic, 365–366
    lattice, 359, 360
    metallic, 220, 365
    molecular, 363–364
    x-ray diffraction, 368–371
    tin, 801
Crystal systems, 359, 360
Cubic centimeter, defined, 17
Cubic decimeter, defined, 17
Cubic structure
  cesium chloride, 365
  closest-packed, 364
  of metallic solids, 365
  of polonium, 820
  unit cells, 360–362
Cupric ion, 854
Cuprite, 850
Curie (unit), 733
Curie, Marie, 164
Curium, 730
Current, units of, 11n, 12, 708
Cyanide, formula, 203
Cyanoacetic acid, 623
Cycloalkanes, 895, 901, 902
  catalytic reforming, 913
  formulas, 901
  from petroleum, 911, 913
Cyclopropane, 501, 901
Cyclotron, 728–729
Cysteine, 943
Cytochromes, 842, 856
Cytosine, 953–954
  base pairing, 955–956
  codon dictionary, 960

δ (charge), 243, 244
Dacron, 924
Dalton, John, 28, 52, 53, 54, 95
Dalton's law of partial pressures, 107–111
Daniell, John Frederick, 683
Daniell cells, 683–685
Dating, radioactive, 738–740

Davison, C., 150
Davy, Humphry, 130n, 571, 768
d-Block transition elements, 179, 201–202, 842–843
de Boisbaudran, Lecoq, 177
de Broglie, Louis, 149, 151
de Broglie relation, 150
de Broglie waves, 149–150
Debyes (unit), 243
Decane, physical properties, 897
Decay constants, 734–735
    and half-life, 736–738
    technetium-99, 736
Deci-, defined, 11, 12
Decimeter, cubic, 17
Decomposition voltage, 705–706
Definite proportions, law of, 60
Degree of ionization, 595–596, 597–601
Dehydration
    hydrate isomer, 867
    with sulfuric acid, 824, 825
    see also Condensation reactions
Deionized water, 486–487
d Electrons, *see d* Orbitals
Delocalized bonding, 218–219
Denaturation, defined, 938–939
Denitrifying bacteria, 448
Densities, of ice and water, 481
Density
    defined, 16
    gases, 104–107, 124–125
    Group IA elements, 767, 768
    Group IIA elements, 774
    Group IIIA elements, 782
    Group IVA elements, 801
    Group VA elements, 813
    Group VIA elements, 820
    Group VIIA elements, 827
    hydrogen, 469
    periodic table and, 177, 178
    physical state and, 39
    transition elements, 844–845
    units of, 18
    water, temperature and, 475, 477
Deoxyribonucleic acid
    codon dictionary, 960
    and genetic code, 956–959
    nucleotides, 953–955
Deoxyribonucleosides, 954
Deoxyribonucleotides, 953, 954, 955
2-Deoxy-D-ribose, 948, 949, 953, 954
Derived units, 16–22
Desalination, water, 486–488
Destructive interference, x-ray, 368, 369, 370
Detection, radiation, 730–733
Detector, mass spectroscope, 137–138
Detergents
    as association colloid, 403–404
    biological molecule denaturation, 939
    phosphorus use in, 817, 818
    water pollution, 484, 818
Deuterium, 729n, 753

Deuterons
    bombardment reaction notations, 727
    defined, 717, 729
    in fusion reactors, 752–753
Developers, photographic, 825
Dew, as phase change, 372, 373
Dextrorotatory, defined, 872
Di-, defined, 34, 225, 226, 899
Diagonal relationships, 766, 774, 785n
Diagrams, orbital, 165–166
Diamagnetism, 876, 877
    cobalt complexes, 876
    defined, 175
Diamond, 800
    crystal structure, 367
    enthalpy of formation, 328, 329
    hardness of, 378
    melting point of, 376
    preparation and uses, 802–803
    properties, 800, 801
    structure, 800
Diatomic molecules, electron configurations of, 261–264
para-Dichlorobenzene, vapor pressure, 354
Dichlorobis(ethylenediamine)cobalt(III), 870–871, 872, 873
Dichloroethylene, *cis–trans* isomers, 256
Dichlorofluoromethane, 375
Dichromate
    formula, 203
    2-propanol oxidation, 921
Diesel fuel, 44
Diethylamine, 926
Diffraction, of light, 142
Diffraction patterns, x-ray diffraction, 204, 239, 368–371
Diffusion, gaseous, 116–118
Dihydrogen phosphate, formula, 203
Dilute, defined, 72
Diluting solutions, and molarity, 74–75
Dimensional analysis, 14–16
Dimethylamine, 926, 603
Dimethylbenzene, isomers of, 907
Dimethylhydrazine, in rocket fuel, 452
2,2-Dimethylpropane, structural formula, 897
Dinitrogen difluoride, 256, 257
Dinitrogen pentoxide, 454–455
    decomposition of, 494
        concentration-time equations, 504–506
        half-life of reaction, 506–507, 508
        kinetic data, 499–500
        mechanisms of, 522–523
        rate law determination, 502–503
        rate of, 497
    oxidation state, 450
Dinitrogen tetroxide, 450, 452
Dinitrogen trioxide, 450, 454
Diopside, 811
Dipeptides, 942, 944, 945
1,2-Diphenylethylene, 235, 256–257

Diphenylmethane, 907
Dipole-dipole forces, 348–349
Dipole moment
    and molecular geometry, 243–246
    water, and solvent properties, 477–478
Dipoles, instantaneous, 348
Diprotic acids
    defined, 286
    hydrogen sulfide, 822
    selenium and telluric, 825
    sulfuric, 823–824, 825
Directional bonding, *see* Molecular geometry
Direction of reaction
    acid–base reactions, relative strengths and, 583
    oxidizing and reducing agent strengths and, 698
    precipitate formation and, 281
    prediction of
        entropies and, 660
        from equilibrium constants, 548–549
    see also Driving of reaction
Discharge tubes, 130, 131–132, 445, 455
Disinfectants
    chlorine use as, 829
    ozone as, 446
    see also Antiseptics
Disodium hydrogen phosphate, 816
D-isomers, 872, 941
Disorder, *see* Entropy
Dispersion forces, 347
Disproportionation, defined, 477
Dissociation
    strong acids and bases, 572
    of water, 477
Dissociation constants
    acids and bases, *see* Ionization constants
    amino acids, 940
    complex ions, 642
Dissolved substances, *see* Solutes
Distillation, 42–44
    fractional, 43–44
        germanium tetrachloride, 804
        liquid air, 440, 457
        petroleum, 902
        silicon tetrachloride, 803
    sea water, 487
Distillation flask, 43, 44
Distorted tetrahedral molecular geometry, 241, 245
Disulfide cross linkage, 946
Disulfides, 914, 927
Divalent states, Group IVA elements, 799
Divanadium pentoxide
    as catalyst, 553, 555, 823
    formation of, 442
Dolomite, 773, 776, 800
d Orbitals
    crystal field theory
        color, 885

d Orbitals, crystal field theory *(cont.)*
high-spin and low-spin complexes, 881–882
tetrahedral field, 883–884
Group VIA elements, 819
octahedral field effects, 879–881
in oxygen release from hemoglobin, 887
second-row elements, 765–766
shapes of, 157
in transition elements, 845, 847
valence bond theory, 873–878
Dosage, radiation, 733–734
Double bonds, 256
bond energies, 333
in fused-ring hydrocarbons, 906
Lewis formula, 209
molecular oxygen, 440
ozone, 444–445
second-row elements, 766
silicon, 806
sulfate ion, 824
Double helix, 956, 958
Double replacement reactions, defined, 276
Downs cell, 769
Driving of reaction
gas formation and, 287
neutralization, 285
precipitate formation and, 281
Dry cells, 687, 688
Dry Ice, 346, 364
d Subshells, 154, 168, 172
Group IIIA elements, 179
ion formation, 200
transition element, 842–843
d-Transition elements, 179, 201–202, 842–843
Ductility, 187, 761
Dumas, Jean-Baptiste Andre, 105
Dumas method, 99, 105, 106
Dyes, chromatography of, 44, 45
Dynamic equilibrium, 536, 537
defined, 273–274
radioactive tracers and, 740
rate law, 523
saturated solutions, 392
solubility, 277–278

Ea (eka-aluminum symbol), 177
EDTA, 859–860
Effective nuclear charge, 182, 845
Efflorescence, defined, 478
Effusion, gaseous, 116–118
Einstein, Albert, 142–143, 145, 146–147, 148
Einstein's equation, 745, 746n
Einstein's theory, 3n, 744
Eka-, defined, 177
Eka-aluminum, 177, 178
Elastic collisions, 113–114
Electrical attraction, *see* Attractive forces; Bonding; Bonds; Charge; Ionic bonds
Electrical conductivity
beryllium compounds, 780

copper, 855
electrolytes, 271–273
graphite, 800
metals vs. nonmetals, 761
tin, 801–802
structure and, 379
see also Semiconductors
Electrical discharge
emission spectra, 455
nitrogen reactions, 455n
ozone generation, 445
Electrical discharge tubes, 130, 131–132, 445, 455
Electrical interactions, and freezing point lowering, 418
Electrical potential, potential difference, 693
Electric current, units of, 11n, 12, 708
Electric fields, *see* Crystal field theory
Electrochemistry
electrochemical cells, 683–686
electrolysis, 704–709. *See also* Electrolysis
voltaic cells
commercial, 687–691
electrode potentials, 695–698
electromotive force, 693–695
emf, concentration and, 700–704
equilibrium constants from emf's, 698–700
notation for, 691–693
Electrode potentials, 695–698, 707
cell emf, 695–697
for nonstandard conditions, 702–703
oxidizing and reducing agent strength, 697–698
table, 696
Electrode reactions
decomposition voltage and, 705–706
sodium hydroxide electrolysis, 769n
Electrodes
cathode ray tube, 130, 131
glass, 704
hydrogen, 691, 692, 703
lead, 805
lithium, 770
mercury cell, 771–772
Electrolysis, 704–705
aluminum production, 783–784
aqueous, 705–707
beryllium chloride, 780
bismuth as by-product, 814
copper purification, 855, 856
defined, 683
Downs cell, 769
Group IA element preparation, 769, 770–771
halogen preparation, 828, 829
magnesium preparation, 775–776
mercury pollutants, 484
as metallurgical technique, 790, 791
perchlorate generation, 833
potassium hydroxide production, 772
seawater, magnesium preparation, 775–776
selenium as by-product, 821

sodium chloride, 771–772
aqueous, 705–707, 714
molten, 704–705, 708
stoichiometry of, 708–709
of water, 438, 468, 469, 472
Electrolytes, 271–276
acid-base concept, Arrhenius theory, 570–573
chemical equilibrium, 273–274
defined, 272
hydrogen ion, 273
strong and weak, 274, 285
Electrolytic cell, 685
defined, 683
structure, 685–686
Electromagnetic radiation, 139
atmosphere and, 428–431
carbon dioxide and, 459–460
electromagnetic spectrum, 141
gamma rays, 718, 719, 724
polarized light, 871–872
Electromotive force, 693–695
cell, calculation of, 695–697
concentration and
electrode potentials for nonstandard conditions, 702–703
Nernst equation, 700–702
pH determination, 703–704
equilibrium constants from, 698–700
halogen oxidation of water, 827–828
Electromotive series, 470, 471
Electron acceptors, beryllium compounds, 780
Electron affinity, 186–187
in electronegativity, 210
Group IVA elements, 801
Group VA elements, 813
Group VIA elements, 820
Group VIIA elements, 827
and hydride ion formation, 473
hydrogen atom, 473
in ionic bonding, 198
Electron capture, 722, 723, 724, 740
Electron configuration, 156, 165–166, 844–845
argon atom, 181
of atoms, table, 169
Aufbau principle, 167–172
bonding and antibonding orbitals, 258–261
of diatomic molecules, 261–264
Group IA elements, 768
Group IIA elements, 774
Group IIIA elements, 782
Group IVA elements, 801
Group VA elements, 813
Group VIA elements, 820
Group VIIA elements, 827
homonuclear diatomic molecule, 263
hydrogen atom, 247
isoelectronic, 206
and periodic table position, 178–180
see also Noble-gas configuration
Electron delocalization, and bond length, 222
Electron distribution, orbital, 156

Electron-dot symbols, *see* Lewis formulas
Electronegativity, 210
  and acid strength, 584–585
  and dipole moment, 244
  Group IA elements, 768
  Group IIA elements, 774
  Group IIIA elements, 782
  Group IVA elements, 801
  Group VA elements, 813
  Group VIA elements, 820
  Group VIIA elements, 827
  and hydrogen bonding, 349, 350
  metallic vs. nonmetallic elements, 760, 761
  and oxidation numbers, 222, 223n
  and oxidation state, 763, 848
  oxygen, 819
  Pauling's method for determining, bond energy in, 335–336
  second-row elements, 765
  transition elements, 844–845
Electronics, *see* Semiconductors
Electronic structure of atoms, 139–157
  atomic orbitals, 153–157
    quantum numbers, 153–155
    shapes of, 156–157
  Aufbau principle, 167–172
  electron spin, 165–166
  Hund's rule, 173–175
  hydrogen atom, Bohr theory, 144–149
    atomic line spectra, 145–146
    Bohr's postulates, 146–149
  light
    particle nature of, 141–144
    photoelectric effect, 142–144
    wave nature of, 139–141, 143
  paramagnetism, 175
  Pauli exclusion principle, 166–167
  quantum effects and photons
    photoelectric effect, 141–144
    photon momentum, 144
    Planck's constant, 142
  quantum mechanics, 149–152
    de Broglie waves, 149–150
    wave functions, 151
  quantum numbers, 153–154
  x rays, atomic numbers, and orbital structures, 172, 173
  see also Periodic table
Electronic structure of molecules
  boranes, 786, 787
  crystal field theory, 878–888
  valence bond theory, 873–878
  see also Bonding; Molecular geometry; Orbitals; *specific theories*
Electron microscope, 150, 151
Electron pairs, *see* Bonding pairs; Electrons, paired
Electron-pair theory, Lewis acids and bases, 587–588
Electron probability distribution, hydrogen molecule, 207
Electron promotion, 248

Electrons, 129
  beryllium compounds as electron acceptors, 780
  beta rays as, 133–134
  bonding, 222
  core, 179n, 185
  defined, 32
  delocalized, 222
  discovery of, 130–132
  in electrochemical cells, 683–686
  equivalents, 303–304
  hydrogen molecule
    formation of, 247
    probability distribution, 207
  inert pair effect, 764n
  ionic bonds, 196–206
  ionization energy, 183–185
  in Lewis formulas, 197, 213–214
  ligands, 858
  molecular oxygen, 440
  in nuclear chemistry, in shell model of nucleus, 720–721
  orbital diagram, 165–166
  orbital distribution, 156. *See also* Orbitals
  and oxidation number, 222–223
  in oxidation-reduction reactions, 293
  paired
    and bond order, 260
    in complex-ion formation, 857
    Lewis acids and bases, 587–588
    ligands, 858
    valence bond theory, *see* Valence bond theory
    VSEPR model, 236, 237–243
  photoelectric effect, 142–143
  properties of, 135, 150
  in rusting, 689–690
  subshells, maximum number in, 166, 167
  Thomson's experiment, 131–132
  unpaired, 208, 216, 845, 876. *See also* Valence bond theory
  valence, 179. *See also* Valence electrons; Valence shells
  water, complex-ion formation, 478
  wave property of, 150
  x-ray interaction, 370
Electron spectroscopy, x-ray, 179n
Electron spin, 165–166. *See also* Spin; Spin complexes
Electron-transfer reactions, *see* Lewis acids; Lewis bases
Electron volts, defined, 183, 728
Electrostatic attraction, ionic bonds, 196–206
Elemental analysis, 61–63
Elementary particles, 129, 719
Elementary reactions, 517–518
Elementary substance, oxidation number of, 223
Elements, 37–38
  Boyle's particle view, 91n
  compound composition, 37–38
  compound composition, empirical

    formula from, 63–64
  defined, 28
  periodic classification of, 176–180. *See also* Periodic table; *specific groups*
  tables
    atomic weights, 54
    symbols and formulas, 30
Emeralds, 773
Emf, *see* Electromotive force
Emissions spectra, noble gases, 455
Empedocles, 467
Empirical formulas, *see* Chemical formulas, empirical
Emulsions, 400, 401
En, defined, *see* Ethylenediamine
Enanthic acid, 919
Enantiomers
  amino acids, 941
  complex ions, 870–872, 873
Endothermic processes, 313, 314–315
  ionic substance dissolution, 398
  phase changes, 374–375
  spontaneity of, 370
-ene, 902, 905n
Energy
  activation, 511, 514–515
  in biological systems, 937–938
    adenosine triphosphate, 669, 961
    carbohydrates and, 952
    lipids and, 962–963
  bonding, 332–336
    intermolecular forces, 347
    ionic, 198–199
    see also Energy, orbital
  chelate formation, 861–862
  in chemical reactions, 3n, 34n
  crystal field theory, crystal field splitting, 881–883, 885
  definitions, 16, 19–20
  electron affinity, 186
  for electron promotion, 248
  enthalpy and, 318, 319
  formation, 665–666, 667, 669, 802
  hydration, and solubility, 395
  ionization, 183–185
  lattice, 336–338
  law of conservation of, 22, 658
  mass and, 3n, 19, 744–746
  mass number and, 748
  nonspontaneous processes and, 657
  nuclear reactions
    breeder reactors, 751–752
    fission, 748–751
    fusion, 752–753
    mass-energy equivalence, 744–746
    nuclear binding energy, 747–748
    reactors, 748–753
  orbital
    atomic vs. molecular hydrogen, 259
    d orbitals, in octahedral field, 880–881
    electron configuration and, 168, 169, 170

Energy, orbital (cont.)
  homonuclear diatomic molecules, 262–263
  hydrogen atom, 155
  orbital interactions, 260–261
  quantum number and, 153, 168
  transitions between levels, 146–148
  see also Energy, bonding
pairing, 882
promotional, 764n
units of, 19–22
water evaporation and condensation, 481
see also Free energy; Free energy changes; Thermodynamics
Energy level postulate, 146
Energy sources, hydrogen as, 474
Energy splitting, d orbital
  in octahedral field, 881
  in tetrahedral field, 883–884
Enrichment, uranium, 118
Enthalpy, 315–317, 656
  in biological systems, 937
  bond energy and, 333, 334–336
  chelate formation, 861–862
  and energy, 318, 319
  of formation, 316, 328–332
  lattice energy determination, 336–338
  phase changes and, 374–376
  of reaction, 316
    from bond energies, 334
    Hess's law, 325–327
  and second law, 660
Enthalpy diagrams
  graphite combustion, 236
  reaction enthalpy from bond energy, 334
Entropy, 657–658
  in biological systems, 937–938
  changes, for reactions, 662–664
  chelate formation, 861–862
  defined, 657
  second law of thermodynamics, 658–660
  and solution formation, 392
  spontaneous processes and, 657–664
  standard, 660–662
  third law of thermodynamics, 660–662
Environment
  acid rain, 444, 553n
  air pollution, 444, 458–459, 553n
  fluoride pollution, 783
  mercury pollution, 772
  ozone depletion, 446
  temperature moderation, atmospheric, 459–460
  water, 444, 480–488, 553n. See also Water, natural
Enzymes, 947–948
  as catalysts, 526
  defined, 937
  in electrodes, 704
  protein structure and, 947

starch-digesting, 952
transition elements in, 856
Enzyme-substrate complex, 947, 948
Equations
  Clausius-Clapeyron, 375–376
  Einstein's, 745, 746n
  Henderson-Hasselbalch, 615
  molar mass, 58n
  Nernst, 700–702
  nuclear, 718–721
  quadratic formulas, 551n, 599, 600
  reaction rates
    concentration-time, 504–506
    elementary reaction, 519–520
Equations, chemical, see Chemical equations
Equatorial directions, 240
Equilibrium
  buffer reaction, 613
  chelates, 862
  chemical, 273–274, 537–539
  defined, 538
  gas formation and, 287
  phase changes at, 659
  precipitate formation and, 281
  radioactive tracers and, 740
  solubility, 392
  standard free energy changes and, 667
  thermal, 313–314
  thermodynamic, see Thermodynamics, and equilibria
Equilibrium, acid-base reactions, 582, 583. See also Acid-base equilibria
Equilibrium, complex-ion, see Complex-ion equilibria
Equilibrium, solubility, 277–278. See also Solubility equilibria
Equilibrium concentrations, calculating, 549–552
Equilibrium-constant expression, 539, 540, 545, 573
Equilibrium constants, 523, 536, 539–540
  and base ionization constant, 603
  catalytic methanation, temperatures and, 560
  definition of, 539–540
  from emf's, 698–700
  equilibrium concentration calculations, 549–552
  heterogenous equilibrium, 544–545
  and hydrolysis constant, 605
  obtaining, 540–543
  partial pressures, 543–544
  predicting direction of reaction, 548–549
  qualitatively interpreting, 546–548
  and solubility product constant, 627
  standard free energy change and, 700
  thermodynamic, calculation of, 671–673
  water autoinization, 573
  weak acid, 595

Equilibrium expressions, 539, 540, 545, 573
  solvents and, 573n
  in terms of partial pressures, 543–544
Equivalence point, defined, 615, 616, 617, 618
Equivalents, 303–304
Erg, 20n
Erlenmeyer flask, 17
Error, systematic, 7n
Escape velocity, 432
Esterification, 922–923
Esters
  formation of, 919, 922–923
  lipids, 961–962
  odors of, 923
  structure of, 914
Ethanal, 917, 918, 920
Ethane
  physical properties, 897
  structural formula, 896
  van der Waals constant, 119
Ethanoic acid, 920
Ethanol, 43n, 916
  combustion of, 666
  as compound, 37–38
  entropy of, 659
  formulas and models, 32
  oxidation of, 920
Ethanolamine, 621
Ethers, 917
Ethyl acetate, 498, 923
Ethylamine, 603, 926
Ethylbenzene, 788
Ethylene, 902, 903
  bonding in, 253–255
  catalytic hydrogenation, 525–526
  Lewis formula, 209
  uses of, 912
Ethylenediamine
  as bidentate ligand, 858–859
  cobalt complexes, 861–862
Ethylenediaminetetraacetate, as ligand, 859–860
Ethylene dibromide, 829
Ethylene glycol, 916
  and freezing point of water, 413
  oxygen use, 441
  polyesters, 924
  structural formula, 57
  and vapor pressure of water, 409–410
Ethylene oxide, 441
Ethyl groups, 898. See also Alkyl groups
3-Ethyl-1-methyl benzene, 907
Exciplexes, 191n
Excited states, defined, 167
Exhaust gases, and smog, 459
Exothermic reactions, 314–315
  hydrogen peroxide, 479
  ionic substance dissolution, 397–398
  potential energy diagram, 515
Experiment
  defined, 5
  reaction rate determination, 498–500

Experiment *(cont.)*
  and theory, 5–6, 7
Explosive reactions
  hydrogen, 188
  hydrogen peroxide, 479
  nitrous oxide preparation, 453
  perchloric acid decomposition, 833
Explosives
  nitric acid use in, 455
  potassium use in, 770
Eye, osmotic pressure in, 417–418

F (Faraday constant), 694
Face-centered cubic structure
  of metallic solids, 365
  unit cell, 361, 362, 364
Face-centered lattices, 366
Fahrenheit scale, 13, 14
Fallout, and background radiation, 734
Faraday, Michael, 401, 708
Faraday constant, 694
Fats, 961–963
Fatty acids, 924, 962, 963, 964
f-Block transition elements, 179, 843
Feldspars, structure, 809, 810
Fermentation, 663, 936
Ferrate ion, preparation of, 852
Ferrochrome, 851
Ferrocyanide ion, 857
Ferromagnetism, defined, 175n
Ferrosilicon, 776, 803
Fertilizers
  nitrogen based, 451
  phosphorus use in, 816, 817
  potassium use in, 770, 772
  sulfur compound uses in, 823, 824
  water pollution, 484
Fibrous proteins, 946
Filtrate, defined, 280
Filtration, in water purification, 485
Fire(s)
  and atmosphere, 458
  as chemical reaction, 34n
  see also Combustion
First ionization energies, 183–185, 187
First-order reactions, 501
  concentration-time equations, 504–505
  graphical plotting, 508–509, 511
  half-life of, 506–507, 508
  vs. second-order reactions, 511
Fischer-Tropsch process, 554–555
Fixative, aluminum hydroxide as, 787
Fixed nitrogen, 448
Fixers, photographic, 825
Flame tests, 129
Flammability, hydrogen, 469–470
Flasks, 17
Flasks, distillation, 43, 44
Flavorings
  aldehydes, 917
  esters, 923
Flerov, Georgi N., 730n
Flotation process, 789, 855
Flue gases, arsenic oxide in, 814

Fluidity, 39, 40
Fluorapatite, 814, 827
Fluorescent tubes, 130n
Fluoridation, water, 833
Fluorides, 225
  alkali-metals, magnesium, and lithium, 766
  noble gas, 457
  pollution, Hall-Heroult process and, 783
  transition elements, 848
Fluorine, 190
  alkane substitution reactions, 908–909
  compounds
    hydrogen fluoride, 829–831
    oxyacids, 832
    see also Fluorides
  enthalpies of formation, 329
  ionization energies, 185
  orbital diagram, 174
  oxidation number, 223
  preparation and uses, 828–829
  properties, 827–828
  reaction with water, 476
  standard entropies, 662
  standard free energies of formation, 667
Fluorite, 629, 827
Fluorocarbons, 829, 831
Foams, 401
Food processing, chlorine use in, 830
Force
  and energy, 20
  and gas pressure, 114
  intermolecular, *see* Intermolecular forces
  units of, 16, 18–19
Forest fires, and atmosphere, 458
Formaldehyde, 61, 474, 917, 918–919
Formate ion, 807
Formation
  enthalpies of, 656
  standard free energies of, 665–666, 667
Formation constants
  defined, 642
  stepwise, 645–646
  table, 643
Formation, free energy of, 665–667
  diamond, 802
  standard emf from, 699
Formation reactions, enthalpy changes for, 328–332
Formic acid, 808
  formulas, 919
  hydrogen bonding in, 349–350
  structure, 221
Formulas, chemical, *see* Chemical formulas
Formulas, quadratic, 551n, 599, 600
Formula units, 33, 418
Formula weights, 55, 56
Forward reaction, in chemical equilibrium, 273–274. *See also* Direction of reaction

Fossil fuels, 44
  and atmospheric carbon dioxide, 460–461
  carbon in, 800
  oxygen consumption, 436
  see also Gasoline; Petroleum
Fractional abundance, 138
Fractional distillation, *see* Distillation, fractional
Fractional precipitation, 637–638
Fractionating column, 43–44
Francium, sources of, 767
Frasch process, 821, 822
Free energy
  changes during reactions, 669–670, 671
  coupling of reactions, 669
  maximum work, 668
  and spontaneity, 664–668
  spontaneity criteria, 666–668, 673
  standard free energy changes, 665–666, 667
  temperature changes and calculation of, 675–676
  spontaneity and, 673–674
Free-energy changes
  in biological systems, 937–938
  chelate formation, 861–862
  from emfs, 699
  and maximum work, 668
  during reactions, 669–670, 671, 672
  standard, 655–666, 667
Freeze drying, 372
Freezing, defined, 372
Freezing point
  depression of, 412–414
    ionic solutions, 418
    solutes and, 390, 391
  determination of, 413
Freezing point constants, 412
Frequency
  electromagnetic spectrum, 141
  of wave, 140
Frequency factor, 515
Frost, as phase change, 372, 373
β-D-Fructofuranose, 951
D-Fructose, 948–949
f Subshells, 154
  ion formation, 200
  transition element, 843
f-Transition elements, 179, 843
Fuel cell, 689
Fuel oils, 44
Fuel rods, nuclear reactor, 750, 828–829
Fuels
  ethanol, 916
  hydrocarbon, 908
  hydrogen as, 474
  rocket, 441, 452, 479, 830
  see also Fossil fuels; Gasoline; Petroleum
Full structural formulas, cycloalkanes, 901
Fumaroles, 433

Functional groups
    alkyl, 898–899, 906–907
    amino acids, 942–944, 946
    defined, 913
    and oxidizing agent reactivity, 922
    table, 914
Furanoses, 950, 951
Fusion, enthalpy of, 375
Fusion, nuclear, 748–752

β-D-Galactopyranose, 951
Galena, 800, 805
Gallium, 189
    from bauxite, 790n
    ground state, 171
    properties of, 782, 783
    radioactive isotopes, 722
    sources of, 782
    uses of, 785
Galvanic cell, defined, 683
Gamma radiation, 133, 723, 724
    defined, 718
    detection of, geiger counters, 731
    effects of, 730–731
    photons, 719
    neutron activation analysis, 742
    relative biological effectiveness, 733
Gangue, 789
Gas chromatography, 45
Gas discharge tube, 130, 457
Gaseous reactions, equilibria in
    addition or removal of reactants or
        products, 555–557
    calculating equilibrium concentra-
        tions, 549–552
    catalysts and, 552–555
    direction of reaction
        forward and reverse, 537–539
        predicting, 548–549
    equilbrium constant, qualitative in-
        terpretation, 546–548
    equilibrium constants, 539–544
    heterogeneous, 544–545
    LeChatelier's principle, 555–557
    optimum conditions, 561–562
    pressure changes and, 558–560
    temperature changes and, 560–561
Gaseous state, 39, 40, 41
    gas laws
        Avogadro's law, 99–100, 101
        Boyle's law, 91–95
        Charles's law, 96–98
        ideal gas law, 101–107
        law of partial pressures, 107–111
        measurement of gas pressure, 89–
            91
    kinetic-molecular theory, 111–120
        diffusion and effusion, 116–118
        ideal gases, 111–115
        molecular speeds, 115–116
        real gases, 118–120
Gases
    coal, 907
    collection over water, 109–111
    critical temperature and pressure,
        381–382

density measurements, 105–107
dissolution of, in solids, 392
enthalpy of vaporization, 375
equilibrium reactions, see Gaseous
    reactions, equilibria in
formation of, from salts, 286–289
ideal, 111–115
miscibility of, 393
molecular substances as, 376
noble, 191. See also Noble gases
in periodic tables, 187–188
phase changes, 371–373, 375
phase diagrams, 380–383
phases, 40, 41
pressure measurement, 89–91
real, 118–120
solubility of
    Henry's law, 397
    pressure and, 398–399
solutions of, 38, 391
vapor phase chromatography, 45
Gas laws, see Gaseous state, gas laws
Gas masks, potassium use in, 771
Gasoline, 44, 459
    combustion of, free energy changes
        during, 669–670
    ethylene dibromide in, 829
    halogen use in, 830
    petroleum refining, 911, 912–913
    synthetic, 554–555
    tetraethyllead, 769–770
    see also Fossil fuels; Petroleum
Gas pressure, measurement of, 89–91
Gay-Lussac, Joseph Louis, 53–54, 95
Gay-Lussac's law, 99n
Geiger counter, 731
Gelatin, 400, 401
Gels, 401
Gems
    beryl, 773
    diamond, see Diamond
    Group IIIA elements, 781, 786–787
    ruby, 362, 781
    silicate structures, 809, 811
Gene, defined, 958
Genetic code
    deoxyribonucleic acid and, 956–959
    ribonucleic acid and, 959–961
Geometric arrangement, hybrid orbit-
    als, 250
Geometric isomerism
    alkenes, 904
    complex ions, 868–870
Geometry, see Isomerism; Molecular
    geometry; Structure
Gerlach, Walther, 165
Germanium, 189
    hydrides, 806
    Mendeleev's prediction of, 178
    minerals, 800
    properties, 800, 801
    purification, 804–805, 806
    zone refining, 806
Germanium fluoride, 238, 239
Germer, L. H., 150
Germicidal lamps, 445

Germicides, see Antiseptics; Disinfec-
    tants
Getters, 771
Ghiorse, Albert, 730n
Gieger, Hans, 134
Glass, 356
    as amorphous solid, 358
    borates in, 786
    hydrofluoric acid and, 831
    silica, 359
    structure, 809
Glass electrodes, 704
Glassware, laboratory, 17
Glauber, J. R., 224
Globular proteins, 946
Glucagon, 940
D-Glucopyranoses, 949, 950, 951, 952
Glucose, 948, 949
    ethanol from, 916
    fermentation, 663
    photosynthesis and, 434–435
Glucose polymers, 952
Glutamate, 943, 944
Glutamine, 943
Glycerol, 359, 916–917
    esters, fatty acids, 924
    surface tension of, 355
    triacylglycerols, 962, 963
    vapor pressure, 353, 354
Glycine, 941, 943
Glycogen, 952
Gold, electromotive series, 471
Gold foil, alpha particle scattering on,
    134
Gold oxide, 438
Goldschmidt process, 785, 851
Gold-silver alloys, 390
Gouy balance, 865–866
Graduated cylinder, 17
Graham, Thomas, 117
Graham's law of effusion, 117
Gram, 11, 15
Gram formula weight, 56
Gram molecular weight, 56
Grape sugar, 936, 948
Graphical plot, reaction order from,
    508–510, 511
Graphite, 800
    carbon dioxide reduction, 561
    combustion of, 322–323, 325–327
    crystal structure, 367
    electrical conductivity of, 379
    enthalpy of formation, 328, 329
    hardness of, 378
    mercury cell, 771–772
    preparation and uses, 802
    properties, 800, 801
    structure, 800
    sulfur and, 560
Gravitation, and atmospheric composi-
    tion, 432
Greenhouse effect, 460
Ground states
    defined, 167
    electron configuration in, 168, 169
    orbital diagrams for, 173–174

Group IA elements (alkali metals), 188, 767–773
  atomic radii, 182
  electron affinities, 187
  hydrolysis, 586
  hydroxides, preparation of, 290–291
  ion formation, 200
  ionization energies, 183
  metal vs. nonmetal character, 760–763
  oxidation number, 223
  oxidation states, 295, 763
  preparation, 768–769
  properties, 767–768
  second-row elements, 765–766
  as strong bases, 572
  uses of, 769–773
  see also Alkali metals
Group IIA elements (alkaline earth metals), 178–179, 188–189, 773–781
  hydrolysis, 586
  ion formation, 200
  metal vs. nonmetal character, 760–763
  oxidation number, 223
  oxidation states, 295, 763, 764
  preparation, 775–776
  properties, 774–775
  second-row elements, 765–766
  as strong bases, 572
  uses of, 776–781
  see also Alkaline earth metals
Group IIIA elements (boron and aluminum family), 178–179, 189, 781–788
  ion formation, 200
  metal vs. nonmetal character, 760–763
  oxidation states, 295, 763–764
  preparation of, 783–784
  properties of, 782–783
  second-row elements, 765–766
  uses of, 784–788
Group IVA elements (carbon family), 189, 799–812
  acid-base behavior, 862
  acids, relative strength of, 584
  compounds, 806–812. *See also* Hydrocarbons
    carbon oxides, 807–808
    hydrides, 806
    silicon dioxide and silicates, 808–809, 810, 811
    tin and lead, 809, 811–812
    uses of, 807
  electron affinities, 187
  ion formation, 200
  oxidation states, 295, 763–764
  preparation and uses, 802–806
  properties of, 799–802
  zone refining, 804–805, 806
Group VA elements (phosphorus family), 189–190, 812–819
  acid-base behavior, 862
  compounds, 815–819

  arsenic, antimony, and bismuth, 818–819
  phosphorus oxides and oxyacids, 815–818
  electron affinities, 187
  electronegativities, 765
  ion formation, 200, 201
  oxidation states, 295, 763–764
  preparation and uses, 814
  properties, 813–814
Group VIA elements (sulfur family), 190, 819–826
  acid-base behavior, 862
  compounds, 821–826
    hydrogen sulfide, selenide, and telluride, 821–822
    selenium and tellurium oxides and oxyacids, 825–826
    sulfur oxides and oxyacids, 822–825
  electron affinities, 186
  electronegativities, 765
  ion formation, 201
  oxidation states, 295, 763–764
  preparation and uses, 821, 822
  properties, 819–821
Group VIIA elements (halogens), 190–191, 826–834
  acid–base behavior, 862
  compounds, 829–834
    hydrogen halides, 829–831
    oxyacids, 831–834
    uses of, 830
    see also Halides; *specific halides*
  electron affinities, 186
  ion formation, 201
  oxidation states, 295, 763–764
  preparation and uses, 828–829, 830
  properties, 827–828
  see also Halogens
Group VIIIA elements, 191. *See also* Noble gases
Group number
  and oxidation state, 763
  and valence electron, 179
Groups, periodic table, 176, 177, 178, 763
g Subshell, 154
Guanine, 953–954
  base pairing, 955–956
  codon dictionary, 960
Guanosine, 954
Gypsum, 773, 819

H, see Enthalpy
Haagen-Smit, Arie J., 459
Haber, Fritz, 448
Haber process, 66–68, 448–449, 536
  catalysts, 555
  optimal conditions, 562
Hafnium, 846
Hahn, Otto, 749
Hair, disulfide cross links, 946n
Half-cells, voltaic cells, 684
Half-life
  of isotopes, 735–738, 767

  of reactions, 506–507, 508, 511
Half-reaction method, oxidation-reduction equation balancing, 297–301
Halic acids, 832
Halides
  covalent bonding, oxidation state and, 765
  of Group IA elements, 768
  of Group IIA elements, 775
  of Group IIIA elements, 783
  of Group IVA elements, 802
  of Group VA elements, 814
  of Group VIA elements, 821
  ligands, bonding strength of, 882
  organic, structure of, 914
  transition elements, 848
  see also specific halides
Halite, 767
Hall-Heroult process, 783, 784
Halogens, 190–191, 473
  acids
    preparation of, 288–289
    relative strengths, 584
  alkane substitution reactions, 908–909
  alkene addition reactions, 909–910
  geometry, 241
  interhalogens, 828
  oxidation number, 223
  as oxidizing agents, relative power, 828
  reaction with water, 476–477
  see also Group VIIA elements; *specific halogens*
Halous acids, oxidation states, 832
Hardness
  Group IIA elements, 773–774
  structure and, 378, 379
  transition elements, 843, 844–845
Hardness, of water, 779
Heart attacks, technetium imaging, 743–744
Heat
  biological molecule denaturation, 939
  and entropy, 658–659
Heat capacity
  of common substances, 475
  defined, 320–321
  of water, 475, 481
Heat content, 316n, 315. *See also* Enthalpy
Heat energy, 3n, 19n, 21–22, 34n. *See also* Thermodynamics
Heat flow, 314
Heat of fusion, 375–376
  of common substances, 475
  water, 475, 476
Heat of vaporization, 375
  calculation, 330, 331
  of common substances, 475
  and entropy, 659
  water, 475, 476
Heats of reaction, *see* Reaction heats
Heat summation, Hess's law of, 325

Heat-transfer agents, 770, 771
Heavy metals, water pollution, 484
Heisenberg, Werner, 145, 151–152
Heitler, W., 207
Helium, 191
    atmospheric, parts per million, 458
    atomic vs. molecular orbital energy, 260
    building-up order, 170
    discovery of, 456
    hexagonal closest-packed crystal, 364
    ionization energies, 185
    orbital diagram, 174
    orbital overlap, 247
    uses of, 457
    van der Waals constant, 119
Helium ions, alpha rays, 133. See also Alpha radiation
Helium nuclei, 721, 747. See also Electron configuration; Noble-gas configuration
Hematite, 850, 852–854
Hemiacetals, 949, 950, 951
Hemiketals, 949
Hemoglobin, 458, 842, 856, 857, 937
    carbon monoxide and, 883
    oxygen release from, 886–887
    quadridentate ligands, 859
    in sickle cell anemia, 395–396
Henderson-Hasselbalch equation, 615
Henry's law, 397
Hepta-, defined, 225, 226
Heptane, physical properties, 897
Hertz, 140
Hess's law, 325–328
Heterogeneous catalysis, 525
Heterogeneous equilibrium, 627
    defined, 544
    gaseous substances, 544–545
Heterogeneous mixtures, defined, 38. See also Mixtures
Heteronuclear diatomic molecules, electron configurations of, 261
Hexa-, defined, 225, 226
Hexagonal structure
    closest-packed, 364
    of metallic solids, 365
    unit cell, 360
Hexane, physical properties, 897
High-spin complexes
    crystal field theory, 881–883
    ions, 876
    oxygen release from hemoglobin, 887
Hindenberg, 470
Histidine, 944
Homogeneous catalysis, 525
Homogeneous equilibrium, defined, 544
Homogeneous mixtures, defined, 38. See also Mixtures
Homologous series, 896
Homonuclear diatomic molecules, electron configurations of, 261
Hooke, Robert, 111–112
Hoopes process, 790, 791

Hormones, 940
Hund, Friedrich, 173
Hund's rule, 173–175, 874, 875, 881
Hybrid orbitals, see Orbitals, hybrid; Valence bond theory
Hydrates, 477–478
    defined, 315n
    isomers, 866–867
Hydration, 56, 394
Hydration complexes, aluminum ions, 787
Hydration energy, and solubility, 395
Hydrazine, 451–452
    formula and ion of, 287
    ionization constant, 603
    oxidation state, 450
Hydrides, 473
    of Group IA elements, 768
    of Group IIA elements, 775
    of Group IIIA elements, boranes, 786, 787
    of Group IVA elements, 806
    lithium, 770
    reaction with water, 477
Hydrocarbons, 44n, 469
    alcohols, 914–917
    aldehydes, 917, 918, 919
    aliphatic, defined, 894
    alkanes, 984–901
        alkane series, 895–897
        cycloalkanes, 901
        methane, 895
        nomenclature, 898–900
        sources of, 901, 902
        substitution reactions, 908–909
    alkenes, 901–904
    alkynes, 905
    amides, 925–926, 927
    amines, 925, 926
    aromatic, 905–908
        benzene derivatives, 906–907
        defined, 894
        source and uses of, 907–908
    association colloids, 403–404
    benzene and derivatives, 905–908
    carbon black from, 802
    carboxylic acids, 919
    chlorination of, hydrogen chloride as by-product, 831
    classification of, 895
    cycloalkanes
        sources of, 901, 902
        substitution reactions, 911
    disulfides, 927
    esters, 919
    ethers, 917
    Fischer-Tropsch process, 554–555
    functional groups, 913, 914
    halogenated, 829, 831
    hydrogen from, 468, 471–472, 536, 544–545
    ketones, 917–918, 919
    mercaptans, 927
    methane, 895
    multiple bonding, 253–257
    nitrogen-containing, 925–926, 927

    oxygen-containing, 807–808
        alcohols and ethers, 914–917
        aldehydes and ketones, 917–919
        carboxylic acids and esters, 919
        esterification and saponification, 922–924
        oxidation-reduction reactions, 920–922
        polyesters, 924
    oxygen reaction with, 443
    oxygen use, 441
    petroleum distillation, 902, 911–913
    and pollution
        atmospheric carbon dioxide, 460–461
        smog, 458–459
    reactions
        addition, 909–910
        esterification and saponification, 922–924
        oxidation, 908
        oxidation-reduction, 920–922
        petroleum refining, 902, 911–913
        polyesters, 924
        substitution, of alkanes, 908–909
        substitution, of aromatic compounds, 911
    saturation, defined, 895
    solubility of, 393
    sulfur containing, 927
    thiols, 927
    see also Biochemistry
Hydrochloric acid, 286
    as electrolyte, 272–273
    hydrogen production from, 109–110, 111
    preparation of, 288
    as strong electrolyte, 274
    titrations
        of ammonia, 618, 619. See also Neutralization, ammonia–hydrochloric acid
        with sodium hydroxide, 615–617
    see also Hydrogen chloride
Hydrofluoric acid, preparation of, 288
Hydrogen, 110, 188, 207
    atmospheric, 433, 458
    atomic
        in atmosphere, 433
        Bohr theory of, 144–149
        bonding and antibonding orbitals, 259
        electron configuration, 247
        electron spin, 165
        energy levels, 146–147, 148
        in ice, 476
        orbital energies, 155
        spectrum of, 146
    catalytic methanation, see Catalytic methanation
    compounds of, 473. See also Hydrocarbons
    discovery of, 467–468
    electromotive series, 471
    elemental analysis, 61–63
    as energy source, 474

Hydrogen *(cont.)*
  enthalpies of formation, 329
  Group IVA element reactivity, 801–
    802
  hydrides, 473
  ion formation, *see* Hydrogen ions
  ionization energies, 185
  metal reduction, 790
  molecular
    bond order, 260
    formation of, 247
    orbital energy, 259
  occurrence of, 469
  orbital diagram, 174
  oxidation number, 223
  percentage composition, determina-
    tion of, 61–63
  physical properties, 469–470
  preparation of, 470–472
    catalysts and reaction conditions,
      555
    Group IIIA elements in, 783
    from metals and acids, 110, 111
    from steam and iron, 544–545
    by water electrolysis, 468, 469
  standard entropies, 662
  standard free energies of formation,
    667
  uses of, 473–474
  van der Waals constant, 119
  water generation, 69–70, 468
Hydrogenation
  catalysts, 555
  catalytic, *see* Catalytic hydrogena-
    tion
  defined, 474
Hydrogen bonds, 349–351
  in biological molecules
    amino acids, 943–944
    base pairing, 955–956
    and conformation, 939
    proteins, 946
  and boiling point, 378, 829
  in hydrogen halides, 829
  and vapor pressure, 353–354
  and viscosity, 357
  water, 475–476, 477
Hydrogen bromide, preparation and
  uses, 829–831
Hydrogen carbonate ion, 772
  formula, 203
  polyprotic acids, 601. *See also* Car-
    bonic acid
Hydrogen chloride
  aqueous solutions, 283–284
  bonding in, 247–248
  dipole-dipole forces, 348–349
  geometry, 239
  preparation and uses, 829–831
  see also Hydrochloric acid
Hydrogen electrodes, 691, 692, 703
Hydrogen fluoride, hydrogen bonding
  in, 350
Hydrogen halides, 829–831. *See also*
  specific compounds
Hydrogen iodide, 551, 557

Hydrogen ion concentration, and pH,
  575–577, 578
Hydrogen ions, 273, 571, 572
  acids and, 283, 284, 286
  amino acid binding of, 940
  and chromium, 850
  formation of, 201
  nomenclature, 226, 227
  in polyprotic acids, 286
Hydrogen peroxide, 438, 452, 478–480
Hydrogen selenide, 377, 822
Hydrogen sulfate, formula, 203
Hydrogen sulfide, 483, 621, 821–822
  boiling point of, 377
  metal ion precipitation, 639–641,
    648–650, 821–822
  oxygen reaction with, 443
  pressure changes and, 560
Hydrogen telluride, 377, 822
Hydrohalic acids, 191, 288–289. *See
  also specific acids*
Hydroiodic acid, preparation and uses,
  829–831
Hydrologic cycle, 480, 481
Hydrolysis, 585–587, 605–608
Hydrolysis constant, 605
Hydronium ion, 273, 283
Hydrophilic interactions, 400–401, 946
Hydrophobic interactions
  colloids, 401–402
  membranes, 963, 964
  proteins, 946, 947
  structure and, 939
Hydrosphere, 469
Hydroxides
  amphoteric, 644, 645
  chromium, 850
  copper, 854
  formula, 203
  of Group IA elements, 768
    preparation of, 290–291, 771–772
    uses of, 770, 771, 772
  of Group IIA elements, 777–778
  of Group IIIA elements, 781
  of Group IVA elements, 809, 811
  iron, 852
  lead, 811
  solubilities, 280
  tin, 809, 811
Hydroxo ions
  of chromium, 850
  of Group IIIA elements, 783
  of Group IVA elements, 802
Hydroxylamine, 452, 621
  formula and ion of, 287
  ionization constant, 603
  oxidation state, 450
Hydroxyl groups
  alcohols, 914–917
  bases and, 283–284
  Group IIA elements, 783
Hygroscopic, defined, 478
Hyphens, in organic nomenclature, 899
Hypo, 825
Hypochlorites, 832–833
  and ammonia, 451–452

  disproportionation reaction, 477
  formula, 203
Hypochlorous acid, formula for, 226n
Hypohalous acids, oxidation states, 832
Hypoiodite ion, 503
Hypothesis, defined, 5

-ic, defined, 227
Ice
  density of, 481
  structure of, 475, 476
Icosahedra, boron, 782
-ide, defined, 34, 227
Ideal gas, kinetic theory of, 111–115
Ideal gas law, 101–107
  calculations using, 103
  density, 104–107
  molecular weight determination,
    104–107
  stoichiometry problems, 103–104
Ilmenite, 850
Impurities, and crystal defects, 362
Incompressibility, 39, 40, 41
Indicators
  acid-base, 284, 577, 578, 616, 617
  defined, 77n
Indium, 782, 783, 785
Inert gases, and noble gases, 191
Inert pair effect, 764n
Initial fast step, mechanisms with, 521–
  524
Initial rate method, 502, 508
Inner transition elements, 179, 843
Inorganic ions, 484
Insoluble, defined, 279
Inspection, balancing by, 36
Instantaneous dipoles, 348
Instantaneous rate of reaction, 496
Insulator, diamond as, 800
Insulin, 744, 940, 945, 946
Interference patterns, in x-ray diffrac-
  tion, 368, 369, 370
Interhalogens, 828
Intermediates, reaction, 517
Intermolecular forces, 113
  and boiling points, 377
  dipole-dipole, 348–349
  hydrogen, 469
  hydrogen bonding, 349–351
  London, 347–348
  and melting point, 376
  and miscibility, 392, 393
  in molecular solids, 363
  and surface tension, 355
  table, 347
  van der Waals, 347–349
  van der Waals constant, 119, 120
  in vaporization, 375
  and vapor pressure, 353, 354–355
  and viscosity, 357
International System units, *see* SI units
Intravenous feeding, 417
Iodides, 225
  copper, 855
  covalent bonding, oxidation state
    and, 765

Iodides *(cont.)*
  ionic radius, 204
  oxidation of, 503
  solubilities, 280
Iodine, 190
  alkane substitution reactions, 908–909
  compounds
    hydrogen iodide, 829–831
    oxyacids, 832
  electron microscope image, 29
  enthalpies of formation, 329
  oxidation number, 223
  periodic table placement, 178
  preparation and uses, 829
  properties, 827–828
  radioactive isotope (I-131)
    half-life, 735, 736
    medical applications, 744n
  standard entropies, 662
  standard free energies of formation, 667
  sublimation, 372, 373
Iodine pentafluoride, geometry, 241
Ion-dipole force, 394
Ion exchange, 486, 840n
Ionic bonds
  in biological molecules
    amino acid side chains and, 943–944
    and conformation, 939
    proteins, 946
  common ions, 199–204
  describing, 197–199
  energy in, 198–199
  formulas of compounds, 202–204
  Group IIIA elements, 781
  ionic radii, 204–206
  Lewis electron-dot symbols, 197–198
  main-group element monatomic ions, 199–201
  nomenclature, *see* Nomenclature
  polyatomic ions, 202, 203
  transition metal ions, 201–202
  unpaired electrons, 208
Ionic dissociation, of water, 477
Ionic equations, 274–276
  net, 275–276
  neutralization, 285–286
Ionic equilibrium, 627
Ionic radii, 204–206
  Group IA elements, 768
  Group IIA elements, 774
  Group IIIA elements, 782
  Group IVA elements, 801
  Group VA elements, 813
  Group VIA elements, 820
  Group VIIA elements, 827
  and lattice energy, 394
  transition elements, 844–845
Ionic solutions, 394–395, 418–419
  electrolytes, 271–276
  equivalents, 303–304
  moles, 56
  normality, 303–305

Ionic substances
  crystalline
    defects in, 362
    hardness of, 378, 379
  hydrides, 473
  melting and boiling points, 374
  metal oxides and peroxides, 441–442
  moles of, 56
  nitrides, 451
  nomenclature, 34, 227
  physical properties, 374, 376, 379
  solids, 365–366, 367
    defined, 357, 358
    melting points of, 376
    water solubility rules, 280
    see also Ionic substances, crystalline
  solubility
    solubility product constant, 626
    temperature and, 397
    in water, rules of, 280
    see also Ionic solutions
  see also Ions
Ionization
  amino acids, 940, 941
  hydrohalic acids, 830
  nuclear radiation, 731
  strong vs. weak electrolytes, 274
  weak acids, 597–601
Ionization chamber, 138
Ionization constants
  acid, 594–601
    amino acids, 940
    buffers, 614
    calculations with, 597–601
    experimental determination of, 595–596
    table, 597
  base, 603–604
    amino acids, 940
    and hydrolysis constant, 605
    table, 603
  complex ions, 642
  see also Dissociation constants
Ionization counters, 731
Ionization energies
  and electronegativity, 210
  Group IA elements, 768
  Group IIA elements, 774
  Group IIIA elements, 782
  Group IVA elements, 801
  Group VA elements, 813
  Group VIA elements, 820
  Group VIIA elements, 827
  and hydride ion formation, 473
  in ionic bonding, 198
  main-group elements, 187, 200
  metallic vs. nonmetallic elements, 760, 761
  periodic patterns, 183–185
  transition elements, 847
Ionization equilibria
  acid, 595–601
  base, 603–604
Ionization isomers, 866
Ionization potential, 183–185

Ion product, defined, 634
Ion-product constant
  conjugate acid base pairs, 606
  for water, 573
Ions, 31–33
  chromium, 849–850
  crystals, 32–33
  defined, 32
  deionization, 486–487
  in electrochemical cells, 683–686
  electrolyte solutions, 271–273
  enthalpies of formation, 331
  equivalents, 303–304
  formula units, 33
  lattice energies, 336–338
  nomenclature, 34
  normality, 303–305
  oxidation number, 223
  peroxide, 441–442
  physical state and, 40, 41
  solubility and precipitation, 277–282
  spectator, 275
  superoxide, 442
  in water, 282, 482, 486–487
  see also Acids; Acid-base equilibria; Bases; Salts; Complex-ion formation; Ionic substances
Ion-sensitive electrode, 704
Iron
  atmospheric oxygen and, 433
  in biological molecules, 562n, 842, 856
    hemoglobin, 859, 887
    proteins, 940
  complex ions, 851, 859, 886–887
    magnetic properties, 881–882
    valence bond theory, 875–876
  coordination numbers, 858
  crystal structure, 365
  EDTA complex, 859
  electromotive series, 471
  ferrochrome, 851
  ferrosilicon alloy, 776, 803
  hemoglobin, 856, 859, 886–887
  hydrogen generation, 468, 544–545
  in nitrogenase, 562n
  ores, 297–301, 850
  oxidation-reduction reactions, 293
  permanganate oxidation, 304
  physical properties, 844–848
  precipitation of, 649
  preparation of, 852–854
  in proteins, 940
  radioactive isotope, medical use of, 744n
  reactions of, 852
  rusting, 433, 669, 689–690
  steel production, 441
  voltaic cells, 690, 691
  x-ray spectra, 172
Iron chloride, hydrates, 478
Iron hydroxide, colloid, 402
Iron-magnesium voltaic cells, 690
Iron pentacarbonyl, 857
Isobutane, 897
Isoelectronic ions, 206

Isoleucine, 943, 944
Isomers
  alkanes, nomenclature, 898–900
  alkenes, 904, 909
  biological
    amino acids, 941
    sugars, 949–950
  chiral, 871
  cis-trans conversion, 256–257
  coordination, 867
  coordination compounds, 865, 866–
    867
  cyclopropane, 507
  defined, 501
  geometric, 256, 868–870
  hydrate, 866–867
  ionization, 866
  linkage, 867
  and octane number, 913
  optical, 870–873
  pentane, 897
  stereoisomerism, 866, 867–873
  structural, 866
  types of, 866–867
Isopropyl groups, 898
Isotope dilution, 741–742
Isotopes
  defined, 52, 137
  radioactive tracers, 740–741
  stable, 721
Isotopic abundance, 138

Jewels, *see* Gems
Joliot-Curie, Frederic, 727
Joule, James P., 112
Joules (unit)
  conversion to calories, 323
  defined, 16, 19, 20–21
  energy changes in nuclear decay, 746
  per kelvin, 658, 659

k (rate constant), 500, 515
K_a, *see* Ionization constants, acid
K_b, *see* Ionization constants, base
K_c, *see* Equilibrium constant
K_d, *see* Dissociation constants
K_f, *see* Formation constants
K_p, 543, 544
K_{sp}, *see* Solubility product constants
Kelvin, 13, 14, 97
Kendrew, John, 368n
Kernite, 781
Kerosene, 44
Ketones, 917–918, 919, 922
  biological molecules, *see* Carbohy-
    drates
  primary alcohol oxidation to, 921
  reduction of, 922
  structure of, 914
Kidney stones, 626
Kilo-, defined, 11, 12
Kilocalorie, defined, 21n
Kilogram
  base unit of mass, 11, 12

derived units, 16
  U.S. unit equivalence, 15
Kilojoule, conversion to kilocalories,
  323
Kilometer, U.S. unit equivalence, 15
Kinetic energy
  defined, 19–20, 115n
  and mass, 744
  molecular, 114
  mole of gas, 116n
Kinetic-molecular theory, 110–120
  diffusion and effusion, 116–118
  of ideal gas, 111–115
  molecular speeds, 115–116
  real gases, 118–120
Kinetics, *see* Reaction mechanisms;
    Reaction rates
Kirschoff, Gustav, 141
Krypton, 191
  atmospheric, parts per million, 458
  compounds of, 457
  discovery of, 456
  p subshell, 170
  radioactive isotope half-life, 738
k Shells, 178, 181

Laboratory equipment
  balances, 3–5
  calorimeter, 319, 320
  glassware, 17
  Gouy balance, 865–866
Lactic acid, 596, 936
Lactose, 951
Lanthanide contraction, 182n, 846
Lanthanides, 179, 843
  coordination numbers, 858n
  in periodic table, 176, 177
Laser fusion reactor, 753
Lasers, 146n
Lattice defects, 362–363. *See also*
    Crystal lattice; Crystal structure
Lattice energies, 336–338
  and melting point, 376
  and solubility, 394–395
Laughing gas, nitrous oxide, 453
Lavoisier, Antoine, 3, 5, 468
Laws, 6
  atomic theory and, 59–60
  combining volumes, 53–54, 99
  conservation of energy, 22, 658
  conservation of mass, 3–4, 5
  Coulomb's, 394
  defined, 5
  gas, *see* Gaseous state, gas laws
  Gay-Lussac's, 53–54
  Graham's, 117
  Henry's, 397
  Hess's, 325–328
  mass action, and equilibrium con-
    stant, 540
  modification of, 6
  partial pressures, 107–111, 113
  periodic, 180
  Raoul's, 410
  rate, *see* Rate laws
  reaction, *see* Reaction laws

of thermodynamics
    first, 318
    second, 658–660
    third, 660–662
Lead, 189
  compounds, 807, 809, 811–812
  covalent bonds, 765
  electromotive series, 471
  enthalpies of formation, 329
  hydrides, 806
  ion formation, 201
  minerals, 800
  oxidation states, 764
  precipitation of, 639–641, 648, 649,
    650
  preparation and uses, 805
  properties, 801, 802
  in radioactive decay series, 721, 726
  standard free energies of formation,
    667
  tetraethyllead, 769–770, 805, 830,
    912
Lead arsenate, solubility, 629
Lead chloride, common-ion effect,
    632–633
Lead chromate, 637
Leaded gasoline
  and catalytic converter, 459
  tetraethyllead, 769–770, 805, 830,
    912
Lead iodide, solubility product con-
    stant, 627, 628
Lead oxide, preparation of, 811
Lead shot, 814
Lead storage cells, 687–688, 811, 814
Lead sulfate, 688
LeChatelier's principle, 398
  addition or removal of products or
    reactants, 555–557
  common-ion effect, 609, 630–631
  complex-ion stability, 647
  pressure changes, 558–559
  temperature changes, 560–561
Leclanché dry cell, 687, 688
Length, units of, 11, 12, 15
Leucine, 943
Leukemia, radiation and, 733
Levorotatory, defined, 872
Lewis, Gilbert Newton, 206, 587
Lewis acids, 587–588
  boric acid, 785–786
  coordination compounds, 857, 858.
    See also Complex-ion formation;
    Coordination compounds
Lewis bases, 587–588
  in complex ions, 642. *See also* Com-
    plex-ion formation
  coordination compounds, 857, 858.
    See also Complex-ion formation;
    Coordination compounds
  ligands, 858
Lewis formulas, 208, 211–215
  bond order, 221, 260n
  covalent bonds, 207–208
  and geometry, 242
  ionic bonds, 197–198

Lewis formulas (cont.)
  molecular shape prediction, 242
  octet rule exception, 215–218
  resonance delocalization, 218–219, 220
  valence-shell electron-pair repulsion model, 236–243
  water, 251
Ligand field theory, 879
Ligands
  coordination compounds, nomenclature, 862–863
  crystal field theory, 878–886
  defined, 642, 858
  high-spin complex ions, 876
  isomerism, 866–867
  nomenclature, 862–863
  polydentate, 858–860
  spectrochemical series, 882–883
  valence bond theory, 873–878
Light
  atmosphere and, 428–431
  chlorophylls and, 938n
  de Broglie waves, 149–150
  electromagnetic spectrum, 141
  as particle, 141–144
  photoelectric effect, 142–144
  polarized, 871–872
  spectra of elements, 129
  speed of, 140
  transition metal complex spectra, 884–886
  Tyndall effect, 400, 401
  wave nature of, 139–141
Light bulbs, argon in, 457
Light energy, in chemical reactions, 34n. See also Photons
Light water nuclear reactor, 750, 751
Ligroin, 44n
Lime, 486, 777–778
Limestone, 626, 800
  acid rain and, 444
  calcium carbonate in, 773
  cave formation, 778–779
  dissolution of, 271, 287–288
  iron production, 852–854
Limiting reactant yield equations, 69–71
Linear arrangement
  and dipole moment, 245
  hybrid orbitals and coordination numbers, 878
  valence-shell electron-pair repulsion model, 236, 237, 238, 239
Line spectrum, 145
Linkage isomers, 867
Lipids, 961–963
  fats and oils, 961–963
  membranes, 963, 964
Liquefaction
  defined, 373
  oxygen, 438
Liquid air machine, 440
Liquid oxygen, 441
Liquids, 39, 40, 41
  boiling points, 372–373

changes of state, see Phase changes
critical temperature and pressure, 381–382
distillation of, 42–44
enthalpy changes, 374–376
filtration of, 41–42
freezing points, 373
immiscibility of, 393
intermolecular forces, 347–351
melting points, 373
phase changes, 371–372. See also Phase changes
phase diagrams, 380–383
phases, 40, 41
properties of
  surface tension, 355–356, 357
  table, 354
  vapor pressure, 351–355
  viscosity, 356–357
solutions of, 391
vapor pressure curve for, 380, 381, 382
L-isomers, 872, 941
Liter, defined, 15, 17
Litharge, 805, 811
Lithium
  atoms
    effective nuclear charge, 182
    electron configuration, 165
  building-up order, 170
  diagonal relationships, 766
  electromotive series, 471
  flame test, 129
  ionization energies, 185
  molecular orbital interaction, 261
  orbital diagram, 174
  preparation of, 769
  properties of, 766, 767, 768
  reactions with nitrogen, 451
  reaction with oxygen, 441
  sources of, 767
  uses of, 770
Lithium aluminum hydride, organic oxide reduction, 922
Lithium fluoride, bonding and antibonding orbitals, 264
Lithium hydroxide, preparation of, 290
Lithium iodide, iodide ion radius, 204
Lithium nitride, oxidation state, 450
Litmus, 284, 577n, 578
Lock-and-key model, enzyme-substrate complex, 947, 948
Logarithms, graphical plotting of reaction rates, 508–510, 511
London, Fritz, 207, 347
London forces, 347–348
  and boiling points, 377
  in covalent network solids, 367
  graphite, 801
  halogens, 827
  and miscibility, 393
  in molecular solids, 363
  and vapor pressure, 353, 354
Lone pairs, water complex-ion formation, 478
Lowry, Thomas M., 578

Low-spin complexes, 876
  crystal field theory, 881–883
  oxygen release from hemoglobin, 887
l Quantum number, 153–154, 168
l Shell
  argon atom electron distribution, 181
  periodic table position and, 178
Lutetium, 843n, 846
Lyman series, energy level transition, 148
Lysine, 944
Lysozyme, 526

Macromolecules, molecular weight determination, 416–417
Magic numbers, 720–721, 730
Magnesite, 773, 776
Magnesium
  compounds, uses of, 777, 779
  diagonal relationships, 766
  electromotive series, 471
  ionization energies, 200
  ores of, 773
  preparation of, 775–776
  properties of, 766, 774, 775
  in proteins, 940
  reactions with nitrogen, 451
  uses of, 776, 777
Magnesium carbonate, 777, 779
Magnesium chloride, electrolysis of, 775–776
Magnesium hydroxide, from sea water, 775–776
Magnesium ion, precipitation of, 635
Magnesium nitride, 447
Magnesium oxide, melting point of, 376
Magnesium perchlorate, 478
Magnetic fields
  cathode ray deflection, 132
  and electrically charged fields, 133
  electron focusing, 150
  mass spectrometry, 137–138
Magnetic properties, 153
  of carbon, 175
  d electrons and, 881–883
  homonuclear diatomic molecule, 263
  liquid oxygen, 258
  paramagnetism, 175
  spin magnetism, 165–166
  types of, 865–866
  valence bond theory, 873–878
Magnetic quantum number, 154
Magnetite, iron production, 852–854
Main-group elements, 187–191
  acid-base behavior, 761–763
  atomic radii in, 183
  defined, 178
  electron affinities of, 186–187
  electron configurations, 179
  Group IA (alkali metals), 767–773
  Group IIA (alkaline earth metals), 773–781
  Group IIIA (boron and aluminum), 781–788
  Group IVA (carbon family), 799–812

Main-group elements *(cont.)*
  Group VA (phosphorus family), 812–819
  Group VIA (sulfur family), 819–826
  Group VIIA (halogens), 826–834
  ionic radius, 205
  metallic-nonmetallic character, 760–763
  metallurgy, 789–791
  monatomic ions of, 199–201
  naming binary compounds, 225
  oxidation states, 295–296, 763–765
  second-row elements, 765–766
  see also specific groups
Malleability
  defined, 167
  of metallic crystals, 378, 379
  metallic vs. nonmetallic elements, 761
Maltose, 951
Manabe, Syukuro, 460
Manganese
  in biological substances, 856
  as catalyst, 525
  in ocean water, 482
  ores, 850
  physical properties, 844–848
  precipitation of, 649
  x-ray spectra, 172
Manganese dioxide, oxygen preparation, 111, 439
Manometer, 90–91
Marble, acid rain and, 444
Markownikoff's rule, 910
Marsden, Ernest, 134
Mass, 106, 135
  atomic, 29, 135
  atomic weights, 52–54
  chemical equation interpretation, 66–67
  chemical formula determination, 60–61
  chromium, 139
  and density, 18
  electron, 132
  elementary particle, 129
  and energy, 19, 744–746
  formula weights, 55
  and gas pressure, 114
  law of conservation of, 3–4, 5, 59–60
  measurement of, 3–4
  and moles, 52–60, 66–71
    atomic weights, 52–54
    chemical equation interpretation, 66–67
    conversion to, 58–59
    definitions, 56, 57
    formula weights, 55
    limiting reactants, 69–71
    molarity of solutions, 72–74
    molar mass, 56–57, 58, 117, 746
    stoichiometry of reactions, 66–69
  nuclear, 745
  in specific heat definition, 21–22
  stoichiometry of reaction heats, 323–324

tables
  atomic weights, 54
  derived units, 16
  nuclear particles, 135, 745
  prefixes, 12
  units, 12, 15, 16
  U.S.-metric system relationships, 15
  units of, 11, 12
Mass action, law of, 540
Mass defect, 747
Mass-energy equivalence, 19, 744–746
Mass number
  and atomic mass units, 136
  chromium, 139
  defined, 136
  nuclear binding energy vs., 747
  in nuclide symbol, 718, 719
  in radioactive decay, 723
  radium, 773–774
Mass percentages
  calculation of, 60–61
  in elemental analysis, 61–63
  of solute, 405
Mass spectrometer, 52, 53, 137–138
Mass spectrum, of neon, 138
Mass-to-charge ratio, 132
Matte, 855
Matter
  Boyle's view, 91n
  classifications of, 37–41
  defined, 3
  physical states of, 40, 41. *See also* Phase changes
Maximum work, 668
Maxwell, James Clerk, 112
Maxwell's distribution, 115, 432
McMillan, E. M., 729
Measurement
  of gas pressure, 89–91
  rounding, 10
  significant figures, 7–10
  see also Units of measurement
Medicine
  bromides, 829
  Group IIA compound uses, 777, 778
  oxygen use, 441
  radioactive isotopes in, 742–744
  technetium application, 717, 718
Mega-, defined, 11, 12
Melting
  defined, 371, 372
  enthalpy change, 374–375
Melting points
  beryllium compounds, 780
  boron allotropes, 782
  of common substances, 475
  Group IA elements, 768
  Group IIA elements, 774
  Group IIIA elements, 782, 783
  Group IVA elements, 800, 801, 809
  Group VA elements, 813
  Group VIA elements, 820
  Group VIIA elements, 827
  hydrogen, 469

  metallic vs. nonmetallic elements, 761
  oxygen, 441
  periodic table and, 177, 178
  solutes and, 390, 391
  straight chain alkanes, 897
  structure and, 376–377, 379
  sulfur, 819–820
  tables, 374, 475
  transition elements, 843, 844–845
Membranes, biological, 940, 963, 746
Mendeleev, Dmitri, 164, 175, 178
Meniscus, surface tension and, 356
Mercaptans, 914, 927
Mercuric oxide, Priestley's experiments, 437
Mercury
  coordination numbers, 858
  electromotive series, 471
  formula, 203
  Lavoisier's experiment, 3–4, 5
  melting point, 376
  precipitation of, 648, 649, 650
  sodium amalgam, 706n, 772
  and transition elements, 843
  vapor pressure, 354
  water pollution, 484
Mercury barometer, 89
Mercury cell, 771–772
Mercury electrodes, 706n, 772
Mescaline, 604
Mesopause, 427, 428
Mesosphere, 427, 428
Messenger RNA, 959–960
Meta-, defined, 907
Metabolic pathways, 937
Metabolism
  defined, 937
  energy in, 937–938
  enzymes, 947–948
  nucleotides in, 961
Metallic bonding, 196, 211n
  in transition elements, 845
  vapor pressure, 354
Metallic-nonmetallic character, 760–763
  beryllium, 774
  Groups IVA to VIIA, 799
  Group VA element transitions, 812
Metalloids, 188
  bonds, 200
  Group VA elements, 812
  Group VIA elements, 819
Metallurgy
  defined, 788
  electrolytic refining, 789, 790. *See also* Electrolysis
  Group IA element uses in, 770
  hydrochloric acid use in, 830, 831
  preliminary treatments, 789–790
  reduction in, 788, 790. *See also* Reduction, metallurgical
Metals, 365
  alkali, 183, 188
  alkaline earth, 188–189

Index

I-25

Metals *(cont.)*
  atomic radii
    defined, 180n
    Group IA elements, 768
    Group IIA elements, 774
    periodic trends, 180–183
    transition metals, 844–846
  boiling points, 374
  bonding, *see* Metallic bonding
  carbides, 799
  complex ions
    nomenclature, 863–864
    valence bond theory, 873–878
    see also Complex-ion formation
  compounds, naming, 224, 225
  crystals
    delocalized bonding, 220
    hardness of, 378, 379
  defined, 357, 358
  electrical conductivity of, 379
  in electrochemical cells, 683–686
  electromotive series, 470, 471
  Group VA elements, 812
  Group VIA elements, 819, 821
  hydrogenation, 474
  hydrogen from, 468, 470–471
  hydrogen reduction of oxides, 474
  ions
    in electrochemical cells, 683–686
    hydrolysis of, 586
    monatomic, 199–201
    precipitation of, 635–641, 648–650, 821–822
  melting points of, 374, 376–377
  monatomic ions of, 199–201
  oxidation states, 295–296
  oxides
    hydrogen and, 474
    oxygen preparation from, 438
  oxygen reactions with, 441–442
  in periodic tables, 187
  physical properties, 379
  production of, *see* Electrolysis; Metallurgy; Reduction, metallurgical
  in proteins, 940
  reactions with nitrogen, 451
  solution, liquid, 391
  sulfides, 635–641, 648–650, 821–822
    nonstoichiometric properties, 821n
    solubility of, pH and, 639–641
  see also Alkali metals; Alkaline earth metals; Group 1A elements; Group IIA elements
Metaphosphoric acids, 817
Metastable nucleus, 724, 742
Metathesis reactions, defined, 276
Meteorites, dating, 740
Meter, base unit of length, 11, 12
Methanation, catalytic, *see* Catalytic methanation
Methane, 37–38, 238, 895
  atmospheric, parts per million, 458
  bonding in, 248–250
  catalytic methanation, *see* Catalytic methanation

clathrate formation, 478
  combustion of, 34–35, 323–324, 547
  in fuel cells, 689
  geometry, 238, 239
  hybrid orbitals in, 250
  hydrogen from, 472
  physical properties, 897
  steam reformation, 471–472, 536, 544–545, 555
Methanol
  hydrogen bond formation, 477
  synthesis of
    catalyst, 555
    hydrogen in, 474
Methionine, 943
Method of successive approximations, 599, 632–633
Methylamine, 621, 926
  formula and ion of, 287
  ionization constant, 603
  preparation of, 290–291
Methylbenzene, structural formula, 906
2-Methylbutane, structural formula, 897
Methyl chloride, 829
Methylene chloride, 828
Methyl groups, 898. *See also* Alkyl groups
Methyl orange, 577, 578
2-Methylpropane, 897
Methyl red, 578, 618, 619
Methyl violet, 578
Metric system, 11–14, 15. *See also* Units of measurement
Meyer, Lother, 175
Mica, 810
Micelles, 403, 939, 963
Micro-, defined, 11, 12
Microorganisms
  and atmosphere, 458
  metabolism in, 938, 939
  Pasteur's studies, 936
  in water, 482, 483
  in water purification, 485
Milk, 402–403
Milk sugar, 936, 948
Milli-, defined, 11, 12
Milligram, defined, 11
Millikan, Robert, 132, 133, 210
Millikan's oil drop experiment, 132, 133
Milliliter, defined, 17
Millimeters of mercury (mmHg), 19
Minerals
  defined, 767
  Group IA elements, 767
  Group IIA elements, 773
  Group IIIA elements, 781
  Group IVA elements, 800
  Group VA elements, 814
  Group VIIA elements, 827
  silicon, 809, 810
  soda-ash from, 772
  sulfur-containing, 819
  see also Ores
Mining, 434

Mirror images, *see* Chiral isomers
Miscibility, defined, 391
Mixtures, 37–38
  defined, 38
  equilibrium, concentration of substances in, 549–550
  gas, *see* Gaseous reactions, equilibria in
  mass percentage calculations, 60–61
  separation of, 41–45
    chromatography, 44–45
    distillation, 42–44
    filtration, 41–42
m_l Quantum number, 154
mmHg, defined, 19, 89, 91
Models
  enzyme-substrate complex, 947, 948
  membrane, 963, 964
  nuclear model of atom, 133–135
  see also Molecular models
Moderator, 750
Molality
  defined, 406
  freezing point depression effects, 412, 413–414
Molar concentration
  gases, 100, 101–102, 115
  and osmotic pressure, 415–416
  and reaction rates, 495–497
  see also Molarity; Moles
Molar gas constant, 101–102, 115
Molar gas volume, 100
Molarity
  defined, 72, 405
  diluting solutions and, 74
  see also Molar concentration; Moles
Molar mass, 56–57, 58
  and effusion, 117
  energy changes for, 746
  formula for, 58n
Molecular crystals, hardness of, 378
Molecular formulas, *see* Chemical formulas, molecular
Molecular geometry
  alkene, 903
  boron allotropes, 782
  crystal field theory
    hemoglobin, oxygen release from, 886–887
    high-spin and low-spin complexes, 881–883
    hybrid orbitals for, 878
    octahedral fields and d orbitals, 879–881
    tetrahedral and square planar complexes, 883–884
    visible spectrum, 884–886
  crystal unit cells, 359, 360
  cycloalkanes, 901
  dipole moment, 243–246
  Lewis formulas and, 212
  multiple bonding, description of, 253–257
  oxygen release from hemoglobin, 887

Molecular geometry *(cont.)*
 valence bond theory
  basic theory, 246–248
  hybrid orbitals, 248–253
  octahedral complexes, 874–877
  tetrahedral and square planar complexes, 877–878
 valence-shell electron-pair repulsion model
  linear arrangement, 237, 238, 239
  octahedral arrangement, 241–242
  summary of, 242–243
  tetrahedral arrangement, 239–240
  trigonal bipyramidal arrangement, 240–241
  trigonal planar arrangement, 238–239
 see also Isomerism; Structure
Molecular interactions, protein structure and, 947. *See also* Bonding
Molecularity, of elementary reactions, 518–519
Molecular models
 defined, 31
 methane, 895
 types of, 32
Molecular orbitals, *see* Orbitals, molecular
Molecular orbital theory, 257–265
 bonding and antibonding orbitals, 258–260, 261
 bond order, 260
 electron configuration of diatomic molecules, 261–264
 molecular oxygen, 440
 orbital interaction, determinations of, 260–261
 orbitals and delocalized bonding, 264–265
 principles
  bonding and antibonding orbitals, 258–260
  factors determining orbital interaction, 260–261
 see also Crystal field theory; Orbitals, molecular
Molecular size, and viscosity, 357
Molecular solutions, 392–394
Molecular speeds, 115–116
 and effusion, 117
 noble gas, 432–433
Molecular structure, *see* Electronic structure; Molecular geometry; Structure
Molecular substances
 as electrolytes, 272–273
 halogens, 827
 melting and boiling points, 374
 physical properties, 379
 solids, 358, 363–364
  crystals, hardness of, 378
  defined, 357, 358
  melting point of, 376
  vapor pressure, 372
 weak bases, ionization of, 274

Molecular weights
 and boiling points, 377
 defined, 55
 determination of
  with colligative properties, 413
  in gases, 104–107
  with osmosis, 416
 freezing point depression and, 414
 gram, defined, 56
 and London forces, 348
 and vapor pressure, 353, 354
Molecules
 defined, 29
 formulas, 30–31
 nuclear radiation, 731
 physical state and, 40, 41
 size of, and viscosity, 357
 skeleton structure, 212
 structures, 213
Mole fractions
 defined, 406–407
 gases, 109
 and vapor pressure lowering, 411
Moles, 55–60
 calculations, 58–59
 during catalytic methanation, 537
 chemical equation interpretation, 66–67
 defined, 12, 56–58
 diluting solutions and, 74–75
 energy changes for, 746
 equivalents and normality, 303–305
 and mass, 66–71
 mass and, 52–60
 in solution, 72–74
 stoichiometry of reaction heats, 323–324
 as unit of concentration, 12
 see also Molar concentration; Molarity
Molino, Mario J., 446
Molybdenite, 850
Molybdenum
 coordination numbers, 858
 deuteron bombardment, 717
 in nitrogenase, 562n
 ores, 850
Momentum
 and gas pressure, 114
 photon, 144
Mond process, 545
Mono-, defined, 34, 225, 226
Monoclinic unit cell, 360
Monohydrogen phosphate, formula, 203
Monoprotic acids, defined, 286
Monosaccharides, 948–950
Mordant, aluminum hydroxide as, 787
Morphine, 604
Morpholine, 926
Mortar, 777, 778
Moseley, Henry G. J., 172, 178
Motion
 molecular, *see* Molecular speeds
 random, 112, 113

M Shell
 electron distribution, 181
 periodic table position and, 178
m_s Quantum number, 154
Multiple bonds, 209, 253–257
 bond energies, 333
 carbon
  addition reactions, 909–911
  Bayer test, 908
  catalytic hydrogenation, 525–526
  hydrocarbons, 901–902. *See also* Alkenes; Alkynes
 Group IVA elements, 806
 second-row elements, 766
Multiple proportions, law of, 60
Munzenberg, Gottfried, 730
Myoglobin, 842, 856
 and hemoglobin, 886–887
 structure of, 947

Naming substances, *see* Nomenclature
Nano-, defined, 11, 12
Nanometer, defined, 11
Nanosecond, 12
Naphthalene, 410
 formula, 906
 sources and uses of, 907
 vapor pressure, 354
Native copper, 850
Natural gas
 carbon black preparation, 802
 carbon monoxide from, 807
 composition of, 109
 in fuel cells, 689
 helium in, 457
 hydrocarbons from, 901
 hydrogen from, 472
 hydrogen sulfide in, 819
 see also Hydrocarbons; Methane
Natural waters, *see* Water, natural
Negative ions
 as anions, 32
 nomenclature, 34
 see also Anions
Negative reaction order, 502
Neon
 atmospheric, parts per million, 458
 discovery of, 456
 frozen, 363, 364
 ionization energies, 185
 isotopes, 137, 138
 London forces, 347–348
 orbital diagram, 174
 p subshell, 170
 uses of, 457
 x-ray photoelectron spectrum of, 173
 see also Electron configuration
Neptunium-238, 729
Nernst, Walther, 701
Nernst equation, 707n
 for cell emf, 700–702
 electrode potential for nonstandard conditions, 702–703
 hydrogen electrode, 703
Net ionic equations, 275–276, 285–286

Network silicates, 808, 809, 810, 811
Neutralization, 285–286
    ammonia-hydrochloric acid, 570, 571
        Arrhenius concept, 572
        Brønsted-Lowry concept, 579, 585, 589
        Lewis concept, 587–588
        titration, 618, 619
    defined, 572
    equivalents and normality, 303–305
    salt preparation, 291
    strong acids and bases, 572–573
Neutron activation analysis, 742
Neutron bombardment
    chain reaction, 749–750
    and fission, 748
    in reactors, 750–751
Neutrons, 129, 135, 750
    band of stability, 721, 722
    bombardment reaction notation, 727
    detection of, geiger counters, 731
    discovery of, 727
    magic numbers, 730
    and mass number, 136
    mass of, 747
    properties of, 135
    in radioactive decay, 723, 724
    in stable isotopes, 721
Neutron to proton ratio, and radioactive decay, 723, 724, 725
Newton (unit), defined, 16, 19
Newton, Isaac, 18, 112, 141
Niacin, 596, 617–618
Nicholson, William, 468
Nickel
    in biological substances, 856
    as catalyst, 555
    complex ions, 643
    electromotive series, 471
    Mond process, 545
    ores, 850
    physical properties, 844–848
    precipitation of, 649
    in stainless steel, 854
    in steam-reforming process, 472
    x-ray spectra, 172
Nickel hydroxide, precipitation of, 280–281
Nicotine, 604
Nicotinic acid, 596, 617–618
Nitrates
    formula, 203
    potassium, 770
    solubilities, 280
    water pollution, 484
Nitric acid, 454–455, 547
    copper oxide oxidation, 301–303
    Group IVA element reactivity, 802
    hydrogen in, 473
    preparation of, 289
    sulfur and, 821
Nitric oxide, 453
    atmospheric, 458
    as catalyst, 525, 526, 527

orbital diagram, 263
    oxidation state, 450
    and ozone, 446
Nitrides, 225, 447
    Group IIA elements, 775
    oxidation state, 450–451
Nitrite, formula, 203
Nitrobenzene, 911
Nitrogen, 189, 190
    bonds, multiple, 209, 766
    compounds
        ammonia and nitrides, 450–451
        dinitrogen pentoxide and nitric acid, 454–455
        dinitrogen trioxide and nitrous acid, 454
        hydrazine, 451–452
        hydroxylamine, 452
        nitric oxide, 453
        nitrogen dioxide, 453
        nitrous oxide, 453
        organic, 925–926
    electrical discharge, 455n
    enthalpies of formation, 329
    fractional distillation of air, 440
    in Haber process, 66–68
    hydrogen reactions with, 473
    ionization energies, 185
    lithium reaction with, 768
    nitric acid formation, equilibrium constants, 547
    nitrogen fixation, 447–450
    orbital diagram, 174
    oxidation states, see Oxidation states, nitrogen compounds
    oxygen reaction with, 443
    properties, 812, 813, 814
    standard free energies of formation, 667
    uses of, 447, 448
Nitrogenase, 562n
Nitrogen bases
    formulas and ions of, 287
    preparation of, 290–291
Nitrogen cycle, 447–448
Nitrogen dioxide, 453, 455n
    atmospheric, 458–459
    decomposition of
        concentration-time equation, 506, 507
        graphical plotting, 510
    oxidation state, 450
Nitrogen fixation, 562n
Nitrous acid, 450, 454
Nitrous oxide, 453
    atmospheric, 458–459
    catalytic converters and, 526
    electron delocalization, 222
    oxidation state, 450
Noble-gas configuration, 179, 197
    ions, 206
    main-group element ions, 200, 201
    valence electrons and, 185
    see also Electron configuration

Noble gases, 170, 191
    atomic radii, 182
    boiling point of, 377
    compounds of, 456–457
    discovery of, 455–456
    ionization energies, 183
    lack of, in atmosphere, 432–433
    molecular solids, 363–364
    in periodic table, 178
    preparation and uses of, 457
Noble metals, oxides, 438
Nomenclature, 7, 34
    acids, 226
    binary compounds, 224–226
    complex ions, 857–858
    coordination compounds, 857–858, 862–864
    ionic substances, 227
    organic compounds
        alcohols, 915–916
        aldehydes and ketones, 916–917
        alkanes, 898–900
        alkenes, 902, 903
        alkynes, 905
        benzene derivatives, 907
        carboxylic acids and esters, 919
        esters, 923
        ethers, 917
    oxidation number, 222–224, 225
Nonane, physical properties, 897
Nonelectrolytes, defined, 272
Nonionic detergents, 404
Nonmetals
    compounds, naming, 224, 225
    Group VA elements, 812
    Group VIA elements, 819
    oxidation states, 295–296
    oxide formation, 442–443, 444
    in periodic table, 187–188
    single bond covalent radii, 221
Nonpolar molecules
    dipole moment and, 245
    side chains, amino acid, 946
Nonspontaneous processes, 657
Nonstoichiometric compounds, 362–363, 821n
Normality, 303–305
Notation, nuclear bombardment reactions, 727
n Quantum number, and orbital size, 182
n Quantum number, 153
    and orbital energy, 168
    period number and, 179
ns Electrons
    inert pair effect, 764n
    transition elements, 848
ν (frequency), defined, 140, 172
Nuclear binding energy, 747–748
Nuclear bombardment reactions, 718, 726–730
    transmutation, 727–729
    transuranium elements, 729–730
Nuclear charge, 168, 169, 182, 718, 845

Nuclear chemistry
  applications, 740–744
    chemical analysis, 740–742
    medical therapy and diagnosis, 742–744
  decay rate, 735–738
  energy of reaction, 744–748
    breeder reactors, 751–752
    fission, 748–751
    fusion, 752–753
    mass-energy calculations, 744–748
    mass-energy equivalence, 744–746
    nuclear binding energy, 747–748
    reactors, 748–753
  half-life, 735–738
  nuclear bombardment reactions, 726
    transmutation, 727–729
    transuranium elements, 729–730
  radiation and matter
    detection, radiation counters, 731–733
    effects, dosage and, 733–734
  radioactive dating, 738–740
  radioactivity
    decay, types of, 723–725
    nuclear stability, 720–723
    nuclear equations, 718–720
    radioactive decay series, 725–726
  synthetic elements, 179
  see also Radioactivity
Nuclear decay
  energy changes, 746
  mass changes, 745–746
  types of, 723–725
Nuclear equations, 718–720
Nuclear fission
  breeder reactors, 751
  defined, 748
  nuclear reactors, 748–751
Nuclear force, 720
Nuclear fusion, 748, 752–753
Nuclear magnetic resonance spectroscopy, 165n
Nuclear model, of atom, 133–135
Nuclear reactors, 748–753
  and background radiation, 734
  beryllium use in, 777
  breeder, 751
  fission reactors, 748–751
  fuel rods, fluorine use in, 828–829
  fusion, 751–752
  Graham's law and, 118
  and thermal pollution, 482, 483
Nuclear spin magnetism, 165n
Nuclear stability, binding energies vs. mass number and, 747–748
Nuclear structure, 135–137
Nuclear weapons
  and background radiation, 734
  beryllium use in, 777
  chain reactions, 749–750
  and ozone, 446
Nucleic acids
  genetic code, DNA and, 956–959

genetic code, RNA and, 959–961
  nucleotides, 953–955, 961
  polynucleotides, 955–956
Nucleons, 719
  binding energies per, 747–748
  defined, 718
  mass of, 747
Nucleosides, 953, 954
Nucleotides, 953–955
  and metabolism, 961
  polynucleotides, 955–956
Nucleus, 129
  defined, 135
  masses, 745
  shell model, 720
  structure of, 135–137
Nucleus of crystallization, and hydrophobic sols, 402
Nuclides
  defined, 718
  stable, 721, 722
  symbol, 718
Numbers
  benzene derivatives, 907
  quantum, *see* Quantum numbers
  significant figures, 8
Nutrition
  potassium in, 772n
  transition elements, 856
Nylon, 455, 926, 927

Octa-, defined, 225, 226
Octahedral holes, 366
Octahedral structure, 236, 241–242
  bidentate ligands and, 870
  and dipole moment, 245
  and d orbitals, 879–881
  hybrid orbitals and coordination numbers, 878
  six-coordinate complexes, 869
  valence bond theory, 874–877
Octane, physical properties, 897
Octane number scale, 912–913
Octet rule
  defined, 209
  exceptions, 215–218
Oil drop experiment, of Millikan, 132, 133
Oil, fuel, 44. *See also* Fossil fuels; Petroleum
Oils, 961–963
Oil wells
  barium sulfate use in, 780
  brines, 827
-ol, 915
Olefins, defined, 902n
Oligosaccharides, 948, 950–951
-one, 918
Optical isomers, 870–872, 873
Optimum conditions, choice of
  gaseous reactions, 561–562
  methanation reaction, 561–562
Orbital diagrams, 165–166, 263
  for ground state, 174
  homonuclear diatomic molecule, 263

Hund's rule for, 174
  nitric oxide, 263
  see also Valence bond theory
Orbital energies, *see* Energy, bonding; Energy, orbital
Orbital interaction, factors determining, 260–261
Orbital overlap
  defined, 246
  graphite, 801
  and orbital interaction, 260
  2p, types of, 262
  sigma and pi bond formation, 254, 255
Orbitals, atomic, 153–157
  diagrams, 165–166
  Group VIA elements, 819
  and maximum number of subshell electrons, 166, 167
  and molecular orbital energy, 262–263
  molecular orbital formation, 260–261
  outermost, size of, 182
  Pauli exclusion principle, 166, 167
  quantum numbers, 153–155, 164
  second-row elements, 765–766
  shapes of, 156–157
  valence, *see* Valence orbitals; Valence shells
  x-ray spectroscopy, 172, 173
Orbitals, molecular
  bonding and antibonding, 258–261
  crystal field theory, 878–886
  hybrid, 248–253
    boranes, 786, 787
    in diamond, 800
    Group IVA elements, 799
    in multiple bonding, 253–255, 256
  in oxygen release from hemoglobin, 887
  peroxides and superoxides, 442
  valence bond theory, 873–878
  see also Molecular orbital theory; Bonding
Ores
  defined, 773
  hydrochloric acid extraction, 831
  preliminary treatment of, 789–790
  transition metal, 849, 850
    chromium, 850, 851
    iron, 852
  see also Metallurgy; Minerals
Organic amines, hydrolysis of, 586
Organic chemistry, *see* Biochemistry; Hydrocarbons
Organic wastes, water pollution, 483, 484, 485, 486
Orientation, orbitals of subshell, 154
Ornithine, 941
Ortho-, defined, 907
Orthoclase, structure of, 810
Orthophosphoric acid, 815–816
Orthorhombic unit cell, 360
Osmium, oxidation state, 848
Osmium tetroxide, 848

Osmosis, 415–418
  defined, 415
  reverse, 487
Osmotic pressure, 415–416
Ostwald, Wilhelm, 418
Ostwald process, 454, 553–554, 555
-ous, defined, 227, 815n
Overall order of reaction, 501–502
Overvoltage, defined, 706
Oxalates, 203, 626, 638–639
Oxalic acid, 621, 626
Oxidation
  at anode, 686
  in biological systems, adenosine tri-
      phosphate, 961
  defined, 293, 920
  Group IVA elements, 802
  of hydrocarbons, 908
  of hydrogen, 69–70
  hydrogen sulfide, 822
  iodide ion, 503
  iron, 852
  see also Combustion; Oxides
Oxidation number method, oxidation-
      reduction equation balancing,
      301–303
Oxidation numbers, 222–224, 693
Oxidation potentials, in emf calcula-
      tions, 697
Oxidation-reduction reactions
  catalysis, 524–525, 526, 527
  defined, 276–277
  direction of, strength of reactants
      and, 697–698
  disproportionation, 832
  equations, balancing, 297–303
      half-reaction method, 297–301
      oxidation number method, 301–
          303
  equations, understanding, 294–297
  equivalents and normality, 303–305
  halogen oxyacids, 831–834
  organic compounds, 920–922
      alcohol, 920–921
      aldehyde, 920, 921–922
      definition, 920
  terminology, 292–294
  transition elements, 843
  see also Oxidation; Oxidizing agents;
      Reducing agents; Reduction
Oxidation states, 222–224, 854
  and acid strength, 584
  biological waste products, 483
  chromium, 850
  copper, 854
  Group I elements, 767
  Group IIA elements, 773
  Group IIIA elements, 781
  Group IVA elements, 799
  Group VA elements, 812. See also
      Oxidation states, nitrogen
      compounds
  Group VIA elements, 819
  Group VIIA elements, 827
  Group VIIA oxyacids, 831–832
  main-group compounds, 763–765

main-group elements, 295–296
and metal oxide pH, 443
in metathesis reactions, 275
nitrogen compounds, 450, 764
  ammonia and nitrides, 450–451
  dinitrogen pentoxide and nitric
      acid, 454–455
  dinitrogen trioxide and nitrous
      acid, 454
  hydrazine, 451–452
  hydroxylamine, 452
  nitric oxide, 453
  nitrogen dioxide, 453
  nitrous oxide, 453
nomenclature, 225
phosphorus oxides, 815
transition elements, 296, 843–844,
      847–849
Oxides
  acid-base behavior, metals vs. non-
      metals, 761–762
  basic and acidic, 443–444
  chromium, 851
  copper, 854
  covalent, 442–443
  formation of, 441
  formula, 203
  of Group IIA elements, 775, 777–
      778, 779
  of Group IIIA elements, 781, 783
  of Group IVA elements, 801
      carbon, 807–808
      lead and tin, 807, 811
      silicon, 807, 808–809, 810, 811
  of Group VA elements, 814
      nitrogen, 450, 451–455
      phosphorus, 815–816, 817
  of Group VIA elements, 820
      selenium and tellurium, 825–826
      sulfur, 822–823
  hydrogen and, 474
  iron, 852
  main-group elements, 187–191
  metallic vs. nonmetallic elements,
      760–761
  oxidation state and, 764
  oxygen preparation from, 438
  periodic tables groups and, 176, 177
  reaction with water, 477
  sulfides and, 443
  tables, 203, 225, 761, 850
  transition elements, 848, 849, 850
Oxidizing agents
  chromium compounds, 851
  defined, 293
  Group VIIA element, 827–828, 831–
      832
  ozone as, 445
  selenium and telluric acids and ox-
      ides, 825
  strength of, 697–698
  sulfuric acid, 824
  xenon difluoride, 457
Oxyacids
  chlorine, 830, 831–834

Group VA element
  metal, 812
  phosphorus, 816–818
Group VIA element
  selenium and tellurium, 825–826
  sulfur, 823–825
Group VIIA element, 830, 831–834
nomenclature, 226, 227
relative strength of, 584
thermal decomposition salts of, 439
Oxyanions
  nomenclature, 226, 227
  nonmetal oxides, 761
  transition elements, 848, 851
Oxygen, 190
  atmospheric, 436–437
      altitude and, 430
      formation of, 434–436
      in primitive atmosphere, 433
      and ultraviolet radiation, 428–431
  atomic
      in atmosphere, 430
      in ice, 476
  biological oxygen demand, 482–483
  bond strength, 437
  compound nomenclature, 224, 225
  discovery of, 3–4
  elemental analysis, 61–63
  enthalpies of formation, 329
  ionization energies, 185
  liquid, paramagnetic properties,
      175n, 258
  nitric acid formation, equilibrium
      constants, 547
  occurrence of, 819
  orbital diagram, 174, 263
  organic compounds containing, 914
      alcohols, 914–917
      aldehyde and ketones, 917–919
      carboxylic acids and esters, 919–
          920
      ethers, 917
      reactions of, 920–924
  oxidation number, 223
  percentage composition, determina-
      tion of, 61–63
  platinum hexafluoride and, 456–457
  preparation of, 111, 437–440
      commercial, 440
      laboratory, 439
      from oxide decomposition, 437,
          438
      from peroxide decomposition, 438–
          439
      from thermal decomposition of
          salts of oxyacids, 439
      water decomposition, 468, 469
  properties, 820
  reactions of
      with basic and acidic oxides, 443–
          444
      with compounds, 443
      with hydrogen, 473
      with metals, 441–442
      with nonmetals, 442–443
  release from hemoglobin, 886–887

Oxygen *(cont.)*
  standard free energies of formation, 667
  structure, valence bonding theory, 257n
  structure and physical properties of, 440–441
  uses of, 441
  van der Waals constant, 119
Oxyhalic acids, preparation of, 289
Oxyhemoglobin, 458
Ozone, 190, 444–446
  atmospheric, parts per million, 458
  properties of, 445–446
  stratospheric, depletion of, 446
  and ultraviolet radiation, 429–431
Ozone generator, 445

Pairing energy, 882
Palladium, 29, 555
Palladium oxide, 438
Paper, 829, 830, 787, 907
Para-, defined, 907
Paramagnetism
  chromium complex ions, 874–875
  cobalt complexes, 876
  defined, 175n, 865
  measurement, 865–866
  oxygen, 258, 440
  valence bond theory, 875, 877
Partial pressures
  equilibrium expression in terms of, 543–544
  law of, 107–111, 113
Particle accelerators, 728–729
Particles, elementary, 129
Particles, light as, 142–144
Particle view of matter, 91n
Particulates, atmospheric, 459–460
Pascal (unit), 16, 19, 90, 91
Paschen series, energy level transitions, 148
Passive form of metal, 783
Pasteur, Louis, 936
Pauli exclusion principle, 166, 236
Pauling, Linus, 210, 395
Pauling's method, of electronegativity determination, 335–336
Pauling's scale, 210–211
Pelargonic acid, formula, 919
Penta-, defined, 34, 225, 226
Pentane, 44n, 897
Pentlandite, 850
Perbromic acid, 834
Percentages, mass, calculation of, 60–61
Percentage yield, defined, 71. *See also* Yield calculations
Percent composition, elemental analysis, 61–63
Percent ionization
  defined, 595–596
  weak acids, 597–601
  see also Ionization constants
Perchlorates, 832, 833
  formula, 203

solubilities, 280
Perchloric acid, 585
Perey, Marguerite, 767
Perhalic acids, 832
Periodic law, 180
Periodic table, 164, 175–191
  definitions, 178
  electronegativity in, 210–211
  main-group elements, 187–191
  melting point, 377
  Mendeleev's, 177
  modern form, 176
  periodic classification of elements, 176–180
  periodic properties, 180–187
    atomic radius, 180–183
    electron affinity, 186–187
    ionization energy, 183–185
    of transition elements, 843–849
  predictions from, 177–178
  see also Main-group elements; *specific groups*
Period number
  and n quantum number, 179
  and oxidation state, 763–764
Permanganate
  Bayer test, 908
  bromide oxidation, 294–295, 296
  formula, 203
  iron oxidation, 297–301, 304
  sulfite oxidation, 300–301
Permutations, 945n
Peroxides
  alkali metals, 766
  barium, 775
  formation of, 441
  formula, 203
  oxygen preparation from, 438–439
  sodium, 770
Perrier, Carlo, 717
Pertechnetate ion, 743
Perutz, Max, 368n
Pesticides, 484, 817
Petroleum
  aromatic hydrocarbons from, 907–908
  carbon black from, 802
  carbon monoxide from, 807
  distillation of
    fractional, 44
    products, 902
  oil well brines, 827
  refining, 474, 911–913
    sodium hydroxide in, 772
    sulfuric acid uses in, 823, 824
  sulfur in, 819
  see also Fossil fuels; Gasoline; Hydrocarbons
Petroleum ether, 44, 45
pH, 575–577, 578
  buffers and, 611–615
  determination of, 703–704
  and solubility
    metal ion separation, by sulfide precipitation, 639–641
    qualitative effects, 638–639

titrations, 615–616, 617, 618
  see also Hydrogen ions
Phase, defined, 40, 41
Phase changes
  boiling point, 372–373. *See also* Boiling points
  condensation, 372
  enthalpy changes, 373–376
  and entropy, 658–659
  equilibrium conditions and, 659
  freezing, 372, 373
  liquefaction, 373
  melting, 371, 372. *See also* Melting points
  phase diagrams, 380–383
    critical temperature and pressure, 381–382
    melting point curve, 380–381
    vapor pressure curves, 380, 381, 382
  sublimation, 372, 373
  table, 372
  vaporization, 372
Phases, in ionic equations, 173
Phenanthracene, formula, 906
Phenolphthalein, 77, 577, 578, 616, 617
Phenylalanine, 943
Phosgene, decomposition of, 498
Phosphate buffers, 612
Phosphates
  alkali metals, magnesium, and lithium, 766
  formula, 203
  solubilities, 280
  water pollution, 484
Phosphide, 225
Phospholipid bilayer, 963, 964
Phosphoric acid, 286, 288, 623
  acid ionization constant, 597, 602
  hydrogen halide preparation, 831
  preparation and uses of, 814, 815
Phosphorous acid, 815
Phosphors, scintillation counters, 732
Phosphorus
  bonding orbitals, 766
  compounds
    halides, 238. *See also* Phosphorus pentachloride
    oxides, 454, 815–816
    oxyacids, 816–818
    uses of, 817
  oxygen reaction with, 443
  preparation and uses of, 814
  properties, 812, 813, 814
  radioactive isotopes
    phosphorus-30, 724, 727
    phosphorus-32, 724, 738, 744n
Phosphorus family, *see* Group VA elements
Phosphorus pentachloride
  equilibrium constant, 552
  geometry, 238, 240
  hybrid orbitals in, 250
  Lewis formula, octet rule exception, 217
Phosphorus trifluoride, 235

Photocathode, scintillation counters, 732
Photochemical smog, 458–459
Photoconductor, selenium as, 820
Photoelectric cells, selenium in, 821
Photoelectric effect, 142–144, 172
Photoelectron spectroscopy, 172, 173
Photography
  halogen uses in, 829, 830
  hydroxylamine in, 452
  silver iodide, 291
  sulfur compound uses, 823, 825
Photolysis, molecular oxygen formation, 434
Photomultipliers, scintillation counters, 732
Photon momentum, 144
Photons, 141–144
  energy of, 429
  gamma, 719
  x-ray spectroscopy, 172, 173
Photosynthesis, 741
  in microorganisms, 937, 938
  oxygen formation, 434–436
Physical adsorption, 525
Physical changes
  defined, 34
  and entropy, 658
  see also Phase changes
Physical properties
  defined, 38
  periodic table and, 177, 178. See also specific main groups
  structure and, 376–379
  of water, 474–476, 477, 481
Physical states of matter, 39–40, 41
Pi (osmotic pressure), 415–416
Pi bonds
  in carbon-carbon double bond, 902
  in cis-trans isomer conversion, 256–257
  in multiple bonding, 254–256, 902
  second-row elements, 766
Pickling, metals, 831
Pico-, defined, 11, 12
Picometer, defined, 11
Picosecond, 12
Pi electrons
  benzene, and substitution reactions, 911
  in covalent network solids, 367
  graphite, 801
  in hydrocarbons, aromatic, 905–906
  see also Pi bonds; Pi orbitals
Pigments
  barium, 777
  chlorophylls, 938n
  chromatography of, 44, 45
  lead, 811
  red selenium, 821
Pi orbitals
  molecular oxygen, 440
  peroxides and superoxides, 442
  second-row elements, 766
  see also Pi bonds; Pi electrons
Pipelines, cathodic protection of, 690

Piperidine, 926
Pipet, 17
Planar geometry, alkene, 903
Planck, Max, 142, 145, 146
Planck's constant, 142, 152
Planck's quantification of energy, 142
Plane-polarized light, 871–872
Plants, carbohydrates in, 938, 948–949, 952
Plasma, defined, 753
Plasticizers, phosphorus use in, 817
Plastics
  chlorine use in, 829, 830
  fluorine use in, 831
  nitric acid use in, 455
  polystyrene, 913
  see also Polymers
Plating, tin, 805
Platinum
  as catalyst, 526, 555
    in catalytic converter, 459
    in contact process, 823
    in Ostwald process, 553–554, 555
    in rocket fuel, 452
  complex ions, 860–861, 867
  coordination numbers, 858
  oxygen reaction with, 442
Platinum hexafluoride, oxygen reaction with, 456–457
Platinum oxide, 438
Plumbite ion, 811
Plutonium-238, 729
Plutonium-239, 729, 751–752
Polar covalent bonds, 209–212
Polar head, membrane, 963, 964
Polarimeter, 872
Polarized light, 871–872
Polar molecules, 695
  and acid strength, 583, 584
  amino acid side chains, 943–944, 946
  in cell membrane, 963, 964
  dipole-dipole forces, 348–349
  dipole moment and, 243–246
  isomers, 869
Pollucite, 767
Pollution
  atmospheric, 458
  catalytic converters and, 526
  fluoride, Hall-Heroult process and, 783
  mercury, 772
  phosphate detergent builders and, 818
Polonium, 190, 722, 819
  as alpha emitter, 725
  oxidation states, 764
  properties, 820
Polyamides, 926, 927
Polyatomic ions, 202, 203
  oxidation number, 223
  structures, 213
  see also Complex-ion formation
Polydentate ligands, 858–860
Polyesters, 916, 924

Polymers
  alkenes, 910
  beryllium compounds, 780
  biological, 938–939. See also Biochemistry
    nucleic acids, 953–961
    polysaccharides, 951–952
    proteins, 940–948
  polyamides, 926, 927
  polyesters, 916, 924
  silicate, 809, 810, 811
  see also Plastics
Polymetaphosphoric acid, 817
Polymorphs, crystal structure, 367
Polynucleotides, 955–956
Polypeptides
  amino acid permutations, 945
  defined, 942
  see also Proteins
Polyphosphoric acids, 816–817
Polyprotic acids
  acid ionization constant, 601–603
  acid strength of, 585
  defined, 286
  phosphoric, 816
Polysaccharides, 948, 951–952
Polystyrene plastic, 913
Polyurethane plastics, nitric acid use in, 455
p Orbitals, 168, 254
  hybrid orbital formation, 248–253
  multiple bonds, 253–257
  oxygen, 819
  second-row elements, 765–766
  shapes of, 157
Positive ions
  as cations, 32
  nomenclature, 34
  see also Cations
Positron, defined, 719
Positron emission, 722
  defined, 723–724
  potassium-40, 740
Postulates of kinetic theory, 112–114
Potassium, 188
  common-ion effect, 631
  electromotive series, 471
  flame test, 129
  oxygen reaction with, 442
  properties of, 767, 768
  radioactive isotope, 724
    and background radiation, 733–734
    dating with, 740
  sodium-potassium solutions, 391
  sources of, 767
  uses of, 770, 771
Potassium chlorate, oxygen preparation from, 111, 439
Potassium chloride, uses of, 772
Potassium chromate, fractional precipitation with, 637
Potassium hydroxide
  production of, 291, 772
  in voltaic cells, 687, 689

Potassium nitrate, oxygen preparation from, 439
Potassium permanganate, *see* Permanganate
Potential difference, 693
Potential energy, defined, 19–20
Potential energy diagrams
    for reactions, 514–515
    sulfur oxide oxidation, 527
Potentiometer, 693, 694
Precipitate, defined, 280
Precipitation reactions, 280–282
    base preparation, 290
    calculations
        completeness, 635–637
        criterion for, 633–635
        fractional, 637–638
    common-ion effect, 629–633
    salt preparation, 291
Precision, 7–10
Prefixes, 225–226
    alkane, 899
    benzene derivatives, 907
    chemical, 34
    coordination compounds, 862–863
    metric, 11, 12
    table of, 12
Pressure
    in Avogadro's law, 99–100, 101
    and boiling point, 373–374
    in Boyle's law, 92, 93
    changes of, and reaction rate, 498, 500, 558–559
    critical, 381–382
    defined, 16
    and equilibrium constant, 543–544
    and heat of reaction, 315, 318, 319
    osmotic, 415–416
    phase diagrams, 380–383
    and solubility, 390, 398–399
    units of, 18–19
Pressure-volume work, and heat of reaction, 318, 319
Priestley, Joseph, 69, 437, 438
Primary alcohols, 916, 921
Primary amines, 925
Primary particles, Boyle's particle view, 91n
Principal quantum number, 146, 153, 182
Principle, defined, 5
Probability distribution, bonding electrons, 210
Probability statements, Heisenberg uncertainty principle, 151–152
Products
    in chemical equations, 35, 36
    and equilibria
        equilibrium constant expression, 545
        equilibrium constant interpretation, 546–548
        removal or addition of, 555–557
    and spontaneity of reaction, 670
    see also Yields
Proinsulin, 946

Proline, 943
Promethium, 722, 849
Promotion, electron, 248
Promotional energy, oxidation state and, 764n
Propane
    burning, equation for, 35–36
    physical properties, 897
    structural formula, 31, 896
2-Propanol, oxidation of, 920–921
Propanone, 917
Properties, chemical and physical, 38
Propionic acid, 621, 622, 919
Proportionality constant, 115
Proportions, laws of, 60
Propylene, uses of, 912
Propyl group, 898. *See also* Alkyl groups
Protactinium-234, 721, 726
Proteins, 940–948
    amino acids, 940–945
    defined, 942
    enzymes, 947–948
    membrane, 963, 964
    solutions, 401
    structure
        primary, 945, 946
        three-dimensional (conformation), 945–947
    synthesis of, 959–961
Protons, 129
    amino acid binding of, 940
    band of stability, 721, 722
    bombardment reactions, 727–729
    hydrogen ion, in aqueous solution, 273
    magic numbers, 730273
    and mass number, 136
    mass of, 747
    properties of, 135
    in radioactive decay, 723, 724
    in shell model of nucleus, 720–721
    spin of, 165n
    in stable isotopes, 721
Proton-transfer reactions
    in acid-base reactions, 286
    Brønsted-Lowry acids and bases, 578–581
Pseudonoble-gas configuration, 200, 201
Ψ (wave function), defined, 152
p Subshell, 154
    and electron affinity, 187
    in noble gases, 170, 178
Pure substance, defined, 38
Purification, by sublimation, 372, 373
Pyranoses, 949–950, 951
    in nucleotides, 953, 954
    polysaccharides, 952
Pyrazinium ion, 858n
Pyrene, formula, 906
Pyridine, 603, 622
Pyridinium ion, 622
Pyrite, 434
Pyrolusite, 68, 850
Pyrosulfuric acid, 823
Pyruvic acid, 601n

q, 314–315. *See also* Reaction heats
Quadratic formula, degree of ionization, 599, 600
Quadridentate ligands, defined, 551n, 859
Qualitative analysis
    chlorine water test for halogen ions, 828
    metal ions, 648–650, 821–822
Quantitation, precipitation reactions, 635–637
Quantum effects, 141–144
Quantum mechanics, 149–152
    de Broglie waves, 149–150
    excited and ground states, 167–168, 169
    wave functions, 151–152
Quantum numbers, 142, 153–155
    in Balmer's formula, 148
    Pauli exclusion principle, 166, 167
    permissible values, 154
    principal, 146
    see also Orbitals; Valence bond theory
Quartz, 359
    crystal structure, 367
    melting point of, 376
    silicon preparation from, 803
Quicklime, 545, 777–778
Quinine, base ionization constant, 604

Rad, defined, 733
Radiation
    background, 733
    in radioactive decay, 723
Radiation counters, 731–733
Radioactive dating, 738–740
Radioactive decay
    rate of, 734–735
    series, 725–726
    types of, 723–725
Radioactive decay constant, 734–735
Radioactivity, 133
    actinides, 843
    curie (unit), 733
    dating, 738–740
    decay, types of, 723–725
    decay series, 725–726
    detection of, 731–733
    discovery of, 133–134
    effects of, 733–734
    half-life, 735–738
    nuclear equations, 718–720
    nuclear stability, 720–723
    polonium, 819
    radium, 773–774
    radon, 456
    transition element, 849
    see also Nuclear chemistry
Radioimmunoassays, 744
Radium, 719
    as alpha emitter, 725
    and background radiation, 733
    decay constant, 734–735
    discovery of, 164
    in uranium ore, 773–774

Radon, 191, 719
  compounds of, 457
  discovery of, 456
  p subshell, 170
  uses of, 457
Ramsay, William, 455
Random motion, 112, 113
Randomness, *see* Entropy
Raoult's law, 410
Rare earths, 843
Rare gases, and noble gases, 191
Rate constant, defined, 500, 515
Rate determining step, 521
Rate equation, for elementary reaction, 519–520
Rate law, 501
  concentration-time equation, 504–506
  defined, 500
  determining, 502–504
  and mechanism of reaction, 520–524
Rate of reaction, 495
Rayleigh, Lord, 455
Reactants
  in chemical equations, 35, 36
  and equilibria
    equilibrium constant expression, 545
    equilibrium constant interpretation, 546–548
    removal or addition of, 555–557
  limiting, yield calculation, 69–71
Reaction coupling, 669
Reaction direction, *see* Direction of reaction; Driving of reaction
Reaction heats, 313–315, 656
  and Arrhenius theory, 572–573
  defined, 314
  measurement of, 322–323
  and second law, 660
  stoichiometry of, 323–325
Reaction intermediate, 517
Reaction laws
  first-order vs. second-order reactions, 511
  graphical plotting, 508
Reaction mechanism
  catalysis, 524–527
  elementary reactions, 517–520
    molecularity, 517–519
    rate equation for, 519–520
  rate law and
    mechanisms with initial fast step, 521–524
    rate determining step, 521
Reaction orders
  defined, 501–502
  graphical plotting, 508–510, 511
  initial rate method, 502
Reaction products, *see* Products; Yields
Reaction quotient, 548
  gaseous reactions, pressure changes and, 559
  precipitation reactions, 633–634
Reaction rates
  Arrhenius equation, 515–517

biological systems
  enzymes and, 948
  regulation of, 937
concentration, dependence on, 500–504
concentration changes with time, 504–511
  equations, 504–506
  first-order and second-order reactions, 511
  graphical plotting, 508–510, 511
  half-life, 506–507, 508
definition of, 495–497
experimental determination, 498–500
factors determining, 494–495
rate law determination, 502–504
reaction order, 501–502
temperature and, 511
  collision theory, 511–513
  potential energy diagrams, 514–515
  transition state theory, 213
Reactions, chemical, 34–35
  defined, 29
  limiting reactant, 69–71
  spontaneity, 670. *See also* Spontaneity
  stoichiometry of, 67–69
Reactivity
  Group IA elements, 767, 768
  Group IIA elements, 773–774
  Group IIIA elements, 782–783
  Group IVA elements, 801–802
  Group VA elements, 814
  Group VIA elements, 820–821
  Group VIIA compounds, 828
  halogens, 830
  hydrocarbons, saturation and, 908, 909
  metallic vs. nonmetallic elements, 760–761
  single vs. double bonds, 221n
Reagent, limiting, yield calculations, 69–71
Real gases, ideal gas law and, 118–120
Receiver, distillation, 43
Rectifiers, selenium in, 821
Red lead, 805, 811
Redox reactions, *see* Oxidation-reduction reactions
Red phosphorus, 813, 814
Red selenium, 821
Reducing agents
  calcium as, 777
  defined, 293
  hydrogen from, 470
  hydrogen peroxide as, 479
  hydrogen sulfide, selenide and telluride, 822
  iron(II) ion, 852
  magnesium and, 776
  sodium as, 770
  stannite ion as, 811
  strength of, 697–698
Reduction
  at cathode, 686

defined, 293, 920. *See also* Oxidation-reduction reactions
  by hydrogen, 474
  metallurgical, 788, 790
    boron preparation, 784
    calcium use in, 777
    chromium preparation, 852
    copper, 855
    Group IA element preparation, 769
    iron ore, 852–853
    lead preparation, 805
  organic compounds, 922
  silicon dioxide, 803
  transition metal compounds
    chromium oxide, 852
    iron oxide, 853
Reduction potentials, 707
  in emf calculations, 697
  table, 696
Refining
  Group VA elements, 814
  metals, 788, 790, 791
  zone, 804–805, 806
Reformation, catalytic, 913
Refractory bricks, 779
Refrigeration
  mechanism of cooling, 375
  trichlorofluoromethane use in, 831
Regulation, biological systems, 937
Relative atomic mass, 52, 53n
Relativity theory, 3n, 744
Rem, defined, 733
Replacement reactions, defined, 276
Reprocessing plants, nuclear fuel, 751
Resistance to flow, 356–357
Resonance, 218–219, 220, 367
Respiration, 435
Retinal, *cis-trans* isomer conversion, 287
Reverse osmosis, 487
Reverse reaction, in chemical equilibrium, *see* Direction of reaction
Rhombic sulfur, 819
Rhombohedral structure
  boron, 782
  unit cell, 360
Ribonucleic acid
  and genetic code, 959–961
  messenger, 959–960
  nucleotides, 953–955
  ribosomal, 959
  transfer, 957, 960–961
Ribonucleosides, 954
Ribonucleotides, 953
D-Ribose, 948, 949, 950, 953, 954
Ribosomes, defined, 959
Ribothymidine, 954
Rigidity, 39, 40
Ring anions, silicate, 809, 810, 811
Roasting, 790
  galena, 805
  Group VA element preparation, 814
  selenium as by-product, 821
Rocket fuel
  hydrazine as, 452, 830
  hydrogen peroxide as, 479

Rocket fuel *(cont.)*
　oxygen as, 441
Rocks, dating, 740
Rock salt, 767
Root-mean-square molecular speed,
　　defined, 116
Rosenberg, Barnett, 2
Rounding, 10
Rowland, F. Sherwood, 446
Rubidium
　oxygen reaction with, 442
　preparation of, 769
　properties of, 768
　sources of, 767
Ruby, 362, 781
Rusting, 852n
　electrochemistry of, 689–690
　energy changes, 669
Rutherford, Daniel, 447
Rutherford, Ernest, 133, 135, 144, 145,
　　172, 726, 727
Rutile, 850

Salicylic acid, 623
Salt bridge, voltaic cell, 684
Salt domes, 819
Salts
　acid, 286
　aluminum, 781, 783, 787–788
　biological molecule denaturation,
　　939
　buffers, 610, 611–615
　defined, 283–284, 571–572
　equivalents and normality, 303–305
　esterification, 923
　hydrolysis, weak acids or bases, 605–
　　608
　ionic bonds, 196, 197
　neutralization, 285–286
　phosphoric acid, 816
　preparation of, 291–292
　reactions of, 286–288
　solubility, pH and, 638–639
　solutions, acid-base properties, 585–
　　587
　table, 283
Sap flow, osmotic pressure and, 417n
Saponification, 923–924, 962
Sapphire, 781
Saturated bonds, 895
　defined, 895
　oxidation of, 908
Saturated solutions
　defined, 278
　solubility product constant, 626
Scandium
　Mendeleev's prediction of, 178
　physical properties, 844–848
Scandium atom, orbital energies, 168,
　　170
Scanning electron microscope, 151
Scattering, alpha particle, 134–135
Scavengers, metallurgical, 770, 777
Scheele, Karl Wilhelm, 437, 438
Scheelite, 850
Schrock, R. R., 220

Schrodinger, Erwin, 145, 151
Scintillation counters, 731–732
Seaborg, Glenn T., 730
Sea shells, 777
Sea water
　alkali metal salts in, 767
　desalination, 486–488
　electrolysis of, magnesium prepara-
　　tion, 775–776
　Group IIA elements in, 773
　Group VIIA elements in, 827
　halogen preparation, 829
　ions in, 482
Second, base unit of time, 12
Secondary alcohols, 916, 921
Secondary amines, 925
Secondary structure, proteins, 946
Second ionization energy, 185
Second law of thermodynamics, 658–
　　660
Second-order equations, concentra-
　　tion-time equation, 506
Second-order reactions, 501
　vs. first-order reactions, 511
　graphical plotting, 508, 510, 511
　half-life of, 507
Second-period elements
　chemical and physical properties,
　　765–766
　Lewis electron dot symbols, 197
　properties of, 826–827
Sedimentary rock, 433, 436
Sedimentation tanks, 485
Seeding, supersaturated solutions, 279
s Electrons, period six elements, 764
Selenides
　hydrogen, 822
　nomenclature, 225
Selenium, 190
　occurrence of, 819
　oxides and oxyacids, 825–826
　periodic table placement, 178
　preparation and uses of, 821
　properties, 820, 821
Self-ionization, of water, 573–574
Semiconductors
　boron as, 781, 782
　gallium, 785
　Group VIA elements, 819
　indium, 785
　main-group elements, 761
　silicon for, 803–804
Semimetals, 188. *See also* Metalloids
Serge, Emilio, 717
Serine, 943, 944
Sewage treatment, 484–485
Shapes, atomic orbitals, 156–157. *See
　　also* Molecular geometry; Struc-
　　ture
Shellfish, calcium carbonate in, 773
Shell model, nucleus, 720
Shells, defined, 153. *See also* Valence
　　shells; *specific shells*
Sickle cell anemia, 395–396
Side chains, amino acids, 942–944, 946
Siderite, 850, 852–854

Sigma bonds (σ)
　in double bond, 256
　in multiple bonding, 254–256
Silica, 831, 911–912
Silica glass, 359
Silicates, 773, 807, 808–809, 810, 811
Silicide, 225
Silicon, 189
　diagonal relationships, 766, 785n
　doubly bonded, 766
　enthalpies of formation, 329
　ferrosilicon alloy, 776
　hydrides, 806
　minerals, 800
　properties, 800, 801, 802
　purification, 803–804
　reaction with water, 476
　standard free energies of formation,
　　667
　zone melting, 806
Silicon carbide, hardness of, 378
Silicon dioxide, 807, 808–809, 810, 811
　uses, 807
　water and, 444
Silicones, chlorine compound use in,
　　829
Silicon hexafluoride, hybrid orbitals in,
　　250
Silver
　building-up principle exceptions,
　　171
　catalysts, 452
　complex ions, 642, 643, 645–648
　crystal structure, 365
　electromotive series, 471
　enthalpies of formation, 329
　hydroxylamine preparation, 452
　precipitation of, 648, 649, 650
　standard entropies, 662
　standard free energies of formation,
　　667
　Tollen's test, 922n
Silver acetate, 633
Silver chloride, 281, 628, 704
Silver electrodes, 686, 687, 704
Silver iodide, 291, 829
Silver-silver chloride electrode, 704
Simple cubic unit cell, 360, 361
Simplest formula, *see* Formulas, empir-
　　ical
Sink, soil as, 458–459
SI units, 11–14
　conversion factors, 15
　derived units, 16–22
　dipole moments, 243
　electrical charge and current, 708
　entropy, 657, 658
　potential difference, 693
　pressure, 90, 91
　tables of, 12, 15
　　derived units, 17
　　pressure, 91
　see also Units of measurement
Six-coordinate complexes, geometry,
　　869
Sizing, paper, 787

Skeleton oxidation-reduction equations, half-reaction method of balancing, 297–300
Skeleton structure
  in Lewis formulas, 213
  of molecule, 212–213
Skutterudite, 850
Slightly soluble, defined, 279
Smog, 458–459
Snow, as phase change, 372, 373
Soap
  as association colloid, 403
  cleaning action of, 404
  from fats, 962
  micelles, 939
  saponification, 923–924
  see also Detergents
Soda ash, 56, 772
Sodium, 28, 34, 188
  alloys, 771
  atomic vs. ionic radii, 205
  atoms
    excited state transitions, 167n
    nucleus of, 135–136
  delocalized bonding, 220
  electromotive series, 471
  enthalpies of formation, 329
  flame test, 129
  and hydride formation, 473
  hydrogen from, 470
  ionization energies, 200
  ions of, 32
  metallic bonding, 196
  in ocean water, 482
  oxidation of, 438
  oxygen reaction with, 442
  preparation of, 769
  properties of, 767, 768
  sodium-potassium solutions, 391
  sources of, 767
  standard entropies, 662
  standard free energies of formation, 667
  uses of, 769–770
Sodium amalgam, 706, 772
Sodium bicarbonate, 623
Sodium carbonate, 56, 224
  and calcium chloride, 275
  hydrolysis, 586
  uses of, 770, 772
Sodium chloride, 28, 34
  Born-Haber cycle, 336–338
  crystal structure, 32–33, 366
  electrolysis of, 704–705, 771–772
  as electrolyte, 271–272
  formation of
    bonding energy, 198–199
    ionic bonds, 196, 197
  formula, 196, 197
  as ionic substance, 31–33
  lattice energy determination, 336–338'
  melting point of, 376
  in sea water, 767
  sodium hydroxide production, 704–705, 771–772

solubility equilibrium, 278
solutions
  electrolysis of, 705–707
  ionic, 418
  uses of, 770, 771
Sodium chromate, chromium compound preparation, 851–852
Sodium citrate, 612
Sodium cyanide, 605–606, 622
Sodium dihydrogen phosphate, 623, 816
Sodium hexafluorosilicate, 833
Sodium hydrogen carbonate, in Solvay process, 772
Sodium hydrogen sulfate, 624
Sodium hydrogen sulfite, 829
Sodium hydroxide, 62n
  aqueous solutions, 283–284
  carbon monoxide reaction with, 807–808
  electrolysis of, 769
  production of, 290, 705–707, 771–772
  titration with, 615–617
  uses of, 770, 771–772
Sodium hypochlorite, 832–833
Sodium lauryl sulfate, 403–404
Sodium nitrate, 454
Sodium peroxide, 438–439, 442
Sodium propionate, 622
Sodium sulfate, 224
Sodium thiosulfate pentahydrate, 825
Sodium triphosphate, 818
Sodium vapor lamps, 130n
Soil, and atmosphere, 458
Solder, 805
Solids
  amorphous, 358–359
  crystalline, see Crystals
  filtration of, 41–42
  intermolecular forces, 347–351
  melting points, 374
  phase, 40, 41
  phase changes, 371–373. See also Phase changes
  phase diagrams, 380–383
  physical properties of, structure and, 376–379
  solutions of, 391, 392
  types of, 357–358
  vapor pressure of, 354, 380, 381, 382
Solid state, 39, 40, 41
Sols, 400, 401, 402
Solubility, 277–282
  alkali metals, magnesium, and lithium salts, 766
  bases, 284
  complex ions and, 646–648
  defined, 392
  factors determining, 392
  Group IIA compounds, 778–779
  Group IIA elements, 783
  Group VIIA elements, 827
  ionic solids, in water, 280
  pressure and, 398–399
  proteins, structure and, 946

temperature and, 397–398
Solubility equilibria, 277–278, 392
  common-ion effect, 629–633
  metal ions, quantitative analysis, 648–650
  pH and solubility, 638–641
  precipitation calculations, 633–638
  solubility product constant, 626–629
  see also Solutions
Solubility product constants, 626–629
  common-ion effect, 629–633
  table, 630
Solubility rules, 279–280
Solutes
  concentration expressions, 404–409
  defined, 72, 391
  mass percentage of, 405
  osmosis, 415–418
  and surface tension, 356
Solutions
  acid-base properties
    hydrolysis, 585–587
    pH, 575–577, 578
    of salts, 585–587
  aqueous, see Aqueous solutions, reactions in
  calculations involving, 71–78
    diluting solutions, 74–75
    molar concentrations, 72–74
    stoichiometry, 76–78
  colligative properties
    boiling point elevation and freezing point lowering, 411–414
    concentration and, 404–409
    vapor pressure, 409–411
  colloids, 399–404
  defined, 390
  gaseous, 38
  as homogeneous mixture, 38
  ionic, 394–395, 418–419
  molecular, 392–394
  osmosis, 415–418
  separation techniques, 42–45
  solubility, factors determining, 392. See also Solubility equilibria
    Henry's law, 398
    pressure and, 398–399
    temperature and, 397–398
  solubility product constant, 626–629, 630
  types of, 391–395
    gaseous, 391
    liquid, 391
    solid, 392
    table, 391
  see also Acid-base equilibria
Solvay process, 772
Solvents
  defined, 72, 391
  osmosis, 415–418
s Orbitals, 168, 254
  hybrid orbital formation, 248–253
  multiple bonding, 253–257
  oxygen, 819
  second-row elements, 765–766
  shapes of, 157

s Orbitals *(cont.)*
  valence bond theory, *see* Valence
    bond theory
Sorensen, S. P. L., 575n
Specific heat, 21–22, 321–322
Specificity, enzyme, 948
Spectator ions, 275
Spectra, 129
  atomic line, 145–146
  Bohr's postulates, 146–149
  of transition metal complexes, 884–
    886
Spectrochemical series, 882
Spectrometry, mass, 52, 53
Spectroscopy
  nuclear magnetic resonance, 165n
  of transition metal complexes, 884–
    886
  x-ray, 172
  x-ray electron, 179n
  x-ray photoelectron, 172, 173
Speed
  defined, 16
  and energy, 19
  molecular, *see* Molecular speed
Sphalerite, 69, 271, 366, 367, 850
Sphygmomanometer, 90n
Spin, 153
  electron, 165–166
  and repulsion of electron pairs, 236
Spin complexes
  crystal field theory, 881–883
  oxygen release from hemoglobin, 887
Spin magnetism, 165–166
Spin orientation, 166
Spin quantum number, 154, 164
Spodumene, 767
Spontaneity
  chelates, 862
  free energy and, 664–668
  oxidizing and reducing agent
    strength and, 698
  standard free energy changes and,
    666–668, 673
  and temperature change, 673–674
Spontaneous reactions
  defined, 657
  free energy changes during, 670
  second law, 660
  see also Entropy, spontaneous proc-
    esses and
sp Orbitals, 250
  ozone bond, 445
  valence bond theory, 873–878
  see also Orbitals, hybrid
Square planar complexes
  crystal field theory, 883–884
  hybrid orbitals and coordination
    numbers, 878
  multiple bonding, 255
  valence bond theory, 877–879
Square planar molecular geometry, 242,
  245
Square pyramidal molecular geometry,
  241, 245
s Subshell, 154

Stability
  band of, 723
  binding energies vs. mass number
    and, 747–748
  boron allotropes, 782
  of chelates, 861–862
  metal ions, 199–200
  nuclear, 720–723
  stable isotopes, 721
Stability constant, 642, 643
Stable isotopes, table, 721
Stainless steel, components of, 854
Standard electrode potential, 695, 696
Standard emf, 696
Standard enthalpies of formation, 328–
  332
Standard entropies, 660–662
Standard free energies of formation,
  665–666
  standard emf from, 699
  table, 667
Standard free energy changes, 665
  as criterion for spontaneity, 666–668,
    673
  from emfs, 699
  and equilibrium constant, 700
  temperature and, calculation of, 675–
    676
Standard potential diagram, chlorine
  and compounds, 831–832
Standard potentials
  electrode, 695, 696
  Group IA elements, 768
  Group IIA elements, 774
  Group IIIA elements, 782
  Group IVA elements, 801
  Group VA elements, 813
  Group VIA elements, 820
  Group VIIA elements, 827
Standards, measurement units, 11n
Standard states, 665
Standard temperature and pressure, 100
Stannate ion, 811, 812
Stannite ion, 811, 812
Starches, 952
State functions, 316, 657
States of matter
  and entropy, 658
  in ionic equations, 173
  physical, 39–40, 41
  solution, 391
  vapor pressure and, 351
  see also Crystal structure; Phase
    changes; *specific states*
Statistical statements, Heisenberg un-
  certainty principle, 151–152
Steam
  Group IVA element reactivity, 801
  hydrogen from, 468, 471–472, 536,
    544–545
Steam-reforming process, 471–472,
    536, 544–545
Stearate micelles, 403, 404
Stearic acid, 924
Steel
  ferrochrome use, 851

  manufacture of, 853–854
  oxygen use, 441
  tin plating, 805
Steel pipeline, cathodic protection of,
  690
Stereoisomerism
  chiral, 871
  defined, 866
  geometric, 868–870
  optical, 870–873
  see also specific types of isomers
Sterilizing lamps, 445
Stern, Otto, 165
Stern-Gerlach experiment, 165
Stock system, nomenclature, 225, 227
Stoichiometry, 66–69
  buffer reactions, 612
  of chemical reactions, 66–69
  defined, 66
  of electrolysis, 708–709
  equivalence point, 615
  with gas volume, 103–104
  ionic reactions, from molecular equa-
    tions, 274–275
  limiting reactant, yields with, 69–71
  molar interpretation of chemical
    equations, 66–67
  mole-mass relations, 66–71
  nonstoichiometric compounds, 362–
    363
  of reaction heats, 323–325
  of solution reactions, 76–78
Storage batteries, *see* Batteries; Voltaic
  cells
Straight chain alkanes, 896
Strassmann, Fritz, 749
Stratopause, 427, 428
Stratosphere, 427, 428
  heating of, 428–431
  ozone depletion, 446
Strong acids and bases, *see* Acids,
  strong; Bases, strong
Strong electrolytes, ionization of,
  274
Strontianite, 773
Strontium
  fractional precipitation, 637–638
  ores of, 773
  oxygen reaction with, 442
  preparation of, 776
  radioactive isotope, 718, 737
Structural formulas, *see* Chemical formulas, structural
Structural isomers, 866–867
  alkanes, 897, 898–900
  defined, 866
Structure
  and acid strength, 583–585
  atomic, *see* Atomic structure; Elec-
    tronic structure of atoms;
    Periodic table
  biological molecules
    amino acid side chains and, 942–
      944
    lipids, 961–963
    membranes, 964

Structure (cont.)
  nucleic acids, 956, 957, 958, 960–961
  nucleotides, 953, 954, 956
  polymer conformations, 938–939
  protein, 945–947
  and boiling point, 377–378, 379
  chlorine oxyacids, 832
  coordination compounds, 865–873
  crystal field theory, see Crystal field theory
  and electrical conductivity, 379
  functional group, 914
  graphite vs. diamond, 800
  and hardness, 378, 379
  of ice, 475, 476
  melting points and, 376–377, 379
  nomenclature, 34
  oxyacids of phosphorus, 815, 816
  phosphorus, elemental, 813
  and physical properties, 376–379
  silicates, 807, 808–809, 810, 811
  skeleton, 212–213
  sulfate ion, 824
  sulfur, 819–820
  sulfur trioxide ions, 823
  tetraborate ion, 785–786
  valence bond theory, see Valence bond theory
  see also Crystals, structure; Molecular geometry
Styrene, 788, 913
Sublimation
  aluminum chloride, 784, 788
  defined, 336, 372, 373
  phosphorus(V) oxide, 815
  triple point and, 381
Subscripts
  in balancing equations, 36
  nuclear equations, 718, 719
Subshells, 153–154
  in building-up order, 168, 170, 171
  and electron affinity, 186, 187
  electrons, maximum number of, 166, 167
  ion formation, 200
  ionization energies, 184
  orbital diagram, 165–166
  periodic table position and, 178
  transition elements, 842–843
  see also Electron configuration
Substitution reactions
  of alkanes, 908–909
  aromatic hydrocarbons, 911
Substrates, enzymes, 947, 948
Successive approximation method, 599, 632–633
Sucrose, 824, 825, 951
Suffixes, 225, 227
  alkyl groups, 899
  chemical, 34
  hydrocarbon, 902
  see also Nomenclature
Sugars, 824, 825, 936
  monosaccharide, 948–950
  in nucleotides, 953–955

oligosaccharides and polysaccharides, 950–952
  see also Glucose
Sulfanilic acid, 621
Sulfates
  Group IIA compounds
    solubilities of, 778
    uses of, 777, 780
  ions, 33, 34
    formula, 203
    quantitation of, 635, 636
  solubilities, 280, 778
Sulfides, 225
  arsenic, 814
  copper, 855
  metal ion precipitation, 635–641, 648–650, 821–822
  nonstoichiometric properties, 821n
  ores
    selenium and tellurium-containing, 819, 821
    transition metal, 849, 850
  oxygen reaction with, 443
  salt preparation from, 291
  solubilities, 280
Sulfites, 823
  bromine reaction with, 297
  formulas, 203, 215
  permanganate oxidation of, 300–301
  salt preparation from, 291
Sulfur, 190
  enthalpies of formation, 329
  gaseous state, 105n
  hydrogen reactions with, 473
  occurrence of, 819
  organic compounds containing, 927
  oxygen reaction with, 442
  preparation and uses of, 821, 822
  properties, 819–820, 821
  in proteins, disulfide cross linkage, 946
  standard entropies, 662
  standard free energies of formation, 667
Sulfur dioxide, 444, 823
  delocalized bonding, 218–219
  formation of, 287, 442
  geometry, 239
  oxidation of, catalysts, 524–525, 526, 527, 553
  thiosulfite decomposition to, 825
Sulfur family, see Group VIA elements
Sulfur hexafluoride, geometry, 241
Sulfuric acid, 288
  hydrate isomer, 867
  hydrogen fluoride preparation, 830
  in lead storage cells, 687, 688
  preparation and uses, 823–824, 825
  sulfur oxides and, 444
Sulfurous acid, 822–823
  acid ionizaton constant, 603
  sulfur oxides and, 444
Sulfur tetrafluoride, geometry, 240
Sulfur trioxide, 444, 823
  catalyst and, 553
  formation of, 442

Supercritical mass, 750
Superoxides
  alkali metals, 766
  formation of, 442–443
  of Group IA elements, 768
  oxidation number, 223
  potassium, 771
Supersaturated solutions, 279, 634
Superscripts, nuclear equations, 718, 719
Supersonic transports, and ozone, 446
Surface area, and reaction rate, 495
Surface catalysts, 526, 527
Surface features, proteins, 947
Surface tension, 354, 355–356, 357
Surroundings, defined, 313
Sylvite, 767
Symbols
  atomic, 29, 30
  electron-dot, see Lewis electron-dot symbols
  nuclear bombardment reactions, 727
  nuclide, 718
Synthesis, vapor phase chromatography in, 45
Synthetic jewels
  diamond, 803
  Group IIIA elements, 786–787
System, defined, 313

Tar, coal, 907
Tartaric acid, isomers, 872
Tartrate buffers, 623
Tassaert, B. M., 860
t-Butyl groups, 898. See also Alkyl groups
Technetium, 717, 722, 724, 849
  decay constant, 736
  discovery of, 729
Technetium pyrophosphate, 743–744
Television tube, 132
Tellurides, 225, 822
Tellurite salts, 825–826
Tellurium, 190
  occurrence of, 819
  oxides and oxyacids, 825–826
  periodic table placement, 178
  preparation and uses of, 821
  properties, 820, 821
Temperature
  absolute, 13
  atmospheric, 427–428
    altitude and, 427
    heating of thermosphere and stratosphere, 428–431
    moderation of, 459–461
    water and, 481
  in Avogadro's law, 99–100, 101
  boiling point, 372–373. See also Boiling points
  in Boyle's law, 92, 93
  and Brownian motion, 114
  and conductivity, 761
  conversion of units, 13
  critical, 381–382

Temperature *(cont.)*
diamond synthesis, 803
enthalpy changes, 373–376
and equilibrium constant, 543–544
free energy changes with
calculations, 675–676
spontaneity and, 673–674
and gaseous reactions, 560–561
and gas volume, 95, 96–97
melting point, 373. *See also* Melting points
and molecular kinetic energy, 114
and nitrogen oxides, 453
and nuclear fusion, 753
phase diagrams, 380–383
and reaction rates, 495
collision theory, 511–513
potential energy diagrams, 514–515
transition state theory, 513
and solubility, 390, 397–398
in specific heat definition, 21–22
and supersaturation, 279
units of, 12, 13–14
and vapor pressure, 110, 353–354
water
and physical properties, 475–476, 477
thermal pollution, 482, 483
and vapor pressure, 110
see also Thermochemistry; Thermodynamics
Termination codons, 959
Terminology, *see* Nomenclature
Termolecular, 518
Tertiary alcohols, 916, 921
Tertiary amines, 925
Tertiary butyl groups, 898
Tertiary structure, proteins, 946, 947
Tetra-
in alkane nomenclature, 899
defined, 34, 225, 226
Tetraaminedichlorocobalt(III) chloride, geometric isomers, 869
Tetraborate ion, structure, 785–786
Tetraethyllead, 769–770, 805, 830, 912
Tetragonal allotropes, boron, 782
Tetragonal unit cell, 360
Tetrahedral arrangement, 236, 238, 239–240
in amino acids, 941
crystal field theory, 883–884
and dipole moment, 245
hemoglobin, 686–687
hybrid orbitals and coordination numbers, 878
silicon, 808, 809, 811
sulfate ion, 824
valence bond theory, 877–879
Tetrahedral bonding, Group IVA elements, 799
Tetrahedral holes, 366, 367
Tetramer, hemoglobin as, 886–887
Tetramethyllead, 829
Textiles
aluminum hydroxide mordants, 787

chlorine use as, 829
sulfur compound uses, 823
Thales of Miletus, 467
Thallium
chlorides, 765
ion formation, 201
oxidation of, 525
oxidation states, 764
properties of, 782, 783
sources of, 782
Theoretical yield, defined, *see* Yield calculations
Theory, 5–6, 7
Thermal equilibrium, 313–314
Thermal pollution, of water, 482
Thermal properties, common substances, 475
Thermite, 785
Thermochemistry
applications of, 332–336
bond energy, 332–336
calorimetry, 319–323
defined, 312
enthalpy, 315–317
enthalpy and energy, 318, 319
enthalpy of formation, 328–332
heat capacity, 320–321
Hess's law, 325–328
lattice energies, 336–338
reaction heat, 313–315
measurements, 322–323
stoichiometry, 323–325
specific heat, 321–322
Thermodynamics, 314
defined, 656
emf measurement and, 699
entropy, 657–658
entropy change for reaction, 662–664
equilibrium constant, 671–673
first law, 318
free energy
changes during reactions, 669–670, 671
coupling of reactions, 669
maximum work, 668
and spontaneity, 664–668
spontaneity criteria, 666–668, 673
standard free energy change, 665–666, 667
free energy, temperature changes and, 673–676
calculation of, 675–676
spontaneity and, 673–674
second law, 658–660
third law, 660–662
Thermodynamic stability, chelates, 862
Thermodynamic system, 314
Thermometers, Amontons' law, 102
Thermosphere, 427, 428–429, 430
Thioacetamide, 649n, 822
Thiols, 914, 927, 946
Thionyl chloride, Lewis formula, 214
Thiosulfates, 203, 824–825
Third law of thermodynamics, 660–662, 701n
Third-order reactions, 501

Third-period elements
ionic radii, 206
Lewis electron-dot symbols, 197
Thomson, George Paget, 150
Thomson, Joseph John, 131, 138
Thorium
abundance of, 726
breeder reactor, 752
Thorium-227, francium decay to, 767
Thorium-232, as alpha emitter, 725
Thorium-234, 721
beta decay, 725
mass losses in, 746
radioactive decay series, 725, 726
Three-center bonds, borane, 786, 787
Threonine, 942, 943
Threshold value, 143
Thymine, 953–954
base pairing, 955–956
structural formula, 953
Thymol blue, 578
Thyroid, iodine-131 and, 735n, 744n
Time, units of, 12
Time-concentration relationship, graphical plotting, 508–510, 511
Time interval, reaction rates, 496
Tin
amphoteric hydroxide, 645
compounds of, 807, 809, 811–812
electromotive series, 471
hydrides, 806
and hydrogen generation from, 468
ion formation, 201, 811, 812
minerals, 800
precipitation of, 649
preparation and uses, 805
properties, 800–801
Tin disease, 801
Titanium
complex ions, visible spectra, 885–886
ores, 850
physical properties, 844–848
x-ray spectra, 172
Titration, 77–78
defined, 77
strong acids and bases, 615–617
weak acids and strong base, 617–618
weak base and strong acid, 618, 619
Tokamak nuclear fusion test reactor, 752, 753
Tollen's test, 922n
Toluene, structural formula, 906
Torr, defined, 19, 91
Torricelli, Evangelista, 89
Toxins, water pollution, 484
Trace constituents, of air, 458–459
Transcription, 959
Transfer RNA, 957, 960–961
Trans isomers, 256–257. See also *Cis-trans* isomers
Transition elements, 179
atomic radii, 844–846
chromium, 849–852
classification of, 842–843

Transition elements (cont.)
  complex ions, see Complex-ion for-
      mation
    crystal field theory, see Crystal
        field theory
    formation and structure, 857–862
    valence bond theory, see Valence
        bond theory
  coordination compounds
    nomenclature, 862–864
    stereoisomerism, 867
    structural isomerism, 866–867
  copper, 853–854
  electron configuration, 179, 844–845
  ionization energies, 847
  iron, 852–854
  melting points, boiling points and
      hardness, 844–845
  metal ions, 201–202
  metal oxides, acidity of, 761n
  ores of, 850
  oxidation states, 296, 847–849
  periodic trends in, 843–849
  see also Chromium; Copper; Iron
Transitions, between energy levels,
    146–148
Transition-state theory, 513, 515n
Translation, 959
Transmutation, 727–729
Transuranium elements, 729–730
Tremolite, 811
Tri-
  in alkane nomenclature, 899
  defined, 34, 225, 226
Triacylglycerols, 961–963
Trichlorofluoromethane, 831
Triclinic unit cell, 360
Triethylamine, 926
Trigonal bipyramidal arrangement,
    240–241, 240, 245
Trigonal planar arrangement, 235, 236,
    238, 239
  and dipole moment, 245
  ozone, 445
Trigonal pyramidal arrangement, 235,
    238, 245
Trilead tetroxide, 811
Trimethylamine, 621, 926
1,3,5-Trimethylbenzene, 907
Tripeptides, 942, 945
Triple bond constants, sigma and pi
    bonds in, 255
Triple bonds
  bond energies, 333
  Lewis formula, 209
  second-row elements, 766
Triple point, 381
Triple superphosphate, 816
Triprotic acids
  defined, 286
  phosphoric, 816
Tritium
  in fusion reactors, 753
  half-life, 737
Trona, 772
Tropopause, 427, 428

Troposphere, 427
Tryptophan, 943
T-shaped molecular geometry, 241,
    245
Tswett, Mikhail, 44–45
Tungsten
  coordination numbers, 858
  hydrogen reduction of oxides, 474
  melting point, 377
  ores, 850
  reduction of, 790
Tyndall effect, 400, 401
Tyrosine, 943

Ultraviolet radiation
  atmosphere and, 428–431
  ozone generation, 445
Uncertainty principle, 151–152
Unidentate ligands, defined, 858
Unimolecular reaction, 518
Unit cells, crystal
  defined, 359
  sodium chloride, 366
  types of, 360–362
Units of measurement
  base, 11, 12
  conversion factors, 15
  of density, 18
  derived, 16–22
  dimensional analysis, 14–16
  electrical, 708
  of energy, 19–22
  entropy, 657, 658
  equivalents and normality, 303–305
  frequency, 140
  ionization energy, 183
  of length, 11, 12
  of mass, 8, 11, 12
  prefixes, 11, 12
  of pressure, 18–19, 91
  radioactive disintegrations, 733
  reaction rates, 495
  significant figures, 7–10
  SI (International system), 11–14. See
      also SI units
  tables
    base units, 12
    derived units, 16
    prefixes, 12
    U.S. and metric units, 15
  of temperature, 13–14
  of time, 12
  of volume, 17–18
Unpaired electrons, 208, 216, 845,
    876
Unsaturated hydrocarbons, 895
  Bayer test, 908
  defined, 901–902
  hydrogenation of, 474, 555
  oxidation of, 908
  see also Alkenes; Alkynes; Multiple
      bonds
Unsaturated solutions, defined, 278
Uracil, 953–954
  base pairing, 955, 956
  codon dictionary, 960

Uranium, 717
  enrichment, 118
  radiation separation, 133
  radioactive decay series, 726
Uranium-235
  breeder reactors, 751–752
  chain reaction, 749–750
  fission of, 748
  in reactor, 750–751
Uranium-238, 718
  as alpha emitter, 725
  breeder reactors, 751–752
  half-life, 736
  mass losses in, 746
  radioactive decay series, 725, 726
Uranium hexafluoride, 828–829
Urea, 603, 656, 663, 671, 672
Urease, 704
Uric acid, 621
U.S. units, metric system relationships,
    15

Vacancies, in crystal lattice, 362
Vacuum tubes, elements used in, 771,
    777
Valence, Group IVA elements, 799
Valence bond theory, 246–253, 873–
    878
  hybrid orbitals, 248–253
  and molecular oxygen structure,
      257n
  octahedral complexes, 874–877
  tetrahedral and square planar com-
      plexes, 877–878
Valence electrons, 179
  in compound formation, 185
  group number and, 179
  ionization energy, 185
  in Lewis formulas, 197, 213, 214
  period 6 elements, 764
Valence-shell electron-pair repulsion
    model, 236–243
  two electron pairs, 237, 238, 239
  three electron pairs, 238, 239
  four electron pairs, 238, 239–240
  five electron pairs, 240–241
  six electron pairs, 241–242
  ozone, 445
Valence shells
  bonding and, 261
  and electron affinity, 187
  Group IA elements, 767
  Group IIIA elements, 781
  Group IVA elements, 799
  main-group elements in, 187–191
Valeric acid, formula, 919
L-Valine, 943
Vanadinite, 850
Vanadium
  in biological substances, 856
  as catalyst, 526, 553, 555
  ores, 850
  physical properties, 844–848
van der Waals, Johannes, 347n
van der Waals constants, 119
van der Waals equation, 119

van der Waals forces
  dipole-dipole, 348–349
  energy of, 347
  London, 347–348
Vanillin, 917
van't Hoff, Jacobus, 418
Vapor, in distillation, 43–44
Vapor density measurements, 105–107
Vaporization
  defined, 351, 372
  enthalpy of, 375
Vaporization heat, of common substances, 475
Vapor phase chromatography, 45
Vapor pressure, 351–355
  defined, 110
  measurement of, 351–352
  of solids, 372
  solute and, 409–411
  table, 354
Vapor-pressure curves, 411–414
Vapor pressure data, heat of vaporization from, 375–376
Vector quantity, dipole moment as, 244
Velocity, and energy, 19
Vinyl chloride, 829
Viscosity, 354, 356–357
Visible spectra, of transition metal complexes, 884–886
Vitamin B$_{12}$, 856, 741–742
Volatile acids, 288
Volatility, defined, 42–43, 354
Volcanic eruptions, 433, 458, 819
Volta, Alessandro, 683
Voltage, 693
  decomposition, 705–706
  and electomotive force, 693
Voltaic cells, 684–685, 693
  alkaline dry cell, 687
  cathodic protection, 690–691
  commercial, 687–691
  defined, 683
  emf calculation, 695–697
  lead storage, 687–688
  mercury cell, 771
  notation for, 691–693
  zinc-carbon, 687, 688
Volts
  defined, 693
  electron, 728
Volume
  defined, 16
  and density, 18
  and gaseous reaction equilibria, 558–559
  and heat of reaction, 318
  units of, 15, 17–18, 72
Volumetric flask, 17
von Laue, Max, 172
VSEPR, *see* Valence-shell electron-pair repulsion model

Washing soda, 772
Waste water treatment, aluminum sulfate, 787

Water, 37–38, 238
  aqua ligands, 863
  basic and acidic oxide products, 443–444
  boiling point of, 377–378
  bonding in, 251–252
  cellulose and, 952
  collecting gases over, 109–111, 439
  complex ion ligand nomenclature, 863
  composition of, 467–468, 469
  deionization, 486–487
  dipole moment, 244–245
  dissociation of, 571–572
  electrolysis, 438, 468, 469, 472, 706
  fluoridation, 833
  formulas and models, 29, 31, 32
  geometry, 239
  Group IA element reactivities, 768
  Group IIA element reactivities, 773–774, 775
  Group IVA element reactivities, 801
  Group VIIA compound reactivities, 827–828
  halogen oxidation of, 827–828
  hydrate isomers, 866–867
  hydrocarbon combustion and, 908
  hydrocarbon solubility in, 393–394
  hydrogen from
    electrolysis, 472
    steam-reforming process, 471–472
  hydrogen oxidation to, 69–70
  hydrolysis, 585–587
  hydrophilic and hydrophobic, defined, 400–402, 939, 946
  ion–dipole force, 394
  metallic vs. nonmetallic elements and, 760–761
  micelles, 939
  natural
    calcium hydrogen carbonate in, 778–779
    desalination, 486–488
    hard, 779
    hydrologic cycle, 480, 481
    limestone dissolution, 271, 287–288
    pollutants, 482–484, 772, 818
    treatment and purification, 484–486, 787
    see also Sea water
  nitrogen oxides in, 453, 454
  oxygen from, 438
  physical states of, 39
    phase diagram for, 380–381
    phases of, 41
  properties of
    chemical, 476–477
    physical, hydrogen bonding and, 475–476, 477
  self-ionization, 573–574
  specific heat of, 321n
  surface tension, 355–356, 357
  triple point for, 381

as unidentate ligand, 858
  vapor pressure, 352, 353, 354
  see also Aqueous solutions, reactions in; Condensation reactions; Solubility equilibria
Water-gas reaction, 471, 807
Water-gas shift reaction, 472, 555
Waterston, John James, 112
Wave functions, 151–152
Wave interference, x-ray, 368, 369, 370
Wavelengths
  de Broglie relation, 149–150
  defined, 139–140
  electromagnetic spectrum, 141
  hydrogen atom spectrum, 146
  transition metal absorption, 885
Wave mechanics, 151
Wave-particle duality, of light, 143
Waves
  de Broglie, 149–150
  light, 139–141
Weak electrolytes
  ionization of, 274
  and neutralization reactions, 285
Weather, 90n, 481. *See also* Atmosphere
Welding, 441, 457
Werner, Alfred, 860–861, 862
Wertherald, Richard, 460
White phosphorus, 813, 814, 815
Witherite, 773
Wohler, Friedrich, 894
Wolframite, 850
Work
  electrical, 693
  maximum, 668
  nonspontaneous processes and, 657
Wurtzite, crystal structure, 366, 367

Xenon, 191
  atmospheric, 458
  clathrate formation, 478
  compounds of, 457
  discovery of, 456
  p subshell, 170
  in troposphere, 431
Xenon difluoride, geometry, 241
Xenon tetrafluoride
  bonding in, 252–253
  geometry, 242
  Lewis formula, 217
X-ray diffraction, 204, 239, 368–371
X-ray electron spectroscopy, 179n
X-ray photoelectron spectroscopy, 172, 173
X rays, 734
  barium sulfate and, 780
  beryllium and, 777
X-ray spectroscopy, 172, 173, 178, 179n, 846

Yields
  calculations, with limiting reactants, 69–71

Table of Atomic Weights and Numbers

Names	Symbol	Atomic number	Atomic weight	Names	Symbol	Atomic number	Atomic weight
Actinium	Ac	89	227.0278	Molybdenum	Mo	42	95.94
Aluminum	Al	13	26.98154	Neodymium	Nd	60	144.24
Americium	Am	95	(243)	Neon	Ne	10	20.179
Antimony	Sb	51	121.75	Neptunium	Np	93	237.0482
Argon	Ar	18	39.948	Nickel	Ni	28	58.69
Arsenic	As	33	74.9216	Niobium	Nb	41	92.9064
Astatine	At	85	(210)	Nitrogen	N	7	14.0067
Barium	Ba	56	137.33	Nobelium	No	102	(259)
Berkelium	Bk	97	(247)	Osmium	Os	76	190.2
Beryllium	Be	4	9.01218	Oxygen	O	8	15.9994
Bismuth	Bi	83	208.9804	Palladium	Pd	46	106.42
Boron	B	5	10.81	Phosphorus	P	15	30.97376
Bromine	Br	35	79.904	Platinum	Pt	78	195.08 ± 3
Cadmium	Cd	48	112.41	Plutonium	Pu	94	(244)
Cesium	Cs	55	132.9054	Polonium	Po	84	(209)
Calcium	Ca	20	40.08	Potassium	K	19	39.0983
Californium	Cf	98	(251)	Praseodymium	Pr	59	140.9077
Carbon	C	6	12.011	Promethium	Pm	61	(145)
Cerium	Ce	58	140.12	Protactinium	Pa	91	231.0359
Chlorine	Cl	17	35.453	Radium	Ra	88	226.0254
Chromium	Cr	24	51.996	Radon	Rn	86	(222)
Cobalt	Co	27	58.9332	Rhenium	Re	75	186.207
Copper	Cu	29	63.546	Rhodium	Rh	45	102.9055
Curium	Cm	96	(247)	Rubidium	Rb	37	85.4678
Dysprosium	Dy	66	162.50	Ruthenium	Ru	44	101.07
Einsteinium	Es	99	(252)	Samarium	Sm	62	150.36
Erbium	Er	68	167.26	Scandium	Sc	21	44.9559
Europium	Eu	63	151.96	Selenium	Se	34	78.96
Fermium	Fm	100	(257)	Silicon	Si	14	28.0855
Fluorine	F	9	18.998403	Silver	Ag	47	107.8682
Francium	Fr	87	(223)	Sodium	Na	11	22.98977
Gadolinium	Gd	64	157.25	Strontium	Sr	38	87.62
Gallium	Ga	31	69.72	Sulfur	S	16	32.06
Germanium	Ge	32	72.59	Tantalum	Ta	73	180.9479
Gold	Au	79	196.9665	Technetium	Tc	43	(98)
Hafnium	Hf	72	178.49	Tellurium	Te	52	127.60
Helium	He	2	4.00260	Terbium	Tb	65	158.9254
Holmium	Ho	67	164.9304	Thallium	Tl	81	204.383
Hydrogen	H	1	1.00794	Thorium	Th	90	232.0381
Indium	In	49	114.82	Thulium	Tm	69	168.9342
Iodine	I	53	126.9045	Tin	Sn	50	118.69
Iridium	Ir	77	192.22	Titanium	Ti	22	47.88
Iron	Fe	26	55.847	Tungsten	W	74	183.85
Krypton	Kr	36	83.80	(Unnilhexium)	(Unh)	106	(263)
Lanthanum	La	57	138.9055	(Unnilpentium)	(Unp)	105	(262)
Lawrencium	Lr	103	(260)	(Unnilquadium)	(Unq)	104	(261)
Lead	Pb	82	207.2	Uranium	U	92	238.0289
Lithium	Li	3	6.941	Vanadium	V	23	50.9415
Lutetium	Lu	71	174.967	Xenon	Xe	54	131.29
Magnesium	Mg	12	24.305	Ytterbium	Yb	70	173.04
Manganese	Mn	25	54.9380	Yttrium	Y	39	88.9059
Mendelevium	Md	101	(258)	Zinc	Zn	30	65.38
Mercury	Hg	80	200.59	Zirconium	Zr	40	91.22

A value in parentheses is the mass number of the isotope of longest half-life.

Values in this table are from the IUPAC report "Atomic Weights of the Elements 1981," *Pure and Applied Chemistry*, Vol. 55, No. 7 (July 1983), pp. 1105–1106.